Handbook of Brain and Behaviour in Human Development

EDITED BY

Alex F. Kalverboer

Department of Developmental and Experimental Psychology,
Faculty of Psychology,
University of Groningen,
The Netherlands

and

Albert Gramsbergen

Department of Medical Physiology,
Faculty of Medical Sciences,
University of Groningen,
The Netherlands

D1728321

KLUWER ACADEMIC PUBLISHERS
DORDRECHT / BOSTON / LONDON

A C.I.P. Catalogue record for this book is available from the Library of Congress

ISBN 0-7923-6943-2

Published by Kluwer Academic Publishers,
P.O. Box 17, 3300 AA Dordrecht, The Netherlands

Sold and distributed in North, Central and South America
by Kluwer Academic Publishers,
101 Philip Drive, Norwell, MA 02061, U.S.A.

In all other countries, sold and distributed
by Kluwer Academic Publishers Group
P.O. Box 322, 3300 AH Dordrecht, The Netherlands.

Printed on acid-free paper

Printed and bound in Great Britain by Antony Rowe Limited.

Dedicated to Minnie

Table of Contents

TABLE OF CONTENTS

III.   MOTOR DEVELOPMENT

TABLE OF CONTENTS

TABLE OF CONTENTS

x

Preface

The study of neurobehavioural development in the human has a relatively short history. Still in the fifties of the 20th century, concepts on the development of brain and behaviour generally were derived from knowledge on human adults suffering from lesions in the brain, or on experiments in animals after transections at various levels in the CNS in order to reduce complexity. Clinical practice in child neurology was often based on notions derived from adult neurology and neuropsychology, which in turn were strongly affected by reflexology-based conceptions and archaic notions on structure–function relationships. The breaking point came when conceptions from ethology on the importance of adaptive functioning in the organism's natural habitat were introduced in theorizing on neurobehavioural development in the human. In this new field of research, which by Heinz Prechtl has been termed "Developmental Neurology", notions were introduced such as "ontogenetic adaptation" and "the primacy of the adaptive function of the brain", based on meticulous investigations into the development of brain–behaviour relationships. The new conceptual framework gave rise to a wealth of detailed observational and experimental studies in the human, firstly in the neonatal period and more recently also in prenatal life. Closely connected to this is research in developing animals seeking for correlates between neurophysiological and neuroanatomical changes and developments in behaviour, and this has offered further perspectives to our present-day thinking on human neurobehavioural development.

In parallel, developments particularly in developmental psychology and ethology, gave further impulses to the development of interdisciplinary research programmes, focusing on human brain–behaviour development. One of the earliest and most creative representatives of this new approach to the study of the adaptive qualities of the young human organism was Hanus Papousek, an "artistic" scientist of exceptional quality who readily agreed to contribute to this handbook. He unfortunately died recently, but we feel it an honour to have his chapter included in this book.

Recently, research into the genetical and molecular aspects of brain development has enjoyed amazingly rapid and vast progress. Undoubtedly the introduction of new techniques has led to an enormous increase in knowledge, and to an onset of insight into the basic mechanisms of brain development. The study of these processes in seemingly primitive organisms such as *C. elegans* and *Drosophila* may have fed the fear of a new wave of reductionism. Recently, it became clear that a surprisingly low amount of genes (estimates in the year 2001 varying between 28,000 and 40,000 genes) constitute the human genome.

Even when appreciating the immense complex interactions which might be thought to occur between genes, and also the intricate timetable of these interactions this (again) points towards the importance of epigenetical factors involved in brain development.

The efforts to disentangle the genetical and epigenetical aspects of developments and properties of the organism, the attempts to further integrate knowledge obtained in research disciplines, the introduction of new theoretical insights such as the dynamic systems theory and the group selection theory, as well as advanced techniques for measuring structures and functions of the developing brain, have led to a rapid accumulation of knowledge on neurobehavioural development, and may substantially contribute to a further understanding of neuro-ontogeny. However, results of these research efforts are still very much scattered in the scientific literature, notwithstanding summaries on restricted fields of research in a number of books and in scientific journals.

Now, at the beginning of the 21st century, the right moment seems to have come to compose this handbook on *Brain and Behaviour in Human Development*. Around a number of core issues, overviews of present-day insights and research are brought together, written by around forty scientists, who are all leading investigators in the field. We aimed at presenting updated knowledge, the newest insights and directions for future research on the development of motor and perceptual systems and on cognitive and emotional and communicative functions, in relation to the neurobiological development of the CNS. We intend to cover the age range from prenatal stages until the onset of adolescence with an emphasis on the earlier phases of human development.

In the first section some basic issues and conceptual problems in the study of brain–behaviour development are discussed. Particularly, attention is paid to some neuropsychological problems inherent to research into early development and to the role of developmental genetics.

The second section consists of chapters on the processes and mechanisms involved in structural development, as well as on some important aspects of morphological development itself. Wherever possible, extrapolations to human development and to other domains research have been made

The third section, on motor development, contains chapters on fetal and postnatal motor development in the human as well as contributions from research in experimental animals and the problem of cross-species extrapolation of developmental data. Next to classical descriptions and interpretations of development, contributions by the dynamical systems approach are also included.

The fourth section on perceptual and cognitive development aims at discussing the development of cognitive functions in relation to perceptual systems and to the development of the brain. Issues such as the development of the temporal organization of behaviour, perception and cognition, memory functions and more complex thought processes are covered.

The fifth section, on the development of communication and emotional development, contains contributions to a growing field of interest. Connections between emotional and cognitive development, as well as the neural basis of these developments, are core topics. Ethological and psychological approaches both to emotional and communicative development are discussed.

Finally, the epilogue comments on some points which are core issues in research on brain–behavior development in the period ahead.

It is our hope that this handbook will help to further integrate knowledge in this most important field of study, and to stimulate further interdisciplinary research efforts.

The editors

Affiliations of the Authors

Armand, J., Laboratoire Mouvement et Perception, Centre National de la Recherche Scientifique & Université de la Mediterranée, Marseille, France

Baron, J., Laboratory of Physiology, University of Oxford, Oxford, UK

Bekoff, A., Department of Environmental, Population and Organismic Biology and Center for Neuroscience, University of Colorado, Boulder, CO, USA

Berger, A., Behavioral Sciences Department and Zlotowski Center for Neuroscience, Ben Gurion University of the Negev, Beer-Sheba, Israel and Sackler Institute Weill Medical College of Cornell University, New York City and University of Oregon, Eugene, OR, USA

Bolhuis, J.J., Behavioural Biology, Institute of Evolutionary and Ecological Sciences, Leiden University, Leiden, The Netherlands

Caldéro, J., Unitat de Neurobiologia Cellular, Department de Ciéncies Mediques Basiques, Facultat de Medicina, Universitat de Lleida, Lleida, Catalonia, España

Caflisch, J.A., University Children's Hospital, Zurich, Switzerland

Carlson, B.X., NeuroScience PharmaBiotec Research Center, Department of Pharmacology, Royal Danish School of Pharmacy, Copenhagen, Denmark

Cazalets, J.-R., UMR 5543, Centre National de la Recherche Scientifique, Université de Bordeaux, Bordeaux, France

De Haan, M., University College London, Institute of Child Health, Cognitive Neuroscience Unit, London, UK and Birkbeck College, Department of Psychology, Centre for Brain and Cognitive Development, London, UK

Dahaene-Lambertz, G., Laboratoire de Sciences Cognitives et Psychoinguistique, Centre National de la Recherche Scientifique, Paris, France, and Service de Neuropediatrie, Centre Hospitalier Universitaire Bicetre, France

Derryberry, D., Department of Psychology, Oregon State University, Corvallis, OR, USA

Doubell, T.P., Laboratory of Physiology, University of Oxford, Oxford, UK

Esquerda, J., Unitat de Neurobiologia Cellular, Department de Ciénces Mediques Basiques, Facultat de Medicina, Universitat de Lleida, Lleida, Catalonia, España

Fagioli, I., Department of Psychology, University of Florence, Florence, Italy

Ficca, G., Department of Psychology, University of Florence, Florence, Italy

Fogel, A., Department of Psychology, University of Utah, Salt Lake City, UT, USA

Forssberg, H., Department of Pediatrics, Karolinska Hospital, Stockholm, Sweden

Giganti, F., Department of Psychology, University of Florence, Florence, Italy

Gordon, A.M., Department of BioBehavioral Sciences, Teachers College, Columbia University, New York, NY, USA and Department of Rehabilitation Medicine, College of Physicians and Surgeons, Columbia University, New York, NY, USA

Gould, T., Department of Neurobiology and Anatomy and the Neuroscience Program, Wake Forest University School of Medicine, Winston-Salem, NC, USA

Gramsbergen, A., Department of Medical Physiology, University of Groningen, Faculty of Medical Sciences, Groningen, The Netherlands

Hadders-Algra, M., Department of Medical Physiology, University of Groningen, Faculty of Medical Sciences, Groningen, The Netherlands

Halit, H., Birkbeck College, Department of Psychology, Centre for Brain and Cognitive Development, London, UK

Hansen, G.H., Institute of Medical Genetics, Panum Institute, University of Copenhagen, Copenhagen, Denmark

Hofer, M.A., Department of Psychiatry, Columbia University, College of Physicians and Surgeons, New York, NY, USA

Hohmann, C.F., Department of Biology, Morgan State University, Baltimore, MD, USA

Hopkins, J.B., Department of Psychology, University of Lancaster, Lancaster, UK

Johnson, S.P., Department of Psychology, Cornell University, Ithaca, NY, USA

Kalverboer, A.F., Department of Developmental and Experimental Psychology, Faculty of Psychology, University of Groningen, Groningen, The Netherlands

Keverne, E.B., Sub-Department of Animal Behaviour, University of Cambridge, High Street, Madingley, Cambridge, UK

King, A.J., Laboratory of Physiology, University of Oxford, Oxford, UK

LaMantia, A.S., Department of Cell and Molecular Physiology, School of Medicine, University of North Carolina, Chapel Hill, NC, USA

Lambert, H.W., Department of Cell Biology and Anatomy, School of Medicine, University of North Carolina, Chapel Hill, NC, USA

Largo, R.H., University Children's Hospital, Zurich, Switzerland

Lauder, J.M., Department of Cell Biology and Anatomy, School of Medicine, University of North Carolina, Chapel Hill, NC, USA

Maness, P.F., Department of Biochemistry, School of Medicine, University of North Carolina, Chapel Hill, NC, USA

Mareschal, D., Centre for Brain and Cognitive Development, Birkbeck College, University of London, London, UK

Mehler, J., Laboratoire de Sciences Cognitives et Psychoinguistique, Centre National de la Recherche Scientifique, Paris, France

Moiseiwitsch, J.R.D., Department of Endodontics, School of Dentistry, University of North Carolina, Chapel Hill, NC, USA

Nelson-Goens, G.C., Institute of Behavioral Genetics, University of Colorado, Bouldon, CO, USA

Oppenheim, R.W., Department of Neurobiology and Anatomy and the Neuroscience Program, Wake Forest University School of Medicine, Winston-Salem, NC, USA

Pantoja, A.P.F., Department of Psychology, California State University, Chico, CA, USA

Papoušek, H., Menchen, Germany (deceased)

Pena, M., Laboratoire de Sciences Cognitives et Psychoinguistique, Centre National de la Recherche Scientifique, Paris, France

Posner, M.I., Sackler Institute Weill Medical College of Cornell University, New York City and Institute of Cognitive and Decision Sciences, University of Oregon, Eugene, OR, USA

Prechtl, H.F.R., Department of Physiology, Karl-Franzens-University of Graz, Graz, Austria

Ribchester, R.R., Department of Neuroscience, University Medical School, Edinburgh, UK

Rios, M., Department of Neuroscience, Tufts University School of Medicine, Boston, MA, USA, and Whitehead Institute, MIT, Cambridge, MA, USA

Roffler-Tarov, S., Department of Neuroscience, Tufts University School of Medicine, Boston, MA, USA and Whitehead Institute, MIT, Cambridge, MA, USA

Rothbart, M.K., Department of Psychology, University of Oregon, Eugene, OR, USA

Salzarulo, P., Department of Psychology, University of Florence, Florence, Italy

Schmid, R.S., Department of Biochemistry, School of Medicine, University of North Carolina, Chapel Hill, NC, USA

Schnupp, J.W.H., Laboratory of Physiology, University of Oxford, Oxford, UK

Schousboe, A., NeuroScience PharmaBiotec Research Center, Department of Pharmacology, Royal Danish School of Pharmacy, Copenhagen, Denmark

Sireteanu, R., Department of Neurophysiology, Max-Planck-Institute for Brain Research, Frankfurt am Main, Germany

Thomas, M.S.C.,

Trevarthen, C., Department of Psychology, the University of Edinburgh, Edinburgh, Scotland, UK

Uylings, H.B.M., Netherlands Institute for Brain Research, Amsterdam and Department of Anatomy, Free University of Amsterdam, Amsterdam, The Netherlands

Welsh, M.C., Department of Psychology, University of Northern Colorado, Greeley, CO, USA

Whitaker-Azmitia, P., Department of Psychology, SUNY at Stony Brook, Stony Brook, NY, USA

I.   BASIC ISSUES TO THE STUDY OF NEUROBEHAVIOURAL DEVELOPMENT

I.1
Brain–Behaviour Relationships in the Human – Core Issues

ALEX F. KALVERBOER and ALBERT GRAMSBERGEN

CROSSING THE BORDERS OF DISCIPLINES

Interest in the development of brain and behaviour in the human has grown explosively in the last decade of the twentieth century, which was labelled the decade of the brain. It has been recognized that functioning of the brain at adult age cannot be understood without accounting for the intrinsic connection during ontogeny between the organism's biological endowment, which is genetically founded, and its environment. This insight implies that we definitively have left behind us the traditional nature–nurture dichotomy, which for so long has ruled discussions in biology. Also, awareness has grown that the traditional segmentation of research and of theorizing in separate disciplines, which has prevailed until recently, seriously hampers the solving of the many outstanding problems which have to be conquered in brain–behaviour research. This for example is indicated by the term *"developmental cognitive neuroscience"*, as coined by Johnson (1997). This field of research covers the growing need to formulate and study problems in such a way that data obtained at different levels of explanation (Rose, 1976) can be theoretically integrated.

The aim of this handbook is to present an overview of present-day knowledge on the structural and functional development of nervous system and adaptive behaviour from this very perspective. Some chapters review interdisciplinary research attempting to integrate knowledge from different domains, and in other chapters results from monodisciplinary studies are discussed which are relevant for understanding particular aspects of brain–behaviour relationships during ontogeny.

Especially in the last two decades in many studies, crossing the borders of disciplines, an attempt has been made to obtain a more integrated insight into brain–behaviour relationships in human development. Such research has often been triggered by new developments in measuring and observation techniques, and by improvements in experimental designs. However, approaches as well as underlying theories and models still largely differ in the various disciplines. Terms and concepts, such as the organization of states, continuity versus

3

A.F. Kalverboer and A. Gramsbergen (eds.), Handbook of Brain and Behaviour in Human Development, 3–10
© *2001 Kluwer Academic Publishers. Printed in Great Britain.*

discontinuity, transitions in development, critical periods, vulnerable periods and others have different connotations in different disciplines. In fact, we are still in a phase in which a wealth of bits and pieces of knowledge on the development of brain and behaviour is available, but which can only incidentally be connected to each other in a meaningful way. Undoubtedly, great progress has been made in studies into the development of motor functions, perception, cognition and emotion, as well as into the understanding of basic mechanisms which govern the structural development of the brain. How these lines of scientific exploration on structure and function will converge in the future, however, depends greatly on the efforts to be invested in the development of multidisciplinary and problem-centred approaches to the study of brain–behaviour relationships in human ontogeny. Such disciplines include developmental neuroanatomy, developmental neurophysiology, developmental neurology, neuropsychology and biopsychology, cognitive psychology, as well as ethology, and these are all represented in this handbook.

On which basis have these approaches been represented here? In other words, which were the main ideas that inspired us in the composition of this handbook?

A DEVELOPMENTAL APPROACH

Ample evidence points to the essential role of early functioning for the structural development of a variety of tissues and also the nervous system (for a review of some of the evidence see e.g., Prechtl, 1981; Oppenheim, 1984). Even more intriguing, however, are the observations indicating that the order in developmental processes follows a seemingly capricious trajectory rather than being governed for example by rostro-caudal or medio-lateral trends in morphological development. Developmental processes should therefore be studied *in their own right* and not merely as a strategy to understand adult morphology or adult functioning. This has been clearly formulated by Prechtl when stating, "During early development the brain does not mature simultaneously in all regions, but only in those parts and structures which are necessary for performing vital functions. These mature selectively and with a higher speed. Such a specific early maturation differs according to the species, depending on the specific ecological circumstances to which the young organisms are phylogenetically adapted" (Prechtl, 1981, p. 199). This implies also that the development of the organism has to be studied in an intrinsic connection to its natural environment, its "ecological niche". This approach characterizes many of the contributions to this handbook, and they all express a direct interest in developmental processes. This implies that aspects such as structural and functional plasticity, continuity versus discontinuity of functional development, compensatory qualities of the developing nervous system, are core issues. Explicitly, as they are in themselves foci of research as reported in many contributions; implicity, as relationships postulated between structure and function are evaluated with reference to the specific characteristics of the developing nervous system.

4

EMPHASIS ON EARLY ONTOGENY

A further characteristic of the approach adhered to in this handbook is the emphasis on ontogenetic processes from the earliest phases of fetal development onward. Many of the contributions concern the first phases of life, including the prenatal period. During these phases the basic circuitry is laid down, the transmitter systems become functional, the muscles form and become innervated. The ultimate set of circuitries in the nervous system is shaped thereafter by selecting those parts of the circuitry which are needed for later functioning, by early perception and functioning of the nervous system itself. The study of these fundamentals of brain–behaviour relationships is therefore a prerequisite for understanding later development. The biological principles most clearly apply to the human species in these early phases of development, and particularly then comparative approaches taking into account results from experimental research in animals are required.

THE NOTION OF ONTOGENETIC ADAPTATIONS

Extremely important has been the notion that the nervous system of a young organism is qualitatively different from that of the adult. This insight was gained in systematic investigations of the neurological development of the fetus and the newborn human infant. In retrospect it seems strange that this insight took so long before it was acknowledged and accepted in clinical neurology and developmental psychology. From birth or shortly thereafter onward, the neural mechanisms for vital functions such as breathing, rooting, sucking and swallowing, crying, spatial orientation, sleeping and waking are fully developed as very complex control mechanisms and not as simple reflexes (Prechtl, 1984). The insight that the nervous system (and the baby as a whole) in a sense is tailored to the environmental possibilities and needs of the moment, and thus differs from that at later ages, obviously has important consequences for the assessment of neurological functioning and the appreciation of the behavioural qualities at early age. Behavioural patterns in human fetuses, in newborn babies and also in later phases of development, have been interpreted in terms of "ontogenetic adaptations". These are adaptations to environmental demands which are specific to particular stages of development (such as the fetal condition at early and later stages and the postnatal stages, when the baby is highly dependent on the caregiver).

Although this view is generally valid, it is important to realize that each species may bear with it the traces of circuitries which have lost their function during evolution or due to long-lasting changes in their habitat. One intriguing example of such lost capabilities during phylogeny are the wing movements in flightless birds (e.g. the penguin). Although the morphology of the sternum and the muscles has changed, and the number of motoneurons to these muscles and their interneurons has decreased, it is likely that the circuitry for this behaviour has persisted despite the loss of its original function (for an essay on this point and related references see Provine, 1988). The question then is how many traces of such circuits still exist in the brain of the species under study, and also which role

5

these circuits may play either during development or at adult age, either in normal undisturbed situations or after brain lesions. Provine specifically questions whether all early behaviour patterns indeed are adaptations, and whether perhaps some of these patterns might be expressions of phylogenetically older patterns which are lost in later life.

THE INTERPLAY BETWEEN GENETIC AND EPIGENETIC FACTORS

This handbook also expresses a core interest in the role of genetic and epigenetic factors and their interplay in brain–behaviour development. In the field of behavioural genetics, particular behaviours have been identified which are mainly genetically determined and others which are mainly determined by environmental factors. In recent neurobiological research, increasing numbers of factors are identified which lead to the expression of genes at earlier or later stages of cellular development and which determine the localization, the fate and the differentiation of neuroblasts and glial cells and also extrinsic cues. The latter cues or signal molecules may bind to membrane-bound receptors which via second messenger systems influence the transcription factors, and which in turn increase or decrease the expression of particular genes. Much of this research has been performed in relatively uncomplicated species such as *Drosophila* and *C. elegans*, but the general opinion is that similar processes have persisted throughout evolution and may also be effective in higher organisms. Many of those who, only two or three decades ago, referred to the impossibility of a few ten-thousands of genes in humans being able to specify the immense complexity of the brain up to the level of individual synapses, are now impressed by the complexity and the intricacy of the various ways in which genes may affect the localization and differentiation of groups of neurons, the circuitry by which they are connected and the nature of the synaptic interactions.

The possibilities of the new technologies (leading to the *human genome project*), the recent and vast increase in knowledge and the optimism with regard to an imminent solving of many problems on brain functioning and brain development held by politicians, pharmaceutical industries, etc. have led to heated debates among neuroscientists aiming at bringing this optimism back to its proper perspectives (see, e.g. the collection of essays in *Alas, poor Darwin*, edited by Rose & Rose, 2000). Several thousands of neurons proliferating each second in the first phases of human development, and the intricacy of the pathways by which the genetic code is translated in the production of proteins, may well prove to be sufficient for laying down the global circuitry of the most complex structure we know of. However, from early development onwards the brain is the organ which senses the environment and ultimately governs, orchestrates and modulates the reactions of the individuum to the unpredictable environment. Epigenetic factors therefore must play a pivotal role in the development of the brain, not only to tune its activity but, as numerous studies have shown, they also take an important part in the shaping of the brain's architecture and circuitry. Indeed, as Oppenheim in his authoritative review on the history of concepts on 'Preformation and epigenesis' (1982) writes at the end: "function and experience, including endogenous neural activity, sensory input, and learning and conditioning do play an important role in neuro-behavioral development".

THE NECESSITY OF ANIMAL STUDIES

New imaging techniques such as MRI and PET scanning enable us to longitudinally study changes in brain structure and to visualize neural processes in the human. It may be expected that the resolution of these techniques will be even further enhanced in the future. However, even then, research into the development of brain structures and behaviour will require animal experimentation in which, by interference with normal development and environmental conditions, causal relationships might be unravelled.

Much of our fundamental knowledge collected so far depends upon observations and experimentation in animals, and many examples of this research are reviewed in this handbook. When considering changes in the developing animal, and when extrapolating such results to human development, an important prerequisite is to scale developmental pace in the species which is considered and to align analogous points in development. It is thus still important to take into consideration the important differences in behavioural competences and the related differences in neuronal circuitry. Shifts in communication between individuals by language or otherwise, the expression of emotions, changes in the processing of, e.g., visual and auditory information and shifts in motor behaviour, among which bipedal locomotion in the human on the one hand and quadrupedal locomotion, swimming or flying in other species on the other hand are the most obvious examples, all have their consequences for the actual neuronal circuitry. These considerations should withhold those who sometimes too readily interpret behavioural patterns or developmental changes in the human on the basis of neurobiological data which were obtained in animal species. On the other hand, many pieces of the puzzles are still missing. The solving of many of these problems will remain dependent on experimentation in animals.

CONTROVERSIES, HOW TO BRIDGE THE GAPS

In particular the research field of the development of brain and behaviour has been the scene of vivid debates on the validity of concepts, on approaches of research, on general lines in explanation. These debates sometimes had a paralysing effect on scientific progress. This was, for example, the case in the study of the development of motor control. Neurobiologists investigating motor development by focusing their research on the development of its components such as neurons, membranes, muscle fibres, etc., and adhering to the neuro-maturational approach, often seem to speak another language than do those who attempt to understand this development by theorizing and experimenting in the context of the dynamic systems approach. In this handbook these approaches, and others aiming at understanding motor development, are all covered, and we consider these different viewpoints as complementary, and badly in need of mutual interaction.

Another controversy exists between, on the one hand, those allocating functions to distributed and specific brain structures and, on the other hand, scientists advocating the existence of a high degree of plasticity in such relationships. Investigators in the fields of developmental neuroanatomy and neuroembryology are often primarily involved in research into the development of brain maps and the highly specific circuitries which later produce certain behaviours or process specific information. Their focus is generally on the development of structure, and brain functions in a sense are derived from this. On the other hand, there is the notion of self-organization in the brain, governed by textures of any kind in the outside world, which requires a high degree of plasticity in the CNS, and almost precludes any rigidity and even perhaps specificity in circuitries. Also here, domains of research have to meet, in order to come to a common understanding and to a further integration of knowledge.

A similar controversy is found in the domain of cognitive neuroscience. Here the insight is growing that the brain does not consist of a collection of circumscribed areas, responsible for highly specific cognitive functions. Rather we should think in terms of "interlocked neuronal (functional) networks", in which for the execution of particular functions certain circumscribed areas may be necessary, but not sufficient. To quote a statement by Lopes da Silva (2000, p. 29), "Even today, a general tendency is to think that because a certain area has a given anatomical name, it should correspond to one function, i.e. one anatomical name = one cognitive function!" In Lopes da Silva's opinion a strict correspondence might be true for a few very simple functions, but this certainly does not hold in general. The study of how structure–function relationships may change during human ontogeny offers the opportunity *par excellence* to challenge simple-minded ideas on this issue.

Rightly enough Panksepp (1998) states that emotional and motivational systems shared by mammalian brains have been largely neglected by mainstream psychology with its almost exclusive focus on differences in higher cognitive functions. There is still a one-sided focus on cognitive processes as indicated by the term cognitive neurosciences, with a lack of interest in the general organizing principles, considered from an evolutionary perspective. According to Panksepp this is largely due to a tendency in the psychology of the last century "to explain everything in human and animal behavior via environmental events that assail organisms in their real-life interactions with the world rather than via the evolutionary skills that are constructured in their brains as genetic birthrights" (Panksepp, 1998, p. 5). However, interest in the biological (the phylogenetic and the genetic) roots of human development has grown considerably in recent decades. The same holds for a growing attention towards emotional and motivational systems. The close connection between emotional and cognitive development is more and more understood. These developments are reflected in this handbook.

THE DIRECT STUDY OF HUMAN DEVELOPMENT

Clearly enough, it is a matter of primary interest to study developmental processes in the human itself. There is still a great lack of dense longitudinal studies on brain–behaviour relationships in the human from the earliest phases onward. Undoubtedly, notions derived from studies on adults or on brain-damaged patients, e.g. in the field of cognitive psychology or neuropsychology, do make valuable contributions to our insight into ontogenetic mechanisms. However, such studies may at best trigger the application of age-adequate approaches to brain–behaviour development, thereby giving suggestions about the effects of dernaged developmental processes.

Finally, there is still a large gap between our knowledge on basic neurological and psychological processes, which are mainly studied in the laboratory, and the way in which such processes play a role in complex adaptive behaviours in the natural habitat. Also from that perspective it is necessary to come to a more integrated approach to the scientific study of brain–behaviour development in the human than that reached hitherto. We hope that this handbook may provide building stones for developing such an approach in order to improve our insight into the ontogeny of neurobehavioural relationships in the human.

References

Johnson, M.H. (1997). *Developmental cognitive neuroscience*. Oxford, UK and Cambridge, USA: Blackwell.

Lopes da Silva, F.H. (2000). *From neuronal networks to consciousness*. Den Haag, The Netherlands: Netherlands Organization for Scientific Research.

Oppenheim, R.W. (1982). Preformation and epigenesis in the origins of the nervous system and behavior: issues, concepts, and their history. In P.P.G. Bateson & P.H. Klopfer (Eds.), *Perspectives in ethology*, vol. 5: *Ontogeny* (pp. 1–100). New York and London, Plenum Press.

Oppenheim, R.W. (1984). Ontogenetic adpatations in neural development: toward a more 'ecological' developmental psychobiology. In H.F.R. Prechtl (Ed.), *Continuity of neural functions from prenatal to postnatal life* (pp. 16–30). Clinics in Developmental Medicine no. 94. Oxford: Blackwell Scientific Publications.

Panksepp J. (1998). *Affective neuroscience: the foundations of human and animal emotions*. New York and Oxford: Oxford University Press.

Prechtl, H.F.R. (1981). The study of neural development as a perspective of clinical problems. In K.J. Connolly & H.F.R. Prechtl (Eds.), *Maturation and development: biological and psychological perspectives* (pp. 198–215). Clinics in Developmental Medicine no. 77/78. London: Heinemann, SIMP.

Prechtl, H.F.R. (1984). Continuity and change in early neural development. In H.F.R. Prechtl (Ed.), *Continuity of neural functions from prenatal to postnatal life* (pp. 1–15). Clinics in Developmental Medicine no. 94, Oxford: Blackwell Scientific Publications.

Provine, R.R. (1988). On the uniqueness of embryos and the difference it makes. In W.P. Smotherman & S.R. Robinson (Eds.), *Behavior of the fetus* (pp. 3–18). Caldwell, NJ: Telford Press.

Rose, S. (1976). *The conscious brain*. London: Penguin.

Rose, H. & Rose, S. (2000). *Alas, poor Darwin. Arguments against evolutionary psychology*. London: Jonathan Cape.

I.2
Ontogeny of Brain and Behaviour: a Neuropsychological Perspective

ANDREA BERGER and MICHAEL I. POSNER

ABSTRACT

The extensive postnatal development of the human brain affords the opportunity for experience to influence its structure and circuitry. However, it has been difficult to study human brain function because of the lack of non-invasive methods for examining changes in the human brain. The past 10 years have allowed increased use of findings from adult neuroimaging studies by the design of tasks that mark the development of specific brain regions. The next 10 years should see more use of methods such as functional magnetic resonance imaging for the non-invasive exploration of the development of brain regions in young children. We review some of these methodological issues in relation to current findings about the development of the brain's attentional control systems.

THE DEVELOPMENTAL CHALLENGE

The human infant has the longest period of dependence upon caregivers of any mammal. While much of the nervous system develops prior to birth, for primates and particularly humans, a long period of postnatal development is needed to complete the basic formation of the brain.

During this period, infants gain control of their behaviour and mental state so that as older children and adults they can explore their sensory world and exercise a degree of central control over their emotions, thoughts and action (see Chapter IV.2 by Sireteanu, in this volume). The attention system (Posner & Petersen, 1990) consists of brain networks through which at least part of this control is exercised. The attention system includes several distinct networks that perform the functions of orienting to sensory events, sustained alertness and self-regulation. The major goal of this chapter is to ask how these networks develop. To reach this goal we first describe some of the methods available for the study

A.F. Kalverboer and A. Gramsbergen (eds.), Handbook of Brain and Behaviour in Human Development, 11–32
© *2001 Kluwer Academic Publishers. Printed in Great Britain.*

of human brain development. Although neuroimaging methods have been applied to adult behaviour for the past decade their use with children is only just beginning. We then describe the developmental course of orienting behaviour during the first year of life and trace the development of attentional control mechanisms in later infancy and childhood. In this chapter we place primary emphasis on the development of the attentional system as being crucial for self-regulation.

METHODS FOR STUDYING THE DEVELOPMENT OF THE NERVOUS SYSTEM

The past quarter-century has produced great improvements in our ability to study the behaviour and cognition of even the youngest infants. However, until recently non-invasive methods for imaging the brain were limited to the use of electrical or magnetic signals collected from outside the skull. In the past decade neuroimaging methods have been used to study regional blood flow. At first these methods were limited to adults, because they involved the use of small amounts of radioactivity (e.g. positron emission tomography, PET). Only recently has the use of functional magnetic imaging (fMRI) allowed approaches to paediatric neuroimaging. Because much of the available neuroimaging data is from adults, we first describe the use of PET and fMRI methods and discuss their relation to electrical recording methods. Next we discuss the recent developments in paediatric neuroimaging. Finally, we consider the use of marker tasks based on adults to study the development of particular brain regions.

Imaging modalities

When neurons are active they change their own local blood supply. This makes it possible to track areas of the brain active during cognitive processes by methods designed to study changes in aspects of the blood supply within the brain. The two most prominent methods for doing this are PET and fMRI (Toga & Mazziotta, 1996). PET uses a radionucleide that gives off positrons. When the positrons are emitted they are annihilated after short distances and exit the skull as photons, which strike detectors at opposite sides of the brain simultaneously. These PET counts can be related to measures of rate of blood flow through particular regions of the brain.

The most commonly used PET method for perceptual and cognitive processes is to examine blood flow with an isotope of oxygen (O^{15}). This is a very short-lived isotope and pictures of activity can be based on sustained activation of less than a minute. Other approaches are also possible using isotopes that allow examination of glucose utilization or which bind to various transmitters. Thus PET can be used to deal with a whole spectrum of questions.

With increased neuronal activity there is also a change in the oxygen content of the blood. It appears that blood flow changes more than oxygen consumption, and since the haemoglobin that carries oxygen is paramagnetic, it is possible to sense these changes and display them as fMRI of brain activity. This is the basis of the most commonly used signal in (fMRI). Because this method is non-invasive it can be used with children, and has become the primary method for examination of cognitive processes in children and adults.

These two major methods are fully complementary and have usually provided converging evidence where they have been applied to cognitive studies. In general, fMRI is less invasive and is replacing PET for most cognitive studies and where both methods can be used. In addition, fMRI has a greater capability of dealing with individual functions. This is partly because its non-invasive nature allows sufficient data to be collected from one individual and the natural connection between images of blood oxygenation (functional MRI) and of brain anatomy (structural MRI) allows for a relatively natural correspondence between individual anatomy and brain activity.

Recording electrical activity from outside the brain (EEG) is a traditional non-invasive way of obtaining information about brain activity, which can be used at any age. However, the method is more useful when employed in conjunction with PET or fMRI as a way of obtaining the activity of functional areas of brain activation in real time (Hillyard & Anllo-Vento, 1998; Toga & Mazziotta, 1996). In the 1960s cognitive psychologists showed that a wide variety of tasks could be described in terms of mental operations (subroutines) that took place over tens to hundreds of milliseconds. The temporal resolution of methods based upon changes in blood flow or chemistry is not fast enough to trace these changes. The integration of electrical and blood flow measures has been achieved in some adult studies (see Hillyard & Anllo-Vento, 1998, for a review). It seems likely that this strategy will be used for children above 6 years old who can be studied with neuroimaging. Below this age EEG may be used alone, or other methods (discussed below) may prove useful.

More methods for imaging human brain activity, for example from external magnetic fields (magnetoencephalography), from light on the surface or through the skull (optical imaging) and from heat sensors, are also becoming available (Toga & Mazziotta, 1996). We can expect more methods to appear in the future. There is already in place a rich set of tools with which to validate and test new methods.

Activation

A frequent design for brain imaging studies involves a comparison of an experimental and a control condition. In most studies one subtracts the data obtained in the control condition from the experimental condition, with the goal of determining what operations are involved (see Posner & Raichle, 1994, for a summary). Consider an effort to discover what parts of the brain are involved in sensing motion. If the experimental condition is a set of moving dots and the control condition comprises the same dots without motion, a subtraction may be thought to involve motion-sensitive areas. This design has been quite successful in localizing areas of the visual system that seem related to motion, colour and form. A quite different design is to use the same stimuli but to require the subject to attend to different aspects of the stimuli. Attention to motion has been shown to activate pretty much the same area as found in the above subtraction.

Things become more complicated when higher-level cognitive processes are involved. It is certainly not possible to say that most experimental and control tasks differ in only one way. For example, consider a task in which one wishes to understand how a word association is generated. As an example, consider an experimental condition that involves generating the use of a word (e.g. pound for the word hammer) and a control condition that involves merely reading the word aloud. When reading is subtracted from generation, it might rid you of the sensory and motor components, but leave intact a number of operations related to differences in attention and effort required by the task and also aspects of semantic processing necessary to generate the word use. This issue has sometimes been called the problem of pure insertion, and it is generally agreed that it is not possible to guarantee successful control of all but one operation. However, a hierarchical experimental design may provide some internal tests of pure insertion. For example, if a passive condition of looking at a fixation point, presentation of visual words, reading aloud and generating word uses are all used as tasks it is possible to ask if subtracting fixation from generation produces the same set of activation as each individual subtraction.

The subtractive method is not the only strategy for analysing functional imaging data (Toga & Mazziotta, 1996). Conjunction designs often attempt to target the same brain area with more than one task which would capture the same mental operation (Friston, 1998). For example, it is possible to target visual motion areas either by comparing a moving and a stationary stimulus or by comparing an instruction to attend to motion with one in which the subject attends to another dimension. These strategies may be called bottom-up or top-down. They have both been shown to activate similar brain areas (Posner & Raichle, 1994).

Sometimes it has been proven useful to use several levels of the same variable. For example, studies of working memory have manipulated the number of items held in store. As they increase there are systematic increases in the strength of activation of brain areas related to storage load (Smith & Jonides, 1997). Since there is often interaction between variables on their effects on particular brain areas, factorial designs are sometimes used. For example, two levels of each variable are examined, such as when fast or slow taping speed is manipulated at the same time as high or low doses of a dopamine agonist. If tapping varies activation of cortical motor areas and parts of the basal ganglia, those parts of the network involving dopamine will probably also be modified by the drug dose.

Resolution

Methods of imaging based on blood flow have some difficulties because there are potential spatial and temporal limitations in a measure that is not directly connected to neuronal activity. The early cognitive studies with PET involved a blur circle surrounding an area of activation of about 18 millimetres. However, even with these devices it was possible to obtain differences between two locations on the retina when 3 mm or less separated them. The development of new scanners and the use of MRI have improved the ability to separate the difference between areas of activation in different tasks to the range of about 1 mm. The earlier cognitive data from PET involved using a standardized brain space and relating activity in each individual brain to a common space. Coregistration of functional and structural MRI scan has provided much precise localization of activity with respect to the varying individual anatomy of human brains.

Many of the PET studies of cognition involve averaging data over a sustained period of 40 s and most fMRI often uses at least 10-s windows. Recently event-related blood flow responses have been obtained by averaging trials that are distributed randomly within a block of trails in a way analogous to time-locked electrical averages of electrical data. This method has been extended to activity that is separated by a second or more within a trial (Rosen *et al.*, 1998). However, blood flow lags behind neuronal activity by a second or more, so that it is not possible to characterize on-going brain activity in real time (e.g. in the millisecond range) by these methods. An approach to this problem has been to relate generator locations developed from PET and fMRI studies to electrical activity through algorithms that relate scalp-recorded voltages in the electroen-cephalogram to brain regions. Studies using these methods have suggested, for example, that attention modulates extrastriate visual responses starting about 100 ms after input (Hillyard & Anllo-Vento, 1998). Because the same brain tissue (e.g. occipital extrastriate visual areas) can be activated directly by visual input or modulated by attention and feedback from higher centres, it is often as important to know when as to know where things are happening.

Pediatric imaging

PET, CT, and SPECT, require exposure to some amount of radiation. Ethical considerations usually prevent this kind of exposure for children unless it is medically justified. However, structural MRI and the development of fMRI techniques have opened the field of MRI to developmental research (see Thomas & Casey, 1999, for a review).

There has been some effort to study the development of brain areas by use of structural MRIs. These studies have provided some normative data on changes of brain size with development (Giedd *et al.*, 1996) and some information on abnormalities in pathological conditions such as attention deficit disorder (Castellanos *et al.*, 1996). Recent use of tensor mapping with MRI may provide better information on the dynamics of growth at particular ages (Thompson *et al.*, 2000). Of importance to this chapter was the finding that in children aged 5.3 to 16 years there is a significant correlation between the volume of the area of the right anterior cingulate and the ability to perform tasks relying upon focal attentional control (Casey *et al.*, 1997a).

The currently published fMRI literature contains very few paediatric studies. This situation is probably about to change in the near future since the studies already conducted seem to show that this is a feasible and preferred method for developmental neuroimaging. Thomas & Casey (1999) reviewed this literature, and summarized the special issues and considerations regarding the application of the fMRI technique for the study of normal infants. Figure 1 illustrates the paediatric fMRI technique, and is kindly contributed by B. J. Casey.

Sedation

Some of the fMRI studies that have been conducted with infants, especially those done with very young infants, sedated their subjects in order to prevent motion artifacts. The use of sedation is problematic regarding interpretation of

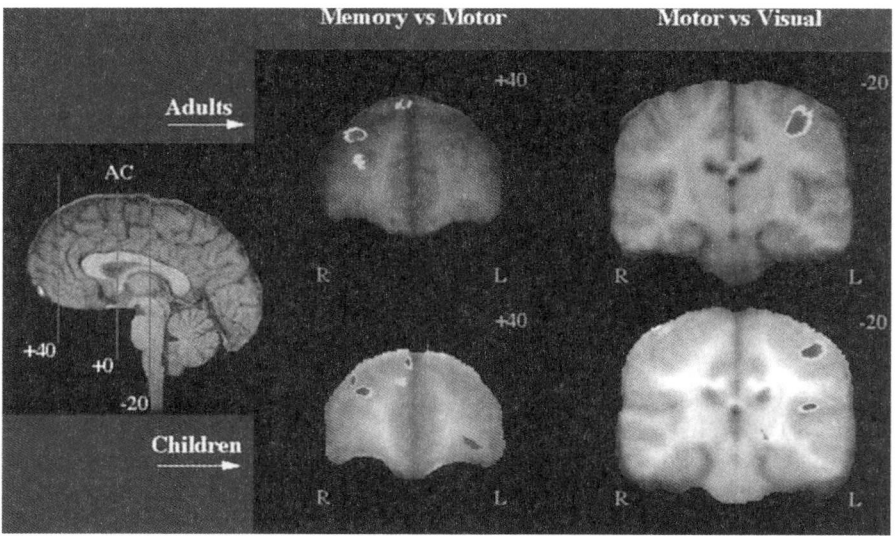

Figure 1 The mid-sagittal image of the left identifies the location of the pictured slices relative to the anterior commissure. Slices indicate activation for adults (top row) and children (bottom row) during the memory and motor tasks of the n-back spatial working memory paradigm respectively. These results are based on a pooled analysis across each age group.

the results, especially since in most of the studies the results of the sedated infants are compared to the ones of non-sedated adults. For example, Joeri *et al.* (1997a) scanned seven sedated children ranging from 4 days after birth to 8 years, and 10 non-sedated adults, during passive visual stimulation. There was no activity in the visual cortex in the newborns, while the older infants showed deactivation of this area, and the 8-year-olds and the adults showed activation. However, these results may reflect the effect of the sedation rather than differences in brain development and maturation. Joeri *et al.* (1997b) showed that the sedated adults showed cortical deactivation of the same areas that were active in the non-sedated condition.

On the other hand, with very young infants, the problems arising when no sedation is used are clearly illustrated by the study of Born *et al.* (1997). In this study, infants younger than 6 weeks old were not sedated. However, 11 of the 13 newborns tested had to be excluded due to motion artifacts.

In the next section we will describe some techniques, other than sedation, that are useful in preventing motion artifacts, and are applicable at least with the older children.

Motion artifacts

The fMRI data are very sensitive to motion artifacts. Even slight movements of the head, hands, etc. can significantly affect the measurement. When dealing with children this problem is especially acute since, for most of them, lying still for long periods of time is extremely difficult. Techniques that were found to be useful include some kind of physical restraint using some combination of a flexible chin-strap, forehead strap, and some kind of padded head clamp. Other methods that are used in adult imaging, and are more successful at preventing movements, e.g. a bite bar and a form-fitting foam, are mostly unappealing to children.

Some laboratories are beginning to use systems of operant conditioning; for example, maintaining a visual display only as long as the child maintains the head still. This kind of device also serves to provide immediate feedback to the subject regarding body position and motion. In addition to the physical restraint techniques, Casey *et al.*, 2000 mention that it is usually necessary to include in the running protocol periodical reminders and instructions to the young subjects. Since motion is never prevented completely, some off-line correction algorithms have been proven to be quite successful (Woods *et al.*, 1992).

Anxiety

Children might experience anxiety and distress upon being in an unknown hospital-like environment, and especially seeing and entering the scanner magnet. Decorating the scanner room, allowing the child to bring blankets or stuffed toys into the magnet, allowing the parent to stay in the room, etc., are simple steps that were found to be extremely helpful in reducing some of these negative reactions. Casey *et al.*, 2000 report using a simulator before the actual scanning, and even using a pre-simulator play tunnel before the simulator experience. In addition to reducing children's anxiety, this procedure provided the opportunity to give children feedback about their head and body motion before they enter the scanner, and provided a realistic measure of how the child might perform during the actual scan.

Paradigms

One of the most critical challenges in the area of paediatric imaging is the development of appropriate behavioural paradigms. Simple adaptation of the paradigms used with adults is not always straightforward. For example, Casey *et al.*, 2000 point out that, when required to press more than one button, children tend to rely on visual input. Adults, in contrast, can more easily rely on the somatosensory feedback to position their fingers on the correct button. In the scanner the child is lying down and typically cannot see his/her fingers, which might be problematic for his/her performance. Most of the paradigms that are in use today rely on passive sensory stimulation and/or very simple manual motor responses. Future development in the field might provide solutions to some of these limitations, by providing the ability to monitor and measure other forms of behavioural responses, such as voice responses and eye movements.

Interpretation

Almost all the imaging studies face some interpretation difficulties due to the usually small sample sizes and large individual variability in brain sizes, and in the localization of anatomical landmarks. In paediatric imaging, differences in brain sizes further complicate the comparison between age groups. Casey *et al.*, 2000 emphasize that morphometric differences must be taken into account. Otherwise, what might seem like smaller regions of activation in one group, for example, might reflect only smaller structures rather than a different use of these brain areas.

Moreover, when comparing between ages, it is extremely important to equate task difficulty across groups; otherwise it is impossible to know if the difference in activity reflects some substantial difference in brain processes, or simply differential effort due to task demands. Some researchers attempted to solve this problem, within studies of spatial working memory, by equating accuracy between the subjects (Orendi *et al.*, 1997; Truwit *et al.*, 1996; Casey *et al.*, 1997b).

Marker tasks

A methodology that has been useful for studying the course of development during early childhood is to use marker tasks. Marker tasks are simple behavioural tasks that have been shown in adult lesion or neuroimaging studies to involve particular brain areas. By studying infants' ability to perform different marker tasks with increasing age, it is possible to make some inferences about the developmental course of these systems and networks.

One clear example is Adele Diamond's work using the "A not B" task and the reaching task. These two marker tasks involve inhibition of action that is strongly elicited by the situation. The "A not B" task involves shifting the location of a hidden object from location A to location B, after retrieving from location A has been previously reinforced as being the correct location, in the previous trials (Diamond, 1988). In the reaching task, visual information about the correct route to a toy is put in conflict with the cues that normally guide reaching. A toy is placed under a transparent box. The opening of the box is on the side (it can be the front side, the back side, etc.), and the subject can reach it only if the tendency to reach directly through the transparent top of the box is inhibited. Important changes in the performance of these tasks are observed from 6 to 12 months. Comparison between monkeys with brain lesions and human infants on the same marker tasks suggests that they are sensitive to the development of the prefrontal cortex. Maturation of this brain area seems to be critical for the development of this form of inhibition (Diamond & Goldman-Rakic, 1989).

We (Posner & Rothbart, 1994; Posner *et al.*, 1997) have used this approach to study the development of the orienting and other attentional networks during infancy, and the approach has also been used to study the development of these same functions in a broader range of ages (Berger *et al.*, in Clohessy *et al.*, 2000; Enns, 1990).

In the next section we review what we have learned regarding the development of attentional functions in infancy. We start with natural behaviour and then turn to data from studies using marker tasks. Only recently, direct evidence from paediatric imaging studies has become available. We believe that this ratio will change in the near future, as new non-invasive paediatric imaging techniques develop.

INFANT ORIENTING BEHAVIOUR

The infant comes into the world with only a small part of the sensory processing systems developed. During the first year of life the visual system undergoes a remarkable development that supports increases in visual acuity (Gwiazda et al., 1989), in the ability to voluntarily control foveation and attention (Johnson et al., 1991) and in recognition of faces and other visual objects (Johnson, 1997).

Visual experience is critically important for the normal maturation of the visual system. For example, if one eye of a cat is closed its representation in the cortex is greatly reduced. Human infants who undergo unilateral removal of cataracts show clear effects of the time of this deprivation period on the development of the extent of the visual field to which they orient (Browning et al., 1996). Thus, the genetic plan for visual system development includes a critical role for exposure to the visual world.

Despite the immature state of the infant sensory systems, the ability to select interesting aspects of the environment and to learn where important stimuli are located begins surprisingly early. It has been possible to trace the development of these skills by careful monitoring of the infant visual system. The developmental course of infants' visual behaviour provides a clear example of the relation between the maturing brain systems and behaviour. At birth only the deeper layers of the visual cortex are well developed. During the next months the more superficial layers and their connection to other brain structures begin to mature. Johnson (1990) argues that the maturation of pathways involving the primary visual cortex controls the development of the infants' orienting. At about 1 month pathways from the basal ganglia to the superior colliculus allow the infant to control fixation. Somewhat later, maturation of the parietal lobe and pathways involving the frontal eye fields allow the infant to voluntarily break fixation and control saccades. These brain developments help explain the behavioural shifts observed during the first months of life described below.

Careful analysis of the infant's visual behaviour shows that newborns' tracking of moving stimuli mostly follows rather than anticipates real-world movements. Saccades tend to be of fixed size, and to fall short of targets. Another characteristic of newborns' visual behaviour is that they have a strong bias to orient with their eyes towards stimuli presented in their temporal visual field (Mauer & Lewis, 1997). At this age infants also tend to visually scan the external contours of stationary objects more than they do their internal features (Salapatek, 1975). All these characteristics reflect the more important role of the retino-tectal pathway in mediating visual behaviour and the relative immaturity of the cortical visual system at this stage. According to Johnson (1990), at this point the functional cortical activity related to the striate cortex is confined mainly to layers 5 and 6, and only serves the purpose of regulatory feedback to the still-maturing lateral geniculate.

Obligatory looking in the 1-month-old

A particularly impressive characteristic of 1-month-old infants is that they tend to maintain fixation at a stimulus for very long times. This so-called "obligatory looking" can also serve to distress the child. Their inability to disengage their covert and overt attention can last from seconds to even minutes. Stechler & Latz (1966), for example, allowed unlimited exposure to drawings of faces and bull's-eyes and observed one infant looking at the stimulus for as long as 35 min. Caregivers can interpret this behaviour as a sign of attachment and love (Ruff & Rothbart, 1996), but it can also lead to distress if a powerful attractor such as a chessboard is used. Johnson (1990) suggested a plausible explanation for the obligatory looking phenomenon based on the time-course of neural development. He suggested that inhibitory projections from the basal ganglia to the superior colliculus mature at this age, inhibiting the orienting towards peripheral stimuli.

The transition at 2–3 months

Ruff & Rothbart (1996) summarized the aspects of looking that begin to change around 2–3 months of age. While, during the first months, infants selected stimuli mainly based on their physical properties and salience, they now become much more likely to select particularly stimuli such as faces or bulls'-eyes, and concentrate their looking at perceptual "figures", regardless of their relative size and brightness (Ruff & Turkewitz, 1975, 1979; Salapatek, 1975). Their scanning of static objects is more likely to include internal areas in addition to the external contours (Salapatek, 1975) and there is an increasing ability to detect stimuli appearing in the nasal visual field. Moreover, pursuit tracking becomes smooth. Conel's analysis of the postnatal growth of the human cortex (Conel, 1939–1967) (although for some of the periods it is rather speculative), suggests that, by the age of 3 months, layers 3, 4, 5 and 6 are already functionally active. When this occurs, pathways that connect the visual cortex to other areas of the brain also become active. Accordingly, the behavioural changes mentioned above are correlated to this development (Johnson, 1990). Smooth pursuit tracking movements, for example, are related to the development of the MT pathway projecting from the primary visual cortex to the middle temporal area and then to the superior colliculus, allowing some cortical modulation of eye movements. The onset of attention towards internal features of patterns and the ability to detect stimuli in the nasal visual field can be explained by the fact that the cortex now not only processes visual input but also influences motor output.

Further development after 3 months from birth

The obligatory looking that characterized the 1–2-month-old infant diminishes around 3–4 months of age. At this time the pathway from the frontal eye fields to the superior colliculus and the parietal cortex mature. This new network allows selectivity to be controlled by more cognitive and experiential factors leading to an increasing coordination between orienting and attention (Posner & Rothbart, 1980). The development of the frontal eye loop also accounts for the emerging ability to track moving targets by faster movements than the target itself, anticipating its future location. Posner & Rothbart (1990) have suggested that the onset of the ability of the infant to make anticipatory eye movements may coincide with the development of covert attention. Prior to this stage the overt "eye movement" system is incapable of marking the future location of a stimulus. Once a covert attention has developed, this can mark a location as a target

20

for overt saccades allowing an "in-advance" preparation of the saccadic motor programme. We turn now to the developments of these attention networks.

THE DEVELOPMENT OF COGNITIVE CONTROL MECHANISMS

Adult imaging studies have revealed three major networks of brain areas involved in attention (Posner & Raichle, 1994). One network involves maintenance of the alert state, another orients to sensory stimuli, and a final executive network is involved in handling conflict and novelty (Posner & Petersen, 1990; Posner & DiGirolamo, 1998). In this section we consider the development of these networks. We deal first with orienting to sensory stimuli and then with higher-level executive control.

Orienting

Luria (1973) distinguished between an early-developing, largely involuntary biological attention system and a later-developing, more voluntary social attention system. A major theme emerging from our work shows that Luria was roughly correct in making a distinction between voluntary and involuntary attention systems, but that, contrary to his original belief, both are shaped by a complex interaction between biology and socialization that together determine their regulatory properties.

Attention to visual stimuli in adults can be either overt or covert, and the key to separation of the two requires the adult to remain fixated, assessing covert attention by improvement in RT, reduction in threshold or increases in electrical activity at the cued location (Posner, 1988). It is impossible to instruct young infants either to remain fixated or to press a key. Thus eye movements are used as the final index of attention.

If eye movements are the response, how then can we know if the attention shift is covert? Several methods have been used. For example, a cue can be presented too briefly to elicit an eye movement, and then a subsequent longer target can be presented at the cued location or on the opposite side (Hood, 1994). There is an early speeding of response to eye movements to targets appearing at the cued location and a later inhibition in response speed if attention is drawn back to fixation. This is also what is found in key presses for adults (Rothbart et al., 1990). However, it is possible that the briefly presented cue programmes the eyes to move in the direction of the subsequent target even if no actual eye movement is involved.

Another effort to examine the relation of eye movement to covert attention in infants involves teaching the infants to perform a counter saccade (Johnson et al., 1994). Infants learn that a target follows a cue presented briefly in one location on the opposite side of the visual field. The infants learn to move their eyes in the direction opposite to the cue. Then on a test trial a target is presented on the same side as the cue. Our idea was that infants first switched attention to the cue and then to the target. If a target was presented shortly after the cue, attention would still be on the cued side and thus a fast eye movement would result. Adults who are given a cue indicating that targets will probably occur on the opposite side (Posner *et al.*, 1982) are faster at the cued location within 150 ms after the cue and faster on the more probable side after 150 ms. Thus infants who have learned to make a counter saccade seem able to prepare to move the eyes in one direction while being fast to a target in the opposite direction.

Attention of this sort appears to undergo a very rapid development between 2 and 4 months. First, we (Johnson *et al.*, 1991) found that the probability of disengaging from a central attractor to process a peripheral target increased very dramatically between 2 and 4 months. This finding might reflect the improvement in acuity that is also found during this period (Gwiazda *et al.*, 1989). However, it turned out that infants of 4 months were also able to learn to use a central cue. When we associated one attractor with a movement to the right and another with a movement to the left, they learned to anticipate the direction of the cue at 4 months but not at 2 months. The common development of peripheral and central cue use may reflect the common mechanisms involved. Both lesion (Posner *et al.*, 1984) and PET (Corbetta *et al.*, 1993) suggest that parietal mechanisms are involved in both central and peripheral cue use. We now suggest that both mechanisms develop between 2 and 4 months, just at the time PET studies suggest a strong increase in metabolic processes in the parietal lobe. If the mechanisms of covert attention to central and peripheral cues involves the same brain structures, it seems less likely the eye movement control of the two forms of cues are really so radically different.

Inhibition of return

One principle of development is that subcortical visual mechanisms tend to develop prior to cortical mechanisms. Inhibition of return is the preference shown for a novel position over the one to which the person had recently oriented (Posner & Cohen, 1984). Since we believed that inhibition of return (IOR) depended upon a collicular mechanism (Posner *et al.*, 1985; Rafal *et al.*, 1989), we expected it to be present at the youngest ages we studied. To examine IOR in infants, we first brought the eyes to a peripheral location and then summoned them away to a central fixation location, after which we gave the infant a choice of two locations, the one to which he/she had recently oriented and another location equally distant from fixation. We were surprised to find that IOR was not present at 3 months, but developed between 4 and 6 months at a time generally overlapping with the cortical covert attention mechanisms we have been describing (Clohessy *et al.*, 1991).

Since our studies, IOR has been demonstrated for newborn infants, presumably reflecting the midbrain mechanisms that are dominant at birth (Valenza *et al.*, 1994). This finding suggests that the basic computations for IOR are present at birth. Their implementation may depend upon whether a stimulus evokes a programmed eye movement or not. We are not sure of the factors that produce so many hypometric eye movements in infants of 3 months. However, these may relate to maturation of the cortical systems (e.g. parietal lobe, frontal eye fields) that appears to occur about this time. These data generally support the idea that the subcortical computations involved in IOR are present earlier than the parietally based ability to disengage from a visual stimulus.

Soothing

In the first 3 months of life most caregivers report trying to soothe their infants by use of rocking and holding. After 3 months caregivers in Western societies begin to use visual orienting to novel stimuli as a method to quiet their infants' crying. Experimental results (Harman *et al.*, 1994) show that 3- and 6-month-old children show a reduction of distress when engaged in orienting towards a novel visual or auditory stimulus. However, once orienting ends, the distress returns at about the same strength as before, even though no new aversive stimulus is presented. These results indicate the importance of orienting as means of early emotional regulation. We believe that caregivers use these methods to train infants to begin to regulate their own distress.

Learning

There is clear evidence that where to orient can be learned at this age (Haith, 1980). We (Clohessy, Posner & Rothbart, 2001) taught infants at 4 months a simple sequence of three visual locations (see Figure 2). The infants learned that, having oriented to stimulus 1, the next stimulus would be 2, etc. They demonstrated learning by showing better-than-chance selection of the next event in anticipation of its being presented. For this simple sequence the results obtained with 4-month-olds were not very different from those older children or adults who learned the sequence when unaware that there was a sequence. Although adults were somewhat better than 4-month-olds, this could be due to partial awareness.

We believe that the infant learning is similar to implicit learning in adults. Studies using PET (Grafton *et al.*, 1995) during sequence learning by adults suggest that basal ganglia and parietal sites are associated with implicit learning of sequences, while frontal and posterior cortical areas are active under conditions when subjects have explicit knowledge of the sequence.

Executive control

Ambiguous sequences

We found that infants as young as 4 months could learn the unambiguous associations, but not until 18 months did they show the ability to learn ambiguous or context-sensitive associations (e.g. 1213). Individual children showed wide differences in their learning abilities, and in our study the ability to learn context-sensitive cues was correlated with the caregivers' report of the child's vocabulary development (Posner *et al.*, 1997).

Conflict

Many behaviours related to higher attention seem to begin maturing around 18 months of age. Around this age important cognitive skills seem to undergo impressive development, including the appearance of multiword sentences (Dore *et al.*, 1976), spontaneous alternation (Vecera *et al.*, 1991), and self-recognition in children's operating on their own images in a mirror (Gallup, 1979). Posner &

23

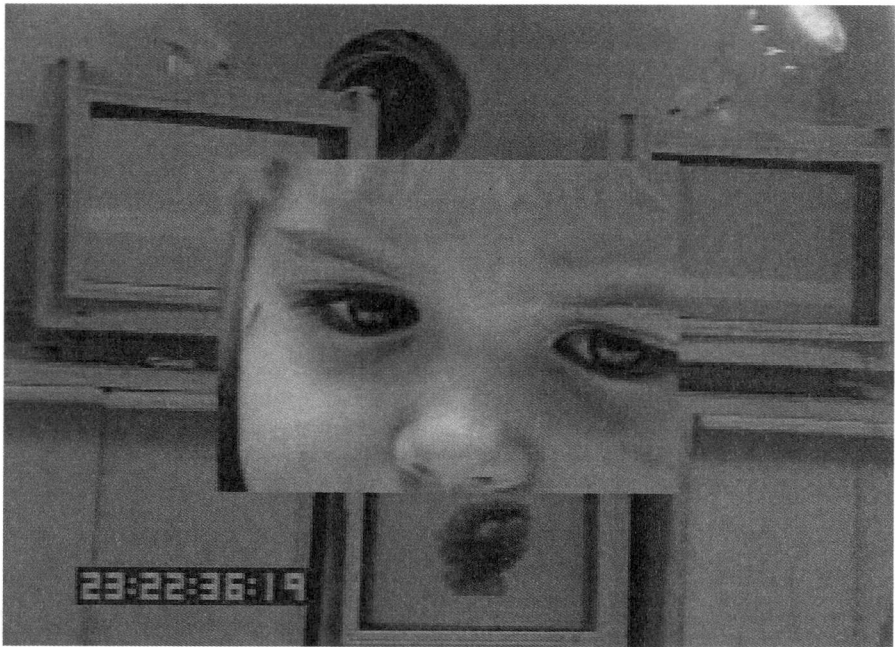

Figure 2 A 24-month-old participating in the visual sequence task. Targets appear one by one, following a constant sequence, in one of the three monitors. Anticipatory looking percentage is measured through a frame-by-frame analysis of the session videotapes.

Rothbart (1994) suggested that these skills rest on the maturation of an executive attention system related to awareness and voluntary control. PET scanning of adult subjects suggests that these skills involve a brain network that includes midline frontal areas, e.g. anterior cingulate gyrus, and areas of the supplementary motor cortex (Posner & DiGirolamo, 1998).

According to the adult cognitive and neuroimaing data (see Bush *et al.*, 1998; Posner & DiGirolamo, 1998) a measure of the development of executive attention might be reflected in the ability to resolve conflict between simultaneous stimulus events as in the Stroop effect. Since children of this age do not read, it is not possible to use the standard Stroop. So instead one can examine the basic visual dimensions of location and identity as an appropriate way to study resolution of conflict in young subjects.

The variant of the Stroop effect we designed to be appropriate for ages 2–3 years involved presenting a simple visual object on one side of a screen in front of the child and requiring the child to respond with a button that matched the stimulus shown (Gerardi-Caulton, 2000). The child had been trained to "pat" the button that matched the stimulus shown. The appropriate button could be either on the side of the stimulus (congruent trial) or on the side opposite the stimulus (incongruent trial). The prepotent response was to press the button on the side of the target irrespective of its identity. However, the task required the child to inhibit that prepotent response and to act instead based on identity. The ability to resolve this conflict is measured by the accuracy and speed of the key press responses.

Results of the study strongly suggest that executive attention undergoes dramatic change during the third year of life. Performance by toddlers at the very beginning of this period was dominated by a tendency to repeat the previous response. Perseveration is associated with frontal dysfunction, and this finding is consistent with the idea that executive attention is still very immature at 24 months.

It was also possible to examine the relationship of our laboratory measures of conflict resolution with a battery of tasks requiring the young child to exercise inhibitory control over his/her behaviour. We found substantial correlations between these two measures. Even more impressive, elements of the laboratory task were significantly correlated with aspects of effortful control and negative affect in parental reports of infant behaviour in their normal environment. Cingulate activity relates to control of distress, and the cognitive measure of conflict resolution also relates to aspects of infant self-control in daily life as reported by the parents.

In the Stroop effect, conflict is introduced between two elements of a single stimulus. We reasoned that an even more difficult conflict might be introduced by the task of executing instructions from one source while inhibiting those from another (Posner & Rothbart, 1998). This conflict task is the basis of the Simon Says game. Previous studies had suggested that the ability to perform this task emerged at about 4 years of age (Reed et al., 1984). In a recent study we asked children of 40–48 months to execute a response when they were given the command by a bear toy, but to inhibit it when it was given by a toy elephant. Children up to 42 months were unable to carry out the instruction at better than a chance level. However, just 2 months later they were virtually perfect. The older children tend to use physical control to inhibit them from executing the commands given by the elephant. It is quite amazing to observe the lengths they go to control their own behaviour.

Cingulate development

There is also some direct evidence that the anterior cingulate is developing during this part of childhood in a way that might support the data on conflict. In children aged 5.3–16 years there is a significant correlation between the volume of the area of the right anterior cingulate and the ability to perform tasks relying upon focal attentional control (Casey et al., 1997a). Moreover, in a recent fMRI study, performance of children aged 7–12 years and adults was studied when performing a go–no-go task in which they were required to withhold pressing to an X while responding to non-Xs. This condition was compared to control tasks where subjects responded to all stimuli and thus never had to withhold a response. Both children and adults showed strong activity in the prefrontal cortex when required to withhold responses. Moreover, the number of false alarms made in the task was significantly correlated with the extent of cingulate activity.

Emotional regulation

The anterior cingulate is the outflow of the limbic system. One may wonder why it should be so closely related to control in cognitive tasks such as the Stroop effect. Elsewhere (Posner & Rothbart, 1998) we have argued that the cingulate is first involved in the regulation of emotion as discussed during the first year of life. Caregivers help to train their infants to regulate their own emotionality. During this period we suggested that the anterior cingulate comes to provide inhibitory control over limbic structures such as those involved in distress. When the toddler begins to be involved in more cognitive functions, this same brain area begins to exert attentional control over the selection of input and output dimensions. In the next section we review some of the methods that might be used to examine this hypothesis.

NEW DIRECTIONS

In this chapter we have tried to outline some of the attentional mechanisms that develop in early childhood. Because paediatric imaging of the human brain is new, we have had to use primarily marker tasks to examine the brain systems that underlie the cognitive life of infants and young children. Even so it has been possible to examine the role of orienting in infancy and to trace its influence on infant visual search, soothing and learning. The maturation of brain systems underlying control of orienting during the first year allows us to observe changes in the infant's visual behaviour. We argue that similar studies in the second year provide information on the maturation of higher-level attentional systems. The systematic application of imaging methods to children should allow more detailed tests and further extension of these ideas.

A useful strategy in paediatric imaging may follow the adult model of relating brain areas obtained from fMRI to high-density electrical recording from scalp electrodes (Abdullaev & Posner, 1998; Hillyard & Anllo-Vento, 1998). So far interpretation of scalp electrical recording in infants has had to rely on comparison of infant data with areas known to be important in adults (Dehaene-Lambertz & Dehaene, 1994). This strategy has the problem that lesion evidence has often suggested that areas of the infant brain that carry out computations may be quite different from the areas used in the adult (Bates *et al.*, 1997).

As an example of this, one study of infant eye movement has argued that infants show the same frontal activity found active with saccades in adults but do not show the parietal activity. Soon fMRI studies with children may help in the interpretation of generators related to this scalp distribution (Csibra *et al.*, 1998).

In addition, new methods will become available for non-invasive examination of human brain development. One possibility is the use of near-infrared optical imaging (Villringer & Chance, 1997). The relatively small amount of bone in infants gives an opportunity for obtaining good optical signals of changes in blood vessels at least from areas of the brain near the surface. Perhaps of even greater significance will be the use of tensor defraction images obtained from MRI (Neil *et al.*, 1998) to indicate the myelination of tracts in the brain of infants. With such a method it might be possible to know when, for example, frontal areas become functionally active and when they develop connections to posterior areas. This information could be of great value in understanding the times at which one might expect particular operations to be possible, and could also be used to validate correlational measures between electrode sites already being used to attempt to understand the transfer of information from one brain area to another. If these coherence measures could be shown to indicate information transfer there would be a strong non-invasive way to assay the connectivity in the human brain. Intriguing results that suggest the importance of the maturation of long-range brain connection (i.e. Srinivasan, 1999; Thacher et al., 1987) await efforts to confirm the correctness of the measures with anatomical methods that indicate myelination.

With methods in place to indicate connectivity between brain areas, it should become possible to mount an attack on the way in which high-level skills become organized in the human brain. For example, in language is the left hemisphere preference for adult language there from the start, or does language begin with a bilateral organization or even a right hemisphere bias, and transfer to the left hemisphere as the skill develops? Must the transfer of information between anterior and posterior language areas await the development of long fibre tracks? If so, when do the appropriate mechanisms that can support transfer of information from posterior word form areas to frontal areas become available?

These methods should make it possible to separate events due to maturation from specific learning experience during brain development. Suppose, as Thacher et al. have suggested from electrical coherence measures (1987), a long pathway comes to connect frontal with posterior areas at about 5 years of age. We should be able to examine the influence of experience in naming visual pictures, letters or words on the rate of development of this pathway. By comparing groups with and without such experience, we could try to separate maturational influences from those due to learning. It may then be possible to compare the influences of different curricula and time of learning in the hope of developing optimal learning experience. Similarly, it should be possible to evaluate the differences between normal and ADHD children in the development and maintenance of connections between remote brain areas. It may then be possible to determine if ADHD is related to a failure of control from frontal to posterior areas and the influence of drugs or other therapies on these connections.

Although neuroimaging has made it possible to examine aspects of the living brain never before seen, there is still much that is missing in our current knowledge. We do not yet have the ability to see how specific learning experience and maturation work in concert to organize the ability of humans to learn a wide variety of skills at the basis of our society. The methods available now, and those that should be added in the near future, can allow a systematic empirical attack on these issues.

References

Abdullaev, Y.G. & Posner, M.I. (1998). Event-related brain potential imaging of semantic encoding during processing single words. *Neuroimage*, **7**, 1–13.

Bates, E., Thal, D., Trauner, D., Fenson, J., Aram, D., Eisele, J. & Nass, R. (1997). From first words to grammar in children with focal brain injury. *Developmental Neuropsychology*, **13**(3), 275–343.

27

Berger, A., Jones, L., Rothbart, M.K. & Posner, M.I. (2000) Computerized games to study the development of attention in childhood. *Behavior research methods, instruments, and computers.* **32**(2), 297–303.

Born, P., Rostrup, E., Larsson, H.B.W., Leth, H., Miranda, M., Peitersen, B. & Lou, H.C. (1997). Infant visual cortex function evaluated by fMRI. *Neuroimage*, **5**(4), S171.

Browning, E.R., Maurer, D., Lewis, T.L., Brent, H.P. & Reidel, P. (1996). The visual field in childhood: normal development and the influence of deprivation. *Developmental Cognitive Neuroscience Technical Report*, **96.I**. London: MRC Cognitive Development Unit.

Bush, G., Whalen, P.J., Rose, B.R., Jenike, M.A., McInerney, S.C. & Rauch, S.L. (1998). The counting Stroop: an interference task specialized for functional neuroimaging – validation study with functional MRI. *Human Brain Mapping*, **6**, 270–282.

Casey, B.J., Thomas, K.M., Welsh, T.F., Livnat, R. & Eccard, C.H. (2000). Cognitive and behavioral probes of developmental landmarks for use in functional neuroimaging. In J.M. Rumsey & M. Ernst (Eds.), *The foundation and future of functional neuroimaging in child psychiatry.* (pp. 155–168). Cambridge, UK: Cambridge University Press.

Casey, B.J., Trainor, R., Giedd, J., Vauss, Y., Vaituzis, C.K., Hamburger, S., Kozuch, P. & Rapoport, J.L. (1997a). The role of the anterior cingulate in automatic and controlled processes: a developmental neuroanatomical study. *Developmental Psychobiology*, **3**, 61–69.

Casey, B.J., Trainor, R.J., Orendi, J.L., Schubert, A.B., Nystrom. L.E., Giedd, J.N., Castellanos, F.X., Haxby, J.V., Noll, D.C., Cohen, J.D., Forman, S.D., Dahl, R.E., & Rapoport, J.L. (1997b). A developmental functional MRI study of prefrontal activation during performance of a go-no-go task. *Journal of Cognitive Neuroscience.* **9**, 835–847.

Castellanos, F.X., Giedd, J.N., Hamburger, S.D., Marsh, W.L. & Rapoport, J.L. (1996). Brain morphometry in Tourette's syndrome: the influence of comorbid attention deficit/hyperactivity disorder. *Neurology*, **47**(6), 1581–1583.

Clohessy, A.B., Posner, M.I., & Rothbart, M.K. (2001). Development of the functional visual field. *Acta Psychologica,* **106**, 51–68.

Clohessy, A.B., Posner, M.I., Rothbart, M.K. & Vecera, S.P. (1991). The development of inhibition of return in early infancy. Journal of Cognitive Neuroscience, **3**(4), 345–350.

Conel, J.L. (1939–1967). *The postnatal development of the human cerebral cortex* (Vols. I–VIII). Cambridge, MA: Harvard University Press.

Corbetta, M., Miezin, F.M., Shulman, G.L. & Petersen, S.E. (1993). A PET study of visuospatial attention. *Journal of Neuroscience*, **13**(3), 1202–1226.

Csibra, G., Tucker, L.A. & Johnson, M.H. (1998). Neural correlates of saccade planning in infants: A high-density ERP study. *International Journal of Psychophysiology*, **29**, 201–215.

Dehaene-Lambertz, G. & Dehaene, S. (1994). Speed and cerebral correlates of syllable discrimination in infants. *Nature*, **370**, 292–295.

Diamond, A. (1988). Abilities and neural mechanisms underlying AB performance. *Child Development*, **59**(2), 523–527.

Diamond, A. & Goldman-Rakic, P. (1989). Comparison of huma infants and rhesus monkeys on Piaget's Anot B Tasks: evidence for dependence on dorsolateral prefrontal cortex. *Experimental Brain Research*, **74**, 24–40.

Dore, J., Franklin, M.B., Miller, R.T. & Ramer, A.L.H. (1976). Transitional phenomena in early language acquisition. *Journal of Child Language*, **3**, 13–27.

Enns, J.T. (1990) Relations between components of visual attention. In J.T. Enns, (Ed.), *The development of attention: Research and theory* (pp. 139–158). Amsterdam: North-Holland, Elsevier.

Friston, K.J. (1998). Imaging neuroscience: principles or maps? *Proceedings of the National Academy of Sciences, USA*, **95**, 796–802.

Gallup, G.G. (1979). Self awareness in primates. *American Scientist*, **67**, 417–422.

Gerardi-Caulton, G. (2000). Sensitivity to spatial conflict and the development of self-regulation in children 24–36 months of age. *Developmental Science*, **3**, 397–404.

Giedd, J.N., Snell, J.W., Lange, N., Rajapakse, J.C., Casey, B.J., Kozuch, P.L., Vaituzis, A.C., Vauss, Y.C., Hamburger, S.D., Kaysen, D. & Rapoport, J.L. (1996). Quantitative magnetic resonance imaging of human brain development: ages 4–18. *Cerebral Cortex*, **6**(4), 551–560.

Grafton, S.T., Hazeltine, E. & Ivry, R. (1995). Functional mapping of sequence learning in normal humans. *Journal of Cognitive Neuroscience*, **7**, 497–510.

Gwiazda, J., Bauer, J. & Held, R. (1989). From visual acuity to hyperacuity: ten year update. *Canadian Journal of Psychology*, **43**, 109–120.

Haith, M.M. (1980). *Rules babies look by*. Hillsdale, NJ: LEA.

Harman, C., Rothbart, M.K. & Posner, M.I. (1994). Distress and attention interactions in early infancy. *Motivation and Emotion*, **21**, 27–43.

Hillyard, S.A. & Anllo-Vento, L. (1998). Event-related potentials in the study of visual selection attention. *Proceedings of the National Academy of Sciences*, **9**, 781–787.

Hood, B.M. (1994). Visual selection attention in infants: a neuroscientific approach. In L. Lipsitt & C. Rovee-Collier (Eds.), *Advances in infancy research*. Hillsdale, NJ: Erlbaum.

Joeri, P., Loenneker, T., Huisman, D., Ekatodramis D., Rumpel, H. & Martin, E. (1997a). fMRI of the visual cortex in infants and children. *Neuroimage*, **3**(3), S279.

Joeri, P., Huisman, D., Loenneker, T., Ekatodramis D., Rumpel, H. & Martin, E. (1997b). Reproducibility of fMRI and effects of pentobarbital sedation on cortical activation during visual stimulation. *Neuroimage*, **3**(3), S280.

Johnson, M.H. (1990). Cortical maturation and the development of visual attention in early infancy. *Journal of Cognitive Neuroscience*, **2**, 81–95.

Johnson, M.H. (1997). *Developmental cognitive neuroscience: an introduction*. Oxford, UK: Blackwell.

Johnson, M.H., Posner, M.I. & Rothbart, M.K. (1991). Components of visual orienting in early infancy: contingency learning, anticipatory looking and disengaging. *Journal of Cognitive Neuroscience*, **3**(4), 335–344.

Johnson, M.H., Posner, M.I. & Rothbart, M.K. (1994). Facilitation of saccades toward a covertly attended locatation in early infancy. *Psychological Science*, (pp. 90–93). Hillsdale, NJ: Erlbaum.

Luria, A.R. (1973). *The Working Brain*. New York: Basic Books.

Mauer, D. & Lewis, T.L. (1997). Overt orienting toward peripheral stimuli: normal development and underlying mechanisms. In J.E. Richards (Ed.), *Cognitive neuroscience of attention: a developmental perspective*. Hillsdale, NJ: Lawrence Erlbaum Associates.

Neil, J.J., Shiran, S.I., McKinstry, R.C., Schefft, G.L., Snyder, A.Z., Almi, C.R., Akbudak, E., Aronvitz, J.A., Miller, J.P., Lee, B.C.P. & Conturo, T.E. (1998). Normal brain in human newborns: apparent diffusion coefficient and diffusion anisotropy measured by using diffusion tensor MR imaging. *Radiology*, **209**(1), 57–66.

Orendi, J.L., Irwin, W. & Ward, R.T. (1997). A fMRI study of cortical activity in children and adults during a spatial working memory task. *Neuroimage*, **5**(4), S603.

Posner, M.I. (1988). Structures and functions of selective attention. In T. Boll and B. Bryant (Eds.), *Master lectures in clinical neuropsychology and brain function: research, measurement, and practice*, (pp. 171–202). Washington DC: American Psychological Association.

Posner, M.I. & Cohen, Y. (1984). Components of attention. In H. Bouma and D. Bowhuis (Eds.), *Attention and performance X* (pp. 531–556). Hillsdale, NJ: Lawrence Erlbaum Associates.

Posner, M.I. & DiGirolamo, G.J. (1998). Executive attention: conflict, target detection and cognitive control. In R. Parasuraman (Ed.), *The Attentive Brain* (pp. 401–423). Cambridge, MA: MIT Press.

Posner, M.I. & Petersen, S.E. (1990). The attention system of the human brain. *Annual Review of Neuroscience*, **13**, 25–42.

Posner, M.I. & Raichle, M.E. (1994). *Images of mind*. New York: Scientific American Books.

Posner, M.I. & Rothbart, M.K. (1980). The development of attentional mechanisms. In J.H. Flowers (Ed), *Nebraska symposium on motivation* (pp. 1–49). Lincoln, NB: Nebraska University Press.

Posner, M.I. & Rothbart, M.K. (1990). The evolution and development of the brain's attention system. *Quarterly Journal of Experimental Psychology*, **42A**, 189–190.

Posner, M.I. & Rothbart, M.K. (1994). Attentional regulation: from mechanism to culture. In P. Bertelson, P. Eelen & G. D'Ydavalle (Eds.), *International perspective on psychological science*, vol. I: *Leading themes* (pp. 41–56). Hillsdale, NJ: LEA Associates.

Posner, M.I. & Rothbart, M.K. (1998). Attention, self-regulation and consciousness. *Transactions of the Philosophical Society of London B*, 1915–1927.

Posner, M.I., Cohen, Y. & Rafal, R.D. (1982). Neural systems control of spatial orienting. *Proceedings of the Royal Society of London*, **298**, 187–199.

Posner, M.I., Rafal, R.D., Choate, L. & Vaughan, J. (1985). Inhibition of return: neural mechanisms and function. *Cognitive Neuropsychology*, **2**, 211–228.

Posner, M.I., Rothbart, M.K. & Thomas-Thrapp, L. (1997). Functions of orienting in early infancy. In P. Lang, M. Balaban & R.F. Simmons, (Eds.), *The study of attention: cognitive perspectives from psychophysiology, reflexology and neuroscience* (pp. 327–346). Hillsdale, NJ: Erlbaum.

Posner, M.I., Walker, J.A., Friedrich, F.J. & Rafal, R.D. (1984). Effects of parietal lobe injury on covert orienting of visual attention. *Journal of Neuroscience*, **4**, 1863–1871.

Rafal, R.D., Calabresi, P.A., Brennan, C.W. & Sciolto, T.K. (1989). Saccade preparation inhibits reorienting to recently attended locations. *Journal of Experimental Psychology: Human Perception & Performance*, **15**(4), 673–685.

Reed, M., Pien, D.L. & Rothbart, M.K. (1984). Inhibitory self-control in preschool children. *Merrill-Palmer Quarterly*, **30**, 131–147.

Rosen, B.R., Buckner, R.L. & Dale, A.M. (1998) Event-related functional MRI: past, present and future. *Proc. Proceedings of the National Academy of Sciences, USA*, **95**, 773–780.

Rothbart, M.K., Posner, M.I. & Boylan, A. (1990). Regulatory mechanisms in infant temperament. In J. Enns (Ed.), *The development of attention: research and theory* (pp. 47–65). Amsterdam: North-Holland.

Ruff, H.A. & Turkewitz, G. (1975). Developmental changes in the effectiveness of stimulus intensity on infant visual attention. *Developmental Psychology*, **11**, 705–10.

Ruff, H.A. & Turkewitz, G. (1979). The changing role of stimulus intensity in infants' visual attention. *Perceptual and Motor Skills*, **48**, 815–826.

Ruff, H.A. & Rothbart, M.K. (1996). *Attention in early development: themes and variations*. New York: Oxford University Press.

Salapatek, P. (1975). Pattern perception in early infancy. In L.B. Cohen & P. Salapatek (Eds.), *Infant perception: from sensation to cognition* (vol. 1, pp. 133–248). New York: Academic Press.

Smith, E.E & Jonides, J. (1997). Working memory: a view from neuroimaging. *Cognitive Psychology*. **33**(1), 5–42.

Srinivasan, R. (1999). Spatial structure of the alpha rhythm: global correlation in adults and local correlation in children. *EEG and Clinical Neurophysiology*. **110**, 1351–1362

Stechler, G. & Latz, E. (1966). Some observations on attention and arousal in the human infant. *Journal of the American Academy of Child Psychiatry*, **5**, 517–525.

Thacher, R.W., Walker, W.A. & Giudice, S. (1987). Human cerebral hemispheres develop at different rates and ages. *Science*, **236**, 1110–1113.

Thomas, K. M. & Casey, B.J. (1999). Functional magnetic resonance imaging in pediatrics. In P. Bandetinni & C. Moonen (Eds.), *Medical radiology: functional magnetic resonance imaging* (pp. 513–523). New York: SpringerVerlag.

Thompson, P.M., Giedd, J.N., Woods, R.P., MacDonald, D., Evans, A.C. & Toga, A.W. (2000). Growth patterns in the developing brain detected by using continuum mechanical tensor maps. *Nature* **404**, 190–193.

Toga, A.W. & Mazziotta, J.C. (Eds.) (1996) *Brain mapping: the methods*. New York: Academic Press.

Truwit, C.L., Le, T.H., Lin, J., Hu, X., Nelson, C.A. & Carver, L. (1996). Functional MR imaging of working memory task activation in children. *Radiology*, **201**, 1326 Suppl S.

Valenza, E., Simion, F. & Umilta, C. (1994). Inhibition of return in newborn infants. *Infant Behavior and Development*, **17**, 293–302.

Vecera, S.P., Rothbart, M.K. & Posner, M.I. (1991). Development of spontaneous alternation in infancy. *Journal of Cognitive Neuroscience*. **3**(4), 351–354.

Villringer, A. & Chance, B. (1997). Non-invasive optical spectroscopy and imaging of human brain function. *Trends in Neurosciences*, **20**(10), 435.

Woods, R.P., Cherry, S.R. & Mazziotta, J.C. (1992). Rapid automated algorithm for aligning and reslicing PET images. *Journal of Computer Assisted Tomography*, **16**, 620–633.

FURTHER READING

1. Johnson, M.H. (1997). *Developmental cognitive neuroscience: an introduction*. Oxford: Blackwell.
2. Parasuraman, R. (Ed.) (1998). *The attentive brain* . Cambridge, MA: MIT Press.

3. Posner, M.I. and Raichle, M.E. (1996). *Images of mind* (revised). New York: Scientific American Library.
4. Thatcher, R.W., Lyon G., Reid, *et al*. (Eds.). (1998). *Developmental neuroimaging: mapping the development of brain and behavior*. San Diego, CA: Academic Press.
5. Toga, A.W. & Mazziotta, J.C. (1996) (Eds.). *Brain mapping: the methods*. New York: Academic Press.

AUTHOR'S NOTE

This work has been partially supported by a Rothschild Foundation fellowship, given to the first author and NSF grant BCS 9907831 to the Sackler Institute. Reprints available from mip2003@mail.med.cornell.edu

I.3
Brain, Behaviour and Developmental Genetics

BARRY KEVERNE

ABSTRACT

In considering the impact of genes on human behaviour, it is especially important to understand the kind of events which have shaped behavioural evolution. The growth of the neocortex, a social lifestyle and the release of behavioural mechanisms from strict hormonal determinants have all played a large part in the evolution of human behaviour. This is particularly notable in the context of primary motivated behaviour. A knowledge of brain evolution is therefore integral to understanding at what kind of level genes might influence human behaviour. In this context genomic imprinting, a process that confers functional differences on parental genomes, has been particularly important. The role of imprinted genes in mammalian brain development and behaviour is first reviewed.

One of the important developments in molecular genetics has been the targeted mutagenesis of known genes in order to examine the functional consequences on the phenotype. The early gene knockout studies which were directed at the brain selected very obvious target genes. Since the consequences were often predictable, the principal revelations were in confirming the validity of the technology rather than providing new insights into how the brain functions. Some of these gene knockout studies were without obvious behavioural consequences, supporting the concept of genetic redundancy or canalization. All neurons have the capacity for learning, and the hippocampal neurons have received considerable molecular genetic attention in the context of spatial learning and memory. The hippocampus was not only the area of behavioural neuroscience first subjected to the molecular genetic approach, but it is also a discipline which has benefited substantially from the gene knockout technologies.

Perhaps the most complex area of behavioural neuroscience is that of psychiatric disorders where the behavioural phenotype may be influenced by a wide range of genetic polymorphisms present on a number

A.F. Kalverboer and A. Gramsbergen (eds.), Handbook of Brain and Behaviour in Human Development, 33–56
© *2001 Kluwer Academic Publishers. Printed in Great Britain.*

of different chromosomes. These are complex disorders, both neuro-logically and neurochemically, and we are only just beginning to scrape the surface of understanding the role of genes in these disorders. However, this is an area of neuroscience which at some future date will certainly benefit from the human genome project. The last section of this review gives a detailed consideration of psychoses, which are a compo-nent of neurological dysfunctions and psychiatric disorders, particularly schizophrenia.

INTRODUCTION

For most living vertebrates the central nervous system coordinates body move-ment and enables the organism to interact with its environment. By means of specialized sensory receptors the brain further registers events that are undergoing change in both the external environment and the body's internal environment. Behaviour, which manifests itself as the motor output of the brain, can therefore be viewed as part of physiological homeostasis, i.e. the behavioural output computed by the brain takes into account the internal bodily needs (via various hormonal messengers and sensory receptors) and integrates these with the supply and demands of the external environment. These general principles can be applied to all motivated behaviour including maternal, sexual and feeding behaviour. The contributions of behaviour to sustaining homeostasis gains additional power with increasing brain size, since the storage of experiences in the form of memory can be integrated into the decision-taking process. Decision-taking remains state-dependent in most mammals, i.e. if the motivational need is hunger then feeding decisions predominate, while similar state-dependent strategies also apply to sexual and parental behaviour.

The very largest of mammalian brains, such as those of primates and especially humans, have to some extent become emancipated from the deterministic influence of the signals originating in the internal and external environments, and have the capacity to make independent decisions that rely principally on past experiences (Keverne, 1992). These decisions can override the motivational needs, i.e. they are not state-dependent, but the process of reasoning and the ability to predict outcomes based on past experiences now ensures the correct decisions are made. Hence primate maternal behaviour can occur without pregnancy and parturition, sexual activity is not confined to discrete periods of fertile oestrus and feeding occurs in anticipation of hunger. Experience in all these contexts is of primary importance in primates. These experiences are not, of course, genetically coded but they do depend upon the brain, how it develops, and when it is constructed, both of which underlie certain behavioural predispositions. As a result of postponing much of the brain's development in humans until the postnatal period (neotany) when most of the brain's connectivities are refined, then developmental behavioural experiences themselves have an impact on organizing such connectivities. This is often referred to as the epigenetic effects on brain development.

In order to understand the kind of impact which genes have on human behaviour it is also important to have an understanding of the kind of events which have shaped behavioural evolution, and a knowledge of what we can and cannot conclude from studies on other species. For example, sexual behaviour is strongly influenced by the environment, but the nature of these environmental influences has changed throughout evolution. In fish the sexes are inter-changeable depending on the social environment (Shapiro, 1992). In reptiles the physical environment determines which sex develops, but behaviour is independent of gonadal secretions. With the development of mammalian oviparity, increased female investment and male competition, behavioural sex differences are marked. These behavioural sex differences are strongly under the influence of gonadal control as a result of early inductive effects, and gonadal hormones are essential for the expression of sexual behaviour in the adult (Keverne, 1985). Even external environmental control of reproductive behaviour by photoperiod or pheromones is via gonadal hormones. Social living and the development of a large brain, characteristic of many primate species, have resulted in the gradual emancipation of sexual behaviour from gonadal influences. Social relationships in primates become the all-important environmental deter-minants influencing sexual behaviour (Keverne, 1992).

The growth of the neocortex, a social lifestyle and the release of behavioural events from strict endocrine determinants have played a large part in the evolution of human behaviour, including parental behaviour. Hence the hormonal priming of the brain during pregnancy, which is critical for a rapid onset of maternal behaviour in most mammals, is not required for the spontaneity of primate parental care, but experimental and cognitive factors are crucial (Keverne, 1995). Thus among monkeys and apes the finding that primiparous mothers give less adequate maternal care than multiparous mothers has been observed among many species (Blaffer-Hrdy, 1976). Caged gorillas have been known to kill their first infant, and captive chimpanzee mothers are often afraid of their firstborn, refusing to touch them or allow them to suckle. In the wild, the case for incompetent care of infants rests on observations of nulliparous juveniles, since primiparous mothers normally give adequate care. The only significant difference between multiparous and primiparous monkeys is the high anxiety and possessiveness of primiparae which contrasts with the firmness in rejection that is accomplished among multiparous mothers. Hence, the impaired maternal care of primiparous captive primates is likely to be a consequence of their lack of prior experience with infants, which in feral primates is a rare event for primiparous females. In free-living social primates most females will have gained some contact with infants prior to motherhood. Maternal behaviour is a highly skilled performance and, since few infants are born to any one female, the loss of an infant through inexperience would be very costly. It is therefore significant that juvenile nulliparous females are frequently seen participating in the care of younger siblings.

Experimentally depriving monkeys of maternal and social contact for the first 8 months of their life has profound effects on their ability to be competent mothers (Holman and Goy, 1995). Their maternal care is at best indifferent, or at worst abusive, requiring intervention in order for the infants to survive. Permitting social contact with peers does procure improvements in subsequent maternal care, but even this is less satisfactory than for feral mothers. Although a socially deprived monkey may show inadequate maternal behaviour to her first offspring, improvements have been found with subsequent offspring. Once adequate care has been displayed to an infant, it is then also likely to be shown to subsequent offspring, whereas if a mother is abusive she is also likely to have received abusive behaviour and be abusive to subsequent infants (Ruppenthal *et al.*, 1976). Hence the impact of brain evolution can be seen particularly in the context of primary motivated behaviour, and a knowledge of this evolution is integral to understanding how genes might influence human behaviour. The additional levels of control that are exhibited by the human brain on behaviour are events which are recapitulated in the influence of genes on brain development. At the cellular level genes are also important in the signalling and transcriptional events that underlie memory formation. It is at these two levels, cellular and development, that a knowledge of gene function is likely to provide the most significant molecular genetic insights into the regulation of human behaviour.

GENOMIC IMPRINTING, BRAIN DEVELOPMENT AND BRAIN FUNCTION

Advances in the field of molecular genetics have recently demonstrated differential roles for maternal and paternal genomes in brain development through genomic imprinting (Keverne *et al.*, 1996; Allen *et al.*, 1995). Genomic imprinting is itself a relatively new finding in mammalian genetics, and confers functional differences on parental genomes such that certain autosomal alleles are expressed only when they originate from father (maternally imprinted and silenced), while others are expressed only on passage through the matriline (paternally imprinted and silenced) (Surani *et al.*, 1984). Maternal and paternal genomes are not, therefore, transcriptionally equivalent, and both sets of autosomal alleles are required for normal development and function.

Imprinting maps of the mouse genome have been constructed which identify chromosomal subregions on a least six different chromosomes (2, 6, 7, 11, 12 and 17) (Beechey and Cattanach, 1995). Mice that carry a parental duplication of these chromosomal regions (isodisomy) show very obvious disturbances in development ranging from foetal death to aberrant postnatal behaviour. In recent years a number of imprinted genes have been identified which are expressed in the brain (SNRPN – ribonucleoprotein which catalyses splicing and processing of mRNA (Leff *et al.*, 1992); KVLQT – a putative potassium channel (Lee *et al.*, 1997); UBE-3A – involved in ubiquitin-mediated protein degradation (Kishino *et al.*, 1997); Peg3 – involved in the TNF signalling pathway and influences hypothalamic development (Li *et al.*, 1999; Relaix *et al.*, 1998); Mest – expressed in the hypothalamus and regulates maternal behaviour (LeFebvre *et al.*, 1998); Grb10 – involved in Silver–Russell syndrome (Miyoshi *et al.*, 1998); NDN – involved in Prader-Willi syndrome (Jay *et al.*, 1997); GNAS1 – regulating learning and memory (Hayward *et al.*, 1998). Although not unique to imprinting, many of the imprinted genes control other genes by acting as oncogenes, tumour suppressors, transcriptional factors, and growth factors, or by regulating the products of other genes through alternative RNA splicing, and protein degradation. Hence their effects are complex and genetic disorders relating to imprinted genes are often pleiotropic (Angelman's syndrome (AS), Prader–Willi syndrome (PWS), Beckwith–Weidemann syndrome (BWS), Silver–Russell syndrome (SRS)).

Genetic imprinting and brain function

Dysfunction of the human brain occurs in a number of genomically imprinted disorders. The best understood are Angelman's syndrome (AS) and Prader–Willi syndrome (PWS). In 1965 Angelman reported three unrelated children with a clinical presentation of mental retardation, absent speech, prognathism and frequent tongue protrusions. The ataxic, jerky limb movements and bouts of inappropriate laughter suggested a superficial resemblance to puppets (Angelman, 1965). Their pathology shows a mild cerebral atrophy but normal gyral development, with a prominent decrease in dendritic aborization in the numbers of dendritic spines in pyramidal cells of layers 3 and 5 (Jay *et al.*, 1991). Genetically, the disorder results from deletions on maternal chromosome 15q 11–13, or from paternal uniparental disomy for chromosome 15 (Lalande, 1997). PWS is the most common form of dysmorphic genetic obesity associated with mild mental retardation. It is characterized by infantile hypotonia and hypothalamic dysfunction causing hypogonadism and hyperphagia. About 60% of cases have a cytological deletion of chromosome 15q 11–13, but in contrast to AS they occur exclusively on the paternal chromosome (Lalande, 1997). Psychoses resembling schizophrenia have also been observed in PWS (Clarke, 1993) and a linkage of a neurophysiological deficit in schizophrenia to chromosome 15q 13–14, overlapping this imprinted region, has recently been reported (Freedman *et al.*, 1996).

The fact that certain of the imprinted loci are clustered on a few chromosomes and replicate asynchronously has prompted suggestions of common regional controls on imprinted genes that can extend their effects over large distances. In the example of AS and PWS on chromosome 15q 11–13, an imprinting centre is believed to regulate this large chromosomal domain (Buiting *et al.*, 1995). Deletions to this "imprinting centre" result in bi-allelic expression of certain genes, but because the deletion itself does not cause the disease phenotype in grandparents, it is believed this centre functions in resetting of the imprint during gametogenesis (Saitoh *et al.*, 1996). Mutation of the "imprinting centre" is thought to result in fixation of an ancestral epigenotype with silent transmission through the matriline but advent of the imprinted phenotype after inheritance through the patriline. However, these views are drawn from human genetics, and await the burden of proof from experimental studies on mice. Recent findings have revealed that mutations of the "imprinting centre" are different in AS and PWS patients. Both are located in non-coding introns and differentially regulate the switch from maternal to paternal or paternal to maternal epigenotype on passage through the germ line (Ferguson-Smith, 1996; Dittrich *et al.*, 1996). Whether or not this has anything in common with the imprinting control centre in BWS and the recently described role of X-ist in the silencing of the X-chromosome remains to be determined (Herzing *et al.*, 1997; Lee and Jaenisch, 1997).

The chromosomal location for Turner's syndrome has been known for many years because it results from the partial or complete deletion of one X chromosome. Females born with Turner's syndrome are of normal intelligence, but depending on the parental origin of their single X chromosome, their social adjustment (as measured by verbal skills and social cognition) varies. Most Turner's syndrome females inherit an X chromosome from their mother (72% are X^M) that is never inactivated, and they exhibit impaired social competence (Skuse et al., 1997). Although of normal intelligence, a large proportion of X^M females have clinically significant social difficulties compared with X^P females, who inherit their single X chromosome from their father (28%). Neuropsychological testing of Turner's syndrome females has revealed a social cognitive impairment whose score is independent of verbal IQ; the impairment is significantly greater in X^M than X^P subjects. A putative imprinted locus has been mapped to the short arm of the paternally derived X chromosome and appears to escape X inactivation. Whether, as the authors suggest, this imprinted locus explains the greater vulnerability of males to a variety of developmental disorders of speech, reading disability and autism is by no means clear, but should certainly focus future attention to this chromosomal domain.

A novel experimental approach to investigating genomic imprinting in the brain has been achieved by the construction of chimaeras in mice. Embryos constructed from a mixture of cells that are parthenogenetic/normal (Pg) or androgenetic/normal (Ag) do survive, but survival requires the total proportion of chimaeric cells not to exceed normal cells (Keverne et al., 1996). The precise locations in the brain to which these chimaeric cells (i.e. those expressing exclusively maternal or paternal genes) participate in development can be determined by the presence of a genetic marker (β globin, or LacZ). Using these techniques a clear and distinct patterning in brain development emerges. At birth, cells that are disomic for the paternal genome (both alleles are from father) contribute substantially to those parts of the brain that are important for primary motivated behaviour (hypothalamus, preoptic area BNST and septum) and are excluded from the developing neocortex and striatum. At the earliest stages of brain development (days 9–10), Ag cells are present in all neural tissues and as gestation progresses they proliferate extensively in the medio-basal forebrain, but at parturition are virtually absent from telencephalic structures. By contrast, parthenogenetic cells (both alleles inherited from mother) are excluded from these medio-basal forebrain areas, but selectively accumulate in those regions where Ag cells are excluded, especially neocortex and striatum. Furthermore, growth of the brain is enhanced by this increased maternally expressed gene dosage, while brains of Ag chimaeras are smaller, both in absolute measurement and especially relative body weight. Not only is it surprising that Pg cells seem to proliferate at the expense of normal cells and produce a larger telencephalon in mouse chimaeras, but this enlarged brain appears anatomically and functionally normal. This is surprising because a large number of genes have been silenced from these cells (all of the imprinted genes which are paternally expressed) and others that are maternally expressed have been duplicated. This would seem to emphasize the importance of maternally expressed alleles in telencephalic development and the lack of importance for paternally expressed imprinted genes in these regions. This is congruent with our knowledge of human imprinted gene syndromes with maternally expressed alleles dysfunctioning in the context of AS (mental retardation and movement disorders) and paternally expressed alleles dysfunctioning in the context of PWS (hyperphagia, obesity and hyposexuality).

Genomic imprinting and brain evolution

The distinct patterning in the distribution of Pg and Ag cells in chimaeras and their differential effects on brain growth suggest genomic imprinting may have been important in forebrain evolution. Allometric scaling of those parts of the brain to which maternally or paternally expressed genes differentially contribute reveals that a remodelling of the brain has occurred during mammalian evolution. On moving across phylogenies from insectivorous mammals to prosimian and then simian primates, it can be seen that neocortex and striatum have significantly increased in size relative to the rest of the brain and body, while the hypothalamus, MPOA and septum have decreased in size (Keverne *et al.*, 1996). Genomic imprinting may thus have facilitated a rapid non-linear expansion of the brain, especially the neocortex and striatum, relative to body size during its development over an evolutionary time scale. However, in order to understand the evolutionary development of the mammalian brain it is important to expand our thinking beyond neurobiology. A larger brain requires a larger skull, and changes in the female pelvic morphology to permit parturition. Brain tissue has one of the highest demands for oxygen and glucose, metabolic needs which can only be developmentally achieved by an adequate vascular supply facilitating placental transfer of nutrients. Interestingly, the larger the neonatal brain the larger and more invasive the placenta for nutrient transfer, features of development also regulated by imprinted genes.

It is still early days in our knowledge of imprinting, but a number of these genes have been shown to regulate the expression of other genes, and are themselves regulated by imprinting centres resulting in the coordination of expression in maternally and paternally derived alleles over large chromosomal domains. Moreover, there is evidence that these epigenetic imprints do not remain stable, but can themselves undergo cell-type specific modifications resulting bi-allelic expression. Hence, further mechanisms must exist not only to erase imprinting patterns, possibly by transcriptional regulators that can induce sequence-specific demethylation, but to provide for cell-type specific imprinting. Taken together, these findings point to a potential for synchronizing and orchestrating huge genetic programmes both within and across tissues.

WHAT DO GENE KNOCKOUT STUDIES IN MICE TELL US ABOUT HUMAN BEHAVIOUR?

At the present time gene knockout studies by targeted mutagenesis in mammals tend to be restricted to mice, and only very few strains of mice unless these are produced by crossing the mutation onto different genetic backgrounds. The reason for this is the availability of mouse embryonic stem cells which eliminate the need for micro-injection of DNA. Moreover, the ability to select among the stem cell cultures for those cells in which the foreign DNA is expressed, increases the likelihood of producing chimaeras with germ-line transmission (Evans, 1996).

The early gene knockout studies with a neurobiological focus were directed at fairly obvious target genes and added very little to our knowledge of behavioural neuroscience. On the contrary, since the behavioural consequences were often predictable, such studies helped to confirm that the genetic deletions had actually been achieved. It was also found that a substantial number of targeted mutagenesis studies of neurally expressed genes were without any obvious behavioural consequences, supporting the concept of genetic canalization and redundancy. Also of importance in gene knockout studies is a knowledge of the level at which a gene is functioning in the expression of a behavioural phenotype. In 1977 Bruce Cattanach and colleagues produced a mutant mouse with a deletion of 33.5 kb in the gene for gonadotrophin-releasing hormone (GnRH), which is normally expressed in hypothalamic neurons. Male mice homozygous for this mutation showed no sexual behaviour (mounting or ejaculation) and were infertile, but the behavioural deficit were secondary to changes in the testis, which in turn was secondary to changes in the brain. Naturally occurring mutations in the human androgen receptor have even more dramatic effects (Bancroft, 1983). Although the genotype is XY and the testes develop and secrete androgens in the normal way, the fetal genitalia are insensitive to this hormone and this insensitivity results in complete regression of the Wolfian ducts and development of female external genitalia. These males are also infertile, but behaviourally they express a feminine gender identity despite their XY genotype. They are attracted to and marry men. I mention these examples for two reasons. They illustrate the kind of complications that can arise in behaviour from a mutation that primarily brings about a change in somatic phenotype and, although in both cases there is an androgen insufficiency, the behavioral consequences differ considerably, depending on the species. Targeting the receptor genes for steroid hormones (androgen, oestrogen and progesterone) and the consequent effects of this on behaviour therefore require careful evaluation.

Learning, memory and cognition

The hippocampus was suspected to play a part in human memory from studies of patients who had undergone bilateral removal of temporal lobe structures in the treatment of intractable epilepsy. Animal models of memory deficit have used spatial learning tasks, since the early work of Olton *et al.* (1979) showed that rats with lesions of the hippocampal connections (fimbria/fornix) experienced working memory loss in a radial maze task. Deleting genes that are expressed in the hippocampus has received considerable attention in the context of spatial learning, and its physiological counterpart of long-term potentiation (LTP) in hippocampal slices. Most, if not all, the gene knockout studies on learning and memory have been directed at genes expressed in the hippocampus, a part of the brain which is important for spatial learning in rodents, and is the anatomical location of "place cells" (O'Keefe, 1976). Behavioural studies using radial and water mazes have shown that bilateral lesions of the hippocampus impair an animal's ability to find its goal (Morris *et al.*, 1982) while similar lesions in humans prevent the formation of new long term memories, but leave past memories intact (Milner *et al.*, 1968). Because individual hippocampal neurons fire nerve impulses when an animal occupies a location it has explored and learnt about, these neurons are assigned specific "places". Of course, individual hippocampal neurons are not hard-wired to this one place, and can participate in more than one location throughout the animal's territory. It is the overall pattern of firing from a population of neurons that is unique to a given location. Hence, while the hippocampus is not the only part of the brain involved in learning and memory, there is a long history of studies that have confirmed the importance of the hippocampus for spatial learning in rodents.

LTP is an important electrophysiological correlate of hippocampal learning, and over the years has become synonymous with learning and memory in general (Bliss and Collingridge, 1993). LTP is also found in other brain structures, including the neocortex, although induction in these regions is usually facilitated by the blockade of intrinsic inhibitory interneurons. LTP can be readily induced in tissue slices of the hippocampus or by "in-vivo" electrical stimulation of the hippocampus. This electrical stimulation is not usually physiological, i.e. of the kind that would occur in the functioning brain, but is of a high frequency (100 Hz) for a short period of time (1 s bursts). The rapidity of this induction on an "in-vitro" tissue slice and its long-lasting effects make LTP a useful synaptic model for learning and memory. It is synapse-specific and can be associative in nature, i.e. synapses that are given a weak stimulus of insufficient intensity to cause LTP, can be potentiated if stimulated at the same time by other stronger inputs to the same neuron.

It was in the context of LTP and spatial learning that an engineered gene knockout was first shown to influence mouse behaviour (Silva et al., 1992a,b). The selection of α-calcium calmodulin kinase (α-CaMKII) was made on the basis of its high concentrations in postsynaptic densities of the hippocampus and cortex. Moreover, this kinase is neuron-specific, which avoided possible somatic confounds, and is mainly expressed postnatally, thereby reducing the possibility of developmental defects sometimes caused by gene knockout in early brain development. The most remarkable feature of the α-CaMKII null mutant mice is the apparent lack of any widespread abnormalities. The mice appeared to function as normally as other laboratory mice, and showed no obvious morphological abnormalities in either the hippocampus or the cortex. Nevertheless, the mutant mice were impaired in spatial learning and were deficient in LTP, despite normal synaptic transmission in the hippocampus.

Following rapidly on this report came the second knockout study which again focused on kinase genes (Grant et al., 1992). The tyrosine kinase genes are expressed in the hippocampus and are part of the cascade leading to the release of retrograde synaptic messengers of importance in learning and memory. Pharmacological inhibitors of tyrosine kinase block the induction of LTP without affecting normal synaptic transmission, but these inhibitors lack pharmacological specificity. To overcome this problem, four tyrosine kinase genes, fyn, src, yes, abl, which are expressed in the hippocampus, were deleted in four separate mouse lines. This study initially examined LTP, which turned out to be impaired in only one of the four variant tyrosine kinase knockouts, the fyn null mutant, and this mutant was therefore tested in behavioural studies. In parallel with the blunted LTP, the fyn knockout mice were impaired on spatial learning when seeking a hidden platform in a water maze, but were also slower at finding the platform made visible by a flag. However, these mice did learn the task following 6 days of training, compared with 2 days for normal mice. In addition to the deficits in spatial learning and LTP, these fyn null mutants had a neuroanatomical phenotype that differed from controls. The arrangement of the various cell types in the hippocampus (dentate granule cells and CA_3 pyramidal cells) was abnormal, possibly arising from a 25% increase in the number of cells in these regions. This suggested a role for the *fyn* gene in the development of the hippocampal neural circuitry, and the possibility that these development defects could themselves be the cause of the physiological and behavioural impairments of the null mutation.

The potential for "compensatory" processes in targeted disruption of single genes should not be ignored, and these early studies revealed this when three of the four different tyrosine kinase knockouts were without effect on hippocampal LTP. Other knockout studies have revealed compensatory effects of other genes. Nitric oxide (NO) has been implicated in hippocampal LTP, but this is normal in mice with a targeted mutation in the neuronal form of NO synthase. LTP is also normal in mice with a targeted mutation in endothelial NO synthase. However, LTP is significantly reduced in mice that are homozygous for the double mutation, suggesting that the neuronal and endothelial forms can compensate for each other (Son *et al.*, 1996). An additional problem of interpretation for gene knockouts in the context of brain and behaviour is the extent to which the rest of the brain may compensate for the function, especially if the gene is regionally expressed. However, this problem of interpretation is not unique to "knockout" studies, since similar questions may be raised against brain lesion studies.

In the initial studies of the fyn and α-CaMKII knockout, and in more recent studies of homozygous null mutation of the γ isoform of protein kinase C, the defects in behaviour (spatial learning) and the physiological defects in LTP were in harmony. Subsequent "gene knockout" studies have succeeded in dissociating these events; spatial learning can occur without LTP, while LTP can be normal but spatial learning impaired. Thus, mice heterozygous at the steel (Sl) locus, a gene which is expressed in neurons that project to and within the hippocampus, have a deficit in spatial learning (Motro *et al.*, 1996). This deficit in hippocampal spatial learning was revealed in the water maze, but no LTP impairment could be detected by electrical stimulation of the projection pathways. There was, however, an electrophysiological deficit in baseline synaptic transmission between the dentate and CA_3. Nevertheless, this study clearly showed that hippocampal learning and hippocampal LTP can be dissociated. The inverse of this dissociation was seen in the *Thy1* knockout mice which exhibited normal spatial learning despite regional inhibition of hippocampal LTP in the dentate. This suppression of LTP in the dentate was paradoxically due to an abnormal gain of function via the $GABA_A$ receptor which enhanced inhibition in the hippocampal dentate (Nosten-Bertrand *et al.*, 1996).

Further knockout studies have revealed a dissociation between LTD and spatial learning in mice that are deficient in the metabotropic glutamate receptor $mGluR_2$ (Yokoi *et al.*, 1996). This subtype of metabotropic receptor is ideal for targeted mutagenesis since it is expressed only in the presynaptic elements of mossy fibres that project from the dentate of the hippocampus to CA_3. Long-term depression (LTD) induced by low-frequency stimulation, was abolished in these mutants but the mice still performed normally in the water maze. Findings such as these have pushed neuroscience to think more carefully about the relationship between certain forms of LTP and hippocampal learning.

These studies point to the importance of phosphorylation initiated by receptor tyrosine kinases and intracellular tyrosine kinases in the functioning of hippocampal learning. However, many cell signalling pathways have been implicated in LTP, including tyrosine kinases, calcium-dependent enzymes and c-AMP-dependent protein kinase (PKA). PKA has a number of regulatory and catalytic subunits (at least six isoforms of the R-regulator, and C-catalytic subunits) which are expressed in different combinations in different parts of the brain including different regions of the hippocampus. Hence, targeting these subunits is helping to refine the understanding of functional specificity of this signalling pathway in hippocampal LTP (Qi *et al.*, 1996).

One of the substrates for PKA is a protein that binds to the c-AMP response element (CREB) and which acts as a transcription factor that in turn activates c-AMP-inducible genes. These are thought to be related to the process of synaptic growth, facilitating transmission. Thus CREB occupies an important step in the c-AMP pathway regulating the proteins for synaptic transmission. Molecular genetic studies and targeted mutations in *Aplysia* have shown that long-term facilitation is regulated by CREB (Bailey *et al.*, 1996), while classical conditioning using olfactory cues in *Drosophila* have shown CREB to be important for the formation of long-term memory (Tully, 1996). Long-term learning requires several trials spaced over intervals, but can be shortened to one trial learning in *Drosophila* by overexpression of CREB activator. Moreover, this same long-term learning can be selectively blocked, leaving short-term memory intact, by inducing expression of a CREB inhibitor (c-AMP response element modulator). CREB became an obvious candidate gene for targeted ablation in mice, but complete knockouts proved to be fatal at birth. The CREB protein has a number of different isoforms (α, β and Δ) and mice lacking only the α and Δ isoforms of the CREB proteins survive. Like the *Drosophila* null mutant, these have impaired long-term memory as revealed from the water maze task, while short-term memory remains intact (Bourtchuladze *et al.*, 1994). These mice show compensatory increases in the levels of the CREB isoform (β) and retain 10–20% of residual CREB activity, which may explain their survival and healthy appearance (Blendy *et al.*, 1996). LTP is also unstable in these CREB α Δ-null mutants, demonstrating that the mutation affects both hippocampal LTP and spatial memory.

CREB α Δ-null mutants are further deficient in a number of learning tasks, including memory for socially transmitted food preferences and Pavlovian conditioning of place preferences. However, all of these learning deficits can be overcome provided the training trials for learning the task are conducted at appropriate intervals. Thus CREB α Δ mice are no different from controls in learning the water maze task providing they are given a 1-h interval between trials (Kogan *et al.*, 1996). If the interval between the trials is shorter (2 or 10 min) they fail to learn the task and, even when the number of trials is increased, they still fail to learn. In the socially transmitted cueing of food preference, CREB α Δ mice are no different from controls in preferring the socially familiar food immediately after the social interaction but, unlike controls, they form no long-term memory to this social information. Again the CREB α Δ mice were equal to controls at long-term social learning when they were given two social interactions with an interval of 1 h between interactions. Likewise, Pavlovian conditioning in the impaired mutants was restored if the animals were given two spaced trials.

KEVERNE

It would appear that additional learning trials provide more CREB activation and the consequent synthesis of proteins involved in memory. However, to be effective these trials must be separated by an interval of 10–60 min. In the mutants, therefore, the machinery that activates CREB is maximally active, but at a reduced level, and requires subsequent reactivation to produce sufficient protein product. It has been suggested that, during the intervals of spaced training, there may be a decrease in the activity of CREB inhibitors and a net increase in the activity of activators, which would allow enhanced transcriptional responses to result from additional trials. While it is comforting to see the cellular conservation of mechanisms subserving memory from *Aplysia* to *Drosophila* to mouse, this is far removed from the understanding of cognition. CREB is the signalling pathway responsive to changes in light (suprachiasmatic nucleus), season (pars tuberalis) and stress (hypothalamus) in other regions of the brain and triggered by different contexts. But it is not just the brain where CREB is active. The immune system, spermatogenesis and the endocrine system all require CREB transcription/phosphorylation in order to function normally. This description of memory at the cellular level may translate to any cellular memory, not just which neuron interacted with the receptor, but which hormone, which drug, or indeed any other ligand which activates this signalling pathway. It is non-specific. It should therefore come as no surprise to learn that there is also a reduction in withdrawal symptoms following morphine abstinence in the CREB knockout. Clearly, a cell cannot become addicted if it has no memory for the addiction.

As mentioned earlier, "*in-vivo*" electrophysiological recordings in the hippocampus reveal that each cell has its own positional-dependent firing rate as the animal moves freely through its environment. The relative location of these "place" receptive fields changes in different environments and must be relearned in each new environment. Simultaneous recording of large numbers of hippocampal neurons, as opposed to single cells, provides a more detailed map of the animal's spatial location, and has enabled us to obtain a more refined picture of the physiology with on-going spontaneous behaviour (Wilson and McNauton, 1993). Combining this electrophysiological refinement with gene knockout in restricted regions of the hippocampus, using cre-lox genetic manipulations, has provided new insights into the functioning of the hippocampus (Wilson and Tonegawa, 1997). Pharmacological studies have reported that NMDA antagonists cause spatial learning defects and prevent LTP induction. It therefore came as some surprise to find that in the NMDA gene knockout mouse stable place fields are still retained by CA1 pyramidal cells without NMDA receptors, but the specificity is weak and the place fields are broader. In these mice there is also a lack of correlation in the firing rates of neurons tuned to similar locations. Since the code for spatial location is carried by activity across a population of hippocampal neurons, then it is essential to have robust covariance of the firing of neurons that have overlapping spatial fields. Without this coordination the hippocampus cannot convey accurate spatial information to other regions of the brain. Further studies directed at understanding how hippocampal "place" cells are established and maintained have been undertaken in transgenic mice that express a form of Ca^2 calmodulin-dependent protein kinase (Ca-MKII) that was mutated to become Ca^2-independent. These mice have normal LTP at high-frequency stimulation but lack LTP in response to low-frequency stimulation (5–10 Hz). In these mice the place cells are less stable and less precise, and the mice are impaired on spatial memory tasks. These low frequencies are in the same range of naturally occurring oscillations in the EEG (theta rhythm) and are generated during exploration. Hence it would appear that preventing theta rhythm interferes with the formation and stability of place cells. These effects can account for the deficits these mice show in spatial memory. Hence there was impaired ability of these knockout mice to navigate the water maze, and find a hidden platform, but they still could exhibit normal performance in the non-spatial visually cued swimming task.

Although these studies were conducted with mice, and some with *Drosophila*, there is every reason to suppose similar mechanisms are operating in the human nervous system. This is because the mutations described have produced deficits in the pathways that regulate cellular memory, and these are conserved from *Drosophila* to mouse, and therefore will certainly include humans. Of course these cellular pathways may be linked to different receptor types in different brain regions, and even receptor types that are uniquely human. Such regional variations in the brain across species may influence the type of memory impairment, but this would be a reflection of the brain region rather than its cellular events, and hence causally related to how and why that group of cells expressed a given receptor type. Answers to the "how and why" questions lie in an understanding of brain development.

BRAIN DEVELOPMENT AND PSYCHIATRIC DISORDERS

In complex psychiatric disorders the behavioural phenotype may be influenced by a wide range of polymorphisms present on a number of different chromosomes (Skuse, 1997). The genotype at any given locus may affect the probability of the disorder developing without fully determining the outcome. Each polymorphic mutation alone may be insufficient to produce a psychiatric phenotype, but in combination they serve to increase susceptibility to disorders. It is also possible that individual phenotypes have not inherited any susceptibility alleles, but manifest the disorder due to environmental causes. Factors such as sex and age may interact with susceptibility alleles as, for example, in autism which first manifests itself in childhood, and schizophrenia which has onset in later adolescence or adulthood, while both are more prevalent in boys. Differential rates of a disorder in monozygotic twins and other family members, compared with rates of the disorder in the general population, have shown that genetic and environmental factors may both contribute to complex psychiatric disorders (Rutter, 1994).

Schizophrenia is characterized by the positive symptoms of delusions or prominent hallucinations occurring in the absence of insight into their pathological nature. Such psychoses also occur as part of other medical conditions including Prader–Willi syndrome, epilepsy, postpartum psychosis, menopausal psychosis and as a result of drug abuse. At the cellular level, GABA-ergic interneurons are a common feature linking psychotic states, and are a principal focus for serotonin and dopamine innervations, as well as playing an important role in cortical development (Keverne, 1999). At the systems level, prefrontal and medial temporal cortices are implicated with activity levels out of synchrony in schizophrenics. How these vast areas of disparately functioning cortical networks are "bound" together to provide coherent conscious experiences is again a function of GABA-ergic interneurons. These interneurons have highly divergent inhibitory projections to large numbers of pyramidal neurons and are themselves synchronized by the ascending dopamine and serotonin innervations. Hence there are a large number of potential features which may have a genetic linkage to psychosis, including 5-HT receptors, dopamine receptors, glutamate transmission, steroid hormones etc., which impinge on GABA-ergic interneurons and influence their interaction with the principal neurons. Underlying all of these is a predisposition that may have arisen during development which is linked to further genes and their dysfunction.

Psychosis – a disorder of neurodevelopment

The view that schizophrenia is a disorder of brain development has received considerable attention in recent years (Harrison, 1997). A number of these reports have pointed to structural changes in the cortex, including neurons being misplaced and asymmetry of the cortex being reduced in schizophrenics (Highley *et al.*, 1999). Moreover children at risk of developing schizophrenia show delayed neurological development, while there is an increased incidence of cerebral anomalies such as agenesis of the corpus collosum in schizophrenics (Lawrie and Abukmeil, 1998). An interesting outcome of a large epidemiological study to evaluate whether events occurring at or around the time of birth contribute to the onset of psychotic illness in adult life, was the finding that children administered vitamin K in the first week of life were more susceptible than controls (Done *et al.*, 1991). Vitamin K is co-factor in the conversion of glutamic acid to γ-carboxyglutamate. Since glutamic acid is also the rate-limiting substrate for GABA synthesis, high doses of vitamin K may have disturbed this biochemical pathway in the developmentally significant process of GABA synthesis.

It stands to reason that if schizophrenia is a disorder of development, and can exhibit a genetic basis, then at least some of the genes that determine schizophrenia must be developmental. The search for susceptibility genes for schizophrenia remains a confusing enterprise with more than a dozen unconfirmed gene loci discovered from linkage studies. These putative genes are distributed across several different chromosomes (1, 2, 3, 6, 8, 9, 11, 12, 15, 17 and 18), but in many cases these findings have been little more than suggestive of linkage or association (McInnes *et al.*, 1998). Nevertheless, it is extremely likely that a number of genetic foci are going to be involved both directly and indirectly. For example, a recent paper linking retinoids with schizophrenia has pointed to loci of known genes within the retinoid signalling pathway or metabolic cascade (Goodman, 1998). A retinoid deficit has been shown to result in symptoms that resemble the schizophrenic phenotype (thought disorder, enlarged ventricles, agenesis of the corpus collosum) including a variety of congenital malformations among which are the craniofacial and digital anomalies. Significantly higher rates of these somatic abnormalities are also found in the extended pedigrees of schizophrenic probands. Among the genes that have been shown to be targets of retinoic acid transcriptional regulation are dopamine receptors and the enzymes that influence dopamine synthesis/metabolism, dopamine β-hydroxylase and tyrosine hydroxylase. Moreover, retinoic acid response elements are found in the promoter region for the dopamine D_2 receptor and in the promoter region of glutamic acid decarboxylase, the enzyme involved in GABA synthesis. Even if a single regulatory gene such as this is causally linked to schizophrenia, then the complexity of the signalling and metabolic cascades will clearly impact on many other genes.

Several studies have identified chromosome 15q 11–14 in the psychotic illness that resembles schizophrenia which may be part of Prader–Willi syndrome (Freedman *et al.*, 1996). Psychosis also occurs in a large French-Canadian kindred that has a recessive demyelination disease linked to markers 15q 14. Moreover, a neurophysiological deficit that regulates auditory stimuli (inhibition of the P50 evoked response to a second stimulus) has been recently linked to 15q 13–14. This defect is common in schizophrenics but is not causally linked to the disorder since the deficit also occurs in non-schizophrenic relatives. However, located at chromosome 15q 13 are the $GABA_A$ receptors (α5, β3 and γ3), which have been used in linkage analysis of schizophrenia (Byerley *et al.*, 1995). These receptors map to cortical regions, especially frontal and temporal cortices. While such diverse studies add strength to this linkage in schizophrenia, it is still not a simple single gene effect, since these will depend on both the location and timing of this gene's effect.

GABA-ergic neurons and cortical development

The cells of the cerebral cortex, both neurons and glia, originate from the ventricular zone and subventricular zone in the developing telencephalon (McKonnell, 1995). The predominant mode of cell migration into the cerebral cortex is both radially and clonally based (Rakic, 1995), but recent studies have revealed a subpopulation of neocortical interneurons which express GABA and migrate from the developing striatum (Rubenstein & Beachy, 1998). This cell migration occurs between the primordia of the basal ganglia and cerebral cortex, suggesting that many of the GABA-ergic neocortical interneurons are generated by the proliferative zone of the basal ganglia. Two members of the Dlx gene family (Dlx1 and Dlx2) are active in cells of the ventricular zones but are switched off once they reach the mantle of the striatum, suggesting a transition from proliferation to differentiation. A mouse containing a null mutation of these two genes produces a phenotype where partially differentiated neurons accumulate in the striatal subventricular zone and fail to migrate into the neocortex (Anderson *et al.*, 1997). Similarly, if the striatum is detached from the neocortical region, neocortical Dlx expression is eliminated from the neocortex and there is a dramatic reduction in the number of cortical interneurons that express GABA. This provides compelling evidence that the GABA-containing neocortical interneurons do not arise uniquely from the neocortical proliferative regions, but a substantial number have their origins in the striatum similar to the medium spiny neurons of the striatum.

These findings are important considerations for the ontogeny of schizophrenia if, as reported, developmental events may contribute a predisposition in the development of the disorders. They are important because the migration of GABA-ergic neurons in the neocortex is orthogonal to the long axis of the radial glia, and hence their path is unlikely to be laid out by the glial scaffold customarily used by migrating cortical neurons. Whatever the molecular regulators of this migration might be, it is clear that the vast evolutionary cortical expansion that has occurred from mouse to humans presents extensive problems of timing and synchronization for such orthogonal migration. This is especially true for the frontal and temporal cortices which represent extreme poles for the migration of GABA-ergic neurons from their origins via the striatal cortical junction. Moreover, developmental patterns in the cytoarchitecture of the human cerebral cortex continue into the postnatal period until age 6, but are most actively differentiated in the first 15 months after birth (Shankle *et al.*, 1998). GABA-ergic neurons themselves may play an important part in this development since they differentiate prior to the principal pyramidal cells and are in a unique situation to modulate differentiation and synaptogenesis of the principal neurons (Ma & Barker, 1995). During the early stages of cortical neurogenesis, GABA depolarizes cells of the rat embryonic cortex and produces increases in intracellular Ca^{2+} (Cherubin *et al.*, 1991). Increases in intracellular Ca^{2+} are essential for neuronal growth and differentiation. Furthermore GABA decreases the number of embryonic cortical cells synthesizing DNA, while $GABA_A$ receptor antagonists increase DNA synthesis indicating an influence of this transmitter on neocortical progenitor cells during the transition from G to S phase of the cell cycle (LoTurco *et al.*, 1995). Cortical GABA-ergic neurons also produce reelin, a protein which is thought to act as a signpost for the final destination of radially migrating cortical neurons (Pesold *et al.*, 1998). Mutations to the "reelin" gene induce disorganization of cortical laminae and many neurons fail to reach their final destination, and their neurites fail to branch appropriately. Moreover, polymorphism in human reelin has been detected in adult brains of schizophrenic family members. This is located in a stretch of genomic DNA important for the regulation of reelin protein secretion. GABA-ergic interneurons in postmortem prefrontal (Brodmann's areas 10 and 46) and temporal cortices (Brodmann's area 22) and hippocampus which express reelin were significantly reduced (approximately 50% for this protein and its mRNA in patients with schizophrenia (Impagnatiello *et al.*, 1998).

The poor penetration of certain genetic abnormalities that may contribute to the cause of schizophrenia suggests that the disorder is not exclusively a multi-gene effect. The absence of symptoms in children who are diagnosed with schizophrenia later in life provides support for other secondary events. Viral illness in the second trimester of pregnancy (Rantakallio *et al.*, 1997), low birth weight, shortened gestation and perinatal brain damage (Jones *et al.*, 1998), could provide vulnerability to schizophrenia. Hormonal changes in neurosteroid synthesis (dehydroepiandrosterone) or pregnancy (allopregnanolone) could provide secondary factors, facilitating excitotoxicity and secondarily precipitating schizophrenia. Reelin insufficiency may, therefore, be seen as an unfavourable background on which the summation of various vulnerability factors operate to produce schizophrenia.

DISCUSSION

This chapter is by no means a comprehensive review of brain, behaviour and developmental genetics. It has, however, selected certain areas of research which have progressed considerably in recent years, and these are used to illustrate the kind of contribution which genomics can make to the understanding of brain and behaviour. Detailed consideration has been given to the very significant role which genomic imprinting plays in brain development and the significance of imprinted genes in human brain developmental disorders. The importance of animal studies in identifying imprinted genes and elucidating their phenotype by targeted mutagenesis is providing novel insights as to how the phenotype develops from gene to cell to neural development and ultimately the behavioural phenotype (Li *et al.*, 1999). Moreover, the construction of chimaeric brains with cells disomic for maternal or for paternal alleles, and containing markers, reveals a reciprocal distribution. Androgenetic cells are virtually excluded from the developing telencephalon but make a substantial contribution to hypothalamic structures and result in a smaller forebrain than controls. Parthenogenetic cells are excluded from the hypothalamus but make a substantial contribution to the cortex and striatum and result in a larger forebrain than controls.

From an evolutionary viewpoint it is noteworthy that there are no differences between control, Ag or Pg chimaeras with respect to brainstem distribution, but it is the phylogenetically recent parts of the prosencephalon where maternal and paternal genomes have most impact. Within mammals, increased neocortical expansion, particularly of the frontal cortex, has been instrumental in the evolution of the primate brain with its capabilities for organizational planning, and in humans creative thinking and language. That imprinted genes, and particularly the maternal genome, could influence the distribution of cells preferentially to this part of the brain has provocative implications for brain evolution. It is noteworthy that evolutionary biologists emphasize the contribution of the matriline to social cohesion in primates (Wrangham, 1987) which in turn has provided an important selection pressure for brain evolution and neocortical enlargement (Humphrey, 1976). That a link could exist between evolution of the brain development, behaviour, and the contribution of parentally imprinted genes to this process is an exciting possibility which may be illuminated by the further characterization of imprinted genes.

An area of neuroscience which at some future date will certainly benefit from the human genome project is the fundamental understanding of psychiatric disorders. These are complex disorders, both neurologically and neurochemically, and are undoubtedly multiple gene disorders. I have tried to illustrate this by a detailed consideration of psychoses, which are a component of many psychiatric and neurological dysfunctions, particularly schizophrenia. However, GABA-ergic neurons are integral to all forms of psychosis, and the way in which the cortex develops may be an important predisposing factor. The migration of GABA-ergic neurons to the neocortex from the striatum orthogonal to the cortical columns of neurons that have their origin in the ventricular zone is especially noteworthy. Since much of the cortical development occurs late in gestation and into the postnatal period, and since GABA-ergic neurons are themselves actively participating in the maturation and synaptic connectivities of the cortex, these are high-risk periods that could explain the correlations of birth problems with subsequent onset of schizophrenia. Moreover, with the evolutionary expansion of the neocortex, the extended migration routes for GABA neurons to the prefrontal and temporal cortices, place these regions under greatest risk, and particularly susceptible to errors in synchronization and timing with respect to the cortical columns. Because GABA itself plays an integral role in cortical development, any perturbation of this neonatal migration may render these most distant areas of the cortex (prefrontal and temporal) most susceptible to connectivity anomalies and a predisposition for schizophrenia. It is especially noteworthy that the cortical areas most distant for the migration of GABA-ergic neurons are Broca's area in the frontal cortex and Wernicke's area in the superior temporal cortex. These areas of the brain are concerned with generating speech (Broca's area) and understanding the spoken word (Wernicke's area). Psychosis is characterized by the subjects experiencing voices speaking about them, and alien thoughts which are not their own intruding or being inserted into their mind. Thoughts are often repeated or echoed and are sufficiently compelling to convince the psychotic subject that these events have a special meaning (Crow, 1997). Of course, it is not known whether hallucinations are dependent on Wernicke's area or if alien thoughts are inserted into Broca's area. However, functional imaging of the brain (fMRI) using regional blood flow of patients with schizophrenia has revealed abnormal patternings in metabolic activity in these regions. Word fluency tests have also produced reduced left frontal activation and increased temporal activity in schizophrenic brains. Language is considered to be structured in relation to the self and has no meaning except in the context of experiencing what is self-generated and distinguishing it from that which is received from significant others. Distinguishing "thoughts to speech" from "speech to understanding" appears to be malfunctioning in psychoses or the nuclear symptoms of schizophrenia (Crow, 1998).

Finally, consideration has been given to targeted mutagenesis of genes known to be expressed in the brain. There is no question that this technology will have a major impact on our understanding of brain development and behaviour. However, it is important not to expect quick simplistic answers of the kind that encourages a deterministic interpretation of genes on behaviour. It is extremely unlikely that any mammalian behaviour will come under the control of a single gene. For example, the imprinted gene (Peg3) has a marked impact on maternal care, but so too does Peg1, fos B, the winged helix gene mf3, and the Dbh gene. Genes are embedded in the genome and we can no more understand their function in isolation than we can understand the function of the brain from electrophysiological recordings of single cells. To complicate interpretations further, the genome also has a capacity to buffer against single gene mutations, a phenomenon long recognized in developmental biology and termed canalization.

Although relatively new to behavioural neuroscience, the gene knockout approach has undergone a rapid learning curve, and the hidden pitfalls of applying this technology to an understanding of behaviour have been clearly flagged. The initial experiments told us more about the validity of the techniques than they did about behaviour, but these were the necessary pioneering studies. The study of learning and memory, more than any other topic of behaviour, has benefited substantially from the application of gene knockout technologies. It is also the area of behavioural neuroscience where this technology was first applied, and where most publications and advances have been made. For this reason there is already an informative history, albeit a relatively brief one, which exemplifies some of the problems, how they have been mastered and hence the progress which this relatively new approach to the brain and behaviour has achieved.

The early learning and memory studies were not without their critics, mainly because the behaviour was directed at hippocampal spatial learning and the gene had been deleted from every cell in the brain, not only in the adult but through-out development. Questions arose as to the functional relevance of gene targeting studies and whether they were truly revealing a gene's function, independent of background effects (Routtenberg, 1995; Gerlai, 1996). Let me add a further caution: brains do not normally undergo the kind of tetanic stimulation required for LTP, and the real world in which mice live never requires swimming in water mazes. Although these techniques have been pivotal to our understanding of the physiology and pharmacology of spatial learning, when considering molecular genetics it is important to take account of the way natural selection has differentially influenced the expression of these genes to benefit a lifestyle and behaviour suitable to the mouse's ecological niche. Not only is there a need to study behaviour in relevant contexts, but it helps if multiple levels of behavioural analysis can be undertaken.

The increasing availability of novel genes through the genome project will undoubtedly further our understanding of brain development and brain dysfunc-tion. However, it should be remembered that behavioural development depends upon action, reaction and interaction, both within the organism and with the changing environment in which the animal lives. Ethologists have recognized this complexity and the importance of both "nature" as well as "nurture", and the need to emphasize the importance of behaviour on both. Nowhere is this more important than in human behavior which is dependent on lifelong experiences. The postponing of much of brain development to the postpartum period ensures that life's early experiences further shape the complex connectivities of the developing brain. Perhaps more than any other species, the complexity of the human brain and the predispositions that are laid down during development owe as much to epigenetic events as they do to genetic programming.

References

Allen, N., Logan, K., Lally, G., Drage, D., Norris, M. & Keverne, E. (1995). Distribution of partheno-genetic cells in the mouse brain and their influence on brain development and behaviour. *Proceedings of the National Academy of Sciences, USA,* **92**, 10782–10786.

Anderson, S.A., Eisenstat, D.D., Shi, L. & Rubenstein, J.L.R. (1997). Interneuron migration from basal forebrain to neocortex: dependence on *Dlx* genes. *Science,* **278**, 474–476.

Angelman, H. (1965). "Puppet"children. A report on three cases. *Developmental Medicine and Child Neurology,* 7, 681–688.

Bailey, C.H., Bartsch, D. & Kandel, E.R. (1996). Toward a molecular definition of long-term memory storage. *Proceedings of the National Academy of Sciences, USA,* **93**, 13445–13452.

Bancroft, J. (1983). Human sexuality and its problems. *Human sexuality and its problems.* Edinburgh: Churchill Livingstone.

Beechey, C.V. & Cattanach, B. (1995). Genetic imprinting map. *Mouse Genome*, **93**, 92.

Blaffer-Hrdy, S. (1976). Care and exploitation of primate infants. *Advances in the Study of Behavior*, **6**, 101–158.

Blendy, J.A., Kaietner, K.H., Schmid, W., Grass, P. & Schutze, G. (1996). Targeting of the CREB gene leads to up-regulation of a novel CREB mRNA isoform. *EMBO Journal*, **15**, 1098–1106.

Bliss, T.V.P. & Collingridge, G.L. (1993). A synaptic model of memory: long-term potentiation in the hippocampus. *Nature*, **361**, 31–39.

Bourtchuladze, R., Fregueilli, B., Blendy, J., Coiffi, D., Schutz, G. & Silva, A. (1994). Deficient long-term memory in mice with a targeted mutation of the cAMP-responsive element binding protein. *Cell*, **79**, 59–68.

Buiting, K., Saitoh, S., Gross, S., Dittric, B., Schwartz, S., Nicholls, R.D. & Horsthemke, B. (1995). Inherited microdeletions in the Angelman and Prader–Willi syndromes define an imprinting centre on human chromosome 15. *Nature Genetics*, **9**, 395–400.

Byerley, W., Bailey, M.E.S., Hicks, A.A., Riley, B.P., Darlison, M.G., Holik, J., Hoff, M., Umar, F., Reimherr, F., Wender, P., Myles-Worsley, M., Waldo, M., Freedman, R., Johnson, K.J. & Coon, H. (1995). Schizophrenia and GABA$_A$ receptor subunit genes. *Psychiatric Genetics*, **5**, 23–29.

Cattanach, B.M., Iddon, C.A., Charlton, H.M., Chiappa, S.A. & Fink, G. (1977). Gonadotropin releasing hormone deficiency in mutant mouse with hypogonadism. *Nature*, **269**, 338–340.

Cherubin, E., Gaiarsa, J.L. & Ben-Ari, Y. (1991). GABA: an excitatory transmitter in early postnatal life. *Trends in Neuroscience*. **14**, 515–519.

Clarke, D.J. (1993). Prader–Willi syndrome and psychoses. *British Journal of Psychiatry*, **163**, 680–684.

Crow T.J. (1998). Nuclear schizophrenia symptoms as a window on the relationship between thought and speech. *British Journal of Psychiatry*, **173**, 303–309.

Crow, T.J. (1997). Is schizophrenia the price that *Homo sapiens* pays for language? *Schizophrenia Research*, **28**, 127–141.

Dittrich, B., Buiting, K., Korn, B., Rickard, S., Buxton, J., Saitoh, S., Nicholls, R.D., Poustka, A., Winterpacht, A., Zabel, B. & Horsthemke, B. (1996). Imprint switching on human chromosome 15 may involve alternative transcripts of th SNRPN gene. *Nature Genetics*, **14**, 163–170.

Done, D.J., Johnstone, E.C., Frith, C.D., Golding, J., Shepherd, P.M. & Crow, T.J. (1991). Complications of pregnancy and delivery in relation to psychosis in adult life: data from the British perinatal mortality survey sample. *British Medical Journal*, **302**, 1576–1580.

Evans, M.J. (1996). The power of embryonic stem cell transgenesis for experimental mammalian genetics. *Endocrinology and Metabolism*, **3**, 45–52.

Ferguson-Smith, A.C. (1996). Imprinting moves to the centre. *Nature Genetics*, **14**, 119–121.

Freedman, R., Coon, H., Myles-Worsley, M., Orr-Urtreger, A., Olincy, A., Davis, A., Polymeropoulos, M., Holik, J., Hopkins, J., Hoff, M., Rosenthall, J., Waldo, M.C., Reimherr, F, Wender, P., Yaw, J., Young, D.A., Breese, C.R., Adams, C., Patterson, D., Adler, L.E., Kruglyak, L., Leonard, S. & Byerley, W. (1996). Linkage of a neurophysiological deficit in schizophrenia to a chromosome 15 locus. *Proceedings of the National Academy of Sciences, USA*, **94**, 587–592.

Gerlai, R. (1996). Gene-targeting studies of mammalian behavior: is it the mutation or the background genotype? *Trends in Neuroscience*, **5**, 177–181.

Goodman, A.B. (1998). Three independent lines of evidence suggest retinoids as causal to schizophrenia. *Proceedings of the National Academy of Sciences, USA*, **95**, 7249–7244.

Grant, S.G.N., O'Dell, T.J., Karl, K.A., Stein, P.L., Soriano, P. & Kandel, E.R. (1992). Impaired long-term potentiation, spatial learning, and hippocampal development in *fyn* mutant mice. *Science*, **258**, 1903–1908.

Harrison, P.J. (1997). Schizophrenia: a disorder of neurodevelopment? *Current Opinion in Neurobiology*, **7**, 285–289.

Hayward, B.E., Moran, V., Strain, L. & Bonthron, D.T. (1998). Bidirectional imprinting of a single gene: *GNAS1* encodes maternally, paternally, and biallelically derived proteins. *Proceedings of the National Academy of Sciences, USA*, **95**, 15475–15480.

Herzing, L.B.K., Romer, J.D., Horn, J.M. & Ashworth, A. (1997). Xist has properties of X-chromosome inactivation centre. *Nature*, **386**, 272–275.

Highley, J.R., McDonald, B., Walker, M.A., Esiri, M.M. & Crow. T.J. (1999). Schizophrenia and temporal lobe asymmetry. *British Journal of Psychiatry*, **175**, 127–134.

Holman, S.E. and Goy, R.W. (1995). Experimental and hormonal correlates of care-giving in rhesus macagues. In C.R. Pryce, R.D. Martin & D. Skuse (Eds.), *Motherhood in human and nonhuman primates* (pp. 87–93). Basel: Karger.

Humphrey, N.K. (1976). The social function of intellect. In: P.P.G. Bateson & R.A. Hinde (Eds.), *Growing points in ethology*. Cambridge, UK: Cambridge University Press.

Impagnatiello, F., Guidotti, A., Pesold, C., Dwiveldi, Y., Caruncho, H., Pisu, M.G., Uzunov, D.P., Smalheiser, N.R., Davis, J.M., Pandey, F.N., Pappas, G.D., Tueting, P., Sharma, R.P. & Costs, E. (1998). A decrease of reelin expression as a putative vulnerability factor in schizophrenia. *Proceedings of the National Academy of Sciences, USA*, **95**, 15718–15723.

Jay, P., Rougeulle, C., Massacrier, A., Moncla, A., Mattei, M.G., Malzac, P., Roeckel, N., Taviaux, S., Lefranc, J-L.B., Cau, P., Lalande, P.B.M. & Muscatelli, F. (1997). The human necdin gene, *NDN*, is maternally imprinted and located in the Prader–Willi syndrome chromosomal region. *Nature Genetics*, **17**, 357–361.

Jay, V., Laurence, E.B., Chan, F-W. & Perry, T.L. (1991). Puppet-like syndrome of Angelman: a pathologic and neurochemical study. *Neurology,* **41**, 416–422.

Jones, P.B., Rantakallio, P., Hartikainen, A.L., Isohanni, M. & Siplia, P. (1998). Schizophrenia as a long-term outcome of pregnancy, delivery, and perinatal complications: a 28-year follow-up of the 1966 North Finland general population birth cohort. *American Journal of Psychiatry*, **155**, 355–364.

Keverne, E.B. (1985). Reproductive behaviour. In C.R. Austin & R.B. Short (Eds.), *Reproductive fitness* (pp. 133–175). Cambridge, UK: Cambridge University Press.

Keverne, E.B. (1992). Primate social relationships: their determinants and consequences. *Advances in the Study of Behavior*, **21**, 1–37.

Keverne, E.B. (1995). Neurochemical changes accompanying the reproductive process: their significance for maternal care in primates and other mammals. In C.R. Pryce, R.D. Martin & D. Skuse (Eds.), *Motherhood in human and nonhuman primates* (pp. 69–77). Basel: Karger.

Keverne, E.B. (1999). GABA-ergic neurons and the neurobiology of schizophrenia and other psychoses. *Brain Research Bulletin*, **48**, 467–473.

Keverne, E.B., Fundele, R., Narashimha, M., Barton, S.C. & Surani, M.A. (1996). Genomic imprinting and the differential roles of parental genomes in brain development. *Developments in Brain Research*, **92**, 91–100.

Keverne, E.B., Martel, F.L. & Nevison, C.M. (1996). Primate brain evolution, genetic and functional considerations. *Proceedings of the Royal Society, B*, **264**, 1 – 8.

Kishino, T., Lalande, M. & Wagstaff, J. (1997). UBE3A/E6-AP mutations cause Angelman syndrome. *Nature Genetics,* **15**, 70–73.

Kogan, J.H., Frankland, P.W., Blendy, J.A., Coblentz, J., Marowitz, Z., Schütz, G. & Silva, A.J. (1996). Spaced training induces normal long-term memory in CREB mutant mice. *Current Biology* **7**, 1–11.

Lalande, M. (1997). Parental imprinting and human disease. *Annual Review of Genetics*, **30**, 173–195.

Lawrie, S. M. & Abukmeil, S.S. (1998). Brain abnormality in schizophrenia. *British Journal of Psychiatry,* **172**, 110–120.

Lee, J.T. & Jaenisch, R.J. (1997). Long-range cis effects of ectopic X-inactivation centres on a mouse autosome. *Nature,* **286**, 275–279.

Lee, M.P., Hu, R-J., Johnson, L.A. & Feinbert, A.P. (1997). Human KVLQT1 gene shows tissue-specific imprinting and encompasses Beckwith–Weidemann syndrome chromosomal rearrangements. *Nature Genetics*, **15**, 181–185.

LeFebvre, L., Viville, S., Barton, S.C., Ishino, F., Keverne, E.B. & Surani, M.A. (1998). Abnormal maternal behaviour and growth retardation associated with loss of the imprinted gene *Mest*. *Nature Genetics*, **20**, 163–169.

Leff, S.E., Branna, C.I., Reed, M.L., Ozcelik, T., Francke, U., Copeland, N.G. & Jenkins, N.A. (1992). Maternal imprinting of the mouse SNRPN gene and conserved linkage homology with the human Prader–Willi syndrome region. *Nature Genetics*, **2**, 259–263.

Li, L-L., Keverne, E.B., Aparicio, S.A., Ishino, F., Barton, S.C. & Surani, M.A. (1999). Regulation of maternal behavior and offspring growth by paternally expressed *Peg3*. *Science*, **284**, 330–333.

LoTurco., J.J., Owens, D.F., Heath, M.J.S., Davies, M.B.E. & Kriegstein, A.R. (1995). GABA and glutamate depolarize cortical progenitor cells and inhibit DNA synthesis. *Neuron*, **15**, 1287–1298.

Ma, W. and Barker, J.L. (1995). Complementary expressions of transcripts encoding GAC(67) and GABA(A) receptor alpha 4, beta-1, and gamma-1 subunits in the proliferative zone of the embryonic rat central-nervous-system. *Journal of Neuroscience*, **15**, 2547–2560.

McInnes, L.A., Reus, V.I. and Nelson, B.F. (1998). Mapping genes for psychiatric disorders and behavioral traits. *Current Opinion in Genetics and Development*, **8**, 287–292.

McKonnell, S.K. (1995). Constructing the cerebral cortex: neurogenesis and fate determination. *Neuron,* **15**, 761–768.

Milner, B., Corkin, S. & Teurber, H.-L. (1968). Further analysis of the hippocampal amnesic syndrome: 14 year follow-up study of HM. *Neurophysioligia,* **6**, 215–234.

Miyoshi, N., Kuroiwa, Y., Kohda, T., Shitara, H., Yonekawa, H., Kawave, T., Hasegawa, H., Barton, S.C., Surani, M.A., Kaneko-Ishino, T. & Ishino, F. (1998). Identification of the *Meg1/Grb10* imprinted gene on mouse proximal chromosome 11, a candidate for the Silver–Russell syndrome gene. *Proceedings of the National Academy of Sciences, USA*, **95**, 1102–1107.

Morris, R.G.N., Garrud, P., Rawlings, J.N.P. & O'Keefe, J. (1982). Place navigation impaired in rats with hippocampal lesions. *Nature,* **297**, 681–683.

Motro, B., Wojtowicz, J.M., Berstein, A. & van der Kooy, D. (1996). Steel mutant mice are deficient in hippocampal learning but not long-term potentiation. *Proceedings of the National Academy of Sciences, USA,* **93**, 1801–1813.

Nosten-Bertrand, M., Errington, M.L., Murphy, K.P.S.J., Tolugawa, Y., Barboni, E., Kozlova, E., Michalovich, D., Morris, R.G.M., Silver, J., Steward, C.L., Bliss, T.V.P. & Morris, R.J. (1996). Normal spatial learning despite regional inhibition of LTP in mice lacking THY-1. *Nature,* **379**, 826–829.

O'Keefe, J. (1976). Place units in the hippocampus of freely moving rats. *Experimental Neurology*, **51**, 78–109.

Olton, D.S., Becker, J.T. & Handelmann, G.E. (1979). Hippocampus, space and memory. *Behavioural and Brain Sciences*, **2**, 313–365.

Pesold, C., Impagnatiello, F., Pisu, M.G., Uzunov, D.P., Costa, E., Guidotti, A. & Caruncho, H.J. (1998). Reelin is preferentially expressed in neurons synthesizing γ-aminobutyric acid in cortex and hippocampus of adult rats. *Proceedings of the National Academy of Sciences, USA*, **95**, 3221– 3226.

Qi, M., Zhuo, M., Skalhegg, B.S., Brandon, E.P., Kandel, E.R. & McKnight, G. (1996). Impaired hippocampal plasticity in mice lacking the $C\beta_1$ catalytic sumunit of cAMP-dependent potein kinase. *Proceedings of the National Academy of Sciences, USA*, **93**, 1571–1576.

Rakic, P. (1995). Radial versus tangenital migration of neuronal clones in the developing cerebral cortex. *Proceedings of the National Academy of Sciences, USA,* **92**, 11323–11327.

Rantakallio, P., Jones, P., Moring, J. & Von Wendt, L. (1997). Association between central nervous system infections during childhood and adult onset schizophrenia and other psychoses: a 28-year follow-up. *International Journal of Epidemiology*, **26**, 837–843.

Relaix, F., Wei, X-j, Wei, X. & Sassoon, D.A. (1998). *Peg3/Pw1* is an imprinted gene involved in the TNF-NFκB signal transduction pathway. *Nature Genetics*, **18**, 287–291.

Routtenberg, A. (1995). Knockout mouse fault lines. *Nature,* **374**, 314–315.

Rubenstein, J.L.R. & Beachy, P.A. (1998). Patterning of the embryonic forebrain. *Current Opininion in Neurobiology*, **8**, 18–26.

Ruppenthal, G.C., Arling, G.L. Harlow, H.F, Sackett, G.P. & Soumi, S.J. (1976). A 10-year perspective of motherless mother monkey behaviour. *Journal of Abnormal Psychology*, **85**, 341–349.

Rutter, M. (1994). Psychiatric genetics: research challenges and pathways forward. *American Journal of Medical Genetics*, **54**, 185–198.

Saitoh, S., Buiting, K., Rogan, P.K., Buxton, J.L., Driscoll, D.J., Arnemann, J., Konig, R., Malcolm, S., Horsthemke, B. & Nicholls, R.D. (1996). Minimal definition of the imprinting center and fixation of a chromosome 15q11-q13 epigenotype by imprinting mutations. *Proceedings of the National Academy of Sciences, USA,* **93**, 7811–7815.

Shankle, W.R., Romney, K.A. Landings, B.H. & Hara, J. (1998). Developmental patterns in the cytoarchitecture of the human cerebral cortex from birth to 6 years examined by correspondence analysis. *Proceedings of the National Academy of Sciences, USA,* **95**, 4023–4028.

Shapiro, D.Y. (1992). Plasticity of gonadal development and protandy in fishes. *Journal of Experimental Zoology,* **261**, 194–203.

Silva, A.J., Paylor, R., Wehner, J.M. & Tonegawa, S. (1992a). Impaired spatial learning in α-calcium-calmodulin kinase II mutant mice. *Science,* **257**, 206–211.

Silva, A.J., Stevens, F.C., Tonegawa, S. & Wang, Y. (1992b). Deficient hippocampal long-term potentiation in a a-calcium-calmodulin kinase II mutant mice. *Science, 257,* 201–206.

Skuse, D.H. (1997). Genetic factors in the etiology of child psychiatric disorders. *Current Opinion in Peadiatrics, 9,* 354–360.

Skuse, D.H., James, R.S., Bishop, D.V.M., Coppin, B., Dalton, P., Aamodt-Leeper, G., Bacarese-Hamilton, M., Creswell, C., McGurk, R. & Jacobs, P.A. (1997). Evidence from Turner's syndrome of an imprinted X-linked locus affecting cognitive function. *Nature, 387,* 705–708.

Son, H., Hawkins, R.D., Martin, K., Kiebler, M., Huang, P.L., Fishman, M.C. & Kandel, E.R. (1996). Long-term potentiation is reduced in mice that are doubly mutant in endothelial and neuronal nitric oxide synthase. *Cell,* **86,** 1015–1023.

Surani, M.A., Barton, S.C. & Norriss, M.L. (1984). Development of reconstituted mouse eggs suggests imprinting of the genome during gametogenesis. *Nature, 308,* 548–550.

Tully, T. (1996). Discovery of genes involved with learning and memory: an experimental synthesis of Hirschian and Benzerian perspectives. *Proceedings of the National Academy of Sciences, USA,* **93,** 13460–13467.

Wilson, M.A. & McNauton, B.L. (1993). Dynamics of hippocampal ensemble code for space. *Science,* **261,** 1055–1058.

Wilson, M.A. & Tonegawa, T. (1997). Synaptic plasticity, place cells and spatial memory: study with second generation knockouts. *Trends in Neuroscience,* **20,** 102–106.

Wrangham, R.W. (1987). Evolution of social structure. In B.B. Smuts *et al.* (Eds.), *Primate Societies* (pp. 282–296). Chicago, IL: University of Chicago Press.

Yokoi, M., Kobayashi, K., Manabe, T., Takahashi, T., Sakaguchi, I., Katsuura, G., Shigemoto, R., Ohishi, H., Nomura, S., Nakamura, K., Nokoa, K., Katsuki, M. & Nakanishi, S. (1996). Impairment of hippocampal mossy fiber LTD in mice lacking mGluR2. *Science,* **273,** 645–647.

SUGGESTIONS FOR FURTHER READING

Crow, T.J. (1998). New schizophrenic symptoms as a window on the relationship between thought and speech. *British Journal of Psychiatry,* **173,** 303–309.

Crow, T.J. (1999). The case for an Xq21.3/Yp homologous locus in the evolution language and the origins of psychosis. *Acta Neuropsychiatrica,* **11,** 54–56.

Impagnatelli, F., Guidotti, A.R., Pesold, C., Swivedi, Y., Caruncho, H., Pisu, M.G., Uzunov, D.P., Smalheiser, N.R., Davis, J.M. Pandey, G.N., Pappas, G.D., Teuting, P., Sharma, R.P. & Costa, E. (1998). A decrease of reelin expression as a putative vulnerability factor in schizophrenia. *Proceedings of the National Academy of Sciences, USA,* **95,** 15718–15723.

Keverne, E.B. (1997). Genomic imprinting in the brain. *Current Opinion in Neurobiology,* **7,** 463–468.

Keverne, E.B. (1997). An evaluation of what the mouse knockout experiments are telling us about mammalian behaviour. *BioEssays,* **19,** 1091–1098.

Keverne, E.B., Martel, F.L. & Nevison, C. (1996). Primate brain evolution: genetic and functional considerations. *Proceedings of the Royal Society of London, B,* **262,** 689–696.

Keverne, E.B., Nevison, C.M. & Martel, F.L. (1997). Early learning and the social bond. In *The Integrative neurobiology of affiliation* (pp. 263–273). Cambridge, MA: MIT Press.

Kogan, J.H., Frankland, P.W., Blendy, J.A., Coblentz, J., Morowitz, Z., Schultz, G. & Silva, A.J. (1996). Spaced training induces normal long-term memory in CREB mutant mice. *Current Biology,* **7,** 1–11.

Lefebvre, L,. Vviville, S., Barton, S.C., Ishino, F., Keverne, E.B. & Surani, M.A. (1998). Abnormal materal behaviour and growth retardation associated with loss of the imprinted gene Mest. *Nature Genetics,* **20,** 163–169.

Li, L.L., Keverne, E.B., Aparicio, S.A., Ishino, F., Barton, S.C. & Surani, M.A. (1999). Regulation of maternal behavior and offspring growth by paternally expressed Peg3. *Science,* **284,** 330–333.

Pearlman, A.L., Faust, P.L., Hatten, M.E. & Brunstrom, J.E. (1998). New directions for neuronal miration. *Current Opinion in Neurobiology,* **8,** 45–54.

Rakic, M. (1995). A small step for the cell, a giant leap for mankind: a hypothesis of neocortical expansion during evolution. *Trends in Neuroscience,* **18,** 383–388.

Rutter, M. (1994). Psychiatric genetics: research challenges and pathways forward. *American Journal of Medical Genetics*, **54**, 185–198.

Skuse, D.H. (1997). Genetic factors in the etiology of child psychiatric disorders. *Current Opinion in Pediatrics*, **9**, 354–360.

Wilson, M.A. & Tonegawa, S. (1997). Synaptic plasticity, place cells and spatial memory: study with second generation knockouts. *Trends in Neuroscience*, **20**, 102–106.

II. DEVELOPMENT OF STRUCTURE

II.1
Development of Structure and Function

JEAN M. LAUDER

INTRODUCTION

This section of the handbook contains chapters by experts in developmental neuroscience and developmental biology, who provide insights into morphological, cellular and molecular mechanisms fundamental to neural and non-neural development in both animals and humans. Important morphological aspects of human cortical development are described in the first chapter and compared to cortical development in other mammals. This is followed by a discussion of early inductive influences leading to development of the rodent forebrain. The next chapters describe roles played by neurotransmitters in development of neural and non-neural cells in the brain, craniofacial region, heart and peripheral nervous system. Neurotransmitters highlighted in these chapters include acetylcholine, serotonin, catecholamines, and the amino acid neurotransmitters, GABA and glutamate. These chapters are followed by a discussion of roles played by cell adhesion molecules in the normal brain and disorders of neural development. The final chapters cover development of descending motor pathways, development and plasticity of neuromuscular innervation by spinal motoneurons, and mechanisms of cell death in the central and peripheral nervous systems.

Based on an extensive collection of human material, **H.B.M. Uylings** discusses morphological aspects of cortical development and regionalization, including generation and migration of neocortical neurons, the role of transient cortical zones, and aspects of cortical neuronal differentiation and neuropil development. Dr Uylings compares cortical development in humans to that of other mammals, and discusses possible deleterious environmental influences on human cortical development.

A.S. LaMantia describes mechanisms of early induction and patterning of the mammalian forebrain, where signalling molecules such as retinoic acid promote patterns of gene expression that lead to formation of dorsal and ventral territories of the telencephalon, predecessor of the forebrain. Evidence presented by Dr LaMantia suggests that early inductive influences from adjacent neural crest-

A.F. Kalverboer and A. Gramsbergen (eds.), Handbook of Brain and Behaviour in Human Development, 59–62
© *2001 Kluwer Academic Publishers. Printed in Great Britain.*

derived mesenchymal cells initiate events culminating in formation of neural circuits in the forebrain, including the primary olfactory pathway. It is hypothesized that early inductive influences promoting patterned gene expression in the forebrain are fundamental developmental processes that may be compromised in human mental disorders such as schizophrenia.

C.F. Hohmann discusses the importance of acetylcholine and its receptors for development of the cerebral cortex in humans and animals. She describes accumulated evidence from animal studies indicating that cholinergic deficits lead to abnormal morphogenesis and connectivity of the cerebral cortex, and discusses underlying cellular and molecular mechanisms. Based on experimental evidence and human disorders where cholinergic neurotransmission is compromised, Dr Hohmann hypothesizes that cholinergic neurotransmission plays an important role in human cognitive development and suggests that prenatal deficits in ontogeny of the cholinergic system may be aetiological in cognitive disabilities in humans.

J.R.D. Moiseiwitsch and colleagues review experimental evidence that serotonin (5-HT) plays important roles in craniofacial and cardiac development, and describe recent evidence for underlying cellular mechanisms. These experimental findings are discussed in light of recent clinical reports that use of selective serotonin reuptake inhibitors (SSRIs) to treat depression in pregnant women is associated with an increased incidence of "groupings of minor birth defects", mainly involving subtle changes in craniofacial features. Together with experimental evidence for serotonergic regulation of craniofacial and cardiac development, these clinical observations raise questions as to whether use of SSRIs or other serotonergic drugs in pregnant women may pose risks to the unborn child. This view is supported in the chapter by P.M. Whitaker-Azmitia, who discusses abnormal brain development and behavioural outcomes in animals prenatally exposed to serotonergic drugs.

Based on the morphological, biochemical and behavioural effects of pharmacological depletion of serotonin (5-HT) in the developing rat brain, **P.M. Whitaker-Azmitia** presents evidence that 5-HT regulates important aspects of neuronal development in immature brain and promotes neuronal plasticity in the adult brain. Results of these studies suggest that one mechanism utilized by 5-HT to regulate neuronal development and plasticity involves 5-HT1A receptors expressed by glial cells. Activation of these receptors causes glial cells to release S-100β, which promotes growth of neurites in developing neurons and stabilizes dendrites in mature cells. Possible cellular mechanisms for these properties of S-100β are discussed, including promotion of neurite outgrowth by inhibition of phosphorylation of the growth-associated protein, GAP-43, and stabilization of dendrites by inhibition of phosphorylation of the microtubule-associated protein, MAP-2. Dr Whitaker-Azmitia concludes with a discussion of possible involvement of altered serotonergic regulation of brain development and plasticity in human developmental disorders such as Down syndrome and autism.

Based on the lethal phenotype of tyrosine hyroxylase (TH) knockout mice, **S. Roffler-Tarlov and M. Rios** discuss the importance of catecholamines, especially norepinephrine, for embryonic survival. Embryonic death occurs prior to establishment of the peripheral nervous system, suggesting that there must be other early source(s) of norepinephrine. Possible sources include neuroblasts of neural crest origin, including those which will form sympathetic ganglia, paraganglia (extramedullary chromaffin cells) as well as neurons in the brain and spinal cord. Norepinephrine would then be released by these cells into the circulation. Another likely source of norepinephrine is the yolk sac, which has been reported to contain catecholamines in early rat embryos (Schlumpf & Lichtensteiger, 1979). Whatever the source of norepinephrine in normal mouse embryos, the reason for the embryonic lethal phenotype in TH knockout mice indicates the absolute requirement for this neurotransmitter for embryonic survival. One reason for this requirement may be that blood-borne norepinephrine regulates heart development. Information gained from studies of TH knockout mice are discussed in relation to humans where, probably because of embryonic lethality, no patients lacking TH have ever been found. However, a few individuals have been identified who lack the next enzyme in the norepinephrine synthetic pathway, dopamine-β-hydroxylase (DβH). Similar to the rare DβH knockout mice that survive to term, these patients have no norepinephrine or epinephrine, but higher levels of dopamine than normal, and many do not survive the perinatal period.

A. Schousboe and colleagues describe experimental evidence that the inhibitory and excitatory neurotransmitters, GABA and glutamate, play important regulatory roles in neuronal survival, differentiation and synaptogenesis. Cellular and molecular mechanisms are described that require appropriate spatiotemporal expression of $GABA_A$ receptors and their subunits, NMDA and non-NMDA glutamate receptors and voltage-gated calcium channels to mediate the trophic functions of GABA and glutamate during critical periods of brain development. Of particular interest is evidence that the neurotrophic effects of the potential excitotoxin, glutamate, are both concentration and age-dependent.

R.S. Schmid and P.F. Maness discuss roles played by two neural cell adhesion molecules, L1 and NCAM, in normal and abnormal brain development, based on comparisons of L1 and NCAM knockout mice and human patients with the CRASH syndrome (*c*orpus callosum dysgenesis/agenesis, *r*etardation, *a*phasia, *s*pastic paraplegia, and *h*ydrocephalus) or schizophrenia, respectively. The authors also describe signal transduction cascades involved in L1 and NCAM signalling in normal development and plasticity. This discussion provides useful insights into possible mechanisms underlying abnormal brain development and behavioural outcomes in animals with mutations in LI and NCAM genes.

J. Armand describes development of medial and lateral descending axonal projections from the brain to the spinal cord in humans and subhuman primates. This information is then related to functional data regarding the developmental emergence of spinal reflexes, and motor responses to sensory stimuli, the latter of which require integration of sensory and motor pathways. Based on comparison of the salient developmental features of descending pathways in humans and other vertebrates, Dr Armand hypothesizes that common ontogenetic aspects of these pathways, such as a retrograde sequence with respect to impulse conduction in spinal reflex pathways, represent general developmental patterns across species. Development of the corticospinal tract (human pyramidal tract) is discussed with respect to the phylogenetically old components of descending pathways terminating on spinal interneurons in most vertebrates studied, compared to the phylogenetically more recent projections to pools of motoneurons innervating the digits in humans and non-human primates.

Next follows a chapter by **R.R. Ribchester** who details principles of development and plasticity of motoneuron innervation of skeletal muscles, as well as formation and morphogenesis of the neuromuscular junction. Included is a discussion of the role of use and disuse in the development and plasticity of neuromuscular innervation, including size–strength plasticity at the adult neuromuscular junction, synapse elimination during development, and the role of motoneuron activity in development of neuromuscular innervation and reinnervation of denervated muscles by motoneurons.

This section of the handbook concludes with a chapter by **R.W. Oppenheim** and colleagues on different types of target-independent programmed cell death in the developing nervous system. The chapter begins with an historical overview of cell death research, which the author uses to illustrate both basic principles and the foresightedness of early researchers in the field. Following are detailed descriptions of different categories of programmed cell death, including death of precursor cells and immature neurons, regulation of glial cell survival by growth and trophic factors, afferent regulation of survival and death of differentiating neurons, and hormonal regulation of neuronal survival. Useful discussions of developmental functions, proximate causes, and possible underlying mechanisms are provided for each type of cell death.

References

Schlumpf, M. & Lichtensteiger, W. (1979). Catecholamines in the yolk sac epithelium of the rat. *Anatomy and Embryology*, **156**, 177–187.

II.2
The Human Cerebral Cortex in Development

ABSTRACT

This chapter describes the generation, migration and differentiation of human neocortical neurons and critical periods in brain development. Neurons in the transient cortical zones (subplate and marginal zone) appear to be essential for a proper migration of cortical neurons and for the correct ingrowth of subcortical axons.

During normal development cell death occurs especially in the group of subplate neurons, the Cajal–Retzius neurons and the subpial granular cells. The differentiation of the layer III and V pyramidal dendritic field takes places mainly in the first few years after birth. This does not mean that the cortex has reached the mature state at that time. Transmitter development and development of peptides show a further maturation during the first decade and until the end of puberty. External, deleterious influences (irradiation, etc.) during prenatal and postnatal cortical development are discussed.

INTRODUCTION

The development of the cerebrum is fascinating because of the multitude and complexity of the processes involved in it. Neurons are generated at particular times and migrate across large distances, through other fibres and cell groups. They differentiate mainly when they have reached their final location or destination and are connected with particular neurons. In this way neuronal systems and cortical area specification arise that are necessary for the execution of many functions (O'Leary, 1989; Rakic, 1988). The general rules of development are genetically programmed rather than every individual connection, which would be impossible because of the incredibly large number of synapses – approximately 10^{14} in the neocortex alone.

A.F. Kalverboer and A. Gramsbergen (eds.), Handbook of Brain and Behaviour in Human Development, 63–80
© 2001 Kluwer Academic Publishers. Printed in Great Britain.

The aim here is to review the prenatal and postnatal development of the human neocortex, with the emphasis on the transient zones that are present mainly in the prenatal phase and their functions, and on the different kinds of critical periods during prenatal and postnatal development.

GENERATION OF NEOCORTICAL NEURONS

The time when most of the neocortical neurons are generated is between approximately 6 and 18 weeks of gestation (Rakic, 1995). With an adult total number of neurons of 27 billion (27×10^9) the average number of neocortical neurons that is generated per minute is 225,000. This figure is likely to be higher due to neuronal cell death that occurs during normal development (e.g. in subplate and marginal zone neurons, etc.). Despite a few reports of a relatively limited neurogenesis in adult cat and rat neocortex (Kaplan, 1981), until 1999 the general view was that neurogenesis in the primate neocortex occurred only in prenatal life (e.g. Rakic, 1995). The only neurogenesis that was known to occur in adulthood took place in the olfactory bulb and hippocampus. This view has been challenged by Gould et al. (1999), who indicated that in adult macaques non-pyramidal neurons are still being generated and migrate into neocortical association areas (prefrontal, inferior temporal and posterior parietal cortex). However, freshly generated neurons were not found in the adult primary visual (striate) cortex. Gould et al. (1999) hypothesize that the newly formed non-pyramidal neurons may play a role in learning and memory, in analogy with the observations that many adult-generated hippocampal neurons die when the mammals are not exposed to complex experiences/environments. This minor group of non-pyramidal neurons generated in adulthood migrate into associational cortex layers I–V, but not into layer VI. Rakic (personal communication) disputes the interpretation of Gould's data and therefore the generation of neocortical neurons in adulthood.

Another classic view is that all neocortical neurons are generated in the "cortical" (sub)ventricular zone of the wall of telencephalic vesicles that were formed from the neural tube (see Figure 1). After the division of stem cells for cortical neurons, one of the two new cells remains a germinal cell in the cortical (sub)ventricular zone of the lateral ventricle. The other one migrates to the outer wall along radial glia fibres. These radial glia fibres span the wall from the inner membrane to the outer pial surface (see Figure 2).

The oldest cortical neurons form the preplate (Supèr et al., 1998), which is separated into a marginal zone and a subplate by the inward migration of new neurons that stop their migration within the preplate (Figure 2). Different types of neurons are marginal zone neurons, i.e. (a) Cajal–Retzius cells, which are Reelin-immunoreactive (Reelin-IR) and calretinin-IR; (b) Reelin-negative neurons, that have axons leaving the marginal zone and are called "pioneer" cells by Meyer et al. (1999); (c) subpial granular layer cells, which are generated not in the neocortical subventricular zone but in a retrobulbous ventricular zone (Gadisseux et al., 1992; Meyer et al., 1999). Most of the early calretinin-negative neurons and the early Reelin-IR and calretinin-IR neurons are likely to be derived from the neocortical (sub)ventricular zone by radial migration. The subpial

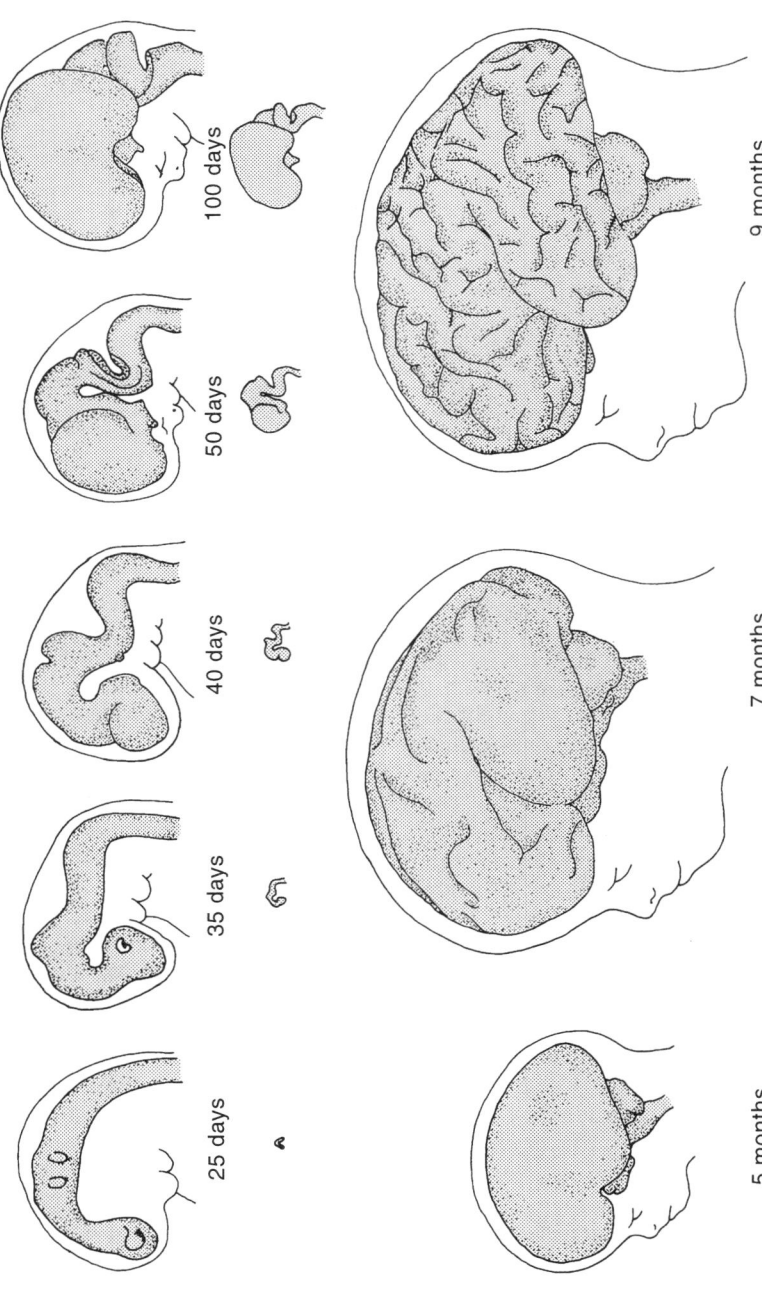

Figure 1 The macroscopic development of the human brain before birth (modified from Cowan, 1979). The drawings are all in relative proportion, while the first five stages have also been enlarged to show the morphological alterations in the neural tube. (Adapted from Uylings et al., *Natuur en Techniek* **56**, 414–423, 1988.)

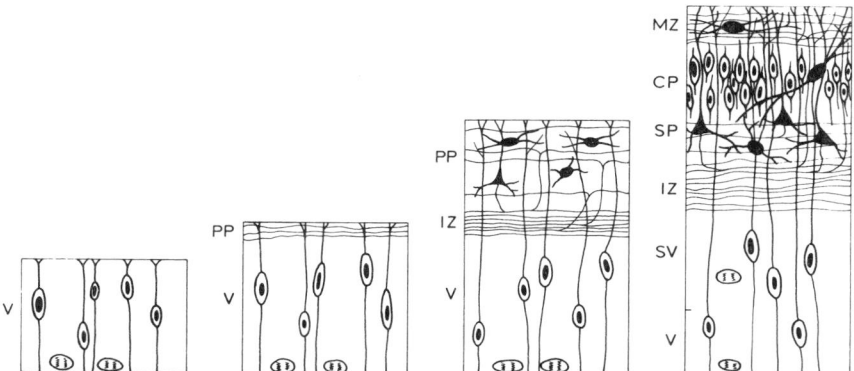

Figure 2 A scheme of four different stages in fetal lamination of the cerebral wall. V = ventricular zone; PP = preplate; this is later subdivided in subplate (SP) and marginal zone (MZ). SV = subventricular zone; IZ = intermediate zone. The Cajal–Retzius cells in the MZ and the subplate neurons in the SP are partly cogenerated. (From Uylings *et al.*, 1994.)

granular layer cells, however, arrived in the marginal zone of the entire neocortex via a tangential migration after a few weeks. They are visible in the marginal zone from the 12th week of gestation (Gadisseux *et al.*, 1992; Uylings *et al.*, unpublished observations). These subpial granular cells are initially Reelin-negative, but also appear to be a precursor pool for Cajal–Retzius cells. Thus at least a number of these cells transform in Reelin-IR neurons with the typical Cajal–Retzius cell forms (Meyer *et al.*, 1999).

Until a few years ago it was thought that the subpial granular cells were exceptional because of their tangential migration. Although various reports have indicated the existence of tangential neuronal migration (e.g. Van Eden *et al.*, 1989; De Diego *et al.*, 1994; O'Rourke *et al.*, 1995; Uylings & Delalle, 1997), it was demonstrated only in recent years that the majority of the cortical non-pyramidal neurons migrate tangentially in the subplate/intermediate zone and the marginal zone (see Parnavelas, 2000, for a review). They are probably generated in the proliferative zone of the medial ganglionic eminence of the ventral telencephalon. The tangential migration is probably guided along axonal, corticofugal bundles from, e.g., the pioneer, calretinin-negative preplate neurons, and later from subplate and marginal zone neurons (Figure 3). Meyer *et al.* (1999) distinguish Cajal–Retzius cells in two cell groups: the earlier Retzius cells and the fusiform, stellate and triangular Cajal cells. It is yet to be demonstrated, however, whether these two groups of cell forms originate as a result of the transformation of Retzius cells or whether they have a different origin (Marín-Padilla, 1990; Uylings & Delalle, 1997; Meyer *et al.*, 1999).

MIGRATION OF NEOCORTICAL NEURONS

The period of major neocortical neuron migration lasts until about 26 weeks of gestation (Marín-Padilla, 1992). Given the fascinating number of neurons

Figure 3 Medial ganglionic eminence (MGE) non-pyramidal neurons migrate to the neocortex in association with corticofugal axons first into the preplate and marginal zone (MZ) (A), somewhat later into the lower intermediate zone (IZ) and subplate (B). Neurons destined for the cortical plate (CP) migrate eventually along radially arranged axons of the MZ (C). IC = internal capsule; LGE = lateral ganglionic eminence. Reprinted from *Trends in Neurosciences,* vol. 23, Parnavelas. The origin and migration of cortical neurones; new vistas, pp. 126–131. Copyright 2000, with kind permission from Elsevier Science. © All rights reserved.

generated (at least 225,000 per minute on average) the correct migration of all these neurons and the correct formation of the right connections (synapses) is a fascinating process. The migration of pyramidal neurons is guided by radial glial fibres. For these radial glia fibres to function properly their interaction with Cajal–Retzius neurons and subpial granular cells in the marginal zone (see Figure 2) is essential (see for review Supèr *et al.*, 1998). By the proper interaction of the terminal branches and end feet of the radial glial fibres with Cajal–Retzius neurons and subpial granular cells, the radial glial fibres are maintained and the migration of cortical neurons follow an *inside-out* rule. This means that the neurons that are the last to be generated, migrate to the peripheral part of the cortical plate just below the marginal zone and pass along the previously generated neurons. The layer II neurons are thus generated later than the layer VI neurons. The majority of non-pyramidal neurons, however, migrate tangentially from the ganglionic eminence (Anderson *et al.*, 1997; Parnavelas, 2000, and see Figure 3). It may be that this tangential migration is guided by glial and axonal fibres, but more research on this topic is needed.

TRANSIENT CORTICAL ZONES

As schematically indicated in Figure 2, the first cortical zone is the preplate (PP). The PP is divided into an outer marginal zone (MZ) and an inner subplate (SP) by the ingrowth of neurons that are generated later, and that form the cortical plate. Still later, the cortical layers VI through II arise from the cortical plate. The MZ and SP are transient and dominant zones in the prenatal period.

Although the subplate (Kostović, 1990) is a transient zone, it is a dominant one during the prenatal period, around 20–27 weeks of gestation. At that time this zone is about five times as thick as the cortical plate (see Figure 4). At birth the subplate is a tiny rim that is not visible in Nissl-stained sections after the age of 1 year. Many subplate neurons disappear by programmed cell death (apoptosis) after they have "fulfilled" their developmental function. Some subplate neurons, however, remain in a zone in the white matter, close to layer VI. However, these cells are scattered in the white matter (Kostović & Rakic, 1980; Mrzljak *et al.*, 1990) and become smaller, especially around 6 months postnatally (Delalle *et al.*, 1997; Uylings & Delalle, 1997). Indeed, in the period of large diminution of the subplate (27 weeks of gestation until about 6 postnatal months), apoptotic cell features have been observed in the subplate during normal development of the human brain. The existence of the subplate became known from Kostović's study of 1976, and it was not until the 1990s that its role was made clearer in studies by Kostović *et al.* (1989), Métin & Godement (1996), Molnár (1998) and Shatz (1992). From primate and rodent studies we can infer that the thalamic fibres and the majority of monoaminergic fibres grow into the subplate and remain in the subplate for some time before they grow into the neocortex (Uylings *et al.*, 1990). The axons of the subplate neurons guide the ingrowth of subcortical fibres into the cortical subplate and are essential for the ingrowth of these subcortical fibres into the correct cortical areas (Shatz, 1992). The growth of thalamic fibres into the neocortex starts 8 weeks after they enter the subplate, i.e. at about 26 weeks of gestation (Kostović & Goldman-Rakic, 1983). This coincides with the start of the

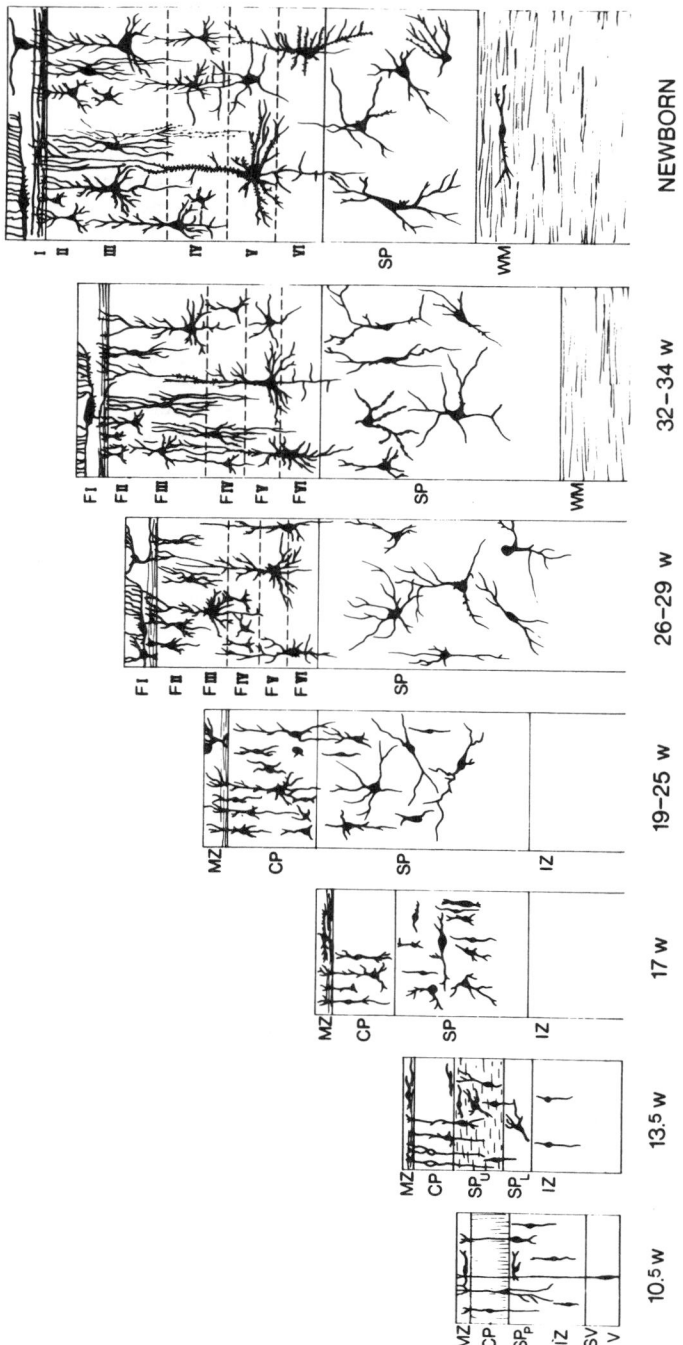

Figure 4 The prenatal development of cortical layers and neurons in human frontal cortex. SPp = first part of SP; SPu = upper part of SP; SP$_L$ = lower part of subplate. FII = fetal layer II. (From Mrzljak *et al.*, 1988.)

thickness reduction of the subplate (Figure 4). As hypothesized elsewhere (Uylings, 1998) the neocortical part of the claustrum below the insula can be considered to be formed from the subplate and thus to be a specialization from the subplate.

The other transient zone that is important during brain development is the marginal zone. This is a second zone where subcortical axonal strata enter. As reviewed elsewhere (Supèr *et al.*, 1998), the marginal zone is the zone for major ingrowth of axonal fibres in the hippocampus, in contrast to the neocortex, where the subplate is the major zone of ingrowth. From this we can predict that in the belt of paralimbic cortical areas (Mesulam, 1985) the relative size of the marginal zone and subplate is around the middle of the scale between the one in neocortex and hippocampus. The data collected by Kostović *et al.* (1993) provide a basis to this hypothesis.

Programmed cell death has been found in the marginal zone from 24 weeks of gestation onwards (Gadisseux *et al.*, 1992; Uylings & Delalle, 1997; Meyer *et al.*, 1999). At 24 weeks of gestation apoptotic characteristics have been reported for some of the subpial granular layer cells (Gadisseux *et al.*, 1992; Uylings & Delalle, 1997). This is disputed by Meyer *et al.* (1999), who suggest that these apoptotic cells are in fact Retzius cells. They claim that all neuronal subpial granular cells transform into other neurons, such as, for instance, Cajal–Retzius cells. Further research on this topic is required to solve this discussion.

At 36 weeks of gestation the subpial granular cells are no longer visible. However, as mentioned above, at least some of these cells have in the meantime been transformed into Cajal–Retzius cells, while others were incorporated into the cortical layer II at a later time. The latter is suggested by the observation that a dissolving of the subpial granular layer coincides with a "migration" of this layer into deeper parts of the marginal zone (Brun, 1965; Gadisseux *et al.*, 1992; Uylings, unpublished observations; but also Meyer & Goffinet, 1998).

Both the subpial granular layer neurons and Cajal–Retzius cells appear to be important for a proper maintenance of the radial glial fibres (see for review Supèr *et al.*, 1998). Pathology studies by Brun (1965) indicate that at those cortical locations where the subpial granular layer has been interrupted, the proper neuronal migration of cortical neurons along the radial glial fibres has also been interrupted. When Cajal–Retzius cells lack Reelin, the proper migration of neurons along radial glial fibres, and thus the proper *inside-out* migration, has been disturbed (see Supèr *et al.*, 1998).

Apoptotic features have been seen in human Cajal–Retzius cells after about 29 weeks of gestation. Several of these Cajal–Retzius cells (the large majority according to Marín-Padilla, 1990) survive and are transformed into smaller nonpyramidal cells. The surviving Cajal–Retzius cells form a relatively small group and are only surviving Cajal cells, according to Meyer *et al.* (1999). In the rodent hippocampus Cajal–Retzius cells degenerate at the arrival of septal fibres after the neuronal migration has been completed (Supèr *et al.*, 1998).

CORTICAL NEURONAL DIFFERENTIATION

The earliest neurons differentiate into the subplate and the calretinin-negative and Cajal–Retzius neurons (Figure 4). The cortical plate neurons differentiate later, in two steps. The layer III and V neurons start to accelerate their outgrowth after the ingrowth of thalamic fibres around 26 weeks of gestation (Mrzljak *et al.*, 1992). The phase of the fastest outgrowth, however, starts at birth and continues during the first 2–3 years (Koenderink *et al.*, 1994; Koenderink & Uylings, 1995) (see Figure 5). This phase, of the fastest dendritic outgrowth, parallels the phase of fast cerebral cortex expansion, which lasts until about 4–5 years of age. Giedd *et al.* (1999) reported that certain human cortical areas, such as the frontal and parietal cortex, keep increasing until the age of about 12–13 years. From that age onwards the volume of these cortices would diminish. Zilles (1978) had already demonstrated, for the tree shrew, that different cortical regions follow their own pattern of development. Each cortex has a particular age at which it reaches a maximum value, after which it can decline. Such a decline is not detectable in the volume of the whole cerebral cortex or whole brain of both humans and tree shrews. We might expect that a decline in a particular cortical volume is paralleled by a decline in dendritic trees present in these cortices. We did not observe this for the dorsolateral prefrontal cortex (see Figure 5) and further research is therefore needed to establish a decline in the human association cortex and the underlying factors.

After 5 years of age a large interindividual variation in dendritic tree size becomes apparent. We may hypothesize on the basis of the animal studies on, e.g., environmental enrichment (Uylings *et al.*, 1978) that this variation is due to genetic and environmental effects.

Although the outgrowth of cortical neurons is at its mature level between 2 and 4 years of age, the chemical, monoaminergic and cholinergic maturation continue until the end of puberty (Goldman-Rakic & Brown, 1982; Kalsbeek *et al.*, 1990; Kostović, 1990; Lewis & Sesack, 1997). For example, the mature level for NPY cortical neurons has been observed from 8 years of age (Delalle *et al.*, 1997; Uylings & Delalle, 1997), while for acetylcholinesterase the mature pattern is reached from 16 to 18 years (Kostović, 1990). The study of chemical transmitter development in postnatal human cortex must be extended before we can arrive at a complete picture.

CRITICAL PERIODS IN CORTICAL DEVELOPMENT

The term "critical period" is used in different definitions. One definition is the period in which a subject is particularly susceptible to damage, i.e. a "period of susceptibility". Another definition of critical period is the period in which recovery or a flexible response occurs, i.e. a "period of plasticity".

Critical "period of susceptibility"

During the period of major neurogenesis of cerebral cortical neurons (8–18 weeks of gestation), the brain is especially susceptible to environmentally inflicted insults, such as irradiation, alcohol use, smoking, drug (e.g. cocaine) abuse and stress. The effects of *irradiation* are age- and dose-dependent. Small doses at ages of major neurogenesis have dramatically greater effect than higher doses in

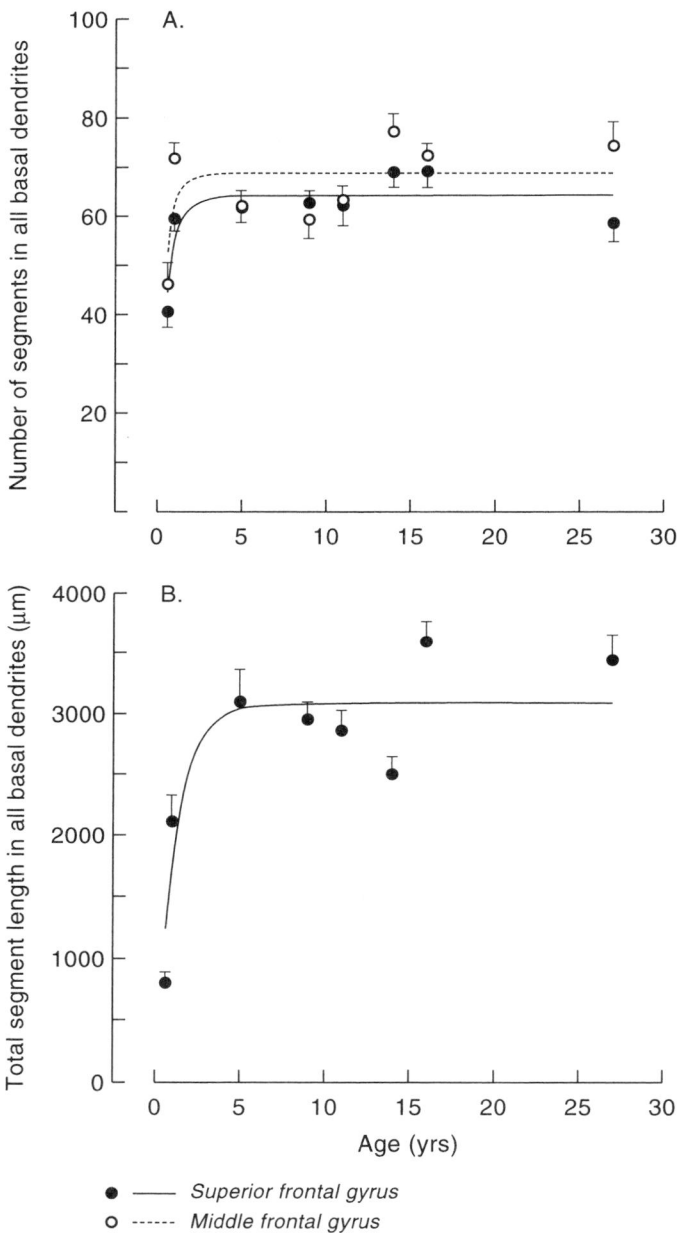

Figure 5 A: the postnatal development of basal dendrites of layer III pyramidal neurons; **B**: the postnatal development of basal dendrites of layer V pyramidal neurons. (From Koenderink *et al.*, 1994; Koenderink & Uylings, 1995.)

adulthood. In the developing rat a dose of 1.5 Gray on the 12th day of gestation (i.e. in the first period of major neocortical neurogenesis) led to massive cell death and cerebral malformation. On postnatal day 7, however, i.e. after the major neocortical neurogenesis and neuronal migration, a dose of 4 Gray did not induce acute cell death (Hicks & D'Amato, 1978). Studies on the effect of atom bombs in Hiroshima and Nagasaki (Shigematsu et al., 1995) also showed that the period of major neocortical neurogenesis (i.e. 7–18 weeks of gestation) is more sensitive to deleterious effects of irradiation. The children who were exposed in utero, before 15 weeks of gestation, and who survived, suffered from significantly more severe mental retardation after an exposure dose of irradiation of 0.7–1.5 Gray.

Drinking one or two glasses of an alcoholic beverage a day in adulthood is thought to be healthy. However, during pregnancy drinking two glasses a day is too much and leads to lower mental capacities (Streissguth et al., 1990). Alcohol affects not only cerebral neurogenesis, but also the correct wiring of neuronal connections (Miller, 1987). All three trimesters of pregnancy have therefore been implicated as vulnerable periods. Concomitant use of other substances can only aggravate the effects of alcohol (Kosofsky, 1999). It is therefore essential that pregnant women temporarily suspend drinking alcohol for the well-being of the fetus. In-utero exposure to alcohol can lead to difficulties at school, as it leads to a reduction in both mental and social capabilities (Kosofsky, 1999).

Smoking, too, affects prenatal development (Slotkin, 1998). Chronic maternal smoking during pregnancy leads to behavioural and cognitive impairment in children and to an increased risk of low birth weight and attention-deficit-hyper-activity disorder (Kline et al., 1989; Weitzman et al., 1992).

Maternal abuse of cocaine leads to impaired fetal growth and altered behaviour in orientation and alertness. Persistent deficits are as yet less obvious than those caused by alcohol, as only a subset of exposed children exhibit clear, lasting deficits (Kosofsky, 1999).

Studies on schizophrenia show that exposure to environmental insults, such as influenza, malnutrition and stress, during the second trimester of pregnancy, leads to an increased incidence of schizophrenia (Huttunen & Niskanen, 1978; Susser et al., 1996).

Fortunately, large malformations due to drug use during the main neurogenesis seldom occur, and are to be considered tragic accidents. Examples are an absent cerebral cortex and hypothalamus (anencephaly) and spina bifida, caused by valproate, a drug against epilepsy; or large malformations of legs and arms, caused by the sleep drug thalidomide (Softenon). Functional teratological effects are, however, less rare (e.g. Swaab et al., 1995). These effects include disturbance of sleep (by, e.g., clonidine, an α_2-receptor agonist used as an antihypertensive that reduces noradrenaline release and interferes with growth hormone secretion), a disturbance of learning ability (by, e.g., lead from the environment), mental retardation (by mercury), increased aggression (by, e.g., progestativa taken during pregnancy or particular artificial food (colour) additives), increased depression (by DES during pregnancy), etc. For a more detailed review on this subject of functional teratology refer to, e.g., Swaab et al. (1995).

In summary, the most "susceptible period" is the period of major neurogenesis and early pathfinding of axons to make the appropriate connections. Damage occurring in this period generally leads to a significant, persistent deficit in the number of neurons and in miswiring of axons.

Critical "period of plasticity"

When the major neurogenesis and the correct timing of making the first set of connections has not been interfered with, damage to the brain is less deleterious. The reason is that the brain becomes less susceptible and can respond flexibly during the next period of development, the "period of plasticity".

The periods in which the brain can be saved from a persisting deficit, or in which it can react with a flexible response, are restricted to certain events, for instance for the inborn error of metabolism (phenylketonuria), for thyroid hormone deficiency due to iodine insufficiency, for gonadal hormones, for restricted lesion by a local infarct around the time of birth, for brain lesion, monocular deprivation, and for the capability to completely master, grammatically, a second language.

Phenylketonuria is a syndrome of severe mental retardation due to elevated levels of phenylalaline (Yudkoff, 1994). Without treatment the majority of the afflicted individuals had an IQ of less than 50 and had to be institutionalized for their entire life (Heminger, 1999). If this inborn error is diagnosed soon after birth, early treatment by restriction of the dietary intake of phenylalanine can generally prevent the mental retardation. Some mild cognitive deficits may remain when the brain damage started prenatally (Heminger, 1999). A similar example is thyroid hormone deficiency due to a diet with insufficient iodine. This leads to severe mental retardation and neurological deficits, which can be largely prevented by adding iodine to the diet soon after birth (Kooistra et al., 1994). Since prenatal thyroid deficiency affects prenatal brain development from about 15 weeks of gestation (Thorpe-Beeston et al., 1991), complete recovery is not possible.

There are also critical periods for persisting effects of testicular hormones on the developing nervous system (e.g. for review Gorski, 2000). The masculine characteristics of brain structure and function are only imposed during the development of the brain by these hormones. These hormones interfere with programmed cell death, in particular nuclei of the hypothalamus and spinal cord, which leads to a sexually dimorphic number of neurons in these nuclei (see, e.g., Hofman & Swaab, 1989; Gorski, 2000). In addition to these influences on neuronal cell number, which are restricted to development, gonadal hormones can also modify particular structural changes such as spines and dendrites in the hippocampus during the oestrous cycle in adulthood (Woolley & McEwen, 1992; McEwen, 1999).

A lesion in the human brain caused by a perinatal infarct can be repaired in the first years of life, as shown by imaging at different ages (Kostović et al., 1989). Studies by Rasmussen & Milner (1977) showed that early unilateral lesions in left cortical language areas did not have a great deal of effect, since the language functions studied shifted to the contralateral side. In all likelihood such lesions are "compensated for" more readily at the youngest ages of postnatal development, due to a transient exuberance of neurons and connections in normal development. This view seems to be in line with the Kennard principle: the earlier the damage, the less severe the functional loss (see discussion in Kolb, 1989).

However, studies performed in the past couple of decades show that this view is too simplistic. Both rat and human studies (e.g. Kolb & Whishaw, 1990, chapters 25 and 26) indicate that earlier is not always better, due to the existence of a critical period of plasticity. Before and after such a period the effect of the same kind of lesions is more severe. In addition, the type of lesion and, especially, the location of the lesion in the brain, are of importance for the existence of a particular critical period of plasticity. Lesions in the left hemisphere before the age of 1 year impair both verbal and performance IQ, while after 1 year left-hemisphere lesions did not affect either kind of IQ compared to controls. Similar kinds of lesions in the right hemisphere before and after the age of 1 year affect only a significantly lower performance IQ (Riva & Cazzaniga, 1986; Woods & Teuber, 1973). Also in rat studies a circumscribed critical period was detected during which frontal lesions did not have an effect on frontal functions such as spatial delayed alternation in later life, whereas lesions in adulthood did have such an effect (Kolb & Tomie, 1988; Kolb et al., 1998).

In addition, a recent paper by the group of Damasio (Anderson et al., 1999) even indicates that very early lesions in the rostromedial and orbitomedial frontal cortex have more severe effects than similar lesions in adulthood. Their report on two cases indicates a functional worsening compared to the effects of similar lesions in adulthood. The two cases with lesions sustained at the age of about 1 year showed not only difficulties with social behaviour, but also a lack of awareness of proper moral reasoning and of the consequences of social misbehaviour.

Critical periods have also been detected for shape and stereoscopic vision. Originally, congenital cataracts (cataracts are opacities of the lens which interfere with vision) were surgically removed between the ages of 10 and 20 years. After studies had demonstrated a permanent impairment in form vision at these ages, these operations appeared to be more beneficial if they were performed earlier, in infancy. Stereoscopic vision develops during the period in which the ocular dominance columns of the primary visual cortex (Huttenlocher, 1994) and the corticocortical connections (Kandel et al., 2000) between the visual and parietal cortex are formed (Hyvärinen, 1982). When stereoscopic vision does not develop in this period of approximately 8 years due to an imbalance in the wiring of the pertinent circuits, it will probably remain permanently impaired.

Critical periods for social isolation or social neglect have been reported. The classic studies of Spitz (1946) in children who lacked proper maternal/parental care in the first year of their life showed that they were withdrawn, that their intellectual performance was below standard and that they were more susceptible to infections, which is indicative of a weakened immune system. For interactions between the immune system and the nervous system, see Ader et al. (1991) and Petitto & Evans (1999). A recent study (Pawbly, 2001) even indicated that maternal postnatal depression during the first 3 months can lead to a considerably lower IQ of sons (group mean value of IQ is 86), due to a shortage of maternal care. The effect on daughters was noticeably smaller. In view of the prevalence of postnatal depression (one in seven mothers), further research on the subject of parental care is of great importance. Harlow (1958) showed, in monkeys, that isolation/maternal deprivation for 6 months during the first 18 months lead to permanent deficits in social behaviour. This did not happen when older monkeys underwent a 6-month period of isolation.

Another critical period appears to exist for most people with respect to second-language acquisition. Until the age of about 7 years the second language can be learned to a level that is grammatically virtually indistinguishable from that of native speakers (Johnson & Newport, 1989). From 8–10 years onwards, however, it becomes increasingly difficult to master a second language completely. This is the conclusion drawn in a study that compared the eventual level of grammatically correct English of immigrants who entered the USA at varying ages with that of native speakers of American English (Johnson & Newport, 1989). Such a critical period may not exist for semantic processing (Neville & Bavelier, 1998).

Furthermore, the period during which the brain responds favourably to environmental enrichment is not restricted. Even in adulthood the brain reacts to the stimulation of environmental enrichment with an increase in spines and a lengthening of dendrites (e.g. Uylings et al., 1978). The extent of the flexible reaction is, however, somewhat smaller than during early postnatal development and becomes smaller with advancing age (Greenough et al., 1990). This was to be expected, as the size of dendrites has generally reached a stable level in adulthood. The environmental enrichment effects are detected in a number of cortical areas, but not in all, and in male rats they are different from those in female rats. The biggest differences are found in the hippocampal dendrites of female rats and in the occipital cortical dendrites of male rats (Juraska, 1991). It is to be expected that, in adulthood, learning also induces morphological alterations in synapses, spines and dendrites, in addition to the above-mentioned hypothesis of Gould et al. (1999), which says that the number of non-pyramidal neurons increases.

Environmental enrichment effects on human brains can be derived from, e.g., a full cross-fostering study on adoptive children (Chapron & Duyme, 1989; Duyme et al., 1999). In these studies environmental enrichment, in addition to the contribution of genetic constitution, appears to increase intellectual capabilities. On the other hand, environmental impoverishment leads to a decrease in synapses, spine numbers and dendrites in the same brain regions that react favourably to environmental enrichment (e.g. Greenough et al., 1990).

In conclusion, critical periods of flexibility coincide with periods of transient exuberance in the number of neurons, axonal projections and synapses. The periods of selective reduction are not identical for each part of the nervous system. After the selective reduction in neuron number, axonal connections and synapses in development, the flexible responses in these brain areas are diminished. This does not mean that the brain cannot change further in adulthood. As indicated above, environmental enrichment remains capable of challenging the brain, even in adulthood.

ACKNOWLEDGEMENTS

I thank Ms W.T.P. Verweij for her secretarial assistance in the preparation of this chapter. I also acknowledge the financial support of the Van den Houten Foundation for parts of the studies described in this chapter.

References

Ader, R., Felten, D.L. & Cohen, N. (Eds.) (1991). *Psychoneuroimmunology*. New York: Academic Press.

Anderson, S.A., Eisenstat, D.D., Shi, L. & Rubenstein, J.L.R. (1997). Interneuron migration from basal forebrain to neocortex: dependence on Dlx genes. *Science,* **278**, 474–476.

Anderson, S.W., Bechara, A., Damasio, H., Tramel, D. & Damasio, A.R. (1999). Impairment of social and moral behavior related to early damage in human prefrontal cortex. *Nature Neuroscience*, **2**, 1032–1036.

Brun, A. (1965). The subpial granular layer of the foetal cerebral cortex in man. Its ontogeny and significance in congenital cortical malformations. *Acta Pathologica et Microbiologica Scandinavica, Suppl.* **179**, 1–98.

Chapron, C. & Duyme, M. (1989). Assessment of effects of socio-economic status of IQ in a full cross-fostering study. *Nature*, **340**, 552–554

Cowan, W.M. (1979). The development of the brain. *Scientific American*, **241**, 113–133.

De Diego, I., Smith-Fernández, A. & Fairén, A. (1994). Cortical cells that migrate beyond area boundaries: characterization of an early neuronal population in the lower intermediate zone of prenatal rats. *European Journal of Neuroscience,* **6**, 983–997.

Delalle, I., Evers, P., Kostović, I. & Uylings, H.B.M. (1997) Laminar distribution of neuropeptide-Y immunoreactive neuron in human prefrontal cortex during development. *Journal of Comparative Neurology,* **379**, 523–540.

Duyme, M., Dumaret, A.C. & Tomkiewicz, S. (1999). How can we boost IQs of 'dull children'?: A late adoption study. *Proceedings of the National Academy of Sciences, USA,* **96**, 8790–8794.

Gadisseux, J.-F., Goffinet, A.M., Lyon, G. & Evrard, Ph. (1992). The human transient subpial granular layer: an optical, immunohistochemical, and ultrastructural analysis. *Journal of Comparative Neurology,* **324**, 94–114.

Giedd, J.N., Blumenthal, J., Jeffries, N.O., Castellanos, F.X, Liu, H., Zijdenbos, A., Paus, T., Evans, A.C. & Rapoport, J.L. (1999). Brain development during childhood and adolescence: a longitudinal MRI study. *Nature Neuroscience*, **2**, 861–863.

Goldman-Rakic, P.S. & Brown, R.M. (1982). Postnatal development of monoamine content and synthesis in the cerebral cortex of rhesus monkeys. *Developmental Brain Research*, **4**, 339–349.

Gorski, R.A. (2000). Sexual differentiation of the nervous system. In E.R. Kandel, J.H. Schwartz & T.M. Jessell (Eds.), *Principles of neural science* (3rd ed.: pp. 1131–1148). New York: McGraw-Hill.

Gould, E., Reeves, A.J., Graziano, M.S.A. & Gross, C.G. (1999). Neurogenesis in the neocortex of adult primates. *Science*, **286**, 548–552.

Greenough, W.T., Withers, G.S. & Wallace, C.S. (1990). Morphological changes in the nervous system arising from behavioral experience: what is evidence that they are involved in learning and memory? In L.R. Squire & E. Lindenlaub (Eds.), *The biology of memory* (Symposia Medica Hoechst, vol. 3, pp. 159–192). Stuttgart: Schattauer.

Harlow, H.F. (1958). The nature of love. *American Psychologist,* **13**, 673–685.

Heminger, G.R. (1999). Special challenges in the investigation of the neurobiology of mental illness. In D.S. Charney, E.J. Nestler & B.S. Bunney (Eds.), *Neurobiology of mental illness* (pp. 89–99). New York: Oxford University Press.

Hicks, S.P. & D'Amato, C.J. (1978). Effects of ionizing radiation on developing brain and behavior. In G. Gottlieb (Ed.), *Studies on the development of behavior and the nervous system* (pp. 36–72). New York: Academic Press.

Hofman, M.A. & Swaab, D.F. (1989). The sexually dimorphic nucleus of the preoptic area in the human brain: a comparative morphometric study. *Journal of Anatomy (London),* **164**, 55–72.

Huttenlocher, P.R. (1994). Synaptogenesis in human cerebral cortex. In G. Dawson & K.W. Fischer (Eds.), *Human behavior and the developing brain* (pp. 137–152). New York: The Guildford Press.

Huttunen, M.O. & Niskanen, P. (1978). Prenatal loss of father and psychiatric disorders. *Archives of General Psychiatry,* **35**, 429–431.

Hyvärinen, J. (1982). *The parietal cortex of monkey and man* (*Studies of Brain Function*, vol. 8). Berlin: Springer Verlag.

Johnson, J.S. & Newport, E.L. (1989). Critical period effects in second language learning: the influence of maturational state on the acquisition of English as a second language. *Cognitive Psychology*, **21**, 60–99.

Juraska, J.M. (1991). Sex differences in "cognitive" regions of the rat brain. *Psychoendocrinology,* **16**, 105–119.

Kalsbeek, A., De Bruin, J.P.C., Feenstra, M.G.P. & Uylings, H.B.M. (1990). Age-dependent effects of lesioning the mesocortical dopamine system upon prefrontal cortex morphometry and PFC-related behaviors. In H.B.M. Uylings, C.G. Van Eden, J.P.C. De Bruin, M.A. Corner & M.G.P. Feenstra (Eds.), *The prefrontal cortex: its structure, function and pathology* (*Progress in Brain Research*, vol. 85, pp. 257–283). Amsterdam: Elsevier.

Kandel, E.R., Jessell, T.M. & Sanes, J.R. (2000). Sensory experience and the fine-tuning of synaptic connections. In E.R. Kandel, J.H. Schwartz & T.M. Jessell, (Eds.), *Principles of neural science* (4th ed.: pp. 1115–1130). New York: McGraw-Hill.

Kaplan, M.S. (1981). Neurogenesis in the 3-month-old rat visual cortex. *Journal of Comparative Neurology,* **195**, 323–338.

Kline, J., Stein, Z. & Susser, M. (1989). *Conception to birth: epidemiology of prenatal development*. New York: Oxford University Press.

Koenderink, M.J.Th. & Uylings, H.B.M. (1995). Postnatal maturation of layer V pyramidal neurons in the human prefrontal cortex. A quantitative Golgi study. *Brain Research*, **678**, 233–243.

Koenderink, M.J.Th., Uylings, H.B.M. & Mrzljak, L. (1994). Postnatal maturation of the layer III pyramidal neurons in the human prefrontal cortex: a quantitative Golgi analysis. *Brain Research*, **653**, 173–182.

Kolb, B. (1989). Brain development, plasticity and behavior. *American Psychologist,* **44**, 1203–1212.

Kolb, B. & Tomie, J.A. (1988). Recovery from early cortical damage in rats. IV. Effects of hemidecortication at 1, 5 or 10 days of age on cerebral anatomy and behavior. *Behavioral Brain Research*, **28**, 259–274.

Kolb, B. & Whishaw, I.Q. (1990). *Fundamentals of human neuropsychology*. (3rd ed.). W.H. Freeman.

Kolb, B., Forgie, M., Gibb, R., Gorny, G. & Rowntree, S. (1998). Age, experience and the changing brain. *Neuroscience Biobehavioral Reviews*, **22**, 143–159.

Kooistra, L., Laane, C., Vulsma, T., Schellekens, J.M.H., Van der Meere, J.J. & Kalverboer, A.F. (1994). Motor and cognitive development in children with congenital hypothyroidism: a long-term evaluation of the effects of neonatal treatment. *Journal of Pediatrics,* **124**, 903–909.

Kosofsky, B.E. (1999). Effect of alcohol and cocaine on brain development. In D.S. Charney, E.J. Nestler & B.S. Bunney (Eds.), *Neurobiology of mental illness* (pp. 601–615). New York: Oxford University Press.

Kostović, I. (1990). Structural and histochemical reorganization of the human prefrontal cortex during perinatal and postnatal life. In H.B.M. Uylings, C.G. Van Eden, J.P.C. De Bruin, M.A. Corner & M.G.P. Feenstra (Eds.), *The prefrontal cortex: its structure, function and pathology*. (*Progress in Brain Research*, vol. 85: pp. 223–240) Amsterdam: Elsevier.

Kostović, I. & Rakic, P. (1980). Cytology and time of origin of interstitial neurons in white matter in infant and adult and monkey telencephalon. *Journal of Neurocytology,* **9**, 219–242.

Kostović, I. & Goldman-Rakic, P.S. (1983). Transient cholinesterase staining in the mediodorsal nucleus of the thalamus and its connections in the developing human and monkey brain. *Journal of Comparative Neurology,* **219**, 431–447.

Kostović, I., Lukinović, N., Judaš, M., Bogdanović, N., Mrzljak, L., Zečević, N. & Kubat, M. (1989). Structural basis of the developmental plasticity in the human cerebral cortex: the role of the transient subplate zone. *Metabolic Brain Disease*, **4**, 17–23.

Kostović, I., Petanjek, Z. & Judaš, M. (1993). Early area differentiation of the human cerebral cortex: entorhinal cortex. *Hippocampus*, **3**, 447–458.

Lewis, D.A. & Sesack, S.R. (1997). Dopamine systems in the primate brain. In A. Björklund, T. Hökfelt & F.E. Bloom (Eds.), *Handbook of clinical neuroanatomy*, vol. 13: *The primate nervous system, part I*. (pp. 263–375). Amsterdam: Elsevier.

Marín-Padilla, M. (1990). Three-dimensional structural organization of layer I of the human cerebral cortex: a Golgi study. *Journal of Comparative Neurology,* **299**, 89–105.

Marín-Padilla, M. (1992). Ontogenesis of the pyramidal cell of the mammalian neocortex and developmental cytoarchitectonics: a unifying theory. *Journal of Comparative Neurology*, **321**, 223–240.

McEwen, B.S. (1999). The effect of stress on structural and functional plasticity in the hippocampus. In D.S. Charney, E.J. Nestler & B.S. Burney (Eds.), *Neurobiology of mental illness* (pp. 475–493). New York: Oxford University Press.

Mesulam, M.-M. (1985). Patterns in behavioral neuroanatomy: association areas, the limbic system and hemispheric specialization. In Mesulam, M.-M. (Ed.), *Principles of behavioral neurology* (pp. 1–70). Philadelphia: F.A. Davis.

Métin, C. & Godement, P. (1996). The ganglionic eminence may be an intermediate target for corticofugal and thalamocortical axons. *Journal of Neuroscience*, **16**, 3219–3235.

Meyer, G. & Goffinet, A.M. (1998). Prenatal development of reelin-immunoreactive neurons in the human neocortex. *Journal of Comparative Neurology*, **397**, 29–40.

Meyer, G., Goffinet, A.M. & Fairén, A. (1999). What is a Cajal–Retzius cell? A reassessment of a classical cell type based on recent observations in the developing neocortex. *Cerebral Cortex*, **9**, 765–775.

Miller, M.W. (1987). Effect of prenatal exposure to alcohol on the distribution and time of corticospinal neurons in the rat. *Journal of Comparative Neurology*, **257**, 372–382.

Molnár, Z. (1998). *Development of thalamocortical connections*. Berlin: Springer Verlag.

Mrzljak, L., Uylings, H.B.M., Kostović, I., Van Eden, C.G. & Judas, M. (1990). Neuronal development in human prefrontal cortex in prenatal and postnatal stages. In H.B.M. Uylings, C.G. van Eden, J.P.C. de Bruin, M.A. Corner & M.G.P. Feenstra (Eds.), *The prefrontal cortex: its structure, function and pathology*, (*Progress in Brain Research*, vol. 85: pp. 185–222). Amsterdam: Elsevier.

Mrzljak, L., Uylings, H.B.M., Kostović, I. & Van Eden, C.G. (1992). The prenatal development of neurons in the human prefrontal cortex: a quantitative Golgi study. *Journal of Comparative Neurology*, **316**, 485–496.

Neville, H.J. & Bavelier, D. (1998). Neural organization and plasticity of language. *Current Opinion in Neurobiology*, **8**, 254–258.

O'Leary, D.D.M. (1989). Do cortical areas emerge from a protocortex? *Trends in Neuroscience*, **12**, 400–406.

O'Rourke, N.A., Sullivan, D.P., Kaznowski, C.E., Jacobs, A.A. & McConnell, S.K. (1995). Tangential migration of neurons in the developing cerebral cortex. *Development*, **121**, 2165–2176

Parnavelas, J.G. (2000). The origin and migration of cortical neurones: new vistas. *Trends in Neuroscience*, **23**, 126–131

Pawbly, S.J. (2001). *Child Psychology and Psychiatry*, (In press).

Petitto, J.M. & Evans, D.L. (1999). Clinical neuroimmunology: understanding the development and pathogenesis of neuropsychiatric and psychosomatic illnesses. In D.S. Charney, E.J. Nestler & B.S. Bunney (Eds.), *Neurobiology of mental illness* (pp. 162–169). New York: Oxford University Press.

Rakic, P. (1988). Specification of cerebral cortical areas. *Science*, **241**, 170–176

Rakic, P. (1995). A small step for the cell, a giant leap for mankind: a hypothesis of neocortical expansion during evolution. *Trends in Neuroscience*, **18**, 383–388.

Rasmussen, T. & Milner, B. (1977). The role of early left-brain injury in determining lateralization of cerebral speech functions. *Annals of the New York Academy of Sciences*, **299**, 355–369.

Riva, D. & Cazzaniga, L. (1986). Late effects of unilateral brain lesions sustained before and after age one. *Neuropsychologia*, **24**, 423–428.

Shatz, C.J. (1992). How are specific connections formed between thalamus and cortex? *Current Opinion in Neurobiology*, **2**, 79–82.

Shigematsu, I., Ito, C., Kamada, N., Akiyama, M. & Sasaki, H. (1995). *Effects of A-bomb radiation on the human body*. Chur: Harwood Academic Publishers.

Slotkin, T.A. (1998). Fetal nicotine or cocaine exposure: which one is worse? *Journal of Pharmacology and Experimental Therapeutics*, **285**, 931–945.

Spitz, R.A. (1946). Hospitalism: a follow-up report on investigation described in volume 1, 1945. *Psychoanalitical Study of Children*, **2**, 113–117.

Streissguth, A.P., Barr, H.M. & Sampson, P.D. (1990). Moderate prenatal alcohol exposure: effects on child IQ and learning problems at age 7½ years. *Alcohol, Clinical and Experimental Research*, **14**, 662–669.

Supèr, H., Soriano, E. & Uylings, H.B.M. (1998). The functions of the preplate in development and evolution of the neocortex and hippocampus. *Brain Research Reviews*, **27**, 40–64.

Susser, E., Neugebauer, R., Hoek, H., Brown, A., Lin, S., Labovitz, D. & Gormna, J. (1996). Schizophrenia after prenatal famine: further evidence. *Archives of General Psychiatry*, **53**, 25–31.

Swaab, D.F., Boer, K. & Mirmiran, M. (1995). Functional teratogenic effects of chemicals on the developing brain. In M.I. Levene, R.J. Lilford, M.J. Bennett & J. Punt (Eds.), *Fetal and neonatal neurology and neurosurgery* (2nd ed.: pp. 263–277). Edinburgh: Churchill Livingstone.

Thorpe-Beeston, J.G., Nicolaides, K.H., Felton, C.V., Butler, J. & McGregor, A.M. (1991). Maturation of the secretion of thyroid hormone and thyroid-stimulating hormone in the fetus. *New England Journal of Medicine, 324*, 532–536.

Uylings, H.B.M. (1998). The (critical) periods of human cortex development. In R. Licht, A. Bouma, W. Slot & W. Kops (Eds.), *Child neuropsychology: reading disability and more* (pp. 17–28). Delft: Eburon.

Uylings, H.B.M. & Delalle, I. (1997). Morphology of neuropeptide Y-immunoreactive neurons and fibers in human prefrontal cortex during prenatal and postnatal development. *Journal of Comparative Neurology, 379*, 523–540.

Uylings, H.B.M., Kuypers, K., Diamond, M.C. & Veltman, W.A.M. (1978). Effects of differential environments on plasticity of dendrites of cortical pyramidal neurons in adult rats. *Experimental Neurology, 62*, 658–677.

Uylings, H.B.M., Mrzljak, L., Kostović, I. & Van Eden, C.G. (1988). Impetuous growth of the brain. *Natuur en Techniek, 56*, 414–423

Uylings, H.B.M., Van Eden, C.G., Parnavelas, J.G. & Kalsbeek, A. (1990). The prenatal and postnatal development of rat cerebral cortex. In B. Kolb & R.C. Tees (Eds.), *The cerebral cortex of the rat* (pp. 35–76). Cambridge, MA: MIT Press.

Uylings, H.B.M., Van Pelt, J., Parnavelas, J.G. & Ruiz-Marcos, A. (1994). Geometrical and topological characteristics in the dendritic development of cortical pyramidal and non-pyramidal neurons. In J. van Pelt, M.A. Corner, H.B.M. Uylings & F.H. Lopes da Silva (Eds.), *The self-organizing brain: from growth cones to functional networks*. (*Progress in Brain Research,* vol. 102: pp. 109–123). Amsterdam: Elsevier.

Van Eden, C.G., Mrzljak, L., Voorn, P. & Uylings, H.B.M. (1989). Prenatal development of GABA-ergic neurons in the neocortex of the rat. *Journal of Comparative Neurology, 289*, 213–227.

Weitzman, M., Gortmaker, S. & Sobol, A. (1992). Maternal smoking and behavior problems of children. *Pediatrics, 90*, 342–349.

Woods, B.T. & Teuber, H.L. (1973). Early onset of complementary specialization of cerebral hemispheres in man. *Transactions of the American Neurological Association, 98*, 113–117.

Woolley, C.S. & McEwen, B.S. (1992). Estradiol mediates fluctuation in hippocampal synapse density during the estrous cycle in the adult rat. *Journal of Neuroscience, 12*, 2549–2554.

Yudkoff, M. (1994) Disorders of amino acid metabolism. In G.J. Siegel, B.W. Agranoff, R.W. Albers & P.B. Molinoff (Eds.), *Basic neurochemistry* (pp. 824–826). New York: Raven Press.

Zilles, K. (1978). Ontogenesis of the visual system. *Advances in Anatomy, Embryology and Cell Biology, 54*, 1–138.

II.3
Induction, Patterning and the Development of Regions and Circuits in the Mammalian Forebrain

ANTHONY-SAMUEL LAMANTIA

ABSTRACT

Recent evidence indicates that the cellular and molecular mechanisms of embryonic induction contribute to the initial development of forebrain regions and pathways. A number of embryological experiments suggest that extrinsic mesenchymal and endodermal cell groups can influence regional differences in gene expression in the developing telencephalon – the embryonic precursor of the forebrain. Furthermore, these cell–cell interactions are mediated by established inductive signalling molecules including retinoic acid (RA), sonic hedgehog (shh), fibroblast growth factors (FGFs) and bone morphogenetic proteins (BMPs). The immediate consequences of these local inductive interactions in the forebrain include patterned expression of several downstream genes that subdivide the telencephalon into dorsal and ventral territories. In addition, these interactions might contribute to the region-specific expression of a number of genes in rudimentary forebrain subdivisions. Finally, there is evidence that inductive interactions can initiate a sequence of events that lead to morphogenesis and differentiation of an entire forebrain pathway: the primary olfactory pathway. Together, these observations establish a cellular and molecular basis for inductive interactions during early forebrain development. It is possible that these interactions represent a target for toxic agents and genetic mutations that disrupt the establishment of normal forebrain organization and function.

INTRODUCTION

Until recently the early differentiation of the forebrain has been among the least well studied and the least well understood steps of early development in mammals.

A.F. Kalverboer and A. Gramsbergen (eds.), Handbook of Brain and Behaviour in Human Development, 81–98
© *2001 Kluwer Academic Publishers. Printed in Great Britain.*

This lack of effort and insight is surprising when one considers the importance of forebrain regions and circuits in the function of the mammalian central nervous system. The mammalian forebrain comprises five major regions: the neocortex, the basal ganglia, the hippocampus, the basal forebrain and the olfactory bulb. The functions that each region mediates – cognition, movement control, learning, memory, emotional and appetitive behaviours, and chemosensation – emphasize the central role of the forebrain in guiding behaviour. Accordingly, an understanding of how forebrain organization emerges during embryonic development seems crucial for a comprehensive understanding of the relationship between early differentiation in the brain and the final circuits and functions that emerge.

The early stages of forebrain development can be divided into a series of related events. The first is induction, the second is patterning, the third is morphogenesis, and the fourth is cellular (and circuit) differentiation. In the forebrain, as in most regions of the embryo, observable differentiation is initiated by a series of morphogenetic changes that apparently lead to the emergence of rudiments of the major subdivisions (Figure 1). The prosencephalon expands into two telencephalic vesicles (plus the diencephalon). Subsequently, the telencephalic neuroepithelium thickens, folds, invaginates or evaginates to form rudiments of the major functional subdivisions. In general, morphogenesis is mediated by tissue–tissue induction, patterned gene expression, differential adhesion, and discontinuities in cell proliferation and differentiation underlie initial development. Despite the apparent morphogenetic process that leads to the establishment of the major forebrain subdivisions, there is little available data on the mechanisms by which forebrain regional development occurs.

There are several challenges for understanding early forebrain development. The first is to identify inductive signals, their sources, their targets, and their influence on the initial patterning of the forebrain. The second is to identify downstream transcriptional regulators whose patterned expression and activity contributes to initial regionalization. The third is to identify the adhesion, recognition, and signalling molecules that guide initial morphogenesis, and the fourth is to identify the distinct aspects of forebrain regions and pathways that these molecules influence. Accordingly, this chapter will focus on current progress in identifying inductive signals, patterning genes, morphogenetic mechanisms, and the resulting differentiation in the developing forebrain. When possible, data or speculation on the relationship between these elements of forebrain development will be discussed so that one might imagine a series of related developmental steps that leads to the initial formation of forebrain regions, pathways and circuits.

THEORETICAL BACKGROUND

Early development relies primarily upon two synergistic mechanisms: induction and patterning. Induction, where one identified population of cells provides signals that influence gene expression and differentiation in a second target population, has been known as a central developmental mechanism for almost a century (reviewed by Gurdon, 1992; Hamburger, 1988; Slack, 1993). Over the

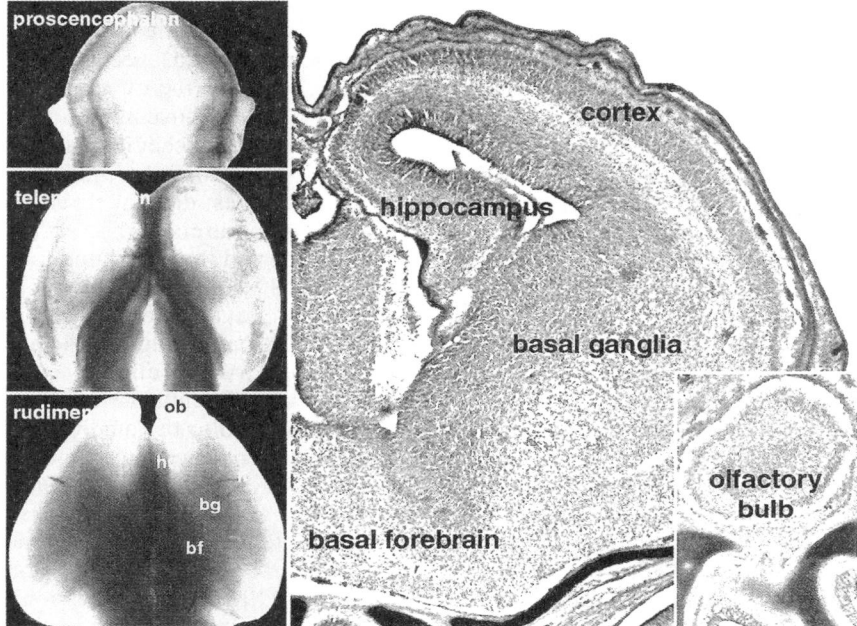

Figure 1 *Left panels*: The progressive morphogenesis of the procencephalic vesicle into the telencephalic vesicles into an immature forebrain with rudiments of the five major subdivisions. Abbreviations: ob: olfactory bulb, nc: neocortex, bg: basal ganglia, bf: basal forebrain, hc: hippocampus. *Right panel*: The cytological differentiation of the major forebrain subdivisions at E16 in the mouse. This nissl-stained coronal section shows the laminated neocortex, hippocampus and olfactory bulb (*inset*), as well as the emerging nuclei of the basal ganglia and basal forebrain.

past 15 years molecular specificity has been added to the experimental embryological descriptions of induction so that there are now a well-defined set of signalling molecules that mediate this process (reviewed by Hammerschmidt *et al.*, 1997; Hogan, 1996; Jessell & Melton, 1992; Linney & LaMantia, 1994; Mason, 1994). Much of this work has been done on the gastrulating embryo to describe the cellular and molecular underpinnings of primary induction (the formation of the germ layers and the establishment of the embryonic axes). The remainder has focused on secondary induction (the process of initiating morphogenesis of distinct body structures) in the developing limbs (reviewed by Johnson & Tabin, 1997; Tickle & Eichele, 1994), face (reviewed by Francis-West *et al.*, 1998; Schneider *et al.*, 1999) or the posterior axial nervous system (reviewed by Lumsden & Krumlauf, 1996). These studies have identified four classes of molecules as essential for induction. These classes are: retinoic acid and its premetabolites, members of the TGFβ family of peptide hormones, members of the acidic FGF family of peptide hormones, and the unique inductive signalling molecule sonic hedgehog. These molecules, along with other not-yet-identified signals, are thought to act in concert at sites of induction to influence a number of cellular responses including post-translational modification of existing proteins and modulation of gene expression.

Inductive signalling is intimately related to the second critical mechanism for early development: patterning. Patterning has two manifestations; the first is morphological and the second is molecular. The morphological indication of patterning is the establishment of distinctly bounded groups of cells in a developing embryo. Thus, embryonic events such as the establishment of the neural plate, the eruption of the limb buds and subsequent formation of digits, the formation of the branchial arches are all indications of patterning in the embryo. Further indications of patterning include the emergence of repeated units such as the metameric segments of the insect embryo (Lewis, 1978; Nusslein-Volhard & Wieschaus, 1980) the somites and axial skeleton of the vertebrate embryo (reviewed by Keynes & Stern, 1988; Yamaguchi, 1997), or the neuromeric units of the vertebrate hindbrain (Lumsden & Keynes, 1989). The molecular indication of patterning is the establishment of bounded domains of gene expression that either precedes or accompanies the morphological subdivision of the embryo into regions or segments. Thus, segmentation in the insect is anticipated by or coincident with the bounded expression of a specific hierarchy of genes (reviewed by Ingham, 1988). Similarly, in the vertebrate, somite formation, axial skeletal differentiation, and some aspects of hindbrain regionalization are matched by a nested expression of a family of genes referred to as Hox genes (reviewed by Gehring, 1993; Wilkinson & Krumlauf, 1990). The expression of many of these genes can be modulated by the inductive signals catalogued above (Marshall et al., 1992; Fan et al., 1994). Thus there is an intimate relationship between the morphogenetic process of patterning during early embryogenesis and the emergence of patterns of gene expression that accompany this process. The question that remains is whether or not this relationship is seen during the initial patterning of the forebrain at later stages of development.

A CRITICAL REVIEW OF THE LITERATURE

Inductive signals in the developing forebrain

Several inductive signalling molecules may play a role in inducing regional development in the forebrain. The role of these molecules has been inferred from localization studies, pharmacological and toxicological manipulation of whole embryos and analysis of mutants that disrupt forebrain development. Retinoic acid (RA) was the first molecule to be identified as a local, endogenous inductive signal (Thaller & Eichele, 1988; Tickle et al., 1982). A role for RA in forebrain development was suggested first by a series of over-exposure studies in rodents. These studies showed that exogenous RA disrupts forebrain regional differentiation (Shenefett, 1972; reviewed by Linney & LaMantia, 1994), particularly that of the olfactory bulb. In a series of experiments from my laboratory we showed that this teratogenesis reflects a normal process of endogenous RA signalling (Anchan et al., 1997; LaMantia et al., 1993, 1995). Our results demonstrated that the forebrain normally receives a localized RA signal from the adjacent frontonasal mesenchyme, and that this signal activates region-specific gene expression in the region of the forebrain that becomes the olfactory bulb. In addition, we showed that RA signalling participates in the initial formation of the olfactory epithelium (LaMantia *et al.*, 1993; Whitesides *et al.*, 1998; LaMantia *et al.*, 2000). Thus, it seemed possible that induction via at least one known signalling molecule, RA, might coordinate the differentiation of at least one major functional pathway of the forebrain – the olfactory pathway.

Subsequently, each of the other classes of inductive signalling molecule has been shown to participate in cell–cell signalling during early forebrain development. A role for TGFβ family members has been suggested primarily by localization studies in embryos. In these studies, several bone morphogenetic proteins (BMPs), including BMP4, have been described in a number of regions of the developing forebrain (Furata *et al.*, 1997). In explant culture experiments purified BMP proteins can elicit specific gene expression in forebrain neuroep-ithelium (Furata *et al.*, 1997). Although there is some experimental evidence that BMP4 and its relative BMP2 can influence craniofacial morphogenesis (Barlow & Francis-West, 1997; Francis-West *et al.*, 1994), there is little currently available data on the biological activity of the BMPs in early embryonic forebrain or related structures *in vivo*. However, in studies of olfactory receptor neuron differentiation in culture, it seems that the BMPs can prevent neuronal differen-tiation (Shou *et al.*, 1999; LaMantia *et al.*, 2000). Thus, BMP signalling might contribute to the development of the peripheral component of one major forebrain pathway, the olfactory pathway. Similarly, it is possible that BMP4 in the forebrain might act to vary the time-course or magnitude of local neuronal differentiation so that distinct regions or pathways emerge. At this time, however, there is no clear indication that this is the case.

Localization studies of the fibroblast growth factor (FGF) family member FGF8 provided an important indication that this particular FGF might play an important role in forebrain inductive signalling. In addition to expression in the limb and branchial arches, FGF8 is expressed in a discrete patch of the ventromedial forebrain (Crossley & Martin, 1995; Heikenheimo *et al.*, 1994). Expression of FGF8 in the forebrain seems to be limited to the junction of the diencephalon with the prospective telencephalon (at this stage referred to as the prosencephalon). Furthermore, FGF8 protein can influence patterned gene expression in forebrain neuroepithelium. When recombinant FGF8 is provided to forebrain target tissue, this protein can modulate the expression of several genes that normally delineate the dorsal and ventral portions of the developing forebrain during the earliest phases of brain development (Shimamura & Rubenstein, 1997). Despite these *in-vitro* observations, no clear function for FGF8 in cellular differentiation in the nervous system has yet been postulated. Null mutations of FGF8 result in embryonic death prior to forebrain differ-entiation (Sun *et al.*, 1999). Thus, the specific function of FGF8 in inducing forebrain regions and pathways remains unclear.

The final class of inductive signalling molecule, sonic hedgehog (shh) has assumed a central role in most inductive signalling processes. The current case for sonic hedgehog signalling in the forebrain is based upon localization studies, *in-vitro* assays, and analysis of shh mutants. A domain of shh expression can be seen in the ventromedial forebrain in a region that overlaps the territory where FGF8 is expressed (Echelard *et al.*, 1993). Since shh and FGF8 apparently interact at other sites of induction including the limbs and the branchial arches

(Helms *et al.*, 1997; Johnson & Tabin 1997), it is tempting to conclude that the same sorts of interactions occur in the forebrain. This conclusion is complicated by the fact that, at these other sites, *shh* and FGF8 expression are segregated in distinct regions and tissues, rather than overlapping in the same neuroepithelial cell population, as is the case for *shh* and FGF8 in the forebrain. Nevertheless, the localization of *shh* in the developing forebrain, and its morphological similarity to the localization of *shh* in the floorplate – a source of inductive signals in the developing spinal cord and hindbrain – makes it likely that this signal plays some role in forebrain induction. *In-vitro* studies show that purified *shh* can drive the differentiation of diencephalic neurons (Dale *et al.*, 1997; Ericson *et al.*, 1995), or the expression of forebrain-specific genes in explanted telencephalic tissue (Kohtz *et al.*, 1998; Shimamura & Rubenstein, 1997). These observations support the hypothesis that *shh* may have a role in ventral forebrain induction in the intact embryo. Finally, in the null mutant of *shh* there are several forebrain anomalies. These include the disruption of the entire olfactory pathway and dramatic malformations of the remaining forebrain that resemble holoprosencephaly (Chiang *et al.*, 1996). It is difficult, however, to be sure that these malformations are a direct consequence of altered activity of *shh* on forebrain neuroepithelial targets, or the indirect effects of altered patterning at earlier times in embryogenesis.

The molecular data on expression or activity of known inductive signals provide an opportunity to consider a more fundamental question about the role of induction in early forebrain development: are there distinct cellular sources for inductive signals for the forebrain? In the spinal cord and hindbrain there is a complex inductive interplay between the notochord, the floorplate, the roofplate and the somites with the neuroepithelium. Similarly, in the limbs a series of interactions between mesenchyme and epithelium provides the cellular locale for molecular transactions that mediate induction (reviewed by Tickle & Eichele, 1994). Comparable local inductive sources and targets have not been defined for the forebrain. Recent work suggests that, for at least the olfactory pathway, the local source of inductive signals might reflect the position of neural-crest-derived frontonasal mesenchyme and its relationship with the adjacent forebrain neuroepithelium and presumptive olfactory epithelium (LaMantia *et al.*, 2000). This local mesenchymal/epithelial-mediated induction in the forebrain has several of the cellular and molecular features associated with mesenchymal/epithelial induction in non-axial structures such as the limbs, the branchial arches, and the heart (reviewed by LaMantia, 1999). Thus, at least one aspect of forebrain induction might be similar to that for non-axial structures such as the limbs, rather than that seen in axial structures such as the spinal cord, hindbrain, and axial skeleton.

Patterned gene expression in the developing forebrain

The molecular hallmark of patterning during early development is the division of embryonic regions into distinct expression domains of developmentally regulated genes. In several cases, most notably the *Drosophila* embryo, as well as the vertebrate limb, and hindbrain, the expression of these genes prefigures differentiation of distinct regions or structures. It seems reasonable to assume that patterned gene expression and its consequences influence early forebrain development. Moreover, since all of the inductive signalling molecules described above have been implicated in the regulation of patterning genes (reviewed by Johnson & Tabin, 1997; Rubenstein & Beachy, 1998) it seems plausible that they might participate in similar functions in the developing forebrain. Indeed, many such genes have now been identified. Their expression is limited to distinct regions of the forebrain (reviewed by Rubenstein *et al.*, 1994). The diversity of their patterns has raised the question of how they might be related to the functional subdivision of the forebrain. Some of these genes appear to divide the dorsal and ventral portions of the telencephalon, others have an apparent relationship to distinct functional regions, and others have no clear correlation with anatomical or functional subdivisions of the forebrain.

There are several patterning genes associated with the early differentiation of the telencephalon. *Otx2* (a homologue of the *Drosophila orthodenticle* gene; Simeone *et al.*, 1992) is found throughout the early mouse forebrain development (around the 9th day of gestation). *Nkx2.1* (a homologue of the *Drosophila ventral nervous system defective* gene; Price *et al.*, 1992) and *Bf1* (a homologue of the *Drosophila forkhead* gene; Tao & Lai, 1992) both characterize the ventral forebrain early in development. *Pax6* (a homologue of the *Drosophila* paired box gene; Walther & Gruss, 1991, 1992) and *Emx1 & 2* (a homologue of the *Drosophila empty spiracles* gene; Simeone *et al.*, 1992) are expressed in the dorsal aspect of the immature forebrain. The expression of at least one of these genes, *Otx2*, can be elicited *in vitro* by apposition of forebrain tissue with anterior mesendoderm (Ang *et al.*, 1994). This accords well with the assumed role of these genes in distinguishing the ventral forebrain early in development. Indeed, when *Bf1* (Xuan *et al.*, 1995), or *Nkx2.1* (Sussel *et al.*, 1999) are inactivated by homologous recombination ("knocked out") ventral telencephalic structures such as the basal ganglia and basal forebrain nuclei show the most severe alterations. These observations suggest that there may be transcriptional regulatory genes whose regional expression influences the early establishment of boundaries and zones in the forebrain.

Many of the genes that distinguish forebrain regions are expressed after rudiments of the forebrain subdivisions can be recognized. The expression patterns and boundaries of these genes have been interpreted as a rough map of forebrain segments or "prosomeres". It is not clear whether these expression domains represent metameric units or a more specific type of distinct developmental field. Nevertheless, several of these regional distinctions in gene expression emerge at the time when rudimentary forebrain regions can be recognized morphologically. Included in the list of genes that distinguish these incipient forebrain subdivisions are the *Dlx* genes (homologues of *Drosophila* distal-less genes; Liu *et al.*, 1997) *Emx* genes, and *Tbr1* (Bulfone *et al.*, 1995). Inactivation of these genes results in either the deletion of subpopulations of interneurons in the cortex or olfactory bulb (*Dlx1,2*, Anderson *et al.*, 1997a,b; Bulfone *et al.*, 1998), projection neurons in the olfactory bulb (*Tbr1*; Bulfone *et al.*, 1998) or entire neuronal fields (CA3 of hippocampus for *Emx2*; Yoshida *et al.*, 1997). The phenotypes of these mutations suggest that these genes contribute to the differentiation of distinct cell classes within already-established forebrain regions. Thus, the role of patterning genes in regional differentiation of the forebrain, versus their role in cellular differentiation of already distinct forebrain regions, remains unclear.

The significance of early and later patterning of gene expression in the forebrain can be interpreted in light of the central role of induction in establishing these patterns. Clearly, in most other embryonic fields, regionally bounded gene expression reflects inductive signaling as a first step in establishing these boundaries. The existence of multiple patterns of gene expression in the developing forebrain – those that distinguish dorsal from ventral as well as those that distinguish individual forebrain regions – suggests that there must be inductive mechanisms that contribute to establishing these patterns. The challenge that remains is making explicit links between identified inductive signaling centers, the molecular signals they employ, and the downstream patterning genes that are influenced by these inductive signals.

A model for induction and regionalization of a forebrain subdivision

Until recently there was no sense that inductive interactions contributed to a sequence of developmental events that lead to the morphogenesis and differentiation of distinct functional pathways in the developing forebrain. Observations in the developing spinal cord (reviewed by Tanabe & Jessell, 1996) and hindbrain (reviewed by Lumsden & Krumlauf, 1996) have demonstrated a primary role for induction in a series of events that constrains the initial development of motor and sensory pathways. It seems reasonable therefore to ask whether inductive interactions associated with early forebrain development have any consequence for the assembly of distinct functional regions and pathways in the developing forebrain.

Work from my laboratory has demonstrated that neural crest-mediated, RA-dependent mesenchymal/epithelial induction in the early forebrain guides the initial formation of the primary olfactory pathway: therefore the olfactory bulb, and its primary source of peripheral innervation, the olfactory epithelium (Figure 2). This induction resembles that which occurs for craniofacial structures, limbs, and the heart. All rely upon local RA signals from neural crest-derived mesenchymal cells. In the developing forebrain RA is locally available from the frontonasal mesenchyme. Furthermore, the production of this ligand apparently relies upon the proper migration and position of a subset of neural crest cells within the frontonasal mesenchyme. The forebrain neuroepithelium and presumptive olfactory epithelium express RA receptors and binding proteins (Anchan et al., 1997; LaMantia et al., 1993, 1995; Whitesides et al., 1998). Thus, the ligand and necessary signal transduction molecules are available in the developing forebrain. RA is not the only inductive signal associated with the nascent olfactory pathway. The related inductive signalling molecules shh, FGF8 and BMP4 are present in and around rudimentary olfactory structures (LaMantia et al., 2000). Accordingly, it is possible that all of these inductive signals, plus others that remain to be identified, mediate the mesenchymal/epithelial interactions that underlie olfactory pathway formation.

We have also demonstrated that RA-mediated gene expression (assessed using an indicator transgenic mouse in which a direct repeat 5-retinoic acid response element (DR5-RARE)/β-galactosidase (β-gal) transgene detects RA-mediated

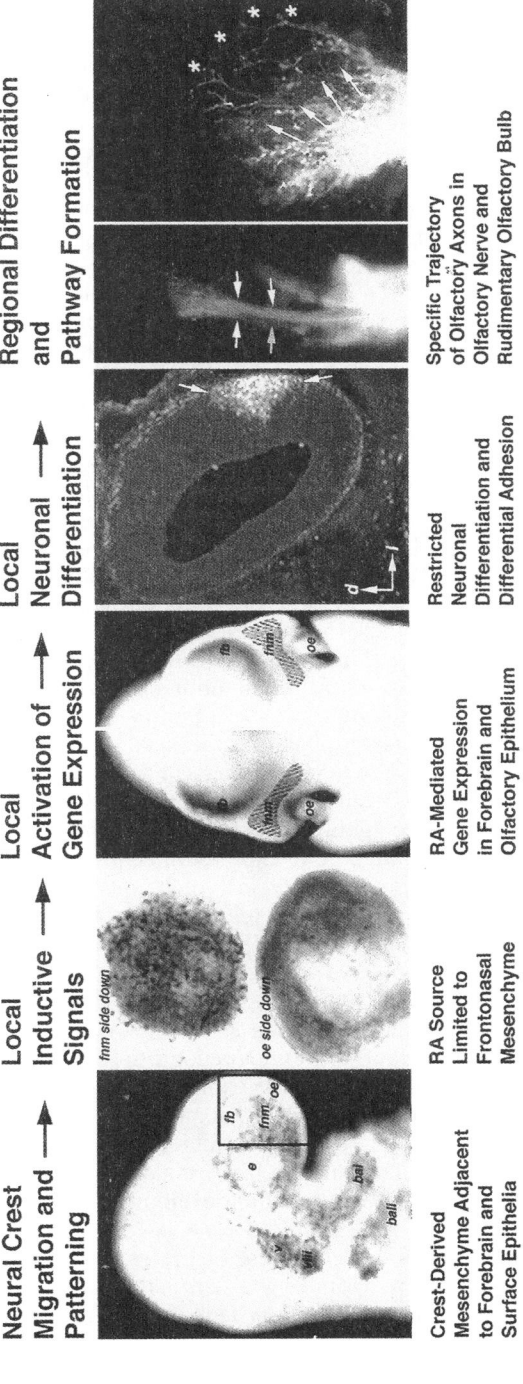

Figure 2 A schematic of the events that underlie the initial morphogenesis and differentiation of the mammalian olfactory pathway during early forebrain development. Our results suggest that one of the earliest steps in this process is the migration of a population of neural crest that is destined for the frontonasal mesenchyme. The position of these cells and the signals they provided then influences the expression and pattern of other inductive signalling molecules and downstream genes. The regions where this differential gene expression occurs then are sites of early neuronal differentiation, differential adhesion and axon ingrowth. This process of cellular differentiation underlies the morphogenesis of the primary olfactory pathway during the period when rudiments of forebrain functional subdivisions emerge. fb: firebrain vesicle, fnm: frontonasal mesenchyme, e: eye, v: triguinal ganglion, vii: spiral ganglion, baI: 1st branchial arch, baII: 2nd branchial arch, oe: olfactory epithelium, d: dorsal, l: lateral.

gene expression; Balkan *et al.*, 1992; Colbert *et al.*, 1993) occurs in distinct domains in the developing forebrain. This RA-dependent transcriptional activation is in register with the position of local RA sources in the frontonasal mesenchyme. Finally, RA-dependent gene expression influences cell adhesion and cellular differentiation in the ventrolateral forebrain, olfactory epithelium, and olfactory nerve (Anchan *et al.*, 1997; LaMantia *et al.*, 1993, 1995; Whitesides & LaMantia, 1995; Whitesides *et al.*, 1998). Thus, all of the available evidence indicates that RA-mediated mesenchymal/epithelial induction plays a role in early forebrain regional and pathway development.

If induction and RA signalling play a role in normal olfactory pathway development then altered induction and RA signalling caused by mutations in relevant genes might compromise the forebrain as well as peripheral structures. We therefore assessed induction and RA-mediated gene expression in mutant mice where regional differentiation in the forebrain plus morphogenesis of the limbs or face are compromised (Figure 3). Homozygous *small eye* mutant mice (a mutation in the murine *Pax6* gene) lack the olfactory bulb in the forebrain as well as the olfactory epithelium and nerve in the periphery (Hogan *et al.*, 1988; Hill *et al.*, 1991; Schmahl *et al.*, 1993; Anchan *et al.*, 1997). This morphogenetic anomaly is seen from the earliest stages. In addition, there are apparent disruptions in cell migration and position in the remaining forebrain subdivisions (Schmahl *et al.*, 1993). These forebrain anomalies are accompanied by small or absent eyes, and a variety of altered craniofacial features (Anchan *et al.*, 1997; Hogan *et al.*, 1988; Quinn *et al.*, 1996). In *extra toes* mutant mice (a mutation in the murine *Gli3* gene) homozygotes lack the olfactory bulb, but not the olfactory epithelium or nerve (Johnson, 1967; Schimmang *et al.*, 1992; Sullivan *et al.*, 1995). In addition, the cytological organization of the remaining forebrain subdivisions is significantly compromised (Schimmang *et al.*, 1992; Figure 3). These forebrain anomalies are accompanied by duplicated digits, craniofacial malformations and cardiovascular anomalies (Johnson, 1967; Schimmang *et al.*, 1992).

We bred the DR5-RARE/β-gal transgene into *small eye* homozygotes to evaluate the integrity of RA-mediated gene expression in these mutant embryos. In *small eye* homozygotes, RA-mediated gene expression cannot be detected in the presumptive olfactory epithelium or forebrain (Anchan *et al.*, 1997; Figure 4). Furthermore, RA production cannot be detected in the frontonasal mesenchyme of these embryos (Anchan *et al.*, 1997; Figure 4, inset). The absence of normal domains of RA-mediated gene expression is not due to a lack of RA responsiveness in *small eye* homozygote embryos or alterations in the expression of several RA receptors and co-factors (Anchan *et al.*, 1997). The widespread expression of RA receptor (RAR)α and retinoid X receptor (RXR) γ is retained in the mutants, and ectopic RA-dependent gene expression is seen throughout the telencephalic vesicle when mutant embryos are exposed to RA via maternal circulation (Anchan *et al.*, 1997; Figure 4). These results suggest that the absence of local RA signals from the frontonasal mesenchyme make a primary contribution to the failure of olfactory bulb and epithelium development in homozygous *small eye* embryos.

The *extra toes* mutation provides a second test of the role of RA-mediated induction in forebrain development. Homozygous *extra toes* embryos lack an olfactory bulb, but have an olfactory epithelium (Figure 3). Accordingly, one

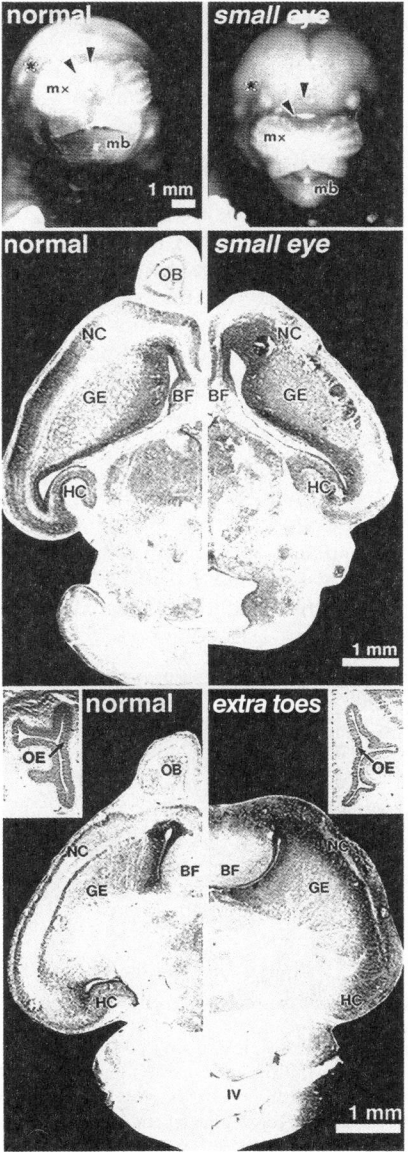

Figure 3 The effects of the *Pax-6* and *Gli-3* mutations on forebrain regional differentiation. *Top panels* show a normal E18 mouse embryo (left) and *small eye* (Pax-6) homozygous mutant littermate (right). mx: maxillary process, mb: mandibular process, arrows indicate nasal processes, which are absent in the small eye mouse. *Middle panels* show the regional organization of the forebrain in normal (left) and homozygous mutant littermates (right) at E18. OB: olfactory bulb, NC : neocortex, GE: ganglionic eminence, BF: basal forebrain, HC: hippocampus. The OB is absent in small eye mice. *Bottom panels* show the organization of the forebrain in a normal E18 mouse (left) and a homozygous extra toes (Gli-3) mutant littermate (right). OE: olfactory epithelium (shown in inset) is present in wild type and mutant extra toes embryo.

would predict that DR5-RARE-dependent RA activation would be absent in the homozygous *extra toes* forebrain, but retained in the presumptive olfactory epithelium. This is indeed the case. RA-mediated gene expression is completely absent in the mutant forebrain; however, RA-mediated gene expression is seen in the presumptive olfactory epithelium (Figure 4). Furthermore, the frontonasal mesenchyme still produces RA (Figure 4, inset). The absence of RA-dependent gene expression reflects an apparent transcriptional insensitivity to RA in the mutant forebrain. Thus, when homozygous *extra toes* embryos are exposed to exogenous RA, it is impossible to drive RA-mediated transgene expression in the forebrain with a wide range of concentrations (effective concentrations of between 10^{-7} and 10^{-4} M; reviewed by Linney & LaMantia, 1994). In these RA-exposed *extra toes* homozygotes, ectopic RA-mediated gene expression persists in other locations including the spinal cord and branchial arches (Figure 4). The forebrain of the *extra toes* homozygote is therefore transcriptionally unresponsive to RA which is normally provided by the mesenchyme. In contrast, the presumptive olfactory epithelium retains its RA responsiveness. This may help to explain why, in homozygote *extra toes* embryos, the olfactory bulb – which is derived from the forebrain vesicle – fails to develop, while the olfactory epithelium and nerve – which differentiate from the placode – are recognizable.

We have recently taken advantage of a single-insertion transgenic "enhancer" or "gene trap" line (Friedrich & Soriano, 1991) to evaluate the role of the neural crest in RA-mediated forebrain induction. This transgenic line (referred to as βgeo-6; originally produced by J. Heemskerk and J. Dodd, Department of

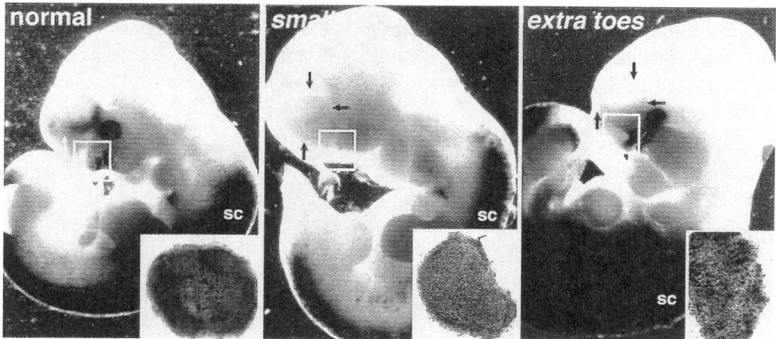

Figure 4 Retinoid-mediated gene expression is disrupted in homozygous mutant *Pax-6* and *Gli-3* mice. The *left panel* shows a normal E10.5 mouse embryo expressing a RA-sensitive transgene that indicates domains of RA-mediated gene expression in the forebrain, presumptive olfactory epithelium, eye and spinal cord. The box indicates the frontonasal mesenchyme and olfactory epithelium. The *inset* shows a RA indicator assay of embryonic frontonasal mesenchyme on a monolayer of cells stably transfected with the same RA-sensitive transgene. Labelled cells (black) indicate regions in the overlying explant that produce RA. The *middle panel* shows that RA signalling is disrupted in both the forebrain and the presumptive olfactory epithelium in *Pax-6* homozygous mice. The box indicates the frontonasal region and the black arrows the forebrain vesicle. The *inset* shows that RA is not produced by the mutant frontonasal mesenchyme. The *right panel* shows that RA signalling is disrupted only in the forebrain (black arrows) in *Gli-3* homozygous mutant mice. The olfactory epithelium (boxed) shows transgene activity that approximates that in the normal (although there are some differences). The *inset* shows that RA is produced by the frontonasal mesenchyme in homozygous *Gli-3* mutant mice.

Physiology, Columbia University College of Physicians and Surgeons) was generated by a single random insertion of a reporter gene behind a cell-specific or tissue-specific regulatory element in the host genome. When a locus is "trapped" enhancer elements in the trapped gene drive expression of a reporter such as β-gal in the relevant cells or tissues. Accordingly, the βgeo-6 enhancer trap identifies a genetic locus expressed uniquely in embryonic neural crest, including neural crest-derived mesenchymal cells. These mesenchymal cells are especially prominent at the interfaces of the mesenchyme with the forebrain and olfactory epithelia as well as in the branchial arches, eye, limb, as well as around the developing heart (Bhasin *et al.*, 2001). Thus they may be appropriately positioned to provide RA to a number of embryonic targets where mesenchymal/epithelial induction occurs, including the forebrain. Our recent work suggests that these cells produce RA. Thus, when βgeo-6-labelled frontonasal mesenchyme is co-cultured on a monolayer of DR5/RARE/green fluorescent protein (GFP) indicator cell line there is precise alignment between the βgeo-6-labelled cells in the mesenchyme of the explant, and the RA-activated, GFP-expressing cells in the indicator monolayer (LaMantia *et al.*, 2000). This implies that neural crest-associated mesenchymal cells or their immediate neighbours produce RA. The registration between these RA-producing neural crest mesenchymal cells and RA-responsive cells in the forebrain neuroepithelium and presumptive olfactory epithelium reinforces this conclusion.

The association of neural crest-derived mesenchymal cells with RA production suggests a critical genetic experiment to test our hypothesis that these mesenchymal cells contribute to forebrain development. If these cells are essential for induction of the olfactory bulb and epithelium, they should be specifically absent from the frontonasal mesenchyme of *small eye* homozygote mutant mice, where RA is not produced. In β-geo6/*small eye* homozygous embryos these cells are absent from the frontonasal mesenchyme, but remain in all other targets of the cranial neural crest (Anchan *et al.*, 1997; LaMantia *et al.*, 2000). Apparently, neural crest-associated, RA-producing mesenchymal cells are necessary for normal forebrain induction and regional differentiation. Thus, cell biological, molecular biological and genetic evidence all points to a primary role for neural crest-derived mesenchymal/ epithelial inductive interactions in the initial development of a major functional subdivision of the forebrain – the olfactory bulb – as well as its primary source of extrinsic innervation – the olfactory epithelium.

DISCUSSION AND CONCLUSION

There is compelling evidence that induction can initiate a sequence of events that leads to the differentiation of distinct forebrain regions, pathways and circuits. Thus, the cellular and molecular mechanisms that underlie forebrain induction might provide a potential target for a number of toxins, infectious agents and genetic mutations that compromise forebrain development or function. Recently, inductive signalling via RA or *shh* has been implicated in developmental syndromes that include compromised forebrain development (Porter *et al.*, 1996; Lanoue *et al.*, 1997; Goodman, 1998; LaMantia, 1999). Although detailed relationships between induction and mature circuitry remain to be

93

determined, there is every reason to believe that disruption of early events can have an impact on the ultimate organization and function of the forebrain.

At present, most knowledge of the function of inductive signals and related patterning genes comes from studies where extensive experimental manipulation or complete genetic deletion has been used to generate extreme and often early lethal phenotypes. While informative, these experiments do not address the impact of inductive signals and patterning genes on shaping mature brain architecture. Thus, experimental strategies are needed to disrupt inductive signalling and patterning in clear but more subtle ways, and then to assess the phenotypic results of these disruptions. Increasingly sophisticated approaches to mutagenesis and gene targeting via homologous recombination offers one avenue to pursue this goal (Woychik *et al.*, 1998; Hrabe de Angelis & Balling, 1998; Evans *et al.*, 1997). In addition, continued efforts to generate and screen mouse mutants for central nervous system deficits may also be useful (Hentges *et al.*, 1999). Finally, it may be possible to take advantage of a new sort of "reverse genetics" using human genetic deletions and mutations as a starting point. Homologues of human genes linked with behavioural and psychiatric disorders may be studied in animals to assess their roles in a number of neurobiological processes – including early development of the forebrain. These molecular and genetic technologies, in combination with careful phenotypic analysis of neural circuits, may provide further insight into the role of induction and patterning in establishing functionally distinct forebrain regions, neurons and circuitry.

References

Anchan, R.M., Drake, D.P., Gerwe, E.A., Haines, C.F. & LaMantia, A.-S. (1997). A failure of retinoid-mediated induction accompanies the loss of the olfactory pathway during mammalian forebrain development. *Journal of Comparative Neurology, 379*, 1–15.

Anderson, S.A., Eisenstat, D.D., Shi, L. & Rubenstein, J.L.R. (1997a). Interneuron migration from the basal forebrain to neocortex: dependence on *Dlx* genes. *Science*, **278**, 474–476.

Anderson, S.A., Qiu, M.S., Bulfone, A., Eisenstat, D.D., Meneses, J., Pedersen, R. & Rubenstein, J.L.R. (1997b). Mutations of the homeobox genes *Dlx-1* and *Dlx-2* disrupt the striatal subventricular zone and differentiation of late born striatal neurons. *Neuron*, **19**, 27–37.

Ang, S.L., Conlon, R.A., Jin, O. & Rossant, J. (1994). Positive and negative signals from mesoderm regulate the expression of mouse Otx2 in ectoderm explants. *Development*, **120**, 2979–2989.

Balkan, W., Colbert, M.C., Bock, C. & Linney, E. (1992). Transgenic indicator mice for studying activated retinoic acid receptors during development. *Proceedings of the National Academy of Sciences, USA*, **89**, 3347–3351.

Barlow, A.J. & Francis-West, P.H. (1997). Ectopic application of recombinant BMP-2 and BMP-4 can change patterning of developing chick facial primordia. *Development*, **124**, 391–398.

Bulfone, A., Smiga, S.M., Shimamura, K., Peterson, A., Puelles, L. & Rubenstein, J.L. (1995). T-brain-1: a homolog of Brachyury whose expression defines molecularly distinct domains within the cerebral cortex. *Neuron*, **15**, 63–78.

Bulfone, A., Wang, F., Hevner, R., Anderson, S., Cutforth, T., Chen, S., Meneses, J., Pedersen, R., Axel, R. & Rubenstein, J.L. (1998). An olfactory sensory map develops in the absence of normal projection neurons or GABAergic interneurons. *Neuron*, **21**, 1273–1282.

Chiang, C., Litingtung, Y., Lee, E., Young, K.E., Corden, J.L., Westphal, H. & Beachy, P.A. (1996). Cyclopia and defective axial patterning in mice lacking *sonic hedgehog* gene function. *Nature*, **383**, 407–413.

Colbert, M.C., Linney, E. & LaMantia, A.-S. (1993). Local sources of retinoic acid coincide with retinoid-mediated transgene activity during embryonic development. *Proceedings of the National Academy of Sciences, USA*, **90**, 657–661.

Crossley, P.H. & Martin, G.R. (1995). The mouse *Fgf8* gene encodes a family of polypeptides and is expressed in regions that direct outgrowth and patterning in the developing embryo. *Development*, **121**, 439–451.

Dale, J.K., Vesque, C., Lints, T.J., Sampath, T.K., Furley, A., Dodd, J. & Placzek, M. (1997). Cooperation of BMP7 and *shh* in the induction of forebrain ventral midline cells by prechordal mesoderm. *Cell*, **90**, 257–269.

Echelard, Y., Epstein, D.J., St-Jacques, B., Shen, L., Mohler, J., McMahon, J.A. & McMahon, A.P. (1993). Sonic hedgehog, a member of a family of putative signaling molecules, is implicated in the regulation of CNS polarity. *Cell*, **75**, 1417–1430.

Ericson, J., Muhr, J., Placzek, M., Lints, T., Jessell, T.M. & Edlund, T. (1995). Sonic hedgehog induces the differentiation of ventral forebrain neurons: a common signal for ventral patterning within the neural tube. *Cell*, **81**, 747–756.

Evans, M.J., Carlton, M.B. & Russ, A.P. (1997). Gene trapping and functional genomics. *Trends in Genetics*, **13**, 370–374.

Fan, C.M. & Tessier-Levigne, M. (1994). Patterning of mammalian somites by surface ectoderm and notochord: evidence for sclerotome induction by a hedgehog homolog. *Cell*, **79**, 1175–1186.

Francis-West, P.H., Tatla, T. & Brickell, P.M. (1994). Expression patterns of the bone morphogenetic protein genes Bmp-4 and Bmp-2 in the developing chick face suggest a role in outgrowth of the primordia. *Developmental Dynamics*, **201**, 168–178.

Francis-West, P., Ladher, R., Barlow, A. & Graveson, A. (1998). Signalling interactions during facial development. *Mechanisms of Development*, **75**, 3–28.

Friedrich, G. & Soriano, P. (1991). Promoter traps in embryonic stem cells: a genetic screen to identify and mutate developmental genes in mice. *Genes and Development*, **5**, 1513–1523.

Furata, Y., Piston, D.W. & Hogan, B.L.M. (1997). Bone morphogenetic proteins (BMPs) as regulators of dorsal forebrain development. *Development*, **124**, 2203–2212.

Gehring, W.J. (1993). Exploring the homeobox. *Gene*, **135**, 215–221.

Goodman, A.B. (1998). Three independent lines of evidence suggest retinoids as causal to schizophrenia. *Proceedings of the National Academy of Sciences, USA*, **95**, 7240–7244.

Gruss, P. & Walther, C. (1992). Pax in development. *Cell*, **69**, 719–722.

Gurdon, J.B. (1992). The generation of diversity and pattern in animal development. *Cell*, **68**, 185–199.

Hamburger, V. (1988). *The heritage of experimental embryology – Hans Spemann and the organizer*. New York: Oxford University Press.

Hammerschmidt, M., Brook, A. & McMahon, A.P. (1997). The world according to hedgehog. *Trends in Genetics*, **13**, 10–17.

Heikinheimo, M., Lawshe, A., Shackleford, G.M., Wilson, D.B. & MacArthur, C.A. (1994). Fgf-8 expression in the post-gastrulation mouse suggests roles in the development of the face, limbs and central nervous system. *Mechanisms of Development*, **48**, 29–38.

Helms, J.A., Kim, C.H., Hu, D., Minkoff, R., Thaller, C. & Eichele, G. (1997). Sonic hedgehog participates in craniofacial morphogenesis and is down-regulated by teratogenic doses of retinoic acid. *Developmental Biology*, **187**, 25–35.

Hentges, K., Thompson, K. & Peterson, A. (1999). The flat-top gene is required for the expansion and regionalization of the telencephalic primordium. *Development*, **126**, 1601–1609.

Hill, R.E., Favor, J., Hogan, B.L.M., Ton, C.C.T., Saunders, G.F., Hanson, I.M., Prosser, J., Jordan, T., Hastie, N.D. & van Heyningen, V. (1991). Mouse *small eye* results from mutations in a paired-like homeobox-containing gene. *Nature*, **354**, 522–525.

Hogan, B.L. (1996). Bone morphogenetic proteins: multifunctional regulators of vertebrate development. *Genes & Development*, **10**, 1580–1594.

Hogan, B.L., Hirst, E.M.A., Horsburgh, G. & Hetherington, C.M. (1988). Small eye (*Pax6[Sey-Neu]*): a mouse model for the genetic analysis of craniofacial abnormalities. *Development*, **103** (Suppl.), 115–119.

Hrabe de Angelis, M. & Balling, R. (1998). Large scale ENU screens in the mouse: genetics meets genomics. *Mutation Research*, **400**, 25–32.

Ingham, P. (1988). The molecular genetics of embryonic pattern formation in *Drosophila*. *Nature*, **335**, 25–34.

Jessell, T.M. & Melton, D.A. (1992). Diffusible factors in vertebrate embryonic induction. *Cell*, **68**, 257–270.

Johnson, D.R. (1967). Extra-toes: a new mutant gene causing multiple abnormalities in the mouse. *Journal of Embryology and Experimental Morphology*, **17**, 543–581.

Johnson, R.L. & Tabin, C.J. (1997). Molecular models for vertebrate limb development. *Cell*, **90**, 979–990.

Keynes, R.J. & Stern, C.D. (1988). Mechanisms of vertebrate segmentation. *Development*, **103**, 413–429.

Kohtz, J.D., Baker, D.P., Corte, G. & Fishell, G. (1998). Regionalization within the mammalian telencephalon is mediated by changes in responsiveness to sonic hedgehog. *Development*, **125**, 5079–5089.

LaMantia, A-S. (1999). Forebrain induction, retinoic acid, and vulnerability to schizophrenia: insights from molecular and genetic analysis in developing mice. *Biological Psychiatry*, **46**, 19–30.

LaMantia, A.-S., Colbert, M.C. & Linney, E. (1993). Retinoic acid induction and regional differentiation prefigure olfactory pathway formation in the mammalian forebrain. *Neuron*, **10**, 1035–1048.

LaMantia, A.-S., Colbert, M.C. & Linney, E. (1995). Induction and the generation of regional and cellular diversity in the developing mammalian brain. In B.H.J. Juurlink *et al*. (Eds.), *Neural cell specification: third annual Altschul symposium proceedings* (pp. 51–65). New York: Plenum Press.

LaMantia, A-S., Bhasin, N., Rhodes, K. & Heemskerk, J. (2000). Mesenchymal/epithelial induction mediates olfactory pathway formation. *Neuron*, **28**, 411–425.

Lanoue, L., Dehart, D.B., Hinsdale, M.E., Maeda, N., Tint, G.S. & Sulik, K.K. (1997). Limb, genital, CNS, and facial malformations result from gene/environment-induced cholesterol deficiency: further evidence for a link to sonic hedgehog. *Americal Journal of Medical Genetics,* **731**, 24–31.

Lewis, E.B. (1978). A gene complex controling segmentation in *Drosophila*. *Nature,* **276**, 565–570.

Linney, E. & LaMantia A.-S. (1994). Retinoid signaling in mouse embryos. *Advances in Developmental Biology,* **3**, 373–114.

Liu, J.K., Ghattas, I., Liu, S., Chen, S. & Rubenstein, J.L. (1997). Dlx genes encode DNA-binding proteins that are expressed in an overlapping and sequential pattern during basal ganglia differentiation. *Developmental Dynamics*, **210**, 498–512.

Lumsden, A. & Keynes, R.J. (1989). Segmental patterns of neuronal development in the chicken hindbrain. *Nature*, **337**, 424–428.

Lumsden, A. & Krumlauf, R. (1996). Patterning the vertebrate neuraxis. *Science*, **274**, 1109–1115.

Marshall, H., Nonchev, S., Sham, M.H., Muchamore, I., Lumsden, A. & Krumlauf, R. (1992). Retinoic acid alters hindbrain Hox code and induces tranformation of rhombomeres 2/ into a 4/5 identity. *Nature*, **360**, 737–741.

Mason, I.J. (1994). The ins and outs of fibroblast growth factors. *Cell*, **78**, 547–552.

Nusslein-Volhard, C. & Wieschaus, E. (1980). Mutations affecting segment number and polarity in *Drosophila*. *Nature*, **287**, 795–801.

Porter, J.A., Young, K.E. & Beachy, P.A. (1996). Cholesterol modification of hedgehog signaling proteins in animal development. *Science*, **274**, 255–259.

Price, M., Lazzaro, D., Pohl, T., Mattei, M-G., Ruther, U., Olivo, J-C., Duboule, D. & DiLauro, R. (1992). Regional expression of the homeobox gene Nkx-2.2 in the developing mammalina forebrain. *Neuron*, **8**, 241–255.

Quinn, J.C., West, J.D. & Hill, R.E. (1996). Multiple functions for *Pax6* in mouse eye and nasal development. *Genes and Development*, **10**, 435–446.

Rubenstein, J.L. & Beachy, P.A. (1998). Patterning of the embryonic forebrain. *Current Opinion in Neurobiology,* **8**, 18–26.

Rubenstein, J.L., Martinez, S., Shimamura, K. & Puelles, L. (1994). The embryonic vertebrate forebrain: the prosomeric model. *Science*, **266**, 578–580.

Schimmang, T., Lemaistre, M., Vortkamp, A. & Ruther, U. (1992). Expression of the zinc finger gene Gli3 is affected in the morphogenetic mouse mutant *extra-toes* (*Xt*). *Development*, **116**, 799–804.

Schmahl, W., Knoedlseder, M., Favor, J. & Davidson, D. (1993). Defects of neuronal migration and the pathogenesis of cortical malformations are associated with *Small eye* (*Sey*) in the mouse, a point mutation at the *Pax-6* locus. *Acta Neuropathologica*, **86**, 126–135.

Schneider, R.A., Hu, D. & Helms, J.A. (1999). From head to toe: conservation of molecular signals regulating limb and craniofaciail morphogenesis. *Cell Tissue Research*, **296**, 103–109.

Shimamura, K. & Rubenstein, J.L.R. (1997). Inductive interactions direct early regionalization of the mouse forebrain. *Development*, **124**, 2709–2718.

Shou, J., Rim, P.C. & Calof, A.L. (1999). BMPs inhibit neurogenesis by a mechanism involving degradation of a transcription factor. *Nature Neuroscience*, **2**, 339–345.

Simeone, A., Acampora, D., Gulisano, M., Stornaiuolo, A. & Boncinelli, E. (1992). Nested expression domains of four homeobox genes in developing rostral brain. *Nature*, **358**, 687–690.

Slack, J.M. (1993). Embryonic induction. *Mechanisms of Development*, **41**, 91–107.

Sullivan, S.L., Bohm, S., Ressler, K.J., Horowitz, L.F & Buck, L.B. (1995). Target-independent pattern specification in the olfactory epithelium. *Neuron*, **15**, 779–789.

Sun, Xm, Meyers, E.N., Lewandowski, M. & Martin, G.R. (1999). Targeted disruption of FGF8 causes failure of cell migration in the gastrulating mouse embryo. *Genes and Development*, **13**, 1834–1846.

Sussel, L., Marin, O., Kimura, S. & Rubenstein, J.L.R. (1999). Loss of Nkx2.1 homeobox gene function results in a ventral to dorsal molecular respecification within the basal telencephalon: evidence for a transformation of the pallidum into the striatum. *Development*, **126**, 3359–3370.

Tanabe, Y. & Jessell, T.M. (1996). Diversity and pattern in the developing spinal cord. *Science*. Nov 15; **274**(5290): 1115–1123.

Tao, W. & Lai, E. (1992). Telencephalon-restricted expression of Bf-1, a new member of the HNF-3/forkhead gene family, in the developing rat brain. *Neuron*, **8**, 957–966.

Thaller, C. & Eichele, G. (1987). Identification and spatial distribution of retinoids in the developing chick limb bud. *Nature*, **327**, 62–628.

Tickle, C. & Eichele, G. (1994). Vertebrate limb development. *Annual Review of Cell Biology*, **10**, 121–152.

Tickle, C., Alberts, B., Wolpert, L. & Lee, J. (1982). Local application of retinoic acid to the limb bud mimics the action of the polarizing region. *Nature*, **296**, 564–565.

Walther, C. & Gruss, P. (1991). *Pax-6*, a murine paired box gene, is expressed in the developing CNS. *Development*, **113**, 1435–1449.

Whitesides, J.G. & LaMantia, A.-S. (1995). Differential adhesion of neurons and neural precursor cells from a distinct domain in the developing mammalian forebrain. *Developmental Biology*, **169**, 229–241.

Whitesides, J.G., Hall, M.E., Anchan, R.M. & LaMantia, A.-S. (1998). Retinoid signaling distinguishes a subpopulation of olfactory receptor neurons throughout life in the mouse olfactory epithelium. *Journal of Comparative Neurology*, **384**, 9–17.

Wilkinson, D.G. & Krumlauf, R. (1990). Molecular approaches to the segmentation of the hindbrain. *Trends in Neuroscience*, **13**, 335–339.

Woychik, R.P., Klebig, M.L., Justice, M.J., Magnuson, T.R. & Avner, E.D. (1998). Functional genomics in the post-genome era. *Mutation Research*, **400**, 3–14

Xuan, S., Baptista, C.A., Balas, G., Tao, W., Soares, V.C. & Lai, E. (1995). Winged helix transcription factor BF-1 is essential for the development of the cerebral hemispheres. *Neuron*, **14**, 1141–1152.

Yamaguchi, T. (1997). New insights into segmentation and patterning during vertebrate somitogenesis. *Current Opinion in Genetics and Development*, **7**, 513–518.

Yoshida, M., Suda, Y., Matsuo, I., Miyamoto, N., Takeda, N., Kurutani, S. & Aizawa, S. (1997). Emx1 and Emx2 function in development of dorsal telencephalon. *Development*, **124**, 101–111.

II.4
Cholinergic Regulation of Cortical Development

CHRISTINE F. HOHMANN

ABSTRACT

Although it is the oldest identified neurotransmitter, the functions and properties of acetylcholine (ACh) and its responding receptors continue to be defined in the central nervous system. Most recently a cholinergic role in cortical development has become apparent, and is now the subject of expanding research. This chapter will review and interpret the experimental literature concerning such an ontogenetic role of the cholinergic system within the context of human cortical development.

To clarify terminologies, as well as anatomical, neurochemical and physiological relationships, a brief overview of the well-characterized aspect of cortical cholinergic neurotransmission, in general, will be provided, followed by a description of the organization of the cholinergic system in the human. A thorough review of the literature on cholinergic development in the human follows, juxtaposed to the developmental literature in the rodent. Comparison of the human and rodent developmental profiles clearly reveals shared features which suggest the involvement of cholinergic signals in cortical pyramidal cell maturation and establishment of appropriate thalamocortical connectivity. In the human the hypothesis of cholinergic modulation of such morphogenetic events remains based on observations of developmental correlates and symptoms associated with certain developmental disorders. However, in the rodent, considerable research now supports a role of acetylcholine in the maturation of cortical neurons and connectivity and its subsequent impact on cognitive function. A detailed review of this research is provided, much of which is composed of animal models of neonatal cortical cholinergic depletion or reconstitution. In addition, molecular experiments, implicating cholinergic receptors and secondary messenger mechanisms in the morphogenetic process, are considered. These molecular mechanisms suggest a possible model for neuronal integration of cholinergic and other developmental signals in the form

A.F. Kalverboer and A. Gramsbergen (eds.), Handbook of Brain and Behaviour in Human Development, 99–138
© *2001 Kluwer Academic Publishers. Printed in Great Britain.*

of local regulation of intracellular calcium currents. Possible implications of this research for human cognitive development, particularly as it may pertain to developmental disorders, are discussed.

INTRODUCTION

In the 1920s acetylcholine (ACh) in the peripheral nervous system became the first identified neurotransmitter (Karczmar, 1993). It would take another 30 years before a role of acetylcholine in central nervous system function became established (Karczmar, 1993). Since then, evidence has been amassed that describes the role of central cholinergic functions in the modulation of sleep and dreaming, breathing and, most prominently, the regulation of cortical cognitive functions including, perhaps, consciousness (Deutsch, 1971; Drachman, 1981; Hobson *et al.*, 1998; Lydic, 1996; Robbins *et al.*, 1998; Sarter & Bruno, 1997; Smythies, 1997; Wenk, 1997a; Woolf, 1997). Only for a little over a decade has attention been directed towards ACh as a regulatory factor in ontogeny (Hohmann & Berger-Sweeney, 1998). While this role is not yet as well established as the role of ACh in adult nervous system function, it is rapidly becoming a very exciting area in which to further our understanding of structure–function relationships in cortical morphogenesis. This line of research has begun to permit insights into the cellular and molecular mechanisms behind the forming cognitive circuitry in normal ontogeny, and offers possibilities for remediation of developmental abnormalities in the future.

The afferent innervation by axons from the deeper brain centers to cerebral neocortex falls into two basic categories. One category conveys *specific* sensory signals or motor sensory integration signals, usually via relay in the thalamus; these types of afferents and their developmental significance are discussed elsewhere in this book. The second category has been termed *modulatory afferents* and, in addition to cholinergic fibres, includes catecholaminergic and serotonergic afferents (Hasselmo, 1995; McGaugh, 1989; Robbins *et al.*, 1998; Smiley, 1996). These modulatory afferents do not carry specific signals to their target neurons but rather determine the level of response their targets will have to the specific signals relayed by fibres of the first category (Hasselmo, 1995; McCormick, 1989; Semba, 1991). Emotion, attention and arousal are functions of modulatory afferents and thus it is no surprise that imbalances in modulatory afferent neurotransmitters are involved in disruptions of cognitive processes and that they also feature in many neuropsychiatric and neurodevelopmental disorders (Berger-Sweeney & Hohmann, 1997; Coyle, 1988; McGaugh, 1989; Robbins *et al.*, 1998; Smythies, 1997).

Cholinergic projections to the cerebral cortex and hippocampus probably comprise the most abundant modulatory afferent innervation and originate in basal forebrain structures including the nucleus basalis of Meynert and closely associated nuclei and the medial septum, respectively (Hohmann *et al.*, 1988; Mesulam, 1995). Afferent fibres travel to the cortical mantle via the external capsule and to the hippocampus predominantly via the fimbria fornix. Once within their target structure, cholinergic afferents distribute widely throughout the cortical and hippocampal neuropil and can influence their target cells using a variety of different receptors which can convey a plethora of divergent signals to the cortical target neurons (Krnjevic, 1988; Levey, 1996; Mesulam, 1995).

The role of the afferent cortical cholinergic innervation in learning and memory has been well established for several decades in both human and animal models (Deutsch, 1971; Drachman, 1981; McGaugh, 1989; Olton et al., 1992; Wenk, 1997a). The specific dimensions (e.g. arousal, sensory, orienting, etc.) which are influenced by cholinergic afferents in the modulation of learning and memory functions are subject to intense ongoing debate among behavioural and cognitive scientists (Sarter & Bruno, 1997). A general consensus is that cholinergic afferents act by modulating the synaptic responsiveness, that is "efficacy", of cortical neurons to sensory and associational information which is conveyed by the above-described "category I" fibres (Jerusalinsky et al., 1997; Juliano, 1998; Krnjevic, 1988; Wenk, 1997a). That is, cholinergic inputs predominantly alter the *impact* of sensory experience and possibly motor processing as well. This insight is backed by extensive animal studies including work in primates. Most recently, a series of elegant studies have conclusively demonstrated this interplay between cholinergic inputs and learning-induced synaptic change in the tonotopic maps of auditory cortex (Kilgard & Merzenich, 1998; Weinberger & Bakin, 1998). These recent experiments stand on the shoulders of a long line of studies demonstrating that cholinergic activity in cortex can impact plastic changes in the physiology and also anatomy of cortical synaptic connections (Bear & Singer, 1986; Donaghue & Carrol, 1987; Hohmann & Berger-Sweeney, 1998; Juliano, 1998; Krnjevic, 1988; McCormick, 1989; Sillito & Murphy, 1987; Singer, 1990; Steriade, 1993; Tremblay et al., 1990).

In view of such cholinergic influences on normal cortical activity, it does not seem surprising that many brain disorders, presenting with disruptions in cognitive performance, also show cholinergic alterations. Most familiar to the reader may be Alzheimer's disease, but cholinergic imbalances also occur in ethanol and lead toxicity, as well as developmental disorders such as Down syndrome, Rett syndrome, periventricular leucomalacia, etc. and schizophrenia (Coyle et al., 1986; Friedman et al., 1999; Johnston et al., 1995; Johnston & Gerring, 1992; Raedler et al., 1998; Sarter & Bruno, 1998; Tandon & Greden, 1989; Wenk, 1997b). Ample evidence that cholinergic afferents have effects in the human brain comparable to those in animal models derives from studies of cortical blood flow, glucose consumption and brain wave measurements, in addition to psychopharmacological assessments and, most recently, functional MRI (Coull, 1998; Dringenberg & Vanderwolf, 1998; Furey et al., 1997; McGaugh & Cahill, 1997; Witte et al., 1997).

Yet, disruptions in cognitive cortical functions may often not directly result from cholinergic imbalances. Many disorders associated with such imbalances, ranging from fetal ethanol and lead poisoning to Down syndrome and schizophrenia, also show signs of altered development of cortical neuromorphology and synaptic connectivity (Beracochea et al., 1989; Coyle et al., 1986; Gould et al., 1989, 1991; Johnston et al., 1995; Lauder, 1983; Miller, 1986; Petit, 1986; Schambra et al., 1990; Takashima et al., 1981; Wenk, 1997a,b). Such pathologies include altered shape, density and position of dendritic trees, and altered spine and synaptic densities. Since the length and branching patterns of dendrites together with their "spines", the sites of synaptic contact on, particularly, cortical pyramidal neurons, determine signal transduction in the cell, these changes in themselves might suffice to generate abnormal cortical function. Abnormal and permanently altered cortical morphogenesis has long been regarded clinically as the substrate for mental retardation.

Abnormal cortical morphogenesis and cholinergic deficits may be closely related causes for abnormal cortical function. Developmental studies in animal models in our laboratory and elsewhere have demonstrated over the past decade that cholinergic deficits during cortical ontogeny will result in altered morphogenesis and connectivity in cortex (reviewed below in detail). Thus, cholinergic afferents appear to sculpt the development of the same cortical neuronal targets whose synaptic function they later modulate. Therewith, acetylcholine is experiencing its third incarnation within the history of neuroscience. Like its fellow modulators, the monoamines discussed elsewhere in this book, acetylcholine has joined the ranks of growth factor-like molecules.

CHOLINERGIC NEUROTRANSMISSION IN THE CORTEX

Cholinergic neurotransmission in the CNS relies on a number of neurochemical processes that deserve to be reviewed here briefly. For a more comprehensive description, I would like to refer the reader to a neurochemistry text. The neurotransmitter ACh is synthesized in the presynaptic terminal by the enzyme choline acetyltransferase (ChAT) from acetylcoenzyme A and choline. Following ACh release, the enzyme acetylcholinesterase (AChE) is responsible for immediate breakdown of the molecule into its original components. The resulting choline is taken back up into the nerve terminals by a choline transporter and recycled into ACh synthesis. The ChAT enzyme is characteristic for cholinergic axons and terminals. In contrast, AChE may appear in both cholinergic and cholinoreceptive neurons. However, in the cerebral cortex of the adult human and other species, the distribution of ChAT immunoreactive fibres and AChE histochemically stained fibres has proven to be so similar that AChE staining has become an accepted method for labelling cortical cholinergic fibres (Kitt *et al.*, 1994; Lysakowski *et al.*, 1986; Mesulam & Geula, 1992). Unlike in many other neurotransmitter synthetic pathways, ChAT activity is not rate-limiting for AChE synthesis; rather, the high-affinity, sodium-dependent choline transporter responsible for choline uptake into the neuron, fulfils this function (Blusztajn & Wurtman, 1983). Choline, of course, is not only used for the purpose of ACh synthesis but also serves as a precursor for membrane synthesis (Blusztajn & Wurtman, 1983). Thus, the cholinergically specific, high-affinity uptake system competes with other uptake pumps for available choline (Blusztajn & Wurtman, 1983; Zeisel, 1987). Assessments of the high-affinity uptake for choline have hence been regarded as the most effective means to assess levels of cholinergic activity in the brain.

Cholinergic afferents to cortical structures originate from a chain of nuclei located in the basal forebrain (for more detailed review see Mesulam, 1995; Hohmann *et al.*, 1988). The nucleus which contributes most of the innervation to neocortical targets is termed the basal nucleus of Meynert. The medial septum, in contrast, provides most of the cholinergic innervation to the hippocampal formation. A variety of additional nuclear groups supply innervation to various other cortical targets including the visual cortex and pyriform cortex. Mesulam and colleagues devised a classification system of the various cholinergic nuclei and their target structures in the primate which was first illustrated by me for the above-mentioned review, and is shown in Figure 1.

Cortical cholinergic receptors have been grouped as either nicotinic or muscarinic cholinergic. Nicotinic receptors are ion-gated channels; that is, they convey their message by directly opening membrane channels which alter intracellular current flow (Montes *et al.*, 1994; Vidal, 1996). The pentameric nicotinic receptors are assembled from a variety of different α and β chains. To date, seven different a and three different β chains have been identified (Boyd, 1997). The exact mechanisms of their assembly are still under study (Boyd, 1997; Conroy & Berg, 1995). Muscarinic receptors, in contrast, are classified as metabotropic receptors (Brown *et al.*, 1997; Hulme *et al.*, 1990; Levey, 1996; Loffelholz, 1996). These type of receptors are coupled to second messenger signalling chains which simultaneously can alter cellular metabolism and gene expression and also change the conductance of separate ion channels to initiate postsynaptic impulse flow. Until a few years ago muscarinic cholinergic receptors were assessed exclusively via ligand-binding techniques, using autoradiography to study patterns of binding distribution (Hulme *et al.*, 1990). Such studies established the presence

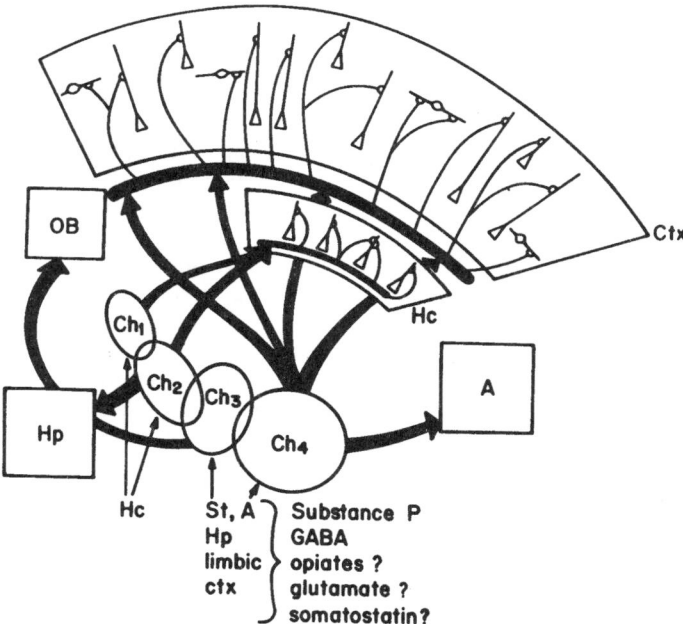

Figure 1 Different cholinergic projection nuclei in the forebrain (Ch₁–Ch₄) in relationship to their cortical targets (according to Mesulam *et al.*, 1985). Ch₁ = septal nucleus; Ch₂ = diagonal band of Broca, vertical limb; Ch₃ = diagonal band of Broca, horizontal limb; Ch₄ = nucleus basalis of Meynert; Ctx = cortex; OB = olfactory bulb; Hc = hippocampus; A = amygdala; Hp = hypothalamus.

of two different affinity types called M1 and M2. Recently, five different muscarinic receptors have been identified, cloned and found to be present in the brain (Bonner et al., 1987; Bonner, 1989; Buckley et al., 1988; Hulme et al., 1990). The muscarinic cholinergic receptor proteins m1, m3 and m5 are associated with a G-protein coupled second messenger system which, via activation of the enzyme phospholipase C (PLC) or PLD, releases downstream the second messenger molecules diazoglycerol (DAG) and inositol-3-phosphate (IP3); the m2 and m4 receptors, on the other hand, are associated with a system that inhibits the synthesis of the messenger molecule c-AMP by the enzyme adenylate cyclase (Brown et al., 1997; Costa, 1994; Hulme et al., 1990; Lauder, 1993; Loffelholz, 1996). Unfortunately, the five different muscarinic receptor proteins, which are independently expressed, do not fit neatly into the M1/M2 affinity model (Buckley et al., 1989). Both nicotinic and muscarinic cholinergic receptors can occur postsynaptically as well as presynaptically, so as to regulate neuro-transmitter release in cerebral cortex (Levey, 1996; Montes et al., 1994; Vidal, 1996). Current evidence suggests that m1 receptors are predominantly associated with cortical pyramidal neurons while many m2 receptors may be found presynaptically on cholinergic as well as thalamocortical afferents (Mrzljak et al., 1998; Rouse et al., 1997).

HUMAN CORTICAL CHOLINERGIC INNERVATION

The adult human cerebral cortex displays a dense network of cholinergic fibers throughout, with differential innervation densities in the various functional areas of cortex (Mesulam et al., 1992). Most of these fibres arrive by way of the external capsule, enter via the deep cortical white matter and, once within cortex, ascend from deep towards superficial layers (Kostovic, 1986; Mesulam et al., 1983). A smaller contingent enters via an anterior route, travelling horizontally in layer I (Kostovic, 1986; Mesulam et al., 1983). As assessed by biochemical measurements of ChAT and AChE activity, as well as by observation of histochemical or immunocytochemical fibre labelling, limbic and paralimbic cortical areas, including the hippocampus and entorhinal cortex, show the highest levels of cholinergic innervation, followed by primary sensory and primary motor areas, with the association cortex displaying the lowest levels (Kostovic, 1986; Mesulam et al., 1992; Mesulam & Geula, 1993). Furthermore, in a given cortical area, fibres have a characteristically patterned distribution. In primary sensory neocortex, plexuses of cholinergic fibres are found predominantly over cortical layers I and II (plus the upper edge of III), the lowest tier of layer IV (IVc) and with lesser intensity in layer VI (Mesulam et al., 1992; Mesulam & Geula, 1992). Both muscarinic and nicotinic cholinergic receptors have been implicated in the modulation of cognitive functions, although nicotinic mechanisms may exert their role predominantly via modulation of ACh release (Levey, 1996; Levin & Simon, 1998).

The adult distribution of muscarinic cholinergic receptors, as assessed by ligand-binding studies, resembles the distribution of cholinergic fibres in cortex. A band of high receptor-binding density can be seen in the superficial layers (II/III); these receptors are predominantly of the M1 affinity type (Zilles et al., 1989). A second, slightly less dense band of binding of mostly M2 affinity type receptors appears in layers VI and V (Zilles *et al.*, 1989). M1 sites are the prdominant affinity type in cerebral cortex but M2 sites are also present in ample numbers (Cortes *et al.*, 1986, 1987). Unfortunately, immunohistochemical or expression studies to reveal the distribution of the muscarinic receptor proteins have not yet become available in the human, but information about likely distribution patterns of m1–m5 may be extracted from non-human primate studies (Mrzljak *et al.*, 1993, 1998). According to such data, both m1 and m2 are associated with likely cholinergic, symmetrical-type synapses as well as with non-cholinergic asymmetrical synapses on the spines and dendrites of pyramidal neurons in cortex. The m2 receptor is found, in addition, presynaptically in axon terminals resembling thalamocortical afferents, as well as in some interneurons and in perisynaptic and extrasynaptic locations; the extra-synaptic m2 distribution suggests that this receptor can respond to "volume transmission" or "en-passant" release of ACh (Mrzljak *et al.*, 1998).

Nicotinic receptor-binding studies reveal that these ion-gated receptors predominate in deep and superficial cortical layers much like muscarinic receptors (Court & Clementi, 1995; Perry *et al.*, 1992; Zilles *et al.*, 1989). Even so, some nicotinic receptor sites also appear to be present in layer IV (Schroder *et al.*, 1989). Binding patterns vary with cortical topography (Court & Clementi, 1995; Perry *et al.*, 1992; Zilles *et al.*, 1989; Schroder *et al.*, 1989). Recent expression studies for the cloned nicotinic receptor proteins essentially confirm the distribution data generated by binding studies (Wevers *et al.*, 1994).

It is still unclear to what extent cholinergic afferents exert their effects via traditional synapses or "en-passant" in a more hormonal manner. Several investigators have attempted to quantify cholinergic synapses by combining immuno-histochemical labelling of cholinergic fibres with ultrastructural, electron microscopic assessment of the tissue, and these studies have counted varying percentages of "classical synaptic sites" but also noticed areas where presynaptic accumulations of cholinergic vesicles were not accompanied by postsynaptic differentiations, suggesting the more general "en-passant" release of ACh (Houser *et al.*, 1985; Mrzljak *et al.*, 1995; Smiley *et al.*, 1997; Umbriaco *et al.*, 1994). The reported percentage of "classical" synapses has been highest for human cortex (Smiley *et al.*, 1997). Even in this study approximately 30% of the presynaptic cholinergic differentiations did not match up with postsynaptic ultrastructural differentiations. The significance of these distinctions lies with the specificity of cholinergic cortical signalling. Neurotransmitter release at a "classical" synapse allows for highly localized signal transduction, whereas "en-passant" release will effect any available receptor within a reasonable diffusion radius, meaning multiple areas on one neuron and even on several neurons simultaneously. We have come to increasingly understand over the past two decades that the cortical cholinergic innervation has a very high level of topography (Mesulam *et al.*, 1992; Woolf, 1997), yet, simultaneously, functional areas might experience a "wholesale" boost of activity following cholinergic stimulation from the basal forebrain (Krnjevic, 1988). The answer may be that this system modulates cortical neuronal activity on both levels; sculpting individual neuronal responses via classical synapses and altering the "gain" in larger areas via additional "en-passant" release. Since all ultrastructural studies to date have focused on the adult brain, the developmental distribution of different synaptic types is an even greater enigma.

DEVELOPMENT OF THE HUMAN CORTICAL CHOLINERGIC INNERVATION

While the adult cholinergic innervation of the human cerebral cortex has received considerable attention, particularly since the early 1980s within the context of Alzheimer's disease, developmental studies in the human cortex are still relatively sparse. In general, developmental studies in the human are fraught with several difficulties such as obtaining normal early postnatal and particularly embryonic tissue, and post-mortem preservation of the tissue. Most of the data we have to date have come from just a handful of laboratories.

Notably Kostovic, in collaboration with colleagues both in Zagreb and the USA, has, over the course of more than a decade, provided much insight into the development of the basal forebrain cholinergic afferents as well as into the development of transient cholinergic enzyme expression within thalamocortical fibres (Kostovic & Racik, 1984; Kostovic, 1986). An overview of the dynamic changes in cholinergic innervation within the context of cortical, thalamocortical and functional development has been provided based on anatomical connections (Kostovic *et al.*, 1991). Acetylcholinesterase, as mentioned above, has proven to be a useful and specific marker for cholinergic fibres in the cerebral cortex of the adult primate, as well as of many other species. In contrast to adult cortex, though, the developing cortex of the human, monkey and rodent display a transient pattern of AChE in non-cholinergic afferent fibres from several thalamic nuclei (Kostovic & Racik, 1984). In addition, cellular AChE staining can be seen in association with non-cholinergic cortical pyramidal cells beginning in the second decade of life (Mesulam & Geula, 1991). We will discuss the significance of this transient appearance of AChE further below, in conjunction with experimental studies.

AChE-stained, presumably cholinergic cell bodies first appear in the area of the future nucleus basalis complex of the human around the 9th week of gestation (GW) and increase considerably by 15 weeks (Kostovic, 1986). ChAT activity in the same area has been detected by one investigator as early as GW8, and was confirmed in the nucleus basalis area by GW12 (Candy *et al.*, 1985; Brooksbank *et al.*, 1978; Gale, 1977). Afferent cholinergic fibres are reported to reach the subplate neurons of the cortex between the GW2 and the GW3, and to enter the cortical plate around GW22 (Kostovic, 1986).

Surprisingly, ChAT activity in the cortical anlage, the very early beginnings of cerebral cortex, has been reported as early as GW12, suggesting that either histochemical reports of fibre ingrowth may have missed some early populations of afferents or that other elements, perhaps members of early subplate neuronal population (see Kostovic & Rakic, 1980; Marin-Padilla, 1978), might express this synthetic enzyme. Furthermore, values for cortical ChAT enzyme activity within the first 12–22 weeks prenatally were reportedly higher than in the mature human (Brooksbank *et al.*, 1978; Candy *et al.*, 1985; Gale, 1977; Perry *et al.*, 1986). The data suggest that there might be several developmental peak periods, a very early peak around GW12 followed by decline and a renewed increase towards the third trimester (Candy *et al.*, 1985; Perry *et al.*, 1986). Another peak apparently occurs in the infant (Court *et al.*, 1993). Unfortunately, ChAT activity has not been monitored systematically in the same study throughout the entire period of cortical maturation, leaving us with discontinuous glimpses during the most important time windows of morphogenesis. In contrast, AChE enzyme activity shows a steady increase throughout the same prenatal time period, as basal forebrain afferents ascend from deep in the cortex towards the superficial layers (Kostovic *et al.*, 1991; Perry *et al.*, 1986). Fibre ingrowth into superficial cortical layers does not occur until the second postnatal year, when the adult pattern of AChE-positive fibre distribution becomes established. According to anatomical studies, peak histochemical intensity of AChE is not reached until young adulthood (Kostovic *et al.*, 1991).

The determination of whether early high ChAT activity measurements or rather the gradual protracted increase of AChE fibres more accurately reflects the maturation of these afferent modulatory projections to cortex will have to await additional studies. Regardless, the early appearance of these cholinergic enzymes in cortex strongly suggests the ontogenetic importance of cholinergic neurotransmission.

The development of cholinergic receptors in cortex proceeds with a similarly early time-course. Nicotinic cholinergic receptors have been revealed by receptor–ligand binding assays and *in-situ* hybridization assays as early as GW8–11 (Agulhon *et al.*, 1998; Hellstrom-Lindahl *et al.*, 1998; Perry *et al.*, 1986), but may be present as early as GW4 when mRNA expression for one of the α subunits was first detected using highly sensitive molecular biology methods (Hellstrom-Lindahl *et al.*, 1998). The highest numbers of nicotinic receptors are seen in mid- to late gestation, but not all investigators agree that there is any correlation between age and receptor numbers (Perry *et al.*, 1986; Agulhon *et al.*, 1998; Hellstrom-Lindahl *et al.*, 1998). Hellstrohm-Lindahl *et al.* furthermore, investigated the distribution of nicotinic α and β subunits and showed that there are differences between the expression patterns of the different subunits in the developing versus mature cortex (Hellstrom-Lindahl *et al.*, 1998). In particular, α5 is expressed very early in cortex, predating even the appearance of synapse formation. Muscarinic cholinergic receptor development still awaits exploration with modern molecular techniques; however, a number of laboratories have investigated the development of muscarinic receptor binding with a variety of different ligands. These studies have shown a developmental onset between GW14–16 in cortex (Aguilar & Lunt, 1985; Bar-Peled *et al.*, 1991; Egozi *et al.*, 1986; Johnston *et al.*, 1985; Perry *et al.*, 1986; Ravikumar & Sastry, 1985). Muscarinic cholinergic receptors gradually increase with cortical maturation to reach peak values around birth which appear to be maintained for a number of postnatal years with significant declines occurring in the second or third postnatal decade of life (Bar-Peled *et al.*, 1991; Johnston *et al.*, 1985; Ravikumar & Sastry, 1985).

Muscarinic cholinergic receptor development displays a dynamic change of binding patterns with age which resembles the changes in basal forebrain afferents (Johnston *et al.*, 1985; Kostovic *et al.*, 1988). That is, muscarinic binding first appears in deep cortical layers and gradually ascends to establish the superficial binding pattern in layers II/III. Regionally, the developmental time-course of muscarinic binding follows the time-course of cortical maturation with early-maturing regions developing receptor populations prior to late-maturing cortical areas (Egozi *et al.*, 1986). M1 and M2 binding appear fairly simultaneously in development and no major changes in affinity kinetics have been reported

(Bar-Peled *et al.*, 1991; Johnston *et al.*, 1985; Perry *et al.*, 1986; Ravikumar & Sastry, 1985). Several receptor-binding studies have pointed out that the developmental profile of muscarinic binding closely parallels that of synaptic overproduction and pruning in cortex (Bar-Peled *et al.*, 1991; Ravikumar & Sastry, 1985). The implication here is that these receptor changes do not simply reflect overall synaptic changes but may themselves be instrumental in bringing about changes in other synapses.

In summary, the data reviewed here demonstrate that cholinergic development accompanies and in several instances predates important milestones in cortical morphogenesis (see Figure 2) (Capone, 1996; Huttenlocher, 1990; Marin-Padilla, 1970; Molliver *et al.*, 1973; Volpe, 1987). Cholinergic fibres appear in neocortex towards the end of neuronal generation and early on in the process of cortical synapse formation and dendritic development. In the human, cholinergic innervation follows the arrival of some thalamic projections from the pulvinar and intralaminar nuclei, but largely overlaps with afferents from the specific thalamic nuclei. The very early appearance of cortical ChAT activity and nicotinic receptors, predating the arrival of afferents from the forebrain, suggests the presence of a transient intrinsic cortical cholinergic mechanism that could be involved in the shaping of the cortical subplate. The transiently present AChE in thalamocortical afferents might interact with such early intrinsic cortical cholinergic chemistry. Muscarinic receptor development, on the other hand, parallels both the ontogeny as well as the "fine tuning" of thalamocortical and cortico-cortical connectivity. Muscarinic receptors as well as ChAT activity decline simultaneously with the pruning of cortical dendrites and synapses, which is widely believed to subserve the fine-tuning of cortical connections allowing for appropriate cognitive function in the adult (see Capone, 1996).

What makes ACh such an appealing candidate to subserve a "trophic" function in early cortical development is that this modulatory system does not have any other clearly apparent function prior to the presence of mature synapses of specific thalamocortical and corticocortical afferents. Furthermore, we already know the cholinergic system in the adult human to impact learning and memory via modulation of cortical neurons (Juliano, 1998; Kilgard & Merzenich, 1998; Krnjevic, 1988; McCormick, 1989; Sillito & Murphy, 1987; Singer, 1990; Tremblay *et al.*, 1990). There is a general consensus that learning and memory processes entail the remodelling of synapses and dendritic spines (Cotman & Nieto-Sampedro, 1982; Greenough *et al.*, 1986; Weiler *et al.*, 1995; Weinberger & Bakin, 1998). Thus, by conjecture, we may assume that cholinergic signal transduction mechanisms have the ability to shape neuronal morphology.

Many of the investigators whose data have been reviewed above have concurred that cholinergic cortical activity is likely to shape morphogenetic events in cortex. However, such impressions in the human have to rely on correlative evidence, including the observations of profound cortical and cognitive abnormalities in certain developmental disorders with early cholinergic deficits (Coyle *et al.*, 1986; Johnston *et al.*, 1995). To test the role of the cholinergic system in cortical ontogenesis and resulting function, one has to resort to animal models which can be experimentally manipulated.

There is now a wealth of literature investigating developmental cholinergic effects in various species. This literature includes behavioural, morphological,

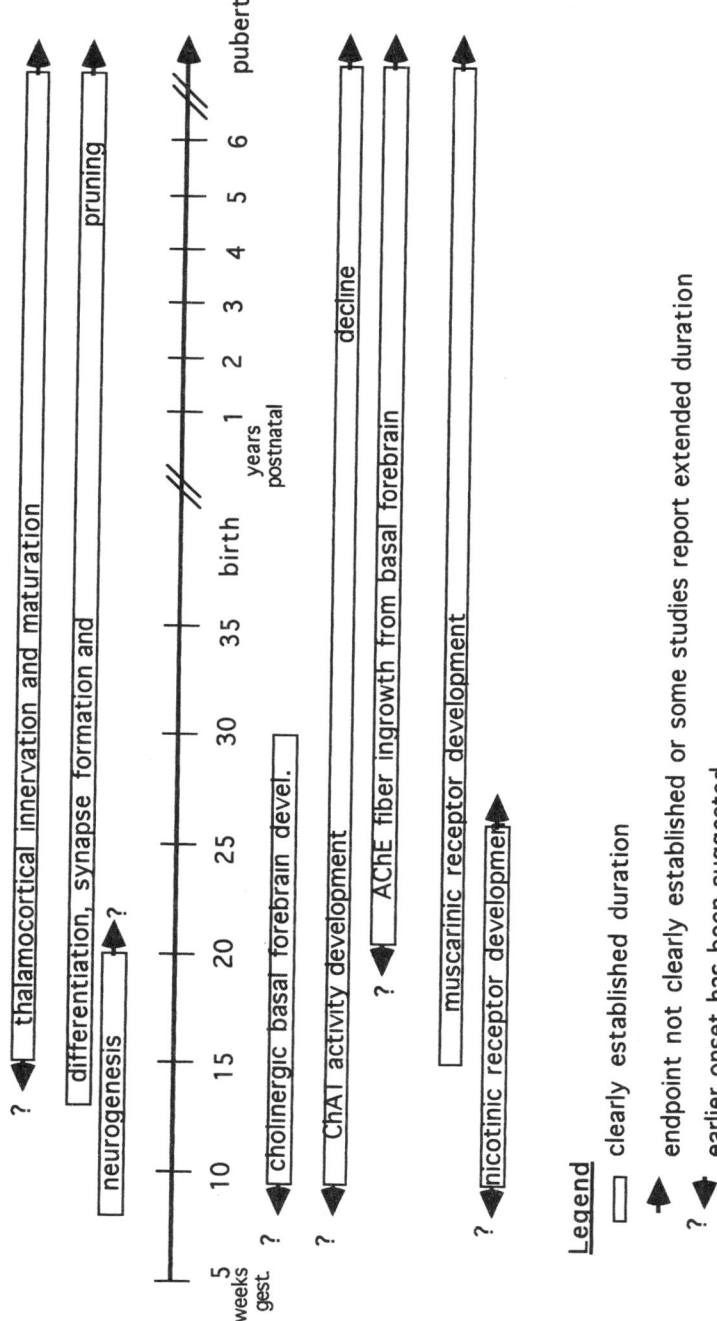

Figure 2 Time relationship in the human between cholinergic development in the forebrain and cortical morphogenesis and formation of cortical connectivity.

neurochemical and molecular studies. We will focus here largely on the literature concerning rat and mouse, as these species have been explored most widely both on the organismal (systems) and cellular/molecular levels within the context of cholinergic development.

DEVELOPMENT OF THE RODENT CORTICAL CHOLINERGIC INNERVATION

The relationships between cortical morphogenesis and afferent cholinergic development are substantially similar in the human and rodent. However, mice and rats obviously have a much shorter lifespan. Thus, a time period that in the human stretches from the GW2 to the end of the second postnatal decade has been condensed to roughly encompass gestational day 10 to postnatal day 60 in mouse or rat. Moreover, since rodents are born substantially more immature than primates, much of the relevant development occurs postnatally. This, of course, has considerable experimental advantages as it allows manipulations of the system without having to resort to *in-utero* pharmacology or surgery.

Figure 3 illustrates cholinergic development in rodents, *vis-à-vis* landmarks in cortical morphogenesis, for comparison with the human charts in Figure 2 (after: Blue & Parnavales, 1983; Catalano *et al.*, 1991; Caviness, 1982; Crandall & Caviness, 1984; Crandall *et al.*, 1985; De Felipe *et al.*, 1997; Jacobson, 1978; Juraska, 1982; Killackey *et al.*, 1995; Koester & O'Leary, 1992; Konig & Marty, 1981; Ivy & Killackey, 1982; McConnell, 1988; Miller, 1981; O'Leary & Stanfield, 1989; Olavarria & Van Sluyters, 1985; Petit *et al.*, 1988; Stanfield *et al.*, 1982; Wise & Jones, 1978). Basal forebrain cholinergic neurons are generated between embryonic day 11 (E11) and E16/17, and mature around the second postnatal week (Brady *et al.*, 1989; Schambra *et al.*, 1989; Sweeney *et al.*, 1989b). AChE-positive fibres enter neocortex on or around the day of birth (there is some variation between rat and mouse and between different strains of mice) (Hohmann & Ebner, 1985; Kristt, 1979). Simultaneously with cholinergic fibre ingrowth into primary sensory areas of cortex, AChE-containing thalamocortical fibres also appear and establish terminal fields in layer IV of primary sensory cortex (Hohmann & Ebner, 1985; Kristt, 1979; Wise & Jones, 1978). These transiently AChE-staining projections are more prominent in rat as compared to mouse neocortex and are clearly non-cholinergic (Hohmann & Ebner, 1985; Kristt, 1979; Robertson *et al.*, 1988; Robertson, 1987). Rat, but not mouse, will subsequently develop permanent intrinsic cortical ChAT positive cells; in this way the mouse more closely resembles the human (Eckenstein & Thoenen, 1983; Kitt *et al.*, 1994; Houser *et al.*, 1985). Cortical ChAT enzyme activity becomes measurable 2–3 days after birth in rodents, peaks around postnatal day (PND) 30, and thereafter declines, reaching adult values by PND60 (Coyle & Yamamura, 1976; Hohmann & Ebner, 1985; Hohmann *et al.*, 1988; Thal *et al.*, 1991; Yamada *et al.*, 1986; Zahalka *et al.*, 1993). The delay in measurable ChAT activity, compared to the appearance of AChE-stained basal forebrain fibres, is most likely a consequence of the measurability threshold. Interestingly, binding to the high-affinity choline transporter does show a prenatal peak (Zahalka *et al.*, 1993); yet it is currently not clear whether these prenatal binding sites are also functionally

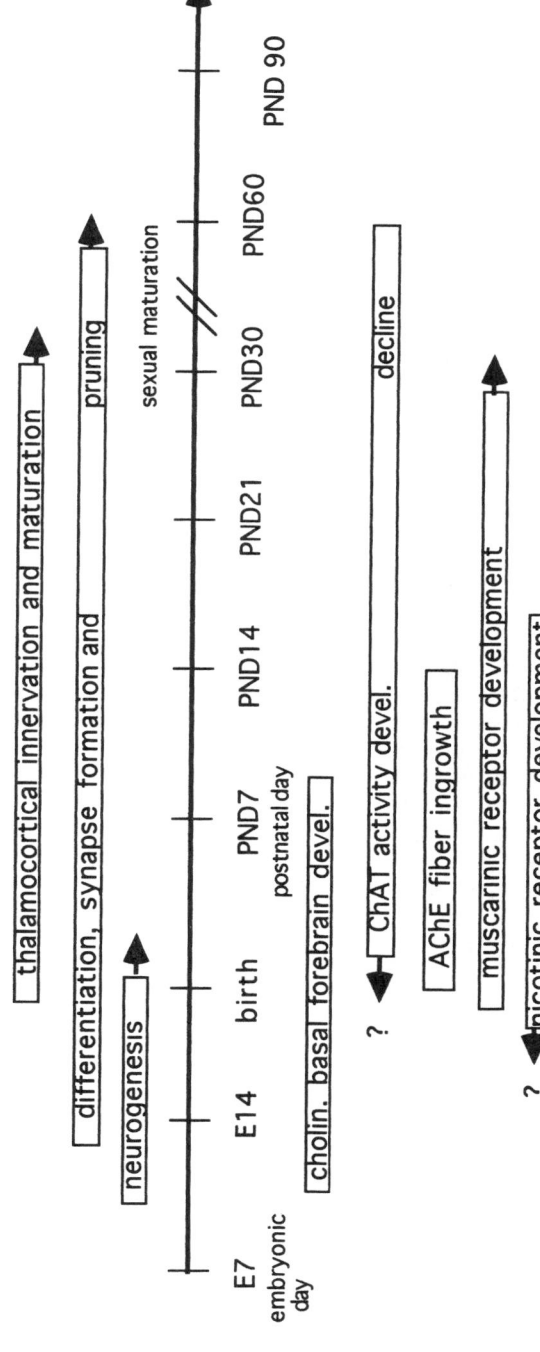

Figure 3 Time relationship in the mouse between cholinergic development in the forebrain and cortical morphogenesis and formation of cortical connectivity.

relevant. Nicotinic cholinergic receptor binding appears in cortex perinatally and increases without substantial change in distribution (Kumar & Schliebs, 1992; Naeff *et al.*, 1992; Prusky *et al.*, 1988; Zhang *et al.*, 1990). In the visual cortex, geniculocortical deafferentation leads to significant losses of these nicotinic binding sites, suggesting that they are presynaptically located on thalamic afferents (Cimino *et al.*, 1995; Prusky *et al.*, 1988). Earlier muscarinic receptor-binding studies and recent studies of the different receptor proteins concur that muscarinic receptors appear in cortex perinatally and undergo considerable pattern changes. The mature distribution is assumed around the peak time of cortical synapse formation (Evans *et al.*, 1985; Kuhar *et al.*, 1980; Hohmann *et al.*, 1985; Miyoshi *et al.*, 1987; Rotter *et al.*, 1979; Tice *et al.*, 1996). I will focus here on the more recent data, showing the developmental distribution of different muscarinic receptor proteins (Hohmann *et al.*, 1995; Lee *et al.*, 1990; Levey *et al.*, 1991; Rossner *et al.*, 1993; Wall *et al.*, 1992). Recent studies in mouse have provided the most detailed picture of the unfolding of the developmental pattern. In the adult mouse somatosensory cortex the distribution of m1 receptors within superficial and deep cortical layers is consistent with the location of this receptor on pyramidal cell dendrites and spines (Hohmann *et al.*, 1995). Yet during the first 2 weeks of cortical development this receptor is also displayed in the whisker barrels of layer IV in mouse in a pattern that corresponds to thalamocortical terminal fields (Hohmann *et al.*, 1995). This particular sensory cortical area has been a focus of plasticity studies for some time (Jeanmonod *et al.*, 1981). The m4 muscarinic receptor protein is also prominently but transiently present in the "barrels" early postnatally. In contrast, m2 is present only in deep cortical layers during the first postnatal week but subsequently shows an intense and permanent presence in the barrels in addition to being found in layer VI and lower V pyramidal layers (Hohmann *et al.*, 1995). Both m1 and m2 proteins are also transiently present in non-cholinergic afferents to cortex and hippocampus respectively, during the first 2 postnatal weeks (Hohmann *et al.*, 1995). One would be hard-pressed to explain this transient receptor distribution in any way other than that it is involved in the shaping of non-cholinergic cortical connectivity. Some distribution differences between mouse and rat cortex may exist, but the presently available literature suggests that similarly dynamic developmental patterns as outlined for mouse are common to all rodents (Buckley *et al.*, 1988; Lee *et al.*, 1990; Levey *et al.*, 1991; Rossner *et al.*, 1993; Schliebs *et al.*, 1994; Wall *et al.*, 1992). A recent quantitative immunoprecipitation study by Tice *et al.* further suggests that the m2 and m4 receptor proteins increase in density until PN90 while m1 peaks around 3 weeks postnatal and declines thereafter (Tice *et al.*, 1996).

Thus, human and rodent development share an early transient AChE expression in thalamic terminal field areas of the primary sensory cortex, a precocious expression of nicotinic receptors, and a substantial redistribution of muscarinic receptors over the course of cortical maturation. These features must have been important enough for cortical ontogeny to remain this well-preserved throughout evolution. They suggest at least two things: (1) the involvement of AChE in the set-up of thalamocortical terminal fields; (2) the involvement of muscarinic cholinergic receptors in pyramidal cell maturation. These may or may not be separate processes. Some evidence has suggested a morphogenetic function of

AChE separate from cholinergic mechanisms (Silver, 1974). In addition to breaking down the bonds between choline and acetyl-CoA, AChE also has peptidase activity which might directly impact neurite growth via these mechanisms (Robertson & Harrell, 1997) or indirectly influence maturation via the modulation of growth-regulating proteins such as amyloid precursor protein (APP) (Robertson & Harrell, 1997). Alternatively, AChE on thalamocortical afferents may well function in the breakdown of ACh, interacting with presynaptic receptors on thalamic afferents. As reviewed above, m1, m4 and nicotinic receptors are expressed, at least transiently, in very high quantities in the terminal zone of these afferents. In the human, plenty of ChAT activity is present early on to ensure active cholinergic transmission. A similar early source for ACh synthesis has not yet been determined clearly in the rodent, but is suggested by the early peak in choline high-affinity transporter binding.

EFFECTS OF EARLY CHOLINERGIC ALTERATIONS ON CORTICAL MORPHOGENESIS

If the early cortical cholinergic innervation indeed has the morphogenetic effects that are suggested by the ontogenesis of its various components, then alteration in this innervation should result in changes in cortical structure and function. A number of experimental approaches have been taken in this direction, and most have resulted in modifications of cortical structure and/or chemistry and behaviour. We will discuss these experiments by the approach that was taken to interfere with normal cholinergic innervation of cortex.

Neonatal basal forebrain lesions

The earliest and to date most comprehensive investigations into neonatal cortical cholinergic depletions have been conducted by this author within the context of several collaborations. In our model, electrical current lesions are aimed at the ventromedial globus pallidus and horizontal limb of the diagonal band of Broca (nBM) area on the day of birth (PND1). These lesions have been shown to cause a transient depletion of ChAT activity in frontoparietal neocortex that was most dramatic during the first postnatal week (85% depletion) and showed full recovery of enzyme activity by the end of the first postnatal month (Hohmann *et al.*, 1988). The disrupted cholinergic innervation, as indicated by acetylcholinesterase (AChE) staining, appears to have completely recovered by adulthood. However, the transient deafferentation of neocortex is accompanied by a delay in the appearance of normal cortical cytodifferentiation in layers II + III through V in all cortical areas that showed cholinergic depletion but spared in areas with undisturbed cholinergic innervation. While cortical differentiation eventually proceeded, permanent alterations of some aspects of cortical cytoarchitecture persisted, such as disrupted boundaries in supragranular layers and altered pattern formation in layer IV of the "barrel fields" in somatosensory cortex (Hohmann *et al.*, 1988). Golgi studies of layer V pyramidal cells have confirmed a delay in cytomorphogenesis ipsilaterally to the neonatal nBM lesion, and indicate additional quantitative as well as qualitative alterations in dendritic branching patterns and spine morphology in sensory motor cortex (Hohmann *et al.*, 1991). The quantitative alterations include significantly decreased soma sizes at both PND7 and PND14, fewer and shorter apical and basal dendritic branches at PND7 and slightly increased apical branching and dendritic length at PND14. Qualitatively, dendritic varicosities and abnormally shaped spines are seen at both ages. Anterograde and retrograde pathway tracing in the adult animal, following unilateral neonatal lesions, demonstrate changes in cortical connectivity, including altered thalamic afferent distribution in the "barrel field" of layer IV and increased efferent connections from layer V to subcortical targets, suggesting perhaps a defect in the pruning process (Hohmann *et al.*, 1991).

The abnormalities in cortical cytomorphology observed following the neonatal electrolytic basal forebrain lesions bear out the hypotheses that cholinergic activity, presumably via muscarinic receptors, is involved in (a) pyramidal cell maturation and (b) the shaping of the terminal fields of thalamocortical afferents. Moreover, the described cortical dysgenesis bears striking resemblance to the pathologies associated with a number of developmental disorders in the human known to result in mental retardation (Bauman & Kemper, 1995; Caviness & Williams, 1979; Coyle *et al.*, 1986; Huttenlocher, 1975; Purpura, 1979; Marin-Padilla, 1975). The latter suggested to us that neonatally nBM lesioned mice might exhibit alterations in cognitive ability as a result of their abnormal cortical morphogenesis and connectivity. Our initial data, using bilateral neonatal nBM lesions in BALB/cByJ mice and testing animals at 2 months of age, showed altered performance on several cognitive tasks, confirming the hypothesis of deficits in memory and learning ability in these animals (Bachman *et al.*, 1994). More importantly, the deficits displayed by the neonatally nBM lesioned mice differed substantially from cognitive deficits observed following lesions to the same forebrain structures in adult animals indicating that they were due to abnormal morphogenesis of cortex and could not be ascribed to cholinergic imbalances at the time of behavioural testing (Sweeney *et al.*, 1988, 1989a).

Recently we have published two comprehensive studies in which a battery of behavioural tests was administered to the same animals that subsequently were assessed either by quantitative morphological procedures or by neurochemistry; moreover, all analyses were conducted by sex of the animals (Arters *et al.*, 1998; Hohmann & Berger-Sweeney, 1998b). These studies show that, by adulthood, the altered width of cortical layers V and IV significantly correlated with pronounced deficits in the acquisition of a spatial swim maze task in male, but not female, neonatally nBM lesioned mice. Some slowing in performance is observable in a non-spatial version of the maze, but is not reflected in open field activity. In contrast, neither task acquisition nor memory is affected in two other simpler, non-spatial tasks (Arters *et al.*, 1998).

The decreased width of layer V following the lesion in both sexes is compatible with previous observations of permanently disrupted cytoarchitecture and decreased size of pyramidal cell somata in this layer (Hohmann *et al.*, 1988, 1991) while the decreased width in layer IV concurs with prior observations of altered afferent thalamocortical projections (Hohmann *et al.*, 1991). Female mice, which did not show comparable deficits in adult behavioural performance following the neonatal nBM lesion, did also display significant decreases in layer IV and V width, but in addition also *decreases* in layers II/III width. Males, in contrast, showed *increases* in width in layers II/III at the time of behavioural testing. The observation of sexually dimorphic responses in layers II + III cells to the neonatal nBM lesion is consistent with prior observations of sexual dimorphism in these layers, but not layer V, in rat pups (Juraska, 1984). It should also be noted that all of these nBM lesion-related morphological changes were measured in the somatosensory-motor cortex, effectively linking the observed cognitive deficits to these primary sensory-motor cortical regions. Cortical ChAT activity as well as AChE histochemistry in the neonatally lesioned mice were the same as littermate control levels at the time of behavioural testing 2 months postnatally, showing that behavioural alterations could not be accounted for by neurotransmitter deficits at the time of behavioural testing (Arters *et al.*, 1998). We have interpreted the structure/function correlates in the male mice to suggest altered sensory-motor programming as a consequence of altered neuronal morphology and connectivity.

Taken together, our studies were able to demonstrate that transient neonatal depletions during the most dynamic phase of cortical morphogenesis will result in altered morphogenesis, which is accompanied by permanent cognitive changes. Moreover, we have demonstrated sex differences in the interaction of the cholinergic system with the developing cerebral cortex, suggesting an impact of sex hormones on these morphogenetic events. Sexual dimorphisms in aspects of the cholinergic innervation and its development have been demonstrated by several studies (Hartnagel *et al.*, 1993; Kornack *et al.*, 1991; Loy & Sheldon, 1987; Luine & McEwen, 1983; McMillian *et al.*, 1996; Ricceri *et al.*, 1997b). The clinical impact that this understanding may have is illustrated by recent research connecting oestrogen to protection from Alzheimer's disease in post-menopausal women (Wickelgren, 1997). Moreover, differences in spatial learning and memory strategies between the sexes have been reported for human and other species (Berger-Sweeney *et al.*, 1995; Gaulin & Hoffman, 1988; Halpern, 1992; Linn & Peterson, 1985; Williams *et al.*, 1990) and have in animal experimentations proven to be particularly vulnerable to cholinergic manipulations (Berger-Sweeney *et al.*, 1995; Hartnagel *et al.*, 1993; Smolen *et al.*, 1987). This suggests the hypothesis that ontogenetic cholinergic mechanisms may be instrumental in establishing the morphological substrates for these dimorphic cognitive performances in males versus females.

One shortcoming of our neonatal nBM lesion paradigm has been that the lesion method is not completely selective for cholinergic afferents even if great pains are taken to eliminate from analysis all cases with obvious lesion encroachment on non-cholinergic basal forebrain structures. Since no other effective lesion method has become available in mouse to date, we have attempted to overcome this shortcoming by implementing control lesions to forebrain areas, such as the medial forebrain bundle, which might have inadvertently sustained damage. We have recently been able to show that neonatal monoaminergic depletions of mouse cortex do not show morphological or behavioural similarities to the nBM lesion effects described above (Berger-Sweeney *et al.*, 1998; Hohmann *et al.*, 1997).

However, in rat, a selective alternative to electrolytic nBM lesions has been available in recent years. An immunotoxin generated from a monoclonal antibody selective to the low-affinity nerve growth factor receptor (p75NGFR) and coupled to a mitochondrial toxin, is selectively taken up only by cholinergic neurons in the basal forebrain projection nuclei to cortex and hippocampus (Wiley *et al.*, 1991). Attempts for production of a similar immunotoxin in mice have all failed to date (Berger-Sweeney *et al.*, 1996; Hohmann & Berger-Sweeney, unpublished observation).

Initial developmental studies using the immunotoxin in neonatal rats were disappointing (Leanza *et al.*, 1996; Pappas *et al.*, 1997). However, it has since become clear that the precise application procedure and timing of early immuno-toxin injections is critical (Ricceri *et al.*, 1997a; Robertson *et al.*, 1998; Zhu & Waite, 1999; Hohmann, Berger-Sweeney and Riccieri, unpublished observations). First, it appears to be important that the cholinergic depletion occur early, within the first postnatal week, and precede the bulk of afferent fibre ingrowth and synapse formation. Using dual injections on the first and third postnatal day, Robertson *et al.* were able to find substantial alterations in pyramidal neuron morphology which substantially resembled those seen with electrolytic lesions in mouse (Robertson *et al.*, 1998). This laboratory recently followed up on their *in-vivo* observations with tissue culture studies in which cortical and basal forebrain cholinergic neurons were co-cultured or grown separately. They were able to show significant trophic influences of basal forebrain neurons on co-cultured cortical pyramidal cells (Ha *et al.*, 1998). In yet another recent study, large single injections of immunotoxin led to effective neonatal cholinergic depletion and resulted in arrested plasticity in the developing barrel field (Zhu & Waite, 1999).

In addition to these well-characterized paradigms in mouse and rat, a number of other approaches have been used to attempt developmental cholinergic depletions in rat. Sengstock *et al.*, using injections of ibotenic acid into the basal forebrain of rat pups, observed increased dendritic branching and volume (Sengstock *et al.*, 1992) in aged adult animals following the neonatal lesions. The effect of cholinergic depletion on cortical and hippocampal development has been assessed also, using the cytotoxic drug ethylcholine mustard aziridinium (AF64A) which reportedly is taken up selectively by cholinergic neurons (Armstrong & Pappas, 1991; Brake & Pappas, 1994; Gaspar *et al.*, 1991; Speiser *et al.*, 1988). Neonatal AF64A treatment resulted in cortical thinning and behavioural deficits in rats (Armstrong & Pappas, 1991; Gaspar *et al.*, 1991; Speiser *et al.*, 1988).

Pharmacological, environmental and molecular manipulations of the developing cholinergic system

Some investigators have attempted to alter cholinergic neurotransmission in ontogeny via inhibition of Acetylcholinesterase (AChE), the degrading enzyme for ACh (Gupta *et al.*, 1985; Castro & Paylor, 1990). Inhibition of AChE increases ACh availability and in the brain of both adult animals and humans and leads to a U-shaped dose–response curve; that is, in low and moderate doses improvements in cognitive functions are observed, but in high doses impairments of cognitive performance are seen in a variety of paradigms (Drachman, 1981; Hohmann *et al.*, 1988; Squire & Davis, 1981). The literature on developmental AChE inhibitor effects appears contradictory at first glance. The AChE inhibitor physostigmine, given to near-weanling age (3–4-week-old) rats, improved performance of a short-term memory task (Castro & Paylor, 1990) by apparently enhancing cholinergic availability at the time of testing. The animals used in this study were evidently past the time window during which cholinergic innervation appears to influence cytomorphogenesis. As such, the data by Castro *et al.* are consistent with the results of the late ontogenetic immunotoxin injections reviewed above. On the other hand, Gupta *et al.* observed impairments in operant behaviour following AChE reductions by methylparathion given to *neonatal* rats (Gupta *et al.*, 1985). Because this group tested several different doses of the esterase inhibitor, and did not see behavioural improvement with any of them, the most likely interpretation of these data is that neonatal AChE inhibition interfered with the ontogenetic role of the cholinergic system in cortex and hippocampus by making too much ACh available at receptor sites.

Attempts to supplement rather than to interfere with cholinergic innervation and function in ontogeny have also been made. Williams, Meck and co-investigators (Meck & Williams, 1987; Williams *et al.*, 1989; Meck & Smith, 1989; Meck *et al.*, 1988) have in recent years explored the effect of perinatal choline supplements on cognitive function and neuronal morphology. In contrast to mostly unsuccessful attempts at cholinergic supplementation via choline in adult and aged individuals (Hohmann *et al.*, 1988), neonates have an increased choline demand, higher blood levels and are more susceptible to dietary choline supplementation (Zeisel, 1987). Thus, when given in ontogeny, choline treatment can increase cholinergic efficacy in cortex and hippocampus and enhance performance on visuospatial memory tasks (Meck *et al.*, 1988; Meck & Smith, 1989; Meck & Williams, 1997; Williams *et al.*, 1989). Simultaneously, basal forebrain neuron morphology and distribution were altered (Williams *et al.*, 1989) and pyramidal cell branching and size were increased in the hippocampus (W. Meck, personal communication). Interestingly, these studies showed differences in the timing of the female and male rats' responses to the choline supplementation. While male rats responded to both prenatal and early postnatal applications of choline, female rats were refractory to postnatal application, suggesting developmental differences in the timing, but perhaps also the function, of the cholinergic afferent innervation in males versus females (Meck *et al.*, 1988; Meck & Smith, 1989; Williams *et al.*, 1989).

Recently, two mouse models of transgenic manipulations of the cholinergic system were developed. In one model, increased expression and presumably activity of AChE were accomplished by introducing the gene for human AChE into the mice (Beeri *et al.*, 1997). These animals exhibited attenuated dendritic branching, enhanced expression of the choline high-affinity transporter and impaired acquisition of the Morris swim maze task. All these effects are compatible with reduced cholinergic transmission resulting from increased breakdown of the neurotransmitter, and resemble the results seen with neonatal cholinergic depletions. Although only data from adult animals were reported in this study, it is safe to presume that the human AChE transgene was expressed throughout development, exerting its effects on cortical ontogenesis. Another group constructed a mouse lacking the m1 muscarinic receptor protein (Hamilton *et al.*, 1997). These animals, in adulthood, no longer displayed the biophysical and physiological responses associated with neurotransmission at these particular muscarinic receptor sites. However, they reportedly did not show morphological or behavioural deficits. This is somewhat surprising, but could be the consequence of compensatory responses from other, perhaps not even cholinergic, receptor sites; alternatively, structural and functional abnormalities may have been too subtle to be assessable by the reported procedures.

Finally, several animal models for environmental toxicities and endocrinological imbalances may also serve as examples for developmentally altered cholinergic innervation. These include lead toxicity, ethanol toxicity and developmental hypothyroidism in rats. All share cholinergic impairments and abnormal cortical cytomorphogenesis (Beracochea *et al.*, 1989; Gould *et al.*, 1989, 1991; Lauder, 1983; Miller, 1986; Petit, 1986; Schambra *et al.*, 1990). In particular, with regard to hypothyroidism, a selective vulnerability of cholinergic basal forebrain neurons to the hormonal deficiency has been established, and forebrain areas developmentally altered by thyroid state are typically those emitting or receiving basal forebrain cholinergic innervation (Gould *et al.*, 1989, 1991; Hayashi & Patel, 1987; Ipina *et al.*, 1987; Lauder, 1983; Ipina *et al.*, 1987; Patel *et al.*, 1987). Furthermore, treatment of neonates with thyroid hormone has been shown to result in increased ChAT activity, increased muscarinic receptor densities and altered cholinergic fibre morphology in male, but not female rats, suggesting interactions between thyroid and gonadal hormones in influencing the maturation and function of the forebrain cholinergic system (Gould *et al.*, 1991). While such complex syndromes are, of course, not suitable to study the isolated effects of cholinergic deficits on cortical development, they make an important clinical point. All of these conditions occur not infrequently in the human, and if cholinergic deficits indeed ranked high among the culprits for cortical dysfunction in these conditions, treatment strategies might be suggested.

In summary, most manipulations of cholinergic efferents to cortex have been shown to alter cortical morphogenesis, cognitive performances or both. A host of diverse approaches has yielded remarkably similar results, such as reduced pyramidal cell soma size, altered pyramidal cell dendrites, shrinkage of cortical width and disruptions of cognitive but not simple motor performance. Sex differences were consistently observed wherever both sexes were examined separately. Thus, the hypothesis that cholinergic afferents to the developing neocortex are necessary for cortical morphogenesis appears to be substantiated by these studies. The next step is to understand how the cholinergic system might exert its influence on shaping cortical neuron morphology and afferent synapses.

MECHANISMS OF ONTOGENETIC CHOLINERGIC EFFECTS

It may be useful here to first review some of the basic concepts regarding the cellular mechanisms that subserve both memory and learning and the sculpting of appropriate connectivity in development of cortical structures. In several places above, in this chapter, I have already invoked the concept of synaptic plasticity as substrate for memory and learning. The idea is that synaptic connectivity, in other words the wiring of cortex, is modified in response to the environmental experience of the organism. It is today widely hypothesized that a determining factor in this synaptic modification is the signal integration carried out by the postsynaptic neuron and subsequent feedback communications to the presynaptic input, signalling strengthening, maintenance or dissolution of the morphological synaptic specialization (Cotman & Nieto-Sampedro, 1982; Jeffery, 1997; Greenough, 1986; Kolb & Wishaw, 1998; Merzenich & Sameshima, 1993; Rauschecker, 1995; Wang et al., 1997; Weiler et al., 1995). The historical concept of the Hebb synapse (Hebb, 1949; Spatz, 1996) postulates that the postsynaptic neuron has a "preferred" input–output relationship it attempts to accomplish and maintain by regulating its inputs. Obviously, each cortical neuron receives hundreds or thousands of inputs from, depending on the neuron, various combinations of afferent and corticocortical axons. At any given point in time a number of these afferents will fire in synergy, eliciting a strong excitatory signal which will result in a postsynaptic action potential; other inputs will be not be capable of exciting the postsynaptic neuron to that point either because their influence is counteracted by inhibitory currents within the same neuron or because their firing was not mediated by a distant stimulus sufficiently strong to recruit a sizable population of afferents. In simple terms the Hebbian idea is that the postsynaptic neuron will, with time, shed these "weak" synapses and consolidate the prior, "strong" synapses which are presumed to correspond to behaviourally meaningful stimuli. These theoretical concepts are supported by a long-estab-lished literature demonstrating that, e.g., sensory deprivation or competition between afferents will result in synaptic atrophy of the weak inputs to cortex, and that during a period of late cortical morphogenesis, a process of "pruning" sculpts synapses to be maintained into adulthood from the overabundance of developmentally established connections (Capone, 1996; Blakemore & Mitchell, 1973; De Felipe et al., 1997; Greenough, 1986; Hubel & Wiesel, 1962; Huttenlocher & Dabholkar, 1997; Ivy & Killackey, 1982; Koester & O'Leary, 1992; McMullen et al., 1988; Olavarria & Van Sluyters, 1985; O'Leary & Stanfield, 1989; Valverde, 1967). It is implied in the concept as laid out here that "strengthening" and "weakening" of afferent connections, permanently or transiently, changes the actual structure of synaptic specializations, dendrites and dendritic spines and perhaps also directs axonal growth and sprouting. The cholinergic neuromodu-latory inputs to cortex are involved in this selective strengthening of specific afferent and corticocortical synapses (Jerusalinsky et al., 1997; Juliano, 1998; Singer, 1990; Woody & Gruen, 1993). It is within this context also that the cholinergic influence on learning and memory has been understood.

The molecular mediators of this rather complex communication process have only recently begun to be elucidated. Most of the specific thalamocortical and corticocortical inputs, presumed to use excitatory amino acid neurotransmitters are equipped with predominantly ion-gated postsynaptic receptor mechanisms. In contrast, most postsynaptic receptors for the cholinergic neuromodulatory inputs have been termed "metabotropic" since they are equipped with second messenger signal transduction chains capable of altering a wide variety of meta-bolic processes as well as gene expression in the cell. Thus, while modulating the immediate physiological effects of specific synapses, these inputs can also modify their future responsiveness. Since cholinergic receptors are present from early on, it is not a real stretch to assume that ontogenetic events may depend on the same molecular mechanisms as learning and memory in the more mature brain. Many of the studies exploring the muscarinic signal transduction effects have been conducted in vitro on immature neurons and neuronal type cell lines

(Ashkenazi *et al.*, 1989; Costa, 1994; Gutkind *et al.*, 1991; Lauder, 1993; Lipton & Kater, 1989; Mattson, 1988; Shapiro, 1973).

Thus, strong evidence points towards muscarinic cholinergic receptors as mediators of maturational cholinergic effects in cortex. As detailed above, these receptors certainly show themselves to be "in the right place at the right time"; moreover, they have the signal transduction mechanisms required to convey the metabolic and gene expression changes necessary for mediating morphogenetic effects. Interestingly, both differentiation- and proliferation-promoting effects have been observed *in vitro*, often following stimulation of the same muscarinic receptor mechanisms in different cells or at different times even in the same cell (Costa, 1994; Lauder, 1993; Shapiro, 1973). However, this may not be too surprising considering that both types of signals must ultimately impact on the cell cycle regulation, and that differentiation cannot be achieved without cessation of proliferation.

Both the IP3/DAG-stimulating pathway, activated via m1 and m3, and the c-AMP-inactivating pathway, activated by m2 and m4, have been shown to influence cell growth *in vitro*. The IP3/DAG pathway has, however, emerged as the more interesting possibility for early cortical growth regulation. Muscarinic receptor coupling to this signal transduction pathway is significantly elevated during the peak period of synapse formation in early cortical ontogeny; hence IP3/DAG turnover is increased at this time (Balduini *et al.*, 1987; Bevilacqua *et al.*, 1995; Rooney & Nahorski, 1989).

The IP3 pathway has been shown to modulate the activity of the growth-associated protein GAP-43. This protein is highly correlated within axonal growth and plasticity in cortex and other brain areas, and has been shown to be expressed in sexually dimorphic fashion in the forebrain (Akers & Routenberg, 1985; Van Hooff *et al.*, 1989a,b; Lustig *et al.*, 1993; Skene, 1989). We have preliminary data suggesting alterations of GAP-43 following the neonatal nBM lesion (Hohmann *et al.*, 1990).

Muscarinic receptors have been shown also to influence MAP-2 protein (Woolf, 1993). Postsynaptically, microtubule-associated proteins, particularly MAP-2, are indicative of normal dendritic development, morphology and plasticity (Johnson & Jope, 1992). We recently observed that MAP-2 expression is altered following the neonatal nBM lesion, as well as in Rett syndrome, a mental retardation paradigm characterized by, among other things, arrested cortical development and cholinergic deficits (Hohmann *et al.*, 1995; Johnston *et al.*, 1995; Kaufmann *et al.*, 1997).

Muscarinic cholinergic activity apparently also modulates the expression of the amyloid precursor protein (APP) which has growth-regulatory functions in development and is associated with pathologies in Alzheimer's disease (AD) and Down syndrome (DS) (Robertson & Harrell, 1997). Our own preliminary data show APP expression to be altered following the neonatal nBM lesion (Hohmann *et al.*, unpublished observations).

Beyond these specific examples of cholinergic regulation of growth- and differentiation-associated proteins, several messenger molecules downstream from the IP3/DAG signal have long been known to effect cellular maturation. The DAG pathway mediates its effects on many metabolic processes and transcriptional activation in cells via the enzyme protein kinase C (PKC), which in turn influences transcriptional activation and other metabolic processes in the neuron (Costa, 1994; Lauder, 1993; McGinty, 1997). Meanwhile, IP3 mobilizes calcium from intracellular stores (Bell & Burns, 1991; Loffelholz, 1996). A rise in intracellular calcium results in regulation of Ca^{2+}-inducible genes, involving Ca^{2+}/calmodulin-dependent kinase II (CaCM KII) (Fields, 1996; Herdegen, 1996; Loffelholz, 1996). This kinase, as well as PKC, phosphorylates nuclear transcription factors which have been directly implicated in dendritic plasticity (Deisseroth et al., 1996; Murphy & Segal, 1997). PKC activity and distribution is significantly affected in a variety of learning and memory paradigms (Olds & Alkon, 1991; Nogues, 1997) and is altered by sensory deprivation in cortex (Kumar & Schliebs, 1992). Likewise, altered CaCM KII has been associated with alterations to cortical plasticity (Soderling, 1995).

The adenylate cyclase-regulating m2 and m4 muscarinic cholinergic receptors are likely to assume their developmental significance later in cortical ontogeny. c-AMP can indirectly regulate the expression of many genes by activating protein kinase A (PKA), which can then phosphorylate transcription factors shown to influence neuronal morphology (Fields, 1996; Herdegen, 1996; Murphy & Segal, 1997). The m2 and m4 receptors are negatively regulating adenylate cyclase (AC); that is, they reduce c-AMP and PKA, and thus the activation of certain metabolic pathways and expression of certain genes in the cells. Some experimental evidence suggests that AC coupling to muscarinic receptors may not be active within the early postnatal time period (Lee et al., 1990). Such later onset and duration of m2-mediated events is certainly consistent with the delayed maturation of this receptor compared to m1 (Tice et al., 1996). Thus, muscarinic receptors coupled to the AC second messenger pathway may be of larger significance in later differentiation processes, including pruning and fine-tuning of connectivity in cortex.

Evidence for nicotinic effects on cortical morphogenesis is somewhat more circumstantial. It has been shown that systemic application of the cholinergic agonist, nicotine, during the perinatal period resulted in a variety of developmental delays in the CNS, including changes of cortical cytomorphology similar to those observed by us after neonatal nBM lesions (Navarro et al., 1989; Roy & Sabherwhal, 1994; Slotkin et al., 1986); yet these effects may well have been transynaptic. The development of cholinergic nBM afferents to cortex and of cortical muscarinic receptors is sensitive to nicotine (Navarro et al., 1989; Zhu et al., 1996); furthermore, nicotinic up-regulation of monoamine release and turnover has been documented (Tani et al., 1997). Thus, perinatal nicotine effects could be the result of *indirect decreases* in cortical muscarinic cholinergic activity. To my knowledge the only evidence for direct nicotinic modulation of neuronal differentiation in the CNS has been reported in retina neurons which respond to nicotinic agonists treatment with enhanced process outgrowth (Lipton et al., 1988). As stated above, nicotinic cholinergic receptors are ligand-gated ion channels; thus their metabolic regulatory impact is limited (Boyd, 1997; Montes et al., 1994). However, nicotinic receptors can regulate intracellular Ca^{2+} concentrations by opening voltage sensitive Ca^{2+} channels and thus impact indirectly on gene expression (McGinty, 1997). In light of their early appearance, particularly in human cortex, nicotinic receptors deserve more careful investigation.

Based on current understanding of the molecular mechanisms involved in plasticity regulation in the CNS, I would like to propose that ACh conveys its

121

main effect on morphogenesis and plasticity via modulation of local calcium concentrations within the neuron. Acetylcholine may thus act as a precisely timed "stop" and "go" signal for dendritic and synaptic growth in cerebral cortex. Calcium fluctuations hold a well-established place in neuronal differentiation (Spitzer, 1994). Relative levels of free Ca^{2+} intracellularly can apparently determine whether a growth cone responds with outgrowth, regression or even cell death (Kater *et al.*, 1988; Mattson, 1988). As suggested by the receptor literature, different muscarinic and even nicotinic receptors might relay the cholinergic message at different ontogenetic times and/or simultaneously, in different neurons and different parts of the same cell, thus sculpting the developing neuron in time and space. As reviewed, all ACh receptors have the ability to modulate postsynaptic Ca^{2+} concentrations either via secondary messengers or, in the case of the nicotinic receptor, via membrane channels. The early appearance of IP3-mediated, Ca^{2+}-increasing cholinergic signals that predominate during the period of rapid synapse formation, and the later maturation of the receptors linked to adenylate cyclase inhibition during the period of synapse pruning, are certainly compatible with such a model.

Calcium modulation as a final common pathway would also explain how cholinergic effects might interact with excitation generated by specific afferent or corticocortical synapses. As stated, most of the specific afferents in cortex are associated with excitatory, ionotropic amino acid receptors (Huntley *et al.*, 1994; McDonald & Johnston, 1990; Salt *et al.*, 1995). Especially early in development, many of these receptors are of the NMDA type. NMDA receptors are associated with Ca^{2+} influx through voltage-gated membrane channels, and regarded as regulators of cortical plasticity in their own right (Huntley *et al.*, 1994; McDonald & Johnston, 1990). Thus, the sum total of cholinergically mediated intracellular Ca^{2+} increase and NMDA-mediated Ca^{2+} influx may determine the growth response of a spine or dendrite in this interaction. However, as NMDA receptors recede in development to make way for AMPA receptors not known to directly affect Ca^{2+} currents, cholinergically mediated Ca^{2+} modulation may still interact with specific excitatory signals. Recent research indicates that back-propagation of action potentials from the soma can reach the dendrites of pyramidal neurons and thus affect synaptic efficacy (Markram *et al.*, 1997). Such back-propagated action potentials impact dendritic Ca^{2+} currents, particularly when a local dendritic EPSP with its own transient Ca^{2+} current recently preceded the action potential (Koester & Sakman, 1998). Such a scenario is supported by our recent observation of developmental changes in a variety of excitatory amino acid receptors following neonatal cholinergic depletions (Hohmann *et al.*, 1999). Both NMDA and AMPA receptors were reduced in a sex-dependent manner by 2 and 4 weeks following the neonatal basal forebrain lesion, indicative of reduced synaptic strength in the cholinergically deprived cortical neurons.

CLINICAL IMPLICATIONS

Throughout this chapter I have advanced the argument that developmental cholinergic innervation is an essential growth-regulatory signal in cortical morphogenesis, maturation and developmental, as well as adult, plasticity. The implications for certain developmental disorders with established cholinergic deficits such as Down syndrome and Rett syndrome are fairly clear. What remains to be established in these complex genetic disorders is not if, but to what extent and exactly how, the cholinergic afferents contribute to cortical abnormalities. In some instances cholinergic enhancement and/or replacement protocols are already under consideration (Capone, 1998). However, the information in this review raises additional issues about cholinergic participation, or rather the possibility of cholinergic deficits in less well-characterized developmental disorders; for example a large variety of unclassified learning disabilities. The alterations in synaptic connectivity that must underlie such disabilities are often too subtle to be noticeable on scans or in pathological examinations. While some might argue that they are of genetic origin, there is in many cases no solid evidence to support this notion. Furthermore, the substantial, recent increase in the diagnosis of learning disabilities, while perhaps in part the result of increased awareness, argues against genetic origins and in favour of environmental reasons. We have described above how a variety of environmental influences and endocrinological alterations including lead, ethanol and hypothyroidism are capable of altering cholinergic neurochemistry and development. This may, by far, not be an inclusive list. Nutritional factors, particularly choline availability in early development, should most certainly be explored further. Stress effects have received very little attention so far with regard to the cholinergic system. An additional reason to suspect cholinergic involvement in learning disabilities is the sex imbalance, with males being substantially more affected (Elliot & Tyler, 1987; Halpern, 1997). As shown, cholinergic effects display a sexual dimorphism in all experimental models where both sexes were studied. In our lesion experiments the males were clearly more affected by cholinergic disruptions than were the females. That some of these developmental sex differences must also exist in the human brain is strongly suggested by recent data concerning oestrogen effects on cholinergic transmission and cognitive performance as relates to Alzheimer's disease. Thus, females may be more protected from developmental deficits in their cholinergic innervation but become more vulnerable in late life.

In conclusion, the studies discussed here concerning cholinergic effects in brain development and behaviour may presently raise more new questions than they have answered; yet such is the yield of most productive research endeavours. It is my hope that some of the basic experimental literature discussed here will generate interest in the investigation of the cholinergic system in normal human cognitive development in relationship to structural brain development. Long impossible, such correlative investigations are becoming increasingly accessible with new non-invasive imaging tools such as functional MRI.

ACKNOWLEDGEMENTS

The author thanks Miss Rhonda Robinson for diligent editorial assistance with the manuscript. The cited research data from the author's laboratory were generated with support from NIH/NIGMS 1S06GM51971, 2P20RR011606 ad HD24448.

References

Aguilar, J.S. & Lunt, G.G. (1985). Muscarinic receptor sites in human foetal brain. *Neurochemistry International*, **7**, 509–514.

Agulhon, C., Charnay, Y., Vallet, P., Bertrand, D. & Malafosse, A. (1998). Distribution of mRNA for the a-4 subunit of the nicotinic acetylcholine receptor in the human fetal brain. *Molecular Brain Research*, **58**, 123–131.

Akers, R. & Routenberg, A. (1985). Protein kinase C phosphorylates a 47 Mr protein (F1) directly related to synaptic plasticity. *Brain Research*, **334**, 147–151.

Armstrong, J.N. & Pappas, B.A. (1991). The histopathological, behavioral and neurochemical effects of intraventricular injections of ethylcholine mustard aziridinium (AF64A) in the developing rat. *Developmental Brain Research*, **64**, 249–257.

Arters, J., Hohmann, C.F., Mills, J,O. & Berger-Sweeney, J. (1998). Sexually dimorphic responses to neonatal basal forebrain lesions in mice: I. Behavior and neurochemistry. *Journal of Neurobiology*, **37**, 582–594.

Ashkenazi, A., Ramachandran, J. & Capon, D.J. (1989). Acetylcholine analogue stimulates DNA synthesis in brain-derived cells via specific muscarinic receptor subtypes. *Nature*, **340**, 146–150.

Bachman, E., Berger-Sweeney, J.E., Coyle, J.T. & Hohmann, C.F. (1994). Developmental regulation of adult cortical morphology and behavior: An animal model for mental retardation. *International Journal of Developmental Neuroscience*, **12**, 239–253.

Balduini, W., Murphy, S.D. & Costa, L.G. (1987). Developmental changes in muscarinic receptor-stimulated phosphoinositide metablism in rat brain. *Journal of Pharmacology and Experimental Therapeutics*, **241**, 421–427.

Bar-Peled, O., Israeli, M., Ben-Hur, H., Hoskins, I., Groner, Y. & Biegon, A. (1991). Developmental pattern of muscarinic receptors in normal and Down's syndrome fetal brain – an autoradiographic study. *Neuroscience Letters*, **133**, 154–158.

Bauman, M.L. & Kemper, T.L. (1995). Pervasive neuroanatomic abnormalities in three cases of Rett syndrome. *Neurology*, **45**, 1581–1586.

Bear, M.F. & Singer, W. (1986). Modulation of visual cortical plasticity by acetylcholine and norepinephrine. *Nature*, **320**, 172–176.

Beeri, R., Le Novere, N., Mervis, R., Huberman, T., Grauer, E., Changeux, J.P. & Soreq, H. (1997). Enhanced hemocholinium binding and attenuated branching in cognitively impaired acetylcholinesterase-transgenic mice. *Journal of Neurochemistry*, **69**, 2441–2451.

Bell, M. & Burns, D.J. (1991). Lipid activation of protein kinase C. *Journal of Biological Chemistry*, **266**, 4661–4664.

Beracochea, D.L., Alaoui-Bouarraqui, I. & Jaffard, R. (1989). Impairment of memory in a delayed non-matching to place task following mamillary body lesions in mice. *Behavioral Brain Research*, **34**, 147–154.

Berger-Sweeney, J. & Hohmann, C.F. (1997). Long-term behavioral consequences of abnormal cortical development in rodent: insights into developmental disabilities. *Behavioral Brain Research*, **86**, 121–142.

Berger-Sweeney, J, Arneld, A, Gabeau, D. & Mills, J (1995). Sex differences in learning and memory in mice: Effects of sequence of testing and cholinergic blockage. *Behavioral Neuroscience*, **109**, 859–873.

Berger-Sweeney, J., Berger, U.V., Lappi, D.A., Hohmann, C.F., Ewusi, A. & Frick, K.M. (1996). Specific cholinergic immunotoxins in mice. *Proceedings of the American Society for Neuroscience*, 1996.

Berger-Sweeney, J., Libbey, M., Arters, J., Junagadhwalla, C. & Hohmann, C.F. (1998). Neonatal monoamine depletions improve performance of a novel odor discrimination task. *Journal of Behavioral Neurobiology*, **112**, 1318–1326.

Bevilacqua, J.A., Downes, C.P. & Lowenstein, P.R. (1995). Transiently selective activation of phosphoinositide turnover in layer V pyramidal neurons after specific mGluR stimulation in rat somatosensory cortex during early postnatal development. *Journal of Neuroscience*, **15**, 7916–7928.

Blakemore, C. & Mitchell, D.E. (1973). Environmental modification of the visual cortex and the neural basis of learning and memory. *Nature*, **241**, 467–468.

Blue, M.E. & Parnavales, J.B. (1983). The formation and maturation of synapses in the visual cortex of the rat. II. Quantitative analysis. *Journal of Neurocytology*, **12**, 697–712.

Blusztajn, J.K. & Wurtman, R.J. (1983). Choline and cholinergic neurons. *Science*, **221**, 614–620.

Bonner, T. (1989). The molecular basis of muscarinic receptor diversity. *TINS*, **12**, 148–151.

Bonner, T.I., Buckley, N.J., Young, A.C. & Brann, M.R. (1987). Identification of a family of muscarinic acetylcholine receptor genes. *Science*, **237**, 527–532.

Boyd, R.T. (1997). The molecular biology of neuronal nicotinic acetylcholine receptors. *Critical Reviews in Toxicology*, **27**, 299–318.

Brady, D.R., Phelps, P.E. & Vaughn, J.E. (1989). Neurogenesis of basal forebrain cholinergic neurons in rat. *Developmental Brain Research*, **47**, 81–92.

Brake, W.G. & Pappas, B.A. (1994). Hemicholinium-3 (HC3) blocks the effect of ethycholine mustard aziridinium ion in the developing rat. *Developmental Brain Research*, **83**, 289–293.

Brooksbank, B.W.L., Martinez, M., Atkinson, D.J. & Balasz, R. (1978). Biochemical development of human brain. I: Some parameters of cholinergic system. *Developmental Neuroscience*, **1**, 267–284.

Brown, D.A., Abogadie, F.D., Allen, T.G., Buckley, N.J., Caulfield, M.P., Delmas, P., Haley, J.E., Lamas, J.A. & Selyanko, A.A. (1997). Muscarinic mechanisms in nerve cells. *Life Sciences*, **60**, 1137–1144.

Buckley, N.J., Bonner, T.I. & Brann, M.R. (1988). Localization of a family of muscarinic receptor mRNAs in rat brain. *Journal of Neuroscience*, **8**, 4646–4652.

Buckley, N.J., Bonner, T.I., Buckley, C.M. & Brann, M.R. (1989). Antagonist binding properties of five cloned muscarinic receptors expressed in CHO-K1 cells. *Molecular Pharmacology*, **35**, 469–476.

Candy, J.M., Perry, E.K., Perry, R.H., Bloxham, C.A., Thompson, J., Johnson, M., Oakley, A.E. & Edwardson, J.A. (1985). Evidence for the early prenatal development of cortical cholinergic afferents from the nucleus of Meynert in the human foetus. *Neuroscience Letters*, **61**, 91–95.

Capone, G.T. (1996). Human brain development. In J. Capute & P. Accardo (Eds.), *Developmetal diagnosis and treatment* (pp. 25–75). Baltimore, MD: Paul Brooks.

Capone, G.T. (1998). Drugs that increase intelligence? Applications for childhood cognitive impairments. *Mental Retardation and Developmental Disabilities Research Reviews*, **4**, 36–49.

Castro, C.A. & Paylor, R. (1990). Cholinergic agent physostigmine enhances short term memory based performance in the developing rat. *Behavioral Neuroscience*, **104**, 390–393.

Catalano, S.M., Robertson, R.T. & Killackey, H.P. (1991). Early ingrowth of thalamocortical afferents to the neocortex of the prenatal rat. *Proceedings of the National Academy of Sciences, USA*, **88**, 2999–3003.

Caviness, V.S. (1982). Neocortical histogenesis in normal and Reeler mice: A developmental study based upon (3H)thymidine autoradiography. *Developmental Brain Research*, **4**, 293–302.

Caviness, V.S. Jr. & Williams, R.S. (1979). Cellular pathology of developing human cortex. In R. Katzman (Ed.), *Congenital and acquired cognitive disorders* (pp. 69–99). New York: Raven Press.

Cimino, M., Marini, P., Colombo, S., Andena, M., Cattabeni, F. & Fornasari, D. (1995). Expression of neuronal acetylcholine nicotinic receptor alpha 4 and beta 2 subunits during postnatal development of the rat brain. *Journal of Neural Transmission, Genetics Section*, **100**, 77–92.

Conroy, W.G. & Berg, D.K. (1995). Neurons can maintain multiple classes of nicotinic acetylcholine receptors distinguished by different subunit compositions. *Journal of Biological Chemistry*, **270**, 4424–4431.

Cortes, R., Probst, A. & Palacios, J.M. (1987). Quantitative light microscopic autoradiographic localization of cholinergic muscarinic receptors in the human brain: forebrain. *Neuroscience*, **20**, 65–107.

Cortes, R., Probst, A., Tobler, H.J. & Palacios, J.M. (1986). Muscarinic cholinergic receptor subtypes in the human brain. II. Quantitative autoradiographic studies. *Brain Research*, **362**, 239–253.

Costa, L. (1994). Signal transduction mechanism in developmental neurotoxicity: the phosphoinositide pathway. *NeuroToxicology*, **15**, 19–28.

Cotman, C.W. & Nieto-Sampedro, M. (1982). Brain function, synapse renewal, and plasticity. *Annual Review of Psychology*, **33**, 371–401.

Coull, J.T. (1998). Neural correlates of attention and arousal: insights from electrophysiological, functional neuroimaging and psychopharmacology. *Progress in Neurobiology*, **55**, 343–361.

Court, J. & Clementi, F. (1995). Distribution of nicotinic subtypes in human brain. *Alzheimer Disease and Associated Disorders*, **9**(Suppl. 2), 6–14.

Court, J.A., Perry, E.K., Johnson, M., Piggott, M.A., Kerwin, J.A., Perry, R.H. & Ince, P.G. (1993). Regional patterns of cholinergic and glutamate activity in the developing and aging human brain. *Developmental Brain Research,* **74**, 73–82.

Coyle, J.T. (1988). Neuroscience and psychiatry. In J.A. Talbot, R.E. Hales & S.C. Yudofsky (Eds.), *Textbook of psychiatry* (pp. 3–32). Washington DC: American Psychiatric Press.

Coyle, J.T. & Yamamura, H.I. (1976). Neurochemical aspects of the ontogenesis of cholinergic neurons in the rat brain. *Brain Research,* **118**, 429–440.

Coyle, J.T., Oster-Granite, M.L. & Gearhart, J.D. (1986). The neurobiologic consequences of Down Syndrome. *Brain Research Bulletin,* **16**, 773–787.

Crandall, J.E. & Caviness, V.S. Jr. (1984). Axon strata of the cerebral wall in embryonic mice. *Developmental Brain Research,* **14**, 185–195.

Crandall, J.E., Whitcomb, J.M. & Caviness, V.S. Jr. (1985). Development of spinal–medullary projections from the mouse barrel field. *Journal of Comparative Neurology,* **239**, 205–215.

De Felipe, J., Marco, P., Fairen, A. & Jones, E.G. (1997). Inhibitory synaptogenesis in mouse somatosensory cortex. *Cerebral Cortex,* **7**, 619–634.

Deisseroth, K., Bito, H. & Tsien, R. (1996). Signaling from synapse to nucleus: postsynaptic CREB Phosphorylation during multiple forms of hippocampal synaptic plasticity. *Neuron,* **16**, 89–101.

Deutsch, J.A. (1971). The cholinergic synapse and the site of memory. *Science,* **174**, 788–794.

Donaghue, J.P. & Carrol, K.L. (1987). Cholinergic modulation of sensory reponses in rat primary somatic sensory cortex. *Brain Research,* **408**, 367–371.

Drachman, D.A. (1981). The cholinergic system, memory and aging. In S.J. Ennor, T., Samoyshi & B. Beer (Eds.), *Brain neurotransmiters and receptors in aging and age related disorders* (pp. 255–268). New York: Raven Press.

Dringenberg, H.C. & Vanderwolf, C.H. (1998). Involvement of direct and indirect pathways in electrocorticographic activation. *Neuroscience and Biobehavioral Research,* **22**, 243–257.

Eckenstein, F. & Thoenen, H. (1983). Cholinergic neurons in the rat cerebral cortex demonstrated by immunohistochemical localization of choline acetyltransferase. *Neuroscience Letters,* **36**, 211–215.

Egozi, Y., Sokolovsky, M., Schejter, E., Blatt, I., Zakut, H., Matzkel, A. & Soreq, H. (1986). Divergent regulation of muscarinic binding sites and acetylcholinesterase in discrete regions of the developing human fetal brain. *Cellular and Molecular Neurobiology,* **6**, 55–70.

Elliot, C.D. & Tyler, S. (1987). Learning disabilities and intelligence test results: a principal component analysis of the British Ability Scales. *British Journal of Psychiatry,* **78**, 325–333.

Evans, R.A., Watson, M., Yamamura, H.I. & Roeske, W.R. (1985). Differential ontogeny of putative M1 and M2 muscarinic receptor binding sites in the murine cerebral cortex and heart. *Journal of Pharmacology and Experimental Therapeutics,* **235**, 612–618.

Fields, R.D. (1996). Signaling from neural impulses to genes. *Neuroscientist,* **2**, 315–325.

Friedman, J.I., Temporini, H. & Davis, K.L. (1999). Pharmacological strategies for augmenting cognitive performance in schizophrenia. *Biological Psychiatry,* **45**, 1–16.

Furey, M.L., Pietrini, P., Haxby, J.V., Alexander, G.E., Lee, H.C., Van Meter, J., Grady, C.L., Shetty, U., Rapaport, S.I., Schapiro, M.B. & Freo, U. (1997). Cholinergic stimulation alters performance and task-specific regional cerebral blood flow during working memory. *Proceedings of the National Academy of Sciences, USA,* **94**, 6512–6516.

Gale, J.S. (1977). Glutamic acid decarboxylase and choline acetyltransferase in human foetal brain. *Brain Research,* **133**, 172–176.

Gaspar, E., Heeringa, M., Markel, E., Luiten, P. & Nyakas, C. (1991). Behavioral and biochemical effects of early postnatal cholinergic lesion in the hippocampus. *Brain Research Bulletin,* **28**, 65–71.

Gaulin, S.J.C. & Hoffman, H.A. (1988). Evolution and development of sex differences in spatial ability. In L. Betzig, M.B. Mulder & P. Turke (Eds.), *Human reproductive behavior: a Darwinian perspective* (pp. 129–152). Cambridge, UK: Cambridge University Press.

Gould, E., Farris, T.W. & Butcher,L.L. (1989). Basal forebrain neurons undergo somatal and dendritic remodeling during postnatal development: a single-section Golgi and choline acetyltransferase analysis. *Developmental Brain Research,* **46**, 297–302.

Gould, E., Wooley, C.S. & McEwen, B.S. (1991). The hippocampal formation: morphological changes induced by thyroid, gonadal and adrenal hormones. *Psychoneuroendocrinology,* **16**, 67–84.

Greenough, W.T. (1986). What's special about development? Thoughts on the bases of experience-sensitive synaptic plasticity. In W.T. Greenaugh & J.M. Juraski (Eds.), *Developmental neuropsychology* (pp. 387–407). Orlando, FL: Academic Press.

Gupta, R.C., Rech, R.H., Lovell, K.L., Welsch, F. & Thornburg, J.E. (1985). Brain cholinergic, behavioral and morphological development in rats exposed *in utero* to methylparathion. *Toxicology and Applied Pharmacology,* **77,** 405–413.

Gutkind, J.S., Novotny, E.A., Brann, M.R. & Robbins, K.C. (1991). Muscarinic acetylcholine receptor subtypes as agonist-dependent oncogenes. *Proceedings of the National Academy of Sciences, USA,* **88,** 4703–4707.

Ha, D.H., Robertson, R.T. & Weiss, J.H. (1998). Distinctive morphological features of a subset of cortical neurons grown in the presence of basal forebrain neurons *in vitro. Journal of Neuroscience,* **18,** 4201–4215.

Halpern, D.F. (1992). *Sex differences in cognitive abilities.* Hillsdale, NJ: Laurence Erlbaum Associates.

Halpern, D.F. (1997). Sex differences in intelligence. Implications for education. *American Psychologist,* **52,** 1091–1102.

Hamilton, S.E., Loose, M.D., Qui, M., Levey, A.I., Hille, B., McKnight, G.S., Idzerda, R.L. & Nathanson, N.M. (1997). Disruption of the m1 receptor gene ablates muscarinic receptor-dependent M current regulation and seizure activity in mice. *Proceedings of the National Academy of Sciences, USA,* **94,** 13311–13316.

Hartnagel, H., Hansen, L., Gisella, K., Schneider, B., Eitamar, A. & Hanin, I. (1993). Sex differences and estrous cycle variations in the AF64A induced cholinergic deficits in the hippocampus of the rat. *Brain Research Bulletin,* **31,** 129–134.

Hasselmo, M.E. (1995). Neuromodulation and cortical function. *Behavioral Brain Research,* **67,** 1–27.

Hayashi, M. & Patel, A.J. (1987). An interaction between thyroid hormone and nerve growth factor in the regulation of choline acetyltransferase activity in neuronal cultures, derived from the septal-diagonal band region of the embryonic rat brain. *Developmental Brain Research,* **36,** 109–120.

Hebb, D.O. (1949). *The organization of behavior: a neurophysiological theory.* New York: John Wiley & Sons.

Hellstrom-Lindahl, E., Gorbounova, O., Seiger, A., Mousavi, M. & Nordberg, A. (1998). Regional distribution of nicotinic receptors during prenatal development of human brain and spinal cord. *Developmental Brain Research,* **108,** 147–160.

Herdegen, T. (1996). Jun, Fos and CREB/ATF transcription factors in the brain: conrol of gene expression under normal and pathophysiological conditions. *Neuroscientist,* **1,** 153–161.

Hobson, J.A., Stickgold, R. & Pace-Schott, E.F (1998). The neuropsychology of REM sleep dreaming. *Neuroreport,* **9,** R1–R14.

Hohmann, C.F. & Berger-Sweeney, J.E. (1998a). Cholinergic regulation of cortical development: new twists on an old story. *Perspectives on Developmental Neurobiolog*y, **5,** 410–425.

Hohmann, C.F. & Berger-Sweeney, J.E. (1998b). Sexually dimorphic responses to neonatal basal forebrain lesions in mice: II. Quantitative assessments of cortical morphology. *Journal of Neurobiology,* **37,** 595–606.

Hohmann, C.F. & Ebner, F.F. (1985). Development of cholinergic markers in mouse forebrain: I. Choline acetyltransferase enzyme activity and acetylcholine histochemistry. *Developmental Brain Research,* **23,** 225–241.

Hohmann, C.F., Antuono, P. & Coyle, J.T. (1988). Basal forebrain cholinergic neurons and Alzheimer's disease. In L.L. Iversen Iversen, D.S. and S.H. Snyder (Eds.), *Handbook of psychopharmacology* (pp. 69–106). New York: Plenum Press.

Hohmann, C.F., Blue, M.E., Kaufmann, WE., Peters, L. & Johnston, M.V. (1995). Cytoskeletal protein and excitatory amino acid receptor development in a potential mouse model for Rett syndrome. *Proceedings of the American Society for Neuroscience.* San Diego.

Hohmann, C.F., Brooks, A.R. & Coyle, J.T. (1988). Neonatal lesions of the basal forebrain cholinergic neurons result in abnormal cortical development. *Developmental Brain Research,* **43,** 253–264.

Hohmann, C.F., Capone, G., Neve, R.L., Bennowitz, L.I. & Coyle, J.T. (1990). Development of GAP-43 expression in the forebrain of normal mice and littermates with neonatal lesions of the basal forebrain.*Proccedings of the American Society for Neuroscience.* St. Louis.

Hohmann, C.F., Kwiterovich, C., Oster-Granite, M.L. & Coyle, J.T. (1991). Newborn basal forebrain lesions disrupt cortical cytodifferentiation as visualized by Rapid Golgi staining. *Cerebral Cortex*, **1**, 143–157.

Hohmann, C.F., Pert, C.C. & Ebner, F.F. (1985). Development of cholinergic markers in mouse forebrain: II. Muscarinic receptor binding in cortex. *Brain Research*, **355**, 243–253.

Hohmann, C.F., Potter, E. & Levey, A.I. (1995). Development of muscarinic receptor subtypes in the forebrain of the mouse. *Journal of Neuroscience*, **358**, 88–101.

Hohmann, C.F., Richardson, C.M., Redding, C., Kaufmann, W., Arters, J. & Berger-Sweeney, J. (1997). Neonatal lesions of cholinergic and monoaminergic afferents to neocortex produce different alterations in cortical morphology. *Proceedings of the American Society for Neuroscience*. New Orleans.

Hohmann, C.F.., Wallace, S., Johnston, M. & Blue, M.E. (1999). Effects of neonatal cholinergic basal forebrain lesions on excitatory amino acid receptors in neocortex. *International Journal of Developmental Neuroscience*, **16**, 645–660.

Hohmann, C.F., Wilson, L. & Coyle, J.T. (1991). Efferent and afferent connections of mouse sensory-motor cortex following transient cholinergic deafferentiation at birth. *Cerebral Cortex*, **1**, 158–172.

Houser, C.R., Crawford, G.D., Salvaterra, P.M. & Vaughn, J.E. (1985). Immunocytochemical localization of choline acetyltransferase in rat cerebral cortex: a study of cholinergic neurons and synapses. *Journal of Comparative Neurology*, **234**, 17–34.

Hubel, D. & Wiesel, T.N. (1962). Receptive fields, binocular interactions and functional architecture in the cat's visual cortex. *Journal of Physiology*, **160**, 106–154.

Hulme, E.C., Birdsall, N.J.M. & Buckley, N.J. (1990). Muscarinic receptor subtypes. *Annual Review of Pharmacology*, **30**, 633–673.

Huntley, G.W., Vickers, J.C. & Morrison, J.H. (1994). Cellular and synaptic localization of NMDA and non-NMDA receptor subunits in neocortex: organizational features related to cortical circuitry, function and diseases. *TINS*, **17**, 536–543.

Huttenlocher, P.R. (1975). Synaptic and dendritic development and mental defect. In N. Buckwaldt & M. Brazier (Eds.) *Brain mechanism in mental retardation* (pp. 123–139). New York: Academic Press.

Huttenlocher, P.R. (1990). Morphometric studies of human cerebral cortex during development. *Neuropsychologica*, **28**, 517–527.

Huttenlocher, P.R. and Dabholkar, A.S. (1997). Regional differences in synaptogenesis in human cerebral cortex. *Journal of Comparative Neurology*, **378**, 167–178.

Ipina, S.L., Ruiz-Marcos, A., Escobar del Rey, F & Morreale de Escobar, G. (1987). Pyramidal cortical cell morphology studied by multivariate analysis: effects of neonatal thyroidectomy, ageing and thyroxine-substitution therapy. *Developmental Brain Research*, **37**, 219–229.

Ivy, G.O. & Killackey, H.P. (1982). Ontogenetic changes in the projections of neocortical neurons. *Journal of Neuroscience*, **2**, 735–743.

Jacobson, M. (1978). *Developmental neurobiology*. New York: Plenum Press.

Jeanmonod, D., Rice, F.L. & Van der Loos, H. (1981). Mouse somatosensory cortex: alterations in the barrel field following receptor injury at different early postnatal ages. *Neuroscience*, **6**, 1503–1535.

Jeffery, K.J. (1997). LTP and spatial learning – where to next? *Hippocampus*, **7**, 95–110.

Jerusalinsky, D., Kornisiuk, E. & Izquierdo, I. (1997). Cholinergic neurotransmission and synaptic plasticity concerning memory processing. *Neurochemical Research*, **22**, 507–515.

Johnson, G.V.W. & Jope, R.S. (1992). The role of microtubule associated protein 2 in neuronal growth, plasticity and degeneration. *Journal of Neuroscience Research*, **33**, 505–512.

Johnston, M.V. & Gerring, J.P. (1992). Head trauma and its sequellae. *Pediatric Annals*, **21**, 362–373.

Johnston, M.V., Hohmann, C.F. & Blue, M.E. (1995). Neurobiology of Rett syndrome. *Neuropediatrics*, **26**, 119–122.

Johnston, M.V., Silverstein, F.S., Reindel, F.O., Penney, Jr. J.B. & Young, A.B. (1985). Muscarinic cholinergic receptors in human infant forebrain: (3H) quinuclidinyl benzilate binding in homogenates and quantitative autoradiography in sections. *Developmental Brain Research*, **19**, 195–203.

Juliano, S.L. (1998). Mapping the sensory mosaic. *Science*, **279**, 1653–1654.

Juraska, J.M. (1982). The development of pyramidal neurons after eye opening in the visual cortex of hooded rats: a quantitative study. *Journal of Comparative Neurology*, **212**, 208–213.

Juraska, J.M. (1984). Sex differences in dendritic response to differential experience in rat visual cortex. *Brain Research*, **295**, 27–34.

Karczmar, A. (1993). Brief presentation of the story and present status of studies of the vertebrate cholinergic system. *Neuropsychopharmacology*, **9**, 181–198.

Kater, S.B., Mattson, M.P., Cohan, C. & Connor, J. (1988). Calcium regulation of the neuronal growth cone. *TINS*, **11**, 315–321.

Kaufmann, W.E., Taylor, C.V., Hohmann, C.F., Sanwall, I.B. & Naidu, S. (1997). Abnormalities in neuronal maturation in Rett syndrome neocortex: preliminary molecular correlates. *European Journal of Childhood and Adolescent Psychiatry*, **6**, 75–77.

Kilgard, M.P. & Merzenich, M.M. (1998), Cortical map reorganization enabled by nucleus basalis activity. *Science*, **279**, 1714–1717.

Killackey, H.P., Rhoodes, W.R. & Bennett-Clarke, C.A. (1995). The formation of a cortical somatotopic map. *TINS*, **18**, 402–407.

Kitt, C.A., Hohmann, C.F., Coyle, J.T. & Price, D.L. (1994). Cholinergic innervation of mouse forebrain structures. *Journal of Comparative Neurology*, **341**, 117–129.

Koester, H.J. & Sakman, B. (1998). Calcium dynamics in single spines during coincident pre-and postsynapstic activity dependent on relative timing of back-propagating action potentials and subthreshhold excitatory postsynaptic potentials. *Proceedings of the National Academy of Sciences, USA*, **95**, 9596–9601.

Koester, S.E. & O'Leary, D.D. (1992). Functional classes of cortical projection neurons develop dendritic distinction by class specific sculpting of an early common pattern. *Journal of Neuroscience*, **12**, 1382–1393.

Kolb, B. & Wishaw, I.Q. (1998). Brain plasticity and behavior. *Annual Review of Psychology*, **49**, 43–64.

Konig, N. & Marty, R. (1981). Early neurogenesis and synaptogenesis in cerebral cortex. *Bibliotheca Anatomica*, **19**, 152–160.

Kornack, D.R., Bai, L & Black, I.B. (1991). Sexually dimorphic expression of the NGF receptor gene in the developing rat brain. *Brain Research*, **542**, 171–174.

Kostovic, I. (1986). Prenatal development of nucleus basalis complex and related fiber systems in man: a histochemical study. *Neuroscience*, **17**, 1047–1077.

Kostovic, I. & Rakic, P. (1980). Cytology and time of origin of interstitial neurons in the white matter in infant and adult human and monkey telencephalon. *Journal of Neurocytology*, **9**, 219–242.

Kostovic, I. & Racik, P. (1984). The development of prestriate visual projections in the monkey and human fetal cerebrum revealed by transient choinesterase staining. *Journal of Neuroscience*, **4**, 25–42.

Kostovic, I., Judas, M., Kostovic-Knezevic, L., Simic, G., Delalle, I., Chudy, D., Sajin, B. & Petanjek, Z. (1991). Zagreb research collection of human brains for developmental neurobiologists and clinical neuroscientists. *International Journal of Developmental Biology*, **35**, 215–230.

Kostovic, I., Skavic, J. & Strinovic, D. (1988). Acetylcholinesterase in the human frontal associative cortex during the period of cognitive development: early laminar shifts and late innervation of pyramidal neurons. *Neuroscience Letters*, **90**, 107–112.

Kristt, D.A. (1979). Development of neuronal circuitry: histochemical localization of AChE in relation to the cell layers of rat somatosensory cortex. *Journal of Comparative Neurology*, **186**, 1–17.

Krnjevic, K. (1988). Acetylcholine as transmitter in the cerebral cortex. In A. Avoli, T.A. Reader, R. Dykes and P. Gloor (Eds.), *Neurotransmitters and brain function* (pp. 227–237). New York: Plenum Press.

Kuhar, J.J., Birdsall, N.J.M., Burgen, A.S.V. & Hulme, E.C. (1980). Ontogeny of muscarinic receptors in rat brain. *Brain Research*, **184**, 375–383.

Kumar, A. & Schliebs, R. (1992). Postnatal laminar development of cholinergic receptors, protein kinase C and dihydropyridine sensitive calcium antagonist binding in rat visual cortex. Effect of visual deprivation. *International Journal of Developmental Neuroscience*, **10**, 491–504.

Lauder, J.M. (1983). Hormonal and humoral influences on brain development. *Psychoneuroendocrinology*, **8**, 121–155.

Lauder, J.M. (1993). Neurotransmitters as growth regulatory signals: role of receptors and second messengers. *TINS*, **16**, 233–243.

Leanza, G., Nilsson, O.G., Nikkhah, G., Wiley, R.G. & Bjorklund, A. (1996). Effects of neonatal lesions on the basal forebrain cholinergic system by 192 immunoglobulin G-saporin: biochemical, behavioral and morphological characterisation. *Neuroscience*, **74**, 119–141.

Lee, W., Nicklaus, K.J., Manning, D.R. & Wolfe, B.B. (1990). Ontogeny of cortical muscarinic receptor subtypes and muscarinic receptor-mediated responses in rat. *Journal of Pharmacology and Experimental Therapeutics,* **252**, 482–490.

Levey, A. (1996). Muscarinic acetylcholine receptor expression in memory circuits: implications for treatment of Alzheimer disease. *Proceedings of the National Academy of Sciences, USA,* **93**, 13541–13546.

Levey, A.I., Kitt, C.A., Simonds, W.R., Price, D.L. & Brann, M.R. (1991). Identification and localization of muscarinic acetylcholine receptor proteins in brain with subtype-specific antibodies. *Journal of Neuroscience,* **11**, 3218–3226.

Levin, E.D. & Simon, B.B. (1998). Nicotinic acetylcholine involvement in cognitive function in animals. *Psychopharmacology,* **138**, 217–230.

Linn, M.C. & Peterson, A.C. (1985). Emergence and characterization of sex differences in spatial ability: a meta analysis. *Child Development,* **56**, 1479–1498.

Lipton, S.A., Frosch, M.P., Phillips, M.D., Tauck, D.L. & Aizenman, E. (1988). Nicotinic antagonists enhance process outgrowth by rat retinal ganglion cells in culture. *Science,* **239**, 1239–1296.

Lipton, S.A. & Kater, S.B. (1989). Neurotransmitter regulation of neuronal outgrowth, plasticity and survival. *TINS,* **12**, 265–270.

Loffelholz, K. (1996). Muscarinic receptors and cell signalling. In J. Klein & K. Loffelholz (Eds.). *Progress in Brain Research,* **109**, 201–208. New York: Elsevier.

Loy, R. & Sheldon, R.A. (1987). Sexually dimorphic development of cholinergic enzymes in the rat septohippocampal system. *Developmental Brain Research,* **34**, 156–160.

Luine, V. N. & McEwen, B. (1983). Sex differences in cholinergic enzymes of the diagonal band nuclei in the rat preoptic area. *Neuroendocrinology,* **36**, 475–482.

Lustig, R.H., Hua, P., Wilson, M. & Federoff, H. (1993). Ontogeny, sex dimorphism and neonatal sex hormone determination of synapse-associated mRNAs in rat brain. *Molecular Brain Research,* **20**, 101–110.

Lydic, R. (1996). Reticular modulation of breathing during sleep and anesthesia. *Current Opinion in Pulmonary Medicine,* **2**, 474–481.

Lysakowski, A., Wainer, B.H., Rye, D.B., Bruce, D. & Hersh, L.B. (1986). Cholinergic innervation displays strikingly different laminar preferences in several cortical areas. *Neuroscience Letters,* **64**, 102–108.

Marin-Padilla, M. (1970). Prenatal and early postnatal ontogenesis of the human motor cortex: a Golgi study. I. The sequential development of the cortical layers. *Brain Research,* **23**, 167–183.

Marin-Padilla, M. (1975). Abnormal neuronal differentiation (functional maturation) in mental retardation. In National Foundation of Birth Defects (Eds.), *Birth defects* (pp. 133–155).

Marin-Padilla, M. (1978). Dual origin of the mammalian neocortex and evolution of the cortical plate. *Anatomy and Embryology,* **152**, 109–126.

Markram, H., Lubke, J., Frotcher, M. & Sakman, B. (1997). Regulation of synaptic efficacy by coincidence of postsynaptic APs and EPSPs. *Science,* **275**, 213–215.

Mattson, M.P. (1988). Neurotransmitters in the regulation of neuronal cytoarchitecture. *Brain Research Review,* **13**, 179–212.

McConnell, S.K. (1988). Development and decision making in the mammalian cerebral cortex. *Brain Research Review,* **13**, 1–23.

McCormick, D.A. (1989). Cholinergic and noradrenergic modulation of thalamocortical processing. *TINS,* **12**, 215–221.

McDonald, J.W. & Johnston, M.V. (1990). Physiological and pathophysiological roles of excitatory amino acids during central nervous system development. *Brain Research Review,* **15**, 41–70.

McGaugh, J.L. (1989). Involvement of hormonal and neuromodulatory systems in the regulaton of memory storage. *Annual Review of Neuroscience,* **12**, 255–287.

McGaugh, J.L. & Cahill, L. (1997). Interaction of neuromodulatory systems in modulating memory storage. *Behavioral Brain Research,* **83**, 31–38.

McGinty, D.D. (1997). Calcium regulation of gene expression: isn't that spatial? *Neuron,* **18**, 183–186.

McMillian, P.J., Singer, C.A. & Dorsa, D.M. (1996). The effects of ovarectomy and estrogen replacement on trkA and choline acetyltransferase mRNA expression in basal forebrain of adult female Sprague Dawley rat. *Journal of Neuroscience,* **16**, 1860–1865.

McMullen, N., Goldberger, B., Suter, C.M. & Glaser, E.M. (1988). Neonatal deafening alters nonpyramidal dendrite orientation in auditory cortex: a computer microscope study in the rabbit. *Journal of Comparative Neurology*, **267**, 92–106.

Meck, W.H. & Smith, R.A. (1989). Organizational changes in cholinergic activity and enhanced visuospatial memory as a function of choline administered prenatally or postnatally or both. *Behavioral Neuroscience*, **103**, 1234–1241.

Meck, W. & Williams, C. (1997). Characterisation of facilitative effects of perinatal choline supplementation on timing and temporal memory. *Neuroreport*, **8**, 2831–2835.

Meck, W.H., Smith, R.A. & Williams, C.L. (1988). Pre- and postnatal choline supplementation produces long-term facilitation of spatial memory. *Developmental Psychobiology*, **21**, 339–353.

Merzenich, M.M. & Sameshima, K. (1993). Cortical plasticity and memory. *Current Opinion in Neurobiology*, **3**, 187–196.

Mesulam, M.M. (1995). Cholinergic pathways and the ascending reticular activating system of the human brain. *Annals of the New York Academy of Sciences*, **134**(2), 169–178.

Mesulam, M.M. and Guela, C. (1991). Acetylcholinesterase-rich neurons of the human cerebral cortex: cytoarchitectonic and ontogenetic patterns of distribution. *Journal of Comparative Neurology*, **306**, 193–220.

Mesulam, M.M. & Geula, C. (1992). Overlap between acetylcholinesterase-rich and choline-acetyltransferase-positive (cholinergic) axons in human cerebral cortex. *Brain Research*, **577**, 112–120.

Mesulam, M.M. & Geula, C. (1993). Chemoarchitectonics of axonal and perikaryal acetylcholinesterase along information processing systems of the human cerebral cortex. *Brain Research Bulletin*, **33**, 137–153.

Mesulam, M.M., Hersh, L.B., Mash, D.C. & Geula, C. (1992). Differential cholinergic innervation within functional subdivisions of the human cerebral cortex: a choline acetyltransferase study. *Journal of Comparative Neurology*, **318**, 316–328.

Mesulam, M.M., Mufson, E.J., Levey, A.I. & Wainer, B.H. (1983). Cholinergic innervation of cortex. *Journal of Comparative Neurology*, **214**, 170–197.

Miller, M. (1981). Maturation of rat visual cortex. A. A quantitative study of Golgi-impregnated pyramidal neurons. *Journal of Neurocytology*, **10**, 859–878.

Miller, M.W. (1986). Effects of alcohol on the generation and migration of cerebral cortical neurons. *Science*, **233**, 1308–1311.

Miyoshi, R., Kito, S., Shimizu, M. & Matzubayashi, H. (1987). Ontogenesis of muscarinic receptors in the rat brain with special emphasis on the differentiation of M1 and M2 subtypes – semiquantitative *in vitro* autoradiography. *Brain Research*, **420**, 302–312.

Molliver, M.E., Kostovic, I. & van der Loos, H. (1973). The development of synapses in cerebral cortex of the human fetus. *Brain Research*, **50**, 403–407.

Montes, J.G., Alkondon, M., Pereira, E.F.R. & Albuquerque, E.X. (1994). Nicotinic acetylcholine receptors of the mammalian central nervous system. In C. Peracchia (Ed.), *Handbook of membrane channels* (pp. 269–286). New York: Academic Press.

Mrzljak, L., Levey, A.I., Belcher, S. & Goldman-Rakic, P.S. (1998). Localization of the m2 muscarinic acetylcholine receptor protein and mRNA in cortical neurons of the normal and cholinergically deafferented rhesus monkey. *Journal of Comparative Neurology*, **390**, 112–132.

Mrzljak, L., Levey, A.I. & Goldman-Rakic, P.S. (1993). Association of m1 and m2 muscarinic receptor proteins with asymmetric synapses in the primate cerebral cortex: morphological evidence for cholinergic modulation of excitatory neurotransmission. *Proceedings of the National Academy of Sciences, USA*, **90**, 5194–5198.

Mrzljak, L., Pappy, M., Leranth, C. & Goldman-Rakic, P. (1995). Cholinergic synaptic circuitry in the macaque prefrontal cortex. *Journal of Comparative Neurology*, **357**, 603–617.

Murphy, D.D. & Segal, M. (1997). Morphological plasticity of dendritic spines in central neurons is mediated by activation of cAMP response element binding protein. *Proceedings of the National Academy of Sciences, USA*, **94**, 1482-1487.

Naeff, B., Schlumpf, M. & Lichtensteiger, W. (1992). Pre- and postnatal development of high affinity (3H)nicotinic binding sites in rat brain regions: an autoradiographic study. *Developmental Brain Research*, **68**, 163–174.

Navarro, H.A., Seidler, F.J., Eylers, J.P., Baker, F.E., Dobbins, S.S., Lappi, S.E. & Slotkin, T.A. (1989). Effects of prenatal nicotine exposure on development of central and peripheral cholinergic neurotransmitter systems. evidence for cholinergic trophic influences in developing brain. *Journal of Pharmacology and Experimental Therapeutics,* **251**, 894–900.

Nogues, X. (1997). Protein kinase C, learning and memory: a circular determinism between physiology and behavior. *Progress in Neuropsychopharmacology and Biological Psychiatry,* **21**, 507–529.

O'Leary, D.D. & Stanfield, B.B. (1989). Selective elimination of axons extended by developing cortical neurons is dependent on regional locale: experiments utilizing fetal cortical transplants. *Journal of Neuroscience,* **9**, 2230–2246.

Olavarria, J. & Van Sluyters, R.C. (1985). Organization and postnatal development of the callosal connections in the visual cortex of the rat. *Journal of Comparative Neurology,* **239**, 1–26.

Olds, J.L. & Alkon, D.L. (1991). A role for protein kinase C in associative learning. *New Biology,* **3**, 27–35.

Olton, D.S., Givens, B.S., Markowska, A.M., Shapiro, M. & Golski, S. (1992). Mneumonic functions of the cholinergic septohippocampal system. In L.R. Squire, N.M. Weinberger, G. Lynch and J.L. McGaugh (Eds.), *Memory: organization and locus of change.* New York: Oxford.

Pappas, B.A., Davidson, C.M., Fortin, T., Park, G.A.S., Mohr, E. & Wiley, R.G. (1997). 192 Ig-Saporin lesion of basal forebrain cholinergic neurons in neonatal rats. *Experimental Brain Research,* **96**, 52–61.

Patel, A.J., Hayashi, M. & Hunt, A. (1987). Selective persistent reduction in choline acetyltransferase activity in basal forebrain of the rat after thyroid deficiency during early life. *Brain Research,* **422**, 182–185.

Perry, E.K., Court, J.A., Johnson, M., Piggott, M.A. & Perry, R.H. (1992). Autoradiographic distribution of (3H) nicotine binding in human cortex: relative abundance in subicular complex. *Journal of Chemical Neuroanatomy,* **5**, 399–405.

Perry, E.K., Smith, C.J., Atack, J.R., Candy, J.M., Johnston, M. & Perry, R.H. (1986). Neocortical cholinergic enzyme and receptor activities in human fetal brain. *Journal of Neurochemistry,* **47**, 1262–1269.

Petit, T.L. (1986). Developmental effects of lead: its mechanism in intellectual functioning and neural plasticity. *Neurotoxicology,* **7**, 483–496.

Petit, T.L., LeBoutillier, J.C., Gregorio, A. & Libstug, H. (1988). The pattern of dendritic development in the cerebral cortex of the rat. *Developmental Brain Research,* **41**, 209–210.

Prusky, G.T., Arbuckle, J.M. & Cynader, M.S. (1988). Transient concordant distribution of nicotinic receptors and acetylcholinesterase activity in infant rat visual cortex. *Developmental Brain Research,* **39**, 154–159.

Purpura, D.P. (1979). *Pathobiology of cortical neurons in metabolic and unclassified amentias.* New York: Raven Press.

Raedler, T.J., Knable, M.B. & Weinberger, D.R. (1998). Schizophrenia as a developmental disorder of the cerebral cortex. *Current Opinion in Neurobiology,* **8**, 157–161.

Rauschecker, J.P. (1995). Developmental plasticity and memory. *Behavioral Brain Research,* **66**, 7–12.

Ravikumar, B.V. & Sastry, P.S. (1985). Muscarinic cholinergic receptors in human foetal brain: Characterization and ontogeny of (3H) quinuclidinyl benzilate binding sites in frontal cortex. *Journal of Neurochemistry,* **44**, 240–246.

Ricceri, L., Calamandrei, G. & Berger-Sweeney, J. (1997a). Different effects of postnatal day 1 vs. 7 192 IgG saporin lesions on learning and exploratory behaviors and neurochemistry in juvenile rats:Insights into the functional maturation of the basal forebrain cholinergic system. *Behavioral Neuroscience,* **111**, 1292–1302.

Ricceri, L., Ewuse, A., Calamandrei, G. & Berger-Sweeney, J. (1997b). Sexually dimorphic effects of anti-NGF treatment in neonatal rats. *Developmental Brain Research,* **101**, 273–276.

Robbins, T.W., Granon, S., Muir, J.L., Durantou, F., Harrison, A. & Everitt, B.J. (1998). Neural systems underlying arousal and attention. *Annals of the New York Academy of Sciences,* **846**, 222–337.

Robertson, M.R. & Harrell, L.E. (1997). Cholinergic activity and amyloid precursor protein metabolism. *Brain Research Reviews,* **25**, 50–69.

Robertson, R.T. (1987). A morphogenetic role for transiently expressed acetylcholinesterase in developing thalamocortical systems. *Neuroscience Letters,* **75**, 259–264.

Robertson, R.T., Gallardo, K.A. & Klaytor, K.J. (1998). Neonatal treatment with 192 IgG Saporin produces long-term forebrain cholinergic deficits and reduces dendritic branching and spine density of neocortical pyramidal neurons. *Cerebral Cortex*, **8**, 142–155.

Robertson, R.T., Hohmann, C.F., Bruce, J.L. & Coyle, J.T. (1988). Neonatal enucleations reduce specific activity of acetylcholinesterase but not choline acetyltransferase in developing rat visual cortex. *Developmental Brain Research*, **39**, 298–302.

Rooney, T.A. & Nahorski, S.R. (1989). Developmental aspects of muscarinic-induced inositol polyphosphate accumulation in rat cerebral cortex. *European Journal of Pharmacology – Molecular Pharmacology Section*, **172**, 425–434.

Rossner, S., Kues, W., Witzeman, V. & Schlieb, R. (1993). Laminar expression of m1, m3 and m4 muscarinic cholinergic receptor genes in the developing rat visual cortex using *in situ* hybridization histochemistry: effects of monocular visual deprivation. *International Journal of Developmental Neuroscience*, **11**, 369–378.

Rotter, A., Field, P.M. & Raisman, G. (1979). Muscarinic receptors in the central nervous system of the rat: Postnatal development of binding of 3H-Propyl benzilycholine mustard. *Brain Research Review*, **1**, 185–205.

Rouse, S.T., Thomas, T.M. & Levey, A.I. (1997). Muscarinic acetylcholine receptor subtype, m2: diverse functional implications of differential synaptic localization. *Life Sciences*, **60**, 1031–1038.

Roy, T.S. & Sabherwhal, U. (1994). Effects of prenatal nicotine exposure on the morphogenesis of somatosensory cortex. *Neurotoxicology and Teratology*, **16**, 411–421.

Salt, T.E., Meier, D.L., Seno, N., Krucker, T. & Herrling, P.L. (1995). Thalamocortical and corticocortical excitatory postsynaptic potentials mediated by excitatory amino acid receptor in the cat cortex *in vivo*. *Neuroscience*, **64**, 433–442.

Sarter, M. & Bruno, J.P. (1997). Cognitive functions and cortical acetylcholine: toward a unifying hypothesis. *Brain Research Reviews*, **23**, 28–46.

Sarter, M. & Bruno, J.P. (1998). Cortical acetylcholine, reality distortion, schizophrenia and lewy body dementia: too much or too little acetylcholine? *Brain Cognition*, **38**, 297–316.

Schambra, U.B., Lauder, J.M., Petruz, P. & Sulik, K.K. (1990). Development of neurotransmitter systems in the mouse embryo following acute ethanol exposure: a histological and immunocytochemical study. *International Journal of Developmental Neuroscience*, **8**, 507–522.

Schambra, U.B., Ulik, K.K., Petrusz, P. & Lauder, J.M. (1989). Ontogeny of cholinergic neurons in the mouse forebrain. *Journal of Comparative Neurology*, **288**, 101–122.

Schliebs, R., Rossner, S., Kumar, A. & Bigl, V. (1994). Muscarinic acetylcholine receptor subtypes in rat visual cortex-a comparative study using quantitative receptor autoradiography and *in situ* hybridization. *Indian Journal of Experimental Biology*, **32**, 25–30.

Schroder, H., Zilles, K., Luiten, P.G., Strosberg, A.D. & Aghchi, A. (1989). Human cortical neurons conrtain both nicotinic and muscarinic acetylcholine receptors: an immunocytochemical double labelling study. *Synapse*, **4**, 319–326.

Semba, K. (1991). The cholinergic basal forebrain: a critical role in cortical arousal. *Advances in Experimetnal Medicine and Biology*, **295**, 197–218.

Sengstock, G.J., Johnson, K.B., Jantzen, P.T., Meyer, E.M., Dunn, A.J. & Arendash, G.W. (1992). Nucleus basalis lesions in neonatal rats induce a selective hypofunction and cognitive deficits during adulthood. *Brain Research*, **90**, 163–174.

Shapiro, D.L. (1973). Morphological and biochemical alterations in rat brain cells cutured in the presence of monobutyryl cAMP. *Nature*, **241**, 203–204.

Sillito, A.M. & Murphy, P.C. (1987). The cholinergic modulation of cortical function. In E.G. Jones & A. Peters (Eds.), *Cerebral cortex* (pp. 161–185). New York: Plenum Press.

Silver, A. (1974). Biology of cholinesterases. In A. Neuberger & E.L. Tortum (Eds.), *Frontiers of biology*. New York: Elsevier.

Singer, W. (1990). Role of acetylcholine in use dependent plasticity of the visual cortex. In M. Steriade & D. Biesold, (Eds.), *Brain cholinergic systems* (pp. 314–336). Oxford: Oxford University Press.

Skene, P. (1989). Axonal growth associated proteins. *Annual Review of Neuroscience*, **12**, 127–156.

Slotkin, T.A., Greer, N., Faust, J., Cho, H. & Seidler, F.J. (1986). Effects of maternal nicotine injections on brain development in the rat: ornithine decarboxylase activity, nucleic acids and proteins in discrete brain regions. *Brain Research Bulletin*, **17**, 41–50.

Smiley, J.F. (1996). Monoamines and acetylcholine in primate cerebral cortex: what anatomy tells us about function. *Rev Brasil Biol*, **56**, 153–164.

Smiley, J.F., Morrell, F. & Mesulam, M.M. (1997). Cholinergic synapses in human cerebral cortex: An Ultrastructural study. *Developmental Neurology*, **144**, 361–368.

Smolen, A., Smolen, T.N., Han, P.C. & Collins, A.C. (1987). Sex differences in the recovery of brain acetylcholinesterase activity following a single exposure of DFP. *Pharmacology, Biochemistry and Behavior*, **26**, 813–820.

Smythies, J. (1997). The functional neuroanatomy of awareness: with a focus on the role of various anatomical systems in the control of intermodal attention. *Consciousness and Cognition*, **6**, 455–481.

Soderling, T.R. (1995). Calcium dependent protein kinases in learning and memory. *Advances in Second Messenger Phosphoprotein Research*, **30**, 175–189.

Spatz, H.C. (1996). Hebb's concept of synaptic plasticity and neuronal cell assemblies. *Behavioral Brain Research*, **78**, 3–7.

Speiser, Z., Amitzi-Sonder, J., Gitter, S. & Cohen, S. (1988). Behavioral differences in the developing rat following postnatal anoxia or postnatal injections of AF64A, a cholinergic neurotoxin. *Behavioral Brain Research*, **30**, 89–94.

Spitzer, N.C. (1994). Spontaneous Ca^{++} spikes and waves in embryonic neurons: signaling systems for differentiation. *TINS*, **17**, 115–118.

Squire, L.R. & Davis, H.P. (1981). The pharmacology of memory: a neurobiological perspective. *Annual Review of Pharmacology and Toxicology*, **21**, 323–356.

Stanfield, B.B., O'Leary, D.D.M. & Fricks, C. (1982). Selective collateral elimination in early postnatal development restricts cortical distribution of rat pyramidal tract neurons. *Nature*, **298**, 371–373.

Steriade, M. (1993). Central core modulation of spontaneous oscillations and sensory transmission in thalamocortical systems. *Current Opinion in Neurobiology*, **3**, 619–625.

Sweeney, J., Hohmann, C.F., Moran, T. and Coyle, J.T. (1988). A long-acting cholinesterase inhibitor reverses spatial memory deficits in mice. *Pharmacology, Biochemistry and Behavior*, **31**, 141–147.

Sweeney, J.E., Hohmann, C.F., Moran, T. & Coyle, J.T. (1989). Galanthamine, an acetylcholinesterase inhibitor: A time course of the effects on performance and neurochemical parameters in mice. *Pharmacology, Biochemistry and Behavior*, **34**, 129–137.

Sweeney, J.E., Hohmann, C.F., Oster-Granite, M.L. and Coyle, J.T. (1989). Neurogenesis of the basal forebrain in euploid and trisomy 16 mice: an animal model for developmental disorders in Down syndrome. *Neuroscience*, **31**, 413–425.

Takashima, S., Becker, L.E., Armstrong, D.L. & Chan, F. (1981). Abnormal neuronal development in the visual cortex of the human fetus and infant with Down's syndrome. A quantitative and qualitative Golgi study. *Brain Research*, **225**, 1–21.

Tandon, R. & Greden, J.F. (1989). Cholinergic hyperactivity and negative schizophrenic symptoms. A model of cholinergic/dopaminergic interactions in schizophrenia. *Archives of General Psychiatry*, **46**, 745–753.

Tani, Y., Saito, K., Tsuneyoshi, A., Imoto, M. & Ohno, T. (1997). Nicotinic acetylcholine receptor agonists – induced changes in brain monoamine turnover in mice. *Psychopharmacology*, **129**, 225–232.

Thal, L.J., Gilbertson, E., Armstrong, D.M. & Gage, F.H. (1991). Development of the basal forebrain cholinergic system: phenotype expression prior to target intervention. *Neurobiology of Aging*, **13**, 62–67.

Tice, M.A.B., Hashemi, T., Taylor, L.A. & McQuade, R.D. (1996). Distribution of muscarinic receptor subtypes in rat brain from postnatal to old age. *Developmental Brain Research*, **92**, 70–76.

Tremblay, N., Warren, R.A. and Dykes, R.W. (1990). Electrophysiological studies of acetylcholine and the role of the basal forebrain in the somatosensory cortex of the cat. II. Cortical neurons excited by somatic stimuli. *Journal of Neurophysiology*, **64**, 1212–1222.

Umbriaco, D., Watkins, K.C., Descarries, L., Cozzari, C. & Hartman, B.K. (1994). Ultrastructural and morphometric features of acetylcholine innervation in adult rat parietal cortex: an electron microscopic study in serial sections. *Journal of Comparative Neurology*, **348**, 351–373.

Valverde, R. (1967). Apical dendritic spines of the visual cortex and light deprivation in the mouse. *Experimental Brain Research*, **3**, 337–352.

Van Hooff, C.O.M., De Graan, P.N.E., Oestreicher, A.B. & Gispen, W.H. (1989a). Muscarinic receptor activation stimulates B-50/GAP 43 phosphorylation in isolated nerve growth cones. *Journal of Neuroscience,* **9**, 3753–3759.

Van Hooff, C.O.M., Oestreicher, A.B., De Graan, P.N.E. & Gispen, W.H. (1989b). Role of the growth cone in neuronal differentiation. *Molecular Neurobiology*, **3**, 101–133.

Vidal, C. (1996). Nicotine receptors in the brain. Molecular biology, function and therapeutics. *Molecular and Chemical Neuropathology,* **28**, 3–11.

Volpe, J.J. (1987). *Neurology of the newborn.* Philadelphia, PA: W.B. Saunders.

Wall, S.J., Yasuda, R.P., Li, M., Ciesla, W. & Wolfe, B.B. (1992). The ontogeny of m1–m5 muscarinic receptor subtypes in rat forebrain. *Developmental Brain Research,* **66**, 181–185.

Wang, J.H., Ko, G.Y. & Kelly, P.T. (1997). Cellular and molecular bases of memory: synaptic and neuronal plasticity. *Journal of Clinical Neurophysiology,* **14**, 264–293.

Weiler, I.J., Hawrylak, N. & Greenough, W.T. (1995). Morphogenesis in memory formation: synaptic and cellular mechanisms. *Behavioral Brain Research,* **66**, 1–6.

Weinberger, N.M. & Bakin, J.S. (1998). Learning -induced physiological memory in adult primary auditory cortex: receptove field plasticity, models and mechanisms. *Audiology and Neurootology,* **3**, 145–167.

Wenk, G.L. (1997a). The nucleus basalis magnocellularis cholinergic system: one hundred years of progress. *Neurobiology of Learning and Memory,* **67**, 85–95.

Wenk, G.L. (1997b). Rett syndrome: neurobiological changes underlying specific symptoms. *Progress in Neurobiology,* **51**, 383–391.

Wevers, A., Jeske, A., Lobron, D., Birtsch, C., Heinemann, S., Maelicke, A., Schroder, R. & Schroder, H. (1994). Cellular distribution of nicotinic acetylcholine receptor subunit mRNAs in the human cortex as reveaqled by non-isotopic *in situ* hybridization. *Molecular Brain Research*, **25**, 122–128.

Wickelgren, I. (1997). Estrogen stakes claim to cognition. *Science*, **276**, 675–678.

Wiley, R.G., Oeltmann, T.N. & Lappi, D.A. (1991). Immunolesioning: selective destruction of neurons using immunotoxin to rat NGF receptors. *Brain Research*, **562**, 149–153.

Williams, C., Barnett, A.M. & Meck, W. (1990). Organizational effects of early gonadal secretion on sexual differentiation in spatial memory. *Behavioral Neuroscience,* **104**, 84–97.

Williams, C.L., Meck, W.H. & Loy, R. (1989). Prenatal choline supplementation produces long-term modifications in the morphology and distribution of cells in the diagonal band region exhibiting NGF receptor immunoreactivity. **15**, no. 110.4.

Wise, S.P. & Jones, E.G. (1978). Developmental studies of thalamocortical and commissural connections in the rat somatic sensory cortex. *Journal of Comparative Neurology,* **178**, 187–208.

Witte, E.A., Davidson, M.C. & Marrocco, R.T. (1997). Effects of altering brain cholinergic activity on covert orienting of attention: comparison of monkey and human performance. *Psychopharmacology*, **132**, 324–334.

Woody, D.D. & Gruen, E. (1993). Cholinergic and glutamatergic effects on neocortical neurons may support rate as well as development of conditioning. *Progress in Brain Research*, **98**, 365–370.

Woolf, N. (1993). Cholinoreceptive cells in rat cerebral cortex: somatodendritic immunoreactivity for muscarinic receptor and cytoskeletal proteins. *Journal of Chemical Neuroanatomy,* **6**, 375–390.

Woolf, N.J. (1997). A possible role for cholinergic neurons of the basal forebrain and pontomesencephalon in consciousness. *Consciousness and Cognition*, **6**, 574–596.

Wu, M., Hohmann, C.F., Coyle, J.T. & Juliano, S. (1989). Lesions of the basal forebrain alter stimulus-evoked metabolic activity in mouse somatosensory cortex. *Journal of Comparative Neurology,* **288**, 414–427.

Yamada, S., Kagawa, Y., Isogai, M., Takayanagi, N. & Hayashi, E. (1986). Ontogenesis of nicotinic acetylcholine receptors and presynaptic cholinergic neurons in mammalian brain. *Life Sciences*, **38**, 637–644.

Zahalka, E.A., Seidler, F.J., Lappi, S.E., Yanai, J. & Slotkin, T.A. (1993). Differential development of cholinergic terminal markers in rat brain regions: implication for nerve terminal density, impulse activity and specific gene expression. *Brain Research*, **601**, 221–229.

Zeisel, S.H. (1987). Choline availability in the neonate. In M.J. Dowdall & J.N. Hawthorne (Eds.), *The cellular and molecular basis of cholinergic function* (pp. 709–719). New York: VCH Weinheim.

Zhang, X., Wahlstrom, G. & Nordberg, A. (1990). Influence of development and aging on nicotinic receptor subtypes in rodent brain. *International Journal of Developmental Neuroscience,* **8**, 715–721.

Zhu, X.O. & Waite, P.M. (1999). Cholinergic depletion reduces plasticity of barrel field cortex. *Cerebral Cortex*, **8**, 63–72.

Zhu, J., Takita, M., Konishi, Y., Sudo, M. & Muramatsu, I. (1996). Chronic nicotine treatment delays the developmental increase in brain muscarinic receptors in rat neonate. *Brain Research*, **732**, 257–260.

Zilles, K., Schroder, H., Schroder, U., Horvath, E., Werner, L., Luiten, P.G., Maelick, A. & Strosbert, A.D. (1989). Distribution of cholinergic receptors in the rat and human neocortex. *Experentia*, **57**, 212–228.

FURTHER READING

Arters, J., Hohmann, C.F., Mills, J.O. & Berger-Sweeney, J. (1998). Sexually dimorphic responses to neonatal basal forebrain lesions in mice: I. Behavior and neurochemistry. *Journal of Neurobiology*, **37**, 582–594.

Candy, J.M., Perry, E.K., Perry, R.H., Bloxham, C.A., Thompson, J., Johnson, M., Oakley, A.E. & Edwardson, J.A. (1985). Evidence for the early prenatal development of cortical cholinergic afferents from the nucleus of Meynert in the human foetus. *Neuroscience Letters*, **61**, 91–95.

Hohmann, C.F. & Berger-Sweeney, J.E. (1998). Cholinergic regulation of cortical development: New twists on an old story. *Perspectives on Developmental Neurobiology*, **5**, 410–425.

Hohmann, C.F. & Berger-Sweeney, J.E. (1998). Sexually dimorphic responses to neonatal basal forebrain lesions in mice: II Quantitative assessments of cortical morphology. *Journal of Neurobiology*, **37**, 595–606.

Juliano, S.L. (1998). Mapping the sensory mosaic. *Science*, **279**, 1653–1654.

Kater, S.B., Mattson, M.P., Cohan, C. & Connor, J. (1988). Calcium regulation of the neuronal growth cone. *TINS*, **11**, 315–321.

Krnjevic, K. (1988). Acetylcholine as transmitter in the cerebral cortex. In A. Avoli, T.A. Reader, R. Dykes & P. Gloor (Eds.), *Neurotransmitters and brain function* (pp. 227–237). New York: Plenum Press.

Kostovic, I., Skavic, J. & Strinovic, D. (1988). Acetylcholinesterase in the human frontal associative cortex during the period of cognitive development: early laminar shifts and late innervation of pyramidal neurons. *Neuroscience Letters*, **90**, 107–112.

Kumar, A. & Schlieb, R. (1992). Postnatal laminar development of cholinergic receptors, protein kinase C and dihydropyridine sensitive calcium antagonist binding in rat visual cortex. Effect of visual deprivation. *International Journal of Developmental Neuroscience*, **10**, 491–504.

Loffelholz, K. (1996). Muscarinic receptors and cell signalling. In J. Klein & K. Loffelholz (Eds.). *Progress in Brain Research*, **109**, 201–208.

Meck, W. & Williams, C. (1997). Characterisation of facilitative effects of perinatal choline supplementation on timing and temporal memory. *Neuroreport*, **8**, 2831–2835.

Mesulam, M.M. (1995). Cholinergic pathways and the ascending reticular activating system of the human brain. *Annals of the New York Academy of Sciences*, 169–178.

Mrzljak, L., Levey, A.I. & Goldman-Rakic, P.S. (1993). Association of m1 and m2 muscarinic receptor proteins with asymetric synapses in the primate cerebral cortex: morphological evidence for cholinergic modulation of excitatory neurotransmission. *Proceedings of the National Academy of Sciences*, **90**, 5194–5198.

Navarro, H.A., Seidler, F.J., Eylers, J.P., Baker, F.E., Dobbins, S.S., Lappi, S.E. & Slotkin, T.A. (1989). Effects of prenatal nicotine exposure on development of central and peripheral cholinergic neurotransmitter systems. Evidence for cholinergic trophic influences in developing brain. *Journal of Pharmacology and Experimental Therapeutics*, **251**, 894–900.

Perry, E.K., Smith, C.J., Atack, J.R., Candy, J.M., Johnston, M. & Perry, R.H. (1986). Neocortical cholinergic enzyme and receptor activities in human fetal brain. *Journal of Neurochemistry*, **47**, 1262–1269.

Ricceri, L., Calamandrei, G. & Berger-Sweeney, J. (1997). Different effects of postnatal day 1 vs. 7 192 IgG saporin lesions on learning and exploratory behaviors and neurochemistry in juvenile rats: insights into the functional maturation of the basal forebrain cholinergic system. *Behavioral Neuroscience*, **111**, 1292–1302.

Robertson, M.R. & Harrell, L.E. (1997). Cholinergic activity and amyloid precursor protein metabolism. *Brain Research Review,* **25**, 50–69.

Robertson, R.T. (1987). A morphogenetic role for transiently expressed acetylcholinesterase in developing thalamocortical systems. *Neuroscience Letters,* **75**, 259–264.

Robertson, R.T., Gallardo, K.A. & Klaytor, K.J. (1998). Neonatal treatment with 192 IgG Saporin produces long-term forebrain cholinergic deficits and reduces dendritic branching and spine density of neocortical pyramidal neurons. *Cerebral Cortex,* **8**, 142–155.

Sarter, M. & Bruno, J.P. (1997). Cognitive functions and cortical acetylcholine: toward a unifying hypothesis. *Brain Research Review,* **23**, 28–46.

Sillito, A.M. & Murphy, P.C. (1987). The cholinergic modulation of cortical function. In E.G. Jones & A. Peters (Eds.), *Cerebral cortex* (pp. 161–185). New York: Plenum Press.

Singer, W. (1990). Role of acetylcholine in use dependent plasticity of the visual cortex. In M. Steriade & D. Biesold (Eds.), *Brain cholinergic systems* (pp. 314–336). Oxford: Oxford University Press.

II.5
Roles for Serotonin in Non-neural Embryonic Development

JULIAN R.D. MOISEIWITSCH, H. WAYNE LAMBERT and
JEAN M. LAUDER

ABSTRACT

Serotonin (5-HT) is a neurotransmitter that is involved in mood disorders such as depression and other psychiatric illnesses. However, it has been demonstrated in animal models that, prior to development of the nervous system, 5-HT also plays important roles in vertebrate craniofacial development. Given the widespread use of selective serotonergic reuptake inhibitors (SSRIs) as antidepressants, and the rising use of 5-HT receptor agonists and antagonists to treat psychiatric disorders, the possibility of teratogenic effects of these drugs if taken in pregnancy should be of real concern. In this review we discuss roles played by 5-HT during mouse craniofacial development, including regulation of cranial neural crest migration, mesenchymal cell proliferation, differentiation and gene expression, as well as dental development. We also discuss possible mechanisms underlying serotonin's morpho-genetic roles in craniofacial development and suggest further avenues of investigation for this controversial topic.

INTRODUCTION AND THEORETICAL BACKGROUND

Serotonin-specific reuptake inhibitors (SSRIs) are presently the most widely prescribed antidepressant drugs in the United States (Pastuszak *et al.*, 1993). The popularity of these drugs stems from their efficacy in controlling symptoms of clinical depression and associated affective disorders, coupled with the much-reduced side-effects compared with previously available antidepressants such as monoamine oxidase inhibitors and tricyclic antidepressants. However, for several years the use of these drugs in pregnant women has been controversial (Cohen & Rosenbaum, 1997; Goldstein *et al.*, 1997; Jones *et al.*, 1997; Roberts, 1996). In

139

A.F. Kalverboer and A. Gramsbergen (eds.), Handbook of Brain and Behaviour in Human Development, 139–152
© 2001 Kluwer Academic Publishers. Printed in Great Britain.

part this controversy has arisen from *in-vitro* findings of craniofacial defects in mouse embryos cultured in the presence of either SSRIs or tricyclic antidepressants (Shuey *et al.*, 1992). At present the pharmaceutical companies that make SSRIs recommend that they not be used during pregnancy. However, they also state that if a patient is on such a drug when she becomes pregnant their therapy should not be stopped (Roberts, 1996). Until recently there was no evidence that SSRIs have any teratogenic effects in either human subjects or animals (Pastuszak *et al.*, 1993; Rosa 1994; Vorhees *et al.*, 1994). While it does appear to be true that no major teratogenic effects are produced by an SSRI taken during pregnancy, a recent study has suggested that a three-fold increase in the incidence of "groupings of minor birth defects" occurs when these drugs are administered early in pregnancy (Chambers *et al.*, 1996). A similar increase in perinatal complications, such as miscarriages, premature delivery, growth abnormalities, colic and respiratory difficulties have been reported when these drugs are administered during the third trimester of pregnancy or in nursing mothers (Baum & Misri, 1996; Chambers *et al.*, 1996). A few clinical reports of such abnormalities might easily be dismissed. However, the correlation between these reports and animal studies reporting widespread sites of 5-HT uptake during craniofacial (Lauder *et al.*, 1988; Lauder & Zimmerman, 1988; Shuey *et al.*, 1992, 1993) and cardiac development (Yaverone *et al.*, 1993a) and in ectoplacental cone and placenta (Yaverone *et al.*, 1993b), together with evidence of severe craniofacial and cardiac malformations in cultured mouse embryos, suggest that these drugs may potentially have more harmful effects than previously thought.

The most widely recognized serotonergic drugs that are in common clinical use are the SSRIs. However, several other classes of serotonergic drugs are also either available for clinical use or are in various stages of testing. Of those presently available, the 5-HT$_3$ receptor antagonists Zofran and Kytril, which are approved as antiemetics, are the most widely used. As they are presently recommended only for chemotherapy patients, particularly those receiving high doses of cisplatin, there is little concern over their teratogenic effects. However, several other serotonergic drugs have been proposed for use as psychoactive drugs, including 5-HT1A receptor agonists. These could have potential use in populations that would be at risk of unplanned pregnancy and hence potential teratogenic side-effects.

CRITICAL REVIEW OF THE LITERATURE

In considering the potential deleterious effects of SSRIs and other serotonergic drugs on prenatal development, it is important to understand the biological basis for such actions. As discussed below, there is accumulating evidence from animal studies that monoamines (see also Chapter II.6, by Roffler-Tarlov and Rios) play critical morphogenetic roles in embryonic development.

Monoamines as morphogens

A "morphogen" is a dose-dependent developmental signal that exerts different types of actions on target cells according to concentration. Concentration gradients of morphogens can be created in developing tissues by the presence of a "source" and a "sink" (where the morphogen is rapidly degraded). A classical example of a morphogen is retinoic acid. In some developmental contexts, neurotransmitters satisfy the operational definition of a morphogen (reviewed by Lauder, 1988). Monoamines (5-HT, catecholamines) are present in the fertilized egg, and appear to regulate early cleavage divisions in both vertebrate and invertebrate embryos (Burden & Lawrence, 1973; Buznikov et al., 1996; Lauder, 1993; Pienkowski, 1977; Sadykova et al., 1992). These neurotransmitters are synthesized by yolk granules and notochord of neurulating chick and frog embryos, and are actively taken up by the neural tube during neurulation (Godin & Gipouloux, 1986; Kirby & Gilmore, 1972; Lawrence & Burden, 1973; Newgreen et al., 1981; Strudel et al., 1977; Wallace, 1982). Monoamines appear to regulate morphogenetic cell movements and cell shape changes during neural tube closure, since exposure of chick embryos to monoamine uptake inhibitors, MAO inhibitors or receptor ligands produces a variety of malformations, incuding neural tube defects (Palen et al., 1979).

Serotonin as a morphogen during craniofacial and cardiac development

Serotonin uptake sites and binding proteins

Studies carried out in mouse embryos have provided evidence that 5-HT acts as morphogen during craniofacial and cardiac development (Lauder et al., 1988; Lauder & Zimmerman, 1988; Shuey et al., 1992, 1993; Yavarone et al., 1993a,b). In these studies, transiently expressed sites of 5-HT uptake and degradation were found in craniofacial epithelia, hindbrain rhombomeres, tooth germs and myocardium (heart). Recently, *in-situ* hybridization studies have provided evidence that 5-HT transporter mRNA is expressed at similar locations in the rat embryo (Hansson et al., 1999). These sites of uptake and degradation could constitute a "sink" for 5-HT according to the morphogen model. The "source" of 5-HT appears to be the maternal-embryonic circulation, where blood-borne 5-HT reaches the embryo from the mother by transport across the placenta (Yavarone et al., 1993a). In addition, serotonin binding proteins (SBPs; Tamir et al., 1994) have been found in craniofacial mesenchyme, the distribution of which becomes progressively more restricted as epithelial 5-HT uptake sites develop, such that by the time of maximal development of epithelial 5-HT uptake sites, SBPs are located only in mesenchyme immediately subadjacent to, and in register with, epithelial 5-HT uptake sites.

When mouse embryos are exposed to 5-HT uptake inhibitors in whole embryo culture, craniofacial malformations occur in structures expressing 5-HT uptake sites (Shuey et al., 1992). In the context of the morphogen model we have hypothesized that these malformations are due to excessive accumulation of 5-HT in mesenchyme as a result of inhibiting 5-HT uptake and degradation by epithelial "sinks". This hypothesis is represented in the model illustrated in Figure 1. Subsequent studies have confirmed many aspects of this model, as discussed below.

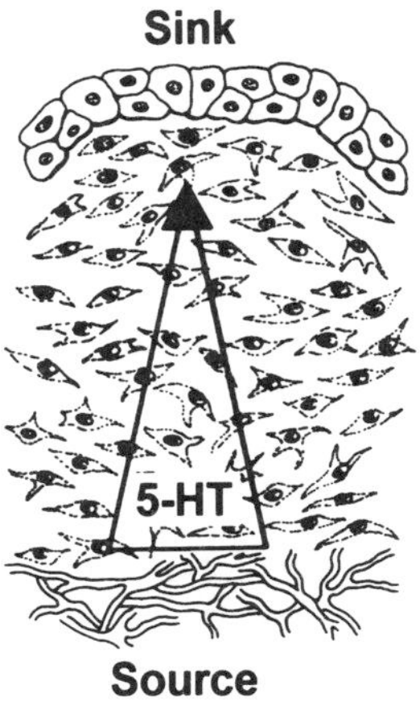

Tissues	Receptors	5-HT & SBP
Epithelium		5-HT uptake
Mesenchyme	5-HT_{1A} $5\text{-HT}_{2A,B,C}$ 5-HT_3 5-HT_4 (?)	SBP
Vascular bed		Circulating 5-HT

Figure 1 Model for 5-HT as a morphogenetic signal during mouse craniofacial development. Blood-borne 5-HT, passed from the mother to the embryo via the maternal–embryonic circulation, acts as a source of 5-HT for differentiating craniofacial mesenchyme. Uptake sites for 5-HT present within the epithelium are thought to form a sink where 5-HT is rapidly degraded. Multiple 5-HT receptor subtypes are expressed by mesenchyme cells located between the source and sink. It is hypothesized that mesenchyme cells are exposed to a 5-HT concentration gradient which regulates their proliferation and differentiation by activation of 5-HT receptors. (Modified from Shuey, 1991 and Buznikov *et al.*, 1996)

In the developing mouse heart, 5-HT is avidly taken up by the myocardium as the heart tube undergoes morphogenesis. This uptake capacity becomes progressively more restricted with age until only the myocardium surrounding developing endocardial cushions in the outflow tract and atrioventricular canal are capable of 5-HT uptake (Yavarone *et al.*, 1993b). Exposure of cultured mouse embryos to 5-HT uptake inhibitors during heart development severely inhibits proliferation of endocardial cushion tissue cells. In the context of a morphogen these effects could result from blocking the "sink" (i.e. myocardial 5-HT uptake and degradation), while not affecting the "source" (the circulation), thereby building up excess levels of 5-HT in developing endocardial cushions (Yavarone, 1991; Yavarone *et al.*, 1993b). Using an *in-vitro* cell migration assay, high doses of 5-HT were found to significantly inhibit migration of cushion tissue cells, whereas low doses of 5-HT tended to stimulate migration (Yavarone *et al.*, 1993b). Therefore, myocardial uptake may help to maintain appropriate levels of 5-HT during development of endocardial cushions.

Taken together, these studies provide evidence that, at least in the mouse embryo, 5-HT acts as a dose-dependent regulatory signal or "morphogen" for important developmental events during both craniofacial and cardiac development. Possible cellular and molecular mechanisms underlying these activities are discussed below.

Serotonin receptors

Exposure of cultured mouse embryos to 5-HT2 and 5-HT3 receptor antagonists (but not antagonists for 5-HT1A or 5-HT4) cause craniofacial malformations similar to those seen with 5-HT uptake inhibitors (see Table 1 and Choi *et al.*, 1997, 1998). The most potent of these antagonists are those targeting 5-HT2B and 5-HT3 receptors, suggesting that these receptor subtypes may be especially important mediators of morphogenetic actions of 5-HT during craniofacial development. Recent studies have revealed expression of 5-HT1A, 5-HT2 and 5-HT3 receptor subtypes in the heart, somites and craniofacial region of developing mouse and rat embryos (Johnson & Heinemann, 1995; Lauder *et al.*, 1998, 2000; Moiseiwitsch & Lauder, 1995; Tecott *et al.*, 1995; Wu *et al.*, 1999). Possible functions of these receptors in specific aspects of craniofacial development are discussed below.

Serotonin and cranial neural crest migration

Cranial neural crest gives rise to most of the mesenchyme of the craniofacial region and contributes to the endocardial cushions of the heart (Kirby & Waldo, 1990; Noden, 1986). If 5-HT acts as a morphogen during craniofacial and cardiac development it might be expected to exert dose-dependent effects on cranial neural crest migration. To test this possibility we investigated effects of 5-HT on migration of cranial neural crest cells isolated from neurulation stage mouse embryos using an *in-vitro* assay (Moiseiwitsch & Lauder, 1995). In this study, 5-HT was found to regulate migration of cranial neural crest cells in an inverse dose-dependent manner, such that lower doses promoted migration, whereas this effect was lost as the dose of 5-HT was increased. The stimulatory effect of 5-HT was prevented by a 5-HT1A receptor antagonist, but not by an antagonist for 5-HT2A/2C receptors. Together with immunocytochemical evidence for the presence of 5-HT1A receptors in neural crest cells and craniofacial mesenchyme at the stage of neural crest migration, these findings provided evidence that these receptors may play an important role in mediating the stimulatory effects of 5-HT on cranial neural crest migration. Recent studies of expression patterns of 5-HT2B and 5-HT3 receptors during neurulation stages of mouse embryogenesis raise the additional possibility that these receptors may also contribute to serotonergic regulation of cranial neural crest migration (Choi *et al.*, 1997; Lauder *et al.*, 2000).

Serotonergic regulation of cell proliferation

Depending on species and cell type, 5-HT can have various effects on proliferating cells (reviewed by Seuwen & Pouyssegur, 1990; Fanburg & Lee, 1997). Serotonin has been reported to increase proliferation of bovine smooth muscle cells (Lee *et al.*, 1991; Nemecek *et al.*, 1986), canine endothelial cells (Pakala *et al.*, 1994), rodent fibroblasts (Seuwen *et al.*, 1988), and rat mesangial cells (Takuwa et al., 1989), as well as cell lines transfected with 5-HT receptors (Garnovskaya *et al.*, 1996; Julius *et al.*, 1989, 1990; Varrault *et al.*, 1992).

Table 1 Teratogenic effects of 5-HT receptor antagonists in whole embryo culture

Treatment	5-HT receptor	μM	n	No. of malformations	Percentage of malformations	NTD	FB/NP	Arches	Eye
Antagonists									
NAN-190	1A	1	27	1	3	+	0	0	0
	1A	10	28	1	4	0	0	+	0
Ketanserin	2A	1	20	2	10	0	+	0	0
	2A	10	20	4	20	0	+	+	0
Mianserin	2A/2C	1	23	2	9	0	+	0	+
	2A/2C	10	21	15	71**	+	+	+	+
Ritanserin	2A/2B/2C	0.1	12	1	8	0	+	0	0
	2A/2B/2C	1	12	12	100**	+	+	+	+
ICS-205	3	1	50	27	54**	+	+	+	+
Zofran	3	1	18	3	17	(+)	+	+	0
Kytril	3	10	23	15	65**	+	+	+	+
	3	1	18	5	28	0	+	+	+
SDZ-205	3	10	25	19	76**	+	+	+	+
	4	10	8	0	0	0	0	0	0
Controls:	–	–	99	12	12	0	(+)	(+)	0

E9 mouse embryos (plug day = E1) were cultured for 48 h in the presence of 5-HT receptor antagonist or vehicle (Sadler, 1979).

Abbreviations: NTD: neural tube defects; FB/NP: forebrain/nasal prominence hypoplasia; Arches: maxillary/mandibular arches; Eye: lens invagination; SDZ-205: SDZ-205.557. (+): occasionally observed, +: frequently observed, 0: not observed

**$p < 0.001$; *$p < 0.01$ (Fisher's exact test).

Specifically, 5-HT2B receptors regulate proliferation of cultured cell lines by activation of the Ras/Raf/MAP kinase pathway (Choi et al., 1997, 1998; Launay et al., 1996), and 5-HT2A and 5-HT2C receptors mediate mitogenic effects of 5-HT when expressed at high density in fibroblasts (Julius et al., 1989, 1990). Serotonin has also been found to synergize with mitogenic growth factors (Lee et al., 1991; Seuwen & Pouyssegur, 1990; Varrault et al., 1992). Serotonin can also inhibit proliferation of some cell types such as human and rat small intestinal cells (Zachrisson & Uribe, 1998), embryonic chick skin cells (de Ridder & Beele, 1988), and bovine smooth muscle cells, when particular 5-HT receptors are activated (Lee et al., 1991).

Recently, we investigated the ability of 5-HT to regulate proliferation of craniofacial mesenchyme cells isolated from embryonic day 12 (E12, plug day = E1) mouse mandible. These studies revealed two different aspects of serotonergic regulation of mesenchymal cell proliferation. The first aspect involves mitogenic effects of 5-HT that appear to be mediated by both 5-HT transport and 5-HT2B receptors, both of which activate the mitogenic Ras/Raf/MAP kinase pathway (Buznikov et al., 2001; Choi et al., 1997, 1998; Fanburg & Lee, 1997; Launay et al., 1996). The second aspect concerns anti-mitogenic effects of 5-HT mediated by receptors that stimulate synthesis of cyclic AMP in mandibular mesenchyme cells, including those of the 5-HT1A and 5-HT4 subtypes (Buznikov et al., 2001; Lambert & Lauder, 1999).

Serotonergic regulation of gene expression

Another important aspect of serotonergic regulation of craniofacial development in mouse embryos is the ability of 5-HT receptors or transporters to differentially regulate expression of other "morphoregulatory molecules" (Edelman & Jones, 1992), including growth factors and extracellular matrix molecules. To date we have found that 5-HT regulates expression of two classes of growth factors (S-100β, insulin-like growth factor I) and extracellular matrix molecules (tenascin, cartilage proteoglycan core protein).

S-100β, tenascin and cartlage proteoglycan core protein
The first indication that 5-HT might have an indirect effect on mouse craniofacial morphogenesis came from evidence that 5-HT receptors were distributed in patterns similar to several previously identified morphoregulatory molecules (Lauder et al., 1994), including, S-100β, a calcium-binding protein, that acts as a growth factor for 5-HT neurons and has been linked to serotonergic regulation of astroglial plasticity (Liu & Lauder, 1992; Whitaker-Azmitia & Azmitia, 1994), and tenascin, an extracellular matrix molecule that plays important roles in cell migration, differentiation and ecto-mesenchymal interactions during embryonic development (Chiquet-Ehrismann et al., 1986; Mackie et al., 1987; Tucker, 1994), and is localized to developing mouse teeth (Tucker et al., 1993).

Using micromass cultures prepared from mandibular mesenchyme (modified from Hassell & Horigan, 1982) we demonstrated that S-100β and tenascin were differentially regulated by 5-HT receptors (Moiseiwitsch & Lauder, 1997). While S-100β mRNA and protein expression were stimulated by the 5-HT2A/2C receptor antagonist mianserin, they were inhibited by the 5-HT3 antagonist, Zofran and unaffected by the 5-HT1A receptor antagonist NAN-190. By contrast, tenascin mRNA and protein expression were stimulated to a varying degree by all three 5-HT receptor antagonists. In an attempt to correlate these findings with a morphological marker we also looked at the extracellular matrix protein, cartilage proteoglycan core protein. Both Zofran and NAN-190 significantly inhibited this cartilage matrix marker.

Insulin-like growth factors
Serotonergic regulation of insulin-like growth factors (IGFs) has recently been investigated in serum-free mandibular micromass cultures from E12 mouse embryos (Lambert & Lauder, 1999, Lambert *et al.*, 2001). Treatment with 5-HT increases levels of IGF-I (but not IGF-II) in an inverse dose-dependent manner where lower doses are more stimulatory than higher doses. Similar effects on IGF-I are produced by agonists for 5-HT1A and 5-HT4 receptors, suggesting that these receptors may mediate the effects of 5-HT. Both agonists activate adenylyl cyclase leading to increased synthesis of cyclic AMP (cAMP). Using a protein kinase A (PKA) inhibitor, we showed that this causes PKA to phosphorylate CREB (cAMP response element binding protein). We hypothesize that this may cause CREB to bind to CRE in the IGF-I promoter, thereby promoting transcription of IGF-I (Buznikov *et al.*, 2001).

Serotonin and dental development

Dental development has frequently been used as a model to study interactions between ectoderm and underlying mesenchyme in developing craniofacial structures. Unlike most other examples of ectomesenchymal interactions, such as hair and glandular development, during dental development there is both an ectodermal and mesenchymal matrix produced. The extremely hard outer surface of teeth, the enamel, is an ectodermal product, while the majority of the tooth is composed of a series of tissues that are all of ecto-mesenchymal origin, including dentin, pulp, cementum and the periodontal ligament.

Dental development is divided morphologically into bud stage, cap stage, early bell stage and late bell stage. Each of these stages involves a series of reciprocal signalling events between mesenchymal and ectodermal tissues responsible for producing the next stage of dental development (Mackenzie *et al.*, 1992). A series of studies in our laboratory has demonstrated that 5-HT plays a key role during several of these stages and that 5-HT is able to stimulate tooth development in mandibular explants from E12-13 mouse embryos when added to the serum free medium (Moiseiwitsch & Lauder, 1996).

During the first stage of dental development, cranial neural crest-derived ectomesenchyme initiates invagination of oral ectoderm to form the dental lamina, which in turn gives rise to the bud stage tooth germ. At this stage, as the dental epithelium invaginates into the underlying mesenchyme it acquires the ability to transport 5-HT (Moiseiwitsch & Lauder, 1996). Simultaneously, 5-HT1A, 5-HT2A, 5-HT2B, 5-HT2C and 5-HT3 receptors are expressed within the developing bud stage tooth germ (Moiseiwitsch *et al.*, 1998; Johnson & Heinemann, 1995; Lauder *et al.*, 2000). During the next stage of development the dental ectoderm proliferates circumferentially to form the cap stage tooth germ. At this time, 5-HT uptake sites are demonstrable in mesenchyme surrounding the dental ectoderm (Lauder & Zimmerman, 1988). This suggests that 5-HT transport may facilitate ecto-mesenchymal interactions that modulate normal dental development (Thesleff & Sharpe, 1997).

Using mandibular explants from E12-13 mouse embryos, we have demonstrated that 5-HT promotes dental development in a dose-dependent manner (Moiseiwitsch & Lauder, 1996). This stimulation is halted at the cap stage by a 5-HT1A receptor antagonist, while a 5-HT3 receptor antagonist prevents development past the bud stage (Moiseiwitsch et al., 1998). The SSRI fluoxetine (Prozac) also prevents dental development from progressing past the early bud stage (Moiseiwitsch & Lauder, 1996). Similarly, 5-HT1A and 5-HT3 receptor antagonists block the stimulatory effects of 5-HT on expression of cartilage proteoglycan core protein, raising the possibility of common pathways in the serotonergic regulation of dental development and chondrogenesis. By the final stage of dental development (bell stage), 5-HT uptake persists in the dental follicle (Lauder & Zimmerman, 1988) in association with expression of 5-HT1A (Moiseiwitsch et al., 1998) and 5-HT3 receptors (Johnson & Heinemann, 1995).

During initial phases of craniofacial development, there appear to be several stages at which 5-HT uptake sites, 5-HT receptors and SBPs all occur in synchrony. It is possible that SBPs may be able to optimize the amount of available 5-HT, ensuring activation of appropriate signal transduction pathways over a wide range of concentrations of 5-HT. Highly specific SBP expression is also seen during tooth germ formation (Lauder et al., 1994; Moiseiwitsch et al., 1998). During the bud stage, condensing dental mesenchyme is surrounded by a zone of non-dental mesenchyme that is highly immunoreactive for SBP. Later, at the bell stage, this encapsulating zone is persistent, but some SBP immunoreactivity is also present in the inner enamel epithelium.

Taken together, these findings raise the possibility that 5-HT may facilitate morphogenesis of the developing tooth germ. At the bud stage, the "source" of 5-HT would be the vasculature, whereas 5-HT uptake sites present in the epithelium would act as a "sink", removing 5-HT by uptake and degradation. At the same time, 5-HT receptors are expressed by the dental epithelium and dental mesenchyme in conjunction with the surrounding zone of SBP-expressing cells. It is postulated that appropriate concentrations of 5-HT may stimulate proliferation of the dental mesenchyme, thereby causing rapid relative enlargement of the enamel organ. In this model (Figure 2A), cells that do not express SBP, located closer to epithelial uptake sites, would proliferate faster than SBP-expressing cells further from 5-HT uptake sites, thereby facilitating the rapid relative increase in size of the enamel organ. As these tissues undergo cytodifferentiation, specific changes in 5-HT receptor expression would occur. By the bell stage (Figure 2B), 5-HT uptake would have switched to the dental follicle. SBP expression persists in the zone of cells surrounding the enamel organ, while expression of 5-HT receptors is localized to the pre-odontoblast and inner-enamel epithelium (Moiseiwitsch et al., 1998). At this time it is postulated that there is a reversal of a 5-HT gradient from the vasculature of the dental papilla toward 5-HT uptake sites in the external layer of the dental mesenchyme. This could facilitate the continued proliferation of dental mesenchyme. Sites of

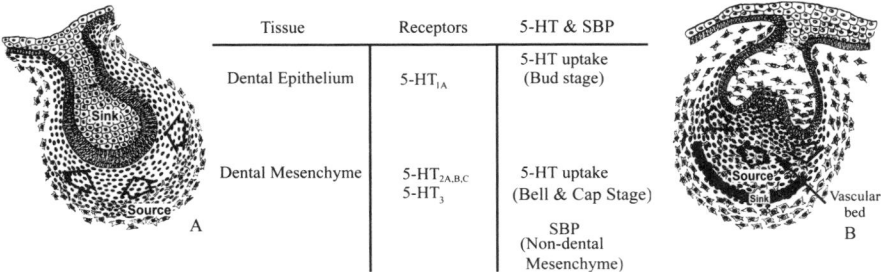

Tissue	Receptors	5-HT & SBP
Dental Epithelium	5-HT$_{1A}$	5-HT uptake (Bud stage)
Dental Mesenchyme	5-HT$_{2A,B,C}$ 5-HT$_3$	5-HT uptake (Bell & Cap Stage) SBP (Non-dental Mesenchyme)

Figure 2 Model for serotonergic regulation of early (**A**) and late (**B**) phases of dental development. (**A**) It is postulated that a 5-HT concentration gradient exists between a vascular source and 5-HT uptake sites present in the invaginating dental epithelium and that the concentration of available 5-HT in the developing tooth germ is regulated by a zone of serotonin-binding protein (SBP) in the surrounding non-dental mesenchyme. The relatively low concentration of serotonin in the dental mesenchyme preferentially stimulates its proliferation and 5-HT receptors transduce the serotonergic signal. Open arrows indicate direction of postulated concentration gradient of 5-HT. (**B**) At later stages of dental development a reversal of the 5-HT concentration gradient occurs. The source of 5-HT remains the rich capillary plexus of the tooth germ. However, by this stage 5-HT uptake sites are present in the peripheral dental mesenchyme (shaded area). The zone of SBP expression in the surrounding non-dental mesenchyme regulates the amount of available 5-HT. With reversal of the 5-HT concentration gradient, cell proliferation occurs preferentially at the periphery of the enamel organ facilitating the down-growth of the enamel epithelium as the bell stage of development progresses. SBP at the site of future cusp development may facilitate formation of molar morphology. Open arrows indicate direction of postulated concentration gradient of 5-HT. (Modified from Moiseiwitsch, 1995.)

SBP expression in the dental epithelium coincide with future cusp tips and could be involved in differential growth necessary for molar tooth morphology. As the bell stage continues to develop, and enamel matrix formation occurs, 5-HT1A receptor expression is co-localized with the G-protein α-subunit Gα$_z$ in both the proximal terminal web complex of the ameloblast and within the odontoblast cell layer (Gilbert *et al.*, 1998), suggesting that 5-HT may continue to play a role in dental development after initiation of amelogenesis.

Dental developmental appears to be particularly sensitive to different concentrations of 5-HT. Therefore it may be important that possible dental anomalies be investigated in children whose mothers have taken SSRIs during pregnancy or nursing.

DISCUSSION

The distribution of 5-HT uptake sites, SBPs and 5-HT receptors in developing mouse embryos, together with the teratogenic effects of 5-HT uptake inhibitors and receptor antagonists in whole embryo culture, may be relevant to craniofacial and cardiac abnormalities found in Down syndrome and trisomy 16 mice (Epstein, 1991; Grausz *et al.*, 1991; Webb *et al.*, 1994). These findings should also be considered in light of previous reports indicating that 5-HT,

L-tryptophan or tricyclic antidepressants can cause malformations of the skull, brain, spinal cord or vertebral column in rodents and humans (Guram *et al.*, 1982; Idänpään-Heikkila & Saxen, 1973; Jurand, 1980; Van Cauteren *et al.*, 1986). Although recent clinical and animal studies suggest that prenatal exposure to highly specific 5-HT uptake inhibitors (SSRIs) such as Prozac (fluoxetine) or Zoloft (sertraline) do not cause major structural malformations in humans or rodents (Rosa, 1994; Pastuszak *et al.*, 1993; Vorhees *et al.*, 1994), the observation of increased miscarriages, minor craniofacial abnormalities and growth abnormalities in children from pregnant women or nursing mothers using SSRIs (Baum & Misri, 1996; Chambers *et al.*, 1996) should raise caution as to the potential dangers of these antidepressants to developing humans, including possible neurobehavioral abnormalities (Baum & Misri, 1996).

References

Baum, A.L. & Misri, S. (1996). Selective serotonin-reuptake inhibitors in pregnancy and lactation. *Harvard Review of Psychiatry*, **4**, 117–125.

Burden, R.W. & Lawrence, I.E. (1973). Presence of biogenic amines in early rat development. *American Journal of Anatomy*, **136**, 251–257.

Buznikov, G.A., Lambert, H.W. & Lauder, J.M. (2001). Serotonin and serotonin-like substances as regulators of early embryogenesis and morphogenesis. *Cell and Tissue Research*, **305**, 177–186.

Buznikov, G.A., Shmukler, Y.B. & Lauder, J.M. (1996). From oocyte to neuron: do neurotransmitters function in the same way throughout development? *Cellular and Molecular Neurobiology*, **16**, 537–559.

Chambers, C.D., Johnson, K.A., Dick, L.M., Felix, R.J. & Jones, K.L. (1996). Birth outcomes in pregnant women taking fluoxetine. *New England Journal of Medicine*, **335**, 1010–1015.

Chiquet-Ehrismann, R., Mackie, E.J., Pearson, C.A. & Sakakura, T. (1986). Tenascin: an extracellular matrix protein involved in tissue interactions during fetal development and oncogenesis. *Cell,* **47**, 131–139.

Choi, D.S., Kellermann, O., Richard, J.F., Colas, J.F., Bolanos-Jimenez, F., Tournois, C., Launay, J.M. & Maroteaux, L. (1998). Mouse 5-HT$_{2B}$ receptor-mediated serotonin trophic functions. *Annals of the New York Academy of Sciences*, **861**, 67–73.

Choi, D.S., Ward, S.J., Messaddeq, N., Launay, J.M. & Maroteaux, L. (1997). 5-HT2B receptor-mediated serotonin morphogenetic functions in mouse cranial neural crest and myocardiac cells. *Development*, **124**, 1745–1755.

Cohen, L.S. & Rosenbaum, J.F. (1997). Birth outcomes in women taking fluoxetine. *New England Journal of Medicine*, **336**, 872.

de Ridder, L. & Beele, H. (1988). Morphological effects of serotonin and ketanserin on embryonic chick skin *in vitro*. *Experientia,* **44**, 603–606.

Edelman, G.M. & Jones, F.S. (1992). Cytotactin: a morphoregulatory molecule and a target for regulation by homeobox gene products. *Trends in Biochemical Science*, **17**, 228–232.

Epstein, C.J. (1991). Aneuploidy and morphogenesis. In C.J. Epstein (Ed.), *The morphogenesis of Down syndrome* (pp. 1–18). New York: Wiley-Liss.

Fanburg, B.L. & Lee, S.L. (1997). A new role for an old molecule: serotonin as a mitogen. *American Journal of Physiology: Lung Cell Molecular Physiology*, **272**, L795–L806.

Garnovskaya, M.N., van Biesen, T., Hawe, B., Casanas Ramos, S., Lefkowitz, R.J. & Raymond, J.R. (1996). Ras-dependent activation of fibroblast mitogen-activated protein kinase by 5-HT1A receptor via a G protein beta gamma-subunit-initiated pathway. *Biochemistry* **35**, 13716–13722.

Gilbert, N. (1998). Signal Transduction Pathways of Bradykinin Receptors and Serotonin Receptors in Rat Molars during Amelogenesis. MS dissertation, University of North Carolina, Chapel Hill.

Godin, I. & Gipouloux, J.D. (1986). Notochoral catecholamines in exogastrulated *Xenopus* embryos. *Development, Growth and Differentiation*, **28**, 137–142.

Goldstein, D.J., Sundell, K.L. & Corbin, L.A. (1997). Birth outcomes in women taking fluoxetine. *New England Journal of Medicine*, **336**, 873.

Grausz, H., Richtsmeier, J.T. & Oster-Granite, M.L. (1991). Morphogenesis of the brain and craniofacial complex in trisomy 16 mice. In C.J. Epstein (Ed.) *The morphogenesis of Down syndrome* (pp. 169–188). New York: Wiley-Liss.

Guram, M.S., Gill, T.S. & Geber, W.F. (1982). Comrataive teratogenicity of chlordiazepoxide, amitriptyline, and a combination of the two compounds in the fetal hamster. *Neurotoxicology,* **3**, 83–90.

Hansson, S.R., Mezey, E. & Hoffman, B.J. (1999). Serotonin transporter messenger RNA expression in neural crest-derived structures and sensory pathways of the developing rat embryo. *Neuroscience,* **89**, 243–265.

Hassell, J.R. & Horigan, E.A. (1982). Chodrogeneis: a model developmental system for measuring teratogenic potential of compounds. *Teratology,* **2**, 325–331.

Idänpään-Heikkila, J. & Saxen, L. (1973). Possible teratogenicity of imipramine/chloropyramine. *Lancet,* **2**, 282–284.

Johnson, D.S. & Heinemann, S.F. (1995). Detection of 5-HT3R-A, a 5-HT3 receptor subunit, in submucosal and myenteric ganglia of rat small intestine using in situ hybridization. *Neuroscience Letters,* **184**, 67–70.

Jones, K.I., Johnson, K.A. & Chambers, C.D. (1997). Birth outcomes in women taking fluoxetine. *New England Journal of Medicine,* **336**, 873.

Julius, D., Huang, K.N., Livelli, T.J., Axel, R. & Jessell, T.M. (1990). The 5HT2 receptor defines a family of structurally distinct but functionally conserved serotonin receptors. *Proceedings of the National Academy of Sciences, USA,* **87**, 928–932.

Julius, D., Livelli, T.J., Jessell, T.M. & Axel, R. (1989). Ectopic expression of the serotonin 1c receptor and the triggering of malignant transformation. *Science,* **244**, 1057–1062.

Jurand, A. (1980). Malformations of the central nervous system induced by neurotropic drugs in mouse embryos. *Developmental Growth and Differentiation,* **22**, 61–78.

Kirby, M.L. & Gilmore, S.A. (1972). A fluorescence study on the ability of the notochord to synthesize and store catecholamines in early chick embryos. *Anatomical Record,* **173**, 469–478.

Kirby, M.L. & Waldo, K.L. (1990). Role of neural crest in congenital heart disease. *Circulation,* **82**, 332–340.

Lambert, H.W. & Lauder, J.M. (1999). Serotonin receptor agonists that increase cyclic AMP positively regulate IGF-I in mouse mandibular mesenchyme cells. *Developmental Neuroscience,* **21**, 105–112.

Lambert, H.W., Weiss, E.R. & Lauder, J.M. (2001). Activation of 5-HT receptors that stimulate the adenylyl cyclase pathway positively regulates IGF-I in cultured craniofacial mesenchymal cells. *Developmental Neuroscience,* **23**, 70–77.

Lauder, J.M. (1988). Neurotransmitters as morphogens. *Progress in Brain Research,* **73**, 365–387.

Lauder, J.M. (1993). Neurotransmitters as growth regulatory signals: role of receptors and second messengers. *Trends in Neuroscience,* **16**, 233–240.

Lauder, J.M. & Zimmerman, E.F. (1988). Sites of serotonin uptake in epithelia of the developing mouse palate, oral cavity, and face: possible role in morphogenesis. *Journal of Craniofacial Genetics and Developmental Biology,* **8**, 265–276.

Lauder, J.M., Moiseiwitsch, J., Liu, J. & Wilkie, M.B. (1994). Serotonin in development and pathophysiology. In H.C. Lou, G. Griesen & J.F. Larsen, Eds., *Brain lesions in the newborn* (pp. 60–72). Copenhagen: Munksgaard.

Lauder, J.M., Tamir, H. & Sadler, T.W. (1988). Serotonin and morphogenesis. I. Sites of serotonin uptake and -binding protein immunoreactivity in the midgestation mouse embryo. *Development,* **102**, 709–20.

Lauder, J.M., Wilkie, M.B., Wu, C. & Singh, S. (2000). Expression of 5-HT$_{2A}$, 5-HT$_{2B}$ and 5-HT$_{2C}$ receptors in the mouse embryo. *International Journal of Developmental Neuroscience,* **18**, 653–662.

Launay, J.M., Birraux, G., Bondoux, D., Callebert, J., Choi, D.S., Loric, S. & Maroteaux, L. (1996). Ras involvement in signal transduction by the serotonin 5-HT2B receptor. *Journal of Biological Chemistry,* **271**, 3141–3147.

Lawrence, I.E., Jr. & Burden, H.W. (1973). Catecholamines and morphogenesis of the chick neural tube and notochord. *American Journal of Anatomy,* **137**, 199–208.

Lee, S.L., Wang, W.W., Moore, B.J. & Fanburg, B.L. (1991). Dual effects of serotonin on growth of bovine pulmonary artery smooth muscle cells in culture. *Circulation Research,* **68**, 1362–1368.

Liu, J. & Lauder, J.M. (1992). S-100β and insulin-like growth factor-II differentially regulate growth of developing serotonin and dopamine neurons *in vitro. Journal of Neuroscience Research,* **33**, 248–256.

Mackenzie, A., Ferguson, M.W.J. & Sharpe, P.T. (1992). Expression patterns of the homeobox gene, *Hox-8*, in the mouse embryo suggests a role in specifying tooth initiation and shape. *Development,* **115**, 403–420.

Mackie, E.J., Thesleff, I. & Chiquet-Ehrismann, R. (1987). Tenascin is associated with chondrogenic and osteogenic differentiation in vivo and promotes chondrogenesis *in vitro. Journal of Cell Biology,* **105**, 2569–2579.

Moiseiwitsch, J.R. & Lauder, J.M. (1995). Serotonin regulates mouse cranial neural crest migration. *Proceedings of the National Academy of Sciences, USA,* **92**, 7182–7186.

Moiseiwitsch, J.R. & Lauder, J.M. (1996). Stimulation of murine tooth development in organotypic culture by the neurotransmitter serotonin. *Archives of Oral Biology,* **41**, 161–165.

Moiseiwitsch, J.R.D. & Lauder, J.M. (1997). Regulation of gene expression in cultured embryonic mouse mandibular mesenchyme by serotonin antagonists. *Anatomical Embryology,* **195**, 71–78.

Moiseiwitsch, J.R.D., Raymond, J.R., Tamir, H. & Lauder, J.M. (1998). Regulation by serotonin of tooth-germ morphogenesis and gene expression in mouse mandibular explant cultures. *Archives of Oral Biology,* **43**, 789–800.

Nemecek, G.M., Coughlin, S.R., Handley, D.A. & Moskowitz, M.A. (1986) Stimulation of aortic smooth muscle cell mitogenesis by serotonin. *Proceedings of the National Academy of Science, USA,* **83**, 674–678.

Newgreen, D.F., Allan, I.J., Young, H.M. & Soutwell, B.R. (1981). Accumulation of exogenous catecholamines in the neural tube and non-neural tissues of the early fowl embryo: Correlation with morphogenetic movements. *Wilhelm Roux's Archiv,* **190**, 320–330.

Noden, D.M. (1986). Origins and patterning of craniofacial mesenchymal tissues. *Journal of Craniofacial Genetics and Developmental Biology,* **2**, 15–31.

Pakala, R., Willerson, J.T. & Benedict, C.R. (1994). Mitogenic effect of serotonin on vascular endothelial cells. *Circulation,* **90**, 1919–1926.

Palen, K., Thorneby, L. & Emanuelsson, H. (1979). Effects of serotonin and serotonin antagonists on chick embryogenesis. *W Roux's Arch,* **187**, 89–103.

Pastuszak, A., Schick-Boschetto, B., Zuber, C., Feldkamp, M., Pinelli, M., Sihn, S., Donnenfeld, A., McCormack, M., Leen-Mitchell, M., Woodland, C., Gardner, A., Hom, M. & Koren, G. (1993). Pregnancy outcome following first-trimester exposure to fluoxetine (Prozac). *Journal of the American Medical Association,* **269**, 2246–2248.

Pienkowski, M.M. (1977). Involvement of biogenic amines in control of development of early mouse embryos. *Anatomical Record,* **189**, 550.

Roberts, E. (1996). Treating depression in pregnancy. *New England Journal of Medicine,* **335**, 1056–1058.

Rosa, F. (1994). Medicaid antidepressant pregnancy outcomes. *Reproductive Toxicology,* **8**, 444.

Sadler, T.W. (1979). Culture of early somite mouse embryos during organogenesis. *Journal of Embryology and Experimental Morphology,* **49**, 17–25.

Sadykova, K.A., Sakharova, N.I. & Marakova, L.N. (1992). The effect of cyclic nucleotides on the sensitivity of early mouse embryos to biogenic monoamine antagonists. *Oncogenes,* **23**, 379–384.

Seuwen, K. & Pouyssegur, J. (1990). Serotonin as a growth factor. *Biochemical Pharmacology,* **39**, 985–990.

Seuwen, K., Magnaldo, I. & Pouyssegur, J. (1988). Serotonin stimulates DNA synthesis in fibroblasts acting through 5-HT1B receptors coupled to a Gi-protein. *Nature,* **335**, 254–256.

Shuey, D.L. (1991). *Serotonergic mechanisms in normal and abnormal craniofacial morphogenesis.* Chapel Hill, NC: Univ. N.C. Sch. Med.

Shuey, D.L., Sadler, T.W. & Lauder, J.M. (1992). Serotonin as a regulator of craniofacial morpho-genesis: site specific malformations following exposure to serotonin uptake inhibitors. *Teratology,* **46**, 367–378.

Shuey, D.L., Sadler, T.W., Tamir, H. & Lauder, J.M. (1993). Serotonin and morphogenesis. Transient expression of serotonin uptake and binding protein during craniofacial morphogenesis in the mouse. *Anatomy and Embryology (Berlin),* **187**, 75–85.

Strudel, G., Recasens, M. & Mandel, P. (1977). Identification de catecholamines et de serotonine dans les chordes d'embryons de poulet. *Comptes Rendus de l'Academie des Sciences, Paris,* **284**, 967–969.

Takuwa, N., Ganz, M., Takuwa, Y., Sterzel, R.B. & Rasmussen, H. (1989). Studies of the mitogenic effect of serotonin in rat renal mesangial cells. *American Journal of Physiology,* **257**, F431–9.

Tamir, H., Liu, K., Hsiung, S., Aldersberg, M. & Gershon, M.D. (1994). Serotonin binding protein: synthesis, secretion, and recycling. *Journal of Neurochemistry*, **63**, 97–107.

Tecott, L., Shtrom, S. & Julius, D. (1995). Expression of a serotonin-gated ion channel in embryonic neural and nonneural tissues. *Molecular and Cellular Neuroscience*, **6**, 43–55.

Thesleff, I. & Sharpe, P. (1997). Signalling networks regulating dental development. *Mechanical Developments*, **67**, 111–123.

Tucker, R.P. (1994). The function of tenascin: hypotheses and current knowledge. *Perspectives in Devevelopmental Neurobiology*, 1–2.

Tucker, R.P., Moiseiwitsch, J.R.D. & Lauder, J.M. (1993). *In situ* localization of tenascin mRNA in developing mouse teeth. *Archives of Oral Biology*, **38**, 1025–1029.

Van Cauteren, H., Vandenberghe, J. & Marsboom, R. (1986). Protective activity of ketanserin against serotonin-induced embryotoxicity and teratogenicity. *Drug Development Research*, **8**, 179–185.

Varrault, A., Bockaert, J. & Waeber, C. (1992). Activation of 5-HT1A receptors expressed in NIH-3T3 cells induces focus formation and potentiates EGF effect on DNA synthesis. *Molecular Biology of The Cell*, **3**, 961–969.

Vorhees, C.V., Acuff-Smith, K.D., Schilling, M.A., Fisher, J.E., Moran, M.S. & Buelke-Sam, J. (1994). A developmental neurotoxicity evaluation of the effects of prenatal exposure to fluoxetine in rats. *Fundamental Applied Toxicology*, **23**, 194–205.

Wallace, J.A. (1982). Monoamines in the early chick embryo: demonstration of serotonin synthesis and the regional distribution of serotonin-containing cells during morphogenesis. *American Journal of Anatomy*, **165**, 261–276.

Webb, S., Anderson, R.A. & Brown, N.A. (1994). Mouse trisomy 16 model of heart defects in Down syndrome: Atrioventricular cushion cells and volumes. *Teratology,* **49**, 373.

Whitaker-Azmitia, P.M. & Azmitia, E.C. (1994). Astroglial 5-HT$_{1A}$ receptors and S-100β in development and plasticity. *Perspectives in Developmental Neurobiology*, **2**, 233–238.

Wu, C., Dias, P., Kumar, S., Lauder, J.M. & Singh, S. (1999). Differential expression of serotonin 5-HT2 receptors during rat embryogenesis. *Developmental Neuroscience*, **21**, 22–28.

Yavarone, M.S. (1991). *Prospective roles for serotonin in heart development*. Chapel Hill, NC: Univ. N.C. Sch. Med.

Yavarone, M.S., Shuey, D.L., Sadler, T.W. & Lauder, J.M. (1993a). Serotonin uptake in the ectoplacental cone and placenta of the mouse. *Placenta,* **14**, 149–161.

Yavarone, M.S., Shuey, D.L., Tamir, H., Sadler, T.W. & Lauder, J.M. (1993b). Serotonin and cardiac morphogenesis in the mouse embryo. *Teratology,* **47**, 573–584.

Zachrisson, K. & Uribe, A. (1998). Serotonin and neuroendocrine peptides influence DNA synthesis in rat and human small intestinal cells *in vitro*. *Acta Physiologica Scandinavica*, 195–200.

Further reading

Buznikov, G.A., Shmukler, Y.B. & Lauder, J.M. (1996). From oocyte to neuron: do neurotransmitters function in the same way throughout development? *Cell Mol Neurobiol*, **16**, 537–559.

Chambers, C.D., Johnson, K.A., Dick, L.M., Felix, R.J. & Jones, K.L. (1996). Birth outcomes in pregnant women taking fluoxetine. *N Engl J Med,* **335**, 1010–1015.

Choi, D.S., Kellermann, O., Richard, J.F., Colas, J.F., Bolanos-Jimenez, F., Tournois, C., Launay, J.M. & Maroteaux, L. (1998). Mouse 5-HT$_{2B}$ receptor-mediated serotonin trophic functions. *Ann N Y Acad Sci,* **861**, 67–73.

Fanburg, B.L., & Lee, S.L. (1997). A new role for an old molecule: Serotonin as a mitogen. *Am. J. Physiol.: Lung Cel Mol Physiol*, **272**, L795–L806.

Lauder, J.M. (1988). Neurotransmitters as morphogens. *Prog Brain Res,* **73**, 365–387.

Lauder, J.M. (1993). Neurotransmitters as growth regulatory signals: role of receptors and second messengers. *Trends Neurosci.* **16**, 233–240.

II.6
Serotonin in Brain Development: Role of the 5-HT1A Receptor and S-100β

PATRICIA M. WHITAKER-AZMITIA

ABSTRACT

Serotonin is one of several neurotransmitters which act both as developmental signals and neurotransmitters. In the younger animal the developmental role predominates. As the animal ages, neurotransmission becomes the larger role. However, both functions of serotonin continue throughout the lifespan. The serotonin receptor particularly involved in development is the serotonin 5-HT1A receptor. This chapter details the effects this receptor has on development, and the effect that development has on the receptor. Finally, the chapter discusses this growth factor role of serotonin as it relates to human illnesses such as Down syndrome and autism, and how these illnesses may some day be treated by using our knowledge of the developmental role of serotonin.

INTRODUCTION

Serotonin is one of the earliest-developing neurotransmitter systems in the mammalian brain (Aitken & Tork, 1988; Lauder, 1990) and eventually becomes one of the mostly widely distributed systems in the brain, contacting most cells of the cortex. Phylogenetically, serotonin is a very old system, being one of the neurotransmitters found in the most simple of organisms. Serotonin is also found in plants where, in combination with the structurally very similar molecule, auxin, it appears to act as a growth-regulating factor by promoting cell elongation. Thus, serotonin develops early enough, and is widespread enough, that it can influence maturation of many other cells in the brain. This role in the mammalian brain recapitulates a role which serotonin has long held in evolution (Lauder & Krebs, 1976, 1978; Turlejski, 1996). The purpose of this chapter, therefore, is to describe the role of serotonin as a growth factor in the developing brain.

The development of serotonin-containing neurons has been extensively studied in a number of animal species, including rat (Lauder and Krebs, 1976,

A.F. Kalverboer and A. Gramsbergen (eds.), Handbook of Brain and Behaviour in Human Development, 153–164
© 2001 Kluwer Academic Publishers. Printed in Great Britain.

1978, 1982, 1985; Lauder, 1990; Lidov & Molliver, 1982; Ivgy-May *et al.*, 1994), chick (Ahmad & Zamenhof, 1978) and human (Hedner *et al.*, 1986). The findings across species are surprisingly consistent – most remarkably that the highest functional status of the serotonin system is reached early in development, and that adult levels of this system are much lower, in terms of receptors, enzymes, transporter molecules and neurons of the serotonin system. It is likely that the two major roles of serotonin, neurotransmission and neurotrophism, coexist throughout life, and the peak activity levels seen early in development are due to the relatively higher levels of trophic activity.

In the human brain, serotonergic neurons are first evident by 5 weeks of gestation (Sundstrom *et al.*, 1993) and increase dramatically by the 10th week of gestation (Kontur *et al.*, 1993; Levallois *et al.*, 1997; Shen *et al.*, 1989). By 15 weeks of gestation the typical organization of serotonin cell bodies into the raphe nuclei can be seen (Takahashi *et al.*, 1986). By birth the highly regulated and important diurnal rhythm of serotonin is already present (Attanasio *et al.*, 1986). Serotonin continues to increase throughout the first 2 years of life and then declines to adult levels by 5 years of age (Toth & Fekete, 1986). This time period coincides with peak periods of synaptogenesis and synaptic remodelling in human brain. Thus, if the animal studies described below can be applied to the human brain, much of the synaptic density and serotonin terminal density of the adult brain has been determined in the first 5 years of life. Understanding the role of serotonin as a trophic factor will have profound implications for understanding how the human brain develops, and what might modulate that process.

THEORETICAL BACKGROUND

Serotonin receptors must be present in the developing brain in order for serotonin to act as a developmental signal. It can be difficult to study receptors in the developing brain, because the pharmacology and cellular responses used to define a receptor can often be quite different in the immature brain (Whitaker-Azmitia, 1991). Nonetheless, a number of receptors have been identified. Principally, our work has focused on the 5-HT1A receptor.

The 5-HT1A receptor plays a number of important roles in the immature brain. First, it directly influences neuronal development. This appears to be by two processes – by a direct influence on process outgrowth from neurons and by release of the trophic factor S-100β. Secondly, the 5-HT1A receptor is what can be termed a programmable receptor – that is adult levels of the receptor are set by events taking place in development. Since this receptor is key in many adult behaviours, early life events influencing it will have profound and permanent effects. Finally, the 5-HT1A receptor also has a direct role to play in neurotransmission in the immature brain.

CRITICAL REVIEW OF THE LITERATURE

The 5-HT1A receptor is one of many specific serotonin receptor subtypes found in the mammalian brain. It is a seven-membrane spanning G-protein coupled receptor, generally associated with inhibition of cyclic AMP production. Interestingly, it is intronless.

Ontogeny of the 5-HT1A receptor

The 5-HT1A receptor clearly shows peak activity early in life. In the rat the 5-HT1A receptor peaks in brainstem by GD 15, and then declines (Hillion *et al.*, 1994). In cultures of serotonin neurons from brainstem, the 5-HT1A receptor is 20-fold higher in immature serotonin neurons than in mature neurons (Eaton *et al.*, 1997) and this has been suggested to indicate a trophic role for this receptor in stimulating development of serotonin neurons. In target regions the levels of serotonin 5-HT1A receptor peak later (Daval *et al.*, 1987: Dyck & Cynader, 1993). The 5-HT1A receptor is also transiently located in developing regions from which it is absent in the adult, such as the cerebellum (Daval *et al.*, 1987). In humans the peak number of 5-HT1A receptors occurs between 16 and 22 weeks of gestation (Bar-Peled *et al.*, 1991). This peak in the human occurs earlier in fetuses with Down syndrome. A recent report has found that an abnormal peak of receptors occurs in adult schizophrenic cerebellum (Slater *et al.*, 1998), at a time when the receptor should have regressed (del Olmo *et al.*, 1994, 1998).

These transiently expressed receptors are probably there for developmental purposes, either direct or indirect. Many of the transiently expressed 5-HT1A receptors can be localized to astroglial cells (Merzak *et al.*, 1996; Whitaker-Azmitia *et al.*, 1993). These receptors are responsible for the release of the astroglial-derived growth factor, S-100β, which in turn may mediate many of serotonin's trophic effects (Whitaker-Azmitia *et al.*, 1990). As the astroglial cell matures, the number of 5-HT1A receptors decreases and thus release of the growth factor is less (Whitaker-Azmitia & Azmitia *et al.*, 1986). However, even in the mature brain these receptors can be found on astroglial cells (Whitaker-Azmitia, *et al.*, 1993), which emphasizes the role which this system continues to play in adult plasticity.

S-100β: Role in Development

S-100β is a member of the family of proteins soluble in 100% ammonium sulphate and thus referred to as S-100 proteins. However, among this family the beta form is unique in having profound effects on brain development and plasticity. S-100β is localized to astroglial cells of the central nervous system, from which it can be released by a number of factors, including the serotonin receptor 5-HT1A (Whitaker-Azmitia *et al.*, 1990). S-100ß causes neurite outgrowth in chick cortical cultures (Kligman & Marshak, 1985) and plays a role in promoting dendritic growth (Whitaker-Azmitia *et al.*, 1997). In blinded kittens the cortical plasticity normally seen is inhibited by an antibody to S-100ß (Muller *et al.*, 1993). As the animal matures, S-100β continues to play a role in neuronal plasticity, in the long-term potentiation model of learning (Lewis & Teyler, 1986) and in maintaining dendrites.

S-100β interacts with and stabilizes microtubule-associated proteins (MAPs), such as tau and MAP-2, possibly through inhibition of phosphorylation (Baudier & Cole, 1988; Baudier *et al.*, 1987; Hesketh & Baudier, 1986).

How does the release of S-100β by the 5-HT1A receptor influence development? Our studies have led to the following model:

As serotonin neurons grow into a target field, the released serotonin acts on the astroglial 5-HT1A receptor. S-100β is then released from the astroglial cell and taken up into the surrounding neurons. Once in neurons, S-100β promotes outgrowth of presynaptic elements by inhibiting the phosphorylation of the growth-associated protein GAP-43. In other neurons S-100β stabilizes dendrites by inhibiting the phosphorylation of microtubule-associated proteins, such as MAP-2. Thus, synaptic contacts are made and stabilized. This model is illustrated in Figure 1.

Much of this model came from studies on the effects on brain development of removing serotonin at specific ages. Depletion of serotonin during synaptogenesis (postnatal day (PND) 10–20 in the rat, or approximately the first 2 years of life in a human) causes loss of dendrites and dendritic spines, which can be prevented by treatment with a 5-HT1A agonist. In our studies, serotonin was depleted by treating animals with the tryptophan hydroxylase inhibitor, parachlorophenylalanine (PCPA) (Mazer et al., 1997). Treating animals from PND 10 to 20 resulted in a significant loss of dendrites in the hippocampus. This loss was evident throughout the hippocampus up to PND 180, but by PND 400 some dendritic density had been restored in CA3. Synaptophysin staining was used as an indicator of the development of presynaptic elements and results showed different changes at different ages. In the youngest animals studied, PND 30 and 60 (or 10 and 40 days after serotonin depletion) there was a loss in presynaptic terminals. At PND 180 it was normal, but by PND 400 it was actually increased compared to controls, especially within the dentate gyrus. There are several possible explanations for this apparent increase. First, it could actually reflect retention of synapses which are normally lost with age. Secondly, it could reflect a "reactive synaptogenesis" secondary to the loss of dendrites. This would raise the possibility that mossy fibres are sprouting and that the animals might be susceptible to seizures. More work needs to be done on this question. However, these results are very exciting because they show that developmental events not only influence development of the brain, but that they also influence how the brain will age.

The loss of synaptic connections from serotonin depletion during synaptogenesis also has behavioural consequences. Principally, the animals showed learning and memory deficits as adolescents and adults. As young animals they showed a decreased ability to reorient to a changing stimulus. Treatment with a 5-HT1A agonist at this time shows opposite effects – i.e. the animals more quickly acquire mature levels of learning performace (Borella et al., 1997). However, there are also studies showing that treatment with a 5-HT1A agonist at this time increases aggression in the adult animal (Albonetti et al., 1996).

Studies of serotonin depletion at earlier times also show a loss of dendritic spines in the hippocampus. This loss gets worse with age, and has been specifically attributed to the 5-HT1A receptor-mediated release of S-100β. Treatment with a 5-HT1A receptor agonist reverses the effects (Yan et al., 1997a,b). Treatment with a 5-HT1A agonist alone, at this time, causes accelerated incisor eruption and eye-opening, a possible consequence of 5HT1A receptor

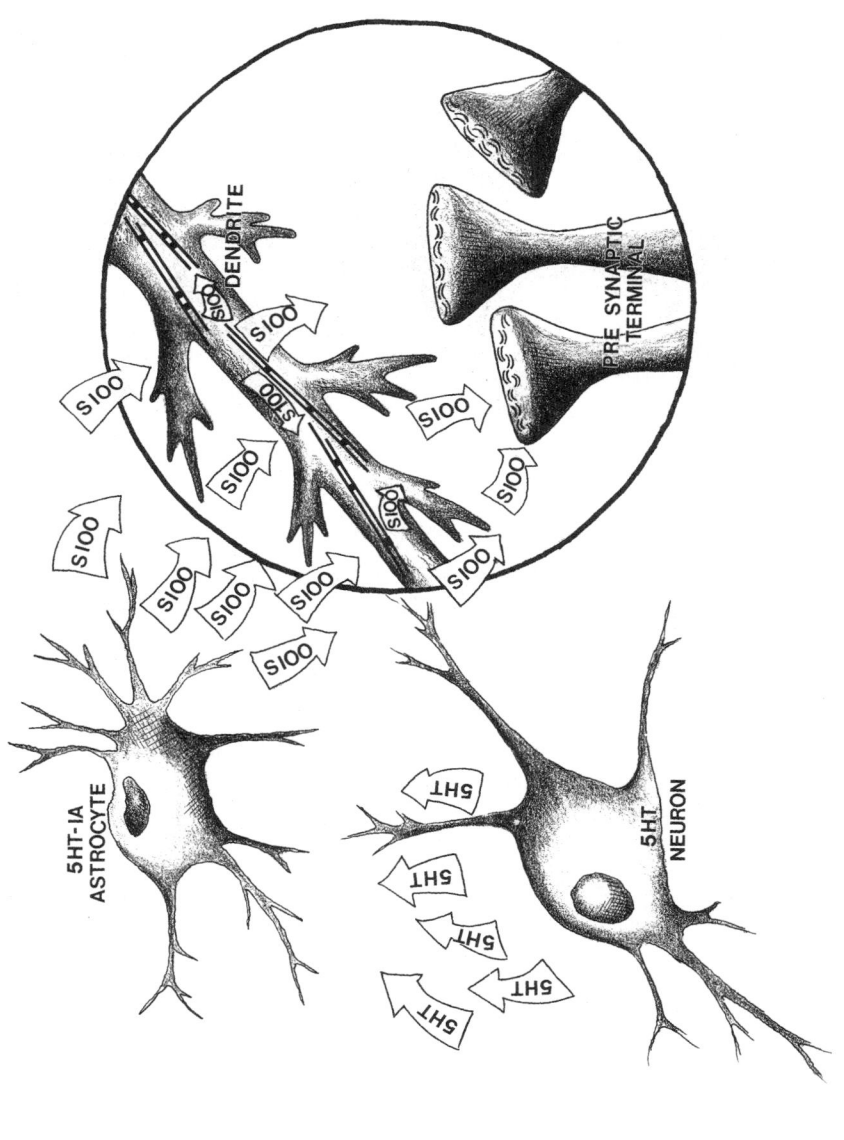

Figure 1 Proposed model of serotonergic regulation of S-100β and synaptogenesis (see text).

interactions with epidermal growth factor (EGF). Behaviourally, the animals are more anxious as adults (Borella *et al.*, 1997).

In general we have found that depletion of the serotonin system itself, after weaning (or PND 21) does not have a profound effect on subsequent brain development. However, we have found that the 5-HT1A receptor continues to be programmable (Borella *et al.*, 1997; J. Hobin, unpublished observations).

Programmable events

As stated above, the 5-HT1A receptor is also a programmable receptor. In studies of environmental enrichment the expression of the 5-HT1A receptor is up-regulated, (Rasmuson *et al.*, 1998) In preliminary studies we housed rats singly (isolated) or in same-sex groups of five (socialized) from weaning (PND 21) to testing and/or sacrifice on PNDs 110–118 and found social isolation results in a loss of 5-HT1A receptors in granule cells of the dentate gyrus (J. Hobin, unpublished observation). The 5-HT1A receptor is also decreased by prenatal malnutrition (Blatt *et al.*, 1994). *In-utero* ethanol impairs the development of 5HT1A receptors in the frontal cortex, parietal cortex, dentate gyrus and lateral septum (Kim *et al.*, 1997).

The 5-HT1A receptor is important in a variety of emotional and cognitive behaviours in the adult. Setting levels of this receptor during development could thus influence the behaviour of the adult.

Behavioral role and neurochemical function

The neurotransmitter and neurochemical functions of this receptor are also evident in the immature brain. Both 5-HT1A and 5-HT1B receptors are present in embryonic tissue and can carry on their usual functions of regulating serotonin release and/or synthesis rates, at the cell body or cell terminal (Hery *et al.*, 1999). As early as 3 days of age a 5-HT1A agonist, such as DPAT, has an anxiolytic effect (Joyce & Carden, 1999). The 5-HT1A receptor plays a role in suckling (Kirstein & Spear, 1988) and calling for the mother (Winslow & Insel, 1991). The ability to induce tolerance to 8-OH-DPAT is not developed before 26 days of age in the rat (Johansson-Wallsten & Meyerson, 1994), suggesting that the receptor predominantly functions in development until that time.

DISCUSSION

Obviously, work with animals has the ultimate goal of being relevant to the human situation. How might our studies be applicable?

During human fetal development maternal serotonin reaches the fetus, and may in fact be concentrated by the fetus. Thus this may be a direct link between the mother's well-being and fetal outcome (Morita *et al.*, 1992). Early life events, such as stress, malnutrition, or a deprived vs enriched environment also have profound effects on the emotional and intellectual outcomes of the child. Could this be through the effects of serotonin? According to our animal studies of the role of serotonin in development, children at risk may have changes in hippocampal dendritic density and expression of the 5-HT1A receptor. This would lead to impaired learning and memory, as well as emotional changes.

There are also a number of specific human developmental diseases which appear to be directly related to serotonin. In particular, we have been working on animal models of autism and Down syndrome.

Autism is a pervasive developmental disorder with profound effects on a variety of functions. It is four times more common in boys than girls, and is not usually diagnosed until 2 years of age when the lack of communication skills first becomes evident. There is lack of imaginative play, repeated body movements, ritualistic behaviours and a paucity of social interactions. Emotional rapport with others is limited. Sensory changes are evident. In some cases there is self-injurious behaviour. In as many as a third of patients there may be seizures. As adults there may be some improvement in language skills, but self-injurious and stereotypic behaviours may be more common (Ballaban-Gil et al., 1996). It is unlikely that a single cause or gene is responsible.

The role of serotonin in autism may be very complex. Neurochemically the most consistent finding in autism is an increase in blood serotonin (Anderson et al., 1990; Cook, 1990; Cook et al., 1993; Naffah-Mazzacoratti et al., 1993; Warren & Singh, 1996). There may be some genetic linkage to the serotonin transporter (Cook et al., 1997) and there are two reports of autistic children having anti-bodies to the 5-HT1A receptor (Singh et al., 1997; Todd & Ciaranello, 1985). Finally, a PET study using the radiolabelled form of a serotonin precursor found decreased serotonin synthesis in cortex and thalamus, although an increase was found in the dentate nucleus (Chugani et al., 1997). In our model, high levels of serotonin during development may be injurious to both the brain and the developing serotonin system itself. Dendritic branching in the postmortem autistic brain is less in CA4 and CA1, suggesting curtailed maturation (Raymond et al., 1996), much as has been seen in our animal studies.

Since the earliest studies on the neurobiology of Down syndrome, a loss of serotonin has been found in postmortem brain (Mann & Yates, 1986; Mann et al., 1985), CSF (Scott et al., 1983) and blood (Boullin & O'Brien, 1971; Tu & Zellweger, 1965; Hayes & Batshaw, 1993). The serotonin 5-HT1A receptor has been shown to peak earlier in developing DS brains than in normals, and to decrease to below normal levels by birth (Bar-Peled et al., 1991). Treatments for DS have long included serotonergic agents, and recent studies suggest that these may be particularly helpful in self-injurious behaviours and in unprovoked aggression (Gedye, 1990, 1991). Changes in the serotonin system are often included as criteria for establishing an animal model of DS. One important clue to the link between serotonin and Down syndrome, is that the gene for S-100β is found on chromosome 21 in what is considered the obligate region for Down syndrome (Allore et al., 1988; Hernandez & Fisher, 1996). Thus, we have been characterizing a mouse model of DS, which overexpresses S-100β. These animals show many of the characteristics of DS, including overdevelopment and early loss of dendrites in the hippocampus (Becker et al., 1986; Takashima et al., 1994). These changes appear to correlate with overexpression followed by early loss of serotonin neurons.

Thus, our work in some way may be useful in understanding these illnesses and ultimately treating them. Treatment options may include cell transplantation or

159

pharmacological agents given at critical times during development. Human fetal serotonin neurons can be grown in culture (Levallois *et al.*, 1997), and an immortalized animal line is available. A great deal is also known of serotonin pharmacology, and animal studies have already shown that 5-HT1A receptor agonists can reverse or prevent the damaging effects of cocaine or alcohol on the developing hippocampus (Akbari *et al.*, 1994; Kim *et al.*, 1997; Tajuddin & Druse, 1993; Whitaker-Azmitia, 1998).

Finally, our studies may also warn against the teratogenic effects of serotonergic agents on the developing brain. To date, most studies in human fetuses exposed to serotonergic agents, such as the selective serotonin re-uptake inhibitors fluoxetine and sertraline, have looked at crude outcome measures such as deformities or fetal loss (Kulin *et al.*, 1998) and have not tested for the effects on emotion and cognition which our studies would predict. In rats, prenatal fluoxetine (Prozac) does decrease serotonin terminal outgrowth (Cabrera-Vera & Battaglia, 1998) and clearly caution should be used when using these drugs in the treatment of pregnant women.

References

Ahmad, G. & Zamenhof, S. (1978). Serotonin as a growth factor for chick embryo brain. *Life Sciences*, **22**(11), 963–970.

Aitken, A.R. & Tork, I. (1988). Early development of serotonin-containing neurons and pathways as seen in wholemount preparations of the fetal rat brain. *Journal of Comparative Neurology*, **274**(1), 32–47.

Akbari, H.M., Whitaker-Azmitia, P.M. & Azmitia, E.C. (1994). Prenatal cocaine decreases the trophic factor S-100β and induced microcephaly: reversal by postnatal 5-HT1a agonist. *Neuroscience Letters*, **170**, 141–144.

Albonetti, E., Gonzalez, M.I., Siddiqui, A., Wilson, C.A. & Farabollini, F. (1996). Involvement of the 5-HT1A subtype receptor in the neonatal organization of agonistic behaviour in the rat. *Pharmacology, Biochemistry and Behavior*, **54**(1), 189–193.

Allore, R., O'Hanlon, D., Price, R., Neilson, K., Willard, H.F., Cox, D.R., Marks, A. & Dunn, R.J. (1988). Gene encoding the beta subunit of S-100 protein is on chromosome 21. Implications for Down's syndrome. *Science*, **239**, 1311.

Anderson, G.M., Horne, W.C., Chatterjee, D. & Cohen, D.J. (1990). The hyperserotonemia of autism. *Annals of the New York Academy of Sciences*, **600**, 331–340.

Attanasio, A., Rager, K. and Gupta, D. (1986). Ontogeny of circadian rhythmicity for melatonin, serotonin, and *N*-acetylserotonin in humans. *Journal of Pineal Research*, **3**(3), 251–256.

Ballaban-Gil, K., Rapin, I., Tuchman, R. & Shinnar, S. (1996). Longitudinal examination of the behavioral, language, and social changes in a population of adolescents and young adults with autistic disorders. *Pediatric Neurology*, **15**(3), 217–223.

Bar-Peled, O., Gross-Isseroff, R., Ben-Hur, H., Hoskins, I., Groner, Y. & Biegon, A. (1991). Fetal human brain exhibits a prenatal peak in the density of serotonin 5-HT1A receptors. *Neuroscience Letters*, **127**(2), 173–176.

Baudier, J. & Cole, R.D. (1988). Interactions between the microtubule associated tau protein and S-100β regulate tau phosphorylation by the Ca++/ calmodulin dependent protein kinase II. *Journal of Biological Chemistry*, **263**, 5876–5883.

Baudier, J., Mochly-Rosen, D., Newton, A., Lee, S.H., Koshland, D.E. & Cole, R.D. (1987). Comparison of S-100b protein with calmodulin: interactions with melittin and microtubule-associated tau proteins and inhibition of phosphorylation of tau proteins by protein kinase C. *Biochemistry*, **26**, 2886–2893.

Becker, L.E., Armstrong, D.L. & Chan, F. (1986). Dendritic atrophy in children with Down's syndrome. *Annals of Neurology*, **20**, 520–526.

Blatt, G.J., Chen, J.C., Rosene, D.L., Volicer, L. & Galler, J.R. (1994). Prenatal protein malnutrition effects on the serotonergic system in the hippocampal formation: an immunocytochemical, ligand binding, and neurochemical study. *Brain Research Bulletin*, **34**(5), 507–518.

Borella, A., Bindra, M. & WhitakerAzmitia, P.M. (1997). Role of the 5HT1A receptor in development of the neonatal rat brain: preliminary behavioral studies. *Neuropharmacology*, **36**(4–5), 445–450.

Boullin, D.J. & O'Brien, R.A. (1971). Abnormalities of 5-hydroxytryptamine uptake and binding by blood platelets from children with Down's syndrome. *Journal of Physiology*, **212**, 287–297.

Cabrera-Vera, T.M. & Battaglia, G. (1998). Prenatal exposure to fluoxetine (Prozac) produces site-specific and age-dependent alterations in brain serotonin transporters in rat progeny: evidence from autoradiographic studies. *Journal of Pharmacology and Experimental Therapeutics*, **286**(3), 1474–1481.

Chugani, D.C., Muzik, O., Rothermel, R., Behen, M., Chakraborty, P., Mangner, T., da Silva, E.A. & Chugani, H.T. (1997). Altered serotonin synthesis in the dentatothalamocortical pathway in autistic boys. *Annals of Neurology*, **42**(4), 666–669.

Cook, E.H. (1990). Autism: review of neurochemical investigation. *Synapse*, **6**(3), 292–308.

Cook, E.H. Jr, Arora, R.C., Anderson, G.M., Berry-Kravis, E.M., Yan, S.Y., Yeoh, H.C., Sklena, P.J., Charak, D.A. & Leventhal, B.L. (1993). Platelet serotonin studies in hyperserotonemic relatives of children with autistic disorder. *Life Sciences*, **52**(25), 2005–2015.

Cook, E.H. Jr, Courchesne, R., Lord, C., Cox, N.J., Yan, S., Lincoln, A., Haas, R., Courchesne, E. & Leventhal, B.L. (1997). Evidence of linkage between the serotonin transporter and autistic disorder. *Molecular Psychiatry*, **2**(3), 247–250.

Daval, G., Verge, D., Becerril, A., Gozlan, H., Spampinato, U. & Hamon, M. (1987). Transient expression of 5-HT1A receptor binding sites in some areas of the rat CNS during postnatal development. *International Journal of Developmental Neuroscience*, **5**(3), 171–180.

Dyck, R.H. & Cynader, M.S. (1993). Autoradiographic localization of serotonin receptor subtypes in cat visual cortex: transient regional, laminar, and columnar distributions during postnatal development. *Journal of Neuroscience*, **13**(10), 4316–4338.

del Olmo, E., Diaz, A., Guirao-Pineyro, M., del Arco, C., Pascual, J. & Pazos, A. (1994). Transient localization of 5-HT1A receptors in human cerebellum during development. *Neuroscience Letters*, **166**(2), 149–152.

del Olmo, E., Lopez-Gimenez, J.F., Vilaro, M.T., Mengod, G., Palacios, J.M. & Pazos, A. (1998). Early localization of mRNA coding for 5-HT1A receptors in human brain during development. *Molecular Brain Research*, **60**(1), 123–126.

Eaton, M.J., Santiago, D.I., Dancausse, H.A. & Whittemore, S.R. (1997). Lumbar transplants of immortalized serotonergic neurons alleviate chronic neuropathic pain. *Pain*, **72**(1–2), 59–69.

Gedye, A. (1990). Dietary increases in serotonin reduce self-injurious behavior in a Down's syndrome adult. *Journal of Mental Deficiency Research*, **34**, 195–203.

Gedye, A. (1991). Serotonergic treatment for aggression in a Down's syndrome adult showing signs of Alzheimer's disease, *Journal of Mental Deficiency Research*, **35**, 247–258.

Hayes, A. & Batshaw, M.L. (1993). Down syndrome. *Pediatric Clinics of North America*, **40**, 523–535.

Hedner, J., Lundell, K.H., Breese, G.R., Mueller, R.A. & Hedner, T. (1986). Developmental variations in CSF monoamine metabolites during childhood. *Biology of the Neonate*, **49**(4), 190–197.

Hernandez, D. & Fisher, E.M. (1996). Down syndrome genetics: unraveling a multifactorial disorder Human. *Molecular Genetics*, **5**, 1411–1416.

Hery, F., Boulenguez, P., Semont, A., Hery, M., Becquet, D., Faudon, M., Deprez, P. & Fache, M.P. (1999). Identification and role of serotonin 5-HT1A and 5-HT1B receptors in primary cultures of rat embryonic rostral raphe nucleus neurons. *Journal of Neurochemistry*, **72**(5), 1791–1801.

Hesketh, J. & Baudier, J. (1986). Evidence that S100 proteins regulate microtubule assembly and stability in rat brain extracts. *International Journal of Biochemistry*, **18**, 691–695.

Hillion, J., Catelon, J., Raid, M., Hamon, M. & De Vitry, F. (1994). Neuronal localization of 5-HT1A receptor mRNA and protein in rat embryonic brain stem cultures. *Developmental Brain Research*, **79**(2), 195–202.

Ivgy-May, N., Tamir, H. & Gershon, M.D. (1994). Synaptic properties of serotonergic growth cones in developing rat brain *Journal of Neuroscience*, **14**(3), 1011–1029.

Johansson-Wallsten, C.E. & Meyerson, B.J. (1994). The ontogeny of tolerance to the 5-HT1A agonist 8-OH-DPAT: a study in the rat. *Neuropharmacology*, **33**(3–4), 325–330.

Joyce, M.P. & Carden, ??. (1999). The effects of 8-OH-DPAT and (+/−)-pindolol on isolation induced ultrasonic vocalizations in 3-, 10-, and 14-day-old rats. *Developmental Psychobiology*, **34**(2), 109–117.

Kim, J.A., Gillespie, R.A. & Druse, M.J. (1997). Effects of maternal ethanol consumption and buspirone treatment on 5-HT1A and 5-HT2A receptors in offspring. *Alcohol Clinical and Experimental Research*, **21**(7), 1169–1178.

Kirstein, C.L. & Spear, L.P. (1988). 5-HT1A, 5-HT1B and 5-HT2 receptor agonists induce differential behavioral responses in neonatal rat pups. *European Journal of Pharmacology*, **150**(3), 339–345.

Kligman, D. & Marshak, D. (1985). Purification and characterization of a neurite extension factor from bovine brain. *Proceedings of the National Academy of Sciences*, **82**, 7136–7142.

Kontur, P.J., Leranth, C., Redmond, D.E. Jr, Roth, R.H. & Robbins, R.J. (1993). Tyrosine hydroxylase immunoreactivity and monoamine and metabolite levels in cryopreserved human fetal ventral mesencephalon. *Experimental Neurology*, **121**(2), 172–180.

Kulin, N.A., Pastuszak, A., Sage, S.R., Schick-Boschetto, B., Spivey, G., Feldkamp, M., Ormond, K., Matsui, D., Stein-Schechman, A.K., Cook, L., Brochu, J., Rieder, M. & Koren, G. (1998). Pregnancy outcome following maternal use of the new selective serotonin reuptake inhibitors: a prospective controlled multicenter study. *Journal of the American Medical Association*, **279**(8), 609–610.

Lauder, J.M. (1990). Ontogeny of the serotonergic system in the rat: serotonin as a developmental signal. *Annals of the New York Academy of Sciences*, **600**, 297–313.

Lauder, J.M. & Krebs, H. (1976). Effects of p-chlorophenylalanine on time of neuronal origin during embyrogenesis in the rat. *Brain Research*, **107**, 638–644.

Lauder, J.M. & Krebs, H. (1978). Serotonin as a differentiation signal in early neurogenesis. *Developmental Neuroscience*, **1**, 15–20.

Levallois, C., Valence, C., Baldet, P. & Privat, A. (1997). Morphological and morphometric analysis of serotonin-containing neurons in primary dissociated cultures of human rhombencephalon: a study of development. *Developmental Brain Research*, **99**(2), 243–252.

Lewis, D. & Teyler, T.J. (1986). Anti-S-100 serum blocks long-term potentiation in the hippocampal slice *Brain Research*, **383**, 159–167.

Lidov, H.G. & Molliver. M.E. (1982). Immunohistochemical study of the development of serotonergic neurons in the rat CNS. *Brain Research Bulletin*, **9**(1–6), 559–604.

Mann, D.M. & Yates, P.O. (1986). Neurotransmitter deficits in Alzheimer's disease and other dementing disorders. *Human Neurobiology*, **5**, 147–158.

Mann, D.M., Yates, P.O., Marcyniuk, B. & Ravindra, C.R. (1985). Pathological deficits in Down's syndrome of middle age. *Journal of Mental Deficiency Research*, **29**, 125–135.

Mazer, C., Muneyyirci, J., Taheny, K., Raio, N., Borella, A & Whitaker-Azmitia, P.M. (1997). Serotonin depletion during synaptogenesis leads to decreased synaptic density and learning deficits in the adult rat. A model of neurodevelopmental disorders with cogntive deficits. *Brain Research*, **760**, 68–73.

Merzak, A., Koochekpour, S., Fillion, M.P., Fillion, G. & Pilkington, G.J. (1996). Expression of serotonin receptors in human fetal astrocytes and glioma cell lines: a possible role in glioma cell proliferation and migration. *Molecular Brain Research*, **41**(1–2), 1–7.

Morita, I., Kawamoto, M. & Yoshida, H. (1992). Difference in the concentration of tryptophan metabolites between maternal and umbilical foetal blood. *Journal of Chromatography*, **576**(2), 334–339.

Muller, C.M., Akhavan, A.C. & Bette, M. (1993). Possible role of S-100 in glia–neuronal signalling involved in activity dependent plasticity in the developing mammalian cortex. *Journal of Chemical Neuroanatomy*, **6**, 215–220.

Naffah-Mazzacoratti, M.G., Rosenberg, R., Fernandes, M.J., Draque, C.M., Silvestrini, W., Calderazzo, L. & Cavalheiro, E.A. (1993). Serum serotonin levels of normal and autistic children. *Brazilian Journal of Medical Biology Research*, **26**(3), 309–317.

Rasmuson, S., Henriksson, B.G., Mohammed, A.H., Kelly, P.A.T., Holmes, M.C. & Seckl, J.R. (1998). Environmental enrichment selectively increases 5-HT1A receptor mRNA expression and binding in the rat hippocampus. *Molecular Brain Research*, **53**, 285–290.

Raymond, G.V., Bauman, M.L. & Kemper, T.L. (1996). Hippocampus in autism: a Golgi analysis. *Acta Neuropathology (Berlin)*, **91**(1), 117–119.

Scott, B.S., Becker, L.E. & Petit, T.L. (1983). Neurobiology of Down's syndrome. *Progress in Neurobiology*, **12**, 199–237.

Shen, W.Z., Luo, Z.B., Zheng, D.R. & Yew, D.T. (1989). Immunohistochemical studies on the development of 5-HT (serotonin) neurons in the nuclei of the reticular formations of human fetuses. *Pediatric Neuroscience*, **15**(6), 291–295.

Singh, V.K., Singh, E.A. & Warren, R.P. (1997). Hyperserotoninemia and serotonin receptor antibodies in children with autism but not mental retardation. *Biology Psychiatry*, **41**(6), 753–755.

Slater, P., Doyle, C.A. & Deakin, J.F. (1998). Abnormal persistence of cerebellar serotonin-1A receptors in schizophrenia suggests failure to regress in neonates. *Journal of Neural Transmission*, **105**(2–3), 305–315.

Sundstrom, E., Kolare, S., Souverbie, F., Samuelsson, E.B., Pschera, H., Lunell, N.O. & Seiger, A. (1993). Neurochemical differentiation of human bulbospinal monoaminergic neurons during the first trimester. *Developmental Brain Research*, **75**(1), 1–12.

Tajuddin, N. & Druse, M.J. (1993). Treatment of pregnant alcohol-consuming rats with buspirone: effects on serotonin and 5-HIAA content in offspring. *Alcohol Clinical and Experimental Research*, **17**, 110.

Takahashi, H., Nakashima, S., Ohama, E., Takeda, S. & Ikuta, F. (1986). Distribution of serotonin-containing cell bodies in the brainstem of the human fetus determined with immunohistochemistry using antiserotonin serum. *Brain Development*, **8**(4), 355–365.

Takashima, S., Iida, K., Mito, T. & Arima, M. (1994). Dendritic and histochemical development and aging in patients with Down's syndrome. *Journal of Intelligence Disease Research,* **38**, 265.

Todd, R.D. & Ciaranello, R.D. (1985). Demonstration of inter- and intraspecies differences in serotonin binding sites by antibodies from an autistic child. *Proceedings of the National Academy of Sciences, USA*, **82**(2), 612–616.

Toth, G. & Fekete, M. (1986). 5-Hydroxyindole acetic excretion in newborns, infants and children. *Acta Pediatrica Hunarica*, **27**(3), 221–226.

Tu, J.B. & Zellweger, H (1965). Blood serotonin deficiency in Down's syndrome. *Lancet*, **2**, 715.

Turlejski, K. (1996). Evolutionary ancient roles of serotonin: long-lasting regulation of activity and development. *Acta Neurobiologica Experimentalis (Warszaw),* **56**(2), 619–636.

Warren, R.P. & Singh, V.K. (1996). Elevated serotonin levels in autism: association with the major histocompatibility complex. *Neuropsychobiology*, **34**(2), 72–75.

Whitaker-Azmitia, P.M. (1991). Role of serotonin and other neurotransmitter receptors in brain development: basis for developmental pharmacology. *Pharmacology Reviews*, **43**, 553–561.

Whitaker-Azmitia, P.M. (1998). Role of the neurotrophic properties of serotonin in the delay of brain maturation induced by cocaine. *Annals of the New York Academy of Sciences*, **846**, 158–164.

Whitaker-Azmitia, P.M. & Azmitia, E.C. (1986). 5-Hydroxytryptamine binding to brain astroglial cells: differences between intact and homogenized preparations and mature and immature cultures. *Journal of Neurochemistry*, **46**, 1186–1189.

Whitaker-Azmitia, P.M. & Azmitia, E.C. (1989). Stimulation of astroglial serotonin receptors produces media which regulates development of serotonergic neurons. *Brain Research*, **497**, 80–85.

Whitaker-Azmitia, P.M., Clarke, C. & Azmitia, E.C. (1993). Localization of 5-HT1a receptors to astroglial cells in adult rats: implications for neuronal-glial interactions and psychoactive drug mechanism of action. *Synapse,* **14**, 201.

Whitaker-Azmitia, P.M., Wingate, M., Borella, A., Gerlai, R., Roder, J. and Azmitia, E.C. (1997). Transgenic mice overexpressing the neurotrophic factor S-100 beta show neuronal cytoskeletal and behavioral signs of altered aging processes: implications for Alzheimer's disease and Down's syndrome. *Brain Research*, **776**(1–2), 51–60.

Whitaker-Azmitia, PM., Murphy, R. & Azmitia, E.C. (1990). S-100 protein is released from astroglial cells by stimulation of 5-HT1a receptors and regulates development of serotonin neurons. *Brain Research*, **528**, 155–160.

Winslow, J.T. & Insel, T.R. (1991). The infant rat separation paradigm: a novel test for novel anxiolytics. *Trends in Pharmacological Sciences*, **12**(11), 402–404.

Yan, W., Wilson, C.C. & Haring, J.H. (1997a). 5-HT1A receptors mediate the effects of 5-HT on developing dentate granule cells. *Developmental Brain Research*, 98, 185–190.

Yan, W., Wilson, C.C. & Haring, J.H. (1997b). Effects of neonatal serotonin depletion on the development of rat dentate granule cells. *Developmental Brain Research*, **98**, 177–184.

II.7
Catecholamines Before
Nervous Activity

SUZANNE ROFFLER-TARLOV and MARIBEL RIOS

ABSTRACT

The lack of catecholamines during prenatal development is fatal. Here we review the evidence that catecholamines, especially norepineph-rine, are essential during mid-gestation. The need for norepinephrine becomes critical soon after the expression of tyrosine hydroxylase (TH), the enzyme that is essential for the production of normal content of the catecholamines (dopamine, norepinephrine, and epinephrine). The evidence compiled here comes mainly from the observations made of several groups of murine mutations that lack the rate-limiting enzyme for catecholamine synthesis, TH, or that lack dopamine β-hydroxylase, the enzyme that converts dopamine to norepinephrine. Other very informative mutants can make norepinephrine but not dopamine. Death of TH-null fetuses occurs during the first developmental phase of sympathetic neuron development, long before the sympathetic or the central catecholaminergic systems are established and are functional. Thus the fatalities occur before nervous activity begins and they have unmasked the critical need for catecholamines during mid-gestation. These mice show that catecholamines are not necessary for the formation of neurons and their projections: they emphasize the unequivocal requirement for catecholamines for embryonic survival.

INTRODUCTION

Mother Nature recycles. The catecholamines are an example of her careful strategy. She uses catecholamines for an as-yet-undefined but essential purpose even before the nervous system is established during the fetal development of mammals. That purpose taken care of, the catecholamines are not discarded but are conserved for their better-understood roles in the sympathetic and central nervous systems. The fact that catecholamines are essential during fetal life has

A.F. Kalverboer and A. Gramsbergen (eds.), Handbook of Brain and Behaviour in Human Development, 165–184

become apparent only recently with the creation of embryonic mice lacking cate-cholamines due to strategies that genetically eliminate the enzymes necessary for the synthesis of catecholamines. This chapter was kindled by the reports of the fact that the lack of catecholamines during fetal life is deadly.

THE CATECHOLAMINES AND THEIR POSTNATAL FUNCTIONS

The catecholamines are the neurotransmitters and neurohormones: dopamine, norepinephrine and epinephrine (see Figure 1). After birth, these small molecules are synthesized in peripheral neurons that are part of the sympathetic nervous system and in the adrenergic and dopaminergic neuronal systems in the brain.

In the periphery the catecholamine-containing cells arise from a common pre-cursor in the neural crest and are closely related in function. The neurotrans-mitter for most postganglionic sympathetic neurons is norepinephrine, while both norepinephrine and epinephrine are released from the adrenal chromaffin cells. Stimulation of the sympathetic nervous system triggers changes in heart rate, blood pressure, sweating, catabolic metabolism, and slowing of functions that are not mobilized during stress. Such stimulation mobilizes the "fight or flight" response originally described by Cannon & Rosenblueth (1937). The mature catecholamine-containing systems may be the best-understood of all neuronal systems having been the focus of synthesis, regulation, localization, and functional studies since the last century. Most studies have been made of the adult, and the functions described have relevance for mature animals.

In the mature brain the catecholamine-containing neurons are clustered in many widely separated cell groups in midbrain, medulla, pons, hypothalamus, retina, and olfactory bulb. Each of the catecholamine-containing groups in mature brain are derived independently, from separate germinal zones in the neural tube. With the exception of those in the olfactory bulb the catecholamine-containing neurons are born during fetal life but most do not function as mature neurons until later. For example, the dopamine-containing cells in the substantia nigra and ventral tegmental nucleus in the midbrain are generated during mid-gestation in mouse (Kawano *et al.*, 1995; Taber Pierce, 1973) but their con-nections and functions are not mature until until about a month after birth (see Roffler-Tarlov & Graybiel, 1987 and references therein).

The functions of the various central catecholamine-containing cell groups are diverse. For example, the dopamine-containing neurons of the substantia nigra in the midbrain modulate motor function, whereas the adjacent dopamine-containing neurons in the ventral tegmental area are thought to influence affect. The dopamine-containing neurons in the olfactory bulb modulate information from primary sensory neurons. The noradrenergic neurons in the lower brain-stem control sleep cycles, coordinate sensory traffic from peripheral organs, and may modify non-catecholamine neurons in the cerebral cortex and cerebellum and, no doubt, a lot else. Probably most of these functions do not become critical until after birth.

A good deal of information regarding catecholamine function has been acquired by attempts to eliminate them and/or the nervous systems in which they serve. Postnatally, catecholamine stores have been reduced through experimental

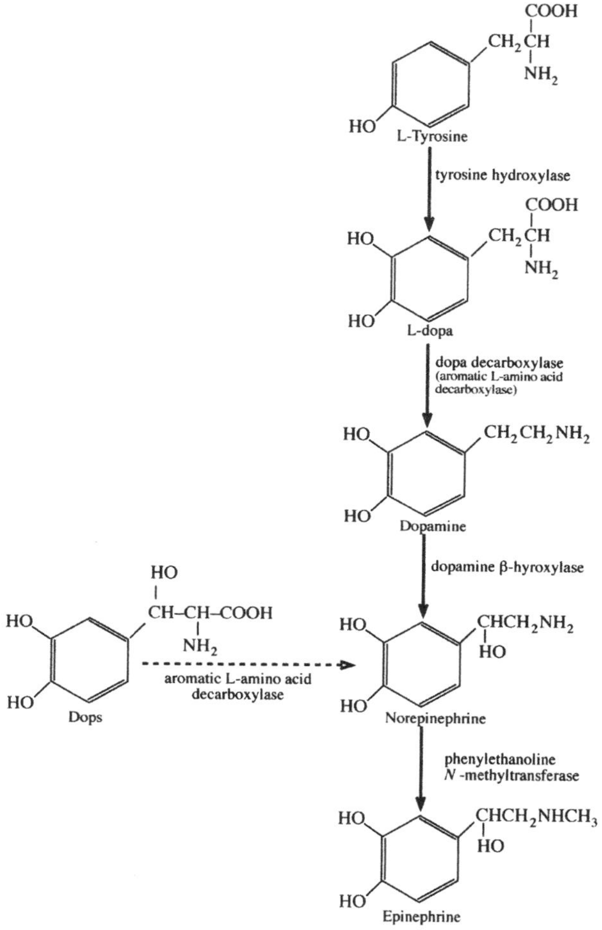

Figure 1 Catecholamine synthesis. The dominant pathway for the biosynthesis of catecholamines (dopamine, norepinephrine, and epinephrine) is indicated by continuous dark arrows. For the rescue of catecholamine-deficient mutants, catecholamines have been supplied to the fetus by the administration of dihydroxyphenyalanine (L-dopa) or dihydroxyphenylserine (dops) to the pregnant dams. Dihydroxyphenylserine supplied to the fetus is converted to norepinephrine; that pathway is indicated by dotted arrows. The decarboxylation of dops by aromatic L-amino acid decarboxylase probably does not occur under normal conditions.

use of various neurotoxins that destroy the catecholamine-containing neurons quite specifically and with the administration of drugs that deplete cate-cholamines from their neuronal stores such as reserpine. The use of the neurotoxins 6-hydroxydopamine and 1-methyl-4-phenyl-1,2,3,6-tetrahydropyridine has provided bountiful information about the central noradrenergic and dopaminergic nervous systems (see Bruno *et al.*, 1984; Koob *et al.*, 1978; Laverty

& Taylor, 1970; Smith *et al.*, 1973; Zigmond *et al.*, 1984) and the peripheral nervous system. One of the lessons learned from neurotoxin-induced destruction of catecholamine neuronal structures is that these systems are remarkably resilient. Sometimes defects caused by the even massive reduction of catecholamine-containing neurons are so rapidly compensated for that they cannot be detected after a short amount of time after neurotoxin treatment unless the poisoned animals are stressed (Zigmond *et al.*, 1984). The administered neurotoxins never destroy the whole of the adrenergic systems or dopaminergic systems, however, and their administration is not fatal to the animal. Prenatally, however, administration of neurotoxins has failed to eliminate the nervous systems because other systems that make them destructive are not yet developed.

THE SYNTHESIS OF THE CATECHOLAMINES

The catecholamines are synthesized from tyrosine in selected central and peripheral neurons and in the adrenal medulla by the sequential action of a series of enzymes in the synthetic pathway illustrated in Figure 1. This is the major pathway for synthesis in the mature nervous systems and there is no evidence that it differs in the fetus. The pathway illustrated in Figure 1 was originally proposed by Blaschko (1939) and was finally demonstrated many years later, after Nagatsu *et al.*, (1964) were able to isolate tyrosine hydroxylase (TH). TH is the first and the rate-limiting enzyme of the pathway catalysing the conversion of tyrosine, acquired from the diet, to L-dihydroxyphenylalanine (L-dopa). In the adult the presence of TH is confined almost entirely to catecholaminergic neurons. L-Dopa is a substrate for dopa-decarboxylase which catalyses its conversion to dopamine. Dopa-decarboxylase, which is aromatic acid decarboxylase, is ubiquitously distributed but is high in monoaminergic neurons. These synthetic steps take place in the cytoplasm of sympathetic neurons and in populations of neurons in the brain where TH is localized. In the sympathetic nervous system dopamine is chiefly a precursor for norepinephrine. In the brain dopamine serves as a neurotransmitter in the neuronal groups that do contain TH, but in which dopamine β–hydroxylase (DβH) is absent. DβH, which is confined to noradrenergic and adrenergic neurons and to chromaffin cells in the adrenal medulla, catalyses the conversion of dopamine to norepinephrine. Norepinephrine is converted to epinephrine by phenyl-*N*-methyl transferase, which is specifically localized in the adrenergic cells of the adrenal gland and in a few clusters of neurons in brain.

FETAL DEATH IN TH-NULL MICE

As TH is thought to be the enzyme essential for the synthesis of catecholamines, absence of TH activity would be expected to deplete the body of catecholamines. The TH-deficient mice discussed here were created by elimination of the TH gene by gene targeting using homologous recombination in embryonic stem (ES) cells (see Capecchi, 1994; Papaioannou & Johnson, 1993, for methods). Gene targeting first requires the cloning of the relevant gene, in this case the mouse

TH gene. Then, one of the endogenous TH alleles was replaced by a faulty copy of the gene that lacks essential coding sequence. This genetic manipulation was done in pluripotent mouse ES cells by insertion of a DNA targeting construct containing the mutant TH into the TH locus. The inserted gene construct contained a marker gene so that the ES cells that had incorporated it could be selected. ES cells heterozygous for the mutant TH allele were injected into a normal host blastocyst which was then populated by a mixture of its own cells and the mutant ES cells. These chimaeric blastocysts were implanted into pseudopregnant females and the pups born were a mixture of normal cells and the TH-deficient cells. The TH-null allele was transmitted to offspring by those chimaeric mice in which the modified ES cells had populated the germ line. The chimaeric mice were crossed with normal mice to create heterozygous carriers which were crossed to produce TH-null animals.

Tyrosine hydroxylase-deficient lines of mice on pigmented hybrid backgrounds have been created in two different countries, in three different laboratories using four different targeting constructs. All of these mutations show the same result: lack of TH is fatal to embryos (Kobayashi *et al.*, 1995; Rios *et al.*, 1999; Zhou *et al.*, 1995). Although, at least in theory, there are several alternative pathways to catecholamines (see Iversen, 1967), no compensatory pathways come to the fore in the knockout mice to eliminate the need for TH. Most of the TH-null mice created are on pigmented backgrounds and thus might be expected to benefit from the activity of tyrosinase, the enzyme which also converts tyrosine to dopa that is part of the melanin pathway. In fact we have found small amounts of norepinephrine in our TH-nulls from the pigmented line. We compared the fate of the pigmented TH-nulls with that of an additional mutant mouse that lacks both TH and tyrosinase. We made these mice by backcrossing heterozygous carriers for the TH-null mutation onto albino ICR mice. The albino mice are homozygous tyrosinase (C-locus)-deficient (Rios *et al.*, 1999). We found that the small amount of catecholamine that we find in our TH-null embryos in our pigmented line confers no advantage for survival; in fact the albino TH-null mice survive longer *in utero*.

Although the mutation is lethal for homozygotes in all the lines produced, the timing of fetal death is somewhat different in these various lines of TH-null mice. Some are a bit more robust than others and genetic background matters. We remind the reader that the gestation period in mice is usually about 18.5 days and that, in calculating embryonic age, the day after the night of conception is counted as embryonic day (E)0.5. The least vigorous of the TH-null embryos appear to be those on a pigmented background created in our laboratory in which no live-born TH-null has ever been found (Rios *et al.*, 1999). The most vigorous may be the TH-null embryo created with the same targeting construct, the albino TH-null. As many as 20–3)% of expected TH-null mice are born from this line, though they do not survive until weaning. Zhou *et al.* reported that 80% of their pigmented TH-null mutants die between E11.5 and E15.5. In their examination of staged embryonic litters Zhou *et al.* found that the TH-null embryos began to die after E11.5 (Zhou *et al.*, 1995). If no lethality had occurred *in utero*, 25% of embryos would be expected to be TH-null; however, at E15.5 only 5% of viable embryos harvested were TH-null. Of the mice born in this line 2.6% were TH-null and none of these survived weaning (Zhou *et al.*, 1995). The pigmented

TH-null mice described by Rios *et al*. were slightly different in terms of vitality in that none survived until birth, as mentioned above, and also the mutation is lethal even earlier (Rios *et al*., 1999). During examinations of embryonic litters in both our pigmented and albino lines a few TH-null embryos have been identified that died at at E9.5, which is only 1 day after TH mRNA can be first detected in mouse embryos (Kobayashi *et al*., 1995; Thomas *et al*., 1995). Demonstrating the earlier vulnerability of the pigmented TH-null line, we found that at E12.5, only 40% of the expected number of pigmented TH-null embryos are present, whereas 84% of expected numbers of the albino TH-nulls are still alive. Koyboyashi *et al*. reported death of their pigmented TH-null embryos to begin at E12.5 and found that at E18.5, in litters that were delivered by caesarean section, 25% of the expected numbers of TH-null fetuses were alive, indicating that in their line many TH-null fetuses live until birth (Kobayashi *et al*., 1995). There is no information concerning whether these TH-null mice can survive birth, which requires a surge of catecholamine (Slotkin & Seidler, 1988).

THE EVIDENCE THAT LACK OF CATECHOLAMINES IS THE CAUSE OF DEATH OF TH-NULL FETUSES

The evidence points directly to the lack of catecholamines as the cause of death of the TH-null fetuses. Mice that have TH but that lack DβH also die *in utero* (Thomas *et al*., 1995). These mutants contain dopamine because of the presence of TH and DβH, whereas they lack norepinephrine and epinephrine due to the absence of DβH. Both types of mutant fetuses, those that lack TH and those that lack DβH, can be rescued until birth by supplying catecholamine precursors to the pregnant dams in their drinking water (Rios *et al*., 1999, Tafari *et al*., 1997; Thomas *et al*., 1995; Zhou *et al*., 1995). This demonstrates that mice without enzymes but with catecholamines will survive. The most efficient survival reported occurs after the administration of L-dopa in the drinking water of the TH–/+ pregnant dams. The administration of L-dopa is a method for restoring at least some of all types of catecholamines to the fetuses (see Figure 1). Zhou *et al*. reported that 100% of expected TH-null pups survive with this supplement, and Rios *et al*. reported survival of 90% of TH-null fetuses treated with L-dopa. The use of dihydroxyphenylserene (dops) which can introduce norepinephrine but not dopamine to the fetuses (see Figure 1) is less effective (Rios *et al*., 1999).

Partial rescue of the catecholamine-deficient mutants can also be accomplished by the administration of noradrenergic agonists supplied in the drinking water of the pregnant dams. Our experience shows that 60–80% of the expected numbers of TH-null fetuses are born when the mothers have received the β–agonist isoproterenol from E8.5 until birth. The combination of α and β agonists may be even more efficient (Thomas & Palmiter, 1998) though the use of the α-adrenergic agonist alone does not rescue. The results of the use of isoproterenol, which alone is very efficient, would indicate that β–receptors may be essential for early fetal survival; however, that does not appear to be the case, for the elimination of both β1 and β2 receptors in double "knockout" mice does not affect their survival (Rohrer *et al*., 1999; Rohrer & Kobilka, 1998).

WHY DO MUTANTS THAT LACK CATECHOLAMINES DIE BEFORE BIRTH?

The cause of fetal death in TH-null mice is unknown, and the causes may be several. Catecholamines are believed to influence cardiovascular and metabolic functions in the fetus. The hearts of TH-null fetuses do not develop normally, according to the findings of Zhou et al. (1995) and Rios et al. (1999). Blood congestion in organs and blood vessels was observed in many of the homozygous mutant embryos from both laboratories, indicating that cardiac function is compromised by the lack of catecholamines. Figure 2 illustrates an example of one of the TH-null fetuses found among our litters with blood congestion in the heart and major blood vessels. About half of our living pigmented TH-null embryos looked like this at mid-gestation. Pools of blood were also present in the liver and lungs of some but not all of these animals. In addition, the hearts of many of the TH-null fetuses had dilated atria, reduced cell density in the ventricles, and thinning of the atrial wall as shown in Figure 3, indicating that lack of catecholamines alters heart development. Similar alterations were found by Zhou et al. (1995).

By contrast, Kobayashi et al. (1995) found no morphological abnormalities in the hearts of their TH mutants examined just before birth. However, when these investigators carefully characterizated cardiac function, monitoring the surface electrocardiograms from five TH-null mice that were taken by caesaerean section on E18.5, they found the heart rate to be decreased significantly in the TH-null pups compared to wild-type (Kobayashi et al., 1995). They found only a slight difference among genotypes in wave morphology. Zhou et al. (1995) also noted, after having counted heart beats of excised hearts in a saline bath, that the TH-nulls suffered from bradycardia.

Catecholamines in cardiac development

Catecholamines have long been implicated in cardiac development. Murine fetal heart tissue has been shown to be responsive to norepinephrine as early as 12 days of gestation. A 10% rise in atrial rate in response to norepinephrine was measured in E12–14 murine hearts: the rate was significantly increased by E15–16 (Wildenthal & Wakeland, 1973). These observations indicate that interaction of catecholamines with cardiac tissue precedes mature sympathetic synaptic function which is not present until after the first postnatal week (Goss, 1938; Seidler & Slotkin, 1985). In-vitro and in-vivo studies have shown that the response of cardiac cells to norepinephrine depends upon the developmental stage of the cell. Cardiac DNA synthesis and cell proliferation are ongoing processes in utero that decrease after birth (Claycomb, 1975). Norepinephrine has been shown to both induce and inhibit proliferation of cardiac myocytes and to induce their hypertrophy once they are post-mitotic. These actions appear to result from the complex regulation of cardiac muscle cell division and differentiation. Neonatal rats that had been exposed to the β-receptor blocker propranolol between 7 days of gestation and birth showed a significant decrease in cardiac DNA levels, pointing to a reduction in cell numbers in the heart (Kudlacz et al., 1990). Basal heart ornithine decarboxylase, an important enzyme for the regulation of cellular

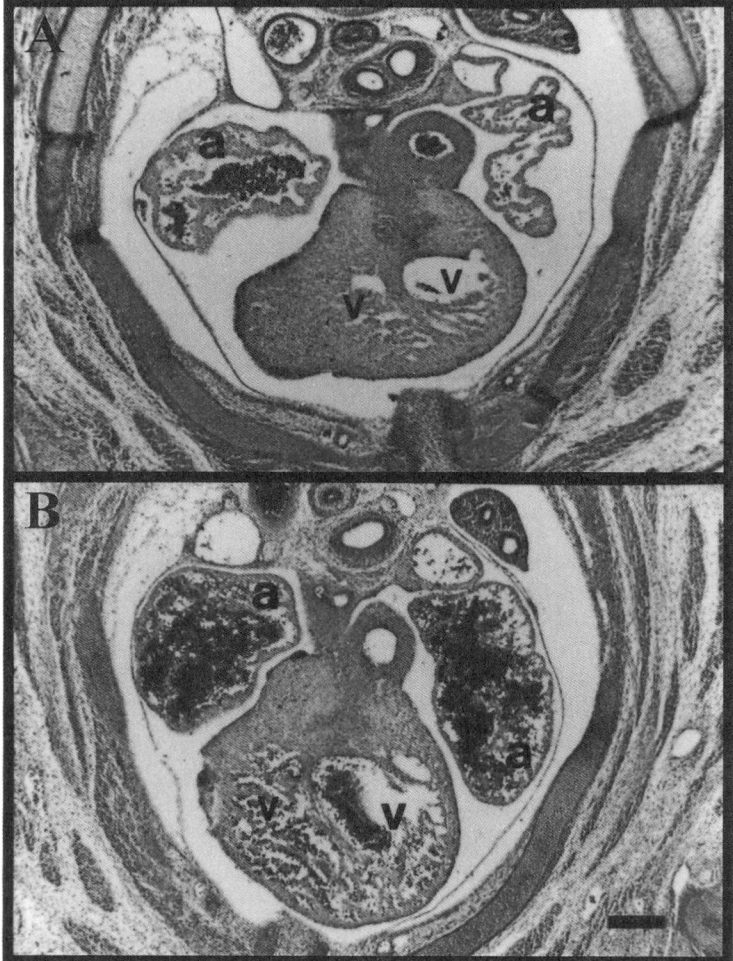

Figure 2 Congestion of blood in the heart of an E13.5 TH-null embryo. The photomicrographs show transverse sections of E13.5 wild-type (panel **A**) and TH-null (panel **B**) hearts stained with haematoxylin and eosin. Pooled blood is present in the heart of the TH-null fetus in addition to dilated atria (a) and reduced cell density in the ventricular (v) tissue. Scale bar = 250 μM.

replication and differentiation, was also altered in the propranolol-treated animals. Noradrenergic induction of cell proliferation in vitro has also been demonstrated in E14 rat fetal heart cells (Marino *et al.*, 1989). In this cell population norepinephrine induced an increase in cell numbers but not in cell size.

On the other hand, the increased cardiac cell proliferation in the absence of norepinephrine has also been reported. Sympathectomy produced by infusion of the neurotoxin 6-hydroxydopamine, during the first postnatal week, resulted in

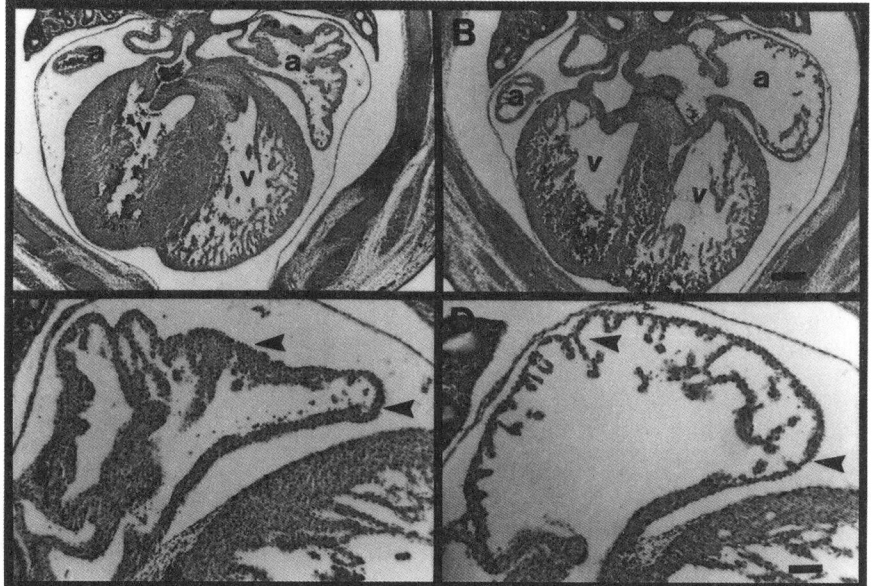

Figure 3 Reduced cell density in the heart of an E13.5 TH-null embryo. The phtotomicrographs show transverse sections of E13.5 wild-type (panels on left, **A** and **C**) and TH-null (panels on right, **B** and **D**) hearts stained with haematoxylin and eosin. Dilation of the atria (a) and reduced cell density in the ventricles (v) and septum can be seen in the mutant fetal heart (**B**). Reduction in the number of cell layers (arrow-heads) is seen in the higher-power view of the TH-null atria (**D**) compared to the wild-type (**C**). Scale bar for top panels= 200 μM; scale bar for bottom panels = 100 μM.

an increase in DNA polymerase activity and cell proliferation in newborn rats (Kugler *et al.*, 1980). Although myocardial cell division normally continues for 10–14 days after birth, it persisted beyond 20 days of age in the sympathec-tomized rats. By contrast, when perinatal heart was exposed to the β-receptor agonist isoproterenol, ongoing cell proliferation was inhibited (Claycomb, 1976). Once cardiac cells become post-mitotic they grow in size in response to catecholamines. Induction of cell hypertrophy has been demonstrated in cultured neonatal rat heart cells through activation of α_1-adrenergic receptors and subsequent induction of protein kinase C (Simpson, 1985).

The capacity of cardiac cells to demonstrate such variable responses to catecholamine alterations probably arises from differential regulation of cate-cholaminergic receptor levels and the ability of the receptors to independently regulate different signalling cascades. Differential regulation of cate-cholaminergic receptors has been demonstrated in newborn rat myocardiocytes *in vitro* and *in vivo* (Rokosh *et al.*, 1996). After chronic stimulation by norepi-nephrine, α_{1B} and α_{1D} mRNA levels were decreased, whereas those of α_{1C} were increased. Thus, the nature of the cardiac response to catecholamines may be correlated with receptor type in the heart tissue.

That cardiac failure may be at least one cause of embryonic lethality in the TH-null mutants is in some ways surprising because the mutant mice die before a functional nervous system has developed. The heart appears to require catecholamines early in development even though the first sympathetic fibres to innervate the atria do not arrive until E13.5, and they are not mature and functional until the first week after birth (Goss, 1938; Seidler & Slotkin, 1985).

Other developmental roles of catecholamines

Although the lethality in TH-null embryos is, at least in part, the result of the cardiovascular defect, other possibilites should be considered. Catecholamines are known to be involved in glucose metabolism through the inhibition of insulin release from the pancreatic islets (Yelich, 1993). They also mediate glucose release into the fetal circulation through the induction of glycogenolysis in the hepatic tissue (Moncany & Plas, 1980). In the absence of catecholamines, hyper-insulinaemia and hypoglycaemia may be induced, conditions which could have deleterious effects in a developing fetus. Reports from clinical studies indicate that infants from mothers treated with propranolol, a β-receptor blocker, during pregnancy are often hypoglycaemic and bradycardic at birth (Cottril et al., 1977; Habib & McCarthy, 1977). The hypoglycaemic state was thought to arise from hyperinsulinaemia, the lack of stimulation of glycogenolysis of skeletal muscle and increased secretion of growth hormone. Because the symptoms resulting from propranolol treatment are caused by a partial block of catecholamine action, more severe effects could be expected in the TH-null mutants that lack catecholamine activity.

TH-null fetuses do not appear to be significantly smaller than their normal littermates in utero. This could be the result of their benefiting from maternal glucose and catecholamines. Thomas et al. (1995) showed that 12% of DβH-deficient mice survived to term and grew to adulthood if they were from heterozygous crosses. However, no DβH-null mutants were born from homozygous DβH-null mothers that did not produce norepinephrine or epinephrine (Thomas et al., 1995). The implication of these observations is that maternal catecholamines present in the fetal circulation contribute to the survival of the affected fetuses, although catecholamines are known to pass the placenta only poorly. A complication for this interpretation is that the heterozygous DβH-null embryos were also vulnerable if carried by the DβH-null mothers.

Even though, in the case of the TH-nulls, maternal catecholamines and glucose are not sufficient to compensate for the lack of endogenous fetal cate-cholamines, they could prevent the mutant fetuses from significantly falling behind in fetal growth. Interestingly, L-dopa and dops-rescued TH-nulls fail to gain weight normally after birth when the drug treatment is terminated and when they no longer have access to maternal catecholamines.

WHICH CATECHOLAMINE(S) ARE NECESSARY BEFORE BIRTH?

Most of the evidence points to norepinephrine as the most crucial catecholamine for survival of the fetus. The observations that support this view come from the fate of a variety of mice which lack either dopamine specifically (Nishii *et al.*, 1998; Zhou & Palmiter, 1995) or which lack norepinephrine and epinephrine (Thomas *et al.*, 1995). The mouse made and described by Thomas *et al.*, in which DβH was eliminated by homologous recombination in embryonic stem cells, were norepinephrine-deficient and were very similar to all of the TH-null mutants in that they died *in utero* in spite of synthesizing dopamine. The DβH-null mutants were successfully rescued until birth by the provision of dops to the pregnant mothers from E8.5 onwards (Thomas *et al.*, 1995).

Like the TH-null mutants, DβH-nulls embryos have abnormal hearts but their cardiac phenotype is less dramatic. Of all DβH-null fetuses examined by Thomas at E10.5, E12.5, E13.5 and E15.5, only one was found to have pooled blood in the liver and large vessels. However, most of the mutants examined appear to have heterogeneity in myocardial cell size and cellular orientation when compared to wild-type littermates. No dilation of atrial or ventricular tissue or thinning of the atrial wall was reported for any of the DβH mutant animals examined (Thomas *et al.*, 1995), and there is no report of a test for bradycardia in these mutants.

The fate of the DβH-null mutants described above is dramatically different from those that lack only dopamine. Mutants that synthesize norepinephrine but that lack dopamine in dopaminergic cells were made using two different manipulations. The first of these creatures reported came from Zhou & Palmiter (1995), who successfully restored norepinephrine in their TH-null mice made earlier (Zhou *et al.*, 1995), by directing TH specifically to the noradrenergic systems by inserting TH in place of one of the endogenous DβH alleles. Through a clever set of manipulations involving gene targeting and breeding schemes, mice were born that were TH –/– and DBH-TH –/+, meaning that the endogenous promoter for DβH drives the TH gene in only the cells where DBH is normally active. The other DβH allele is normal, which allows production of the enzyme. This strategy provides no TH in the dopaminergic cells. Although the resulting mice are heterozygous for DβH in the adrenergic cells, they produce normal amounts of norepinephrine. At birth, all of the expected numbers of mutants are present, showing that availability of norepinephrine but not dopamine (except as a precursor to norepinephrine), has successfully rescued the mutants until birth. After birth they fail to thrive (Zhou & Palmiter, 1995).

Nishii *et al.* (1998) created a second type of norepinephrine-producing and dopamine-deficient mutant mouse. This mouse was made in a different way. These scientists who had already created the TH-null mutant discussed above (Kobayashi *et al.*, 1995) used the DβH promoter to drive a TH transgene in TH-null mice. These investigators found that the presence of the transgene effected the almost complete rescue of the mutant embryos. They also found that the presence of norepinephrine reversed the slowing of the heart which occurred in the TH-null embryos (Kobayashi *et al.*, 1995; Zhou *et al.*, 1995) and specifically, the heart rate of the TH-nulls was 230 ± 40 compared to a rate of 300 ± 30 in the wild-type mice. The mutants that expressed TH driven by the DβH promoter had a heart rate of 310 ± 30 beats per minute. The heart beats were measured in hearts taken at E18.5 by caesarean section. Quantitative measurements of catecholamines extracted from tissues taken from the perinatal mutants showed completely normal content of norepinephrine and epinephrine in brain heart and adrenal (Nishii *et al.*, 1998). Dopamine was not reported in the peripheral tissues and was about 25% of normal values in the brain of the mutants. This line of mutants, like the one from Zhou, perished during the second and third weeks after birth, indicating that life cannot perist on norepinephrine alone. The need demonstrated for norepinephrine by the early embryo probably does not apply to epinephrine also because PNMT, the enzyme necessary for the conversion of norepinephrine to epinephrine, does not become active until after the initial wave of embryonic death.

There are a couple of findings that seem inconsistent with the idea that norepinephrine is the most essential catecholamine before birth. The first is that, in our experience, TH-null embryos are more successfully rescued by L-Dopa than by dops supplements supplied at the same concentration in the pregnant mother's drinking water. L-Dopa is a precursor for all catecholamines, whereas dops would be expected to specifically supply norepinephrine and epinephrine not influencing the content of dopamine in the embryo (see Figure 1). Thus, the greater efficiency of L-dopa would indicate that dopamine also influences prenatal survival, whereas the *in-vivo* experiments described above show that embryos survive without dopamine if they have norepinephrine. It would be complicated to make conclusions about the relative survival value of catecholamines based on the rescue experiments, because only a fraction of the normal content of cate-cholamine can be furnished in this way. Also the regulation of the supply must be very abnormal as it is available only when the mother drinks. Another abnormality involved in the production of catecholamines through the adminis-tration of precursors is that the decarboxylation of dops and L-dopa may occur in many ectopic sites because of the widespread distribution of aromatic amino acid decarboxylase.

The second possibly inconsistent finding is the lack of correlation that we find between amount of catecholamine present in the fetus and its survival. We have made a comparison of catecholamine content between TH-null fetuses from our pigmented and our albino lines. We found that the TH-null fetuses from the pigmented line die earlier that those from the albino line. On E12.5, 84% of expected TH-nulls were alive in the albino line, whereas only 40% of expected TH-nulls were alive in the pigmented line. Yet the TH-null fetuses in the pigmented line contained small amounts of norepinephrine and those from the albinos line had no detectable norepinephrine. The greater vitality of the albino TH-nulls is connected with something else, and the small amount of norepinephrine found in the pigmented TH-null fetuses during mid-gestation just is not enough.

THE SOURCE OF CATECHOLAMINES IN THE NORMAL FETUS

As discussed above, fetuses that carry two copies of the TH-null allele begin to die as early as E9.5, just 1 day after TH mRNA is detectable in a normal fetus (Kobayashi et al., 1995; Zhou et al., 1995). In our pigmented line 60% of the expected number of TH-null fetuses are dead at E12.5. Thus, given that the TH-null fetuses die because they lack catecholamines, this would indicate that, in a normal fetus, TH is active soon after its expression, and that the complex choreography for catecholamine synthesis, including the activity of the other key synthetic enzymes for norepinephrine (DDC and DβH), is in place. It also makes apparent that the fetus needs catecholamines soon after TH is expressed. Some catecholamine may be available to the fetus by maternal transfer, but clearly this does not suffice because the TH-null embryos are exposed to the maternally derived catecholamine to the same extent as their heterozygous and wild-type littermates. A key question then is where the catecholamines are produced and released in normal embryos during mid-gestation.

The source of catecholamines cannot be the sympathetic nervous system because there is no sympathetic nervous system in operation at the time during which the the vitality of the fetus begins to depend upon catecholamines. Catecholamine dependence occurs well in advance of the development of the sympathetic nervous system and of the development of the noradrenergic and dopaminergic transmitter systems in the brain. The sympathetic innervation of the heart, for example, occurs much later; the first sympathetic fibres to reach the atria do not appear until E13.5 and they do not function until the first week after birth. Like the sympathetic innervation, the adrenal gland is also an unlikely source because it also forms later (see Tischler & Coupland, 1994, for review). Like the sympathetic ganglia, the adrenal medulla is formed by neural crest-derived progenitors that migrate ventrally from the thoracic region. The invasion of the adrenal cortex *anlage* by the first of the adrenal medullary cells occurs at about embryonic day 13 in the rat and a little earlier in the mouse.

If the sympathetic nervous system is not a source of fetal norepinephrine, what is? As noted above, TH expression is first detected in mouse embryos at E8.5 by reverse transcriptase-PCR (Kobayashi *et al.*, 19956; Zhou *et al.*, 1995) and the TH-null mutations have demonstrated that catecholamines are needed very soon thereafter. A possible source of catecholamines in the fetuses are neuroblasts in which the TH-message can be detected early. Neuroblasts destined to become catecholamine-containing neurons in the sympathetic system and in the central nervous system express TH (Cochard *et al.*, 1978; Rothman *et al.*, 1980; Specht *et al.*, 1981a). These are the recently born cells of neural crest derivation that migrate to form sympathetic structures in the peripheral nervous system and those derivatives of the neural tube that migrate from their place of birth in the brain to form the nuclei from which the central catecholamine-releasing pathways originate. In addition there are neuroblasts that transiently express TH during early development (see Cochard *et al.*, 1978; Francis & Landis, 1999). These cells may release catecholamines into the circulation long before synaptic connections are established: there is no evidence, however, that they do.

Information about the possible source of catecholamines during prenatal development is derived from the immunocytochemical localization of TH and of localization of catecholamines detected by histofluorescence in rat and mouse embryos during various intervals during embryogenesis (see Cochard *et al.*, 1978; Jonakait *et al.*, 1989; Schimmel *et al.*, 1999; Specht *et al.*, 1981a; Teitelman *et al.*, 1979, 1981). This discussion will focus mainly on a recent study that includes information concerning anatomical location of TH immunoreactivity just after expression of TH mRNA in the fetus until birth in the normal mouse (Schimmel *et al.*, 1999). It was demonstrated by Schimmel *et al.* (1999), and is seen in the *Chemoarchitectonic atlas of the developing mouse brain* (Jacobwitz & Abbot, 1998) that the earliest expression pattern for TH during mid-gestation is different from that found during late gestation and in postnatal animals. Thus, the catecholamines may have many embryonic sources, some of which disappear in the adult.

The earliest histochemical detection of TH in the embryonic mice studied was on E10.5 when the most robust expression was found in the ventral lateral spinal cord (Figure 4A). In this region the anterior horn cells, the motor neurons for the somatic motor system, develop. Anterior horn cells do not express TH in mature animals; their neurotransmitter is acetylcholine. Yet, for just a few days in the embryonic mouse, between E10.5 and about E12, TH is present in the region of the motor neurons and in fibres that project towards the somites (Figure 4B). During this window, TH expression is found along the entire rostral to caudal length of the spinal cord and extends into the brainstem. Whether the transient TH in the anterior horn is part of a catecholamine-producing and releasing pathway is not known. In fact, a report by Foster *et al*. (1985) describes detection of TH without detectable DβH, aromatic amino acid decarboxylase, or catecholamine histofluorescence in the anterior horn of the embryonic rat spinal cord between E11 and E13. On the other hand, a reporter gene can be directed by the DβH promoter to these locations in the embryonic mouse spinal cord, indicating that the catecholamine pathway may be expressed here during this developmental period (Kapur *et al*., 1991).

Following the demise of TH expression in the spinal cord, TH is most pronounced immunocytochemically in the sympathetic ganglia. TH was detectable in developing sympathetic ganglia in the E12–E13 embryos and could be observed in two chains flanking the aorta at E13.5. Certainly these cells do produce catecholamine, which can be seen by glyoxylic acid-induced histofluorescence in thoracic ganglia at this time in normal mice (for example see Rios *et al*., 1999). Whether it is released from the cell bodies is not known, but the ganglia are a potential source of norepinephrine.

TH immunoreactivity is also evident within a subset of cells in the dorsal root ganglia at this time. The speckled pattern of TH staining in the dorsal root

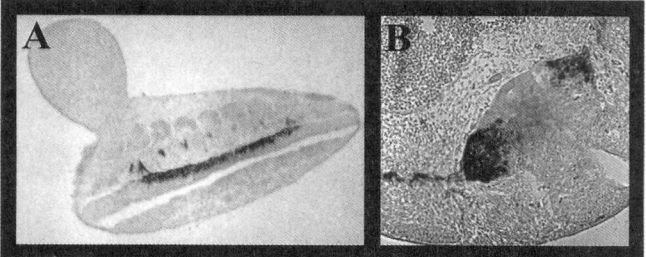

Figure 4 TH immunoreactivity during mid-gestation in normal mouse embryos. **A**: Longitudinal section of E10.5 embryo shows TH staining in the developing spinal cord. This was the earliest unambiguous staining seen in the study of Schimmel *et al*. (1999), from which the figure is taken. **B**: TH immunoreactivity in a transverse section of an E11.5 spinal cord that shows TH staining in the ventrolateral region of the spinal cord and TH-positive fibres projecting laterally. This is the district in which the anterior horn cells develop.

ganglia is present also in cranial sensory ganglia, including the trigeminal, nodose and petrosal ganglia, in parasympathetic ganglia, and in enteric ganglia. Some of these are locations of transient TH expression in which TH disappears by about E13 in mouse. Expression in the trigeminal ganglia and dorsal root ganglia, however, extended through later periods. These sites of transient TH expression have been noted in both rat and mouse, and DβH has also been found in some of them (Baetge et al., 1990; Baetge & Gordon, 1989; Jonakait et al., 1984; Katz et al., 1983; Lawrence & Burden, 1973; Leblanc & Landis, 1989; Price, 1983; Teitelman et al., 1981).

After E14 TH, expression in the central nervous system resembles, to a great extent, the pattern seen in the adult rodent (Specht et al., 1981b). TH immunoreactivity is evident in the dopamine-containing cell groups in the ventral midbrain, in hypothalamic nuclei, in the olfactory bulb, in the presumptive noradrenergic brainstem nuclei. In the periphery, TH staining appears in a subset of cells within sensory ganglia and is intense in the ganglia of the sympathetic chain. Immunoreactivity also appears in the enteric ganglia and the adrenal primordium. TH staining was observed in cells in the outer layer of the duodenum and in the pancreatic primordium (Baetge et al., 1990; Teitelman & Lee, 1987).

An additional source of embryonic catecholamines may be a transient population of TH-positive cells known as the paraganglia or extramedullary chromaffin cells. Extramedullary chromaffin cells of neural crest origin are distributed along the sympathetic and parasympathetic chains (Tischler, 1995). The sympathetic paraganglia are found in connective tissue in or near the walls of pelvic organs, in the mesenteric artery, the aortic bifurcation, and the organ of Zuckerkandl, the only macroscopic sympathetic paraganglion. Parasympathetic paraganglia are located along the cranial and thoracic branches of the glossopharyngeal and vagus nerves in the carotid bodies, the middle ear, near the bases of the great vessels of the heart and in the interatrial septum (Coupland, 1960). Paraganglia have been detected between the atria of birds (Kose, 1902) and among myocardiocyte bundles (Gobbi et al., 1991).

The distribution and number of paraganglionic cell groups changes as development progresses. Examples of this are the organ of Zuckerkandl, which in various mammalian species enlarges during and until the end of fetal life, but regresses after birth (De Gallardo et al., 1974) and the carotid bodies, which increase in size between infancy and adult life (Zak & Lawson, 1983). Even though it is known that paraganglia are present in fetal tissue, the time when they first appear has not been examined carefully. The physiological role of paraganglionic cells also remains unclear, but they are known to produce dopamine, norepinephrine, and epinephrine and neuropeptides such as enkephalin (Zak & Lawson, 1983). Because paraganglia synthesize catecholamines and express neuroendocrine markers (Gobbi et al., 1991), they are thought to have an endocrine role in the developing fetus. As is the case with other endocrine tissues, fetal paraganglia have been found to be associated with well-developed capillary sinusoidal networks . Higher catecholamine content has been detected in the organ of Zuckerkandl compared to the adrenal medulla in fetal humans, rhesus monkey, dog, cat, rabbit, and guinea pig (Niemineva & Pekkarinen, 1953; Shepherd & West, 1952; West et al., 1953). Thus, catecholamines necessary for fetal development are perhaps released into the circulation by the paraganglionic tissue. However, in his study of TH distribution in the embryonic mouse, Schimmel did not see a lot of such tissue, and found no evidence of paraganglia in the embryonic mouse heart.

OF MICE AND MEN

Catecholamines, so necessary for embryonic life, have no back-up plan. There is no redundancy, no alternative pathway that can produce enough catecholamine to sustain survival. It is not surprising, then, that no humans have been found who lack TH. It is remarkable, though, that a handful of people have been reported who lack DβH activity (Man in't Veld *et al.*, 1987; Robertson *et al.*, 1986, 1991). Like the rare DβH-null mice that survive birth, these patients have no norepinephrine and epinephrine, and have greater than normal dopamine which, in the case of the humans, has been measured in plasma, cerebrospinal fluid and urine (Man in't Veld *et al.*, 1987; Robertson *et al.*, 1986). The perinatal period has been recorded as being particularly perilous for these patients (Man in't Veld *et al.*, 1987) as would be expected given the great dependence upon catecholamines during the period around birth (Slotkin & Seidler, 1988).

The lives of human neonates lacking DβH are complicated by hypotension, hypoglycaemia, hypothermia, often by vomiting, and by some seizures. Later any exertion becomes too taxing, and symptoms tend to become more severe in late adolescence and early adulthood. Patients must adopt inventive strategies to maintain an upright posture. The most effective therapy for DβH-deficient patients is the administration of dops, which shows evidence of providing norepinephrine for the neuronal stores of these people (Robertson *et al.*, 1991). Robertson and colleagues found that the administration of dopa increased plasma and urinary levels of norepinephrine. The restoration of norepinephrine was correlated with increase in mean arterial blood pressure and ability to maintain upright posture (Biaggioni *et al.*, 1987).

References

Baetge, G. & Gordon, M.D. (1989). Transient catecholaminergic (TC) cells in the vagus nerves and bowel of fetal mice: relationship to the development of enteric neurons. *Developmental Biology,* **132**, 189–211.

Baetge, G., Pintar, J. & Gershon, M. (1990). Transient catecholaminergic cells in the bowel of the fetal rat: precursors of noncatecholaminergic enteric neurons. *Developmental Biology,* **141**, 353–380.

Biaggioni, I., Hollister, A.S. & Robertson, D. (1987). Dopamine in dopamine-beta-hydroxylase deficiency. *New England Journal of Medicine,* **317**, 1415.

Blaschko, H. (1939). The specific action of L-dopa decarboxylase. *Journal of Physiology (London),* **96**, 50–51.

Bruno, J.P., Synder, A.M. & Stricker, E.M. (1984). Effect of dopamine-depleting brain lesions on suckling and weaning in rats. *Neuroscience,* **98**, 165–168.

Cannon, W. & Rosenblueth, T. (1937). *Autonomic neuro-effector systems, XIV.* New York: Macmillan.

Capecchi, M.R. (1994). Targeted gene replacement. *Scientific American,* **207**, 52–59.

Claycomb, W. (1975). Biochemical aspects of cardiac muscle differentiation. Deoxyribonucleic acid synthesis and nuclear and cytoplasmic deoxyribonucleic acid polymerase activity. *Journal of Biological Chemistry,* **250**, 3229–3235.

Claycomb, W. (1976). Biochemical aspects of cardiac muscle differentiation. Possible control of deoxyribonucleic acid synthesis and cell differentiation by adrenergic innervation and cyclic adenosine 3':5'-monophosphate. *Journal of Biological Chemistry,* **251**, 6082–6089.

Cochard, P., Goldstein, M. & Black, I.B. (1978). Ontogenetic appearance and disappearance of tyrosine hydroxylase and catecholamines in the rat embryo. *Proceedings of the National Academy of Sciences, USA,* **75**, 2986–2990.

Cottrill, C., McAllister, R., Gettes, L. & Noonan, J. (1977). Propanolol therapy during pregnancy, labor, and delivery: evidence for transplacental drug transfer and impaired neonatal drug disposition. *Journal of Pediatrics,* **91**, 812–814.

Coupland, R. (1960). The post-natal distribution of the abdominal chromaffin tissue in the guinea pig, mouse and white rat. *Journal of Anatomy,* **94**, 244–256.

De Gallardo, M., Freire, F. & Tramezzani, J. (1974). The organ of Zuckerkandl of the newborn rat and its postnatal involution. *Acta Physiologic Latinoamericana,* **24**, 290–304.

Foster, G.A., Schultzberg, M., Dahl, D., Goldstein, M. & Verhofstad, A.A.J. (1985). Ephemeral existence of a single catecholamine synthetic enzyme in the olfactory placode and the spinal cord of the embryonic rat. *International Journal of Developmental Neuroscience,* **3**, 597–608.

Francis, N.J. & Landis, S.C. (1999). Cellular and molecular determinants of sympathetic neuron development. *Annual Review of Neuroscience,* **22**, 541–566.

Gobbi, H., Barbosa, A., Teixeira, V. & Almeida, H. (1991). Immunocytochemical identification of neurondocrine markers in human cardiac paraganglion-like structures. *Histochemistry,* **95**, 337–340.

Goss, C.M. (1938). The first contractions of the heart in rat embryos. *Anatomical Record,* **70**, 505–524.

Habib, A. & McCarthy, J. (1977). Effects on the neonate of propranolol administered during pregnancy. *Journal of Pediatrics,* **91**, 808–811.

Hernoven, A. (1971). Development of catecholamine storing cells in human fetal paraganglia and adrenal medulla. *Acta Physiologica Scandinavica,* **368**, 3.

Iversen, L. (1967). *The uptake and storage of noradrenaline in sympathic nerves.* Cambridge, UK: Cambridge University Press.

Jacobwitz, D.M. & Abbot, L.C. (1998). *Chemoarchitectonic atlas of the developing mouse brain.* New York: CRC Press.

Jonakait, G.M., Markey, K.A. Goldstein, M. & Black, I.B. (1984). Transient expression of selected catecholaminergic traits in cranial sensory and dorsal root ganglia of the embryonic rat. *Developmental Biology,* **101**, 51–60.

Jonakait, G.M., Rosenthal, M. & Morrell, J.I. (1989). Regulation of tyrosine hydroxylase mRNA in catecholaminergic cells of embryonic rat: analysis by in situ hydridization. *Journal of Histochemistry and Cytochemistry,* **37**, 1–5.

Kapur, R.P., Hoyle, G.W., Mercer, E.H., Brinster, R.L. & Palmiter, R.D. (1991). Some neuronal cell populations express human dopamine B-hydroxylase-lacZ transgenes transiently during embryonic development. *Neuron,* **7**, 717–727.

Katz, D.M., Markey, K.A., Goldstein, M. & Black, I.B. (1983). Expression of catecholaminergic characteristics by primary sensory neurons in the normal adult rat *in vivo. Proceedings of the National Academy of Sciences, USA,* **80**, 3526–3530.

Kawano, H., Ohyama, K., Kawamura, K. & Nagatsu, I. (1995). Migration of dopaminergic neurons in the embryonic mesencephalon of mice. *Developmental Brain Research,* **86**, 101–113.

Kobayashi, K., Morita, S., Sawada, H., Mizuguchi, T., Yamada, K., Nagatsu, I., Hata, T., Watanabe, Y., Fujita, K. & Nagatsu, T. (1995). Targeted disruption of the tyrosine hydroxylase locus results in severe catecholamine depletion and perinatal lethality in mice. *Journal of Biological Chemistry,* **270**, 27235–27243.

Koob, G.F., Riley, S.F., Smith, S.C. & Robbins, T.W. (1978). Effects of 6-hydroxydopamine lesions of the nucleus accumbens septi and olfactory tubercle on feeding, locomotor activity, and amphetamine anorexia in the rat. *Physiological Psychology,* **92**, 917–927.

Kose, W. (1902). Uber das Vorkommen einer "Carotisdrose" und der "chromaffinen Zellen" bei Vogel. Nebst Bemerkungen uber die Kiemenspaltenderivative. *Anatomischer Anzeiger,* **22**, 162–170.

Kudlacz, E., Navarro, H., Kavlock R. & Slotkin, T. (1990). Regulation of postnatal beta-adrenergic receptor/adenylate cyclase development by prenatal agonist stimulation and steroids: alterations in rat kidney and lung after exposure to terbutaline or dexamethasone. *Journal of Developmental Physiology,* **14**, 273–281.

Kugler, J., Gillette, P., Graham, S., Garson, A., Goldstein, M. & Thompson, H. (1980). Effect of chemical sympathectomy on myocardial cell division in the newborn rat. *Pediatric Research*, **14**, 881–884.

Laverty, R. & Taylor, K.M. (1970). Effects of intraventricular 2,4,5-trihydroxyphenylethylanine (6-hydroxydopamine) on rat behavior and brain catecholamine metabolism. *Research Communications in Chemistry and Pathology*, **40**, 836–846.

Lawrence, I.E. & Burden, H.W. (1973). Catecholamines and morphogenesis of the chick neural tube and notochord. *American Journal of Anatomy*, **137**, 199–208.

Leblanc, G.G. & Landis, S.C. (1989). Differentiation of noradrenergic traits in the principal neurons and small intensely fluorescent cells of the parasympathetic sphenopalatine ganglion of the rat. *Developmental Biology*, **131**, 44–59.

Man in't Veld, A.J., Boomsma, F., Moleman, P. & Schalekamp, M.A.D.H. (1987). Congenital dopamine-B-hydroxylase deficiency: a novel orthostatic syndrome. *Lancet*, **1**, 183–187.

Marino, T., Walter, R., D'Ambra, K. & Mercer, W. (1989). Effects of catecholamines on fetal rat cardiocytes *in vitro*. *American Journal of Anatomy*, **186**, 127–132.

Moncany, M. & Plas, C. (1980). Interaction of glucagon and epinephrine in the regulation of adenosine 3'5'-monophosphate-dependent glycogenolysis in the cultured fetal hepatocyte. *Endocrinology*, **107**, 1667–1675.

Nagatsu, T., Levitt, M. & Udenfriend, S. (1964). Tyrosine hydroxylase: the initial step in norepinephrine biosynthesis. *Journal of Biological Chemistry*, **239**, 2910–2917.

Niemineva, K. & Pekkarinen, A. (1953). Determination of adrenaline and noradrenalin in human fetal adrenals and aortic bodies. *Nature*, **171**, 436.

Nishii, K., Matsushita, N., Sawada, H., Sano, H., Noda, Y., Mamiya, T., Nabeshima, T., Nagatsu, I., Hata, T., Kiuchi, K., Yoshizata, H., Nakashima, K., Nagatsu, T. & Kobayashi, K. (1998). Motor and learning dysfunction during postnatal development in mice defective in dopamine neuronal transmission. *Journal of Neuroscience Research*, **54**, 450–464.

Papaionnou, V. & Johnson, R. (1993). Production of chimeras and genetically defined offspring from targeted ES cells. In A. Joyner (Ed.), *Gene targeting: a practical approach*. New York: Oxford University Press.

Price, J. (1983). A subpopulation of rat dorsal root ganglion neurones is catecholaminergic. *Nature*, **301**, 241–243.

Rios, M., Habecker, B., Sasaoka, T., Eisenhofer, G., Tain, H., Landis, S., Chikaraishi, D. & Roffler-Tarlov, S. (1999). Catecholamine synthesis is mediated by tyrosinase in the absence of tyrosine hydroxylase. *Journal of Neuroscience*, **19**, 3519–3526.

Robertson, D., Goldberg, M.R., Hollister, A.S., Onrot, J., Wiley, R., Thompson, J.G. & Robertson, R.M. (1986). Isolated failure of autonomic noradrenergic neurotransmission: evidence for impaired beta-hydroxylation of dopamine. *New England Journal of Medicine*, **314**, 1494–1497.

Robertson, D., Haile, V., Perry, S.E., Robertson, R.M., Phillips, J.A. III & Biaggioni, I. (1991). Dopamine β-hydroxylase deficiency: a genetic disorder of cardiovascular regulation. *Hytertension*, **18**, 1–8.

Roffler-Tarlov, S. & Graybiel, A. (1987). The postnatal development of the dopamine-containing innervation of dorsal and ventral striatum: effects of the weaver gene. *Journal of Neuroscience*, **7**, 2364–2372.

Rohrer, D.K. & Kobilka, B.K. (1998). Insights from *in vivo* modification of adrenergic receptor gene expression. *Annual Review of Pharmacology and Toxicology*, **38**, 351–373.

Rohrer, D.K., Chruscinski, A., Schauble, E.H., Bernstein, D. & Kobilka B.K. (1999). Cardiovascular and metabolic alterations in mice lacking both in B1- and B2-adrenergic receptors. *Journal of Biological Chemistry*, **274**, 16701–16708.

Rokosh, D., Stewart, A., Chang, K., Bailey, B., Karliner, J., Carmacho, S., Long, C. & Simpson, P. (1996). Alpha1-adrenergic receptor subtype mRNAs are differentially regulated by alpha1-adrenergic and other hypertrophic stimuli in cardiac myocytes in culture and *in vivo*. Repression of alpha1B and alpha1D but induction of alpha1C. *Journal of Biological Chemistry*, **271**, 5839–5843.

Rothman, T., Specht, L., Gershon, M., Joh, T., Teitelman, G., Pickel, V. & Reis, D. (1980). Catecholamine biosynthetic enzymes are expressed in replicating cells of the peripheral but not the central nervous system. *Proceedings of the National Academy of Sciences, USA*, **77**, 6221–6225.

Schimmel, J.J., Crews, L., Roffler-Tarlov, S. & Chikaraishi, D.M. (1999). 4.5 KB of the rat tyrosine hydroxylase 5′ flanking sequence directs tissue specific expression during development and contains consensus sites for multiple transcription factors. *Molecular Brain Research*, **74**, 1–14.

Seidler, F.J. & Slotkin, T.A. (1985). Adrenomedullary function in the neonatal rat: responses to acute hypoxia. *Journal of Physiology (London),* **358**, 1–16.

Shepherd, D. & West. G. (1952). Noradrenaline and accessory chromaffin tissue. *Nature,* **170**, 42.

Simpson, P. (1985). Stimulation of hypertrophy of cultured neonatal rat heart cells through an alpha 1-adrenergic receptor and induction of beating through an alpha 1- and beta 1-adrenergic receptor interaction. Evidence for independent regulation of growth and beating. *Circulation Research,* **56**, 884–894.

Slotkin, T.A. & Seidler, F.J. (1988). Adrenomedullary catecholamine release in the fetus and newborn: secretory mechanisms and their role in stress and survival. *Journal of Developmental Physiology,* **10**, 1–16.

Smith, R.D., Cooper, B.R. & Breese, G.R. (1973). Growth and behavioral changes in developing rats treated intracisternally with 6-hydroxydopamine: evidence for involvement of brain dopamine. *Journal of Pharmacology and Experimental Therapeutics,* **185**, 609–619.

Specht, L.A., Pickel, V.M., Joh, T.H. & Reis, D.J. (1981a). Light-microscopic immunocytochemical localization of tyrosine hydroxylase in prenatal rat brain. I. Early ontogeny. *Journal of Comparative Neurology,* **199**, 233–253.

Specht, L.A., Pickel, V.M., Joh, T.H. & Reis, D.J. (1981b). Light-microscopic immunocytochemical localization of tyrosine hydroxylase in prenatal rat brain. II. Late ontogeny. *Journal of Comparative Neurology,* **199**, 255–276.

Taber Pierce, E. (1973). Time of origin of neurons in the brain stem of the mouse. *Progress in Brain Research,* **40**, 53–65.

Tafari, A.T., Thomas, S.A. & Palmiter, R.D. (1997). Norepinephrine facilitates the development of the murine sweat response but is not essential. *Journal of Neuroscience,* **17**, 4275–4281.

Teitelman, G. & Lee, J. (1987). Cell lineage analysis of pancreatic islet cells development: glucagon and insulin cells arise from catecholaminergic precursors present in the pancreatic duct. *Developmental Biology,* **121**, 454–466.

Teitelman, G., Baker, H., Joh, T.H. & Reis, D.J. (1979). Appearance of catecholamine-synthesizing enzymes during development of rat sympathetic nervous system: possible role of tissue environment. *Proceedings of the National Academy of Sciences, USA,* **76**, 509–513.

Teitelman, G., Gershon, M.D., Rothman, T.P., Joh, T.H. & Reis, D.J. (1981). Proliferation and distribution of cells that transiently express a catecholaminergic phenotype during development in mice and rats. *Developmental Biology,* **86**, 348–355.

Thomas, S.A. & Palmiter, R.D. (1998). Examining adrenergic roles in development, physiology, and behavior through targeted disruption of the mouse dopamine beta-hydroxylase gene. *Advances in Pharmacology,* **42**, 57–60.

Thomas, S., Matsumoto, A. & Palmiter, R. (1995). Noradrenaline is esential for mouse fetal development. *Nature,* **374**, 643–646.

Tischler, A. (1995). Triple immunohistochemical staining for bromodeoxyuridine and catecholamine biosynthetic enzymes using microwave antigen retrieval. *Journal of Histochemistry and Cytochemistry,* **43**, 1–4.

Tischler, A.S. & Coupland, R.E. (1994). Changes in structure and function of the adrenal medulla. *Pathobiology of the Aging Rat,* **2**, 245–268.

West, G., Shepherd, D. & Hunter, R. (1953). The function of the organs of Zuckerkandl. *Clinical Science,* **12**, 317.

Wildenthal, K. & Wakeland, J. (1973). Differential effects of acetylcholine, norepinephrine, theophylline, tyramine, glucagon, and dibutyril cyclic AMP on atrial in hearts of fetal mice. *Journal of Clinical Investigation,* **52**, 2250–2258.

Yelich, M. (1993). The effect of epinephrine on glucagon and insulin secretion from the endotoxin rat pancreas. *Pancreas,* **8**, 450–458.

Zak, F. & Lawson, W. (1983). *The paraganglionic chemoreceptor system. Physiology, pathology, and clinical medicine.* New York: Springer-Verlag.

Zhou, Q. & Palmiter, R. (1995). Dopamine-deficient mice are severely hypoactive, adipsic, and aphagic. *Cell,* **83**, 1197–1209.

Zhou, Q., Quaife, C. & Palmiter, R. (1995). Targeted disruption of the tyrosine hydroxylase gene reveals that catecholamines are required for mouse fetal development. *Nature,* **374**, 640–643.

Zigmond, M.J., Acheson, A.L., Stachowiak, M.K. & Strickerm, E.M. (1984). Neurochemical compensation after nigrostriatal bundle injury in an animal model of preclinical parkinsonism. *Archives of Neurology,* **41**, 856–861.

II.8
Amino Acid Neurotransmitters as Developmental Signals

ARNE SCHOUSBOE, GERT H. HANSEN and BERIT X. CARLSON

ABSTRACT

Neuronal development and differentiation including synaptogenesis is influenced by environmental signals. A large body of experimental evidence suggests that the inhibitory and excitatory neurotransmitters, gamma aminobutyric acid (GABA) and glutamate, play important roles in this signalling. The exact mechanisms have yet to be fully elucidated, but intracellular Ca^{2+} homeostasis, and possibly that of polyamines, may be involved. It has been well established that Ca^{2+} is an important factor in intracellular events, such as signal transduction, gene expression, and enzymatic pathways. Likewise during neuronal development, promotion/regulation of Ca^{2+} intracellular levels would indeed be an essential component to cell survival/differentiation. This review will discuss how the neurotransmitters, GABA and glutamate, and activation of their receptors promote the necessary cues for the proper maturation of the CNS.

INTRODUCTION

Neuronal growth, differentiation, and specialization into specific functional phenotypes is governed by a large number of internal and external stimuli. Some events are activity-dependent and others are dependent on the switching on and off of genes, which are specifically involved in these processes (Allendoerfer & Shatz, 1994). For example, the organization of neurons in functional columns in the cerebral cortex and the cerebellum, as described by Rakic (1988), must clearly be the result of a concerted action of such cues as this structural organization is not present at very early, embryonic stages. It should be noted that this developmental process requires an interaction between neurons and glial cells, a specialized type of astrocyte (Rakic, 1988). Moreover, it involves the interactions of a large number of cell surface adhesion molecules, the expression of which is

A.F. Kalverboer and A. Gramsbergen (eds.), Handbook of Brain and Behaviour in Human Development, 185–198
© 2001 Kluwer Academic Publishers. Printed in Great Britain.

also influenced by a number of epigenetic factors (see Maar *et al.* (1988) for references). Studies of the regulation of these developmental processes have provided evidence that a variety of classical neurotransmitters, in addition to their function as such, play important roles in neuronal growth and differentiation (Lauder, 1983, 1988; Kater & Haydon, 1987; Redburn & Schousboe, 1987; Lipton & Kater, 1989). The present review will be particularly engaged in a discussion of the extent to which the presence of GABA and glutamate is important for these processes.

GABA AS A NEUROTROPHIC AGENT

In a series of elegant experiments in which GABA could be continuously applied to rat superior cervical ganglia and cerebral cortex, GABA facilitated synapto-genesis during early development (Wolff *et al.*, 1978, 1979a,b). In subsequent studies, using different preparations of neurons in culture, it was shown that the addition of GABA to the culture medium led to the enhancement of neurite outgrowth, as well as functional differentiation of cells, including synaptogenesis (Spoerri & Wolf, 1981; Eins *et al.*, 1983; Hansen *et al.*, 1984; Spoerri, 1988). It was additionally demonstrated that, in exposing cerebellar granule neurons in culture to GABA during the first week of postnatal development, the cytoplas-mic density of ultrastructural organelles (see Figure 1) involved in protein synthesis (such as the Golgi apparatus, neurotubules, rough endoplasmis reticulum and vesicles) was increased (Hansen *et al.*, 1984; Meier *et al.*, 1985; Spoerri, 1988). This effect of GABA appeared to be selective for the organelles involved in protein synthesis as the density of mitochondria was unaffected by the exposure of the neurons to GABA (Table 1). These findings may be interesting in the light of observations that elevated GABA levels have been reported to increase the biosynthesis of specific proteins during early postnatal development *in vivo*, as well as during development of neurons in culture (Meier & Jørgensen, 1986; Meier *et al.*, 1987; Schousboe *et al.*, 1988). The overall protein synthesis in the brain, however, does not seem to be affected by elevated GABA levels (Toth & Lajtha, 1984).

From studies of nerve cell development *in vivo*, as well as in culture, it appears that the neurodifferentiative or neurotrophic activity of GABA is most pro-nounced during an early developmental period (Hansen *et al.*, 1987, 1988; Meier *et al.*, 1987), and the effects on the ultrastructural organelles are seen very rapidly after the exposure to GABA. This effect occurs at a time period preceding the effects of GABA on the expression of certain membrane-associated proteins (Hansen *et al.*, 1987). This finding is compatible with the observation that GABA, *in vivo,* is synthesized and released early in the developing central nervous system. Thus, in both cerebral cortex and retina, GABA is present at early developmental stages and in some cases in a population of neurons which is eliminated during later stages of development (Madtes & Redburn, 1983a,b; Schnitzer & Rusoff, 1984; Redburn & Madtes, 1986; Ghosh *et al.*, 1990). GABA synthesis may take place either from glutamate via glutamate decarboxylase or via polyamines utilizing ornithine as the precursor (Seiler, 1980; Martin & Rimval, 1993; Waagepetersen *et al.*, 1999). Since vesicular release mechanisms

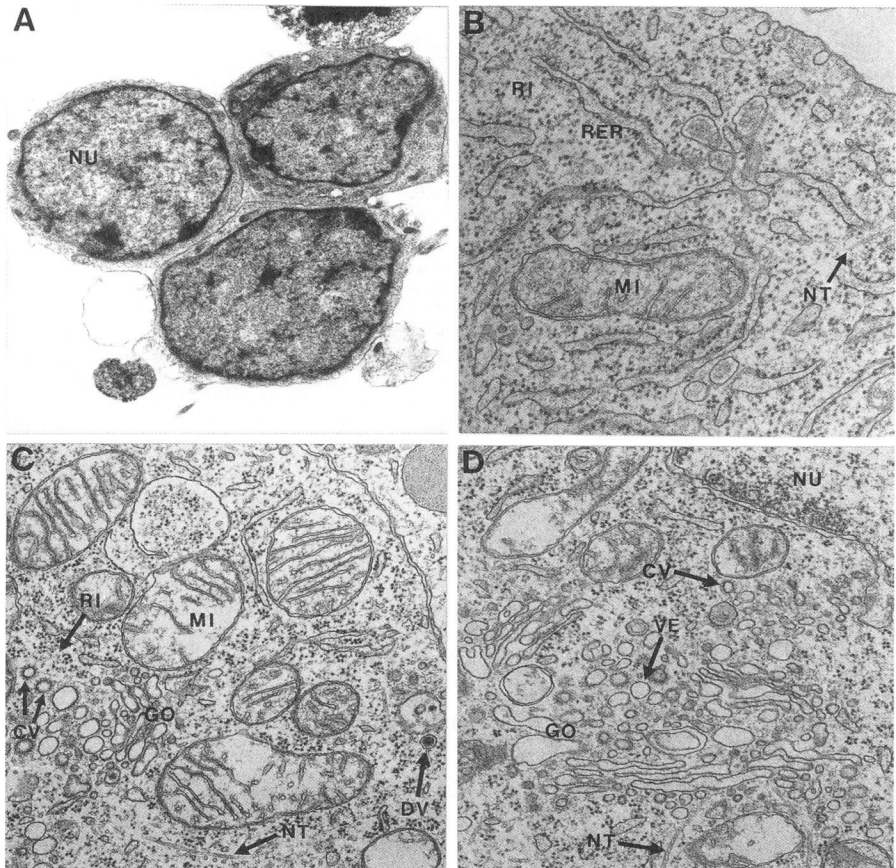

Figure 1 Electron micrographs illustrating the different subcellular structures in cultured cerebellar granule neurons (7 DIV). The data presented in Table 1 are derived from countings of the density of these structures. NU: nucleus; MI: mitochondria; RER: rough endoplasmic reticulum; RI: ribosomes; NT: neurotubules; GO: Golgi apparatus; CV: coated vesicles; VE: vesicles; DV: dense core vesicles.

are not fully developed at these early stages of development (Redburn *et al.*, 1978), it is most likely that GABA release primarily occurs via reversal of the plasma membrane-associated GABA carriers which are present during early development (Redburn *et al.*, 1978). As the efficiency of these carriers, determined by the stoichiometry between Na^+, Cl^- and GABA, is increased during development, it is likely that the extracellular concentration of GABA may be relatively high during this early developmental period (Redburn *et al.*, 1978; Schousboe, 1981). At this stage of development it is thought that the actions of GABA are neurotrophic. Indeed, this release of GABA occurs via reversal of the GABA carriers in a Ca^{2+}-independent manner, as has been demonstrated in a variety of brain cell preparations (Gordon-Weeks *et al.*, 1984; Schwartz, 1987; Pin & Bochaert, 1989; O'Malley *et al.*, 1992; Belhage *et al.*, 1993).

Table 1 Effect of GABA on the density of ultrastructural organelles in granule neurons

Organelle	Effect of GABA	
Mitochondria	None,	–
Golgi apparatus	Increased,	$p < 0.0005$
Neurotubules	Increased,	$p < 0.025$
Ribosomes	None,	–
Rough endoplasmic reticulum	Increased,	$p < 0.025$
Smooth endoplasmic reticulum	None,	–
Coated vesicles	Increased,	$p < 0.05$

Cerebellar granule cells were cultured for 1–7 days in the absence or presence of 50 µM GABA and the temporal development of the cytoplasmic densities of the different organelles was followed. Electron micrographs (cf. Figure 1) were taken at random, and the densities of the structures were determined double-blind using the point method for morphometric analysis. Summary from Hansen *et al.* (1984).

The neurotrophic actions of GABA as described above appear to be closely correlated with activation of $GABA_A$ receptors. Thus, the effect of GABA in cultured cerebellar granule neurons is mimicked by the specific $GABA_A$ receptor partial agonist, THIP (4,5,6,7-tetrahydroisoxazolo[5,4-c]pyridin-3-ol) and blocked by the $GABA_A$ receptor antagonist, bicuculline (Curtis *et al.*, 1971; Krogsgaard-Larsen, 1983; Meier *et al.*, 1985). This observation is interesting in light of the demonstrations that GABA or $GABA_A$ receptor agonists not only exert effects on neuronal structural development and differentiation, but $GABA_A$ receptor agonists also stimulate expression and alter the molecular properties of its own receptors, as seen *in vivo* and *in vitro* (Madtes & Redburn, 1983a,b; Meier *et al.*, 1983, 1984; Sykes *et al.*, 1984; Beart *et al.*, 1985; Sykes & Horton, 1986; Belhage *et al.*, 1986; Madtes & Bashir-Elahi, 1986; Ito *et al.*, 1988; Hansen *et al.*, 1991; Kim *et al.*, 1993; Elster *et al.*, 1995; Carlson *et al.*, 1997). In cultured, cerebellar granule neurons exposure during early postnatal development will lead to an increase in the number of low-affinity $GABA_A$ receptors (Meier *et al.*, 1984; Belhage *et al.*, 1986, 1988). The function and subunit composition of these low-affinity $GABA_A$ receptors during early development has yet to be fully elucidated; however, it has been proposed that this low-affinity $GABA_A$ receptor is perhaps related to the ρ subunit-containing $GABA_C$ receptor (Martina *et al.*, 1997). Furthermore, expression of selected subunits of the $GABA_A$ receptor family, α_1, $\beta_{2/3}$ and α_6, has been shown to increase dramatically in the presence of $GABA_A$ agonists (Elster *et al.*, 1995; Carlson *et al.*, 1997). *In vivo* in the brain, the expression of these as well as other subunits changes during postnatal development (Laurie *et al.*, 1992; Fritschy *et al.*, 1994). Such switches in the $GABA_A$ receptor subunit composition may have profound consequences for the kinetic and pharmacological properties of the receptor and thus for GABAergic function (Carlson *et al.*, 1998).

As mentioned above, the neurodifferentiative action of GABA requires the activation of $GABA_A$ receptors, the exact subunit composition of which is not yet known. How such activation of $GABA_A$ receptors is translated into a signal, which triggers intracellular events, that lead to the observed increase in mRNA and protein synthesis of GABA$_A$ receptor expression, has yet to be elucidated (Belhage *et al.*, 1990; Kim *et al.*, 1993; Elster *et al.*, 1995; Carlson *et al.*, 1997). Nevertheless, it is evident that GABA-mediated hyperpolarizations are essential for proper neuronal development since hyperpolarizing agents, such as bromide and valinomycin, can mimic the trophic actions of GABA (Joo *et al.*, 1980; Belhage *et al.*, 1990). Furthermore, it may be of interest that inhibition of ornithine decarboxylase, which is essential for synthesis of polyamines (Seiler, 1980), has been shown to prevent the neurotrophic action of GABA (Abraham *et al.*, 1993). This finding may be compatible with demonstrations that these compounds are important for growth and maturation of the brain as well as other tissues (Heby, 1981; Seiler, 1982). This involvement of polyamines is, however, likely to be a more general effect and not specific for the neurotrophic action of GABA, as these compounds are crucial for the correct function of ribosomes (Rudkin *et al.*, 1984; Hölttä, 1985).

GABA AS AN EXCITATORY TRANSMITTER DURING EARLY DEVELOPMENT

Calcium ions represent an extremely important intracellular messenger involved in the regulation of enzymatic pathways related to cell growth, differentiation, and ultimately cell death (Kater *et al.*, 1989). Also, it appears that GABA-mediated intracellular Ca^{2+} rises is a mechanism by which this traditionally inhibitory neurotransmitter exerts its neurotrophic actions during early development. Indeed, it has been documented that GABA can elicit depolarizations by activating GABA$_A$ receptors in a bicuculline-sensitive manner in various brain regions, including granule neurons in immature cerebellar explant cultures (Connor *et al.*, 1987), as well as in a variety of neuronal preparations from hypothalamus, cortex, striatum, hippocampus, colliculus, medulla, and spinal cord, dorsal horn neurons (Cherubini *et al.*, 1990; Fiszman *et al.*, 1990; Segal, 1993; Walton *et al.*, 1993; Obrietan & van de Pol, 1995). It is interesting to note that GABA$_A$ receptors have been shown to be co-localized with L-type voltage-gated Ca^{2+} channels (Hansen *et al.*, 1992). Thus, due to high intracellular Cl^- concentrations, activation of GABA$_A$ receptors allows for the efflux of Cl^- ions and consequently a membrane potential depolarization which triggers the activation of voltage-sensitive Ca^{2+} channels. Likewise, GABA-mediated intracellular rises in Ca^{2+} have been shown to be blocked by voltage-gated Ca^{2+} channel antagonists, and more specifically by the L-type variety (Obrietan & van de Pol, 1995; Borodinsky & Fiszman, 1999). It should be noted that not all cases of GABA$_A$ receptor-mediated increases in intracellular Ca^{2+} were related to depolarizations. For example, as discussed by Segal (1993) it may be the result of Ca^{2+} entry as well as its release from intracellular Ca^{2+} stores. Furthermore, in cortical slices, exposure to GABA can result in an increase in IP_3 production, which would set in motion a cascade of events leading to an increase of intracellular free Ca^{2+} (Li *et al.*, 1990). Nevertheless, this excitatory effect of GABA appears as early as embryonic day 15 in hypothalamic cultures (Obrietan & van de Pol, 1995) and, as seen in hippocampal cells in culture, the depolarizing effect observed in early developmental stages changes to a mixed hyperpolarizing and depolarizing effect at later stages (Fiszman *et al.*, 1990). This switch from GABA$_A$ receptor-mediated depolarizations to hyperpolarizations could be a consequence of the developmental induction of the Cl⁻ extruding K⁺/Cl⁻ co-transporter, KCC2 (Rivera *et al.*, 1999). The expression of the KCC2 co-transporter is essential for establishing the proper Cl⁻ electrochemical gradient needed for GABA$_A$-mediated hyperpolarizations (Rivera *et al.*, 1999).

Although developing neurons can show a greater Ca²⁺ cytosolic rise via GABA than to glutamate (Obrietan & van de Pol, 1995), GABA-mediated depolarizations, and thus the resulting trophic actions of GABA, do not occur unchecked. The c-AMP signal transduction pathway has been shown to modulate GABAergic activity at the presynaptic and postsynaptic level, by inhibiting GABA release or optimizing GABA$_A$ receptor activity via phosphorylation, respectively (Obrietan & van de Pol, 1997). Moreover, in developing neurons, adenylate cyclase-coupled receptors, such as metabotropic glutamate and GABA$_B$ receptors, have been shown to be involved in tonic inhibition of GABA$_A$ depolarizations pre- and postsynaptically (Obrietan & van de Pol, 1998; van de Pol *et al.*, 1998). In a sense a reversal of traditional neurotransmitter roles can occur during early neuronal development, where glutamate, an excitatory transmitter acting at metabotropic receptors, can inhibit GABA$_A$-mediated depolarizations. Although during early neuronal development GABA$_A$ receptors promote depolarizations, GABA can also act as an inhibitory transmitter, as mentioned, via GABA$_B$ receptors, either by inhibiting GABA$_A$- or glutamate-activated Ca²⁺ transients (Obrietan & van de Pol, 1998, 1999). This inhibitory effect may correlate with the finding that, in slices of hippocampal pyramidal neurons, application of GABA induces hyperpolarization in the soma and depolarization in dendrites (Zhang & Jackson, 1993). In all, GABA-mediated depolarizations via Ca²⁺ transients are a critical component to proper neuronal development, but it should not be overlooked that these critical developmental signals are also mediated by another neurotransmitter with neurotrophic/neurodifferentiative capabilities namely, glutamate.

GLUTAMATE AS A NEUROTROPHIC AGENT

It is well documented that certain neurons require a mildly depolarizing signal such as exposure to 25 mM KCl during early development in order to survive (Scott & Fisher, 1970; Lasher & Zagon, 1972; Gallo *et al.*, 1987; Balazs *et al.*, 1988c; Collins & Lile, 1989; Damgaard *et al.*, 1996). This depolarizing signal enhances the functional maturation of neurons that survive, as seen by the stimulation of biochemical markers for glutamatergic neurons (Balazs *et al.*, 1988c; Moran & Patel, 1989; Peng *et al.*, 1991; Damgaard *et al.*, 1996). Other depolarizing agents, such as excitatory amino acids, have for decades been considered neurotoxic substances (Hayashi, 1954; Olney *et al.*, 1971; Choi, 1988; Schousboe & Frandsen, 1995). In the light of this fact it was somewhat unexpected that Balazs *et al.* (1988a,b) reported that activation of *N*-methyl-D-aspartate (NMDA) receptors in cerebellar granule neurons during early development in culture could replace 25 mM K⁺ as a depolarizing signal, preventing the cell loss which would occur in a serum-containing culture medium with a non-depolarizing concentration of KCl (5–10 mM). Moreover, it was also shown by Moran & Patel (1989) that the glutamatergic property of these neurons could be enhanced by replacing 25 mM KC1 with NMDA. It was subsequently shown that not only activation of the NMDA receptors, but also activation of the kainate subtype of the glutamate receptor superfamily, could rescue and stimulate differentiation of these neurons kept in a culture medium with 5 mM KCl (Balazs *et al.*, 1990; Damgaard *et al.*, 1996). For example, in a series of elegant experiments in which cultured hippocampal neurons were exposed to different glutamate concentrations (Mattson *et al.*, 1988), neurite outgrowth from these neurons was enhanced at low glutamate concentrations while at higher glutamate concentrations neurites were pruned and ultimately died (as illustrated in Figure 2). Most likely, glutamate mediates these dual effects via intracellular rises in Ca^{2+}. For instance, it has been documented that glutamate produced an optimal growth and motility of the growth cones by inducing an intracellular Ca^{2+} concentration of 200 nM, while at higher Ca^{2+} concentrations cells were damaged (Kater *et al.*, 1988). Thus during early neuronal development it appears that there is a fine line between the actions of glutamate being neurotrophic or neurotoxic.

THE ROLE OF Ca^{2+} IN GLUTAMATE DEPOLARIZATIONS

The neurotrophic role of glutamate appears to be related to the fact that it has profound effects on the intracellular level of Ca^{2+}, and this effect appears to coincide with the expression of Ca^{2+}-permeable, non-NMDA glutamate receptors (i.e. ionotropic glutamate receptors), which are gated by AMPA (i.e. alpha-amino-3-hydroxy-5-methyl-4-isoxazoleproprionate) and/or kainate, early in neuronal development, i.e. embryonic day 15 (van de Pol *et al.*, 1995, reviewed in Bardoul *et al.*, 1998). This gating of Ca^{2+} by non-NMDA receptors is interesting in light of the fact that they primarily gate Na^+ ions in adult brain[1]. At 2–4 days later the expression of NMDA receptors and metabotropic receptors is evident, and can also play a role in mediating the intracellular rises of Ca^{2+} induced by glutamate (van de Pol *et al.*, 1995, 1998). It is interesting to note that this rise in Ca^{2+} via non-NMDA glutamate receptors occurs before chemical synapses have been formed (reviewed in Bardoul *et al.*, 1998).

As mentioned above, the neurotrophic effect of glutamate is concentration-dependent. Likewise, it is age-dependent. In cultured neocortical and hippocampal neurons, exposure to glutamate in younger cultures led to a much lower increase in intracellular Ca^{2+} than in older cultures treated the same way, and hence less neurotoxic consequences (Wahl *et al.*, 1989; Marks *et al.*, 1996). This effect correlates well with the demonstration that, in similar cultures, glutamate is not toxic in young cultures but will kill these neurons at later stages of development (Frandsen & Schousboe, 1990).

Although many of the neurotoxic effects of high intracellular Ca^{2+}, seen later in development, are mediated via metabotropic glutamate receptors (mGluR) and

---

[1]Ionotropic, non-NMDA glutamate receptors which do not contain the edited GluR2 subunit are capable of gating Ca^{2+} (Burnashev *et al.*, 1992). Therefore, it is assumed that there is little or no expression of GluR2 in early-developing neurons.

Glutamate concentration

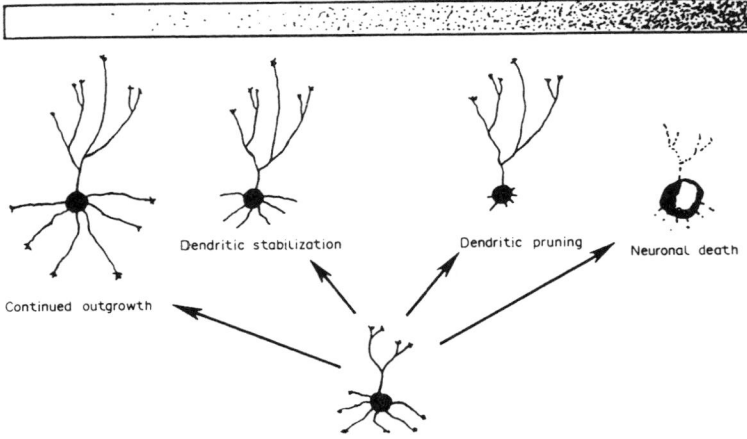

Figure 2 Regulation of hippocampal pyramidal neuron dendritic architecture by glutamate. A growing neuron may encounter different levels of glutamate within the proximate environment. Very low levels do not alter the course of outgrowth (left); moderate levels lead to dendritic stabilization, while further increases in glutamate concentration cause reduction in dendritic length. At very high levels, glutamate is toxic. From Mattson *et al.* (1988) with permission.

NMDA receptors (Marks *et al.*, 1996), it has been shown that the mGluR5 may be of importance in regulating the intracellular dynamics of Ca^{2+} rises, and hence developmental benefits (Flint *et al.*, 1999). Flint and colleagues (1999) have shown that the mGluR5 receptor is highly expressed during embryonic and early postnatal neurons, and activation of this specific receptor subtype induces intracellular Ca^{2+} oscillations via a cyclical process of phosphorylation and dephosphorylation of the receptor. It is thought that these oscillations of intracellular Ca^{2+} are crucial during development for the proper signalling needed to regulate gene expression, such as Ca^{2+}-dependent protein kinase activity, which regulates the trafficking of ionotropic glutamate receptors to their final postsynaptic destination, as well as its maintenance at these postsynaptic sites (Flint *et al.*, 1999; Rongo & Kaplan, 1999). To this end it appears evident that a concerted action between ionotropic and metabotropic glutamate receptors is inherent for optimizing intracellular Ca^{2+} levels in developing neurons.

Further reading with regard to the role of GABA and glutamate in neuronal development can be found in the following citations: Bardoul *et al.*, 1998; Cherubini *et al.*, 1998; Leinekugel *et al.*, 1999.

ACKNOWLEDGEMENTS

The expert secretarial assistance by Ms Hanne Danø is cordially acknowledged. The work has received financial support from the Danish Medical Research Council (9700761) and the NOVO Nordisk, Lundbeck and Alfred Benzon Foundations.

References

Abraham, J.H., Hansen, G.H., Seiler, N. & Schousboe, A. (1993). Depletion of polyamines prevents the neurotrophic activity of the GABA-agonist THIP in cultured rat cerebellar granule cells. *Neurochemical Research,* **18,** 153–158.

Allendoerfer, K.L. & Shatz, C.J. (1994). The subplate, a transient neocortical structure: its role in the development of connections between thalamus and cortex. *Annual Review of Neuroscience,* **17,** 185–218.

Balázs, R., Jørgensen, O.S. & Hack, N. (1988a). Stimulation of the *N*-methyl-D-aspartate receptor has a trophic effect on differentiating cerebellar granule cells. *Neuroscience Letters,* **87,** 80–86.

Balazs, R., Hack, N. & Jørgensen, O.S. (1988b). *N*-methyl-D-aspartate promotes the survival of cerebellar granule cells in culture. *Neuroscience,* **27,** 437–451.

Balazs, R., Gallo, V. & Kingsbury, A. (1988c). Effect of depolarization on the maturation of cerebellar granule cells in culture. *Developmental Brain Research,* **40,** 269–279.

Balazs, R,. Hack, N. & Jørgensen, O.S. (1990). Selective stimulations of excitatory amino acid receptor subtypes and the survival of cerebellar granule cells in culture: effect of kainic acid. *Neuroscience,* **37,** 251–258.

Bardoul, M., Drian, M-J. & König, N. (1998). Modulation of intracellular calcium in early neural cells by non-NMDA ionotropic glutamate receptors. *Perspectives in Developmental Neurobiology,* **5,** 353–371.

Beart, P.M., Scatton, B. & Lloyd, K.G. (1985). Subchronic administration of GABAergic agonist elevates ^{3}H-GABA binding and produces tolerance in striatal dopamine catabolism. *Brain Research,* **335,** 169–173.

Belhage, B., Meier, E. & Schousboe, A. (1986). GABA-agonists induce the formation of low-affinity GABA-receptors on cultured cerebellar granule cells via preexisting high affinity GABA receptors. *Neurochemical Research,* **11,** 599–606.

Belhage, B., Hansen, G.H., Schousboe, A. & Meier, E. (1988). GABA agonist promoted formation of low affinity GABA receptors on cerebellar granule cells in restricted to early development. *International Journal of Developmental Neuroscience,* **6,** 125–128.

Belhage, B., Hansen, G.H., Meier, E. & Schousboe, A. (1990). Effects of inhibitors of protein synthesis and intracellular transport on the γ-aminobutyric acid agonist-induced functional differentiation of cultured cerebellar granule cells. *Journal of Neurochemistry,* **55,** 1107–1113.

Belhage, B., Hansen, G.H. & Schousboe, A. (1993). Depolarization by K+ and glutamate activates different neurotransmitter release mechanisms in GABAergic neurons: vesicular versus non-vesicular release of GABA. *Neuroscience,* **54,** 1019–1034.

Borodinsky, L.N. & Fiszman, M.L. (1999). Neural activity affects cultured cerebellar granule cells morphology. *Journal of Neurochemistry,* 73(Suppl.), S124.

Burnashev, N., Monyer, H., Seeburg, P.H. & Sakmann, B. (1992). Divalent ion permeability of AMPA receptor channels is dominated by the edited form of a single subunit. *Neuron,* **8,** 189–198.

Carlson, B.X., Belhage, B., Hansen, G.H., Olsen, R.W. & Schousboe, A. (1997). Expression of the GABA$_A$ receptor α$_6$ subunit in cultured cerebellar granule cells is developmentally regulated by activation of GABA$_A$ receptors. *Journal of Neuroscience Research,* **50,** 1053–1062.

Carlson, B.X., Elster, L. & Schousboe, A. (1998). Pharmacological and functional implications of developmentally-regulated changes in GABA$_A$ receptor subunit expression in the cerebellum. *European Journal of Pharmacology,* **352,** 1–14.

Cherubini, E., Martina, M., Sciancalepore, M. & Strata, F. (1998). GABA excites immature CA3 pyramidal cells through bicuculline-sensitive and -insensitive chloride-dependent receptors. *Perspectives in Developmental Biology,* **5,** 289–304.

Cherubini, E., Rovira, C., Gaiarsa, J.L., Corradetti, R. & Ben-Ari, Y. (1990). GABA mediates excitation in immature rat CA3 hippocampal neurons. *International Journal of Developmental Neuroscience,* **8,** 481–490.

Choi, D.W. (1988). Glutamate neurotoxicity and diseases of the nervous system. *Neuron,* **1,** 623–634.

Collins, F. & Lile, J.D. (1989). The role of dihydropyridine-sensitive voltage-gated calcium channels in potassium-mediated neuronal survival. *Brain Research,* **502,** 99–108.

Connor, J.A., Tseng, H.Y. & Hockberger, P.E. (1987). Depolarization- and transmitter-induced changes in intracellular Ca^{2+} of rat cerebellar granule cells in explant cultures. *Journal of Neuroscience,* **7,** 1384–1400.

Curtis, D.R., Duggan, A.W., Felix, A., Johnston, G.A.R. & McLennan, H. (1971). Antagonism between bicuculline and GABA in cat brain. *Brain Research,* **31**, 53–73.

Damgaard, I., Trenkner, E., Sturman, J.A. & Schousboe, A. (1996). Effect of K⁺- and kainate-mediated depolarization on survival and functional maturation of GABAergic and glutamatergic neurons in cultures of dissociated mouse cerebellum. *Neurochemical Research,* **21**, 267–275.

Eins, E., Spoerri, P.E. & Heyder, E. (1983). GABA or sodium bromide-induced plasticity of neurites of mouse neuroblastoma cells in cultures. A quantitative study. *Cell and Tissue Research,* **229**, 457–460.

Elster, L., Hansen, G.H., Belhage, B., Fritschy, J.M., Möhler, H. & Schousboe, A. (1995). Differential distribution of GABA_A receptor subunits in soma and processes of cerebellar granule cells: effects of maturation and a GABA agonist. *International Journal of Developmental Neuroscience,* **13**, 417–428.

Fiszman, M.L., Novotny, E.A., Lange, G.D. & Barker, J.L. (1990). Embryonic and early postnatal hippocampal cells respond to nanomolar concentrations of muscimol. *Developmental Brain Research,* **53**, 186–193.

Flint, A.C., Dammerman, R.S. & Kriegsten, A.R. (1999). Endogenous activation of metabotropic glutamate receptors in neocortical developmental causes neuronal calcium oscillations. *Proceedings of the National Academy of Sciences, USA,* **96**, 12144–12149.

Frandsen, A.A. and Schousboe, A. (1990). Development of excitatory amino acid induced cytotoxicity in cultured neurons. *International Journal of Developmental Neuroscience,* **8**, 209–216.

Fristchy, J.M., Paysan, J., Enna, A. & Möhler, H. (1994). Switch in the expression of rat GABA_A-receptor subtypes during postnatal development: an immunohistochemical study. *Journal of Neuroscience,* **14**, 5302–5324.

Gallo, V., Kingsbury, A., Balázs, R. & Jørgensen, O.S. (1987). The role of depolarization in the survival and differentiation of cerebellar granule cells in culture. *Journal of Neuroscience,* **7**, 2203–2213.

Ghosh, A., Antonin, A., McConnel, S.K. & Shatz, C.J. (1990). Requirement for subplate neurons in the formation of thalamocortical connections. *Nature,* **347**, 179–181.

Gordon-Weeks, P.R., Lockerbie, R.O. & Pearce, B.R. (1984). Uptake and release of [³H]GABA by growth cones isolated from neonatal rat brain. *Neuroscience Letters,* **52**, 205–210.

Hansen, G.H., Meier, E. & Schousboe, A. (1984). GABA influences the ultrastructure composition of cerebellar granule cells during development in culture. *International Journal of Developmental Neuroscience,* **2**, 247–257.

Hansen, G.H., Belhage, B., Schousboe, A. & Meier, E. (1987). Temporal development of GABA agonist induced alterations in ultrastructure and GABA receptor expression in cultured cerebellar granule cells. *International Journal of Developmental Neuroscience,* **5**, 263–269.

Hansen, G.H., Belhage, B., Schousboe, A. & Meier, E. (1988). γ-Aminobutyric acid agonist-induced alterations in the ultrastructure of culture cerebellar granule cells is restricted to early development. *Journal of Neurochemistry,* **51**, 243–245.

Hansen, G.H., Belhage, B. & Schousboe, A. (1991). Effect of a GABA-agonist on the expression and distribution of GABA_A-receptors in the plasma membrane of cultured cerebellar granule cells: an immunocytochemical study. *Neuroscience Letters,* **124**, 162–165.

Hansen, G.H., Belhage, B. & Schousboe, A. (1992). First direct electron microscopic visualization of a tight spatial coupling between GABA_A-receptors and voltage sensitive calcium channels. *Neuroscience Letters,* **137**, 14–18.

Hayashi, T. (1954). Effects of sodium glutamate on the nervous system. *Keio Journal of Medicine,* **3**, 183-192.

Heby, O. (1981). Role of polyamines in the control of cell proliferation and differentiation. *Differentiation,* **19**, 1–20.

Höltta, E. (1985). Polyamine requirement for polyribosome formation and protein synthesis in human lymphocytes. In L. Selmeci, M.E. Brosnan & N. Seiler (Eds.), *Recent progress in polyamine research* (pp. 137–150). Budapest: Akadémiai Kiadó.

Ito, Y., Lim, D.K., Hoskins, B. & Ho, I.K. (1988). Bicuculline upregulation of GABA_A receptors in rat brain. *Journal of Neurochemistry,* **51**, 145–152.

Joo, F., Dames, W. & Wolff, J.R. (1980). Effect of prolonged sodium bromide administration of the fine structure of dendrites in the superior cervical ganglion of adult rat. *Progress in Brain Research,* **51**, 109–115.

Kater, S.B. & Haydon, P.G. (1987). Multifunctional roles for neurotransmitters: the regulation of neurite outgrowth, growth-cone motility and synaptogenesis. In A. Vernadakis, A. Privat, J.M. Lauder, P.S. Timiras & E. Giacobini (Eds.), *Model systems of development and aging of the nervous system* (pp. 239–255). Boston, MA: Martinus Nijhoff.

Kater, S.B., Mattson, M.P. & Guthrie, P.B. (1989). Calcium-induced neuronal degeneration: a normal growth cone regulating signal gone awry(?) *Annals of the New York Academy of Sciences,* **568,** 252–261.

Kater, S.B., Mattson, M.P., Cohan, C. & Connor, J. (1988). Calcium regulation of the neuronal growth cone. *Trends in Neuroscience,* **11,** 315–321.

Kim, H.Y., Sapp, D.W., Olsen, R.W. & Tobin, A.J. (1993). GABA alters GABA$_A$ receptor mRNAs and increases ligand binding. *Journal of Neurochemistry,* **62,** 2334–2337.

Krogsgaard-Larsen, P. (1983). GABA-agonists: structural, pharmacological and clinical aspects. In L. Hertz, E.G. McGeer & A. Schousboe (Eds.), *Glutamine, glutamate, and GABA in the central nervous system* (pp. 537–599). New York: Alan R. Liss.

Lasher, R.S. & Zagon, I.S. (1972). The effect of potassium on neuronal differentiation in cultures of dissociated newborn rat cerebellum. *Brain Research,* **41,** 482–488.

Lauder, J.M. (1983). Hormonal and humoral influences on brain development. *Psychoneuroendocrinology,* **8,** 121–155.

Lauder, J.M. (1988). Neurotransmitters as morphogens. *Progress in Brain Research,* **73,** 365–387.

Laurie, D.J., Seeburg, P.H. & Wisden, W. (1992). The distribution of 13 GABA$_A$ receptor subunit mRNAs in the rat brain. II. Olfactory bulb and cerebellum. *Journal of Neuroscience,* **12,** 1063–1076.

Leinekugel, X., Khalilov, I., McLean, H., Caillard, O., Gaiarsa, J.L., Ben Ari, Y. & Khazipov, R. (1999). GABA is the principal fast-acting excitatory transmitter in the neonatal brain. *Advances in Neurology,* **79,** 189–201.

Li, X.H., Song. L. & Jope, R.S. (1990). Modulation of phosphoinositide metabolism in rat brain slices by excitatory amino acids, arachidonic acid, and GABA. *Neurochemistry Research,* **15,** 725–738.

Lipton, S.A. & Kater, S.B. (1989). Neurotransmitter regulation of neuronal outgrowth, plasticity and survival. *Trends in Neuroscience,* **12,** 265–270.

Maar, T.E., Lund, T.M., Gegelashvili, G., Hartmann-Petersen, R., Moran, J., Pasantes-Morales, H., Berezin, V., Bock, E. & Schousboe, A. (1998). Effects of taurine depletion on cell migration and NCAM expression in cultures of dissociated mouse cerebellum and N2A cells. *Amino Acids,* **15,** 77–88.

Madtes, P. & Bashir-Elahi, R. (1986). GABA receptor binding site "induction" in the rat retina after nipecotic acid treatment: changes during postnatal development. *Neurochemical Research,* **11,** 55–61.

Madtes, P.C. & Redburn, D.A. (1983a). Synaptic interactions in the GABA system during postnatal development in retina. *Brain Research Bulletin,* **10,** 741–745.

Madtes, P.C. & Redburn, D.A. (1983b). GABA as a trophic factor during development. *Life Sciences,* **33,** 979–984.

Marks, J.D., Friedman, J.E. & Haddad, G.G. (1996). Vulnerability of CA1 neurons to glutamate is developmentally regulated. *Developmental Brain Research,* **97,** 194–206.

Martin, D.L. & Rimval, K. (1993). Regulation of γ-aminobutyric acid synthesis in the brain. *Journal of Neurochemistry,* **60,** 395–407.

Martina, M., Virginio, C. & Cherubini, E. (1997). Functionally distinct chloride-mediated GABA responses in rat cerebellar cells cultured in a low-potassium medium. *Journal of Neurophysiology,* **77,** 507–510.

Mattson, M.P., Dou, P. & Kater, S.B. (1988). Outgrowth-regulating actions of glutamate in isolated hippocampal pyramidal neurons. *Journal of Neuroscience,* **8,** 2087–2100.

Meier, E. & Jørgensen, O.S. (1986). Gamma-aminobutyric acid affects the developmental expression of neuron-associated proteins in cerebellar granule cell cultures. *Journal of Neurochemistry,* **46,** 1256–1262.

Meier, E., Drejer, J. & Schousboe, A. (1983). Trophic actions of GABA on the development of physiologically active GABA receptors. In P. Mandel & F.V. DeFeudis (Eds.), *CNS-receptors from molecular pharmacology to behavior* (pp. 47–58). New York: Raven Press.

Meier, E., Drejer, J. & Schousboe, A. (1984). GABA induces functionally active low-affinity GABA receptors on cultured cerebellar granule cells. *Journal of Neurochemistry,* **43,** 1737–1744.

Meier, E., Hansen, G.H. & Schousboe, A. (1985). The trophic effect of GABA on cerebellar granule cells is mediated by GABA-receptors. *International Journal of Developmental Neuroscience, 3,* 401–407.

Meier, E., Jørgensen, O.S. & Schousboe, A. (1987). Effect of repeated THIP treatment on postnatal neural development in rats. *Journal of Neurochemistry, 49,* 1462–1470.

Moran, J. & Patel, A.J. (1989). Stimulation of the *N*-methyl-D-aspartate receptor promotes the biochemical differentiation of cerebellar granule neurons and not astrocytes. *Brain Research, 486,* 14–25.

Obrietan, K. & van de Pol, A.N. (1995). GABA neurotransmission in the hypothalamus: developmental transition from Ca^{2+} elevating to depressing. *Journal of Neuroscience, 15,* 5065–5077.

Obrietan, K. & van de Pol, A.N. (1997). GABA activity mediating cytosolic Ca^{2+} rises in developing neurons is modulated by cAMP-dependent signal transduction. *Journal of Neuroscience, 17,* 4785–4799.

Obrietan, K. & van de Pol, A.N. (1998). $GABA_B$ receptor-mediated inhibition of $GABA_A$ receptor calcium elevations in developing hypothalamic neurons. *Journal of Neurophysiology, 79,* 1360–1370.

Obrietan, K. & van de Pol, A.N. (1999) $GABA_B$ receptor-mediated regulation of glutamate-activated calcium transients in hypothalamic and cortical neuron development. *Journal of Neurophysiology, 81,* 94–102.

Olney, J.W., Ho, O.L. & Rhee, V. (1971). Cytotoxic effects of acidic and sulphur containing amino acids on the infant mouse central nervous system. *Experimental Brain Research, 14,* 61–76.

O'Malley, D.M., Sandell, J.H. & Masland, R.H. (1992). Co-release of acetylcholine and GABA by the starburst amacrine cells. *Journal of Neuroscience, 12,* 1394–1408.

Peng, L., Juurlink, B.H.J. & Hertz, L. (1991). Difference in transmitter release, morphology, and ischemia-induced cell injury between cerebellar granule cell cultures developing in the presence of and in the absence of a depolarizing potassium concentration. *Developmental Brain Research, 63,* 1–12.

Pin, J.P. & Bockaert, J. (1989). Two distinct mechanisms, differentially affected by excitatory amino acids, trigger GABA release from fetal mouse striatal neurons in primary culture. *Journal of Neuroscience, 9,* 648–656.

Rakic, P. (1988). Specification of cerebral cortical areas. *Science, 241,* 170–176.

Redburn, D.A. & Madtes, P. (1986). Postnatal development of ^{3}H-GABA accumulating cells in rabbit retina. *Journal of Comparative Neurology, 243,* 41–57.

Redburn, D.A. & Schousboe, A. (1987). *Neurotrophic activity of GABA during development.* New York: Alan R. Liss.

Redburn, D.A., Broome, D., Ferkany, J. & Enna, S.J. (1978). Development of rat brain uptake and calcium-dependent release of GABA. *Brain Research, 152,* 511–519.

Rivera, C., Voipio, J., Payne, J.A., Ruusuvuori, E., Lahtinen, H. Lamsa, K., Pirvola, U., Saarma, M. & Kaila, K. (1999). The K^+/Cl^- co-transporter KCC2 renders GABA hyperpolarizing during neuronal maturation. *Nature, 397,* 251–255.

Rongo, C. & Kaplan, J.M. (1999). CaMKII regulated the density of central glutamatergic synapses *in vivo. Nature, 402,* 195–199.

Rudkin, B.B., Mamont, P.S. & Seiler, N. (1984). Decreased protein-synthetic activity is an early consequence of spermidine depletion in rat hepatoma tissue-culture cells. *Biochemical Journal, 217,* 731–741.

Schnitzer, J. & Rusoff, A.C. (1984). Horizontal cells of the mouse retina contains glutamic acid decarboxylase-like immunoreactivity during early developmental stages. *Journal of Neuroscience, 4,* 2948–2955.

Schousboe, A. (1981). Transport and metabolism of glutamate and GABA in neurons and glial cells. *International Review of Neurobiology, 22,* 1–45.

Schousboe, A. & Frandsen, A.A. (1995). Glutamate receptors and neurotoxicity. In T.W. Stone (Ed.), *CNS neurotransmitters and neuromodulators: glutamate* (pp. 239–251). Boca Raton, FL: CRC Press.

Schousboe, A., Belhage, B., Meier, E., Hammerschlag, R. & Hansen, G.H. (1988). Gamma-aminobutyric acid as a neurotrophic agent. In G. Huether (Ed.), *Amino acid availability and brain function in health and disease* (pp. 449–456). Berlin: Springer-Verlag.

Schwartz, E.A. (1987). Depolarization without calcium can release GABA from a retinal neuron. *Science, 238,* 350–355.

Scott, B.S. & Fisher, K.C. (1970). Potassium concentration and number of neurons in cultures of dissociated ganglia. *Experimental Neurology*, **27**, 16–22.

Segal, M. (1993). GABA induces a unique rise of $[Ca]_i$ in cultured rat hippocampal neurons. *Hippocampus*, **3**, 229–238.

Seiler, N. (1980). On the role of GABA in vertebrate polyamine metabolism. *Physiological Chemistry and Physics*, **12**, 411–429.

Seiler, N. (1982). Polyamines. In A. Lajtha (Ed.), *Handbook of neurochemistry*, vol. 1 (pp. 223–255). New York: Plenum Press.

Spoerri, P.E. (1988). Neurotrophic effects of GABA in cultures of embryonic chick brain and retina. *Synapse*, **2**, 11–22.

Spoerri, P.E. & Wolff, J.R. (1981). Effect of GABA-administration on murine neuroblastoma cells in culture. *Cell and Tissue Research*, **218**, 567–579.

Sykes, C.C. & Horton, R.W. (1986). Development of the gamma-aminobutyric acid neurotransmitter system in the rat cerebral cortex during repeated administration of the GABA-transaminase inhibitor ethanolamine *O*-sulphate. *Journal of Neurochemistry*, **46**, 213–217.

Sykes, C.C., Prestwich, S. & Horton, R. (1984). Chronic administration of the GABA-transaminase inhibitor ethanolamine *O*-sulphate leads to up-regulation of GABA binding sites. *Biochemical Pharmacology*, **33**, 387–393.

Toth, E. & Lajtha, A. (1984). Brain protein synthesis rates are not sensitive to elevated GABA, taurine or glycine. *Neurochemical Research*, **9**, 173–179.

Van de Pol, A.N., Obrietan, K., Cao, V. & Trombley, P.Q. (1995). Embryonic hypothalamic expression of functional glutamate receptors. *Neuroscience*, **67**, 419–439.

Van de Pol, A.N., Gao, X.B., Patrylo, P.R., Ghosh, P.K. & Obrietan, K. (1998). Glutamate inhibits GABA excitatory activity in developing neurons. *Journal of Neurosciences*, **18**, 10749–10761.

Waagepetersen, H.S., Sonnewald, U. & Schousboe, A. (1999). The GABA paradox: Multiple roles as metabolite, neurotransmitter, and neurodifferentiative agent. *Journal of Neurochemistry*, **73**, 1335–1342.

Wahl, P., Schousboe, A., Honoré, T. & Drejer, J. (1989). Glutamate induced increase in intracellular Ca^{2+} in cerebral cortex neurons is transient in immature cells but permanent in mature cells. *Journal of Neurochemistry*, **53**, 1316–1319.

Walton, M.K., Schaffner, A.E. & Barker, J.L. (1993). Sodium channels, $GABA_A$ receptors, and glutamate receptors develop sequentially on embryonic rat spinal cord cells. *Journal of Neuroscience*, **13**, 2068–2084.

Wolff, J.R., Foo, F. & Dames, W. (1978). Plasticity in dendrites shown by continuous GABA administration in superior cervical ganglion of adult rat. *Nature, London*, **274**, 72–74.

Wolff, J.R., Joo, F., Dames, W. & Feher, O. (1979a). Induction and maintenance of free postsynaptic membrane thickenings in the adult superior cervical ganglion. *Journal of Neurocytology*, **8**, 549–563.

Wolff, J.R., Rickmann, M. & Chronwall, B.M. (1979b). Axoglial synapses and GABA-accumulating glial cells in the embryonic neocortex of the rat. *Cell and Tissue Research*, **201**, 239–248.

Zhang, S.J. & Jackson, M.B. (1993). GABA-activated chloride channels in secretory nerve endings. *Science*, **259**, 531–534.

II.9
Cell Recognition Molecules and Disorders of Neurodevelopment

RALF-STEFFEN SCHMID and PATRICIA F. MANESS

ABSTRACT

Neural cell adhesion molecules of the immunoglobulin superfamily (L1 and NCAM) mediate cell adhesion, neuronal process outgrowth, and axon fasciculation necessary for proper development of synaptic connections. New evidence also implicates these adhesion molecules in aspects of synaptic plasticity in the adult nervous system. L1 and NCAM function at the cell membrane through homophilic and heterophilic binding mechanisms that involve specific domains of their extracellular regions. L1 and NCAM binding activates distinct intracellular signalling cascades composed of tyrosine kinases and phosphatases, culminating in phosphorylation of nuclear transcription factors.

Mutation of the L1 gene on the human X chromosome leads to a complex mental retardation disorder termed the CRASH syndrome (*c*orpus callosum dysgenesis/agenesis, *r*etardation, *a*phasia, *s*pastic paraplegia, and *h*ydrocephalus). Over 70 mutations in L1 have been observed clinically, and their location within the molecule dictates the severity of the disease. NCAM defects are linked with neurodevelopmental abnormalities that may contribute to schizophrenia. Gene knockout mice offer new animal models for the study of neurodevelopmental defects associated with CRASH and schizophrenia. The mouse L1 gene knockout displays a phenotype resembling the CRASH syndrome, and the NCAM knockout mouse shows dilated lateral ventricles and sensorimotor gaiting impairments characteristic of schizophrenia. These and second-generation animal models offer new routes for assessing the contribution of faulty neuronal migration and cell adhesion to human neurological disease.

A.F. Kalverboer and A. Gramsbergen (eds.), Handbook of Brain and Behaviour in Human Development, 199–218
© *2001 Kluwer Academic Publishers. Printed in Great Britain.*

L1

Structure and function of L1

The neural cell adhesion molecule L1 belongs to a class of related vertebrate and invertebrate members of the immunoglobulin superfamily: L1, neurofascin, NrCAM, NgCAM, CHL1(close homologue of L1) and neuroglian (*Drosophila*). The extracellular region of L1 contains six immunoglobulin (Ig)-like and five fibronectin type III (FN) domains, followed by a transmembrane region and a short cytoplasmic tail (Schachner, 1991) (Figure 1). L1 plays important roles in cell adhesion, neurite outgrowth, and axon bundling (Persohn & Schachner, 1990), and is implicated in survival of dopaminergic neurons (Hulley *et al.*, 1998). L1 may also contribute to learning and memory, since antibodies to L1 perturb the induction of long-term potentiation (LTP) in rat hippocampal slices (Lüthi *et al.*, 1994) and prevent passive avoidance learning in chicken (Scholey *et al.*, 1993).

During development of the nervous system, L1 is expressed in postmitotic, differentiating neurons throughout the brain. In the mouse neocortex, immuno-staining studies have revealed L1 expression in neuronal cell bodies in the intermediate zone but not in mitotic neural progenitors of the ventricular ger-minal matrix (Demyanenko *et al.*, 1999; Fushiki & Schachner, 1986). At the time of axonogenesis L1 is localized on axonal shafts and growth cones (Demyanenko *et al.*, 1999; Fushiki & Schachner, 1986; Persohn & Schachner, 1990; Stallcup & Beasley, 1985) but upon myelination it declines dramatically (Bartsch *et al.*, 1989; Martini *et al.*, 1994). In the embryonic mouse neocortex L1 is seen on apical dendrites of pyramidal neurons, diminishing with maturation and synaptogenesis (Demyanenko *et al.*, 1999). L1 continues to be expressed on unmyelinated axons in the mature brain, and on Schwann cells, lymphocytes, and melanocytes (Takeda *et al.*, 1996). Although only low levels of L1 remain on neuronal processes in the adult cerebral cortex, it persists at slightly elevated levels in process-rich layers of regions noted for lasting plasticity such as the hippocampus (Demyanenko *et al.*, 1999; Miller *et al.*, 1993; Persohn & Schachner, 1990) and cerebellum (Persohn & Schachner, 1987).

Human L1 is encoded by a single gene consisting of 28 exons located on the X chromosome (Xq28) and has three different splice variants (Figure 1). In neuronal cells the sequence RSLE encoded by exon 27 is inserted into the L1 cytoplasmic domain (Miura *et al.*, 1991; Reid & Hemperly, 1992). This insert is necessary for correct sorting of L1 to axonal growth cones (Kamiguchi & Lemmon, 1998). In Schwann cells and haematopoietic cells the RSLE sequence is not present in the mature protein (Miura *et al.*, 1991). In B lymphocytes a unique sequence encoded by exon 2 is inserted into the second Ig-like domain (Jouet *et al.*, 1995). The effect of this splice variation on binding properties of L1 is not known.

Homophilic binding of L1 occurs between L1 molecules on the cell surfaces of apposing cells. The major homophilic interaction site was mapped to a portion of the second Ig-like domain (Jouet *et al.*, 1995; Zhao *et al.*, 1998); however, for maximal neurite outgrowth, *trans* interaction of Ig 1–2, Ig 5–6 and FN 1–2 domains of L1 is necessary (Appel *et al.*, 1993). Heterophilic binding occurs between the RGD-binding motif on the Ig6 domain of L1 and certain integrins ($\alpha_v\beta_3$, $\alpha_v\beta_1$, $\alpha_5\beta_1$), an interaction that medicates cell attachment and spreading

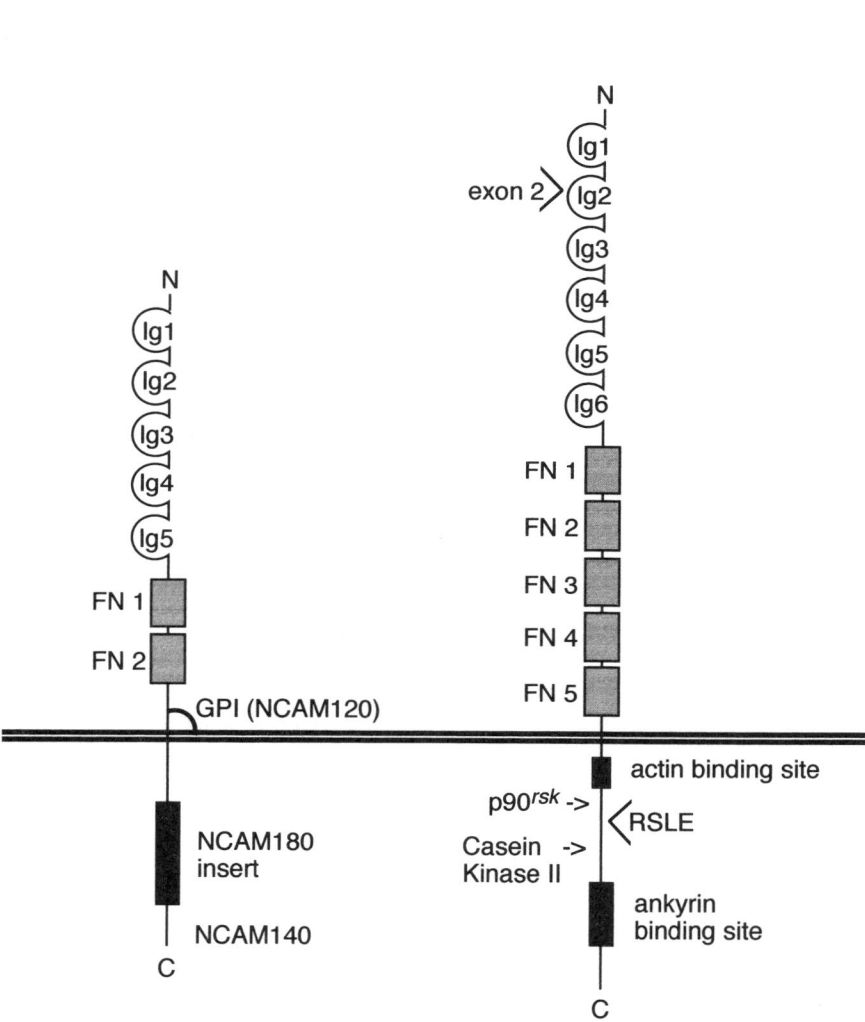

Figure 1 Schematic representation of NCAM and L1 structure. NCAM (120, 140 and 180 isoforms) and L1 belong to the Ig supe family. L1 contains several potential binding and phosphorylation sites on its intracellular domain.

(Felding-Habermann *et al.*, 1997; Montgomery *et al.*, 1996; Ruppert *et al.*, 1995). L1 has also been reported to bind laminin (Grumet *et al.*, 1993). Other binding partners of L1 include glycophosphatidyl inositol-linked axonin-1/TAG-1 (Felsenfeld *et al.*, 1994; Kuhn *et al.*, 1991), F3/F11/contactin (Brummendorf *et al.*, 1993), neurocan (Friedlander *et al.*, 1994), and phosphacan (Milev *et al.*, 1994).

L1 and TAG-1 show a strong *cis* interaction in the plane of the membrane when expressed in *Drosophila* S2 cells (Malhotra *et al.*, 1998).

The cytoplasmic domain of L1 is able to interact with molecules involved in actin cytoarchitecture (Figure 1). Recently, it has been shown that the cytoplasmic domain of L1 contains a juxtamembrane actin interaction site, which could provide direct anchorage to the actin cytoskeleton (Dahlin-Huppe *et al.*, 1997). Importantly, a conserved sequence near the carboxyterminal tail of L1 family members binds to the cytoskeletal adaptor protein ankyrin (Davis & Bennett, 1994). Ankyrin is known to bind to spectrin, which associates directly with actin filaments in the subcortical region of the cell (Letourneau & Shattuck, 1989). The ankyrin–spectrin–actin complex may anchor L1 in the plasma membrane, providing a base for effective L1-dependent cell adhesion and axon fasciculation (Stoeckli & Landmesser, 1995). In *Drosophila* S2 cells, L1- or neuroglian-directed homophilic adhesion induces the recruitment of ankyrin to sites of cell–cell contact (Dubreuil *et al.*, 1996; Malhotra *et al.*, 1998). Tyrosine 1229 within the FIGQY motif of the ankyrin binding site of the L1-related neurofascin molecule is phosphorylated in a developmentally regulated manner (Garver *et al.*, 1997). Ankyrin binding is abolished and lateral motility of neurofascin is increased within the plasma membrane when tyrosine 1229 is phosphorylated (Garver *et al.*, 1997; Tuvia *et al.*, 1997). Mutations of the same conserved tyrosine residue in *Drosophila* neuroglian similarly reduces ankyrin binding (Dubreuil *et al.*, 1996; Hortsch *et al.*, 1998). This suggests that tyrosine phosphorylation of FIGQY by a yet-unidentified kinase is able to regulate interactions of L1 family members with the cytoskeleton. Tyrosine phosphorylation may be necessary for weak or transient substrate binding, such as during migration of the growth cone, while dephosphorylation may favour stable adhesion interactions such as in fasciculating axons.

Serine and threonine-specific protein kinases are also able to phosphorylate the cytoplasmic tail of L1, and may regulate L1 function in axonal growth. Casein kinase II can phosphorylate serine 1181 adjacent to the RSLE insert *in vitro* (Sadoul *et al.*, 1989; Wong *et al.*, 1996a). The serine-threonine kinase p90^{rsk} can phosphorylate the cytoplasmic tail of L1 at serine 1152, which occurs in neurons during neurite extension (Wong *et al.*, 1996b). Inhibitory peptides spanning the serine 1152 site reduce outgrowth of neurites in neurons plated on L1, suggesting that p90^{rsk} is important in regulating L1-dependent neurite outgrowth (Wong *et al.*, 1996b).

L1 signal transduction

L1 peptides or L1 antibodies that are able to mimic homophilic binding activate intracellular signal transduction cascades (Figure 2) involving tyrosine kinases and phosphatases essential for neurite outgrowth in culture (Atashi *et al.*, 1992; Beggs *et al.*, 1997; Ignelzi *et al.*, 1994; Klinz *et al.*, 1995; Saffell *et al.*, 1997). Phosphoinositide turnover, intracellular Ca^{2+} and pH changes occur subsequently in cultured neuronal cells (Schuch *et al.*, 1989; von Bohlen und Halbach *et al.*, 1992). The signal transduction pathways activated by L1 stimulation are beginning to be understood. Neurite outgrowth on purified L1 is regulated by the non-receptor tyrosine kinase pp60^{c-src}. The rate of neurite extension on L1 is reduced in cerebellar neurons from *src*-minus mice but not in neurons from

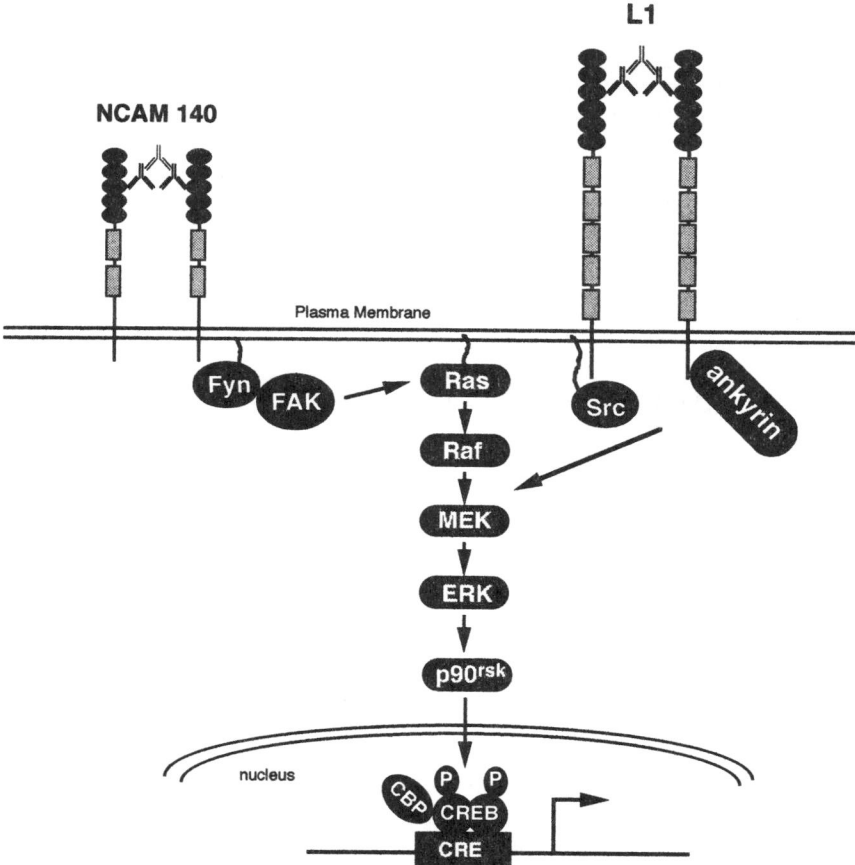

Figure 2 Pathway diagram for NCAM and L1 signal transduction. NCAM140 and L1 both activate the MAP kinase ERK; however, by different proximal signalling pathways. Eventually, gene transciption necessary for neurite outgrowth and neuron survival may be activated.

p59^{fyn} or pp62^{yes} knockout mice (Ignelzi *et al.*, 1994). This suggests that pp60$^{c\text{-}src}$ specifically regulates neurite outgrowth on L1 and that Src-related tyrosine kinases do not serve redundant functions in L1-mediated neurite outgrowth (Ignelzi *et al.*, 1994). *In vivo*, p59^{fyn} and pp60$^{c\text{-}src}$ are both required for proper axon guidance in the mouse olfactory nerve, perhaps due to their separate involvement in NCAM and L1-dependent axon growth (Morse *et al.*, 1998). It is not yet clear whether ankyrin recruitment to L1 is required for activation of signal transduction.

One substrate of pp60$^{c\text{-}src}$ is the cytoskeletal protein tubulin (Atashi *et al.*, 1992; Aubry & Maness, 1988; Matten *et al.*, 1990; Simon *et al.*, 1998), which is phosphorylated on tyrosine in developing brain and primary neuronal cultures

(Matten *et al.*, 1990). Tubulin isoforms are also tyrosine phosphorylated upon differentiation of PC12 cells induced by activated *src* or by treatment with nerve growth factor (Cox & Maness, 1991, 1993). Direct tyrosine phosphorylation of purified tubulin by pp60^{c-src} does not affect polymerization of microtubules *in vitro*, but does produce a small increase in microtubule dynamics in extracts from fetal rat brain (Simon *et al.*, 1998). Other potential targets of pp60^{c-src} are proteins associated with the actin cytoskeleton, i.e. tensin, cortactin, and paxillin. Actin cytoarchitecture seems to be affected by L1 signalling, as growth cones assume a filopodia-rich shape on an L1 substrate but different morphologies on other cell adhesion molecules (Burden-Gulley *et al.*, 1995; Payne *et al.*, 1992).

Selective chemical inhibitors of mitogen-activated protein kinase (MAPK) phosphorylation reduce L1-dependent neurite outgrowth of cultured cerebellar neurons by more than 50%, suggesting the requirement of an intact MAPK pathway for neurite outgrowth on L1 (Schmid *et al.*, 2000). MAPK plays an important role in cell survival, and is required for induction of LTP (English & Sweatt, 1997). Clustering of L1 activates MAPK phosphorylation in rat B35 neuroblastoma cells and cultured cerebellar neurons (Schmid *et al.*, 1998, 2000) (Figure 2). These clusters may become internalized during L1 activation (Schmid *et al.*, 2000), which may be needed for growth cones to migrate (Kamiguchi & Lemmon, 1998).

L1 activation causes rapid Ca^{2+} influx in neuronal cells through N- and/or L-type calcium channels, and this Ca^{2+} influx depends on tyrosine kinase activity (Schuch *et al.*, 1989; Williams *et al.*, 1994b). Pharmacological experiments implicate L1 in another cascade that regulates neurite outgrowth: fibroblast growth factor receptor (FGFR) activation leads to phospholipase Cγ stimulation, resulting in the production of arachidonic acid and subsequent opening of voltage-gated calcium channels (Williams *et al.*, 1994b). It is also possible that full L1 activation requires association of other cell-recognition molecules such as NCAM, F11 or CD9 to L1 (Heiland *et al.*, 1998; Schmidt *et al.*, 1996).

The CRASH syndrome

Mutations in the L1 gene on the human X chromosome lead to a complex mental retardation disorder referred to as CRASH syndrome, with an incidence of approximately 1 in 20,000 births (Fransen *et al.*, 1995; Jouet *et al.*, 1994; Kamiguchi *et al.*, 1998). The term "CRASH" summarizes the major characteristic symptoms associated with the human L1 syndrome: *c*orpus callosum hypoplasia, *r*etardation, *a*phasia, *s*pastic paraplegia, and *h*ydrocephalus (Wong *et al.*, 1995). CRASH syndrome patients display a range of infirmity, from slight disabilities to severe mental dysfunction. Patients may show one or more of the following symptoms: adducted thumbs, micro- or macrocephaly, delayed speech development, shuffling gait, spasticity of lower limbs, and agenesis or dysgenesis of the corpus callosum. Because symptoms vary among affected family members and between families, additional modifier genes are likely to contribute to the disease. Female heterozygotes are unaffected, or in some cases are mildly retarded, while homozygous males are more severely affected (Kaepernick *et al.*, 1994). Many of the symptoms associated with the CRASH syndrome overlap with those in fetal alcohol syndrome (FAS). Maternal alcohol consumption may perturb L1-mediated cellular responses during fetal development, because it has been demonstrated in cell cultures that alcohol can diminish L1-dependent adhesion of certain cell types (Charness *et al.*, 1994; Ramanathan *et al.*, 1996; West & Goodlett, 1990; Wilkemeyer & Charness, 1998), and L1-directed neurite outgrowth (Bearer *et al.*, 1999).

Over 70 mutations of L1 have been observed clinically, and these can be divided into three classes based on their location within the molecule: (a) missense mutations in the extracellular region, (b) truncation mutations resulting in a soluble L1, and (c) missense and truncation mutations in the cytoplasmic domain. Mutations in the extracellular domain may interfere with L1 homophilic or heterophilic binding, producing severe CRASH defects. A cluster of mutation sites on the second Ig-like domain, which contains the homophilic binding site (Rao *et al.*, 1993), has been observed in some CRASH families. A mutation in exon 2 in humans is associated with severe hydrocephalus (Jouet *et al.*, 1995). Some truncation mutations produce L1 extracellular fragments which, if secreted intact, may affect cell adhesion and axonal growth. Secreted L1 fragments might compete with endogenous L1 for homophilic or heterophilic adhesion.

Cytoplasmic mutations of L1 generally show milder symptoms than extracellular domain mutations. Perhaps cytoplasmic mutations alter only the signalling properties of L1, whereas extracellular mutations may additionally disrupt adhesion (Kamiguchi *et al.*, 1998). CRASH-associated point mutations in the cytoplasmic tail of L1 are also found within actin and ankyrin binding sites. The exchange of a single amino acid may disrupt binding of kinases or cytoskeletal molecules that play an essential role in L1 function. Frameshift mutations in the cytoplasmic sequence of the tail can also lead to truncated L1 cytoplasmic domains, reducing or abrogating its function. It has yet to be determined in which way different cytoplasmic L1 mutations induce the clinical symptoms mentioned above.

L1 knockout mice as CRASH models

Two different groups have used different strategies of homologous recombination in embryonic stem cells to obtain L1 knockout mice strains (Cohen *et al.*, 1997; Dahme *et al.*, 1997). The L1-minus mice of Soriano display errors in axon guidance in the corticospinal tract (Cohen *et al.*, 1997) and have dilated brain ventricles characteristic of mild hydrocephalus (Demyanenko *et al.*, 1999; Fransen *et al.*, 1997). These mice perform poorly in a spatial learning paradigm and exhibit reduced social exploration and stereotypical circling behaviour (Fransen *et al.*, 1997). The L1-minus mouse strain of the Schachner group (Dahme *et al.*, 1997) displays a reduced size of the corticospinal tract and decreased axonal association with non-myelinating Schwann cells. Ventricular enlargement in these mice depends on the genetic background of the mice, suggesting that modifier genes strongly influence the L1 phenotype. Agenesis or dysgenesis of the corpus callosum, the principal interhemispheric commissure, is often observed in CRASH patients. Dysgenesis of the corpus callosum has been observed in L1 knockout mice in the Sv129SvJae inbred strain (Demyanenko *et al.*, 1999). Additionally, the hippocampus of these mice is reduced in size with approximately 30% fewer pyramidal and granular neurons (Demyanenko *et al.*, 1999). Earlier experiments using L1 antibody and peptide perturbation resulted in inhibition of LTP in rat hippocampal slices, suggesting that hippocampal L1 plays some role in learning responses (Lüthi *et al.*, 1994). The hippocampal defects of L1 knockout mice may contribute to the impaired spatial LTP and memory of L1 knockout mice, as well as their poor performance in other behavioural tests (Fransen *et al.*, 1997). Axon guidance defects have also been noted in flies in which the *Drosophila* L1 homologue, neuroglian, is mutated (Bieber *et al.*, 1989; Hall & Bieber, 1997).

L1-minus mice display atypical pyramidal neurons in layer V of the cerebral cortex (Demyanenko *et al.*, 1999). Many apical dendrites of these neurons fail to reach their normal targets in layer I, lack apical tufts and display a tortuous morphology. Adhesive contact during cortical development may occur normally between the dendritic growth cones of the apical dendritic tuft and L1-containing axonal components of layer I, disruption of which could alter synaptic connectivity and cause learning and other cognitive defects.

Outside the nervous system the effects of L1 gene deletion have not yet been analysed, but L1 could be important for interactions between leucocytes, between epithelial crypt cells of the small intestine (Kadmon & Altevogt, 1997), and in generation and maintenance of the collecting duct system in kidneys (Debiec *et al.*, 1998).

Discussion

The L1 gene has gained particular attention since the CRASH syndrome has been linked to its mutation. It is presently not understood how the L1 mutations cause different symptoms and degree of severity of the CRASH syndrome. To better understand the underlying mechanisms of the CRASH syndrome, two major lines of investigation are being pursued. Biochemical methods are aimed at elucidating the steps of L1 signal transduction that may be defective in patients with certain types of L1 mutations. It will be interesting to determine the effects of cytoplasmic domain mutations that disrupt ankyrin and actin binding to L1 with regard to adhesion, fasciculation and neurite outgrowth. Second, L1 knockout mice will be useful both as models and as recipients for transfer of mutant versions of the L1 gene to further investigate the role of L1 in development, learning and memory. Evidence is compelling that mutated L1 plays a crucial role in correct outgrowth of neuronal processes in the developing brain. It is less clear if L1 has a synaptic function in the hippocampus or cerebral cortex. How L1 defects may lead to complex cognitive defects in affected individuals is another open question. Most likely the CRASH syndrome will prove to be a multi-genetic disease, influenced by many genetic modifiers. Identification of other genetic contributions, such as *src* family members, ankyrin, other signalling molecules and ion channels will help understand the wide variation in symptoms displayed by affected CRASH families.

NCAM

Developmental function and structural features of NCAM

NCAM is a related membrane-associated glycoprotein of the Ig superfamily that controls a diversity of cellular events important in nervous system development and function (Cunningham *et al.*, 1987). NCAM is expressed on the surface of developing and mature neurons throughout the brain, including regions implicated in schizophrenia: the hippocampus, cingulate, prefrontal and temporal cortex (Rutishauser & Jessell, 1988). It is also a component of myogenic cells and some glia. NCAM is localized on the surface of nerve growth cones, dendrites, axons, and cell bodies (Persohn & Schachner, 1990), and has a complex isoform-specific expression pattern. The NCAM gene in humans is located on chromosome region 11q23, and a novel related gene (NCAM2) maps to 21q21 (Paoloni-Giacobino *et al.*, 1997).

NCAM regulates aspects of neural migration during development such as outgrowth of axons and dendrites, axon fasciculation and branching. These cellular responses are mediated through interactions of its extracellular cell recognition region, which consists of five Ig-like and two fibronectin type III domains. The third Ig-like domain mediates *trans* homophilic binding to the same domain of another NCAM molecule (Rao *et al.*, 1993) (Figure 1). Heterophilic binding occurs between the second Ig domain of NCAM and the extracellular matrix protein agrin (Burg *et al.*, 1995; Cole & Glaser, 1986; Storms & Rutishauser, 1998) and other cell surface proteoglycans, such as neurocan (Friedländer *et al.*, 1994). NCAM can also bind L1 in a carbohydrate-dependent *cis* interaction within the plane of the membrane, facilitating L1 homophilic binding (Kadmon *et al.*, 1990a,b). The fourth Ig domain of NCAM contains the carbohydrate recognition site (Horstkorte *et al.*, 1993). During embryogenesis homophilic and heterophilic interactions between NCAM molecules in the axon and substrate may assist in promoting the growth of axons towards appropriate targets (Stoeckli & Landmesser, 1995; Thanos *et al.*, 1984).

Variable mRNA splicing of the single NCAM gene generates a multiplicity of isoforms (Figure 1), the principal ones of which are 180- and 140-kDa transmembrane isoforms, and a 120-kDa glycophosphatidyl inositol (GPI)-linked form (Murray *et al.*, 1986). The 180 kDa isoform (NCAM180) differs from the 140 kDa form (NCAM140) in having an additional 261-amino acid sequence inserted into the cytoplasmic domain (Murray *et al.*, 1986). A "variable alternatively spliced exon", termed VASE, encodes a 10-amino acid sequence, which when inserted into the fourth Ig-like domain causes down-regulation of neurite growth (Doherty *et al.*, 1992). Another alternatively spliced variant is a secreted form of NCAM consisting of most of the extracellular region (Gower *et al.*, 1988). The functional relevance of the different principal isoforms is not well understood but there are differences in their developmental patterns of expression, subcellular localization, and ability to activate intracellular signalling pathways.

All three principal isoforms of NCAM undergo posttranslational addition of α-2, 8-sialic acid to extracellular domains during development with desialylation generally occurring postnatally in many brain regions (Hekmat *et al.*, 1990; Rutishauser, 1994). Polysialylation of NCAM reduces the adhesive strength of NCAM homophilic binding with the effect of decreasing fasciculation and promoting neuronal process extension (Rutishauser, 1994; Tang *et al.*, 1994). Interestingly, sites noted for plasticity retain elevated levels of polysialylated NCAM (PSA-NCAM) in the adult, perhaps serving to enhance the growth of new synaptic connections (Rutishauser, 1994; Seki & Rutishauser, 1998).

Role of NCAM in learning and memory

Several lines of evidence implicate NCAM in mature synaptic function in addition to its role in development. NCAM is present in vertebrate brain synaptic membranes (Persohn et al., 1989; Pollerberg et al., 1986) and becomes enriched at the neuromuscular junction in an activity-dependent manner (Covault & Sanes, 1985). Several lines of evidence implicate NCAM in LTP and learning. Antibodies to NCAM perturb the induction of LTP in rat hippocampal slices (Arami et al., 1996; Lüthi et al., 1994) and reduce learning capability in the chick (Scholey et al., 1993). Mice with a gene knockout that deletes all forms of NCAM display deficits in spatial learning and hippocampal LTP (Cremer et al., 1994). Possibly contributing to this impairment, the hippocampus of NCAM mutant mice shows alterations in fasciculation, pathfinding, and distribution of mossy fibre terminals (Cremer et al., 1997). These mice are also more anxious and aggressive than wild-type mice (Stork et al., 1997).

It is not yet clear whether NCAM-dependent effects on LTP occur at the pre- or postsynaptic side of nerve terminals. The 140 kDa NCAM isoform (NCAM140) is present both pre- and postsynaptically, whereas the NCAM180 isoform is restricted to postsynaptic densities (Persohn et al., 1989). NCAM180 may become localized to the postsynaptic density through its unique ability to interact with brain spectrin (fodrin), which restricts its lateral mobility in the plasma membrane (Pollerberg et al., 1986). In invertebrates a decrease in NCAM-related molecules at presynaptic sites appears to facilitate new synaptic connections by promoting defasciculation. Down-regulation of the NCAM-related protein, fasciclin II, at the Drosophila neuromuscular junction has been shown to be both necessary and sufficient for activity-dependent presynaptic sprouting (Schuster et al., 1996a,b). Similarly, in Aplysia the NCAM-like adhesion molecule, apCAM, becomes internalized in sensory neurons after repeated serotonin stimulation, leading to increased connectivity (Bailey et al., 1997; Zhu et al., 1994).

Polysialylation of NCAM during LTP may facilitate sprouting of terminals by decreasing adhesion and fasciculation, and thus could be analogous to activity-dependent down-regulation of NCAM-like molecules in invertebrates. Stimulation of LTP in the normal rodent hippocampus causes an increase in cell surface levels of PSA-NCAM, and this up-regulation is necessary for both LTP and long-term depression (Muller et al., 1996). Enzymatic removal of polysialic acid from NCAM results in spatial memory defects and inhibits the formation of LTP in rats (Becker et al., 1996). Activity-dependent up-regulation of surface PSA-NCAM can be the consequence of its redistribution of PSA-NCAM from secretory vesicles to the cell surface as shown in cultured cortical neurons (Kiss et al., 1994). Surprisingly, enzymatic removal of polysialic acid in mice induces collateral sprouting of mossy fibres and ectopic synapses in the pyramidal layer of the hippocampus, but the changes might arise through a different mechanism of inhibition of process retraction (Seki & Rutishauser, 1998).

Activation of intracellular signalling pathways by NCAM

Homophilic binding interactions of the NCAM extracellular region activate intracellular signal transduction pathways governed by protein tyrosine kinases and phosphatases (Figure 2). NCAM peptides and antibodies that bind appropriate extracellular domains of NCAM elicit rapid changes in the activity of tyrosine kinases and phosphatases (Atashi et al., 1992; Beggs et al., 1994; Beggs et al., 1997; Klinz et al., 1995), and modulate intracellular Ca^{2+} levels, pH, and phosphoinositide turnover (Doherty & Walsh, 1992; Schuch et al., 1989). The non-receptor tyrosine kinase $p59^{fyn}$ has been demonstrated to carry out a critical function in NCAM-dependent axonal growth in neuronal cultures from fyn knockout mice (Beggs et al., 1994). Like NCAM knockout mice, fyn null mutants display impaired LTP and spatial memory as expected of the role of $p59^{fyn}$ in NCAM signalling (Grant et al., 1992). Recently, a signalling pathway has been defined leading from NCAM through $p59^{fyn}$ to activation of mitogen-activated protein kinases (MAPKs) (Beggs et al., 1997; Schmid et al., 1999). This pathway is activated by stimulation of the NCAM140 isoform but not by NCAM180 or NCAM120. Clustering of NCAM140 in the plasma membrane induces the recruitment of the focal adhesion kinase $p125^{fak}$ to a membrane-bound complex consisting of NCAM140 and $p59^{fyn}$, resulting in rapid and transient tyrosine phosphorylation of both $p125^{fak}$ and $p59^{fyn}$ (Beggs et al., 1997). It is not known if NCAM140 associates directly with these kinases or whether a linker molecule such as an integrin acts as a coreceptor.

Subsequently, the MAPKs "extracellular-regulated kinases 1 and 2" (ERK1, 2) become phosphorylated and activated in a cascade that depends critically on $p125^{fak}$, Ras, and MEK (MAP kinase kinase) (Schmid et al., 1999) (Figure 2). MEK inhibitors diminish neurite outgrowth on NCAM, demonstrating that an intact MAPK cascade is needed for NCAM-dependent neurite outgrowth (Schmid et al., 1999). The small G protein Rho, a regulator of the actin cytoskeleton, also functions in NCAM140-stimulated MAPK activation.

The pathway culminates in phosphorylation of the c-AMP response-element binding protein (CREB) on serine 133 (Schmid et al., 1998), a modification that is known to activate the ability of CREB to act as a transcription factor in conjunction with calcium-dependent binding of the accessory transcription factor CBP (CREB binding protein) (Goldman et al., 1997).

CREB gene knockout mice have severely impaired long-term memory (Bourtchuladze et al., 1994) and CREB2 in Aplysia participates in long-term facilitation (Bartsch et al., 1995), thus NCAM140 binding has the potential to induce the expression of genes important in memory. Since glutamate receptor stimulation also induces CREB phosphorylation on serine 133 (Pende et al., 1997), NCAM140 may synergize with other signalling pathways set in motion by neurotransmitter receptor activation to stimulate expression of genes that enhance synaptic responses.

Another NCAM signalling pathway has been described in which FGF receptor stimulates neurite outgrowth through phospholipase Cγ and diacylglycerol lipase (Doherty & Walsh, 1994; Saffell et al., 1997; Williams et al., 1994a). This may be a more general signaling mechanism than the NCAM140–MAPK–CREB cascade, because it is invoked by a variety of substrates (NCAM, L1, N-cadherin) and does not show isoform specificity for NCAM in responding neurons (Saffell et al., 1997). Physical association of the FGF receptor with these cell adhesion molecules has not been demonstrated; thus the nature of the initial interaction between cell adhesion molecules and FGF receptor is not clear. Because of their different signalling components it will be important to determine if the two pathways mediate distinct functions in axon growth and synaptic plasticity.

Phosphorylation of p125^{fak}, p59^{fyn}, MAPK and CREB induced by NCAM140 stimulation is rapid and transient, suggesting that protein phosphatases are coordinately activated. Indeed, NCAM and L1 stimulate tyrosine and serine/threonine phosphatase activity in growth cone-enriched membranes from fetal rat brain (Klinz et al., 1995). Candidate tyrosine phosphatases expressed in rat brain during the major period of axon growth include PTPδ, LAR, LAR-PTP2, and LRP/PTPδ (Shock et al., 1995). These tyrosine phosphatases, like NCAM, bear Ig-like and fibronectin III domains in their extracellular regions, raising the possibility that they may be involved in cell–cell recognition.

Aberrant NCAM expression in schizophrenia

Schizophrenia is a brain disease manifested by psychotic symptoms of, hallucinations (often acoustic), thought disorganization, memory impairment, and delusions. Evidence suggests that schizophrenia is a disorder of neurodevelopment even though symptoms often appear at adolescence (Egan & Weinberger, 1997). A number of studies document the occurrence of displaced or misoriented neurons in affected regions of the cerebral cortex and the hippocampus, suggesting possible defects in neuronal migration and neuronal process extension in the aetiology of the brain.

Recent findings link NCAM with aspects of maldevelopment of the schizophrenic brain. Mice in which the NCAM180 isoform is specifically deleted through gene knockout technology display two features characteristic of schizophrenia: increased size of brain lateral ventricles and sensorimotor gaiting impairment manifested by a reduced prepulse inhibition of acoustic startle (Wood et al., 1998). NCAM180 knockout mice also have features reflecting aberrant neuronal migration, and show alterations in the structure of the hippocampus, cerebellum, and olfactory bulb (Tomasiewicz et al., 1993). Of particular relevance to schizophrenia, diminished levels of PSA-NCAM are observed in NCAM180 minus-mice and in affected regions of the human schizophrenic brain (Barbeau et al., 1995).

A striking feature of the human schizophrenic brain is the presence of elevated levels of a soluble form of NCAM consisting of its extracellular domain (Forss-Petter et al., 1990; van Kammen et al., 1998; Vawter et al., 1998). NCAM extracellular fragments are elevated in soluble fractions from affected brain regions (hippocampus, cingulate, prefrontal and temporal cortex), and are also found in cerebrospinal fluid from individuals diagnosed with schizophrenia. Increased levels of the soluble form of NCAM are not found in brains of patients with bipolar disorder or suicidal history. Soluble NCAM could arise from overexpression of the alternatively spliced isoform comprising the NCAM extracellular region (Gower et al., 1988). Alternatively, it may be the consequence of aberrant proteolytic cleavage of transmembrane NCAM isoforms or release of GPI-linked NCAM120 from the cell surface. Regardless of how it is generated, overabundance of a soluble form of NCAM has the potential to function as a dominant mutation, for example by eliciting constitutive intracellular signalling through homo- or heterophilic binding.

Discussion

A number of outstanding questions remain concerning the molecular mechanism of NCAM and its roles in neurodevelopment, adult learning and memory, and schizophrenia. What are the unique functions of each of the NCAM isoforms? Specifically, what are the immediate molecular binding partners for NCAM140, and does the insert into the cytoplasmic domain of NCAM180 disrupt this binding? Do NCAM interactions in the growth cone produce local changes in actin cytoarchitecture necessary for axon guidance or synaptogenesis, or change the form of synapses during plasticity? What genes are activated by NCAM signalling pathways and how do the products of these genes modulate neurite outgrowth or synaptic properties? Does the soluble NCAM variant found in schizophrenic brain and cerebrospinal fluid contribute to the pathology of the disease? These questions may be accessed through studies with knockout mice. Other than the NCAM180 knockout mouse, only three animal models (in rat) are currently used for studies of schizophrenia: neonatal hippocampal lesions, isolation rearing, and maternal deprivation. These models lack a molecular/ genetic basis and are not specific for schizophrenia (Ellenbroek *et al.*, 1998). A high priority will be to generate a transgenic NCAM mutant mouse in which neurons secrete the extracellular domain of NCAM in regions of the cortex affected in schizophrenia. NCAM mouse models will also allow evaluation of contributory factors to schizophrenia such as deficiencies in neurotransmitter and receptor systems (dopamine, serotonin, glutamate), maternal deprivation, stress, hormones of puberty, and social isolation.

ACKNOWLEDGEMENTS

This work was supported by NIH grants NS26620 and HD35170, a Howard Hughes Medical Institute Pilot Grant, and a NARSAD Distinguished Investigator Award (PFM).

References

Appel, F., Holm, J., Conscience, J.-F. & Schachner, M. (1993). Several extracellular domains of the neural cell adhesion molecule L1 are involved in neurite outgrowth and cell body adhesion. *Journal of Neuroscience*, **13**, 4764–4775.

Arami, S., Jucker, M., Schachner, M. & Welzl, H. (1996). The effect of continuous intraventricular infusion of L1 and NCAM antibodies on spatial learning in rats. *Behaviour and Brain Research*, **81**, 81–87.

Atashi, J.R., Klinz, S.G., Ingraham, C.A., Matten, W.T., Schachner, M. & Maness, P.F. (1992). Neural cell adhesion molecules modulate tyrosine phosphorylation of tubulin in nerve growth cone membranes. *Neuron*, **8**, 1–20.

Aubry, M. & Maness, P.F. (1988). Developmental regulation of protein tyrosine phosphorylation in rat brain. *Journal of Neuroscience Research*, **21**, 473–479.

Bailey, C.H., Kaang, B.-K., Chen, M., Martin, K.C., Lim, C.-S., Casadio, A. & Kandel, E.R. (1997). Mutation in the phosphorylation sites of MAP kinases blocks learning-related internalization of apCAM in *Aplisia* sensory neurons. *Neuron*, **18**, 913–924.

Barbeau, D., Liang, J.J., Robitalille, Y., Quirion, R. & Srivastava, L.K. (1995). Decreased expression of the embryonic form of the neural cell adhesion molecule in schizophrenic brains. *Proceedings of the National Academy of Sciences, USA*, **92**, 2785–2789.

Bartsch, D., Ghirardi, M., Skehel, P.A., Karl, K.A., Herder, S.P., Chen, M., Bailey, C.H. & Kandel, E.R. (1995). *Aplysia* CREB2 represses long-term facilitation: relief of repression converts transient facilitation into long-term functional and structural change. *Cell*, **83**, 979–992.

Bartsch, U., Kirchhoff, F. & Schachner, M. (1989). Immunohistological localization of the adhesion molecules L1, N-CAM, and MAG in the developing and adult optic nerve of mice. *Journal of Comparative Neurology*, **284**, 451–462.

Bearer, C.F., Swick, A.R., O'Riordan, M.A. & Cheng, G. (1999). Ethanol inhibits L1-mediated neurite outgrowth in postnatal rat cerebellar granule cells. *Journal of Cell Biology*, **274**, 13264–13270.

Becker, C.G., Artola, A., Gerardy-Schahn, R., Becker, T., Welzl, H. & Schachner, M. (1996). The polysialic acid modification of the neural cell adhesion molecule is involved in spatial learning and hippocampal long-term potentiation. *Journal of Neuroscience Research*, **45**, 143–152.

Beggs, H.E., Soriano, P. & Maness, P.F. (1994). NCAM-dependent neurite outgrowth is inhibited in neurons from *fyn*-minus mice. *Journal of Biological Chemistry*, **127**, 825–833.

Beggs, H.E., Baragona, S.C., Hemperly, J.J. & Maness, P.F. (1997). NCAM-140 interacts with the focal adhesion kinase p125^{fak} and the src-related tyrosine kinase p59^{fyn}. *Journal of Biological Chemistry*, **272**, 8310–8319.

Bieber, A.J., Snow, P.M., Hortsch, M., Patel, N.H., Jacobs, J.R., Traquina, Z.R., Schilling, J. & Goodman, C.S. (1989). *Drosophila* neuroglian: a member of the immunoglobulin superfamily with extensive homology to the vertebrate neural cell adhesion molecule L1. *Cell*, **59**, 447–460.

Bourtchuladze, R., Frenguelli, B., Blendy, J., Cioffi, D., Schutz, G. & Silva, A.J. (1994). Deficient long-term memory in mice with a targeted mutation of the cAMP-responsive element-binding protein. *Cell*, **79**, 59–68.

Brümmendorf, T., Hubert, M., Treubert, U., Leuschner, R., Tarnok, A. & Rathjen, F.G. (1993). The axonal recognition molecule F11 is a multifunctional protein: specific domains mediate interactions with Ng-CAM and restrictin. *Neuron*, **10**, 711–727.

Burden-Gulley, S.M., Payne, H.R. & Lemmon, V. (1995). Growth cones are actively influenced by substrate-bound adhesion molecules. *Journal of Neuroscience*, **15**, 4370–4381.

Burg, M.A., Halfter, W. & Cole, G.J. (1995). Analysis of proteoglycan expression in developing chicken brain: characterization of a heparan sulfate proteoglycan that interacts with the neural cell adhesion molecule. *Journal of Neuroscience Research*, **41**, 49–64.

Charness, M.E., Safran, R.M. & Perides, G. (1994). Ethanol inhibits neural cell–cell adhesion. *Journal of Biological Chemistry*, **269**, 9304–9309.

Cohen, N.R., Taylor, J.S.H., Scott, L.B., Guillery, R.W., Soriano, P. & Furley, A.J.W. (1997). Errors in corticospinal axon guidance in mice lacking the neural cell adhesion molecule L1. *Current Biology*, **8**, 26–33.

Cole, G.J. & Glaser, L. (1986). A heparin-binding domain from N-CAM is involved in neural cell-substratum adhesion. *Journal of Cell Biology*, **102**, 403–412.

Covault, J. & Sanes, J.R. (1985). Neural cell adhesion molecule (N-CAM) accumulates in denervated and paralyzed skeletal muscles. *Proceedings of the National Academy of Sciences, USA*, **82**, 4544–4548.

Cox, M.E. & Maness, P.F. (1991). Neurite extension and protein tyrosine phosphorylation elicited by inducible expression of the v-src oncogene in a PC12 cell line. *Experimental Cell Research*, **195**, 423–431.

Cox, M.E. & Maness, P.F. (1993). Tyrosine phosphorylation of a-tubulin is an early response to NGF and pp60^{v-src} in PC12 cells. *Journal of Molecular Neuroscience*, **4**, 63–72.

Cremer, H., Lange, R., Christoph, A., Plomann, M., Vopper, G., Roes, J., Brown, R., Baldwin, S., Barthels, D., Rajewsky, K. & Wille, W. (1994). Inactivation of the N-CAM gene in mice results in size reduction of the olfactory bulb and deficits in spatial learning. *Nature*, **367**, 455–459.

Cremer, H., Chazal, G., Goridis, C. & Represa, A. (1997). NCAM is essential for axonal growth and fasciculation in the hippocampus. *Molecular and Cellular Neuroscience*, **8**, 323–335.

Cunningham, B.A., Hemperly, J.J., Murray, B.A., Prediger, E.A., Brackenbı ry, R. & Edelman, G.A. (1987). Neural cell adhesion molecule: structure, immunoglobulin-like domains, cell surface modulation, and alternative RNA splicing. *Science*, **236**, 799–806.

Dahlin-Huppe, K., Berglund, E.O., Ranscht, B. & Stallcup, W.B. (1997). Mutational analysis of the L1 neuronal cell adhesion molecule identifies membrane-proximal amino acids of the cytoplasmic domain that are required for cytoskeletal anchorage. *Molecular and Cellular Neuroscience*, **9**, 144–156.

Dahme, M., Bartsch, U., Martini, R., Anliker, B., Schachner, M. & Mantei, N. (1997). Disruption of the mouse L1 gene leads to malformations of the nervous system. *Nature Genetics*, **17**, 346–349.

Davis, J.Q. & Bennett, V. (1994). Ankyrin binding activity shared by the neurofascin/L1/NrCAM family of cell adhesion molecules. *Journal of Biological Chemistry*, **269**, 27163–27166.

Debiec, H., Christensen, E.I. & Ronco, P.M. (1998). The cell adhesion molecule L1 is developmentally regulated in the renal epithelium and is involved in kidney branching morphogenesis. *Journal of Cell Biology*, **143**, 2067–2079.

Demyanenko, G., Tsai, A. & Maness, P.F. (1999). Abnormalities in neuronal process extension, hippocampal development, and the ventricular system of L1 knockout mice. *Journal of Neuroscience*, **19**, 4907–4920.

Doherty, P. & Walsh, F.S. (1992). Cell adhesion molecules, second messengers and axonal growth. *Current Opinion in Neurobiology*, **2**, 595–601.

Doherty, P. & Walsh, F.S. (1994). Signal transduction events underlying neurite outgrowth stimulated by cell adhesion molecules. *Current Opinion in Neurobiology*, **4**, 49–55.

Doherty, P., Moolenaar, C.E.C.K., Ashhton, S.V., Michalides, R.J.A.M. & Walsh, F.S. (1992). The VASE exon downregulates the neurite growth-promoting activity of NCAM 140. *Nature*, **356**, 791–793.

Dubreuil, R.R., MacGivar, G., Dissanayake, S., Liu, C., Homer, D. & Hortsch, M. (1996). Neuroglian-mediated cell adhesion induces assembly of the membrane skeleton at cell contact sites. *Journal of Cell Biology*, **133**, 647–655.

Egan, M.F. & Weinberger, D.R. (1997). Neurobiology of schizophrenia. *Current Opinion in Neurobiology*, **7**, 701–707.

Ellenbroek, B.A., van den Kroonenberg, P.T. & Cools, A.R. (1998). The effects of an early stressful life event on sensorimotor gating in adult rats. *Schizophrenia Research*, **30**, 251–260.

English, J.D. & Sweatt, J.D. (1997). A requirement for the mitogen-activated protein kinase cascade in hippocampal long term potentiation. *Journal of Biological Chemistry*, **272**, 19103–19106.

Felding-Habermann, B., Silletti, S., Mei, F., Siu, C.H., Yip, P.M., Brooks, P.C., Cheresh, D.A., O'Toole, T.E., Ginsberg, M.H. & Montgomery, A.M. (1997). A single immunoglobulin-like domain of the human neural cell adhesion molecule L1 supports adhesion by multiple vascular and platelet integrins. *Journal of Cell Biology*, **139**, 1567–1581.

Felsenfeld, D., Hynes, M.A., Skoler, K.M., Furley, A.J. & Jessell, T.M. (1994). TAG-1 can mediate homophilic binding, but neurite outgrowth on TAG-1 requires an L1-like molecule and b1 integrins. *Neuron*, **12**, 675–690.

Forss-Petter, S., Danielson, P.E., Catsicas, S., Battenberg, E., Price, J., Nerenberg, M. & Sutcliffe, J.G. (1990). Transgenic mice expressing beta-galactosidase in mature neurons under neuron-specific enolase promoter control. *Neurology*, **5**, 187–197.

Fransen, E., Lemmon, V., Van Camp, G., Vits, L., Coucke, P. & Willems, P.J. (1995). CRASH syndrome: clinical spectrum of corpus callosum hypoplasia, retardation, adducted thumbs, spastic paralysis and hydrocephalus due to mutation in one single gene, L1. *European Journal of Human Genetics*, **3**, 273–284.

Fransen, E., VanCamp, G., Vits, L. & Willems, P.J. (1997). L1-associated diseases: clinical geneticists divide, molecular geneticists unite. *Human Molecular Genetics*, **6**, 1625–1632.

Friedländer, D.R., Milev, P., Karthikeyan, L., Margolis, R.K., Margolis, R.U. & Grumet, M. (1994). The neuronal chondroitin sulfate proteoglycan Neurocan binds to the neural cell adhesion molecules Ng-CAM/L1/NILE and N-CAM, and inhibits neuronal adhesion and neurite outgrowth. *Journal of Cell Biology*, **125**, 669–680.

Fushiki, S. & Schachner, M. (1986). Immunocytological localization of cell adhesion molecules L1 and N-CAM and the shared carbohydrate epitope L2 during development of the mouse neocortex. *Brain Research*, **389**, 153–167.

Garver, T.D., Ren, Q., Tuvia, S. & Bennett, V. (1997). Tyrosine phosphorylation at a site highly conserved in the L1 family of cell adhesion molecules abolishes ankyrin binding and increases lateral mobility of neurofascin. *Journal of Cell Biology*, **137**, 703–714.

Goldman, P.S., Tran, V.K. & Goodman, R.H. (1997). The multifunctional role of the co-activator CBP in transcriptional regulation. *Recent Progress in Hormone Research*, **52**, 103–120.

Gower, H.J., Barton, C.H., Elsom, V.L., Thompson, J., Moore, S.E., Dickson, G. & Walsh, F.S. (1988). Alternative splicing generates a secreted form of N-CAM in muscle and brain. *Cell*, **55**, 955–964.

Grant, S.G.N., O'Dell, T.J., Karl, K.A., Stein, P.L., Soriano, P. & Kandel, E.R. (1992). Impaired long-term potentiation, spatial learning, and hippocampal development in *fyn* mutant mice. *Science*, **258**, 1903–1910.

Grumet, M., Friedlander, D.R. & Edelman, G.M. (1993). Evidence for the binding of Ng-CAM to laminin. *Cell Adhesion Communications*, **1**, 177–190.

Hall, S.G. & Bieber, A.J. (1997). Mutations in the *Drosophila* neuroglian cell adhesion molecule affect motor neuron pathfinding and peripheral nervous system patterning. *Journal of Neurobiology*, **32**, 325–340.

Heiland, P.C., Griffith, L.S., Lange, R., Schachner, M., Hertlein, B., Traub, O. & Schmitz, B. (1998). Tyrosine and serine phosphorylation of the neural cell adhesion molecule L1 is implicated in its oligomannosidic glycan dependent association with NCAM and neurite outgrowth. *European Journal of Cell Biology*, **75**, 97–106.

Hekmat, A., Bitter-Suermann, D. & Schachner, M. (1990). Immunocytological localization of the highly polysialylated form of the neural cell adhesion molecule during development of the murine cerebellar cortex. *Journal of Comparative Neurology*, **291**, 457–467.

Horstkorte, R., Schachner, M., Magyar, J.P., Vorherr, T. & Schmitz, B. (1993). The fourth immunoglobulin-like domain of NCAM contains a carbohydrate recognition domain for oligomannosidic glycans implicated in association with L1 and neurite outgrowth. *Journal of Cell Biology*, **121**, 1409–1421.

Hortsch, M., Homer, D., Malhotra, J.D., Chang, S., Frankel, J., Jefford, G. & Dubreuil, R.R. (1998). Structural requirements for outside-in and inside-out signalling by *Drosophila* neuroglian, a member of the L1 family of cell adhesion molecules. *Journal of Cell Biology*, **142**, 251–261.

Hulley, P., Schachner, M. & Lubbert, H. (1998). L1 neural cell adhesion molecule is a survival factor for fetal dopaminergic neurons. *Journal of Neuroscience Research*, **53**, 129–134.

Ignelzi, M.A., Miller, D.R., Soriano, P. & Maness, P.F. (1994). Impaired neurite outgrowth of src-minus cerebellar neurons on the cell adhesion molecule L1. *Neuron*, **12**, 873–884.

Jouet, M., Rosenthal, A., Armstrong, G., MacFarlane, J., Stevenson, R., Paterson, J., Metzenberg, A., Ionasescu, V., Temple, K. & Kenwrick, S. (1994). X-linked spastic paraplegia (SPG1), MASA syndrome and X-linked hydrocephalus result from mutations in the L1 gene. *Nature Genetics*, **7**, 402–407.

Jouet, M., Rosenthal, A. & Kenwrick, S. (1995). Exon 2 of the gene for neual cell adhesion molecule L1 is alternatively spliced in B cells. *Molecular Brain Research*, **30**, 378–380.

Kadmon, G. & Altevogt, P. (1997). The cell adhesion molecule L1: species- and cell-type-dependent multiple binding mechanisms. *Differentiation*, **61**, 143–150.

Kadmon, G., Kowitz, A., Altevogt, P. & Schachner, M. (1990a). Functional cooperation between the neural adhesion molecules L1 and N-CAM is carbohydrate dependent. *Journal of Cell Biology*, **110**, 209–218.

Kadmon, G., Kowitz, A., Altevogt, P. & Schachner, M. (1990b). The neural cell adhesion molecule N-CAM enhances L1-dependent cell-cell interactions. *Journal of Cell Biology*, **110**, 193–208.

Kaepernick, L., Legius, E., Higgins, J. & Kapur, S. (1994). Clinical aspects of the MASA syndrome in a large family, including expressing females. *Clinical Genetics*, **45**, 181–185.

Kamiguchi, H. & Lemmon, V. (1998). A neuronal form of the cell adhesion molecule L1 contains a tyrosine-based signal required for sorting to the axonal growth cone. *Journal of Neuroscience*, **18**, 3749–3756.

Kamiguchi, H., Hlavin, M.L., Yamasaki, M. & Lemmon, V. (1998). Adhesion molecules and inherited diseases of the human nervous system. *Annual Review of Neurosciences*, **21**, 97–125.

Kiss, J.Z., Wang, C., Olive, S., Rougon, G., Lang, J., Baetens, D., Harry, D. & Pralong, W.F. (1994). Activity-dependent mobilization of the adhesion molecule polysialic NCAM to the cell surface of neurons and endocrine cells. *EMBO Journal*, **13**, 5284–5292.

Klinz, S.G., Schachner, M. & Maness, P.F. (1995). L1 and NCAM antibodies trigger protein phosphatase activity in growth cone-enriched membranes. *Journal of Neurochemistry*, **65**, 84–95.

Kuhn, T.B., Stoeckli, E.T., Condrau, M.A., Rathjen, F.G. & Sonderegger, P. (1991). Neurite outgrowth on immobilized axonin-1 is mediated by a heterophilic interaction with L1(G4). *Journal of Cell Biology*, **115**, 1113–1126.

Letourneau, P.C. & Shattuck, T.A. (1989). Distribution and possible interactions of actin-associated proteins and cell adhesion molecules of nerve growth cones. *Development*, **105**, 505–519.

Lüthi, A., Laurent, J.P., Figurov, A., Muller, D. & Schachner, M. (1994). Hippocampal long-term potentiation and neural cell adhesion molecules L1 and NCAM. *Nature*, **372**, 777–779.

Malhotra, J.D., Tsiotra, P., Karagogeos, D. & Hortsch, M. (1998). *Cis*-activation of L1-mediated ankyrin recruitment by TAG-1 homophilic cell adhesion. *Journal of Biological Chemistry*, **273**, 33354–33359.

Martini, R., Xin, Y. & Schachner, M. (1994). Restricted localization of L1 and N-CAM at sites of contact between Schwann cells and neurites in culture. *Glia*, **10**, 70–74.

Matten, W.T., Aubry, M., West, J. & Maness, P.F. (1990). Tubulin is phosphorylated at tyrosine by pp60c-src in nerve growth cone membranes. *Journal of Cell Biology*, **111**, 1959–1970.

Milev, P., Friedländer, D.R., Sakurai, T., Karthikeyan, L., Flad, M., Margolis, R.K., Grumet, M. & Margolis, R.U. (1994). Interactions of the chondroitin sulfate proteoglycan phosphacan, the extracellular domain of a receptor-type protein tyrosine phosphatase, with neurons, glia, and neural cell adhesion molecules. *Journal of Cell Biology*, **127**, 1703–1715.

Miller, D.R., Lee, G.M. & Maness, P.F. (1993). Increased neurite outgrowth induced by inhibition of protein tyrosine kinase activity in PC12 pheochromocytoma cells. *Journal of Neurochemistry*, **60**, 2134–2144.

Miura, M., Kobayashi, M., Asou, H. & Uyemura, K. (1991). Molecular cloning of cDNA encoding the rat neural cell adhesion molecule L1 - two L1 isoforms in the cytoplasmic region are produced by differential splicing. *FEBS Letters*, **289**, 91–95.

Montgomery, A.M.P., Becker, J.C., Siu, C., Lemmon, V.P., Cheresh, D.A., Pancook, J.D., Zhao, X. & Reisfeld, R.A. (1996). Human neural cell adhesion molecule L1 and rat homologue NILE are ligands for integrin $\alpha_v\beta_3$. *Journal of Cell Biology*, **132**, 475–485.

Morse, W.R., Whitesides III, J.G., LaMantia, A.-S. & Maness, P.F. (1998). p59^{fyn} and pp60^{c-src} modulate axonal guidance in the developing mouse olfactory pathway. *Journal of Neurobiology*, **36**, 53–63.

Müller, D., Wang, C., Skibo, G., Toni, N., Cremer, H., Calaora, V., Rougon, G. & Kiss, J.Z. (1996). PSA-NCAM is required for activity-induced synaptic plasticity. *Neuron*, **17**, 413–422.

Murray, B.A., Hemperly, J.J., Pridiger, E.A., Edelman, G.M. & Cunningham, B.A. (1986). Alternatively spliced mRNAs code for different polypeptide chains of the chicken neural cell adhesion molecule (N-CAM). *Journal of Cell Biology*, **102**, 189–193.

Paoloni-Giacobino, A., Chen, H. & Antonarakis, S.E. (1997). Cloning of a novel human neural cell adhesion molecule gene (NCAM2) that maps to chromosome region 21q21 and is potentially involved in Down syndrome. *Genomics*, **43**, 43–51.

Payne, H.R., Burden, S.M. & Lemmon, V. (1992). Modulation of growth cone morphology by substrate-bound adhesion molecules. *Cell Motility and Cytology*, **21**, 65–73.

Pende, M., Fisher, T.L., Simpson, P.B., Russell, J.T., Blenis, J. & Gallo, V. (1997). Neurotransmitter- and growth factor-induced cAMP response element binding protein phosphorylation in glial cell progenitors: role of calcium ions, protein kinase C, and mitogen-activated protein kinase/ribosomal S6 kinase pathway. *Journal of Neuroscience*, **17**, 1291–1301.

Persohn, E. & Schachner, M. (1987). Immunoelectron microscope localization of the neural cell adhesion molecules L1 and N-CAM during postnatal development of the mouse cerebellum. *Journal of Cell Biology*, **105**, 569–576.

Persohn, E. & Schachner, M. (1990). Immunohistological localization of the neural cell adhesion molecule L1 and NCAM in the developing hippocampus of the mouse. *Neurocytology*, **19**, 807–819.

Persohn, E., Pollerberg, G.W. & Schachner, M. (1989). Immunoelectron-microscopic localization of the 180kDa component of the neural cell adhesion molecule N-CAM in postsynaptic membranes. *Journal of Comparative Neurology*, **288**, 92–100.

Pollerberg, G.E., Schachner, M. & Davoust, J. (1986). Differentiation state-dependent surface mobilities of two forms of the neural cell adhesion molecule. *Nature*, **324**, 462–465.

Ramanathan, R., Wilkemeyer, M.F., Mittal, B., Perides, G. & Charness, M.E. (1996). Alcohol inhibits cell-cell adhesion mediated by human L1. *Journal of Cell Biology*, **133**, 381–390.

Rao, Y., Wu, X.-F., Yip, P., Gariepy, J. & Siu, C.-H. (1993). Structural characterization of a homophilic binding site in the neural cell adhesion molecule. *Journal of Biological Chemistry*, **268**, 20630–20638.

Reid, R.A. & Hemperly, J.J. (1992). Variants of human L1 cell adhesion molecule arise through alternate splicing of RNA. *Journal of Molecular Neuroscience*, **3**, 127–135.

Ruppert, M., Aigner, S., Hubbe, M., Yagita, H. & Altevogt, P. (1995). The L1 adhesion molecule is a cellular ligand for VLA-5. *Journal of Cell Biology*, **131**, 1881–1891.

Rutishauser, U. (1994). Adhesion molecules of the nervous system. *Current Opinion in Neurobiology*, **3**, 709–715.

Rutishauser, U. (1998). Polysialic acid at the cell surface: biophysics in service of cell interactions and tissue plasticity. *Journal of Cellular Biology*, **70**, 304–312.

Rutishauser, U. & Jessell, T.M. (1988). Cell adhesion molecules in vertebrate neural development. *Physiology Reviews*, **68**, 819–857.

Sadoul, R., Kirchhoff, F. & Schachner, M. (1989). A protein kinase activity is associated with and specifically phosphorylates the neural cell adhesion molecule L1. *Journal of Neurochemistry*, **53**, 1471–1478.

Saffell, J.L., Williams, E.J., Mason, I.J., Walsh, F.S. & Doherty, P. (1997). Expression of a dominant negative FGF receptor inhibits axonal growth and FGF receptor phosphorylation stimulated by CAMs. *Neuron*, **18**, 231–242.

Schachner, M. (1991). Neural recognition molecules and their influence on cellular functions. In P.C. Letourneau, S.B. Kater, & E.R. Macagno (Eds.), *The nerve growth cone* (pp. 237–254). New York: Raven Press.

Schmid, R.-S., Graff, R. & Maness, P.F. (1998). The neural cell adhesion molecule L1 activates MAP kinase and CREB signaling pathways. *Society of Neuroscience Abstracts*, **24**, 117.2.

Schmid, R.S., Graff, R., Schaller, M.D., Chen, S., Schachner, M., Hemperley, J.J. & Maness, P.F. (1999). NCAM stimulates the Ras-MAPK pathway and CREB phosphorylation in neuronal cells. *Journal of Neurobiology*, **38**, 542–555.

Schmid, R.-S., Pruitt, W.M. & Maness, P.F. (2000). A MAP kinase signalling pathway mediates and requires Src-dependent endocytosis neurite outgrowth on L1. *Journal of Neuroscience*, **11**, 4177–4188.

Schmidt, C., Kunemund, V., Wintergerst, E.S., Schmitz, B. & Schachner, M. (1996). CD9 of mouse brain is implicated in neurite outgrowth and cell migration in vitro and is associated with the alpha 6/beta 1 integrin and the neural adhesion molecule L1. *Journal of Neuroscience Research*, **43**, 12–31.

Scholey, A.B., Rose, S.P., Zamani, M.R., Bock, E. & Schachner, M. (1993). A role for the neural cell adhesion molecule in a late, consolidating phase of glycoprotein synthesis six hours following passive avoidance training of the young chick. *Neuroscience*, **55**, 499–509.

Schuch, U., Lohse, M.J. & Schachner, M. (1989). Neural cell adhesion molecules influence second messenger systems. *Neuron*, **3**, 13–20.

Schuster, C.M., Davis, G.W., Fetter, R.D. & Goodman, C.S. (1996a). Genetic dissection of structural and functional components of synaptic plasticity. I. Fasciclin II controls synaptic stabilization and growth. *Neuron*, **17**, 641–654.

Schuster, C.M., Davis, G.W., Fetter, R.D. & Goodman, C.S. (1996b). Genetic dissection of structural and functional components of synaptic plasticity. II. Fasciclin II controls presynaptic structural plasticity. *Neuron*, **17**, 655–667.

Seki, T. & Rutishauser, U. (1998). Removal of polysialic acid–neural cell adhesion molecule induces aberrant mossy fiber innervation and ectopic synaptogenesis in the hippocampus. *Journal of Neuroscience*, **18**, 3757–3766.

Shock, L.P., Bare, D.J., Klinz, S.G. & Maness, P.F. (1995). Protein tyrosine phosphatases expressed in developing brain and retinal Müller glia. *Molecular Brain Research*, **28**, 110–116.

Simon, J.R., Graff, R.D. & Manes, P.F. (1998). Microtubule dynamics in a cytosolic extract of fetal rat brain. *Journal of Neurocytology*, **27**, 119–126.

Stallcup, W.B. & Beasley, L. (1985). Involvement of the nerve growth factor-inducible large external glycoportein (NILE) in neurite fasciculation in primary culutres of rat brain. *Proceedings of the National Academy of Sciences, USA*, **82**, 1276–1280.

Stoeckli, E.T. & Landmesser, L.T. (1995). Axonin-1, Nr-CAM, and Ng-CAM play different roles in the *in vivo* guidance of chick commisural neurons. *Neuron*, **14**, 1165–1179.

Stork, O., Welzl, H., Cremer, H. & Schachner, M. (1997). Increased intermale aggression and neuroendocrine response in mice deficient for the neural cell adhesiuon molecules. *European Journal of Neuroscience*, **9**, 424–434.

Storms, S.D. & Rutishauser, U. (1998). A role for polysialic acid in neural cell adhesion molecule heterophilic binding to proteoglycans. *Journal of Biological Chemistry*, **273**, 27124–27129.

Takeda, Y., Asou, H., Murakami, Y., Miura, M., Kobayashi, M. & Uyemura, K. (1996). A non-neuronal isoform of cell adhesion molecule L1: tissue-specific expression and functional analysis. *Journal of Neurochemistry*, **66**, 2338–2349

Tang, J., Rutishauser, U. & Landmesser, L. (1994). Polysialic acid regulates growth cone behavior during sorting of motor axons in the plexus region. *Neuron*, **13**, 405–414.

Thanos, S., Bonhoeffer, F. & Rutishauser, U. (1984). Fiber-fiber interaction and tectal cues influence the development of the chicken retinotectal projection. *Proceedings of the National Academy of Sciences, USA*, **81**, 1906–1910.

Tomasiewicz, H., Ono, K., Yee, D., Thompson, C., Goridis, C., Rutishauser, U. & Magnuson, T. (1993). Genetic deletion of a neural cell adhesion molecule variant (N-CAM-180) produces distinct defects in the central nervous system. *Neuron*, **11**, 1163–1174.

Tuvia, S., Garver, T.D. & Bennett, V. (1997). The phosphorylation state of the FIGQY tyrosine of neurofascin determines ankyrin-binding activity and patterns of cell segregation. *Proceedings of the National Academy of Sciences, USA*, **94**, 129957–12962.

van Kammen, D.P., Poltorak, M., Kelley, M.E., Yao, J.K., Gurklis, J.A., Peters, J.L., Hemperly, J.J., Wright, R.D. & Freed, W.J. (1998). Further studies of elevated cerebrospinal fluid neuroral cell adhesion molecule in schizophrenia. *Biological Psychiatry*, **43**, 680–686.

Vawter, M.P., Hemperly, J.J., Freed, W.J. & Garver, D.L. (1998). CSF N-CAM in neuroleptic-naive first-episode patients with schizophrenia. *Schizophrenia Research*, **34**, 123–131.

von Bohlen und Halbach, F., Taylor, J. & Schachner, M. (1992). Cell type-specific effects of the neural adhesion molecules L1 and N-CAM on diverse second messenger systems. *European Journal of Neuroscience*, **4**, 896–909.

West, J.R. & Goodlett, C.R. (1990). Teratogenic effects of alcohol on brain development. *Annals of Medicine*, **22**, 319–326.

Wilkemeyer, M.F. & Charness, M.E. (1998). Characterization of ethanol-sensitive and insensitive fibroblast cell lines expressing human L1. *Journal of Neurochemistry*, **71**, 2382–2391.

Williams, E.J., Furness, J., Walsh, F.S. & Doherty, P. (1994a). Activation of the FGF receptor underlies neurite outgrowth stimulated by L1, N-CAM, and N-cadherin. *Neuron*, **13**, 583–594.

Williams, E.J., Walsh, F.S. & Doherty, P. (1994b). The production of arachidonic acid can account for calcium channel activation in the second messenger pathway underlying neurite outgrowth stimulated by NCAM, N-Cadherin, and L1. *Journal of Neurochemistry*, **62**, 1231–1234.

Wong, E.V., Kenwrick, S., Willems, P. & Lemmon, V. (1995). Mutations in the cell adhesion molecule L1 cause mental retardation. *Trends in Neuroscience*, **18**, 168–172.

Wong, E.V., Schaefer, A.W., Landreth, G. & Lemmon, V. (1996a). Casein kinase II phosphorylates the neural cell adhesion moleculae L1. *Journal of Neurochemistry*, **66**, 779–786.

Wong, E.V., Schaefer, A.W., Landreth, G. & Lemmon, V. (1996b). Involvement of p90rsk in neurite outgrowth mediated by the cell adhesion molecule L1. *Journal of Cell Biology*, **271**(30), 18217–18223.

Wood, G.K., Tomasiewicz, H., Rutishauser, U., Magnuson, T., Quirion, R., Rochford, J. & Srivastava, L.K. (1998). NCAM-180 knockout mice display increased lateral ventricle size and reduced prepulse inhibition of startle. *Neuroreport*, **9**, 461–466.

Zhao, X., Yip, P.M. & Siu, C.H. (1998). Identification of a homophilic binding site in immunoglob-ulin-like domain 2 of the cell adhesion molecule L1. *Journal of Neurochemistry*, **71**, 960–971.

Zhu, H., Wu, F. & Schacher, S. (1994). Aplysia cell adhesion molecules and serotonin regulate sensory cell–motor cell interactions during early stages of synapse formation *in vitro*. *Journal of Neuroscience*, **14**, 6886–6900.

Further Reading

Demyanenko, G., Tsai, A. & Maness, P.F. (1999). Abnormalities in neuronal process extension, hippocampal development, and the ventricular system of L1 knockout mice. *Journal of Neuroscience*, **19**, 4907–4920.

Kamiguchi, H. & Lemmon, V. (1998). A neuronal form of the cell adhesion molecule L1 contains a tyrosine-based signal required for sorting to the axonal growth cone. *Journal of Neurosciece*, **18**, 3749–3756.

Rutishauser, U. & Jessell, T.M. (1988). Cell adhesion molecules in vertebrate neural development. *Physiology Reviews*, **68**, 819–857.

II.10
The Development of Descending Projections

JEAN ARMAND

ABSTRACT

In order to study the development of descending projections we will begin with an overview, presented from below, that is from the final motor common pathway, namely the motoneurons projecting to the various muscles. The medially and laterally descending brainstem pathways could thus be viewed as controlling the axio-proximal and the distal musculature, respectively. In addition, descending brainstem, aminergic pathways, terminating in the dorsal horn and directly on the motoneurons, are part of the "emotional motor system". Finally, the corticospinal pathway belongs to both the medial and lateral systems, allowing the cortex to control the axio-proximo-distal musculature. From a phylogenetical point of view these latter projections have an increasing area of spinal terminations, from the dorsal horn, to the intermediate zone and finally directly on the motoneurons.

In some way the timetable of maturation of descending projections in humans follow the above sequence, from motoneurons to medial and lateral brainstem projections, and finally to corticospinal projections.

The motoneurons are the first neuroblasts to differentiate and to connect their target muscles, long before dorsal root ganglion neurons enter the spinal grey matter and contact spinal interneurons and motoneurons. In the meantime, medially descending pathways arrive in the spinal cord and, by way of interneurons, can subserve axial movements. During the first prenatal weeks there are three critical periods of behavioural development that correspond to three critical periods of synapse formation: the onset of reflex activities at 5.5 weeks, the appearance of local reflex activities at 7.5–8.5 weeks, and the emergence of integrated motor responses evoked by combined stimuli. The sequential production of spinal neurons in the human embryo as a "retrograde" sequence with respect to impulse conduction in the "reflex pathway" appears to be a general developmental process, which is recognized

A.F. Kalverboer and A. Gramsbergen (eds.), Handbook of Brain and Behaviour in Human Development, 219–260
© *2001 Kluwer Academic Publishers. Printed in Great Britain.*

from amphibians to mammals, as exemplified in rats. It is suggested that coordinated, regional, spontaneous motility may have a morphological substrate in the spinal cord before the reflex circuit is closed.

In the development of descending projections, brainstem pathways are the first to invade the spinal cord, and finally have access to the final motor command path. This is the case in human development; it is also the case for more ancient vertebrates, where medial descending brainstem pathways, controlling axial and proximal muscles, develop earlier than lateral descending pathways, controlling distal limb muscles. This temporal sequence corresponds to the phylogenetic sequence of appearance of the brainstem pathways. In order to gain a better insight into human development the well-documented data obtained in amphibians and rodents will be summarized. The corticospinal tract, the phylogenetically most recent motor pathway, develops last in the human embryo, and here also the comparison with subprimates corticospinal development will give a better insight into primate corticospinal development. Current knowledge on the downward growth, the myelination and the conduction velocity of the human pyramidal tract is discussed.

The human pyramidal tract is not only the last-maturing motor pathway during development, it also contains phylogenetically ancient projections that terminate on spinal interneurons, and phylogenetically more recent ones, namely the corticomotoneuronal connections. The development of these two kinds of corticospinal projections has been extensively studied in subprimates and non-human primates, respectively. The protracted expansion of corticomotoneuronal projections to hand muscles motor nuclei in primates and humans is in marked contrast to the retraction of exuberant projections that characterizes the development of corticospinal projections, and further of other descending pathways, to spinal interneurons.

INTRODUCTION

To better understand motor control, one can observe the development of descending projections in humans. The first step is the early maturation of the final common motor pathway, where the motoneurons project to the various muscles. The second step is the downward growth of descending brainstem pathways that steer the axial musculature, as in the limbless vertebrate ancestor, where swimming was the only locomotor pattern. The third step is the arrival in the spinal cord of the brainstem pathways that drive limb movements. With the increasing development of the cerebral cortex, the last-maturing descending pathways, the corticospinal projections, begin to appear. These first cortical projections extend to the motoneurons via the spinal interneurons, as is also seen in subprimates. However, later direct corticomotoneuronal connections are established, which provide in humans, as in some non-human primates, a monosynaptic linkage between the motor cortex and the spinal motoneurons.

To understand the development and function of the descending projections in humans, it is thus necessary to adopt a phylogenetic perspective. We will begin with an overview of the descending projections to the spinal cord, followed by some milestones in human brain development. Current knowledge on the cytogenetic principles of spinal cord ontogeny will be exemplified in humans as a common sequence shared by other vertebrates. The development of medially descending brainstem pathways in humans, that preferentially control axial and proximal limb muscles, will be compared to that observed in amphibians, where a limbless tadpole stage is observed. Then the development of laterally descending brainstem pathways in humans, that preferentially control limb muscles, will be compared to that observed in quadrupedal rodents. Finally, the development of corticospinal projections in humans will be compared to the different types of development observed in various mammals. The distinction between the phylogenetically more ancient corticospinal projections to spinal interneurons, and the more recent corticomotoneuronal connections will also be discussed.

DESCENDING PROJECTIONS TO THE SPINAL CORD: AN OVERVIEW

The descending projections from the cerebral cortex and the brainstem to the spinal interneurons and motoneurons represent the way in which the brain regulates movements of the body and limbs, which is commonly known as the motor system.

Until the 1960s, the motor system tended to be identified with the motor cortex (Walshe, 1942). This was probably due to the early observations that, after cortical damage in human patients, a severe motor defect known as hemiplegia was observed (for a review see Armand, 1982). In the early 1960s Kuypers high-lighted the fact that the descending cortical and brainstem projections do not derive their motor capacities from their connections with the motor cortex, but from their direct or indirect access to the final common motor pathway, namely the motoneurons (Kuypers, 1964, 1981; Lawrence & Kuypers, 1968a,b). "In order to unravel the anatomical and functional organization of the motor system it should therefore be viewed from below, i.e., from the level of the interneurons and motoneurons rather than from above, e.g. from the level of the motor cortex" (Kuypers, 1982).

The somatic motoneurons are concentrated into medial and lateral longitudinal groups, which are located in the spinal ventral horn (Figure 1). Whereas the motoneurons of the medial group innervate axial muscles, the motoneurons of the lateral group innervate the remainder of the body and limb muscles. Within the lateral motoneuronal cell group a detailed somatotopic organization is observed in both the cervical and lumbosacral enlargements. Thus, the ventromedial motoneuronal cell columns innervate girdle and proximal limb muscles, whereas the dorsolateral motoneurons innervate the muscles intrinsic to the arm and leg. In addition the columns of motoneurons which innervate the most distal extremity muscles, namely the small hand and foot muscles, are located in the most dorsolateral cell groups of the most caudal segments of the cervical (C8–T1; Fritz et al., 1986a,b) and lumbosacral (L7–S1) enlargement, respectively.

The main afferent to the motoneurons is composed of spinal interneurons, the cells of which are concentrated in the intermediate zone (Rexed laminae V–VIII;

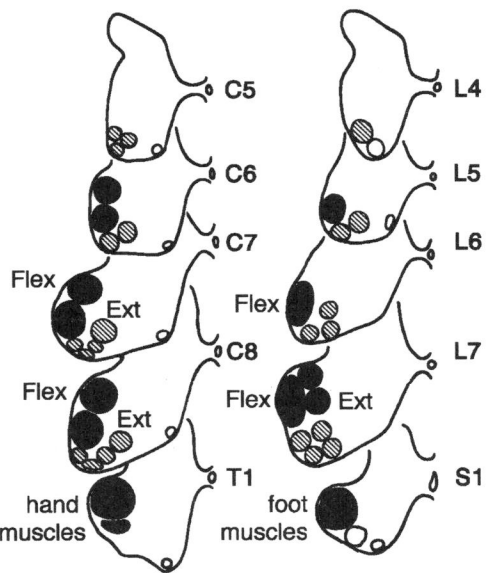

Figure 1 Diagram of the distribution of the different motoneuronal cell groups in the cervical (left column: C5–T1) and lumbosacral (right column: L4–S1) enlargements innervating the different limb muscles in the cat (adapted from Kuypers, 1973). The medial motoneuronal cell group (white) innervate axial muscles, the ventromedial (hatched), girdle and proximal extremity muscles, the dorsolateral (black), the muscles intrinsic to the arm and leg. Within this latter group the ventrally located motoneurons innervate the extensor muscles (Ext.) on the dorsal aspect of the arm, while the dorsally located motoneurons innervate the flexor muscles (Flex.) of the ventral aspect of the arm. Groups of motoneurons, which innervate the most distal extremity muscles (small hand and foot muscles) are situated in the most caudal part of the enlargements.

Rexed, 1952). The spinal interneurons also display a certain somatotopic organization. The relatively short propriospinal neurons, located in the dorsal and lateral parts of the intermediate zone, project, by way of the dorsolateral funiculus, preferentially ipsilaterally, to motoneurons of distal extremity muscles. The relatively long propriospinal neurons, located in the ventromedial part of the intermediate zone, project by way of the ventral funiculus, to some extent bilaterally and throughout the length of the spinal cord, to motoneurons of axial muscles. The central part of the intermediate zone contains short propriospinal neurons, as well as neurons of intermediate length, which axons project, by way of the ventral part of the lateral funiculus, to motoneurons of girdle and proximal limb muscles.

On the basis of their terminal distribution in the spinal intermediate zone and motoneuronal cell groups, the descending brainstem projections may be subdivided into a medial and lateral system (Figure 2).

The medial system of brainstem pathways. This characteristically terminates in the ventromedial parts of the intermediate zone, though some of its fibres also terminate in the dorsal and lateral parts (tectospinal and medullary reticulospinal

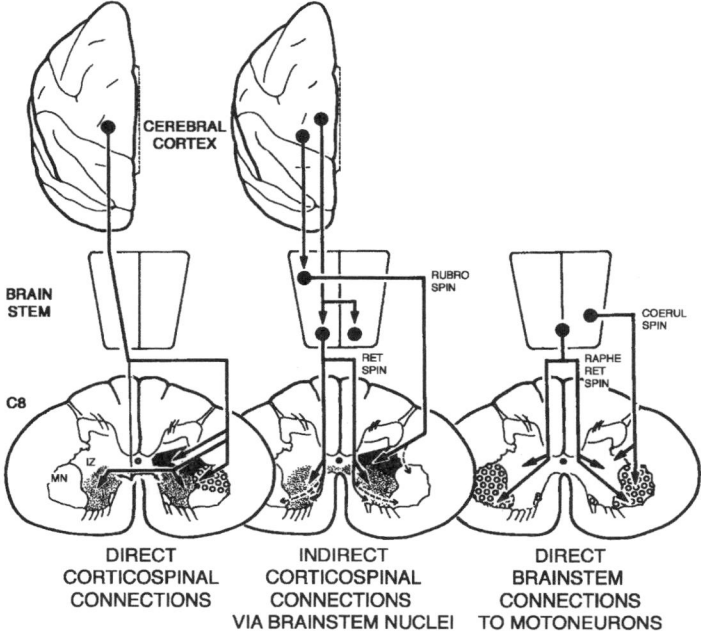

Figure 2 Schematic representation of the descending cortical and brainstem pathways to the spinal intermediate zone (IZ, dorsolateral part: black; ventromedial part: stippled) and motoneuronal cell groups (MN, open circles) in monkey C8 spinal segments (adapted from Kuypers, 1982). Abbreviations: coerul. spin.: coerulospinal tract; raphe ret. spin.: raphe and reticulospinal tracts; ret. spin.: reticulospinal tract; rubro spin.: rubrospinal tract.)

tracts). Interstitiospinal, mesencephalic, pontine and medullary reticulospinal, tectospinal and vestibulospinal pathways descend by way of the ventral and ventrolateral spinal funiculi and terminate on long propriospinal neurons or neurons of intermediate length. The medial brainstem system, steering preferentially axial and proximal limb muscles, represents the basic system by which the brain controls movements. This control is mainly concerned with body and integrated limb–body movements and with synergistic movements of the whole limb.

The lateral system of brainstem pathways. This characteristically terminates in the dorsolateral part of the intermediate zone. The rubrospinal tract, certain reticulospinal fibres and raphespinal fibres, originating from rostral parts of the raphe nucleus, descend by way of the dorsolateral funiculus, and terminate on short propriospinal neurons. The lateral brainstem system, and in particular the rubrospinal tract, adds some resolution to the control of the whole limb devoted to the medial system. It provides the capacity to execute independent movements of the individual limbs, especially their distal parts.

In non-mammalian quadrupeds the presence of a rubrospinal tract is related to the presence of limbs (ten Donkelaar, 1988); however, the bulk of the supraspinal projections is formed by reticulospinal projections (ten Donkelaar,

2000). These projections are not only part of the classical medial and lateral brainstem pathways, but also comprise *aminergic projections terminating in the dorsal horn and in the motoneuronal cell groups* (Figure 2). Brainstem projections to sensory cell groups, such as the dorsal horn and the lateral cervical nucleus, are derived from the ventrolateral tegmentum, from the nucleus coeruleus and sub-coeruleus (noradrenergic projections), from the ventral portion of the medial reticular formation and from the raphe nuclei in the rostral medulla oblongata (serotonergic projections). These projections could thus subserve control of pain transmission, as indicated by physiological and behavioural findings (Holstege & Kuypers, 1982). Brainstem projections to both the medial and lateral somatic motoneuronal cell groups are mainly derived from the caudal medial reticular formation and the raphe nuclei (serotonergic projections), but also from the dorsolateral tegmentum comprising the area of nucleus coeruleus and sub-coeruleus (noradrenergic projections). These aminergic projections to motoneurons may therefore represent gain-setting systems which determine the responsiveness of the final common pathway (Holstege & Kuypers, 1982). Kuypers considered these extensive noradrenergic and serotonergic brainstem projections as a third component of the motor system (Kuypers, 1982). The brainstem areas at the origin of these spinal projections are under powerful limbic control (Holstege & Kuypers, 1987; Holstege, 1991). This third component of the motor system might therefore be instrumental in providing motivational drive in the execution of movements. This "emotional motor system", which is present from the most ancient non-mammalian species until primates, can exert a powerful influence on all regions of the spinal cord, and becomes a determinant drive of motor behaviour (Holstege, 1991, 1992).

The descending cortical pathways include *the corticospinal tract* which passes directly to the spinal cord and cortical projections to the cells of origin of descending brainstem pathways, and thus establish indirect connections to the spinal cord (Figure 2). The cortex is thus concerned with a direct and indirect control of the axio-proximo-distal musculature. With regard to its cortical origin, the corticospinal tract has an increasingly expanding area among mammals, from a sensorimotor amalgam in marsupials, such as the North American opossum (Martin & Fisher, 1968), to distinct motor and sensory cortices, such as in carnivores (Van Crevel & Verhaart, 1963a,b; Armand & Kuypers, 1980) and to sensory, primary and non-primary motor cortices (premotor, supplementary motor and cingulate motor areas) in non-human primates and humans (Dum & Strick, 1991; Minckler *et al.*, 1944). As regards its spinal terminations within the intermediate zone, the corticospinal pathway belongs to both the medial and lateral systems, but also terminates in the dorsal horn, and in some species in the motoneuronal cell groups (Kuypers & Brinkman, 1970). The projections to the dorsal horn seem to represent the primordial component of the corticospinal tract. This is mainly derived from the granular somatosensory cortex, and probably subserves sensory control. The components projecting to the intermediate zone and the motoneuornal cell groups are derived from the agranular motor cortex and mainly subserve motor control (for a review see Porter & Lemon, 1993). The spinal distribution of these different corticospinal components differs in four different groups of mammals, more or less in parallel with their motor capacity (Kuypers, 1981; Armand, 1982):

1. Mammals with corticospinal fibres extending only to cervical or mid-thoracic segments and terminating in the dorsal horn, such as marsupials (opossum: Martin & Fisher, 1968; Martin *et al.*, 1975);
2. Mammals with corticospinal fibres extending throughout the spinal cord and terminating in the dorsal horn and the intermediate zone, such as carnivores (cat: Niimi *et al.*, 1963; Nyberg-Hansen & Brodal, 1963; Armand & Kuypers, 1980; Armand *et al.*, 1985);
3. Mammals with corticospinal fibres extending throughout the spinal cord and terminating in the dorsal horn, intermediate zone, and dorsolateral parts of the lateral motoneuronal cell groups, such as most non-human primates (macaque monkey: Kuypers, 1960; Kuypers & Brinkman, 1970; Bortoff & Strick, 1993);
4. Mammals with corticospinal fibres extending throughout the spinal cord and terminating in dorsal horn, intermediate zone, and dorsolateral as well as ventromedial parts of the lateral motoneuronal cell groups, such as some non-human primates and humans (Kuypers, 1964; Petras, 1968; Schoen, 1964, 1969).

Fine cortical control of movements is thus achieved in most mammals by phylogenetically ancient corticospinal projections onto the different categories of spinal interneurons of the intermediate zone, and in some mammals by phylogenetically more recent projections directly onto motoneurons, namely the corticomotoneuronal connections.

In some way the timetable of the development of descending projections in humans follows the above sequence, from motoneurons to medial and lateral brainstem projections, and finally to corticospinal projections.

SOME MILESTONES IN HUMAN BRAIN DEVELOPMENT

Staging problems

An exact comparison between the data reported in the literature concerning human brain development is difficult, because there is always uncertainty about the age of a human embryo (Windle, 1970). The stated time of last menstruation is notoriously inaccurate. The patient's recollection of the dates of intercourse can also be erroneous but, even if known, the time of fertilization may occur later. The size of the embryo is also a poor criterion of age. Measurements taken before fixation differ from those afterwards. Furthermore, size measurements vary with the kind of fixative solution used. The best one can do in most instances is to estimate embryonic age from the state of development of certain clearly evident structures.

The Carnegie system for staging primate embryos was first defined by Streeter (Heuser & Streeter, 1941; Streeter, 1951), and complemented for the earlier stages by Heuser & Corner (1957) and O'Rahilly (1973). This staging system is based on the external form and internal structures of human embryos. The beginning of the ossification of the cartilaginous precursor in the humerus was arbitrarily defined as the end of the embryonic and the beginning of the fetal period. More recently the transition from embryo to fetus was defined as the time of closure of the secondary palate (Wilson, 1973). As to the embryonic period, 23 "horizons" were distinguished. The horizons were each 2 days apart, and numbered by Roman numerals. The original Streeter's horizons were later named "stages" and numbered by Arabic numerals (O'Rahilly & Müller, 1987). Although it is worthwhile to adopt a staging based upon the state of development of certain clearly evident structures, it is thus necessary to better correlate the two methods based on age and length for a given horizon (Olivier & Pineau, 1962). As a consequence we will adopt the postovulation age along with the crown-rump length (CRL) of the embryo, and the Carnegie stage.

The Carnegie system for staging human embryos has been applied to other primates (rhesus monkey: Hendrickx & Sawyer, 1975; Gribnau & Geijsberts, 1981, 1985), and has also been adapted to rodents, up to stage 17 (ten Donkelaar *et al.,* 1979). From studies in different mammals, mainly in rodents, but also in carnivores, and in primates, the development of the different organs in general, and of the central nervous system in particular, follows the same chronological sequence (ten Donkelaar, 2000). All vertebrates pass through stages resembling the primitive three and five brain vesicles of lower forms; but in many ways the development of the human brain thereafter is intrinsically different from that of a fish or an amphibian. Some developmental processes are faster in some species, but slower in others. Ontogeny does not recapitulate phylogeny very closely (Windle, 1970).

General human brain development

Prenatal human development is subdivided into two periods. The embryonic period lasting 8 weeks (until stage 23; embryonic day 56 = E56; O'Rahilly & Müller, 1987, 1994) is followed by the fetal period that continues until the time of delivery at 260–280 days. After the time of implantation at E5–6 the neural groove appears at E18. The three primary brain vesicles are visible at E20 (3 weeks ovulation age), before the closure of the neural tube. Closure of the rostral and caudal neuropore occur at E24 and E26 (3.5 weeks), respectively. At that age limb buds are formed, forelimbs at E26, hindlimbs at E28. Secondary brain vesicles, particularly cerebral hemispheres, become visible at E32 (4 weeks, stage 14). At that age the spinal cord grey matter begins to differentiate in ventral horn, dorsal horn, and intermediate zone. Hand and foot plates develop at E33 and E37 (5 weeks), followed by fingers and toes at E41 and E44 (6 weeks), respectively. Finally at E56–60 the fetal period replaces the embryonic period.

CYTOGENETIC PRINCIPLES OF SPINAL CORD ONTOGENY

Neuronal development in the spinal cord is relatively well known in humans, as in other vertebrates. We will first present the data obtained in humans, including some interesting aspects of synaptogenesis. We will then extend to more general cytogenetic principles of spinal cord ontogeny in vertebrates.

Neuronal development in the human spinal cord

Primary efferent neuroblasts – primordial motoneurons – are among the first elements of the human brain to be differentiated (Windle & Fitzgerald, 1937).

Neuroblasts, primordial neurons, are recognizable (using pyridine-silver methods) when neurofibrils appear in the cytoplasm and neuronal processes begin to sprout (Windle, 1970). In a 4-week (ovulation age) embryo the primordial motoneurons occupy a medial position on either side of the floor plate of the neural tube, forming an interrupted ventral column from the lower rhomben-cephalon (posterior cerebral vesicle) to the upper spinal cord (Figure 3).

Differentiation of the primary efferent neuroblasts slightly precedes that of primary afferent and second-order elements. In a late 4-week embryo only a few neuroblasts of primary afferents – primordial sensory ganglion neurons – have differentiated in the cranial ganglia but none in the spinal ganglia. Moreover, very few axons leave the ganglia, and none enters the neural tube. These axons do not send axonal collaterals into the primordial of the spinal neuropil until the end of the 7th week, completing reflex connections in the brachial region during the 8th week.

During the second half of the embryonic period (4–8 weeks) the neuronal development of the human spinal cord proceeds in a retrograde fashion with res-pect to the direction of flow of nerve impulses, namely from the ventral horn to the intermediate zone, and finally the dorsal horn (Okado et al., 1979; Okado, 1980).

At light microscopic level (2 μm thick spinal cord transverse sections stained with toluidine blue), the motor nucleus forms a bulge in the ventrolateral aspect of the spinal cord, whereas the dorsal horn and the intermediate zone are still undifferentiated, at 4 weeks (8 mm CRL, stage 14). Nerve fibres of the ventral root are seen to pass to the periphery as a spinal nerve, and the dorsal root fibres derived from the spinal ganglion to the spinal cord.

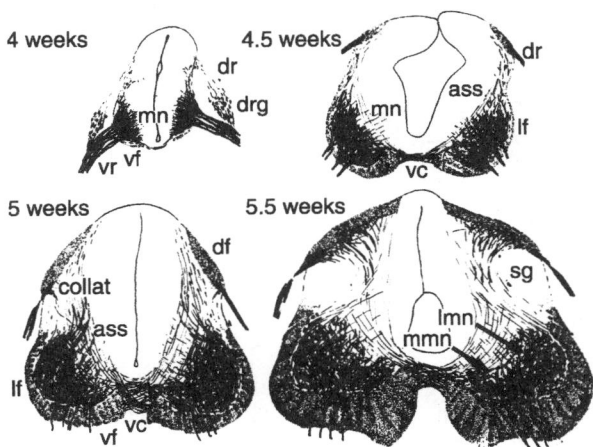

Figure 3 Camera lucida drawings of C7 spinal cord sections stained by the pyridine silver method, of human embryos of 4, 4.5, 5 and 5.5 weeks ovulation age (adapted from Windle & Fitzgerald, 1937 and the age equivalent Table 2 of Okado et al., 1979). (ass: association neurons, collat: collaterals of dorsal root fibres, df: dorsal funiculus, dr: dorsal root, drg: dorsal root ganglion, lf: lateral funiculus, lmn: lateral motoneurons, mn: motoneurons, mmn: medial motoneurons, sg: gelatinous substance, vc: ventral commissure, vf: ventral funiculus, vr: ventral root.)

At 5 weeks the ventral horn develops two motor nuclei, medial and lateral, in a 14 mm CRL embryo (stage 17; Figure 4), which appear well separated in a 16 mm CRL embryo (stage 18; Figure 5). At that age the intermediate zone is more developed, but the dorsal horn remains undifferentiated. The white matter develops into dorsal, lateral and ventral funiculi. From 5.5 to 7 weeks the different parts of the spinal grey matter progressively differentiate. Thus, at 5.5 weeks (18 mm CRL, stage 19), the ventral horn and the intermediate zone are modeately developed. At 6 weeks (22 mm CRL, stage 20), the dorsal horn and intermediate zone develop rapidly. Finally, at 7 weeks (29 mm CRL), the total area of the motor neuropil and the distance separating the motoneuron somata increase.

The neuronal development in the human spinal cord, as observed at the light microscopic level, is further emphasized by the synaptogenesis, as observed at the ultrastructural level. The synaptogenesis in the cervical spinal cord of human embryos revealed that, at 4 weeks, no synapses are found in any region of the cervical enlargement (Okado *et al.*, 1979). Synapses first appear at 4.5 weeks in the medial and lateral motor neuropil (Okado, 1981). Most of these synaptic contacts are found to form axodendritic synapses, but an occasional axosomatic synapse is also encountered. By this embryonic stage the dorsal root fibres have not yet entered into the spinal grey matter. As a consequence the first synaptic contact in the cervical spinal cord may represent connections between spinal interneurons and motoneurons. It is only at 5 weeks ovulation age that synapses were found outside the motor nucleus, namely in the dorsal marginal layer, where primary afferent fibres contacted spinal interneurons (Okado, 1981).

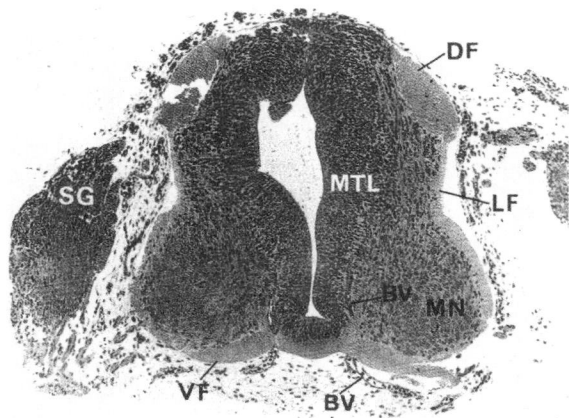

Figure 4 Low magnification (× 70) of a 2-μm thick, epon-embedded, toluidine-blue-stained transverse section through the cervical cord of a 5-week (ovulation age) human embryo (14 mm crown–rump length). The marginal layer develops to the dorsal (DF), the lateral (LF) and the ventral funiculi (VF). The central canal extends in the dorsoventral direction and is surrounded by a relatively thick matrix layer (MTL). The motor nucleus (MN) is markedly increasing in size and occupies the mantle layer of the basal plate. Note the blood vessels (BV) that are penetrating into the spinal cord along the interface between the ventral matrix and the mantle layer. Ventral root fibres are seen to pass to the periphery and dorsal root fibres derived from the spinal ganglion (SG) to the spinal cord. (Reproduced with permission from Okado *et al.*, 1979.)

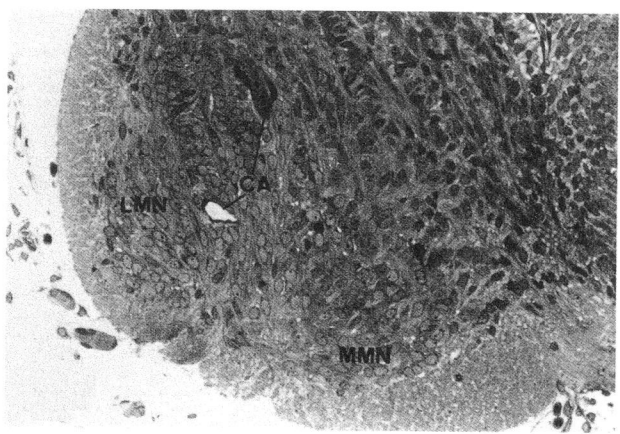

Figure 5 High magnification (×220) of the left ventral quadrant of the cervical spinal cord of a human embryo of 5-week ovulation age (16 mm crown–rump length). The motor nucleus is well separated into the medial (MMN) and the lateral (LMN) nuclei, with an increasing number of motoneurons having large and pale cell nuclei. Penetration of capillaries (CA) into the motor neuropil is seen. (Reproduced with permission, Okado *et al.*, 1979.)

This is in keeping with the development of human fetal reflexes. Fitzgerald & Windle (1942) have shown that the first cutaneous reflex activities can be elicited at 5.5 weeks (Hooker, 1952). However, at the onset time of spinal cutaneous reflex only a small number of synapses were observed in the motor neuropil, and mainly axodendritic synapses (Figure 6). Subsequently there are two other critical periods of behavioural development that coincide with two critical periods of synapse formation in the human spinal cord: the appearance of local reflex activities between 7.5 and 8.5 weeks, corresponding to the first increase in the density of axodendritic (Figure 6) and axosomatic synapses (Figure 7).

Humphrey (1964) reported that local reflexes begin to appear at 7.5 weeks, responding to sensory trigeminal stimuli. She further reported that the partial closure of all fingers can be elicited between 8 and 8.5 weeks. It is at the end of the 8th week that motoneurons become multipolar in shape and their dendrites extend distally (Okado, 1980; Windle, 1970). Therefore the receiving area of axodendritic synapses undergoes an extraordinary increase. In addition, the range of the area of the motoneuron somata becomes broader and the mean value increases gradually (Figure 8). Therefore the receiving area of axosomatic synapses also increases markedly (Okado, 1980).

The second important change in the synapse formation is the marked increase in the covering ratio of axosomatic synapses between 11 and 13.5 weeks (Figure 7). This is in keeping with the changes in the reflex pattern occurring between 11.5 and 12.5 weeks (Humphrey & Hooker, 1959). At 11.5 weeks reflexes are elicited only at the proximal part of the body by the simultaneous stimuli applied to different reflexogenic sites, whereas in fetuses older than 12.5 weeks reflexes can be elicited at both distal and proximal sites.

In summary, during the first 19 prenatal weeks there are three critical periods of behavioural development that correspond to three critical periods of synapse

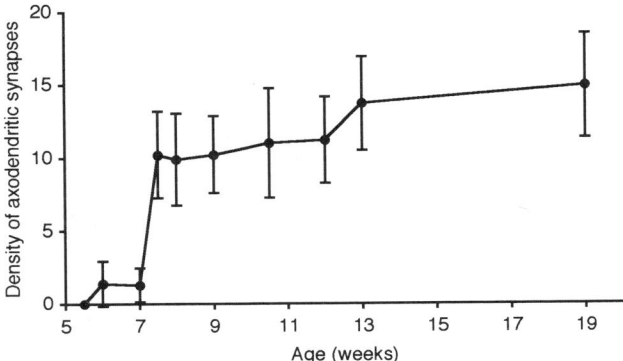

Figure 6 Developmental changes in the density of axodendritic synapses per 200 μm² unit area or the lateral motor column neuropil (the mean and the standard deviation of a 10 unit area for each specimen, postovulation age is in weeks). A significant increase ($p < 0.01$) in the density of axodendritic synapses occurs between 7 and 7.5 weeks, with no significant increase occurring afterwards. (Adapted from Okado, 1980.)

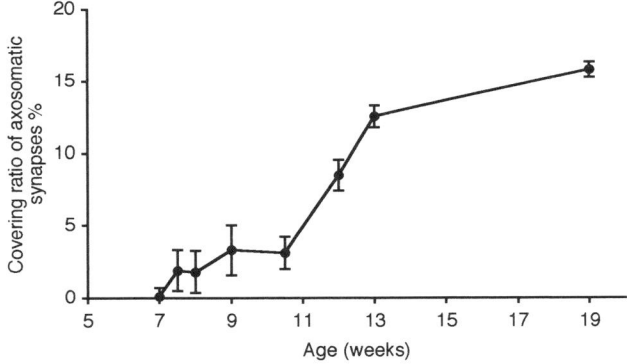

Figure 7 Developmental changes in the covering ratio of axosomatic synapses. The ratio of the circumference of 13 lateral motor column neuron somata to the surface length covered by axosomatic boutons with synaptic contacts were calculated for each specimen (means and standard deviations, postovulation age is in weeks). A significant increase ($p < 0.05$) in the covering ratio of axosomatic boutons occurs between 7 and 7.5 weeks, which corresponds to the increase in the density of axodendritic synapses (see Figure 6). Furthermore, there is a marked increase ($p < 0.01$) in the bouton-covering ratio between 11 and 13.5 weeks, followed by a gradual increase until the end of the embryonic period. (Adapted from Okado, 1980.)

formation: the onset of reflex activities (5.5 weeks), corresponding to the synapse formation outside of the motor neuropil; the appearance of local reflex activities (7.5–8.5 weeks), corresponding to the first increase in the density of axodendritic and axosomatic synapses in the motor neuropil; and the emergence of responses elicited by simultaneously applied double stimuli (11.5–12.5 weeks), corresponding to the second increase in number of axosomatic synapses (Okado, 1980).

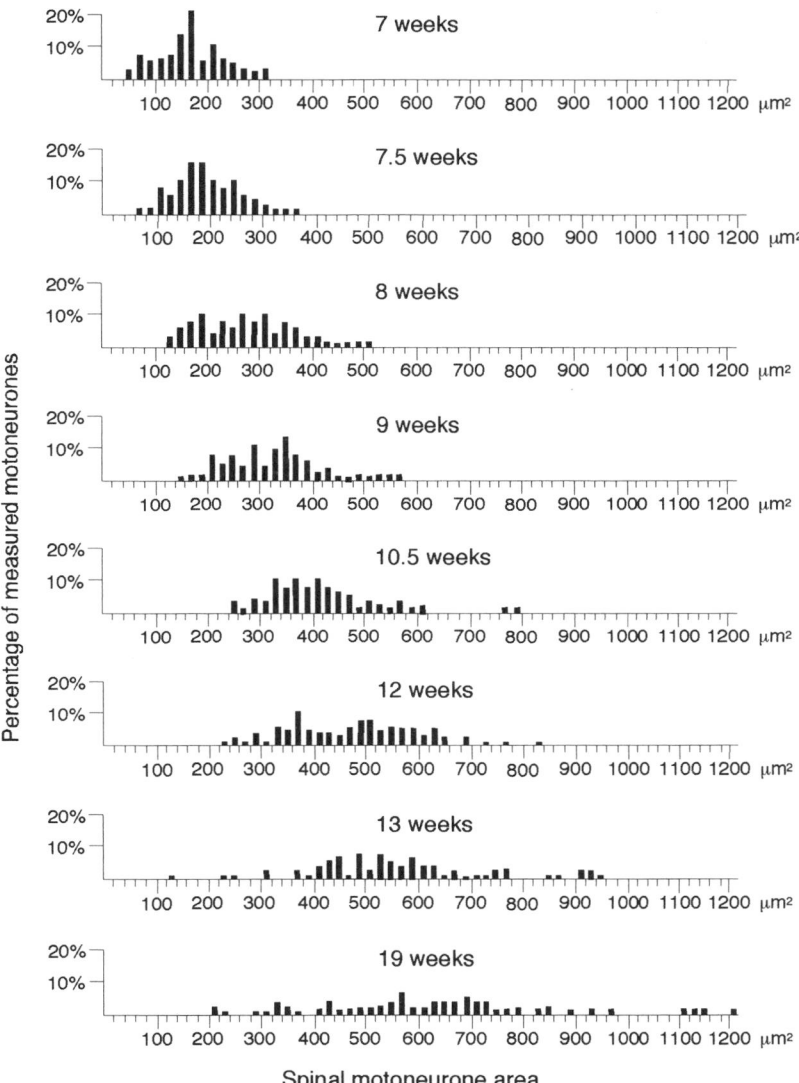

Figure 8 Comparison of the histograms of the area of spinal motoneurons at different embryonic ages (postovulation age in weeks). The cross-sectional area (μm^2) of lateral motor column neuron somata cut through the nucleolar plane, in 2 μm thick toluidine blue-stained epon sections, was measured at a final magnification of $\times$ 1000. In each specimen the number of motoneurons measured was 72–106. At increasing embryonic ages the range of the area of the motoneuron somata becomes broader, the mean value and standard deviation increases gradually (79 ± 64 at 7 weeks, 200 ± 58 at 7.5 weeks, 276 ± 89 at 8 weeks, 328 ± 94 at 9 weeks, 411 ± 108 at 10.5 weeks, 470 ± 123 at 12 weeks, 551 ± 177 at 13 weeks, 608 ± 199 at 19 weeks), whereas the coefficient of variance is approximately the same for fetuses older than 7 weeks. (Adapted from Okado, 1980.)

231

Neuronal development in the vertebrate spinal cord

The sequential production of spinal neurons in human embryo as a "retrograde" sequence with respect to impulse conduction in the "reflex pathway" (Okado *et al.*, 1979; Okado, 1980) appears to be a general developmental process. As emphasized by Altman & Bayer (1984), the ventral-to-dorsal neurogenetic gradient represents an early developmental phenomenon as an expression of certain cytogenetic and organizational principles of spinal cord ontogeny, and the subsequent distortion of this gradient is due to ensuing morphogenetic events in different species.

In anuran amphibians, as in avians, after the establishment of contacts between motoneurons and their target muscles these motoneurons receive connections first from interneurons, then from descending supraspinal fibres, and finally from dorsal root fibres, indirectly by way of spinal interneurons or directly (Okado & Oppenheim, 1985; ten Donkelaar, 2000, for a review).

In mammals this dorsal-to-ventral neurogenetic sequence has been extensively studied in rodents, and is exemplified in rat (Nornes & Carry, 1978; Ozaki & Snider, 1997; Sims & Vaughn, 1979; Wentworth, 1984a,b; for a review see Altmann & Bayer, 1984). Motoneurons are first differentiated from the basal plate to form the ventral horn, then spinal interneurons in the intermediate zone; finally the alar plate differentiates the interneurons of the dorsal horn (Figure 9).

The early-generated motoneurons appear at E11, and their production is maximum at E12 at the cervical level and E14 at thoracic and lumbar levels. From these motoneurons the outgrowth of peripheral motor fibres is apparent at E13 (Figure 9). The early-generated interneurons of the intermediate zone appear at E13. The first interneurons to be generated project their axons contralaterally, forming the ventral commissure and the ventral funiculi at E14. The next interneurons to be generated, migrating laterally, project their axons ipsilaterally forming the lateral funiculus. Finally at E15, interneurons in the dorsal horn are generated. Although dorsal root ganglion cells are first observed within the ganglion at E12 (large ganglion cells are generated before small ones), dorsal root fibres do not enter the cervical cord before E15. Before entering the dorsal horn grey matter, dorsal root fibres distribute different axonal collaterals according to their final destination. The earliest-bifurcating dorsal root fibres (E14) have essentially an intrasegmental distribution. Then (E15) the ascending and descending intersegmental collaterals form the dorsal funiculus propriospinal zone. Finally (E17) suprasegmental dorsal root collaterals form the dorsal funiculus ascending zone. In the meantime intrasegmental axonal collaterals enter the dorsal horn and contact the dorsal horn interneurons. The ingrowth of the dorsal root fibres within the spinal grey matter, and the outgrowth of cervical motoneuron dendrites, have been shown to occur between E15 and E17, with the use of carbocynanine dye (Snider *et al.*, 1992). Finally, between E17 and E19 dense terminal arborization and synaptogenesis of muscle Ia afferents occur on the motoneurons.

In summary, in rats the earliest propriospinal and ascending supraspinal projections appear to develop around E12, that is at the age of peak production of spinal motoneurons and of ingrowth of descending supraspinal projections in the spinal cord. Dorsal root fibres do not enter the spinal grey matter before

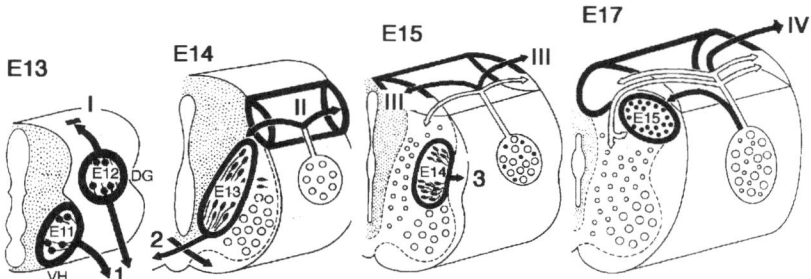

Figure 9 Major developmental events (heavy lines) in the rat cervical spinal cord on embryonic days E13-E17 (adapted from Altman & Bayer, 1984). **At E13** (1): onset of growth of peripheral motor fibres from the early-produced motoneurons of the ventral horn (E11) and of sensory fibres from the early-produced spinal ganglion cells (E12); (I): arrival of the earliest root fibres at the site of the dorsal root entrance zone. **At E14**, (2): formation of the ventral commissure and ventral funiculus by the earliest axons of the contralaterally projecting interneurons that are generated predominantly on day E13; (II): formation of the dorsal root bifurcation zone by the growing intrasegmental dorsal root collaterals. **At E15**, (3): lateral migration of the ipsilaterally projecting interneurons that are generated predominantly on day E14; (III): formation of the dorsal funiculus propriospinal zone by the growing intersegmental (propriospinal) dorsal root collaterals, and production of the small dorsal root ganglion cells (filled circles). **At E17**, (IV): formation of the dorsal funiculus ascending zone by the growing suprasegmental dorsal root collaterals. Settling of the interneurons generated on day E15 in the dorsal horn. Presumed arrival of the small-calibre dorsal root afferents dorsally, and rotation of the large-calibre dorsal root afferents from a ventral direction towards the dorsal horn (white arrow.)

E15. The inverted sequence of neurogenesis within the central reflex pathway is not restricted to the spinal cord. Altman & Bayer (1982) have shown that the motoneurons of the trigeminal, facial, and retrofacial nuclei are generated before the neurons of the sensory portion of the trigeminal nucleus, and the motoneurons of the dorsal vagal nucleus before the neurons of sensory solitary nucleus. Although the issue of the embryonic development of motility cannot be resolved by anatomical evidence, at least with regard to the maturation of the afferent link of spinal reflexes, intrasegmental "coordination" should precede intersegmental "coordination", and the development of brainstem suprasegmental "integration" should be the last event in spinal cord development (Hamburger, 1968).

The importance of the issue of whether the reflex circuit is closed in a retrograde or anterograde sequence is related to the controversy of behavioural embryologists, in the 1930s and 1940s, concerning whether the earliest form of motility is "spontaneous" or "reflexogenic" in nature. In other words, is the developmental organiz ation of motor control due to the integration of initially spontaneous random twitches and jerks, or is it due to initially independent but coordinated local reflexes? If spontaneous movements appear much before elicited or reflexive movements, then it follows that some organization is present in the spinal cord before the reflex circuit is functionally closed. However, investigators who focused their attention on behavioural development in mammalian embryos (Windle, 1944) downplayed the importance of the early random jerks and twitches, and argued that the organization of motility begins with local reflexes, and that these reflexes are the building blocks of behaviour. In contrast, it is suggested that coordinated, regional, spontaneous motility may have a morphological substrate in the spinal cord before the reflex circuit is closed. In E14 rats both the ipsilaterally and contralaterally projecting interneurons are present with axons that can be traced into the ventral and lateral funiculi, respectively (Figure 9). The synapses that have been observed in the ventral horn by this time, as early as E13.5, could very well belong to these interneurons (Vaughn & Grieshaber, 1973). While reflexogenic movements cannot be elicited in the rat before E15.5, the anatomical evidence predicts that spontaneous motility *in utero* appear as early as day E14 (Angulo y Gonzalez, 1932).

In the development of descending projections brainstem pathways are the first to invade the spinal cord, and finally have access to the final motor common path. This is the case in human development, it is also the case for more ancient mammals, in which medial descending brainstem pathways, steering axial and proximal muscles, develop earlier than lateral descending pathways, steering distal limb muscles. This temporal sequence corresponds to the phylogenetic sequence of appearance of the brainstem pathways. In order to gain a better insight into human development we will thus summarize the well-documented data obtained in anuran amphibians and rodents. The corticospinal tract, the phylogenetically most recent motor pathway, develops last in the human embryo; here also the comparison with subprimates' corticospinal development will give a better insight into primate corticospinal development.

DEVELOPMENT OF DESCENDING BRAINSTEM PATHWAYS IN AMPHIBIANS

Anuran amphibians present a unique opportunity to study the development of descending brainstem pathways (see ten Donkelaar, 2000, for review). Their transition from aquatic, limbless tadpoles to juvenile tetrapod animals occurs over a protracted period of time, during which the animal is accessible for experimental studies. Moreover, changes in the locomotor pattern are observed during metamorphosis, when the tail and gills disappear and limbs and lungs develop.

The clawed toad's (*Xenopus laevis*) development has been divided into 66 stages (Nieuwkoop & Faber, 1967). After hatching (around stage 35/38), during all the larval period (until stage 58), locomotion (swimming) consists of coordinated, alternate contractions of the axial muscles on each side of the body. At the end of this larval period hindlimb (stage 46/48) and forelimb (stage 50/54) buds appear. After this limb production (from stage 63) locomotion is gradually taken over by the extremities and swimming is accomplished solely with the limbs. These behavioural developmental changes parallel the development of descending brainstem pathways to the spinal cord.

The earliest-descending brainstem projections arrive in a rather immature spinal cord, very early in development, before hatching (Table 1). The cells of origin of the first descending supraspinal fibres have been identified in the hindbrain reticular formation, namely the nucleus reticularis inferior and medius, by using the retrograde transport of horseradish peroxidase (HRP crystal) applied in the rostral spinal cord (Figure 10). These fibres penetrate the spinal cord only

Table 1 Development of supraspinal projections in anuran amphibian *Xenopus laevis* at different stages (st). [a]: Gonzalez *et al.*, 1994a; [b]: Nordlander *et al.*, 1985; [c]: ten Donkelaar & de Boer-van Huizen, 1982; [d]: ten Donkelaar *et al.*, 1991; [e]: van Mier & ten Donkelaar, 1984; [f]: van Mier *et al.*, 1986 (adapted from ten Donkelaar, 2000).

Nuclei	Innervation of the spinal cord	
	Rostral cord[e]	Tail cord[b]
Reticular formation		
Inferior	st. 28	st. 37
Medius	st. 28	st. 37
Superior	st. 35/36	st. 37
Nucleus reticularis isthimi	st. 43/44	st. 43
Interstitial nucleus of the MLF	st. 30/31	st. 39
Raphe nuclei	< st. 35/36	st. 41
Setotonergic projections[f]	st. 32	
Mauthner cells	st. 30/31	st. 37
Vestibular nuclei		
Lateral nucleus	st. 35/36	st. 39
Medial and inferior nuclei	< st. 50[c]	
Locus coeruleus		
Coeruleospinal projections	st. 43?	?
Noradrenergic projections[a]	st. 41	
Red nucleus	st. 48[d]	st. 58
Hypothalamus	st. 57[c]	

30 hours after fertilization, when the neural tube has just closed (stage 28) (van Mier & ten Donkelaar, 1984). At this stage of development only the rostral part of the body musculature is able to contract in a head flexure movement. As behaviour develops, and uncoordinated swimming occurs (stage 30/31), contacts between ingrowing reticulospinal axons and primary motoneurons are seen (van Mier *et al.*, 1989). In addition, more brainstem centres project to the rostral spinal cord, namely the interstitial nucleus of the medial longitudinal fasciculus (MLF) and the transitory Mauthner cells. After hatching, the developing larva becomes free-swimming (stage 35/36), with more rostrally arising reticulospinal projections, from the nucleus reticularis superior, as well as vestibulospinal projections, from the lateral vestibular nucleus, also reaching the rostral spinal cord. At that age (stage 32) the first serotonergic raphespinal projections also arrive in the rostral spinal cord, and the inflexible embryonic spinal locomotor network is transformed into a potentially more flexible adult-like one (Sillar *et al.*, 1992, 1995; van Mier *et al.*, 1986). All these early-descending brainstem projections to the rostral spinal cord reach the tail spinal cord (at stage 37), by way of the ventral funiculi, before limb buds appear. When hindlimb buds appear, reticulospinal projections are comparable with the adult pattern. These include the projections from the nucleus reticularis isthmi, and the putative coerulospinal projections, but exclude the rubrospinal and hypothalamospinal projections (Gonzalez *et al.*, 1994a,b).

Shortly after hindlimb buds arise (stage 46/48), lumbar motoneurons establish contacts with their target muscles (van Mier & ten Donkelaar, 1988) and the

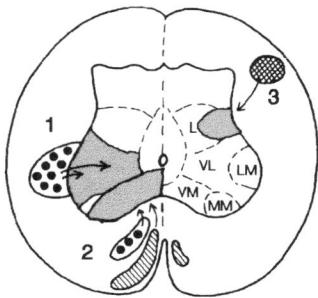

Figure 10 Progressive cervical grey matter invasion of supraspinal projections in anuran amphibian (adapted from ten Donkelaar, 1982). (**1**) The uncrossed lateral tract, containing ipsilateral, reticulospinal projections (large dots) terminates (stages 28/43) in the ventrolateral field (VL), but also in the lateral group of motoneurons (LM), as distinguished by Ebbeson (1976). (**2**) The ventral tract, containing ipsilateral reticulospinal projections (large dots) and bilateral vestibulospinal projections (hatched) terminates (stages 35/50) in the ventromedial field (VM, comparable to the ventromedial part of laminae VII and VIII in other vertebrates), but also in the medial group of motoneurons (MM). (**3**) The crossed dorsolateral tract, containing rubrospinal projections (crosshatched) terminates (stage 48) in the lateral field (L).

rubrospinal fibres start invading the rostral spinal cord (stage 48, ten Donkelaar *et al.*, 1991). Soon after (stage 50), monoaminergic coerulospinal projections increase, and vestibulospinal projections, derived from the ventral vestibular nucleus, arrive in the rostral spinal cord (ten Donkelaar & de Boer-van Huizen, 1982). Finally, hypothalamospinal projections are the last supraspinal projections to arrive in the rostral spinal cord (at stage 57). Rubrospinal projections reach the lumbosacral cord when hindlimbs are used for locomotion (stage 58, ten Donkelaar *et al.*, 1991).

In summary, in anuran amphibians, as in other vertebrates, medially descending brainstem pathways, namely the reticulospinal pathways, steering axial and proximal muscles, differentiate very early, whereas the laterally descending brainstem pathways, namely the anuran homologue of the rubrospinal tract, steering limb muscles, are present notably later in development (Figure 10). In lower vertebrates the reticular formation represents a "final common supraspinal pathway" by way of which higher nervous centres can influence the spinal motoneurons (Nieuwenhuys, 1998).

DEVELOPMENT OF DESCENDING SUPRASPINAL PROJECTIONS IN RODENTS

The development of descending supraspinal projections in rodents has been extensively studied by means of a great variety of retrograde or anterograde tract-tracing and immunohistochemical techniques. The early development of brainstem–spinal cord projections was recently studied in fixed rat embryos or isolated brain-spinal cord *in-vitro* preparation (De Boer-van Huizen & ten Donkelaar, 1999). By applying a small crystal of carbocyanine dye (DiI:1,1′-

dioctoadecyl-3,3,3'3'-tetramethylindocarbocyanine perchlorate), or byotinylated dextran amine (BDA) to the rostral spinal cord, labelled cells of origin of descending brainstem–spinal cord projections were observed as early as embryonic day 12 (E12). As in amphibians, reticulospinal projections are the first to arrive in the spinal cord, and are derived from the interstitial nucleus of the MLF, and from various parts of the reticular formation – mesencephalic, pontine, and medullary (Figure 11, E12).

In addition, just below the cerebellum a conspicuous small group of labelled cells is found in a position reminiscent of the locus coeruleus. On E13 and E14 almost similar observations were made (Figure 11, E14). In addition some raphespinal and putative coerulospinal projections are present. The small number of vestibulospinal neurons observed can be explained by the fact that in these experiments the marker applied did not reach the most ventral part of the spinal cord where vestibulospinal fibres are supposed to pass (Figure 11). Moreover, at that stage of development the vestibular nuclear complex is difficult to distinguish from the medullary reticular formation. The arrival of these early brainstem–spinal cord projections to high cervical level was complemented by their gradual descent to midcervical (Auclair *et al.*, 1993), low thoracic (Kudo *et al.* 1993) and lumbosacral (Lakke, 1997) level (Table 2).

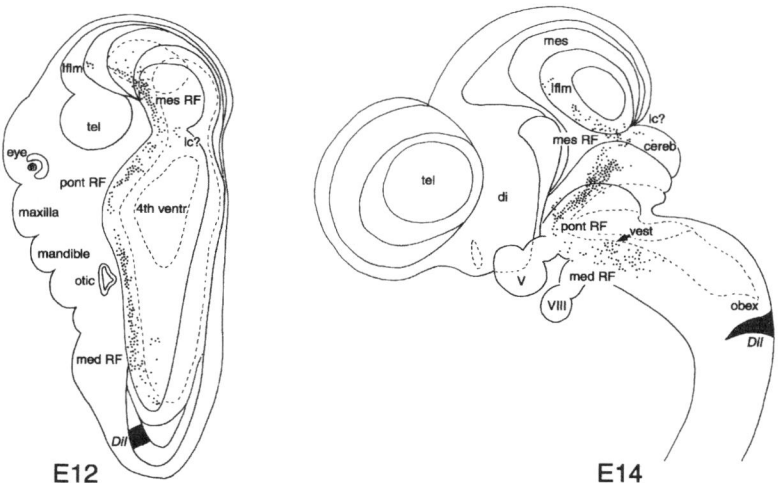

E12   E14

Figure 11 Distribution of labelled brainstem neurons (dots) in an E12 (embryonic day 12) and an E14 rat embryo following the application of a DiI crystal just behind the obex shown in the reconstruction of a nearly horizontally sectioned (E12) and a sagittally sectioned (E14) brain (adapted from de Boer-van Huizen & ten Donkelaar, 1999). cereb: cerebellum, di: diencephalon, eye: eye vesicle, Iflm: interstitial nucleus of the fasciculus longitudinalis medialis, lc?: putative locus coeruleus, maxilla: maxillary process, mandible: mandibular process, medRF: medullary reticular formation, mes: mesencephalon, mesRF: mesencephalic reticular formation, otic: otic vesicle, pontRF: pontine reticular formation, tel: telencephalon, vest: vestibular nuclear complex, 4th ventr: 4th ventricle, V: trigeminal ganglion, VII: vestibulocochlear ganglion.

Table 2 Development of supraspinal projections in rats. [a]: de Boer-van Huizen & ten Donkelaar, 1999; [b]: Auclair *et al.*, 1993; [c]: Kudo *et al.*, 1993; [d]: Lakke, 1997; [e]: Rajaofetra *et al.*, 1989; [f]: Rajaofetra *et al.*, 1992; [g]: Lakke & Marani, 1991; [h]: Gribnau *et al.*, 1986.

Nuclei	*Innervation of spinal cord*			
	High cervical[a]	*Midcervical[b]*	*Low thoracic[c]*	*Lumbosacral[d]*
Reticular formation				
Medullary	E12	E14	E15	E17–E19
Pontine	E12	E13	E14	E18
Mesencephalic	E12	?	E14	E18
Interstitial nucleus of the MLF	E12	E13	E14	E18
Raphe nuclei	E14	E14	E15	E17
Serotonergic projections[e]	E14	E14	E15	E17
Vestibular nuclei				
Lateral nucleus	E13	E13	E14	E17
Medial and inferior nuclei	?	?	E20	E18–E21
Locus coeruleus				
Coerulospinal projections	E12?	E13?	E15	E20
Noradrenergic projections[f]	< E16	< E16	E16–E17	E17–E18
Red nucleus[g]	E17	E18	E19	E21
Hypothalamus				
Paraventricular nucleus			E20.5	P1
Corticospinal projections[h]	P0	P1	P4–P5	P7–P9

The descent of serotonergic and noradrenergic fibres, studied by means of immunohistochemical techniques, is in line with the tract-tracing data (Rajaofetra *et al.*, 1989, 1992). Serotonergic projections have reached the cervical level at E14 and the lumbar level at E17, whereas the noradrenergic projections reach the lumbar level at E17/18. The age at which descending brainstem projections arrive at the lumbosacral level was further established using the retrograde transport of WGA-HRP (Lakke, 1997): at E17 projections from the lateral vestibular nucleus, the raphe magnus nucleus and the gigantocellular nucleus; at E18 projections from the interstitial nucleus of the MLF, the mesencephalic, caudal pontine and ventral medullary reticular formation, the inferior vestibular nucleus, the nucleus raphe obscurus, the subcoerulean nucleus; at E19 projections from the oral pontine, parvocellular and ventral gigantocellular reticular formation; at E20 projections from the locus coeruleus, the raphe pallidus nucleus, and several reticular nuclei. Last to arrive in the lumbar cord during the prenatal period, at E21, are projections from the medial vestibular nucleus, the solitary nucleus and the red nucleus (Lakke & Marani, 1991). Fibres from the paraventricular hyphothalamic nucleus and the lateral hypothalamic area arrive in the lumbosacral cord only on the 1st postnatal day (P1). Finally, the descent of corticospinal projections in the spinal cord is entirely postnatal, arriving in the high- and mid-cervical cord at P0 and P1, respectively, then in the low thoracic cord at P4–P5, and in the lumbosacral cord at P7–P9 (Gribnau *et al.*, 1986).

In summary, the gradual arrival of descending fibres in the rat lumbosacral cord (Table 2) suggests that the supraspinal control at birth is exerted primarily on proximal muscles, by means of medially descending brainstem pathways, whereas the control over distal muscles increases during the first days after birth, by means of laterally descending brainstem pathways (Brocard *et al.*, 1999), and finally control over the axio-proximo-distal musculature is overcome during the first postnatal weeks by means of the corticospinal projections.

DEVELOPMENT OF DESCENDING BRAINSTEM PATHWAYS IN HUMANS

In humans the temporal sequence of development of medially and laterally descending brainstem projections followed by corticospinal projections is also observed. However, in humans the differentiation of the cells' origin of these various pathways is not so well correlated to their downward growth in the spinal cord, as shown in animal experiments, because tract-tracing methods cannot be used. As a consequence we will focus first on the cells of origin of the descending brainstem pathways. During the embryonic period (from day 18 to day 56, stages 8–23), the development of the human brain was extensively studied by R. O'Rahilly & F. Müller in a series of 15 articles based on the examination of 340 serially sectioned embryos, including 51 silver impregnations and graphic reconstructions from 89 brains.

In human embryos the first neuroblasts at the origin of *the reticulospinal projections* differentiate at 28 postovulatory days (4–5 mm length, stage 13). At that age the brain for the first time is part of a closed system, called the neural tube (the rostral neuropore closed in stage 11, the caudal one in stage 12; Rhines & Windle, 1941). Soon after that age (early 5th week), the axons of these neuroblasts sprout and form the MLF (Figure 12). The MLF thus arise from cells located in the interstitial nucleus and related Darkschevitch nucleus, which lie at the junction of the diencephalon (D2) and the mesencephalon (M1) (Müller & O'Rahilly, 1988a; Rhines & Windle, 1941; Windle, 1970). The MLF has been considered to represent "the oldest longitudinal pathway" (Mesdag, 1909). This group of fibres forms a homolateral descending pathway which, at that stage, can reach the isthmus rhombencephali, in between the mesencephalon and the rhombencephalon proper (Rhines & Windle, 1941). Whereas the fibres of the MLF, at 28 days (stage 13) are purely descending, some of those present at later stages (*ca.* 41 days, stage 17) are ascending fibres from the vestibular area, comparable to those in the adult (Müller & O'Rahilly, 1988a).

The distinction between the MLF and the ventral longitudinal fasciculus (VLF) has been clarified by a comparative study between rat, cat and humans (Rhines & Windle, 1941). The MLF, arising from the interstitial nucleus, descends (stages 13–17), whereas the VLF ascends from neurons of the rhombencephalon, in the region of the trigeminal, glossopharyngeal, vagus and accessory nerves. Rostrally directed growth cones were seen in the VLF (Rhines & Windle, 1941). In humans the two join and overlap at approximately 32 postovulatory days (stage 14; Figure 12). In addition, the lateral longitudinal fasciculus looks like a rather scattered layer of fibres located in the lateral wall of the rhombencephalon, and can be traced to the mesencephalon, but reaches probably more rostral regions (Müller & O'Rahilly, 1988a). This lateral longitudinal fasciculus also contains descending fibres which sprout from neuroblasts located at the

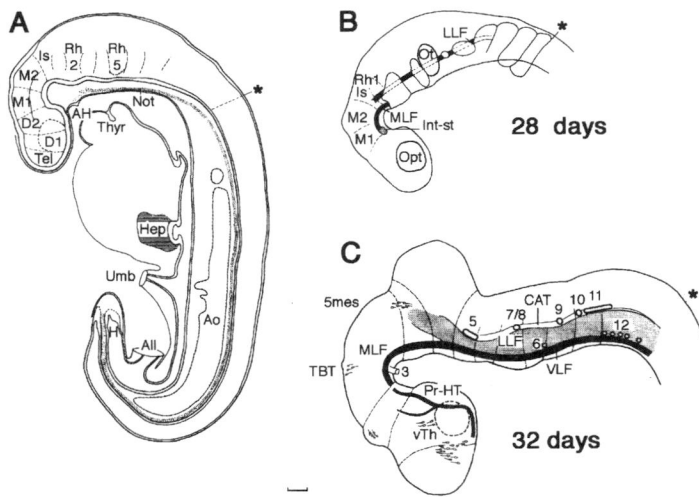

Figure 12 Tract development in human embryos of 28 (**A, B**) and 32 (**C**) postovulatory days (adapted from Müller & O'Rahilly, 1988a,b) **A:** Median section of a human embryo of 28 postovulatory days (stage 13, 5 mm length). At the mesencephalic flexure the head is bent towards the cardiac eminence. **B:** Brain reconstructed from transverse sections showing the longitudinal tracts. The three brain vesicles can be subdivided in several neuromeres (Müller & O'Rahilly, 1997). The forebrain can be subdivided into three parts of approximately equal length: Tel: telencephalon, D1 and D2: diencephalon. The midbrain or mesencephalon has two segments (M1 and M2). The hindbrain or rhombencephalon is composed of eight rhombomeres (Rh 1–8), and an isthmic segment (Is) between the mesencephalon and rhombencephalon. Asterisk indicates borderline between brain and spinal cord. **C:** Thirty-two postovulatory day embryo (stage 14, 5–7 mm length). The medial longitudinal fasciculus (MLF, black) joins the ventral longitudinal fasciculus (VLF) of the rhombencephalon. The lateral longitudinal fasciculus (LLF, stippled) is spread out on the lateral wall of the rhombencephalon; it can be followed to the mesencephalon but probably reaches further rostrally. The exits of cranial nerves 6 and 12 are ventral, whereas all the others are at the level of the sulcus limitans, where all afferent fibres enter in the common afferent tract (CAT, white). Apart from those main tracts, some developing short fibres indicate the medial tectobulbar tract (TBT), and the mesencephalic root of the cranial nerve 5 (5 mes). The bar represents 0.2 mm. *Abbreviations*: A–H: adenohypophyseal pocket; All: allantoic primordium; Ao: aorta; CAT: common afferent tract; H: hindgut; Hep: hepatic primordium; Int-st: interstitial nucleus; LLF: lateral longitudinal fasciculus; MLF: medial longitudinal fasciculus; Not: notochord; Opt: optic vesicle; Ot: otic vesicle; Pr-HT: preopticohypothalamotegmental tract; TBT: medial tectobulbar tract; Thyr: thyroid primordium; Umb: umbilical stalk, VLF: ventral longitudinal fasciculus; vTh: ventral tahlamus; numbers 3–12: cranial nerves.

lower rhombencephalon, the primordium of medullary reticular formation (Windle, 1970). These reticulospinal fibres are the first descending fibres to reach the spinal cord, at around 34–36 days (10–12 mm, stage 14–15; Table 3); however, they do not course more than one or two segments, and become incorporated into the lateral spinal funiculus (Windle & Fitzgerald, 1937).

Tectobulbar and tectospinal projections are concerned with ear muscle and platysma motoneurones (Courville, 1966; Kume *et al.*, 1978; Panneton & Martin, 1979), on the one hand, and on the other hand with indirect excitation of

Table 3 Developmet of supraspinal projections identified in human embryos. Adapted from Humphrey, 1960; Windle, 1970; Müller & O'Rahilly, 1988b, 1989a, 1990a; Puelles & Verney, 1998.

Nuclei and pathway	Innervation of the spinal cord (postovulatory days)
Reticular formation	
Interstitiospinal	32
Medullary reticulospinal	32–35
Tectum	
Medial tectospinal	37
Locus coeruleus	
Coerulospinal	
Noradrenergic projections	44
Vestibular nuclei	
Lateral vestibulospinal	44
Medial and inferior	Before end of embryonic period
Vestibulospinal	
Red nucleus	
Rubrospinal	?
Sensorimotor cortex	Early fetal period
Corticospinal	57

contralateral neck muscle motoneurons (Anderson *et al.*, 1972). The medial tectobulbar tract arises from the rostral part of the mesencephalic tectum as early as 32 days (5–7 mm length, stage 14; Müller & O'Rahilly, 1988b). At 37 days (11–14 mm, stage 16) these fibres, after decussating in the dorsal tegmental commissure, join a particularly voluminous MLF (Müller & O'Rahilly, 1989a; O'Rahilly *et al.*, 1987; Windle, 1970). The lateral tectobulbar tract originates from the caudal part of the mesencephalic tectum, at 41 days (11–14 mm, stage 17), and joins the lateral longitudinal fasciculus.

Cells of origin of rubrospinal projections can be seen for the first time at around 41 days, lateral to the oculomotor nucleus (Cooper, 1946; Morrell, 1985; Müller & O'Rahilly, 1989b). The red nucleus further shifts at 44–51 days, in its lateral position (Müller & O'Rahilly, 1990a; O'Rahilly *et al.*, 1988).

Cells of origin of coerulospinal projections are recognized at around 35 days (8 mm, stage 14), in the primordium of the locus coeruleus, ventral to the region which will form the cerebellum (Windle & Fitzgerald, 1942). The fibres originating from this presumptive locus coeruleus join the lateral longitudinal fasciculus. At around 41 days the locus coeruleus is well distinguishable as unilayer of particularly tall cells at the junction between the mesencephalon and the rhombencephalon (isthmic neuromere; Müller & O'Rahilly, 1989b, 1997). The presence of catecholamine or serotonin in the brainstem was first detected at 8 weeks, and catecholaminergic medullary–spinal projections at 10 weeks (Olson *et al.*, 1973). Recently, using antibodies against enzymes of the catecholaminergic pathway, noradrenergic neurons were visualized as early as 6 weeks in the medulla and the locus coeruleus (Verney *et al.*, 1991; Zecevic & Verney, 1995). More recently, tyrosine-hydroxylase immunoreactive neurons were already observed at 4.5 weeks (10 mm, stage 15), at the junction between the mesencephalon and rhombencephalon (isthmic neuromere; Puelles & Verney, 1998), and descending catecholaminergic fibres were found in the spinal cord at 6 weeks (stage 18).

Cells of origin of vestibulospinal projections are detectable between 41 and 44 days, and fibres from the more medially located nucleus can be seen to reach the MLF (Müller & O'Rahilly, 1989b, 1990a,b). From 52 days (stage 21) to the end of the embryonic period (at about 57 days, stage 23) additional fibres from all four vestibular nuclei can be followed to the MLF (Figure 13). In particular, the lateral vestibular nucleus, the clearest of the four vestibular nuclei, is at the origin of the vestibulospinal tract.

DEVELOPMENT OF CORTICOSPINAL PROJECTIONS IN MAMMALS

In humans corticospinal fibres are recognizable for the first time at 57 days (31 mm, stage 23), on both sides of the rhombencephalon as two small ventral

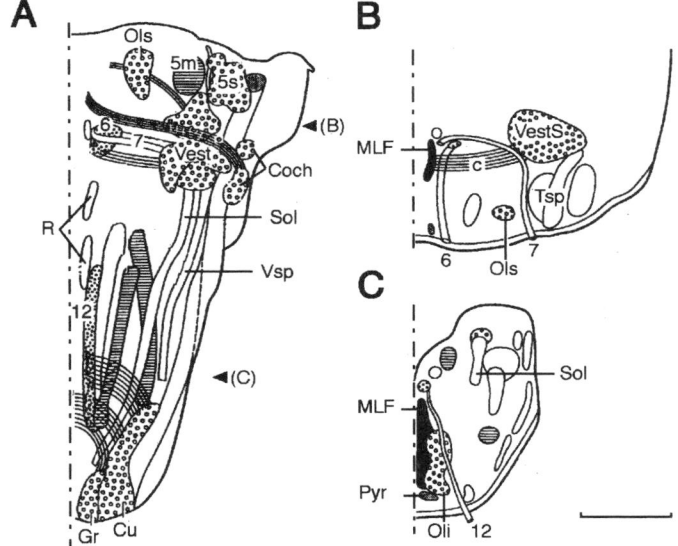

Figure 13 Tracts and nuclei in the human rhombencephalon at the end of the embryonic period (8 weeks postovulation; adapted from Müller & O'Rahilly, 1990b). **A:** Diagrammatic reconstruction of a dorsal view of the right half of the rhombencephalon showing tracts and nuclei. **B** and **C:** Reconstructed coronal slices at the level marked by arrowheads in **A. B:** Level of abducens (6) and facial (7) nerves, showing the superior olivary nucleus (Ols), the superior vestibular nucleus (VestS) and its fibres (c) to the MLF. **C:** Level of inferior olivary nucleus (Oli), showing the small bundle of pyramidal fibers (Pyr). Sensory nuclei are shown by black circles, somatomotor nuclei in bold stippling, visceral efferent nuclei are hatched. The bar represents 1 mm. *Abbreviations:* c: vestibular fibres to the MLF; Coch: area of cochlear nuclei; Cu: cuneate nucleus; Gr: gracile nucleus; MLF: medial longitudinal fasciculus; Oli: inferior olivary nucleus; Ols: superior olivary nucleus; Pyr: pyramidal tract; R: raphe nuclei; Sol: solitary tract; Tsp: trigeminospinal tract; Vest: vestibular nuclei; VestS: superior vestibular nucleus; Vsp: vestibulospinal tract. Arabic numbers: cranial nerves (5m: motor portion, 5s: sensory portion of the trigeminal nerve.)

bundles, which can be traced in the caudal part of the medulla as decussating fibres, and in the cervical spinal cord (Humphrey, 1960; Müller & O'Rahilly, 1990b). As suggested by Altman & Bayer (1984), in order to form this early pyramidal decussation "the axons of layer V pyramidal cells have grown to a considerable length while their cell bodies are still migrating to the cortical plate".

At about 48 days (stage 19) the thalamostriatal tract ("Stammbündel des Thalamus" of His, 1904) represents the fibres around which the internal capsule will continue to develop. These fibres, however, stop abruptly and do not yet continue to the pallium. At that stage of development in human embryo the internal capsule represents a mammalian neoencephalic component of the submammalian lateral forebrain bundle which contains ascending as well as descending fibres (Kuhlenbeck, 1977; O'Rahilly et al., 1988). The first neopallial fibres can be traced to an area that is comparable to the later central gyrus (Müller & O'Rahilly, 1990a). The cortical plate appears at about 52 days (22 mm, stage 21) in the area of the future insula, and at 57 days (31 mm, stage 23) it reaches over the main neocortical surfaces of the hemispheres (Müller & O'Rahilly, 1990a)

The growth of the pyramidal tract was investigated by Humphrey (1960) in human fetuses and in one newborn (Figure 14). She found that the corticospinal fibres appear in the cerebral peduncle at an estimated age of 7 postovulatory weeks (for the correspondence between estimated postovulatory ages and menstrual ages of this study, see Wozniak & O'Rahilly, 1982). At about 1 week later these fibres reach rostral levels of the medulla. At about 14 or 15 weeks the pyramidal decussation is completed, and corticospinal fibres are present down the mid-thoracic spinal level. The caudal limits of the corticospinal fibres at 27 weeks and at term are comparable to those for the adult.

Alternate periods of downward growth and of maturation characterize the development of the corticospinal tracts. Concurrent with the rapid downward growth into mid-thoracic cord levels between 14 and 15 weeks, there is an extremely large increase in area occupied by corticospinal fibres in both the pons and caudal medulla levels. In contrast, after 15 weeks the area occupied by the corticospinal fibres in the pons remains about the same, for a time at least. However, at all levels, particularly caudal to the cervical enlargement, the pyramidal tract areas increase throughout fetal life. Some general trends, along with some significant variations, have been described in the patterns of decussation of medullary pyramids and the distribution of pyramidal tracts on the two sides of the spinal cord (Yakovlev & Rakic, 1966). In more than two-thirds of the medullary pryamids (100 specimens) and spinal cords (130 specimens) of fetuses and neonates, the fibres of the left pyramid hold the "right-of-way", and cross to the right side of the spinal cord at higher levels in the decussation than the fibres of the right pyramid which cross just below. Furthermore, in 87% of the samples more fibres of the left pyramid crossed to the right side of the spinal cord than fibres of the right pyramid that crossed to the left side. This asymmetry in the distribution of corticospinal fibres on the two sides of the spinal cord is enhanced by the fact that more fibres of the right pyramid descend uncrossed into the right side of the spinal cord (17.7%) than fibres of the left pyramid that descend towards the left side (3%). As a consequence the right side of the spinal cord, at least in the cervical region, receives more pyramidal tract fibres from both cerebral

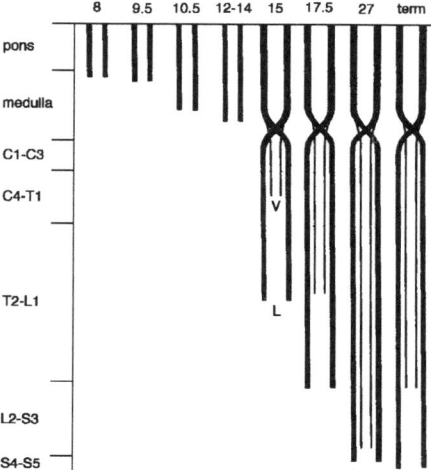

Figure 14 Progressive downward growth of the human corticospinal tracts through the brainstem and spinal cord at different postovulatory weeks and at term (adapted from Humphrey, 1960; Nissl stains, various silver techniques and Weigert's haematoxylin method were used). The lengths allocated to the different brainstem and spinal cord regions are based on those of a 16.5-week-old fetus. The postovulatory ages are approximately 2 weeks less than the menstrual ages (for the correspondence between these two ages, see Wozniak & O'Rahilly, 1982). L: lateral corticospinal tracts; V: ventral corticospinal tracts.

hemispheres than the left side, and with regard to corticospinal innervation the right side of the spinal cord appears to be the preferred or dominant side.

Myelination of corticospinal fibres

The end of the descent of corticospinal fibres within the spinal cord at term does not mean that this pathway is fully mature. As suggested by Yakovlev at the symposium on Regional Development of the Brain in Early Life: "I think of myelination as a morphological criterion of the functional maturity of a conduction path" (Yakovlev & Lecours, 1967). In this classical study, using the Loyez method for myelin sheaths, a synchronization was established between the cycles of myelination of the specific and non-specific projection fibres to the respective cortical areas and the cycles of myelination of the corticofugal fibres from these areas, in which the respective afferents and efferents conjugate as limbs of an arc. This general rule seems to apply particularly well to the specific thalamic afferents to pre- and postcentral cortex and the pyramidal tracts. The specific thalamo-cortico-pyramidal arc has a relatively short cycle of myelination, beginning at about 6–10 fetal months and being completed at about the second year of life (Yakovlev & Lecours, 1967; their figure 1). Among the descending pathways the corticospinal tract is thus the last one to achieve its myelogenetic cycle in humans (Flechsig, 1920; Langworthy, 1933).

Recent studies using antibodies against myelin-associated protein, and particularly myelin basic protein (MBP) have established that the first descending projections are the first to be myelinated (Bodhireddy *et al.*, 1994; Brody *et al.*, 1987; Kinney *et al.*, 1988; Weidenheim *et al.*, 1992, 1993, 1996). Myelination of the MLF begins at 8 postovulatory weeks in the brainstem, and in the vestibulospinal tract at about 24 weeks. Myelination of the different spinal funiculi has recently been studied during the fetal period, from 10 to 24 gestational weeks, using antibodies against myelin-associated proteins. MBP represents approximately 30% of total myelin protein, and has proven to be an extremely sensitive marker for tracing the progress of myelination of human fetal spinal cord (Bodhireddy *et al.*, 1994). This myelin marker, along with later-expressed myelin proteins (around 2 weeks later), such as proteolipid proteins (PLP, representing approximately 50–60% of total protein) and myelin-associated glycoproteins (MAG, 1%), are expressed in the ventral to dorsal and rostral to caudal gradient (Weidenheim *et al.*, 1992, 1996). As early as 10 weeks, light MBP-positive cells and/or processes are detected in the ventral funiculus of the cervical spinal cord. Between 10 and 12 weeks a similarly light MBP expression is observed not only in the cervical ventral, ventrolateral and dorsal funiculi, but also in the thoracic and lumbosacral ventral and ventrolateral funiculi. Between 12 and 16 weeks a moderately intense MBP is expressed at cervical level in all three funiculi. Between 16 and 24 weeks the intensity of the reaction product increases at all levels. Finally, at 24 weeks MBP expression is intense and uniform in the ventral, ventrolateral and dorsal funiculi, except for the region of the ventral and lateral corticospinal tracts, which myelinate only after birth.

The postnatal myelination of the corticospinal tract has been studied in postmortem brains of a large number of infants (162 specimens, from birth to 2 years), by using the Luxol fast blue technique when the amount of myelin is sufficient to produce an easily visible blue colour (Brody *et al.*, 1987). From the internal capsule to the lumbar spinal cord a mature pattern of myelination (degree 3 of staining intensity) is reached according to a rostro-caudal sequence which develops from birth for more than 2 years (Figure 15). In this sample 50% of the brains showed a mature pattern of myelination. By about 15 postnatal months 50% of the brains of the age group showed a mature pattern of myelination at the medullary pyramid level, and by about 20 months at the cervical level.

This protracted myelination process is in agreement with the ultrastructural observation of the pyramidal tract at the level of the decussation in a 23-week-old fetus (postovulation age, 220 mm CRL, Wozniak & O'Rahilly, 1982). In the lower part of the medulla the pyramids consist of bundles of unmyelinated axons (0.2–1.5 µm in diameter) and scattered myelinated ones (0.8–2 µm). Both the myelinated and unmyelinated fibres cross the median plane, passing to the opposite side of the spinal cord. Below the decussation the anterior and lateral corticospinal tracts are composed mainly of unmyelinated fibres, although some myelinated fibers are also present.

The conduction velocity is a good indicator for the axon size and degree of myelination. This assumption is supported by the fairly constant ratio between the conduction velocity of the largest corticospinal axons and their diameter in the medullary pyramid in the developing monkey (Olivier *et al.*, 1997). In human neonates the conduction velocity of corticospinal axons in the spinal cord has

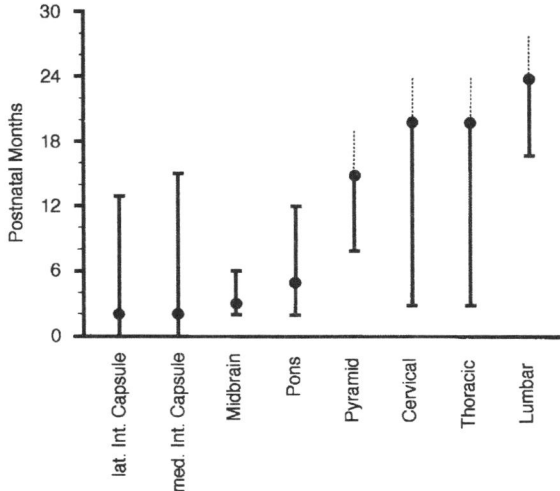

Figure 15 Rostrocaudal gradient of myelin maturation of the human corticospinal tract at different postnatal months (adapted from Brody *et al.*, 1987). Filled circles indicate median values of the age at which infants of a given age group attain the degree of "mature myelination" (degree 3 of the grading system used in the National Collaborative Perinatal Project, Gilles *et al.*, 1983) for different levels of the corticospinal tract (lat. int. capsule and med. int. capsule: lateral and medial half of posterior limb of internal capsule at midthalamic level, respectively; midbrain: central portion of middle 3/5 of crus pedunculi; pons: descending tract anywhere in the basis pontis; pyramid: any level; cervical, thoracic, lumbar: centre of lateral spinal funiculus at respective levels). The vertical bars give the 10th and 90th percentile values as estimated by the Ayer method (Ayer *et al.*, 1955). The vertical dotted lines indicate that the exact 90th percentile is not known but is greater than 24 months.

been estimated at about 10 m/s, compared to about 50–70 m/s in adults (Boyd *et al.*, 1986; Herdmann *et al.*, 1991; Inghilleri *et al.*, 1989; Kather-Boidin & Duron, 1991). This is in keeping with the low conduction velocity of the fastest corticospinal axons in the spinal cord of the neonatal monkey (5-day-old), estimated at about 8 m/s, that is 10 times slower than the mean value found in the adult monkeys (80 m/s; Olivier *et al.*, 1997). After a rapid increase in conduction velocity from birth to 5 months there is a protracted increase that probably lasts for another year or more. This is in keeping with changes in the diameter of the largest corticospinal axons in the pyramid, which increases from 1 to 7 µm during the first 2 years (Verhaart, 1950). Further, in the monkey as in humans, the conduction velocity of the fastest corticospinal axons, and therefore their diameter, may continue to increase slightly with age, long after the rapid changes observed during the first postnatal months (Eyre *et al.*, 1991, 2000; Olivier *et al.*, 1997). In addition, a rostro-caudal gradient of myelination of monkey corticospinal axons was indicated by the earlier maturation of the conduction velocity in their cranial rather than their spinal course. This is also in keeping with a comparable rostro-caudal gradient of myelination suggested by anatomical studies in humans (Niebroj-Dobosz *et al.*, 1980; Brody *et al.*, 1987). As stated by Lemon *et al.* (1997):

Recent research into the development of the corticospinal system have highlighted the protracted nature of the structural and functional changes involved. Rather than there being a sudden change at a particular age, there are gradual changes which last for several years (Eyre *et al.*, 1991; Müller & Hömberg, 1992; Armand *et al.*, 1994; Müller *et al.*, 1994). It is evident that the capacity to perform a precision grip is present long before the conduction velocity of the fastest fibres reaches an adult value, and this means that the corticospinal system is functional long before full myelination of the axons is complete. These longer and slower changes are of course paralleled by gradual and protracted improvements in motor skill (Halverson, 1943; Forssberg *et al.*, 1991; Müller & Hömberg, 1992). Such changes may be important in determining the fine temporal structure of the motor programmes, leading to smoother coordination as well as better anticipatory control of reafferent sensory inputs generated by finger movement.

Development of the different components of the corticospinal projections

The human pyramidal tract is not only the last-maturing descending pathway during development, it also contains phylogenetically ancient corticospinal projections that terminate on spinal interneurons and phylogenetically more recent ones, namely the corticomotoneuronal projections (Armand, 1982; Heffner & Masterton, 1975, 1983; Kuypers, 1981; Nudo & Masterton, 1988, 1990a,b; Nudo *et al.*, 1995). The development of these two kinds of corticospinal projections has been extensively studied in subprimates and non-human primates, respectively. It appears that these two kinds of projections have two different modes of development (for a review see Armand *et al.*, 1996; Lemon *et al.*, 1997).

The development of corticospinal projections that terminate on spinal interneurones in subprimates is characterized by exuberant cortical areas of origin, transient corticospinal axons and aberrant spinal projections. Thus, in the developing marsupial, rodent and carnivore, exuberant corticospinal axons originate from a more widespead cortical area of origin than in the adult (Cabana & Martin, 1984; D'Amato & Hicks, 1978; Kalil, 1985; Leong, 1983; Martin *et al.*, 1982; Meissirel *et al.*, 1993; Reh & Kalil, 1982; Schreyer & Jones, 1988a). Moreover, transient corticospinal axonal collaterals are made to supraspinal structures and to different parts of the spinal grey matter (Adams *et al.*, 1983; Cabana & Martin, 1985; Curfs *et al.*, 1994; Girard *et al.*, 1993; Joosten *et al.*, 1987; Mihailoff *et al.*, 1984; Reh & Kalil, 1981; Schreyer & Jones, 1982, 1988b; Stanfield *et al.*, 1982; Stanfield & O'Leary, 1985). For instance in the cat, diffuse bilateral corticospinal projections were observed in all parts of the spinal grey matter in kittens of 4–5 postnatal weeks, i.e. in the dorsal horn and in the dorsolateral and ventromedial parts of the intermediate zone (Alisky *et al.*, 1992; Theriault & Tatton, 1989). By 6–7 weeks there was selective elimination of the transient ipsilateral projections to the dorsal horn and the dorsolateral part of the intermediate zone, leaving an adult-like pattern of termination. Exuberant corticospinal projections during development have also been found in non-human primates. At E108 the cortical distribution of retrogradely labelled neurons, after injection of fluorescent tracer into the macaque spinal cord just below the pyramidal decussation, extended rostro-caudally from behind the frontal pole to the posterior lip of the lateral sulcus, and over most of the lateral and medial surface of the neocortex (Killackey *et al.*, 1997). However, at this relatively early developmental stage some regional specialization was already detected. The density of labelled neurons within the presumptive motor cortex was 10 times greater than in the peripheral areas. In the newborn and adult macaque the patterns of corticospinal projections from the different cortical regions are "strikingly similar" (Darian-Smith *et al.*, 1996; Galea & Darian-Smith, 1995). However, substantial maturational changes occur during the 8th month after birth. These included a halving of the area of cerebral cortex from which the contralateral corticospinal projection originated and a threefold reduction in the number of labelled corticospinal neurons projecting to all segments of the cord. These authors suggested that spinal collateral elimination rather than neuronal cell death was the likely mechanism for this reduction in the population and areal extent of corticospinal neurons during the first 6–8 months of postnatal development. By 5 months exuberant cortical areas could no longer be dected by Biber *et al.* (1978; Kuypers, 1962; Sloper *et al.*, 1983). With regard to their site of spinal terminations, Armand *et al.* (1994, 1997), using anterograde tracing methods, have found that in neonatal and adult macaque monkeys, corticospinal fibres from the motor cortex hand area terminate in the same regions of the grey matter intermediate zone, where spinal interneurons are located. The relative densities of terminal labelling in the different parts of the spinal intermediate zone were comparable at the different postnatal ages, including a 5-day-old infant monkey. However, there was a big increase in the overall density of labelling from birth to 2.5 months, with smaller increases until adulthood.

As far as corticospinal termination developmental changes are concerned, the contrast in the results from primates and non-primates may be due to the very considerable differences in the relative stage of development achieved by the different species at birth (Passingham, 1985). As stated by Lemon et al. (1997):

> The corticospinal system is so much more advanced in the neonatal primate; in man, for example, the fibres are thought to have reached sacral levels by a gestational age of 29 weeks (Humphrey, 1960). Thus it is probable that the early fetal stages of human corticospinal development are also exuberant in character, but that there has been a great deal of selection and elimination by birth. This is supported by the observation that such exuberance is most pronounced in prematurely born macaques (Galea & Darian-Smith, 1995).

The development of corticomotoneuronal projections requires particular attention, because their integrity appears to be an essential perequisite for skilled finger movements (Lawrence & Hopkins, 1976; Passingham *et al.*, 1983). Kuypers (1962) was the first to study the development of the corticospinal tract in the macaque, by using the Nauta technique to visualize the terminals of degenerating corticospinal axons after large lesions in the motor cortex. In a neonate he found that these fibres had reached all levels of the spinal cord white matter and that terminals were already present within the spinal intermediate zone, but not in the motor nuclei. The only exception to this was the presence of a few terminals among the dorsal margins of the hand muscle motor nuclei. He also reported an "almost adult pattern" of terminal labelling in the hand motor nuclei, at 8 months of age. Armand *et al.* (1994, 1997) reinvestigated corticospinal development in the macaque, by using anterograde transport of wheat-germ agglutinin conjugated to horseradish peroxidase (WGA-HRP), injected in the hand region of the primary motor cortex at different stages of development, and concentrated on the pattern of termination among the dorsolateral motoneurons at the C8–T1 levels which supply hand and finger muscles (Jenny & Inukai, 1983). In the adult monkey terminal labelling was present throughout the dorsolateral group of motoneurones (Figure 16). In contrast there was no labelling in the vicinity of the ventral motoneurons, which supply axial and proximal arm muscles (Bortoff & Strick, 1993; Kuypers, 1981). By birth, although the corticospinal fibres have reached all levels of the spinal cord white matter, the penetration of the spinal grey matter is far from complete. In a 5-day-old monkey only weak labelling was observed. Faint labelling was present among the dorsolateral motor nuclei, and somewhat densest along their dorsal margins (Stanfield & Asanuma, 1993). No

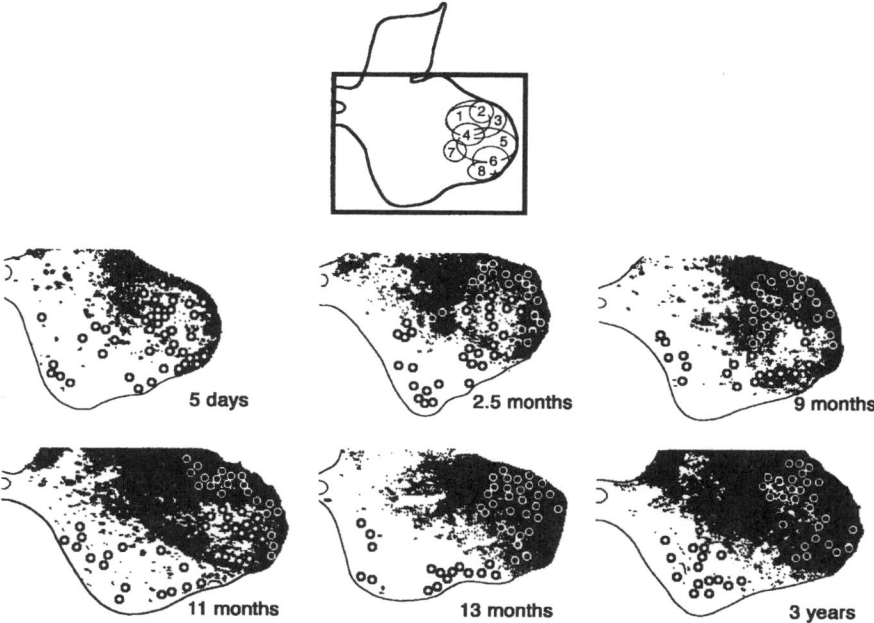

Figure 16 Postnatal changes in corticospinal projections at the C8–T1 junction in the macaque monkey (adapted from Armand *et al.*, 1997). The inset in the top diagram demarcates the region of grey matter represented below and also shows the distribution of motor nuclei innervating eight selected hand muscles. (**1**), First dorsal interosseus; (**2**), lateral lumbrical; (**3**), adductor pollicis; (**4**), abductor and flexor pollicis brevis; (**5**), flexor digitorum profundus and superficialis; (**6**), extensor digitorum communis, abductor and extensor pollicis longus; (**7**), flexor carpi ulnaris; (**8**), extensor carpi ulnaris (Jenny & Inukai, 1983). After injection of WGA-HRP in the primary motor cortex hand area, the distribution of corticospinal terminal labelling (black stipple) and the location of motoneurons (circles) were obtained from digitized paratungstate–tetramethyl benzidine-reacted sections counterstained with neutral red.

labelling was observed among the ventral motoneurons. In the 2.5-month-old monkey there was heavy labelling in the spinal intermediate zone, which formed a distinct ring around the dorsolateral motor nuclei. In the 9- and 11-month-old monkeys labelling was similar to the 2.5-month-old animal, with further encroachment of fibres into the dorsolateral group of motoneurons. In the 13-month-old monkey the characteristic ring around the dorsal motoneurons was completely in-filled. However, the labelling was still less extensive and intense than in the adult case. These data suggest a much more protracted encroachment of the motor nuclei by the corticospinal fibres than previously thought, and that takes at least a year. In addition, these corticomotoneuronal projections seem to be directly addressed to their adult target motoneurons. No aberrant projections to the ipsilateral motoneurons were observed, nor any early innervation of the more ventral motoneurons, which might subsequently have been withdrawn.

It has been recently claimed that, in humans, functional corticospinal innervation, including corticomotoneuronal connections, are established before birth (Eyre *et al.*, 2000). This morphological study is based on the immunoreactivity to growth-associated protein 43 (GAP 43), a protein expressed by growth cones (Benowitz & Routtenberg, 1997). By 33 weeks post-conceptional age strong and diffuse immunoreactivity was described in the whole grey matter, including the dorsal horn, intermediate zone and all motoneuronal nuclei. Unfortunately, corticospinal axons were not identified as such. Moreover, GAP 43 continues to be expressed by corticospinal axons long after corticospinal synaptogenesis has taken place (Fitzgerald *et al.*, 1991). Finally, it is not clear whether GAP 43 expression represents functional connections.

CONCLUDING REMARKS

The development of descending brainstem and cortical projections to the spinal cord occurs according to a temporal sequence which is comparable in humans and other vertebrates. The phylogenetically most ancient medially descending brainstem pathways are the first to develop. The medial longitudinal fasciculus (MLF) is the first supraspinal projection to invade the spinal cord. This is composed of interstitiospinal and reticulospinal fibres, to which medial vestibulospinal fibres later join. The lateral vestibulospinal tract descends separately through the brainstem. The phylogenetically more recent laterally descending brainstem pathways, such as the rubrospinal tract, reach the spinal cord much later. Finally, the phylogenetically most recent corticospinal tract, with its various components, develops the latest, with its connections to spinal interneurons taking place prior to the corticomotoneuronal connections.

The first supraspinal projections arrive in a rather immature spinal cord. The earliest synapses in the spinal cord are axodendritic synapses between spinal interneurons and motoneurons. This is well before the formation of synapses between dorsal root collaterals and interneurons or motoneurons. This is, however, the time at which fibres derived from the interstitial nucleus of the MLF, the reticular formation and the vestibular nuclear complex have reached the spinal cord. By this time, around 5.5 weeks (postovulation age), the first discernible spontaneous movement occurs.

The phylogenetic perspective is essential to shed some light on the development of descending projections in humans. The process of development of these different pathways, however, is not unique. This is exemplified by the corticospinal projections onto spinal interneurons and motoneurons. The protracted postnatal expansion of corticomotoneuronal projections to hand muscle motor nuclei in primates is in marked contrast to the retraction of exuberant projections that characterizes the development of corticospinal projections, and further of other descending pathways, to spinal interneurons.

ACKNOWLEDGEMENTS

The author thanks the Wellcome Trust and the CNRS-Royal Society Exchange Scheme for financial support. I am most grateful to Roger Lemon, Steve Edgley and Etienne Olivier for their collaboration on the primate developmental project.

References

Adams, C.E., Mihailoff, G.A. & Woodward, D.J. (1983). A transient component of the developing corticospinal tract arises in visual cortex. *Neuroscience Letters,* **36**, 243–248.

Alisky, J.M., Swink, T.D. & Tolbert, D.L. (1992). The postnatal spatial and temporal development of corticospinal projections in cats. *Brain Research*, **88**, 265–276.

Altman, J. & Bayer, S.A. (1982). Development of the cranial nerve ganglia and nuclei in the rat. *Advances in Anatomy, Embryology and Cell Biology*, **74**, 1–90.

Altman, J. & Bayer, S.A. (1984). The development of the rat spinal cord. *Advances in Anatomy, Embryology and Cell Biology*, **85**, 1–164.

Anderson, M.E.M., Yoshida, M. & Wilson, V.J. (1972). Tectal and tegmental influences on cat forelimb and hindlimb motoneurons. *Journal of Neurophysiology,* **35**, 462–470.

Angulo Y Gonzalez, A.W. (1932). The prenatal development of behavior in the albino rat. *Journal of Comparative Neurology,* **55**, 395–442.

Armand, J. (1982). The origin, course and terminations of corticospinal fibers in various mammals. In H.G.J.M. Kuypers & G.F. Martin (Eds.), A*natomy of descending pathways to the spinal cord* (*Progress in brain research*, vol. 57, pp. 329–360). Amsterdam: Elsevier.

Armand, J. & Kuypers, H.G.J.M. (1980). Cells of origin of crossed and uncrossed corticospinal fibers in the cat. A quantitative horseradish peroxidase study. *Experimental Brain Research,* **40**, 23–34.

Armand, J., Edgley, S.A., Lemon, R.N. & Olivier, E. (1994). Protracted postnatal development of corticospinal projections from the primary motor cortex to hand motoneurones in the macaque monkey. *Experimental Brain Research,* **101**, 178–182.

Armand, J., Holstege, G. & Kuypers, H.G.J.M. (1985). Differential corticospinal projections in the cat. An autoradiographic tracing study. *Brain Research*, **343**, 351–355.

Armand, J., Olivier, E., Edgley, S.A. & Lemon, R.N. (1996). The structure and function of the developing corticospinal tract: some key issues. In A.H. Wing, P. Haggard & J.R. Flanagan (Eds.), *Hand and brain: the neurophysiology and psychology of hand movements* (pp. 125–145). San Diego, CA: Academic Press.

Armand, J., Olivier, E., Edgley, S.A. & Lemon, R.N. (1997). Postnatal development of corticospinal projections from motor cortex to the cervical enlargement in the Macaque monkey. *Journal of Neuroscience,* **17**, 251–266.

Auclair, F., Bélanger, M.-C. & Marchand, R. (1993). Ontogenetic study of early brain stem projections to the spinal cord in the rat. *Brain Research Bulletin*, **30**, 281–289.

Ayer, M., Brunk, H.D., Ewing, G.M., Reid, W.T. & Silverman, E. (1955). An empirical distribution function for sampling with incomplete information. *Annals of Mathematical Statistics,* **26**, 641–647.

Benowitz, L.I. & Routtenberg, A. (1997). GAP-43: An intrinsic determinant of neuronal development and plasticity. *Trends in Neuroscience,* **20**, 84–91.

Biber, M.P., Kneisley, L.W. & LaVail, J.H. (1978). Cortical neurons projecting to the cervical and lumbar enlargements of the spinal cord in young and adult monkeys. *Experimental Neurology,* **59**, 492–508.

Bodhireddy, S.R., Lyman, W.D., Rashbaum, W.K. & Weidenheim, K.M. (1994). Immunohistochemical detection of myelin basic protein is a sensitive marker of myelination in second trimester human fetal spinal cord. *Journal of Neuropathology and Experimental Neurology,* **53**, 144–149.

Bortoff, G.A. & Strick, P.L. (1993). Corticospinal terminations in two New World primates: further evidence that corticomotoneuronal connections provide part of the neural substrate for manual dexterity. *Journal of Neuroscience,* **13**, 5105–5118.

Boyd, S.G., Rothwell, J.C., Cowan, J.M.A., Webb, P.J., Morley, T., Asselman, P. & Marsden, C.D. (1986). A method of monitoring function in corticospinal pathways during scoliosis surgery with a note on motor conduction velocities. *Journal of Neurology, Neurosurgery and Psychiatry,* **49**, 251–257.

Brocard, F., Vinay, L. & Clarac, F. (1999). Gradual development of the ventral funiculus input to lumbar motoneurons in the neonatal rat. *Neuroscience,* **90**, 1543–1554.

Brody B.A., Kinney, H.C., Kloman, A.S. & Gilles, F.H. (1987) Sequence of central nervous system myelination in human infancy. I. an autopsy study of myelination. *Journal of Neuropathology and Experimental Neurology,* **46**, 283–301.

Cabana, T. & Martin, G.F. (1984). Developmental sequence in the origin of descending spinal pathways. Studies using retrograde transport techniques in the North American Opossum (*Didelphis virginiana*). *Developmental Brain Research,* **15**, 247–263.

Cabana, T. & Martin, G.F. (1985). Corticospinal development in the North-American Opossum: Evidence for a sequence in the growth of cortical axons in the spinal cord and for transient projections. *Developmental Brain Research,* **23**, 69–80.

Cooper, E.R.A. (1946). The development of the human red nucleus and corpus striatum. *Brain,* **69**, 34–44.

Courville, J. (1966). Rubrobulbar fibres to the facial nucleus and the lateral reticular nucleus (nucleus of the lateral funiculus). An experimental study in the cat with silver impregnation methods. *Brain Research,* **1**, 317–337.

Curfs, M.H.J.M., Gribnau, A.A.M. & Dederen, P.J.W.C. (1994). Selective elimination of transient corticospinal projections in the rat cervical spinal cord gray matter. *Developmental Brain Research,* **78**, 182–190.

D'Amato, C.J. & Hicks, S.P. (1978). Normal development and post-traumatic plasticity of corticospinal neurons in rats. *Experimental Neurology,* **60**, 557–569.

Darian-Smith, I., Galea, M.P., Darian-Smith, C., Sugitani, M., Tan, A. & Burman, K. (1996). The anatomy of manual dexterity. The new connectivity of the primate sensorimotor thalamus and cerebral cortex. *Advances in Anatomy, Embryology and Cell Biology,* **133**, 1–140.

De Boer-van Huizen, R.T. & ten Donkelaar, H.J. (1999). Early development of descending supraspinal pathways: a tracing study in fixed and isolated rat embryos. *Anatomy and Embryology,* **199**, 539–547.

Dum, R.P. & Strick, P.L. (1991). The origin of corticospinal projections from the premotor areas in the frontal lobe. *Journal of Neuroscience,* **11**, 667–689.

Ebbesson, S.O.E. (1976). Morphology of the spinal cord. In R. Llinás & W. Precht (Eds.), *Frog Neurobiology* (pp. 679–706). Heidelberg: Springer.

Eyre, J.A., Miller, S., Clowry, G.J., Conway, E.A. & Watts, C. (2000). Functional corticospinal projections are established prenatally in the human fœtus permitting involvement in the development of spinal motor centres. *Brain,* **123**, 51–64.

Eyre, J.A., Miller, S. & Ramesh, V. (1991). Constancy of central conduction delays during development in man: investigation of motor and somatosensory pathways. *Journal of Physiology (London),* **434**, 441–452.

Fitzgerald, J.E. & Windle, W.F. (1942). Some observations on early human fetal movements. *Journal of Comparative Neurology,* **76**, 159–167.

Fitzgerald, M., Reynolds, M.L. & Benowitz, L.I. (1991). GAP-43 expression in the developing rat lumbar spinal cord. *Neuroscience,* **41**, 187–199.

Flechsig, P. (1920). *Anatomie des menschlichen Gehirns und Rückenmarks auf myelogenetischer Grundlage*. Leipzig: Georg Thieme.

Forssberg. H., Eliasson, A.C., Kinoshita, H., Johansson, R.S. & Westling, G. (1991). Development of human precision grip. I. Basic coordination of force. *Experimental Brain Research,* **85**, 451–457.

Fritz, N., Illert, M. & Saggau, P. (1986a). Location of motoneurones projecting to the cat distal forelimb. I. Deep radial motornuclei. *Journal of Comparative Neurology,* **244**, 286–301.

Fritz, N., Illert, M. & Saggau, P. (1986b). Location of motoneurones projecting to the cat distal forelimb. II. Median and ulnar motornuclei. *Journal of Comparative Neurology,* **244**, 302–312.

Galea, M.P. & Darian-Smith, I. (1995). Postnatal maturation of the direct corticospinal projections in the Macaque monkey. *Cerebral Cortex,* **5**, 518–540.

Gilles, F.H., Shankle, W. & Dooling, E.C. (1983). Myelination tracts: Growth patterns. In F.H. Gilles, A. Leviton & E.C. Dooling (Eds.), *The developing human brain: growth and epidemiologic neuropathology* (pp. 117–183). Boston, MA: Wright.

Girard, L., Bleicher, C. & Cabana, T. (1993). Development of skilled locomotion in the kitten: A comparison with the development of the corticospinal tract. *Society of Neuroscience Abstracts,* **19**, 63.16.

Gonzalez, A., Marin, O. Tuinhof, R. & Smeets, W.J.A.J. (1994a). Ontogeny of catecholamine systems in the central nervous system of anuran amphibians: An immunohistochemical study with antibodies against tyrosine hydroxylase and dopamine. *Journal of Comparative Neurology,* **346**, 63–79.

Gonzalez, A., Marin, O. Tuinhof, R. & Smeets, W.J.A.J. (1994b). Developmental aspects of catecholamine systems in the brain of anuran amphibians. In W.J.A.J. Smeets & A. Reiner (Eds.), *Phylogeny and development of catecholamine systems in the CNS of vertebrates* (pp. 343–360). Cambridge, UK: Cambridge University Press.

Gribnau, A.A.M., de Kort, E.J.M., Dederen, P.J.W.C. & Nieuwenhuys, R. (1986). On the development of the pyramidal tract in the rat. II. An anterograde tracer study of the outgrowth of the corticospinal fibers. *Anatomy and Embryology,* **175**, 101–110.

Gribnau, A.A.M. & Geijsberts, L.G.M. (1981). Developmental stages in the rhesus monkey (Macaca mulatta). *Advances in Anatomy, Embryology and Cell Biology,* **68**, 1–84.

Gribnau, A.A.M. & Geijsberts, L.G.M. (1985). Morphogenesis of the brain in stage Rhesus monkey embryos. *Advances in Anatomy, Embryology and Cell Biology,* **91**, 1–69.

Halverson, H.M. (1943). The development of prehension in infants. In R.G. Barker, J.S., Kounin & H.F., Wright (Eds.), *Child behavior and development* (pp. 49–65). New York: McGraw Hill.

Hamburger, V. (1968). Emergence of nervous coordinations: origins of integrated behavior. *Developmental Biology Supplement,* **2**, 251–271 (cited by Altman & Bayer, 1984).

Heffner, R. & Masterton, B. (1975). Variation in form of the pyramidal tract and its relationship to digital dexterity. *Brain, Behavior and Evolution,* **12**, 161–200.

Heffner, R. & Masterton, B. (1983). The role of the corticospinal tract in the evolution of human digital dexterity. *Brain, Behavior and Evolution,* **23**, 165–183.

Hendrickx, A.G. & Sawyer, R.H. (1975). Embryology of the rhesus monkey. In G.H. Bourne (Ed.), *The rhesus monkey,* vol. II: *Management, reproduction and pathology* (pp. 141–169). New York: Academic Press.

Herdmann, J., Dvorak, J., Rathmer, L., Theiler, R., Peuschel, K.N., Zenker, W. & Lumenta, C.B. (1991). Conduction velocities of pyramidal tract fibres and lumbar motor nerve roots: Normal values. *Zentralblatt für Neurochirugie,* **52**, 197–199.

Heuser, C.H. & Corner, G.W. (1957). Developmental horizons in human embryos. Description of age group X, 4 to 12 somites. *Carnegie Institute, Washington, Contributions in Embryology,* **36**, 29–39.

Heuser, C.H. & Streeter, G.L. (1941). Development of the macaque embryo. *Carnegie Institute, Washington, Contributions in Embryology,* **29**, 15–55.

His, W. (1904). Die Entwicklung des menschlichen Gehirns während der ersten Monate. Untersuchungsergebnisse. Leipzig: Hirzel (cited by Müller & O'Rahilly, 1990b).

Holstege, G. (1991). Descending motor pathways and the spinal motor system: Limbic and non-limbic components. *Progress in Brain Research,* **87**, 307–421.

Holstege, G. (1992). The emotional motor system. *European Journal of Morphology,* **30**, 67–79.

Holstege, G. & Kuypers, H.G.J.M. (1982). The anatomy of brain stem pathways to the spinal cord in cat. A labeled amino acid tracing study. In H.G.J.M. Kuypers & G.F. Martin (Eds.), *Anatomy of descending pathways to the spinal cord (Progress in brain research,* Vol. 57: pp. 145–175). Amsterdam: Elsevier.

Holstege, G. & Kuypers, H.G.J.M. (1987). Brainstem projections to spinal motoneurons: an update. *Neuroscience*, **23**, 809–821.

Hooker, D. (1952). The prenatal origin of behavior. Eighteenth Porter Lecture. University of Kansas Press, Lawrence, Kansas.

Hooker, D. (1954). Early human fetal behavior, with a preliminary note on double simultaneous fetal stimulation. *Research Publication, Association for Research into Nervous and Mental Disease,* **33**, 98–113.

Humphrey, T. (1960). The development of the pyramidal tracts in human fetuses, correlated with cortical differentiation. In D.B. Tower & J.P. Schadé (Eds.), *Structure and function of the cerebral cortex* (pp. 93–103). Amsterdam: Elsevier.

Humphrey, T. (1964). Some correlations between the appearance of human fetal reflexes and the development of the nervous system. In D.P. Purpura & J.P. Schadé (Eds.), *Growth and maturation of the brain* (*Progress in brain research*, vol. 4: pp. 93–135). Amsterdam: Elsevier.

Humphrey, T. & Hooker, D. (1959). Double simultaneous stimulation of human fetuses and the anatomical patterns underlying the reflexes elicited. *Journal of Comparative Neurology,* **112**, 75–102.

Inghilleri, M., Berardelli, A., Cruccu, G., Priori, A. & Manfredi, M. (1989). Corticospinal potentials after transcranial stimulation in humans. *Journal of Neurology, Neurosurgery and Psychiatry,* **52**, 970–974.

Jenny, A.B. & Inukai, J. (1983). Principles of motor organization of the monkey cervical spinal cord. *Journal of Neuroscience,* **3**, 567–575.

Joosten, E.A.J., Gribnau, A.A.M. & Dederen, P.J.W.C. (1987). An anterograde tracer study of the developing corticospinal tract in the rat: three components. *Developmental Brain Research,* **36**, 121–130.

Kalil, K. (1985). Development and plasticity of the sensorimotor cortex and pyramidal tract. In *Development, organization and processing in somatosensory pathways*. (pp. 87–96). Alan R. Liss.

Kather-Boidin, J. & Duron, B. (1991). Postnatal development of descending motor pathways studied in man by percutaneous stimulation of the motor cortex and the spinal cord. *International Journal of Developmental Neuroscience,* **9**, 15–26.

Killackey, H.P., Dehay, C., Giroud, P., Berland, M. & Kennedy, H. (1997). Distribution of corticospinal projection neurons in the neocortex of the fetal Macaque monkey. *Society of Neuroscience Abstracts,* 359.10.

Kinney, H.C., Brody, B.A., Kloman, A.S. & Gilles, F.H. (1988). Sequence of central nervous system myelination in human infancy. II. Patterns of myelination in autopsied infants. *Journal of Neuropathology and Experimental Neurology,* **47**, 217–234.

Kudo, N., Furukawa, F. & Okado, N. (1993). Development of descending fibers to the rat embryonic spinal cord. *Neuroscience Research,* **16**, 131–141.

Kuhlenbeck, H. (1977). Derivatives of the prosencephalon: Diencephalon and telencephalon. In *The central nervous system of vertebrates* (vol. 5, Part I: pp. 461–888). Basel: Karger (cited by Müller & O'Rahilly, 1990b).

Kume, M., Uemura, M., Matsuda, K., Matsushima, R. & Mizuno, N. (1978). Topographical representation of peripheral branches of the facial nerve within the facial nucleus. A HRP study in the cat. *Neuroscience Letters,* **8**, 5–8.

Kuypers, H.G.J.M. (1960). Central cortical projections to motor and somatosensory cell group. *Brain,* **83**, 161–184.

Kuypers, H.G.J.M. (1962). Corticospinal connections: postnatal development in the rhesus monkey. *Science,* **138**, 678–680.

Kuypers, H.G.J.M. (1964). The descending pathways to the spinal cord, their anatomy and function. In J.C. Eccles & J.P. Schadé (Eds.), *Organization of the spinal cord* (*Progress in brain research*, vol. 11: pp. 178–202). Amsterdam: Elsevier.

Kuypers, H.G.J.M. (1973). The anatomical organization of the descending pathways and their contributions to motor control especially in primates. In J.E. Desmedt (Ed.). *New developments in electromyography and clinical neurophysiology* (pp. 38–68). Basel: Karger.

Kuypers, H.G.J.M. (1981). Anatomy of the descending pathways. In V.B. Brooks, J.M. Brookhart & V.B. Mountcastle (Eds.). *Handbook of physiology – the nervous system*, vol. 2: *Motor systems* (pp. 597–666). Bethesda, MD: American Physiological Society.

Kuypers, H.G.J.M. (1982). A new look at the organization of the motor system. In H.G.J.M. Kuypers & G.F. Martin (Eds.), *Anatomy of descending pathways to the spinal cord* (*Progress in brain research*, vol. 57: pp. 381–403). Amsterdam: Elsevier.

Kuypers, H.G.J.M. & Brinkman, J. (1970). Precentral projections to different parts of the spinal intermediate zone in the rhesus monkey. *Brain Research*, **24**, 29–48.

Lakke, E.A.J. (1997). The projections to the spinal cord of the rat during development; a time-table of descent. *Advances in Anatomy, Embryology and Cell Biology*, **135**, 1–143.

Lakke, E.A.J. & Marani, E. (1991). Prenatal descent of rubrospinal fibers through the spinal cord of the rat. *Journal of Comparative Neurology*, **314**, 67–78.

Langworthy, O. (1933). Development of behaviour patterns and myelination of the nervous system in the human fetus and infant. *Contributions in Embryology*, **139**, 1–57.

Lawrence, D.G. & Hopkins, D.A. (1976). The development of motor control in the rhesus monkey: evidence concerning the role of corticomotoneuronal connections. *Brain*, **99**, 235–254.

Lawrence, D.G. & Kuypers, H.G.J.M. (1968a). The functional organization of the motor system in the monkey. I. The effects of bilateral pyramidal lesions. *Brain*, **91**, 1–14.

Lawrence, D.G. & Kuypers, H.G.J.M. (1968b). The functional organization of the motor system in the monkey. II. The effects of lesions of the descending brain-stem pathways. *Brain*, **91**, 15–33.

Lemon, R.N., Armand, J., Olivier, E. & Edgley, S.A. (1997). Skilled action and the development of the corticospinal tract in primates. In K.J. Connolly & H. Forssberg (Eds.), *Neurophysiology and neuropsychology of motor development* (*Clinics in developmental medicine*, no. 143/144: pp. 162–176). Cambridge, UK: Cambridge University Press.

Leong, S.K. (1983). Localizing the corticospinal neurons in neonatal, developing and mature albino rat. *Brain Research*, **265**, 1–9.

Martin, G.F. & Fisher, A.M. (1968). A further evaluation of the origin, the course and the termination of the opossum corticospinal tract. *Journal of Neurological Science*, **7**, 177–187.

Martin, G.F., Beattie, M.S., Bresnahan, J.C., Henkel, C.K. & Hughes, H.C. (1975). Cortical and brainstem projections to the spinal cord of the North American opossum, *Didelphis marsupialis virginiana*. *Brain Behavior and Evolution*, **12**, 270–310.

Martin, G.F., Cabana, T., DiTirro, F.J., Ho, R.H. & Humberston, A.O. (1982). The development of descending spinal connections. Studies using the North American opossum. In H.G.J.M. Kuypers & G.F. Martin (Eds.), *Anatomy of descending pathways to the spinal cord* (*Progress in brain research*, vol. 57: pp. 131–144). Amsterdam: Elsevier.

Mesdag, M.T. (1909). Bijdrage tot de ontwikkelingsgeschiedens van de structuur fer hersenen bij het Kipembryo. M. De Waal, Groningen (cited by Rhines & Windle, 1941).

Meissirel, C., Dehay, C. & Kennedy, H. (1993). Transient cortical pathways in the pyramidal tract of the neonatal ferret. *Journal of Comparative Neurology*, **338**, 193–213.

Mihailoff, G.A., Adams, C.E. & Woodward, D.J. (1984). An autoradiographic study of the postnatal development of sensorimotor and visual components of the corticopontine system. *Journal of Comparative Neurology*, **222**, 116–127.

Minckler, J., Klemme, R.M. & Minckler D. (1944). The course of efferent fibers from the human premotor cortex. *Journal of Comparative Neurology*, **81**, 259–277.

Morrell, N.W. (1985). The development of the human midbrain tegmentum with particular reference to the red nucleus. *Journal of Anatomy*, **140**, 544.

Müller, F. & O'Rahilly, R. (1988a). The development of the human brain from a closed neural tube at stage 13. *Anatomy and Embryology*, **177**, 203–224.

Müller, F. & O'Rahilly, R. (1988b). The first appearance of the future cerebral hemispheres in the human embryo at stage 14. *Anatomy and Embryology*, **177**, 495–511.

Müller, F. & O'Rahilly, R. (1989a). The human brain at stage 16, including the initial evagination of the neurohypophysis. *Anatomy and Embryology*, **179**, 551–569.

Müller, F. & O'Rahilly, R. (1989b). The human brain at stage 17, including the appearance of the future olfactory bulb and the first amygdaloid nuclei. *Anatomy and Embryology*, **180**, 353–369.

Müller, F. & O'Rahilly, R. (1990a). The human brain at stages 18–20, including the choroid plexuses and the amygdaloid and septal nuclei. *Anatomy and Embryology*, **182**, 285–306.

Müller, F & O'Rahilly, R. (1990b). The human brain at stages 21–23, with particular reference to the cerebral cortical plate and to the development of the cerebellum. *Anatomy and Embryology*, **182**, 375–400.

Müller, F. & O'Rahilly, R. (1997). The timing of appearance of neuromeres and their derivatives in stage human embryos. *Acta Anatomica*, **158**, 83–99.

Müller, K. & Hömberg, V. (1992). Development of speed of repetitive movements in children is determined by structural changes in corticospinal efferents. *Neuroscience Letters,* **144**, 57–60.

Müller, K., Ebner, B. & Hömberg, V. (1994). Maturation of fastest afferent and efferent central and peripheral pathways: No evidence for a constancy of central conduction delays. *Neuroscience Letters,* **166**, 9–12.

Niebroj-Dobosz, I. Fizianska, A., Rafalowksa, J. & Sawicka, E. (1980). Correlative biochemical and morphological studies of myelination in human ontogenesis. I. Myelination of the spinal cord. *Acta Neuropathologica (Berlin)*, **49**, 145–152.

Nieuwenhuys, R. (1998). Morphogenesis and general structure. In R. Nieuwenhuys, H.J. ten Donkelaar & C. Nicholson (Eds.), *The central nervous system of vertebrates* (pp. 159–228). Berlin: Springer. (cited by ten Donkelaar, 2000).

Nieuwkoop, P.D. & Faber, J. (1967). *Normal table of* Xenopus laevis *(Daudin)*. (2nd ed.) Amsterdam: North-Holland.

Niimi, K., Kishi, S., Miki, M. & Fujita, S. (1963). An experimental study of the course and termination of the projection fibers from cortical areas 4 and 6 in the cat. *Folia Psychiatrica et Neurologica Japonica,* **17**, 167–216.

Nordlander, R.H., Baden, S.T. & Ryba, T. (1985). Development of early brainstem projections to the tail spinal cord of *Xenopus*. *Journal of Comparative Neurology,* **231**, 519–529.

Nornes, H.O., & Carry, M. (1978). Neurogenesis in spinal cord of mouse: an autoradiographic analysis. *Brain Research*, **159**, 1–16.

Nudo, R.J. & Masterton, R.B. (1988). Descending pathways to the spinal cord: A comparative study of 22 mammals. *Journal of Comparative Neurology,* **277**, 53–79.

Nudo, R.J. & Masterton, R.B. (1990a). Descending pathways to the spinal cord. III. Sites of origin of the corticospinal tract. *Journal of Comparative Neurology,* **296**, 559–583.

Nudo, R.J. & Masterton, R.B. (1990b). Descending pathways to the spinal cord. IV. Some factors related to the amount of cortex devoted to the corticospinal tract. *Journal of Comparative Neurology,* **296**, 584–597.

Nudo, R.J., Sutherland, D.P. & Masterton, R.B. (1995). Variation and evolution of mammalian corticospinal somata with special reference to primates. *Journal of Comparative Neurology,* **358**, 181–205.

Nyberg-Hansen, R. & Brodal, A. (1963). Sites of termination of corticospinal fibers in the cat. An experimental study with silver impregnation methods. *Journal of Comparative Neurology,* **120**, 369–391.

Okado, N. (1980). Development of the human cervical spinal cord with reference to synapse formation in the motor nucleus. *Journal of Comparative Neurology,* **191**, 495–513.

Okado, N. (1981). Onset of synapse formation in the human spinal cord. *Journal of Comparative Neurology,* **201**, 211–219.

Okado, N., Kakimi, S. & Kojima, T. (1979). Synaptogenesis in the cervical cord of the human embryo: Sequence of synapse formation in a spinal reflex pathway. *Journal of Comparative Neurology,* **184**, 491–518.

Okado, N. & Oppenheim, R.W. (1985). The onset and development of descending pathways to the spinal cord in the chick embryo. *Journal of Comparative Neurology,* **232**, 143–161.

Olivier, E., Edgley, S.A., Armand, J. & Lemon, R.N. (1997). An electrophysiological study of the postnatal development of the corticospinal system in the Macaque monkey. *Journal of Neuroscience,* **17**, 267–276.

Olivier, G. & Pineau, H. (1962). Horizons de Streeter et âge embryonnaire. *Comptes Rendus de l'Association des Anatomistes*, **47**, 573–576.

Olson, L., Boreus, L.O. & Seiger, A. (1973). Histochemical demonstration and mapping of 5-hydroxytryptamine and catecholamine containing neuron systems in the human fetal brain. *Zeitschrift für Anatomie und entwicklungsgeschichte*, **139**, 259–282.

O'Rahilly, R. (1973). The early development of the hypophysis cerebri in staged human embryos. *Anatomical Record*, **175**, 511.

O'Rahilly, R. & Müller, F. (1987). Developmental stages in human embryos including a revision of Streeter's "horizons" and a survey of the Carnegie collection. *Carnegie Institution, Washington*, Publication 637.

O'Rahilly, R. & Müller F. (1994). *The embryonic human brain. an atlas of developmental stages*. New York: Wiley-Liss.

O'Rahilly, R., Müller, F., Hutchins, G.M. & Moore, G.W. (1987). Computer ranking of the sequence of appearance of 73 features of the brain and related structures in staged human embryos during the sixth week of development. *American Journal of Anatomy*, **180**, 69–86.

O'Rahilly, R., Müller, F., Hutchins, G.M. & Moore, G.W. (1988). Computer ranking of the sequence of appearance of 40 features of the brain and related structures in staged human embryos during the seventh week of development. *American Journal of Anatomy*, **182**, 295–317.

Ozaki, S. & Snider, W.D. (1997). Initial trajectories of sensory axons toward laminar targets in the developing mouse spinal cord. *Journal of Comparative Neurology*, **380**, 215–229.

Panneton, W.M. & Martin, G.F. (1979). Midbrain projections to the trigeminal, facial and hypoglossal nuclei in the opossum. A study using axonal transport techniques. *Brain Research*, **168**, 493–511.

Passingham, R.E. (1985). Rates of brain development in mammals including man. *Brain, Behavior and Evolution*, **26**, 167–175.

Passingham, R.E., Perry, V.H. & Wilkinson, F. (1983). The long-term effects of removal of sensori-motor cortex in infant and adult rhesus monkeys. *Brain*, **106**, 675–705.

Petras, J.M. (1968). Corticospinal fibers in New World and Old World simians. *Brain Research*, **8**, 206–208.

Porter, R., & Lemon R. (1993). *Corticospinal function and voluntary movement*. Monographs of the Physiological Society 45. Oxford: Oxford University Press.

Puelles, L. & Verney, C. (1998). Early neuromeric distribution of tyrosine-hydroxylase-immunoreactive neurons in human embryos. *Journal of Comparative Neurology*, **394**, 283–308.

Rajaofetra, N., Sandillon, F., Geffard, M. & Privat, A. (1989). Pre- and postnatal ontogeny of serotonergic projections to the rat spinal cord. *Journal of Neuroscience Research*, **22**, 305–321.

Rajaofetra, N., Poulat, P., Marlier, L. Geffard, M. & Privat, A. (1992). Pre- and postnatal development of noradrenergic projections to the rat spinal cord: an immunocytochemical study. *Developmental Brain Research*, **67**, 237–246.

Reh, T. & Kalil, K. (1981). Development of the pyramidal tract in the hamster. I. A light microscopic study. *Journal of Comparative Neurology*, **200**, 55–67.

Reh, T. & Kalil, K. (1982). Development of the pyramidal tract in the hamster. II. An electron microscopic study. *Journal of Comparative Neurology*, **205**, 77–88.

Rexed, B. (1952). The cytoarchitectonic organization of the spinal cord in the cat. *Journal of Comparative Neurology*, **96**, 415–496.

Rhines, R. & Windle, W.F. (1941). The early development of the fasciculus longitudinalis medialis and associated secondary neurons in the rat, cat and man. *Journal of Comparative Neurology*, **75**, 165–189.

Schoen, J.H.R. (1964). Comparative aspects of the descending fiber systems in the spinal cord. In J.C. Eccles & J.P. Schadé (Eds.), *Organization of the spinal cord* (*Progress in brain research*, vol. 11: pp. 203–222). Amsterdam: Elsevier.

Schoen, J.H.R. (1969). The corticofugal projection in the brain stem and spinal cord in man. *Psychiatria, Neurologica, Neurochirurgica*, **72**, 121–128.

Schreyer, D.J. & Jones, E.G. (1982). Growth and target finding by axons of the corticospinal tract in prenatal and postnatal rats. *Neuroscience*, **7**, 1837–1853.

Schreyer, D.J. & Jones, E.G. (1988a). Topographic sequence of outgrowth of corticospinal axons in the rat: a study using retrograde axonal labeling with fast blue. *Developmental Brain Research*, **38**, 89–101.

Schreyer, D.J. & Jones, E.G. (1988b). Axon elimination in the developing corticospinal tract of the rat. *Developmental Brain Research*, **38**, 103–119.

Sillar, K.T., Wedderburn, J.F. & Simmers, A.J. (1992). Modulation of swimming rhythmicity by 5-hydroxytryptamine during post-embryonic development in *Xenopus laevis*. *Proceedings of the Royal Society of London (Biology)*, **250**, 107–114.

Sillar, K.T., Woolston, A.M. & Wedderburn, J.F. (1995). Involvement of brainstem serotonergic interneurons in the development of a vertebrate spinal locomotor circuit. *Proceedings of the Royal Society of London (Biology)*, **259**, 65–70.

Sims, T.J. & Vaughn, J.E. (1979). The generation of neurons involved in an early reflex pathway of embryonic mouse spinal cord. *Journal of Comparative Neurology*, **183**, 707–720.

Sloper, J.J., Brodal, P. & Powell, T.P.S. (1983). An anatomical study of the effects of unilateral removal of sensorimotor cortex in infant monkeys on the subcortical projections of the controlateral sensorimotor cortex. *Brain*, **106**, 707–716.

Snider, W.D., Zhang, L., Yusoof, S., Gorukanti, N. & Tsering C. (1992). Interactions between dorsal root axons and their target motor neurons in developing mammalian spinal cord. *Journal of Neuroscience,* **12**, 3494–3508.

Stanfield, B.B. & Asanuma, C. (1993). The distribution of corticospinal axons within the spinal gray of infant rhesus monkeys. *Society of Neuroscience Abstracts,* 19.673.

Stanfield, B.B. & O'Leary, D.D.M. (1985). The transient corticospinal projection from the occipital cortex during the postnatal development of the rat. *Journal of Comparative Neurology,* **238**, 236–248.

Stanfield, B.B., O'Leary, D.D.M. & Fricks, C. (1982). Selective collateral elimination in early postnatal development restricts cortical distribution of rat pyramidal tract neurones. *Nature*, **298**, 371–373.

Streeter, G.L. (1951). Developmental horizons in human embryos (horizons xi to xxiii). Carnegie Institution, Washington, Publication 592, *Contributions in Embryology,* **34**, 165–196.

ten Donkelaar, H.J. (1982). Organization of descending pathways in the spinal cord in amphibians and reptiles. In H.G.J.M. Kuypers & G.F. Martin (Eds.), *Anatomy of descending pathways to the spinal cord* (*Progress in brain research*, vol. 57: pp. 25–67). Amsterdam: Elsevier.

ten Donkelaar, H.J. (1988). Evolution of the red nucleus and rubrospinal tract. *Behavioral Brain Research*, **28**, 9–20.

ten Donkelaar, H.J. (2000). Development and regenerative capacity of descending supraspinal pathways in tetrapods: a comparative approach. *Advances in Anatomy, Embryology and Cell Biology,* **154**, 1–145.

ten Donkelaar, H.J. & de Boer-van Huizen, R. (1982). Observations on the development of descending pathways in the clawed toad, *Xenopus laevis. Anatomy and Embryology,* **163**, 461–473.

ten Donkelaar, H.J., de Boer-van Huizen, R. & van der Linden, J.A.M. (1991). Early development of rubrospinal and cerebellorubral projections in *Xenopus laevis. Developmental Brain Research*, **58**, 297–300.

ten Donkelaar, H.J., Geijsberts, L.G.M. & Dederen, P.J.W. (1979). Stages in the prenatal development of the Chinese hamster (*Cricetulus griseus*). *Anatomy and Embryology*, **156**, 1–28.

Theriault, E. & Tatton, W.G. (1989). Postnatal redistribution of pericruciate motor cortical projections within the kitten spinal cord. *Developmental Brain Research*, **45**, 219–237.

Van Crevel, H. & Verhaart, W.J.C. (1963a). The rate of secondary degeneration in the central nervous system. I. The pyramidal tract of the cat. *Journal of Anatomy (London)*, **97**, 429–449.

Van Crevel, H. & Verhaart, W.J.C. (1963b). The rate of secondary degeneration in the central nervous system. II. The optic nerve of the cat. *Journal of Anatomy (London)*, **97**, 451–464.

Van Mier, P. & ten Donkelaar, H.J. (1984). Early development of descending pathways from the brain stem to the spinal cord in *Xenopus laevis. Anatomy and Embryology*, **170**, 295–306.

Van Mier, P. & ten Donkelaar, H.J. (1988). The development of primary afferents to the lumbar spinal cord in *Xenopus laevis. Neuroscience Letters,* **84**, 35–40.

Van Mier, P., Armstrong, J. & Roberts, A. (1989). Development of early swimming in *Xenopus laevis* embryos: Myotomal musculature, its innervation and activation. *Neuroscience*, **32**, 113–126.

Van Mier, P., Joosten, H.W.J., Van Rheden, R. & ten Donkelaar, H.J. (1986). The development of serotonergic raphespinal projections in *Xenopus laevis. International Journal of Developmental Neuroscience,* **4**, 465–475.

Vaughn, J.E. & Grieshaber, J.A. (1973). A morphological investigation of an early reflex pathway in developing rat spinal cord. *Journal of Comparative Neurology,* **148**, 177–210.

Verhaart, W. (1950). Hypertrophy of pes pedunculi and pyramid as result of degeneration of contralateral corticofugal fiber tracts. *Journal of Comparative Neurology,* **92**, 1–15.

Verney, C., Zecevic, N., Nikolic, B., Alvarez, C. & Berger, B. (1991). Early evidence of catecholaminergic cell groups in 5- and 6-week-old human embryos using tyrosine hydroxylase and dopamine-β-hydroxylase immunocytochemistry. *Neuroscience Letters,* **131**, 121–124.

Walshe, F.M.R. (1942). The giant cells of Betz, the motor cortex and the pyramidal tract: a critical review. *Brain*, **65**, 409–461.

Weidenheim, K.M., Bodhireddy, S.R., Rashbaum W.K. & Lyman W.D. (1996). Temporal and spatial expression of major myelin proteins in the human fetal spinal cord during the second trimester. *Journal of Neuropathology and Experimental Neurology,* **55**, 734–745.

Weidenheim, K.M., Epshteyn, I. Rashbaum W.K. & Lyman W.D. (1993). Neuroanatomical localization of myelin basic protein in the late first and early second trimester human fœtal spinal cord and brainstem. *Journal of Neurocytology, 22,* 507–516.

Weidenheim, K.M., Kress, Y., Epshteyn, I. Rashbaum W.K. & Lyman W.D. (1992). Early myelination in the human fetal lumbosacral spinal cord: characterization by light and electron microscopy. *Journal of Neuropathology and Experimental Neurology, 51,* 142–149.

Wentworth, L.E. (1984a). The development of the cervical spinal cord of the mouse embryo. I. A Golgi analysis of ventral root neuron differentiation. *Journal of Comparative Neurology, 222,* 81–95.

Wentworth, L.E. (1984b). The development of the cervical spinal cord of the mouse embryo. II. A Golgi analysis of sensory, commissural, and association cell differentiation. *Journal of Comparative Neurology, 222,* 96–115.

Wilson, J.G. (1973). *Environment and birth defects.* New York: Academic Press.

Windle, W.F. (1944). Genesis of somatic motor function in mammalian embryos: a synthesizing article. *Physiological Zoology, 17,* 247–260 (cited by Altman & Bayer, 1984).

Windle, W.F. (1970). Development of neural elements in human embryos of four to seven weeks gestation. *Experimental Neurology Supplement, 5,* 44–83.

Windle, W.F. & Fitzgerald, J.E. (1937). Development of the spinal reflex mechanism in human embryos. *Journal of Comparative Neurology, 67,* 493–509.

Windle, W.F. & Fitzgerald, J.E. (1942). Development of the human mesencephalic trigeminal root and related neurons. *Journal of Comparative Neurology, 77,* 597–608.

Wozniak, W. & O'Rahilly, R. (1982). An electron microscopic study of myelination of pyramidal fibers at the level of the pyramidal decussation. *Journal für Hirnforschung, 23,* 331–342.

Yakovlev, P.I. & Lecours A.R., (1967). The myelogenetic cycles of regional maturation of the brain. In A. Minkowski (Ed.), *Regional development of the brain in early life* (pp. 3–70). Oxford: Blackwell.

Yakovlev, P.I., & Rakic, P. (1966). Patterns of decussation of bulbar pyramids and distribution of pyramidal tracts on two sides of the spinal cord. *Transactions of the American Neurological Association, 91,* 366–367.

Zecevic, N. & Verney, C. (1995). Development of the catecholamine neurons in human embryos and fetuses, with special emphasis on the innervation of the cerebral cortex. *Journal of Comparative Neurology, 351,* 509–535.

Further Reading

Altman, J. & Bayer, S.A. (1984). The development of the rat spinal cord. *Advances in Anatomy, Embryology and Cell Biology, 85,* 000–000.

Armand, J. (1982). The origin, course and terminations of corticospinal fibers in various mammals. In H.G.J.M. Kuypers & G.F. Martin, (Eds.), *Anatomy of descending pathways to the spinal cord* (*Progress in brain research*, vol. 57: pp. 329–360). Amsterdam: Elsevier.

Armand, J., Olivier, E., Edgley, S.A. & Lemon, R.N. (1996). The structure and function of the developing corticospinal tract: Some key issues. In A.H. Wing, P. Haggard & J.R. Flanagan (Eds.), *Hand and brain: the neurophysiology and psychology of hand movements* (pp. 125–145). San Diego, CA: Academic Press.

Kuypers, H.G.J.M. (1981). Anatomy of the descending pathways. In V.B. Brooks, J.M. Brookhart & V.B. Mountcastle (Eds.), *Handbook of physiology – the nervous system*, vol. 2: *Motor systems* (pp. 597–666). Bethesda, MD: American Physiological Society.

Kuypers, H.G.J.M. (1982). A new look at the organization of the motor system. In H.G.J.M. Kuypers & G.F. Martin (Eds.), *Anatomy of descending pathways to the spinal cord* (*Progress in brain research*, vol. 57: pp. 381–403). Amsterdam: Elsevier.

Lemon, R.N., Armand, J., Olivier, E. & Edgley, S.A. (1997). Skilled action and the development of the corticospinal tract in primates. In K.J. Connolly & H. Forssberg (Eds.), *Neurophysiology and neuropsychology of motor development* (*Clinics in Developmental Medicine*, no. 143/144: pp. 162–176). Cambridge, UK: Cambridge University Press.

Martin, G.F., Cabana, T., DiTirro, F.J., Ho, R.H. & Humberston, A.O. (1982). The development of descending spinal connections. Studies using the North American opossum. In H.G.J.M. Kuypers

& G.F. Martin (Eds.), *Anatomy of descending pathways to the spinal cord* (*Progress in brain research*, vol. 57: pp. 131–144). Amsterdam: Elsevier.

Porter, R. & Lemon R. (1993). Corticospinal function and voluntary movement. Monographs of the Physiological Society 45. Oxford: Oxford University Press.

ten Donkelaar, H.J. (2000). Development and regenerative capacity of descending supraspinal pathways in tetrapods: a comparative approach. *Advances in Anatomy, Embryology and Cell Biology,* **154** (In press).

II.11
Development and Plasticity of Neuromuscular Innervation

RICHARD R. RIBCHESTER

ABSTRACT

Neuromuscular junctions comprise the last synapses in the chain mediating voluntary and reflex commands that under behaviour. Access to general mechanisms of synaptic development and plasticity is facilitated at neuromuscular junctions by the stereotyped patterns of motoneuron innervation of muscle, and by the large size of the postsynaptic structures at motor end-plates on individual muscle fibres. These features permit detailed descriptions at systems, cellular and molecular levels of the organization of neuromuscular connections, and enable experiments to be designed which probe and quantify the role of activity in stabilization and elimination of synaptic connections. Recent experiments suggest that activity is influential, but not decisive, in these processes.

1. INTRODUCTION

Professor Sir Bernard Katz recounted one motivation behind his pioneering work in neuromuscular synaptic physiology, during the Fenn Lecture to the International Union of Physiological Sciences held in Glasgow in 1993: namely that "the neuromuscular junction is an experimentally favourable object whose study could throw light on mechanisms of synaptic transmission elsewhere" (Katz, 1996). Three years later Professor Mu-Ming Poo remarked, during a presentation at a symposium on Mechanisms of Synaptic Plasticity held in Edinburgh, that the neuromuscular junction may be regarded as "the ultimate synapse". Such anecdotes propel an argument that neuromuscular synapses are important; not merely because we literally cannot live without them, but because their properties are representative of synapses in general; and these properties remain more accessible to experimental investigation using combinations of

A.F. Kalverboer and A. Gramsbergen (eds.), Handbook of Brain and Behaviour in Human Development, 261–342
© 2001 Kluwer Academic Publishers. Printed in Great Britain.

techniques that are not always possible or easy to apply to central synapses *in situ*. There is now little reason to doubt the validity of Katz's hypothesis, since it has been so amply supported by subsequent analysis of synaptic transmission in the central nervous system. And Poo was only half-joking: it is of course literally true that neuromuscular junctions are the final connection in the system that mediates wilful behaviour via neural activity. It seems plausible to suggest (and this is implicit in Poo's metaphor) that knowledge and understanding we acquire by studying neuromuscular development and plasticity will also eventually be extended to presynaptic and postsynaptic cell relationships elsewhere in the nervous system. This ultimately promises insights into the role of synaptic development and plasticity in all higher brain functions, including environmental adaptation; memory and learning; as well as the planning and execution of complex movements.

The anatomical patterning of connections between motoneurones and skeletal muscle in adult mammals is highly stereotyped (Figure 1). Each motoneuron, with few exceptions, innervates just one anatomically defined muscle; but within that muscle the motor axon branches many times. Remarkably, each of the thousands of muscle fibres contained in a muscle is targeted by only one axon collateral, derived from the small contingent of motoneurons uniquely supplying it. Each motor axon terminal is constrained to make functional synaptic contacts on a small, circumscribed area on its target muscle fibre: the motor endplate. Here the terminal may remain throughout life, often unaltered but with the potential to undergo adaptive remodelling of both function and form.

This beautiful phenomenon raises obvious questions: how are neuromuscular synaptic connections formed? What processes bring about selective innervation of a muscle fibre by only one motoneuron? What properties constrain the extent of an individual motoneuron's axonal arbour? How are the size and strength of a neuromuscular synapse regulated once the stable connection has formed? What processes and factors govern the reaction to nerve injury and bring about nerve sprouting and regeneration of neuromuscular synapses?

Perhaps the most beguiling questions, however, are those that relate to the importance of neural activity and function in the formation, maintenance and repair of neuromuscular innervation and gene expression. At a systems level, axons supplying muscle fibres are made and broken in a fashion that appears to optimize the potential of the neuromuscular system to adapt to changing functional requirements during development, or to recover function after injury. At a cellular level, neural activity modifies the size and strength of individual synaptic connections and the organization of subcellular components that comprise them. At a molecular level, activity up- or down-regulates the state of individual signalling proteins, and alters the expression of genes that provide the protein scaffolds required to modify and maintain overall neuromuscular synaptic form and function. How does use or disuse bring about such changes? Above all, is activity both sufficient and necessary for the induction and expression of neuromuscular synaptic plasticity?

My aim in this chapter is to provide first, as a foundation, a straightforward overview of the salient features of the mammalian neuromuscular system; some of the cellular and molecular relationships at neuromuscular synapses; and the formation of these connections during development. The references in this

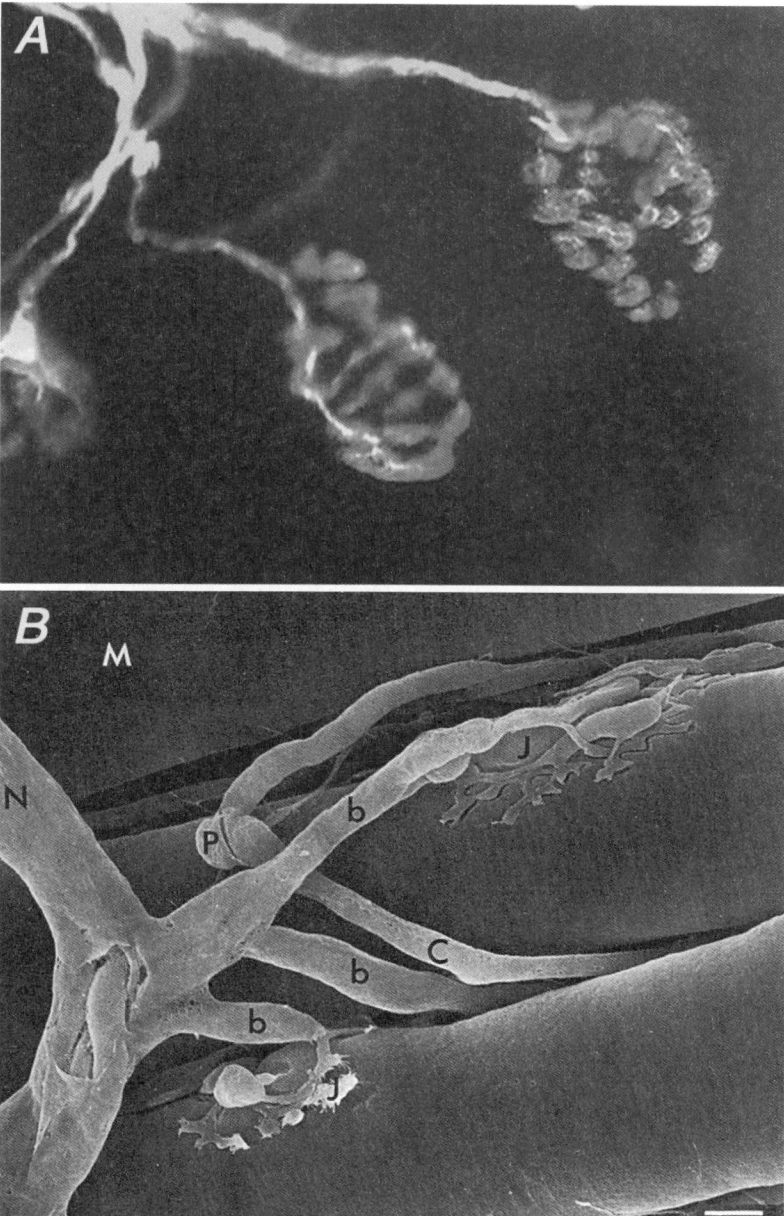

Figure 1 Images of mammalian neuromuscular junctions. **A**: Fluorescence micrograph of motor endplates, terminals and axons in a mouse muscle stained with TRITC-α-bungarotoxin and immunocytochemically with antibodies against SV2 and neurofilament (R.R. Ribchester & D. Thompson, unpublished). **B**: Scanning electron micrograph of neuromuscular junctions in hamster muscle. (From Desaki & Uehara, 1981, with permission.) Calibration: **A**, 10 μm; **B**, 5 μm.

section are not intended to be extensive, representing only a sample of those relating to the facts presented. Next, I have attempted a more critical appraisal on three issues: plasticity of relationships between synaptic structure and function in adult muscles; elimination of polyneuronal innervation during development; and the synapse elimination that accompanies competitive reinnervation of muscle after nerve injury and axon regeneration. These are critical issues in the search for understanding the importance of activity in the development and plasticity of synaptic structure and function. The conclusion that emerges is that activity is not absolutely necessary for the induction or expression of any of these phenomena in skeletal muscle, though it is sufficient for some of them.

The context of this handbook largely precludes an extensive review of the literature relevant to the topics discussed here, however. For example, I have side-stepped most of the literature concerned with valuable studies of neuromuscular synapses in non-mammalian vertebrates and invertebrates. Grinnell (1995) provides excellent coverage of important findings acquired from these species up to relatively recent times. Though relevant, I shall also not refer in detail to the relationships between neuromuscular synaptic development and plasticity, and similar phenomena elsewhere in the nervous system. Jansen & Fladby (1990), Purves (1994) and Snider & Lichtman (1996) provide outstanding essays attempting to integrate these topics. Finally, the reader must look elsewhere for descriptions of the development and plasticity of the complex sensorimotor innervation of muscle receptors (Patak *et al.*, 1992; Patten & Ovalle, 1991; Rees *et al.*, 1994; Walro & Kucera, 1999); the autonomic innervation of skeletal, cardiac and smooth muscle (Burnstock & Hoyle, 1992; Gabella, 1995); and the reflex innervation of spinal motoneurons by peripheral and central afferents and interneurons (Kalb & Hockfield, 1992; Konstantinidou *et al.*, 1995; Lin *et al.*, 1998); all of which are beyond the remit and scope of this chapter.

2. BACKGROUND AND OVERVIEW

2.1. The neuromuscular system

2.1.1. *Organization of adult motoneurons and their connections*

Spinal motoneurons supplying each skeletal muscle are loosely clustered dorsoventrally, mediolaterally and rostrocaudally in sausage-shaped "pools" in the ventral horn horn of spinal cord grey matter (Figure 2). These usually extend over one to three spinal segments (Crockett *et al.*, 1987; Lance-Jones & Landmesser, 1980). The location and extent of a motoneuron pool can be determined experimentally by mapping the extracellular field potential with an extracellular microelectrode in response to antidromic stimulation of muscle nerves; or by applying fluorescent or enzymic retrograde tracer molecules to the muscle nerves and locating the motoneuron cell bodies in sections of spinal cord, using fluorescence microscopy or histochemistry (Clowry & Vrbova, 1991; Lev-Tov & O'Donovan, 1995; Mesulam, 1982). Such methods have demonstrated that motoneurons supplying muscle groups that are either synergistic or antagonistic in function – flexors and extensors in the limb, for example – though spatially quite discrete, nonetheless have their cell bodies, dendritic trees and afferent inputs intermingled (A.G. Brown, 1981). Given this refined chaos it is remarkable that each motoneuron projects a single axon out of one ventral root unerringly – without branching – to one specific, anatomically distinct muscle or compartment. The distance each motor axon extends to its specific target varies from a few microns, in a mouse embryo for example, to a few metres – for instance from spine to toes in a fully mature giraffe. But after reaching their targets during early embryogenesis, maintenance of contact as the limbs elongate is probably controlled by different mechanisms from the initial selection (Bray, 1984).

The specificity and extent of motor axon projection, though impressive, pales before the astonishing selectivity of innervation of the motoneurons themselves by primary sensory afferents, intraspinal interneurons and descending connections from higher motor centres (Figure 2C). It is truly awe-inspiring to consider that primary sensory axons supplying muscle spindles have been contrived by natural selection to make their monosynaptic central connections onto homonymous motoneurons, in far greater numbers and strengths than to those motoneurons supplying muscles that are synergistic or antagonistic in function. This occurs during development in spite of the cell bodies and dendritic arbors of both synergists and antagonists becoming close neighbours within the same spinal segment (Eide et al., 1982). Considering the apparent opportunities for random connectivity, the developmental ordering of ascending and descending connections to motoneurons are equally astounding phenomena (Armand et al., 1997; Lemon, 1998). Experimental analysis of spinal cord development in embryonic chickens and frogs suggest that in some cases the central connections are specified after peripheral motor and sensory connectivity is laid down, implying that peripheral-to-central signalling, and activity-dependent reinforcement of central synaptic connections are involved (Seebach & Ziskind-Conhaim, 1994; Wenner & Frank, 1995). However, it is unlikely that such orthograde and retrograde influences are ultimately responsible for all the nuances of central connectivity, and at least the broad pattern of selective connections is probably mapped out by a programmed distribution of recognition molecules (Purves, 1988, 1994).

There are many fewer motoneurons within a specific motoneuron pool than muscle fibres in the target muscle to which they project. This is only partly a consequence of extensive cell death among the population of motoneurons, which reduces some motoneuron pools by more than 50% during mid-embryogenesis (see Oppenheim et al., this handbook). Thus, once confined to its target muscle compartment, each motor axon branches many times, and each collateral branch ultimately supplies a neuromuscular junction on just one muscle fibre: a state referred to as "mononeuronal innervation". The number of muscle fibres innervated by an individual motoneuron defines the motor unit size. The numbers may vary over two orders of magnitude – from tens to thousands – but within a specific adult muscle the coefficient of variation of motor unit size is usually about 30–50%. Different muscles have characteristic distributions of motor unit size: in some the number varies approximately symmetrically about a mean, as in the mouse soleus muscle for example; in others the distribution has a positive skew, as in the mouse extensor digitorum longus. These size distributions are characteristic for the types of nerves that supply the muscles. Thus cross-reinnervation of muscle confers a motor unit size distribution characteristic of the nerve implant, rather than the muscle host (Taxt, 1983a).

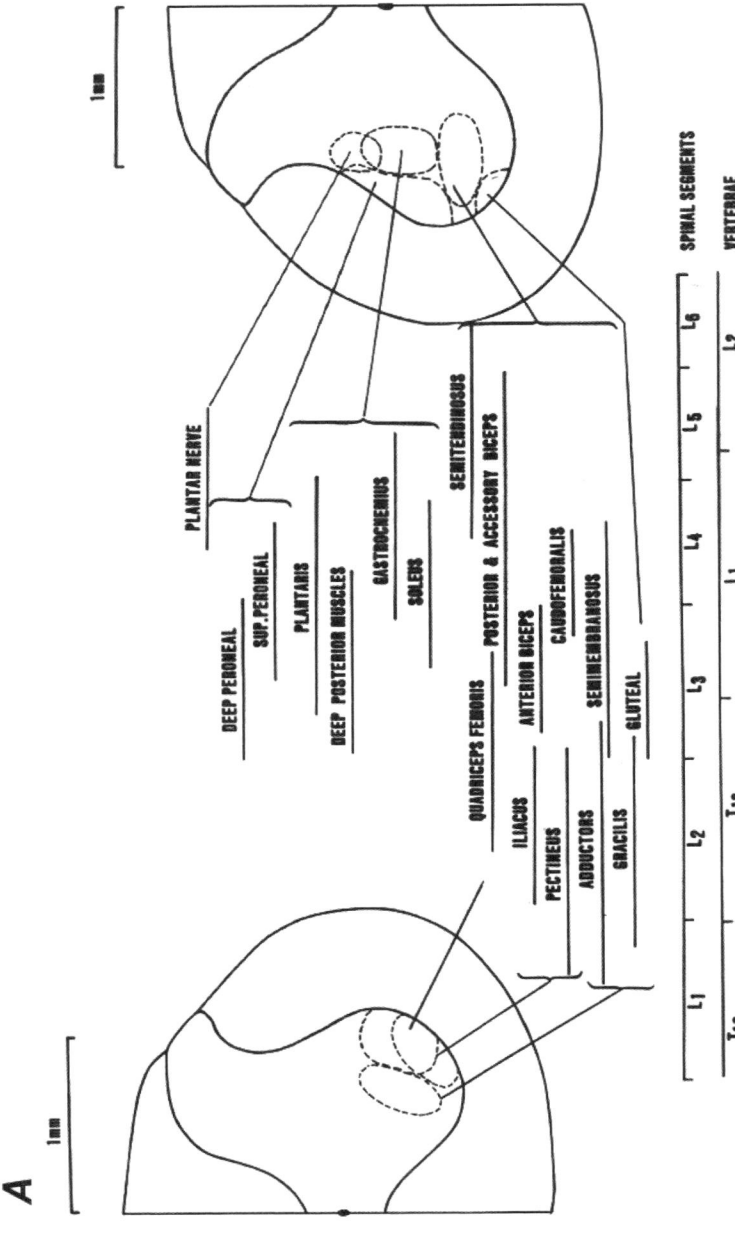

Figure 2 (*see also facing page*) Motoneurones localized by uptake (**A**, **B**) or injection (**C**) of horseradish peroxidase. **A**: Diagram indicating the location of motoneuron pools in the lumbar spinal cord of the adult rat. **B**: Each spot shows the location of a motoneuron whose axon passes to the foot via the medial and lateral plantar nerves or sural nerve. **C**: Single motoneuron in the ventral horn of the cat spinal cord (lower right) and the afferent arbor from a single group 1a afferent arising from a muscle spindle in the same muscle. (From: A, Nicoloupoulos-Stournos & Iles1983; **B**, Crockett *et al.*, 1987; **C**, A.G. Brown, 1981, with permission.)

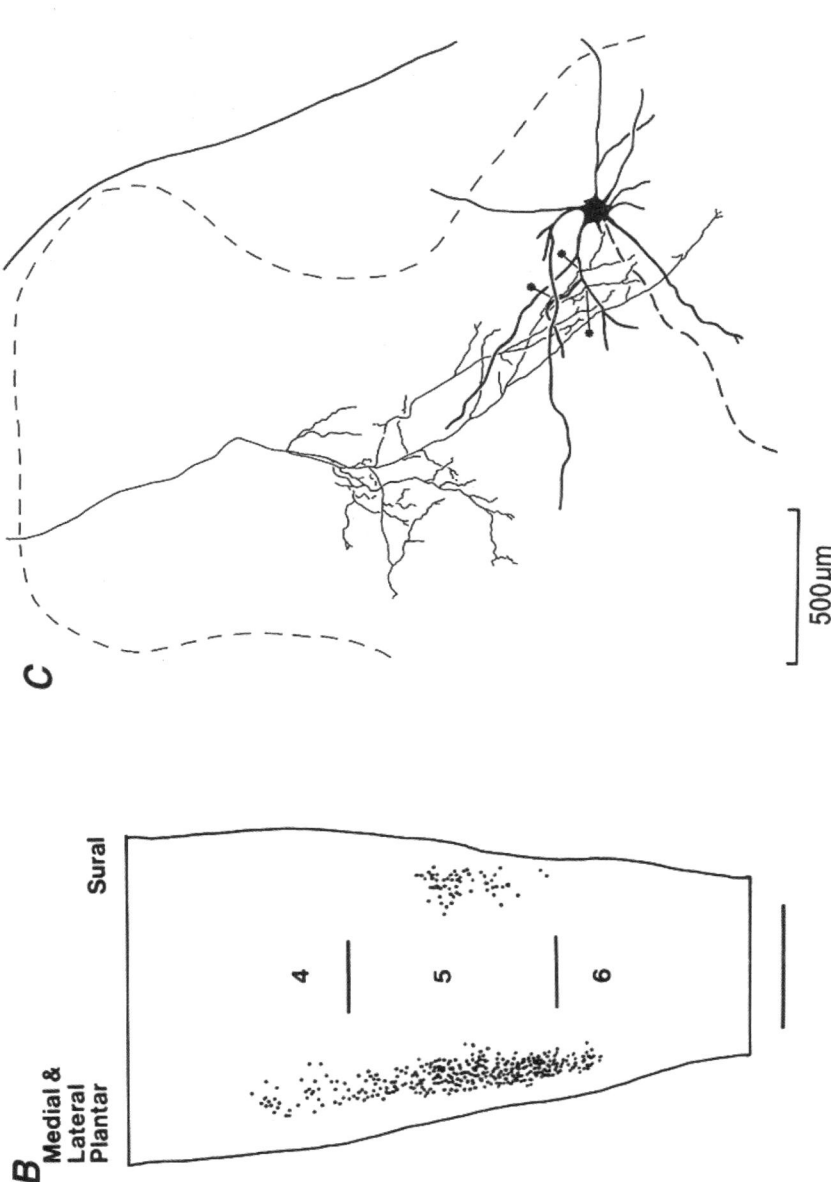

Motoneurons supplying different muscle show very distinctive firing patterns in their day-to-day use, and this is determined partly by the activity of cortical connections that connect with them (e.g. Bennett & Lemon, 1996) and partly by their intrinsic firing capabilities (adaptive range; Westgaard & Lomo, 1988). In the rat, motoneurons supplying phasic muscles engage intermittent firing patterns, at about 80 Hz for very brief periods; whereas those supplying postural muscle are more tonically active, firing for longer periods at around 20 Hz (Hennig & Lomo, 1985). These patterns may play a role in specifying muscle fibre types and synaptic efficacies at neuromuscular junctions (see below). Variable motor unit numbers and sizes, activity patterns, and mononeuronal innervation are crucial adaptaptions that optimize the range of forces a muscle is required to produce for its overall function (this handbook; Kernell, 1998).

2.1.2. Organization of adult muscles and muscle compartments

Skeletal muscles are anatomically discrete, defined by the patterns created by migration, segmentation and coalescence of somite mesoderm cells during development. Each adult muscle is bounded by tissue fascia. Many muscles are subdivided into compartments which comprise major anatomical subdivisions, and these are largely selectively innervated by discrete populations of neurons. Each muscle compartment is further subdivided into internal fascicles by similar connective tissue and microvascular barriers. For example, the extensor digitorum longus muscle of the rat is divided into at least two major compartments (Balice-Gordon & Thompson, 1988a). Although there is some overlap in their innervation up to the first few days after birth, the nerve supply is restricted to motoneurons which supply only one compartment or the other (Figure 3). Similar compartmental boundaries are evident in muscles with a segmental nerve supply such as the thoraco-abdominal muscles, or the gluteal muscles. The distribution of motoneuron collaterals within compartments and between fascicles is quite promiscuous, however. For example, a fascicle comprising about a dozen muscle fibres can be supplied in some cases by as many motoneurons; and an individual motoneuron can extend collateral branches to as many fascicles as there are present in a muscle compartment.

Muscle fibres may be "typed" according to a system of classification based on their contractile properties, isoforms of myosin and histochemical profile of enzymes they express. Each fascicle normally comprises a mixture of muscle fibre types, although one type or other may predominate in different, anatomically distinct muscles. The soleus muscle fibres of rodents, for example, are predominantly "slow twitch" – that is, the rise time of isometric force development is relatively long, about 20–30 ms; and soleus muscles comprise mostly type I fibres. These are characterized by high levels of expression of "slow" myosin; an acid-stable ATPase; and enzymes important in oxidative metabolism. The fibres of extensor digitorum longus, by contrast, are predominantly type II: expressing "fast" isoforms of myosin; alkali-stable ATPase; and enzymes important in glycolytic metabolism. Isometric twitch rise times of these muscles are of the order of 10–20 ms at normal body temperature.

Muscle fibre types may be further subdivided on the basis of a combination of physiological and morphological critieria: type II fibres, for example are

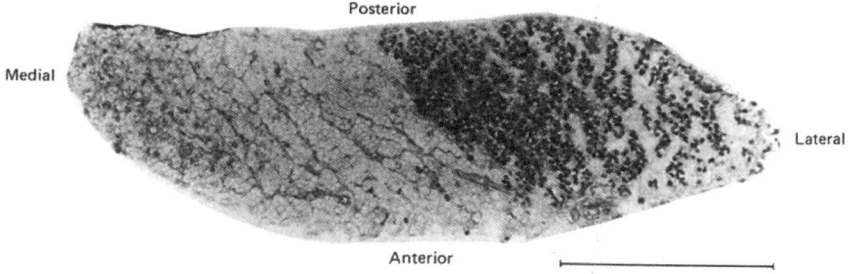

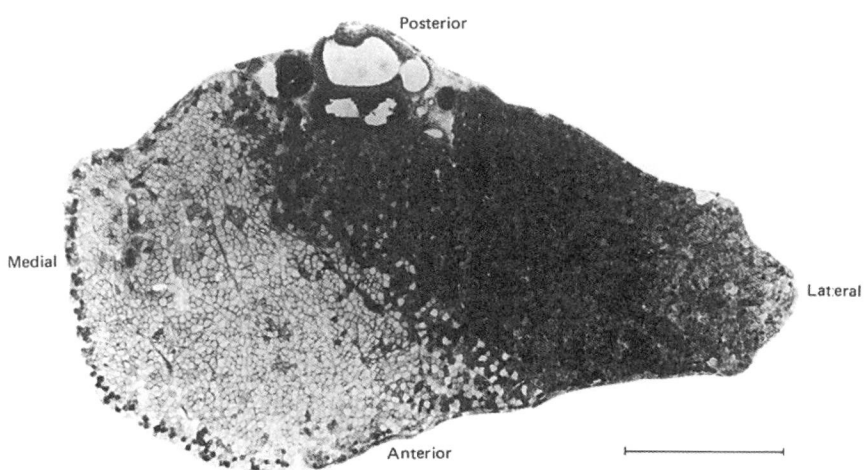

Figure 3 Transverse section of extensor digitorum longus muscles in newborn (**A**) and 14-day-old (**B**) rats. Muscle fibres in the medial compartment have been depleted of glycogen by tetanic stimulation of the compartmental nerve. Note that there is no overt difference in the segregation of muscle fibres in the two compartments with age. Calibrations: 500 μm (From Balice-Gordon & Thompson, 1988, with permission.)

classified into at least three subtypes (IIA, B; and type IIX – also called type IID by some investigators). These subtypes are either fast or intermediate in their contraction speed; fatiguable or fatigue-resistant; and express either glycolytic, or a mixture of glycolytic and oxidative enzymes. In addition, some muscle fibres (most in the case of the rat fourth deep lumbrical muscle) bind antibodies to both type I and type IIA antibodies and are described as type IIC (Ridge, 1989; see Figure 4). Even these subcategories may represent the tip of an iceberg in muscle fibre phenotyping: combinations of myosin isoform expression generate a large continuum of subtly different muscle fibre subtypes (Pette & Staron, 1993; Pette & Vrbova, 1992).

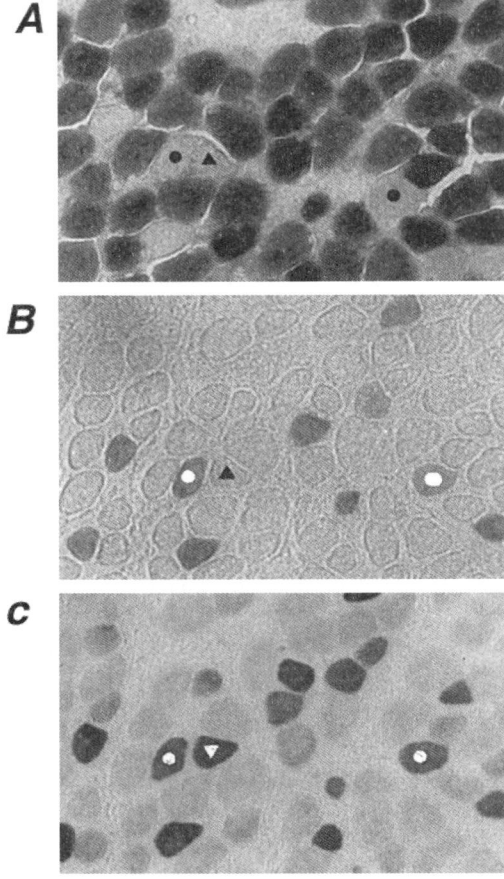

Figure 4 Transverse section of an adult rat lumbrical muscle following glycogen depletion of a single motor unit (**A**) and immunocytochemical staining for slow (type I) and fast (type II) myosin antibodies respectively (**B,C**). The symbols denote three fibres in the same unit (all glycogen-depleted) and motor unit is therefore a mixed IIC/IIA unit because two of the fibres stain positively with slow myosin antibody (**B**) and all three stain positive with fast myosin antibody (**C**). Calibration: 100 μm. (From Gates *et al.*, 1991, with permission.)

2.1.3. Matching of motoneuron and muscle fibre types

Motoneurons appear normally to be matched to specific types of muscle fibre. Thus, although muscles normally contain a mixed population of fibre types, with one or two exceptions (see below) the muscle fibres within a motor unit are homogeneous with respect to type. This is thought to be primarily due to imposition or matching of muscle fibre properties to motoneuron activity. For example, muscle fibre subtype characteristics are transformed by imposing activity via electrical stimulation. Chronic low-frequency stimulation of dissociated adult skeletal muscle fibres in culture up-regulates the expression of β-myosin heavy chain within 6 days (Liu & Schneider, 1998). Low-frequency stimulation of rat EDL muscle *in vivo* for up to 4 months converts almost all type IIB and type IIX fibres they normally contain to type IIA and then to type I (Windisch *et al.*, 1998). Conversely, stimulation of rat soleus muscles with high-frequency phasic patterns converts type I fibres to type IIB and IIX (Hamalainen & Pette, 1996). However, sometimes motor units are of mixed fibre type; for example, motor units containing a mixture of type IIA fibres and type IIC occur in adult rat lumbrical muscles (Gates *et al.*, 1991). This has a bearing on the role of activity in muscle fibre type development and determination (Section 2.3.2 below), and on the role of nerve terminal-muscle fibre recognition in competitive synapse elimination (Section 3.2.2).

Muscle fibres in muscles with characteristically different fibre type compositions and the motoneurons that supply them differ significantly in their membrane electrophysiological properties, including resting membrane potential, input resistance and action potential characteristics (e.g. Bakels & Kernell, 1993; Harris & Luff, 1970; Wood & Slater, 1995). Remarkably little seems to be known about the specificity and regulation of the links between activity, membrane excitability, and intracellular characteristics that distinguish different muscle fibre types.

2.1.4. Summary

The muscle fibres that comprise each motor unit – that is those supplied by a single motoneuron – may be dispersed among many fascicles, but are normally confined to a discrete compartment (when the anatomically defined muscle comprises multiple compartments). Each motor unit normally, though not necessarily, comprises muscle fibres of the same histochemical type. One of the most striking features of the organization of the system of motor units is the apparent matching of a motoneuron's firing properties to the type of motor unit it supplies. This matching appears normally to be appropriate to the postural or phasic function of the motor units concerned. An important goal of developmental neuroscience research into the organization of the motor system is to explain how motoneurons become matched in specific muscle compartments to functionally appropriate muscle fibres in a fascicular mosaic.

2.2. Cellular and molecular biology of the neuromuscular junction

2.2.1. Structure and function of neuromuscular synpases

Most of the fundamental structural and functional characteristics of neuromuscular junctions were discovered and measured by members of Bernard Katz's school over the decades from 1950 to 1980 (see, for example, Aidley, 1998; Martin, 1977). Work during this period, surely a golden era of cellular neuroscience, established that neuromuscular synaptic transmission is a process of coupling membrane depolarization of a nerve terminal to that of the muscle fibre via a chemical intermediary, acetylcholine (ACh); and that this process is "quantized", probably by exocytosis causing instantaneous release of the neurotransmitter molecules constituting each quantum (Figure 5). The random, spontaneous release of transmitter is explosively accelerated when an action potential invades the nerve terminal. Following an influx of Ca^{2+} ions the amount of transmitter released is proportional to the fourth power of the Ca^{2+} ion concentration. In addition to quantized release of transmitter there is a constant trickle of ACh from nerve terminals via membrane transporters. Molecules of ACh subsequently act on receptors concentrated in the motor endplate membrane. Binding of ACh to postsynaptic receptors gates a small inward current due to the transmembrane flux of cations, mainly Na^+ and K^+ also Ca^{2+}. These summate, giving rise to spontaneous miniature endplate current (MEPCs; each peaking about a nanoamp) and nerve-evoked endplate currents (EPCS; several nanoamps). Non-quantal release of ACh adds a steady backround current amounting to about 100 pA. The action of ACh is terminated by its catalytic hydrolysis mediated by acetylcholinesterase, an enzyme also specifically localized to the neuromuscular junction. One of the breakdown products, choline, is transported back into nerve terminals and used as a substrate for resynthesis of ACh (mediated by cholineacetyltransferase) and recycled. The motor endplates of mammalian muscle are also the source of an endogenous DC current, mediated by non-uniform distribution of chloride ion channels (Betz *et al.*, 1980, 1984).

The elegant and mathematically rigorous studies of the biophysics and biochemistry of synaptic transmission at the neuromuscular junctions, that established the above description, were contemporaneously matched by careful ultrastructural studies (Heuser, 1989). The neuromuscular junction was thus revealed to be a close apposition between a muscle fibre and a highly differentiated presynaptic nerve terminal, containing mitochondria and richly endowed with many thousands of 50 nm spherical synaptic vesicles (Figure 5). Synaptic vesicles fuse with active zones in the presynaptic membrane, releasing their transmitter contents by exocytosis. ACh continues to be released via transporter incorporated into vesicular membranes exposed to the synapse, accounting for non-quantal release (Nikolsky *et al.*, 1994). Recycling of vesicular membrane occurs via clathrin-mediated endocytosis, either directly restoring vesicles or indirectly via internal membrane stores. The presynaptic terminal is apposed to an elevated, thickened postsynaptic membrane that is thrown into a labyrinth of folds; a complex referred to as the motor endplate. The presynaptic terminal is capped by a small number of Schwann cells. Interspersed between the Schwann cells, nerve terminals and motor endplate membrane is a conspicuous electron-opaque fuzz: the basal lamina.

A number of ingeneous experiments demonstrated the fusion of vesicles with active zones in the presynaptic membrane; and measurements of the concentration of ACh in single synaptic vesicles. Together with the cloning and three-dimensional reconstruction of the structure of the ACh receptor; the correlation of its ion channel properties with fundamental features of synaptic transmission; and the biochemical characterization of acetylcholinesterase in the basal lamina and cholineacetyltransferase presynaptically; the main framework for a satisfactory and quantitative explanation for the electrophyiological phenomena

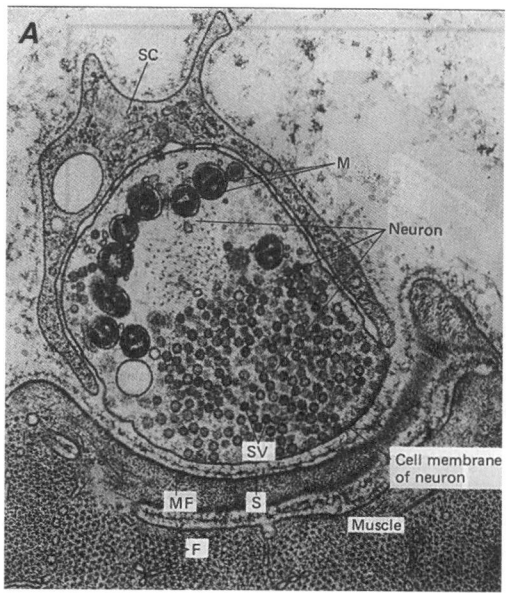

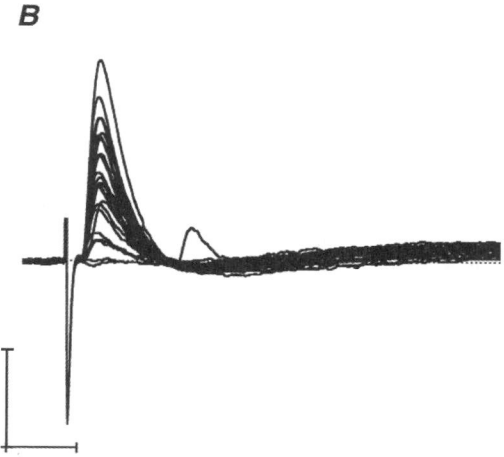

Figure 5 A: Electron micrograph of the neuromuscular junction of a frog muscle, showing the motor nerve terminal containing synaptic vesicles (V) and mitochondria (M); capped by a terminal Schwann cell (SC). The terminal forms a synapse (S) with the muscle fibre (MF) at the motor endplate, which is thrown into junctional folds (F). The basal lamina is visible as an electron-dense fuzz between the terminal and muscle fibre plasma membranes. (From Heuser, 1989, with permission.) **B**: Intracellular recordings from a mouse neuromuscular junction in which transmitter release was partially suppressed. Endplate potentials evoked by nerve stimulation at constant intensity, fluctuate in a "quantized" fashion, by steps equivalent in magnitude to spontaneous MEPPs, one of which has occurred spontaneously, to the right of the evoked responses (R.R. Ribchester & D. Thomson, unpublished recordings) . Calibrations: horizontal : **A**, 200 nm; **B**, 5 ms; vertical: 5 mV.

recorded postsynaptically from the motor endplate was complete by about 1980 (see Aidley, 1998).

A complete review of the structure and function of adult neuromuscular synapses is beyond the scope of this chapter but a few more recent findings, of relevance to studies of development and plasticity, may be highlighted. The process of vesicle exocytosis and recycling at neuromuscular junctions has been visualized with fluorescent stytryl dyes, revealing that recycled vesicles remix randomly with the intraterminal pool (Betz & Bewick, 1992; Cochilla *et al.*, 1999). Synaptic vesicles are recycled at different rates in motor nerve terminals innervating fast- and slow-twitch muscle fibres (Reid *et al.*, 1999). Studies combining membrane capacitance measurements, voltammetry, pharmacology, immunocytochemistry, and molecular biology in other systems have generated the 'SNARE' hypothesis, a molecular mechanism for docking and fusion of a synaptic vesicle with the presynaptic membrane (Angleson & Betz, 1997; Fernandez-Chacon & Sudhof, 1999; Robinson & Martin, 1998). Short-term synaptic plasticity, represented by facilitation, depression, augmentation or post-tetanic potentiation, is associated with persistent changes in intracellular Ca (Tang & Zucker, 1997): these are mediated by mitichondria and a multitude of ion channels and ion exchangers in presynaptic membranes (Rahamimoff *et al.*, 1999); and presynaptic receptors for neurotransmitters and purines (Ribeiro *et al.*, 1996; Tian *et al.*, 1997). Recently, evidence for a pivotal role for Schwann cells in short-term synaptic depression has been demonstrated, employing combined recording from Schwann cells and muscle fibres (Robitaille, 1998).

It may also be of importance that neurotransmitter is released by exocytosis from some postsynaptic membranes (Girod *et al.*, 1995). Following surgical denervation, motor endplates in some muscles also exhibit conspicuous endo-cytosis, in association with a loss of postsynaptic folding (Lawoko & Tagerud, 1995). It is well established that denervated or paralysed muscle fibres re-express a neonatal form of ACh receptors (containing a γ-subunit substituted for the ε-subunit), and that this response is reversed by reinnervation or by imposing muscle activity (Lomo & Westgaard, 1975). Denervation or paralysis also accelerates the turnover, although not the isoform, of junctional ACh receptor expression, due at least in part to phosphorylation of the β-subunits (Meir & Wallace, 1998; Stiles & Salpeter, 1997; Witzemann *et al.*, 1991).

2.2.2. *Molecular constitution of neuromuscular junctions*

The ultrastructure of motor nerve terminals, motor endplates, Schwann cells and the extracellular matrix between them are quite different from elsewhere along the muscle fibre or motor axon. This could either arise because distinctive molecules are expressed at these sites, or by virtue of a special arrangement of molecules that are expressed ubiquitously. The molecular analysis of neuromuscular junctions, and the consequences for neuromuscular development of knocking-out synapse-specific proteins, have recently been reviewed by Sanes *et al.* (1998) and by Sanes & Lichtman (1999; Figure 6).

The best-characterized molecule whose expression is restricted to the neuro-muscular junction is the nicotinic ACh receptor (Unwin, 1998), concentrated at

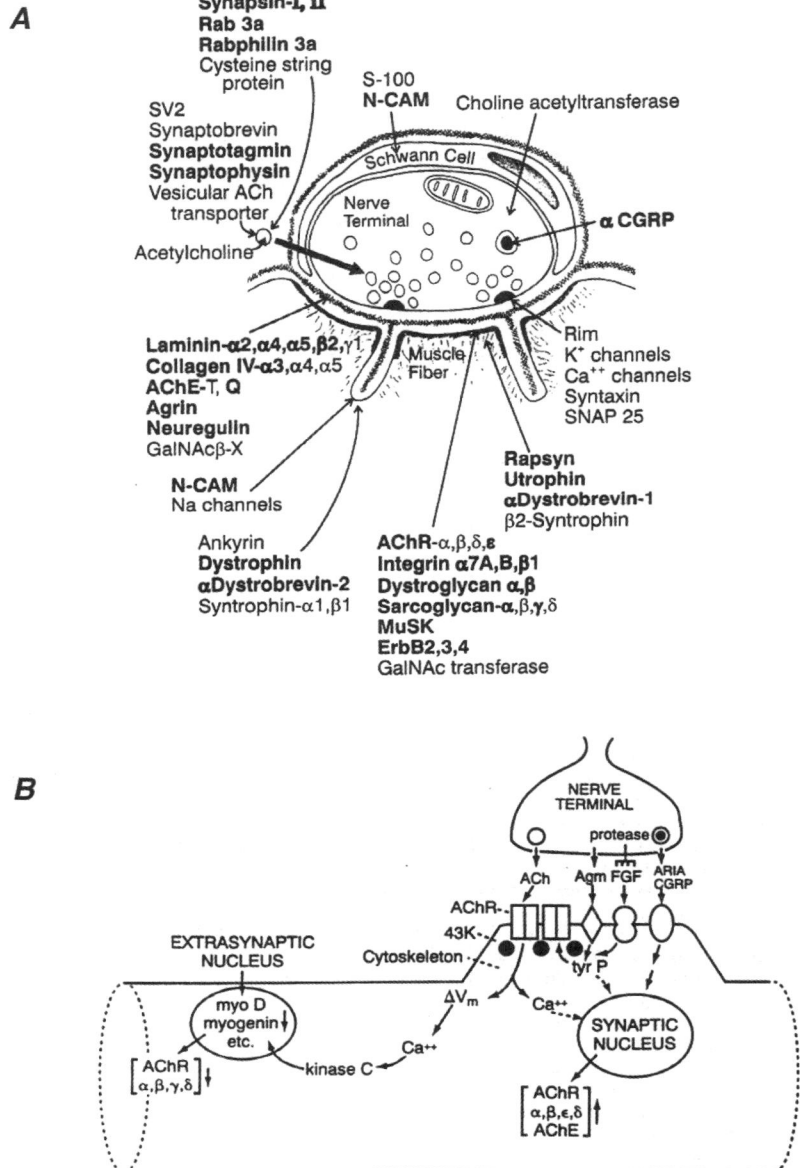

Figure 6 Schematic diagrams illustrating some of the specific protein molecules found concentrated at neuromuscular junctions (**A**); and the pathways from orthograde factors released by motor nerve terminals leading to regulation of muscle fibre gene expression and localization of acetylcholine receptors in synaptic and extrasynaptic membranes. (From: **A**, Sanes *et al.*, 1998, **B**, Edgerton *et al.*, 1996; with permission.)

the crests of the junctional folds in the motor endplate membrane. Voltage-gated ion channels are also concentrated at neuromuscular synapses, including sodium, potassium and calcium channels expressed pre- and postsynaptically (Braga *et al.*, 1992; Katz *et al.*, 1996; Lupa *et al.*, 1993). Somewhat less is understood about the distribution of metabotropic receptors on either side of the synaptic cleft, or ion exchangers; for example, the sodium–calcium exchanger (Mulkey & Zucker, 1992), sodium–hydrogen exchanger, and sodium–potassium pump (Zahler *et al.*, 1996). Specific isoforms of neural cell adhesion molecules, including NCAM itself, and others including cadherin and integrins, are localized to neuromuscular synapses (Chen & Grinnell, 1995). Proteins involved in exocytosis, endocytosis and recycling of synaptic vesicles are receiving increasing attention, though most of the characterization of these is being done in brain rather than peripherally. Molecules localized to extracellular matrix include acetylcholinesterase, synaptic forms of laminin and other distinct β-GalNAc-containing carbohydrates (Martin *et al.*, 1999). Molecules affiliated with the membrane proteins and ion channels that directly mediate neuromuscular function have been identified. These include rapsyn (formerly called 43K protein), spectrin, dystrophin, utrophin, and a muscle-specific protein kinase (MuSK) which acts as a physiological receptor for agrin (Bewick *et al.*, 1996; Wood & Slater, 1998). Neuromuscular junctions are also sites of selective expression of specific serine proteases and protease inhibitors, protein kinases and phosphatases (Akaaboune *et al.*, 1998; Dai & Peng, 1998; Kim *et al.*, 1998; Plomp & Molenaar, 1996). Finally, molecules that direct and control the morphogenesis of the endplate by regulating or localizing structural protein expression have been identified (Figure 6B): these include orthograde, presynaptic neurotrophic factors including agrin, neuregulin (formerly known as ARIA) that probably play a role at all neuromuscular junctions (Loeb *et al.*, 1999); and molecules such as calcitonin gene-related neuropeptide (CGRP) that are probably expressed at only some junctions, or only in response to nerve injury (Sala *et al.*, 1995). Knockout mice that fail to express either CGRP, FGF-5, tenascin C or NCAM show either minor or no abnormality of neuromuscular structure and development (Moscoso *et al.*, 1998). Muscle and Schwann cell-derived factors have been found that may mitigate motoneuron death and synaptic development and plasticity. Some are as yet uncharacterized; others include neurotrophins such as BDNF and NT4; insulin-like growth factors (IGF); glial cell-line derived neurotrophic factor (GDNF); and leucocyte inhibitory factor (LIF). Other factors such as NGF, CNTF and GAP-43 may be expressed and/or secreted in responses to denervation or muscle injury (Caroni, 1997; McAllister *et al.*, 1999).

2.2.3. Summary

The motor unit functions in a unitary fashion, because ea·h adult neuromuscular junction normally transmits action potentials from the presynaptic motor nerve terminal to the muscle fibre with exceptionally high fidelity and within a large safety margin. This specialization is accomplished by the concentration of synapse-specific molecules that take part either directly in synaptic transmission and cell–cell adhesion; or indirectly, where their roles as ancillary proteins include localization of the functional proteins to these sites; or homeostatic maintenance of the local environment of the nerve terminals and motor endplates.

2.3. Muscle development

2.3.1. Formation of muscle

Vertebrate muscles are derived from individual somites that generate two distinct populations of myoblasts: those that give rise to myotomal muscle, forming axial and trunk musculature; and a migratory population that colonizes developing limbs and gives rise to the limb musculature. The latter arises from two muscle masses on either side of the limb, dorsal and ventral, that then partition into different muscle groups (Purves & Lichtman, 1985). Muscle cell differentiation involves activation of specific transcription factors (see Section 2.3.2 below), myoblast cell proliferation, cell-cycle arrest, fusion of myoblasts to form myotubes, and expression of structural muscle proteins (Ludolph & Konieczny, 1995).

Cell–cell adhesion, mediated at least in part by NCAM and N-cadherin, is a precursor to myoblast fusion; but the role of these adhesion molecules in myogenesis is unclear. Transgenic mice overexpressing NCAM show morphological abnormalities in the generation of secondary but not primary myotubes (Fazeli et al., 1996). Polysialation reduces the adhesiveness of NCAM, and this property may underlie myotube separation during secondary myogenesis, as well as defasciculation of axons when they reach their target muscle (Allan & Greer, 1998).

The myotubes that are generated first in mammalian embryos comprise a set called "primary" myotubes (Kelly & Zacks, 1969a; Rubinstein & Kelly, 1981). These multinucleated syncytia form by near-synchronous, longitudinal fusion of tens to thousands of myoblasts. In rats the primary myotubes develop autonomously, between embryonic days 13 and 17 depending on the proximal–distal location of the muscles. These myotubes act as a scaffold around which a secondary population of myotubes are clustered, but myoblasts join with the secondary myotubes at random positions along their length (A.J. Harris et al., 1989a). Secondary myogenesis is complete before birth in some muscles, but continues for about a week after birth in the most distal muscles of the rat hind limb (Betz et al., 1979; Duxson, 1992; Duxson & Sheard, 1995; Ross et al., 1987). Primary and secondary muscle fibres remain electrically coupled to one another via gap junctions until about a week after birth (Fladby, 1987; Schmalbruch, 1982). Shortly after fusion, myotubes express the protease inhibitor nexin I on their surfaces; and the myotubes express nitric oxide synthase. These may play a later role in the stabilization of growth cones after they contact the myotube surface (Chao et al., 1997; Verdiere-Sahuque et al., 1996).

Primary myotubes form at about the same time as motor axons grow into a muscle, but their development does not require innervation, because primary myotubes will develop in the absence of a nerve supply. By birth these aneural muscles have a similar fibre type composition in terms of the myosin isoforms they express as innervated muscles, suggesting that this property is also independent of a nerve supply (Condon et al., 1990b). Secondary myogenesis is also not strictly dependent on a nerve supply as it also is still initiated in aneural limbs (Condon *et al.*, 1990b). Maturation and survival of secondary myotubes are strongly dependent on an intact nerve supply, however, since either pre- or postnatal denervation rapidly leads to profound diminution in the number of secondary muscle fibres (Betz *et al.*, 1980a; Ross *et al.*, 1987). On the other hand, fully differentiated synaptic contacts are not required for this regulation, since muscle fibre numbers develop normally in knockout mice lacking agrin, despite the absence of properly formed neuromuscular junctions in these mice (Gautam *et al.*, 1996).

2.3.2. *Molecular mechanisms of muscle formation*

Distinct signals are necessary for formation of distinct myoblast populations in the vertebrate somite (Christ *et al.*, 1998; Hughes & Salinas, 1999). Several helix–loop–helix muscle-specific transcription factors have been identified; MyoD, myogenin, Myf-5 and MRF4 (also called herculin or MRF-6) being the most studied. MyoD and myf-5 are essential genes that commit somitic cells to myogenesis, because myoblasts and myofibres are entirely absent in double-knockout mouse embryos that lack both Myf-5 and MyoD (Kablar *et al.*, 1999). Mice lacking myogenin are deficient in skeletal muscle, with only a few muscle fibres present at birth. Mice lacking MRF4 are viable and have skeletal muscle; but they up-regulate myogenin expression and this could compensate for the absence of MRF4 (Rawls *et al.*, 1998). Other transcription factors upstream of the Myf-5/ MyoD have been identified. Pax genes 1–7, and other early genes, including sonic-hedgehog, Wnt's and Bmp's have been implicated in initiating myogenesis and controlling the fate of myoblasts. Migrating somitic cells express Pax-3 before other muscle transcription factors and Pax-3 mutations result in disruption of muscle organization (Tremblay *et al.*, 1998). Experiments with retroviruses and myotoxins (the latter in studies of regenerating adult muscle) have suggested a plausible sequence: Wnt and sonic hedgehog signals induce expression of Pax-3 and Pax-7, concomitantly with Myf-5. This then triggers MyoD, followed by myogenin and MRF4 expression (Maroto *et al.*, 1997; Mendler *et al.*, 1998). These genes then induce expression of structural genes for the contractile proteins.

The pattern of myosin isoform expression in both primary and secondary myotubes is strongly influenced by the nerve supply. All primary myotubes initially express two isoforms of myosin: an "embryonic" isoform and a "slow" isoform. Later some primary tubes switch to expression of a "neonatal" myosin isoform. All secondary myotubes initially express neonatal myosin, but some later switch to expression of the slow isoform (Condon *et al.*, 1990a,b). Postnatally – by about 2 weeks after birth – those fibres still expressing neonatal myosin switch to expression of fast isoforms (Dhoot, 1992). Clones of myoblast cells analysed in culture express forms of myosin heavy chain that are type-specific for the muscle from which they are derived, suggesting a form of lineage determination of muscle fibre types. Embryonic and neonatal forms of myosin heavy chain are expressed by distinct lineages of myoblasts (Cho *et al.*, 1993). However, when clones of these different lineages were retrovirus-labelled and injected into different muscles they contributed by differentiation to a variety of muscle fibre types (Hughes & Blau, 1992). The distribution of isoforms, and the switches between them are most likely determined by interactions with the motoneurons that supply these muscle fibres. For example, MyoD levels are higher in fast-twitch (type IIB/IIX) fibres than in slow-twitch fibres, and disruption of MyoD expression shifts the fibre type composition of muscle to slower types (Hughes *et al.*, 1997).

2.3.3. Summary

The profile of muscle fibre types in an adult muscle is ultimately determined by the muscle environment, and the nerve supply probably plays an instructive role in this fibre-type determination. Part of the influence may be over levels of muscle-specific transcription factors. Nonetheless, the fact that some fibre types can develop independently of a nerve supply raises the possibility that these types become selectively innervated by motoneurons of complementary type. The implications of this for the subsequent refinement of the innervation pattern are considered in Section 3.2.2.

2.4. Neuromuscular development

2.4.1. Formation of neuromuscular junctions

The generation of neuromuscular contacts in vertebrates has been very elegantly visualized in zebra fish embryos, exploiting the relative transparency of their skin in combination with time-lapse fluorescence microscopy. Neuromuscular contacts are formed within 24 h of fertilization in these rapidly developing creatures (Liu & Westerfield, 1990). Physiological studies of neurite growth and synapse formation in cultures of embryonic *Xenopus* neurons and myocytes suggest that growth cones release neurotransmitter and myocytes express ACh receptors prior to formation of synapses between them, and that spontaneous and evoked synaptic transmission can be recorded within seconds of mutual contact (Evers *et al.*, 1989). It rapidly accelerates over minutes, and continues to increase substantially throughout pre- and postnatal development, as the motor nerve terminals enlarge. Motor nerve terminals accumulate large numbers of synaptic vesicles, and release of neurotransmitter by exocytosis is further potentiated by the differentiation of specific active zones in the nerve terminal membrane.

Descriptions of neuromuscular synapse formation in mammals have been, of necessity, more indirect. In rodents, neuromuscular connections begin to form at around embryonic day 14 (E14), only 1-week before birth. By comparison, neuromuscular junctions form in human muscle during the 8th–9th week of gestation (Hesselmans *et al.*, 1993); that is, about one-fourth the way through a normal pregnancy. This is about the time most expectant mothers first experience active fetal movements.

Neuromuscular synapse formation has been systematically described in studies of thoracic (intercostal, diaphragm) muscles of embryonic rats. Axon bundles grow into the intercostal muscle region of rat embryos by E13, about two-thirds through gestation in the rat. Axon collaterals are visible less than a day later (E14). Spontaneous or evoked synaptic potentials can be recorded from myotubes at this time. The site where the first successful synaptic contact is made remains considerably labile to other ingrowing motoneuron growth cones, resulting in polyneuronal innervation of neuromuscular junctions. By E17 most fibres are maximally supplied by functional synapses from several motoneurons. These convergent inputs are almost always focused on single motor endplate sites (Dennis *et al.*, 1981). At the level of the electron microscope, rudimentary neuromuscular appositions first appear in intercostal muscles at E16. These immature synapses are already capped by Schwann cells. By E18 the multiple axon terminal sprouts that converge on an endplate are well differentiated and contain numerous synaptic vesicles. Postsynaptic folding and accumulation of junctional myonuclei are apparent by birth (E21; Kelly & Zacks, 1969a,b). Studies of the rat hemidiaphragm muscle have offered a tidy explanation for why endplates appear to be constrained in some adult muscles to a tight band across the muscle surface: the first neuromuscular contacts appear to be formed by motoneuron growth cones randomly on the surface of the immature muscle fibres, which then increase their length by adding sarcomeres to their ends (Bennett & Pettigrew, 1974). A transitional stage in the innervation of some primary and secondary myotubes is that two immature fibres may be transiently innervated by a single synaptic terminal, implying that in some instances secondary myotubes may acquire their initial nerve supply by progressive transfer of synaptic terminals from primary to secondary myotubes (Betz *et al.*, 1979; Duxson *et al.*, 1986).

Motor endplate differentiation has been revealed by fluorescent staining of ACh receptors, and by scanning electron microscopy (Figure 7). At birth, rat endplates have the form of a smooth, cup-like impressions about 20 μm^2 in area, stamped into the muscle fibre surface. The variegated structure characteristic of mammalian motor endplates appears over the course of the following 1–2 weeks (Desaki & Uehara, 1987; Marques *et al.*, 2000; Slater, 1982a). Once established, there is some remodelling over the course of the animal's life in some muscles; others show little change in endplate form over years (Balice-Gordon *et al.*, 1990; Barker & Ip, 1966; Tuffery, 1971; Wigston, 1989). The variegations of the endplate form may be determined at least in part by selective removal of ACh receptors from parts of the initial plaque, preceding (or at least coincident with) elimination of overlying synaptic boutons (Sanes & Lichtman, 1999). Scanning and transmission EM studies show that endplate gutters and ridges form between P5 and P10, and only by P13–P15 are these are infiltrated with a pattern of numerous slit-like secondary folds (Desaki & Uehara, 1987). By adulthood, rat endplates are about 300 μm^2 in area.

Complete denervation leads to atrophy of existing muscle fibres, as in adults. Axotomy- induced degeneration of motor nerve terminals causes endplates and their receptors to expand (Slater, 1982b), and this coincides with redistribution of a number of cytoskeletal proteins, including rapsyn, spectrin, dystrophin, and utrophin (Bewick *et al.*, 1996). Neonatal muscle fibres appear to be transiently refractory to reinnervation (Dennis & Harris, 1980). It has recently been shown that Schwann cells at neonatal endplates rapidly undergo programmed cell death by apoptosis in response to axotomy; but if this is prevented by administering the neuregulin, glial growth factor II, successful reinnervation can occur in neonates (Trachtenberg & Thompson, 1996). One of the functions of synaptic forms of

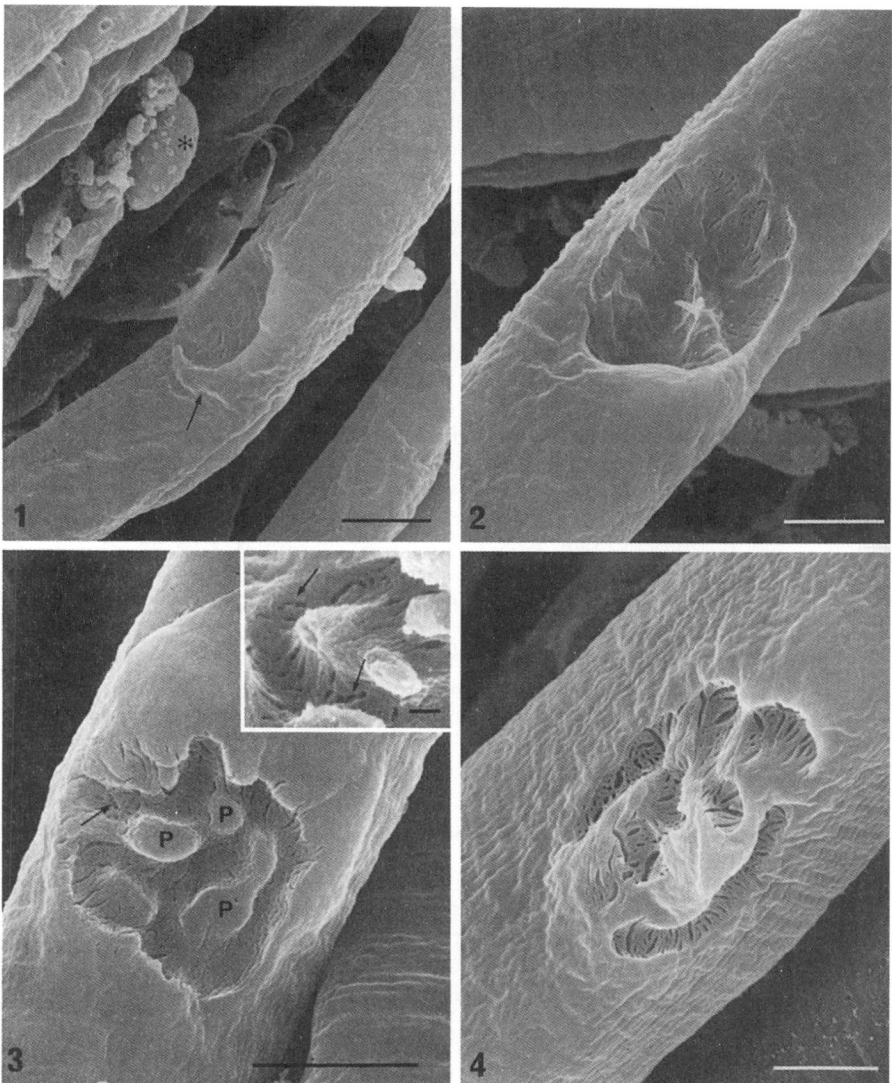

Figure 7 Scanning electron micrographs of motor endplates in neonatal rat muscle fibres at different stages of maturation. **1**: Birth: cup-like depression; **2**: 5 days postnatal: formation of ridges within the endplate trough; **3**: 15 days postnatal: distinct nodal protusions (P) and gutters; **4**: 30 days postnatal: formation of secondary junctional folds. Calibration: 5 μm. (From Desaki & Uehara, 1987, with permission.)

laminin may be to deter engulfment of motor terminals by overlying, perisynaptic Schwann cells (Patton *et al.*, 1998).

2.4.2. Molecular mechanisms of neuromuscular synapse formation

Postsynaptic differentiation

Agrin is responsible for initiating the organization of postsynaptic receptors and the other membrane and extracellular features that characterize the mature synapse. Agrin may also be viewed as an initiator of presynaptic differentiation, although this probably occurs indirectly via retrograde actions of molecules derived from the myotubes or Schwann cells. Agrin is released from growth cones and motor nerve terminals and becomes bound and localized in extracellular matrix. This bound agrin is subsequently sufficient and necessary for initiating aggregation of ACh receptors and other molecules in the muscle. The receptor complex for agrin includes a muscle-specific tyrosine kinase (MuSK). Agrin does not bind directly to this, however; rather, it requires a co-factor that has not clearly been identified, but which may be either α- or β-dystroglycan. Binding of agrin to its receptor complex leads to phosphorylation of MuSK, and also phosphorylation of the β-subunits of the ACh receptors. It also triggers the aggregation of the 43 kDa protein rapsyn, a molecule that is always associated via RATL with clustering of ACh receptors. Clustering of rapsyn is in fact required for ACh receptor aggregation; for synthesis of ACh receptors and other synapse-specific membrane and cytoskeletal proteins (including sodium channels, dystrophin, utrophin, spectrin-like molecules and NCAM); and for secretion of other specific proteins into the synaptic basal lamina (including synaptic isoforms of laminin and acetylcholinesterase). Rapsyn-knockout mice fail to aggregate ACh receptors during *de-novo* synapse formation and the mice therefore die of paralysis (Sanes & Lichtman, 1999).

The synthesis of synapse-specific molecules expressed in motor endplate membranes is regulated by neuregulin, also synthesized by motoneurons and released from motor nerve terminals, and probably by terminal Schwann cells also. Neuregulin acts on erbB3 and erbB4 receptors in the muscle fibre membrane, resulting in subsequent transcriptional regulation of gene expression in the nuclei located nearest the sites of synaptic contact (Fischbach & Rosen, 1997).

Presynaptic differentiation

Presynaptic differentiation may be a response to factors secreted into the basal lamina, or to diffusible neurotrophic molecules released by the myotubes at the sites of contact. Small molecules such as nitric oxide, and derivatives of arachidonic acid, notably 5HPETE, may also be important (Fitzsimonds & Poo, 1998; Harish & Poo, 1992). The identity of the agents reponsible for presynaptic differentiation is not yet established, but it is mimicked and neurotransmitter release is enhanced by neurotrophins (such as BDNF, NT3 and NT4) and neurotrophic cytokines (such as CNTF, GDNF, LIF and IGF). Some of these molecules have local effects on synaptic function; others require signalling via the motoneuron cell body. Membrane depolarization, elicited by activity, rapidly and specifically increased the levels of NT-3 mRNA in developing *Xenopus* muscle cells in culture (Xie *et al.*, 1997). A recent study suggests that the inductive effects of neurotrophic factor on transmitter release are potentiated by presynaptic depolarization (Boulanger & Poo, 1999).

2.4.3. Summary

Molecular and cellular differentiation of postsynaptic and presynaptic specializations of the neuromuscular junction occur more or less synchronously, beginning within seconds or minutes of the first contacts by a motoneuron growth cone at a random site on a myotube surface. The axon stops elongating, and the growth cone develops the presynaptic specializations that facilitate synaptic transmission. At the same time postsynaptic differentiation begins, probably in response to signalling by extracellular agrin and neuregulins released from the developing motor nerve terminals into the extracellular matrix. Receptors and sodium channels accumulate in the nascent motor endplate. As myotubes begin to expand in circumference the motor nerve terminal grows in parity, increasing its synaptic strength (see Section 3.1). Motor endplates initially acquire additional inputs from several motoneurons; this is a transient feature of neonatal muscle innervation and synapse elimination subsequently establishes the mononeuronal innervation pattern to each muscle fibre (see Section 3.2). Polyneuronal innervation and synapse elimination are restored in adult muscle after nerve injury and regeneration (see Section 3.3).

3. SPECIAL TOPICS: ROLE OF USE AND DISUSE IN THE DEVELOPMENT AND PLASTICITY OF CONNECTIONS

This section addresses the issues of necessity and sufficiency of muscle activity in the formation, maintenance and repair of neuromuscular structure and function. Plasticity of size–strength relationships at neuromuscular junctions will be considered first. Motor nerve terminal and endplate size vary in proportion to muscle fibre size, so it is interesting to consider the physiological consequences of this size relationship for the efficacy (strength) of neuromuscular junctions, and how it might be regulated. Next, the role of use and disuse in the elimination of polyneuronal innervation in neonatal muscles will be discussed. Synapse elimination brings about a profound change in the pattern of muscle innervation, with obvious functional consequences for the contribution and relative strengths of motor units to muscle use. Finally, similar transformations in innervation pattern that occur in reinnervated adult muscles will be considered. Motor axon sprouting and regeneration are the principal mechanisms responsible for compensation and restoration of motor unit function (respectively) after nerve injury, and it is worthwhile to compare and contrast the role of activity in these processes, with developmental synapse formation and elimination.

3.1. Size–strength plasticity at the adult neuromuscular junction

The average amounts of evoked neurotransmitter release increase as muscle grow with postnatal age (Kelly & Robbins, 1983). Repeated observations of the same motor terminals at different ages in re-anaesthetized mice reveal that the general form of an endplate does not change radically as its muscle fibre grows throughout most of the animal's life, but the overall size does (Balice-Gordon *et al.*, 1990; Lichtman *et al.*, 1987). In mature muscles motor nerve terminal size is strongly correlated with muscle fibre size (Figure 8). Taking these observations together, it comes as no surprise that transmitter release varies in proportion to the size of the nerve terminal (Betz *et al.*, 1993; Harris & Ribchester, 1979; Kuno *et al.*, 1971; Tsujimoto *et al.*, 1990; see Figure 8C). The principal basis of this appears to be that synaptic vesicle densities and active zones – the specific sites of transmitter exocytosis – are distributed periodically in motor nerve terminal membranes (Betz *et al.*, 1992; Pawson *et al.*, 1998a), and the bigger the terminal,

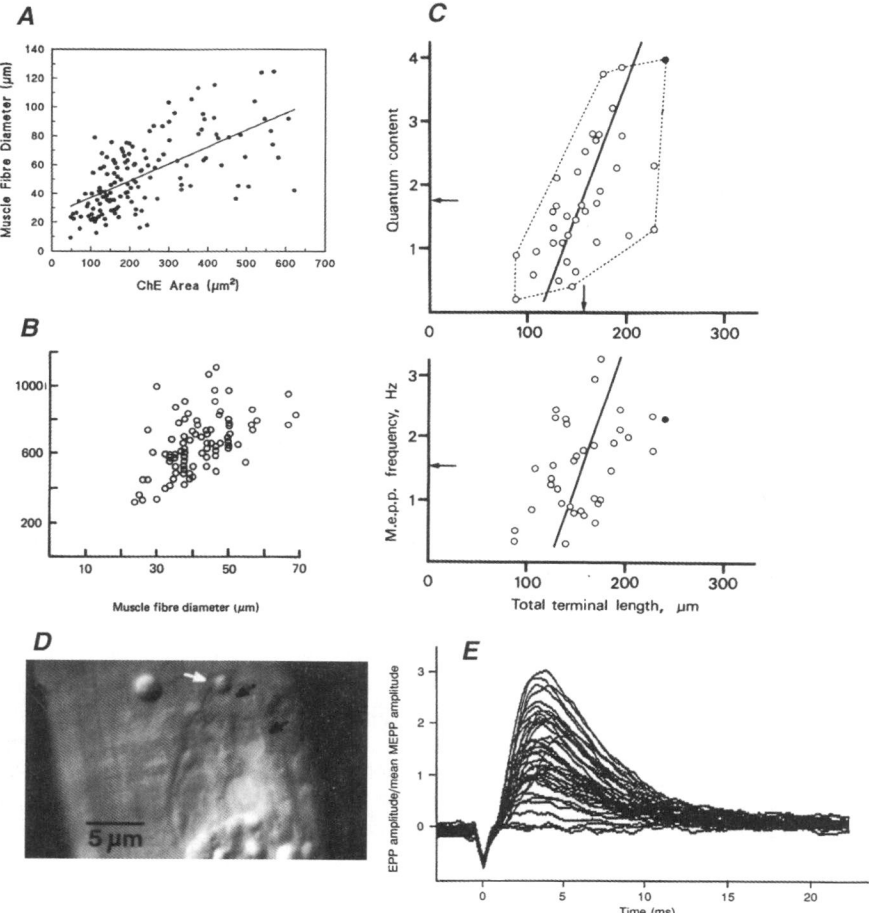

Figure 8 Relationships between size and strength of neuromuscular junctions. **A,B**: Fibre diameter/endplate area correlation in human (**A**) and mouse (**B**). C: transmitter release/nerve terminal length correlations for evoked (upper) and spontaneous (lower) transmitter release. **D**: Isolation and relocation of single synaptic boutons at neuromuscular junctions in the snake, with **E**: synaptic responses from a single relocated synaptic bouton. (From: **A**, Slater *et al.*, 1992; **B**, Harris & Ribchester, 1979; **C**, Tsujimoto *et al.*, 1990; **D,E**, Wilkinson *et al.*, 1996, with permission.)

the more vesicles and active zones are present . Thus, according to the simple binomial model of neuromuscular transmission ($m = n \cdot p$; Katz, 1969) mean quantal content (m) increases in proportion to the increased number (n) of available quanta or release sites, assuming no change in the unit probability of release (p); hence large terminals tend to release more transmitter than small terminals, other factors being equal.

Measurements of the amount of neurotransmitter release per unit area have been elegantly obtained by Wilkinson and his colleagues in studies of synaptic transmission in isolated preparations of garter snake muscle (Wilkinson *et al.*, 1996; Figure 8D,E). They enzymically detached individual synaptic boutons from endplates and micromanipulated them back onto the same or adjacent denervated endplates. Estimates of quantal content per unit area varied between about 0.1 and 2.0 quanta/μm^2 using this method, consistent with estimates based on overall nerve terminal size and quantal content. At frog neuromuscular junctions, quantal contents of 200 (vesicles) are typical, and at mammalian neuromuscular junctions quantal contents of around 70–100 are the norm. The critical quantal content required to depolarize muscle fibres from a resting potential of –75 mV to the firing threshold of –63 mV is about 13–20 quanta (Wood & Slater, 1995; 1997). Thus, taking these estimates of the amount of transmitter required to depolarize motor endplate membranes to threshold together with morphological data from different species including humans (Martin, 1994; Slater *et al.*, 1992), it would appear that most motor nerve terminals are between two and 10 times the size they are required to be, in order to be effective in triggering muscle fibre action potentials.

3.1.1. Function of size–strength plasticity

One possible function of a size-strength relationship is to provide compensation for the smaller synaptic potentials that accrue in larger-diameter muscle fibres; or conversely, the reduced requirement for large amounts of transmitter to be released from a terminal innervating a small-diameter muscle fibre. Muscle fibres of differing diameter show inverse variation with an electrophysiological measure: "input resistance". This is defined empirically as the ratio of steady-state membrane depolarization to current injected into a muscle fibre. Large-diameter fibres have a higher axial conductance per unit length and so current escapes more readily from the locus of current injection. Also, current leaks more readily across the surface membrane of a large-diameter fibre, because there is a greater surface area per unit length over which that current may also escape. For muscle fibres with a constant specific membrane and specific axial resistance, it is thus expected that, the larger the muscle fibre, the lower the input resistance of the fibre (Katz & Thesleff, 1957), and the smaller the depolarization of the endplate to each unit of depolarizing current provided by each exocytotic release of neurotransmitter from the nerve terminal. It follows from this that the critical quantal content required to reach action potential threshold in a muscle fibre will be greater for large-diameter than for small-diameter muscle fibres. A plausible hypothesis is that the size–strength relationship is maintained by a muscle-activity-dependent, negative feedback loop. Thus, if a synaptic response dips below threshold – for instance with a decline in input resistance as a muscle fibre grows – then this could stimulate production of growth-promoting or synaptic strength-promoting molecules that act on nerve terminals to restore functional synaptic transmission to an adequate level.

However, this hypothesis cannot be an entirely satisfactory one, on the basis of several reasons and experimental observations. These would tend to suggest that muscle activity is not necessary for inducing or maintaining the muscle fibre/motor nerve terminal size–transmitter release relationship.

First, as indicated above, the safety factor for transmission is normally much greater than unity. Thus we can expect that muscle fibre action potentials are nearly always produced, even if mean quantal content were to be reduced by a factor of two or more. Even tetanic stimulation fails to reduce the quantal content of EPPs below about 50% of its initial value (Elmqvist & Quastel, 1965). Secondly, there are non-linearities in the relationship between nerve terminal size and transmitter release, at least at frog neuromuscular junctions, where transmitter release per unit length declines with distance from the nerve entry point. This feature could favour the hypothesis above, because it is likely to impair synaptic transmission most in large muscle fibres with large terminals (Grinnell, 1995; Nudell & Grinnell, 1982). However, the amount of depolarization produced by single quanta as a function of muscle fibre diameter is also non-linear: MEPPs and endplate currents measured in large fibres are substantially larger than predicted by Ohm's Law (Wilkinson et al., 1992), suggesting that single quantal efficacy is probably regulated postsynaptically. Finally, the safety margins for synaptic transmission in diverse species are roughly the same, in spite of differences in their geometry.

Motor nerve terminals in human skeletal muscle are relatively small, and have low quantal content as a consequence, in relation to muscle fibre size; whereas the ratio of rodent, frog or chicken motor terminal size to muscle fibre diameter is relatively large (Figure 9). The ratio of endplate area:muscle fibre diameter in humans is about $200:50 \ \mu m$ (4:1), whereas in mice it is about $600:30 \ \mu m$ (20:1): about a five-fold difference (Harris & Ribchester, 1979; Slater et al., 1992) and yet the safety factor for transmission is high in both species. The principal explanation may lie in the degree of junctional folding: in humans the primary and secondary folds are extensive; in chickens they are virtually non-existent. Rodents have junctional folds that are intermediate in their density and complexity. Junctional folds contain high densities of sodium channels, and theoretical calculations suggest that the high densities of these voltage-gated channels coupled with the fold geometry substantially enhance the current density and lower the excitation threshold of the endplate (Martin, 1994).

Although muscle action potentials may not play a necessary role in the neuromuscular size–strength relationship, activity could still be sufficient to regulate it. Changes in muscle fibre size stimulated by training bring about increases in nerve terminal size and neurotransmitter release in some muscles but not others (Deschenes et al., 1993; Waerhaug et al., 1992; Figure 10). Paradoxically, however, disuse of terminals also brings about increases in transmitter release, and growth of motor nerve terminals by reactive sprouting. Initial investigations suggested that increases in transmitter release were a passive consequence of increases in nerve terminal size (Snider & Harris, 1979), but it now appears that transmitter release is regulated independently from nerve terminal size. For example,

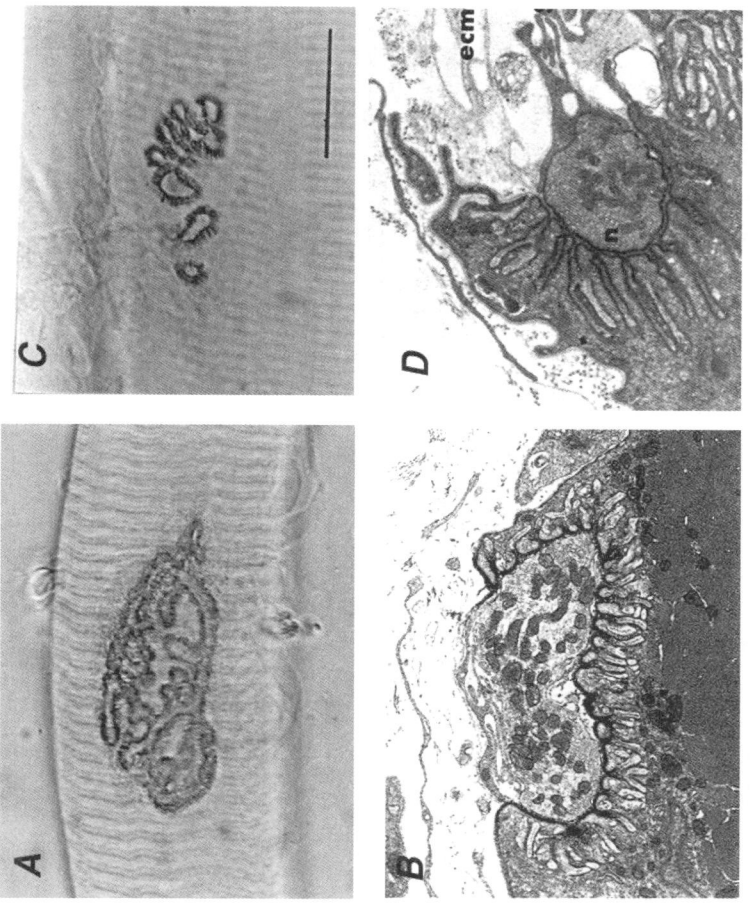

Figure 9 Differences in relative size of motor endplates and extent of junctional folds in rat (**A,B**) and human (**C,D**) neuromuscular junctions. Rat junctions, shown in **A,C** with cholinesterase staining, are much larger in proportion to muscle fibre diameter than human junctions, but the length and extent of junctional folds is much greater in humans. Calibration: **A,C**, 20 μm; **B,D**, 2 μm. (From: **A,B**, Wood & Slater, 1997; **C,D**, Slater *et al.*, 1992; with permission.)

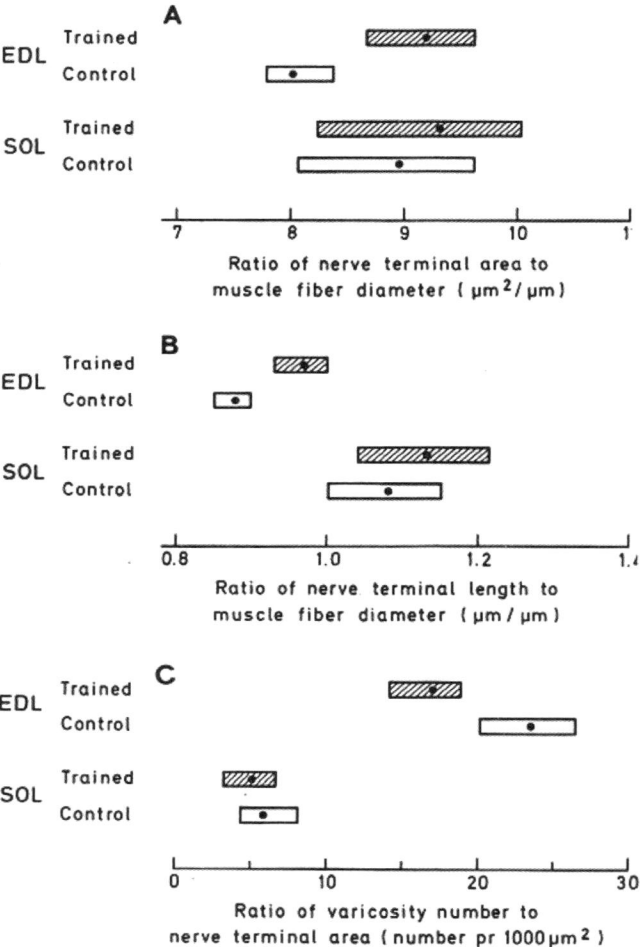

Figure 10 Effect of training on the growth of rat EDL and soleus motor nerve terminals. Enforced wheel running for 6 weeks caused a significant increase in the size, but not complexity, of terminals in EDL muscles, but not soleus. (From Waerhaug *et al.*, 1992, with permission.)

treadmill training of mice leads to a significant increase in evoked quantal content at neuromuscular junctions without significant changes in muscle fibre diameter (Dorlochter *et al.*, 1991). Conversely, Tsujimoto *et al.* (1990) showed that muscle disuse, brought on by tetrodotoxin block of nerve conduction, produced increases in transmitter release within 24–48 h; about a day before any increase in motor nerve terminal size by sprouting is observed. Inactivity also rapidly induces increased turnover of ACh receptors (Akaaboune *et al.*, 1999). Recent genetic manipulations in *Drosophila* suggest that plastic changes in quantal size and quantal content in relation to muscle fibre size and innervation are regulated by independent molecular mechanisms (Davis & Goodman, 1998; Petersen et al., 1997; Sigrist et al., 2000).

3.1.2. Mechanism of size–strength plasticity

Muscle action potentials may not comprise necessary stimuli for nerve terminal size–strength homeostasis, but the sufficiency of activity does not discount the possibility that terminal size and strength are regulated by a negative-feedback, trophic relationship between the terminal and its motor endplate as the muscle fibre grows. For example, Balice-Gordon et al. (1990), using a technique of repeated visualization of neuromuscular junctions over months, demonstrated that correlation between terminal size and muscle fibre size was maintained in mice treated with testosterone to increase muscle fibre diameter in muscles with receptors for this hormone. Evidence has been obtained that adhesion molecules or neurotrophic factors may regulate nerve sprouting in response to disuse (Booth et al., 1990; Gurney et al., 1992). Neurotrophins, cytokines and arachidonic acid derivatives also enhance transmitter release in both cell cultured and intact muscle preparations (Harish & Poo, 1992; Lohof et al., 1993; Ribchester et al., 1998; Stoop & Poo, 1995). Transmitter release is up-regulated in rat muscle during chronic, partial neuromuscular block induced by administration of α-bungarotoxin. This up-regulation is blocked by protein kinase inhibitors, particularly those which affect protein kinase C (Plomp et al., 1994; Plomp & Molenaar, 1996). Endurance training in rats leads to increases in the level of CGRP in some motoneurons (Gharakhanlou et al., 1999).

If muscle activity directly influences the production of molecules that regulate nerve growth then this would imply that the role of activity is instructive. However, the effects of activity could be merely permissive, an indirect consequence of its general effects on muscle growth. For example, growth-induced swelling of muscle fibres might cause mechanical deformation of terminals, that could then stimulate the physiological mechanisms that regulate nerve terminal form and function. Mechanically distorting muscle fibres stretches endplates and motor nerve terminals in much the same way as a logo on balloon as it is blown up. Increases in spontaneous transmitter release induced by muscle stretch are blocked by administration of RGD peptides, suggesting that this response is mediated by integrins (Chen & Grinnell, 1995, 1997). It is conceivable that deformation could also signal, in similar fashion, the intercalation of membrane components that would render such deformations permanent.

3.1.3. Summary

The functional significance of the size–strength relationships between motor terminals and muscle fibres remains obscure. Though some insights into the physiological regulation of transmitter release have been obtained, we still do not know how these are related to regulation of nerve terminal size and form. The idea that negative feedback based on the critical firing threshold of muscle fibres might provide a necessary stimulus to nerve terminal growth is not plausible, although the role of activity may be permissive in this regard. Physiological changes in transmitter release are evidently regulated independently from overt morphological growth of terminals, which may expand either in response to use or disuse (though perhaps via different mechanisms). Hormones and neuro-trophic factors may act, via membrane receptors, on specific protein kinases and phosphatases to regulate transmitter release; but how these processes are trans-lated into regulated growth of terminals is still not known. It would be interesting to establish, for example, whether induction of nerve terminal growth by training is also inhibited by blocking presynaptic integrins; or by blocking specific protein kinases or phosphatases.

3.2. Elimination of polyneuronal innervation during development

Polyneuronal innervation was first observed in neonatal muscles using silver staining methods by Tello (1917) and confirmed by Boeke (1921). Interest in this phenomenon was revived with its demonstration using physiological techniques, beginning with the study by Redfern (1970), who based his measurements on intracellular recording from motor endplates in the rat diaphragm. Graded stimulation of motor axons produced stepwise increments in the size of the motor endplate potential, whereas evoked synaptic responses in adult muscle fibres show no stepwise increments. The simplest explanation is that convergent axons with distinct electrical thresholds convergently innervate most muscle fibres in neonates (Figure 11A). This pattern of innervation is lost over the first 2–3 postnatal weeks by a process now normally referred to as "synapse elimi-nation". Tetanic muscle tension and/or histological measurements later revealed polyneuronal innervation in kitten, neonatal mouse and neonatal rabbit muscle (Bagust *et al.*, 1973; Brown *et al.*, 1982a,b; Gordon & Van Essen, 1983). Polyneuronal innervation in human muscles has been demonstrated histologically in fetuses (Figure 11B) and most synapse elimination occurs prenatally, by about the 25th week of gestation (Figure 12); but there appears to be some uncertainty about how much synapse elimination occurs postnatally in humans (Gramsbergen *et al.*, 1997; Hesselmans *et al.*, 1993).

In a study that represents the watershed in this field, Brown *et al.* (1976) demonstrated that the elimination of polyneuronal innervation cannot be adequately accounted for by loss of entire motor units through cell death, because the number of functional motor units is similar in neonates and in adults. Rather, most motoneurons in neonatal muscles innervate more muscle fibres than they do in adults. The adult pattern (mononeuronal innervation) of muscle fibres is achieved through a process of withdrawal of many, and in some cases most, of the collateral branches each motoneuron supplies to individual muscle fibres. Brown and colleagues also carried out partial denervation experiments that identi-fied a role for competitive and non-competitive components to synapse elimination (see below); and cross-reinnervation experiments showing that distance between synaptic sites is a mitigating feature of synapse elimination. Thus, the process of neonatal synapse elimination defines the final motor unit size distribution in the normal adult muscle. Since mononeuronal innervation of muscle fibres optimizes the control of muscle force by combinatorial activation of motor units, synapse elimination may be viewed as a process of fundamental importance in the development and maturation of function in the neuromuscular

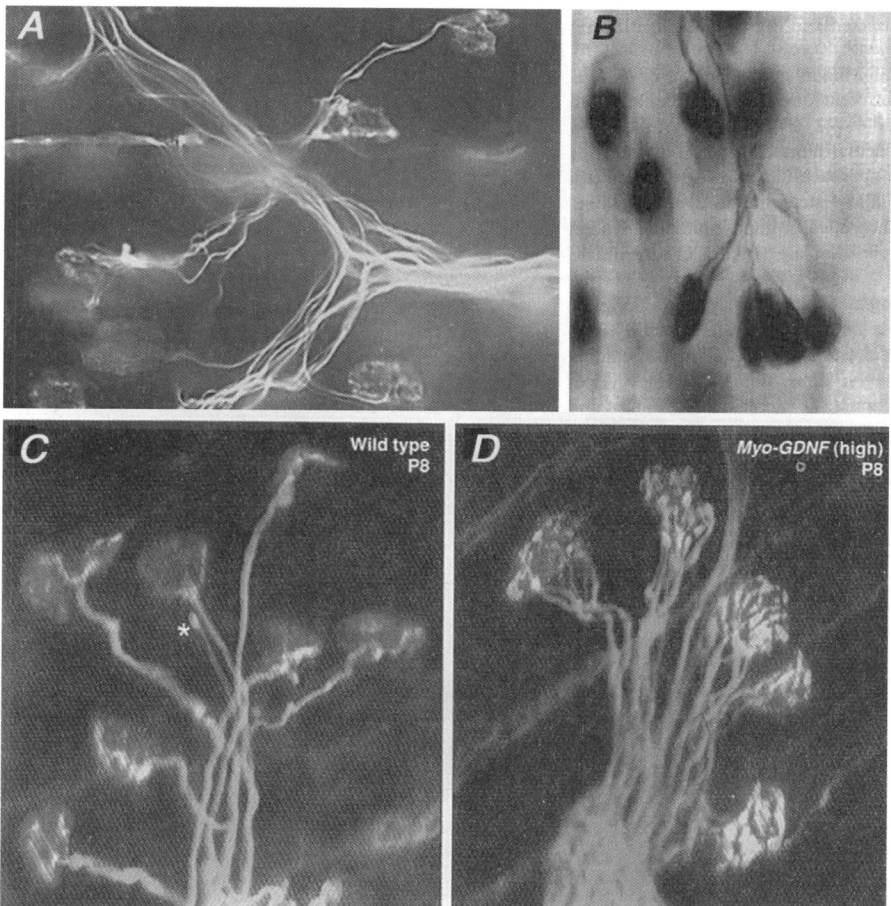

Figure 11 Examples of polyneuronal and mononeuronal innervation in fetal and neonatal muscle. **A:** Neonatal rat muscle, 6 days postnatal, axons and synaptic terminals stained with FITC-conjugated antibodies against neurofilament and SV2, and receptors stained with TRITC-bungarotoxin. Most of the junctions are polyneuronally innervated by two or more axons (R.R. Ribchester and R. Panteri, unpublished). **B:** Combined silver/cholinesterase stain of a polyinnervated junction in human psoas muscle from a fetus at 28 weeks gestation. (From Gramsbergen *et al.*, 1997, with permission.) **C:** monoinnervated junctions stained immunocytochemically in a wild-type mouse muscle, littermate control to **D**, a transgenic mouse overexpressing GDNF under the control of the myogenin promoter. Note the extensive convergent innervation of endplates in the mutant. (From Nguyen *et al.*, 1998, with permission.) Calibration: **A**, ca. 3 μm; **B**, ca. 5 μm; **C,D**, 10 μm.

system. The interactions that result in mononeuronal innervation take place within the confines of the motor endplate.

The possibility that significant numbers of adult skeletal muscle fibres sustain polyneuronal innervation was intially suggested by Hunt & Kuffler (1954), but

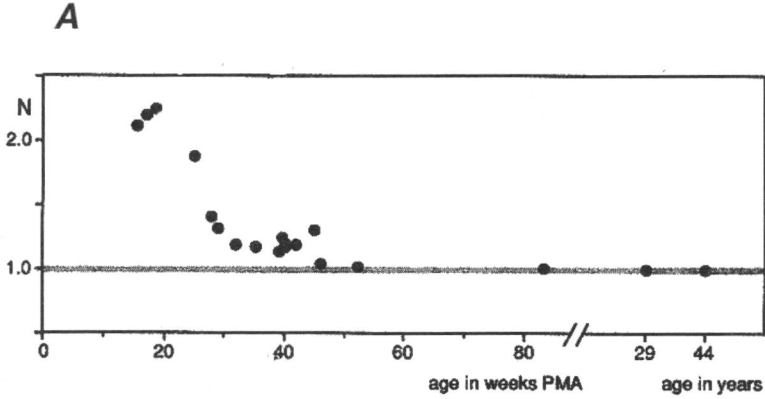

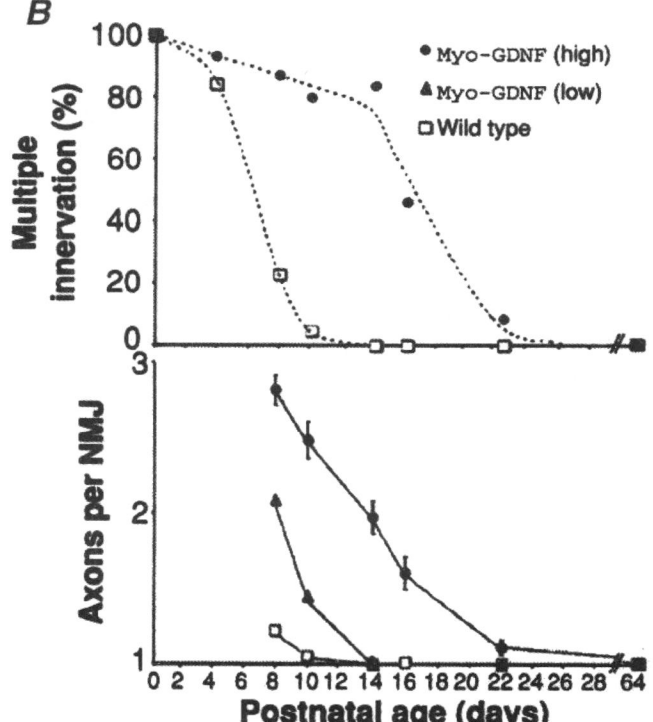

Figure 12 Time course of elimination of polyneuronal innervation in **A**, human; **B**, mouse muscle; **B**, shows data from wild-type and transgenic myo-GDNF mice overexpressing the trophic factor to different degrees. Elimination is largely complete in human neonates by birth, but occurs more rapidly in the postnatal period in rodents. (From **A**, Gramsbergen *et al.*, 1997; **B**, Nguyen *et al.*, 1998; with permission.)

this was discounted by Brown & Matthews (1962), who carefully illustrated the pitfalls associated with reliance on isometric twitch tension recordings (see Section 3.4). The debate lingers, to some extent. Taxt (1983b) reported small amounts of persistent polyneuronal innervation in adult rat lumbrical muscles, an infrequent phenomenon also noted by Barry & Ribchester (1995). However, Chamberlain & Ridge (1989) failed to find any consistent evidence for it in their studies of the same muscle. Soha et al. (1987) attributed apparent instances of polyinnervation in adult muscles to an artefact of intracellular recording techniques, particularly when recording from deep muscle fibres. This can lead to pickup of synaptic potentials generated at endplates on nearby fibres. On the other hand, anatomical studies have provided unequivocal evidence of persistent polyneuronal innervation in about 14% of frog twitch muscles; about 30% of muscle fibres in tensor fasciae latae in adult mouse muscles following neonatal paralysis with botulinum toxin; and a similar percentage of fibres in rat levator ani muscles following treatment with androgenic steroids (Brown et al., 1982b; Herrera, 1984; Lubischer et al., 1992).

The principal structural characteristics of polyneuronally innervated neuromuscular junctions and their elimination from neonatal mammalian muscles are as follows (Figure 13):

1. The converging terminals synapse within the same, circumscribed motor endplate region (Bixby, 1981; Brown et al., 1976; O'Brien et al., 1978);
2. The synaptic boutons of the converging axons are initially intermingled, but subsequently undergo some segregation. Eventually all the boutons belonging to all but one of the axons are withdrawn, and the axon collateral – tipped by a "retraction bulb" – is rapidly sequestered by the parent axon (Balice-Gordon et al., 1993; Gan & Lichtman, 1998; Riley, 1977, 1981).
3. ACh receptors are initially distributed uniformly in motor endplates supplied by the multiple axons, reflecting a simple plaque-like structure to the motor endplates. Subsequently, the form of postjunctional apparatus adopts a variegated pattern at the surface of the muscle fibre, conferring a unique shape to each neuromuscular junction (Desaki & Uehara, 1987; Marques et al., 2000; Slater, 1982a,b).
4. The formation of the variegated adult structure is accompanied by the redistribution and/or removal of ACh receptors and other extracellular and intracellular proteins including cytoskeletal proteins, membrane-bound cell adhesion molecules, and molecules in the extracellular matrix (Sanes & Lichtman, 1999).
5. The reorganization of postsynaptic structure coincides with, and sometimes evidently precedes, the withdrawal of overlying synaptic boutons (Balice-Gordon & Lichtman, 1993, 1994; Rich & Lichtman, 1989a).
6. Both polyneuronally innervated and mononeuronally innervated neuromuscular junctions are normally capped by between one and six perijunctional (terminal) Schwann cells. These Schwann cell numbers are regulated by their nerve supply; and dispersal of Schwann cells defocuses the innervation of muscle fibres (Love & Thompson, 1998; Trachtenberg & Thompson, 1997).
7. Transmitter release per unit area and postsynaptic sensitivity are initially uniform at sites of synaptic contact. As synapse removal and rearrangement

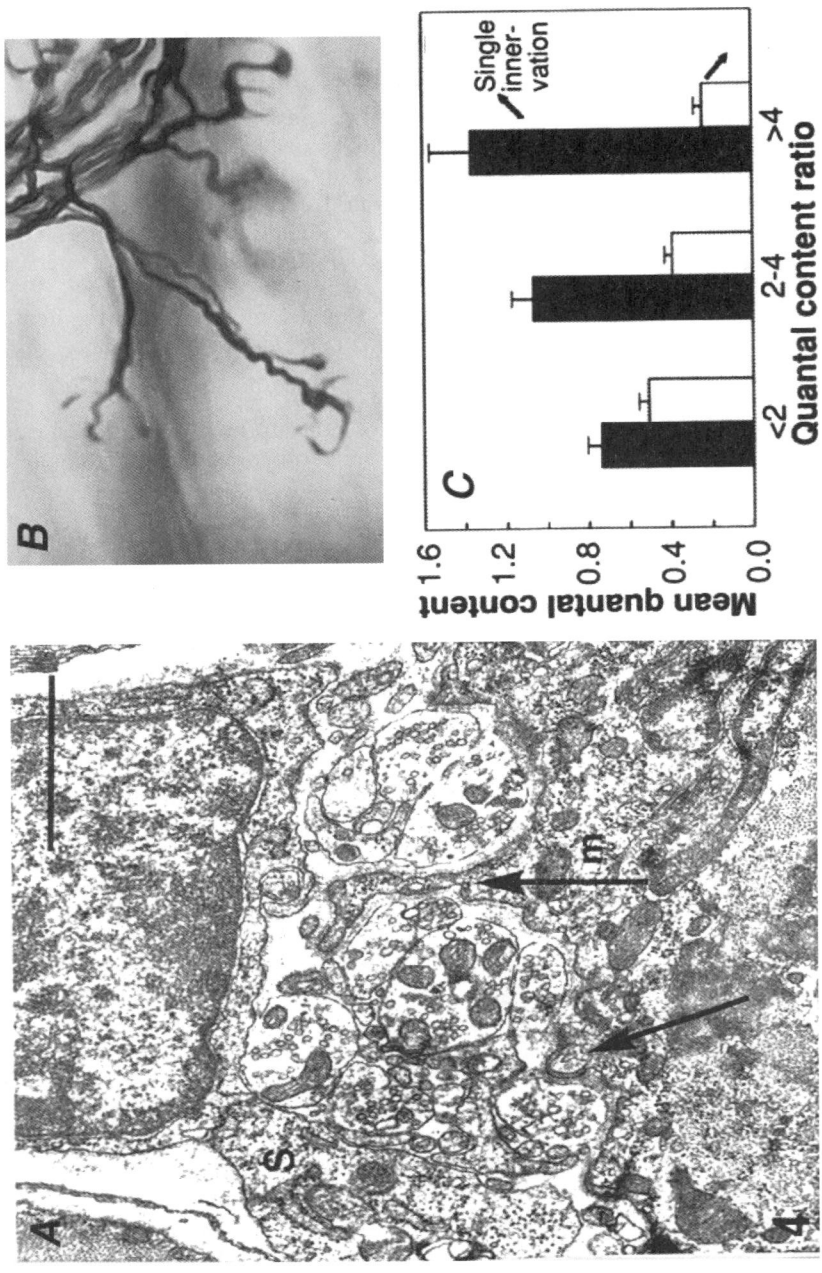

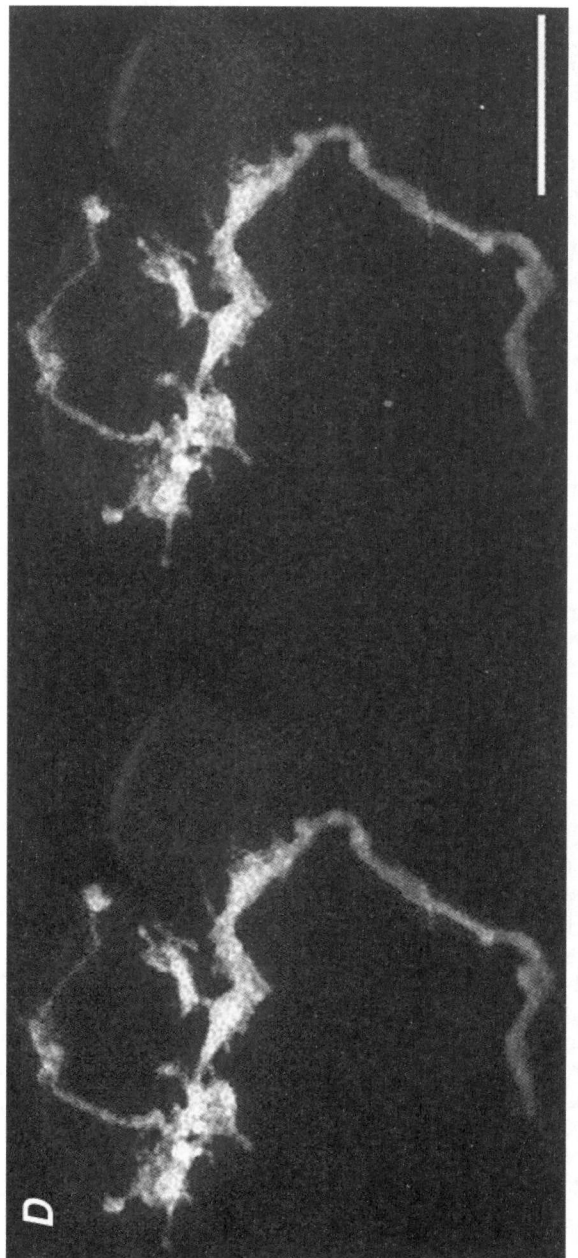

Figure 13 Structure and function of polyneuronally innervated neuromuscular junctions in rodents. **A**: Electron micrograph showing several closely apposed synaptic boutons at a single neuromuscular junction. Some of the boutons are stacked upon one another. **B**: Silver stain of "retraction bulbs" (arrows). **C**: Quantal contents of synaptic potentials evoked by converging boutons diverge during synapse elimination. **D**: Stereo pair fluorescence micrograph. One of at least two converging axon terminals has been labelled with carbocyanine dye. A small retraction connected by fine axonal filament is detaching from the endplate. Calibrations: **A**, 1 μm; **B**, 10 μm; **D**, 10 μm. (From: **A**, Korneliussen & Jansen, 1976; **B**, R.R. Ribchester, unpublished; **C**, Colman et al., 1997; **D**, Gan & Lichtman, 1999, with permission.)

continue, synaptic efficacies of the converging inputs diverge (Colman *et al.*, 1997; Kopp *et al.*, 2000).

8. The surviving motor nerve terminal continues to expand and increase its levels of transmitter release in proportion, matching the circumferential dimensions of the muscle fibre it supplies (see Section 3.1). The basic form of most endplates and their terminals may not change significantly over the remaining course of the animal's life, however (Balice-Gordon *et al.*, 1990; Harris & Ribchester, 1979; Lichtman *et al.*, 1987).

9. In normal animals, synapse elimination is coincident with increased postnatal use of muscles. Experimental alteration of neuromuscular activity influences the rate, and – at least partly – the outcome of the competition between converging synapses on the endplate (Callaway *et al.*, 1987; O'Brien *et al.*, 1978; Ridge & Betz, 1984; Thompson, 1983a).

10. In addition to the activity-dependent, competitive mechanisms of synapse elimination, there is evidence for activity-independent competitive and non-competitive withdrawal of mammalian neuromuscular synapses (Betz *et al.*, 1980a; Fladby & Jansen, 1987; Gates & Ridge, 1992).

Parallels have been drawn between the transformation of the polyneuronal to the mononeuronal state and the transformation of patterns of innervation in the autonomic and central nervous system, including the cerebral cortex. The most studied paradigms, other than the neuromuscular junction, are the elimination of thalamocortical connections in the visual system and the elimination of multiple climbing fibre inputs to cerebellar Purkinje cells (Jansen & Fladby, 1990; Lohof *et al.*, 1996). All show activity-dependence, although some recent studies suggest that, at least in the visual system, visual experience and/or activity may not be as decisive in patterning retinothalamocortical connections as presumed hitherto (Crowley & Katz, 1999).

3.2.1. Competition based on activity

Activity is not required for the formation or differentiation of neuromuscular junctions at sites of growth-cone:muscle fibre apposition, since they will do so in culture or in paralysed embryos (Cohen, 1972; Ding *et al.*, 1983; Giacobini *et al.*, 1973; Liu & Westerfield, 1990; Verhage *et al.*, 2000; Westerfield *et al.*, 1990) or after regenerating into paralysed adult muscles (Ribchester, 1988a; Taxt, 1983b). However, since synapse elimination results from a removal of inputs at a time when muscles are becoming active, it is natural to enquire whether activity plays a decisive role in this remodelling process. Two fundamental questions have been addressed. First, does the overall level of activity affect the rate of synapse elimination? Second, do differences in the activity of converging inputs affect the outcome of synaptic competition?

Answers to the first question have been quite clear-cut. In a nutshell, blocking activity profoundly delays or prevents synapse elimination; whilst stimulating muscles accelerates it. There are some nuances to these findings, however. For example, Thompson *et al.* (1979) showed that the effect of muscle paralysis induced by a nerve conduction block kicked in only after a couple of days; thus, synapse elimination continued for a day or so after the onset of nerve block, but then went into reverse and levels of polyinnervation of the rat soleus muscles they studied climbed back to almost 100% of muscle fibres. Paralysis in this study was induced by implanting plastic pellets impregnated with tetrodotoxin sub-epineurially. Blocking synaptic transmission presynaptically, using botulinum toxin, or postsynaptically using α-bungarotoxin, also preserves polyneuronal innervation (Brown et al., 1982a,b; Callaway & Van Essen, 1989; Duxson, 1982), although the effect with α-bungarotoxin is not as great as with either botulinum toxin, or tetrodotoxin. Greensmith & Vrbova (1996) have reported that α-bungarotoxin block in early postnatal development transiently accelerates synapse elimination, then delays it. The reasons are not understood, but perhaps they are related to an as-yet-unestablished role for spontaneous transmitter release and action. Synapse elimination is also delayed in the *paralysé* mutant mouse (Blondet et al., 1989), and by paralysing muscles indirectly, via tenotomy or by surgical isolation of peripheral and descending inputs to spinal motoneurons (Benoit & Changeux, 1975; Caldwell & Ridge, 1983). Although the neonatal elimination of synapses is associated with, or even preceded by, a reduction in ACh receptor density (Balice-Gordon & Lichtman, 1993; Colman et al., 1997), ACh receptors *per se* are unlikely to play a pivotal role in the rate of developmental synapse elimination. Synapses continue to be withdrawn apparently in accordance with a developmental schedule in mutant zebra fish which lack ACh receptors (Liu & Westerfield, 1990; Westerfield et al., 1990). Finally, it was found in a recent study of knockout mice which fail to express the adult (ε-subunit) form of the receptor, that synapse elimination occurred at a normal rate, even while levels of the neonatal (γ-subunit) form of the receptor were declining (Missias et al., 1997; Schwarz et al., 2000).

The effects of stimulation are not entirely uniform either. O'Brien et al. (1978) first reported that imposing a daily regime of electrical stimulation of nerve and muscle in neonatal rats accelerates synapse elimination. However, the greater effects were observed from the physiological data based on intracellular recording of EPPs, compared with the histological measure based on combined silver/cholinesterase staining. Later, Thompson (1983a) showed that phasic stimulation at 100 Hz was effective in promoting synapse elimination in the neonatal rat soleus muscle; but continuous stimulation at 1 Hz did not affect the rate of elimination. Other, more indirect methods of increasing muscle activity are reported to increase the rate of elimination. These include, for example, prolonging the action of ACh by inhibiting acetylcholinesterase (Duxson & Vrbova, 1985); incubating muscles in solutions containing elevated concentrations of potassium ions (Vrbova et al., 1988); calcium ions (Zhu & Vrbova, 1992); and increasing spontaneous, non-quantal release of ACh (Vyskocil & Vrbova, 1993). These latter studies report effects on levels of polyneuronal innervation over the course of a few hours at room temperature *in vitro*, in otherwise plain mammalian physiological saline. This is very surprising, because it suggests that the dismantling of motor nerve terminals has a Q_{10} that is very low. However, recent estimates suggest that, once initiated, elimination of a synaptic input by the normal physiological mechanisms takes about a day (Gan & Lichtman, 1998), so it is plausible that overt pharmacological effects on innervation pattern should be observed after treatments lasting only a few hours.

The effects of selective activation of motor axons converging on polyneu-ronally innervated neuromuscular junctions are even less straightforward. The experiments are technically difficult to conduct, because they require surgical or semi-acute experiments using preparations with an accessible dual nerve supply. These constraints have also meant that experiments designed to probe the role of activity in the outcome of synaptic competition (in terms of which terminals persist and which are eliminated) have been concentrated in adult models (see Section 3.3 below). Nonetheless, some experiments on neonates have been performed, and there are additionally a number of studies on nerve and muscle cells in tissue culture.

Ridge & Betz (1984) subjected neonatal rat lumbrical muscles to selective stimulation in a daily regime, and reported that the stimulated motoneurons acquired exclusive innervation of more muscle fibres, at the expense of unstim-ulated units. Their conclusions were based on isometric twitch tension measure-ments, and the inherent variability in the motor unit sizes in the lumbrical muscle meant that the effect became readily apparent only when the data were normalized. Callaway et al. (1987) carried out the complementary experiment, selectively blocking activity in some of the axons supplying the neonatal rabbit soleus muscle. However they reported the opposite finding: that the inactive motor units were the ones that acquired a competitive advantage and became larger. The data obtained by these authors also required considerable processing to tease out their conclusion, and this study was also based on an indirect, twitch-tension method (see Ribchester, 1988a; and Section 3.3 below).

Tissue culture studies have produced results that are equally fraught with difficulty of interpretation. For example, Magchielse & Meeter (1986) reported that selective stimulation of cholinergic ciliary ganglion neurons, innervating chick myotubes in culture, provoked selective elimination of the unstimulated inputs. However, Nelson et al. (1993) report that either selective or non-selective stimulation of mammalian preparations produced non-selective elimination of the stimulated inputs. More recent reports from this group suggest that selective activity-dependent elimination does occur, but only in the presence of activators of the thrombin receptor of protein kinase C (Jia et al., 1999). One of the problems with these tissue culture models is that they do not properly reproduce one of the key structural features of polyneuronally innervated mammalian neuromuscular junctions in vivo: namely, the convergent innervation of motor terminals belonging to different motoneurons on the same postsynaptic, motor endplate site. Distance between competing synaptic inputs has been shown to be an important additional variable in synapse elimination (Brown et al., 1976; Kuffler et al., 1980), so this important constraint needs to be considered in the design of tissue culture studies. Moreover, most of the culture paradigms used to study synapse elimination are devoid of Schwann cells, which are now known to have a role in regulating ion concentrations in the synaptic cleft and synaptic function (Attwell & Iles, 1979; Robitaille, 1998). Increasing attention is being focused on the possible role of Schwann cells in the regulation of synaptic competition (Culican et al., 1998; Hirata et al., 1997; Koenig et al., 1998; Parson et al., 1998; Patton et al., 1998; Trachtenberg & Thompson, 1997).

Attempts to influence the physiology of synaptic transmission by selective stimulation – considered to be a likely precursor to structural, heterosynaptic elimination – have also been carried out in isolated nerve–muscle preparations and in culture. Betz *et al*. (1989) found that paired stimulation of inputs to polyinnervated muscle fibre in isolated rat lumbrical muscles produced a heterosynaptic suppression of the endplate current (EPC). The effect lasted a few tens of milliseconds and was fully decayed within 100 ms. A pair of elegant experiments by Poo and his colleagues have demonstrated similar effects in a culture system of frog myocytes and neural tube cells. Lo & Poo (1991) found that stimulating separate inputs to spherical myocytes caused suppression of quantal content from the unstimulated input. Furthermore, if the stimulated input was initially relatively weak, its quantal content became stronger; but if it was initially strong then there was no homosynaptic effect. These authors also showed, by comparing the effects using elongated rather than spherical myocytes, that heterosynaptic suppression did not occur if the converging synapses were separated longitudinally by more than 75 μm. In a subsequent study, Dan & Poo (1992) showed that heterosynaptic suppression could be produced by substituting the stimulated neuron with an ACh-filled iontophoretic micropipette. The unstimulated (real) neuronal synapses were suppressed for about 60 ms from the time of administration of ACh. As in the study in the rat by Betz *et al*. (1989), these selective effects of stimulation on heterosynaptic neurotransmitter release were shown, by voltage-clamping, to depend not on muscle fibre membrane depolarisation *per se*; but rather on postsynaptic increases in intracellular calcium ions. Thus the heterosynaptic effects were blocked by injecting the calcium chelator BAPTA into the myocytes. Interestingly, the heterosynaptic suppression of transmitter release was not accompanied by any changes in ACh sensitivity at either the stimulated or unstimulated sites. This suggests that if heterosynaptic suppression is a precursor to structural elimination, it is not *necessarily* a consequence of ACh receptor loss, even though reduction and redistribution of ACh receptors accompanies synapse elimination at mammalian neuromuscular junctions (Colman *et al*., 1997; Slater, 1982a,b).

In sum, there is little doubt that experimentally altering neuromuscular activity patterns can measurably influence the rate and outcome of both acute and chronic heterosynaptic interactions that lead to synapse elimination. It remains uncertain as to how decisive activity is in this regard. The issue of the necessity and sufficiency of activity to drive elimination of polyneuronal innervation is discussed further in Section 3.3.2, in connection with competitive synapse elimination in reinnervated adult muscle.

3.2.2. Activity-independent competition

Competition based on specificity of matching of motoneuron type to muscle fibre type has been debated by a number of investigators. The possibilities that have been explored are: competition based on segmental origin of motoneurons; competition based on organization of muscle compartments; competition based on inherent fast–slow characteristics of skeletal muscles.

Studies on the role of the rostrocaudal placement of motoneurons in the spinal cord suggest that segmental interactions are important in establishing the innervation pattern of some muscles. For example, although segmentally organized muscles such as the gluteal muscles receive input predominantly from the appropriate spinal segment from the outset, selective elimination of projections from neighbouring segments may play a role in establishing the final innervation pattern (Brown & Booth, 1983). During reinnervation there is also evidence for preferential synapse regeneration in segmentally innervated muscles by the original spinal segment. Other studies suggested that segments supplying muscles whose fibres are not segmentally organized are also selectively eliminated with respect to the rostrocaudal location of their motoneurons in the spinal cord (Miyata & Yoshioka, 1980). However, subsequent studies carried out in different laboratories failed to replicate this finding (Gordon & Van Essen, 1983; Thompson, 1983b). More recently, competition favouring axons of the appropriate rostrocaudal position has been demonstrated during reinnervation of neonatal serratus anterior muscles (Laskowski et al., 1998).

The organization of motor units between distinct muscle compartments has also been scrutinized and debated. The emerging conclusion is that muscles organized into anatomically distinct compartments are selectively innervated by different groups of motoneurons from birth. The rat EDL muscle falls into this category, for example (Balice-Gordon & Thompson, 1988a,b). These compartmental motoneurons are sometimes organized segmentally, for example L4 and L5 motoneurons are unequally distributed to rat gastrocnemius muscles (Bennett et al., 1986). But some aberrant projections between compartments do exist in neonates (Dennis et al., 1981, Donahue & English, 1987) and these are subsequently withdrawn, suggesting that compartmental identity is a mitigating factor in competition. Segmental denervation, removing some of the competitors, stabilizes at least some of these aberrant projections (Gatesy & English, 1993).

Competition based on matching of motoneuron type to inherent fast-versus-slow characteristics of muscle fibres has also generated a stimulating debate. At issue is whether muscle fibres become selectively innervated at an early stage in synapse formation/elimination; or whether muscle fibres are initially non-selectively innervated and the selective pattern emerges only as a consequence of synapse elimination. Studies by Fladby & Jansen (1987), Soha et al. (1987), and by Thompson et al. (1990) support the notion that muscle fibres are selectively innervated from the outset, based on the high proportions of muscle fibres of a particular immunocytochemical or histochemical profile within motor units at the onset of synapse elimination. However, studies by Jones et al. (1987a) suggested that muscle fibres in rat lumbrical muscles are initially indiscriminately innervated. A compelling study by Gates & Ridge (1992) shows that, following partial denervation of lumbrical muscles at birth, adult muscles comprising a single motor unit contain a mixture of muscle fibre types in similar proportion to the composition of neonatal units, whereas most motor units in normal adult muscles are homogeneous with respect to fibre type. They thus conclude that competition is at least in part based on "mismatch withdrawal". Ribchester & Barry (1994) have proposed that one way of accommodating such findings is if there were an "induced fit" between surface markers or adhesion molecules on motor nerve terminals and muscle fibres, but this hypothesis has yet to withstand experimental test.

3.2.3. Molecular basis of competition

Competition in biological systems is based on consumption or control of access to finite resources. In a neurobiological context, synaptic competition can be based either on consumption of neurotrophic resources, or on access to space at the motor endplate (Ribchester & Barry, 1994; Van Essen et al., 1990). The winners in this process may express superior capacity to exploit the relevant resources (as in a drinking race) or by interference (as in a boxing match). Arguably, competition can be established only once the relevant resources have been identified (Keddy, 1989), which has not yet been achieved in the context of neuromuscular development. Thus, the evidence that the adult innervation pattern arises through competition remains indirect, though highly plausible. Attempts to identify molecules that might define a spatial competition have so far drawn a blank. For example, it is remarkable how normally developmental elimination of synapses occurs in various transgenic animals in which expression of cell surface or extracellular matrix molecules has been disrupted (NCAM, AChR etc.; Sanes et al., 1998). On the other hand, there are a number of reported effects of diffusible neurotrophic factors on synapse elimination and these factors are candidates for a consumptive mode of competition based on exploitation of the neurotrophic resources. Alternatively (or in addition) there is evidence that neuromuscular synapses might engage in competition based on interference, via secretion of proteases and protease inhibitors.

Neurotrophic factors
Experiments involving pharmacological or transgenic regulation of neurotrophic factors produce small or transient effects on elimination of polyneuronal innervation (English & Schwarz, 1995; Jordan, 1996a,b; Kwon et al., 1995; Kwon & Gurney, 1996). However, a recent study by Nguyen et al. (1998b) of transgenic mice overexpressing GDNF under the control of the myogenin promoter, revealed extensive polyneuronal innervation at a relatively late postnatal stage in these mice (Figures 11C,D, 12). Mononeuronal innervation was eventually established, but about 2 weeks later than normal. It seems unlikely based on more recent evidence (Nguyen et al., 1998a; J.W. Lichtman and colleagues, personal communication) that GDNF is the endogenous mediator of neuromuscular synaptic competition. On the other hand, GDNF is strongly expressed in human muscle (Lie & Weis, 1998; Suzuki et al., 1998). Either way, the *principle* of an interaction between the transient expression of a neurotrophic molecule and the asynchronous regulation of its receptors expressed presynaptically is given a considerable boost by the findings of Nguyen et al. (1998b). Thus, a role for GDNF as an endogenous "synaptotrophin" cannot be ruled out at present (Snider & Lichtman, 1996).

Proteases and protease inhibitors
Inspection of neuromuscular junctions at different stages in the the transition from polyneuronal to mononeuronal innervation has revealed that boutons of the terminal that will ultimately be eliminated are withdrawn in stepwise, piecemeal fashion (Balice-Gordon et al., 1993; Gan & Lichtman, 1998). Whether this process is completely different from the process of degeneration, of the type

induced by nerve injury, has been debated. Rosenthal & Taraskevich (1977) presented electrophysiological and ultrastructural evidence in support of this idea, but most investigators in this field now accept instead the conclusion made initially by Korneliussen & Jansen (1976) and later, in an exhaustive electron microscope study, by Bixby (1981), that synapse withdrawal during elimination is quite distinct from Wallerian degeneration. This conclusion was endorsed by the electrophysiological analysis presented by Colman *et al.* (1997), which demonstrated gradual suppression of eliminating synapses. This contrasts with the abrupt cessation of synaptic transmission that accompanies degeneration of nerve terminals following nerve section (Miledi & Slater, 1968, 1970)

Notwithstanding the evident distinction between synapse elimination and degeneration, a role for proteases and protease inhibitors has been sought in the case of both, but the discussion here will be restricted to synapse elimination. The search for a physiological role for proteases and their inhibitors may in part be motivated by the idea (and the evidence for it) that growing axons and their growth cones are invasive cellular proboscices, which must clear a path through a dense cellular and extracellular matrix in order to reach their targets (Fawcett & Housden, 1990; Seeds *et al.*, 1997). Once motor axon growth cones reach their targets they become transformed from a "growing state" to a "transmitting state" (Greensmith & Vrbova, 1996; Watson, 1974). The first evidence that a balance between the levels of proteases and protease inhibitors might play a role in synapse stabilization and subsequent elimination was presented by O'Brien *et al.* (1978), who suggested that muscle fibres are the source of proteases at the neuromuscular junction. Earlier, Betz & Sakman (1973) had shown that treating frog neuromuscular junctions with a cocktail of proteases gently releases synaptic terminals from muscle fibres. Subsequently, G. Vrbova and her colleagues have investigated the possible role of specific proteases by administering selective inhibitors via impregnated plastic implants placed in the vicinity of neonatal muscles (Connold *et al.*, 1986; Connold & Vrbova, 1994; O'Brien *et al.*, 1984; Vrbova *et al.*, 1988). The protease baton has since been taken up by P.G. Nelson and B. Festoff and their colleagues (Liu *et al.*, 1994). The serine protease inhibitor nexin I is concentrated at neuromuscular junctions (Akaaboune *et al.*, 1998) and mRNAs for this inhibitor, and for prothrombin and thrombin receptor, are all expressed in mouse mouse (Kim *et al.*, 1998). Levels of all three are elevated at birth but decline – though not at the same rate – by day 20; by two orders of magnitude in the case of prothrombin and thrombin receptor; and by a factor of 4–5 in the case of nexin I. Inhibition of thrombin with another selective protease inhibitor, hirudin, delays or prevents synapse elimination in a cell culture model (Zoubine *et al.*, 1996), and removes the selective effect of nerve stimulation on heterosynaptic elimination – also in culture (Jia *et al.*, 1999).

Thus there is growing evidence that proteases and their inhibitors could play a significant physiological role in neonatal synapse elimination (Chang & Balice-Gordon, 1997). Indeed a balance between adhesion of nerve terminal boutons to muscle fibres, and retractile, motor forces exerting tension within growth cones or their postsynaptic endplate sites (Bray, 1987; Crick, 1982) could readily be tipped in favour of detachment and withdrawal, versus strong adhesion and persistence, by disturbances in levels of proteases or their inhibitors. Tension–adhesion forces within axons and their targets are thought to play an important role elsewhere in the nervous system, for example in synaptic plasticity (Fischer *et al.*, 1998) and in sculpting the form of the cerebral cortex during development (Van Essen, 1997).

3.2.4. Non-competitive synapse elimination

Non-competitive elimination of neuromuscular synapses has been the subject of an intriguing dialogue, that has yet to be definitevely resolved. Brown *et al.* (1976) suggested, on the basis of motor unit size measurements made in adult rat soleus muscles following partial denervation at birth, that motoneurons withdraw some of the connections in a programmed fashion irrespective of the presence or absence of competing motor terminals supplied by different motoneurons to the same endplates. This possibility is commonly referred to as "intrinsic withdrawal". It was supported by Thompson & Jansen (1977), who carried out similar partial denervation experiments in a strain of rats in which the soleus muscle receives a dual nerve supply. However, the design of the experiments left open the possibility that partial denervation left intact motoneurons competing for innervation of the same muscle fibres, and that this could account for the subsequent reduction in motor unit size. Betz *et al.* (1980a) attempted to resolve this issue by studying lumbrical muscles partially denervated at birth so as to leave a proportion innervated only by a single remaining motoneuron. They concluded that these motoneurons underwent no reduction in the numbers of muscle fibres they innervated (Figure 14A). However, the interpretation was complicated by the fact that new muscle fibres are added to rat lumbrical muscles postnatally, and this failed to occur after partial denervation at birth. Thus tension measurements used to establish motor unit size in neonates overestimate the number of muscle fibres receiving suprathreshold input from motoneurons unless the number of muscle fibres is explicitly determined.

The issue of intrinsic withdrawal was revisited by Fladby & Jansen (1987) in a study of synapse elimination in partially denervated mouse soleus muscle. These muscles have their adult complement of muscle fibres established by birth. Some of the soleus motor units provide subthreshold inputs to muscle fibres (Fladby, 1987), as in the rat lumbrical muscle (Jones *et al.*, 1987b), so muscle tension measurements underestimate the numbers of neuromuscular connections supported by motor axons at this stage. Moreover, by comparing the observed adult motor unit sizes following neonatal partial denervation with the predictions of a simple stochastic model defining the initial innervation pattern, Fladby & Jansen (1987) showed a consistent shortfall in the numbers of innervated fibres in the experimental muscles (Figure 14B). Intrinsic remodelling of terminal arbors, independent of activity or competitors, has also been described in normal and mutant zebra fish (Liu & Westerfield, 1990). Activity-independent competition has also been demonstrated in reinnervated muscle (see Section 3.3). Thus a significant body of evidence favours the hypothesis that neonatal synapse elimination is controlled in part by an unknown, intrinsic property of motoneurons.

It is interesting to note, in passing, that in mutant Wld[s] mice with slow or absent Wallerian degeneration, axotomy induces a form of synapse withdrawal also characterized by piecemeal removal of synaptic boutons and gradual

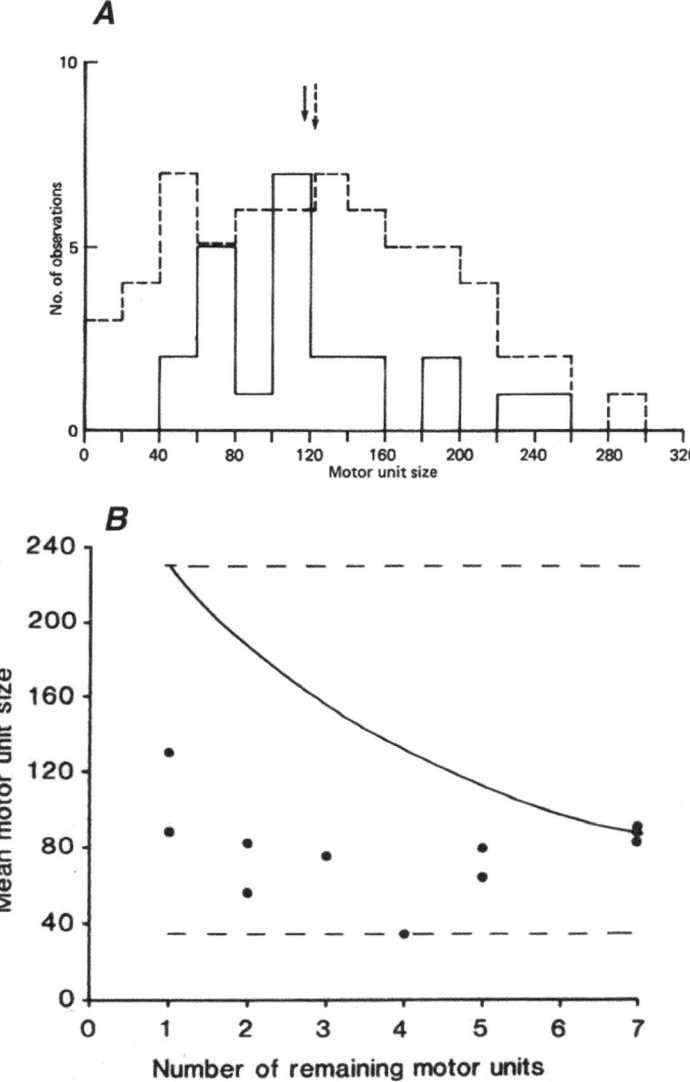

Figure 14 Evidence for competition (**A**) and intrinsic withdrawal (**B**) of terminals in neonatal muscle. **A**: Histograms of motor unit size in newborn rat lumbrical muscle (dotted line) and in adult muscle following partial denervation at birth (solid line). In contrast to normal adult motor units there was no reduction in the size of the units in partially denervated muscles. (From Betz *et al.*, 1980.) **B**: Motor unit size in mouse soleus muscles after partial denervation at birth, as a function of the number of remaining motor units. All the points fall below the solid curved line, which represents the expected motor unit size if competition alone accounted for synapse elimination. The upper dotted line represents average motor unit size at birth and the lower dotted line average motor unit size in adults. Thus, about half the synapse elimination occurs by a non-competitive mechanism. (From Fladby & Jansen, 1987, with permission.)

suppression of transmitter release. Overall the axotomy-induced synapse withdrawal in Wld^s mice shows considerable morphological and physiological resemblance to neonatal synapse elimination in wild-type animals (Ribchester *et al.*, 1995; Parson *et al.*, 1997; T. Gillingwater and R.R. Ribchester, 2001). An intriguing possibility is that this may be the consequence of disrupted orthograde trafficking of intrinsic factors that are responsible for non-competitive withdrawal of motor axon collaterals during normal development.

3.2.5. Computational models of synapse elimination

Computational models of neonatal synapse elimination have attempted to describe the experimental findings, but in only a few cases have the models predicted fundamental properties of the neurons that are accessible to further experiments. Many propose a role for neurotrophic factors secreted into the local environment of motor terminals, coupled with presynaptic factors that limit the maximum number of terminals a motoneuron can support at any given stage of development (Bennett & Robinson, 1989; Jean-Pretre *et al.*, 1996). Autocatalytic biochemical reactions in presynaptic terminals have been invoked to generate the positive feedback required to selectively eliminate disadvantaged terminals (Gouze *et al.*, 1983). Some models posit a balance between forces that strengthen the remaining synapses in a motor unit as others are eliminated, by competition for presynaptic and postsynaptic resources (Rasmussen & Willshaw, 1993; Van Ooyen & Willshaw, 1999a; Willshaw, 1981). Others model the emerging patterns of innervation in terms of competition for substances versus competition for space (Van Essen *et al.*, 1990).

Perhaps the most challenging recent model has been proposed by Van Ooyen & Willshaw (1999b), who have computed the outcome of a wide range of experimental interventions, based on action of neurotrophic factors and the induction of presynaptic receptors for these factors, in a fashion that strikes a chord with the experimental findings of Nguyen *et al.* (1998b). A specific prediction of Van Oojen & Willshaw's model is that levels of presynaptic receptors for growth factors are up- or down-regulated in a fashion that indicates which terminals will persist and which will be eliminated. The measurements required to test this hypothesis are feasible, but they have not yet been made.

3.2.6. Summary

Synapse elimination at polyneuronally innervated junctions transforms not only the pattern and degree of convergence but also the shape of individual neuromuscular junctions. Growing evidence suggests that, although competition and activity strongly influence the time course and outcome of this process, non-competitive and activity-independent mechanisms must also play a significant role. The challenge over the next few years is to identify the nature of the (fixed versus diffusible) resources underpinning the competitive mechanisms and whether non-competitive withdrawal is due to selective innervation of muscle fibre or trafficking of intracellular maintenance factors along specific motor axon collaterals.

3.3. Competition during muscle reinnervation in adults

The effectiveness of regeneration after complete crush or section of peripheral nerves is well known. For example, after sciatic nerve crush in young adult rodents even the most distal muscle fibres become fully reinnervated within about a month. Many of the reinnervated muscle fibres acquire transient poly-neuronal innervation, as in postnatal development. Polyneuronal innervation in reinnervated mammalian skeletal muscle was first recorded in 1916 by J. Boeke (Figure 15) and his actually represents the first description of polyneuronal innervation of skeletal muscle fibres under any circumstances. (J.F. Tello failed to spot this phenomenon in his seminal study of reinnervated muscle published in 1907, although he was the first to espy it in neonatal muscle.) Boeke's silver stains and text descriptions of reinnervated muscles in intercostal muscles of adult hedgehogs appear to show a number of instances where muscle fibres receive convergent innervation from different motor axons. Boeke was evidently undecided whether the converging axons were really independent, however, or whether his stains depicted a single motor nerve terminal forming *en passant* by an axon that looped back into the peripheral nerve. Subsequent reanalysis using intracellular recording, histology and vital staining (Barry & Ribchester, 1995; McArdle, 1975; Gorio *et al.*, 1983) now leave no room for doubt about the robust nature of this phenomenon (Figure 15). Costanzo *et al* (1999, 2000) recently utilized the abbreviation "π-junction" to refer to polyneuronally innervated neuromuscular junctions in reinnervated muscles. It has aesthetic appeal, and serves just as well to describe the same phenomenon in neonatal muscles.

The relationship of polyneuronal innervation and its subsequent elimination in reinnervated muscle to synapse elimination in neonates is secured by four main observations: the convergent inputs are located at the same motor endplate site (Gorio *et al.*, 1983); elimination of polyneuronal innervation in reinnervated muscle is delayed or arrested by muscle paralysis (Barry & Ribchester, 1995; Taxt, 1983b); synaptic boutons are eliminated in piecemeal fashion (Rich & Lichtman, 1989b); and, in at least some instances, selective removal of ACh receptor plaques precedes nerve terminal withdrawal (Culican *et al.*, 1998); all as in development. On the other hand, regenerating axons reoccupy motor end-plates that are already present and differentiated before nerve injury, rather than inducing these postsynaptic sites. Thus, we must exercise caution when interpreting findings made about synapse stabilization and elimination in reinnervated muscles, and when comparing them with the analogous developmental phenomena.

The identity of the molecules responsible for competitive elimination of polyneu-ronal innervation in reinnervated adult muscles is unknown, as in development. However, denervated adult muscle fibres express several intracellular and membrane characteristics in common with fetal or neonatal muscle. Such "denervation changes" include reactivation of myogenin (Witzemann & Sakmann, 1991), expression of extrajunctional (γ-subunit form) ACh receptors and molecules associated with them, such as rapsyn, RATL and MuSK (Sanes & Lichtman, 1999; Witzemann *et al.*, 1991), TTX-resistant sodium channels (Bambrick & Gordon, 1987; Lupa *et al.*, 1995; Trimmer *et al.*, 1990) adhesion molecules including N-CAM (Cashman *et al.*, 1987; Covault & Sanes, 1986;

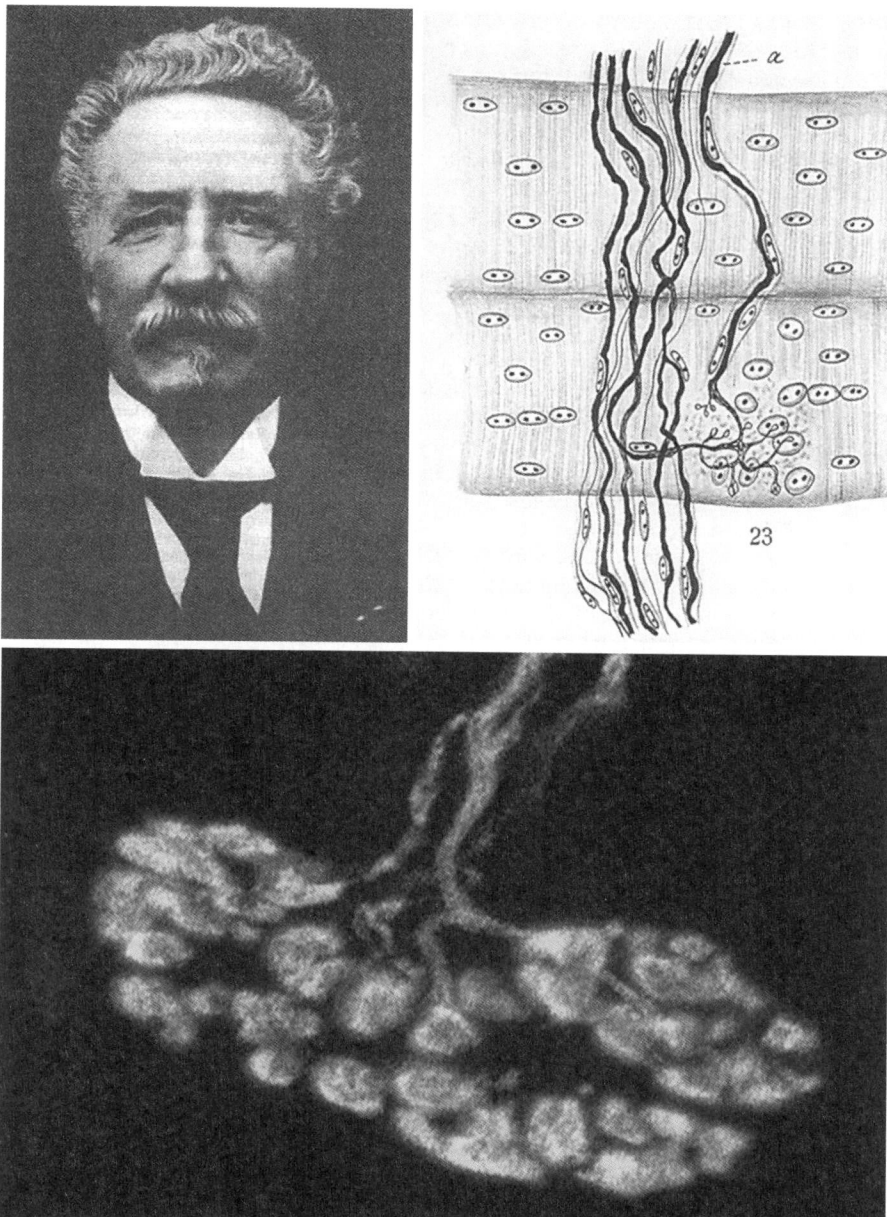

Figure 15 Polyneuronal innervation in reinnervated adult muscles. Left, J. Boeke who, in 1917, first described polyneuronal innervation in reinnervated intercostal muscle fibres of hedgehogs (right). Bottom, confocal micrograph of a reinnervated motor endplate in rat lumbrical muscle, stained immunocytochemically. Three axons converge supplying unique sets of synaptic boutons. Each bouton is about 2 μm in diameter.

Sanes *et al.*, 1986), neurotrophic factors (Caroni *et al.*, 1994; Funakoshi *et al.*, 1995; Lie & Weis, 1998; Rassendren *et al.*, 1992), and myosin isoforms (Hamalainen & Pette, 1996; Windisch *et al.*, 1998). These "neonatal" properties are reversed and muscle fibres re-express their normal, mature patterns of gene expression following reinnervation.

The remainder of this section will focus on studies of sprouting and competitive reinnervation of *partially* denervated muscles, an appealing model with a number of distinctive features appropriate to experiments designed to probe mechanisms of competitive synapse elimination.

3.3.1. Motor nerve sprouting

Collateral sprouts arise both from nerve terminals (terminal, or "ultraterminal" sprouts) and intramuscular nodes of Ranvier (nodal sprouts). Both forms of sprouting take place from intact axons following crush or section of a fraction of the motor nerve supply to a muscle. Most contemporary studies of nerve sprouting in response to partial denervation stem from the studies of Edds (1953) and Hoffman (1950, 1953). The excellent review by M.C. Brown and colleagues (Brown *et al.*, 1981a) documents the field over the following 25 years. Competitive reinnervation of partially denervated muscles was studied using tension measurements, intracellular recording and anatomical methods in two important, contemporaneous reports by Brown & Ironton (1978) and Thompson (1978).

Intact axons react within about 2–3 days of injury to their near neighbours, and the sprouts are directed towards the endplates of nearby denervated fibres. During sprouting, some intact terminals withdraw synaptic boutons from their endplates at the same time as they generate sprouts. This weakens synaptic transmission at the intact endplates (Lubischer & Thompson, 1999; Weiss & Edds, 1946). With time, however, the sprouted terminals are strengthened, such that most are capable of robust synaptic transmission. Extensive partial denervation results in maximal sprouting of intact motor units, which become between two and five times their initial size (Brown & Ironton, 1978; Gordon *et al.*, 1993; Rafuse *et al.*, 1992; Ribchester, 1988b; Thompson, 1978). Motor nerve sprouting is primarily a phenomenon of mature muscle, however. Neither can it be induced by partial denervation in neonatal muscles: the earliest sprouting reactions occur about a week after neonatal synapse elimination is complete (Brown *et al.*, 1982a; Lubischer & Thompson, 1999).

Extensive sprouting can also be induced by paralysing muscle, either with botulinum toxin, α-bungarotoxin or via a chronic nerve conduction block (Brown *et al.*, 1981b). The sprouting that occurs with all these treatments is generally restricted to the nerve terminals rather than axons. Ultraterminal sprouting is very profuse after botulinum toxin injection, but less so after bungarotoxin or TTX block of nerve conduction. In the latter cases few sprouts form functional contact on adjacent muscle fibres (Betz *et al.*, 1980b; Brown *et al.*, 1982b). Electrical stimulation of muscles blocks terminal sprouting (Brown *et al.*, 1980). Terminal sprouting (induced by botulinum toxin) is also blocked by administration of antibodies against the cell-adhesion molecule N-CAM (Booth *et al.*, 1990). No treatments to date have been identified that block nodal sprouting after partial denervation.

Both terminal and nodal sprouting can also be stimulated by "products of nerve degeneration". For example, when only muscle afferents are injured – by dorsal rhizotomy – motor axons in the deafferented muscles sprout within a few days (Brown et al., 1978). Sprouting is also evoked by tissue damage or inflammation. Myotoxic or mechanical injury to muscle fibres triggers withdrawal then directional sprouting of axon terminals (Grubb et al., 1991; Rich & Lichtman, 1989b; Van Mier & Lichtman, 1994). Local administration of neurotrophic molecules (neurotrophins, cytokines) also triggers a mild sprouting reaction but it is unclear at present whether this is a direct effect of the growth factors on presynaptic receptors, or a secondary consequence of a more generalized inflammatory reaction to injection of the factors (Caroni et al., 1994; Funakoshi et al., 1995; Gurney et al., 1992).

It is becoming clear from a number of studies that neuroglial cells play a crucial role in motor nerve sprouting (Figure 16). One of the earliest responses to partial denervation or paralysis is reactive differentiation of Schwann cells associated with the denervated muscle fibres, and this can be detected by immunostaining with the 4E2 monoclonal antibody or antibodies against nestin (Kopp & Thompson, 1998; Son et al., 1996). Schwann cells also proliferate and then generate their own cellular sprouts. These are directed towards endplates on denervated muscle fibres, forming bridges between them and the innervated endplates (Love & Thompson, 1998). Axonal sprouts appear to be passively directed along these Schwann cell bridges and onto the denervated muscle fibres (Reynolds & Woolf, 1992; Son & Thompson, 1995a,b). Recent data suggest that either complete paralysis or muscle stimulation blocks formation of Schwann cell bridges in partially denervated muscles, and thereby suppresses directional terminal sprouting. This suggests that some kind of gradient between active, innervated muscle fibres and nearby denervated fibres is necessary to produce directed Schwann cell and motor nerve sprouting (F. Love and W.J. Thompson, personal communication). The nature of this gradient remains unidentified, but it could be generated either by diffusible neurotrophic molecules or by fixed molecules in extracellular matrix. The physiological regulator is unlikely to be the glial cell-line-derived neurotrophic factor, GDNF, however. Injecting plasmids containing code for GDNF into adult muscle fibres leads to local synthesis and release of this cytokine, but mature motor nerve terminals do not react to its presence (Bernstein et al., 1998). Schwann cell sprouting is itself preceded by secretion of specific components of the synaptic basal lamina (Chen & Ko, 1994; Ko & Chen, 1996; Astrow et al., 1997). Synaptic forms of laminin inhibit sprouting of Schwann cells (Patton et al., 1998).

The overall picture that emerges with regard to the state of partially denervated muscle is that many (or all) of the motor endplates vacated by injured axons become taken over by sprouts, as a consequence of an activity-dependent formation of Schwann cell bridges and directed growth from motor nerve terminals and preterminal nodes of Ranvier. The occupancy of these endplates effectively constitutes an energy barrier which regenerating axons must overcome in order to reform their connections and restore function. Some thought-provoking findings have been made concerning, the effectiveness of regeneration and the inevitability of synapse elimination; and the necessity and sufficiency of activity in bringing it about.

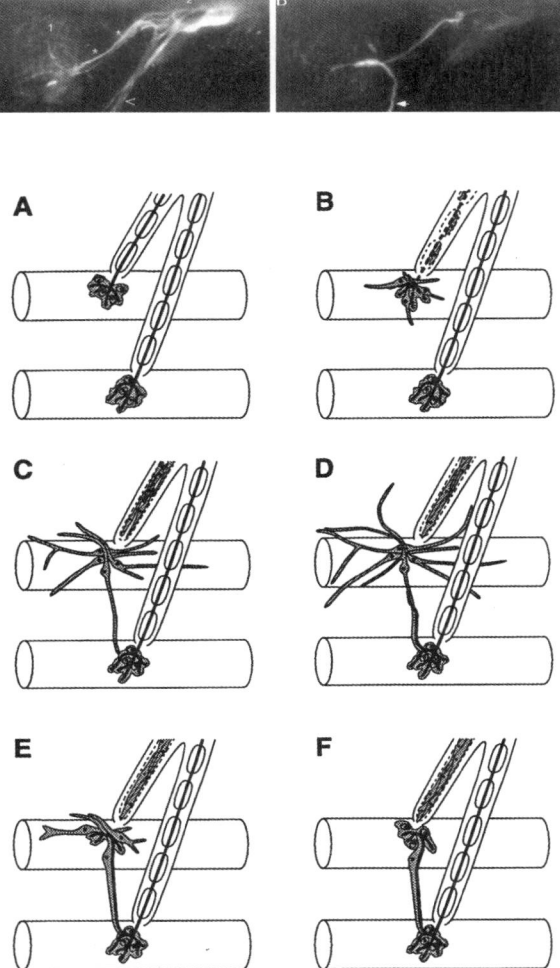

Figure 16 Sprouting of Schwann cells and axons following partial denervation. Top: fluorescence micrographs of terminal Schwann cells (left) and axon sprouts (right) stained immunocytochemically with monoclonal antibodies. Cartoons A–F show stages in the degeneration of an injured axon, the reactive proliferation and sprouting of Schwann cells at the denervated endplate, and the formation of a bridging sprout by an intact axon terminal on a neighbouring muscle fibre. (From Son & Thompson, 1995, with permission.)

3.3.2. Reinnervation of partially denervated muscles

Functional recovery by the regenerating axons, measured from the isometric tension, is accompanied by regression of the intact motor units (Hoffman, 1953; Rafuse & Gordon, 1998; Ribchester, 1986, 1988b; Weiss & Edds, 1946). It is rare

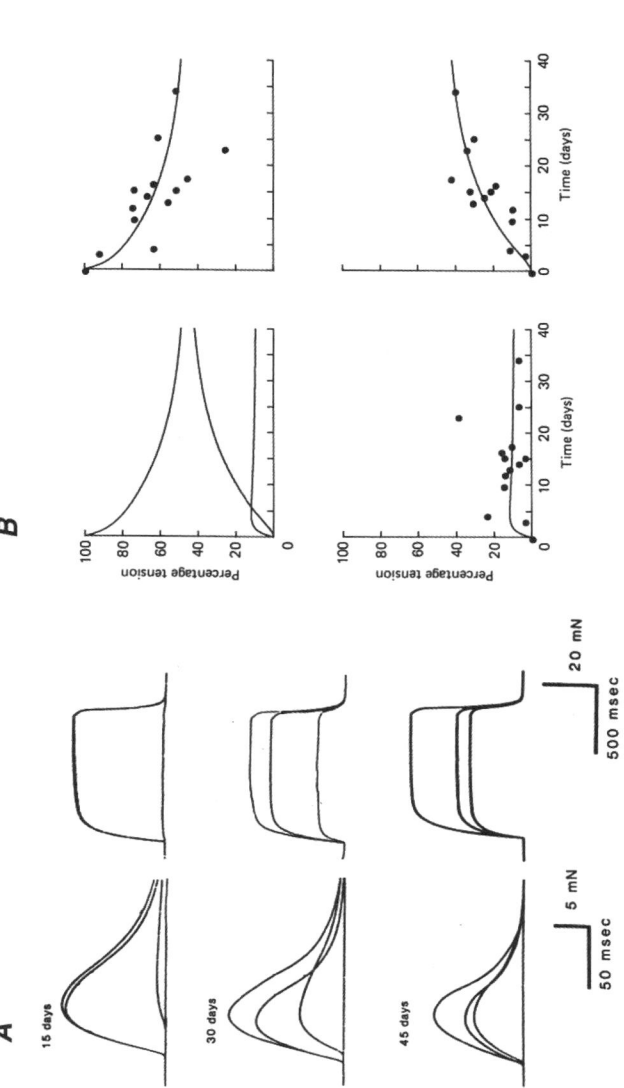

Figure 17 Reinnervation of partially denervated rat lumbrical muscle following nerve crush to most of the axons supplying the muscles. **A**: Twitch (left) and tetanic tension recordings from different muscles after partial denervation and regeneration of injured axons. Lower traces in each case are from the regenerating nerve; middle traces, intact nerve; upper traces, combined response to stimulation of both nerves. Before injury the regenerating axons innervated about 70% of the muscle. By 15 days after injury, sprouts from the intact nerve provided functional innervation to all the denervated muscle fibres. By 30 days, significant recovery of muscle fibres by regenerating axons has occurred, with concomitant decline in the innervation supplied by the intact nerve. The levels stabilize by 45 days. **B**: Graphs showing data and results of a simulation based on a simple, reversible kinetic model of competitive reinnervation. Top left graph: model only; upper trace, decline in innervation by the intact nerve; middle trace, slight increase then decline to a stable level of polyneuronal innervation; lower trace, progressive rise in reinnervation by regenerating nerve. Top right graph: data on regression of sprouts; lower left, levels of polyneuronal innervation; lower right, levels of recovery of tension by regenerating axons. (From Ribchester, 1988.)

to find that regenerating axons fully restore their original motor unit sizes, however (Figure 17). The longer the delay in regeneration, the less the recovery of function (Fu & Gordon, 1995; Grow *et al.*, 1995; Thompson, 1978). Moreover, the smaller the number of axons axotomized by nerve injury (minor partial denervation) the less effective are the injured axons in recapturing the fibres they once innervated (Ribchester, 1988a, 1993; Ribchester & Taxt, 1984). Put another way, the less extensive the requirement for sprouting by intact units, the more capable the sprouts are in retaining their new synaptic sites in the face of competition by regenerating axons, and the less successful those regenerating fibres become in recovering exclusive control of the motor endplates they previously supplied.[1]

Remarkably, however, many axons regenerating into partially denervated muscle do recover exclusive innervation of motor endplates, and this requires competitive elimination of the occupying terminals formed by the sprouts (Brown & Ironton, 1978; Thompson, 1978). This recovery takes place even if all the endplates are occupied by intact terminals and their sprouts. Intracellular recordings and selective staining with the activity-dependent vital dyes FM1-43 and RH414, show that synaptic boutons belonging to sprouted and regenerated axons become intermingled (Barry & Ribchester, 1995; Betz *et al.*, 1992; Brown & Ironton, 1978; Thompson, 1978). It is likely, though not proven, that the regenerating axons also vie for endplates that were always occupied by the uninjured, intact axons (Ribchester & Taxt, 1984). An extreme instance of regenerating axons competing successfully with intact ones was demonstrated by implanting a foreign nerve into the endplate region of the fully innervated soleus muscle (Bixby & Van Essen, 1979).

3.3.3. The role of activity in competitive reinnervation

There is no doubt that the ability of regenerating axons to reform and sustain connections in partially denervated muscle is strongly influenced by muscle activity. Thus, levels of polyneuronal innervation and the size of regenerating motor units are significantly increased when intact motor units are paralysed during reinnervation (Barry & Ribchester, 1995; Ribchester & Taxt, 1984). The effects of chronic electrical stimulation on the rate of synapse elimination from reinnervated motor endplates remain unknown. The prediction from comparable experiments carried out in neonatal muscles would be that fast patterns of chronic stimulation should accelerate synapse elimination and that

---

[1]In this respect the reinnervation of partially denervated skeletal muscle bears some resemblance to regeneration in the central nervous system, after spinal injury or stroke, for example. Owing to the myriad connections that survive injury, and the compensatory sprouting that results, CNS lesions nearly always constitute a very minor partial denervation, from the viewpoint of individual postsynaptic neurons. The failure of regenerating axons to reconnect under these circumstances may only be quantitatively different from the failure of axons regenerating into partially denervated muscles to restore all their connections. Partially denervated muscles may therefore constitute an effective paradigm in which to address general issues raised by the failure of regenerating axons to fully recover and restore functional innervation after nerve lesions.

selective stimulation should give an advantage to the active motor units (O'Brien *et al.*, 1978; Ridge & Betz, 1984; Thompson, 1983a). Recent studies by Cangiano and colleagues (Bussetto *et al.*, 2000) suggest that synchronised nerve stimulation preserves polyneuronal innervation rather than accelerates synapse elimination. However, these experiments were carried out on muscles where the regenerating axons were forced to make ectopic synapses on muscle fibres, rather than competing for reinnervation of original motor endplates. Nonetheless, this novel finding, taken together with other observations described below, raises questions not only about the mechanism by which use and disuse might stabilize synaptic connections, but also whether activity is both sufficient and necessary for synapse elimination.

Activity is influential, but not decisive
A tempting conclusion from the majority of studies on the effects of manipulating neuromuscular activity is that action potentials are required to induce and/or execute synapse elimination (Sanes & Lichtman, 1999). The first suggestion that activity may not play a decisive role in competitive elimination of synapses emerged from a study of partially denervated muscles in which the regenerating axons – rather than the intact ones – were rendered inactive (Ribchester, 1988a). Whilst fewer of the regenerating axons were able to reinnervate muscle fibres, nevertheless there were many fibres that became exclusively reinnervated by these axons. The experimental design of this study left open the possibility that the regenerating axons merely occupied endplates that had failed to attract sprouts from the intact nerve. A more definitive experimental design, in which all activity was blocked after minor partial denervation, ruled out this possibility. The outcome and inference were the same. Electrophysiological analysis of innervation patterns after reinnervation of completely paralysed muscles established two features: first, paralysed muscles provide a conducive environment for regenerating axons to manufacture and restore functional synaptic connections. Second, inactive terminals are capable of competitively displacing other terminals even when these are also inactive (Ribchester, 1993; Figure 18 A,B).

Recently, Costanzo *et al.* (2000) carried out a critical test of the activity hypothesis, using vital dyes to selectively stain and distinguish regenerating and intact (or sprouted) terminals that either mononeuronally or polyneuronally innervated muscle fibres. In this study synapse elimination occurred from some junctions in completely paralysed muscles (by TTX superfusion of peripheral nerves and daily intramuscular injection of muscles with α-bungarotoxin). In addition to fibres that became exclusively innervated by regenerating axons, most of the polyneuronally innervated fibres became innervated by synaptic boutons supplied mostly by the inactive regenerating nerve, rather than the inactive, intact nerve (Figure 18C–E). Control experiments established that more than 99% of endplates were innervated; almost all endplates were fully occupied by sprouts; and both presynaptic and postsynaptic activity blocks were patent immediately prior to and during the return of the regenerating axons. Thus, elimination of silent synapses by other silent synapses occurred after the nerve regeneration. We believe this finding constitutes compelling evidence that activity is not necessary for competitive synapse elimination.

313

Other experiments challenge the sufficiency of activity in promoting synapse elimination once it is restored. Barry & Ribchester (1995) observed that following recovery from chronic nerve conduction block, many reinnervated muscle fibres in partially denervated muscles retained polyneuronal innervation, for at least 8 weeks after resumption of activity. The influence of activity on synapse elimination was confirmed in two ways: first, levels of polyneuronal innervation were significantly elevated in the paralysed and reinnervated muscles before nerve conduction block wore off. Second, significant amounts of synapse elimination occurred within the first 2 weeks of resumption of activity. Thereafter, levels of polyneuronal innervation remained about three times higher than in control muscles that were not paralysed during nerve regeneration. Recent data suggest that the synapses provided to an endplate by converging terminals need not be large, or evoke suprathreshold responses in order to become stable (Figure 19); and that the synapses belonging to stable converging inputs are of similar strength per unit area whatever their relative size (Costanzo et al., 1999).

These findings accord with others suggesting that synapses are not necessarily eliminated as a consequence of muscle activity. First, some polyneuronal innervation, normally less than about 10% of muscle fibres, persists for long periods after reinnervation of partially denervated muscle (Brown & Ironton, 1978; Weiss & Edds, 1946; Werle & Herrera, 1987, 1991). In fact Hoffman (1953) clearly illustrated this persistence ("hyperneurotization") as long as a year after partial denervation and reinnervation of rat gastrocnemius and soleus muscles (Figure 20). Second, when some neonatal muscles are paralysed with botulinum toxin, then allowed to recover, levels of polyneuronal innervation remain elevated through to adulthood (Brown et al., 1982b; Figure 20). Third, treatment of other neonatal muscles with testosterone delays synapse elimination and about 30% of fibres remain polyneuronally innervated for a considerable period after testosterone therapy has been withdrawn (Lubischer et al., 1992). Finally, in transgenic mice engineered to overexpress GDNF secretion from muscle in neonates, the extent of polyneuronal innervation is profound and synapse elimination is delayed for at least 2 additional weeks, despite a significant excess of muscle activity in these mice (Nguyen et al., 1998b).

The above findings contrast with other studies of synapses undergoing competitive elimination in adults, however. For example, Rich & Lichtman (1989a) reported that synapse elimination always runs to completion in reinnervated adult mouse sternomastoid muscle; and that synapse elimination begins with elimination of small patches of ACh receptors followed by removal of the overlying synaptic boutons. In an elegant extension of this finding, Balice-Gordon & Lichtman (1994) blocked small patches of receptors in adult innervated neuromuscular junctions by topical administration of α-bungarotoxin in vivo, rendering the synapses overlying them functionally ineffective. They then repeatedly visualized the same sites over several weeks (Figure 21). Based on this technical tour-de-force the authors were able to conclude that, provided the areas of local inactivity were small, the synaptic boutons overlying them were almost guaranteed to become eliminated. Areas of paralysis extending to more than about 40% of an endplate normally did not lead to elimination of either receptor patches or overlying boutons, however. These findings, taken together

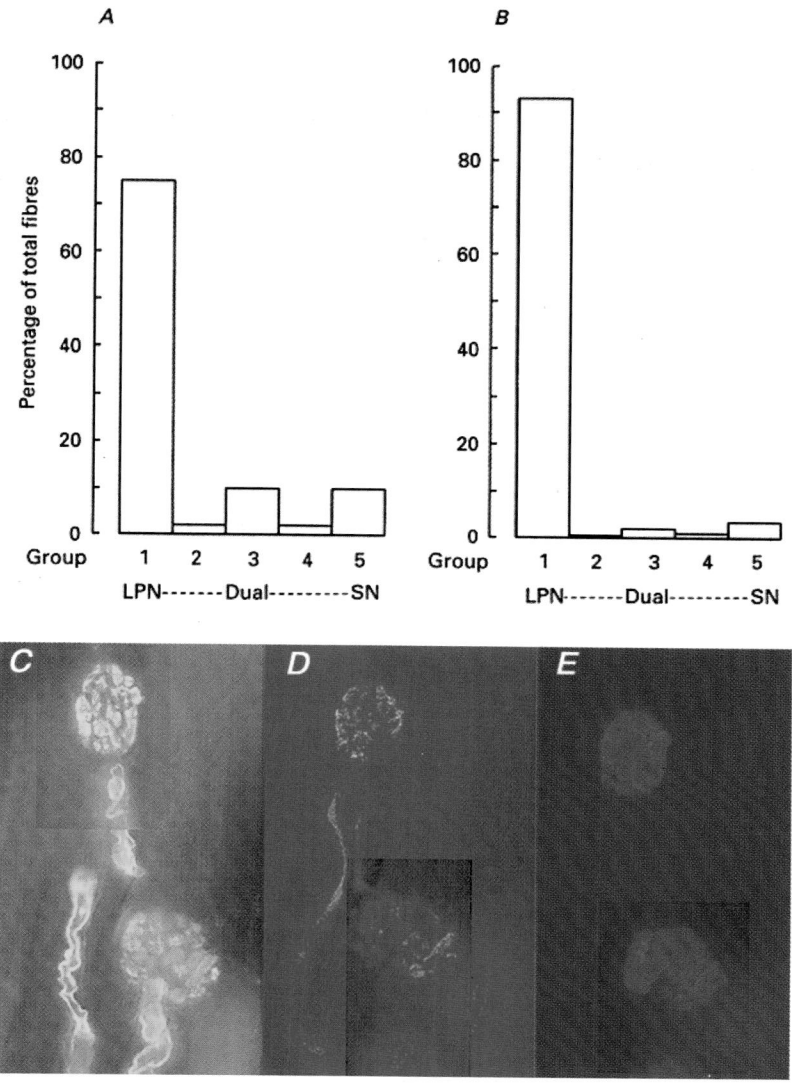

Figure 18 Levels of mononeuronal and polyneuronal innervation in reinnervated rat lumbrical muscle following minor partial denervation and nerve conduction block during reinnervation (**A**) compared with contralateral reinnervated controls (**B**). The presence of mononeuronal (SN) fibres indicates that, at at least some reinnervated endplates, activity is not necessary for synapse elimination to occur. Staining with styryl dyes to distinguish intact (orange) from regenerating (yellow/green) synaptic boutons (**C**) and with subsequent immunostaining (**D**) in completely paralysed muscle confirmed that some endplates (**E**, rhodamine bungarotoxin stain) become mostly or completely occupied by regenerating terminals, and thus that competition can occur at silent neuromuscular junctions. (From Ribchester, 1993 and Costanzo et al., 2000.)

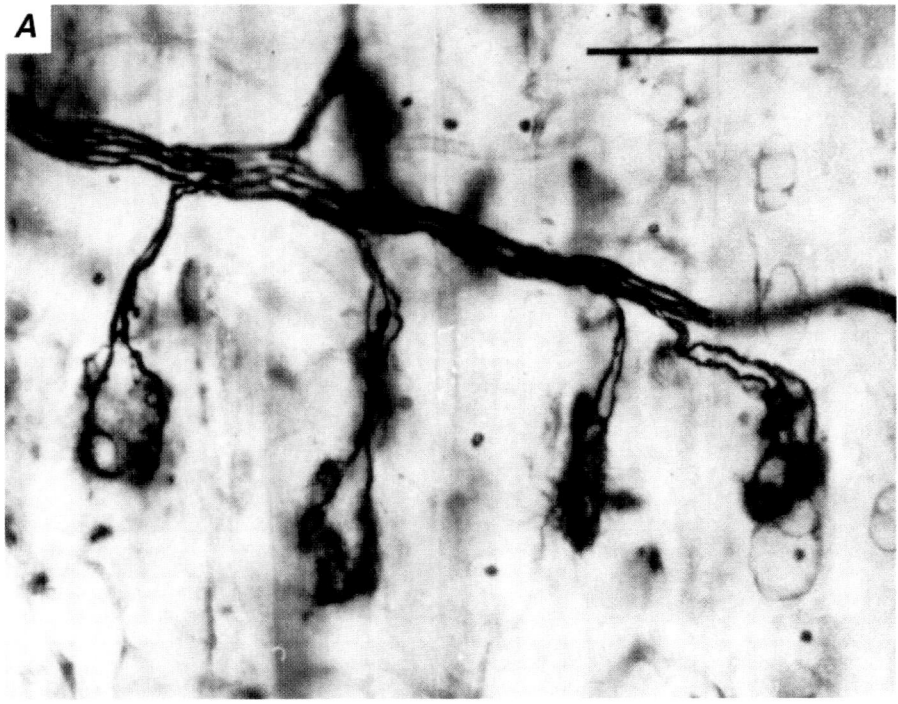

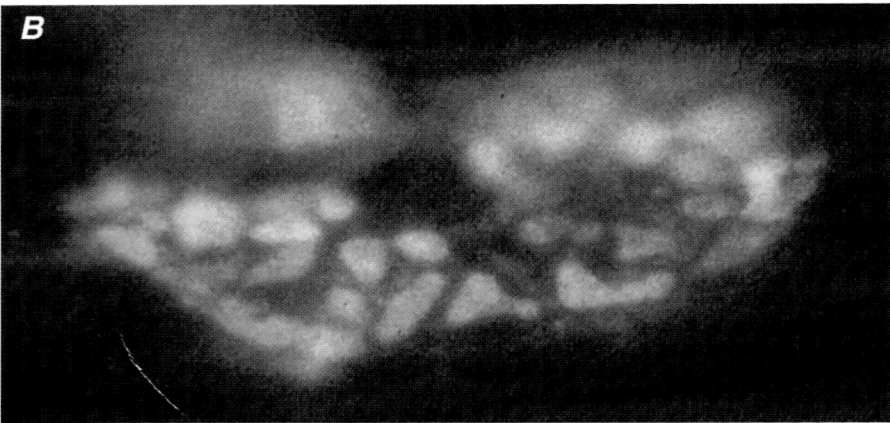

Figure 20 Persistent polyneuronal innervation in adult rodent muscles. **A**: Mouse tensor fasciae latae muscle following recovery of activity after neonatal injection of botulinum toxin. Silver stains show clear evidence of convergence of different axons. (From Brown *et al.*, 1982b, with permission). **B**: Endplate in a rat fourth deep lumbrical muscle stained with vital styryl dyes to reveal synaptic boutons from different converging axons (green/orange), 8 weeks after recovery from chronic nerve conduction block during reinnervation. (From Barry & Ribchester, 1995). Calibrations: **A**, 30 µm; **B**, 5 µm.

A   *B*

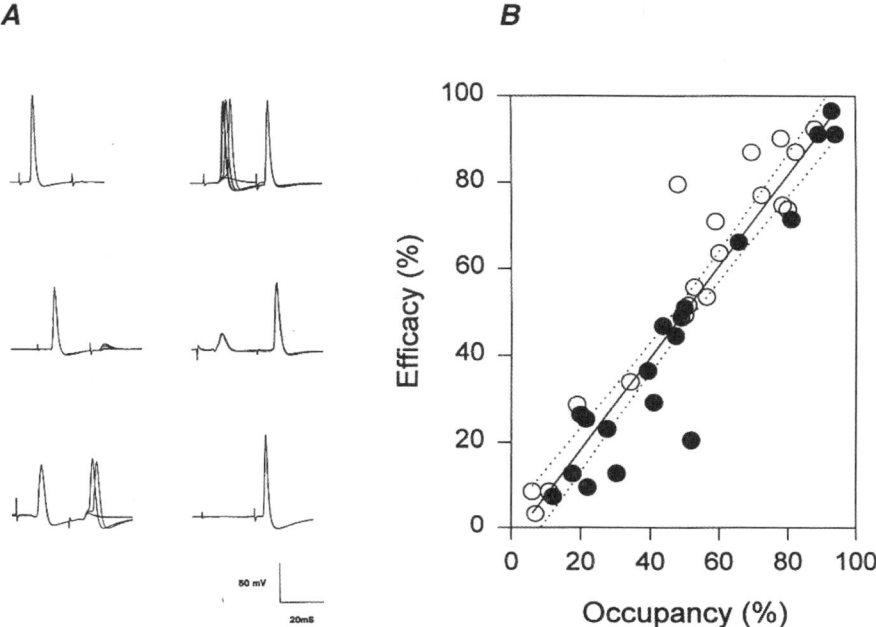

Figure 19 Persistent polyneuronal innervation in reinnervated rat lumbrical muscles 4–8 weeks after recovery from chronic nerve conduction block. **A**: Examples of monoinnervated (top left and bottom right traces) and polyneuronally innervated junctions. **B**: Relationship between relative synaptic strength (endplate potential amplitude) and fractional occupancy of polyneuronally innervated endplates. The strong linear correlation indicates that neither small nor large inputs are disproportionately weak at these stable polyinnervated junctions. The persistence of stable polyinnervated junctions for weeks, months and years after resumption of activity indicates that the presence of activity is not sufficient to bring about synapse elimination. (From Barry & Ribchester, 1995; and Costanzo *et al.*, 1999.)

with descriptions of physiological and morphological properties of synapses undergoing elimination in neonates (Colman *et al.*, 1997), have led to the idea that there is an inevitability to the elimination of small, relatively ineffective synaptic boutons at motor endplates, and that this is inexorably driven by differences in the activity of the majority over the minority inputs to the endplate (Frank, 1997; Jennings, 1994).

It is not clear how the idea of a decisive role of activity during synapse elimination favoured by Lichtman and colleagues will be reconciled with observations of persistent polyneuronal innervation in partially denervated and reinnervated adult muscle, or conversely the elimination of synapses from paralysed muscle fibres (Ribchester, 1993; Barry & Ribchester, 1995; Costanzo *et al.*, 1999, 2000). One testable hypothesis is that paralysed muscle releases neurotrophic factors to which regenerating axons are more responsive than intact axons and their terminals, by virtue of enhanced expression of the appropriate receptors in their presynaptic membranes. This may enable the levels of

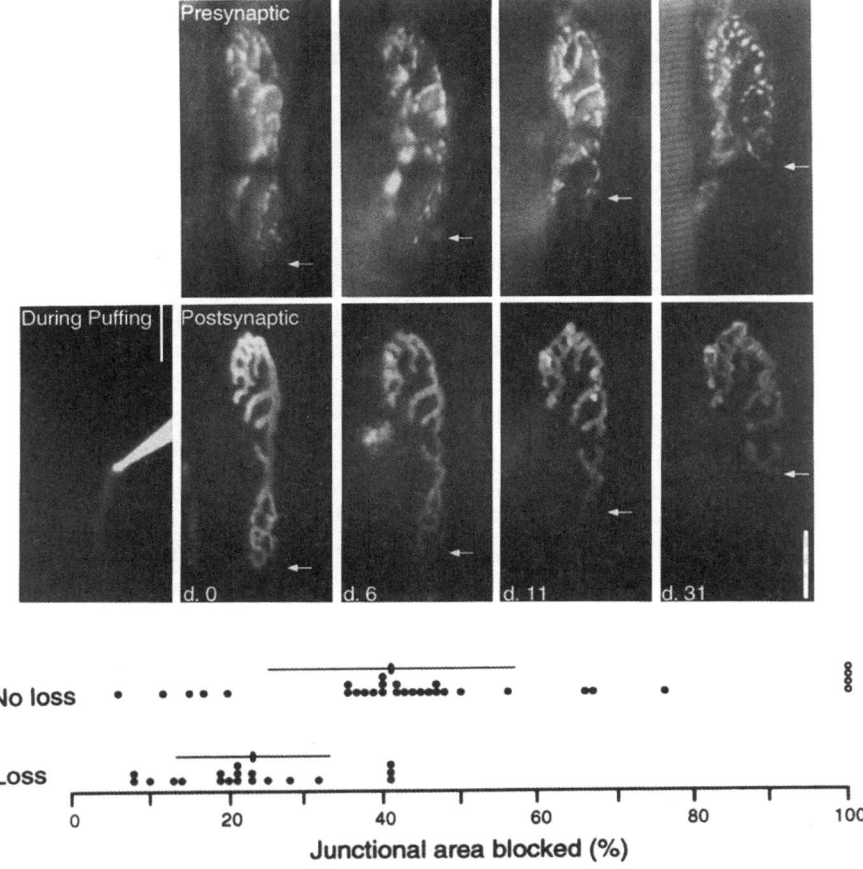

Figure 21 Elimination of synaptic boutons in an adult mouse neuromuscular junction following focal blockade of a small fraction of the endplate by local superfusioin with α-bungarotoxin. The endplate has been repeatedly visualized *in vivo* over the course of a month using vital fluorescent dyes. Both receptors and overlying synaptic boutons are eliminated from the inactivated area. Graph below summarizes data from junctions with different extents of focal nerve block. Elimination occurred consistently only from regions where less than 30% of the endplate was blocked. Calibration: 20 μm. (From Balice-Gordon & Lichtman, 1994, with permission.)

neurotrophic factor to promote the consolidation of the regenerating over the intact or sprouted terminal more effectively, whilst the activity of the terminal *per se* need not play a decisive role. Recent evidence supports the hypothesis that regenerating terminals are more responsive than intact terminals to the neurotrophic cytokines CNTF and GDNF (Bernstein *et al.*, 1998; Ulenkate *et al.*, 1994). Enhanced release of proteases is a further possibility, although this would have to be highly selective, since components of the extracellular matrix are not altered prior to synapse elimination (Culican *et al.*, 1998; Rich & Lichtman, 1989a).

3.3.4. Summary

Polyneuronal innervation recurs after complete or partial nerve injury in adult mammals and in several respects this resembles the developmental phenomenon (Section 3.2). In contrast to development, however, regenerating axons normally reinnervate preformed synaptic sites; and synapse elimination does not always completely remove all the synaptic boutons from supernumerary motor axons converging on the endplate. With respect to which boutons persist or are eliminated, activity appears to be a mitigating factor, rather than a decisive one. Further experiments are required to establish the molecular and biophysical characteristics of polyneuronal versus mononeuronally innervated muscle fibres; and whether different molecular mechanisms are responsible for synapse elimination in neonatal and reinnervated adult muscles.

3.4. A technical note on evaluating polyneuronal innervation

All methods for estimating the amounts of polyneuronal innervation in muscle are individually flawed. Isometric tension measurements may overestimate levels of polyneuronal innervation owing to non-saturation of muscle series compliance (Brown & Matthews, 1962); or underestimate it owing to the ineffectiveness of subthreshold inputs (Fladby, 1987). Intracellular recording is a reasonably reliable method for distinguishing poly- from mononeuronally innervated fibres. However, this method also has pitfalls; for example, sampling errors due to clustering of π-junctions. In addition, evoked EPPs vary in amplitude with repeated stimulation at the same suprathreshold level of excitation, owing to random variation in the number of synaptic vesicles undergoing exocytosis from stimulus to stimulus. This random variability becomes worse when mean quantal content is reduced (Boyd & Martin, 1956), and this can make it difficult to assess whether a muscle fibre receives weak input from one or from several axons (see for example the recording from a *mononeuronally* innervated (μ) junction in Figure 5B). Moreover, if a weak input is contributed by an axon with an electrical threshold higher than a strong input, it may not be easily detected when the stimulus strength is maximal; and non-linear summation of synaptic potentials exacerbates this problem. Selective stimulation of separate motor nerve supplies to a muscle, for example as in the fourth deep lumbrical muscle (Betz *et al.*, 1979), overcomes this problem to some extent. However, in the limit, only stimulation of all the motor axons individually – by splitting ventral root filaments – will give the most definitive measurements of polyneuronal innervation (Brown *et al.*, 1976). On the other hand, such preparations are technically difficult to make – especially in neonates – and damage to one or more ventral root filaments will therefore lead to underestimates of polyneuronal innervation. Anatomical methods – based on silver staining or immunocytochemistry – give no clue as to the origin and physiological effectiveness of a visible input (O'Brien *et al.*, 1978). Selective staining of convergent terminals with different carbocyanine dyes can be achieved in neonates, using fixed preparations (Balice-Gordon *et al.*, 1993; Gan & Lichtman, 1998; Laskowski *et al.*, 1998); but these methods do not work well on adult preparations. It is possible to unequivocally identify polyneuronally innervated junctions in adult muscles using selective vital staining with styryl dyes that stain recycling synaptic vesicles in an activity-dependent fashion (Barry & Ribchester, 1995; Betz *et al.*, 1992; Costanzo *et al.*, 1999, 2000; Lichtman *et al.*, 1985; Lichtman & Wilkinson, 1987). However, for reasons which are not understood, these dyes do not work very well on neonatal mammalian preparations (although they are effective on synapses between neurons in culture; see for example Forti *et al.*, 1997). Moreover, they rely on maintained transmission during prolonged stimulation to ensure dye uptake.

The important point is that, if conclusions are to be made with confidence about the pattern of innervation of muscle, multiple methods should always be applied, particularly when attempting to interpret the outcome of experiments designed to probe mechanisms of synapse elimination (Ribchester, 1988c).

4. CONCLUSIONS AND DISCUSSION

Three main points summarizing knowledge and understanding of the role played by activity in formation, maintenance and repair of neuromuscular connections can be made. Maintenance of form and function has been considered first, because issues raised there provide a context for the relationships between form and function during postnatal development, or after nerve injury in adults.

1. Once neuromuscular synapses have formed and the adult pattern of innervation is established, mature terminals continue to grow throughout life, but the general form of a neuromuscular junction does not change substantially. The size, shape and strength of neuromuscular connections can be altered as a consequence of extreme use or disuse, however. Thus, both muscle paralysis and muscle stimulation or excessive exercise increase the amount of transmitter release and the size of nerve terminals, and the changes in synaptic strength (efficacy; quantal content) precede any morphological changes, suggesting they are regulated independently. Some of the changes may be a consequence of mechanical distortion, mediated by tension-sensing receptors (such as the integrins) in nerve terminal membranes. Changes brought on by excessive use tend to result in uniform expansion of terminals, matching the effects of activity to increases in muscle fibre diameter. Such changes may be adaptive, controlled by negative feedback to ensure that muscle fibres reliably respond to presynaptic action potentials. But militating against this is the high safety factor for neuromuscular transmission that normally prevails, together with evidence that activity is not necessary for boosting transmitter release or synaptic area. Changes brought on by paralysis normally do so by inducing terminal sprouting, disrupting the normal size–strength relationship. This may be mediated by trophic factors released by inactive muscle or Schwann cells; or by the re-expression of adhesion molecules in the muscle fibre membranes or extracellular matrix.

2. Following initial synapse formation during fetal development, muscle fibres acquire inputs from several motoneurons. Postnatally, all but one of these inputs is eliminated, synaptic bouton-by-bouton in piecemeal fashion. This synapse elimination is not due to degeneration or death of entire motoneurons, but rather to withdrawal of a significant fraction of each motoneuron's

intramuscular nerve branches, producing a fall in motor unit size. Synapse elimination appears to be driven by both competitive and non-competitive mechanisms. Though coincident with the onset of activity and selective use of muscles, there are activity-dependent and activity-independent components to both competitive elimination and non-competitive synapse withdrawal. The molecular mechanisms are unknown. A number of molecules expressed at the neuromuscular junction (such as the adult form of the acetylcholine receptor) have been ruled out as mediators of synapse elimination, but there is appealing evidence that neurotrophic factors and their receptors regulate its rate and outcome. There is growing evidence that terminal Schwann cells and glial growth factors (neuregulins) may also play a role. Sequestration of the detached axon and its retraction bulb occurs quite rapidly, possibly as a consequence of the activity of motor proteins in the axon which might gener- ate significant intra-neuritic traction. One view is that synapse elimination is fundamentally a consequence of breakdown in cell–cell adhesion. The search for evidence for a specific role of known adhesion molecules has not yet produced a compelling outcome. However, proteases and their inhibitors have been implicated and these could play a pivotal role in regulating adhe- sion or intra-neuritic tension and withdrawal during synapse elimination.

3. After nerve injury, distal axons and their terminals normally degenerate rapidly by an active mechanism that is still poorly understood. Intact axons and terminals and Schwann cells that remain in a muscle react to the presence of denervated fibres and sprout, providing compensatory, collateral reinnervation. There is compelling evidence that motor nerve sprouts, and regenerating axons, are guided passively by Schwann cell sprouts that form bridges between intact and denervated endplates. Regenerating axons often polyneuronally innervate muscle fibres, as in development. Subsequently many, but not all, of the polyneuronal inputs are removed, via activity- dependent and activity-independent mechanisms that may resemble those controlling developmental synapse elimination. The molecular mechanisms are unknown, but sprouting can be induced by paralysis or by neurotrophic molecules, and inhibited by antibodies against neural cell adhesion mole- cules. The inputs that remain at persistently polyneuronally innervated junctions have equivalent synaptic strengths that are scaled up or down together; and it is not necessary for an input to occupy large fractions of the endplate, or to generate suprathreshold synaptic responses in order to become stable. Thus activity is neither sufficient nor necessary to promote synapse elimination at all neuromuscular junctions in reinnervated muscles.

A conventional view is that polyneuronal innervation of mammalian muscle fibres is an inherently unstable state (Frank, 1997; Jennings, 1994). Disparities in the relative size, synaptic strength and/or activity are presumed in these models to lead inexorably and inevitably to the withdrawal of the weaker competitors until only one axon terminal remains. However, the number of instances in which muscle fibres remain polyneuronally innervated for considerable periods, together with evidence for intrinsic and activity-independent synapse elimination, suggests that this model is oversimplified. Muscle fibres can sustain a persistent polyneuronal innervation in a number of different circumstances, indicating, *ipso facto*, that polyneuronal innervation is not an inherently unstable state. Figure 22 summarizes the possible relationships between different stable and unstable polyneuronally innervated states. According to this view, stable mononeuronal and polyneuronal innervation represent two alternative endpoints: low-energy states in which the synapses all have equivalent synaptic strengths. These alternative endpoints can be reached from an unstable starting condition via low-threshold switches that carry synapses along one or other of the two trajectories to their stable endpoints. Which switch predominates under any given set of circumstances is unknown; but we might view development as a circumstance in which the trajectory leading to mononeuronal innervation is preferrred, while the other trajectory is preferred following reinnervation of a paralysed adult muscle, for example. According to this model it should be possible for a neuro-muscular junction to leave a stable polyinnervated state and become mono-innervated, provided sufficient energy (trophic factor?) is injected into the system, to fuel the transition from one of the low-energy stable states to the high-energy state represented by unstable polyinnervation.

A number of questions for future investigations arise (Figure 23). Do converging terminals compete for access to space, or consumption of substances localized at endplates? Are there unique molecules, or biophysical properties associated with stable monoinnervated versus polyinnervated junctions? Are increases in the size and strength of synaptic connection – induced by either

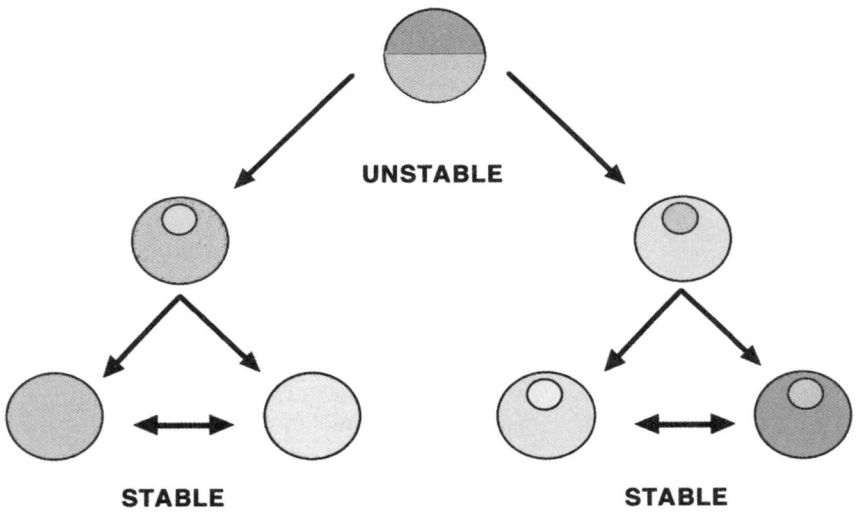

UNSTABLE

STABLE   **STABLE**

Figure 22 Alternative trajectories from an initially unstable pattern of innervation, based on data on the size, strength and pattern of innervation of immature and adult neuromuscular junctions. Normally, polyneuronal innervation is eliminated during development, leaving monoinnervated junctions in which synaptic strength can be up- or down-regulated by use or disuse. Sometimes, after recovery from paralysis for example, muscle fibres sustain a persistent, stable polyinnervation in which the strengths of converging inputs are also co-regulated. (From Costanzo *et al.*, 1999.)

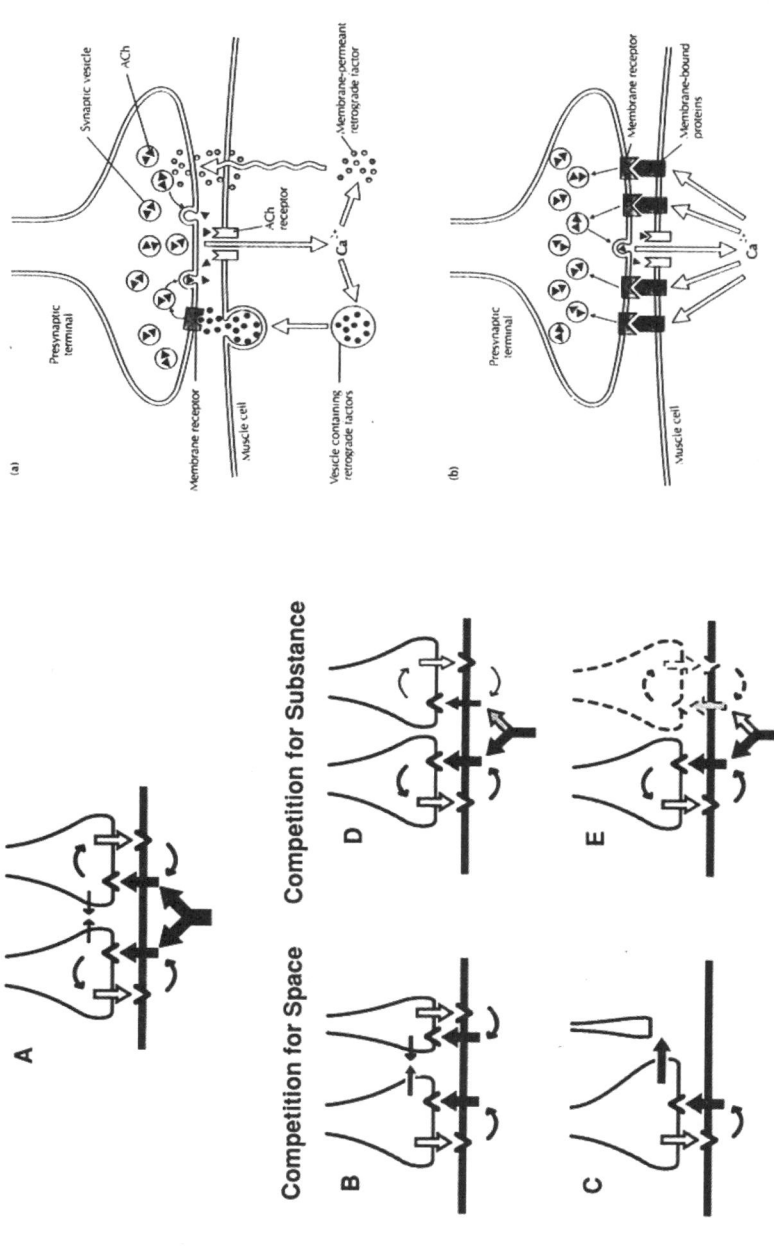

Figure 23 Cartoons illustrating left: possible modes of competition, either direct or indirectly mediated, and alternatively for substance or space at motor endplates. (From Van Essen *et al.*, 1990, with permission.) Right: possible molecular mechanisms of feedback between muscle fibres and motor nerve terminals that could regulate their structure and function. (From Dan & Poo, 1994, with permission.)

activity or inactivity – mediated by common signalling pathways? How might the strengths of synaptic boutons supplied by the same or different axons be co-regulated? How important are cell–cell recognition, orthograde myotrophic factors, retrograde neurotrophic factors, and local neuroglial factors in formation, maintenance and plasticity of neuromuscular synapses? Does "intrinsic withdrawal" of axon collaterals result from perturbed cell–cell interactions, or to selective trafficking of materials (nucleic acid and/or protein) in motor axon collaterals?

Whilst the diversity of stable and unstable states of muscle fibre innervation suggested here may initially surprise, given the paucity of polyneuronal innervation in most normal adult muscles, it should not be considered astonishing when we consider that most other kinds of muscle – including for example the intrafusal muscle fibres of skeletal muscle – and virtually all neurons elsewhere in the nervous system, receive convergent innervation from different sources throughout life. Interestingly, several recent studies have questioned the adequacy of the Hebbian synaptic model (the requirement for coincident presynaptic and postsynaptic activity) to account for homosynaptic potentiation, heterosynaptic depression or the fine-tuning of cortical circuitry in the central nervous system (Crair et al., 1998; Crowley & Katz, 1999; Engert & Bonhoeffer, 1997; Fitzsimonds et al., 1997; Zhang et al., 1998). Taken together, such observations vindicate the choice of the mammalian neuromuscular junctions as a paradigm for investigating general mechanisms of synaptic plasticity, beyond their mere accessibility to experimental investigation. Ultimately, analysis of neuromuscular synaptic development and plasticity will doubtless have significant impact on our understanding of the control of neuronal convergence and divergence, not only in skeletal muscle but elsewhere in the nervous system as well.

ACKNOWLEDGEMENTS

I thank all my colleagues and friends, past and present, for their influence, valuable discussions and insights; especially: John Harris, Clarke Slater, Bill Betz, John Caldwell, Bob Martin, Jan Jansen, Torfinn Taxt, Wes Thompson, Richard Morris, Simon Parson, Jacki Barry, and Ellen Costanzo. Derek Thomson has contributed outstanding technical assistance in our recent work. Research in my laboratory has been possible thanks to grants from the Medical Research Council, the Wellcome Trust, Action Research and the Royal Society.

References

Aidley, D.J. (1988). The physiology of excitable cells (4th ed.) Cambridge: Cambridge University Press.
Akaaboune, M., Culican, S.M., Turney, S.G. & Lichtman, J.W. (1999). Rapid and reversible effects of activity on acetylcholine receptor density at the neuromuscular junction in vivo. Science, **286**, 503–507.

Akaaboune, M., Hantai, D., Smirnova, I., Lachkar, S., Verdieresahuque, M. & Festoff, B.W. (1998). Developmental regulation of the serpin, protease nexin I: localization during activity-dependent polyneuronal synapse elimination in mouse skeletal muscle. *Journal of Comparative Neurology*, **397**, 572–579.

Allan, D.W. & Greer, J. (1998). Polysialylated ncam expression during motor axon outgrowth and myogenesis in the fetal rat. *Journal of Comparative Neurology*, **391**, 275–292.

Angleson, J.K. & Betz, W.J. (1997). Monitoring secretion in real time: capacitance, amperometry and fluorescence compared. *Trends in Neurosciences*, 20, 281–287.

Armand, J., Olivier, E., Edgley, S.A. & Lemon, R.N. (1997). Postnatal development of corticospinal projections from motor cortex to the cervical enlargment in the macaque monkey. *Journal of Neuroscience*, **17**, 251–266.

Astrow, S.H., Tyner, T.R., Nguyen, M.T.T. & Ko, C.P. (1997). Schwann cell matrix component of neuromuscular junctions and peripheral nerves. *Journal of Neurocytology*, **26**, 63–75.

Attwell, D. & Iles, J.F. (1979). Synaptic transmission: ion concentration changes in the synaptic cleft. *Proceedings of the Royal Society of London, Series B: Biological Sciences*, **206**, 115–131.

Bagust, J., Lewis, D.M. & Westerman, R.A. (1973). Polyneuronal innervation of kitten skeletal muscle. *Journal of Physiology, London*, **229**, 241–255.

Bakels, R. & Kernell, D. (1993). Matching between motoneuron and muscle unit properties in rat medial gastrocnemius. *Journal of Physiology, London*, **463**, 307–324.

Balice-Gordon, R.J. & Lichtman, J.W. (1990). *In vivo* visualization of the growth of presynaptic and post-synaptic elements of neuromuscular junctions in the mouse. *Journal of Neuroscience*, **10**, 894–908.

Balice-Gordon, R.J. & Lichtman, J.W. (1993). *In vivo* observations of presynaptic and postsynaptic changes during the transition from multiple to single innervation at developing neuromuscular junctions. *Journal of Neuroscience*, **13**, 834–855.

Balice-Gordon, R.J. & Lichtman, J.W. (1994). Long-term synapse less induced by focal blockade of postsynaptic receptors. *Nature*, **372**, 519–524.

Balice-Gordon, R.J. & Thompson, W.J. (1988a). The organization and development of compartmentalized innervation in rat extensor digitorum longus muscle. *Journal of Physiology, London*, **398**, 211–231.

Balice-Gordon, R.J. & Thompson, W.J. (1988b). Synaptic rearrangements and alterations in motor unit properties in neonatal rat extensor digitorum longus muscle. *Journal of Physiology, London*, **398**, 191–210.

Balice-Gordon, R.J., Breedlove, S.M., Bernstein, S. & Lichtman, J.W. (1990). Neuromuscular junctions shrink and expand as muscle fiber size is manipulated – *in vivo* observations in the androgen-sensitive bulbocavernosus muscle of mice. *Journal of Neuroscience*, **19**, 2660–2671.

Balice-Gordon, R.J., Chua, C.K., Nelson, C.C. & Lichtman, J.W. (1993). Gradual loss of synaptic cartels precedes axon withdrawal at developing neuromuscular junctions. *Neuron*, **11**, 801–805.

Bambrick, L. & Gordon, T. (1987). Acetylcholine receptors and sodium channels in denervated and botulinum-toxin-treated adult rat muscle. *Journal of Physiology, London*, **382**, 69–86.

Barker, D. & Ip, M.C. (1966). Sprouting and degeneration of mammalian motor axons in normal and de-afferentated skeletal muscle. *Proceedings of the Royal Society of London, Series B: Biological Sciences*, **163**, 538–554.

Barry, J.A. & Ribchester, R.R. (1995). Persistent polyneuronal innervation in partially denervated rat muscles after reinnervation and recovery from prolonged nerve-conduction block. *Journal of Neuroscience*, **15**, 6327–6339.

Bennett, K.M.B. & Lemon, R.N. (1996). Corticomotoneuronal contribution to the fractionation of muscle activity during precision grip in the monkey. *Journal of Neurophysiology*, **75**, 1826–1842.

Bennett, M.R. & Pettigrew, A.G. (1974). The formation of synapses in striated muscle during development. *Journal of Physiology, London*, **241**, 515–545.

Bennett, M.R. & Robinson, J. (1989). Growth and elimination of nerve terminals at synaptic sites during polyneuronal innervation of uscle cells – a trophic hypothesis. *Proceedings of the Royal Society of London, Series B: Biological Sciences*, **235**, 299–320.

Bennett, M., Ho, S. & Lavidis, N. (1986). Competition between segmental nerves at end-plates in rat gastrocnemius muscle during loss of polyneuronal innervation. *Journal of Physiology, London*, **381**, 351–376.

Benoit, P. & Changeux, J.P. (1975). Consequences of tenotomy on the evolution of multiinnervation in developing rat soleus muscle. *Brain Research*, **99**, 354–358.

Bernstein, M.L., Parsadanian, A.S., Keller-Peck, C.R., Snider, W.D. & Lichtman, J.W. (1998). Axotomy induces both gdnf sensitivity in regenerating axons and gdnf expression in schwann cells. *Society for Neuroscience Abstracts*, **24**, 413.3.

Betz, W.J. & Bewick, G.S. (1992). Optical analysis of synaptic vesicle recycling at the frog neuromuscular junction. *Science*, **255**, 200–203.

Betz, W. & Sakmann, B. (1973). Effects of proteolytic enzymes on function and structure of frog neuromuscular junctions. *Journal of Physiology, London*, **230**, 673–688.

Betz, W.J., Caldwell, J.H. & Kinnamon, S.C. (1984). Physiological basis of a steady endogenous current in rat lumbrical muscle. *Journal of General Physiology*, **83**, 175–192.

Betz, W.J., Caldwell, J.H. & Ribchester, R.R. (1979). The size of motor units during post-natal development of rat lumbrical muscle. *Journal of Physiology, London*, **297**, 463–478.

Betz, W.J., Caldwell, J.H. & Ribchester, R.R. (1980a). The effects of partial denervation at birth on the development of muscle fibres and motor units in rat lumbrical muscle. *Journal of Physiology, London*, **303**, 265–279.

Betz, W.J., Caldwell, J.H. & Ribchester, R.R. (1980b). Sprouting of active nerve terminals in partially inactive muscles of the rat. *Journal of Physiology, London*, **303**, 281–297.

Betz, W.J., Caldwell, J.H. Ribchester, R.R., Robinson, K.R. & Stump, R.F. (1980). Endogenous electric field around muscle fibres depends on the Na$^+$ K$^+$ pump. *Nature*, **287**, 235–237.

Betz, W.J., Chua, M. & Ridge, R.M.A.P. (1989). Inhibitory interactions between motoneuron terminals in neonatal rat lumbrical muscle. *Journal of Physiology, London*, **418**, 25–51.

Betz, W.J., Mao, F. & Bewick, G.S. (1992). Activity-dependent fluorescent staining and destaining of living vertebrate motor-nerve terminals. *Journal of Neuroscience*, **12**, 363–375.

Betz, W.J., Ridge, R.M.A.P. & Bewick, G.S. (1993). Comparison of FM1-43 staining patterns and electrophysiological measures of transmitter release at the frog neuromuscular-junction. *Journal of Physiology, Paris*, **87**, 193–202.

Bewick, G.S., Young, C. & Slater, C.R. (1996). Spatial relationships of utrophin, dystrophin, beta-dystroglycan and beta-spectrin to acetylcholine receptor clusters during postnatal maturation of the rat neuromuscular junction. *Journal of Neurocytology*, **25**, 367–379.

Bixby, J.L. (1981). Ultrastructural observations on synapse elimination in neonatal rabbit skeletal muscle. *Journal of Neurocytology*, **10**, 81–100.

Bixby, J.L. & Van Essen, D.C. (1979). Competition between foreign and original nerves in adult mammalian skeletal muscle. *Nature*, **282**, 726–728.

Blondet, B., Duxson, M.J., Harris, A.J., Melki, J., Guenet, J.L., Pinconraymond, M. & Rieger, F. (1989). Nerve and muscle development in paralyze mutant mice. *Developmental Biology*, **132**, 153–166.

Boeke, J. (1916). Studien zur nervenregeneration. i di regeneration der motorischen nervenelemente und die regeneration der nerven der muskelspindeln. *Verhandlingen Koninkonen Akadamie Wetenschaffen (Amsterdam)*, **18**, 1–120.

Boeke, J. (1921). The innervation of striped muscle fibres and Langley's receptive substance. *Brain*, **44**, 1–22.

Booth, C.M., Kemplay, S.K. & Brown, M.C. (1990). An antibody to neural cell-adhesion molecule impairs motor-nerve terminal sprouting in a mouse muscle locally paralyzed with botulinum toxin. *Neuroscience*, **35**, 85–91.

Boulganer, L. & Poo, M.-M. (1999). Presynaptic depolarization facilitates neurotrophin-induced synaptic potentiation. *Nature Neuroscience*, **2**, 346–351.

Boyd, I.A. & Martin, A.R. (1956). The end-plate potential in mammalian muscle. *Journal of Physiology, London*, **132**, 74–91.

Braga, M.F.M., Anderson, A.J., Harvey, A.L. & Rowane, E.G. (1992). Apparent block of k+ currents in mouse motor-nerve terminals by tetrodotoxin, mu-conotoxin and reduced external sodium. *British Journal of Pharmacology*, **106**, 91–94.

Bray, D. (1984). Axonal growth in response to experimentally applied mechanical tension. *Developmental Biology*, **102**, 379–389.

Bray, D. (1987). Growth cones – do they pull or are they pushed? *Trends in Neuroscience*, **10**, 431–434.

Brown, A.G. (1981). *Organization in the spinal cord*. Berlin: Springer Verlag.

Brown, M.C. & Booth, C.M. (1983). Segregation of motor nerves on a segemental basis during synapse elimination in neonatal muscles. *Brain Research*, **273**, 188–190.

Brown, M.C. & Ironton, R. (1978). Sprouting and regression of neuromuscular synapses in partially denervated mammalian muscles. *Journal of Physiology, London*, **278**, 325–348.

Brown, M.C. & Matthews, P.B.C. (1962). An investigation into the possible existence of polyneuronal innervation of individual skeletal muscle fibres in certain hind-limb muscles of the cat. *Journal of Physiology, London*, **151**, 436–457.

Brown, M.C., Holland, R.L. & Hopkins, W.G. (1981a). Motor nerve sprouting. *Annual Reviews of Neuroscience*, **4**, 17–42.

Brown, M.C., Holland, R.L. & Hopkins, W.G. (1981b). Restoration of focal multiple innervation in rat muscles by transmission block during a critical stage of development. *Journal of Physiology, London*, 318, 355–364.

Brown, M.C., Holland, R.L. & Ironton, R. (1980). Nodal and terminal sprouting from motor nerves in fast and slow muscles of the mouse. *Journal of Physiology, London*, **306**, 4493–4510.

Brown, M.C., Holland, R.L. & Ironton, R. (1978). Degenerating nerve products affect innervated muscle fibres. *Nature*, **275**, 652–654.

Brown, M.C., Hopkins, W.G. & Keynes, R.J. (1982a). Comparison of effects of denervation and botulinum toxin paralysis on muscle properties in mice. *Journal of Physiology, London*, **327**, 29–37.

Brown, M.C., Hopkins, W.G. & Keynes, R.J. (1982b). Short-term and long-term effects of paralysis on the motor innervation of 2 different neonatal mouse muscles. *Journal of Physiology, London*, **329**, 439–?.

Brown, M.C., Jansen, J.K.S. & Van Essen, D. (1976). Polyneuronal innervation of skeletal muscle in newborn rats and its elimination during maturation. *Journal of Physiology, London*, **261**, 387–422.

Burnstock, G. & Hoyle, C.H.V. (1992). *Autonomic neuroeffector mechanisms*. Philadelphia, PA: Harwood Academic Publishers.

Busetto, G., Buffelli, M., Tognana, E., Bellico, F. & Cangiano, A. (2000). Hebbian mechanisms revealed by electrical stimulation at developing rat neuromuscular junctions. *Journal of Neuroscience*, **20**, 685–695.

Caldwell, J.H. & Ridge, R.M.A.P. (1983). The effects of deafferentation and spinal-cord transection on synapse elimination in developing rat muscles. *Journal of Physiology, London*, **339**, 145–159.

Callaway, E.M. & Van Essen, D.C. (1989). Slowing of synapse elimination by alpha-bungarotoxin superfusion of the neonatal rabbit soleus muscle. *Developmental Biology*, **131**, 356–365.

Callaway, E.M., Soha, J.M. & Van Essen, D.C. (1987). Competition favoring inactive over active motor neurons during synapse elimination. *Nature*, **328**, 422–426.

Caroni, P. (1997). Intrinsic neuronal determinants that promote axonal sprouting and elongation. *Bioessays*, **19**, 767–775.

Caroni, P., Schneider, C., Kiefer, M.C. & Zapf, J. (1994). Role of muscle insulin-like growth-factors in nerve sprouting – suppression of terminal sprouting in paralyzed muscle by IGF-binding protein-4. *Journal of Cell Biology*, **125**, 893–902.

Cashman, N.R., Covault, J., Wollman, R.L. & Sanes, J.R. (1987). Neural cell-adhesion molecule in normal denervated, and myopathic human-muscle. *Annals of Neurology*, **21**, 481–489.

Chamberlain, S. & Ridge, R.M.A.P. (1989). Adult rat lumbrical muscle fibers are not polyneuronally innervated. *Quarterly Journal of Experimental Physiology*, **74**, 375–378.

Chang, Q. & Balice-Gordon, R.J. (1997). Nip and tuck at the neuromuscular junction: a role for proteases in developmental synapse elimination. *Bioessays*, **19**, 271–275.

Chao, D.S., Silvagno, F., Xia, H., Cornwell, T.L., Lincoln, T.M. & Bredt, D.S. (1997). Nitric oxide synthase and cyclic gmp-dependent protein kinase concentrated at the neuromuscular endplate. *Neuroscience*, **76**, 665–672.

Chen, B.M. & Grinnell, A.D. (1995). Integrins and modulation of transmitter release from motor-nerve terminals by stretch. *Science*, **269**, 1578–1580.

Chen, B.M. & Grinnell, A.D. (1997). Kinetics, Ca^{2+} dependence, and biophysical properties of integrin-mediated mechanical modulation of transmitter release from frog motor nerve terminals. *Journal of Neuroscience*, **17**, 904–916.

Chen, L.L. & Ko, C.P. (1994). Extension of synaptic extracellular matrix during nerve terminal sprouting in living frog neuromuscular-junctions. *Journal of Neuroscience*, **14**, 796–808.

Cho, M., Webster, S.G. & Blau, H.M. (1993). Evidence for myoblast extrinsic regulation of slow myosin heavy-chain expression during muscle fiber formation in embryonic development. *Journal of Cell Biology*, **121**, 795–810.

Christ, B., Schmidt, C., Huang, R.J., Wiltshire, J. & Brandsaberi, B. (1998). Segmentation of the vertebrate body. *Anatomy and Embryology*, **197**, 1–8.

Clowry, G.J. & Vrbova, G. (1991). Embryonic motoneurons grafted into the spinal cord of an adult rat can innervate a muscle. *Restorative Neurology and Neuroscience*, **2**, 299–302.

Cochilla, A.J., Angleson, J.K. & Betz, W.J. (1999). Monitoring secretory membrane with fm1-43 fluorescence. *Annual Review of Neuroscience*, **22**, 1–10.

Cohen, M.W. (1972). The development of neuromuscular connexions in the presence of d-tubocurarine. *Brain Research*, **41**, 457–463.

Colman, H., Nabekura, J. & Lichtman, J.W. (1997). Alterations in synaptic strength preceding axon withdrawal. *Science*, **275**, 356–361.

Condon, K., Silberstein, L., Blau, H.M. & Thompson, W.J. (1990a). Differentiation of fiber types in aneural musculature of the prenatal rat hindlimb. *Developmental Biology*, **138**, 275–295.

Condon, K., Silberstein, L., Blau, H.M. & Thompson, W.J. (1990b). Development of muscle fiber types in the prenatal rat hindlimb. *Developmental Biology*, **138**, 256–274.

Connold, A.L. & Vrbova, G. (1994). Neuromuscular contacts of expanded motor units in rat soleus muscles are rescued by leupeptin. *Neuroscience,* **63**, 327–338.

Connold, A.L., Everse, J.V. & Vrbova, G. (1986). Effect of low calcium and protease inhibitors on synapse elimination during postnatal-development in the rat soleus muscle. *Developmental Brain Research,* **28**, 99–107.

Costanzo, E.M., Barry, J.A. & Ribchester, R.R. (1999). Co-regulation of synaptic efficacy and occupancy at stable polyneuronally innervated neuromuscular junctions in reinnervated rat muscle. *Journal of Physiology*, **521**, 365–374.

Costanzo, E.M., Barry, J.A. & Ribchester, R.R. (2000). Competition at silent synapse in reinnervated muscle. *Nature Neuroscience*, **3**, 694–700.

Covault, J. & Sanes, J.R. (1986). Distribution of N-CAM in synaptic and extrasynaptic portions of developing and adult skeletal muscle. *Journal of Cell Biology*, **102**, 716–730.

Crair, M.C., Gillespie, D.C. & Stryker, M.P. (1998). The role of visual experience in the development of columns in cat visual cortex. *Science*, **279**, 566–570.

Crick, F. (1982). Do dendritic spines twitch? *Trends in Neuroscience*, **5**, 44–46.

Crockett, D.P., Harris, S.L. & Egger, M.D. (1987). Plantar motoneuron columns in the rat. *Journal of Comparative Neurology*, **265**, 109–118.

Crowley, J.C. & Katz, L.C. (1999). Development of ocular dominance columns in the absence of retinal input. *Nature Neuroscience*, **2**, 1125–1130.

Culican, S.M., Nelson, C.C. & Lichtman, J.W. (1998). Axon withdrawal during synapse elimination at the neuromuscular junction is accompanied by disassembly of the postsynaptic specialization and withdrawal of Schwann cell processes. *Journal of Neuroscience*, **18**, 4953–4965.

Dai, Z.S. & Peng, H.B. (1998). Role of tyrosine phosphatase in acetylcholine receptor cluster dispersal and formation. *Journal of Cell Biology*, **141**, 1613–1624.

Dan, Y. & Poo, M.M. (1992). Hebbian depression of isolated neuromuscular synapses *in vitro*. *Science*, **256**, 1570–1573.

Dan, Y. & Poo, M.M. (1994). Retrograde interactions during formation and elimination of neuromuscular synapses. *Current Biology*, **4**, 95–100.

Davis, G.W. & Goodman, C.S. (1998). Synapse-specific control of synaptic efficacy at the terminals of a single neuron. *Nature*, **392**, 82–86.

Dennis, M.J. & Harris, A.J. (1980). Transient inability of neonatal motoneurones to reinnervate muscle. *Developmental Biology*, **74**, 173–183.

Dennis, M.J., Ziskindconhaim, L. & Harris, A.J. (1981). Development of neuromuscular junctions in rat embryos. *Developmental Biology*, **81**, 266–279.

Desaki, J. & Uehara, Y. (1981). The overall morphology of neuromuscular junctions as revealed by scanning electron microscopy. *Journal of Neurocytology*, **10**, 101–110.

Desaki, J. & Uehara, Y. (1987). Formation and maturation of subneural apparatuses at neuromuscular junctions in postnatal rats – a scanning and transmission electron microscopic study. *Developmental Biology*, **119**, 390–401.

Deschenes, M.R., Maresh, C.M., Crivello, J.F., Armstrong, L.E., Kraemer, W.J. & Covault, J. (1993). The effects of exercise training of different intensities on neuromuscular-junction morphology. *Journal of Neurocytology*, **22**, 603–615.

Dhoot, G.K. (1992). Neural regulation of differentiation of rat skeletal muscle cell types. *Histochemistry*, **97**, 479–486.

Ding, R., Jansen, J.K.S., Laing, N.G. & Tonnesen, H. (1983). The innervation of skeletal muscles in chickens curarized during early development. *Journal of Neurocytology*, **12**, 887–919.

Donahue, S.P. & English, A.W. (1987). The role of synapse elimination in the establishment of neuromuscular compartments. *Developmental Biology*, **124**, 481–489.

Dorlochter, M., Irintchev, A., Brinkers, M. & Wernig, A. (1991). Effects of enhanced activity on synaptic transmission in mouse extensor digitorum longus muscle. *Journal of Physiology, London*, **436**, 283–292.

Duxson, M.J. (1982). The effect of postsynaptic block on development of the neuromuscular-junction in postnatal rats. *Journal of Neurocytology*, **11**, 395–408.

Duxson, M.J. (1992). The relationship of nerve to myoblasts and newly-formed secondary myotubes in the 4th lumbrical muscle of the rat fetus. *Journal of Neurocytology*, **21**, 574–588.

Duxson, M.J. & Sheard, P.W. (1995). Formation of new myotubes occurs exclusively at the multiple innervation zones of an embryonic large muscle. *Developmental Dynamics*, **204**, 391–405.

Duxson, M.J. & Vrbova, G. (1985). Inhibition of acetylcholinesterase accelerates axon terminal withdrawal at the developing rat neuromuscular junction. *Journal of Neurocytology*, **14**, 337–363.

Duxson, M.J., Ross, J.J. & Harris, A.J. (1986). Transfer of differentiated synaptic terminals from primary myotubes to newly-formed muscle cells during embryonic development in the rat. *Neuroscience Letters*, **71**, 147–152.

Duxson, M.J., Usson, Y. & Harris, A.J. (1989). The origin of secondary myotubes in mammalian skeletal muscles – ultrastructural studies. *Development*, **107**, 743–750.

Edds, M.V. (1953). Collateral nerve regeneration. *Quarterly Reviews in Biology*, **28**60–276.

Edgerton, V.R., Bodine-Fowler, S., Roy, R.R., Ishihara, A. & Hodgson, J.A. (1996). Neuromuscular adaptation. In L.B. Rowell & J.T. Shepherd (Eds.), *Handbook of physiology Section 12*, (pp. 54–88). New York, Oxford: Oxford University Press.

Eide, A.L., Jansen, J.K.S. & Ribchester, R.R. (1982). The effect of lesions in the neural crest on the formation of synaptic connections in the embryonic chick spinal cord. *Journal of Physiology, London*, **324**, 453–478.

Elmqvist, D. & Quastel, D.M.J. (1965). A quantitative study of endplate potentials in isolated human muscle. *Journal of Physiology, London*, **178**, 505–529.

Engert, F. & Bonhoeffer, T. (1997). Synapse specificity of long-term potentiation breaks down at short distances. *Nature*, **388**, 279–284.

English, A.W. & Schwartz, G. (1995). Both basic fibroblast growth factor and ciliary neurotrophic factor promote the retention of polyneuronal innervation of developing skeletal muscle fibers. *Developmental Biology*, **169**, 57–64.

Evers, J., Laser, M., Sun, Y.A., Xie, Z.P. & Poo, M.M. (1989). Studies of nerve–muscle interactions in *Xenopus* cell culture – analysis of early synaptic currents. *Journal of Neuroscience*, **9**, 1523–1539.

Fatt, P. & Katz, B. (1952). Spontaneous subthreshold activity at motor nerve endings. *Journal of Physiology, London*, **117**, 109–128.

Fawcett, J.W. & Housden, E. (1990). The effects of protease inhibitors on axon growth through astrocytes. *Development*, **109**, 59–66.

Fazeli, S., Wells, D.J., Hobbs, C. & Walsh, F.S. (1996). Altered secondary myogenesis in transgenic animals expressing the neural cell adhesion molecule under the control of a skeletal muscle alpha-actin promoter. *Journal of Cell Biology*, **135**, 241–251.

Fernandez-Chacon, R. & Sudhof, T.C. (1999). Genetics of synaptic vesicle function, toward the complete functional anatomy of an organelle. *Annual Review of Physiology*, **61**, 753–776.

Fischbach, G.D. & Rosen, K.M. (1997). ARIA, A neuromuscular junction neuregulin. *Annual Review of Neuroscience*, **20**, 429–458.

Fischer, M., Kaech, S., Knutti, D. & Matus, A. (1998). Rapid actin-based plasticity in dendritic spines. *Neuron*, **20**, 847–854.

Fitzsimonds, R.M. & Poo, M.M. (1998). Retrograde signaling in the development and modification of synapses. *Physiological Reviews*, **78**, 143–170.

Fitzsimonds, R.M., Song, H.J. & Poo, M.M. (1997). Propagation of activity-dependent synaptic depression in simple neural networks. *Nature*, **388**, 439–448.

Fladby, T. (1987). Postnatal loss of synaptic terminals in the normal mouse soleus muscle. *Acta Physiologica Scandinavica*, **129**, 229–238.

Fladby, T. & Jansen, J.K.S. (1987). Postnatal loss of synaptic terminals in the partially denervated mouse soleus muscle. *Acta Physiologica Scandinavica*, **129**, 239–246.

Fladby, T. & Jansen, J.K.S. (1988). Selective innervation of neonatal fast and slow muscle fibers before net loss of synaptic terminals in the mouse soleus muscle. *Acta Physiological Scandinavica*, **134**, 561–562.

Fladby, T. & Jansen, J.K.S. (1990). Development of homogeneous fast and slow motor units in the neonatal mouse soleus muscle. *Development*, **109**, 723–732.

Forti, L., Bossi, M., Bergamaschi, A., Villa, A. & Malgaroli, A. (1997). Loose-patch recordings of single quanta at individual hippocampal synapses. *Nature*, **388**, 874–878.

Frank, E. (1997). Synapse elimination: for nerves it's all or nothing. *Science*, **275**, 324–325.

Fu, S.Y. & Gordon, T. (1995). Contributing factors to poor functional recovery after delayed nerve repair – prolonged denervation. *Journal of Neuroscience*, **15**, 3886–3895.

Funakoshi, H., Belluardo, N., Arenas, E., Yamamoto, Y., Casabona, A., Persson, H. & Ibanez, C.F. (1995). Muscle-derived neurotrophin-4 as an activity-dependent trophic signal for adult motor-neurons. *Science*, **268**, 1495–1499.

Gabella, G. (1995). The structural relations between nerve fibers and muscle cells in the urinary bladder of the rat. *Journal of Neurocytology*, **24**, 159–187.

Gan, W.B. & Lichtman, J.W. (1998). Synaptic segregation at the developing neuromuscular junction. *Science*, **282**, 1508–1511.

Gan, W.-B., Bishop, D., Turney, S.G. & Lichtman, J.W. (1988). Observing naturally occurring synaptic withdrawal at developing neuromuscular junctions at high spatial and temporal resolution. *Society for Neuroscience Abstracts*, **24**, 111.12.

Gates, H.J. & Ridge, R.M.A.P. (1992). The importance of competition between motoneurons in developing rat muscle – effects of partial denervation at birth. *Journal of Physiology, London*, **445**, 457–472.

Gates, H.J., Ridge, R.M.A.P. & Rowlerson, A. (1991). Motor units of the 4th deep lumbrical muscle of the adult-rat – isometric contractions and fiber type compositions. *Journal of Physiology, London*, **443**, 193–215.

Gatesy, S.M. & English, A.W. (1993). Evidence for compartmental identity in the development of the rat lateral gastrocnemius muscle. *Developmental Dynamics*, **196**, 174–182.

Gautam, M., Noakes, P.G., Moscoso, L., Rupp, F., Scheller, R.H., Merlie, J.P. & Sanes, J.R. (1996). Defective neuromuscular synaptogenesis in agrin-deficient mutant mice. *Cell*, **85**, 525–535.

Gharakhanlou, R., Chadan, S. & Gardiner, P. (1999). Increased activity in the form of endurance training increases calcitonin gene-related peptide content in lumbar motoneuron cell bodies and in sciatic nerve in the rat. *Neuroscience*, **89**, 1229–1239.

Giacobini, G., Filogamo, G., Weber, M., Boquet, P. & Changeux, J.P. (1973). Effects of a snake alpha neurotoxin on the development of innervated skeletal muscles in chick embryo. *Proceedings of the National Academy of Sciences, USA*, **70**, 1708–1712.

Girod, R., Popov, S., Alder, J., Zheng, J.Q., Lohof, A. & Poo, M.M. (1995). Spontaneous quantal transmitter secretion from myocytes and fibroblasts – comparison with neuronal secretion. *Journal of Neuroscience*, **15**, 2826–2838.

Gordon H. & Van Essen, D.C. (1983). The relation of neuromuscular synapse elimination to spinal position of rabbit and rat soleus motoneurones. *Journal of Physiology, London*, **339**, 591–597.

Gordon, T., Yang, J.F., Ayer, K., Stein, R.B. & Tyreman, N. (1993). Recovery potential of muscle after partial denervation – a comparison between rats and humans. *Brain Research Bulletin*, **30**, 477–482.

Gorio, A., Carmignoto G., Finesso, M., Polato, P. & Nunzi, M.G. (1983). Muscle reinnervation 2. sprouting, synapse formation and repression. *Neuroscience*, **8**, 403–416.

Gouze, J.L., Lasry, J.M. & Changeux, J.P. (1983). Selective stabilization of muscle innervation during development: a mathematical model. *Biological Cybernetics*, **46**, 207–215.

Gramsbergen, A., Ijkemapaassen, J., Nikkels, P.G.J. & Haddersalgra, M. (1997). Regression of polyneural innervation in the human psoas muscle. *Early Human Development*, **49**, 49–61.

Greensmith, L. & Vrbova, G. (1996). Motoneurone survival: a functional approach. *Trends in Neurosciences*, **19**, 450–455.

Greensmith, L., Dick, J., Emanuel, A.O. & Vrbova, G. (1996). Induction of transmitter release at the neuromuscular junction prevents motoneuron death after axotomy in neonatal rats. *Neuroscience*, **71**, 213–220.

Grinnell, A.D. (1995). Dynamics of nerve–muscle interaction in developing and mature neuromuscular junctions. *Physiological Reviews*, **75**, 789–834.

Grow, W.A., Kendallwassmuth, E., Ulibarri, C. & Laskowski, M.B. (1995). Differential delay of reinnervating axons alters specificity in the rat serratus anterior muscle. *Journal of Neurobiology*, **26**, 553–562.

Grubb, B.D., Harris, J.B. & Schofield, I.S. (1991). Neuromuscular transmission at newly formed neuromuscular junctions in the regenerating soleus muscle of the rat. *Journal of Physiology, London*, **441**, 405–421.

Gurney, M.E., Yamamoto, H. & Kwon, Y. (1992). Induction of motor-neuron sprouting *in vivo* by ciliary neurotrophic factor and basic fibroblast growth factor. *Journal of Neuroscience*, **12**, 3241–3247.

Hamalainen, N. & Pette, D. (1996). Slow-to-fast transitions in myosin expression of rat soleus muscle by phasic high-frequency stimulation. *FEBS Letters*, **399**, 220–222.

Harish, O.E. & Poo, M.M. (1992). Retrograde modulation at developing neuromuscular synapses – involvement of g protein and arachidonic acid cascade. *Neuron*, **9**, 1201–1209.

Harris, A.J., Duxson, M.J., Fitzsimons, R.B. & Rieger, F. (1989a). Myonuclear birthdates distinguish the origins of primary and secondary myotubes in embryonic mammalian skeletal-muscles. *Development*, **107**, 771–784.

Harris, A.J., Fitzsimons, R.B. & McEwan, J.C. (1989b). Neural control of the sequence of expression of myosin heavy-chain isoforms in fetal mammalian muscles. *Development*, **107**, 751–769.

Harris, J.B. & Luff, A.R. (1970). The resting membrane potentials of fast and slow skeletal muscle fibres in the developing mouse. *Comparative Biochemistry and Physiology*, **33**, 923–931.

Harris, J.B. & Ribchester, R.R. (1979). The relationship between end-plate size and transmitter release in normal and dystrophic muscles of the mouse. *Journal of Physiology, London*, **296**, 245–265.

Hennig, R. & Lomo, T. (1985). Firing patterns of motor units in normal rats. *Nature*, **314**, 164–166.

Herrera, A.A. (1984). Polyneuronal innervation and quantal transmitter release in formamide-treated frog sartorius muscles. *Journal of Physiology, London*, **355**, 267–280.

Hesselmans, L.F.G.M., Jennekens, F.G.I., Vandenoord, C.J.M., Veldman, H. & Vincent, A. (1993). Development of innervation of skeletal-muscle fibers in man – relation to acetylcholine receptors. *Anatomical Record*, **236**, 553–5621.

Heuser, J.E. (1989). Review of electron-microscopic evidence favoring vesicle exocytosis as the structural basis for quantal release during synaptic transmission. *Quarterly Journal of Experimental Physiology*, **74**, 1051–1069.

Hirata, K., Zhou, C., Nakamura, K. & Kawabuchi, M. (1997). Postnatal development of Schwann cells at neuromuscular junctions, with special reference to synapse elimination. *Journal of Neurocytology*, **26**, 799–809

Hoffman, H. (1950). Local reinnervation in partially denervated muscle: a histophysiological study. *Australian Journal of Experimental Biology and Medical Sciences*, **28**, 383–397.

Hoffman, H. (1953). The persistence of hyperneurotized end-plates in mammalian muscles. *Journal of Comparative Neurology*, **99**, 331–345.

Holland, R.L. & Brown, M.C. (1981). Nerve growth in botulinum toxin poisoned muscles. *Neuroscience*, **6**, 1167–1179.

Hopkins, W.G., Brown, M.C. & Keynes, R.J. (1985). Postnatal growth of motor nerve terminals in muscles of the mouse. *Journal of Neurocytology*, **14**, 525–540.

Hughes, S.M. & Blau, H.M. (1992). Mucle-fiber pattern is independent of cell lineage in postnatal rodent development. *Cell*, **68**, 659–671.

Hughes, S.M. & Salinas, P.C. (1999). Control of muscle fibre and motoneuron diversification. *Current Opinion in Neurobiology*, **9**, 54–64.

Hughes, S.M., Koishi, K., Rudnicki, M. & Maggs, A.M. (1997). Myod protein is differentially accumulated in fast and slow skeletal muscle fibres and required for normal fibre type balance in rodents. *Mechanisms of Development*, **61**, 151–163.

Hunt, C.C. & Kuffler, S.W. (1954). Motor innervation of skeletal muscle: multiple innervation of individual muscle fibres and motor unit function. *Journal of Physiology, London*, **126**, 293–303.

Ishihara, A., Roy, R.R. & Edgerton, V.R. (1995). Succinate dehydrogenase activity and soma size of motoneurons innervating different portions of the rat tibialis anterior. *Neuroscience*, **68**, 813–822.

Jansen, J.K.S. & Fladby, T. (1990). The perinatal reorganization of the innervation of skeletal-muscle in mammals. *Progress in Neurobiology*, **34**, 39–90.

Jean-Pretre, N., Clarke, P.G.H. & Gabriel, J.P. (1996). Competitive exclusion between axons dependent on a single trophic substance: a mathematical analysis. *Mathematical Biosciences*, **135**, 23–54.

Jennings, C. (1994). Developmental neurobiology – death of a synapse. *Nature*, **372**, 498–499.

Jia, M., Li, M., Dunlap, V. & Nelson, P.G. (1999). The thrombin receptor mediates functional activity-dependent neuromuscular synapse reductions via protein kinase c activation *in vitro*. *Journal of Neurobiology*, **38**, 369–381.

Jones, S.P. & Ridge, R.M.A.P. (1987). Motor units in a skeletal muscle of neonatal rat – mechanical properties and weak neuromuscular transmission. *Journal of Physiology, London*, **386**, 355–375.

Jones, S.P., Ridge, R.M.A.P. & Rowlerson, A. (1987a). The nonselective innervation of muscle-fibers and mixed composition of motor units in a muscle of neonatal rat. *Journal of Physiology, London*, **386**, 377–394.

Jones, S.P., Ridge, R.M.A.P. & Rowlerson, A. (1987b). Rat muscle during postnatal-development – evidence in favor of no interconversion between fast-twitch and slow-twitch fibers. *Journal of Physiology, London*, **386**, 395–406.

Jordan, C.L. (1996a). Morphological effects of ciliary neurotrophic factor treatment during neuromuscular synapse elimination. *Journal of Neurobiology*, **31**, 29–40.

Jordan, C.L. (1996b). Ciliary neurotrophic factor may act in target musculature to regulate developmental synapse elimination. *Developmental Neuroscience*, **18**, 185–198.

Kablar, B., Krastel, K., Ying, C.Y., Tapscott, S.J., Goldhamer, D.J. & Rudnicki, M.A. (1999). Myogenic determination occurs independently in somites and limb buds. *Developmental Biology*, **206**, 219–231.

Kalb, R.G. & Hockfield, S. (1992). Activity-dependent development of spinal-cord motor neurons. *Brain Research Reviews*, **17**, 283–289.

Katz, B. (1969). *The release of neural transmitter substances*. Liverpool: Liverpool University Press.

Katz, B. (1996). Neural transmitter release: from quantal secretion to exocytosis and beyond. The Fenn Lecture. *Journal of Neurocytology*, **25**, 677–686.

Katz, B. & Thesleff, S. (1957). On the factors which determined the amplitude of the miniature endplate potential. *Journal of Physiology, London*, **137**, 267–278.

Katz, E., Ferro, P.A., Weisz, G. & Uchitel, O.D. (1996). Calcium channels involved in synaptic transmission at the mature and regenerating mouse neuromuscular junction. *Journal of Physiology, London*, **497**, 687–697.

Keddy, P. (1989). *Competition*. London: Chapman & Hall.

Kelly, A.M. & Zacks, S.I. (1969a). The histogenesis of rat intercostal muscle. *Journal of Cell Biology*, **42**, 135–153.

Kelly, A.M. & Zacks, S.I. (1969b). The fine structure of motor endplate morphogenesis. *Journal of Cell Biology*, **42**, 154–169.

Kelly, S.S. & Robbins, N. (1983). Progression of age changes in synaptic transmission at mouse neuromuscular junctions. *Journal of Physiology, London*, **343**, 375–383.

Kerenell, D. (1998). The final common pathway in postural control – developmental perspective. *Neuroscience and Biobehavioral Reviews*, **22**, 479–484.

Kim, S., Buonanno, A. & Nelson, P.G. (1998). Regulation of prothrombin, thrombin receptor, and protease nexin-1 expression during development and after denervation in muscle. *Journal of Neuroscience Research*, **53**, 304–311.

Ko, C.P. & Chen, L.L. (1996). Synaptic remodeling revealed by repeated *in vivo* observations and electron microscopy of identified frog neuromuscular junctions. *Journal of Neuroscience*, **16**, 1780–1790.

Koenig, J., Delaforte, S. & Chapron, J. (1998). The Schwann cell at the neuromuscular junction. *Journal of Physiology, Paris*, **92**, 153–155.

Konstantinidou, A.D., Silossantiago, I., Flaris, N. & Snider, W.D. (1995). Development of the primary afferent projection in human spinal cord. *Journal of Comparative Neurology*, **354**, 1–12.

Kopp, D.M. & Thompson, W.J. (1998). Innervation-dependent expression of nestin at the mammalian neuromuscular junction. *Society for Neuroscience, Abstracts*, **24**, 415.6

Kopp, D.M., Perkel, D.J. & Balice-Gordon, R.J. (2000). Disparity in neurotransmitter release probability among competing inputs during neuromuscular synapse elimination. *Journal of Neuroscience*, **20**, 8771–8779.

Kopp, D.M., Trachternberg, J.T. & Thompson, W.J. (1997). Glial growth factor rescues Schwann cells of mechanoreceptors from denervation-induced apoptosis. *Journal of Neuroscience*, **17**, 6697–6706.

Korneliussen, H. & Jansen, J.K. (1976). Morphological aspects of the elimination of polyneuronal innervation of skeletal muscle fibres in newborn rats. *Journal of Neurocytology*, **5**, 591–604.

Kuffler, D.P., Thompson, W. & Jansen, J.K. (1980). The fate of foreign endplates in cross-innervated rat soleus muscle. *Proceedings of the Royal Society of London, Series B, Biological Sciences*, **208**, 189–222

Kuno, M., Turkanis, S.A. & Weakly, J.N. (1971). Correlation between nerve terminal size and transmitter release at the neuromuscular junction of the frog. *Journal of Physiology, London*, **213**, 545–446.

Kwon, Y.W. & Gurney, M.E. (1994). Systemic injections of ciliary neurotrophic factor induce sprouting by adult motor neurons. *Neuroreport*, **5**, 789–792.

Kwon, Y.W. & Gurney, M.E. (1996). Brain-derived neurotrophic factor transiently stabilizes silent synapses on developing neuromuscular junctions. *Journal of Neurobiology*, **29**, 503–516.

Kwon, Y.W., Abbondanzo, S.J., Stewart, C.L. & Gurney, M.E. (1995). Leukemia inhibitor factor influences the timing of programmed synapse withdrawal from neonatal muscles. *Journal of Neurobiology*, **28**, 35–50.

Lance-Jones, C. & Landmesser, L. (1980). Motoneurone projection patterns in the chick hindlimb following early partial reversals of the spinal cord. *Journal of Physiology, London*, **302**, 581–602.

Laskowski, M.B., Colman, H., Nelson, C. & Lichtman, J.W. (1998). Synaptic competition during the reformation of a neuromuscular map. *Journal of Neuroscience*, **18**, 7328–7335.

Lawoko, G. & Tagerud, S. (1995). High endocytic activity occurs periodically in the end-plate region of denervated mouse striated muscle fibers. *Experimental Cell Research*, **219**, 598–603.

Lemon, R.N. (1998). Pathways and mechanisms subserving cortical control of the hand in primates. *Current Science*, **75**, 458–463.

Levtov, A. & O'Donovan, M.J. (1995). Calcium imaging of motoneuron activity in the end-bloc spinal cord preparation of the neonatal rat. *Journal of Neurophysiology*, **74**, 1324–1334.

Lichtman, J.W. & Wilkinson, R.S. (1987). Properties of motor units in the transversus abdominis muscle of the garter snake. *Journal of Physiology, London*, **393**, 355–374.

Lichtman, J.W., Magrassi, L. & Purves, D. (1987). Visualization of neuromuscular junctions over periods of several months in living mice. *Journal of Neuroscience*, **7**, 1215–1222.

Lichtman, J.W., Wilkinson, R.S. & Rich, M.M. (1985). Multiple innervation of tonic endplates revealed by activity-dependent uptake of fluorescent-probes. *Nature*, **314**, 357–359.

Lie, D.C. & Weis, J. (1998). GDNF expression is increased in denervated human skeletal muscle. *Neuroscience Letters*, **250**, 87–90.

Lin, J.H., Saito, T., Anderson, D.J., Lancejones, C., Jessell, T.M. & Arber, S. (1998). Functionally related motor neuron pool and muscle sensory afferent subtypes defined by coordinate ets gene expression. *Cell*, **95**, 393–407.

Liu, D.W.C. & Westerfield, M. (1990). The formation of terminal fields in the absence of competitive interactions among primary motoneurons in the zebrafish. *Journal of Neuroscience*, **10**, 3947–3959.

Liu, Y., Fields, R.D., Fitzgerald, S., Festoff, B.W. & Nelson, P.G. (1994). Proteolytic activity, synapse elimination, and the Hebb synapse. *Journal of Neurobiology*, **25**, 325–335.

Liu, Y.W. & Schneider, M.F. (1998). Fibre type-specific gene expression activated by chronic electrical stimulation of adult mouse skeletal muscle fibres in cultures. *Journal of Physiology, London*, **512**, 337–344.

Lo, Y.J. & Poo, M.M. (1991). Activity-dependent synaptic competition *in vitro* – heterosynaptic suppression of developing synapses. *Science*, **254**, 1019–1022.

Loeb, J.A., Khurana, T.S., Robbins, J.T., Yee, A.G. & Fischbach, G.D. (1999). Expression patterns of transmembrane and released forms of neuregulin during spinal cord and neuromuscular synapse development. *Development*, **126**, 781–791.

Lohof, A.M., Delhayebouchaud, N. & Mariani, J. (1996). Synapse elimination in the central nervous system: functional significance and cellular mechanisms. *Reviews in the Neurosciences*, **7**, 85–101.

Lohof, A.M., Ip, N.Y. & Poo, M.M. (1993). Potentiation of developing neuromuscular synapses by the neurotrophins NT-3 and BDNF. *Nature*, **363**, 350–353.

Lomo, T. & Westgaard, R.H. (1975). Further studies on the control of ACh sensitivity by muscle activity in the rat. *Journal of Physiology, London*, **252**, 603–626.

Love, F.M. & Thompson, W.J. (1998). Schwann cells proliferate at rat neuromuscular junctions during development and regeneration. *Journal of Neuroscience*, **18**, 9376–9385.

Lubischer, J.L. & Thompson, W.J. (1999). Neonatal partial denervation results in nodal but not terminal sprouting and a decrease in efficacy of remaining neuromuscular junctions in rat soleus muscle. *Journal of Neuroscience*, **19**, 8931–8944.

Lubischer, J.L., Jordan, C.L. & Arnold, A.P. (1992). Transient and permanent effects of androgen during synapse elimination in the levator ani muscle of the rat. *Journal of Neurobiology*, **23**, 1–9.

Ludolph, D.C. & Konieczny, S.F. (1995). Transcription factor families: muscling in on the myogenic program. *FASEB Journal*, **9**, 1595–1604.

Lupa, M.T., Krzemien, D.M., Schaller, K.L. & Caldwell, J.H. (1993). Aggregation of sodium-channels during development and maturation of the neuromuscular-junction. *Journal of Neuroscience*, **13**, 1326–1336.

Lupa, M.T., Krzemien, D.M., Schaller, K.L. & Caldwell, J.H. (1995). Expression and distribution of sodium-channels in short-term and long-term denervated rodent skeletal muscles. *Journal of Physiology, London*, **483**, 109–118.

Magchielse, T. & Meeter, E. (1986). The effect of neuronal activity on the competitive elimination of neuromuscular junctions in tissue culture. *Developmental Brain Research*, **25**, 211–220.

Maroto, M., Reshef, R., Munsterberg, A.E., Koester, S., Goulding, M.& Lassar, A.B. (1997). Ectopic pax-3 activates myod and myf-5 expression in embryonic mesoderm and neural tissue. *Cell*, **89**, 139–148.

Marques, M.J., Conchello, J.A. & Lichtman, J.W. (2000). From plaque to pretzel: fold formation and acetylcholine receptor loss at the developing neuromuscular junction. *Journal of Neuroscience*, **20**, 3663–3675.

Martin, A.R. (1977). Junctional transmission. II. Presynaptic mechanisms. In E. Kandel (Ed.), *Handbook of the nervous system* (vol. 1, pp. 329–335). Baltimore, MD: American Physiological Society.

Martin, A.R. (1994). Amplification of neuromuscular transmission by postjunctional folds. *Proceedings of the Royal Society of London, Series B, Biological Sciences*, **258**, 321–326.

Martin, P.T., Scott, L.J.C., Porter, B.E. & Sanes, J.R. (1999). Distinct structures and functions of related pre- and postsynaptic carbohydrates at the mammalian neuromuscular junction. *Molecular and Cellular Neuroscience*, **13**, 105–118.

Matsuda, Y., Oki, S., Kitaoka, K., Nagano, Y., Nojima, M. & Desaki, J. (1988). Scanning electron-microscopic study of denervated and reinnervated neuromuscular-junction. *Muscle & Nerve*, **11**, 1266–1271.

McAllister, A.K., Katz, L.C. & Lo, D.C. (1999). Neurotrophins and synaptic plasticity. *Annual Review of Neuroscience*, **22**, 295–318.

McArdle, J.J. (1975). Complex end-plate potentials at the regenerating neuromuscular junction of the rat. *Experimental Neurology*, **49**, 629–638.

Meier, T. & Wallace, B.G. (1998). Formation of the neuromuscular junction: molecules and mechanisms. *Bioassays*, **20**, 819–829.

Mendler, L., Zador, E., Dux, L. & Wuytack, F. (1998). MRNA levels of myogenic regulatory factors in rat slow and fast muscles regenerating from notexin-induced necrosis. *Neuromuscular Disorders*, **8**, 533–541.

Mesulam, M-M. (1982). *Tracing neuronal connections with horseradish peroxidase*. New York: Wiley & Sons.

Miledi, R. & Slater, C.R. (1968). Electrophysiology and electron microscopy of rat neuromuscular junctions after nerve degeneration. *Proceedings of the Royal Society of London, Series B: Biological Sciences*, **169**, 289–306.

Miledi, R. & Slater, C.R. (1970). On the degeneration of rat neuromuscular junctions after nerve section. *Journal of Physiology, London*, **207**, 507–528.

Missias, A.C., Mudd, J., Cunningham, J.M., Steinbach, J.H., Merlie, J.P. & Sanes, J.R. (1997). Deficient development and maintenance of postsynaptic specializations in mutant mice lacking an "adult" acetylcholine receptor subunit. *Development*, **124**, 5075–5086.

Miyata, Y. & Yoshioka, K. (1980). Selective elimination of motor nerve terminals in the rat soleus muscle during development. *Journal of Physiology, London*, **309**, 631–646.

Moscoso, L.M., Cremer, H. & Sanes, J.R. (1998). Organization and reorganization of neuromuscular junctions in mice lacking neural cell adhesion molecule, tenascin-c, or fibroblast growth factor-5. *Journal of Neuroscience*, **18**, 1465–1477.

Mulkey, R.M. & Zucker, R.S. (1992). Posttetanic potentiation at the crayfish neuromuscular junction is dependent on both intracellular calcium and sodium-ion accumulation. *Journal of Neuroscience*, **12**, 4327–4336.

Nelson, P.G., Fields, R.D., Yu, C. & Liu, Y. (1993). Synapse elimination from the mouse neuromuscular junction *in vitro*: a non-Hebbian activity-dependent process. *Journal of Neurobiology*, **24**, 1517–1530.

Nguyen, Q.T., Keller-Peck, C.R., Lichtman, J.W., Silos-Santiago, L. & Snider, W.D. (1998a). GDNF expression is restricted to a subset of muscles and muscle fibers. *Society for Neuroscience Abstracts*, **24**, 800.11.

Nguyen, Q.T., Parsadanian, A.S., Snider, W.D. & Lichtman, J.W. (1998b). Hyperinnervation of neuromuscular junctions caused by GDNF overexpression in muscle. *Science*, **279**, 1725–1729.

Nicoloupoulos-Stournos, S. & Iles, J.F. (1983). Motor neuron columns in the lumbar spinal cord of the rat. *Journal of Comparative Neurology*, **217**, 75–85.

Nikolsky, E.E., Zemkova, H., Voronin, V.A. & Vyskocil, F. (1994). Role of nonquantal acetylcholine-release in surplus polarization of mouse diaphragm fibers at the end-plate zone. *Journal of Physiology, London*, **477**, 497–502.

Nudell, B.M. & Grinnell, A.D. (1982). Inverse relationship between transmitter release and terminal length in synapses on frog-muscle fibers of uniform input resistance. *Journal of Neuroscience*, **2**, 216–224.

Nussinovitch, I. & Rahamimoff, R. (1988). Ionic basis of tetanic and post-tetanic potentiation at a mammalian neuromuscular junction. *Journal of Physiology, London*, **396**, 435–455.

O'Brien, R.A., Ostberg, A.J. & Vrbova, G. (1978). Observations on the elimination of polyneuronal innervation in developing mammalian skeletal muscle. *Journal of Physiology, London*, **282**, 571–582.

O'Brien, R.A.D., Ostberg, A.J.C. & Vrbova, G. (1984). Protease inhibitors reduce the loss of nerve terminals induced by activity and calcium in developing rat soleus muscles *in vitro*. *Neuroscience*, **12**, 637–646.

O'Malley, J.P., Waran, M.T. & Balice-Gordon, R.J. (1999). *In vivo* observations of terminal Schwann cells at normal, denervated, and reinnervated mouse neuromuscular junctions. *Journal of Neurobiology*, **38**, 270–286.

Parson, S.H., Dilley, J., Gandhi, N., Gillingwater, T.H. & Ribchester, R.R. (1998). Schwann cell responses at disconnected nerve terminals in organ cultures of slds mouse neuromuscular junctions. *Society for Neuroscience Abstracts*, **24**, 413.17.

Parson, S.H., Mackintosh, C.L. & Ribchester, R.R. (1997). Elimination of motor nerve terminals in neonatal mice expressing a gene for slow wallerian degeneration (C57BL/WLD(S)). *European Journal of Neuroscience*, **9**, 1586–1592.

Patak, A., Proske, U., Turner, H. & Gregory, J.E. (1992). Development of the sensory innervation of muscle spindles in the kitten. *International Journal of Developmental Neuroscience*, **10**, 81–92.

Patten, R.M. & Ovalle, W.K. (1991). Muscle-spindle ultrastructure revealed by conventional and high-resolution scanning electron-microscopy. *Anatomical Record*, **230**, 183–198.

Patton, B.L., Chiu, A.Y. & Sanes, J.R. (1998). Synaptic laminin prevents glial entry into the synaptic cleft. *Nature*, **393**, 698–701.

Pawson, P.A., Grinnell, A.D. & Wolowske, B. (1998a). Quantitative freeze–fracture analysis of the frog neuromuscular junction synapse. II. Proximal–distal measurements. *Journal of Neurocytology*, **27**, 379–391.

Perry, V.H., Lunn, E.R., Brown, M.C., Cahusac, S. & Gordon, S. (1990). Evidence that the rate of Wallerian degeneration is controlled by a single autosomal dominant gene. *European Journal of Neuroscience*, **2**, 408–413.

Petersen, S.A., Fetter, R.D., Noordermeer, J.N., Goodman, C.S. & DiAntonio, A. (1997). Genetic analysis of glutamate receptors in Drosophila reveals a retrograde signal regulating presynaptic transmitter release. *Neuron*, **19**, 1237–1248.

Pette, D. & Staron, R.S. (1993). The molecular diversity of mammalian muscle fibers. *News in Physiological Sciences*, **8**, 153–157.

Pette D. & Vrbova, G. (1992). Adaptation of mammalian skeletal muscle fibers to chronic electrical stimulation. *Reviews of Physiology, Biochemistry and Pharmacology*, **120**, 115–202.

Plomp, J.J. & Molenaar, P.C. (1996). Involvement of protein kinases in the upregulation of acetyl-choline release at endplates of alpha-bugarotoxin-treated rats. *Journal of Physiology, London*, **493**, 175–186.

Plomp, J.J., Vankempen, G.T.H. & Molenaar, P.C. (1994). The up-regulation of acetylcholine-release at end-plates of alpha-bungarotoxin-treated rats – its dependency on calcium. *Journal of Physiology, London*, **478**, 125–136.

Purves, D. (1988). *Body and brain: a trophic theory of neural connections.* Cambridge, MA: Harvard University Press.

Purves, D. (1994). *Neural activity and the growth of the brain.* Cambridge, UK: Cambridge University Press.

Purves, D. & Lichtman, J.W. (1985). *Principles of neural development.* Sunderland, Massachusetts, Sinauer Associates Inc.

Rafuse, V.F. & Gordon, T. (1996a). Self-reinnervated cat medial gastrocnemius muscles. 1. Comparisons of the capacity for regenerating nerves to form enlarged motor units after extensive peripheral nerve injuries. *Journal of Neurophysiology*, **75**, 268–281.

Rafuse, V.F. & Gordon, T. (1996b). Self-reinnervated cat medial gastrocnemius muscles. 2. Analysis of the mechanisms and significance of fiber type grouping in reinnervated muscles. *Journal of Neurophysiology*, **75**, 282–297.

Rafuse, V.F. & Gordon, T. (1998). Incomplete rematching of nerve and muscle properties in motor units after extensive nerve injuries in cat hindlimb muscle. *Journal of Physiology, London*, **509**, 909–926.

Rafuse, V.F., Gordon, T. & Orozco, R. (1992). Propertional enlargement of motor units after partial denervation of cat triceps surae muscles. *Journal of Neurophysiology*, **68**, 1261–1276.

Rahamimoff, R., Butkevich, A., Duridanova, D., Ahdut, R., Harari, E. & Kachalsky, S.G. (1999). Multitude of ion channels in the regulation of transmitter release. *Philosophical Transactions of the Royal Society of London, Series B. Biological Sciences*, **354**, 281–288.

Ramon Y Cajal, S. (1911). *Histologie du systeme nerveux de l'homme et des verterbres.* Madrid: Inst. Ramon Y Cajal.

Rasmussen, C.E. & Willshaw, D.J. (1993). Presynaptic and postsynaptic competition in models for the development of neuromuscular connections. *Biological Cybernetics*, **68**, 409–419.

Rassendren, F.A., Blochgallego, E., Tanaka, H. & Henderson, C.E. (1992). Levels of messenger-RNA coding for motoneuron growth-promoting factors are increased in denervated muscle. *Proceedings of the National Academy of Sciences, USA*, **89**, 7194–7198.

Rawls, A., Valdez, M.R., Zhang, W., Richardson, J., Klein, W.H. & Olson, E.N. (1998). Overlapping functions of the myogenic bhlh genes mrf4 and myod revealed in double mutant mice. *Development*, **125**, 2349–2358.

Redfern, P.A. (1970). Neuromuscular transmission in new-born rats. *Journal of Physiology, London*, **209**, 701–709.

Rees, S., Rawson, J., Nitsos, I. & Brumley, C. (1994). The structural and functional development of muscle spindles and their connections in fetal sheep. *Brain Research*, **642**, 185–198.

Reid, B., Slater, C.R. & Bewick, G.S. (1999). Synaptic vesicle dynamics in rat fast and slow motor nerve terminals. *Journal of Neuroscience*, **19**, 2511–2521. Reynolds, M.L. & Woolf, C.J. (1992). Terminal Schwann cells elaborate extensive processes following denervation of the motor end-plate. *Journal of Neurocytology*, **21**, 50–66.

Ribchester, R.R. (1986). Neural activity and the reorganisation of motor units in reinnervated skeletal muscle. In W.A. Nix & G. Vrbov (Eds.), *Electrical stimulation and neuromuscular disorders*, (pp. 201–206). Berlin: Springer.

Ribchester, R.R. (1988a). Activity-dependent and activity-independent synaptic interactions during reinnervation of partially denervated rat muscle. *Journal of Physiology, London*, **401**, 53–75.

Ribchester, R.R. (1988b). Models, mechanisms and kinetics of neuromuscular synapse elimination in reinnervated adult skeletal muscle. In H. Flohr (Ed.), *Post-lesion neural plasticity*, (pp. 11–23). Berlin: Springer.

Ribchester, R.R. (1988c). Competitive elimination of neuromuscular synapses. *Nature*, **331**, 21.

Ribchester, R.R. (1993). Coexistence and elimination of convergent motor-nerve terminals in reinnervated and paralyzed adult rat skeletal muscle. *Journal of Physiology, London*, **466**, 421–441.

Ribchester, R.R. & Barry, J.A. (1994). Spatial versus consumptive competition at polyneuronally innervated neuromuscular-junctions. *Experimental Physiology*, **79**, 465–494.

Ribchester, R.R. & Taxt, T. (1983). Motor unit size and synaptic competition in rat lumbrical muscles reinnervated by active and inactive motor axons. *Journal of Physiology, London*, **344**, 89–111.

Ribchester, R.R. & Taxt, T. (1984). Repression of inactive motor-nerve terminals in partially denervated rat muscle after regeneration of active motor axons. *Journal of Physiology, London*, **347**, 497–511.

Ribchester, R.R., Mao, F. & Betz, W.J. (1994). Optical measurements of activity-dependent membrane recycling in motor-nerve terminals of mammalian skeletal-muscle. *Proceedings of the Royal Society of London, Series B: Biological Sciences*, **255**, 61–66.

Ribchester, R.R., Thomson, D., Haddow, L.J. & Ushkaryov, Y.A. (1998). Enhancement of spontaneous transmitter release at neonatal mouse neuromuscular junctions by the glial cell line-derived neurotrophic factor (GDNF). *Journal of Physiology, London*, **512**, 635–641.

Ribchester, R.R., Tsao, J.W., Barry, J.A., Asgarijirhandeh, N., Perry, V.H. & Brown, M.C. (1995). Persistence of neuromuscular junctions after axotomy in mice with slow Wallerian degeneration (C57BL/WLD(S)). *European Journal of Neuroscience*, **7**, 1641–1650.

Ribeiro, J.A., Cunha, R.A., Correiadesa, P. & Sebastiao, A.M. (1996). Purinergic regulation of acetylcholine release. *Progress in Brain Research*, **109**, 31–241.

Rich, M. & Lichtman, J.W. (1989a). Motor-nerve terminal loss from degenerating muscle fibers. *Neuron*, **3**, 677–688.

Rich, M.M. & Lichtman, J.W. (1989b). *In vivo* visualization of presynaptic and postsynaptic changes during synapse elimination in reinnervated mouse muscle. *Journal of Neuroscience*, **9**, 1781–1805.

Ridge, R..M.A.P. (1989). Motor unit organization in developing muscle. *Comparative Biochemistry and Physiology A, Physiology*, **93**, 115–123.

Ridge, R.M.A.P. & Betz, W.J. (1984). The effect of selective, chronic stimulation on motor unit size in developing rat muscle. *Journal of Neuroscience*, **4**, 2614–2620.

Ridge, R.M.A.P. & Rowlerson, A. (1990). Sprouting motoneurons break as well as make contacts in partially denervated rat muscle. *Journal of Physiology, London*, **425P**, 15P.

Ridge, R.M.A.P. & Rowlerson, A. (1996). Motor units of juvenile rat lumbrical muscles and fibre type compositions of the glycogen-depleted component. *Journal of Physiology, London*, **497**, 199–210.

Riley, D.A. (1977). Spontaneous elimination of nerve terminals from the endplates of developing skeletal myofibers. *Brain Research*, **134**, 279–285.

Riley, D.A. (1981). Ultrastructural evidence for axon retraction during the spontaneous elimination of polyneuronal innervation of the rat soleus muscle. *Journal of Neurocytology*, **10**, 425–440.

Robinson, L.J. & Martin, T.F.J. (1998). Docking and fusion in neurosecretion. *Current Opinion in Cell Biology*, **10**, 483–492.

Robitaille, R. (1998). Modulation of synaptic efficacy and synaptic depression by glial cells at the frog neuromuscular junction. *Neuron*, **21**, 847–855.

Rosenthal, J.L. & Taraskevich, P.S. (1977). Reduction of multiaxonal innervation at the neuromuscular junction of the rat during development. *Journal of Physiology, London*, **270**, 299–310.

Ross, J.J., Duxson, M.J. & Harris, A.J. (1987). Neural determination of muscle fiber numbers in embryonic rat lumbrical muscles. *Development*, **100**, 395–409.

Rubinstein, N.A. & Kelly, A.M. (1981). Development of muscle fiber specialization in the rat hindlimb. *Journal of Cell Biology*, **90**, 128–144.

Sala, C., Andreose, J.S., Fumagalli, G. & Lomo, T. (1995). Calcitonin-gene-related peptide – possible role in formation and maintenance of neuromuscular junctions. *Journal of Neuroscience*, **15**, 520–528.

Salim, S.Y., Dezaki, K., Tsuneki, H., Abdelzaher, A.O. & Kimura, I. (1998). Calcitonin gene-related peptide potentiates nicotinic acetylcholine receptor-operated slow Ca^{2+} mobilization at mouse muscle endplates. *British Journal of Pharmacology*, **125**, 277–282.

Sandrock, A.W., Dryer, S.E., Rosen, K.M., Gozani, S.N., Kramer, R., Theill, L.E. & Fischbach, G.D. (1997). Maintenance of acetylcholine receptor number by neuregulins at the neuromuscular junction *in vivo*. *Science*, **276**, 599–603.

337

Sanes, J.R. & Lichtman, J.W. (1999). Development of the vertebrate neuromuscular junction. *Annual Review of Neuroscience*, **22**, 389–442.

Sanes, J.R., Apel, E.D., Burgess, R.W., Emerson, R.B., Feng, G., Gautam, M., Glass, D., Grady, R.M., Krejci, E., Lichtman, J.W., Lu, J.T., Massoulie, J., Miner, J.H., Moscoso, L.M., Nguyen, Q., Nichol, M., Noakes, P.G., Patton, B.L., Son, Y.J., Yancopoulos, G.D. & Zhou, H (1998). Development of the neuromuscular junction: genetic analysis in mice. *Journal of Physiology, Paris*, **92**, 167–172.

Sanes, J.R., Schachner, M. & Covault, J. (1986). Expression of several adhesive macromolecules (N-CAM, L1, J1, nile, uvomorulin, laminin, fibronectin, and a heparan-sulfate proteoglycan) in embryonic, adult, and denervated adult skeletal-muscle. *Journal of Cell Biology*, **102**, 420–431.

Schmalbruch, H. (1982). Skeletal-muscle fibers of newborn rats are coupled by gap junctions. *Developmental Biology*, **91**, 485–490.

Schwarz, H., Giese, G., Muller, H., Koenen, M. & Witzemann, V. (2000). Different functions of fetal and adult AChR subtypes for the formation and maintenance of neuromuscular synapses revealed in epsilon-subunit-deficient mice. *European Journal of Neuroscience*, **12**, 3107–3116.

Seebach, B.S. & Ziskind-Conhaim, L. (1994). Formation of transient inappropriate sensorimotor synapses in developing rat spinal-cords. *Journal of Neuroscience*, **14**, 4520–4528.

Seeds, N.W., Siconolfi, L.B. & Haffke, S.P. (1997). Neuronal extracellular proteases facilitate cell migration, axonal growth, and pathfinding. *Cell & Tissue Research*, **290**, 367–370.

Sigrist, S.J., Thiel, P.R., Reiff, D.F., Lachance, P.E.D., Lasko, P. & Schuster, C.M. (2000). Postsynaptic translation affects the efficacy and morphology of neuromuscular junctions. *Nature*, **405**, 1062–1065.

Slater, C.R. (1982a). Postnatal maturation of nerve muscle junctions in hindlimb muscles of the mouse. *Developmental Biology*, **94**, 11–22.

Slater, C.R. (1982b). Neural influence on the postnatal changes in acetylcholine-receptor distribution at nerve–muscle junctions in the mouse. *Developmental Biology*, **94**, 23–30.

Slater, C.R., Lyons, P.R., Walls, T.J., Fawcett, P.R.W. & Young, C. (1992). Structure and function of neuromuscular junctions in the vastus lateralis of man – a motor point biopsy study of 2 groups of patients. *Brain*, **115**, 451–478.

Snider, W.D. & Harris, G.L. (1979). A physiological correlate of disuse-induced sprouting at the neuromuscular junction. *Nature*, **281**, 69–71.

Snider, W.D. & Lichtman, J.W. (1996). Are neurotrophins synaptotrophins? *Molecular and Cellular Neuroscience*, **7**, 433–442.

Soha, J.M., Yo, C. & Vanessen, D.C. (1987). Synapse elimination by fiber type and maturational state in rabbit soleus muscle. *Developmental Biology*, **123**, 136–1441.

Soileau, L.C., Silberstein, L., Blau, H.M. & Thompson, W.J. (1987). Reinnervation of muscle-fiber types in the newborn rat soleus. *Journal of Neuroscience*, **7**, 4176–4194.

Son, Y.J. & Thompson, W.J. (1995a). Schwann-cell processes guide regeneration of peripheral axons. *Neuron*, **14**, 125–132.

Son, Y.J. & Thompson, W.J. (1995b). Nerve sprouting in muscle is induced and guided by processes extended by Schwann cells. *Neuron*, **14**, 133–141.

Son, Y.J., Trachtenberg, J.T. & Thompson W.J. (1996). Schwann cells induce and guide sprouting and reinnervation of neuromuscular junctions. *Trends in Neurosciences*, **19**, 280–285.

Stiles, J.R. & Salpeter, M.M. (1997). Absence of nerve-dependent conversion of rapidly degrading to stable acetylcholine receptors at adult innervated endplates. *Neuroscience*, **78**, 895–901.

Stoop, R. & Poo, M.M. (1995). Potentiation of transmitter release by ciliary neurotrophic factor requires somatic signaling. *Science*, **267**, 695–699.

Suzuki, H., Hase, A., Miyata, Y., Arahata, K. & Akazawa, C. (1998). Prominent expression of glial cell line-derived neurotrophic factor in human skeletal muscle. *Journal of Comparative Neurology*, **402**, 303–312.

Tang, Y.G. & Zucker, R.S. (1997). Mitochondrial involvement in post-tetanic potentiation of synaptic transmission. *Neuron*, **18**, 483–491.

Taxt, T. (1983a). Cross-innervation of fast and slow-twitch muscles by motor axons of the sural nerve in the mouse. *Acta Physiologica Scandinavica*, **117**, 331–341.

Taxt, T. (1983b). Local and systemic effects of tetrodotoxin on the formation and elimination of synapses in reinnervated adult-rat muscle. *Journal of Physiology, London*, **340**, 175–194.

Taxt, T., Ding, R. & Jansen, J.K.S. (1983). A note on the elimination of polyneuronal innervation of skeletal muscles in neonatal rats. *Acta Physiologica Scandinavica*, **117**, 557–560.

Tello, J.F. (1907). Dégéneration et régéneration des plaques motrices aprés la section des nerfs. *Travaux du Laboratoire de Recherches Biologique*, **5**, 117–149.

Tello, J.F. (1917). Genesis de la terminaciones nerviosas motrices y sensitivas. *Trabathos do Laboratorio Invest. Biol. Univ. Madrid*, **15**, 101–199.

Thompson, W. (1978). Reinnervation of partially denervated rat soleus muscle. *Acta Physiologica Scandinavica*, **103**, 81–91.

Thompson, W. (1983a). Synapse elimination in neonatal rat muscle is sensitive to pattern of muscle use. *Nature*, **302**, 614–616.

Thompson, W.J. (1983b). Lack of segmental selectivity in elimination of synapses from soleus muscle of newborn rats. *Journal of Physiology, London*, **335**, 343–352.

Thompson, W.J. (1986). Changes in the innervation of mammalian skeletal muscle fibers during postnatal development. *Trends in Neurosciences*, **9**, 25–28.

Thompson, W. & Jansen, J.K. (1977). The extent of sprouting of remaining motor units is partly denervated immature and adult rat soleus muscle. *Neuroscience*, **2**, 523–535.

Thompson, W.J., Condon, K. & Astrow, S.H. (1990). The origin and selective innervation of early muscle-fiber types in the rat. *Journal of Neurobiology*, **21**, 212–222.

Thompson, W., Kuffler, D.P. & Jansen, J.K. (1979). The effect of prolonged, reversible block of nerve impulses on the elimination of polyneuronal innervation of new-born rat skeletal muscle fibers. *Neuroscience*, **4**, 271–281.

Thompson, W.J., Soileau, L.C., Balice-Gordon, R.J. & Sutton, L.A. (1987). Selective innervation of types of fibers in developing rat muscle. *Journal of Experimental Biology*, **132**, 249–263.

Thompson, W.J., Sutton, L.A. & Riley, D.A. (1984). Fiber type composition of single motor units during synapse elimination in neonatal rat soleus muscle. *Nature*, **309**, 709–711.

Tian, L.J., Prior, C., Dempster, J. & Marshall, I.G. (1997). Hexamethonium- and methyllycaconitine-induced changes in acetylcholine release from rat motor nerve terminals. *British Journal of Pharmacology*, **122**, 1025–1034.

Trachtenberg, J.T. & Thompson, W.J. (1996). Schwann cell apoptosis at developing neuromuscular junctions is regulated by glial growth factor. *Nature*, **379**, 174–177.

Trachtenberg, J.T. & Thompson, W.J. (1997). Nerve terminal withdrawal from rat neuromuscular junctions induced by neuregulin and Schwann cells. *Journal of Neuroscience*, **17**, 6243–6255.

Tremblay, P., Dietrich, S., Mericskay, M., Schubert, F.R., Li, Z.L. & Paulin, D. (1998). A crucial role for pax3 in the development of the hypaxial musculature and the long-range migration of muscle precursors. *Developmental Biology*, **203**, 49–61.

Trimmer, J.S., Cooperman, S.S., Agnew, W.S. & Mandel, G. (1990). Regulation of muscle sodium-channel transcripts during development and in response to denervation. *Developmental Biology*, **142**, 360–367.

Tsujimoto, T., Umemiya, M. & Kuno, M. (1990). Terminal sprouting is not responsible for enhanced transmitter release at disused neuromuscular-junctions of the rat. *Journal of Neuroscience*, **10**, 2059–2065.

Tuffery, A.R. (1971). Growth and degeneration of motor end-plates in normal cat hind limb muscles. *Journal of Anatomy*, **110**, 221–247.

Ulenkate, H.J.L.M., Kaal, E.C.A., Gispen, W.H. & Jennekens, F.G.I. (1994). Ciliary neurotrophic factor improves muscle-fiber reinnervation after facial-nerve crush in young rats. *Acta Neuropathologica*, **88**, 558–564.

Unwin, N. (1998). The nicotinic acetylcholine receptor of the torpedo electric ray. *Journal of Structural Biology*, **121**, 181–190.

Van Essen, D.C. (1997). A tension-based theory of morphogenesis and compact wiring in the central nervous system. *Nature*, **385**, 313–318.

Van Essen, D.C., Gordon, H., Soha, J.M. & Fraser, S.E. (1990). Synaptic dynamics at the neuromuscular junction: mechanisms and models. *Journal of Neurobiology*, **21**, 223–249.

Van Mier, P. & Lichtman, J.W. (1994). Regenerating muscle fibers induce directional sprouting from nearby nerve-terminals – studies in living mice. *Journal of Neuroscience*, **14**, 5672–5686.

Van Ooyen, A. & Wilshaw, D.J. (1999a). Poly- and mononeuronal innervation in a model for the development of neuromuscular connections. *Journal of Theoretical Biology*, **196**, 495–511.

Van Ooyen, A. & Wilshaw, D.J. (1999b). Competition for neurotrophic factor in the development of nerve connections. *Proceedings of the Royal Society of London, Series B*, **266**, 883–892.

Verdiere-Sahuque, M., Akaaboune, M., Lachkar, S., Festoff, B.W., Jandrotperrus, M., Garcia, L., Barlovatzmeimon, G. & Hantai, D. (1996). Myoblast fusion promotes the appearance of active protease nexin I on human muscle cell surfaces. *Experimental Cell Research*, **222**, 70–76.

Verhage, M., Maia, A.S., Plomp, J.J., Brussard, A.B., Heeroma, J.H., Vermeer, H., Toonen, R.F., Hammer, R.E., van den Berg, T.K., Missler, M., Geuze, H.J. & Sudhof, T.C. (2000). Synaptic assembly of the brain in the absence of neurotransmitter secretion. *Science*, **287**, 864–869.

Vrbova, G., Lowrie, M.B. & Evers, J. (1988). Reorganization of synaptic inputs to developing skeletal-muscle fibers. *Ciba Foundation Symposia*, **138**, 131–151.

Vyskocil, F. & Vrbova, G. (1993). Nonquantal release of acetylcholine affects polyneuronal innervation on developing rat muscle fibers. *European Journal of Neuroscience*, **5**, 1677–1683.

Waerhaug, O., Dahl, H.A. & Kardel, K. (1992). Different effects of physical-training on the morphology of motor-nerve terminals in the rat extensor digitorum longus and soleus muscles. *Anatomy and Embryology*, **186**, 125–128.

Walro, J.M. & Kucera, J. (1999). Why adult mammalian intrafusal and extrafusal fibers contain different myosin heavy-chain isoforms. *Trends in Neurosciences*, **22**, 180–184.

Watson, W.E. (1974). *Cell biology of brain*. London: Chapman & Hall.

Weiss, P. & Edds, M.V. (1946). Spontaneous recovery of muscle following partial denervation. *American Journal of Physiology*, **145**, 587–607.

Wenner, P. & Frank, E. (1995). Peripheral target specification of synaptic connectivity of muscle spindle sensory neurons with spinal motoneurons. *Journal of Neuroscience*, **15**, 8191–8198.

Werle, M.J. & Herrera, A.A. (1987). Synaptic competition and the persistence of polyneuronal innervation at frog neuromuscular junctions. *Journal of Neurobiology*, **18**, 375–389.

Werle, M.J. & Herrera, A.A. (1991). Elevated levels of polyneuronal innervation persist for as long as 2 years is reinnervated frog neuromuscular junctions. *Journal of Neurobiology*, **22**, 97–103.

Westerfield, M., Liu, D.W., Kimmel, C.B. & Walker, C. (1990). Pathfinding and synapse formation in a zebrafish mutant lacking functional acetylcholine receptors. *Neuron*, **4**, 867–874.

Westgaard, R.H. & Lomo, T. (1988). Control of contractile properties within adaptive ranges by patterns of impulse activity in the rat. *Journal of Neuroscience*, **8**, 4415–4426.

Wigston, D.J. (1989). Remodeling of neuromuscular-junctions in adult-mouse soleus. *Journal of Neuroscience*, **9**, 639–647.

Wilkinson, R.S., Lunin, S.D. & Stevermer, J.J. (1992). Regulation of single quantal efficacy at the snake neuromuscular-junction. *Journal of Physiology, London*, **448**, 413–436.

Wilkinson, R.S., Son, Y.J. & Lunin, S.D. (1996). Release properties of isolated neuromuscular boutons of the garter snake. *Journal of Physiology, London*, **495**, 503–514.

Willshaw, D.J. (1981). The establishment and the subsequent elimination of polyneuronal innervation of developing muscle: theoretical considerations. *Proceedings of the Royal Society of London, Series B: Biological Sciences*, **212**, 233–252.

Windisch, A., Gundersen, K., Szabolcs, M.J., Gruber, H. & Lomo, T. (1998). Fast to slow transformation of denervated and electrically stimulated rat muscle. *Journal of Physiology, London*, **510**, 623–632.

Witzemann, V. & Sakmann, B. (1991). Differential regulation of myod and myogenin messenger-RNA levels by nerve induced muscle-activity. *FEBS Letters*, **282**, 259–264.

Witzemann, V., Brenner, H.R. & Sakmann, B. (1991). Neural factors regulate achr subunit messenger-RNAs at rat neuromuscular synapses. *Journal of Cell Biology*, **114**, 125–141.

Wood, S.J. & Slater, C.R. (1995). Action-potential generation in rat slow-twitch and fast-twitch muscles. *Journal of Physiology, London*, **486**, 401–410.

Wood, S.J. & Slater, C.R. (1997). The contribution of postsynaptic folds to the safety factor for neuromuscular transmission in rat fast- and slow-twitch muscles. *Journal of Physiology, London*, **500**, 165–176.

Wood, S.J. & Slater, C.R. (1998). Bet-spectrin is colocalized with both voltage-gated sodium channels and ankyrin(g) at the adult rat neuromuscular junction. *Journal of Cell Biology*, **140**, 675–684.

Xie, K.W., Wang, T., Olafsson, P., Mizuno, K. & Lu, B. (1997). Activity-dependent expression of nt-3 in muscle cells in culture: implications in the development of neuromuscular junctions. *Journal of Neuroscience*, **17**, 2947–2958.

Young, S.H. & Poo, M.M. (1983). Spontaneous release of transmitter from growth cones of embryonic neurons. *Nature*, **305**, 634–637.

Zahler, R., Sun, W., Ardito, T., Zhang, Z.T., Kocsis, J.D. & Kashgarian, M. (1996). The alpha 3 isoform protein of the Na$^+$, K$^+$-ATPase is associated with the sites of cardiac and neuromuscular impulse transmission. *Circulation Research*, **78**, 870–879.

Zhang, L.I., Tao, H.W., Holt, C.E., Harris, W.A. & Poo, M.M. (1998). A critical window for cooperation and competition among developing retinotectal synapses. *Nature*, **395**, 37–44.

Zhu, P.H. & Vrbova, G. (1992). The role of Ca-$^{2+}$ in the elimination of polyneuronal innervation of rat soleus muscle-fibers. *European Journal of Neuroscience*, **4**, 433–437.

Zoubine, M.N., Ma, J.Y., Smirnova, I.V., Citron, B.A. & Festoff, B.W. (1996). A molecular mechanism for synapse elimination: novel inhibition of locally generated thrombin delays synapse loss in neonatal mouse muscle. *Developmental Biology*, **179**, 447–457.

II.12
Target-independent Programmed Cell Death in the Developing Nervous System

RONALD W. OPPENHEIM, JORDI CALDERÓ, JOSEP ESQUERDA and THOMAS W. GOULD

ABSTRACT

Previous studies and reviews of programmed cell death in the developing nervous system have focused mainly on the target-dependent death of post-mitotic differentiating neurons. In the present review, by contrast, we discuss target-independent forms of neuronal death, including the death of precursor cells and immature post-mitotic neurons, as well as the death of glial cells in the central and peripheral nervous system. A major effort here is to understand the extrinsic mechanisms that regulate death and survival and to examine the adaptive significance of the various categories of neuronal and glial cell death. As a part of this goal, we also review the literature in both vertebrate and invertebrate species regarding the role of hormones in neuronal survival. From all of this, we conclude that a narrow focus on target-dependent programmed cell death does not adequately reflect the diversity, richness or complexity of either the mechanisms or the evolutionary significance of developmental cell death in the nervous system.

INTRODUCTION

The degeneration of large numbers of cells during the normal development of vertebrates has long been recognized to be an important event during most of embryogenesis, beginning in the late blastula stage and extending through fetal and early postnatal stages, and involving virtually all germ cell layers, tissues and organs (Glücksmänn, 1930, 1951; Clarke & Clarke, 1996; Sanders & Wride, 1995; Saunders, 1966; Källén, 1965; Hensey & Gautier, 1998; Jacobson *et al.*, 1997; Ernst, 1926). Cell death during the development of invertebrates has also been

A.F. Kalverboer and A. Gramsbergen (eds.), Handbook of Brain and Behaviour in Human Development, 343–408
© 2001 Kluwer Academic Publishers. Printed in Great Britain.

recognized for many years (Truman, 1984; Sanders & Wride, 1995; Clarke & Clarke, 1996; Ellis *et al.*, 1991; Steller & Grether, 1994). In both vertebrates and invertebrates cell death in the developing nervous system is especially conspicuous and has been the subject of research for more than 100 years (Clarke & Clarke, 1996). Historically, most of the early studies of developmental neuronal cell death were descriptive, mainly documenting the occurrence and the spatial–temporal pattern of cell loss, with few attempts to address the functional significance (i.e. the adaptive role) of cell death, and with an almost total neglect of the proximate cell interactions or biochemical and molecular mechanisms involved.

Ernst (1926) and Glücksmann (1930, 1951) provided the first comprehensive reviews of the occurrence of cell death during normal development, and Glücksmann, in particular, attempted the first crude classification of cell death according to its developmental function. He distinguished three categories of cell death: (1) morphogenetic (e.g. interdigital cell death for creating digits); (2) histogenetic (e.g. cell death involved in creating cell layers in the retina); and (3) phylogenetic (e.g. loss of the pronephros during kidney development in higher vertebrates). Historically this classification represents an important and valiant attempt to understand the biological utility of developmental cell death, but viewed in retrospect it falls short of a comprehensive conceptualization of the many currently recognized roles of developmental cell death (e.g. Ellis *et al.*, 1991: Sanders & Wride, 1999; Jacobson *et al.*, 1997; Oppenheim, 1999). Although, in the absence of empirical data, they were forced to speculate on the proximate causes of cell death, to their credit both Ernst and Glücksmann discussed possible exogenous and endogenous causes for why cells die. For example, they suggested that impaired nutrition, loss of metabolic activity, loss of proliferative signals, mechanical factors (physical stress) and reduced vascularization may all account for individual cell deaths.

A significant landmark in the history of this field was the thoughtful review by the embryologist John Saunders published in *Science* in 1966. Not only did he summarize much of the experimental literature on the functional significance and proximate causes of developmental cell death, but his review was notable for identifying and addressing virtually all of the major issues that have been the focus of investigators in this field right up to the present time. For example, in anticipation of current views regarding the genetic control of cell death, Saunders stated that:

> It is abundantly clear that regressive phenomena during embryonic development and metamorphosis are programmed to occur in a highly predictable manner. Thus, degenerative changes in development would seem to proceed under genetic control, even as the progressive ones do; and as might be expected, there are mutant genes that achieve phenotypic expression through extending or contracting the normal pattern of cellular death, or by bringing about extensive death of tissues or organs in which degradative changes normally do not occur (1966, p. 607).

In this context Saunders also considered whether developmental cell death occurs by "suicide" or "assassination". That is, whether "cells are capable of effecting their own demise – by releasing autolytic enzymes or alternatively whether cells scheduled for death might be slain by extrinsic factors" (1966, p. 608). The explosion of research during the past 10 years that has documented in considerable detail the active, regulated nature of programmed cell death (PCD) has fully vindicated Saunders' visionary, albeit at the time counter-intuitive, ideas about the genetic regulation of PCD.

In view of the remarkable increase in information concerning PCD that has appeared since the reviews of Glücksmann (1951) and Saunders (1966), it is not surprising that the classification of developmental PCD has been considerably expanded beyond the three general categories of morphogenetic, histogenetic and phylogenetic as explanations for the biological utility of cell loss during development (Ellis *et al.*, 1991; Jacobson *et al.*, 1997). As summarized in Table 1, for most species, tissues and organs, the evolutionary question of why cell death occurs has several potential answers. Consideration of these different possibilities, together with the relatively sparse evidence on the proximate mechanisms involved in each one in the context of the developing nervous system, is the major focus of this review. Because, historically, most of the attention on cell death in the nervous system has focused on target-dependent regulation of cell death and survival, and because this form of cell death has been extensively reviewed recently (Burek & Oppenheim, 1999; Oppenheim, 1981, 1991, 1999; Pettmann & Henderson, 1998), we wish to focus our review on non-target-dependent forms of cell death in the developing nervous system. This means focusing on cell death prior to the formation of synaptic connectivity, as well as on survival signals that may be provided by afferents, glia and hormones rather

Table 1 Some possible reasons for the occurrence of developmental cell death

Reasons	Examples
1. Cytoplasmic inheritance predetermines death	Lineage related cell death in *C. elegans*
2. Death of excess cells provides metabolites, energy sources or inductive signals for surviving cells	No available evidence
3. Morphogenetic sculpting of structures	Formation of digits
4. Creation of sexual dimorphism	Loss of the Müllerian ducts in males and the Wolffian ducts in females
5. Removal of vestigial structures	Loss of the pronephros in mammals
6. Negative selection of abnormal, redundant or misplaced cells	Deletion of lymphocytes with self-reactive receptors; creation of segment-specific ganglia in insect nervous system
7. Removal of cells or structures that subserve a transient function	Death of many structures (e.g. the tadpole tail, larval muscles) during metamorphosis
8. Control of cell numbers	Target-dependent cell death in the nervous system
9. Deletion of harmful cells	Death of cells with gene or chromosomal abnormalities
10. Producing differentiated cells without organelles	Red blood cells; keratotinocytes

than solely from target-derived sources. We also include here an assessment of the regulation of survival of non-neuronal glial cells in the peripheral and central nervous system.

PRECURSOR CELLS AND IMMATURE POSTMITOTIC NEURONS

Because the emerging recognition of massive cell death in the developing nervous system in the 20th century occurred historically within the context of studies of neuron–target interactions, such as occurs between motoneurons and their muscle targets or between retinal ganglion cells and their CNS targets (Oppenheim, 1981, 1991; Hamburger, 1992), this form of target-dependent cell death, which occurs relatively late during neuronal differentiation, has received an inordinate amount of attention. Accordingly, although other forms of neuronal cell death that occur earlier in development were known (Glücksmann, 1930, 1951), they have frequently been neglected, overlooked or forgotten.

Target-dependent neuronal death is considered to be a means of controlling neuronal numbers for the establishment of optimal connectivity and innervation of target populations, and is thought to be controlled by competition for target-derived trophic molecules. In this context the decision to live or die requires interactions between neurons and their targets mediated via axonal projections. However, if neurons die prior to this stage of differentiation, it is obvious that the death is occurring for other reasons and must be mediated by mechanisms distinct from those involved in target-dependent cell death. Using the categorization provided in Table 1 we shall review the evidence for the different types of target-independent cell death in the developing nervous system and consider the proximate mechanisms that promote survival or death in each situation.

Cytoplasmic inheritance

The inheritance of maternally derived cytoplasmic components that are selectively distributed to dividing embryonic cells, and act to control cell fate, is a common strategy during invertebrate and, at least early stages of, vertebrate development (Gilbert, 1997). In the nematode worm *Caenorhabditis elegans*, most dying cells in the embryo (and most of these are neurons or neuronal precursors) appear to be induced to die by cytoplasmic factors that they inherit from the maternal egg. These cells are predetermined to die at birth based strictly on their lineage, although the specific proximate factors in the cytoplasm that they inherit from ancestors that specifies their fate are not well understood (Gilbert, 1997; Ellis *et al.*, 1991). For most of these cells, cell interactions do not appear to play any role in their fate; rather the cell deaths occur cell autonomously. Although the death-initiating, lineage-based, cytoplasmic factors are unknown many of the downstream steps that specify and execute the death signal are well characterized (Oppenheim, 1999; Conradt & Horvitz, 1998). The function or utility of this form of predetermined cell death is not well understood, although in both *C. elegans* and insects some deaths may be required for generating differences in cell numbers in specific segmental ganglia (Sulston & Horvitz, 1977; Goodman & Bate, 1981).

Although the strategy of cytoplasmic inheritance for generating phenotypic diversity is not entirely lost in vertebrate embryos, its role appears to be restricted to the earliest stages of development and is primarily involved in germ cell determination. In vertebrates its role in phenotypic fate determination is replaced in large measure by cell interactions and, accordingly, we are unaware of any evidence for this type of cytoplasmic predetermined cell death in the developing vertebrate nervous system. In some vertebrates (e.g. frogs) there is an analogous process in which a maternally derived death programme is set up at fertilization but is then activated at the mid-blastula stage only if the embryo is injured (Hensey & Gautier, 1997). However, in this case all of the cells of the embryo, not just select cells, undergo cell death. The similarity to invertebrates then is only that the potential for undergoing cell death is maternally derived, presumably via cytoplasmic determinants.

Constituents of dying cells are utilized by surviving cells

Although it is conceivable that the degradative products of dying cells could be recycled and utilized by surviving cells, even if this were shown to occur this is not likely to be a plausible rationale for the evolution of developmental cell death. At best, it could be thought of as a kind of secondary co-opting of a type of cell death that arose for other reasons. We are unaware of any direct evidence that dying cells in any tissue are recycled in this way. However, it is interesting that cell deaths in many embryonic tissues, including the nervous system, are often phagocytosed by adjacent surviving cells of the same or different phenotype rather than by professional macrophages (e.g. Chu-Wang & Oppenheim, 1978; García-Porrero & Santander, 1979). Recent demonstrations of the prevention of normal neuronal cell death may provide a means for directly testing the validity of this idea (White *et al.*, 1998; Kuida *et al.*, 1996; Hakem *et al.*, 1998).

Morphogenetic sculpting of tissues

Most of the classic examples of this form of cell death involve non-neuronal structures and tissues (e.g. Glücksmann, 1951; Saunders, 1966; Jacobson *et al.*, 1997; Sanders & Wride, 1995). Nonetheless, many apparent examples of this form of cell death also exist in the developing nervous system (Table 2). We use the term "apparent" here because, although it is relatively easy to ascribe such a putative role to a particular instance of cell death, proving that it is actually required for the morphogenetic event in question is not always so easy. Cell death could be the result rather than the cause of a morphogenetic event. Some examples of putative morphogenetic cell deaths in the nervous system are provided in Table 2 and additional examples can be found in the cited references. Here we wish to focus on just a few of these for which experimental evidence exists for the morphogenetic requirement of cell death, or for which the mechanisms regulating death or survival have begun to be examined.

Several of the examples provided in Table 2 involve major morphogenetic events such as invagination, delamination/detachment and fusion of cells and cell layers. Although it remains to be shown that the cell deaths that accompany these specific events are required for their occurrence (but see Silver, 1978), in other

Table 2 Morphogenetic cell death in the nervous system

Example	Putative function
Neurulation	Detachment from ectoderm; neural tube closure[1-4]
Secondary neurulation	Cavitation of caudal neural tube[5]
Optic vesicle/optic stalk	Formation of optic cup; creation of space for axon growth[6-11]
Otic vesicle	Invagination and detachment from ectoderm[1]
Inner ear	Formation of semicircular canals[12]
Olfactory placode/vesicle	Invagination and detachment from ectoderm[1]
Cerebral cortex	Differential thickness of layers[13]
Telencephalon	Formation of choroid plexus[14]
Retina	Creation of layers; space for axon growth[15-18]
Cerebellar medulla	Space for axon growth[19]
Olfactory epithelium	Space for axon growth/cell migration[20-21]
Corpus callosum ("glial sling")	Fusion of cerebral hemispheres; axon growth; creation of cavum septi pellucidi[22]
Neural crest in rhombomeres 3 and 5	Craniofacial development[23]
Spinal cord	Space for axon growth[3,24]

[1]Glücksmänn, 1951
[2]Weil et al., 1997
[3]Homma et al., 1994
[4]Oppenheim et al., 1999
[5]Findeis & Milligan, 1995
[6]Garcia-Porrero et al., 1984
[7]Garcia-Porrero & Santander, 1979
[8]Knabe & Kuhn, 1998
[9]Luemle et al., 1999
[10]Van den Eijnde et al., 1999
[11]Frago et al., 1998
[12]Fekete et al., 1997

[13]Finlay, 1992
[14]Furuta et al., 1997
[15]Ulshafer & Clavert, 1979
[16]Cuadros & Rios, 1988
[17]Frade et al., 1996
[18]Cook et al., 1998
[19]Ashwell, 1990
[20]Pellier & Astic, 1994
[21]Carr & Farbman, 1993
[22]Hankin et al., 1998
[23]Graham et al., 1996
[24]Frade & Barde, 1999

situations integrin-mediated cell–matrix interactions are thought to be critically involved in promoting survival since the loss of cell attachments results in cell death and aberrant morphogenesis (Ruoslahti & Reed, 1994; Coucouvanis & Martin, 1995). It seems plausible, therefore, that the loss of integrin-mediated signalling may be a more general mechanism that triggers the death of intervening cells and thereby contributes to the detachment or delamination of tissue layers. Inhibition of normal cell death in the neural folds can prevent their fusion and results in the failure of neural tube closure (Weil et al., 1997; Oppenheim et al., 1999). In this situation cell death may be necessary for promoting cell–cell or cell–matrix attachments required for neuronal tube closure. Cell death in the neural folds in the hindbrain also appears to be required for neural tube closure (Kuan *et al.*, 1999).

Secondary neurulation in the caudal spinal cord occurs by cavitation and involves cell death (Findeis & Milligan, 1995). Although the mechanisms involved here are not known, in an analogous process of cavitation involving cell death in the blastula–gastrula stage of mouse embryos for creating the proamniotic cavity, cell interactions (attachments) with the extracellular matrix (ECM) were shown to provide survival signals, whereas the loss of these attachments resulted in cell death/cavitation (Coucouvanis & Martin, 1995). Either (or both) the loss of a survival signal provided by the ECM or the active participation of a death-inducing receptor, such as a Fas-type receptor (Abbas, 1996; Nagata, 1997), could explain the selective death in such situations.

The survival and differentiation of the neural tube caudal to the thoracic region is dependent upon signals derived from notochord/floor plate. Excision of a specific region of Hensen's node at the 5–6 somite stage in the chick embryo results in the loss of notochord and floor plate, the absence of sonic hedgehog expression and, ultimately, massive PCD of all tissues, including the previously formed neural tube caudal to the brachial/thoracic region, such that by E5 surviving embryos are truncated at the forelimb level (Charrier *et al.*, 1999). Thus, cells derived from the excised region of Hensen's node are required for promoting the survival of caudal tissues including neural tube.

In the chick embryo the selective death of many neural crest cells in rhombomeres 3 and 5 of the hindbrain is thought to be required for normal craniofacial patterning (Graham *et al.*, 1996). This rhombomere selective cell death helps to segregate streams of migrating crest cells that are destined to make distinct contributions to skeletal and connective tissue components of the face and head. Although perturbations of rhombomeric neural crest cell death are reported to result in craniofacial malformations (e.g. Takahashi *et al.*, 1998), we are unaware of any direct evidence showing that cell death in rhombomeres 3 and 5 (r3 and r5) is specifically required for normal craniofacial morphology. Assuming that this will eventually be shown, however, this provides a paradigm example of morphogenetic cell death in the nervous system. This example is also especially informative in that the proximate mechanisms involved are reasonably well understood. The induction of cell death in r3 and r5 requires signals derived from adjacent rhombomeres. Although the molecular nature of the signals is unknown, in their absence crest cells in r3 and r5 survive. Two genes, *BMP 4* and *msx-2*, are selectively expressed in crest cells in r3 and r5, and both are required for cell death (Graham *et al.*, 1996). Mis-expression of high levels of *msx-2* is sufficient to induce the ectopic death of crest cells in even-numbed rhombomeres (Takahashi *et al.*, 1998). BMPs and msx have also been implicated in interdigital cell death in the limbs (Zou & Niswander, 1996) and in morphogenetic cell death in the forebrain (Furuta *et al.*, 1997). In all of these cases, BMPs appear to act by an autocrine pathway as death-inducing molecules, analogous to the Fas system in lymphocytes. However, at least in the case of crest cells in the hindbrain, signals from adjacent rhombomeres are required for inducing the BMP/msx death pathway in r3 and r5.

In some regions of the developing nervous system the early death of cells is associated with the growth of axons, suggesting that cell death creates physical spaces required for the passage of projecting axons (Horsburgh & Sefton, 1986, Cuadros & Rios, 1988; Navascues et al., 1988; Homma et al., 1994; Silver & Robb, 1979; Frade & Barde, 1999). If correct, this would be another compelling example of morphogenetic cell death in the nervous system. However, we are unaware of any evidence that such spaces are required for normal axonal growth. In fact, in one case involving the death of cells in the path of growing spinal commissural axons in the chick embryo, the prevention of cell death had no apparent effect on axon growth (Homma & Oppenheim, 1991). Because we consider it more likely that this type of cell death serves other functions, we will consider some examples as well as the proximate mechanisms involved in the section below on negative selection.

Removal of vestigial structures

This is the type of cell death categorized as phylogenetic cell death by Glücksmann (1951). Many examples of this form of cell death have been described in non-neuronal tissues (e.g. the pronephros, the mesonephros and the tail bud). However, we are unaware of any *bona-fide* examples of this form of cell death in the developing vertebrate nervous system, although the lineage-related death of segmental neurons in *C. elegans* and some insects may fit this classification (Ellis et al., 1991).

In the chick embryo, Levi-Montalcini (1950) suggested that a transient population of motoneurons in the cervical neural tube that degenerate, represent vestigial sympathetic preganglionic visceral motoneurons that persist in the adults of some cold-blooded ancestral vertebrates. However, a recent re-examination of this population has shown that, although their death is in some respects unique, they are in fact somatic motoneurons that project to skeletal muscle prior to their death, and therefore are not vestigial preganglionic cells (Yaginuma et al., 1996).

Cells that serve a transient function

This type of cell death is common among organisms such as insects and amphibians that undergo dramatic structural changes during metamorphosis that often involves wholesale reorganization of the nervous system, including the death of many parts of the embryonic and larval nervous system (Ellis et al., 1991; Truman, 1984). The utility of this form of cell death is obvious and because hormones are often involved in promoting both cell survival and death we will discuss this type of cell death in the section below on hormones and neuronal cell death.

In addition to neuronal cell death during metamorphosis, however, there are other examples of cells in the nervous system that serve a transient function and are then removed by cell death. One prominent example is neuronal or glial cells that are used to guide growing axons along selective pathways towards their targets, and which then undergo cell death. Although most such cases involve invertebrate species (Stewart et al., 1987; Awad & Truman, 1997; Sonnenfeld & Jacobs, 1995; Kutsch & Bentley, 1987; Bate et al., 1981), at least one extensively studied example is known in mammalian cerebral cortical development. This involves the cortical subplate in which neurons appear to serve a transient function as intermediate targets for thalamic afferents as well as for pioneering pathways for cortical axonal projections, after which subplate cells then undergo cell death (Ferrer et al., 1992). Although the proximate mechanisms responsible for the death of subplate neurons are not entirely known, one possibility is that, as has been shown for many other developing neurons (Linden, 1994), the loss of thalamic afferents following their projection beyond the subplate removes a required transient source of orthograde trophic support, resulting in the degeneration of the subplate cells (Price & Lotto, 1996). Recent evidence suggests that blockade of NMDA receptors, thereby mimicking loss of afferents, increases cell death of subplate cells, and this appears to be mediated by rises in intracellular calcium via voltage-dependent channels (McKellar et al., 1999).

Deletion of harmful cells

Although it seems highly likely that some of the cell deaths that occur normally in the developing nervous system are of this type, few well-established examples exist in the literature. The most compelling example of this type of cell death involves not the nervous system but rather the immune system. During normal development, for example, many thymocytes are generated with T-cell receptors capable of recognizing and attacking an individual's own tissues (autoimmune diseases). To prevent this before these cells mature and leave the thymus, they are selectively induced to undergo cell death by cell–cell interactions involving Fas ligand and Fas receptors (Nagata, 1997; Abbas, 1996). Virally infected cells in many tissues, including neurons, also have the capacity to trigger their own cell death as a means for limiting the spread of infection (Ameisen, 1996). Another general process that probably also occurs in most developing tissues is the selective elimination of abnormal cells that contain lethal gene mutations, chromosomal abnormalities, aberrant cell cycle or meiotic events, defective biogenesis or other types of damage. Presumably, if allowed to proliferate and continue their development, these cells would reduce the chances of survival of the embryo or future progeny (e.g., germ cells). Damaged or aberrant cells appear to activate intrinsic death programmes (Hensey & Gautier, 1997; Vaux et al., 1994). At least some of the cell death reported to occur in proliferating and immature precursor cells in the nervous system may fit this category (but see below). An interesting recent example from outside the immune system involves wing development in insects. Experimental perturbation of the expression of a wing morphogen can result in aberrant wing morphogenesis by disrupting proximal–distal positional information. However, the embryo may avoid this by activating a selective cell death programme in the affected regions of the wing, that removes only those wing cells with abnormal morphogen expression and, thereby restoring normal wing morphogenesis (Adahi-Yamada et al., 1999).

Finally, it could be argued that neurons with aberrant synaptic connections that selectively undergo developmental cell death (Clark & Cowan, 1976; O'Leary et al., 1986; Catsicas et al., 1987) represent harmful cells. At present this may be the only *bona-fide* example of this kind of cell death in the nervous system. However, even here it is not known whether the retention of these cells in the nervous system would in fact be maladaptive. Two recent variations on this theme are quite interesting. Studies of the repulsive guidance molecule Semaphorin III suggest that it may act also as a death-inducing factor selectively eliminating peripheral sensory neurons whose axons are misguided and stray away from the main nerve fascicle and into a region expressing high levels of Semaphorin III (Gagliardini & Frankhauser, 1999). In conceptually related studies of spinal commissural neurons it was shown that the survival of these cells depends upon trophic signals derived from an intermediate target along their pathway, the floor plate (Wang & Tessier-Lavigne, 1999). As with Semaphorin III, this provides a mechanism for eliminating misprojecting neurons and preventing the formation of aberrant connections during development.

Cells without organelles

The process by which lens cells, skin keratinocytes and mammalian red blood cells lose their nucleus and other organelles during development has been suggested to be a modified form of cell death (Jacobson *et al.*, 1997; Wride *et al.*, 1999), in that the intracellular death machinery may have been co-opted to mediate this kind of cellular reorganization. Although the modified lens and red blood cells continue to survive intact, the skin cells do, in fact, undergo an abortive cell death-like transition in which the cell corpses are not phagocytosed but rather are retained and utilized to form the squamous layer of the skin surface. The transition of keratinocytes into organelle-less squamous cells is thought to involve the activation of pro-apoptotic proteases (caspases) that have been shown to be important for the complete degeneration of other cell types (Weil *et al.*, 1999). In the case of the lens, the removal of nuclei and organelles also involves caspases and serves to provide a transparent region in the centre of the lens (Wride *et al.*, 1999). There are no known examples of this abortive form of "cell death" in the nervous system. However, a related kind of degradation occurs in individual myonuclei of striated muscle cells. In normal ageing or denervated muscle, individual nuclei undergo apoptotic degeneration and elimination without the loss of other myonuclei and in the absence of myofibre degeneration (Borisov *et al.*, 2000). This may be a means of eliminating unneeded subsynaptic myonuclei in the process of synaptic remodelling of the neuromuscular junction.

Control of cell numbers

Overproduction of cells in many developing tissues by proliferation appears to be a common strategy which, together with cell death, provides a two-step process for optimizing cell numbers to attain functional efficiency. The well-studied case of target-dependent neuronal death is a paradigm example of this type of cell death involving postmitotic cells (Oppenheim, 1981, 1991, 1999; Burek & Oppenheim, 1999; Pettmann & Henderson, 1998). Neuronal interactions with targets and other cells, as well as glial–neuron interactions (e.g. glia–axon interactions), involve a competition for trophic support that, if successful, promotes the survival of postmitotic neuronal and glial cells and ultimately therefore controls the regulation of optimal cell numbers (see section on regulation of glial cell survival, below).

As discussed above, the developmental death of mitotically active cells may serve to remove damaged and potentially harmful cells. However, cell death among proliferating populations of stem or precursor cells may be another way of controlling final cell numbers that is distinct from both proliferation itself (and cell cycle duration) and the more common death of excess postmitotic cells. That is, cell death during this stage of neurogenesis may be a means of controlling the size of the pool of progenitor cells. Although this potential role for cell death during neurogenesis is often ignored (e.g. Williams & Herrup, 1988; Williams *et al.*, 1998; Ross, 1996), the normal cell death of mitotically active (or newly postmitotic) cells in the nervous system has frequently been reported (Carr & Simpson, 1982; Barres *et al.*, 1992a; Wood *et al.*, 1993; Carr & Farbman, 1993; Homma *et al.*, 1994; Blaschke *et al.*, 1996, 1998; Furuta *et al.*, 1997; Lefcourt *et al.*, 1996; Liebl *et al.*, 1997; Thomaidou *et al.*, 1997; El Shamy *et al.*, 1998). In some of these cases local autocrine or paracrine availability of neurotrophic factors appears to regulate the survival of proliferating precursor cells (e.g., El Shamy *et al.*, 1998; Ockel *et al.*, 1996). For example, proliferating precursor cells in the dorsal root ganglion that are deprived of NT-3 fail to complete the S phase of the cell cycle and undergo cell death by what appears to be a ligand-induced cell death process. In this way cell death acts to control gangliogenesis and cell number prior to target innervation, by regulating how many mitotically active precursor cells are available.

Negative selection

The classic example of this form of cell death involves the ligand-induced death of potentially harmful lymphocytes in the immune system (Abbas, 1996; Nagata, 1997), a process by which specific cellular phenotypes are recognized and eliminated. Although it has been argued that the immune system is unique in this respect (Abbas, 1996), in fact other tissues, including the nervous system, appear to utilize a similar process for actively removing cells that are misplaced or otherwise aberrant based on their phenotype. We included this type of cell death in the immune system under the category of deletion of harmful cells, but such cell death may be part of a more general strategy of negative phenotypic selection that also contributes to the patterning or morphogenesis of other tissues. However, whether this type of cell death is categorized as negative selection, removal of harmful cells or morphogenetic will have to be decided on a case-by-case basis and, in the final analysis, may be more a matter of semantics than of substance.

In the nervous system negative selection by cell death has generally been postulated to play a role in mitotically active and newly postmitotic precursor cells (Homma *et al.*, 1994; Blaschke *et al.*, 1998; Oppenheim *et al.*, 1999). However, in principle, even some forms of target-dependent cell death could be included in this category. For example, if neurons that fail to compete successfully for trophic support from targets (and on that basis are considered aberrant), are then selectively removed by ligand-induced cell death, then this would also be a form of negative selection. Recent studies suggest that such a mechanism may be involved in the normal death of cortical neurons (Cheema *et al.*, 1999), spinal motoneurons (Raoul *et al.*, 1999) and olfactory neurons (Farbman *et al.*, 1999).

As discussed in the previous section, the death of precursor cells has been repeatedly documented in many parts of the developing nervous system. However, the full extent of cell death in proliferating and newly postmitotic cells has only recently been appreciated (Morshead & Van der Kooy, 1992; Homma *et al.*, 1994; Blaschke *et al.*, 1996, 1998; Thomaidou *et al.*, 1997) and remains controversial (Gilmore *et al.*, 1999). In the most detailed study that is presently available, it was estimated that the percentage of mitotically active ventricular zone (VZ) cells in rat cerebral cortex that die before becoming part of the postmitotic population is: 1.7% at E14; 8.8% at E16; 17.5% at E19; and 37% of the subventricular zone (SVZ) cells at birth (Thomaidou *et al.*, 1997). Because many newly postmitotic cells in the VZ and SVZ also undergo cell death, it is clear that the death of precursor cells in the nervous system is a quantitatively significant event, though perhaps not nearly as massive as once suggested (cf. Blaschke *et al.*, 1996 with Gilmore *et al.*, 2000).

Although in the present review we have chosen not to discuss the extensive literature on intracellular pathways that mediate cell death, it is nonetheless of considerable interest that the death of cortical precursor cells appears to be mediated by a pathway distinct from that of target-dependent postmitotic neurons. Whereas genetic deletion of the pro-apoptotic genes caspase-3, caspase-9 and Apaf1 (Ced-4) prevents the death of cortical neuronal precursor cells (Kuida *et al.*, 1996, 1998; Cecconi *et al.*, 1998) these mice have apparent normal numbers of postmitotic target-dependent neurons. By contrast, genetic deletion of the pro-apoptotic gene *Bax* affects the survival of target-dependent postmitotic neurons but not cortical precursor cells (White *et al.*, 1998).

As discussed above (also see Voyvodic, 1996), there are at least three plausible reasons for the death of neuronal precursor cells: (1) removal of harmful or damaged cells, (2) regulation of cell numbers, and (3) the negative selection of inappropriate phenotypes as a constructive process in patterning the developing nervous system. A fourth possibility, namely that cell death at this stage is required for creating physical pathways for axon growth, cannot be excluded but seems unlikely (see section on morphogenetic sculpting, above). Although there is presently a notable paucity of experimental evidence that would allow one to argue in favour of one or more of these possibilities, we tend to favour the role of negative selection.

Precursor cells appear to become committed to die during or shortly following S phase (Thomaidou *et al.*, 1997), a stage when postmitotic cell fate is determined in many cells (McConnell & Kaznoski, 1991). Perturbations of cell fate in the spinal cord neural tube result in increased cell death of precursor cells, at least some of which express an aberrant phenotype relative to their position along the dorsal–ventral axis of the spinal cord (Oppenheim *et al.*, 1999). In the developing immune system chromosomal breaks and recombination events that are mediated by specific genes involved in repairing DNA double-strand breaks are used as a normal strategy to generate diversity in antigen receptor expression in lymphocytes. Targeted mutations in these genes result in defective lymphogenesis and embryonic lethality (Gao *et al.*, 1998). Surprisingly, these mutant mouse embryos also exhibit extensive cell death of early postmitotic neurons throughout the central and peripheral nervous system between E10 and E16, and this death occurs in a pattern associated with the onset and cessation of neurogenesis in each region. One explanation of the neuronal death in this situation is that, as in the immune system, neuronal chromosomes are normally cleaved and rejoined as part of a selection process that deletes those cells making aberrant recombination products (Chun & Shatz, 1999). The aberrant cells may then be removed by ligand-induced cell death similar to Fas-induced death in the immune system. Although this is in principle an interesting idea, there are several reasons for caution, not the least of which is the apparent absence of a single molecule in neurons that, like immunoglobulins in the immune system, could be the target of recombination (Gilmore et al., 2000).

Although most cell death in the nervous system is thought to result from trophic factor deprivation (Pettmann & Henderson, 1998; Raff et al., 1993), ligand-induced cell death is also known to occur (Cassaccia-Bonnefil et al., 1998). In the developing spinal cord and retina newly postmitotic cells undergo a normal phase of ligand-induced cell death involving NGF acting via the neurotrophin p75 receptor (Frade et al., 1996; Frade & Barde, 1999). NGF derived from microglia in the retina is required for cell death to occur (Frade & Barde, 1998). This represents a type of direct paracrine-mediated cell death in which the target cells express the "death" receptor. Fas-mediated cell death is a similar means of removing cells by PCD, and although Fas-mediated death is best known from the immune system it also appears to function in the nervous system (see above). Although we are unaware of any similar evidence for neurons, still another form of indirect paracrine-mediated cell death has been described in reproductive tract development. Regression of the Müllerian duct in males is mediated by a death-inducing TGF-β family member, Müllerian inhibiting substance (MIS). Surprisingly, however, the Müllerian duct cells that die (epithelium) are not those that express the MIS receptor. Rather, the mesenchymal cell layer expresses this receptor and evidence suggests that MIS induces death by acting on the mesenchyme to alter mesenchymal–epithelial interactions which then indirectly results in epithelial cell death (Roberts et al., 1999).

Because NGF-induced cell death in the retina occurs prior to target innervation and involves early postmitotic neurons, it may represent the negative selection of inappropriate cellular phenotypes. In the future, more detailed studies of the phenotype of neurons rescued from this early form of cell death (Kuida et al., 1996; Motoyama et al., 1995; Hakem et al., 1998) may provide a means to assess the validity of negative selection as a strategy for patterning the developing nervous system.

Summary and conclusions

Although the target-dependent death of developing postmitotic neurons has been more extensively studied, and is better understood regarding its utility and proximate mechanisms, it is nonetheless clear that the death of precursor cells prior to synaptic interactions is also a significant event during nervous system development and for that reason deserves more attention in order to reveal its role(s) in the generation of the mature nervous system. The functions of early cell death in the nervous system appear to be diverse, although experimental evidence for many of these is lacking. The mechanisms that regulate this phase of cell death also appear to be diverse and include ligand-induced death, cell–matrix interaction-regulated death and autocrine/paracrine loops in which trophic factor availability is the deciding factor.

REGULATION OF GLIAL CELL SURVIVAL

Until recently developmentally regulated PCD within the nervous system has been primarily recognized and characterized in neurons. However, it is now clear that normal PCD of developing glial cells also occurs in both oligodendrocytes and Schwann cells, and to a lesser extent in astrocytes. In the case of oligodendrocytes, studies in the rat optic nerve show that about 50% of generated oligodendrocytes die normally within 2–3 days after they develop (reviewed in Barres et al., 1992a, Barres & Raff, 1994). Axons decisively influence oligodendrocyte development. It is known that electrical activity in axons stimulates the proliferation of oligodendrocyte precursor cells, probably by the release of growth factors (Barres & Raff, 1993). In addition, oligodendrocytes compete with each other for limiting amounts of neuronally derived survival factors, such that in the absence of signals derived from axons oligodendrocytes undergo PCD, whereas PCD is reduced in the face of more than normal numbers of optic axons (Burns et al., 1996). Oligodendrocyte progenitor cells express PDGF alpha-receptors (Hall et al., 1996), and PDGF appears to be the main known trophic factor for newly formed oligodendrocytes, although IGF-1, IGF-2 and bFGF also have trophic actions on these cells (Barres et al., 1992a, Yasuda et al., 1995), and neurotrophins (e.g. BDNF and NT-3) can also enhance survival and proliferation of oligodendrocyte precursors by acting through functional trk receptors (McTigue et al., 1998).

Developing Schwann cells (Schwann cell precursors) can be maintained in vitro only in presence of signals derived from neurons (Jessen et al., 1994). In peripheral nerves a special relationship is established between axons and their ensheathing myelinating Schwann cells, such that 1:1 cell matching is attained. Studies on developing nerves of chick suggest that, like oligodendrocytes, the numerical adjustment of the Schwann cell population is achieved by eliminating excess Schwann cell precursors by means of normal PCD (Ciutat et al., 1996). A well-defined period of normal Schwann cell PCD has been described in the chick embryo (Figure 1) and the experimental reduction of axons results in a dramatic increase in Schwann cell apoptosis (Ciutat et al., 1996), whereas increasing motoneuron numbers and their axons reduces normal PCD in Schwann cells (Calderó et al., 1996; Esquerda et al., 1999). Schwann cell susceptibility to undergo apoptosis following the loss of axon-derived signals is transitory. It is well known that mature Schwann cells survive and proliferate, rather than die by apoptosis, in response to axonal deprivation (for example after peripheral nerve lesion) (Abercrombie & Johnson, 1946; Meier et al., 1999; Payer, 1979). In the chick embryo the vulnerability of Schwann cells to undergo apoptosis following axon loss disappears after E14 (Esquerda et al., 1999) whereas in the rat this period of vulnerability ends during the early postnatal period (Grinspan et al., 1996; Meier et al., 1999). Recently, it has been found that the transcription factors Krox-20 and SCIP (also named Oct-6) have critical roles in the differentiation of Schwann cells. At the stage preceding myelination, SCIP is transiently expressed

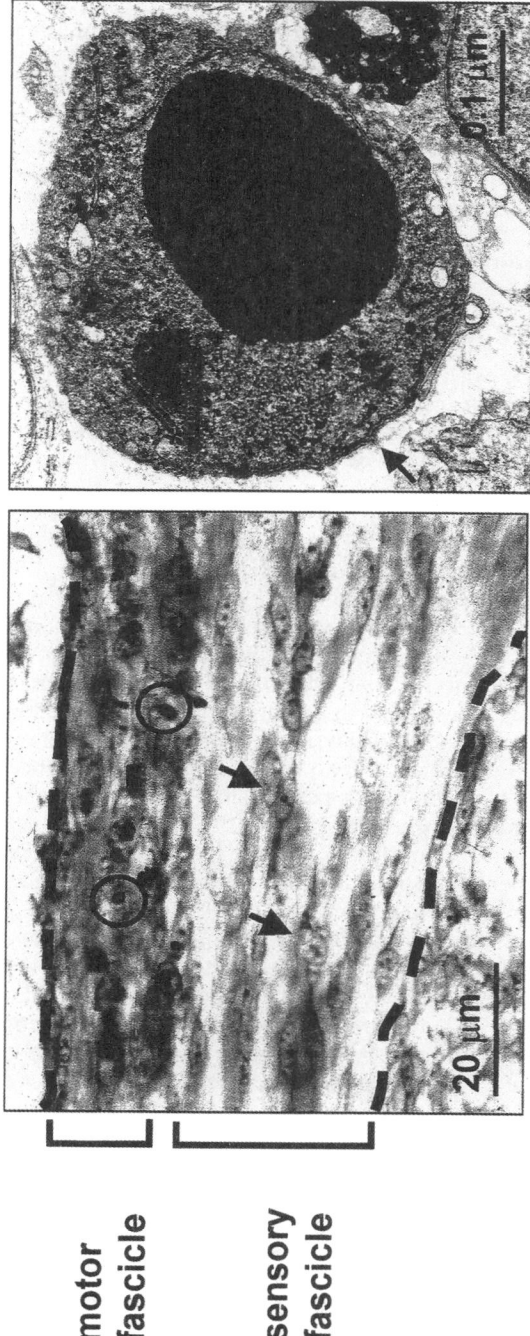

Figure 1 Schwann cell apoptosis in developing nerve following axonal degeneration. *Left*: a longitudinal section of a spinal cord nerve in the chick embryo after treatment with an excitotoxin that induces a massive degeneration of motor axons, secondary to selective motoneuron death. As shown, a high number of pyknotic Schwann cells are seen (circles) in the motor fascicle of nerve. Undamaged sensory fascicles contain only healthy Schwann cells (arrows). *Right*: ultrastructure of a dying Schwann cell displaying an apoptotic morphology; note the chromatin condensation, the densification of the cytoplasm and the presence of a so-called ribosome crystal (*), a structure that has been described in several apoptotic cell types (the arrow indicates the basal lamina).

in Schwann cells at the time they proliferate and are susceptible to undergo apoptosis; by contrast, *Krox-20* appears after SCIP in the myelinating Schwann cell lineage coinciding with both the loss of their susceptibility to die by apoptosis and further progression of Schwann cell differentiation (Jaegle *et al.*, 1996; Zorick *et al.*, 1999).

Neuregulins are the best-established neuronally derived factors that promote the survival and proliferation of Schwann cell precursors in vitro (Dong *et al.*, 1995; Syroid *et al.*, 1996) and neuregulins have also been shown to rescue Schwann cells from normal and induced PCD *in vivo* (Grinspan *et al.*, 1996; Trachtenberg & Thompson, 1996; Esquerda *et al.*, 1999). Members of the ErbB family of receptor tyrosine kinases mediate the survival response of Schwann cells to neuregulins (Carraway & Burden, 1995). Three transmembrane neuregulin receptors (ErbBs) have been identified, ErbB2, ErbB3 and ErbB4. Neuregulins directly bind with low affinity to ErbB3 and with high affinity to ErbB4. On the other hand, ErbB2 and ErbB4, but not ErbB3, display intrinsic high tyrosine kinase activity, and receptor heterodimerization allows neuregulin-dependent activation of ErbB2 (Lemke, 1996). ErbB2 and ErbB3 are the major neuregulin signalling receptors in Schwann cells (Meyer & Birchmeier, 1995; Grinspan *et al.*, 1996; Vartanian *et al.*, 1997). Consistent with the important role of neuregulin signalling in Schwann cell development, Schwann cells are absent in mouse mutants lacking either neuregulin (Meyer & Birchmeier, 1995) or the ErbB3 neuregulin receptor (Riethmacher *et al.*, 1997). In addition, it appears that neuregulins are a major component of the diffusible factors produced by the neural tube that induce *Krox-20* expression in developing Schwann cell precursors and that promote both survival and Schwann cell differentiation (Murphy *et al.*, 1996). Other growth factors that can promote Schwann cell survival *in vitro* include IGF, PDGF, NT-3 and FGF (Gavrilovic *et al.*, 1995; Syroid *et al.*, 1999). Recent studies indicate that IGF, PDGF and NT-3 constitute an autocrine circuit which postnatal and adult Schwann cells use for survival in the absence of axon-derived factors such as neuregulin (Meier *et al.*, 1999; Syroid *et al.*, 1999). This autocrine loop is absent from embryonic Schwann cell precursors and its presence later in development (and in the adult) ensures that Schwann cells will survive following peripheral nerve lesions.

At present little information is available concerning normal PCD in astrocytes. It is thought that, at least in certain regions, astrocyte numbers are mainly regulated by proliferation rather than by cell death. In contrast to oligodendrocytes, astrocytes do not undergo PCD in immature rodent optic nerve during development (Barres *et al.*, 1992b) or after nerve transection (Barres *et al.*, 1993). However, in the developing postnatal cerebellum, large numbers of astrocytes are normally eliminated by apoptosis (Krueger *et al.*, 1995). In this case astrocyte cell death peaks at the time when oligodendrocytes begin to migrate and differentiate in the white matter. It has been suggested that astrocytes might be a transient source of trophic support for axons before oligodendrocytes differentiate and myelinate the axons. Because it is thought that glial cells require neuronally derived survival factors, the arrival of oligodendrocytes would force astrocytes to compete with them for limiting amounts of axonally derived factors for survival. It is possible that normal PCD of astrocytes occurs only in regions of the nervous system in which similar competitive interactions occur.

TARGET-INDEPENDENT REGULATION OF DIFFERENTIATING POSTMITOTIC NEURONS

There is a large body of evidence indicating that the normal programmed cell death (PCD) of developing neurons is influenced by signals derived from targets. This is especially clear in the peripheral nervous system, because of the ability to experimentally alter neuron–target interactions. Target elimination results in a massive death of innervating neurons, during the period of normal PCD (Hamburger, 1958; Oppenheim et al., 1978, Phelan & Hollyday, 1991; Grieshammer et al., 1998). Conversely, target enlargement promotes the survival of neurons that otherwise would die (Hollyday & Hamburger, 1976). Exogenous administration of target-derived neurotrophic factors, such as NGF or tissue extracts, can mimic target-survival promoting effects in dorsal root ganglia or spinal cord motoneurons, respectively (Hamburger & Yip, 1984; Calderó et al., 1998a). The understanding of trophic neuron–target interactions began with the pioneering studies of Hamburger and Levi-Montalcini that led to the discovery of nerve growth factor (NGF) and the postulation of the neurotrophic hypothesis (Oppenheim, 1996). In its initial formulation this theory stated that neurons depend on and compete for neurotrophic factors released in limiting amounts by their targets (Oppenheim, 1989). Thus, only neurons that obtain sufficient amounts of neurotrophic factors would survive the period of normal PCD.

During the past two decades an increasing number of trophic molecules interacting with distinct receptors with survival-promoting activity on restricted neuronal populations have been discovered (Henderson, 1996). These trophic factors are not always present in targets but, presumably, are delivered to neurons from other sources such as neuronal afferent projections, glial cells or even through non-synaptic autocrine/paracrine mechanisms (Korsching, 1993; Davies, 1996; Oppenheim, 1996). How non-target-derived factors cooperate with target-derived ones in maintaining neurons is not well understood. Additionally, the physiology of neurotrophic interactions is complicated by the fact that the pattern of temporal and spatial expression of neurotrophic factors and their receptors, as well as the neurotrophic requirements of cells, change during the development of neurons (Korsching, 1993; Davies, 1994). Additionally, the neurotrophic factor requirements of cells can be modulated by neuronal activity, in that, for example, depolarization favours neuronal survival even in the absence of neurotrophic supply (Johnson et al., 1992). Finally, neuronal activity can regulate the expression of neurotrophic factors and their receptors (Zafra et al., 1990; Bessho et al., 1993; see also Bonhoeffer, 1996). Thus, it is evident that neurotrophic interactions are more complex than originally assumed by the neurotrophic theory, in which neurotrophic support was considered to be restricted to target influences. It is in the context of this modified framework that we will focus on the role of non-target-derived trophic mechanisms involved in regulating the survival and death of developing neurons.

Role of afferents

In contrast to the reasonably well-studied role of target influences on neuronal survival, the role of afferent inputs in regulating PCD is not as well understood.

Studies to address this issue have used several experimental strategies to explore how PCD is altered after: (a) suppression or enlargement of afferent inputs, (b) modification of electrical activity derived from afferent inputs and (c) exogenous administration of trophic agents presumably delivered by afferents to their targets by means of anterograde axonal transport and release from terminals and receptor-mediated uptake by postsynaptic cells.

Experimental alteration of afferent innervation

While the complete ablation of targets is a feasible and reproducible experimental model, especially in the peripheral nervous system, experimental deafferentation by contrast usually only allows one to partially remove afferent inputs. Furthermore, the experimental enlargement of afferent innervation is difficult and has been reported in only a few specific experimental paradigms. It is known that removal of peripheral sense organs induces neuronal death in their developing central targets, as shown in the avian cochlear nucleus after otocyst removal (Levi-Montalcini, 1949; Parks, 1979) and in the mammalian lateral geniculate nucleus after eye enucleation (Finlay et al., 1986). Eye removal at birth also results in neuronal loss in the superior colliculus. Moreover, in this system attempts have been made to examine whether increased optic afferents could rescue cells from PCD in the superior colliculus (Cunningham, 1982). Since the implantation of a supernumerary eye is not feasible in mammals, increasing afferent inputs was achieved by lesioning the lateral geniculate nucleus, which is a major target for optic axons; in its absence optic axons increase their projections to the superior colliculus and the nucleus of the optic tract, thereby providing additional inputs to these populations. In this situation there are increased neuronal numbers in both nuclei (Cunningham et al., 1979), presumably as a result of reduced normal PCD.

Afferents are also required to maintain neuronal survival even after the cessation of normal PCD, as shown for instance in post-hatching or postnatal cochlear nucleus neurons (Hashisaki & Rubel, 1989). Studies using this model have provided new insights concerning the cellular mechanisms involved in afferent control of neuronal survival. The chicken brainstem cochlear nucleus, the nucleus magnocellularis (NM), undergoes a 25–30% neuronal loss when the activity evoked by its normal afferent input (VIIIth nerve) is suppressed by either cochlear removal or functional silencing with tetrodotoxin (TTX, Kelley et al., 1997). Neuronal death in NM is preceded by a rapid increase in protein degradation and also by a failure in protein synthesis and disruption of ribosomes (Garden et al., 1995; Kelley et al., 1997). Removing the inputs to NM neurons also results in a rapid increase in their intracellular calcium concentration $[Ca^{2+}]_i$ that reach lethal levels, which can be prevented by orthodromic stimulation of afferents. Afferent-evoked lowering of Ca^{2+} within NM neurons can be blocked by metabotropic glutamate receptor (mGluR) antagonists but not by ionotropic glutamate receptor antagonists (Zirpel & Rubel, 1996). The mechanism by which mGluR activation attenuates the Ca^{2+} influx involves cAMP-dependent decrease in Ca^{2+} influx through voltage-dependent Ca^{2+} channels as well as protein kinase A and C dependent $[Ca^{2+}]_i$ regulatory mechanisms (Lachica et al., 1995; Zirpel et al., 1998).

Afferents also appear to be involved in controlling the survival of spinal cord motoneurons in the chick during normal development. Extensive elimination of afferent inputs to motoneurons arising from supraspinal and/or propriospinal sources, combined with the elimination of primary sensory inputs from dorsal root ganglia, results in a significant loss of motoneurons during the major period of normal PCD. A more restricted elimination of afferents, affecting only one of these sources, also reduces motoneuron survival, but in this case the loss occurs mainly during the terminal stage of normal PCD (Okado & Oppenheim, 1984; Qin-Wei et al., 1994). Increasing the DRG primary afferent inputs to the spinal cord does not appear to modify the number of spinal motoneurons. This was shown in experiments in which the number of DRG neurons was markedly increased by treatment with NGF (Oppenheim et al., 1982).

The chick isthmo-optic nucleus (ION) projects to the contralateral retina and receives afferent inputs from the ipsilateral tectum. In addition to the role of target-derived retrograde signals in regulating ION neuronal survival during development, anterograde signals via afferents have also been shown to be important (Clarke, 1985). Unilateral lesions of the source of ION afferents in the optic tectum results in a dramatic increase in ION neuronal death ipsilaterally during the second half of the normal period of PCD. Neuronal death following afferent deprivation of the ION displays morphological changes which can be distinguished from that induced by target deprivation (Clarke, 1992).

Another interesting experimental model for examining the role of afferents in PCD is the chick ciliary ganglion. The avian ciliary ganglion is composed of parasympathetic autonomic neurons that project to smooth and striated musculature in the eye. This ganglion receives afferents from a single source, the accessory oculomotor nucleus (AON, homologous to the Edinger–Westphal nucleus of mammals) and requires normal target-dependent trophic signals for neuronal survival; target removal results in the loss of 80–90% of ciliary ganglion neurons compared to only a 50% loss in normal embryos (Landmesser & Pilar, 1974). However, the destruction of AON before the onset of normal PCD of ciliary ganglion cells also leads to a massive (85–90%) death of these neurons at the end of the normal PCD period (Furber et al., 1987). Removal of both targets and afferents results in a complete loss of all ciliary ganglion neurons. The amount of neuronal death induced by deafferentation in this model is much higher than that observed in the other experimental paradigms cited above. One likely reason for this is that the avian ciliary ganglion neurons can be completely deafferented by lesioning the AON and, therefore, target neurons do not receive compensatory inputs from persisting afferents after partial deafferentation, as occurs in other cases such as spinal motoneurons.

The role of afferent neuronal activity

It is well established that the trophic factor-dependent survival of cultured neurons is minimized when they are permanently depolarized by the presence of high [K^+] in the media. The survival-promoting effects of K^+ appear to be mediated by Ca^{2+} influx through voltage-dependent Ca^{2+} channels (VSCC), since this effect can be inhibited by VSCC blockers or by removal of extracellular Ca^{2+} (Koike et al., 1989). Based on these findings, a "calcium set-point hypothesis" for neurotrophic factor dependence was proposed. Because $[Ca^{2+}]_i$ may be either toxic or trophic for neurons, this hypothesis establishes distinct steady-step levels or set-points for $[Ca^{2+}]_i$ actions on cell survival (Franklin & Johnson, 1992). When $[Ca^{2+}]_i$ is either very low or very high neuronal survival would be seriously compromised even in the presence of trophic factors. When $[Ca^{2+}]_i$ is in the range of normal resting levels, trophic factors are needed for survival, whereas moderate increases in $[Ca^{2+}]_i$ promote neuronal survival in the absence of neurotrophic factors. Although the survival-promoting effects of K^+ depolarization are robust and have been found in diverse neuronal types *in vitro*, it is not clear whether or how Ca^{2+} entry associated with physiological depolarization *in vivo* induced by afferent synaptic activity is required for neuronal survival. It has been suggested that $[Ca^{2+}]_i$ is necessary for regulating the survival of neurons before they become dependent on neurotrophic factors at early stages of development (Larmet *et al.*, 1992). Although it is possible that Ca^{2+} entry through VSCC during physiological firing of afferents may modulate the neurotrophic factor dependence for neuronal survival, there are only a few reported cases in which the suppression of electrical activity *in vivo* by means of the Na^+ channel blocker tetrodotoxin (TTX) increases neuronal death at later stages. In the optic tecta of chick embryos an attempt was made to determine whether the survival-promoting effects of afferents from the retina on tectal cells are dependent on electrical activity versus axonally transported factors (Catsicas *et al.*, 1992). When action potentials were blocked by means of TTX a rapid induction (9 hours) of neuronal death was found in a defined neuronal population located in the deep layers of the tectum, whereas blockade of axoplasmic transport by colchicine resulted in a slower (13 hours) death of only the superficial tectal neurons. Comparable results were obtained in the rodent retinotectal system after blocking spontaneous firing of retinal ganglion cells by TTX, but in this case increased numbers of dying cells were observed already after 1 hour of afferent blockade (Galli-Resta *et al.*, 1993). An enhancement of naturally occurring PCD in sympathetic and ciliary ganglion cells has also been seen following blockade of ganglionic transmission (Wright, 1981; Maderdrut *et al.*, 1988). However, it is possible that other activity-mediated survival signals different from Ca^{2+} entry through VSCC may also be affected by activity blockade such as the release of neurotransmitters or the regulation of neurotrophic factors. For example, in a recent striking example, prevention of neurotransmitter release in the developing brain by genetic deletion of a single protein, Munc 18-1, that is required for neurotransmitter release throughout the brain, results in widespread PCD following the normal formation of connectivity (Verhage *et al.*, 2000).

It appears that $[Ca^{2+}]_i$ is critically dependent on activation of postsynaptic receptors, particularly Ca^{2+}-permeable ionotropic glutamate receptors. Additionally, changes in membrane potential due to the activation of non-Ca^{2+}-permeable ionotropic receptors can indirectly affect Ca^{2+} permeability through VSCC. There are many examples in which the neurotrophic effects of high $[K^+]$ in cultured neurons can be mimicked by the pharmacological activation of Ca^{2+}-permeable NMDA-subtype glutamate receptors (see for example Balázs *et al.*, 1988; Burgoyne *et al.*, 1993). However, glutamate receptor overactivation is more commonly associated with several forms of excitotoxic neurodegeneration. In these cases, Ca^{2+} overload triggers a cascade of events that lead to a catastrophic disruption of cellular structure. Excitotoxic degeneration associated with pathological glutamate overload is postulated to be involved in several forms of neurodegenerative diseases, but it is not yet clear whether during normal development neurons also undergo PCD by an excitotoxic-mediated glutamate mechanism. From the available literature one can infer that endogenous glutamate may either promote or inhibit PCD depending on the experimental situation. It has been reported that glutamate receptor antagonists prevent the normal and deafferentation-induced PCD in chick embryo auditory neurons (Solum et al., 1997) and the normal PCD of chick spinal motoneurons (Calderó et al., 1997). Moreover, experimentally induced synaptic release of endogenous excitotoxins can produce either apoptotic or necrotic neuronal damage in adult brain (Mitchell et al., 1994, Sloviter et al., 1996). By contrast, it has recently been reported that the blockade of NMDA receptors at specific stages of development, coinciding with the period in which normal PCD takes place, triggers a wave of apoptotic neurodegeneration in the developing rat brain (Ikonomidou et al., 1999). Similarly, in avian motoneurons chronic stimulation of NMDA receptors during a critical developmental period, beginning just before the onset of normal PCD, promotes survival (Calderó et al., 1998b; Lladó et al., 1999) and the same treatment also promotes the survival of motoneurons after early target (limb) ablation, indicating that, in the absence of target-derived neurotrophic support, neuronal survival is promoted by activation of NMDA receptors (Calderó et al., 1998b, Lladó et al., 1999).

During development, many neurons undergo a period in which glutamate receptors are transiently highly expressed and consequently neurons show an increased vulnerability to excitotoxic damage when these receptors are activated (Ikonomidou et al., 1989; Kalb et al., 1992; Greensmith & Vrbová, 1996). It is noteworthy that in rat brain this period of hypersensitivity to glutamate coincides with both the main period of normal PCD and – paradoxically – with the period in which NMDA antagonists are most effective in triggering apoptosis in neurons (Ikonomidou et al., 1999). This suggests that, at least during certain stages of development, that are coincident with glutamate receptor hypersensitivity, increases in $[Ca^{2+}]_i$ elicited by the activation of NMDA receptors should be in a range capable of promoting neuronal survival instead of inducing excitotoxicity. However, it remains to be determined whether increases in $[Ca^{2+}]_i$ induced by endogenous glutamate may be toxic to developing neurons when higher levels are attained during intervals of glutamate receptor hypersensitivity, as seems to occur in the adult (Sloviter et al., 1996). Thus, the extent to which glutamate receptors are involved in regulating normal or induced neuronal death may be specific and differ in each particular neuronal population, and the situation may be further complicated by developmental regulation of both glutamate and its receptors. To complicate matters even further, there are additional feedback cues that regulate synaptic receptor development that must be taken into account. For instance, an activity-dependent mechanism regulates the number of physiologically active glutamate receptors, as shown for both NMDA (Rao & Craig, 1997) and AMPA (O'Brien et al., 1998) receptors. In both cases overall reductions in activity result in increased synaptic receptor density; conversely, increased excitatory synaptic activity produces a compensatory reduction in excitatory synaptic responses. These modulatory responses are generally thought to be important in the regulation of synaptic function and plasticity, but they may also be relevant for controlling activity-dependent neuronal survival. Thus, it has been shown that the rescue of developing motoneurons from chronic NMDA treatment at one age is associated with a down-regulation of the obligatory NMDA receptor subunit NR1 by NMDA agonists at an earlier age. A decrease in NMDA receptors may underlie the shift from a toxic to trophic response following NMDA receptor overstimulation, by maintaining NMDA-induced Ca^{2+} influx in a range in which motoneurons can be kept alive even in the absence of target-derived trophic support (Lladó *et al.*, 1999). Moreover, it is known that blockade of NMDA receptors induces a rapid up-regulation of NMDA receptors in the brain and increases neuronal sensitivity to NMDA-mediated injury. In this situation NMDA receptor antagonists may paradoxically make neurons more, *not less*, vulnerable to excitotoxic lesions (McDonald *et al.*, 1990).

One intracellular signalling pathway that mediates the survival effects of modest increases in $[Ca^{2+}]_i$ involves calmodulin (Soler *et al.*, 1998) and Ca^{2+}/calmodulin-dependent protein kinase kinase (CaM-KK) stimulation. These factors in turn activate protein kinase B (PKB, also known as Akt) resulting in the phosphorylation of the proapoptotic Bcl-2 family member BAD which protects cells from apoptosis (Yano *et al.*, 1998). Normally, BAD promotes cell death by heterodimerizing with Bcl-2 or Bcl-X_L in the mitochondrial membrane, thereby neutralizing the survival-promoting activity of Bcl-2 or Bcl-X_L. By contrast, when BAD is phosphorylated it no longer has access to the mitochondrial membrane and remains sequestered in the cytosol bound to 14-3-3 protein (Kroemer, 1997). The potential downstream targets following more marked changes in $[Ca^{2+}]_i$ homeostasis induced by glutamate overstimulation include: phospholipases, nitric oxide synthase (NOS), mitochondria, proteases, phosphatases, and kinases. It is believed that the activation of these downstream targets by Ca^{2+} leads to a non-specific catastrophic breakdown of cellular structure and function (Leist & Nicotera, 1997).

The role of trophic factors/signals provided or modulated by afferents

Synaptic afferents can influence target neurons not only by the release of neurotransmitters but also via trophic factors (e.g. neurotrophins) thought to be transported to, and released from, presynaptic terminals. Some neurotrophic factors, such as brain-derived neurotrophic factor (BDNF) have been shown to be anterogradely transported to nerve terminals and preterminal axons (Conner *et al.*, 1997; Yan *et al.*, 1997). This is particularly evident in several brain areas where neurotrophic factor protein is present in absence of its mRNA, as in the striatum, in which BDNF is derived from extrinsic afferents coming from cortex or substantia nigra (Altar *et al.*, 1997). The anterogradely transported neurotrophins may reach postsynaptic targets by means of axodendritic transfer, as shown in the case of exogenous NT-3 in the developing avian visual system (von Bartheld *et al.*, 1996). Putative afferent-derived neurotrophic factors experimentally delivered *in ovo* to the chick embryo prevent motoneuron loss induced by surgical deafferentation of the spinal cord (Qin-Wei *et al.*, 1994) and can resume avian songbird cortical neurons from cell death following deafferentation (Johnson *et al.,* 1997).

Activity of presynaptic neurons may also regulate their own expression of trophic factors. For example, after pharmacological stimulation, AMPA receptors can act as both ion channel and cell-surface signal transducers through their interaction with the Src-family non-receptor protein tyrosine kinase Lyn (Hayashi *et al.*, 1999). Activation of Lyn results in expression of the mitogen-activated protein kinase signalling pathway and increases the expression of BDNF mRNA in an activity-dependent manner. However, the activation of postsynaptic target neurons may also dynamically modulate their sensitivity to neurotrophic factors by regulating receptor expression. It has been shown that the expression of TrkA is permanently increased by depolarization and cAMP elevation (Birren *et al.*, 1992) and that a rapid regulation of the levels of cell surface TrkB, by translocation of intracellular membrane pools to the plasma membrane, occurs in central nervous system neurons by a mechanism involving depolarization and cAMP (Meyer-Franke *et al.*, 1998). Activity can also promote neuronal survival by stimulating the synthesis and release of trophic factors, as shown for cerebellar granule cells (Marini *et al.*, 1998). In this model, NMDA-induced activity contributes to neuronal survival by the activation of a BDNF autocrine loop.

Glial-derived trophic signals involved in neuronal survival

Schwann cells and neuronal survival

As reviewed above, immature Schwann cells are critically dependent on axonal-derived signals for survival. However, Schwann cells can also potentially influence neuronal survival by their ability to synthesize many neurotrophic factors, including: insulin-like growth factor (IGF); NGF; brain-derived neurotrophic factor (BDNF); neurotrophin-3 (NT-3); ciliary neurotrophic factor (CNTF); leukaemia inhibitory factor (LIF); and glial cell line-derived neurotrophic factor (GDNF). Schwann cells may provide a major source of trophic support for developing motor and sensory neurons before they establish synaptic contacts with peripheral targets and become dependent on target-derived trophic factors for survival (Davies, 1998). Additionally, Schwann cell-derived and target-derived trophic factors may also cooperate in promoting survival after target innervation. In the absence of neuregulin ErbB3 receptors and the resulting loss of Schwann cells, dorsal root ganglion neurons and spinal cord motoneurons initially differentiate and develop axons, but then many of them later die during the normal time of PCD, presumably due to loss of Schwann cell-derived survival signals (Riethmacher *et al.*, 1997).

GDNF is a highly potent neuronal survival factor for developing and adult motoneurons when applied *in vivo* (Oppenheim *et al.*, 1995; Yan *et al.*, 1995; Matheson *et al.*, 1997). GDNF can be detected on Schwann cells ensheathing peripheral nerves early in development (Henderson *et al.*, 1994) and GDNF receptors are present on embryonic motoneurons (Pachnis, 1993; Treanor *et al.*, 1996). *In vitro* studies on cultured rat motoneurons have shown that media conditioned (CM) by muscle and Schwann cell lines have potent synergistic effects on motoneuronal survival (Arce *et al.*, 1998). By using blocking antibodies against GDNF and CT-1 the survival-promoting activity of Schwann cell CM or muscle cell CM can be respectively inhibited in a selective way (Arce *et al.*, 1998). It appears

Table 3 Various populations of neurons whose survival during development is regulated by hormones

Population	Animal	Time	Hormone	Effect	Hormone action
A_3–A_5 pupal abdominal MNs*	Fruitfly, moth	Adult ecdysis	Ecdysterone	Decline in endogenous hormone causes death	Direct but may require other activities
A_3–A_6 larval proleg MNs	Fruitfly, moth	Pupal ecdysis	Ecdysterone	Peak in endogenous hormone causes death	Appears direct
Mauthner neuron	Bullfrog, clawed frog	Metamorphosis	Thyroid hormone	Decline in endogenous hormone causes death	Appears direct
Lumbar spinal MNs	Bullfrog, clawed frog	Metamorphosis	Thyroid hormone	Peak in endogenous hormone causes death	Direct but may require other activities
IMAN and RA telencephalic nuclei	Canary, zebra finch	~P20–P40	Oestrogen	Exogenous hormone reduces death	Appears indirect
SDN-POA hypothalamic nucleus	Rat	~E18–P5	Oestrogen or androgen	Exogenous hormone reduces death	Appears indirect
SNB spinal MNs	Rat	~P1–P10	Androgen	Exogenous hormone reduces death	Indirect
Laryngeal cranial MNs	Clawed frog	Metamorphosis	Androgen and thyroid hormone	Exogenous hormone reduces death	Appears direct
Cranial, spinal MNs	Rat, chicken	After embryonic or neonatal injury	Androgen	Exogenous hormone reduces death	Appears direct

*For references, see text.

MN = motoneuron; IMAN = lateral magnocellular nucleus of the anterior neostriatum; RA = robust nucleus of the archistriatum; P = postnatal day; SDN-POA = sexually dimorphic nucleus of the preoptic area; E = embryonic day; SNB = spinal nucleus of the bulbocavernosus.

that this type of synergy may also occur during motoneuron development *in vivo*, since the combined *in ovo* application of either Schwann cell and muscle cell CM or of GDNF and CT-1 results in similar synergistic effects on chick embryo motoneurons (Prevette *et al.*, 1999). The fact that, in mice lacking Schwann cells, massive motoneuron death occurs (about 80% – Riethmacher *et al.*, 1997) could indicate that target (muscle) is normally only a minor source of trophic support for developing motoneurons. However, specific ablation of muscle cells leaving Schwann cells intact also results in extensive motoneuron death (90% – Grieshammer *et al.*, 1998), suggesting that both sources of trophic support are important. Thus, motoneurons are likely to integrate signals from target and non-target sources for optimal survival.

Another highly enriched protein in neurons and Schwann cells is S100, which has neuronal survival-promoting activity on chicken motoneurons when administered *in ovo* (Bhattacharyya *et al.*, 1992). However, the function of S100 as a specific physiologically relevant Schwann cell-derived trophic factor remains to be established. CNTF is also a protein abundant in the cytosol of mature Schwann cells, but it cannot be secreted, due to the lack of hydrophobic leader sequence. Although CNTF mRNA expression in peripheral nerves is extremely low during the period of normal neuronal PCD (Stöckli *et al.*, 1991), CNTF in Schwann cells increases progressively during the postnatal period in which neurons become less vulnerable to axotomy-induced cell death (Sendtner *et al.*, 1994). Exogenous administration of CNTF has a potent protective action against motoneuronal death in normal development (Oppenheim *et al.*, 1991) and following early neonatal axotomy (Sendtner *et al.*, 1990).

Astrocytes and neuronal survival

In the central nervous system, neurons are closely associated with astroglial cells, which are thought to regulate several fundamental aspects of neuronal development, such us neuronal survival, neuronal migration and axonal growth and guidance. Most of these actions are mediated by a variety of cytokines and growth factors secreted by astrocytes (Banker, 1980; Schwartz & Nishiyama, 1994). Furthermore, after injury, locally secreted inflammatory cytokines and growth factors induce an astrocytic activation that is directly involved in central nervous system repair mechanisms (Eddleston & Mucke, 1993). For example, fibroblast growth factors (FGF) can stimulate the synthesis and secretion of NGF by astrocytes (Yoshida & Gage, 1991). Microenvironmental ionic imbalance at sites of brain injury leading to increases in extracellular K^+ may induce secretion of NGF by astrocytes, which in turn might help to promote survival of neurons damaged by ischaemia (Abiru *et al.*, 1998).

Radial glia are transient elongated cells that, at early stages of development, span the full thickness of the brain and spinal cord, providing a substrate for migrating neurons and axonal growth. When neurogenesis ends, radial neuroglial cells are transformed into astrocytes, oligodendrocytes and ependymal cells (Schmechel & Rakic, 1979; Choi *et al.*, 1983). Several members of the transforming growth factor-β (TGF-β) superfamily, including bone morphogenetic proteins (BMPs), are expressed in radial glial cells and may play a role in promoting the survival of early-developing neurons before they interact with their targets (Schluesener &

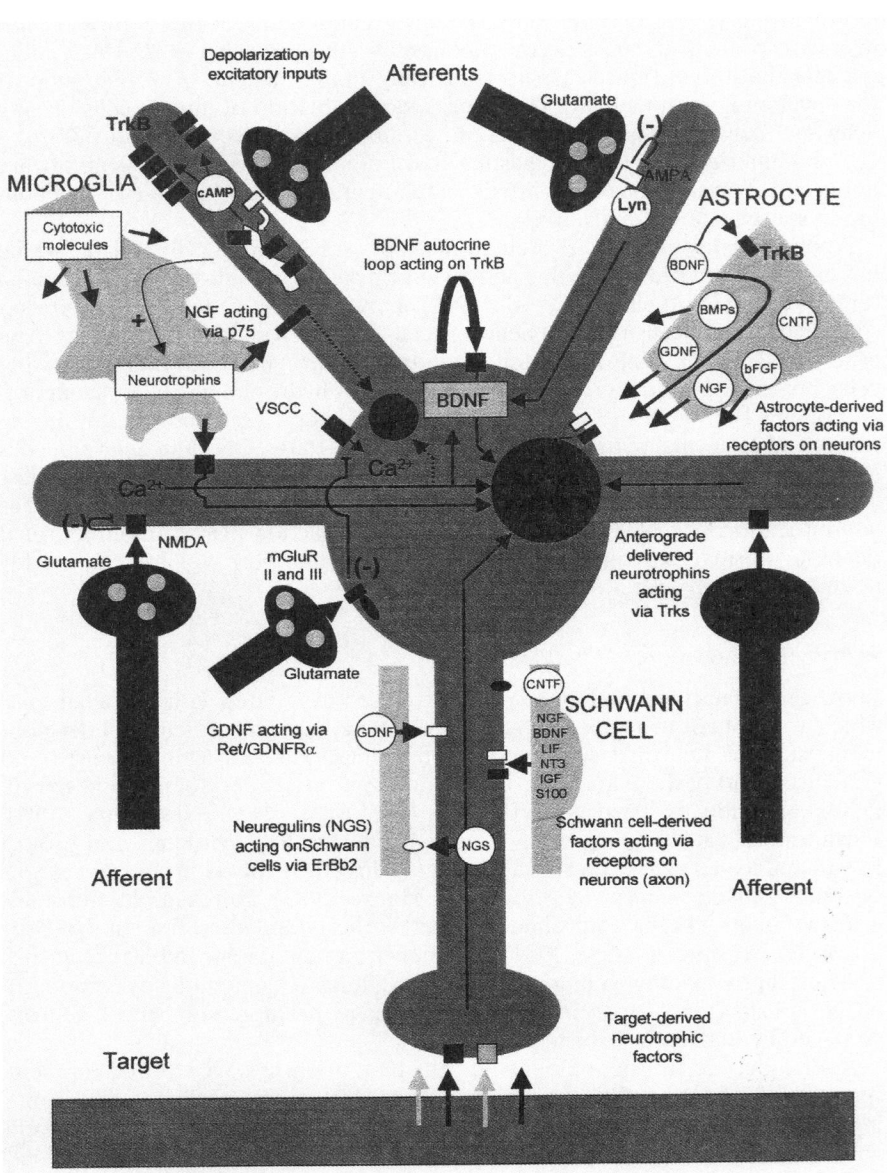

Meyermann, 1994). BMPs may also have an indirect neurotrophic action by stimulating the release of other neurotrophic factors from astrocytes (e.g. Jordan *et al.*, 1997). Astrocytes express the functional neurotrophin receptor trkB as well as non-catalytic truncated forms of trkB (McKeon *et al.*, 1997; Rudge *et al.* 1994). Although it appears that astrocytes do not synthesize BDNF mRNA or protein (Rudge *et al.*, 1992), they may, nonetheless, secrete this neurotrophin by being able to bind, internalize (and store without degradation) and then later release BDNF (Rubio, 1997). CNTF is present at very low levels in the normal adult central nervous system, but is highly increased in astrocytes after central nervous system lesions, and in this way may contribute to the neurotrophic activity of CNTF found in lesion sites (Ip *et al.*, 1993). Other relevant factors contributing to the neurotrophic action of astrocytes are GDNF family members, such as GDNF and persephin (Schaar *et al.*, 1993; Moretto *et al.*, 1996, Jaszai *et al.*, 1998). Finally, neurotrophic factors such as basic fibroblast growth factor (bFGF) and

Figure 2 (*Opposite*) Putative target and non-target cellular interactions regulating PCD in neurons. Most of the putative signalling pathways derived from targets, afferent inputs, glia and autocrine loops which are described in the text are shown. Neurotrophins and other neurotrophic factors such as cardiotrophin-1 (CT-1) derived from targets interact with their specific receptors at nerve terminals. Receptors bound to their ligands are internalized by endocytosis and activate an intracellular survival-promoting signal transduction pathway. Neurotrophic factors can also be delivered by terminal afferents or glial cells. In peripherally projecting neurons, Schwann cells are an important source of survival-promoting factors for neurons. Among these the most well-defined are glial-derived neurotrophic factor (GDNF) and ciliary neurotrophic factor (CNTF). It is likely that CNTF can gain access to neurons only if Schwann cells are lesioned. Other putative neurotrophic factors derived from Schwann cells are indicated in the box. Neurons can, at the same time, provide trophic signals for Schwann cells by, for example, the release of neuregulins (NGS) acting on tyrosine kinase ErbB2 receptors on the Schwann cell membrane. In the case of neurons projecting intrinsically within the CNS, axons are ensheathed by oligodendrocytes instead of Schwann cells. No neuronal survival- promoting factors have been described in oligodendrocytes. Astrocytes are also the source of a plethora of survival-promoting factors for neurons, and they express neurotrophin receptor trkB, as indicated. Although astrocytes can release brain-derived neurotrophic factor (BDNF), they lack mRNA for BDNF; thus, released BDNF is probably sequestered from the extracellular compartment. Microglia can have either neurotoxic or neurotrophic actions. The former are mainly mediated by cytokines and other inflammatory factors secreted in pathological conditions; microglia can also release neurotrophins that have a survival-promoting activity or, conversely, activate a ligand-induced death pathway through NGF acting via the low-affinity (non-trkA) NGF receptor (as indicated by dotted arrow). Adjacent neurons can positively influence the neurotrophic action of microglia. Intracellular Ca^{2+} levels modulate the requirements of neurotrophic factors for survival but high levels of $[Ca^{2+}]_i$ are toxic for neurons (dotted arrow). Moderate rises in $[Ca^{2+}]_i$ through voltage-sensitive Ca^{2+} channels (VSCC) or Ca^{2+}-permeable glutamate receptors promote survival via a signal transduction pathway that is presumably independent of neurotrophic factor receptor activation. Moreover, activation of type II and III metabotropic glutamate receptors inhibits calcium entry through VSCC. Additionally, survival-promoting effects of $[Ca^{2+}]_i$ can be potentiated by Ca^{2+} activation of an autocrine neurotrophic loop involving BDNF synthesis. Synthesis of BDNF can be stimulated by the activation of AMPA glutamate receptors through a pathway involving cell surface signal transduction by means of interactions with the Src-family non-receptor protein tyrosine kinase Lyn, that is physically associated with AMPA receptor. Activity regulates the levels of surface receptors for neurotransmitters and neurotrophic factors. NMDA and non-NMDA type of glutamate receptors are down-regulated by their sustained activation. Depolarization by excitatory inputs induces a rapid insertion of TrkB in the plasma membrane by its translocation from intracellular pools through a cAMP-dependent mechanism (see text for references).

CNTF, that are released by astrocytes, may also be involved in synaptic plasticity and sprouting that follow deafferentation (Gómez-Pinilla et al., 1992; Guthrie et al., 1997).

Oligodendrocytes and neuronal survival

Oligodendrocytes are located primarily in the white matter and provide myelin sheath to axons in the CNS of vertebrates. In contrast to Schwann cells and astrocytes, oligodendrocytes are terminally differentiated postmitotic cells. They arise from proliferating precursor cells migrating from distinct germinal zones of the ventral neuroepithelium into developing white matter (Raff et al., 1983; Hall et al., 1996). However, at present there is no evidence that oligodendrodrocytes can directly supply trophic agents to promote neuronal survival in the same way as Schwann cells or astrocytes.

Microglia and neuronal survival

Microglia, a class of mononuclear phagocytes derived from blood but intrinsic to the central nervous system, are the principal immune effector elements of the brain (Giulian, 1987). Activated microglia are a source of neurotoxic factors including short-lived free radicals and cytokines, that are expressed after traumatic, ischaemic or inflammatory injury within the central nervous system. Neurotoxicity induced by active microglia may be counterbalanced by trophic factors derived from reactive astrocytes at later phases of wound repair (Giulian, 1995). Cellular responses after injury also involve astrocyte activation mediated by factors released from microglia. Although there is extensive evidence for the participation of microglia in cytotoxicity underlying neuropathological conditions, much less attention has been given to the potential beneficial action of these cells in promoting neuronal survival. Because microglia are also the source of growth factors they may play a role in the trophic support of developing or mature neurons (Rappolee et al., 1989; Elkabes et al., 1996; Batchelor et al., 1999). There is evidence that activated microglia can either promote neuronal survival or exacerbate neuronal death, depending on several factors: the environment, which molecules they secrete, the vulnerability of neurons, and the activity of the surrounding macroglial cells (Zietlow et al., 1999). A change in microglial action from toxic to protective can be induced by molecules secreted by neurons and astrocytes. However, when these influences are not present, microglia with an activated macrophage phenotype secrete toxic molecules.

Neurotrophins may act on microglia by an autocrine feedback loop since they secrete and respond to NT-3, which has both mitogenic and trophic actions on microglia (Elkabes et al., 1996). Microglia are also important during developmental periods when neuronal death or synaptic elimination occurs by mediating phagocytic activity necessary for clearance of cellular debris (Ferrer et al., 1990). It is interesting to note that microglia may also play a primary role in the induction of naturally occurring neuronal death in certain regions by means of the secretion of NGF acting on neurons expressing p75[NTR] but lacking trkA (Frade & Barde, 1998; see section above on precursor cells and immature postmitotic neurons).

Summary and conclusions

The role of targets in regulating neuronal PCD is at present well established and supported by a large body of experimental studies. Thus, it is generally assumed that targets are the most important source of trophic signals that support the survival of developing neurons. However, trophic signals derived from both afferent inputs and non-neuronal cells, such as central and peripheral glia, also appear to participate in controlling cell survival (Figure 2). Afferents may influence target neurons by the release of neurotransmitters as well as via trophic factors anterogradely transported to the nerve terminals. Moreover, electrical activity of target neurons mediated by their afferent inputs can modulate the expression of trophic factors and their receptors. On the other hand, moderate increases in $[Ca^{2+}]_i$ evoked by electrical activity may restrict the dependence of neurons on trophic factors. Complex bidirectional trophic interactions are established between neurons and glial cells. The optimal number of glial cells is attained by elimination of the excess cells by means of PCD, as observed in oligodendrocytes, astrocytes and Schwann cells. Glia synthesize many trophic factors that can act on neurons, but they also require neuronally derived trophic factors for their own survival. The elimination of axons at specific stages of development results in a massive death of glial cells and conversely the targeted elimination of glial cells (Schwann cells) induces a massive loss of sensory neurons and motoneurons. Target-independent sources of trophic support are potentially important for the survival of many populations of developing central and peripheral neurons.

HORMONAL REGULATION OF NEURONAL SURVIVAL

Hormones are another source of developmental signals that may have both neurotrophic-like or neurotoxic-like (i.e. death-inducing) effects on neurons, and that often act independently of classic neuron–target interactions (Lewis *et al.*, 1976; Kollros, 1981; Truman & Schwartz, 1984; Weeks & Truman, 1985; Konishi & Akutagawa, 1985; Nordeen *et al.*, 1985; Davis *et al.*, 1996; Kay *et al.*, 1999). Some of the earliest descriptions of PCD during development were in metamorphosing insects and amphibians, which we now know is mediated by changing titres of systemic hormones (Clarke & Clarke, 1996). In this section we will consider neuronal death mediated by hormones as distinct from that regulated by the target, despite the caveat that in certain instances hormones may influence neuronal survival in part through their actions on targets. Nonetheless, as hormones are usually synthesized by specific glands distant from rather than within the target, they clearly represent a type and mode of trophic signal which differs from classic target-produced trophic support. We will concentrate on those hormones which exert their effects on neuronal survival through the nuclear hormone receptor family, which include members of the steroid, thyroid and retinoid hormone superfamily. These receptors are known as ligand-activated transcription factors because they possess discrete hormone-binding and DNA-binding domains (Yamamoto, 1985; Evans, 1988). These two domains are widely conserved across phylogeny for a specific receptor (Berg, 1989) and therefore suggest similar mechanisms mediating their biological activity.

Hormones – definitions, sources and modes of action

The definition of a hormone generally includes those factors which are synthesized by a particular gland, organ or tissue and which exert their biological effects in another location. This is achieved because the endocrine cells which synthesize hormones secrete them into the bloodstream, from which they gain systemic access to their target cells. As their molecular weight is usually less than 300, steroid hormones can easily cross the endothelial and astroglial blood–brain barrier and subsequently diffuse through the plasma membrane of both glia and neurons. Within the cell these hormones bind their receptor, which exists in an unoccupied state within the cytoplasm (Gorski et al., 1968). Previous reports provided evidence that hormone receptors exist exclusively within the nucleus of target cells (King & Greene, 1984; Welshons et al., 1984), but more recent reports detect unoccupied receptors throughout the cell. For example, the presence of oestrogen receptors (OR) in the absence of ligand has been shown in neuronal cytoplasm, dendrites and axon terminals in the guinea pig (Blaustein & Turcotte, 1989; Blaustein et al., 1992). Similarly, the neuronal distribution of free androgen receptors (AR) changes from predominantly nuclear to cytoplasmic upon gonadectomy of both male and female hamsters (Wood & Newman, 1993, 1999), suggesting that the degree of nuclear localization depends upon the level of bound receptor. Once bound, the hormone–receptor complex migrates to chromatin within the nucleus and activates transcription through an interaction between the receptor's DNA-binding domain and hormone response elements (HRE) within the genome (Yamamoto, 1985; Evans, 1988; Beato, 1989; Carson-Jurica et al., 1990; Truss & Beato, 1993; Pfahl, 1993). A growing body of research, however, indicates that hormones may mediate signalling through non-genomic means. For example, the presence of OR (Blaustein et al., 1992) and aromatase (Naftolin et al., 1996) at synaptic terminals suggests that both the synthesis of oestradiol (E_2) and the activation of OR can occur distal to the nucleus, and suggests that hormones may act as neurotransmitters (Matsumoto, 1991).

Hormonally regulated transcription has been extensively investigated in the fruitfly, Drosophila melanogaster. In the salivary gland, translocation of the activated ecdysterone receptor (EcR) complex to the nucleus results in a hierarchical pattern of chromosomal puffing and gene expression which directly controls the histolysis of specific cells and tissues during metamorphosis (Ashburner et al., 1974; Richards, 1976; Robinow et al., 1993; Woodard et al., 1994; Jiang et al., 1997). One of the targets of ecdysterone signalling in Drosophila larvae is the EcR receptor itself (Karim & Thummel, 1992). Similarly, thyroxine stimulates the expression of thyroid hormone receptor (TR) during metamorphosis of the African clawed frog, Xenopus laevis (Yaoita & Brown, 1990; Kawahara et al., 1991). Finally, autoinduction of AR expression by testosterone (T) occurs in the testes and kidney of adult male canaries and in the laryngeal motor nucleus of juvenile frogs (Nastiuk & Clayton, 1994; Pérez et al., 1996). The effects of steroid and thyroid hormones, therefore, appear to depend on both systemic availability and receptor distribution, which itself is hormonally regulated. The following sections describe several of the best-characterized examples of the hormonal regulation of neuronal death during development (Table 3). Particular attention will be given to the role of hormones as agents which not only sculpt neuronal structures through modulating death but also create dramatic sexually distinct differences in function and behaviour as a result of these neuroanatomical changes (i.e. sexual dimorphism).

Invertebrate metamorphosis: multiple roles of ecdysterone

The tobacco hornworm *Manduca sexta* hatches from the egg as a first-instar larva and sheds its cuticle in a process called moulting. The caterpillar moults four times in a period of 2 weeks under the influence of the peptide hormone prothoracicotropin (PTTH), which stimulates release of the steroid hormone ecdysone from the prothoracic glands of the first thoracic segment (Carrow *et al.*, 1981). Ecdysone is hydroxylated by mitochondria within peripheral tissue to the active hormone, 20-hydroxyecdysone, which is referred to either as 20-HE or ecdysterone (Hoffmeister, 1966). The effects of ecdysterone on each larval moulting are accomplished in two pulses and modified by the presence of juvenile hormone (JH), which is released from the corpora allata, a neurohaemal organ (Riddiford & Truman, 1972). The disappearance of detectable JH levels in the fifth-instar larva is followed by a small "commitment pulse" of ecdysterone, which triggers pupal differentiation (Riddiford, 1980). Pupal ecdysis is accompanied by a surge in ecdysterone known as the prepupal peak that can be inhibited by removal of the prothoracic glands (Dominick & Truman, 1985). During the next 18–21 days of adult development the larval musculature degenerates, with the exception of the internal intersegmental muscles (ISMs) that line abdominal segments A_3–A_5 (Finlayson, 1975; Lockshin & Williams, 1965). Concurrently, new external muscles differentiate from myoblasts to form the adult musculature (Stocker & Nuesch, 1975). At the end of this period, as ecdysterone levels decline, the adult animal ecdyses its pupal case, digs to the surface after being buried in the ground and emerges using the ISMs, which then degenerate during the first 36 hours of adult life (Bollenbacher *et al.*, 1981; Finlayson, 1975; Lockshin & Williams, 1965; Schwartz, 1992).

Although most of the motoneurons (MNs) which innervate larval musculature in *Manduca* remain alive and acquire new muscle targets as they differentiate during adult development, about half of the A_3–A_5 abdominal MNs and interneurons (INs) die after adult ecdysis (Taylor & Truman, 1974). The temporal pattern and hormonal modulation of these INs and MNs within the fourth abdominal ganglion have been extensively investigated (Truman, 1983; Truman & Schwartz, 1984). These cells die in a strikingly stereotypic sequence, with the first INs dying just before adult ecdysis and the first MNs dying 10 hours later (Truman & Schwartz, 1980). Individually identified MNs die at specific times after ecdysis (Truman & Schwartz, 1982) and in a similar sequence in isolated abdominal ganglia *in vitro* (Bennett & Truman, 1985).

Ecdysterone also regulates the survival of both the abdominal ISMs and MNs which die after adult ecdysis. A surgical clamp of the abdominal–thoracic junction prevents these cells from gaining access to systemic ecdysterone and accelerates their death, whereas injections of ecdysterone on the 14th or 16th day of adult development block or attenuate ISM and MN death (Schwartz & Truman, 1983). Studies on the silkworm *Antharaea polyphemus* suggest that ecdysterone separately regulates the survival of the ISMs and the abdominal MNs which innervate them. Removal of the abdominal nerve segments after pupal ecdysis results in the same temporal pattern of death of the ISMs, whereas injections of ecdysterone after the onset of ISM degeneration, but before MN death, maintain these MNs despite the death of their target (Truman & Schwartz, 1984). Furthermore, MNs in isolated abdominal ganglia survive in the presence of ecdysterone and die in its absence (Bennett & Truman, 1985).

Evidence that ecdysterone directly regulates the survival of A_3–A_5 ISM-innervating MNs is also provided by autoradiographic studies in *Manduca*, which demonstrate a high level of tritiated ponasterone A binding in these MNs near the end of adult development (Bidmon *et al.*, 1988; Fahrbach & Truman, 1989). Finally, the cloning of *Drosophila* EcR has allowed the generation of molecular antibodies against two of its three isoforms (Koelle *et al.*, 1991; Talbot *et al.*, 1993). Strikingly high levels of EcR-A immunoreactivity are detected within the 300 neurons in the abdominal nerve cord which are fated to die after emergence (Robinow *et al.*, 1993). The identity of these cells as ISM-innervating MNs is supported by their location within the ganglia, their time of death and their ability to be rescued from death by exogenous ecdysterone (Robinow *et al.*, 1993). Furthermore, expression of the *Drosophila* cell death executor genes *grim* and *reaper* is induced within these MNs prior to their death (Robinow *et al.*, 1997). Although it is unclear whether the high level of EcR-A expression in *Drosophila* in these abdominal MNs reflects or specifies their fate to die, it may create an ecdysterone-dependent identity for cells which are programmed to die upon the combined stimuli of hormonal deprivation and target degeneration. Indeed, evidence for the requirement of a death-inducing signal other than simply low ecdysterone titre is provided by the finding in *Manduca* that removal of afferent input to the ISM-innervating MN-12 cells prevents them from dying in response to declining ecdysterone (Fahrbach & Truman, 1987).

In contrast to these A_3–A_5 ISM-innervating MNs, other MNs which innervate the transient larval proleg musculature in segments A_3–A_6 of *Manduca* die during the prepupal peak of ecdysterone (Weeks & Truman, 1985). The competence of these MNs to die in response to ecdysterone is independent of signals mediated by interactions with targets, sensory afferents or other ganglia (Lubischer & Weeks, 1996; Weeks & Davidson, 1994). In addition, only proleg-innervating MNs die in response to ecdysterone when isolated *in vitro* (Streichert *et al.*, 1997). Although these neurons bind ponasterone A (Fahrbach & Truman, 1989), the lack of probes or antibodies against EcR in *Manduca* precludes a detailed analysis of the distribution of various EcR isoforms within this species. In *Drosophila*, however, the EcR expression pattern distinguishes MNs which live throughout metamorphosis as their targets change versus those which die after emergence (Robinow *et al.*, 1993; Truman *et al.*, 1994). The fact that surviving MNs express EcR-A even before pupal ecdysis argues against the idea that the expression of this isoform by itself is sufficient to cause death (see above; Truman, 1996).

Taken together, these studies demonstrate the necessity and sufficiency of ecdysterone to directly activate neuronal-specific programmes to live or die. This hormonal regulation of MN survival underlies the ontogeny of specific behaviours involved in metamorphosis. For example, MNs innervating muscles involved in digging out of the ground during emergence die before MNs innervating muscles involved in wing expansion, which occurs only after the animal has reached the surface (Truman & Schwartz, 1982). The molecular mechanisms by which the disappearance of ecdysterone mediates such a temporally specific pattern of death are becoming elucidated (Robinow et al., 1993, 1997) and illustrate just one of the many specific examples of hormone-dependent anatomical reorganization that is required for the insect to progress through metamorphosis.

Effects of thyroxine – amphibian metamorphosis and mammalian CNS development

A similarly diverse array of effects are exerted on amphibians by hormones within the thyroid gland, the extracts of which have long been known to induce premature metamorphosis in tadpoles (Gudernatsch, 1912; Huxley, 1920). Within the anuran nervous system, thyroid hormone (TH) is responsible for the atrophy of Mauthner's neuron (Stefanelli, 1951), the death or remodelling of spinal MNs (Race, 1961), spinal (Rohon-Beard) and cranial sensory neurons (Lamborghini, 1980; Kollros & McMurray, 1956), the maturation of the cerebellum (Gona, 1972), the differentiation of retinal neurons (Hoskins & Grobstein, 1985) and the proliferation of cells in the olfactory epithelium (Burd, 1990). In the developing rat cerebellum, TH regulates the survival of neurons within the internal granule layer (IGL) and the maturation of Purkinje cells (Lewis et al., 1976; Legrand, 1979). Although such heterogeneity in response to TH has yet to be explained by molecular analysis, striking similarities between the regulation by this hormone of neuronal survival within amphibians and that by ecdysterone within insects suggest that similar phylogenetic programmes may underlie the hormonal control of nervous system development within these species.

The tyrosine derivative thyroxine (T_4) is synthesized by the amphibian thyroid gland under the control of the hyothalamus–pituitary–thyroid axis (Etkin, 1968). Within premetamorphic anuran larvae (tadpoles), prolactin secreted by the pituitary gland stimulates the growth of the organism, including the hypothalamus (Etkin & Gona, 1967). As this nucleus develops it secretes signals which induce the production of T_4 as well as inhibiting the synthesis of prolactin (White & Nicoll, 1981). Rising levels of T_4 during prometamorphosis initiate the period of climax, during which the major events of metamorphosis take place, and after which T_4 levels decline. Triiodothyronine (T_3), which is converted from T_4 within responsive cells, is thought to be the active hormone, as it induces metamorphosis at significantly lower concentrations than T_4 when administered to thyroid-deficient tadpoles (Kistler et al., 1977). In order to avoid confusion, this hormone (T_3) will be referred to as TH. The developmental control of TH synthesis by prolactin is analogous to the regulation of ecdysterone production by JH in the larval insect. Furthermore, the similarity in structure between thyroid and steroid hormone receptors indicates that, although ecdysterone and TH are unrelated in structure or biosynthesis, they exert their biological effects through similar ligand-activated transcription factors (Evans, 1988).

Early studies investigated the role of TH in regulating the survival of the Mauthner neuron (M cell), a large cell on either side of the medulla which innervates the tail and mediates swimming movements in both the bullfrog, Rana pipiens and in Xenopus (Stefanelli, 1951). This neuron grows during larval periods, atrophies after metamorphosis but maintains a projection to the lumbar spinal cord, although a significant proportion of its afferent input and target MNs have degenerated (Kollros, 1981; Davis & Farel, 1990). Loss of the peripheral target, through tail removal or spinal cord transection, fails to change the size of M cells up to 29 days after the operation (Weiss & Rosetti, 1951). Thyroidectomized *Rana* larvae exhibit smaller M cells, however, and healthy larvae treated for 9 days with TH display larger M cells than unoperated or untreated controls (Pesetsky, 1962). Based on these observations it has been suggested that high levels of TH during prometamorphosis create a TH-dependence within M cells, which respond by increasing their size even as their target degenerates. The decline in TH titre after metamorphic climax then induces their subsequent atrophy (Pesetsky, 1962). Although M cells appear to survive after metamorphosis, their hormonal dependence appears analogous to that exhibited by the abdominal MNs, which die upon the disappearance of ecdysterone in emerging insects.

One neuronal population in anurans which undergoes significant TH-dependent degeneration during metamorphosis is the lateral motor column (LMC), whose MNs innervate the developing limbs (Reynolds, 1963; Prestige & Wilson, 1972). In lumbosacral regions these MNs proliferate during larval development and migrate into the LMC, which contains 10,000 MNs in *Rana* before the prometamorphic peak of TH secretion (Reynolds, 1963). While the limbs are rapidly growing in response to this peak, the number of MNs within the lumbar LMC is dramatically reduced to 2000 by the time of metamorphic climax (Beaudoin, 1955; Reynolds, 1963). The relative contribution of the limbs and of TH to LMC formation and degeneration has been investigated. Thyroid-deficient tadpoles generate a normal number of LMC MNs but retain them longer, suggesting that the initial recruitment of MNs into the LMC is TH-independent (Race, 1961). By contrast, tadpoles given exogenous TH exhibit MN degeneration earlier than controls (Beaudoin, 1956; Race, 1961). Finally, limb amputation appears to delay MN degeneration, both in the presence and absence of elevated TH (Race, 1961). Therefore, although the prometamorphic peak of TH is required for LMC MNs to die during hindlimb innervation, a signal from the differentiating limb is also necessary for this death to occur on schedule. More recent studies have questioned the requirement of target-derived signals for LMC MN maturation (Farel, 1989). Limb differentiation does appear capable of mediating some of the effects of TH on the LMC, however, since the expression of TR mRNA in the hindlimb bud of *Xenopus* occurs at the onset of metamorphosis (stage 55; Kawahara *et al.*, 1991; Yaoita & Brown, 1990). Alternatively, high expression of TR mRNA in the spinal cord during the same period indicates that TH could cause spinal MN death directly.

Although TH exerts many other effects on neuronal development of amphibian (reviewed by Kollros, 1981) the best-studied examples of its role in neuronal atrophy and death have been the M cells and the LMC MNs. One other population of neurons which undergoes TH-regulated death during metamorphosis in both *Rana* and *Xenopus* is the mesencephalic fifth nucleus (M-V), which contains jaw-innervating sensory neurons (Kollros & McMurray, 1956). About 40–45% of the neurons die within this nucleus, which displays pyknotic cells even before the onset of larval jaw muscle degeneration, suggesting a direct effect of TH on the death of these neurons (Kollros, 1984; Kollros & Thiesse, 1985). The remaining neurons in M-V change their innervation target from larval to adult jaw musculature (Omerza & Alley, 1992).

Finally, TH has also been shown to positively regulate neuronal survival in the cerebellum of the developing rat. Thyroidectomized rats display a marked increase in pyknotic nuclei within the IGL (Lewis *et al.*, 1976). In addition, Purkinje cells, whose dendrites serve as targets for developing granule cells, exhibit a marked delay in maturation in these animals (Legrand, 1979), suggesting that TH may cause granule cell death indirectly through target interactions. The expression of TR mRNA in granule but not in Purkinje cell layers of rats cerebellum, however, argues for a direct neurotrophic role of TH in granule cell development (Bradley *et al.*, 1989).

Vertebrate sexual differentiation – neurotrophic role of sex steroids

Similar to the role of metamorphic hormones used for establishing the profound anatomical reorganization in insects and amphibians which permits their transition from aquatic to terrestrial life, the secretion of sex steroids by vertebrates creates anatomical differences between the sexes which are indispensable for sex-specific behaviour and reproduction. In both instances it appears that the heterogeneity of responses to hormones by various tissues is determined before the actual availability of these ligands (Tata, 1994; Arnold, 1997). In principle this could be achieved by hormone-independent expression or activation of proteins which affect the synthesis, release, systemic availability (i.e. binding proteins), degradation or transduction of the hormone. In the frog *Xenopus laevis*, for example, the expression of TR mRNA in neuronal tissues precedes the prometamorphic peak of TH secretion (Kawahara *et al.*, 1991). Similarly, the expression of AR mRNA is sexually dimorphic in the higher vocal control (HVc) nucleus of the zebra finch before gonadal steroids establish a difference in its size (Gahr & Metzdorf, 1999). This ability of hormone-independent mechanisms to create a template for hormone-dependent changes has been a particular focus of several recent reviews of sexual differentiation (Balthazart & Ball, 1995; Arnold, 1997; Schlinger, 1998; Gahr & Metzdorf, 1999).

In striking contrast to the potent ability of ecdysterone and TH to induce neuronal death, sex steroids positively regulate neuronal survival during development (but see Arai *et al.*, 1996). The neurotrophic-like effects of these hormones on specific neuronal populations are necessary for the emergence of specific sexually dimorphic behaviours. The establishment of these sexual differences in adult behaviour depends on steroid exposure during a critical period of development, and is therefore considered *organizational* (Phoenix *et al.*, 1959). In contrast, the differences in systemic hormone levels that trigger sexually dimorphic behaviour in the adult are referred to as *activational*. We will focus here on the organizational role of sex steroids in creating sexually dimorphic numbers of neurons within four areas of the developing nervous system: (1) the song nuclei of the zebra finch telencephalon, (2) the sexually dimorphic nucleus of the preoptic area (SDN-POA) of the rodent hypothalamus, (3) the spinal nucleus of the bulbocavernosus (SNB) of the rat spinal cord and (4) the laryngeal motor nucleus of the anuran brainstem. However, steroid hormones also appear to be involved in the establishment of sexual dimorphisms of neuronal number within the dorsal root ganglion (Mills & Sengelaub, 1993), superior cervical ganglion (Wright & Smolen, 1987) and hippocampus (Wimer & Wimer, 1988; Sloviter *et al.*, 1996) of the rat and mouse.

Effects of sex steroids on avian song nuclei

The first sexual dimorphism to be identified in the nervous system was within the telencephalon of the adult canary (*Serinus canarius*) and zebra finch (*Poephila guttata*), two species of songbird in which the males sing and females do not (Nottebohm & Arnold, 1976). The regions within the brain which mediate singing can be separated into two circuits. The first mediates the motor commands of song and is composed of the robust nucleus of the archistriatum (RA), which receives input from the HVc and innervates the hypoglossal MNs which control vocalization through movement of the syrinx (Nottebohm *et al.*, 1976, 1982). The second circuit appears necessary for song acquisition rather than production and includes a projection from HVc to area X, from area X to thalamus, from thalamus to the lateral magnocellular nucleus of the anterior neostriatum (lMAN), and from lMAN back to the RA (Bottjer *et al.*, 1984, 1989; Arnold, 1991). The volumes of the HVc, the RA and area X are as much as five times greater in adult males than in females (Nottebohm & Arnold, 1976; Nottebohm *et al.*, 1976). Males also possess higher amounts of androgen-accumulating neurons in the HVc and in the lMAN (Arnold & Saltiel, 1979) as well as more neurons overall within the HVc, the RA and presumably also the lMAN (Gurney, 1981; Konishi & Akutagawa, 1985; Bottjer *et al.*, 1986; Nordeen *et al.*, 1987). Finally, the size of somata is larger within male HVc, RA and lMAN neurons (Arnold, 1980; Konishi & Akutagawa, 1985).

The ontogeny of neuronal number within these sexually dimorphic songbird nuclei has been extensively investigated (Bottjer *et al.*, 1985, 1986; Nordeen & Nordeen, 1988). Two different patterns emerge in the differentiation of these nuclei (Arnold, 1991). Neuron numbers within the HVc and area X are low in both sexes at P12, whereas by P53 a massive incorporation of new neurons into male but not female nuclei accounts for their sexual dimorphism (Bottjer *et al.*, 1985). By contrast, neuronal number within the RA and lMAN is higher in both sexes at P12–P15 and decreases dramatically in the female by P45–P53 (Konishi & Akutagawa, 1985; Bottjer *et al.*, 1985; Nordeen *et al.*, 1987). Although cell death had been suspected to contribute to sexual dimorphisms in the number of HVc neurons (Kirn & DeVoogd, 1989), subsequent studies showed this not to be the case (Burek *et al.*, 1997). Cell death has been shown to play a major role, however, in reducing the number of neurons within the female lMAN and RA nuclei (Bottjer & Sengelaub, 1989; Kirn & DeVoogd, 1989).

The effects of steroid hormones, innervation targets and afferent projections on the survival of neurons within the RA and lMAN nuclei have been examined (Konishi & Akutagawa, 1985; Nordeen *et al.*, 1987; Herrman & Arnold, 1991; Johnson & Bottjer, 1994). Subcutaneous administration of E_2 to female zebra finches during development masculinizes many of the anatomical markers associated with songbird telencephalic nuclei (Gurney & Konishi, 1980) and induces these birds to sing (Gurney, 1982; Pohl-Appel & Sossinka, 1984; Simpson &

Vicario, 1991). In contrast, exogenous T given to developing females is able to masculinize the volume and somata size but not the number of neurons in RA (Grisham & Arnold, 1995). In addition, hatchling females treated with T do not sing (Gurney & Konishi, 1980), suggesting that the effects of masculinization are exerted by ER- and not AR-mediated signaling (Schlinger, 1998).

Exogenous delivery of E_2 to female zebra finches masculinizes the number of androgen-accumulating neurons within RA and lMAN (Gurney, 1981; Nordeen et al., 1987). An analysis of the extent of neuronal cell death has been undertaken in the RA, in which a significant increase in pyknotic figures in the female relative to the male is observed beginning at P30–P35 (Kirn & DeVoogd, 1989). Interestingly, this timepoint coincides with the arrival of a dense HVc projection to the RA, which occurs in males but not in females (Konishi & Akutagawa, 1985). Additionally, HVc lesions in hatchling females prevent exogenous E_2 from masculinizing the RA (Herrmann & Arnold, 1991). Furthermore, these lesions induce additional degeneration of the RA in females not treated with E_2. Finally, lesions of the lMAN, another source of afferent support to the RA whose axons arrive at P15, results in the loss of 40% of the neurons in the male RA (Johnson & Bottjer, 1994). These results provide evidence for the regulation of RA survival by afferents both in the presence (males and E_2-treated females) or absence (untreated females) of E_2. Afferent regulation of RA survival could either be direct, via presynaptic–postsynaptic interactions, or indirect. Recent evidence indicates that the number of glial cells is less in the female RA and that this difference precedes the increased loss of neurons in females (Nordeen et al., 1998). By increasing glial proliferation with FGF-2 in females the number of degenerating RA neurons was reduced. These data are consistent with an indirect mode of action of afferents via glial cells, and raise the question of whether the sexual dimorphism in glial development is directly mediated by steroid hormones.

The temporal pattern of cell death in lMAN of the developing male zebra finch has also been examined (Bottjer & Sengelaub, 1989). This analysis is not possible in the adult female lMAN due to the difficulty in describing its boundaries. The peak of cell death in this nucleus occurs between P20 and P35 and includes neurons which send projections to RA as well as those which do not (Korsia & Bottjer, 1989). The timing of this death also suggests that HVc afferents which innervate RA between P30 and P35 may compete with lMAN afferents for trophic support from RA. Because lMAN neurons concentrate androgens, and because E_2 regulates their survival, it would be interesting to determine if: (1) the 40% loss of RA neurons in males induced by lesions of lMAN is exacerbated in females; (2) lesions in HVc affect survival of lMAN neurons; (3) dual lesions of lMAN and HVc reduce RA survival more than either lesion alone; and (4) lesions of RA influence survival of lMAN or HVc. Answers to these questions would shed light on the complex interactions between targets, afferents and hormones in the regulation of RA or lMAN survival.

It appears unlikely, however, that E_2 exerts a direct neurotrophic effect on lMAN and RA nuclei in zebra finches, or indeed that endogenous E_2 is required at all for the organization of the sexual dimorphic songbird nuclei (Arnold, 1997; Schlinger, 1998; Gahr & Metzdorf, 1999). Concerns have been primarily raised on the basis of four findings: (1) although an initial report showed higher levels of circulating E_2 in young males than females (Hutchison et al., 1984), subsequent reports failed to replicate this difference (Adkins-Regan *et al.*, 1990; Schlinger & Arnold, 1992); (2) the distribution of brain aromatase expression is not sexually dimorphic (Schlinger & Arnold, 1992); (3) ER mRNA distribution is neither sexually dimorphic during the E_2-sensitive period nor predominant within song control regions (Konishi & Akutagawa, 1988; Gahr, 1996); and (4) treatment of male zebra finches with well-characterized aromatase inhibitors fails to inhibit masculinization (Balthazart *et al.*, 1994; Wade *et al.*, 1994, 1996). These findings have led to an altered conceptualization of how steroid hormones achieve sexual differentiation within the brain. Instead of determining genetic differences only at the level of the gonad, sex-determining factors on the Y chromosome may also specify a masculine fate within male neural primordia, permitting the construction of a brain whose competence to respond to steroids is already sexually dimorphic (Berta *et al.*, 1990; Arnold, 1997; Gahr & Metzdorf, 1999).

Finally, although somewhat tangential to our focus here on development, it is nonetheless of considerable interest that in at least one of the sexually dimorphic nuclei of songbirds (HVc), neurogenesis in males continues into adulthood (Alvarez-Buyella & Kirn, 1997). This continued neurogenesis is seasonal, related to the yearly timing of song production, and appears to be regulated by T which modulates proliferation, differentiation and survival. Additionally, the effects of T on the survival of these newly generated neurons that are destined to populate the adult HVc is mediated by the neurotrophin BDNF (Rasika *et al.*, 1999). Although questions regarding the functional significance of the newly added song neurons remain, this model nonetheless provides an opportunity to examine the cellular and molecular mechanisms that control the survival of such neurons in adult vertebrates. For example, recent studies in both birds and mammals suggest that neurogenesis in another class of neurons that continue to be produced into adulthood, the dentate gyrus of the hippocampus, is influenced by learning and related experiential events (Barnea & Nottebohm, 1996; Gould *et al.*, 1999; vanPraag *et al.*, 1999). Although not a sexually dimorphic phenomenon, experiential regulation of hippocampal neurogenesis appears to occur primarily at the level of enhanced survival of the newly generated neurons.

Effects of sex steroids on the rodent hypothalamus

The medial preoptic area of the rat hypothalamus includes six nuclei (Simerly *et al.*, 1984), the largest of which is referred to as the medial preoptic nucleus (MPN) and has been implicated in the control of male copulatory behaviour (Hansen *et al.*, 1982) and pituitary gonadotropin secretion (Docke *et al.*, 1984). On the basis of cytoarchitectonics and neurotransmitter phenotype, the MPN can be divided into central, lateral and medial components (Simerly *et al.*, 1986; Bloch & Gorski, 1988). Widespread projections from the forebrain and brainstem innervate the MPN, including dense inputs from other hypothalamic nuclei (Simerly & Swanson, 1986). Many of these afferents, like the MPN itself, concentrate steroids (Pfaff & Keiner, 1973; Sar & Stumpf, 1975). Similarly, the MPN projects to a wide variety of regions within the brainstem and forebrain, most of which project back to the MPN (Simerly & Swanson, 1988). One of the strongest projections to the central portion of the MPN (MPNc) arises from a sexually dimorphic component of the bed nucleus of the stria terminalis (Hines *et al.*, 1985).

A significant difference in the synaptic organization of the MPN between male and female rats was initially observed by Raisman & Field (1973), followed by the discovery that part of this nucleus is also sexually dimorphic in volume (Gorski et al., 1978). Sexual dimorphisms within this portion of the hypothalamus have also been reported in the guinea pig, gerbil, ferret, and human (Bleier et al., 1982; Commins & Yahr, 1984; Tobet et al., 1986; Swaab & Fliers, 1985; Allen et al., 1989). Lesions of this nucleus disrupt some aspects of male sexual behaviour, although they do not appear to disrupt copulatory behaviour itself (Arendash & Gorski, 1983; De Jonge et al., 1989). The number of neurons within the SDN-POA, which includes the MPNc and portions of the MPNm, is equal per unit area in both sexes (Gorski et al., 1980; Jacobson et al., 1980), implying that the larger male SDN-POA contains more neurons. The overall volume and neuronal density of the male SDN-POA, which becomes sexually dimorphic on P1, increases for 10 days in the male while remaining the same in females (Jacobson et al., 1980). Sexually dimorphic neuronal numbers in the rat SDN-POA do not appear to evolve from differential neurogenesis or migration (Jacobson et al., 1985; Dodson et al., 1988). In the cell-dense region of the MPNc within the SDN-POA, however, the number of neurons decreases in females from P4 to P10 (Dodson & Gorski, 1993). Similarly, increased numbers of dying neurons during this time are observed within the hypothalamus of female rats and gerbils but not female ferrets (Davis et al., 1996; Holman et al., 1996; Park et al., 1998; Yoshida et al., 2000).

The formation of the SDN-POA is dependent on gonadal steroids, as postnatal treatment of females with T or castration of males on P1 increases or decreases its volume, respectively (Gorski et al., 1978). Treatment of castrated males with T restores the SDN-POA, and prolonged perinatal administration of T or a synthetic oestrogen to females completely sex-reverses SDN-POA volume (Döhler et al., 1982a,b). Perinatal steroid treatment is effective in influencing SDN-POA development only during the period extending from E18 to P5 (Rhees et al., 1990). E_2 probably mediates these effects, because postnatal treatment with the OR antagonist tamoxifen reduces both male and female SDN-POA volume and because male rats with mutant ARs still exhibit masculine SDN-POA volume (Döhler et al., 1984; Naess et al., 1976; Gorski & Jacobson, 1981; Yarbrough et al., 1990).

E_2 reportedly does not promote sexual dimorphism of the SDN-POA through differential neurogenesis, as injections of this steroid still cause a sex-reversal even after the last presumptive neurons of this nucleus have become postmitotic (Dodson et al., 1988). Similarly, a common pattern of migration of a subset of SDN-POA neurons is observed in males and females and is unchanged in castrated males (Jacobson et al., 1985). However, within the MPNc the number of dying cells is significantly higher in females and in castrated males than in normal males between P7 and P10 (Davis et al., 1996), whereas castrated males or normal females treated with T or E_2 display significantly fewer dying neurons within the MPNc than controls treated with oil (Dodson & Gorski, 1993; Davis et al., 1996; Arai et al., 1996).

Similar to the situation in the songbird telencephalon, the mechanism by which gonadal steroids regulate neuronal survival in the rat SDN-POA is unclear. First, systemic E2 levels are similar in neonatal males and females (Ojeda et al., 1975; Weisz & Ward, 1980). Secondly, although expression of OR mRNA is sexually dimorphic in the SDN-POA, its presence in females fails to support a model in which E_2-signalling directly creates sexual differences within this nucleus (DonCarlos & Handa, 1994). Finally, the diverse array of SDN-POA inputs and targets which accumulate androgen complicates the analysis of hormonal action *in vivo*. For example, the fact that one source of input to the MPNc is itself sexually dimorphic illustrates the difficulty of attributing direct neurotrophic actions to any region of the hypothalamus (Hines *et al.*, 1985).

Testosterone regulates SNB survival in the male rat

Although the sexual dimorphisms in the rodent SDN-POA and songbird telencephalon appear to be mediated by E_2, sexual differences in the survival of both the perineal muscles at the base of the rat penis and the SNB MNs which innervate them are dependent on T and not on E_2 (Cihak *et al.*, 1970; Breedlove *et al.*, 1982; Breedlove & Arnold, 1983a; Goldstein & Sengelaub, 1990). Adult male and female rats exhibit a striking disparity in the number of MNs present in the ventral–medial region of the fifth and sixth lumbar spinal cord segments (L_5–L_6), which in males contains about 200 and in females 60 MNs, respectively (Breedlove & Arnold, 1980, 1981). This nucleus is referred to as the SNB on the basis of its peripheral striated muscle targets, the bulbocavernosus (BC) and the levator ani (LA) (Breedlove & Arnold, 1980), which mediate erection and copulation in males and are absent in adult females (Hayes, 1965; Hart & Melese-d'Hospital, 1983; Sachs, 1982). The proliferation of SNB MNs begins at E12 and is completed by P1, resulting in equal numbers of male and female MNs within the SNB at E18 (Breedlove *et al.*, 1983; Nordeen *et al.*, 1985). More MNs exist within the male than female SNB at E20–E22, however, suggesting sexual differences in the process of SNB MN specification or migration (Nordeen *et al.*, 1985; Sengelaub & Arnold, 1986). Evidence which argues against the former of these two alternatives is provided by the finding that the SNB of adult females treated with tritiated thymidine at E12 and T at P1 is composed exclusively of densely tritium-labelled MN nuclei, indicating that the additional SNB MNs in these females differentiate into MNs before androgen treatment (Breedlove, 1986). In addition to the original difference in number of SNB MNs noted at E22, a dramatic decrease in surviving and an increase in pyknotic cells in the female SNB between P1 and P10 has been reported (Nordeen *et al.*, 1985). MNs are also reduced within the male SNB during this period, suggesting that a period of naturally occurring cell death (PCD) regulates SNB MN number in both sexes (Nordeen *et al.*, 1985).

SNB MNs and their muscle targets accumulate androgens and require them to survive (Jung & Baulieu, 1972; Breedlove & Arnold, 1981, 1983a,b). The SNB is demasculinized in rats treated with the AR antagonist flutamide and in rats with a mutation in the AR receptor (Breedlove & Arnold, 1981, 1983a; Yarbrough *et al.*, 1990). Exogenous T fails to alter the process of SNB MN generation or differentiation but prevents the PCD of these MNs and their muscle targets in females and castrated males (Breedlove *et al.*, 1983; Breedlove, 1986; Nordeen *et al.*, 1985). Circulating levels of androgen, which first become detectable at E15, are sexually dimorphic by E18, which coincides with the beginning of the critical period during which T can regulate SNB MN survival (Picon, 1976; Weisz & Ward, 1980; Breedlove & Arnold, 1983b). Although adult male SNB MNs accumulate androgen and express AR (Breedlove & Arnold, 1981; Freeman *et al.*, 1995), neonatal rats fail to accumulate T until after this critical period (Fishman *et al.*, 1990; Jordan *et al.*, 1991).

Several lines of evidence indicate that T rescues SNB MNs indirectly by acting on the BC/LA muscle targets, both during development and after injury: (1) T rescues BC/LA muscles from degeneration even in the absence of the lumbar spinal cord (Fishman & Breedlove, 1985a); (2) androgen blockade in the muscle prevents T from rescuing SNB MNs (Fishman & Breedlove, 1987); (3) axotomy of the pudendal nerve on P14, after the critical period, results in the degeneration of 50% of SNB MNs, suggesting that even after the onset of AR expression in SNB MNs, T is capable of regulating SNB MN survival indirectly through its target (Lubischer & Arnold, 1995); and (4) the size of SNB MNs which are axotomized in the adult, although still capable of responding to T, remains small, even in the presence of T, when they are forced to reinnervate soleus rather than perineal muscles (Araki *et al.*, 1991). Finally, the role of afferent projections in the regulation of SNB MN survival appears to be excluded based on the finding that T prevents MN death in neonatal females even when the thoracic spinal cord is transected before androgen treatment (Fishman & Breedlove, 1985b).

The mechanisms by which T mediates these effects via the muscle have also been examined, particularly because female SNB MNs die despite initially making functional contact with their appropriate target, the transient BC/LA muscles (Sengelaub & Arnold, 1986; Rand & Breedlove, 1987). The fact that T regulates the strength and number of synaptic contacts between SNB MNs and their targets (Jordan *et al.*, 1989, 1992) suggests that androgens may interact with classic target-derived neurotrophic signalling pathways. Treatment of female rat pups with ciliary neurotrophic factor (CNTF), for example, masculinizes the number of MNs within the SNB, and targeted deletion of the signal-transducing component of its receptor abolishes the sexual dimorphism of the SNB (Forger *et al.*, 1993, 1997). Furthermore, the expression of this CNTF receptor within SNB MNs is regulated by androgens, suggesting that T regulates CNTF trophic signalling between BC/LA muscles and SNB MNs (Forger *et al.*, 1998). A similar example of hormone-mediated neurotrophic signalling has been reported in the adult canary, which undergoes seasonal T-dependent neuronal addition into the HVc (Kirn & Nottebohm, 1993). Survival of these new neurons has been shown to be dependent upon BDNF signalling, as survival is increased by the exogenous addition of BDNF to adult females and decreased by infusion of neutralizing BDNF antibodies in males (Rasika *et al.*, 1999).

These studies demonstrate that androgens most probably regulate the survival of neonatal SNB MNs and their perineal muscle targets by indirect actions at the target. The finding that CNTF can mediate androgen-regulated neuronal survival should aid future research into the relationship between hormone receptor activation and the up-regulation of neurotrophic signalling between sexually dimorphic populations of neurons and their efferent targets. For example, the reciprocal regulation by E_2 and NGF of their receptors demonstrates the capacity for these different ligand-activated pathways to cooperatively regulate neuronal survival (Sohrabji *et al.*, 1994). This effect has been reported in the excitotoxic-induced death of rat embryonic cortical neurons, whose rescue by E_2 involves the activation of a well-characterized neurotrophic pathway (Singh *et al.*, 1999).

Effects of androgen on laryngeal MNs in *Xenopus*

The laryngeal neuromuscular system of *Xenopus laevis*, which controls sexually dimorphic vocalization during mating (Kelley & Tobias, 1998), is dependent upon both thyroid and androgenic hormones for its establishment (Watson *et al.*, 1993; Robertson *et al.*, 1994; Robertson & Kelley, 1996). This system consists of the brainstem nucleus of cranial nerve IX–X, (N. IX–X) and its innervation target, the larynx, whose regulated contraction produces vocalization in males (Tobias & Kelley, 1987). Mate-calling behaviour itself is also dependent upon androgens (Kelley & Pfaff, 1976; Wetzel & Kelley, 1983). The establishment of this sexually dimorphic behaviour and its neurobiological correlate has been extensively investigated by Darcy Kelley and her colleagues.

The number of neurons and axons in N. IX–X and in the laryngeal nerve, and the number and type of muscle fibres present within the larynx, are sexually dimorphic in the adult frog (Hannigan & Kelley, 1981; Kelley & Dennison, 1990; Sassoon & Kelley, 1986; Tobias *et al.*, 1991). The generation of presumptive N. IX–X MNs, which occurs between stages 11 and 56, appears similar between sexes and is followed by a period of differential cell loss which starts at stage 54 and ends by postmetamorphic day 1 (PM1) (Gorlick & Kelley, 1987; Kay *et al.*, 1999). In contrast, the number of N. IX–X MN axons is similar in males and females at stage 56 (the end of proliferation) but then becomes greater in males by stage 62 (Kelley & Dennison, 1990). This difference persists into adulthood and reflects both an increased production and a decreased loss of laryngeal MN axons in males relative to females (Kelley & Dennison, 1990). Finally, the number and type of laryngeal muscle fibres is the same in males and females at stage 62 but increases in males by PM1 (Marin *et al.*, 1990). Therefore, sexual differentiation in the number of laryngeal MNs and MN axons appears to precede that of muscle fibres in the larynx.

Differentiation of the gonads (stage 56) and secretion of TH (stage 54) occur just before the appearance of laryngeal MN dimorphism, suggesting the possible dependence of such changes on hormones. Indeed, both laryngeal MNs and muscle cells concentrate androgens, and express AR protein and mRNA (Kelley *et al.*, 1975, 1989; Kelley, 1980, 1989; Cohen & Kelley, 1994; Fischer *et al.*, 1995; Pérez *et al.*, 1996). In addition, TR is expressed by the larynx as early as stage 52 (Cohen & Kelley, 1996). The role of androgen in regulating these dimorphisms has been tested by the castration of males, which retards laryngeal muscle fibre addition and transformation, and by the exogenous delivery of DHT to females, which masculinizes laryngeal muscle fibre number and type (Marin *et al.*, 1990; Tobias *et al.*, 1991). DHT-treated females and normal males also exhibit equal numbers of laryngeal MN axons and MNs, suggesting that androgens normally establish sexual dimorphism of these MNs by preventing PCD (the proliferative period for these MNs ends before the onset of androgen secretion) (Gorlick & Kelley, 1987; Robertson *et al.*, 1994; Kay *et al.*, 1999). In contrast, although TH secretion fails to regulate differentiation of the gonad, manipulations in TH avail-ability have shown that this hormone controls the onset of androgen responsivity in the larynx and therefore regulates androgen-dependent sexual dimorphisms (Robertson & Kelley, 1996; Cohen & Kelley, 1996).

The fact that the sexual dimorphism in the number of laryngeal MN axons is created before dimorphism of the larynx suggests that androgens directly regulate the survival of these MNs. Although no studies have been performed to address this issue, the opposite experiment, in which the development of muscle is exam-ined after denervation, demonstrates that the nerve contributes to, but is not necessary for, laryngeal muscle fibre responsivity to androgen (Tobias *et al.*, 1993). Other experiments have tested the ability of laryngeal MNs in juvenile frogs to respond to androgen after axotomy (Pérez & Kelley, 1996, 1997). Because the critical period during which androgen regulates laryngeal MN survival is over by PM1, the relative contribution of target muscle and DHT to survival could not be evaluated in these studies. However, the ability of DHT to rescue laryngeal MNs from axotomy-induced cell death, together with the fact that these MNs up-regulate AR expression in response to this injury (and normally express AR earlier during metamorphosis) supports the possibility that androgen directly regulates MN survival during development (Kelley *et al.*, 1989; Pérez & Kelley, 1996)

Steroid hormone effects after neuronal injury

Although the effects of exogenous androgen treatment on neuronal survival have been largely confined to the previously mentioned groups of sexually dimorphic neurons, considerable evidence suggests that these hormones can regulate the survival of other populations of non-dimorphic neurons after injury, including cranial and spinal MNs of the rat and hamster (Yu, 1989; Kujawa *et al.*, 1991). Additionally, we have shown that both male and female chick lumbar spinal MNs, which concentrate androgens and express AR, are rescued after axotomy at E12 by treatment with dihydrotestosterone (DHT) from E12 to E15 (Sar & Stumpf, 1977; Weill, 1986; Lumbroso *et al.*, 1996; Gould *et al.*, 1999). This effect is reduced by coadministration of the androgen antagonist flutamide, although flutamide alone fails to affect neuronal survival, suggesting that endogenous androgens are not involved in this response (Gould *et al.*, 1999). The fact that axotomy in the chick embryo requires hindlimb amputation suggests that exoge-nous DHT exerts these survival-promoting effects by acting directly on AR-expressing MNs rather than indirectly through the target. This conclusion is also consistent with the finding that cultured embryonic mouse spinal MNs are rescued by treatment with androgen even in the absence of muscle or muscle extract (Hauser & Toran-Allerand, 1989). Furthermore, because spinal MNs do not appear to concentrate oestrogen or express OR, and because DHT cannot be aromatized to E_2, the attenuation of injury-induced neuronal death appears to be androgen-specific. However, treatment of adult-injured facial MNs in the hamster with either T or E_2 increases their rate of regeneration, an interesting result considering that cranial MNs also fail to express OR as adults (Tanzer & Jones, 1997). Although these MNs do express OR transiently during development (Yokosuka & Hayashi, 1992), adult axotomy fails to induce the expression of OR (Tanzer *et al.*, 1999). The cellular mechanism, therefore, by which steroids regulate neuronal survival after injury has yet to be elucidated.

Summary and conclusions

The studies reviewed here identify thyroid and steroid hormones as key regulators of neuronal survival during development. Interestingly, it appears as though the net effect on neuronal survival of thyroxine and ecdysterone secretion is negative or "pro-apoptotic". For example, the presence of ecdysterone either stimulates the death of target cells, as in the case of proleg-innervating MNs during the prepupal peak of ecdysterone, or death occurs in its absence, as in the case of the MNs involved in emergence, which express high levels of EcR during the disappearance of ecdysterone. This creation of a state of hormone "addiction" has been articulated by Pesetsky (1962) in his description of the hormonal dependence of the Mauthner neuron, which atrophies due to disappearing TH. Perhaps this method of cell death is a phylogenetically conserved programme which permits the destruction of certain cell populations. By contrast, in vertebrates the sex steroid hormones appear to mediate neuronal survival during development. In each case, studies have described the ability of sex steroids to create the neuronal differences which underlie a particular sexually dimorphic behaviour, and in some cases these steroid effects are mediated via classic neurotrophic factors such as CNTF and BDNF. Equally as interesting would be the discovery of androgenic or oestrogenic neurotrophic effects on non-sexually dimorphic populations of neurons during development. The roles of sex steroids in regulating the survival of such neurons after injury are becoming increasingly revealed, and promise to serve as a model for examining the molecular mechanisms underlying this effect. Finally, these studies have demonstrated the diversity of cellular mechanisms by which hormones can influence a particular neuron's survival. For example, the ability of these factors to prevent the death of neurons which do not express the appropriate receptor indicates that hormones either can act through non-hormone receptor-mediated pathways (e.g. neurotrophic factors) or that their effect is transduced through intermediary neuronal populations (i.e. targets, afferents, glia, etc.).

References

Abbas, A.K. (1996). Die and let live: eliminating dangerous lymphocytes. *Cell*, **84**, 655–657.

Abercrombie, M. & Johnson, M.L. (1946). Quantitative histology of Wallerian degeneration. I. Nuclear population in rabbit sciatic nerve. *Journal of Anatomy*, **80**, 37–50.

Abiru, Y., Katho-Semba, R., Nishio, C. & Hatanaka, H. (1998). High potassium enhances secretion of neurotrophic factors from cultured astrocytes. *Brain Research*, **809**, 115–126.

Adachi-Yamada, Y., Fujimura-Kamada, K., Nishida, Y. & Matsumoto, K. (1999). Distortion of proximodistal information causes JNK-dependent apoptosis in *Drosophila* wing. *Nature*, **400**, 166–169.

Adkins-Regan, E., Abdelnabi, M., Mobarak, M. & Ottinger, M.A. (1990). Sex steroid levels in developing and adult male and female zebra finches. *General and Comparative Endocrinology*, **78**, 93–109.

Allen, L.S., Hines, M., Shryne, J.E. & Gorski, R.A. (1989). Two sexually dimorphic cell groups in the human brain. *Journal of Neuroscience, 9*, 497–506.

Alley, K.E. & Omerza, F.F. (1998). Reutilization of trigeminal motoneurons during amphibian metamorphosis. *Brain Research, 813*, 187–190.

Altar, C.N., Cai, N., Bliven,T., Juhasz, M., Conner, J.M., Acheson, A.L., Lindsay, R.M. & Wiegand, S.J. (1997) Anterograde transport of brain-derived neurotrophic factor and its role in the brain. *Nature, 389*, 856–860.

Alvarez-Buyella, A. & Kirn, J. (1997). Birth, migration, incorporation and death of vocal control neurons in adult songbirds. *Journal of Neurobiology, 33*, 585–601.

Ameisen, J.C. (1996). The origin of programmed cell death. *Science, 272*, 1278–1279.

Arai, Y., Sekine, Y. & Murakamai, S. (1996). Estrogen and apoptosis in the developing sexually dimorphic preoptic area in female rats. *Neuroscience Research, 25*, 403–407.

Araki, I., Harada, Y. & Kuno, M. (1991). Target-dependent hormonal control of neuron size in the rat spinal nucleus of the bulbocavernosus. *Journal of Neuroscience, 11*, 3025–3033.

Arce, V., Pollock, R.A., Philippe, J.M., Pennica, D., Henderson, C.E. & deLapeyriere, O. (1998). Synergistic effects of Schwann- and muscle-derived factors on motoneuron survival involve GDNF and cardiotrophin-1 (CT-1). *Journal of Neuroscience, 18*, 1440–1448.

Arendash, G.W. & Gorski, R.A. (1983). Effects of discrete lesions of the sexually dimorphic nucleus of the preoptic area or other medial preoptic regions on the sexual behavior of male rats. *Brain Research Bulletin, 10*, 147–154.

Arnold, A.P. (1991). Developmental plasticity in neural circuits controlling birdsong: sexual differentiation and the neural basis of learning. *Journal of Neurobiology, 23*, 1506–1528.

Arnold, A.P. (1997). Sexual differentiation of the zebra finch song system: positive evidence, negative evidence, null hypotheses, and a paradigm shift. *Journal of Neurobiology, 33*, 572–584.

Arnold, A.P. & Saltiel, A. (1979). Sexual difference in pattern of hormone accumulation in the brain of a songbird. *Science, 295*, 702–705.

Arnold, A.P. (1980). Sexual differences within the brain. *American Scientist, 68*, 165–173.

Ashburner, M., Chihara, C., Meltzer, P. & Richards, G. (1974). Temporal control of puffing activity in polytene chromosomes. *Cold Spring Harbor Symposia in Quantitative Biology, 38*, 655–662.

Ashwell, K. (1990). Microglia and cell death in the developing mouse cerebellum. *Developmental Brain Research, 55*, 219–230.

Awad, T.A. & Truman, J. (1997). Postembryonic development of the midline glia in the CNS of *Drosophila*: proliferation, programmed cell death and endocrine regulation. *Developmental Biology, 187*, 283–297.

Balázs, R., Jørgensen, O.S. & Hack, N. (1988). N-methyl-D-aspartate promotes the survival of cerebellar granule cells in culture. *Neuroscience, 27*, 437–451.

Balthazart, J. & Ball, G.F. (1995). Sexual differentiation of brain and behavior in birds. *Trends in Endocrinology and Metabolism, 6*, 21–29.

Balthazart, J., Absil, P., Fiasse, V. & Ball, G.F. (1994). Effects of the aromatase inhibitor R76713 on sexual differentiation of brain and behavior in zebra finches. *Behavior, 131*, 225–260.

Banker, G.A. (1980). Trophic interactions between astroglial cells and hippocampal neurons in culture. *Science, 209*, 809–810.

Barnea, A. & Nottebohm, F. (1996). Recruitment and replacement of hippocampal neurons in young and adult chickadees: an addition to the theory of hippocampal learning. *Proceedings of the National Academy of Sciences, USA, 93*, 714–718.

Barres, B.A. & Raff, M.C. (1993). Proliferation of oligodendrocyte precursor cells depends on electrical activity in axons. *Nature, 361*, 258–260.

Barres, B.A. & Raff, M.C. (1994). Control of oligodendrocyte number in the developing rat optic nerve. *Neuron, 12*, 935–942.

Barres, B.A., Hart, I.K., Coles, H.S.R., Burne, J.F., Voyvodic, J.T., Richardson, W.D. & Raff, M.C. (1992a). Cell death in the oligodendrocyte lineage. *Journal of Neurobiology, 23*, 1221–1230.

Barres, B.A., Hart, I.K., Coles, H.S.R., Burne, J.F., Voyvodic, J.T., Richardson, W.D. & Raff, M.C. (1992b). Cell death and control of cell survival in the oligodendrocyte lineage. *Cell, 70*, 31–46.

Barres, B.A., Jacobson, M.D., Schmid, R., Sendtner, M. & Raff, M.C. (1993). Does oligodendrocyte survival depend on axons? *Current Biology, 3*, 489–497.

Batchelor, P.E., Liberatore, G.T., Wong, J.Y., Porrit, M.J., Frerichs, F., Donnan, G.A. & Howells, D.W. (1999). Activated macrophages and microglia induce dopaminergic sprouting in the injured striatum and express brain-derived neurotrophic factor and glial cell line-derived neurotrophic factor. *Journal of Neuroscience,* **19**, 1708–1716.

Bate, M., Goodman, C.S. & Spitzer, N.C. (1981). Embryonic development of identified neurons: segment-specific differences in the H cell homologies. *Journal of Neuroscience,* **1**, 103–106.

Beato, M. (1989). Gene regulation by steroid hormones. *Cell,* **56**, 336–344.

Beaudoin, A.R. (1955). The development of lateral motor column cells in the lumbosacral cord in *Rana pipiens*. I. Normal development and development following unilateral limb ablation. *Anatomical Record,* **121**, 81–96.

Beaudoin, A.R. (1956). The development of lateral motor column cells in the lumbosacral cord in *Rana pipiens*. II. Development under the influence of thyroxine. *Anatomical Record,* **125**, 247–259.

Bennett, K.L. & Truman, J.W. (1985). Steroid-dependent survival of identifiable neurons in cultured ganglia of the moth *Manduca sexta*. *Science,* **229**, 58–60.

Berg, J.M. (1989). DNA-binding specificity of steroid receptors. *Cell,* **57**, 1065–1068.

Berta, P., Hawkins, J.R., Sinclair, A.H., Taylor, A., Griffiths, B.L. & Goodfellow, P.N. (1990). Genetic evidence equating SRY and the testis-determining factor. *Nature,* **348**, 448–450.

Bessho, Y., Nakanishi, S. & Nawa, H. (1993). Glutamate receptor agonists enhance the expression of BDNF mRNA in cultured cerebellar granule cells. *Molecular Brain Research,* **18**, 201–208.

Bhattacharyya, A., Oppenheim, R.W., Prevette, D., Moore, B.W., Brackenbury, R. & Ratner, N. (1992). S100 is present in developing chicken neurons and Schwann cells and promotes motor neuron survival *in vivo*. *Journal of Neurobiology,* **23**, 451–466.

Bidmon, H.J., Oettling, G., Koolman, J., Granger, N.A. & Stumpf, W.E. (1988). Autoradiological localization of ecdysteroid receptors in last instar larvae of *Calliphora vicina* and *Manduca sexta*. *American Zoologist,* **28**, 83A.

Birren, S.J., Verdi, J.M. & Anderson, D.J. (1992). Membrane depolarization induces p140trk and NGF responsiveness but not p75LNGFR in MAH cells. *Science,* **257**, 395–397.

Blaschke, A.J., Staley, K. & Chun, J. (1996). Widespread programmed cell death in proliferative and postmitotic regons of the fetal cerebral cortex. *Development,* **122**, 1165–1174.

Blaschke, A.J., Weiner, J.A. & Chun, J. (1998). Programmed cell death is a universal feature of embryonic and postnatal neuroproliferative regions throughout the central nervous system. *Journal of Comparative Neurology,* **396**, 39–50.

Blaustein, J.D. & Turcotte, J.C. (1989). Estrogen receptor-immunostaining of neuronal cytoplasmic processes as well as cell nuclei in guinea pig brain. *Brain Research,* **495**, 75–82.

Blaustein, J.D., Lehman, M.N., Turcotte, J.C. & Greene G. (1992). Estrogen receptors in dendrites and axon terminals in the guinea pig hypothalamus. *Endocrinology,* **131**, 281–290.

Bleier, R., Byne, W. & Siggelkow, I. (1982). Cytoarchitectonic and sexual dimorphisms of the medial pre-optic and anterior hypothalamic area in guinea pig, rat, hamster and mouse. *Journal of Comparative Neurology,* **212**, 118–130.

Bloch, G.J. & Gorski, R.A. (1988). Cytoarchitectonic analysis of the SDN-POA of the intact and gonadectomized rat. *Journal of Comparative Neurology,* **275**, 604–612.

Bollenbacher, W.E., Smith, S.L., Goodman, W. & Gilbert, L.I. (1981). Ecdysteroid titer during the larval-pupal-adult development of the tobacco hornworm, *Manduca sexta*. *General and Comparative Endocrinology,* **44**, 302–306.

Bonhoeffer, T. (1996). Neurotrophins and activity-dependent development of the neocortex. *Current Opinion in Neurobiology,* **6**, 119–126.

Borisov, A.E., Dedkov, E.I. & Carlson, B.M. (2000). Apoptotic elimination of individual subsynaptic and extrasynaptic nuclei in denervated and aging muscle fibers: remodeling of nuclear domains following the loss of neuromuscular junctions? *Journal of Neurobiology* (In press).

Bottjer, S.W. & Sengelaub, D.R. (1989). Cell death during development of a forebrain nucleus involved with vocal learning in zebra finches. *Journal of Neurobiology,* **20**, 609–618.

Bottjer, S.W., Glaessner, S.L. & Arnold, A.P. (1985). Ontogeny of brain nuclei controlling song learning in zebra finches. *Journal of Neuroscience,* **5**, 1556–1562.

Bottjer, S.W., Halsema, K.A., Brown, S.A. & Miesner, E.A. (1989). Axonal connections of a forebrain nucleus involved with vocal learning in zebra finches. *Journal of Comparative Neurology,* **279**, 312–316.

Bottjer, S.W., Miesner, E.A. & Arnold, A.P. (1984). Forebrain lesions disrupt development but not maintenance of song in passerine birds. *Science*, **224**, 901–903.

Bottjer, S.W., Miesner, E.A. & Arnold, A.P. (1986). Changes in neuronal number, density and size account for increases in volume of song-control nuclei during song development in zebra finches. *Neuroscience Letters*, **67**, 263–268.

Bradley, D.J., Young III, W.S. & Weinberger, C. (1989). Differential expression of and thyroid hormone receptor genes in rat brain and pituitary. *Proceedings of the National Academy of Sciences, USA*, **86**, 7250–7254.

Breedlove, S.M. (1986). Cellular analyses of hormone influence on motoneuronal development and function. *Journal of Neurobiology*, **17**, 157–176.

Breedlove, S.M. & Arnold, A.P. (1980). Hormone accumulation in a sexually dimorphic nucleus of the rat spinal cord. *Science*, **210**, 564–566.

Breedlove, S.M. & Arnold, A.P. (1981). Sexually dimorphic motor nucleus in the rat lumbar spinal cord: response to adult hormone manipulation, absence in androgen-insensitive rats. *Brain Research*, **225**, 297–307.

Breedlove, S.M. & Arnold, A.P. (1983a). Hormonal control of a developing neuromuscular system: I. Complete demasculinization of the spinal nucleus of the bulbocavernosus in male rats using the anti-androgen flutamide. *Journal of Neuroscience*, **3**, 414–423.

Breedlove, S.M. & Arnold, A.P. (1983b). Hormonal control of a developing neuromuscular system: II. Sensitive periods for the androgen-induced masculinization of the rat spinal nucleus of the bulblocavernosus. *Journal of Neuroscience*, **3**, 424–432.

Breedlove, S.M., Jacobson, C.D., Gorski, R.A. & Arnold, A.P. (1982). Masculinization of the female rat spinal cord following a single neonatal injection of testosterone propionate but not estradiol benzoate. *Brain Research*, **237**, 173–181.

Breedlove, S.M., Jordan, C.L. & Arnold, A.P. (1983). Neurogenesis in the sexually dimorphic spinal nucleus of the bulbocavernosus in rats. *Developmental Brain Research*, **9**, 39–43.

Burd, G.D. (1990). Role of thyroxine in neural development of the olfactory system. In K.B. Doving (Ed.), *Proceedings of the tenth international symposium on olfaction and taste* (pp. 196–205). Oslo: GCS AS.

Burek, M.J. & Oppenheim, R.W. (1999). Cellular interactions that regulate programmed cell death in the developing vertebrate nervous system. In V.E. Koliatosos & R.R. Ratan (Eds.), *Cell death and disease of the nervous system* (pp. 145–179). Totowa, NJ: Humana Press.

Burek, M.J., Nordeen, K.W. & Nordeen, E.J. (1997). Sexually dimorphic neuron addition to an avian song-control region is not accounted for by sex differences in cell death. *Journal of Neurobiology*, **33**, 61–71.

Burgoyne, R.D., Graham, M.E. & Cambray-Deakin, M. (1993). Neurotrophic effects of NMDA receptor activation on developing cerebellar granule cells. *Journal of Neurocytology*, **22**, 689–695.

Burns, J.F., Staple, J.K. & Raff, M.C. (1996). Glial cells are increased proportionately in transgenic optic nerves with increased numbers of axons. *Journal of Neuroscience*, **16**, 2064–2073.

Calderó, J., Ciutat, D., Oppenheim, R.W., Ribera, J., Casanovas, A., Tarabal, O., Lladó, J. & Esquerda, J.E. (1996). Regulation of Schwann cell death in developing spinal cord nerve roots. *Society of Neuroscience Abstracts*, **22**, 44.

Calderó, J., Ciutat D., Lladó, J., Castán, E., Oppeneheim, R.W. & Esquerda J.E. (1997). Effects of excitatory amino acids on neuromuscular development in the chick embryo. *Journal of Comparative Neurology*, **387**, 73–95.

Calderó, J., Prevette, D., Mei, X., Oakley, R.A., Li, L., Milligan, C., Houenou, L., Burek, M. & Oppenheim, R.W. (1998a). Peripheral target regulation of the development and survival of spinal sensory and motor neurons in the chick embryo. *Journal of Neuroscience*, **18**, 356–370.

Calderó, J., Lladó, J., Ribera, J., Tarabal, O., Ciutat, D., Casanovas A., Ayala, V., Casas, C., Serrando, M., Oppenheim, R.W. & Esquerda, J.E. (1998b). Prevention of normal and induced programmed cell death by chronic administration of N-methyl-D-aspartate (NMDA). *Society of Neuroscience Abstracts*, **24**, 1790.

Carr, V.M. & Farbman, A.I. (1993). The dynamics of cell death in the olfactory epithelium. *Experimental Neurology*, **124**, 308–314.

Carr, V.M. & Simpson, Jr, S.B. (1982). Rapid appearance of labeled degenerating cells in the dorsal root ganglia after exposure of chick embryos to tritiated thymidine. *Developmental Brain Research,* **2**, 157–162.

Carraway III, K.L. & Burden, S.J. (1995). Neuregulins and their receptors. *Current Opinion in Neurobiology,* **5**, 606–612.

Carrow, G., Calabrese, R.L. & Williams, C.M. (1981). Spontaneous and evoked release of prothoracicotropin from multiple neurohaemal organs of the tobacco hornworm, *Manduca sexta*. *Proceedings of the National Academy of Sciences, USA,* **78**, 5866–5870.

Carson-Jurica, M.A., Schrader, W.T & O'Malley, B.W. (1990). Steroid receptor family: structure and functions. *Endocrinology Reviews,* **11**, 201–220.

Cassaccia-Bonnefil, P., Kong, H. & Chao, M.V. (1998). Neurotrophins: the biological paradox of survival factors eliciting apoptosis. *Cell Death and Differentiation,* **5**, 357–364.

Catsicas, M., Péquignot, Y. & Clarke, P.G.H. (1992). Rapid onset of neuronal death induced by blockade of either axoplasmic transport or action potentials in afferent fibers during brain development. *Journal of Neuroscience,* **12**, 4642–4650.

Catsicas, S., Thamos, S. & Clarke, P.G.H. (1987). Major role for neuronal death during brain development: refinement of topographical connections. *Proceedings of the National Academy of Sciences, USA,* **84**, 8165–8168.

Cecconi, F., Alvarez-Bolado, G., Meyer, B.I. Roth, K.A. & Gruss, P. (1998). Apaf1 (CED-4 homolog) regulates programmed cell death in mammalian development. *Cell,* **94**, 727–737.

Charrier, J.-B., Teillet, M.-A., Lapointe, F. & LeDouarin, N.M. (1999). Defining subregions of Hensen's node essential for caudalward movement, midline development and cell survival. *Development,* **126**, 4l77–4183.

Cheema, Z.F., Wade, S.B., Sata, M., Walsch, K., Sohrabji, F. & Miranda, R.C. (1999). Fas/apo [apoptosis]-1 and associated proteins in the differentiating cerebral cortex: Induction of caspase– dependent cell death and activation of NF-κB. *Journal of Neuroscience,* **19**, 1754–1770.

Choi, B.H., Kim, R.C. & Lapham, L. (1983). Do radial glia give rise to both astroglia and oligodendroglial cells? *Developmental Brain Research,* **8**, 119–130.

Chu-Wang, I.W. & Oppenheim, R.W. (1978). Cell death of motoneurons in the chick embryo spinal cord. *Journal of Comparative Neurology,* **177**, 33–86.

Chun, J. & Schatz, D.G. (1999). Rearranging views on neurogenesis: neuronal death in the absence of DNA end-joining proteins. *Neuron,* **22**, 7–10.

Cihak, R., Gutmann, E. & Hanzlikova, V. (1970). Involution and hormone-induced persistence of the muscle sphincter (levator) ani in female rats. *Hormonal Behavior,* **10**, 40–53.

Ciutat, D., Calderó, J., Oppenheim, R.W. & Esquerda, J.E. (1996). Schwann cell apoptosis during normal development and after axonal degeneration induced by neurotoxins in the chick embryo. *Journal of Neuroscience,* **16**, 3979–3990.

Clarke, P.G.H. (1985). Neuronal death during development in the isthmo-optic nucleus of the chick: sustaining role of afferents from the tectum. *Journal of Comparative Neurology,* **234**, 365–379.

Clarke, P.G.H. (1992). Neuron death in the developing avian isthmo-optic nucleus, and its regulation in the establishment of functional circuitry. *Journal of Neurobiology,* **23**, 1140–1158.

Clarke, P.G.H. & Clarke, S. (1996). Nineteenth-century research on naturally occurring cell death and related phenomena. *Anatomy and Embryology,* **193**, 81–99.

Clarke, P.G.H. & Cowan. W.M. (1976). The development of the isthmo-optic tract in the chick, with special reference to the occurrence and correction of developmental errors in the location and connections of isthmo-optic neurons. *Journal of Comparative Neurology,* **167**, 143–164.

Cohen, M.A. & Kelley, D.B. (1994). Thyroxine exposure governs the onset of androgen sensitivity in the larynx of *Xenopus laevis*. *Developmental Biology,* **163**, 548a.

Cohen, M.A. & Kelley, D.B. (1996). Androgen-induced proliferation in the developing larynx of *Xenopus laevis* is regulated by steroid hormone. *Developmental Biology,* **178**, 113–123.

Commins, D. & Yahr, P. (1984). Adult testosterone levels influence the morphology of a sexually dimorphic area in the Mongolian gerbil brain. *Journal of Comparative Neurology,* **224**, 132–140.

Conner, J.M., Lauterborn, J.C., Yan Q., Gall, C.M. & Varon, S. (1997). Distribution of BDNF protein and mRNA in the normal adult rat CNS: evidence for anterograde transport. *Journal of Neuroscience,* **17**, 2295–2313.

Conradt, B. & Horvitz, H.R. (1998). The *C. elegans* protein EGL-1 is required for programmed cell death and interacts with the Bcl-2-like protein CED-9. *Cell*, **93**, 519–529.

Cook, B., Portera-Cailliau, C. & Adler, R. (1998). Developmental neuronal death is not a universal phenomenon among cell types in the chick embryo retina. *Journal of Comparative Neurology*, **396**, 12–19.

Coucouvanis, E. & Martin, G.R. (1995). Signals for death and survival: a two-step mechanism for cavitation in the vertebrate embryo. *Cell*, **83**, 279–287.

Cuadros, M.A. & Rios, A. (1988). Spatial and temporal correlation between early nerve fiber growth and neuroepithelial cell death in the chick embryo retina. *Anatomy and Embryology*, **178**, 543–551.

Cunningham, T.J. (1982). Naturally occurring neuron death and its regulation by developing neural pathways. *International Reviews of Cytology*, **74**, 163–185.

Cunningham, T.J., Huddelston, C. & Murray, M. (1979). Modification of neuron numbers in the visual system of the rat. *Journal of Comparative Neurology*, **184**, 423–434.

Davies, A.M. (1994). Switching neurotrophin dependence. *Current Biology*, **4**, 273–276.

Davies, A.M. (1996). Paracrine and autocrine actions of neurotrophic factors. *Neurochemistry Research*, **21**, 749–753.

Davies, A.M. (1998). Neuronal survival: early dependence on Schwann cells. *Current Biology*, **8**, R15–R18.

Davis, E.C., Popper, P. & Gorski, R.A. (1996). The role of apoptosis in sexual differentiation of the rat sexually dimorphic nucleus of the preoptic area. *Brain Research*, **734**, 10–18.

Davis Jr, G.R. & Farel, P.B. (1990). Mauthner cells maintain their lumbar projections in adult frog. *Neuroscience Letters*, **113**, 139–143.

De Jonge, F.H., Louwerse, A.L., Ooms, M.P., Evers, P., Endert, E. & Van de Poll, N.E. (1989). Lesions of the SDN-POA inhibit sexual behavior of male Wistar rats. *Brain Research Bulletin*, **23**, 483–492.

Docke, F., Rohde, W., Gerber, P., Chaoui, R. & Dorner, G. (1984). Varying sensitivity to negative oestrogen feedback during the ovarian cycle of female rats: evidence for the involvement of oestrogen and the medial preoptic area. *Journal of Endocrinology*, **102**, 287–294.

Dodson, R.E. & Gorski, R.A. (1993). Testosterone propionate administration prevents the loss of neurons within the central part of the medial preoptic nucleus. *Journal of Neurobiology*, **24**, 80–88.

Dodson, R.E., Shryne, J.E. & Gorski, R.A. (1988). Hormonal modification of the number of total and late-arising neurons in the central part of the medial preoptic nucleus of the rat. *Journal of Comparative Neurology*, **275**, 623–629.

Döhler, K-D., Coquelin, A., Davis, F., Hines, M., Shryne, J.E. & Gorski, R.A. (1982a). Differentiation of the sexually dimorphic nucleus in the preoptic area of the rat brain is determined by its perinatal hormone environment. *Neuroscience Letters*, **33**, 295–298.

Döhler, K-D., Hines, M., Coquelin, A., Davis, F., Shryne, J.E. & Gorski, R.A. (1982.) Pre- and postnatal influence of diethylstilbestrol on differentiation of the sexually dimorphic nucleus in the preoptic area of the female rat brain. *Neuroendocrinology Letters*, **4**, 361–365.

Döhler, K-D., Srivastava, S.S., Shryne, J.E., Jarzab, B., Sipos, A. & Gorski, R.A. (1984). Differentiation of the sexually dimorphic nucleus in the preoptic area of the rat brain is inhibited by postnatal treatment with an estrogen antagonist. *Neuroendocrinology*, **38**, 297–301.

Dominick, O.S. & Truman, J.W. (1985). The physiology of wandering behavior in *Manduca sexta*. II. The endocrine control of wandering behavior. *Journal of Experimental Biology*, **117**, 45–68.

donCarlos, L.L. & Handa, R.J. (1994). Developmental profile of estrogen receptor mRNA in the preoptic area of male and female neonatal rats. *Developmental Brain Research*, **79**, 283–289.

Dong, Z., Brennan, A., Liu, N., Yarden, Y., Lefkowitz, G., Mirsky, R. & Jessen, K.R. (1995). Neu differentiation factor is a neuron-glia signal and regulates survival, proliferation and maturation of rat Schwann cell precursors. *Neuron*, **15**, 585–596.

Eddleston, M. & Mucke, L. (1993). Molecular profile of reactive astrocytes: implications for their role in neurologic diseases. *Neuroscience*, **54**, 15–36.

Elkabes, S., DiCicco-Bloom, E.M. & Black, I.B. (1996). Brain microglia/macrophages express neurotrophins that selectively regulate microglial proliferation and function. *Journal of Neuroscience*, **16**, 2508–2521.

Ellis, R.E., Yuan, J. & Horvitz, H.R. (1991). Mechanisms and function of cell death. *Annual Review of Cell Biology*, **7**, 663–698.

El Shamy, W.M., Fridvall, L.K. & Ernfors, P. (1998). Growth arrest failure, G1 restriction point override, and S phase death of sensory precursor cells in the absence of neurotrophin-3. *Neuron*, **21**, 1003–1015.

Ernst, M. (1926). Ueber Untergang von Zellen während der normalen Entwicklung bei Wirbeiteren. *Zeitschrift für Anatomie und Entwicklungsgeschichtz*, **79**, 228–262.

Esquerda, J.E., Calderó, J., Ciutat, D., Prevette, D., Scott, S., Wang, G. & Oppenheim, R.W. (1999). Regulation of programmed cell death of Schwann cells in the chick embryo by growth factors and axonally derived signals. *Society of Neuroscience Abstracts*, **607**, 11.

Etkin, W. (1968). Hormonal control of amphibian metamorphosis. In W. Etkin & L.I. Gilbert (Eds.), *Metamorphosis: a problem in developmental biology* (pp. 313–348). New York: Appleton-Century-Crofts.

Etkin, W. & Gona, A.G. (1967). Antagonism between prolactin and thyroid hormone in amphibian development. *Journal of Experimental Zoology*, **165**, 249–258.

Evans, R.M. (1988). The steroid and thyroid hormone superfamily. *Science*, **240**, 889–895.

Fahrbach, S.E. & Truman, J.W. (1987). Possible interactions of a steroid hormone and neural inputs in controlling the death of an identified neuron in the moth *Manduca sexta*. *Journal of Neurobiology*, **18**, 497–508.

Fahrbach, S.E. & Truman, J.W. (1989). Autoradiographic identification of ecdysteroid-binding cells in the nervous system of the moth *Manduca sexta*. *Journal of Neurobiology*, **20**, 681–702.

Farbman, A.I., Buchholz, J.A., Suzuki, Y., Coines, A. & Speert, D. (1999). A molecular basis of cell death in olfactory epithelium. *Journal of Comparative Neurology*, **414**, 306–314.

Farel, P.B. (1989). Naturally occurring cell death and differentiation of developing spinal motoneurons following axotomy. *Journal of Neurosciene*, **9**, 2103–2113.

Fekete, D.M., Homburger, S.A., Waring, M.T., Riedl, A.E. & Garcia, L.F. (1997). Involvement of programmed cell death in morphogenesis of the vertebrate inner ear. *Development*, **124**, 2451–2461.

Ferrer, I., Berent, E., Soriano, E., del Rio, T. & Fonseca, M. (1990). Naturally occurring cell death in the cerebral cortex of the rat and removal of dead cells by transitory phagocytes. *Neuroscience*, **39**, 451–458.

Ferrer, I., Soriano, E., DelRio, J.A., Alcántara, S. & Auladell, C. (1992). Cell death and removal in the cerebral cortex during development. *Progress in Neurobiology*, **39**, 1–43.

Findeis, E.K. & Milligan, C.E. (1995). Secondary neurulation in teleosts occurs by apoptosis. *Society of Neuroscience Abstracts*, **21**, 1531.

Finlay, B.L. (1992). Cell death and the creation of regional differences in neuronal numbers. *Journal of Neurobiology*, **9**, 115–1171.

Finlay, B.I. & Slattery, M. (1983). Local differences in the amount of early cell death in neocortex predict adult local specialization. *Science*, **219**, 1349–1351.

Finlay, B.L., Sengelaub, D.R. & Berian, C.A. (1986). Control of cell number in the developing visual system. I. Effects of monocular enucleation. *Developmental Brain Research*, **28**, 1–10.

Finlayson, L.H. (1975). Development and differentiation. In P.N.R. Usherwood (Ed.), *Insect muscle* (pp. 75–149). New York: Academic Press.

Fischer, L.M., Catz, D. & Kelley, D.B. (1995). Androgen-directed development of the *Xenopus laevis* larynx: control of androgen receptor expression and tissue differentiation. *Developmental Biology*, **170**, 115–126.

Fishman, R.B. & Breedlove, S.M. (1985a). Neonatal androgen maintains sexually dimorphic perineal muscles in the absence of innervation. *Society of Neuroscience Abstracts*, **11**, 530.

Fishman, R.B. & Breedlove, S.M. (1985b). The androgenic induction of spinal sexual dimorphism is independent of supraspinal afferents. *Developmental Brain Research*, **23**, 255–258.

Fishman, R.B. & Breedlove, S.M. (1987). Androgen blockade of bulbocavernosus muscle inhibits testosterone-dependent masculinization of spinal motoneurons in newborn female rats. *Society of Neuroscience Abstracts*, **13**, 1520.

Fishman, R.B., Chism, L., Firestone, G.L. & Breedlove, S.M. (1990). Evidence for androgen receptors in sexually dimorphic perineal muscles of neonatal male rats. Absence of androgen accumulation by the perineal motoneurons. *Journal of Neurobiology*, **21**, 694–705.

Forger, N.G., Howell, M.L., Bengston, L., MacKenzie, L., DeChiara, T.M. & Yancopoulos, G.D. (1997). Sexual dimorphism in the spinal cord is absent in mice lacking the ciliary neurotrophic factor receptor. *Journal of Neuroscience*, **17**, 9605–9612.

Forger, N.G., Roberts, S.L., Wong, V. & Breedlove, S.M. (1993). Ciliary neurotrophic factor maintains motoneurons and their target muscles in developing rats. *Journal of Neuroscience,* **13**, 4720–4726.

Forger, N.G., Wagner, C.K., Contois, M., Bengston, L. & MacLennan, A.J. (1998). Ciliary neurotrophic factor receptor alpha in spinal motoneurons is regulated by gonadal hormones. *Journal of Neuroscience,* **18**, 8720–8729.

Frade, J.M. & Barde, Y.A. (1998). Microglia-derived nerve growth factor causes cell death in the developing retina. *Neuron,* **20**, 35–41.

Frade, J.M. & Barde, Y-A. (1999). Genetic evidence for cell death mediated by nerve growth factor and the neurotrophin receptor p75 in the developing mouse retina and spinal cord. *Development,* **126**, 683–690.

Frade, J.M. & Rodriguez-Tébar & Barde, Y-A (1996). Induction of cell death by endogenous nerve growth factor through its p75 receptor. *Nature,* **383**, 166–168.

Frago, L.M., León, Y., de la Rosa, E.J., Muñoz, A.G. & Varela-Nieto, I. (1998). Nerve growth factor and ceramides modulate cell death in the early developing inner ear. *Journal of Cell Science,* **111**, 549–556.

Franklin, J.L. & Johnson, Jr, E.M. (1992). Suppression of programmed neuronal death by sustained elevation of cytoplasmic calcium. *Trends in Neuroscience,* **15**, 501–507.

Freeman, L.M., Padgett, B.A., Prins, G.S. & Breedlove, S.M. (1995). Distribution of androgen receptor immunoreactivity in the spinal cord of wild-type, androgen-insensitive and gonadectomized male rats. *Journal of Neurobiology,* **27**, 51–59.

Furber, S., Oppenheim, R.W. & Prevette, D. (1987). Naturally-occurring neuron death in the ciliary ganglion of the chick embryo following removal of preganglionic input: evidence for the role of afferents in ganglion cell survival. *Journal of Neuroscience,* **7**, 1816–1832.

Furuta, Y., Piston, D.W. & Hogan, B.L.M. (1997). Bone morphogenetic proteins (BMPs) as regulators of dorsal forebrain development. *Development,* **124**, 2203–2212.

Gagliardini, V. & Frankhauser, C. (1999). Semaphorin III can induce death in sensory neurons. *Molecular and Cellular Neuroscience,* **14**, 301–316.

Gahr, M. (1996). Developmental changes in the distribution of estrogen receptor mRNA expressing cells in the forebrain of female, male and masculinized female zebra finches. *Neuroreport,* **7**, 2469–2473.

Gahr, M. & Metzdorf, R. (1999). The sexually dimorphic expression of androgen receptors in the song nucleus hyperstriatalis ventrale pars caudale of the zebra finch develops independently of gonadal steroids. *Journal of Neuroscience,* **19**, 2628–2636.

Galli-Resta, L., Ensini, M., Fusco, E., Gravina, A. & Margheritti, B. (1993). Afferent spontaneus electrical activity promotes the survival of target cells in developing retinotectal system of the rat. *Journal of Neuroscience,* **13**, 243–250.

Gao, Y., Sun, Y., Frank, K.M., Dikkes, P., Fujiwara, Y., Seidl, K.J., Sekiguchi, J.M., Rathbun, G.A., Swat, W., Wang, J., Bronson, R.T., Malynn, B.A., Bryans, M., Zhu, C., Chaudhuri, J., Davidson, L., Ferrini, R., Stamato, T., Orkin, S. H., Greenberg, M.E. & Alt, F.W. (1998). A critical role for DNA end-joining proteins in both lymphogenesis and neurogenesis. *Cell,* **95**, 891–902.

García-Porrero, J.A. & Santander, J.L.O. (1979). Cell death and phagocytosis in the neuroepithelium of the developing retina. A TEM and SEM study. *Separatum Experientia,* **35**, 375–376.

García-Porrero, J.A., Colvée, E. & Ojeda, J.L. (1984). Cell death in the dorsal part of the chick optic cup. Evidence for a new necrotic area. *Journal of Embryology and Experimental Morphology,* **80**, 241–249.

Garden, G.A., Redeker-De Wulf, V. & Rubel, E.W. (1995). Afferent influences on brainstem auditory nuclei of the chicken: regulation of transcriptional activity following cochlea removal. *Journal of Comparative Neurology,* **359**, 412–423.

Gavrilovic, H., Brennan, A., Mirsky, R. & Jessen, K.R. (1995). Fibroblast growth factors and insulin growth factors combine to promote survival of rat Schwann cell precursors without induction of DNA synthesis. *European Journal of Neuroscience,* **7**, 77–85.

Gilbert, S.F. (1997). *Developmental biology.* Sunderlund, MA: Sinauer.

Gilmore, E.C., Nowakowski, R.S., Caviness, V.S. & Herrup, K. (2000). Cell birth, cell death, cell diversity and DNA breaks: How do they all fit together? *Trends in Neuroscience,* **23**, 100–105.

Giulian, D. (1987). Ameboid microglia as effectors of inflammation in the central nervous system. *Journal of Neuroscience Research,* **18**, 155–171.

Giulian, D. (1995). Microglia and neuronal dysfunction. In H. Kettenmann & B.R. Ransom (Eds.), *Neuroglia* (pp. 671–684). New York: Oxford University Press.

Glücksman, A. (1930). Ueber die Bedentung Von Zellvörgangen für di Formbildung epitheliarer Organe *Zeitschrift für Anatomie und Entwicklungsgeschichte,* **93**, 35–92.

Glücksmann, A. (1951). Cell deaths in normal vertebrate ontogeny. *Biology Reviews,* **26**, 59–86.

Goldstein, L.A. & Sengelaub, D.R. (1990). Hormonal control of neuron number in sexually dimorphic spinal nuclei of the rat: IV. Masculinization of the spinal nucleus of the bulbocavernosus with testosterone metabolites. *Journal of Neurobiology,* **21**, 719–730.

Gómez-Pinilla, F., Li, W.K. & Cotmann, C.W. (1992). Basic FGF in adult rat brain: Cellular distribution and response to entorhinal lesion and fimbria-fornix transection. *Journal of Neuroscience,* **12**, 345–355.

Gona, A.G. (1972). Morphogenesis of the cerebellum of the frog tadpole during spontaneous metamorphosis. *Journal of Comparative Neurology,* **146**, 133–142.

Goodman, C.S. & Bate, M. (1981). Neuronal development in the grasshopper. *Trends in Neuroscience,* **4**, 163–169.

Gorlick, D. & Kelley, D.B. (1987). Neurogenesis in vocalization pathway of *Xenopus laevis*. *Journal of Comparative Neurology,* **257**, 614–627.

Gorski, J., Toft, D., Shyamala, G., Smith, D. & Notides, A. (1968). Hormone receptors: studies on the interaction of estrogen with the uterus. *Recent Progress in Hormone Research,* **24**, 45–80.

Gorski, R.A. & Jacobson, C.D. (1981). Sexual differentiation of the brain. In S.J. Kogan & E.S.E. Hafez (Eds.), *Clinics in andrology: pediatric andrology*. (Vol. VII, pp. 109–134). Boston: Martinus Nijhoff.

Gorski, R.A., Gordon, J.H., Shryne, J.E. & Southam, A.M. (1978). Evidence for a morphological sex difference within the medial preoptic area of the rat brain. *Brain Research,* **148**, 333–346.

Gorski, R.A., Harlan, R.E., Jacobson, C.D., Shryne, J.E. & Southam, A.M. (1980). Evidence for the existence of a sexually dimorphic nucleus in the preoptic area of the rat. *Journal of Comparative Neurology,* **193**, 529–539.

Gould, E., Beylin, A., Tanapat, P., Reeves, A. & Shors, T.J. (1999). Learning enhances adult neurogenesis in the hippocampal formation. *Nature Neuroscience,* **2**, 260–265.

Gould, T.W., Burek, M.J., Ishihara, R., Lo, A.C., Prevette, D. & Oppenheim, R.W. (1999). Androgens rescue avian embryonic lumbar spinal motoneurons from injury-induced but not naturally-occurring cell death. *Journal of Neurobiology,* **41**, 585–595.

Graham, A., Koentges, G. & Lumsden, A. (1996). Neural crest apoptosis and the establishment of craniofacial pattern: an honorable death. *Molecular and Cellular Neuroscience,* **8**, 76–83.

Greensmith, L. & Vrbová, G. (1996). Motoneurone survival: a functional approach. *Trends in Neuroscience*, **19**, 450–455.

Grieshammer, U., Lewandoski, M., Prevette, D., Oppenheim, R.W. & Martin, G.R. (1998). Muscle-specific cell ablation conditional upon Cre-mediated DNA recombination in transgenic mice leads to massive spinal and cranial motoneuron loss. *Developmental Biology,* **197**, 234–247.

Grinspan, J.B., Marchionni, M.A., Reeves, M., Coulaloglou, M. & Scherer, S.S. (1996). Axonal interactions regulate Schwann cell apoptosis in developing peripheral nerve: neuregulin receptors and the role of neuregulins. *Journal of Neuroscience,* **16**, 6107–6118.

Grisham, W. & Arnold, A.P. (1995). A direct comparison of the masculinizing effects of testosterone, androstenedione, estrogen and progesterone on the development of the zebra finch song system. *Journal of Neurobiology,* **26**, 163–170.

Gudernatsch, J.F. (1912). Feeding experiments on tadpoles. I. The influence of specific organs given as food on growth and differentiation. A contribution to the knowledge of organs with internal secretion. *Wilhelm Roux Archiv für Entwicklungsmechanik der Organism,* **35**, 457–483.

Gurney, M.E. (1981). Hormonal control of cell form and number in the zebra finch song system. *Journal of Neuroscience,* **1**, 658–673.

Gurney, M.E. (1982). Behavioral correlates of sexual differentiation in the zebra finch song system. *Brain Research,* **231**, 153–173.

Gurney, M.E. & Konishi, M. (1980). Hormone-induced sexual differentiation of brain and behavior in zebra finches. *Science,* **208**, 1380–1383.

Guthrie, K.M., Nguyen, T. & Gall, C.M. (1995). Insulin-like growth factor-1 mRNA is increased in deafferented hippocampus: Spatiotemporal correspondence of atrophic event with axon sprouting. *Journal of Comparative Neurology,* **352**, 147–160.

Guthrie, K.M., Woods, A.G., Nguyen, T. & Gall, C.M. (1997). Astroglial ciliary neurotrophic factor mRNA expression is increased in fields of axonal sprouting in deafferented hippocampus. *Journal of Comparative Neurology*, **386**, 137–148.

Hakem, R., Hakem, A., Duncan, G.D., Henderson, J.T., Woo, M., Soengas, M.S., Elia, A., de la Pompa, J.L., Kagi, D., Khoo, W., Potter, J., Yoshida, R., Kaufman, S.A., Lowe, S.W., Penninger, J.M. & Mak, T. W. (1998). Differential requirements for caspase 9 in apoptotic pathways *in vivo*. *Cell*, **94**, 339–352.

Hall, A., Giese, N.A. & Richardson, W.B. (1996). Spinal cord oligodendrocytes develop from ventrally derived progenitor cells that express PDGF alpha-receptors. *Development*, **122**, 4085–4094.

Hamburger, V. (1958). Regression versus peripheral control of differentiation in motor hypoplasia. *American Journal of Anatomy*, **102**, 365–409.

Hamburger, V. (1992). History of the discovery of neuronal death in embryos. *Journal of Neurobiology*, **9**, 1116–1123.

Hamburger, V & Yip, J.W. (1984). Reduction of experimentally induced neuronal death in spinal ganglia of the chick embryo by nerve growth factor. *Journal of Neuroscience*, **4**, 767–774.

Hankin, M.H., Schneider, B.F. & Silver, J. (1988). Death of the subcallosal glial sling is correlated with formation of the cavum septi pellucidi. *Journal of Comparative Neurology*, **272**, 191–202.

Hannigan, P. & Kelley, D.B. (1981). Male and female laryngeal motor neurons in *Xenopus laevis*. *Society of Neuroscience Abstracts*, **7**, 269.

Hansen, S., Köhler, C.H., Goldstein, M. & Steinbusch, H.V.M. (1982). Effects of ibotenic acid-induced neuronal degeneration in the medial preoptic area and the lateral hypothalamic area on sexual behavior in the male. *Brain Research*, **239**, 213–232.

Hart, B.L. & Melese-d'Hospital, Y. (1983). Penile mechanisms and the role of the striated penile muscles in penile reflexes. *Physiological Behavior*, **31**, 807–813.

Hashisaki, G.T. & Rubel, E.W. (1989). Effects of unilateral cochlea removal on antero-ventral cochlear nucleus neurons in developing gerbils. *Journal of Comparative Neurology*, **283**, 465–473.

Hauser, K.F. & Toran-Allerand, C.D. (1989). Androgen increases the number of cells in fetal mouse spinal cord cultures: implications for motoneuron survival. *Brain Research*, **485**, 157–164.

Hayashi, T., Umemori, H., Mishina, M. & Yamamoto, T. (1999). The AMPA receptor interacts with and signals through the protein tyrosine kinase Lyn. *Nature*, **397**, 72–76.

Hayes, K.J. (1965). The so-called levator ani of the rat. *Acta Endocrinologica*, **48**, 337–347.

Henderson, C.E. (1996). Role of neurotrophic factors in neuronal development. *Current Opinion in Neurobiology*, **6**, 64–70.

Henderson, C.E., Phillips, H.S., Pollock, R.A., Davies, A.M., Lemeulle, C., Armani, M., Simpson, L.C., Moffet, B., Vandlen, R.A., Koliatsos, V.E. & Rosenthal, A. (1994). GDNF: a potent survival factor for motoneurons present in peripheral nerve and muscle. *Science*, **266**, 1062–1064.

Hensey, C. & Gautier, J. (1997). A developmental timer that regulates apoptosis at the onset of gastrulation. *Mechanical Developments*, **69**, 183–195.

Hensey, C. & Gautier, J. (1998). Programmed cell death during *Xenopus* development: a spatio-temporal analysis. *Developmental Biology*, **203**, 36–48.

Herrman, K. & Arnold, A.P. (1991). Lesions of HVC block the developmental masculinizing effects of estradiol in the female zebra finch song system. *Journal of Neurobiology*, **22**, 29–39.

Hines, M., Davis, F.C., Coquelin, A., Goy, R.W. & Gorski, R.A. (1985). Sexually dimorphic regions in the medial preoptic area and the bed nucleus of the stria terminalis of the guinea pigbrain: a description and an investigation of their relationship to gonadal steroids in adulthood. *Journal of Neuroscience*, **5**, 40–47.

Hoffmeister, H. (1966). Ecdysterone, a new metamorphosis hormone of insects. *Angewandte Chemié, International Edition*, **2**, 248–249.

Hollyday, M. & Hamburger, V. (1976). Reduction of naturally occurring motor neuron loss by enlargement of the periphery. *Journal of Comparative Neurology*, **170**, 311–320.

Holman, S.D., Collado, P., Skepper, J.N. & Rice, A. (1996). Postnatal development of a sexually dimorphic, hypothalamic nucleus in gerbils: a stereological study of neuronal number and apoptosis. *Journal of Comparative Neurology*, **376**, 315–325.

Homma, S. & Oppenheim, R.W. (1991). The prevention of early cell death during the earliest stages of spinal cord development in the chick embryo. *Society of Neuroscience Abstracts*, **17**, 229.

Homma, S., Yaginuma, H. & Oppenheim, R.W. (1994). Programmed cell death during the earliest stages of spinal cord development in the chick embryo: a possible means of early phenotypic selection. *Journal of Comparative Neurology,* **345**, 377–395.

Horsburgh, G.M. & Sefton, A.J. (1986). The early development of the optic nerve and chiasm in embryonic rat. *Journal of Comparative Neurology,* **243**, 547–560.

Hoskins, S.G. & Grobstein, P. (1985). Development of the ipsilateral retinothalamic projection in the frog *Xenopus laevis.* III. The role of thyroxine. *Journal of Neuroscience,* **5**, 930–940.

Hutchison, J.B., Wingfield, J.C. & Hutchison, R.E. (1984). Sex differences in plasma concentrations of steroids during the sensitive period of brain differentiation. *Journal of Endocrinology,* **103**, 363–369.

Huxley, J. (1920). Metamorphosis of axolotl caused by thyroid feeding. *Nature,* **104**, 436.

Ikonomidou, C., Bosch, F., Miksa, M., Bittigau, P., Vöckler, J., Dikranian, K., Tenkova, T., Stefovska, V., Turski, L. & Olney, J.W. (1999). Blockade of NMDA receptors and apoptotic neurodegeneration in the developing brain. *Science,* **283**, 70–74.

Ikonomidou, C., Mosinger, J.L., Shahid Salles, K., Labruyere, J. & Olney, J.W. (1989). Sensitivity of the developing rat brain to hypobaric/ischemic damage parallels sensitivity to *N*-methyl-aspartate neurotoxicity. *Journal of Neuroscience,* **9**, 2809–2818.

Ip, N.Y., Wiegand, S.J., Morse, J. & Rudge, J.S. (1993). Injury-induced regulation of ciliary neurotrophic factor mRNA in the adult rat brain. *European Journal of Neuroscience,* **5**, 25–33.

Jacobson, C.D., Davis, F.C. & Gorski, R.A. (1985). Formation of the sexually dimorphic nucleus of the preoptic area: neuronal growth, migration and changes in cell number. *Developmental Brain Research,* **21**, 7–18.

Jacobson, C.D., Shryne, J.E., Shapiro, F. & Gorski, R.A. (1980). Ontogeny of the sexually dimorphic nucleus of the preoptic area. *Journal of Comparative Neurology,* **193**, 541–548.

Jacobson, M.D., Weil, M. & Raff, M.C. (1997). Programmed cell death in animal development. *Cell,* **88**, 347–354.

Jaegle, M., Mandemakers, W., Broos, L., Zwart, R., Karis, A., Visser, A., Grosveld, F. & Meijer, D. (1996). The POU factor Oct-6 and Schwann cell differentiation. *Science,* **273**, 507–510.

Jaszai, J., Farkas, L., Galter, D., Reuss B., Strelau, J., Unsicker, K. & Krieglstein, K. (1998). GDNF-related factor persephin is widely distributed throughout the nervous system. *Journal of Neuroscience Research,* **53**, 494–501.

Jessen, K.R., Brennan, A., Morgan, L., Mirsky, R., Kent, A., Hashimoto, Y. & Gavrilovic, J. (1994). The Schwann cell precursor and its fate: a study of cell death and differentiation during gliogenesis in rat embryonic nerves. *Neuron,* **12**, 509–527.

Jiang, C., Baehrecke, E.H. & Thummel, C.S. (1997). Steroid-regulated programmed cell death during *Drosophila* metamorphosis. *Development,* **124**, 4673–4683.

Johnson, Jr, E.M., Koike, T. & Franklin, J. (1992). A "calcium set-point hypothesis" of neuronal dependence on neurotrophic factor. *Experimental Neurology,* **115**, 163–166.

Johnson, F. & Bottjer, S.W. (1994). Afferent influences on cell death and birth during development of a cortical nucleus necessary for learned vocal behavior in zebra finches. *Development,* **120**, 13–24.

Johnson, F., Hohmann, S.E., Di Stefano, P.S. & Bottjer, S.W. (1997). Neurotrophins suppress apoptosis induced by deafferentation of an avian motor-cortical region. *Journal of Neuroscience,* **17**, 2101–2111.

Jordan, C.L., Breedlove, S.M. & Arnold, A.P. (1991). Ontogeny of steroid accumulation in spinal lumbar motoneurons of the rat: implications for androgen's site of action during synapse elimination. *Journal of Comparative Neurology,* **312**, 1–8.

Jordan, C.L., Letinsky, M.S. & Arnold, A.P. (1989). The role of gonadal hormones in neuromuscular synapse elimination in rats. I. Androgen delays the loss of multiple innervation in the levator ani muscle. *Journal of Neuroscience,* **9**, 229–238.

Jordan, C.L., Pawson, P.A., Arnold, A.P. & Grinell, A.D. (1992). Hormonal regulation of motor unit size and synaptic strength during synapse elimination in the rat levator ani muscle. *Journal of Neuroscience,* **12**, 4447–4459.

Jordan, J., Bottner, M., Schluesenes, H.J., Unsickev, K. & Kriegelstein, K. (1997). Bone morphogenetic proteins: neurotrophic roles for midbrain dopaminergic neurons and implications for astroglial cells. *European Journal of Neuroscience,* **9**, 1699–1709.

Jung, I. & Baulieu, E.E. (1972). Testosterone cytosol "receptor" in the rat levator ani muscle. *Nature,* **237**, 24–26.

Kalb, R.G., Lidow, M.S., Halsted, M.J. & Hockfield, S. (1992). *N*-methyl-D-aspartate receptors are transiently expressed in developing spinal cord ventral horn. *Proceedings of the National Academy of Sciences, USA*, **89**, 8502–8506.

Källén, B. (1965). Degeneration and regeneration in the vertebrate central nervous system during embryogenesis. *Progress in Brain Research*, **14**, 77–96.

Karim, F.D. & Thummel, C.S. (1992). Temporal coordination of regulatory gene expression by the steroid hormone ecdysone. *EMBO Journal*, **38**, 4083–4093.

Kawahara, A., Baker, B. & Tata, J.R. (1991). Developmental and regional expression of thyroid hormone receptor genes during *Xenopus* metamorphosis. *Development*, **112**, 933–943.

Kay, J.N., Hannigan, P. & Kelley, D.B. (1999). Trophic effects of androgen: Development and hormonal regulation of neuron number in a sexually dimorphic vocal motor nucleus. *Journal of Neurobiology*, **40**, 375–385.

Kelley, D.B. (1980). Auditory and vocal nuclei in the frog brain concentrate sex hormones. *Science*, **207**, 553–555.

Kelley, D.B. & Dennison, J. (1990). The vocal motor neurons of *Xenopus laevis*: development of sex differences in axon number. *Journal of Neurobiology*, **21**, 869–882.

Kelley, D.B. & Pfaff, D.W. (1976). Hormone effects on male sex behavior in adult South African clawed frogs, *Xenopus laevis*. *Hormonal Behavior*, **7**, 159–182.

Kelley, D.B. & Tobias, M.L. (1998). Vocal communication in *Xenopus laevis*. In M. Hauser & M. Konishi (Eds.), *Neural mechanisms of communication* (pp. 9–35). Cambridge, MA: MIT Press.

Kelley, D.B., Morrell, J.I. & Pfaff, D.W. (1975). Autoradiographic localization of hormone-concentrating cells in the brain of the amphibian, *Xenopus laevis*. I. Testosterone. *Journal of Comparative Neurology*, **164**, 47–62.

Kelley, M.S., Lurie, D.I. & Rubel, E.W. (1997). Rapid regulation of cytoskeletal proteins and their mRNAs following afferent deprivation in the avian cochlear nucleus. *Journal of Comparative Neurology*, **389**, 469–483.

Kelley, D.B., Sassoon, D., Segil, N. & Scudder, M. (1989). Development and hormone regulation of androgen receptor levels in the sexually dimorphic larynx of *Xenopus laevis*. *Developmental Biology*, **131**, 111–118.

King, W.J. & Greene, G.L. (1984). Monoclonal antibodies localize estrogen receptor in the nuclei of target cells. *Nature*, **307**, 745–747.

Kirn, J.R. & DeVoogd, T.J. (1989). The genesis and death of vocal control neurons during sexual differentiation in the zebra finch. *Journal of Neuroscience*, **9**, 3176–3187.

Kirn, J.R. & Nottebohm, F. (1993). Direct evidence for loss and replacement of projection neurons in adult canary brain. *Journal of Neuroscience*, **13**, 1654–1663.

Kistler, A., Yoshizato, K. & Frieden, E. (1977). Preferential binding of tri-substituted thyronine analogs by bullfrog tadpole tail fin cytosol. *Endocrinology*, **100**, 134–137.

Knabe, W. & Kuhn, H.J. (1998). Pattern of cell death during optic cup formation in the tree shrew *Tupaia belangeri*. *Journal of Comparative Neurology*, **401**, 352–366.

Koelle, M.R., Talbot, W.S., Segraves, W.A., Bender, W.A., Cherbas, P. & Hogness, D.S. (1991). The *Drosophila* EcR gene encodes an ecdysone receptor, a new member of the steroid receptor superfamily. *Cell*, **67**, 59–78.

Koike, T., Martin, D.P. & Johnson, Jr, E.M. (1989). Role of calcium channels in the ability of membrane depolarization to prevent neuronal death induced by trophic factor deprivation: evidence that levels of Ca^{2+} determine nerve growth factor dependence of sympathethic ganglion cells. *Proceedings of the National Academy of Sciences, USA*, **86**, 6421–6425.

Kollros, J.J. (1981). Transitions in the nervous system during amphibian metamorphosis. In L.I. Gilbert & E. Frieden (Eds.), *Metamorphosis: a problem in developmental biology* (pp. 445–449). New York: Plenum Press.

Kollros, J.J. (1984). Growth and death of cells of the mesencephalic fifth nucleus in *Rana pipiens* larvae. *Journal of Comparative Neurology*, **102**, 47–63.

Kollros, J.J. & McMurray, V.M. (1956). The mesencephalic V nucleus in anurans. II. The influence of thyroid hormone on cell size and cell number. *Journal of Experimental Zoology*, **131**, 1–26.

Kollros, J.J. & Thiesse, M.L. (1985). Growth and death of cells of the mesencephalic fifth nucleus of *Xenopus laevis*. *Journal of Comparative Neurology*, **233**, 481–489.

Konishi, M. & Akutagawa, E. (1985). Neuronal growth, atrophy and death in a sexually dimorphic song nucleus in the zebra finch brain. *Nature*, **315**, 145–147.

Konishi, M. & Akutagawa, E. (1988). A critical period for estrogen action on neurons of the song control system in the zebra finch. *Proceedings of the National Academy of Sciences, USA*, **85**, 7006–7007.

Korsching, S. (1993). The neurotrophic factor concept: a reexamination. *Journal of Neuroscience*, **13**, 2739–2748.

Korsia, S. & Bottjer, S.W. (1989). Developmental changes in the cellular composition of a brain nucleus involved with song learning in zebra finches. *Neuron*, **3**, 451–460.

Kroemer, G. (1997). The proto-oncogene Bcl-2 and its role in regulating apoptosis. *Nature Medicine*, **3**, 614–620.

Krueger, B.K., Burne, J.F. & Raff, M.C. (1995). Evidence for large-scale astrocyte death in the developing cerebellum. *Journal of Neuroscience*, **15**, 3366–3374.

Kuan, C-Y., Yang, D.D., Roy, D.R.S., Davis, R.J., Rakic, P. & Flavell, R.A. (1999). The JnK1 and JnK2 protein kinases are required for regional specific apoptosis during early brain development. *Neuron*, **22**, 667–676.

Kuida, K., Zheng, T.S., Na, S., Kuan, C-Y., Yang, D., Karasuyama, H., Rakic, P. & Flavell, R.A. (1996). Decreased apoptosis in the brain and premature lethality in CPP32-deficient mice. *Nature*, **384**, 368–372.

Kuida, K., Haydar, T.F., Kuan, C.Y., Gu, Y., Taya, C., Karasuyama, H., Su, M.S., Rakic, P. & Flavell, R.A. (1998). Reduced apoptosis and cytochrome c-mediated caspase activation in mice lacking caspase-9. *Cell*, **94**, 325–337.

Kujawa, K.A., Emeric, E. & Jones, K.J. (1991). Testosterone differentially regulates the regenerative properties of injured hamster facial motoneurons. *Journal of Neuroscience*, **11**, 3898–3906.

Kutsch, W. & Bentley, D. (1987). Programmed death of peripheral pioneer neurons in the grasshopper embryo. *Developmental Biology*, **123**, 517–525.

Lachica, E.A., Rubsamen, R., Zirpel, L. & Rubel, E.W. (1995). Glutamatergic inhibition of voltage-operated calcium channels in the avian cochlear nucleus. *Journal of Neuroscience*, **15**, 1724–1734.

Laemle, L.K., Puszkarczuk, M. & Feinberg, R.N. (1999). Apoptosis in early ocular morphogenesis in the mouse. *Developmental Brain Research*, **112**, 129–133.

Lamborghini, J.E. (1980). Rohon–Beard cells and other large neurons in *Xenopus* embryos originate during gastrulation. *Journal of Comparative Neurology*, **189**, 323–333.

Landmesser, L. & Pilar, G. (1974). Synapse formation during embryogenesis on ganglion cells lacking a periphery. *Journal of Physiology, London*, **247**, 715–736.

Larmet, Y., Dolphin, A.C. & Davies, A.M. (1992). Intracellular calcium regulates the survival of early sensory neurons before they become dependent on neurotrophic factors. *Neuron*, **9**, 563–574.

Lefcourt, F., Clary, D.O., Rusoff, A.C. & Reichardt, L.F. (1996). Inhibition of the NT-3 receptor trkC, early in chick embryogenesis, results in severe reduction in multiple neuronal sub-populations in the dorsal root ganglia. *Journal of Neuroscience*, **16**, 3704–3713.

Legrand, J. (1979). Morphogenetic actions of thyroid hormones. *Trends in Neuroscience*, **2**, 234–236.

Leist, M. & Nicotera, P. (1997). Calcium and neuronal death. *Reviews in Physiology, Pharmacology and Biochemistry*, **132**, 79–125.

Lemke, G. (1996). Neuregulins in development. *Molecular and Cellular Neuroscience*, **7**, 247–262.

Levi-Montalcini, R. (1949). The development of the acoustico-vestibular centers in the chick embryo in the absence of the afferent root fibers and of descending fiber tracts. *Journal of Comparative Neurology*, **91**, 209–242.

Lewis, P.D., Patel, A.J., Johnson, A.L. & Balazs, R. (1976). Effect of thyroid deficiency on cell acquisition in the postnatal rat brain: a quantitative histological study. *Brain Research*, **104**, 49–62.

Liebl, D.J., Tessarollo, L., Palko, M.E. & Parada, L.F. (1997). Absence of sensory neurons before target innervation in brain-derived neurotrophic factor-, neurotrophin 3- and trkC-deficient embryonic mice. *Journal of Neuroscience*, **17**, 9113–9121.

Linden, R. (1994). The survival of developing neurons: a review of afferent control. *Neuroscience*, **58**, 671–682.

Lladó, J., Calderó, J., Ribera, J., Tarabal, O., Oppenheim, R.W. & Esquerda, J.E. (1999). Opposing effects of excitatory amino acids on chick embryo spinal cord motoneurons: excitotoxic degeneration or prevention of programmed cell death. *Journal of Neuroscience*, **19**, 10803–10812.

Lockshin, R.A. & Williams, C.M. (1965). Programmed cell death. III. Neuronal control of the breakdown of the intersegmental muscles of silkmoths. *Journal of Insect Physiology,* **11**, 601–610.

Lubischer, J.L. & Arnold, A.P. (1995). Effects of axotomy during early postnatal life on androgen-sensitive motoneurons: evidence for target regulation of the development of steroid-sensitivity. *Developmental Neuroscience,* **17**, 106–117.

Lubischer, J.L. & Weeks, J.C. (1996). Target muscles and sensory afferents do not influence steroid-regulated, segment-specific death of identified motoneurons in *Manduca sexta*. *Journal of Neurobiology,* **31**, 449–460.

Lumbroso, S., Sandillon, F., Georget, V., Lobaccaro, J.M., Brinkmann, A.O., Privat, A. & Sultan, C. (1996). Immunohistochemical localization and immunoblotting of androgen receptor in spinal neurons of male and female rats. *European Journal of Endocrinology,* **134**, 626–632.

Maderdrut, J.L., Oppenheim, R.W. & Prevette, D. (1988). Enhancement of naturally occurring cell death in the sympathetic ganglia of the chicken embryo following blockade of ganglionic transmission. *Brain Research*, **444**, 189–194.

Marin, M.L., Tobias, M.L. & Kelley, D.B. (1990). Hormone-sensitive stages in the sexual differentiation of laryngeal muscle fiber number in *Xenopus laevis*. *Development*, **110**, 703–712.

Marini, A.M., Rabin, S.J., Lipsky, R.H. & Mocchetti, I. (1998). Activity-dependent release of brian-derived neurotrophic factor underlies the neuroprotective effect of *N*-methyl-D-aspartate. *Journal of Biological Chemistry,* **273**, 29394–29399.

Matheson, C.R., Carnahan, J., Urich, J.L., Bocangel, D., Zhang, T.J &, Yan, Q. (1997). Glial cell line-derived neurotrophic factor (GDNF) is a neurotrophic factor for sensory neurons: comparison with the effects of the neurotrophins. *Journal of Neurobiology*, **32**, 22–32.

Matsumoto, A. (1991). Synaptogenic action of sex steroids in developing and adult neuroendocrine brain. *Psychoneuroendocrinology*, **16**, 25–40.

McConnell, S.K. & Kaznoski, C.E. (1991). Cell cycle dependence of laminar determination in the developing neocortex. *Science*, **254**, 282–285.

McDonald, J.W., Silverstein, F.S. & Johnston, M.V. (1990). MK-801 pretreatment enhances *N*-methyl-D-aspartate-mediated brain injury and increases brain *N*-methyl-D-aspartate recognition site binding in rats. *Neuroscience*, **38**, 103–113.

McKellar, C.E., DeFreitas, M.F. & Shatz, C.J. (1999). Activity-dependent survival of subplate neurons. *Society of Neuroscience Abstracts,* **25**, 1777.

McKeon, R.J., Silver, J. & Large, T.H. (1997). Expression of full-length trkB receptors by reactive astrocytes after chronic CNS injury. *Experimental Neurology,* **148**, 558–567.

McTigue, D.M., Horner, P.J., Stokes, B.T. & Gage, F.H. (1998). Neurotrophin-3 and brain-derived neurotrophic factor induce oligodendrocyte proliferation and myelination of regenerating axons in the contused adult rat spinal cord. *Journal of Neuroscience,* **18**, 5354–5365.

Meier, C., Parmantier, E., Brennan, A., Mirsky, R. & Jessen, K.R. (1999). Developing Schwann cells acquire the ability to survive without axons by establishing an autocrine circuit involving insulin-like growth factor, neurotrophin-3, and platelet-derived growth factor. *Journal of Neuroscience*, **19**, 3847–3859.

Meyer, D. & Birchmeier, C. (1995). Multiple essential functions of neuregulin in development. *Nature*, **378**, 386–390.

Meyer-Franke, A., Wilkinson, G.A., Kruttgen, A., Hu, M., Munro, E., Hanson, Jr, M.G., Reichardt, L.F. & Barres, B.A. (1998). Depolarization and cAMP elevation rapidly recruit TrkB to the plasma membrane of CNS neurons. *Neuron*, **21**, 681–693.

Mills, A.C. & Sengelaub, D.R. (1993). Sexually dimorphic neuron number in lumbosacral dorsal root ganglia of the rat: development and steroid regulation. *Journal of Neurobiology,* **24**, 1543–1553.

Mitchell, I.J., Lawson, S., Moser, B., Laidlaw, S.M., Cooper, A.J., Walkinshaw, G. & Waters, C.M. (1994). Glutamate-induced apoptosis results in a loss of striatal neurons in the parkinsonian rat. *Neuroscience*, **63**, 1–5.

Moretto, G., Walker, D.G., Lanteri, P., Taioli, F., Zaffagnini, S., Xu, R.Y. & Rizzuto, N. (1996). Expression and regulation of glial-cell-line-derived neurotrophic factor (GDNF) mRNA in human astrocytes *in vitro*. *Cell Tissue Research*, **286**, 257–262.

Morshead, C.M. & van der Kooy, D. (1992). Postmitotic death is the fate of constitutively prolif-erating cells in the subependymal layer of the adult mouse brain. *Journal of Neuroscience,* **12**, 249–256.

Motoyama, N., Wang, F. Roth, K.A., Sawa, H., Nakayam, K., Negishi, I. Senju, S. Zhang, Q. & Fuji, S. (1995). Massive cell death of immature hematopoietic cells and neurons in Bcl-x-deficient mice. *Science*, **267**, 1506–1510.

Murphy, P., Topilko, P., Schneider-Maunoury, S., Seitanidou, T., Baron-van Evercooren, A. & Charnay, P. (1996). The regulation of Krox-20 expression reveals important steps in the control of peripheral glial cell development. *Development*, **122**, 2847–2857.

Naess, O., Haug, E., Attramadal, A., Aakvaag, A., Hansson, V. & French, F. (1976). Androgen receptors in the anterior pituitary and central nervous system of the androgen "insensitive" (Tfm) rat: correlation between receptor binding and effects of androgens on gonadotropic secretion. *Endocrinology*, **99**, 1295–1303.

Naftolin, F., Horvath, T.L., Jakab, R. L., Leranth, C., Harada, N. & Balthazart, J. (1996). Aromatase immunoreactivity in axon terminals of the vertebrate brain. An immunocytochemical study on quail, rat, monkey and human tissues. *Neuroendocrinology*, **63**, 191–195.

Nagata, S. (1997). Apoptosis by death factor. *Cell*, **88**, 355–365.

Nastiuk, K.L. & Clayton, D.F. (1994). Seasonal and tissue-specific regulation of canary androgen receptor mRNA. *Endocrinology*, **134**, 640–649.

Navascues, J., Martin-Partido, I., Avarez, S. & Rodrigues-Gallardo, L. (1988). Cell death in sub-optic necrotic centers of chick embryo diencephalon and their topographic relationship with the earliest optic fiber fasicles. *Journal of Comparative Neurology*, **278**, 34–46.

Nordeen, K.W. & Nordeen, E.J. (1988). Projection neurons within a vocal motor pathway are born during song learning in zebra finches. *Nature*, **334**, 149–151.

Nordeen, E.J., Nordeen, K.W. & Arnold, A.P. (1987). Sexual differentiation of androgen accumulation within the zebra finch brain through selective cell loss and addition. *Journal of Comparative Neurology*, **25**, 393–399.

Nordeen, E.J., Nordeen, K.W., Sengelaub, D.R. & Arnold, A.P. (1985). Androgens prevent normally occurring cell death in a sexually dimorphic spinal nucleus. *Science*, **229**, 671–673.

Nordeen, E.J., Voelkel, L. & Nordeen, K.W. (1998). Fibroblast growth factor-2 stimulates cell proliferation and decreases sexually dimorphic cell death in an avian song control nucleus. *Journal of Neurobiology*, **37**, 573–581.

Nottebohm, F. & Arnold, A.P. (1976). Sexual dimorphism in vocal control areas of the songbird brain. *Science*, **194**, 211–213.

Nottebohm, F., Kelley, D.B. & Paton, J.A. (1982). Connections of vocal control nuclei in the canary telencephalon. *Journal of Comparative Neurology*, **207**, 344–357.

Nottebohm, F., Stokes, T.M. & Leonard, C.M. (1976). Central control of song in the canary *Serinus canarius*. *Journal of Comparative Neurology*, **165**, 457–486.

Ockel, M. Lewin, G. & Barde, Y.A. (1996). *In vivo* effects of neurotrophin-3 during sensory neurogenesis. *Development*, **122**, 301–307.

Ojeda, S.R., Kalra, P.S. & McCann, S.M. (1975). Further studies on the maturation of the estrogen negative feedback on gonadotropin release in the female rat. *Neuroendocrinology*, **18**, 242–255.

Okado, N. & Oppenheim, R.W. (1984). Cell death of motoneurons in the chick embryo spinal cord. IX. The loss of motoneurons following removal of afferent inputs. *Journal of Neuroscience*, **4**, 1639–1652.

Omerza, F.F. & Alley, K.E. (1992). Redeployment of trigeminal motor axons during metamorphosis. *Journal of Comparative Neurology*, **325**, 124–134.

Oppenheim, R.W. (1981). Neuronal cell death and some related regressive phenomena during neurogenesis: a selective historical review and progress report. In W.M. Cowan (Ed.), *Studies in developmental neurobiology: essays in honor of Viktor Hamburger* (pp. 73–133). New York: Oxford University Press.

Oppenheim, R.W. (1989). The neurotrophic theory and naturally occurring motoneuron death. *Trends in Neuroscience*, **12**, 252–255.

Oppenheim, R.W. (1991). Cell death during development of the nervous system. *Annual Review of Neuroscience*, **14**, 453–501.

Oppenheim, R.W. (1996). Neurotrophic survival molecules for motoneurons: an embarrassment of riches. *Neuron*, **17**, 195–197.

Oppenheim, R.W. (1999). Programmed cell death. In M. Zigmond, F. Bloom, S. Landis, J. Roberts & L. Squire (Eds.), *Fundamental neuroscience* (pp. 581–609). New York: Academic Press.

Oppenheim, R.W., Chu-Wang, I.-W. & Maderdrut, J.L. (1978). Cell death of motoneurons in the chick embryo spinal cord. III. The differentiation of motoneurons prior to their induced degeneration following limb-bud removal. *Journal of Comparative Neurology,* **177**, 87–112.

Oppenheim, R.W., Homma, S., Marti, E., Prevette, D., Wang, S., Yaginuma, H. & McMahon, A.O. (1999). Modulation of early but not later stages of programmed cell death in avian spinal cord by sonic hedgehog. *Molecular and Cellular Neuroscience,* **13**, 348–361.

Oppenheim, R.W., Houenou, L.J., Johnson, J.E., Lin, L.-F.H., Li, L., Lo, A.C., Newsome, A.L., Prevette, D.M. & Wang, S. (1995). Developing motor neurons rescued from programmed and axotomy-induced cell death by GDNF. *Nature,* **373**, 344–346.

Oppenheim, R.W., Maderdrut, J.L. & Wells, D.J. (1982). Cell death of motoneurons in the chick embryo spinal cord. VI. Reduction of naturally occurring cell death in the thoracolumbar column of Terni by nerve growth factor. *Journal of Comparative Neurology,* **210**, 174–189.

Oppenheim, R.W., Prevette, D., Qin-Wei, Y., Collins, F. & MacDonald, J. (1991). Control of embryonic motoneuron survival *in vivo* by ciliary neurotrophic factor. *Science,* **251**, 1616–1618.

O´Brien, R.J., Kamboj, S., Ehlers, M.D., Rosen, K.R., Fischbach, G.D. & Huganir, R.L. (1998). Activity-dependent modulation of synaptic AMPA receptor accumulation. *Neuron,* **21**, 1067–1078.

O'Leary, D.D.M., Fawcett, J.W. & Cowan, W.M. (1986). Topographic targeting errors in the retinocollicular projection and their elimination by selective ganglion cell death. *Journal of Neuroscience,* **6**, 3692–3705.

Pachnis, V., Mankoo, B. & Constantini, F. (1993). Expression of c-ret protooncogene during mouse embryogenesis. *Development,* **119**, 1005–1017.

Park, J-J., Tobet, S.A. & Baum, M.J. (1998). Cell death in the sexually dimorphic dorsal preoptic area anterior hypothalamus of perinatal male and female ferrets. *Journal of Neurobiology,* **34**, 242–252.

Parks, T.N. (1979). Afferent influences on the development of the brain stem auditory nuclei of the chicken: otocyst ablation. *Journal of Comparative Neurology,* **183**, 665–678.

Payer, A.F. (1979). An ultrastructural study of Schwann cell response to axonal degeneration. *Journal of Comparative Neurology,* **183**, 365–384.

Pellier, V. & Astic, L. (1994). Cell death in the developing olfactory epithelium of rat embryos. *Developmental Brain Research,* **79**, 307–315.

Pérez, J. & Kelley, D.B. (1996). Trophic effects of androgens: receptor expression and the survival of laryngeal motoneurons after axotomy. *Journal of Neuroscience,* **16**, 6625–6633.

Pérez, J. & Kelley, D.B. (1997). Androgen prevents axotomy-induced decreases in calbindin expression in motor neurons. *Journal of Neuroscience,* **17**, 7396–7403.

Pérez, J., Cohen, M. & Kelley, D.B. (1996). Androgen receptor mRNA expression in Xenopus laevis CNS: sexual dimorphism and regulation in the laryngeal motor nucleus. *Journal of Neurobiology,* **30**, 556–568.

Pesetsky, I. (1962). The thyroxine-stimulated enlargement of Mauthner's neuron in anurans. *General and Comparative Endocrinology,* **2**, 229–235.

Pettmann, B. & Henderson, C.E. (1998). Neuronal cell death. *Neuron,* **20**, 633–647.

Pfaff, D.W. & Keiner, M. (1973). Atlas of estradiol-containing cells in the central nervous system of the female rat. *Journal of Comparative Neurology,* **151**, 121–158.

Pfahl, M. (1993). Nuclear receptor AP-1 interaction. *Endocrine Reviews,* **14**, 651–658.

Phelan, K.A. & Hollyday, M. (1991). Embryonic development and survival of brachial motoneurons projecting to muscleless chick wings. *Journal of Comparative Neurology,* **311**, 313–320.

Phoenix, C.H., Goy, R.W., Gerall, A.A. & Young, W.C. (1959). Organizing action of prenatally administered testosterone propionate on the tissues mediating mating behavior in the female guinea pig. *Endocrinology,* **65**, 369–382.

Picon, F. (1976). Testosterone secretion by fetal rat testes *in vitro. Journal of Endocrinology,* **71**, 231–237.

Pohl-Appel, G. & Sossinka, R. (1984). Hormonal determination of song capacity in females of the zebra finch: critical phase of treatment. *Zetischrift für Tierpsychologie,* **64**, 330–336.

Prestige, M.C. & Wilson, M.A. (1972). Loss of axons from ventral nerve roots during development. *Brain Research,* **41**, 467–470.

Prevette, D., Arce, V., Henderson, C., Wang, S., MacDonald, M., deLapeyiere, O. & Oppenheim, R.W. (1999). Synergistic effects of Schwann cell and muscle cell derived factors on avian motoneuron survival may involve GDNF and CT-1. *Society of Neuroscience Abstracts,* **25**, 1526.

Price, D.J. & Lotto, R.B. (1996). Influences of the thalamus on the survival of subplate and cortical plate cells in cultured embryonic mouse brain. *Journal of Neuroscience,* **16**, 3247–3255.

Qin-Wei, Y., Johnson, J., Prevette, D. & Oppenheim, R.W. (1994). Cell death of spinal motoneurons in the chick embryo following deafferentation: rescue effects of tissue extracts, soluble proteins, and neurotrophic agents. *Journal of Neuroscience,* **14**, 7629–7640.

Race, J. (1961). Thyroid hormone control of development of lateral motor column cells in the lumbosacral cord in hypophysectomized *Rana pipiens. General and Comparative Endocrinology,* **I**, 323–331.

Raff, M.C., Barres, B.A., Burne, J.F., Coles, H.S., Ishizaki, Y. & Jacobson, M.D. (1993). Programmed cell death and the control of cell survival: lessons from the nervous system. *Science,* **262**, 695–700.

Raff, M.C., Miller, R.H. & Noble, M. (1983). A glial progenitor cell that develops *in vitro* into an astrocyte or an oligodendrocyte depending on culture medium. *Nature,* **303**, 390–396.

Raisman, G. & Field, P.M. (1973). Sexual dimorphism in the neuropil of the preoptic area of the rat and its dependence on neonatal androgen. *Brain Research,* **54**, 1–29.

Rand, M.N. & Breedlove, S.M. (1987). Ontogeny of functional innervation of bulbocavernosus muscles in male and female rats. *Developmental Brain Research,* **33**, 150–152.

Rao, A. & Craig, A.M. (1997). Activity regulates the synaptic localization of the NMDA receptor in hippocampal neurons. *Neuron,* **19**, 801–812.

Raoul, C., Yamamoto, Y. Henderson, C.E. & Pettmann, B. (1999). Fas triggers programmed cell death of embryonic motoneurons independently of neurotrophic support. *Neuron* (In press).

Rappolee, D.A., Mark, D., Banda, M.J. & Werb, Z. (1989). Wound macrophages express TGF-alpha and other growth factors *in vivo*: analysis by mRNA phenotyping. *Science,* **241**, 708–712.

Rasika, S., Alvarez-Buylla, A. & Nottebohm, F. (1999). BDNF mediates the effects of testosterone on the survival of new neurons in an adult brain. *Neuron,* **22**, 53–62.

Reynolds, W.A. (1963). The effects of thyroxine upon the initial formation of the lateral motor column and differentiation of motor neurons in *Rana pipiens. Journal of Experimental Zoology,* **153**, 237–249.

Rhees, R.W., Shryne, J.E. & Gorski, R.A. (1990). Termination of the hormone-sensitive period for differentiation of the sexually dimorphic nucleus of the preoptic area in male and female rats. *Developmental Brain Research,* **52**, 17–23.

Richards, A.G. (1976). Sequential gene activation by ecdysone in polytene chromosomes of *Drosophila melanogaster.* V. The late prepupal puffs. *Developmental Biology,* **54**, 256–263.

Riddiford, L.M. (1980). Interaction of ecdysteroids and juvenile hormone in the regulation of larval growth and metamorphosis of the tobacco hornworm. In J.A. Hoffman (Ed.), *Progress in ecdysone research* (pp. 409–430). Amsterdam: Elsevier/North Holland Biomedical.

Riddiford, L.M. & Truman, J.W. (1972). Delayed effects of juvenile hormone on insect metamorphosis are mediated by the corpus allota. *Nature,* **237**, 458.

Riethmacher, D., Sonnenberg-Riethmacher, E., Brinkmann, V., Yamaai, T., Lewin, G.R. & Birchmeier, C. (1997). Severe neuropathies in mice with targeted mutations in the ErbB3 receptor. *Nature,* **16**, 725–730.

Roberts, L.M., Hirokawa, Y., Nachtigal, M.W. & Ingraham, H.A. (1999). Paracrine-mediated apoptosis in reproductive tract development. *Developmental Biology,* **208**, 110–122.

Robertson, J.C. & Kelley, D.B. (1996). Thyroid hormone controls the onset of androgen sensitivity in the developing *Xenopus laevis. Developmental Biology,* **176**, 108–123.

Robertson, J.C., Watson, J.T. & Kelley, D.B. (1994). Androgen directs sexual differentiation of laryngeal innervation in developing *Xenopus laevis. Journal of Neurobiology,* **25**, 1625–1636.

Robinow, S., Talbot, W.S., Hogness, D.S. & Truman, J.W. (1993). Programmed cell death in the *Drosophila* CNS is ecdysone-related and coupled with a specific ecdysone receptor isoform. *Development,* **119**, 1251–1259.

Robinow, S., Draizen, T.A. & Truman, J.W. (1997). Genes that induce apoptosis: transcriptional regulation in identified, doomed neurons of the *Drosophila* CNS. *Developmental Biology,* **190**, 206–213.

Ross, M.E. (1996). Cell division and the nervous system: regulating the cycle from neural differentiation to death. *Trends in Neuroscience,* **19**, 62–68.

Rubio, N. (1997). Mouse astrocytes store and deliver brain-derived neurotrophic factor using the non-catalytic gp95[trkB] receptor. *European Journal of Neuroscience,* **9**, 1847–1853.

Rudge, J.S., Alderson, R.F., Pasnikowsky, E., McClain, J., Ip, N.Y. & Lindsay, R.M. (1992). Expression of ciliary neurotrophic factor and the neurotrophins-NGF, BDNF, and NT-3 in cultured rat hippocampal astrocytes. *European Journal of Neuroscience,* **4**, 459–471.

Rudge, J.S., Li, Y., Pasnikowsky, E.M., Mattsson, K., Pan, L., Yankopoulos, G.D., Wiegand, S.J., Lindsay, R.M. & Ip, N.Y. (1994). Neurotrophic factor receptors and their signal transduction capabilities in rat astrocytes. *European Journal of Neuroscience,* **6**, 693–705.

Ruoslahti, E. & Reed, J.C. (1994). Anchorage dependence, integrins, and apoptosis. *Cell,* **77**, 477–478.

Sachs, B.D. (1982). Role of striated penile muscles in penile reflexes, copulation, and induction of pregnancy in the rat. *Journal of Reproduction and Fertility,* **66**, 433–443.

Sanders, E.J. & Wride, M.A. (1995). Programmed cell death in development. *International Review of Cytology,* **163**, 105–173.

Sar, M. & Stumpf, W.E. (1975). Distribution of androgen-concentrating neurons in rat brain. In W.E. Stumpf & L.P. Grant (Eds.), *Anatomical neuroendocrinology* (pp. 120–133). Basel: Karger.

Sar, M. & Stumpf, W.E. (1977). Androgen concentration in motor neurons of cranial nerves and spinal cord. *Science,* **197**, 77–79.

Sassoon, D. & Kelley, D.B. (1986). The sexually dimorphic larynx of *Xenopus laevis*: development and androgen regulation. *American Journal of Anatomy,* **177**, 457–472.

Saunders, Jr, J.W. (1966). Death in embryonic systems. *Science,* **154**, 604–612.

Schaar, D.G., Sieber, B.A., Dreyfus, C.F. & Black, I.B. (1993). Regional and cell specific expression of GDNF in rat brain. *Experimental Neurology,* **124**, 368–371.

Schlinger, B.A. (1998). Sexual differentiation of avian brain and behavior: current views on gonadal hormone-dependent and independent mechanisms. *Annual Review of Physiology,* **60**, 407–429.

Schlinger, B.A. & Arnold, A.P. (1992). Plasma sex steroids and tissue aromatization in hatchling zebra finches: implications for the sexual differentiation of singing behavior. *Endocrinology,* **130**, 289–299.

Schluensener, H.J. & Meyermann, R. (1994). Expression of BMP-6, a TGF-β related morphogenetic cytokine in rat radial glial cells. *Glia,* **12**, 161–164.

Schmechel, D.E. & Rakic, P. (1979). A Golgi study of radial glial cells in developing monkey telencephalon: morphogenesis and transformation into astrocytes. *Anatomy and Embryology,* **156**, 115–152.

Schwartz, J.P. & Nishiyama, N. (1994). Neurotrophic factor gene expression in astrocytes during development and following injury. *Brain Research Bulletin,* **35**, 403–407.

Schwartz, L.M. (1992). Insect muscle as a model for programmed cell death. *Journal of Neurobiology,* **9**, 1312–1326.

Schwartz, L.M. & Truman, J.W. (1983). Hormonal control of rates of metamorphic development in the tobacco hornworm, *Manduca sexta*. *Developmental Biology,* **99**, 103–114.

Sendtner, M., Carroll, P., Holtmann, B., Hughes, R.A. & Thoenen, H. (1994). Ciliary neurotrophic factor. *Journal of Neurobiology,* **25**, 1436–1453.

Sendtner, M., Kreutzberg, G.W. & Thoenen, H. (1990). Ciliary neurotrophic factor prevents the degeneration of motor neurons after axotomy. *Nature,* **345**, 440–441.

Sengelaub, D.R. & Arnold, A.P. (1986). Development and loss of early projections in a sexually dimorphic rat spinal nucleus. *Journal of Neuroscience,* **6**, 1613–1620.

Silver, J. (1978). Cell death during development of the nervous system. In M. Jacobson (Ed.), *Handbook of sensory physiology*, Volume IX: *Development of sensory systems* (pp. 419–436). Berlin: Springer-Verlag.

Silver, J. & Robb, R.M. (1979). Studies on the development of the eye cup and optic nerve in normal mice and in mutants with congenital optic nerve aplasia. *Developmental Biology,* **68**, 175–190.

Simerly, R.B. & Swanson, L.W. (1986), The organization of neural inputs to the medial preoptic nucleus of the rat. *Journal of Comparative Neurology,* **246**, 312–342.

Simerly, R.B. & Swanson, L.W. (1988). Projections of the medial preoptic nucleus: a *Phaseolus vulgaris* leucoagglutinin anterograde tract-tracing study in the rat. *Journal of Comparative Neurology,* **270**, 209–242.

Simerly, R.B., Gorski, R.A. & Swanson, L.W. (1986). The neurotransmitter specificity of cells and fibers in the medial preoptic nucleus: an immunohistochemical study in the rat. *Journal of Comparative Neurology,* **246**, 364–381.

Simerly, R.B., Swanson, L.W. & Gorski, R.A. (1984). The cells of origin of a sexually dimorphic serotonergic input to the medial preoptic nucleus of the rat. *Brain Research,* **324**, 185–189.

Simpson, H.B. & Vicario, D.S. (1991). Early estrogen treatment alone causes female zebra finches to produce male-like vocalizations. *Journal of Neurobiology,* **22**, 755–776.

Singh, M., Sétálo, G., Guan, X., Warren, M. & Toran-Allerand, C.D. (1999). Estrogen-induced activation of mitogen-activated protein kinase in cerebral cortical explants: convergence of estrogen and neurotrophin signaling pathways. *Journal of Neuroscience, 19,* 1179–1188.

Sloviter, R.S., Dean, E., Sollas, A.L. & Goodman, J.H. (1996). Apoptosis and necrosis induced in different hippocampal neuron populations by repetitive perforant path stimulation in the rat. *Journal of Comparative Neurology, 366,* 516–533.

Sohrabji, F., Greene, L.A., Miranda, R.C. & Toran-Allerand, C.D. (1994). Reciprocal regulation of estrogen and NGF receptors by their ligands in PC12 cells. *Journal of Neurobiology, 25,* 974–988.

Soler, R.M., Egea, J., Mintening, G.M., Sanz-Rodriguez, C., Iglesias, M. & Comella, J.X. (1998). Calmodulin is involved in membrane depolarization-mediated survival of motoneurons by phosphatidylinositol-3-kinase- and MAPK-independent pathways. *Journal of Neuroscience, 18,* 1230–1239.

Solum, D., Hughes, D., Major, M.S. & Parks, T.N. (1997). Prevention of normally occurring and deafferentation-induced neuronal death in chick brainstem auditory neurons by periodic blockade of AMPA/kainate receptors. *Journal of Neuroscience, 17,* 4744–4751.

Sonnenfeld, M.J. & Jacobs, J.R. (1995). Apoptosis of the midline glia during *Drosophila* embryogenesis: a correlation with axon contact. *Development, 121,* 569–578.

Stefanelli, A. (1951). The Mauthnerian apparatus in the Ichthyopsida; its nature and function and correlated problems of histogenesis. *Quarterly Reviews of Biology, 26,* 17–34.

Steller, H. & Grether, M.E. (1994). Programmed cell death in *Drosophila*. *Neuron, 13,* 1269–1274.

Stewart, R.R., Gao, W.-Q., Peinado, A., Zipser, B. & Macagno, E.R. (1987). Cell death during gangliogenesis in the leech: Bipolar cells appear and then degenerate in all ganglia. *Journal of Neuroscience, 7,* 1919–1927.

Stocker, R.F. & Nuesch, H. (1975). Ultrastructural studies on neuromuscular contacts and the formation of junctions in the flight muscles of *Antheraea polyphemus*. I. Normal adult development. *Cell Tissue Research, 159,* 245–266.

Stöckli, K.A., Lillien, L.E., Näher-Noe, M., Breitfeld, G., Hughes, R.A., Thoenen, H. & Sendtner, M. (1991). Regional distribution, developmental changes, and cellular localization of CNTF mRNA and protein in the rat brain. *Journal of Cell Biology, 115,* 447–459.

Streichert, L.C., Pierce, J.T., Nelson, J.A. & Weeks, J.C. (1997). Steroid hormones act directly to trigger segment-specific programmed cell death of identified motoneurons *in vitro*. *Developmental Biology, 183,* 95–107.

Sulston, J.E. & Horvitz, H.R. (1977). Postembryonic cell lineages of the nematode *C. elegans*. *Developmental Biology, 56,* 110–156.

Swaab, D.F. & Fliers, E. (1985). A sexually dimorphic nucleus in the human brain. *Science, 228,* 1112–1114.

Syroid, D.E., Maycox, P.R., Burrola, P.O., Liu, N., Wen, D., Lee, K.-F., Lemke, G. & Kilpatrick T.J. (1996). Cell death in the Schwann cell lineage and its regulation by neuregulin. *Proceedings of the National Academy of Sciences, USA, 93,* 9229–9234.

Syroid, D.E., Zorick, T.S., Arbet-Engels, C., Kilpatrick, T.J., Eckhart, W. & Lemke, G. (1999). A role of insulin-like growth factor-I in the regulation of Schwann cell survival. *Journal of Neuroscience, 19,* 2059–2068.

Takahashi, K., Nuckolis, G.H., Tanaka, O., Semba, I., Takahashi, I., Dashner, R., Shum, L.O. & Slavkin, H.C. (1998). Adenovirus-mediated ectopic expression of Msx2 in even-numbered rhombomeres induces apoptotic elimination of cranial neural crest cells *in ovo*. *Development, 125,* 1627–1635.

Talbot, W.S., Swyryd, E.A. & Hogness, D.S. (1993). *Drosophila* tissues with different metamorphic responses to ecdysone express different ecdysone receptor isoforms. *Cell, 73,* 1323–1337.

Tanzer, L. & Jones, K.J. (1997). Gonadal steroid regulation of hamster facial nerve regeneration: effects of dihydrotestosterone and estradiol. *Experimental Neurology, 146,* 258–264.

Tanzer, L., Sengelaub, D.R. & Jones, K.J. (1999). Estrogen receptor expression in the facial nucleus of adult hamsters: does axotomy recapitulate development? *Journal of Neurobiology, 39,* 438–446.

Tata, J.R. (1994). Hormonal regulation of programmed cell death during amphibian metamorphosis. *Biochemistry and Cell Biology, 72,* 581–588.

Taylor, H.M. & Truman, J.W. (1974). Metamorphosis of the abdominal ganglia of the tobacco hornworm, *Manduca sexta*. *Journal of Comparative Neurology, 90,* 367–388.

Thomaidou, D., Mione, M.C., Cavanagh, J.F.R. & Parnavelas, J.G. (1997). Apoptosis and its relation to the cell cycle in the developing cerebral cortex. *Journal of Neuroscience, 17,* 1075–1085.

Tobet, S.A., Zahniser, D.J. & Baum, M.J. (1986). Sexual dimorphism in the preoptic/anterior hypothalamic area of ferrets: effects of adult exposure to sex steroids. *Brain Research*, **364**, 249–257.
Tobias, M.L. & Kelley, D.B. (1987). Vocalizations of a sexually dimorphic isolated larynx: peripheral constraints on behavioral expression. *Journal of Neuroscience,* **7**, 3191–3197.
Tobias, M.L., Marin, M.L. & Kelley, D.B. (1991). Development of functional sex differences in the larynx of *Xenopus laevis*. *Developmental Biology,* **147**, 251–259.
Tobias, M.L., Marin, M. & Kelley, D.B. (1993). The roles of sex, innervation and androgen in laryngeal muscle fibers of *Xenopus laevis*. *Journal of Neuroscience,* **13**, 324–333.
Trachtenberg, J.T. & Thompson, W.J. (1996). Schwann cell apoptosis at developing neuromuscular junctions is regulated by glial growth factor. *Nature*, **379**, 174–177.
Treanor, J.J., Goodman, L., de Sauvage, F., Stone, D.M., Poulsen, K.T., Beck, C.D., Gray, C., Armanini, M.P., Pollock, R.A., Hefti, F., Phillips, H.S., Goddard, A., Moore, M.W., Buj-Bello, A., Davies, A.M., Asai, N., Takahashi, M., Vandlen, R., Henderson, C. & Rosenthal, A. (1996). Characterization of a multifunctional component receptor for GDNF encoded by the c-ret proto-oncogene. *Nature*, **381**, 785–788.
Truman, J.W. (1984). Cell death in invertebrate nervous systems. *Annual Review of Neurosciences,* **7**, 171–188.
Truman, J.W. (1983). Programmed cell death in the nervous system of an adult insect. *Journal of Comparative Neurology,* **216**, 445–452.
Truman, J.W. (1996). Steroid receptors and nervous system metamorphosis in insects. *Developmental Neuroscience,* **18**, 87–101.
Truman, J.W & Schwartz, L.M. (1982). Programmed death in the nervous system of a moth. *Trends in Neuroscience*, **5**, 270–273.
Truman, J.W. & Schwartz, L.M. (1984). Steroid regulation of neuronal death in the moth nervous system. *Journal of Neuroscience,* **4**, 274–280.
Truman, J.W. & Schwartz, L.M. (1980). Peptide hormone regulation of programmed cell death of neurons and muscle in an insect. In F.E. Bloom (Ed.), *Peptides: integrators of cell and tissue function* (pp. 55–67). New York: Raven Press.
Truman, J.W., Talbot, W.S., Fahrbach, S.E. & Hogness, D.S. (1994). Ecdysone receptor expression in the CNS correlates with stage-specific responses to ecdysteroids during *Drosophila* and *Manduca* development. *Development*, **120**, 219–234.
Truss, M. & Beato, M. (1993). Steroid hormone receptors: Interaction with deoxyribonucleic acid and transcription factors. *Endocrinology Review,* **14**, 459–479.
Ulshafer, R.J. & Clavert, A. (1979). Cell death and optic fiber penetration in the optic stalk of the chick. *Journal of Morphology,* **162**, 67–75.
van den Eijnde, S. M., Lips, J., Boshart, L., Vermeij-Keers, C., Marani, E., Reutelingsperger, C.P.M. & De Zeeuw, C.I. (1999). Spatiotemporal distribution of dying neurons during early mouse development. *European Journal of Neuroscience,* **11**, 712–724.
vanPraag, H., Kempermann, G. & Gage, F.H. (1999). Running increases cell proliferation and neurogenesis in the adult mouse dentate gyrus. *Nature Neuroscience*, **2**, 266–270.
Vartanian, T., Goodearl, A., Viehöver, A. & Fischbach, G. (1997). Axonal neuregulin signals cells of the oligodendrocyte lineage through activation of HER4 and Schwann cells through HER2 and HER3. *Journal of Cell Biology,* **137**, 211–220.
Vaux, D.L., Haecker, G. & Strasser, A. (1994). An evolutionary perspective on apoptosis. *Cell,* **76**, 777–779.
Verhage, M., Maia, A.S., Plomp, J.J., Brussaard, A.B., Heeroma, J.H., Vermeer, H., Toonen, R.F., Hammer, R.E., van den Berg, T.K., Missler, M. Geuze, H.J. & Südhof, T.C. (2000). Synaptic assembly of the brain in the absence of neurotransmitter secretion. *Science*, **287**, 864–869.
von Bartheld, C.S., Byers, M.R., Williams, R. & Bothwell, M. (1996). Anterograde transport of neurotrophins and axodendritic transfer in the developing visual system. *Nature*, **379**, 830–833.
Voyvodic, J.T. (1996). Cell death in cortical development: How much? why? so what? *Neuron*, **16**, 693–696.
Wade, J., Schlinger, B.A., Hodges, L. & Arnold, A.P. (1994.) Fadrozole: a potent and specific inhibitor of aromatase in the zebra finch brain. *General and Comparative Endocrinology,* **94**, 53–61.
Wade, J., Springer, M.L., Wingfield, J.C. & Arnold, A.P. (1996). Neither testicular androgens nor embryonic aromatase activity alter morphology of the neural song system in zebra finches. *Biology of Reproduction,* **33**, 1126–1132.

Wang, H. & Tessier-Lavigne, M. (1999). *En passant* neurotrophic action of an intermediate axonal target in the developing mammalian CNS. *Nature,* **401,** 765–769.

Watson, J.T., Robertson, J., Sachdev, U. & Kelley, D.B. (1993). Laryngeal muscle and motor neuron plasticity in *Xenopus laevis*: Testicular masculinization of a developing neuromuscular system. *Journal of Neurobiology,* **24,** 1615–1625.

Weeks, J.C. & Davidson, S.K. (1994). Influence of interganglionic interactions on steroid-mediated dendritic reorganization and death of motor neurons in the tobacco hornworm, *Manduca sexta. Journal of Neurobiology,* **24,** 125–140.

Weeks, J.C. & Truman, J.W. (1985). Independent steroid control of the fates of motoneurons and their muscles during insect metamorphosis. *Journal of Neuroscience,* **5,** 2290–2300.

Weil, M., Jacobson, M.D. & Raff, M.C. (1997). Is programmed cell death required for neural tube closure? *Current Biology,* **7,** 281–284.

Weil, M., Raff, M.C. & Braga, V.M.M. (1999). Caspase activation in the terminal differentiation of human epidermal keratinocytes. *Current Biology,* **9,** 361–364.

Weill, C.L. (1986). Characterization of androgen receptors in embryonic chick spinal cord. *Developmental Brain Research,* **24,** 127–132.

Weiss, P. & Rosetti, F. (1951). Growth responses of opposite sign among different neuron types exposed to thyroid hormone. *Proceedings of the National Academy of Sciences, USA,* **37,** 540–556.

Weisz, J. & Ward, I.L. (1980). Plasma testosterone and progesterone titers of the pregnant rat, their male and female fetuses, and neonatal offspring. *Endocrinology,* **106,** 306–316.

Welshons, W.V., Lieberman, M.E. & Gorski, R.A. (1984). Nuclear localization of unoccupied oestrogen receptors. *Nature,* **307,** 747–749.

Wetzel, D. & Kelley, D.B. (1983). Androgen and gonadotropin control of the mate calls of male South African clawed frogs, *Xenopus laevis. Hormonal Behavior,* **17,** 388–404.

White, B.H. & Nicoll, C.S. (1981). Hormonal control of amphibian metamorphosis. In L.I. Gilbert & E. Frieden (Eds.), *Metamorphosis: a problem in developmental biology* (pp. 363–396). New York: Plenum Press.

White, F.A., Keller-Peck, C.R., Knudson, C.M. Korsmeyer, S.J. & Snider, W.D. (1998). Widespread elimination of naturally occurring neuronal death in Bax-deficient mice. *Journal of Neuroscience,* **18,** 1428–1439.

Williams, R.W. & Herrup, K. (1988). The control of neuron number. *Annual Review of Neuroscience,* **11,** 1–33.

Williams, R.W., Strom, R.C. & Goldowitz, D. (1998). Natural variation in neuron number in mice is linked to a major quantitative trait locus on chr 11. *Journal of Neuroscience,* **18,** 138–146.

Wimer, R.E. & Wimer, C.C. (1988). On the development of strain and sex differences in granule cell number in the area dentata of house mice. *Developmental Brain Research,* **42,** 191–197.

Wood, K.A., Dipasquale, B. & Youle, R.J. (1993). *In situ* labeling of granule cells for apoptosis-associated DNA fragmentation reveals different mechanisms of cell loss in developing cerebellum. *Neuron,* **11,** 621–632.

Wood, R.I. & Newman, S.W. (1993). Intracellular partitioning of androgen receptor immunoreactivity in the brain of the male Syrian hamster: effects of castration and steroid replacement. *Journal of Neurobiology,* **24,** 925–938.

Wood, R.I. & Newman, S.W. (1999). Androgen receptor immunoreactivity in the male and female syrian hamster brain. *Journal of Neurobiology,* **39,** 359–369.

Woodard, C.T., Baehrecke, E.H. & Thummel, C.S. (1994). A molecular mechanism for stage specificity of the *Drosophila* prepupal genetic response to ecdysone. *Cell,* **79,** 607–615.

Wride, M.A., Parker, E. & Sanders, E.J. (1999). Members of the Bcl-2 and caspase families regulate nuclear degeneration during chick lens differentiation. *Developmental Biology,* **213,** 142–156.

Wright, L. (1981). Cell survival in chick embryo ciliary ganglion is reduced by chronic ganglionic blockade. *Brain Research,* **227,** 283–286.

Wright, L.L. & Smolen, A.J. (1987). The role of neuron death in the development of the gender difference in the number of neurons in the rat superior cervical ganglion. *International Journal of Developmental Neuroscience,* **5,** 305–311.

Yaginuma, H., Tomita, M., Takashita, N., McKay, S.E., Cardwell, C., Yin Q.-W. & Oppenheim, R.W. (1996). A novel type of programmed neuronal death in the cervical spinal cord of the chick embryo. *Journal of Neuroscience,* **16,** 3685–3703.

Yamamoto, K.R. (1985). Steroid receptor regulated transcription of specific genes and gene networks. *Annual Review of Genetics*, **19**, 209–252.

Yan, Q., Matheson, C. & Lopez, O.T. (1995). *In vivo* neurotrophic effects of GDNF on neonatal and adult facial motor neurons. *Nature*, **373**, 341–344.

Yan, Q., Rosenfeld, R.D., Matheson, C.R., Hawkins, N., Lopez, O.T., Bennet, L. & Welcher, A.A. (1997). Expression of brain-derived neurotrophic factor (BDNF) protein in the adult rat central nervous system. *Neuroscience*, **78**, 431–448.

Yano, S., Tokumitsu, H. & Soderling, T.R. (1998). Calcium promotes cell survival through CaM-K kinase activation of the protein-kinase-B-pathway. *Nature*, **396**, 584–587.

Yaoita, Y. & Brown, D.D. (1990). A correlation of thyroid hormone receptor gene expression with amphibian metamorphosis. *Genes and Development*, **4**, 1917–1924.

Yarbrough, W.G., Quarmby, V.E., Simental, J.A,, Joseph, D.R., Sar, M., Lubahn, D.B., Olsen, K.L., French, F.S. & Wilson, E.M. (1990). A single base mutation in the androgen receptor gene causes androgen insensitivity in the testicular feminized rat. *Journal of Biological Chemistry*, **265**, 8893–8900.

Yasuda, T., Grinspan, J., Stern, J., Franceschini, B., Banneman, P. & Pleasure, D. (1995). Apoptosis occurs in the oligodendroglial lineage, and is prevented by basic fibroblast growth factor. *Journal of Neuroscience Research*, **40**, 306–317.

Yokosuka, M. & Hayashi, S. (1992). Transient expression of estrogen receptor-like immunoreactivity (ER-LI) in the facial nucleus of the neonatal rat. *Neuroscience Research*, **15**, 90–95.

Yoshida, K. & Gage, F.H. (1991). Fibroblast growth factors stimulate nerve growth factor synthesis and secretion by astrocytes. *Brain Research*, **538**, 118–126.

Yu, W-H.A. (1989). Administration of testosterone attenuates neuronal loss following axotomy in the brain-stem motor nuclei of female rats. *Journal of Neuroscience*, **9**, 3908–3914.

Zafra, F., Hengerer, B., Leibrock, J., Thoenen, H. & Lindholm, D. (1990). Activity dependent regulation of BDNF and NGF mRNAs in the rat hippocampus is mediated by non-NMDA glutamate receptors. *EMBO Journal*, **9**, 3545–3550.

Zietlow, R., Dunnett, S.B. & Fawcett, J.W. (1999). The effect of microglia on embryonic neuronal survival in vitro: diffusible signals from neurons and glia change microglia from neurotoxic to neuroprotective. *European Journal of Neuroscience*, **11**, 1657–1667.

Zirpel, L. & Rubel, E.W. (1996). Eighth nerve activity regulates intracellular calcium concentration of avian cochlear nucleus via a metabotropic glutamate receptor. *Journal of Neurophysiology*, **76**, 4127–4139.

Zirpel, L., Lippe, W.R. & Rubel, E.W. (1998). Activity-dependent regulation of $[Ca^{2+}]_i$ in avian cochlear nucleus neurons: roles of protein kinases A and C and relation to cell death. *Journal of Neurophysiology*, **79**, 2288–2302.

Zorick, T.S., Syroid, D.E., Brown, A., Gridley, T. & Lemke, G. (1999). Krox-20 controls SCIP expression, cell cycle exit, and susceptibility to apoptosis in developing myelinating Schwann cells. *Development*, **126**, 1397–1406.

Zou, H. & Niswander, L. (1996). Requirement for BMP signaling in interdigital apoptosis and scale formation. *Science*, **272**, 738–741.

III. MOTOR DEVELOPMENT

III.1
Function and Structure in Motor Development

ALBERT GRAMSBERGEN and HANS FORSSBERG

The early stages of motor behavior in relation to morphological and neuro-physiological development have been the topic of many investigations and heated debate in the past century. The individuation–integration controversy, the problem of the autonomous emergence of motility, the discussions on the epigenetical or preformational determination of brain development are of historical interest, but they all have importantly contributed to current definitions of concepts on development.

Research in experimental animals, involving meticulous descriptions of the neuroanatomical development in relation to motor behaviour, studies into correlations between neurophysiological processes and changes in motor output, as well as studies into the effects of experimental interferences with normal neuro-ontogeny, have provided an important body of knowledge and a rich source for extrapolations to early phases of brain–behaviour relationships in the human.

Our knowledge on motor development in human fetuses was until fairly recently based mainly upon indirect evidence or observations in exteriorized fetuses. Since then, the advent of high-resolution ultrasound scanning of movements in the recent past has opened up possibilities for following the development of movements of undisturbed human fetuses longitudinally, and this development is described in the chapter by **H.F.R. Prechtl**, the pioneer in the field of behavioural studies both in human fetuses and in newborn infants. Movements in normal human fetuses as depicted by ultrasound scanning are characterized by elegance, fluency and a large degree of variability, and in these qualities these movements differ drastically from the movements and reactions to stimuli as recorded on film by Davenport Hooker. Recordings of fetal movement patterns revealed that the repertoire of movements and movement patterns which develops from around 7 weeks of gestation, remains more or less unchanged until after the second month after birth. From then onwards new behaviours and movement patterns develop. On the one hand this developmental course underscores the continuity of motor development from prenatal stages onwards, and on the other hand it raises the issue of the timing of birth; points which are amply discussed by Dr Prechtl. General movements, the most

A.F. Kalverboer and A. Gramsbergen (eds.), Handbook of Brain and Behaviour in Human Development, 411–414
© 2001 Kluwer Academic Publishers. Printed in Great Britain.

frequently occurring movement pattern, gradually change in their phenomenology, from complex before term age (i.e. in the prematurely born baby), via the writhing type until about 2 months and the fidgety type thereafter. Particularly the shift from the writhing type into the fidgety type has attracted attention, as the absence of this change indicates a compromised brain development which only at later age is made apparent by motor handicaps (see also the chapter by Dr Hadders-Algra).

The new knowledge on fetal movement patterns in the human now also gives a firmer ground for extrapolations from animal research. The early stages of motor development in chick embryos and rat fetuses seem to be less different, compared with those in the human, than was suspected until recently, and this gives new momentum to investigations into neurobiological substrates. Studying the chick embryo offers the unique possibilities to follow the development of the brain in relation to the development of movement patterns throughout the embryonic period and beyond. In her chapter, **A. Bekoff** points to the earliest movement patterns being organized from the beginning, as was demonstrated by kinematic analysis of these movements. This points strongly to neural circuitries which produce these movements being present from the beginning. At later stages a larger variability in movements develops, indicating a secondary reorganization of circuitry. In chick embryos the earliest movements occur prior to the establishment of the afferent connections to the spinal cord, and this raises the question of what influence sensory input might have on later motor development. Brain damage in chick embryos leads to qualitative rather than quantitative changes of behaviour, and this shows important parallels with data in the human fetus (see the chapter by Dr Prechtl).

The neural circuits producing rhythmic motor patterns are the subject of the chapter by **J.-R. Cazalets**. Dr Cazalets discusses the early development of the central pattern generator (CPG) and the supraspinal modulation of its activity in rats. Only a few days after the motoneurons have arrived at their localization synchronous bursts of activity can be recorded, and this is before any motor output is observed and before sensory fibres have reached the dorsal horns of the spinal cord. The intricate interplay between excitatory amino acids (EAAs) and serotonin (5-HT) in the activation and modulation of the rhythmic activity of the CPG and the inhibitory transmitter GABA (and possibly glycine) in the fine-tuning of this activity were studied in ingenious experiments on the isolated spinal cord of newborn rats. This knowledge is obviously highly relevant to interpreting observations on rhythmic leg movements in human fetuses before birth, the infantile stepping response in newborn babies (see the chapter by Dr Prechtl) and the development of walking movements later (see the chapter by Dr Hadders-Algra).

Extrapolation to the human of neural and behavioural developments in experimental animals depends upon the identification of analogous processes in brain development. This issue is discussed in the chapter on behavioural and neural development in the rat by **A. Gramsbergen**. An important difference between rat fetuses (and probably the human fetus) on the one hand and chick embryos on the other seems to be the timing of the arrival of sensory fibres in the spinal cord. In chick embryos this arrival is later compared to that in rats. Investigations into the postnatal development of motor behaviour indicate that the development of extremity movements and the accompanying postural adjustments in rats initially develop independently, and only at a later stage do they become interconnected, thereby enabling feed-forward programming of fluent movements (see also the chapter by B. Hopkins). The development of postural control being the limiting factor for these new behaviour patterns (e.g. the adult-like type of fluent walking) to develop was demonstrated in experiments in which postural development selectively was retarded. This relatively late development of postural control in rats has also been described and studied in human infants (see the chapters by Drs Hadders-Algra and Gordon).

The development of manual skills in human infants is discussed by **A. Gordon**. Arm movements occur shortly after birth, and these movements are often followed visually, but "voluntary" and controlled reaching movements emerge only after 6 weeks of age. The adult-like trajectories of such movements, however, are reached only after 2 years. Finger movements in grasping occur spontaneously or upon exteroceptive stimulation of the hand-palm. In the following months voluntary grasping develops, and after 10–12 months individual and voluntary finger movements occur, probably by virtue of the development of monosynaptic connections between corticospinal axons and motoneurons of the distal arm and hand muscles (see Armand, in chapter II.12 of this handbook). The further refinement in manipulative skills and reaching movements generally takes another 10 years. This chapter discusses studies on prehensile forces during object manipulation and the formation of appropriate synergies which seem to simplify motor control. These studies suggest an interplay between the developing nervous system and the environment in the development of skilled manipulations (see also the chapter by Hopkins, for a discussion on the direction of this influencing).

The development of gross motor behaviour in human infants after birth is discussed by **M. Hadders-Algra**. General movements, consisting of a series of gross body movements in which all parts of the body participate, are the most frequently occurring movement pattern until the 3rd or 4th month, but in parallel a repertoire of goal-directed movements develops, such as reaching movements and locomotor activities. Adequate postural control is the limiting factor for succesfully reaching, sitting, standing and walking to develop, but feed-forward control in these patterns becomes apparent only from about 15 months. The further refinement of this control probably continues until adulthood. An interesting possibility raised in this chapter is that behavioural development starts with primary repertoires from which, by sensory information, adaptive motor patterns are selected according to the neuronal group selection theory, as formulated by Edelman.

The neuromotor development from Kindergarten age to puberty or adolescence has been a relatively neglected field of research. The chapter by **R.H. Largo** and **J.A. Caflisch** fills some of the gaps. The Zurich Neuromotor Test Assessment has been developed to test motor abilities in children from 5 to 18 years of age, and this chapter reports both on qualitative assessments and quantitative measures of specific movements. Speeds and durations of repetitive movements in fingers, hands or feet, the occurrence of associated movements, handedness and gender differences were studied and, interestingly, possible relations between deviations in measures and disorders in emotional stability, and in lingual development and learning disabilities are discussed.

The chapter by **B. Hopkins** introduces the dynamical systems thinking on the motor development of infants. There are three interrelated approaches: the *ecological approach* which deals with the problem of how information from the outside world is continuously adjusted to developmental changes in skeleto-muscular dynamics; the *perception–action approach*, treating the mutuality of perception and action and particularly stressing the successive reorganizations in their interdependencies (not merely by changes in perceptual abilities or changing facilities in acting) as well as the *dynamical approach*. The latter approach, which groups a number of related theories, is the main instrument for the description and interpretation of changes between successive stable states in the natural–physical approach (the umbrella term for the three approaches mentioned). Particularly during development, the transitions from the one stable mode to another (e.g. from the scissor grasp to the pincer grasp), the instabilities during the transitions and the nature of the transitions (a disappearance of the earlier pattern before the newer one occurs, or a coexistence of both patterns for a certain period) allows us to study the nature of developmental changes and to reveal the organizational principles. The natural–physical approach acknowledges that the brain ultimately controls and coordinates the development of actions, but changes in these actions from this point of view are not initiated from the brain or controlled by it. Rather, the brain adapts to changes in the dimensions and dynamics of the body parts. In a broader sense the brain only globally specifies the dynamics for maintaining stability against external perturbations or for the shifting of the system into other coordinative modes.

Research in recent decades has seen important increases in our insights into the development of motor behaviour. Many gaps in our knowledge still exist, such as the nature of transitions in motor behaviour, the interrelation between posture and movements during development, motor development at later ages, and the compensational capacity of the human brain after damage. Many of these issues are addressed in several chapters of this section. The identification and outlining of these and other problems, as well as the availability of advanced technologies in the fields of movement research and in related disciplines, now provide a firm basis for further investigation of these problems.

III.2
Prenatal and Early Postnatal Development of Human Motor Behaviour

H.F.R. PRECHTL

ABSTRACT

Neonates born at term have a rich repertoire of spontaneous move-ments as well as of responses and reflexes to sensory stimulation. The spontaneous movements of the newborn infant have a long prenatal history. From 8 weeks onwards (postmenstrual age) the fetus moves in distinct motor patterns. There is no period of amorphic and random movements. The patterns are immediately recognizable, as all of them can be seen after birth. Ultrasound observations of the moving fetus revealed a number of new findings which had not been known from the reflex studies on exteriorized fetuses. Particularly, the frequent occur-rence of endogenously generated movements is striking.

There are hardly any changes in the form and pattern of the spon-taneous fetal movements after birth, despite the profound changes in the environmental condition. During the first 2 months post-term the human neonate demonstrates a continuum of neural functions from prenatal to postnatal life. Around the third month of life a major transformation of many motor and sensory patterns occurs. This makes the infant more fit to meet the requirements of the extrauterine envi-ronment. This delay can be explained by the constraints in maternal energy metabolism in humans which led to insufficient lengthening of the pregnancy duration during the evolution of hominids.

The developmental course of spontaneous movements during the first 20 weeks post-term shows the emergence and disappearance of various forms of movements. One particular pattern, the so-called general movements, deserve special interest as they are in their altered quality a most reliable indicator of brain dysfunction with a specific prediction of later developing cerebral palsy.

A.F. Kalverboer and A. Gramsbergen (eds.), Handbook of Brain and Behaviour in Human Development, 415–428
© 2001 Kluwer Academic Publishers. Printed in Great Britain.

INTRODUCTION

The rich repertoire of spontaneous movements in the newborn infant has a long prenatal history. This discovery was only possible due to the rapid development of ultrasound equipment with sufficient resolution and dynamics to see the fetal movements sufficiently clearly.

Before commencing systematic observation of fetal movements I had 35 years experience in developmental neurology of young infants. In addition, during the 1970s a specific pilot study on spontaneous movement patterns was carried out in carefully selected low-risk preterm infants (Prechtl *et al.*, 1979). I had hoped to be able to recognize the various motor patterns of the preterm infant also in the fetus when ultrasound equipment became good enough for fetal observations. This hope was underestimated, as the fetus performed movement patterns identical to those previously seen in the preterm. In the light of the profound environmental changes from intra- to extrauterine life this observation came as a surprise. In addition, it turned out that all the different motor patterns of the fetus can also be seen after birth, most of them shortly after birth; a few only weeks after birth; all in all we can speak of a continuum of neural functions from prenatal to postnatal life. The reason for this will be discussed later in this chapter.

The systematic observations of normal and abnormal fetuses have led to many changes in ideas on fetal behaviour. Previous studies on fetal movements were carried out on exteriorized fetuses. The survival was limited to a few minutes, during which the fetus was stimulated with tactile stimuli (Minkovski, 1928; Hooker, 1952; Humphrey, 1978). These studies remained strictly in the tradition of reflexology and behaviourism, and it is not surprising that the endogenously generated and thus spontaneous activity was totally overlooked or wrongly interpreted. To be fair it must be mentioned that it may not have been possible to see spontaneous movements during the short survival time. Today, we know from non-invasive ultrasound observations that these previous studies described rather abnormal movement patterns of dying fetuses. Moreover, the fetus responded to artificial tactile stimuli, which are never present in the natural situation.

THE TIMETABLE OF MOTOR DEVELOPMENT IN THE FETUS

There is a time sequence in the emergence of the various motor patterns. This became possible to detect in close collaboration between the Departments of Developmental Neurology and Obstetrics at the Groningen University Hospital, the Netherlands. Weekly repeated 1-hour continuous ultrasound recordings of fetal movements, including videotaping of the recordings, were the basis for these studies (de Vries *et al.*, 1982, 1984, 1985, 1987, 1988; Prechtl 1989a, 1997a, 1999).

The first movements to occur are sideward bendings of the head. These are first seen at 7½–8 weeks postmenstrual age (counted from the first day of the last menstruation before the amenorrhoea). These first movements can clearly be seen by transvaginal transducers but are poorly detected by transabdominal ultrasound. Time and movement patterns are actually the same as in Davenport Hooker's (1952) observation after perioral stimulation with a Frey's hair. This trigeminal stimulation elicited the first movement which was due to the newly formed connections of the cervical motor neurons with the neck muscles. Thus, spontaneously generated movements follow the same line.

At 9–10 weeks postmenstrual age complex and generalized movements occur. These are the so-called general movements (Prechtl *et al.,* 1979) and the startles. Both include the whole body, but the general movements are slower and have a complex sequence of involved body parts, while the startle is a quick, phasic movement of all limbs and trunk and neck. It should be mentioned here that general movements became of extreme importance for the early diagnosis of brain dysfunction and the prediction of later neurological outcome.

To our great surprise local and isolated movements of one arm or leg emerge only 1 week later than the generalized movements. It may be surmised that isolated movements are more difficult to produce by the very young nervous system than global motor activity. There was another unexpected finding. Traditionally it is accepted that the early ontogenetic process goes from cranial to caudal. Although this was primarily based on stimulation experiments (Hooker, 1952) the motor system does not follow that rule. Isolated arm and isolated leg movements emerge at the same time, at 9 weeks. It is, however, true that isolated arm movements occur more frequently than isolated leg movements, and this might previously have been overlooked in short-lasting recordings.

These observations led to changes of paradigms: In the traditional literature it was assumed that in ontogeny jerky movements precede slow and tonic movements. This is certainly not the case. By the same token the study of early movement patterns revealed that tonic and phasic movements emerge at the same age. It might have been the case that the traditionally short recordings overestimated the more frequent jerky movements, and thus led to the wrong conclusion. In the same way it is not true that early movements are random and amorphic, followed only later by specific and distinct movements. All early and later fetal movement patterns are differentiated, and this is true from their very first appearance onwards.

The next pattern to emerge at 9–10 weeks is the hiccup, caused by a mostly repetitive short contraction of the diaphragm. Such episodes of hiccups may last for several minutes. They can be so forceful that the whole fetus is passively moved in the amniotic cavity.

Around 11 weeks postmenstrual age the head of the fetus becomes very mobile. Head retroflexion, even anteflexion and particularly head rotations from one to the other side are common events. Together with these head movements, arm movements may occur and produce hand–face contacts. Most of these are obviously accidental, and it is not at all likely that such contacts are intentional.

Breathing movements emerge at around 11 weeks and appear episodically. Interestingly enough their rate of occurrence is related to maternal glucose level. Hence, they are most easily observed after the mother has had a meal. It is important to notice that fetal breathing movements are an exception in this respect while the occurrence of all other fetal movements is independent of maternal glucose level. Moreover, breathing movements do not lead to an influx of amniotic fluid into the fetal lungs.

A very interesting phenomenon in fetal motor development is the early emergence of stretches and yawns. Both are complex movements, and the most interesting aspect is their maintenance throughout the whole of life, without changing their form or pattern. This is a rather unique event while all other early movement patterns disappear or later change their pattern.

It is shortly after the 12th week that the fetus starts to drink amniotic fluid with rhythmical sucking movements and swallowing. At the end of pregnancy the fetus drinks about 1 litre of fluid per 24 hours.

A very important aspect of fetal movements is the change of fetal position *in utero*. It is not quite understood why this is so significant, but all animals studied so far move in the egg or the uterus. This universal phenomenon must thus have great significance.

In the human fetus positional changes are frequent, and may run up to 25 changes per hour during the first half of pregnancy (de Vries *et al.,* 1985). Later these positional changes become rarer, but are still present. It is important that several specific movement patterns are essential for these changes. Trunk rotations, general movements and alternating leg movements, leading to a somersault if proper contact of the feet can be made with the uterine wall – all these produce changes in intrauterine position. These motor patterns are obviously an ontogenetic adaptation and have an effective function during prenatal life. The alternating leg movements outlive the duration of pregnancy and are known as newborn stepping.

Other movement patterns are anticipating later functions, becoming effective only during postnatal life. To this group belong breathing movements, as well as sucking and swallowing movements. The later already have a significant intrauterine function of regulating the amount of amniotic fluid.

Relatively late are the slow eye movements, added at 20 weeks, and rapid eye movements emerging at about 22 weeks (Birnholz, 1981; Inoue *et al.,* 1986). Fetal eye movements were first discovered by Bots *et al.* (1981) and later confirmed by Birnholz (1981). The eye movements are another example of an anticipatory motor pattern which much precedes its ultimate function. All these examples are evidence for the primacy of the motor system. Such movement patterns develop earlier in ontogeny and are ready for their actual function after birth. A similar example is the smiling pattern which can be easily seen in the preterm infant and with ultrasound also *in utero*. This may occur half-sided or bilaterally. Despite this early onset the real function of this motor pattern does not start before about 6 weeks post-term age, when social smiling emerges.

The majority of fetal movement patterns develop during the first half of pregnancy and continue not only until term but also after birth.

THE MOVEMENT REPERTOIRE AFTER BIRTH

Only after birth are a number of new functions added. Prenatally it was not possible to elicit any vestibular responses in the fetus, when the mother had been adequately moved by the experimenter and the fetus simultaneously observed by ultrasound (Prechtl, 1997a). However, after birth vestibular responses such as the vestibular–ocular response (von Bernuth and Prechtl, 1969) and the Moro response are clearly present.

There are other aspects arising after birth which I will describe using the example of sucking. Before birth the fetus can drink amniotic fluid at any time during a 24-hour period. After birth, however, sucking is needed for food intake only if the feeding situation is guaranteed by the caretaker. Hence, rooting and sucking have gradually to come under afferent control to be elicited in the biologically adequate situation. The sensory trigger mechanism of an otherwise spontaneous movement pattern becomes mandatory in the postnatal adaptation of the newborn infant.

New in the motor repertoire after birth are functions depending on the newly installed lung ventilation. Reflexes for protection of the airway, such as sneezing and coughing, as well as the communication signal of crying, are seen only after birth.

By and large, however, there is an amazing continuation of the fetal repertoire during about the first 2 months after birth at term (Prechtl, 1984). Needless to say, in healthy preterm infants this continuation lasts until the same postmenstrual age as in infants born at term (i.e. the corrected age for preterm birth).

Although psychologists often speak about the "competent newborn", it is amazing how many neural functions are delayed in the human infant, compared with neonates of infrahuman primates, in particular with chimpanzees and other apes. There seems to be a very special delay in the ontogenetic developmental course in the human species.

THE THIRD MONTH: MAJOR TRANSFORMATION OF MANY NEURAL FUNCTIONS

Although there are a few vital adaptations which occur shortly after birth, there are very few behavioural changes during the first 2 months post-term. At about 8–10 weeks, however, many motor and sensory systems change their properties and improve the infant's fitness and adaptation to the extrauterine environment. The specific changes are:

1. The previously body-oriented posture is becoming space-oriented (Prechtl, 1989b).
2. The infant's head is no longer in a side position but is now held centred in the midline.
3. The muscle power is rapidly increasing and can more easily overcome the force of gravity, including proper head control.
4. The sucking pattern with peristaltic waves of the tongue changes to a new pattern (Iwayama & Eishima, 1997).
5. The form of general movements gradually lose their writhing character and a new pattern of general movements emerges: fidgety movements (small circular movements of moderate speed, going in all directions and, in the awake infant, in a continual manner).
6. Visual attention evolves.
7. Manipulation movements of the hands on the baby's clothing or other objects can be observed.
8. The voluntary modulation of the respiration makes cooing and other vocalization possible.

This transformation of so many motor and sensory functions occurs within a relatively short period of a few weeks. There are, as always in early development, large inter-individual differences. The timetable of these changes may be different even in individuals of the same population. An excellent example of this variation was provided by Touwen's study, *Neurological development in infancy* (1976) carried out on 50 healthy infants in 4-weekly repeated examinations until the children could walk freely. The author demonstrated that the different functions (i.e. postures, spontaneous movements, responses to sensory stimulation and reflexes) surprisingly develop independent of each other. This study is still unique in quality and detailed insight into functional neural development during infancy. It is still unparalleled, and was preceded only by Gesell's *The embryology of behavior* (1945) and by Myrtle McGraw's (1943) *The neuromuscular maturation of the human infant*. Both books describing a life's work were written in theoretical context different from the Touwen study, which was within the theoretical concept of developmental neurology.

The 3-month transformation has many consequences for the infant's behaviour and for the child–mother interaction (van Wulfften-Palthe and Hopkins, 1984). The newly improved social responsiveness of the infant is greatly rewarding for the mother and the caretaker.

WHY IS THIS BEHAVIOURAL ADAPTATION TO THE EXTRAUTERINE ENVIRONMENT SO LATE AND SPECIFIC FOR THE HUMAN?

As far as is known from studies on infrahuman primate species, such a late adaptation to the extrauterine environment has not yet been reported. What could be the reason for this delay in the developmental course of so many neural functions in the human infant? The traditional idea of anthropologists was, firstly, the development of the upright body posture leading to an adaptation of the small pelvis, and, secondly, the later-occurring increase in the brain size of the baby to be born. These all happened in the evolution of the hominids. If this concept was correct, and the selective pressure had been concentrated on the small pelvis size and the enlarged brain, two things should have happened. In the first place a clear sex-dimorphism of the small pelvis size should have occurred, and secondly those infants with small brains would have been more successful in surviving, leading to a relative microcephaly in infancy. Neither phenomenon occurred in the evolution of the hominids, and the theory thus can be refuted.

An alternative concept is based on allometric measurements of body size, brain size, basal metabolism, maternal–neonatal body weight relation and longevity among primates. In this context it became clear that the human pregnancy of 40 weeks is relatively short compared to other primates (Prechtl, 1986). This may not be too surprising, as there are two factors specific to the human. One is the relatively rapid brain growth of a very differentiated brain, which puts a high demand on the energy metabolism of the maternal organism. The second point is the phenomenon, unique among primates, of subcutaneous white fat in the human. Obviously, as compensation for the loss of fur during evolution, the thermoregulation became insured by this insolating white fat. This is in the order of 13% of neonatal body weight, and has to be produced by the maternal metabolism. This is nonexistent in any of the infrahuman species. Similar to the human neonate, infrahuman primates at birth have 3% of the body weight as brown fat for chemical thermoregulation. The production of the white subcutaneous fat is an additional burden on the energy metabolism of the expectant human mother. For these reasons it can be conjectured that, in the evolution of hominids, the duration of gestation was not accordingly prolonged as would have been expected from the aforementioned allometric values with the increase in human body weight and brain weight, metabolic rate, etc. In conclusion, it is primarily a metabolic constraint for human pregnancy duration, and less likely to be the features proposed earlier by anthropologists.

THE PROBLEM OF ENDOGENOUSLY GENERATED ACTIVITY VERSUS REFLEX AND RESPONSE

Naturalistic observations of infant movement patterns and behaviour had a tradition during the 19th century (e.g. Preyer, 1884). A new era of observations of unstimulated infants, focusing on spontaneous motor behaviour, started with the work of Peter Wolff (1959, 1966) and Prechtl (1958). While Piaget interpreted neonatal behaviour during the first 4 weeks post-term as a stage of reflex behaviour, naturalistic observations led to the conclusion of the dominance of spontaneous behaviour – i.e. behaviours not generated by sensory stimulation. Although this point is still a matter of debate, results from developmental neurobiology are rapidly accumulating convincing evidence that the classical neurophysiological studies on experimentally "brain-damaged" or anaesthetized animals have historically resulted in a distorted concept of the functions of the intact nervous system. Particularly the young nervous system is, under normal conditions, an eminently active organ. The endogenously generated motor activity in many different species has now been studied in detail, even with single-neuron recordings and manipulation of the involved transmitters and membrane receptors (for review see Prechtl, 1997b).

In the observation of infants, interest has changed from analysis of the capacities to respond to a manifold of sensory stimulations to observation of the unstimulated infant. The above-mentioned fetal studies with ultrasound have certainly promoted this approach. This change in paradigm concerning the properties of the young nervous system and the new methodology of research on infants has produced a great scientific and clinical payoff.

POST-TERM DEVELOPMENT OF SPONTANEOUS MOVEMENTS

While there is a wealth of descriptions in the literature on reflex and response studies (e.g. Carmichael, 1946; Peiper, 1961), reports on the ontogeny of postnatal spontaneous behaviour are still scarce. The first systematic description of the post-term development of the spontaneous movement repertoire during the first 18 weeks was given by Hopkins & Prechtl (1984), and was preceded only by similar studies on fetuses and preterm infants (see above). The classification of the various movement patterns was as follows:

1. *Wiggling–oscillating movements*. Irregular, oscillatory, waving-like movements, which are most noticeable in partially or fully extended arms, where they have a frequency of 2–3 Hz. Movements of this quality are generally of small amplitude and moderate speed. This quality should be clearly distinguished from tremulous movements, which are less smooth in appearance and have a more regular rhythm.
2. *Saccadic movements*. Jerky, zig-zag movements which continually vary in direction; these movements are most noticeable in partially or fully extended arms. Movements of this quality are generally of moderate to large amplitude and moderate speed. In some infants these movements overshoot, i.e. there is a partial return in trajectory of the movement from its maximal excursion.
3. *Swiping movements*. Movements with a sudden but fluid onset and smooth offset. These have a ballistic-like appearance and can go in downward or upward direction. These movements are most noticeable in extended arms but can sometimes occur in partially or fully extended legs. Movements of this quality are always of large amplitude and high speed. This movement quality should be clearly distinguished from elliptical movements, which occur only until about 6 weeks.
4. *Mutual manipulation of fingers*. The two hands are brought together in the midline and the fingers of both hands repetitively touch, stroke or grasp each other.
5. *Manipulation of clothing*. The fingers of one or both hands repetitively touch, stroke or grasp some article of clothing.
6. *Reaching and touching*. One or both arms extend to some object in the immediate environment. The fingers contact the surface of the object.
7. *Legs lift, extension at knees*. Both legs lift vertically upward, with partial or full extension at the knees. The hips are slightly tilted upward. Typically this movement involves one or both hands touching or grasping the knees. Sometimes it involves anteflexion of the head. This movement should be clearly distinguished from legs lifting without extension, which can occur at all ages under consideration.
8. *Trunk rotation*. As a result of the soles of the feet pushing down on the lying surface, one side of the hips is lifted and rotated.
9. *Axial rolling*. The whole body is turned from supine to prone lying in a movement started by the head. Sometimes the infant returns to prone lying.

A recent definition of the *general movements* is added (Prechtl *et al.*, 1997): While during preterm age we merely call them general movements, at term age until about 6–9 weeks post-term age they are called writhing general movements. The definition is: they are gross movements, involving the whole body. They may last from a few seconds to several minutes or longer. What is particular about them is the variable sequence of arm, leg, neck and trunk movements. They wax and wane in intensity, force and speed, and they have a gradual beginning and ending. The majority of sequences of extension and flexion movements of arms and legs is complex, with superimposed rotations and often slight changes in the direction of the movement. These additional components make the movement fluent and elegant, and create the impression of complexity and variability.

At the time of the major transformation a new type of general movements appears. These are called general movements of fidgety character. *Fidgety movements* are circular movements of small amplitude, moderate speed, and variable acceleration of neck, trunk, and limbs in all directions. They are continual in the awake infant, except during focused attention, fussing and crying. They may be concurrent with other movements. Fidgety movements may be seen as early as 6 weeks, but usually occur around 9 weeks and are present until 20 weeks or even a bit longer.

The developmental sequence of the above-described motor patterns was provided by Hopkins & Prechtl (1984) for 12 low-risk full-term infants. This study was replicated on another group of 10 full-term infants (Cioni *et al.*, 1989) and on 10 preterm infants (Cioni & Prechtl, 1990). All three studies have been based on video replay analysis of 1-hour recordings, repeated every week during the preterm period and afterwards every 3 weeks until the 18th week post-term.

In a further study on 22 full-term infants the developmental course of the post-term repertoire from 2 to 18 weeks was analysed again (Hadders-Algra & Prechtl, 1992). These recordings were made at 2-week intervals. As a rather wide scatter between the individuals was prominent, it became clear that even an age correction for term-born infants (38–42 weeks of gestation) is necessary for this kind of study. After correction for postmenstrual age the inter-individual scatter became remarkably reduced. A clear temporal sequence in the development of the various spontaneous (i.e. endogenously generated) motor activity became apparent. Table 1 provides the developmental sequence of the various patterns in detail.

GENERAL MOVEMENTS ARE A WINDOW INTO THE BRAIN

After the change in paradigm from reflex to spontaneous motor activity had been achieved the way was paved for the discovery of the importance of the quality of general movements for the assessment of the integrity of the young nervous system. Studies on the relationship between documented brain pathology and changes in general movements have shown that the quantity of occurrence of

Table 1 Timetable of the emergence and disappearance of the various spontaneous movements pattern during the first 20 weeks post-term

	Emergence	*Disappearance*
Writhing general movements	At term	8 weeks
Fidgety general movements	6 weeks	20 weeks
Wiggling–oscillating movements	6 weeks	12–14 weeks
Saccadic movements	6 weeks	15 weeks
Swiping movements	8 weeks	20 weeks
Hand–hand manipulation	12–15 weeks	⟶
Manipulation of clothing	15 weeks	⟶
Reaching, touching	12–18 weeks	⟶
Legs lifted, hand–knee contact	15 weeks	⟶
Trunk rotation	12–15 weeks	⟶
Axial rolling	18–20 weeks	⟶

general movements and also of other spontaneous movements is not altered (Prechtl & Nolte, 1984; Ferrari *et al.*, 1990; Bos *et al.*, 1997a). This was an unexpected finding. However, we noticed that babies with brain damage move differently from those with intact brains (Prechtl & Nolte, 1984). This view was supported by previous studies on fetuses with problems, and particularly by a study on anencephalic fetuses (Visser *et al.*, 1985). It became clear that malformations of the forebrain and the diencephalon dramatically change the expression of general movements into a kind of chaotic movement pattern. This was later confirmed by Ferrari *et al.* (1997) on various forms of brain malformations after birth. From all these studies it can be concluded that an intact brain is a prerequisite for the normal quality of general movements. Hence, general movements are a window into brain integrity. A series of studies has again and again confirmed this statement (Bekedam *et al.*, 1985; Sival *et al.*, 1992; Geerdink & Hopkins, 1993; Prechtl *et al.*, 1993; Einspieler, 1994; Albers & Jorch, 1994; Bos *et al.*, 1997b; Cioni *et al.*, 1997a,b; Kainer *et al.*, 1997; Bos, 1998; Bos *et al.*, 1998a,b).

What is particular about this finding is the fact that only with the qualitative assessment of general movements is a specific prediction of later development of cerebral palsy possible. So far all neurological examinations (Prechtl, 1977 or Dubowitz & Dubowitz, 1981) have failed to provide any specific prediction in this respect.

There are two patterns which reliably predict the later neurological outcome of cerebral palsy:

1. A persistent pattern of cramped–synchronized general movements. These are general movements which appear rigid and lack the normal smooth and fluent character. All limb and trunk muscles contract and relax almost simultaneously (Einspieler *et al.*, 1997). If this pattern exists over several weeks during preterm and term age, a cerebral palsy will develop at later age (Ferrari *et al.*, 1990).
2. The second predictor concerning the quality of general movements is the absence of fidgety movements (Prechtl *et al.*, 1997). This has a sensitivity of 95% and a specificity of 96%.

This breakthrough in diagnostic procedure we owe to the investigation of spontaneous movements. The fact that general movements are the most important item to be selected for this method is based on their frequent occurrence, their complexity, and their relatively long duration. These properties make them the ideal movement pattern for functional neurological assessment. One of the great advantages of this method is the fact that it can be applied from fetal life until about 20 weeks post-term using the same criteria (Prechtl & Einspieler, 1997).

The neural mechanisms of general movements consist of central pattern generators, somewhere in the brainstem. A separate generator must be assumed for the fidgety movements, as general movements of writhing and fidgety character can occur simultaneously during a period of overlap of both movements patterns (Einspieler *et al.*, 1994, Prechtl, 1997b).

Why, however, is the quality of general movements changed if lesions in the brain are localized in the periventricular area of the cerebral hemispheres? This area includes the connection from the cortex to the lower brainstem and spinal cord, the corticospinal tract. This neural system does not function before at least the end of the third month post-term age. Despite this fact an influence of these structures on the more caudally situated structures is beyond doubt. These neural connections are formed rather early during pregnancy, but reach functional effect rather late. Despite this discrepancy a lesion of this neural system has an effect on the quality, but not the quantity, of the production of the central pattern generators of general movements. Only in the case of fidgety movements is the quantity reduced to zero, and this is the worst sign for neurological outcome.

The assessment of general movements, based on detailed study of spontaneous motor activity during fetal life and early post-term life, has provided an insight into brain–behaviour relations which was not possible from reflex and response studies, including all efforts to investigate tonus and posture of infants. In this sense the general movements are indeed a window into the brain.

ACKNOWLEDGEMENTS

My sincere thanks go to Professor Christa Einspieler for her invaluable help in preparing this chapter.

References

lbers, S. & Jorch, G. (1994). Prognostic significance of spontaneous motility in very immature preterm infants under intensive care treatment. *Biology of the Neonate*, **66**, 182–187.

Bekedam, D.J., Visser, G.H.A., de Vries, J.I.P. & Prechtl H.F.R. (1985). Motor behaviour in the growth retarded fetus. *Early Human Development*, **12**, 155–166.

von Bernuth H. & Prechtl, H.F.R. (1969). Vestibular-ocular response and its state dependency in newborn infants. *Neuropaediatrie*, **1**, 11–24.

Birnholz, J.C. (1981). The development of human fetal eye movement patterns. *Science*, **213**, 679–681.

Bos, A.F. (1998). Analysis of movement quality in preterm infants. *European Journal of Obstetrics, Gynecology and Reproductive Biology*, **76**, 117–119.

Bos, A.F, van Loon, A.J., Martijn, A., van Asperen, R.M., Okken, A. & Prechtl, H.F.R. (1997a). Spontaneous motor behaviour in preterm, small for gestational age infants. I. Quantitative aspects. *Early Human Development*, **50**, 115–129.

Bos, A.F., van Loon, A.J., Hadders-Algra, M., Martijn, A., Okken, A. & Prechtl, H.F.R. (1997b). Spontaneous motor behaviour in preterm, small for gestational age infants. II. Qualitative aspects. *Early Human Development*, **50**, 115–129.

Bos, A.F., Martijn, A., van Asperen, R.M., Hadders-Algra, M., Okken, A. & Prechtl, H.F.R. (1998a). Qualitative assessment of general movements in high risk preterm infants with chronic lung disease requiring dexamethasone therapy. *Journal of Pediatrics*, **132**, 300–306.

Bos, A.F., Martijn, A., Okken, A. & Prechtl, H.F.R. (1998b). Quality of general movements in preterm infants with transient periventricular echodensities. *Acta Paediatrica*, **87**, 328–335.

Bots, R.S.G.M., Nijhuis, J.G., Martin, C.B. Jr. & Prechtl, H.F.R. (1981). Human fetal eye movements: detection in utero by ultrasonography. *Early Human Development*, **5**, 87–94.

Carmichael, L. (1946). *Manual of child psychology*. New York: Wiley.

Cioni, G. & Prechtl, H.F.R. (1990). Preterm and early postterm motor behaviour in low-risk premature infants. *Early Human Development*, **23**, 159–193.

Cioni, G., Ferrari, F., Einspieler, C., Paolicelli, P.B., Barbani, M.T. & Prechtl, H.F.R. (1997a). Comparison between observation of spontaneous movements and neurological examination in preterm infants. *Journal of Pediatrics*, **130**, 704–711.

Cioni, G., Ferrari, F. & Prechtl, H.F.R. (1989). Posture and spontaneous motility in fullterm infants. *Early Human Development*, **7**, 247–262.

Cioni, G., Prechtl, H.F.R., Ferrari, F., Paolicelli, P.B., Einspieler, C. & Roversi, M.F. (1997b). Which better predicts later outcome in fullterm infants: quality of general movements or neurological examination? *Early Human Development*, **50**, 71–85.

Dubowitz, L.M.S. & Dubowitz, V. (1981). *The neurological assessment of the preterm and fullterm newborn infant*. Clin. Dev. Med. vol. 79, London: Heinemann.

Einspieler, C. (1994). Abnormal spontaneous movements in infants with repeated sleep apneas. *Early Human Development*, **36**, 31–49.

Einspieler, C., Prechtl, H.F.R., van Eykern, L. & de Roos B. (1994). Observation of movements during sleep in ALTE and apnoeic infants. *Early Human Development*, **40**, 39–50.

Einspieler, C., Prechtl, H.F.R., Ferrari, F., Cioni, G. & Bos, A.F. (1997). The qualitative assessment of general movements in preterm, term and young infants – review of the methodology. *Early Human Development*, **50**, 47–60.

Ferrari, F., Cioni, G. & Prechtl, H.F.R. (1990). Qualitative changes of general movements in preterm infants with brain lesions. *Early Human Development*, **23**, 193–233.

Ferrari, F., Prechtl, H.F.R., Cioni, G., Roversi, M.F., Einspieler, C., Gallo, C, Paolicelli, P.B. & Cavazutti, G.B. (1997). Posture, behavioural state organization and spontaneous movements in infants affected by brain malformation. *Early Human Development*, **50**, 87–113.

Geerdink, J.J. & Hopkins, B. (1993). Qualitative changes in general movements and their prognostic value in preterm infants. *European Journal of Pediatrics*, **152**, 362–367.

Gesell, A. & Amatruda, C.S. (1945). *The embryology of behaviour*. New York: Harper.

Hadders-Algra, M., & Prechtl, H.F.R. (1992). Developmental course of general movements in early infancy. I. Descriptive analysis of change in form. *Early Human Development*, **28**, 201–213.

Hooker, D. (1952). *The prenatal origin of behavior*. Lawrence, KA: University of Kansas Press.

Hopkins, B. & Prechtl, H.F.R. (1984). A qualitative approach to the development of movements during early infancy. In H.F.R. Prechtl (Ed.), *Continuity of neural functions from prenatal to postnatal life* (pp. 179–197). Clin. Dev. Med., vol. 94. Oxford: Blackwell Scientific Publications.

Humphrey, T. (1978). Function of the nervous system during prenatal life. In U. Stave (Ed.), *Perinatal physiology* (pp. 651–683). New York: Plenum.

Inoue, M., Koyanagi, T., Nakahara, H., Hara, K., Hori, E. & Nakano, H. (1986). Functional development of human eye movements *in utero* assessed quantitatively with real-time ultrasound. *American Journal of Obstetrics and Gynecology*, **155**, 170–174.

Iwayama, K. & Eishima M. (1997). Neonatal sucking behaviour and its development until 14 months. *Early Human Development*, **47**, 1–9.

Kainer, F., Prechtl, H.F.R., Engele, H., & Einspieler, C. (1997). Prenatal and postnatal assessment of the quality of general movements in infants of women with type-I-diabetes mellitus. *Early Human Development*, **50**, 13–25.

McGraw, M.B. (1943). *The neuromuscular maturation of the human infant*. New York: Columbia University Press.

Minkowsky, M. (1928). *Neurobiologische Studien am menschlichen Fötus*. Handbuch der biologischen Arbeitsmethoden, Abtl. 5, Teil 5B, pp. 511–618.

Peiper, A. (1961). *Die Eigenart der kindlichen Hirntätigkeit*. Leipzig: Georg Thieme.

Prechtl, H.F.R. (1958). The directed head turning response and allied movements of the human body. *Behaviour*, **8**, 212–242.

Prechtl, H.F.R. (1977). *The neurological examination of the full-term newborn infant*. (2nd ed.). Clin. Dev. Med. no. 63, London: Heinemann.

Prechtl, H.F.R. (1984). *Continuity of neural functions from prenatal to postnatal life*. Clin. Dev. Med. no. 94. Oxford: Blackwell.

Prechtl, H.F.R. (1986). New perspectives in early human development. *European Journal of Obstetrics, Gynecology and Reproductive Biology*, **21**, 347–355.

Prechtl, H.F.R. (1989a). Fetal behaviour. In A. Hill, J. Volpe (Eds.). *Fetal neurology* (pp. 1–16). New York: Raven Press.

Prechtl, H.F.R. (1989b). Development of postural control in infancy. In C. von Euler, H. Forssberg & H. Lagercrantz (Eds.), *Neurobiology of early infant behaviour* (pp. 598–68). Wenner-Gren International Symposium Series, vol. 55. London: MacMillan.

Prechtl, H.F.R. (1997a).The importance of fetal movements. In K.J. Connolly, H. Forssberg (Eds.), *Neurophysiology and psychology of motor development* (pp. 42–53). Clin. Dev. Med. nos. 143/144. Cambridge University Press.

Prechtl, H.F.R. (1997b). State of the art of a new functional assessment of the young nervous system. An early predictor of cerebral palsy. *Early Human Development*, **50**, 1–11.

Prechtl, H.F.R. (1999). How can we assess the integrity of the fetal nervous system? In P. Arbeille, D. Manlik, R.N. Laurini (Eds.), *Fetal hypoxia* (pp. 109–115). New York: Parthenon.

Prechtl, H.F.R. & Einspieler, C. (1997). Is neurological assessment of the fetus possible? *European Journal of Obstetrics, Gynecology and Reproductive Biology,* **75**, 81–84.

Prechtl, H.F.R. & Nolte, R. (1984). Motor behaviour of preterm infants. In H.F.R. Prechtl (Ed.), *Continuity of neural functions from prenatal to postnatal life* (pp. 79–93) Clin. Dev. Med. no. 94. Cambridge University Press.

Prechtl, H.F.R., Einspieler, C., Cioni, G., Bos, A.F., Ferrari, F. & Sontheimer, D. (1997). An early marker of developing neurological deficits after perinatal brain lesions. *Lancet*, **349**, 1361–1363.

Prechtl, H.F.R., Fargel, J.W., Weinmann, H.M. & Bakke, H.H. (1979). Postures, motility and respiration of low-risk preterm infants. *Developmental Medicine and Child Neurology,* **21**, 3–27.

Prechtl, H.F.R., Ferrari, F. & Cioni, G. (1993). Predictive value of general movements in asphyxiated fullterm infants. *Early Human Development*, **35**, 91–120.

Preyer, W. (1884). *Die Seele des Kindes*. Leipzig Grieben.

Sival, D., Visser, G.H.A. & Prechtl, H.F.R. (1992). The effect of intrauterine growth retardation on the quality of general movements in the human fetus. *Early Human Development*, **28**, 119–132.

Touwen, B.C.L. (1979). *Neurological development in infancy*. Clin. Dev. Med. no. 58, London: Heinemann.

van Wulfften-Palthe, T. & Hopkins, B. (1984). Development of the infant's social competence during early face to face interaction. A longitudinal study. In H.F.R. Prechtl (Ed.), *Continuity of neural functions from prenatal to postnatal life* (pp. 198–220). Clin. Dev. Med. no. 94. Cambridge University Press.

Visser, G.H.A., Laurini, R.N., de Vries, J.I.P. & Prechtl, H.F.R. (1985). Abnormal motor behaviour in anencephalic fetuses. *Early Human Development*, **11**, 221–229.

de Vries, J.I.P., Visser, G.H.A. & Prechtl, H.F.R. (1982). The emergence of fetal behaviour. I. Qualitative aspects. *Early Human Development*, **7**, 301–322.

de Vries, J.I.P., Visser, G.H.A. & Prechtl, H.F.R. (1984). Fetal motility in the first half of pregnancy. In H.F.R. Prechtl (Ed.), *Continuity of neural functions from prenatal to postnatal life* (pp. 46–64). Clin. Dev. Med. no. 94. Cambridge University Press.

de Vries, J.I.P., Visser, G.H.A. & Prechtl, H.F.R. (1985). The emergence of fetal behaviour. II. Quantitative aspects. *Early Human Development*, **12**, 99–120.

de Vries, J.I.P., Visser, G.H.A., Mulder, E.J.H. & Prechtl, H.F.R. (1987). Diurnal and other variations in fetal movement and heart rate patterns at 20 to 22 weeks. *Early Human Development*, **15**, 333–348.

de Vries, J.I.P., Visser, G.H.A. & Prechtl, H.F.R. (1988). The emergence of fetal behaviour. III. Individual differences and consistencies. *Early Human Development*, **16**, 85–103.

Wolff, P.H. (1959). Observations on newborn infants. *Psychosomatic Medicine*, **21**, 110–118.

Wolff, P.H. (1966). *The causes, controls and organization of behavior in the neonate*. Psychological Issues Monograph Series, vol. 5, no. 1. New York: International University Press.

III.3
Development of Motor Behaviour in Chick Embryos

ANNE BEKOFF

ABSTRACT

Spontaneous movements occur throughout embryonic and fetal ontogeny in a wide variety of organisms, including chicks and humans. Chick embryos provide an ideal model system for using experimental techniques to examine the neural mechanisms underlying embryonic movements. Kinematic and electromyographic (EMG) analyses show that the basic neural circuitry develops, and that embryonic movements can be produced, in the absence of sensory input. Nevertheless chick embryos are responsive to exteroceptive stimuli and it remains possible that there are subtle changes in the movement patterns that have not been detected in the studies carried out to date. Sensory signals have been shown to be crucial in initation and production of later, goal-directed behaviours, such as hatching and walking.

INTRODUCTION

Embryonic motor behaviour is particularly interesting because it is highly variable both from moment to moment and between developmental stages. While this makes it more challenging to characterize than regular, rhythmic postnatal motor behaviours such as walking (Bekoff, 1992), the function and source of this variability is intriguing (see also Hadders-Algra, this volume). The role of embryonic motor behaviour in neural development remains largely unknown. However, it is necessary for normal anatomical development since interfering with fetal movement has been shown to result in severe joint malformations and muscle atrophy in chick embryos (Drachman & Sokoloff, 1966). Furthermore, analysis of fetal movement in humans has recently been shown to provide useful diagnostic information about fetal brain function (Prechtl & Einspieler, 1997; Prechtl, this volume). The source of the variability and transitions is also largely unknown; whether it is due to instability of the pattern-generating circuitry, or is related to

429

A.F. Kalverboer and A. Gramsbergen (eds.), Handbook of Brain and Behaviour in Human Development, 429–446
© *2001 Kluwer Academic Publishers. Printed in Great Britain.*

the fact that descending and sensory inputs are being established during the embryonic period, is currently a subject of great interest and will be discussed in detail below.

The chick embryo provides a unique opportunity for studying these issues, as well as unique possibilities for relating functional and morphological development, because of its accessibility throughout embryonic life. In contrast to human embryos, which develop inside the mother's uterus, each chick embryo is encased in a hard eggshell, an environment that is independent of the mother. This makes it possible for us to make a hole in the shell at any stage of development and make observations or carry out experiments while the embryo remains relatively undisturbed in its normal fetal environment. This advantage can be exploited to allow detailed, high-resolution, analysis of movements from the earliest stages through the bird's equivalent of birth – hatching. The ability to look in detail at both movements and motor output patterns, and to manipulate sensory input and input descending from the brain, has provided important information on the mechanisms that are used to produce the distinctive patterns of movements that characterize embryonic motor behaviours.

Despite the differences between *in ovo* and *in utero* development, there are many parallels between chick and human embryonic motor behaviour. For example, in both cases embryonic movements are endogenously generated (Hamburger, 1963; Prechtl, 1997). Because alterations in general movements in human fetuses have clinical significance (Prechtl, 1997; Prechtl & Einspieler, 1997), the availability of an animal model in which the neural mechanisms underlying the production of such movements can be experimentally studied is of considerable interest.

THEORETICAL BACKGROUND

The nature and function of early embryonic movements has long intrigued scientists (e.g. Preyer, 1885; Windle, 1940; Hamburger, 1963; Oppenheim, 1981). While later embryonic movements may resemble goal-directed postnatal behaviours such as face-wiping, sucking or swallowing (Smotherman & Robinson, 1988; Prechtl, 1997; this volume), the early embryonic movements (called Type 1 motility in chicks; Bekoff, 1981; Hamburger, 1963) bear little resemblance to goal-directed postnatal behaviours. In humans, born with very immature nervous systems, similar movements, called general movements, may continue for the first few months after birth (Cioni & Prechtl, 1990). However, in chicks, which are relatively more mature when they emerge from the egg, embryonic-type behaviour ends at hatching.

It is widely accepted that embryonic behaviour is neurogenic and produced by the spinal cord (Bekoff, 1981). However, early investigators argued about whether the embryonic movements were "spontaneous", generated by the central nervous system in the absence of sensory or descending input, or "reflexogenic," initiated and organized by sensory input (for reviews see Bekoff, 1981; Hamburger, 1963). It is now generally accepted that embryonic movements are spontaneous, in the sense that they can continue in the absence of sensory or descending input.

However, the role played by sensory and/or descending input is not yet fully understood. This issue will be explored in the following section.

CRITICAL REVIEW OF THE LITERATURE

Spontaneous embryonic motility

Embryonic motor behaviour has been extensively studied in recent years. In part because of its variability, it appears strikingly different from adult behaviours. For example, like the general movements in human fetuses (Prechtl, 1997), embryonic movements in chicks show extensive variation in both speed and amplitude (Watson & Bekoff, 1990). In addition, intra- and inter-individual differences are great (Prechtl, 1997; Sharp *et al.*, 1999. It is only fairly recently that technical advances in the use of non-invasive techniques, such as ultrasonography, have allowed a fascinating view of the human fetus *in situ*, showing a remarkable diversity of movements (Cioni *et al.*, 1997; deVries *et al.*, 1982; Prechtl, this volume). These studies have suggested strong similarities to other mammalian fetuses, as well as to avian embryos. Furthermore, Smotherman & Robinson (1988; Robinson & Smotherman, 1992) have developed techniques for studying rat fetal movements *in utero,* and these have been used to extend earlier behavioural observations (Bekoff & Lau, 1980; Hamburger, 1975), as well as to carry out experimental manipulations (e.g., Robinson & Smotherman, 1995; Ronca & Alberts, 1995). Additional studies have been carried out in sheep fetuses (Robertson, 1988; Robertson & Bacher, 1995).

However, the most detailed observations and experimental studies of fetal movements have been carried out on chick embryos. Hamburger and his colleagues were responsible for the renewed interest in embryonic behaviour that emerged in the 1960s (e.g. Hamburger, 1963; Hamburger & Balaban, 1963; Hamburger & Oppenheim, 1967). This group emphasized the spontaneous and centrally generated nature of embryonic movements (Hamburger *et al.*, 1966) and they distinguished three types of motility: Types I, II and III (Hamburger & Oppenheim, 1967). From its initiation on day 3 the embryonic motor activity is episodic and the majority of embryonic movements are jerky and uncoordinated in appearance. Within an episode of activity, any or all body parts may move. This behaviour, called Type I embryonic motility, is seen throughout embryonic life. Type II motility, consisting of "startles" or "wriggles", was first seen at about 11 days of incubation and was much rarer. Note that a recent study by Bradley (1999) suggests that Type II behaviour can be identified in kinematic records as early as embryonic day 9. The embryos gradually become more active until, on day 13, they are active nearly continuously. The amount of activity decreases again near the time of hatching on day 21. The first Type III motility appears on day 17. This behaviour is tucking, the motor pattern used by the embryo to fold itself into the hatching position. Type III motility involves movements that are smoother and more coordinated in appearance than Types I and II. Finally, hatching appears on day 21. In the discussion to follow *embryonic motility* will be used to refer to Type I embryonic motility. When Type II or III motility is intended, this will be stated.

431

Initial studies using multi-unit recordings characterized the neural activity of the embryonic spinal cord and showed that embryonic movements are correlated with neural activity (Provine *et al.*, 1971; Provine, 1971; Ripley & Provine, 1972). The motor patterns underlying embryonic motility have also been examined in detail by recording electromyograms (EMGs) from leg muscles in spontaneously behaving chick embryos (Bekoff *et al.*, 1975; Bekoff, 1976; Bradley & Bekoff, 1990; Landmesser & O'Donovan, 1984). These studies showed that, despite the jerky and disorganized appearance of Type I embryonic motility, organized neural circuitry is present. This circuitry produces a coordinated sequence of muscle activation that can be recognized and quantified. This motor pattern consists of coactivation of extensor muscles, coactivation of flexor muscles and alternation of these two synergies. Furthermore, this coordinated pattern is seen as early as day 7, prior to the establishment of functional sensory input (Bekoff *et al.*, 1975). Based on EMG recordings from chick embryos on days 9 and 10, we have made the suggestion that a simple, symmetrical, basic circuit for leg movements is built during embryonic stages in the chick, and that this circuit is then modulated to produce a variety of different leg motor patterns in the post-hatching chick, including the asymmetric pattern typical of walking (Bradley & Bekoff, 1990).

Initially, the EMG results, which showed organized patterns of leg muscle contractions on days 9 and 10, seemed to be somewhat at odds with the behavioural observations, which indicated that embryonic movements were disorganized and uncoordinated. A kinematic study by Watson & Bekoff (1990) helped integrate these findings. They found that leg movements in 9- and 10-day embryos were, in fact, organized in that hip, knee and ankle typically moved at the same time and in the same direction. That is, when all three joints moved, they extended and flexed together. However, in some movements only one or two of the three joints moved. In addition, the durations of the movements varied greatly, from short "jerks" to longer, smoother movements. These latter two features are apparently sufficient to make the behaviour appear disorganized. These results were confirmed and extended by Bradley and colleagues (Bradley, 1997, 1999; Chambers *et al.*, 1995) in studies in which they considered interlimb coordination between leg and wing in addition to interjoint coordination within the leg and the wing. Bradley (1999) has suggested that the identifying characteristics of type I embryonic motility are: nearly synchronous initiation of movement in leg and wing at the onset of an episode, a pause in all joints prior to the last one or two cycles and nearly-simultaneous termination of movement in all joints at the end of an episode.

Bradley (1999) has suggested that the kinematic correlate of Type II motility is an abrupt movement in which a joint rapidly flexes or extends and then, usually, abruptly returns to the initial position. These short "jerks" appear quite frequently at day 12 (Bradley, 1999) and, although they are less frequent at day 9, they do occur then (Bradley, 1999; Watson & Bekoff, 1990). Thus, the kinematic analysis is able to pick up the occurrence of Type II motility somewhat earlier (day 9) than the behavioural observations (day 11). It has been suggested that these short jerks may be the product of the short-duration, synchronous bursts that are seen in EMG records (Bekoff *et al.*, 1976; Landmesser & O'Donovan, 1984; Watson & Bekoff, 1990). However, synchronized EMG and kinematic recordings will be needed to evaluate this suggestion.

Another puzzle introduced by the EMG results is the fact that the EMG pattern, which is well organized at day 9, appears less well organized at day 13, then becomes more organized again by day 17 (Bekoff, 1976). Recently, two kinematic studies have begun to address this by examining embryos at mid-incubation stages (days 11–13) and comparing the results to day 9 (Bradley, 1999; Sharp *et al.*, 1999). Both found evidence for a decrease in regular cyclical activity and changes in coordination. Using a method for determining the percentage of each episode in which all three, two or no leg joints moved synchronously in the same direction, Sharp and colleagues (1999) showed that coordination among all three leg joints decreases between 9 and 13 days, while coordination of two joints (hip and knee) increases. Thus, the ankle movements become less tightly coupled to movements of the other leg joints. This study suggests that this is an adaptive response to the spatial constraints of the egg as the embryo increases in size. That is, as the embryo gets larger, extending all three joints at once would be increasingly likely to result in the feet hitting the shell and potentially damaging the extraembryonic membranes. In contrast, moving the ankle independently of the knee and hip allows the ankle to move through a large excursion without pressing the feet against the shell.

Looking at the correlation of peak excursions in knee and ankle, Bradley's (1999) extensive kinematic study of day 12 embryos showed an increase rather than a decrease in interjoint coordination. The differences between the two studies most likely indicate that, while major excursions become more tightly coupled, the fine structure of the coupling of ankle with the other joints becomes more variable. Bradley (1999) also examined the correlation of the ipsilateral elbow and ankle peak excursions and found a significant decrease in wing–leg interlimb coordination, indicating that the coupling between limbs decreases. As Bradley (1999) points out, this uncoupling of wing and leg may occur in anticipation of later behaviours in which legs and wings must perform different functions. A similar conclusion was reached by Provine (1980) based on recording how often the wing and leg moved at the same time. However, he did not consider other aspects of coordination, such as whether concurrent movements were in the same direction.

Several laboratories are currently providing exciting new results to add to our understanding of the neural basis of spontaneous embryonic motility in chicks. For example, O'Donovan and his colleagues have used an isolated spinal cord preparation to make significant progress in understanding some of the cellular mechanisms involved in the production of organized motor patterns in chick embryos from 5½ to 12 days of incubation (e.g. Ho & O'Donovan, 1993; O'Donovan, 1999; Ritter *et al.*, 1999). They have used optical imaging as well as electrophysiological recordings and pharmacological manipulations to investigate spontaneous activity in developing spinal networks. Their data suggest that rhythm generation and pattern generation are generated by different mechanisms (Ho & O'Donovan, 1993). Key features of the current model for rhythm generation include: (1) a population of ventral interneurons that is rhythmically active, interconnected by recurrent excitatory synaptic connections, and projects to motoneurons and (2) the presence of activity-induced transient depression of network excitability (O'Donovan, 1999).

433

Role of sensory input in embryonic motility

There are a variety of roles that sensory input could play in embryonic development. One is that it could be required to produce the coordinated patterns that are observed. A second is that a coordinated pattern could be produced centrally, but sensory input could serve to modulate this "basic" pattern. This modulation could take the form of simply reinforcing the basic pattern to make it more stable, or could actively modify the pattern. Developing sensory input could also disrupt organized patterns; or, of course, it could have no role until later stages.

Availability of sensory input

The earliest limb movements in chick embryos occur prior to the establishment of functional reflex arcs. For example, EMG recordings show that coordinated activation of extensor and flexor leg muscles occurs as early as day 7 (Bekoff *et al.*, 1975). This is prior to the time at which lumbosacral sensory afferents have made functional contacts in the spinal cord (Lee & O'Donovan, 1991; Lee *et al.*, 1988; Davis *et al.*, 1989). Therefore sensory input is not required either for the development or the production of this pattern. Nevertheless, by days 9–10 functional sensory input is potentially available. Both homonymous and synergist motoneurons have been monosynaptically contacted by Ia afferents (Lee & O'Donovan, 1991; Lee *et al.*, 1988). Furthermore, collateral Ia axons have extended over an area of up to 20 spinal segments by day 10 and are beginning to retract to their final more restricted range (Eide & Glover, 1995).

Evidence against the use of proprioceptive sensory input in day 9 embryonic motility

Despite the potential availability of proprioceptive sensory input by day 9–10 there are several reasons to think that it may not play a substantial role in the production of the "basic pattern" seen at this age, or in the development of the underlying neural circuitry. For example, Hamburger and colleagues (1966) removed both sensory and descending input at 2 days of incubation and observed little change in the amount or organization of motility in embryos up to 15 days of incubation, when compared to controls with only descending input removed. Furthermore, Oppenheim (1972) has shown that embryos fail to increase their motor activity in response to proprioceptive sensory stimulation. That is, they appear to adapt rapidly to these stimuli and stop responding after the first or second stimulus. Using another approach, Landmesser & Szente (1986) found that, after blocking movement-induced sensory input beginning at 6 days by using a neuromuscular blocker, alternation of flexors and extensors still occurred at 9–12 days. Further evidence is provided by the observation that alternation of extensors and flexors is also seen in the isolated spinal cord preparation, which lacks sensory input (Landmesser & O'Donovan, 1984; O'Donovan & Landmesser, 1987). Finally, deafferentation experiments in post-hatching chicks show that the walking motor pattern, when deprived of sensory input, becomes very similar to 9-day embryonic motility (Bekoff *et al.*, 1987). Together, these experimental results suggest that production of the 9-day embryonic motor patterns is not dependent on sensory input.

These results are in line with the idea that embryos, floating in the amniotic fluid, are effectively buffered from many sources of sensory input, both internal as discussed here, and external as discussed by Reynolds (1962). Perhaps too much sensory input, or sensory input arriving too early, would be disruptive to normal motor development. Evidence in favour of this idea is provided by Sleigh & Lickliter (1996, 1998), who have altered the timing or amount of prenatal auditory stimulation and observed significant effects on postnatal auditory and visual responsiveness in bobwhite quail chicks. Although these studies did not examine the effects of altered sensory stimulation on motor behaviour, they do suggest that regulating the timing and levels of sensory stimulation during embryonic development may be important.

It is important to note that there is at least one argument against interpreting the results of the deafferentation and paralysis experiments described above as evidence that sensory input is not used at all in the production of the motor behaviour at 9 days. This is that the studies to date have only examined the motor activity of the experimental embryos for general features: amount of movement, general appearance of movements, presence of alternation of antagonist muscle activity. The motor patterns have not yet been evaluated using kinematic or more detailed EMG analyses that might detect more subtle deficits.

Evidence for effects of exteroceptive sensory input on day 9 embryonic motility

Early studies showed no evidence for quantitative changes in numbers of movements in embryos following manipulations of exteroceptive stimuli. For example, Oppenheim (1972) showed that embryos fail to increase their motor activity in response to tactile sensory stimulation. They also do not respond to changes in contraction frequency of the amnion (Oppenheim, 1966; Oppenheim & Levin, 1975).

However, more detailed kinematic analyses have found changes. For example, Bradley (1997) has found that reducing buoyancy by removing amniotic fluid has significant effects on motor patterns of chick embryos on day 9. Under these conditions embryos were more active and showed increased interjoint, but decreased interlimb, coordination. Bradley suggests that these results support the idea that the transitions in motility during the embryonic period may be, at least in part, due to changes in environmental cues as the embryo increases in size within the unchanging spatial constraints of the eggshell. This is also supported by a study by Sharp and colleagues (1999; see below). The finding that qualitative changes are more prominent than quantitative changes has an interesting parallel in human fetal movements: qualitative, but not quantitative, alterations in general movements are seen following brain damage (Prechtl, 1997).

Role of sensory input at later embryonic stages

Even less is known about the use of sensory input between days 9–10 and the onset of hatching. Several of the studies cited above followed embryos to days 12 or 15 and still saw little evidence that sensory input plays an important role in the generation of embryonic motility (Hamburger et al., 1966; Landmesser & Szente,

1986; O'Donovan & Landmesser, 1987; Oppenheim, 1972). Nevertheless, one intriguing finding is that the rhythmic pattern that is so clearly established in 9-day embryos becomes much less well organized in 12- and 13-day embryos (Bekoff, 1976; Bradley, 1999; Sharp et al., 1999). A possible interpretation of this is that the developing sensory and/or descending (see below) input has a disruptive effect on the motor behaviour. One suggestive piece of evidence is the observation that isolated spinal cord preparations appear to have more stable rhythmic activity than is seen in intact embryos (Bradley, 1999). Alternatively, other developmental changes in the pattern-generating network may be responsible (Bradley, 1999; Chub & O'Donovan, 1998; O'Donovan & Chub, 1997).

Role of descending input in embryonic motility

The time of arrival of descending inputs to the spinal cord has been determined in chick embryos (e.g. Okado & Oppenheim, 1985; Glover, 1993). For example, reticulospinal and vestibulospinal projections reach the spinal cord at day 5 and day 8, respectively, and their distributions are adult-like by day 10 (Okado & Oppenheim, 1985). Serotonergic input from the raphe nuclei arrives in the lumbosacral cord by day 8 (Okado et al., 1992). Work from Steeves' laboratory suggests that most brainstem–spinal neurons have reached the lumbar spinal cord by day 13 (e.g. Hasan et al, 1993).

A study by Sholomenko & O'Donovan (1995) showed that brainstem stimulation in the isolated spinal cord preparation could elicit motor activity as early as embryonic day 6. The neurons responsible for these results lie in the reticular formation, most likely in cells of the ventral pontine and medullary regions, which send long descending axons to the lumbosacral spinal cord.

Nevertheless, removal of descending input appears to have a limited effect on the production of embryonic motility. Oppenheim (1975) found that there were alterations in the amount and distribution of activity after chronic spinal transection, but that the movements did not differ in appearance from controls. Provine & Rogers (1977) found similar patterns of rhythmic neural activity in control and chronically transected chicks at day 6 and day 13. Furthermore, typical patterns of extensor and flexor motor activity were seen in acutely transected embryos on day 8 to day 10 (Landmesser & O'Donovan, 1984). In a detailed analysis of EMG activity in day 9 and 10 embryos with chronic spinal transections, Bradley & Bekoff (1992) confirmed the results of Oppenheim (1975), finding shortened cycle period duration. This study also showed that the alternation of flexors and extensors was similar to controls. These results are in keeping with studies of post-hatching chicks in which cervical spinal transection had less evident effects on hatching and walking than did lumbosacral deafferentation (Bekoff et al., 1987; Bekoff et al., 1989).

Embryonic development of neurotransmitter systems

An extensive literature is available on the development of neurotransmitter systems in chick embryos. For the most part it is not yet possible to relate this information directly to the development of motor behaviour; therefore it will not be considered in detail. Nevertheless, this is likely to become important in the future so a few references are provided (e.g. *glycine and GABA*: Berki *et al.*, 1995; Chub & O'Donovan, 1998; Milner & Landmesser, 1999; Reitzel *et al.*, 1979; Reitzel & Oppenheim, 1980; *serotonin*: Muramoto *et al.*, 1996; Okado *et al.*, 1992; Wallace *et al.*, 1986; *substance P*: Du *et al.*, 1987; *enkephalin*: Du & Dubois, 1988; *calcitonin gene-related peptide*: Carr & Wenner, 1998; *glutamate*: Chubb & O'Donovan, 1998; *acetylcholine*: Chubb & O'Donovan, 1998; Milner & Landmesser, 1999).

Prehatching and hatching behaviours

Tucking and pipping

Beginning about day 17 the chick begins to produce prehatching behaviours that are strikingly different from type I and type II embryonic motility. Hamburger & Oppenheim (1967) named these type III behaviours. In contrast to the jerky and disorganized appearance of the earlier behaviours, type III behaviours are characterized by smooth and obviously coordinated movements. In addition, their goal is apparent. For example, on day 17 the chick folds itself into the hatching position using a type III behaviour called tucking. In the hatching position the legs are tightly folded, with the ankles extending into the pointed end of the shell and the toes near the head. The head, at the blunt end of the shell near the air space, is tightly bent around to the right so that the beak is oriented over the back pointing upwards into the air space. To achieve this the embryo pushes its beak under the right wing over and over using rotational movements until finally, some time on day 17, the head stays there. At some point after attaining the hatching position, and prior to the initiation of hatching, the embryo produces another type III behaviour: pipping. In this behaviour, the beak makes a small hole through the shell. Pipping is usually completed several hours or even days prior to the initiation of hatching.

Nothing is known about the mechanisms involved in the initiation of tucking or pipping. In fact, beyond the behavioural descriptions supplied by Hamburger & Oppenheim (1967), these prehatching behaviours have not been extensively studied. One study, which analysed EMG recordings of the leg muscle activity during tucking, found coactivity in antagonist muscles (Bekoff, 1976). In this study it was found that tucking did not simply replace embryonic motility. Nor were the two types of behaviour intermingled. Instead, periods of tucking behaviour lasting approximately 15 minutes alternated with periods of embryonic motility lasting similar lengths of time. Embryonic motility on day 17, as at earlier stages, was characterized by rhythmic alternation of antagonist leg muscle activity. However, the mean cycle period of embryonic motility at 17 days (326 milliseconds) is much shorter than that seen in day 9 embryonic motility (1269 milliseconds).

Hatching

Hatching takes place at the end of the incubation period, on day 21, several days after the embryo has achieved the characteristic hatching position (Hamburger & Oppenheim 1967). The hatching process takes approximately 1 hour in chick embryos, from the time the embryo starts cracking the shell, until it has cracked around the circumference of the shell far enough to push off the top and escape from the egg. During that time it produces episodes of hatching behaviour that last about 1–3 seconds, with inter-episode intervals of about 20–30 seconds (Bekoff & Kauer, 1982, 1984). Filmed records show that the most obvious part of the behaviour is the backward thrusting of the head that causes the beak to progressively break the shell in a counterclockwise direction (Bakhuis, 1974). Another essential component of hatching is the thrusting of the legs against the shell. The wings also extend forwards as the beak contacts the shell. During a hatching episode the head, neck, wings and legs all extend to enable the beak to crack the shell. Both film (Bakhuis, 1974) and EMG (Bekoff & Kauer, 1984) recordings show that the legs thrust against the shell synchronously. These leg thrusts appear to supply the force that causes the embryo to rotate in the shell. This rotation is essential. Without it the beak would hit over and over at the same location, instead of progressing around the shell.

Initiation of hatching

Hatching is unlike embryonic motility in that it occurs only once, at the end of the 21-day incubation period. It is essential that hatching not be initiated until after the lungs and other organs are mature enough to support postnatal survival, but it must be completed before yolk supplies are exhausted or the extraembryonic membranes dry out. This effectively means that hatching in chickens must occur late on day 20 or on day 21.

Among the mechanisms that have been suggested as likely candidates for involvement in the initiation of hatching are: changes in O_2/CO_2 tension, vocalizations of other embryos, hormones, and sensory signals. O_2/CO_2 tension does not appear to affect the time of hatching (Visschedijk, 1968). Vocalizations of neighbouring embryos have been shown to be an effective mechanism for either advancing or delaying the time of hatching (Vince, 1969), but they are not required for the initiation of hatching since isolated embryos will hatch successfully. Hormones, or other neurochemical signals, are almost certainly involved. This seems likely because the onset of hatching movements must be coordinated with the uptake of the yolk sac and clamping off of the extraembryonic circulation, and a neurochemical signal seems well suited for coordinating such disparate events. There is evidence that corticosterone (Cornwall et al., 1998) and thyroxine (reviewed by Oppenheim, 1973) do appear to play roles in hatching, because manipulating their levels several days prior to hatching can change the time of hatching. Nevertheless, no hormone or other neurochemical signal has yet been identified as a trigger for the immediate initiation of hatching, including behaviour, yolk sac withdrawal and clamping off of the extraembryonic circulation.

Neck bending

However, a specific sensory signal that is involved in triggering the initiation of hatching behaviour has been identified: asymmetric bending of the chick embryo's neck. A study in which chicks were suspended from rings glued to their backs, leaving all body parts free to move, provides the most convincing evidence that the bent neck is a specific signal for the initiation of hatching (Bekoff & Kauer,

1982). Specific body parts (neck, legs or wings), were taped in the position they would normally be in during hatching. Restraining the legs or wings failed to elicit hatching. However, bending the neck laterally and firmly taping it into this position resulted in the initiation of hatching movements. Note that hatching was not elicited by bending the neck downwards toward the breast, or by immobilizing it in a straight position. Therefore neither symmetrical bending nor simple immobilization was sufficient. Despite the fact that in the normal hatching position the head and neck are always bent to the right, this study found that bending the neck either to the right or to the left elicited hatching. This indicates that asymmetrical bending of the neck is a key factor.

In another study the source of the sensory signal within the neck was examined (Bekoff & Sabichi, 1987). Injections of the local anaesthetic, lidocaine, into either the skin alone or into the neck muscles, were used to determine whether the signal was due to cutaneous input from the neck region, or to proprioceptive input. Only the injections into the neck muscles were effective in preventing hatching. Taken together, the results from the neck-bending experiments and the anaesthetic injections suggest that the initiation signal for hatching depends on the asymmetric stretching of the muscles of the chick embryo's neck.

Further evidence suggests that the termination signal for hatching is the straightening of the neck. If the head is released from the egg so that the neck straightens, hatching ceases (Bekoff & Kauer, 1982; Bekoff, 1992). In contrast, if the legs are released but the neck remains bent, hatching movements continue. Moreover, if the egg shell is taped in such a way as to prevent the chick from pushing the top off so that it can get its head out and straighten its neck, it continues to rotate around, completing as many as three rotations before stopping (Provine, 1972).

Circuitry

As soon as hatching is completed, and the chick has emerged from the egg, hatching behaviour ceases and the newly hatched chick's behaviour takes on characteristics more appropriate to postnatal life. For example, the leg movements immediately switch from the synchronous pattern of coordination that is typical of hatching to alternation, which is characteristic of postnatal walking (Bekoff & Nicholl, 1992). The fact that the rhythm is unchanged while the pattern switches from synchronous to alternating is evidence for separate rhythm- and pattern-generating circuitry. The switch is triggered by the unbending of the neck since re-bending the neck and placing chicks in artificial glass eggs will re-elicit hatching in chicks up until at least 61 days after hatching (Bekoff & Kauer, 1984). The glass egg experiments show clearly that the circuitry underlying the initiation and production of hatching behaviour is still present and functional long after hatching behaviour is normally completed. Thus the hatching circuitry remains intact in the adult chicken. Similarities in the EMG patterns underlying the leg movements of hatching and walking, have led to the suggestion that the same, or at least elements of the same, circuitry is used for both behaviours (Bekoff et al., 1987).

Furthermore, both hatching and walking appear to share elements of the so-called "basic circuit" used in the production of the leg movements of embryonic motility at 9 days of incubation (Bradley & Bekoff, 1990; Bekoff, 1992). Therefore, it seems likely that at least some of the circuitry used for hatching develops very early in embryonic life. It is then reconfigured or modulated to produce the characteristic hatching behaviour by a hormonal trigger and a sensory signal from the bent neck at the end of the incubation period.

DISCUSSION

Clearly we have learned a great deal about the development of motor behaviour in chicks. The appearance of type I and type II embryonic motility and the organization of limb movements have been well characterized through mid-incubation (days 12–13) using kinematic and/or EMG techniques. More limited information is available about motor behaviour in later embryos (Bekoff et al., 1976; Hamburger & Oppenheim, 1967), until the time of hatching, which has been well characterized (for review see Bekoff, 1992). We also know a substantial amount about the development of sensory input in chicks. For example we know when various afferent inputs arrive and become functional (Lee & O'Donovan, 1991; Lee et al., 1988; Davis et al., 1989). However, we know relatively little about what sensory input contributes to motor development. It is clear that normal development of motor behaviour is not completely dependent on sensory input. However, it does not appear to be entirely independent. The real issue is not whether it is completely dependent or independent, but what does sensory input contribute and when?

To address this issue we need to close a significant gap in our understanding of the effects of deafferentation or paralysis on embryonic motor patterns. To date the studies that have been carried out have used behavioural measures or EMG analyses to examine relatively gross aspects of the motor patterns. More detailed analyses are needed to determine whether there are subtle deficits that have not previously been detected. For example, in adult animals, including post-hatching chicks, a coordinated stepping pattern, consisting of alternation of antagonist muscles and synchronous activation of synergists, occurs following deafferenta-tion (Bekoff et al., 1987; Grillner & Zangger, 1984). This is analogous to the results that have been obtained so far in deafferented or paralysed embryos. However, detailed EMG analysis shows that many of the complexities that normally characterize the walking motor pattern are lost in deafferented adults. This level of analysis has not yet been carried out in deafferented or paralysed embryos. Of course, it is also possible that removal of sensory information has no impact on any aspect of motor development at early embryonic stages. If so, at what stage does elimination of sensory input first result in deficits?

A related issue is to determine what modalities are important during embryonic stages in which sensory information is needed. For example, is proprioception (muscle spindles, Golgi tendon organs, joint receptors) needed earlier than cutaneous input? Do they play different roles in the embryo? This information is not easy to obtain. However, methods with the potential for selectively elimi-nating specific classes of sensory neurons are currently being developed (Sharp et al., 1998). Another approach is to apply specific sensory perturbations during ongoing movement, similar to methods that have been used successfully during locomotion in adults (e.g. Pearson *et al.*, 1998). The major difficulty with these experiments in embryos is that the natural variability in motor patterns makes the interpretation of the results of the perturbations problematic (Sharp *et al.*, 1997). Small size of the embryos and immature anatomy also contribute to the difficulty. Nevertheless, as our ability to characterize the embryonic motor patterns becomes more sophisticated, we may be able to overcome these obstacles. Another approach would be to take advantage of a semi-isolated chick spinal cord preparation in which a limb remains attached (Landmesser & O'Donovan, 1984). In this case such variables as the limb position, muscle stretch, or tactile stimulation could be manipulated and the effects on the motor patterns monitored.

The studies of hatching show that asymmetric sensory input from the bent neck plays a key role in initiating hatching at 21 days. However, the neck is initially bent into the hatching position during tucking on day 17. A study of why hatching is not elicited until several days later would be useful. Some possibilities include immaturity or inhibition of proprioceptive afferents in the neck, requirement for hormonal signals, or insufficient stretching of the neck muscles.

Finally, evidence is accumulating that chick embryos are responsive to aspects of their environment such as the amount of amniotic fluid (Bradley, 1997) or the size of the embryonic environment (Sharp *et al.*, 1999). Further manipulations of environmental variables are likely to prove interesting.

ACKNOWLEDGEMENTS

I am grateful for the generous funding from the National Science Foundation and the National Institutes of Health that has made the research reported here possible. My current work is funded by NSF IBN-963049.

References

Bakhuis, W.L. (1974). Observations on hatching movements in the chick (*Gallus domesticus*). *Journal of Comparative Physiology and Psychology, 87*, 997–1003.

Bekoff, A. (1976). Ontogeny of leg motor output in the chick embryo: a neural analysis. *Brain Research, 106*, 271–291.

Bekoff, A. (1981). Embryonic development of the neural circuitry underlying motor coordination. In M.W. Cowan (Ed.), *Studies in developmental neurobiology. Essays in honor of Viktor Hamburger* (pp. 134–170). Oxford: Oxford University Press.

Bekoff, A. (1992). Neuroethological approaches to the study of motor development in chicks: achievements and challenges. *Journal of Neurobiology, 23*, 1486–1505.

Bekoff, A. & Kauer, J.A. (1982). Neuronal control of hatching: role of neck position in turning on hatching leg movements in post-hatching chicks. *Journal of Comparative Physiology, 145*, 497–504.

Bekoff, A. & Kauer, J.A. (1984). Neural control of hatching: fate of the pattern generator for the leg movements of hatching in post-hatching chicks. *Journal of Neuroscience, 4*, 2659–2666.

Bekoff, A., Kauer, J.A., Fulstone, A. & Summers, T.R. (1989). Neural control of limb coordination. II. Hatching and walking motor output patterns in the absence of input from the brain. *Experimental Brain Research, 74*, 609–617.

Bekoff, A. & Lau, B. (1980). Interlimb coordination in 20-day old rat fetuses. *Journal of Experimental Zoology, 214*, 173–175.

Bekoff, A. & Nicholl, D. L. (1992). Timing vs. pattern in cyclic motor patterns in chicks. *Society for Neuroscience Abstracts, 18*, 1278.

Bekoff, A. & Sabichi A.L. (1987). Sensory control of the initiation of hatching in chicks: effects of a local anesthetic injected into the neck. *Developmental Psychobiology, 20*, 489–49.

Bekoff, A., Nusbaum, M.P., Sabichi, A.L. & Clifford, M. (1987). Neural control of limb coordination. I. Comparison of hatching and walking leg motor output patterns in normal and deafferented chicks. *Journal of Neuroscience, 7*, 2320–2330.

Bekoff, A., Stein, P.S.G. & Hamburger, V. (1975). Coordinated motor output in the hindlimb of the 7-day chick embryo. *Proceedings of the National Academy of Sciences, USA, 72*, 1245–1248.

Berki, A.C.S., O'Donovan, M.J. & Antal, M. (1995). Developmental expression of glycine immuno-reactivity and its colocalization with GABA in the embryonic chick lumbosacral spinal cord. *Journal of Comparative Neurology, 362*, 583–596.

Bradley, N.S. (1997). Reduction in buoyancy alters parameters of motility in E9 chick embryos. *Physiology and Behavior, 62*, 591–595.

Bradley, N.S. (1999). Transformations in embryonic motility in chick: kinematic correlates of Type I and II motility at E9 and E12. *Journal of Neurophysiology, 81*, 1486–1494.

Bradley, N.S. & Bekoff, A. (1990). Development of coordinated movement in chicks: I. Temporal analysis of hindlimb synergies at embryonic days 9 and 10. *Developmental Psychobiology, 23*, 763–782.

Bradley, N.S. & Bekoff, A. (1992). Development of coordinated movement in chicks: II. Temporal analysis of hindlimb muscle synergies at embryonic day 10 in embryos with spinal gap transections. *Journal of Neurobiology, 23*, 420–432.

Carr, P.A. & Wenner, P. (1998). Calcitonin gene-related peptide: distribution and effects on spon-taneous rhythmic activity in embryonic spinal cord. *Developmental Brain Research, 106*, 47–55.

Chambers, S.H., Bradley, N.S. & Orosz, M.D. (1995). Kinematic analysis of wing and leg movements for type I motility in E9 chick embryos. *Experimental Brain Research, 103*, 218–226.

Chub, N. & O'Donovan, M.J. (1998). Blockade and recovery of spontaneous rhythmic activity after application of neurotransmitter antagonists to spinal networks of the chick embryo. *Journal of Neuroscience, 18*, 294–306.

Cioni, G. & Prechtl, H.F.R. (1990). Preterm and early postterm behaviours in low-risk premature infants. *Early Human Development, 23*, 159–191.

Cioni, G., Prechtl, H.F.R., Ferrari, F., Paolicelli, P.B., Einspieler, C. & Roversi, M.F. (1997). Which better predicts later outcome in fullterm infants: quality of general movements or neuromuscular examination? *Early Human Development, 50*, 71–85.

Cornwall, G., White, E., Spencer, R. & Bekoff, A. (1998). Fetal corticosterone appears to be involved in the initiation of hatching in the domestic chicken. *Society for Neuroscience Abstracts, 24*, 118.

Davis, B.M., Frank, E., Johnson, F.A. & Scott, S.A. (1989). Development of central projections of lumbosacral sensory neurons in the chick. *Journal of Comparative Neurology, 279*, 556–566.

DeVries, J.I.P., Visser, G.H.A. & Prechtl, H.F.R. (1982). The emergence of fetal behavior. I. Qualitative aspects. *Early Human Development, 7*, 301–322.

Drachman, D.B. & Sokoloff, L. (1966). The role of movement in embryonic joint development. *Developmental Biology, 14*, 401–420.

Du, F. & Dubois, P.M. (1988). Development and distribution of enkephalin-immunoreactive elements in the chicken spinal cord. *Neuroscience, 27*, 251–266.

Du, F., Charnay, Y. & Dubois, P. (1987). Development and distribution of substance P in the spinal cord and ganglia of embryonic and newly hatched chick: an immunofluoresence study. *Journal of Comparative Neurology, 263*, 436–454.

Eide, A.L. & Glover, J.C. (1995). Development of the longitudinal projection patterns of lumbar primary sensory afferents in the chicken embryo. *Journal of Comparative Neurology, 353*, 247–259.

Glover, J.C. (1993). The development of brain stem projections to the spinal cord in the chicken embryo. *Brain Research Bulletin, 30*, 265–271.

Glover, J.C. & Petursdottir, G. (1988). Pathway specificity of reticulospinal and vestibulospinal projections in the 11-day chicken embryo. *Journal of Comparative Neurology, 270*, 25–38.

Grillner, S. & Zangger, P. (1984). The effect of dorsal root transection on the efferent motor pattern in the cat's hindlimb during locomotion. *Acta Physiologica Scandinavica, 120*, 393–405.

Hamburger, V. (1963). Some aspects of the embryology of behavior. *Quarterly Review of Biology, 38*, 342–365.

Hamburger, V. (1975). Fetal behavior. In E.F.E. Hafez (Ed.), *The mammalian fetus* (pp. 68–81). Springfield, IL: Charles C. Thomas.

Hamburger, V. & Balaban, M. (1963). Observations and experiments on spontaneous rhythmical behavior in the chick embryo. *Developmental Biology, 7*, 533–545.

Hamburger, V. & Oppenheim, R. (1967). Prehatching motility and hatching behavior in the chick. *Journal of Experimental Zoology, 166*, 171–203.

Hamburger, V., Wenger, E. & R. Oppenheim, R. (1966). Motility in the absence of sensory input. *Journal of Experimental Zoology, 162*, 133–160.

Hasan, S.J., Keirstead, H.S., Muir, G.D. & Steeves, J.D. (1993). Axonal regeneration contributes to repair of injured brain-stem spinal neurons in embryonic chick. *Journal of Neuroscience, 13*, 492–507.

Ho, S. & O'Donovan, M.J. (1993). Regionalization and intersegmental coordination of rhythm-generating networks in the spinal cord of the chick embryo. *Journal of Neuroscience, 13*, 1354–1371.

Landauer, W. (1967). The hatchability of chicken eggs as influenced by environment and heredity. *Storrs Agricultural Bulletin, 1*.

Landmesser, L.T. & O'Donovan, M.J. (1984). Activation patterns of embryonic hindlimb muscles recorded in ovo and in an isolated spinal cord preparation. *Journal of Physiology (London), 347*, 189–204.

Landmesser, L.T. & Szente, M. (1986). Activation patterns of embryonic chick hindlimb muscles following blockade of activity and motoneurone cell death. *Journal of Physiology (London), 380*, 157–174.

Lee, M.T. & O'Donovan, M.J. (1991). Organization of hindlimb muscle afferent projections to lumbosacral motoneurons in the chick embryo. *Journal of Neuroscience, 11*, 2564–2573.

Lee, M.T., Koebbe, M.J. & O'Donovan, M.J. (1988). The development of sensorimotor synaptic connections in the lumbosacral cord of the chick embryo. *Journal of Neuroscience, 8*, 2530–2543.

Milner, L.D. & Landmesser, L.T. (1999). Cholinergic and GABAergic inputs drive patterned spontaneous motoneuron activity before target contact. *Journal of Neuroscience, 19*, 3007–3022.

Muramoto, T., Mendelson, B., Phelan, K.D., Garcia-Rill, E., Skinner, R.D. & Puskarich-May, C. (1996). Developmental changes in the effects of serotonin and *N*-methyl-D-aspartate on intrinsic membrane properties of embryonic chick motoneurons. *Neuroscience, 75*, 607–618.

O'Donovan, M.J. (1999). The origin of spontaneous activity in developing networks of the vertebrate nervous system. *Current Opinion in Neurobiology, 9*, 94–104.

O'Donovan, M.J. & Chubb, N. (1997). Population behavior and self-organization in the genesis of spontaneous rhythmic activity by developing spinal networks. *Seminars in Cell and Developmental Biology, 8*, 21–28.

O'Donovan, M.J. & Landmesser, L.T. (1987). The development of hindlimb motor activity studied in the isolated spinal cord of the chick embryo. *Journal of Neuroscience, 7*, 3256–3264.

Okado, N. & Oppenheim, R.W. (1985). The onset and development of descending pathways to the spinal cord in the chick embryo. *Journal of Comparative Neurology, 232*, 143–161.

Okado, N., Sako, H., Homma, S. & Ishikawa, K. (1992). Development of serotonergic system in the brain and spinal cord of the chick. *Progress in Neurobiology, 38*, 93–123.

Oppenheim, R.W. (1966). Amniotic contraction and embryonic motility in the chick embryo. *Science, 152*, 528–529.

Oppenheim, R.W. (1972). Embryology of behavior in birds: a critical review of the role of sensory stimulation in embryonic movement. *Proceedings of the International Ornithology Conference, 15*, 283–302.

Oppenheim, R.W. (1973). Prehatching and hatching behavior: a comparative and physiological consideration. In G. Gottlieb (Ed.), *Behavioral embryology* (pp. 163–244). New York: Academic Press.

Oppenheim, R.W. (1975). The role of supraspinal input in embryonic motility: a re-examination in the chick. *Journal of Comparative Neurology, 160*, 37–50.

Oppenheim, R.W. (1981). Ontogenetic adaptations and retrogressive processes in the development of the nervous system and behaviour: a neuroembryological perspective. In K.J. Connolly & H.F.R. Prechtl (Eds.), *Maturation and development: biological and psychological perspectives* (pp. 73–109). Philadelphia, PA: J.P. Lippincott.

Oppenheim, R.W. & Levin, H.L. (1975). Short-term changes in incubation temperature–behavioral and physiological effects in the chick embryo from 6 to 20 days. *Developmental Psychobiology, 8*, 103–115.

Pearson, K.G., Misiaszek, J.E. & Fouad, K. (1998). Enhancement and resetting of locomotor activity by muscle afferents. *Annals of the New York Academy of Sciences, 860*, 203–215.

Prechtl, H.F.R. (1997). The importance of fetal movements. In: K.J. Connolly & H. Forssberg (Eds.), *Neurophysiology and neuropsychology of motor development* (pp. 42–53). Cambridge, UK: Cambridge University Press.

Prechtl, H.F.R. & Einspieler, C. (1997). Is neurological assessment of the fetus possible? *European Journal of Obstetrics, Gynecology and Reproductive Biology,* **75**, 81–84.

Preyer, W. (1885). *Specielle physiologie des embryo.* Leipzig: Grieben.

Provine, R.R. (1971). Embryonic spinal cord: synchrony and spatial distribution of polyneuronal burst discharges. *Brain Research,* **29**, 155–158.

Provine, R.R. (1972). Hatching behavior of the chick (*Gallus domesticus*): plasticity of the rotatory component. *Psychonomic Science,* **2**, 27–28.

Provine, R.R. (1980). Development of between-limb synchronization in the chick embryo. *Developmental Psychobiology,* **13**, 151–163.

Provine, R.R. & Rogers, L. (1977). Development of spinal-cord bioelectric activity in spinal chick embryos and its behavioral implications. *Journal of Neurobiology,* **8**, 217–228.

Provine, R.R., Sharma, S.C., Sandel, T.C. & Hamburger, V. (1971). Electrical activity in the spinal cord of the chick embryo in situ. *Proceedings of the National Academy of Sciences, USA,* **65**, 508–515.

Reitzel, J.L. & Oppenheim, R.W. (1980). Ontogeny of behavioral sensitivity to glycine in the chick embryo. *Developmental Psychobiology,* **13**, 455–461.

Reitzel, J.L., Maderdrut, J.L. & Oppenheim, R.W. (1979). Behavioral and biochemical analysis of GABA-mediated inhibition in the early chick embryo. *Brain Research,* **172**, 487–504.

Reynolds, S.R.M. (1962). Nature of fetal adaptation to the uterine environment: a problem of sensory deprivation. *American Journal of Obstetrics and Gynecology,* **83**, 800–808.

Ripley, K.L. & Provine, R.R. (1972). Neural correlates of embryonic motility in the chick. *Brain Research,* **45**, 127–134.

Ritter, A., Wenner, P., Ho, S., Whelan, P.J. & O'Donovan, M.J. (1999). Activity patterns and synaptic organization of ventrally located interneurons in the embryonic chick spinal cord. *Journal of Neuroscience,* **19**, 3457–3471.

Robertson, S.S. (1988). Mechanism and function of cyclicity in spontaneous movement. In W.P. Smotherman & S.R. Robinson (Eds.), *Behavior of the fetus* (pp. 77–94). Caldwell, NJ: Telford Press.

Robertson, S.S. & Bacher, L.F. (1995). Oscillation and chaos in fetal motor activity. In J.-P. Lacanuet, W.P. Fifer, N.A. Krasnegor & W.P. Smotherman (Eds.), *Fetal development* (pp. 169–189). Hillsdale, NJ: Erlbaum.

Robinson, S.R. & Smotherman, W.P. (1992). Fundamental motor patterns of the mammalian fetus. *Journal of Neurobiology,* **23**, 1574–1600.

Robinson, S.R. & Smotherman, W.P. (1995). Habituation and classical conditioning in the rat fetus: opioid involvements. In J.-P. Lacanuet, W.P. Fifer, N.A. Krasnegor & W.P. Smotherman (Eds), *Fetal development* (pp. 295–314). Hillsdale, NJ: Erlbaum.

Ronca, A.E. & Alberts, J.R. (1995). Maternal contributions to fetal experience and the transition from prenatal to postnatal life. In J.-P. Lacanuet, W.P. Fifer, N.A. Krasnegor & W.P. Smotherman (Eds), *Fetal development* (pp. 331–350). Hillsdale, NJ: Erlbaum.

Sharp, A.A., Ma, E. & Bekoff, A. (1997). Kinematic analysis of embryonic motility in 11-day chick embryos. *Society for Neuroscience Abstracts,* **23**, 636.

Sharp, A.A., Ma, E. & Bekoff, A. (1999). Developmental changes in leg coordination of the chick at embryonic days 9, 11 and 13: uncoupling of ankle movements. *Journal of Neurophysiology,* **82**, 2406–2414.

Sholomenko, G.N. & O'Donovan, M.J. (1995). Development and characterization of pathways descending to the spinal cord in the embryonic chick. *Journal of Neurophysiology,* **73**, 1223–1233.

Sleigh, M.J. & Lickliter, R. (1996). Differential amounts of specific prenatal stimulation alter postnatal responsiveness to maternal cues in bobwhite quail chicks. *Developmental Neurobiology,* **29**, 78.

Sleigh, M.J. & Lickliter, R. (1998). Timing of presentation of prenatal auditory stimulation alters auditory and visual responsiveness in bobwhite quail chicks. *Journal of Comparative Neurology,* **112**, 153–160.

Smotherman, W.P. & Robinson, S.R. (1988). Dimensions of fetal investigation. In W.P. Smotherman & S.R. Robinson (Eds.), *Behavior of the fetus* (pp. 19–34). Caldwell, NJ: Telford Press.

Vince, M.A. (1969). Embryonic communication, respiration and the synchronization of hatching. In R.A. Hinde (Ed.), *Bird vocalizations* (pp. 233–260). Cambridge, UK: Cambridge University Press.

Visschedijk, A.H.J. (1968). The airspace and embryonic respiration. 2. The times of pipping and hatching as influenced by artificially changed permeability of the shell over the airspace. *British Poultry Science,* **9**, 185–196.

Wallace, J.A., Allgood, P.C., Hoffman, T.J., Mondragon, R.M. & Maez, R.R. (1986). Analysis of the change in number of serotonergic neurons in the chick spinal cord during embryonic development. *Brain Research Bulletin,* **17**, 297–305.

Watson, S.J. & Bekoff, A. (1990). A kinematic analysis of hindlimb motility in 9- and 10-day old chick embryos. *Journal of Neurobiology,* **21**, 651–660.

Windle, W.F. (1940). *Physiology of the fetus: origin and extent of function in prenatal life.* Philadelphia, PA: Saunders.

Further Reading

The following references provide useful reviews of the literature on embryonic chick motor patterns: Bekoff (1992); Hamburger (1963); Oppenheim (1981); O'Donovan (1999).

III.4
Development of the Neural Correlates of Locomotion in Rats

JEAN-RENÉ CAZALETS

ABSTRACT

The rat is an extremely suitable model for studying ontogenesis, due to the rapid stereotyped changes that occur during this animal's development. This allows us to acquire an overview of neuro-ontogeny within a limited period of time. During the embryogenic stages, movements occur from embryonic day 15 (E15) onwards, and synchrony between leg movements can be observed from E18. In the same period of development, sensory-motor connections are established. At birth neonates cannot spontaneously perform locomotor movements: these have to be evoked in swimming or air-stepping experiments. It has been established that the rats' inability to perform locomotor movements during the early stages is due to postural constraints and to the weakness of the postural muscle tone. This conclusion has been confirmed by in-vitro experiments showing that the locomotor networks are in fact expressed early on in the fetus. In parallel, the primary sensory afferents are installed by E16.5 and the descending pathways to the lumbar spinal cord are already present. The neurochemical control of locomotor activity has been extensively studied in the isolated in-vitro spinal cord from P0 onwards. Various neuromodulatory mechanisms are already at work at birth. Excitatory amino acids, serotonin and dopamine can initiate a locomotor-like activity, which can be modulated by peptides (subs-P) and GABA. The organization of the locomotor network has been studied from the day of birth (P0). The interneuronal network, which is mainly located in the rostral lumbar cord (L1/L2 segments), sends excitatory and inhibitory inputs successively to the motoneuron during each locomotor cycle. The data available so far on all these mechanisms are reviewed here, and it emerges that, despite their apparent immaturity during the first week of life, most of the mechanisms which are characteristic of adult locomotor system are already present in the newborn rat.

A.F. Kalverboer and A. Gramsbergen (eds.), Handbook of Brain and Behaviour in Human Development, 447–466
© 2001 Kluwer Academic Publishers. Printed in Great Britain.

INTRODUCTION

Studies on a motor system during the early stages of development focus on a critical period during which the neuronal networks go through transient states involving important changes. The first basic idea which underlies ontogenetic studies is that one can approach the functioning of the system more easily in neonates than in adult animals, since the internal loops are presumably less complex at birth than those which are set up during successive learning processes. Secondly, by analysing a rapidly changing system, it is possible to study the gradual build-up, and the stabilization into the adult configuration. From this point of view the rat appears to be a suitable model, since its motor development is fast and reproducible. The gestational age can be determined precisely and the animal rapidly evolves from an extremely immature state at birth to adulthood within about 3 weeks. In addition, studies on the fetus can be performed with relative ease. An additional advantage of using the rat as a model for development is that data can be collected on these animals at the cellular level. In this chapter some of the characteristics will be reviewed of the events which occur during the development of spinal motor circuits, with special emphasis on the locomotor networks. The chapter will successively review how the various components of the sensory-motor apparatus and its control mechanisms are developing until the adult circuits are acquired.

GENERATION OF LOCOMOTOR ACTIVITY IN MAMMALS

Since the pioneering studies by Brown (1914) at the beginning of the century, the concept of the central pattern generator has gradually emerged. The use of spinal animals has made it possible to unambiguously demonstrate that the spinal cord has the intrinsic ability of generating locomotor activity. In the early 1970s research on invertebrates definitely established the idea that the central nervous system can autonomously generate rhythmic motor activity. During locomotion, tens of muscles are activated per locomotor cycle, each according to a specific temporal pattern. Subsequent studies have demonstrated that the central programme is not only limited to the simple alternating activation of flexor and extensor units, as proposed by Engberg & Lundberg (1969), who assumed that the complexity of the final motor output might be due to the sensory afferents. It has been reported, however, that in spinal animals which were curarized to block phasic sensory feedbacks, a complex locomotor pattern could be generated comparable to that recorded in the intact animal (for reviews see Grillner, 1981; Rossignol, 1996). Further evidence was provided by experiments on the isolated *in-vitro* spinal cord, in which fictive locomotion can be recorded. These *in-vitro* preparations have also been used to study the functioning of motor networks and to identify their components at the cellular level under restrictive conditions. This has been achieved in lower vertebrates, in which the cellular organization of the networks responsible for swimming is now beginning to be well understood (Grillner & Matsushima, 1991; Roberts *et al.*, 1997). The recent use of an isolated *in-vitro* spinal cord preparation from newborn rat has led to significant progress being made in this field, and the general functional pattern of the locomotor network in neonatal rat is now beginning to emerge.

EARLY DEVELOPMENT OF LOCOMOTOR ACTIVITY IN THE RAT

Behavioural data

A prerequisite of ontogenetic studies is the availability of a reference state, which is generally that of the adult animal. With regard to studies on the locomotor mechanisms and the initiation of locomotion in intact animals, numerous data are available on the adult cat. However, much less research has been performed on the adult rat. During either galloping or swimming it has been established that the minimum locomotor period is around 250 ms (Cazalets *et al.*, 1990; Cohen & Gans, 1975; De Leon *et al.*, 1994; Gruner & Altman, 1980; Ménard *et al.*, 1991). In a decerebrate rat, Nicolopoulos-Stournaras & Iles (1984) studied the hindlimb muscle activity occurring during spontaneous sequences of locomotion, or during the stimulation of the mesencephalic locomotor region (MLR). The muscle activity observed was found to be very similar to that occurring in the cat. Similar conclusions were reached using a decerebrate rat during hypothalamic stimulation (Goudard *et al.*, 1992; Sinnamon, 1993). It was therefore concluded that the mechanisms involved in the generation of locomotor activity are very conservative in mammals.

On the other hand, the development of locomotor activity from the embryonic to weanling stages has been thoroughly documented in the rat. As a preliminary remark one should keep in mind that motoneurons appear in the lumbar spinal cord at embryonic day 12–13 (E12–13; Nornes & Das, 1974; gestation time 21 days). Shortly afterwards, at E16, spontaneous movements of the head, trunk, legs and less frequently the tail are first observed in the rat fetus (Narayanan *et al.*, 1971). At E16 the forelimbs are more active than the hindlimbs, which only sporadically perform weak movements. These movements reach a peak at E18 and then tend to decline. At E18 Narayanan *et al.* (1971) have described synchronous movements occurring in right and left legs. This finding was confirmed by Bekoff & Lau (1980), who observed short sequences of interappendicular coordination at E20. Robertson & Smotherman (1990) also reported that in the fetus (E20), the lumbar cord is able to produce spontaneous hindlimb movements following a section at the thoracic level. Surprisingly, however, despite the fact that the neuronal circuitry is capable of mediating coordinated movements in the fetus, the neonate does not express any spontaneous locomotor activity. Locomotor activity is observed only when the animal is free from gravitational constraints. This is the case, for example, during swimming (Figure 1; Bekoff & Trainer, 1979; Cazalets *et al.*, 1990) or air-stepping (Van Hartesveldt *et al.*, 1991; Fady *et al.*, 1998). Interestingly, it was noted that the motor pattern expressed during development in swimming and L-DOPA-induced air-stepping experiments exhibits the same changes (Van Hartesveldt *et al.*, 1991), since there exists a rostrocaudal gradient in the motor activity. At postnatal day 0–1 (P0–1), only the forelimbs are involved. From P2–3 the hindlimbs also begin to be active, whereas from P12–14 onwards only the hindlimbs participate in swimming or air-stepping.

In all the cases studied an antiphase pattern was observed between the right and left limbs and the ipsilateral forelimbs and hindlimbs. It was also found that, in the course of development, there was a decrease in the variability of the motor pattern (Figure 1C; Cazalets *et al.*, 1990), and that the rhythmic leg movements became faster with age (Figure 1B, C; Ménard *et al.*, 1991).

The development of walking activity is more controversial, however. This is due to the fact that, in the early stages, rats do not support their body weight, and they perform what is called crawling. Therefore, the problem arises whether crawling and locomotion are different types of behaviour, and whether a switch occurs from the one form of activity to the other. The slight discrepancies between the data obtained in this context probably result from the use of different protocols, and from the fact that it is difficult to clearly distinguish between these two types of behaviour. The authors of the first reports on crawling stated that this behaviour occurs up to postnatal day 7 (Altman & Sudarshan, 1975). Recently, however, it was observed that the shift from crawling to walking occurs earlier (Jamon & Clarac, 1998). In an experiment in which the pups received olfactory stimulation provided by the litter material, these authors elicited an early form of stepping at P3–4 (Jamon & Clarac, 1998). In contrast, spontaneous walking

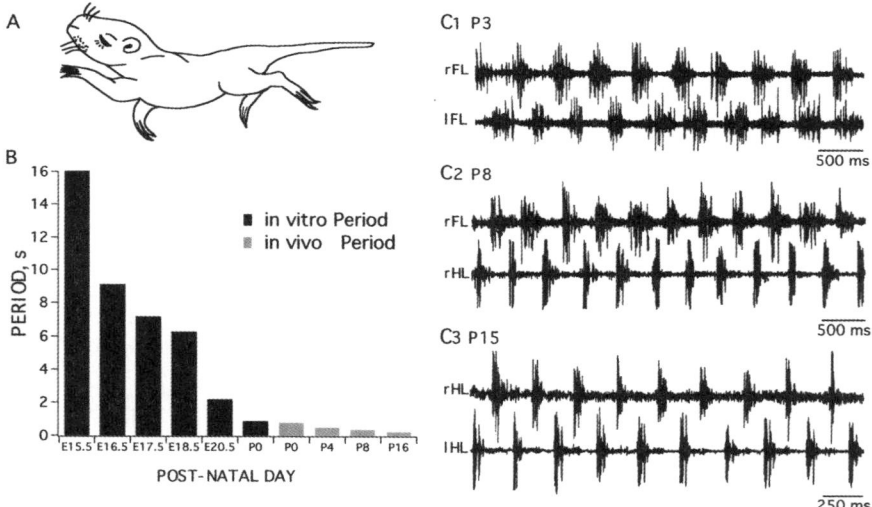

Figure 1 Development of locomotor activity in rat. **A**: Drawing of a rat pup showing the antiphase pattern between legs. During swimming experiments, rhythmic locomotor patterns can be elicited within the first hours after birth. **B**: Changes in the motor period during development. The black bars represent the motor period recorded from isolated fetal spinal cords during bath-application of NMA (adapted from Ozaki *et al.*, 1996) and the grey bars the motor period recorded from intact swimming newborn rats (adapted from Ménard *et al.*, 1991). **C**: Pattern of electromyographic activity at three different stages. Note the progressive shift in phase relationships occurring at P3 (**C1**) and the changes in motor period (adapted from Cazalets *et al.*, 1990). Abbreviations: FL, forelimb; HL, hindlimb; l, left; P, postnatal day; r, right.

normally starts around P10, although the movements are still rather clumsy and unstable; the motor pattern suddenly undergoes major changes at around P15, and begins to develop towards the adult pattern (Westerga & Gramsbergen, 1990, 1993; see also Gramsbergen, this handbook). This set of behavioural data suggests that the locomotor networks are functional very early in life, although they are active only occasionally. In order to be activated, however, the animal needs to be either released from the postural constraints, as occurs during swimming (Bekoff & Trainer, 1979; Cazalets *et al.*, 1990), or exposed to a strong stimulation inducing olfactory-guided behaviour (Fady *et al.*, 1998; Jamon & Clarac, 1998).

In-vitro data

Since the expression of locomotor behaviour appears to be largely dependent on external limits, the experiments performed so far on intact animals do not provide sufficient information regarding the real locomotor capabilities of the spinal cord itself during the embryonic and early postnatal stages. In the spinal cord the emergence of the electrical and chemical excitability was initially studied by performing optical recordings on dissociated cells (Mandler *et al.*, 1990). It was suggested that functionally excitable Na^+ channels may emerge at E13. Extracellular recordings on ventral roots at day 13.5 of gestation (E13.5; Saïto, 1979) and intracellular recordings on embryonic motoneurons *in vitro* (E14; Ziskind-Conhaim, 1988), also demonstrated that the action potential is carried by Na^+ ions. Saïto (1979) reported that synaptic responses (whether spontaneous or elicited by applying electrical stimulation to the cord) were detected in motoneurons at E14.5 but not at E13.5. At the same time the neuronal responses to various transmitters begin to develop. The first detectable responses to GABA have been observed at E12, although not consistently, while at E14, GABA and the $GABA_A$ agonist muscimol elicited depolarizing responses via Cl^- channels (Mandler *et al.*, 1990). These depolarizing responses persisted up to E18. Similar responses were also obtained with glycine, starting 1 day later. Kainate starts depolarizing cells at E14, while glutamate responses are detected only at E15.

In parallel to this pattern of development of the neuronal responses, the spinal circuits are progressively built up. The use of isolated spinal cord preparations has yielded some important insights into the onset of spinal network activity and the development of rhythmogenic properties during ontogeny. Using an isolated *in-vitro* spinal cord, Nishimaru *et al.* (1996) recorded synchronous bursts of action potentials from E14.5. This is 2 days before any spontaneous movements are observed in the fetus, and this could be due to the delayed development of the nerve–muscle innervation. This spontaneous activity, which is synaptically mediated, declines around E18, which is in agreement with the behavioural observations of Narayanan *et al.* (1971). The glycinergic receptor blocker strychnine totally abolishes the spontaneous bursts between E14.5 and 16.5, and then becomes inefficient, which suggests that the effects are mediated by glycinergic connections (Nishimaru *et al.*, 1996). These authors suggested that, from E14.5 to E16.5, glycine may have a depolarizing action, and may then be supplanted by glutamate. This idea is supported by the fact that, at E18–19, glycinergic inputs tend to become inhibitory (Wu *et al.*, 1992). These results are also in agreement with previous studies showing that spinal embryonic neurons *in vitro* first exhibit responses to GABA and glycine and later to glutamate (Mandler *et al.*, 1990).

Besides this spontaneous bursting, an organized motor pattern can be elicited at an early age (E15.5) by bath-applying glutamatergic agonists or serotonin (5-HT; Ozaki *et al.*, 1996; Iizuka *et al.*, 1998). Adding *N*-methyl-D-aspartate (NMDA) or 5-HT to the saline elicits synchronous bursts of action potentials on contralateral lumbar ventral roots, with a 10 s period at E16.5. A right and left alternating activity appears only at E18.5 (Kudo *et al.*, 1991; Ozaki *et al.*, 1996), concomitant with the development of inhibitory glycinergic connections (Ozaki *et al.*, 1996; Wu *et al.*, 1992). The cycling period diminishes quickly with age and is around 1 s at birth. It has also been observed that alternating activity between flexor and extensor groups starts to occur at E20.5 (Iizuka *et al.*, 1998). All these results therefore show that a functional locomotor circuitry has developed already *in utero*.

It is from birth, however, that the *in-vitro* spinal cord has been extensively used for studying locomotor processes. Kiehn & Kjaerulff (1996) have investigated the patterns of muscle activity generated by the isolated spinal cord with one intact limb attached, from P0 to P4. They simultaneously recorded several muscles (up to six) during locomotor-like activity induced by serotonin or dopamine. They observed that the relative timing of the various muscle units showed many similarities with the adult motor output. These findings demonstrate and confirm therefore that: (1) very early in life, the locomotor circuitry exhibits complex adult-like characteristics; and that (2) the *in-vitro* spinal cord is a relevant preparation for studying locomotor mechanisms.

ONTOGENY OF SENSORIMOTOR CONNECTIONS

The development of sensory connections in the spinal cord has been investigated both anatomically and electrophysiologically. In the rat fetus Narayanan *et al.* (1971) have observed the appearance of motor responses to cutaneous stimulation of the hindlimb at E17. In an *in-vitro* spinal cord, however, Saïto (1979) detected synaptic activity in motoneurons at E14.5, while the first motor responses to the dorsal root stimulation were observed at E15.5. These initial reflex responses were polysynaptic. Saïto (1979) reported that he was able to identify initial monosynaptic responses as early as E17.5. This, however, was queried in subsequent studies by Kudo & Yamada (1985, 1987a), since, in response to dorsal root stimulation, they recorded the first postsynaptic potential (PSP) in the motoneurons between E16 and E17. They specifically studied the monosynaptic reflex by directly stretching the triceps surae muscle. They could therefore clearly identify both the polysynaptic reflex (E17.5, after cutaneous stimulations) and the monosynaptic reflex (E19.5) at a later stage (Kudo & Yamada, 1985; see also Ziskind-Conhaim, 1990). In addition, they performed retrograde labelling by horseradish peroxidase of dorsal roots in order to study the growth of sensory afferents from E15.5 to birth (Kudo & Yamada, 1987a) and they observed that the first fibres enter the motoneuronal area (lamina IX) at E17.5. Fitzgerald (1987) analysed the properties of primary afferent neurons in the fetus. As early as E17 she identified various types of primary afferents, as defined by their receptive fields and their responses to mechanical stimulation. Jahr & Yoshioka (1986) reported that the Ia afferent (muscle spindle afferent) response is mediated by excitatory amino acids in newborn rats, and subsequently Ziskind-Conhaim (1990) observed that, in the fetus, the motoneuronal responses to dorsal root stimulation are mostly mediated via the NMDA receptor activation at E16, and become sensitive to CNQX (a non-NMDA receptor blocker) only at E17. The PSP amplitude increases during embryogenesis and reaches the action potential firing threshold at E17–18. A parallel can be established from developmental data in premature babies (Hakamada *et al.*, 1988). By examining the H reflex in triceps surae these authors found that the monosynaptic reflex was functional at the post-conceptional age of 25 weeks. In the preterm period until term age there is an increase in the size of the H reflex comparable to that observed in rat fetuses by Kudo & Yamada (1987a) suggesting that similar mechanisms are at work during ontogeny.

In addition to the segmental spinal sensory pathways, it was also demonstrated that, at birth, complex sensory motor loops are already installed (Vinay *et al.*, 1995). These polysynaptic pathways play an important role in the adaptation of the animals' responses to changes in the environment (Noga *et al.*, 1988; see also olfactory guided locomotion, Jamon & Clarac, 1998). It was established in this study that stimulating the trigeminal nerve elicits PSPs in the lumbar motoneurons. These involve connections from both the brainstem and the spinal cord (Vinay *et al.*, 1995). It has not yet been ascertained whether this may also be the case in the embryo.

DESCENDING PATHWAYS AND CONTROL OF LOCOMOTOR ACTIVITY IN THE NEONATAL RAT

As we have seen, the spinal motor networks and the sensory afferents have established before birth. A third main component involved in locomotor processes are influences from supraspinal centres which control the activity of the spinal circuitry. In adult rats electrical stimulation of the mesencephalic locomotor region (MLR, Nicolopoulos-Stournaras & Iles, 1984) or to various hypothalamic sites (Sinnamon, 1993; Goudard *et al.*, 1992) are known to elicit locomotor-like activity in the decerebrate rat.

Likewise, in the neonatal rat from P0 onwards, it has been demonstrated that two regions of the brainstem, the MLR and the medioventral medulla, can give rise to locomotor-like activity in the *in-vitro* brainstem spinal cord with the hindlimb attached (Atsuta *et al.*, 1990). The electromyographic pattern was found to be comparable to that observed during stepping in adult rats. It was also observed that electrical stimulation of the ventrolateral funiculus can induce locomotor-like activity *in-vitro* (Magnusson & Trinder, 1997). In addition, locomotor-like movements of the hindlimbs, in a semi-isolated preparation in which the limbs are retained, are evoked when direct stimulation is applied to the lumbar enlargement (Iwahara *et al.*, 1992). In some other experiments chemical stimulation was applied to the brainstem to induce locomotor-like activity. A partitioned brainstem/spinal cord preparation was used to superfuse the two structures independently. The bath application of excitatory amino acids (Cazalets *et al,* 1994) or bicuculline (a GABA$_A$ receptor antagonist; Atsuta *et al.*, 1991; Smith *et al.*, 1988) induced rhythmic and alternating activity. The development of descending ventral pathways was recently investigated by stimulating the ventral funiculus. The responses (excitation and inhibition) evoked in the lumbar motoneurons were found to increase in amplitude from P0 to P6 (Brocard *et al.*, 1999). It was suggested here that the progressive arrival of the developing nerve endings may be responsible for the progressive acquisition of postural control (Brocard *et al.*, 1999).

The origin of the descending pathways has been investigated in the rat embryo (Kudo *et al.*, 1993) and the neonatal rat (Shieh *et al.*, 1983; Leong *et al.*, 1984). After injecting retrogradely transported horseradish peroxidase in the low thoracic segments, it has been noted that the first cells are labelled in the brainstem at E14.5. Cells in the red nucleus are first labelled at E16.5. By birth the major supraspinal inputs, with the exception of the corticospinal fibres, have reached the lower spinal cord (Kudo *et al.*, 1993; Leong *et al.*, 1984). All these findings converge to show that various functional pathways which originate in the brainstem and are involved in the control of lumbar locomotor networks are present at birth.

In the absence of any stimulation, the *in-vitro* isolated spinal cord is generally silent. Therefore the challenge is to establish how it can be adequately stimulated in order to promote the locomotor-like activity. In the neonatal rat (Atsuta *et al.*, 1990) have demonstrated that electrical stimulation of the mesencephalic locomotor region can initiate locomotion (as in the adult cat, see review by Shik & Orlovsky, 1976), and therefore that descending pathways are functional at birth. These stimulation experiments are not easy to perform, however, and they are also difficult to reproduce from one trial to another. For this reason pharmacological approaches have been used to study the involvement of the supra segmental centres in the control of locomotor networks. Since descending fibres are able to elicit locomotor behaviour by releasing various neurotransmitters, the aim here was to mimic the action of these specific pathways, by directly applying the neurotransmitters or their agonists to the spinal cord. These substances have been widely tested on the isolated spinal cord. As the blood–brain barrier is absent in this preparation, it is possible to bath-apply all the neurochemicals. In addition, specific targets in the spinal cord can be superfused. It emerges from these studies that locomotor activity can be triggered by various types of neurotransmitters in the neonatal rat spinal cord, and I will now briefly review some of these mechanisms and their ontogeny.

THE NEUROTRANSMITTERS CONTROLLING THE SPINAL LOCOMOTOR NETWORKS DURING DEVELOPMENT

Ontogeny of neurotransmitter innervation

Along with the emerging spinal motor circuitry the various systems which are involved in controlling these circuits also undergo some considerable structural and functional changes. One can make the following rough distinction between two types of transmitters: (1) those which are extrinsic to the spinal cord, such as the amines (serotonin, dopamine, noradrenaline) and which originate in the brainstem; (2) those which are intrinsic to the spinal cord. Both intrinsic and extrinsic innervation might be present at early stages. This is the case, for example, for GABA (Barber *et al.*, 1982; Holstege, 1991) and a similar type of coexistence is likely to occur in the case of other transmitters such as excitatory amino acids (EAAs).

It has been possible in several cases to establish the ontogeny of neurotransmitters in the spinal cord. GABA is the first transmitter to be present in the spinal cord (Lauder *et al.*, 1986; Ma *et al.*, 1992). It has been detected at the cervical level at E13, when no cell bodies are immunoreactive in the spinal cord, which suggests that these initial GABA fibres are of extrinsic origin. The first GABA-immunoreactive somata to be detected and which presumably are motoneurons, are located in the ventral horn at E14, while labelling begins to occur in the dorsal horn at E17 (Ma *et al.*, 1992). At E18–19 the GABA reaches its peak distribution, and then at E20 the labelling in the ventral regions greatly decreases. One week after birth GABA has also disappeared from the motoneurons in the lumbar region.

The development of the serotoninergic (5-HT; which arises in the raphe nucleus) and noradrenergic (NA) innervation has also been studied using immunocytochemical procedures. These two aminergic systems show comparable changes during development (Bregman, 1987; Rajaofetra *et al.*, 1989, 1992). 5-HT appears at E14 in the upper spinal cord slightly before noradrenaline (NA; E16), and reaches the lumbosacral cord by E16–17 (E18 for NA). Both the dorsal and ventral horn are progressively innervated. At birth the innervation is already dense and it continues to develop until adulthood.

Initiation of locomotor activity by excitatory amino acids and serotonin

The use of the *in-vitro* spinal cord preparation has made it possible to establish the effects of various transmitters on the locomotor network (Figure 2). We have already mentioned that the NMDA and 5-HT initiate a rhythmic coordinated motor pattern in the lumbar ventral roots in the fetal period (see Ozaki *et al.*, 1996; Iizuka *et al.*, 1998). It was only from P0, however, that a detailed analysis of the effects of various neurotransmitters has been performed.

In several studies it has been established that the bath application of 5-HT (Figure 2C) or EAAs (Figure 2B), induces a rhythmic motor pattern (Cazalets *et al.*, 1992; Kiehn & Kjaerulff, 1996; Kudo & Yamada, 1987b; Smith *et al.*, 1988). This treatment elicits bursts of spikes alternating between the left and right sides of each lumbar segment and between the flexor and extensor units (Cazalets *et al.*, 1992). 5-HT-induced bursting was blocked by ketanserin and cyproheptadine but was not mimicked by any of the 5-HT agonists tested. The receptor involved, therefore, might be of the 5-HT2_C type (Cazalets *et al.*, 1992; see Wallis, 1994).

EAAs have also been demonstrated to play a crucial role in the activation of the locomotor CPGs in mammals. Kudo & Yamada (1987b) and Smith *et al.* (1988) provided the initial evidence for the involvement of NMDA receptors in the emergence of locomotion in newborn rats. We confirmed that NMDA receptors initiate locomotor-like activity but also that non-NMDA receptors,

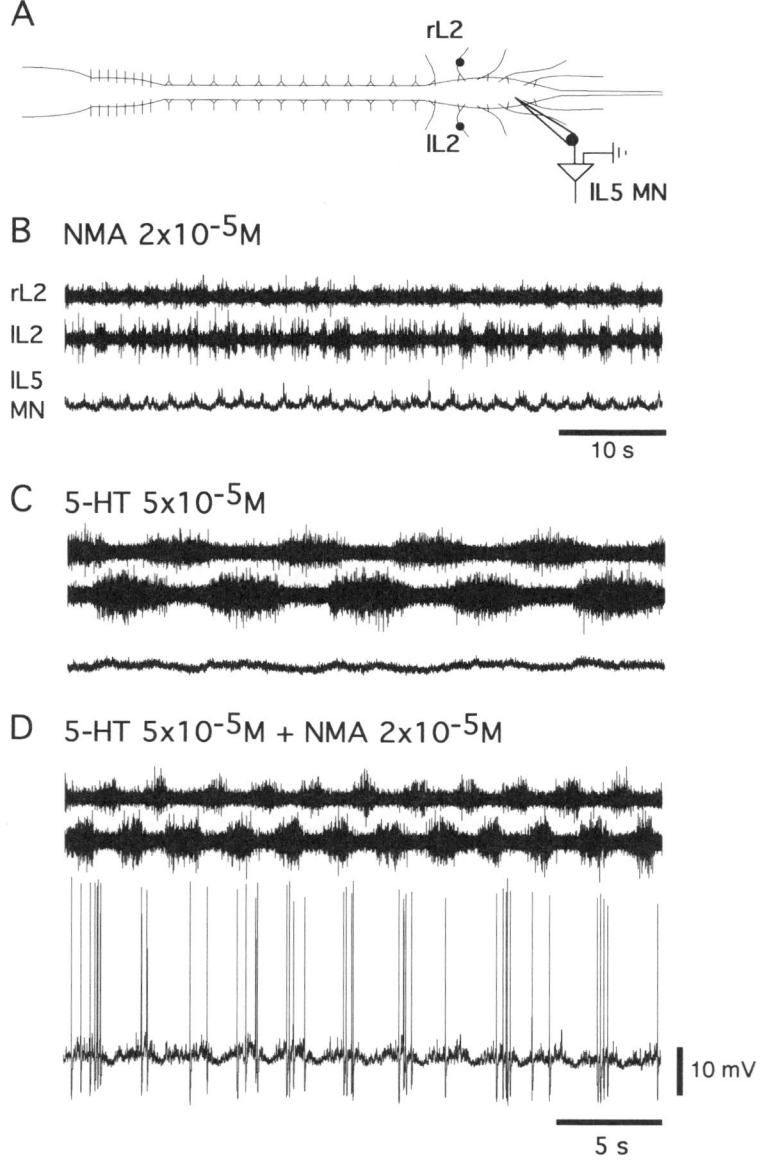

Figure 2 Locomotor-like activity in the brainstem/spinal cord preparation of neonatal rat. **A:** Schematic drawing of the isolated preparation. The fictive locomotion is monitored from extracellular recordings of the lumbar ventral roots (L2) and intracellularly from a motoneurone located in the L5 segment. The dots indicate the recording sites. **B–D:** The locomotor activity is successively induced by bath-application of an agonist of the NMDA receptor (NMA, **B**), serotonin (5–HT, **C**) or a mixture of the two compounds (**D**). Note the changes in the motor period and in the locomotor drive amplitude. All the recordings are from the same experiment.

which were activated by kainate but not by AMPA, were capable of triggering locomotor-like activity (Cazalets *et al.,* 1992; Sqalli-Houssaini *et al.*, 1993). With all the EAAs tested, as well as with 5-HT, their effects were found to be dose-dependent and the antagonists gradually decreased the rhythm before it ceased completely. Recently, we have found that the various allosteric mechanisms controlling the NMDA receptor are already at work at birth, although they differ from what is found in the adult (Bertrand & Cazalets, 1999). It was suggested that these differences between the adult and the neonate reflected an adaptation to developmental constraints. All these findings therefore demonstrated that, like 5-HT, the EAAs act directly on the rhythmic component of the CPG.

When locomotor-like activity is experimentally induced as described above, only one compound at a time is tested in its effects to trigger rhythmic activity. It is unlikely, however, that in a stepping animal only one pathway is activated. Rather it seems that several pathways are acting together to continuously adjust the locomotor behaviour. To approach this problem the interactions between 5-HT and EAAs have been studied by mixing them (Sqalli-Houssaini *et al.*, 1993). A mixture of serotonin (5-HT) and excitatory amino acids (EAAs) elicited fictive locomotion in experiments where the compounds alone, even at higher concentrations, did not evoke any rhythmic activity. NMDA and non-NMDA receptors were involved in this type of co-modulatory effects. In experiments in which EAAs initiated locomotor-like activity, 5-HT added to the NMA saline modulated the motor pattern (Figure 2D). The period of the EAA/5-HT motor pattern was set at an intermediate value and the burst amplitude and duration increased (Figure 2D; Sqalli-Houssaini *et al.*, 1993). The variability of all these parameters decreased under 5-HT/NMA conditions. It was concluded that the simultaneous presence of the two compounds is necessary to initiate a stable and regular motor pattern. The bath-application of a mixture of 5-HT and EAA is the most powerful means of eliciting long-lasting sequences of locomotor-like activity, and it is now widely used (Sqalli-Houssaini *et al.*, 1993; Kjaerulff *et al.*, 1994; Magnusson & Trinder, 1997; Kremer & Lev-Tov, 1997; Cowley & Schmidt, 1997).

In conclusion, although it is clear that the two compounds have synergistic effects from a qualitative point of view, they have conflicting effects on the period, and it is therefore possible that the speed of the working CPG is modulated by varying their respective concentrations.

In addition to excitatory amino acids and serotonin, other transmitters have been found to induce or to modulate locomotor-like activity at birth. Dopamine induced slower and more irregular rhythmic activity than 5-HT, and some characteristic phase relationships were found to exist between the various muscle groups (Kiehn & Kjaerulff, 1996). Substance P bath-applied alone either elicits a slow rhythmic non-locomotor activity, or increases the locomotor frequency and the duration of the bursts of a NMDA-induced activity (Barthe & Clarac, 1997).

GABAergic control of the lumbar locomotor system

Another transmitter which plays an important role in the early control of motor systems is GABA. In the mature spinal cord it has an inhibitory action, together with glycine (Curtis *et al.*, 1959). It has been particularly implicated as the mediator of presynaptic inhibition of the primary afferent terminals (Eccles, 1964). It has been proposed that, during embryonic life, GABA has a depolarizing rather than a hyperpolarizing effect (Wu *et al.*, 1992). It should be noted, however, that this property is not restricted to immature systems, since presynaptic inhibition in the adult rat is mediated by GABA-induced depolarization of primary afferents (Eccles, 1964), due to a reversal in the Cl gradient. We recently observed that GABA plays other functional roles in the spinal cord of the neonatal rat (Cazalets *et al.*, 1994; Bertrand & Cazalets, 1999). We established that, in the *in-vitro* spinal cord preparation, GABA can finely tune the motor pattern, since it has access at all stages to the spinal locomotor system. We established that endogenous GABA has a direct action on the premotoneuronal network and that the GABAergic inputs may mask the expression of this network. The bath-application of GABA can either slow down or suppress ongoing activity (Cazalets *et al.*, 1994). Secondly, we noted that GABA acts presynaptically at the premotoneuronal level on the connections that the locomotor network establishes with the motoneurons, thus adjusting the level of motoneuronal activity (Bertrand & Cazalets, 1999). Thirdly, GABA can modify the excitability of motoneurons, and in this way regulate their ability to respond to premotoneuronal inputs (Bertrand & Cazalets, 1999). The existence of the dual control exerted by activating and inactivating systems acting in parallel is probably of major importance to motor behaviour, since it is likely that the level of activity of the locomotor system is based on the effects of all the modulatory pathways at these three different stages. In addition, these data again demonstrate that complex regulatory mechanisms are already at work very early in life.

THE ORGANIZATION OF SPINAL LOCOMOTOR NETWORKS AT BIRTH

As we have seen above, as early as E15.5 a rhythmic activity with right and left coordination can be evoked in the isolated spinal cord under bath-application of EAAs or 5-HT (Iizuka *et al.*, 1998; Ozaki *et al.*, 1996), which suggests that the locomotor circuits are functional from a very early stage. A detailed analysis of the structure of these locomotor networks has been performed from P1 to P4. The aims were: (1) to map the locomotor circuits if possible; (2) to precisely define the organization of the interneuronal network; (3) to define the relationships between the locomotor network and the output from the system, i.e. the motoneurons. In order to reveal the rhythmogenic properties of the various parts of the spinal cord we used a technique which makes it possible to compartmentalize the lumbar cord without injuring it (Cazalets *et al.*, 1995, 1996; Cazalets, 1999). First, a Vaseline wall was built at the T12–T13 level in order to set a higher limit in the superfusion area. Other walls were built at several other lumbar levels. It was established that, when the wall was built between the L2 and L3 segments, bath-application of the mixture of 5-HT and EAAs to the rostral (T13–L2) segments induced a locomotor-like activity which was similar to that evoked upon bath-applying the neurotransmitters to the whole lumbar cord. This locomotor-like activity was not only recorded extracellularly in the rostral L1–L2 segments but was also evoked in the other lumbar segments, as revealed by intracellular recordings of motoneurons from the caudal region. In the L3–L6 segments the motoneurons exhibit a rhythmic activity which is correlated with that recorded in the L2 ventral roots. When, on the contrary, the neurotransmitters were bath-applied to the caudal region, only tonic activity was elicited in the caudal motoneurons and no activity was elicited in the rostral segments. Using the same protocol we also demonstrated that when the segments L1–L2 were pharmacologically isolated by superfusing the other part of the spinal cord with tetrodotoxin (a fast Na^+ channel blocker which suppresses spiking activity), they were able by themselves to generate the same locomotor-like activity without any changes in the motor period or the phase relationships (Cazalets *et al.*, 1995). We also observed that, after a longitudinal section in the sagittal plane from the caudal end of the cord to the L2 segments, the right and left as well as the flexor and extensor alternation persisted at the L3, L4 and L5 levels, despite the disappearance of the physical connections. It was therefore concluded that the cross-inhibitory connections originate, at least partly, in the rostral T13–L2 segments (Cazalets *et al.*, 1995). It was subsequently established that cross-inhibitory connections, although they are weaker, are also present in lower segments (Kjaerulff & Kiehn, 1996; Kremer & Lev-Tov, 1997). Later complementary studies have also shown that when the caudal segments were sectioned to isolate them, they can also express rhythmic activities in response to bath-applied neurotransmitters (Cowley & Schmidt, 1997; Kjaerulff & Kiehn, 1996; Kremer & Lev-Tov, 1997). In all the cases studied these bursting activities were much slower and more irregular than those recorded in the L1/L2 segments, and it is agreed that the key elements for generating locomotion in the neonatal rat are located in this thoracolumbar area (Cowley & Schmidt, 1997; Kjaerulff & Kiehn, 1994; for an extensive review of this problem see Cazalets, 1999). It is not known, however, whether, from P4 onwards, changes still continue to occur in the course of development. From a study by Ho & O'Donovan (1993), who analyzed the changes in the rhythmogenic capacity occurring during development in the chick spinal lumbar cord, it appears that from day 8 to day 13 the caudal segments lose 80% of their bursting capacity in comparison with the rostral segments. The possibility cannot be ruled out that the decrease in the bursting capacity may continue and, therefore, that the rhythmogenesis involves only the rostral segments in adults. It seems likely that similar processes occur in the rat, and that the rostralization of the network proceeds with age.

The locomotor drive received by the motoneurons is complex since, during one stride, a muscle unit (and therefore the corresponding motoneuronal pool) is inhibited by its antagonists on the ipsilateral side and by its contralateral homologue (for an alternating gait). Therefore, the observed inhibitory drive will be the result of a mixture of both ipsilateral and contralateral synaptic influences. Using the isolated spinal cord it has been possible to separate, at least partly, these various elements in the neonatal rat. In our former study we established that the interneuronal locomotor network located in L1/L2 synaptically drives the motoneuronal activity through monosynaptic connections (Cazalets *et al.*, 1995). In a second step we took advantage of this, in order to identify the components of the synaptic drive that the motoneurons receive during locomotor activity. By using the experimental protocol defined in Figure 2 it becomes possible to physically separate the L1–L2 network from its motoneuronal target located in the caudal segments. In this way one can intervene independently at each level and study the characteristics of the locomotor drive, by modifying the synaptic functioning with various drugs in the caudal compartment, without affecting the locomotor-like activity induced by bath-application to the rostral pool. We found that the motoneuronal drive is biphasic, consisting within a locomotor cycle of successive bursts of IPSPs and EPSPs (Cazalets *et al.*, 1996; Cazalets, 1999). By using selective blockers of the EAAs we found that the depolarizing excitatory phase consisted of two components: (1) the NMDA part, sensitive to 2-amino-5-phosphonovaleric acid (AP-5); and (2) the non-NMDA part, sensitive to 6-cyano-7-nitroquinoxaline-2-3-dione (CNQX). The remaining inhibitory synaptic volley was reversed at a membrane potential of -60 mV and it was mediated by glycine (since it was sensitive to strychnine), probably through the opening of Cl$^-$ channels (Cazalets *et al.*, 1996). Using a comparable procedure, Kjaerulff & Kiehn (1997) directly identified the cross-inhibitory connections. For this purpose the barrier was built longitudinally along the midline. They pharmacologically activated an hemispinal cord and recorded intracellularly the motoneurons on the opposite side. The tested motoneurons received a rhythmic glycinergic inhibitory input sensitive to strychnine, and which was in most cases conveyed via a polysynaptic pathway (Kjaerulff & Kiehn, 1997). It was also found that the lumbar spinal cord of the neonatal rat contains substantial numbers of commissural interneurons with axon projections and collateral ranges spanning several segments, and that theses neurons projecting rostrally versus caudally have different distributions in the transverse plane (Eide *et al.*, 1999). To date it is not known if these synaptic characteristics are similar to those in the adult rat. However, based on the data collected in the cat, it is likely that the same processes occur (Orsal *et al.*, 1986).

COMPARISON WITH HUMAN DATA

One question that can be asked at this point is how relevant this information may be when addressing the issue of human locomotion and development. The first point is to know whether the notion of a CPG can be extended to humans. Although this problem is still under debate, there is an increasing amount of evidence showing that comparable mechanisms exist in humans (for a review see Duysens & Van de Crommert, 1998). Since no direct experimental demonstrations are possible in this case, the most useful data available come from clinical findings on patients with a complete spinal cord section. First, it has been observed in human paraplegics that applying electrical stimulation to the flexor reflex afferents (FRA) induces a late flexion reflex similar to that observed in acute spinal cats with L-DOPA (Roby-Brami & Bussel, 1987). The same research group observed that a patient with a cervical spinal cord transection still produced rhythmic movements of the trunk and lower limbs, which shows that the spinal cord deprived of its supraspinal centres still keeps the ability to generate rhythmic movements (Bussel *et al.*, 1988, 1996). Further convincing evidence on these lines has recently been obtained. The locomotor ability of patients with complete spinal cord section was investigated by performing direct epidural stimulation of the spinal cord at various levels from T10 (thoracic level 10) to S1 (sacral level 1). The authors reported that the electrical stimulation elicited locomotor-like EMG patterns and stereotyped stepping movements when it was delivered exactly at the level of the second lumbar segment (Dimitrijevic *et al.*, 1998). Other stimulation sites were ineffective in eliciting stepping activity. This finding is comparable to data in the rat (Cazalets *et al.*, 1995, 1996; see also the previous section), in which the lumbar segments L1–L2 have been identified as the key area for generating locomotor-activity. If these results obtained on humans as well as on the rat are confirmed, this indicates a striking similarity between the spinal structures of these two species.

Another point is the relevance of animal data to human ontogeny. A common feature is the early appearance of spontaneous movements during fetal life since they are observed at 7–8 weeks after conception in humans (de Vries *et al.*, 1982; see Prechtl, 1997, for review). As in newborn animals, there is a continuum in the development of locomotor activity in humans. Newborn infants exhibit locomotor-like behaviour when supported (Forssberg, 1985; for review see Forssberg, 1999; Forssberg & Dietz, 1997). This suggests that, as observed in newborn rats, postural constraints limit the expression of existing locomotor networks. Like neonatal rats, however, the young infant has an immature locomotor pattern lacking the plantigrad foot placing (Forssberg, 1985). The harmonious development of the fluent locomotor behaviour both in adult humans and in adult rats results from a close interaction between postural and locomotor activities (Gramsbergen, 1998; Gramsbergen, this handbook).

CONCLUSION

The first concluding remark which has to be made is that the circuits underlying hindlimb locomotor-related movements in the rat are present from very early in life. This is several weeks before they are actually used for locomotion under "natural" conditions. Another striking characteristic of the development of these motor circuits is the almost perfect simultaneity in the development of all the requisite components. This shows that this development does not result from a linear sequence of successive phenomena, but from a parallel development of superimposed processes. The various sensorimotor components are not successively installed, but they all are simultaneously combined. In addition, the results discussed here show the high speed with which all these events take place during development in the rat.

The question may arise what the functional significance may be of this early development. Overall, these results demonstrate that the neuronal circuitry which underlies locomotor activity is established very early during development, although it is expressed only later in the intact animal, due to the postural constraints involved. Another point that has not been discussed here in great detail is the trophic role of the neurotransmitters. We have seen above how the various transmitter systems are installed at an early developmental stage, before the onset of specific functions, and how they subsequently control the locomotor networks. The actions of these various transmitters are multiple, however. In addition to, and prior to, their function in the regulation of interneuronal communications, they also play a role as trophic factors in the development and maturation of the nervous system (see chapters in Section II of this handbook).

References

Altman, J. & Sudarshan, K. (1975). Postnatal development of locomotion in the laboratory rat. *Animal Behaviour*, **23**, 896–920.

Atsuta Y., Garcia-Rill E. & Skinner, R.D. (1990). Characteristics of electrically induced locomotion in rat in vitro brain stem-spinal cord preparation. *Journal of Neurophysiology*, **64**, 727–735.

Atsuta, Y., Abraham, P., Iwahara, T., Garcia-Rill, E. & Skinner, R.D. (1991). Control of locomotion *in vitro*: II. Chemical stimulation. *Somatosensory Motor Research*, **8**, 55–63.

Barber, R.P., Vaughn, J.E. & Roberts, E. (1982). The cytoarchitecture of GABAergic neurons in rat spinal cord. *Brain Research*, **238**, 305–328.

Barthe, J-Y. & Clarac, F. (1997). Modulation of the spinal network for locomotion by substance P in the neonatal rat. *Experimental Brain Research*, **115**, 485–492.

Bekoff, A. & Lau, B. (1980). Interlimb coordination in 20-days-old rat fetuses. *Journal of Experimental Zoology*, **214**, 173–175.

Bekoff, A. & Trainer, W. (1979). The development of interlimb coordination during swimming in postnatal rats. *Journal of Experimental Biology*, **83**, 1–11.

Bertrand, S. & Cazalets, J-R. (1999). Presynaptic GABAergic control of the locomotor drive in the isolated spinal cord of neonatal rats. *European Journal of Neuroscience*, **11**, 583–592.

Bregman, B.S. (1987). Development of serotonin immunoreactivity in the rat spinal cord and its plasticity after neonatal spinal cord lesions. *Developmental Brain Research*, **34**, 245–263.

Brocard, F., Vinay, L. & Clarac, F. (1999). Gradual development of the ventral funiculus input to lumbar motoneurons in the neonatal rat. *Neuroscience*, **90**, 1543–1554

Brown, T.G. (1914). On the nature of the fundamental activity of the nervous centres, together with an analysis of the conditioning of rhythmic activity in progression, and a theory of evolution of function in the nervous system. *Journal of Physiology (London)*, **48**, 18–46.

Bussel, B., Roby-Brami, A., Azouvi, P., Biraben, A., Yakovleff, A. & Held, J.P. (1988). Myoclonus in a patient with spinal cord transection. Possible involvement of the spinal stepping generator. *Brain*, **111**, 1235–1245.

Bussel, B., Roby-Brami, A., Neris, O.R. & Yakovleff, A. (1996). Evidence for a spinal stepping generator in man. Electrophysiological study. *Acta Neurobiologiae Experimentalis (Warsaw)*, **56**, 465–468.

Cazalets, J-R. (1999). Organization of the spinal locomotor network in neonatal rat. In R. Kalb & S.M. Stritmatter (Eds.) *Neurobiology of spinal cord injury* (pp. 89–111). Totowa, NJ: Humana Press.

Cazalets, J-R., Borde, M. & Clarac, F. (1995). Localization of the central pattern generator for hindlimb locomotion in newborn rat. *Journal of Neuroscience*, **15**, 4943–4951.

Cazalets, J-R., Borde, M. & Clarac, F. (1996). The synaptic drive from the locomotor network to the motoneurons in newborn rat. *Journal of Neuroscience*, **16**, 298–306.

Cazalets, J.R., Sqalli-Houssaini, Y. & Clarac, F. (1992). Activation of the central pattern generators for locomotion by serotonin and excitatory amino acids in neonatal rat. *Journal of Physiology (London)*, **455**, 187–204.

Cazalets, J.R., Sqalli-Houssaini, Y. & Clarac, F. (1994). GABAergic inactivation of the central pattern generators for locomotion in isolated neonatal rat spinal cord. *Journal of Physiology (London)*, **474**, 173–181.

Cazalets, J.R., Ménard, I., Crémieux, J. & Clarac, F. (1990). Variability as a characteristic of immature motor systems: an electromyographic study of swimming in the newborn rat. *Behavioural Brain Research*, **40**, 215–225.

Cohen, A.H. & Gans, C. (1975). Muscle activity in rat locomotion: movement analysis and electromyography of the flexors and extensors of the elbow. *Journal of Morphology*, **146**, 177–196.

Cowley, K.C. & Schmidt, B.J. (1997). Regional distribution of the locomotor pattern-generating network in the neonatal rat spinal cord. *Journal of Neurophysiology*, **77**, 247–259.

Curtis, D.R., Phillis, J.W. & Watkins, J.C. (1959). The depression of spinal neurons by γ-amino-n-butyric acid and β-alanine. *Journal of Physiology (London)*, **146**, 185–203.

deLeon R., Hodgson, J.A, Roy, R.R. & Edgerton, V.R. (1994). Extensor- and flexor-like modulation within motor pools of the rat hindlimb during treadmill locomotion and swimming. *Brain Research*, **654**, 241–250.

de Vries, J.I., Visser, G.H. & Prechtl, H.F. (1982). The emergence of fetal behaviour. I. Qualitative aspects. *Early Human Development*, **7**, 301–322.

Dimitrijevic, M.R., Gerasimenko, Y. & Pinter, M.M. (1998). Evidence for a spinal central pattern generator in humans. *Annals of the New York Academy of Sciences*, **860**, 360–376.

Duysens J. & Van de Crommert, H.W. (1998). Neural control of locomotion; The central pattern generator from cats to humans. *Gait and Posture*, **7**, 131–141.

Eccles, J.C. (1964). *The physiology of synapses*. Berlin: Springer-Verlag.

Engberg, I. and Lundberg, A. (1969). An electromyographic analysis of muscular activity in the hindlimb of the cat during unrestrained locomotion. *Acta Physiologica Scandinavica*, **75**, 614–630.

Eide A.L., Glover J., Kjaerulff O. & Kiehn O. (1999). Characterization of commissural interneurons in the lumbar region of the neonatal rat spinal cord. *Journal of Comparative Neurology*, **403**, 332–345.

Fady, J-C., Jamon, M. & Clarac, F. (1998). Early olfactory-induced rhythmic limb activity in the newborn rat. *Developmental Brain Research*, **108**, 111–123.

Fitzgerald, M. (1987). Spontaneous and evoked activity of fetal primary afferents *in vivo*. *Nature*, **326**, 603–605.

Forssberg, H. (1985). Ontogeny of human locomotor control. I. Infant stepping, supported locomotion and transition to independent locomotion. *Experimental Brain Research*, **57**, 480–493.

Forssberg, H. (1999). Neural control of human motor development. *Current Opinion in Neurobiology*, **9**, 76–82.

Forssberg, H. & Dietz, V. (1997). Neurobiology of normal and impaired locomotor development, in K.J. Connolly & H. Forssberg (Eds.), *Neurophysiology and neuropsychology of motor development* (pp. 78–100). London: MacKeith Press.

Goudard, I., Orsal, D. & Cabelguen, J-M. (1992). An electromyographic study of the hindlimb locomotor movements in the acute thalamic rat. *European Journal of Neuroscience*, **4**, 1130–1139.

Gramsbergen, A. (1998). Posture and locomotion in the rat: independent or interdependent development? *Neuroscience and Biobehavioural Reviews*, **22**, 547–553.

Grillner, S. (1981). Control of locomotion in bipeds, tetrapods and fish. In J.M.Brookhart & V.B. Mountcastle (Eds.) *Handbook of physiology. The nervous system II* (pp. 1179–1236). Maryland: American Physiological Society.

Grillner, S. & Matsushima, T. (1991). The neural network underlying locomotion in lamprey-synaptic and cellular mechanisms. *Neuron*, **7**, 1-15.

Gruner, J.A. & Altman, J. (1980). Swimming in the rat: analysis of locomotor performance in comparison to stepping. *Experimental Brain Research*, **40**, 374–382.

Hakamada, S., Hayakawa, F., Kuno, K. & Tanaka, R. (1988). Development of the monosynaptic reflex pathway in the human spinal cord. *Brain Research*, **470**, 239–246.

Ho, S. & O'Donovan, M.J. (1993). Regionalization and intersegmental coordination of rhythm-generating networks in the spinal cord of the chick embryo. *Journal of Neuroscience*, **13**, 1354–1371.

Holstege, J.C. (1991). Ultrastructural evidence for GABAergic brain stem projections to spinal motoneurons in the rat. *Journal of Neuroscience*, **11**, 159–167.

Iizuka, M., Nishimaru, H. & Kudo, N. (1998). Development of the spatial pattern of 5-HT-induced locomotor rhythm in the lumbar spinal cord of rat fetuses *in vitro*. *Neuroscience Research*, **31**, 107–111.

Iwahara, T., Atsuta, Y., Garcia-Rill, E. & Skinner, R.D. (1992). Spinal cord stimulation-induced locomotion in the adult cat. *Brian Research Bulletin*, **28**, 99–105.

Jahr, C.E. & Yoshioka, K. (1986). Ia afferent excitation of motoneurones in the in-vitro newborn rat spinal cord is selectively antagonized by kynurenate. *Journal of Physiology (London)*, **370**, 515–530.

Jamon, M. & Clarac, F. (1998). Early walking in the neonatal rat: a kinematic study. *Behavioural Neuroscience*, **112**, 1218–1228.

Kiehn, O. & Kjaerulff, O. (1996). Spatiotemporal characteristics of 5-HT and dopamine-induced rhythmic hindlimb activity in the *in vitro* neonatal rat. *Journal of Neurophysiology*, **75**, 1472–1482.

Kjaerulff, O. & Kiehn, O. (1994). Localization of the central pattern generator for hindlimb locomotion in the neonatal rat *in vitro*. A lesion study. *Society of Neuroscience Abstracts*. 1757.

Kjaerulff, O. & Kiehn, O. (1996). Distribution of networks generating and coordinating locomotor activity in the neonatal rat spinal cord *in vitro*: a lesion study. *Journal of Neuroscience*, **16**, 5777–5794.

Kjaerulff, O. & Kiehn, O. (1997). Crossed rhythmic synaptic input to motoneurons during selective activation of the contralateral spinal locomotor network. *Journal of Neuroscience*, **17**, 9433–9447.

Kjaerulff, O., Barajon, I. & Kiehn, O. (1994). Sulphorhodamine-labelled cells in the neonatal rat spinal cord following chemically induced locomotor activity *in vitro*. *Journal of Physiology (London)*, **478**, 265–273.

Kudo, N. & Yamada, T. (1985). Development of the monosynaptic stretch reflex in the rat: an in vitro study. *Journal of Physiology (London)*, **369**, 127–144.

Kudo, N. & Yamada, T. (1987a). Morphological and physiological studies of development of the monosynaptic reflex pathway in the rat lumbar spinal cord. *Journal of Physiology (London)*, **389**, 441–459.

Kudo, N. & Yamada, T. (1987b). *N*-Methyl-D,L-aspartate-induced locomotor activity in a spinal cord–hindlimb muscles preparation of the newborn rat studied *in vitro*. *Neuroscience Letters*, **75**, 43–48.

Kudo, N., Ozaki, S. & Yamada, T. (1991). Ontogeny of rhythmic activity in the spinal cord of the rat. In M. Shimamura, S. Grillner & V. R. Edgerton (Eds.), *Neurobiological basis of human locomotion* (pp. 127–136). Tokyo: Japan Scientific Societies Press.

Kudo, N., Furukawa, F. & Okado, N. (1993). Development of descending fibers to the rat embryonic spinal cord. *Neuroscience Research*, **16**, 131–141.

Kremer, E., & Lev-Tov, A. (1997). Localization of the spinal network associated with generation of hindlimb locomotion in the neonatal rat and organization of its transverse coupling system. *Journal of Neurophysiology*, **77**, 1155–1170.

Lauder, J.M., Han, V.K., Henderson, P., Verdoorn, T. & Towle, A.C. (1986). Prenatal ontogeny of the GABAergic system in the rat brain: an immunocytochemical study. *Neuroscience*, **19**, 465–493.

Leong, S.K., Shieh, J.Y. & Wong, W.C. (1984). Localizing spinal-cord-projecting neurons in neonatal and immature albino rats. *Journal of Comparative Neurology*, **228**, 18–23.

Ma, W., Behar, T. & Barker, J.L. (1992). Transient expression of GABA immunoreactivity in the developing rat spinal cord. *Journal of Comparative Neurology*, **325**, 271–290.

Magnuson, D.S. & Trinder, T.C. (1997). Locomotor rhythm evoked by ventrolateral funiculus stimulation in the neonatal rat spinal cord *in vitro*. *Journal of Neurophysiology*, **77**, 200–206.

Mandler, R.N., Schaffner, A.E., Novotny, E.A., Lang, G.D., Smith & S.V., Barker, J.L. (1990). Electrical and chemical excitability appear one week before birth in the embryonic rat spinal cord. *Brain Research*, **522**, 46–54.

Ménard, I., Cremieux, J. & Cazalets, J-R. (1991). Evolution non linéaire de la fréquence des mouvements de nage pendant l'ontogenèse chez le rat. *Comptes Rendus de l'Académie des Sciences*, **312**, 233–240.

Narayanan, C.H., Fox, M.W. & Hamburger, V. (1971). Prenatal development of spontaneous and evoked activity in the rat (*Rattus norvegicus albinus*). *Behaviour*, **40**, 100–134.

Nicolopoulos-Stournaras, S. & Iles, J.F. (1984). Hindlimb muscle activity during locomotion in the rat (*Rattus norvegicus*) (Rodentia: Muridae). *Journal of Zoology (London)*, **203**, 427–440.

Nishimaru, H., Iizuka, M., Ozaki, S. & Kudo, N. (1996). Spontaneous motoneuronal activity mediated by glycine and GABA in the spinal cord of rat fetuses *in vitro*. *Journal of Physiology (London)*, **497**, 131–143.

Noga, B.R., Kettler, J. & Jordan, L.M. (1988). Locomotion produced in mesencephalic cats by injections of putative transmitter substances and antagonists into the medial reticular formation and the pontomedullary locomotor strip. *Journal of Neuroscience*, **8**, 2074–2086.

Nornes, H.O. & Das, G.D. (1974). Temporal pattern of neurogenesis in spinal cord of rat. I. An autoradiographic study – time and sites of origin and migration and settling patterns of neuroblasts. *Brain Research*, **73**, 121–138.

Orsal, D., Perret, C. & Cabelguen, J.M. (1986). Evidence of rhythmic inhibitory synaptic influences in hindlimb motoneurons during fictive locomotion in the thalamic cat. *Experimental Brain Research*, **64**, 217–224.

Ozaki, S., Yamada,T., Iizuka, M., Nishimaru, H. & Kudo, N. (1996). Development of locomotor activity induced by NMDA receptor activation in the lumbar spinal cord of the rat fetus studied *in vitro*. *Developmental Brain Research*, **97**, 118–125.

Prechtl, H.F.R. (1997). The importance of fetal movements. In K.J. Connolly & H. Forssberg (Eds.), *Neurophysiology and neuropsychology of motor development* (pp. 42–53). London: MacKeith Press.

Rajaofetra, N., Sandillon, F., Geffard & M., Privat, A. (1989). Pre- and post-natal ontogeny of serotonergic projections to the rat spinal cord. *Journal of Neuroscience Research*, **22**, 305–321.

Rajaofetra, N., Poulat, P., Marlier, L., Geffard, M. & Privat, A. (1992). Pre- and postnatal development of noradrenergic projections to the rat spinal cord: an immunocytochemical study. *Developmental Brain Research*, **67**, 237–246.

Roberts, A., Soffe, S.R. & Perrins, R. (1997). Spinal networks controlling swimming in hatchling *Xenopus* tadpoles. In P.S.G. Stein, S. Grillner, A.I. Selverston & G.G.Stuart (Eds.). *Neurons, networks, and motor behavior* (pp. 83–89). Cambridge, MA: Bradford Book MIT Press.

Robertson, S.S. & Smotherman, W.P. (1990). The neural control of cyclic motor activity in the fetal rat (*Rattus norvegicus*). *Physiology and Behaviour*, **47**, 121–126.

Roby-Brami, A. & Bussel, B. (1987). Long-latency spinal reflex in man after flexor reflex afferent stimulation. *Brain,* **110**, 707–725.

Rossignol, S. (1996). Neural control of stereotypic limb movements. In B. Rowell & J.T. Sheperd (Eds.), Handbook of physiology, section 12: *Exercise: Regulation and integration of multiple systems* (pp. 173–216). Oxford: American Physiological Society.

Saito, K. (1979). Development of spinal reflexes in the rat fetus studied *in vitro*. *Journal of Physiology (London)*, **294**, 581–594.

Shieh, J.Y., Leong, S.K. & Wong, W.C. (1983). Origin of the rubrospinal tract in neonatal, developing, and mature rats. *Journal of Comparative Neurology*, **214**, 79–86.

Shik, M.L. & Orlovsky, G.N. (1976). Neurophysiology of locomotor automatism. *Physiology Reviews*, **56**, 465–501.

Sinnamon, H.M. (1993). Preoptic and hypothalamic neurons and the initiation of locomotion in the anesthetized rat. *Progress in Neurobiology*, **41**, 323–344

Smith, J.C., Feldman, J.L. & Schmidt, B.J. (1988). Neural mechanisms generating locomotion studied in mammalian brain stem-spinal cord in vitro. *FASEB Journal*, **2**, 2283–2288.

Sqalli-Houssaini, Y., Cazalets J.R. & Clarac, F. (1993). Oscillatory properties of the central pattern generator for locomotion in neonatal rats. *Journal of Neurophysiology*, **70**, 803–813.

Van Hartesveldt, C., Sickles, A.E., Porter, J.D. & Stehouwer, D.J. (1991). L-dopa-induced air-stepping in developing rats. *Developmental Brain Research*, **58**, 251–255.

Vinay, L., Cazalets, J-R. & Clarac F. (1995). Evidence for the existence of a functional polysynaptic pathway from trigeminal afferents to lumbar motoneurones in the neonatal rat. *European Journal of Neuroscience*, **7**, 143–151.

Wallis, D.I. (1994). 5-HT receptors involved in initiation or modulation of motor patterns: opportunities for drug development. *Trends in Pharmacology Sciences*, **15**, 288–292.

Westerga, J. & Gramsbergen, A. (1990). The development of locomotion in the rat. *Developmental Brain Research*, **57**, 163–174.

Westerga, J. & Gramsbergen, A. (1993). Changes in the electromyogram of two major hindlimb muscles during locomotor development in the rat. *Experimental Brain Research*, **92**, 479–88.

Wu, W.L., Ziskind-Conhaim, L. & Sweet, M.A. (1992). Early development of glycine- and GABA-mediated synapses in rat spinal cord. *Journal of Neuroscience*, **12**, 3935–3945.

Ziskind-Conhaim, L. (1988). Electrical properties of motoneurons in the spinal cord of rat embryos. *Developmental Biology*, **28**, 21–29.

Ziskind-Conhaim, L. (1990). NMDA receptors mediate poly- and monosynaptic potentials in motoneurons of rat embryos. *Journal of Neuroscience*, **10**, 125–135.

III.5
Neuro-ontogeny of Motor Behaviour in the Rat

Experimental studies in developing rats for understanding human motor development

ALBERT GRAMSBERGEN

ABSTRACT

The investigation of motor behaviour during the rat's development in relation to the neuroanatomy and neurophysiology of its motor systems and neuromuscular interactions has provided important insights into their mutual relations. In this chapter investigations into the fetal development of movement patterns in the rat are reviewed and many aspects of this development show striking parallels to those in the human. After birth the development of movements, particularly at later stages, is strongly dependent upon postural development and this also holds similarly for the development of reaching, standing and walking in the human. The neural and muscular systems involved in postural control have a different evolutionary history and during development they initially develop independently. Only at later stages does an intimate interdependence between movements and postural control emerge, which then allows feed-forward control of movements.

The similarities in many aspects of the development of motor behaviour in the rat and the human offer promising possibilities for extrapolating results from neurophysiological and neuroanatomical investigations, provided that analogous developmental stages are compared and differences in the temporal aspects of processes are taken into account.

INTRODUCTION

Our knowledge on the early stages of behavioural development in the human has increased enormously in recent decades. New technologies such as ultrasound

467

A.F. Kalverboer and A. Gramsbergen (eds.), Handbook of Brain and Behaviour in Human Development, 467–512
© 2001 Kluwer Academic Publishers. Printed in Great Britain.

scanning of fetal movements; videorecording of movements in the postnatal period which allows the replay of movements at normal and slow speeds; kinematic and kinesiological recording and analysis techniques have revealed details of this development which were obscured until a few decades ago. This development was paralleled by the new insights into neural circuitries in the nervous system which were obtained with neuroanatomical tracers and immunological techniques in experimental animals. More recently, details on molecular aspects of signal transmission and modulation of cellular interactions, as well as genetical aspects of early development, were revealed, and these again have drastically influenced our concepts on neural functioning over the past 10 years.

The nature of the methodologies which are involved in neuroanatomical research and in molecular biology do not allow us to study developmental aspects of circuitry and cellular interactions in the human. Until recently, structure–function relations in human neuro-ontogeny depended mainly on extrapolations from post-mortem studies. Although the visualization of brain lesions and metabolic processes with modern imaging techniques and PET scans now provides important new insight, the interpretation of these data largely depends on research in animals. In animal experiments, behavioural development can be related to cross-sectional data on neuroanatomy or neurochemistry, and such experiments also allow us to interfere experimentally with normal development under standardized conditions. However, extrapolation of the results to humans poses important problems.

The first problem is to what extent behaviours and neural structures in humans are analogous to those in animals. The ecological biotope and behavioural competence are mutually interconnected with neural organization. The behavioural repertoire in humans and animals differs importantly, and this consequently is paralleled by important differences in the nervous system. The relation between manipulative skills and the projection pattern of the cortiocospinal tract is a case in point. In higher primates, monosynaptic connections exist between the motor cortex and spinal motoneurones and these probably enable monkeys, apes and humans to move their fingers independently and to manipulate objects (Kuypers, 1962; Lawrence & Hopkins, 1976; Lemon et al., 1997; Armand, this handbook). Rabbits, on the other hand, are unable to manipulate objects by means of independent movements of the toes of the foreleg. The corticospinal tract in these animals descends only as far as the upper cervical segments and does not reach the cervical intumescence, in which the foreleg motoneurons are localized (Hobbelen et al., 1992). This implies that the cortex in these animals can only indirectly be involved in foreleg and finger movements, and this becomes apparent in the nature of their paw movements.

Another problem of extrapolation, which particularly applies to neuro-ontogenetic research, is the timing of birth in relation to neural development. The length of the fetal period and the stage of brain development at birth differ importantly between animal species and man (Dobbing, 1981) and this implies, for example, that changes in sensory input and nutrition related to birth take place at different phases of neural development. Most importantly in the perspective of brain development, however, "*stages, not ages* should be compared" (J. Dobbing, personal communication).

Investigations into the normal and experimentally disturbed development of the nervous system have often been performed in rats. The main reason for this is that rats are born at an early stage of brain development, and this allows us to study important apects of the early development in the postnatal period. A wealth of data has been collected with regard to fundamental aspects of the development of neural and muscular development in this species, and also on neuroanatomy and neurophysiology at adult age. The investigation of structure–function relationships during motor development therefore offers many perspectives for unravelling its principles in this species.

In this chapter I will review the development of motor behaviour in rats in relation to morphological and neurophysiological changes. This development will be discussed in the perspective of human neuro-ontogeny.

THEORETICAL BACKGROUND

History of movement research

Movements indicate life, and the onset of movements has therefore intrigued scientists over a long period of time. The relation between form and function, the animation of form, and motion as expression of the soul were key notions in the thinking of Aristotle (384–322 BC), and his authority was appreciated until the 16th century. A few centuries later, Galen (AD 130–203) investigated the prenatal development of chickens and mammals, and incidentally reference was made to their motility at early stages of development (see Adelman's foreword (1967) in Fabricius' *De Formatione Ovi et Pulli* and *De Formato Foetu*).

Interest in the significance of the nervous system in governing movements originates from more recent times. The distinction between voluntary and involuntary movements, and the parts of the nervous system involved in both these types of movements, were important issues in debates from the end of the 17th century. Voluntary movements were considered to be controlled by the brain, in contrast to involuntary movements and reflexes, which were thought to be produced by the spinal cord. Much later, Hall (1850) still considered the spinal cord and the spinal nerves as a system, acting independently from the brain. Hence, reflexes were a category of movements distinct from voluntary movements. Later, the concept of brain reflexes (next to the "absolute" or spinal reflexes) emerged (Pavlov, 1927). The brain according to Pavlov takes an active part in conditioned reflexes and the animal adapts to its environment by becoming sensitive to the stimuli which are most effective, not only in so-called vital functions but also in the domain of voluntary movements. Sherrington (1906), in contrast, considered reflexes as basic elements of neural functioning. Simple reflexes were integrated into a coordinated action. Not the environment (as Pavlov theorized) but the brain is the ordering force in behaviour (for extended accounts on this history see Jeannerod, 1985; Brazier, 1988; Jacobson, 1993). This history of neurophysiological thinking provides a perspective for understanding the controversies which emerged during the study of the early development of movements.

Scientific interest into the development of the nervous system started only at the end of the 19th century (see Hamburger, 1988). The experimental neuroembryology as it emerged was directed both towards the developmental process in its own right and to the relationship between brain and behaviour during neuro-ontogeny. As the relatively simple organization of the developing nervous system increases in its complexity, and simultaneously the functional repertoire expands, this offers opportunities to unravel their mutual relationships. Both meticulous descriptions of normal development and investigation into the effects of experimentally disturbed development have vastly expanded our knowledge in the 20th century (Hamburger, 1988).

Concepts on the development of movements

Behavioural embryology in the first half of the 20th century was overshadowed by two problems which to a certain extent were interrelated. The first was whether movements occur spontaneously or whether they are elicited. Preyer (1885), an obstetrician working in Vienna, concluded on the basis of his observations in human pregnant females and also in chick embryos that movements emerge spontaneously, and that they are not induced by external stimuli. This conclusion was neglected for a long time. The conception that the central nervous system spontaneously and autonomously generates activity conflicted with the *Zeitgeist* of the early decades of the 20th century. In that period behaviourism prevailed, which in turn was a reaction to intuitive interpretations of brain functions as they were proposed in the 19th century. It is generally accepted now that motility in the behavioural development of mammals may emerge spontaneously and without any external stimuli (Hamburger, 1963, 1973; Oppenheim, 1974; Prechtl, 1984).

The other problem was whether movements from the outset are part of patterns, or whether they are strictly localized and perhaps reflexes. Angulo y Gonzalez (1932) studied fetal motility in rats, and reported that the first movements basically are mass patterns, involving larger parts of the body. These results were in line with earlier results from investigations by Coghill in the salamander (for a summary of these results see Coghill, 1929). Coghill had observed that already the first movements in these species involve greater parts of the body while local movements in extremities only develop at later stages. In Coghill's opinion even these local movements involve a pattern, consisting of the activation of certain parts and simultaneously the inhibition of other parts of the body (see also Oppenheim, 1978). On the other hand Windle, on the basis of his studies in cats (and other mammals including the human) concluded that the first movements in mammals are local movements (e.g. Windle & Griffin, 1931; see also Windle, 1940). From these observations the controversy emerged between Coghill and others, stating that behaviour was integrated from the first moment onwards, and Windle and others, who shared the opinion that integration was a secondary process (Windle, 1940; for a full account of this controversy see Oppenheim, 1978). These differing observations set the stage for the individuation–integration controversy which continued at least until he 1940s and which involved many investigators such as Kuo (1932, 1963) studying chick embryos, Barcroft & Barron (1939), fetal sheep, and Hooker (1952), human fetuses. Hamburger (1963) has pointed out that the differences between salamanders on the one hand and mammals on the other may largely explain the differences in the results. A distinguishing aspect between mammals and lower vertebrates is probably differences in the intersegmental connectivity by propriospinal inter-neurons (Holstege, 1988).

FETAL DEVELOPMENT OF MOVEMENTS

Gestational period and time of birth

Fetal ages are counted from the day the female rat is pregnant as indicated by the presence of sperm or the closure of the vagina by a plug. Alternatively, vaginal cells in smears are inspected histologically for the rat being in oestrus. The day of conception is referred to as day zero of pregnancy (Narayanan *et al.*, 1971), day zero of gestation (Smotherman & Robinson, 1988) and in this chapter as embryonic day 0 (E0). Parturition in rats takes place around the end of E21 and the beginning of E22 (Gramsbergen & Mulder, 1998; Lakke, 1997; Smotherman & Robinson, 1988a).

Methods to observe fetal movements

Most of the studies on the development of motility in fetal rats were based upon direct observations with a dissecting miscroscope under cold-light illumination; in later studies movements were also recorded on videotape. In these studies the uterus was exteriorized through the abdominal wall and the female's body and uterine horns were immersed in a thermostatically controlled bath of physio-logical saline at about 37°C. Either the fetuses were observed through the uterine wall or the uterus was opened while leaving the placental circulation intact. In the latter situation minute details of activity could be observed, but a side-effect was that the level of fetal activity increased (Narayanan *et al.*, 1971). This may point towards the influence of physical contraints on fetal activity when the fetuses remain within the uterus.

As anaesthetics pass the placental barrier the anaesthesia of the pregnant female is an important variable. General narcosis of the pregnant female makes it possible to close the abdominal wound after the observations and also to test the pups postnatally (e.g. Stickrod *et al.*, 1982), but fetal activity in this procedure is depressed (Kirby & Holtzman, 1982). In other methods the female's spinal cord is transected at lower thoracic levels (e.g. Kirby & Holtzman, 1982; Narayanan *et al.*, 1971; Smotherman & Robinson, 1984), or blocked chemically by a 100% ethyl-alcohol injection (Smotherman & Robinson, 1984). More recently it has been proven possible to reversibly block the spinal cord by lidocaine injections at lower levels (Smotherman *et al.*, 1986). Smotherman & Robinson (1986) observed fetal rat movements using an implanted endoscope. This method seems particu-larly suited for following quantitative aspects of fetal motility.

Motor development before birth

Angulo y Gonzalez (1932), Windle *et al.* (1935) and Windle & Baxter (1936) studied fetal behaviour in rats, particularly in reaction to tactile stimuli; in addition they described the spontaneous occurrence of movements. Narayanan and co-workers (1971), and more recently Smotherman & Robinson (1988) were primarily interested in qualitative and quantitative aspects of spontaneously occurring movements; additionally they also studied the development of extero-ceptive reflexes. Reports by Swenson (in the early 1930s; see Oppenheim, 1978), Carmichael (cited in Narayanan et al., 1971) and others, presently are difficult to obtain and were not considered.

Narayanan et al. (1971) categorized movements into total movements involving the whole body, regional movements in parts of the body as well as local movements in limb-segments, the face, the mouth, etc. They emphasized that all three categories were observed from the earliest stages of motor development onwards. Smotherman's group (Smotherman & Robinson, 1988) distinguished seven categories, movements of the head, mouth, foreleg and hindleg as well as flexion of the trunk ("curling"), extension ("stretch") and twitches. Complex movements were also considered, consisting of movements in two or more parts of the body.

The first movements are barely discernible and they occur late on E15 (Angulo y Gonzalez, 1932). From E16 a repertoire of movements in trunk, head, forelegs and hindlegs develops (Narayanan et al., 1971; Smotherman & Robinson, 1988). Movements at this stage mainly comprise lateral flexions of the head, uncoordinated wriggling movements of the trunk and movements of the forelegs. Movements of the arm are by far the most frequent on E16 and the following days, and Smotherman & Robinson calulated that, throughout pregnancy, over 45% of all movements occur in the forelegs. On E17 the head is moved in all directions and even rotatory head movements may be observed, as well as bouts of mouth and tongue movements. The foreleg movements are more vigorous at this age and hindleg movements may also now be observed, although the latter are distinctly less vigorous compared to movements in the forelegs. Trunk movements consist of irregular wriggles and no cephalocaudal waves of activity in the trunk occur. Movements at E18 and E19 are performed even more vigorously, and from E20 onwards they are noticeable for their smooth execution. At E20 intersegmental coordination within extremities has been demonstrated to occur (Bekoff & Lau, 1980) but interlimb coordination has not been observed prior to birth (Narayanan et al., 1971). Both Narayanan and co-workers and Smotherman & Robinson (1988) observed that mouth movements are most frequent on E20 and this, as well as their late emergence (on E17), indicates that the development of fetal movements does not follow a rostrocaudal gradient. Narayanan and co-workers concluded their studies at E20 but Smotherman & Robinson (1988) continued until E22 and during the last days of pregnancy they observed eye blinking, wiping of the face with one or both forelegs, and foreleg–mouth contacts. These patterns, however, occur infrequently.

Reflexes in the fetal period

Narayanan et al. (1971) studied the effects of reactions to tactile stimulations. At E16, stimulation of the vibrissae and the palmar surface of the forepaw leads to reactions such as head extension and flexion of the limbs. At E17 and E18 reactions to tactile stimulation of the face area have increased in intensity and then they often comprise generalized mass actions. This is in line with earlier descriptions by Angulo y Gonzalez (1932). It is interesting in this respect that electrical stimulation of trigeminal afferents in the neonatal rat even evoked excitatory responses in lumbar motoneurons (Vinay *et al.*, 1995). These responses are thought to be mediated by polysynaptic input through reticulospinal neurons located in the pontomedullary locomotor strip. Behaviourally, from E19 and particularly from E20, the reactions gradually become more localized and then they also decrease in vigour. The reactions to tactile stimulation roughly follow a cephalocaudal trend (with the forelegs being responsive before the hindlegs) but reactions to stimulation of the oral region develop relatively late.

Temporal organization of activity

Both Narayanan *et al.* (1971) and Smotherman & Robinson (1988) reported that fetal activity in general increases until E18 and decreases again thereafter. In both studies the temporal organization of motility was also analysed quantitatively. Narayanan and co-workers analysed 15-min periods (divided into 20-s bouts) and they found no indication for a clustering of activity with development. Smotherman & Robinson performed a similar statistical analysis but they based this on the actual duration of intervals. Their results indicate that from E21 a clustering of activity emerges (and this implies that intervals of silence also develop). No obvious precursor of a rest–activity cycle or a sleep–wakefulness cycle develops in the rat's fetal period. This lack of behavioural organization continues until after birth. Cycles of stable sleep and wakefulness states, characterized by specific sets of parameters (such as absence or presence of gross body movements, regular or irregular respiration) in rats develop only several days after birth (Gramsbergen *et al.*, 1970; Jouvet-Mounier *et al*, 1970).

Continuity of behaviour from fetal to postnatal life

Narayanan and co-workers (1971) raised the question of whether antecedents of postnatal behaviour patterns occur in the fetal period. Spontaneously occurring leg movements resembling locomotion in the postnatal period were not observed, but tactile stimulation could evoke withdrawal reactions as well as mouthing, face wiping and hindlimb scratching, and these movements are similar to those after birth. Smotherman & Robinson (1987) could elicit a variety of specific patterns such as the grasp reflex and face wiping during the last days of gestation (E21 and E22). As similar patterns also can be evoked in the postnatal period they took this to indicate a continuity in the development of such behaviours before and after birth.

Significance of fetal movements

The problem of the significance of fetuses moving before birth has been raised by several authors. One of the possibilities is that movements play a role in the adaptation of the fetus to the uterine environment. It has been suggested that movements may prevent adhesions of the fetus with the uterine wall (Oppenheim, 1981b). Strong evidence in the human exists that leg movements play an important role in attaining a vertex position in the last trimester of pregnancy (Prechtl, 1965). Breech positions may result from diminished space because of decreases in amniotic fluid or neurological abnormalities (Hytten, 1982; Tompkins, 1946). Leg and trunk movements in the rat are possibly also important in the birth process, as death shortly before or during birth severely hampers parturition (personal observation).

Another possibility is that certain movements are anticipatory to postnatal behaviour patterns. Human fetuses make breathing movements, and such movements have also been described in a variety of mammals such as fetal lambs (e.g. Dawes et al., 1972) and chick embryos (see Bekoff, this handbook). Such movements might be an exercise for the pattern which postnatally is of vital importance. In a similar vein, trunk and extremity movements in the fetal period might be important as a preparation for postnatal movements.

Strong evidence exists for neural and muscular activity at early stages playing an essential role in histogenesis and morphogenesis. As soon as developing neurons are integrated in neural circuits, action potentials are generated (Corner & Crain, 1964; Crain, 1964, 1966). It has been suggested that this endogenous activity might play a role in neurogenesis and the establishment of circuitry (Crain, 1966; Gottlieb 1976; Oppenheim, 1974) although a blockade of neural activity in other cases may still lead to a seemingly normal behavioural development (see Harrison, 1904). Muscular contractions have also been demonstrated to stimulate the formation of additional sarcomeres, to enhance muscular differentiation (Drachman, 1968) and to stimulate shaft bone growth (Moessinger, 1988). The blockade of movements by curare leads to malformed joints or even ankylosis (Drachman & Sokoloff, 1966). Neural activity at later stages of development has been shown to play an important role in the process of apoptosis (see Oppenheim, 1981a, 1991; Oppenheim, this handbook), in the regression of polyneural innervation of muscles (Ribchester, this handbook) and in dendritic arborization (O'Hanlon & Lowrie, 1994b).

In summary, it has been suggested that fetal movements are adaptive to the fetal environment and anticipatory to postnatal movement patterns. Evidence is accumulating that movements at early stages of development are essential for histogenesis and morphogenesis.

MOTOR DEVELOPMENT IN THE POSTNATAL PERIOD

Methodology

The postnatal development of movements and reflexes in rats has been studied repeatedly in the past century (e.g. Almli & Fisher, 1977; Altman & Sudarshan, 1975; Blanck et al., 1967; Bolles & Woods, 1964; Geisler et al., 1993; Gramsbergen et al., 1970; Kretschmer & Schwartze, 1974; Petrosini et al., 1990; Small, 1899; Smart & Dobbing, 1971; Tilney, 1934; Westerga & Gramsbergen, 1990). The aims of these investigations, and consequently the scope of the investigators, varied, however, and this hampers a straightforward comparison of the results. Small studied motor development as an aspect of a "comparative embryology of the soul" in the tradition of the 19th-century vitalism. His study, based on the diaries of five pups, is mainly concerned with sensory development, but in addition he briefly touches on the development of locomotion and equilibrium.

Most of the more recent studies have been performed in order to serve as a basis for comparisons with studies on motor development after experimental inter-ference (such as undernutrition, cerebellar lesions, X-irradiation or vestibular deprivation). As the effects of such treatments become apparent at different ages, or only in particular skills, the focus on the developmental stage and the rat's competence shifted accordingly. The results of these studies when considered collectively, however, provide a comprehensive description of postnatal motor development. Slight differences in the timetable of developmental transitions are probably related to differences in strain, breeding conditions and possibly the nutritional status of laboratory rats (the quality of rat food has increased con-siderably in the past century; J.L. Smart, personal communication).

In a few of these studies the rats were observed while remaining in their litter (e.g. Bolles & Woods, 1964) and in most they were studied when alone in a cage. As rats are poikilothermic in the early postnatal period, the ambient temperature, particularly when they are observed alone, is an important variable. Ambient temperatures sometimes were not mentioned, and in other studies the temperature was adjusted at 21–24.5°C (Altman & Sudarshan, 1975) or the temperature varied (Gramsbergen *et al.*, 1970), from 33°C at early ages and linearly decreasing to 26°C at older ages, which is about 2°C below the age-specific neutral temperature of the rats at these respective ages (Taylor, 1960).

The description of motor development in the next section is primarily based upon our own studies in rats of the white and black hooded Lister strain. We observed the rats when alone in a walking alley, at a few degrees below the age-specific neutral temperature. Their behaviour was recorded on videotape. In a few studies markers on their joints enabled kinematic analysis of movements and in other studies EMG recordings were made, in synchrony with videorecordings.

Postnatal development of motor behaviour

On the first postnatal day (P1) rats lift their head for short periods, and at P2 they make horizontal head movements (or rooting movements) and this lasts until between P4 and P7 when head movements in the vertical plane may also be observed (Altman & Sudarshan, 1975; Geisler *et al.*, 1993). In the first days they make alternating crawling movements with the forelegs but these movements are generally not effective in changing the rat's position. Coordination of all four legs has been observed at neonatal age during swimming in a waterbath (Bekoff & Trainer, 1979). As the trunk in this situation is supported by upward pressure of the water, this might indicate that immature postural control inhibits movements of the extremities and particularly of the hindlegs. However, Jamon & Clarac (1998) recently demonstrated that allowing neonatal pups to smell a cardboard cylinder filled with nesting material from the maternal litter induces vigorous walking movements, and even lifting of the ventral body surface from the floor. This result demonstrates that olfactory stimulation is a potent factor in activating trunk and extremity muscles, and that this stimulation, even in the first days of life, is able to induce walking movements.

In overground situations the hindpaws are involved in crawling only from P4, and this leads to so-called pivoting, in which the rat's body in prone position is turning in various orientations (Gramsbergen *et al.*, 1970; Gramsbergen &

Mulder, 1998). From P5 to P8 rats are able to stand with their ventral body surface off the floor (Geisler *et al.*, 1993). A few staggering steps are sometimes made but gradually these bouts of "free walking" increase in length. Westerga & Gramsbergen (1990), in black and white hooded Lister rats, and Bolles & Woods (1964) in Sprague-Dawley rats, observed longer bouts of walking from P11, and Altman & Sudarshan (1975) found this to occur from around P13 to P14. When rats scan their environment (at this age mainly with the olfactory and auditory systems and by tactile stimulation of their whiskers) they stand still and then move their heads in ventrodorsal and lateral directions.

From P15, in the course of 1 or 2 days, the immature pattern of locomotion is replaced by a faster and strikingly different walking pattern (Figure 1). A similar shift has also been reported by other investigators. Bolles & Woods (1964) reported that running occurs from P14 to P15, and Altman & Sudarshan observed fast locomotion from P16; these results probably refer to the same transition. After this shift rats walk in an adult-like fashion, they run smoothly, at higher speeds (or they walk at remarkably slow speed – Bolles & Woods, 1964) and they may suddenly vary the speed of walking (Westerga & Gramsbergen, 1990). Kinematic analysis shows that joint angle-traces which were not fluent before P15 have now become smooth (Westerga & Gramsbergen, 1990). Before P15 the hindlegs rotate outwardly during the stance phase but after P16 they remain adducted (Figure 2). From this age they are able to stand on their hindpaws (rearing) for extended periods of time, and head movements may occur independently from ongoing movements or postures. After this transition the walking speed increases still further and rats may continue walking or running for longer periods but, apart from this, no obvious changes in walking patterns can be noted. Clarke & Williams (1994) showed, however, that the pattern of foot placing still shows immature traits until P22 and that the adult foot-fall pattern develops only in the weeks thereafter (see also Clarke & Parker, 1986).

The transition in the walking pattern around P15–P16 is accompanied by changes in the EMG recorded from hindlimb muscles (Westerga & Gramsbergen, 1993, 1994). Recordings from the gastrocnemius, the soleus and the tibialis anterior muscles until P14 show an irregular EMG during walking with co-contractions in antagonistic muscles (Figure 3). Similar patterns of co-contraction in antagonistic leg muscles at early ages have also been observed in newborn human babies during the so-called infantile stepping response (Forssberg, 1985; see Gordon, this handbook). From P15 to P16 the EMG changes into an interference pattern, bursts are clearly delineated from then onwards with a sudden onset and a clear-cut end of the burst and without co-contractions in antagonistic muscles.

The shift in walking pattern is accompanied by important changes in feeding behaviour. Until the end of the second week of life rats are fed by their mother, and transportation to and from the litter is mainly effected by the mother carrying them at their neck. From P17 the pups start to chew solid rat food and from then they often leave their litter (Bolles & Woods, 1964; Smart & Dobbing, 1971).

The question is what induces the transition in walking pattern around P15. It has been suggested that the development of the smooth walking pattern is related to the opening of the eyes. Rats open their eyes around P15 (between 13.9 and

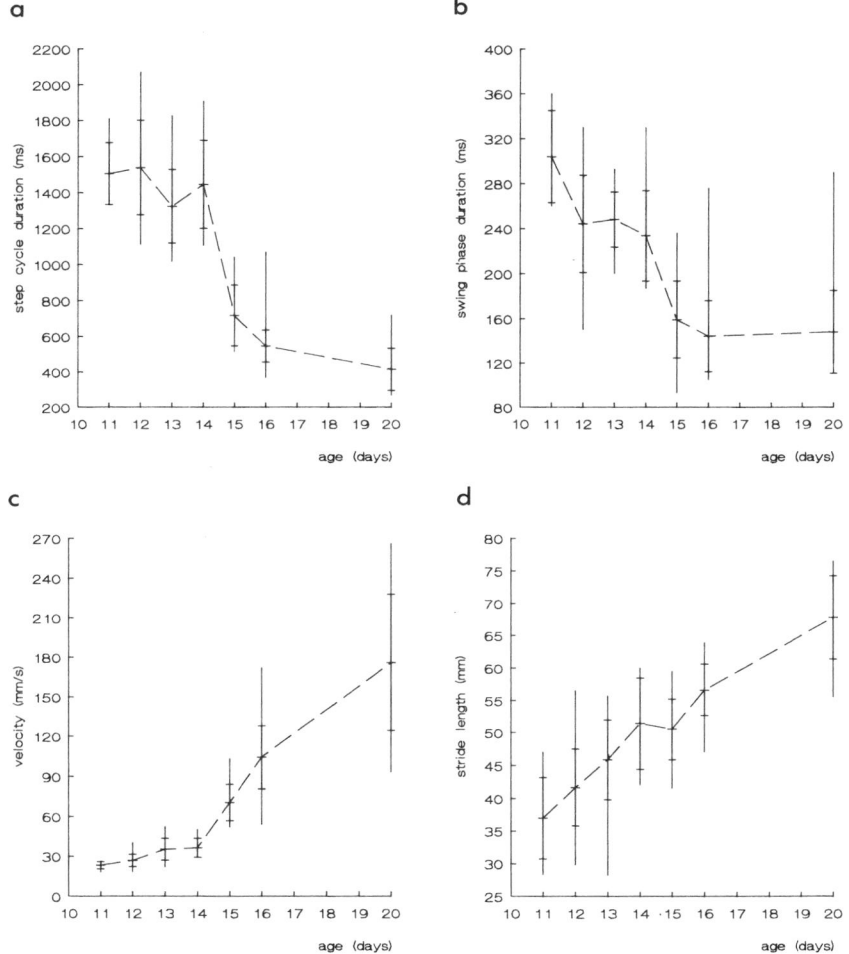

Figure 1 Mean, standard deviations and ranges of various parameters of locomotion at different ages: step cycle duration (**a**), swing phase duration (**b**), velocity (**c**) and stride length (**d**). Note that the durations of the step cycle and swing phase, as well as the velocity, change abruptly at P15, whereas the stride length increases more gradually. (Reproduced by courtesy of Elsevier.)

15.9 days; Gramsbergen & Mulder, 1998; see also Smart & Dobbing, 1971). In recent experiments we sutured the eyelids from P9 and in addition the rats were reared in a dark environment. Recordings of walking movements from P10 until P30 occurred in dim light conditions. This visual deprivation, however, did not postpone the emergence of the adult-like walking pattern or postural development, although it influenced the organization of motor behaviour (Gramsbergen, 2001b).

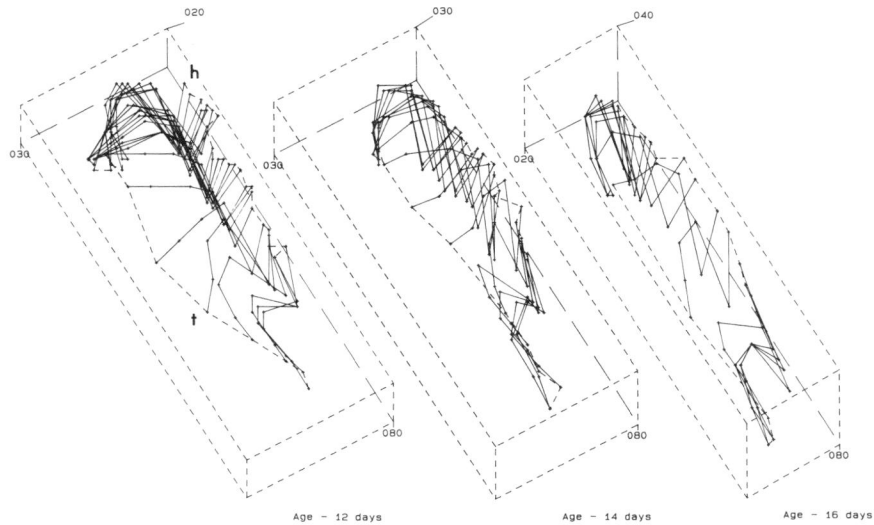

Figure 2 Three-dimensional reconstruction of movements of the right hindlimb of a rat at P12, P14 and P16. "h" represents the hip joint, and "t" the trajectory of the toe. Note the abduction of the hindlimb, particularly at P12 and still at P14, and the adducted hindlimb at P16. (Reproduced by courtesy of Elsevier.)

Another possibility is that this transition is effected by a sudden increase in the force which can be produced by muscles. Bennett *et al.* (1986) measured maximal tetanic force which can be produced by the lateral head of the gastrocnemius muscle around P10, and Close (1964) measured this force in the soleus muscle. These measurements indicate that muscle force at this age might be insufficient for bearing body weight for longer periods (note, however, that Jamon & Clarac, 1998, after olfactory stimulation, observed neonatal rats walking with the ventral body surface free from the floor). Thereafter the force which can be produced by muscles increases considerably. In the gastrocnemius muscle the maximal tetanic force has been shown to increase substantially from P10 and, theoretically, this force would be sufficient to bear body weight (Huiing *et al.*, unpublished results). These data point to the muscle force which is available around the end of the second week not being the limiting factor for the adult-like walking pattern to occur.

Still another factor might be changes in the properties of motoneurons. No data are available on membrane properties of motoneurons around P15. The data in the literature on neurons of neonatal rats (Fulton & Walton, 1987) and older rats suggest, however, that motoneuronal properties at all developmental ages are matched to muscle properties (for review see Kernell, 1998). This suggests that changes in motoneurons are not the limiting factor for producing tonic discharges, which are obviously needed for activation of the antigravity muscles.

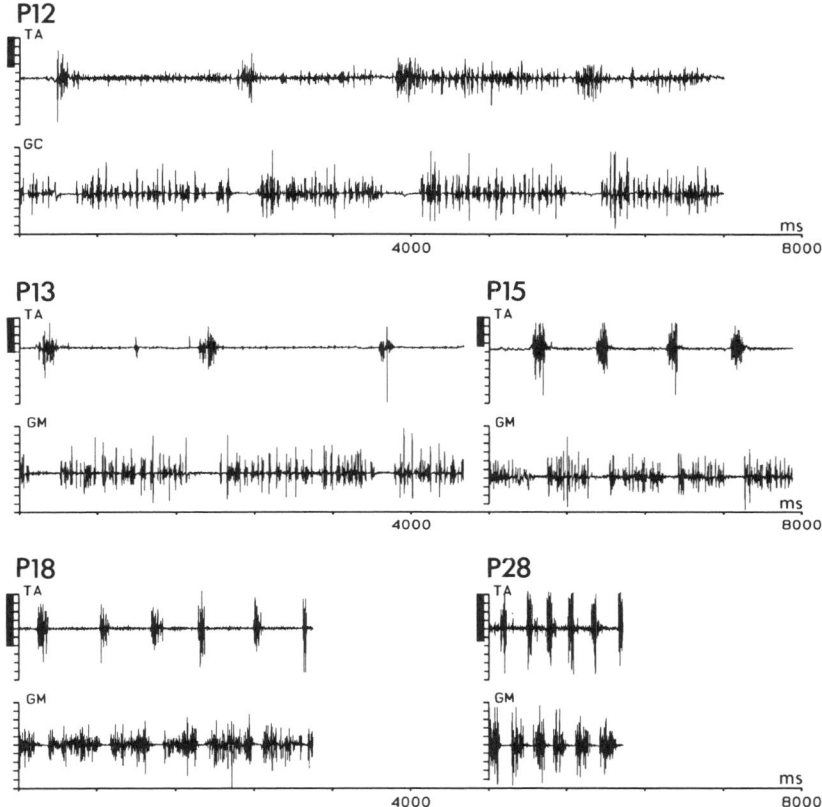

Figure 3 Representative EMGs of tibialis anterior (TA) and medial gastrocnemius (GM) muscles at various ages. The time scale is the same in all graphs. (Reproduced by courtesy of Springer Verlag).

We hypothesized that the sudden shift into the fluent walking pattern in rats is related to developmental changes in postural control (Geisler *et al.*, 1993; Gramsbergen, 1998). In order to check this possibility we recorded the EMG from postural muscles in the back, the medially located multifidus muscle and the laterally located longissimus muscles. Until P14–P15 the muscles are activated irregularly and without a consistent relation with the step cycle of the fore- or hindpaws but from P16 EMG bursts in the back muscles become linked to the paw movements during walking (Geisler *et al.*, 1996b). These results lend some support to the hypothesis of changes in postural control being pivotal in the transition in walking pattern. Greater support was gained by experiments into the effects of vestibular deprivation. In rats at P5 the semicircular canals were plugged (Geisler *et al.*, 1996a) and this prevents the circulation of endolymph in the semicircular canals and stimulation of the ciliae. This interference leads to a retardation in the development of postural development as indicated, e.g., by a

5-day retardation in the emergence of unsupported rearing. Pertinent to our hypothesis is that the adult-like type of smooth walking was also delayed, by about 3 days. In a subsequent study we demonstrated that vestibular deprivation also retards the development of EMG patterning in the longissimus muscles (Geisler & Gramsbergen, 1998). These results, in sum, strongly support the hypothesis that the development of postural control is one of the main factors in the transition from the immature pattern of locomotion into the adult-like type of walking.

Another transition occurs around the end of the third week of life. EMG recordings of leg and trunk muscles indicate that, from P16, the longissimus muscle is activated twice during the step cycle, i.e. both during the stance phase at the ipsilateral side and at the contralateral side. From P16 until P20 the bursts with the highest amplitudes are phase linked to the stance phase in the *contralateral* hindleg (Figure 4). Remarkably, however, after a transitional period of 1 or 2 days, the strongest activity is coupled to the stance phase in the *ipsilateral* hindleg, and this pattern is basically maintained until adulthood (Geisler *et al.*, 1996b; Gramsbergen *et al.*, 1999). This shift in the coupling between trunk and hindleg muscles around the end of the third week of life indicates a fundamental change in the interaction between postural control and locomotion. Human babies during their first year, when crawling on a flat surface, show a pattern which is similar to that in young rats and this pattern also strongly resembles the pattern which can be observed, for example, in adult amphibiae and lizards during locomotion. We interpret this result as follows. The contraction of the longissimus muscle and the extension of the contralateral hindleg before P20 indicate the active involvement of the (contralateral) trunk muscles in propulsion of the body After P21 the function of the trunk muscles has shifted towards the postural stabilization of the trunk, and locomotion is mainly effected by extremity movements. This shift is paralleled by a contraction of the ipsilateral trunk muscles from then onwards (for further discussion see Gramsbergen, 1998; Gramsbergen *et al.*, 1999).

In adult humans, when walking upright, the strongest activity in the longissimus muscle shortly precedes the stance phase of the contralateral leg (Thorstensson *et al.*, 1982). Similar results were obtained by Shapiro & Jungers (1994) in humans, apes and monkeys. This indicates that bipeds and quadrupeds differ in the nature of coupling of leg movements and the activation of the trunk muscles.

Research into the locomotion of adult rats has concentrated upon characteristics of the step cycle at higher speeds and foot-fall patterns during the different types of walking, trot and gallop. A detailed discussion of this literature is beyond the scope of this chapter (for reviews see Gramsbergen, 2001a; Grillner, 1981; Grillner *et al.*, 1991).

ASPECTS OF NEURAL AND MUSCULAR DEVELOPMENT IN THE RAT

In the following paragraphs a few selected aspects of the histogenesis of the rat's nervous and muscular system are discussed, the primary aim being to temporally relate these structural changes, particularly in the spinal cord and in muscles, to motor development. For detailed essays on neural development the reader is

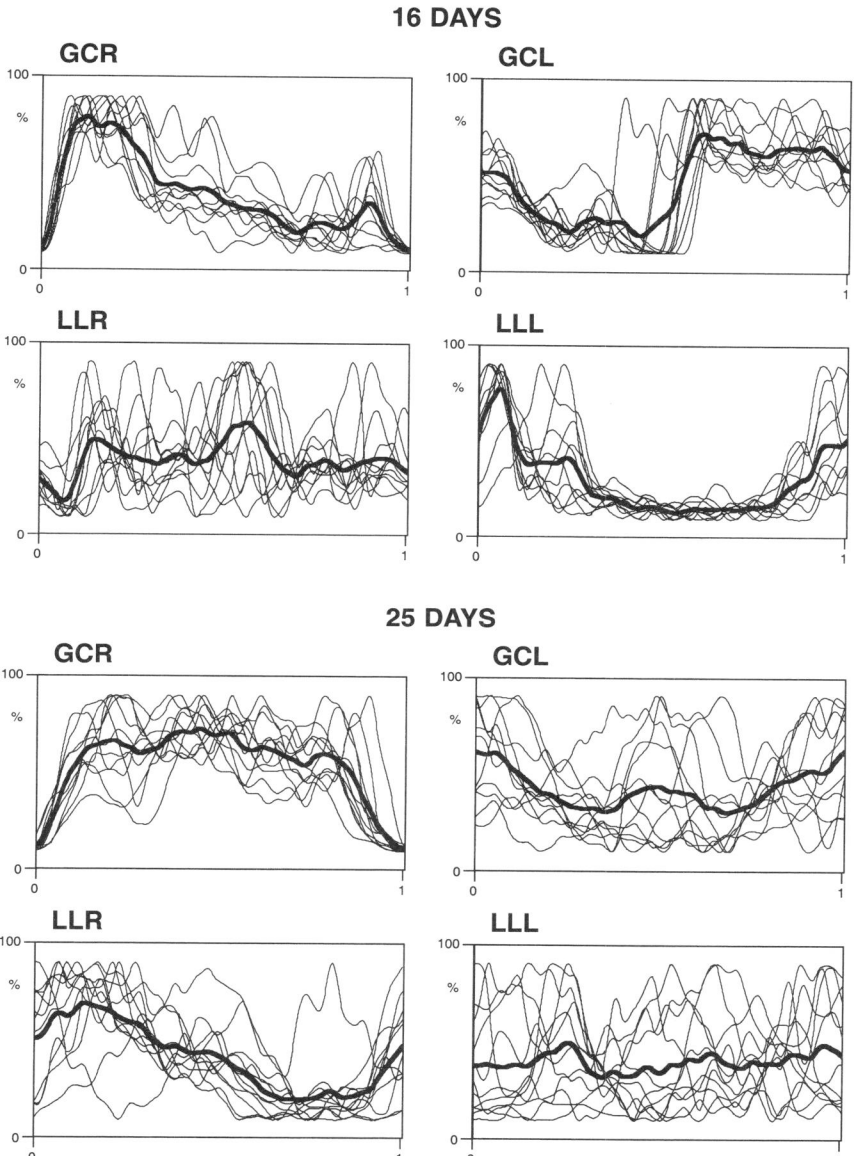

Figure 4 Superimposed plots of EMG bursts ($n = 11$) in the gastrocnemius muscle (GC) at the right side, the tibialis muscle (TA) at the right side and the longissimus muscles (LL) at a caudal level at the right and the left side. Plots depict EMG activity from the onset of the EMG burst in the GC muscle onto the onset of the next EMG burst. Ages of the rats, 16 days (upper panel) and 25 days (lower panel).

referred to the literature cited and the chapters in the section II of this handbook.

Proliferation and migration

The large motoneurons in the ventral parts of the ependymal layer in the spinal cord are generated between E11 and E13. Remarkably, the motoneurons, in the laterally located motoneuronal pools, which later innervate muscles in the extremities, are generated on E11 and E12, while the cells in the medially located pools which will innervate the filogenetically older axial muscles are produced 1 day later. The neuroblasts for the intermediate grey and the preganglionic motorneurons proliferate between E12 and E15 and the future dorsal horn neurons develop between E14 and E16 (Altman & Bayer, 1984; Nornes & Das, 1974). After proliferation the neurons migrate to their final destination, along radially oriented glial cells as has been shown in the mouse spinal cord (Henrikson & Vaughn, 1974). The results indicate a ventrodorsal gradient in the proliferation of neuroblasts. However, a rostrocaudal gradient in the production of motoneurons is less clear. The motoneurons for head and forelimbs develop early and before those of the hindlegs, but motoneurons for extra-ocular and facial muscles develop relatively late. This temporal order in proliferation precedes a similar order in the development of movements, those in the trunk and extremities emerging before those of the mouth and perioral area (see above).

The migration and differentiation of the neuroblasts in the spinal cord, on the other hand, do follow a rostrocaudal gradient. Neuroblasts in the cervical spinal cord start to differentiate on E12. In the lumbar intumescence this occurs on E14, and at that age those in sacral areas only start to migrate (Altman & Bayer, 1984). The motoneurons in the cervical spinal cord have reached their final and specific destinations already at E13.5 (Smith & Hollyday, 1983). As this occurs before neuromuscular junctions can be detected in the intercostal muscles (Dennis et al., 1981), this suggests that motoneurons and their axons are specified before their muscles (see also Altman & Bayer, 1984), rather than that a specification of motoneurons occurs by their peripheral connections (as has been postulated, e.g., by Romanes, 1941).

Outgrowth of motoneuronal axons

The first roots emanating from the ventral horn have been demonstrated at E12, which is shortly after the migration of the motoneurons to the ventral horn (Altman & Bayer, 1984). The axons growing out firstly are from motoneurons located ventrally in the ventral horn and these head towards groups of mesenchymal cells in the ventral plate. These muscles later develop into extensor muscles. The dorsally located motoneurons somewhat later reach groups of mesenchymal cells in the dorsal plate which are the precursors of flexor muscles (as has been shown in rat fetuses – Altman & Bayer, 1984; and in chick embryos – Bennett et al., 1980; Hollyday, 1985). The medioventrally located axons head for the mesenchymal cells in the segmentally arranged somites. These mesenchymal cells later give rise to the axial muscles.

482

Altman & Bayer (1984) detected in some regions of the spinal cord a lateral-to-medial gradient in the maturation of the spinal cord, and this seems to point towards an independent developmental course of two initially separate neuromuscular systems – the medial tier of motoneurons, subserving the future axial muscles and the lateral tier of motor columns related to the extremity muscles.

Recent years have seen an explosion of experimental data on principles of axonal pathfinding both in the central and in the peripheral nervous system. Now it is known that the outgrowth of the fibres is governed by an intricate interplay between attracting as well as repulsive factors, acting over relatively long ranges (netrins and semaphorins). Related to these are factors which lead to growth along common trajectories (fasciculation) and those that permit axons to divert from this path (defasciculation). Near their neuronal or muscular target the growth cone at the tip of the axon is attracted by cell adhesion molecules (CAMs) of the immunoglobulin superfamily (see Edelman, 1993) and of the cadherin superfamily (see Takeichi, 1995) which are located either on the cellular membrane or in the extracellular medium, but they may also be repelled, for example, by Eph's which belong to the receptor protein tyrosine kinases (Goodman & Tessier-Lavigne, 1997). The basic principles of axonal pathfinding were discovered mainly in insects, but evidence exists that these principles have been maintained during evolution (Goodman & Tessier-Lavigne, 1997; for further details see the chapters in section I of this handbook). It is remarkable that motoneuronal axons grow out and are directed when only the first condensations of shaft bones can be observed and before the stage when future muscles have started to develop. This suggests that the direction of axonal outgrowth, at least initially, is governed by tropic factors, other than those released by muscles (Altman & Bayer, 1984). Results from experiments on axonal pathfinding in chick embryos in which additional limb segments were transplanted onto extremities seem to suggest an increased complexity in vertebrates (Hollyday, 1985).

Development of muscles

The trunk and neck muscles and the muscles on the head develop from the segmentally arranged myotomes, while those in the extremities develop from limb buds. Formation of the forelimb bud is in progress by E12, and by E13 the first motor fibres have reached their destination. Outgrowing motor axons penetrate limb buds before the mesenchymal cells have condensed into groups of future muscles, and this suggests that they might play a role in their initial development.

The development of the axial muscles, which are older from an evolutionary point of view, starts somewhat later (as do the neurons innervating axial muscles, see above). Myotomes at cervical levels can first be recognized on E12. On the following days groups of cells migrate to form the axial muscles (Altman & Bayer, 1984). The motor axons to these muscles grow out as early as E13 and reach the myotomes on the same day, but only E15 separate muscles can be distinguished and then the axons penetrate the muscle mass.

Cytologically, the mesenchymal cells which are the precursors of the later muscle cells elongate and long myofilaments emerge in these ellipsoid cells.

From that stage onwards the cells are termed myoblasts. Helix–loop–helix proteins such as myogenin and MyoD play an important role in the expression of the specific contractile proteins in muscle cells (for review see Buckingham *et al.*, 1992) and this differentiation is further regulated in a complex way by different growth factors such as fibroblast growth factor (FGF), epidermal growth factor (EGF), insulin-like growth factor-I (IGFI) (for review see Florini *et al.*, 1991a–c). The myoblasts become oriented longitudinally and subsequently they fuse into polynuclear primitive myocytes or *primary myotubes*, a process which in the rat takes place from about E12 (for reviews see Kelly, 1983; Kelly & Rubinstein, 1986; Wakelam, 1985). At first the primary myotubes are electrotonically connected via gap junctions, but later they separate and develop into independent units (Dennis *et al.*, 1981; Kelly & Zacks, 1969). This separation coincides wih the formation of a next generation of *secondary myotubes* from E13 to E14 developing from undifferentiated mesenchymal cells, and which are first enclosed with the primary myotubes within the same basal lamina. The addition of new fibres continues for some time and this ends at around birth (Kelly & Zacks, 1969) or at about the time when N-CAM expression diminishes (in the rat around the end of the first postnatal week; Moore & Walsh, 1985; see also Betz *et al.*, 1979). On the basis of the latter results it has been suggested that the number of muscle fibres is regulated, in part, by the transient expression of N-CAMs during development (Kelly & Rubinstein, 1986).

At later stages a phenotypic diversity develops between tonic type I fibres and phasic type IIa and IIb fibres. The early phases of this differentiation may proceed autonomously without neural innervation (Condon *et al.*, 1990; Drachman, 1968; Hoh, 1991; McLennan, 1994) and possibly this is related to the primary myotubes generating predominantly slow type I fibres and the secondary myotubes the fast type II muscle fibres. Evidence for this hypothesis was obtained, for example, from experiments in severely undernourished rats (Wilson *et al.,* 1988). Undernutrition from early gestation onwards did not affect the number of primary myotubes, but the number of secondary myotubes had decreased drastically. At a later age the muscles showed a selective decrease of type II muscle fibres but no changes in the numbers of type I muscle fibres. This result was interpreted as indicating first that the number of primary myotubes is genetically determined, and secondly that these fibres are the precursors of type I muscle fibres. The development of the secondary myotubes would be influenced by epigenetic factors such as hormonal status and undernutrition. Consequently, factors such as undernutrition would specifically affect the numbers of type II muscle fibres. However, in other research it has been shown that transitions between the myotube type and muscle fibre type do occur. The soleus muscle of the neonatal rat contains about equal numbers of primary and secondary fibres (Rubinstein & Kelly, 1981) but, as in the mature rat about 80% of the muscle fibres are of type I (e.g. Gramsbergen *et al.*, 1996), this indicates that many secondary myotubes are transformed into slow-twitch muscle fibres (Kelly & Rubinstein, 1986). In other muscles, such as the extensor digitorum longus muscle which is a predominantly fast muscle, a reverse development has been noted. In the C57 strain of mice about 150 primary myotubes were counted at birth but as at adult age many fewer type I muscle fibres occur, this implies that in this muscle a shift from primary myotubes into type II muscle fibres must have occurred. Interestingly, the extent of this shift appeared to be strain-specific (Leforavich & Kelly, cited in Kelly & Rubinstein, 1986). Such transitions are probably effected by alterations in motoneuronal firing patterns (Kugelberg, 1976). More recent results applying immunohistochemical techniques indicate that the primary myotubes contain precursors of both light and heavy chain myosins. As these proteins are the constituents of the future slow-twitch type I and fast-twitch type II muscle fibres this indicates that primary myotubes indeed might not be the exclusive origin of type I muscle fibres (McLennan, 1994).

Neuromuscular connections

The first synaptic contacts between motoneuronal axons and muscles in the rat have been demonstrated at E14 in the intercostal muscles, and from that age muscle contractions can be elicited (Dennis et al., 1981). At these early stages, however, the transmitter release is generally low and this only leads to rates of miniature endplate potentials (MEPPs) of only up to 1 per minute (Dennis et al., 1981). Strikingly, the rate of spontaneous transmitter release begins to increase at about the time when the multiple innervation of muscle fibres has been eliminated (Diamond & Miledi, 1962; see below).

In tissue cultures of Xenopus laevis it has been demonstrated that the muscle cell induces the release of acetylcholine (ACh). From that moment, further outgrowth of the axon is arrested, protein molecules (cadherins) connecting the axon to the target cell develop, and in the following 15 min the adhesion strength increases dramatically (Sanes & Poo, 1988). The first identifiable contacts between axons and muscles in rats have been detected electron microscopically only at E16 (Kelly & Zacks, 1969). Further differentiation of the neuromuscular junction takes up to 3 weeks, and this involves a bidirectional exchange of signals between presynaptic and postsynaptic elements (for reviews see Hall & Sanes, 1993; Sanes & Scheller, 1997).

Development of afferents

Cells in the dorsal root ganglia are generated from the neural crest (or possibly from the somites) between E12 and E15 (Altman & Bayer, 1984) and from E13 the first sensory fibres were observed to join the motor nerves and penetrate the limb bud. Afferents at cervical levels grow into the dorsal horn at E14.5 (Vaughn & Grieshaber, 1973; Windle & Baxter, 1936) and in the thoracic cord at E15.5 (Smith, 1983). They bifurcate in an ascending and a descending branch to form the dorsal funiculus. On E17 the suprasegmentally ascending collaterals grow out, and this is the very day when the cells in the dorsal column nuclei can be recognized.

Tactile stimulation of the hindlegs induces movements from E15 to E16 (Angulo y Gonzalez, 1932; Narayanan et al., 1971). In two out of nine rats electrical stimulation of the dorsal roots of the L4 segment demonstrated that the first motoneuronal discharges could be elicited already from E15.5. Stimulation caused a slowly rising positivity and this is indicative for polysynaptically induced depolarizations of motoneurones (Kudo & Yamada, 1987). (Saito, 1979), reported synaptic responses even at E14.5.) Proprioceptive reflexes, on the other hand, which at least in part are mediated by monosynaptic connections between afferent fibres and motoneurons, develop a few days later. Kudo & Yamada (1985) could elicit the myotatic reflex of the triceps surae muscle only from E19.5. They detected from E18.5 a steep rise in motoneuronal activation after stimulation of the dorsal roots, and this is indicative for a monosynaptic driving (Kudo & Yamada, 1987). They also demonstrated, at this same day, afferent fibres from the dorsal ganglia starting to make direct synaptic contacts with motoneurons.

It is interesting that the peak in prenatal neuronal death in the L4 dorsal ganglion (at P17–P19) correlates well with the establishment of the central connections of the neurons. A second phase of cell death, which ends at about P5, seems to be particularly influenced by peripheral target factors (Coggeshall *et al.*, 1994).

Fitzgerald (1987) raised the point that the establishment of afferent input might be an important factor in reorganizing fetal movements in rats. Although discontinuities in the course of movement development have not been addressed specifically, this interesting possibility should be pursued further.

Single-unit recordings from L4 dorsal ganglion cells indicated that all major cutaneous receptor types, including nociceptors, are present at birth in the rat (Fitzgerald, 1987). Results from this study suggest that further changes in the cutaneous reflexes after birth are mainly caused by the maturation of the end-organs and myelination. Descending inhibition by suprasegmental structures through fibres passing the dorsolateral funiculus is effective from P10 to P12, although these fibres could be demonstrated histologically as early as P6 (Fitzgerald & Koltzenburg, 1986).

Muscle spindles were demonstrated before birth, but their maturation is by no means ended at that point. At birth the nuclear bags still have a primitive structure and the intrafusal muscle fibres become innervated by gamma motoneurons only in the course of the first week (Milburn, 1973). Nuclear chain organs appear at the end of the first week and the adult morphology of muscle spindles is reached only at the end of the second postnatal week. These results are in agreement with neurophysiological data obtained by Vejsada *et al*. (1985b), who demonstrated that stimulation leads to adult-like reactions only from the 18th postnatal day.

Dendrites

Dendritic arborization starts only after the axon has reached its target (Barron, 1943; Yawa, 1987) and it has been suggested that a retrogradely transported trophic agent from the axon induces dendritic maintenance. Ramon y Cajal (1909/1911) noted that dendrites initially develop in great abundancy, while at later phases the dendritic tree decreases in size. An important part of dendritic development and the subsequent pruning of the dendritic trees of spinal motoneurons occurs in the rat postnatally and until P30 (Cummings & Stelzner, 1984; Curfs *et al.*, 1993). These changes are obviously related to changes in the distribution of receptors on the neuronal surface and linked to trends in synaptic connectivity (Dekkers *et al.*, 1994). Experiments in which neuromuscular inter-action was blocked, or in which the hindlimb nerves were crushed during early development, showed abnormal dendritic arborization, indicating that both efferent and afferent connections of motoneurons interfere with normal development (O'Hanlon & Lowrie, 1994a).

A particular reorganization of dendrites into dendrite bundles occurs in the pools of the long back muscles, the abdominal muscles and specific extensor muscles in the extremities (Gramsbergen *et al.*, 1996; Roney *et al.*, 1979; Scheibel & Scheibel, 1970). These muscles have in common that they fulfil important tasks in postural control. In rats it has been shown that this reorganization (in the pools of the long back muscles from P6 to P9 and in the pool of the soleus muscle from P14 to P16) coincides with changes in postural control (Geisler *et al.*, 2000; Westerga & Gramsbergen, 1992) (Figure 5). Such bundles are possibly the sites

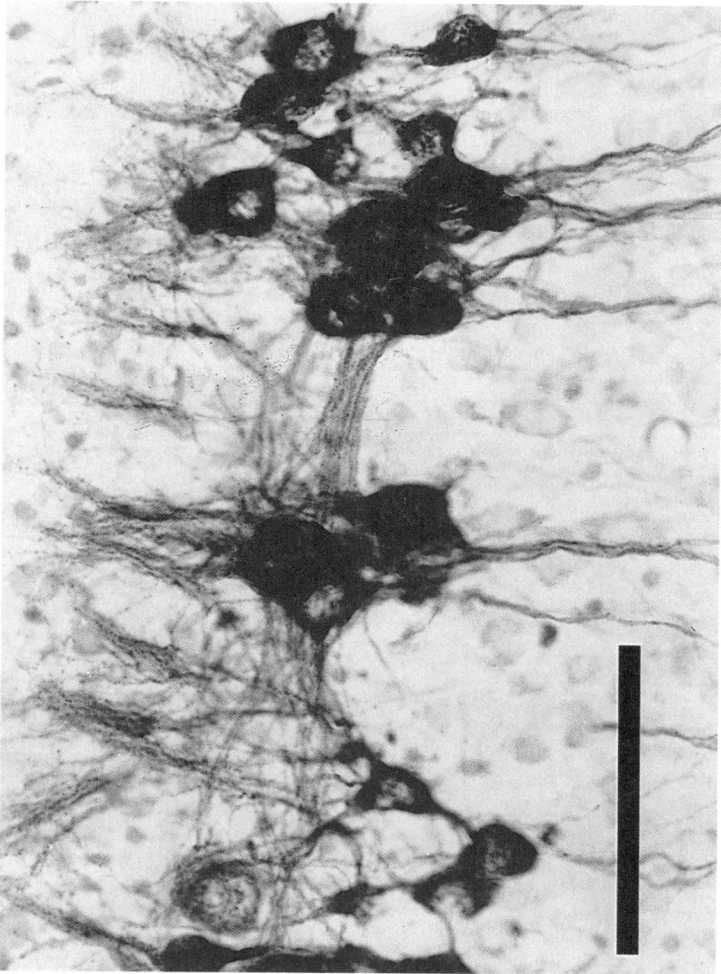

Figure 5 Motoneuronal pool of the multifidus muscle at P9. Note that nearly all dendrites run in transversal or longitudinally oriented dendritic bundles. Scaling bar (medially in the spinal cord): 100 μm.

of multiple electrotonic gap junctions between dendrites (Van der Want *et al.*, 1998) (Figure 6) and their function might be to synchronize the activity of motoneurons located in several spinal cord segments (Gramsbergen, 1998). Dendrite bundles have also been demonstrated in the spinal cord of human fetuses (Schoenen, 1982) but details of their further development are unknown.

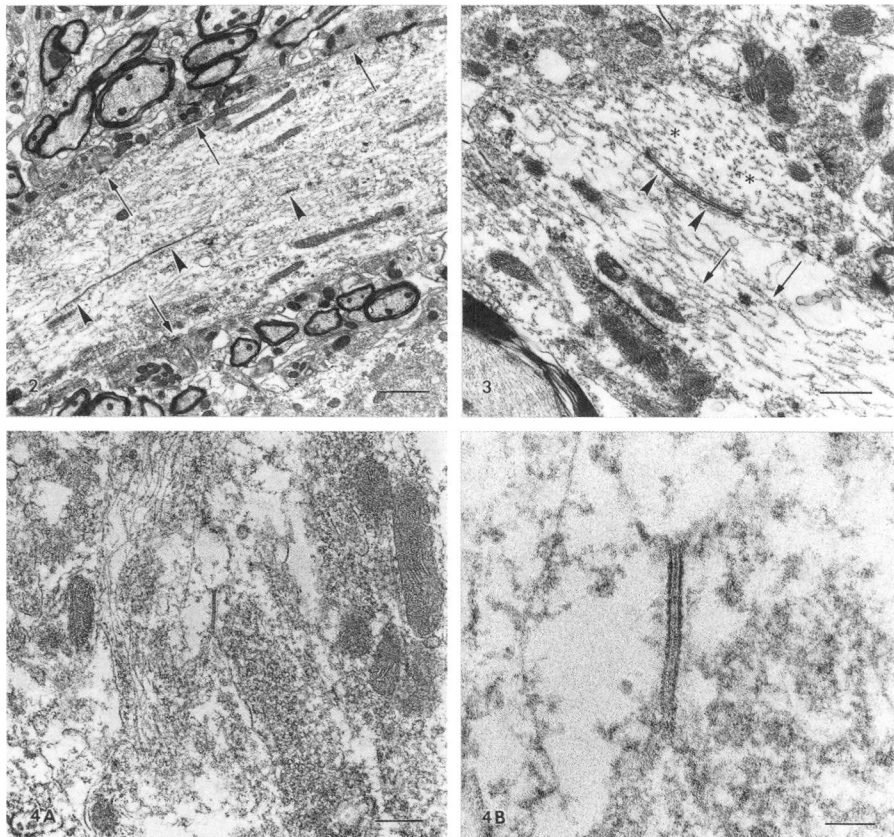

Figure 6 Electron micrograph of two CTB-labelled dendrites with extended gap junctional complexes (arrowheads) in the midline. The dendrites are surrounded by myelinated and non-myelinated axonal profiles, at the borderline of the dendritic bundles some terminal boutons form synaptic contacts (arrows). The dendrites predominantly contain filamentous material, mitochondria and clusters with rough endoplasmic reticulum. Note that in the vicinity of the gap junctions organelles such as mitochondria, endoplasmic reticulum or lamellar bodies are absent. Bar = 1.0 µm. (Reproduced by courtesy of Elsevier.)

Descending projections

Monoaminergic projections

Spinal projections of the limbic motor system (Kuypers, 1982; also called the emotional motor system by Holstege (1991, 1995)) are among the first to descend and to make connections at segmental levels. At adult age this system emerges from medial portions of the hypothalamus and the mesencephalon, the raphe nuclei and the locus coeruleus and subcoeruleus complex. Its medial components influence the excitatory state of interneurons and motoneurons via diffusely projecting monoaminergic fibres. The modulation of proprioceptive and extero-ceptive reflexes (but not the nociceptive reflexes) in different behavioural states in rats (Kretschmer & Schwartze, 1974) and in human babies (Prechtl, 1974) most probably depends on behavioural state-dependent "tuning" by this motor system.

In rat fetuses the cells of origin of the locus coeruleus are generated at E12 and E13 (Wallace & Lauder, 1983). The spinally projecting noradrenergic fibres have reached lumbar levels already at E18 and sacral levels at E19, and their descent continues until P2 (Bregman, 1987; Bregman & Bernstein-Goral, 1991; Rajaofetra et al., 1992). Serotonergic projections from the raphe nuclei have reached lumbosacral levels even at E16–E17. Rajaofetra et al. (1989) studied the 5-HT containing synapses electron microscopically. As the terminals are initially widespread, but later are only in the dorsal and ventral horn, Rajaofetra hypoth-esized that these 5-HT containing fibres at early stages might play a role in the stabilization of the innervation patterns of other projections in the spinal cord. Adult projection patterns are reached only around P21.

Reticulospinal and vestibulospinal projections

Descending projections from reticular nuclei, both of the ventral tier and the dorsal tier, reach the cervical spinal cord around E14–E16 and the lumbar spinal cord around the second postnatal day (Auclair et al., 1993; Kudo et al., 1993; Leong et al., 1984). More recent studies by Lakke (1997) demonstrated that fibres from a few reticular nuclei have reached lumbar levels even at E17. Neurons of the lateral, medial and spinal vestibular nuclei are generated around E12–E14 and their spinal projections have reached lumbar levels between E18 and P4 (Auclair, 1993; Kudo et al., 1993; Lakke, 1997; Leong et al., 1984). The reticulospinal and vestibulospinal projections are particularly important for the innervation of the axial muscles and the proximal muscles of the extremities. Electrical stimulation at the lower brainstem indicated that these fibres are already functional on the first postnatal day, although not in an adult-like fashion (Floeter & Lev-Tov, 1993, Vinay & Clarac, 1999).

Rubrospinal tract and the cerebellum

With some oversimplification the role of the cerebello-rubrospinal system for governing movements in lower vertebrates such as quadrupeds and birds may be considered as important as the role of the corticospinal system in apes and humans. For this reason a few lines will be devoted to cerebellar development. (For a comprehensive review on cerebellar development in the rat and other mammals the reader is referred to Altman & Bayer, 1996.) It is important to remember that the first two stages in cerebellar development, the production of the so-called deep neurons and Purkinje cells (stage "I") and the formation of the deep nuclei and of the primitive cerebellar cortex (stage "II") is for the greater part completed before birth. On the other hand, the formation of the circuitry of the cerebellar cortex develops only after birth (stage "III"). From E17 until birth the extra granular layer migrates and diffuses over the cerebellar surface and, from birth, the proliferative cells in this layer generate the future granular cells, basket cells and stellate cells. Cerebellar circuitry is mature around P21.

The rubrospinal tract, which is the descending tract of this system, is particularly important for motor behaviour in quadrupeds. The nucleus ruber receives a prominent projection from the cerebellum via the deep cerebellar nuclei.

Purkinje cells, which channel the output from the cerebellum, are spontaneously active from P1 (Woodward et al., 1969) and from P3 they can be driven by electrical stimulation. The nature of this activation suggested that at this age this is induced by climbing fibres (Puro & Woodward, 1977a), which emanate from the inferior olive. The development of the other main afferent system to the cerebellar cortex, the parallel fibre system, occurs much later. From P4 the future granule cells migrate through the Purkinje cell layer (see above) and synaptic transmission from the parallel fibre system was detected only from P7 (Puro & Woodward, 1977b). In view of this late development it is intriguing that the first cerebellorubral connections have been demonstrated already from E16 (Cholley et al., 1989) and rubrospinal projections occur from E17 (Lakke & Marani, 1991).

A cerebellar hemispherectomy in rats at adult age leads to serious motor handicaps. Such lesions at P5 or P10 become apparent only in disturbed functioning from P14 to P15 (Gramsbergen, 1982) and this indicates that the cerebello-rubrospinal system starts to play an important role in motor functioning from about the moment when the adult-like type of walking normally develops. Possibly the delay between the descent of the rubrospinal tract and the functional consequences of cerebellar lesions points towards a waiting period between this descent and the establishment of synaptic connections at segmental levels. Similar evidence was obtained, for example, with regard to descending inhibitory pathways in the dorsolateral funiculus of the newborn rat (Fitzgerald & Koltzenburg, 1986), to several descending projections in the opossum (Cassidy & Cabana, 1993), and to the descent of the corticospinal tract in monkeys (see Armand, this Handbook).

Corticospinal tract

The corticospinal tract in the rat contacts motoneurons mainly via interneurons although direct connections have been demonstrated (Liang et al., 1991, Ugolini et al., 1989). Lesioning of the corticospinal tract affects fine digital flexion movements in rats (Castro, 1972). Its main role, however, might be to modulate sensory input (Porter & Lemon, 1993). The tract courses in the ventral part of the dorsal funiculus and contains relatively thin fibres, many of which are myelinated but some are unmyelinated (Gorgels *et al.*, 1989).

Outgrowth of the pyramidal tract starts at E16 from the cortex, and at the day of birth the first pioneer fibres have descended as far as the cervical intumescence (Gribnau *et al.*, 1986; Joosten *et al.*, 1987; Schreyer & Jones, 1982). Between P7 and P10 they have descended along the extent of the spinal cord. Interestingly, the number of fibres in this tract increases to about 400,000 at P4, but thereafter the number decreases, first rapidly and more slowly thereafter to 150,000 fibres at adult age (Gorgels *et al.*, 1989; Schreyer & Jones, 1988). This decrease is largely dependent on the withdrawal of collaterals, e.g. from the occipital cortex (Stanfield *et al.*, 1982; Stanfield & O'Leary, 1985). At adult age the pyramidal tract originates in the sensorimotor cortex, as has been shown both by tracer studies (e.g. Leong *et al.*, 1984) and electrophysiological mapping studies (Donoghue & Wise, 1982; Neafsey *et al.*, 1986). Myelination of the corticospinal fibres starts only after P10, and is largely completed by P28 (Joosten *et al.*, 1989).

Development of ascending projections

Relatively few studies were involved in the development of ascending projections and this might be due to the methodological problems encountered in such studies (such as the specification of the target structures in the young brain or the impossibility of selectively injecting the neurons of origin). The birth date of ascending tract neurons was established by a combination of injecting tritiated thymidine in rat fetuses at E13–E15 and by retrogradely labelling with Fluoro-Gold from the C3 spinal cord segment at adult age (Nandi *et al.*, 1990). This study indicated that the bulk of neurons giving rise to ascending tracts proliferates around E13, although a small percentage is stillborn at E15. It has been shown by tritiated thymidine labelling at embryonic ages, and by retrogradely labelling from either the dorsal thalamus or the cerebellum, that the neurons which later project to either of these structures are already separated from the start (Beal & Bice, 1994).

Spinocerebellar connections were studied by biocytin labelling, and it has been demonstrated that by E13–E14 the first fibres have reached the cerebellar region. At E18–E19 they invade and branch into the deep cerebellar nuclei (Grishkat & Eisenman, 1995) and thereafter they reach the central white matter. After a waiting period these afferent fibres invade the granular layer around P5 and the adult patterns have developed by P7 (Arsenio-Nunes & Sotelo, 1985).

The neurons projecting to the thalamic nuclei are produced around E13–E15 (Beal & Bice, 1994) and by the day of birth the spinothalamic fibres have already reached their target nuclei where they start arborization. However, at that age some fibres extend beyond their terminal field and some have even invaded the internal capsule. Such erroneous projections are eliminated and the mature pattern of terminal branching is reached by P30.

Motoneurones and apoptosis

During normal development large proportions of the neurons (and among them motoneurons) degenerate in a process which has been termed programmed cell death or apoptosis (for reviews see Oppenheim, 1981a, 1991; Oppenheim *et al.*, this handbook). Considerable evidence exists that neuronal cell death in post-mitotic cells partly is controlled by the interaction of neurons with their efferent targets and with afferent input. In the lumbar spinal cord of the rat this target-dependent neuronal cell death ends around birth (Oppenheim, 1986). Oppenheim and co-workers demonstrated in the chick embryo that administration of a target-derived neurotrophic factor (prepared from hindlimb muscles) led to an increase in surviving motoneurons, and this was also observed in chick embryos in which the hindlimb was amputated (Oppenheim & Haverkamp, 1988). Similar results were reported for other neurotrophic factors, such as nerve growth factor (NGF), brain derived neurotrophic factor (BDNF), neurotrophin 3 (NT-3) , NT-4 and the ciliary neurotrophic factor (CNTF), but the artificial administration of such growth factors seems to protect motoneurons only for a short period. The survival of neurons is also dependent upon their afferent input, as well as upon hormonal influences (for an extensive review on these aspects see Oppenheim *et al.*, this handbook).

Much recent research is devoted to elucidating the aspects of the molecular and genetic factors which are involved in this cell death. Basically, it appears that neurons are doomed to die unless they are saved by extracellular survival factors, such as neurotrophins which bind to tyrosine kinase receptors. Such binding may lead to the up-regulated production of proteins that protect cells from cellular programmes for apoptosis in which caspases play a role (Wiese *et al.*, 1999). Other factors are now known to promote neuronal death by binding to receptors, such as the P75[NTR] peptide and steroid hormones, as well as oxidative stress. Research, particularly in *C. elegans* has identified the *ced-3* and *ced-4* genes which encode killer activities. The *ced-3* gene encodes for the CED-3 protein which in part is identical to a mammalian cysteine protease, interleukin-1β-converting enzyme (ICE) which plays a role in inflammatory processes. It has been sug-gested that CED-3 might have similar proteolytic properties and therefore could play a role in the induction of cell death. The expression of another gene in *C. elegans*, *ced-9*, on the other hand seems to protect cells against death. This gene encodes for a homologue of the mammalian proto-oncogene *bcl-2*. Expression of *bcl-2* has been demonstrated to protect cells from apoptosis (for further details see review by Agapite & Steller, 1997, and references in that review). Although these new and exciting findings have unravelled details on how apoptosis might be induced and regulated, the key question remains why cell death hits the one neuron and not the other. Greensmith & Vrbova (1996) suggested that the morphological changes in the succesful nerve endings after having made contact with their muscle target (rendering them into a "transmitting" neuron) are crucial for their survival (see also Oppenheim *et al.*, in this handbook).

Regression of polyneural innervation in the rat

Soon after the myotubes have formed, they are contacted by a multitude of axons, and this multiple innervation of muscle fibres remains even after their differentiation into type I and type II muscle fibres (for reviews see Jansen & Fladby, 1990; Ribchester, this handbook). Polyneural innervation was described initially in the soleus muscle of young rats (Redfern, 1970). Thereafter, it has been demonstrated to be a general feature of the neuromuscular innervation of developing muscles. Polyneural innervation appears to vary considerably in different muscles, ranging from maximally two axons per fibre in lumbrical muscles (Betz et al., 1979) to six axons in the soleus muscle (Bennett & Pettigrew, 1974) and it has been suggested that these differences might be related to the proportion of type I muscle fibres. Investigations by Brown et al. (1976) implied that motor units in the soleus muscle at neonatal age are roughly four to five times as large as those at young adult age. Theoretically this should have consequences for the recruitment patterns of the soleus and other muscles at early age, but possibly these are masked by a different viscosity and greater elasticity of the muscle tissue at early age.

In the further course of development the supernumerary nerve endings are retracted, and this leads to mononeural innervation. In some muscles of the rat this regression occurs around birth (e.g. in the intercostal muscles – Dennis et al., 1981), while in the gastrocnemius muscle this takes place only between P8 and P12 (Bennett et al., 1986). The soleus muscle is mononeurally innervated from around P16 (Brown et al., 1976; O'Brien et al., 1978) and the psoas muscle from P21 (IJkema-Paasen & Gramsbergen, 1998). For reviews and details on the mechanisms involved see Greensmith & Vrbova (1991) and Ribchester, this handbook.

Myelination

The conduction velocity of peripheral nerves in rats increases dramatically during development. In the rat's tibial nerve conduction increases from an average value of 1.4 m/s at P1 to 35 m/s at P30, and further to 60–84 m/s in adult rats (Vejsada et al., 1985a). This increase is in a complex way related to myelination and fibre size. For a further discussion on glial cell and Schwann cell development, as well as myelination in the central and peripheral nervous system, see Jacobson, 1991.

Central pattern generators and postural control

Strong evidence exists showing that rhythmic hindleg movements in rats during locomotion are generated by an assembly of neurons in the spinal cord, termed a central pattern generator (CPG; see Cazalets, this handbook). It has been demonstrated, in a variety of experimental animals, that such CPGs, even in the absence of afferent feedback or supraspinal influences, are able to produce orderly activation patterns in extremity muscles (Grillner & Zangger, 1975; for review see Grillner et al., 1991). Recently Cazalets et al. (1995), on the basis of research in an isolated spinal cord preparation of neonatal rats, demonstrated that the CPG for hindlimb movements in the rat is localized in the first lumbar segment, L1 (for details and further references see Cazalets, this handbook).

The problem arises as to whether this CPG also governs the rhythmic activation of the back muscles during locomotion. Zomlefer and co-workers (1984) studied this problem in cats which were spinalized at Th13. Bursts in the back muscles were recorded, which were in phase with locomotor movements, and the interpretation was that back muscles are steered by the CPG for locomotion. In experiments in freely moving rats we, however, detected a large variability in the latencies between bursts in the back muscles and in the gastrocnemius muscles (which are active from the onset of the stance phase). This has led us to suppose that the activation of the hindleg muscles at older ages is relatively independent from the CPG for hindleg movements, and that afferent feedback and supra-spinal influences (particularly from the cerebellum, via the rubrospinal tract and from the lateral vestibular nucleus, via the lateral vestibular tract) play a prominent role in this coupling. The supraspinal systems probably adjust postural control continuously during locomotion (see also Gramsbergen, 2001a; Grillner, 1975).

Development of movements in relation to structural changes in rats

In this section a few of the unsolved problems are listed with regard to the development of movements in relation to structural changes.

A rostrocaudal gradient?

The proliferation of neurons in the rat's spinal cord takes place in 2–3 days. The chronological differences between the production of neurons at rostral and caudal levels are only small, and the differences within a spinal cord segment are of the same order of magnitude. A rostrocaudal gradient is much more pronounced in the migration and further differentiation of the neurons. The emergence of motility in the trunk and in extremities follows this pattern, but movements of the face and of the perioral region appear at a later stage of development (contractions in the facial area are obviously produced by activity of motoneurons in the brainstem). Anokhin (1964, 1967) suggested that the neural systems which are involved in functions such as feeding, breathing or locomotion develop more or less independently, and only later become integrated. He further hypothesized that the significance of neural systems for vital functions determines their priority in the maturation. This priority, in turn, was suggested to be related to evolutionary, selective mechanisms.

Alternative explanations for the later emergence of movements in the head region are that the complexity of the neural control of facial and perioral muscles requires a longer period to develop, or that the gradient in development proceeds from a pole at brainstem level both in a rostral and a caudal direction.

Motor primacy?

Comparing the results of Angulo y Gonzalez (1932), Narayanan et al. (1971) and Smotherman & Robinson (1988) indicate that the first reactions to tactile stimuli and the onset of spontaneous motility occur almost simultaneously. Histological data, on the other hand, indicate that the neurons in the ventral half of the neuraxis (the motor and the premotor neurons) are produced before those at the dorsal side (Altman & Bayer, 1984). Illustrative in this respect is that the motoneurons of the trigeminal nerve are generated before the neurons of the sensory portion of this nerve, an order which was termed by the authors an "inverted sequence" (Altman & Bayer, 1982). Connections between the afferents entering the dorsal horn and interneurons and motoneurons occur a few days after spontaneous movements could be observed (Kudo & Yamada, 1987)

494

but, on the other hand, electrical stimulation of the dorsal root in *in-vitro* experiments suggested that such connections might be functional even before the axons of the motoneurons have reached the muscle mass (Saito, 1979). The temporal differences in these studies, however, might well be related to the use of different animal species or the methodologies involved. An investigation into the development of efferent and afferent connections in the same strain of rats might lead to an unequivocal time scale of events.

Motor development related to postural development?

Research into the development of locomotion and postural control has suggested that both developments in the first stages proceed independently, but that they become interdependent at a later stage. In the rat the axial muscles in the neck and trunk play an important role in postural maintenance. Altman & Bayer (1984) observed that the motoneurons in the medial tier, innervating the axial muscles, develop about a day later than those in the lateral tier, which at later stages innervate the extremity muscles. Remarkably, the maturation of the axial muscles precedes that of the muscles in the limb buds, and Altman & Bayer suggested that developmental relations between motoneurons and muscles differ in the two systems. Observations of fetal motor development, on the other hand, showed that movements of the rostral portions of the trunk precede those in the rostral extremities. These paradoxes in developmental order in neural and muscular development, as well as in motor development, have not been solved, but the data strongly point towards different maturational trajectories of the systems involved in posture, on the one hand, and movements of extremities on the other. At later stages of maturation an independent development of postural control and movement control was also noted (see above). As deficits in postural control mechanisms might be a causal factor in movement disorders in the human it is of great importance to investigate details of their mutual relationships further.

DISCUSSION

Research into the development of the nervous system traditionally had two important roots. The investigation of early and less complex stages of the developing nervous system would first help to gain insight into the immense complexity of the mature brain. Neuroanatomists, for example, have chosen this strategy, and this approach eventually led to the formulation of the *neuron theory* (see Jacobson, 1993). The other root was to study the developmental process itself. Wilhelm Roux (1881) might be considered as the founder of this field of research, which was particularly directed towards the relationships between the successive changes in the organism and the relation between these stages and the environment.

Neuro-ontogenetic research has made clear that, even at early stages of development, structure–function relationships are by no means as simple as was often hoped for. Transient structures emerge and disappear (e.g. the Cajal–Retzius cells in the most superficial layer of the cortex, to mention just one example) and transient behavioural patterns develop and disappear (hatching movements in the chick embryo, see Bekoff, this handbook, and rooting and sucking behaviour in the human baby, see Prechtl, this handbook and Hadders-Algra, this handbook). Obviously these structures and behaviours fulfil a pivotal role only during relatively short periods in development. They illustrate that neuro-ontogeny is not characterized by a monotonous increase in complexity. Neuro-ontogenetic studies in the human pose specific problems, as inferences on structural development rely largely upon extrapolations from animal research. In this section certain parallels in human and rat development will be discussed, as well as anchoring points for extrapolation.

Early development

The older literature on human motor development was based on direct observations in fetuses after abortion (e.g. Erbkam, 1837; Minkowski,1923; Pflueger, 1877). Observations in four human embryos with the placental connections intact were made by Fitzgerald & Windle (1942) and they reported that strong stimuli (pinching or heavy stroking) at 7.5 or 8 weeks could elicit local contractions, particularly of trunk muscles. Hooker and his colleagues (see Hooker, 1952) were the first to systematically study motor development and reactions to stimuli in an impressive series of 131 human fetuses with menstrual ages varying from 6.5 weeks to 45 weeks. It is only fair to note that they themselves pointed out that the progressive anoxia of the fetus and the anaesthesia of the mother interfere with normal movements. This also holds, they claimed, for manipulations of the fetus and the amniotic cavity. It will be remembered that similar considerations were put forward with regard to research in fetal rats.

Obviously such influences are avoided when non-invasively studying the motility of fetuses in the uterine environment. Real-time ultrasound scanning of fetal movements, which has been developed recently, allows us to visualize the fetus in great detail, and this also enables us to study motor development longitudinally (Birnholz et al., 1978; Prechtl, 1984; Reinold, 1976; de Vries et al., 1982; Prechtl, this handbook). Ultrasound scanning of fetal movements in individual rats is not yet feasible due to the limited resolutional capacity of scanners (attempts of our own).

The first movements in human fetuses can be observed from the onset of the 7th week of postmenstrual age (PMA; De Vries et al., 1982) and this is at about the same age when Hooker could for the first time elicit neck and trunk movements by tactile stimulation (between the middle of the 7th week and the onset of the 8th week). These "just discernible movements" initially involve only minor head and neck movements, but a few days later also trunk and extremity movements occur (De Vries et al., 1982) in so-called general movements (Prechtl, this handbook). In the weeks thereafter a repertoire develops of arm and leg movements, trunk and head movements, as well as specific movement patterns such as sucking and breathing movements. From the 14th to 16th week until the 2nd month after term birth, the repertoire of movement patterns remains largely unchanged (for details see Prechtl, 1984; de Vries et al., 1982; Prechtl, this handbook).

The trends in motor development in the human fetus and in the rat show striking similarities. Both in the human and in the rat, movements from the onset are part of patterns in which several muscle groups are involved. Both in the human and in the rat, movements start in the neck region and later the trunk and the arms are involved, while movements of the mouth and the tongue develop relatively late. In the human the first movements emerge (de Vries *et al.*, 1982) shortly before reactions can be elicited by tactile stimulation (Hooker, 1952). In the rat, movements could be elicited (Angulo y Gonzalez, 1932) half a day before the onset of spontaneous motility (Narayanan *et al.*, 1971; Smotherman & Robertson, 1988). The differences in phasing might be caused by different methodologies, and it seems that the emergence of spontaneous motility almost coincides with the possibility to elicit movements by tactile stimulation.

In the rat, stimulation of the trigeminal area and the forepaws leads to movements in widespread areas of the body, but later reactions are restricted to the stimulated area. In the human, stimulation of the perioral skin area from the 7.5th week leads to neck and trunk flexion at the contralateral side and startle-like movements. In the following days and weeks the movements involve increasingly larger areas of the body, but thereafter they are replaced by local activities (Hooker, 1952).

Another similarity is the depression of fetal activity towards the end of pregnancy. In the human such a depression was described by Edwards & Edwards (1970) and a similar trend was also observed in rats (Narayanan *et al.*, 1971, Smotherman & Roberts, 1988). As birth in the human and in the rat takes place at different stages of development (see below) it seems that this depression is caused by external influences.

Interestingly, it appears that fetal movements in rats are not eliminated by transecting the spinal cord or by decapitation (Hooker, 1930; Narayanan *et al.*, 1971) and this indicates that movements at early stages of development are generated autonomously by circuits in the spinal cord. Fetal movements also occur in anencephalic human fetuses, but Preyer (1885) reported that these movements are distinctly abnormal. Ultra-sound scanning of movements in seven anencephalic human fetuses with gestational ages varying from 16 to 35 weeks indicated that such movements lack the fluency and variability of those movements in normal fetuses (Visser *et al.*, 1985). Another interesting finding of this investigation was that the absence of the pontine region is accompanied by a disturbed temporal distribution of motility. This indicates that, at least from the 16th week PMA, brainstem projections modulate fetal activity in the human fetus.

The stage of brain development at birth

In order to be able to extrapolate data from the rat's development to the human, it is important to consider similar stages of brain development. Dobbing & Sands (1979) compared trends in total brain weight of seven different mammalian species and related this to birth. This study indicated that the peak in the brain growth spurt in the human occurs at term age, and in the rat around 7 days after birth. This implies that, with respect to this parameter, a rat at birth could be compared to a human fetus of about 24–28 weeks of gestation. When also considering other parameters, however, it appears that this comparison is not valid for the development of the brain as a whole. The development of electrical activity in the rat's cerebral cortex at P10–P11, for example, is similar to that of a human baby at term age. Both in rats of that age and in human babies at term age the EEG during so-called quiet sleep or "state 1" shows a waxing and waning of amplitudes which was termed a "tracé-alternant"(Figure 7) (in the rat: Gramsbergen et al., 1970, Gramsbergen, 1976; Jouvet-Mounier et al., 1970; in the baby, Dreyfus-Brisac et al., 1955; Parmelee et al., 1968; Prechtl, 1984; see also Salzarulo et al., this handbook).

Romijn et al. (1991), in considering several parameters of brain development (numerical data on synapse formation, development of decarboxylase activity, the development of choline acetyltransferase activity and EEG development), concluded that the stage of brain development in rats between P10 and P13 is similar to that in human babies at term birth. It should be noted, however, that this comparison refers only to the timing of birth at term age in the human baby. When including other landmarks in neural development, such as the onset of motility (P16 in the rat; 7 weeks PMA in the human), the emergence of free walking (P8–P16 in the rat, around 15 months in the human) or the end of polyneural innervation of the psoas muscle (P20 in the rat, in babies at 3 months term age; Figure 8) this indicates that the time scale of neuro-ontogenetic development in the human is not a scaled-up version of that development in the rat.

General movements in rats?

General movements in human fetuses and newborn babies have been defined as a "series of gross movements of variable speed and amplitude which involve all parts of the body with a duration from a few seconds to several minutes" (for details see Prechtl, 1990 and Prechtl, this handbook). These movements occur from the 6th to 7th week postmenstrual age until a few months after birth. It is assumed that general movements, at least initially, are generated by spinal circuits (Prechtl, 1997). As an abnormal quality of these movements in the first postnatal months reliably predicts the emergence of serious neurological abnormalities at a later age caused by lesions in the brain, this indicates that suprasegmental projections are involved in their generation (see Hadders-Algra, this handbook). The nature of these influences and the projections involved are unknown; therefore it would be of interest to study behavioural and neurobiological aspects of general movements in experimental animals. Movement patterns which involve trunk and extremities have been shown in fetal rats by Narayanan et al., (1971, "total movements") and Smotherman & Robertson (1988, "complex movements"). These movement patterns occur with a certain regularity and they are possibly analogous to general movements in human fetuses and babies (R.W. Oppenheim, personal communication). During the first 10 postnatal days, similar movement patterns occur particularly when they are quietly asleep ("state 1", characterized by regular respiration). An important difference with general movements in human babies, however, is that rats show these movements only when in a prone position, Supine orientations for longer periods occur only when pups are fed by their mother. When observed alone, and when put in a supine position, rats struggle vigorously to return to a prone position. A definite answer with regard to the occurrence of general movements

STATE 1

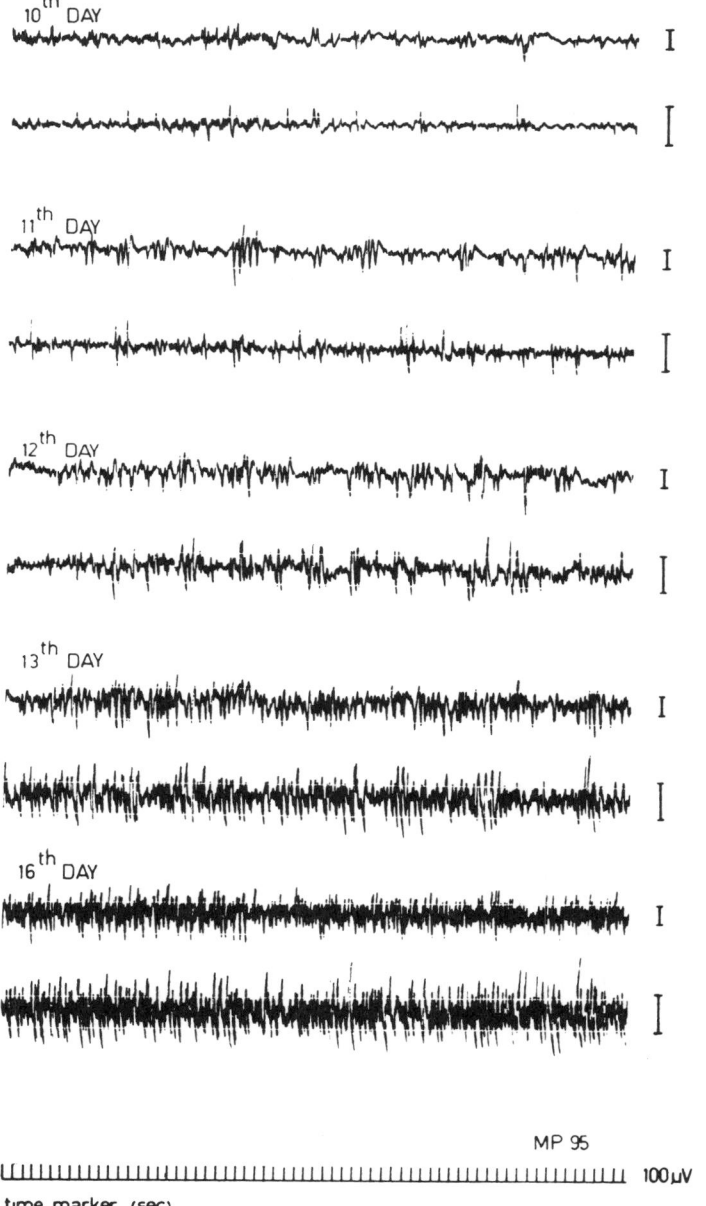

MP 95

time marker (sec)

Figure 7 Trace alternant in the EEG from a single rat at P11, P12, P14, and P16. Upper and lower trace: EEG from the sensory motor cortex and visual cortex, Time constant, 1 s.

in rats and, eventually, the occurrence of abnormalities after lesions in central areas, awaits further research (for a discussion on this point see also Hadders-Algra, this handbook).

Development of postural control

The behavioural repertoire of human babies expands importantly, particularly from the 2nd month onwards (Prechtl, 1984). Reactions to the caregiver change

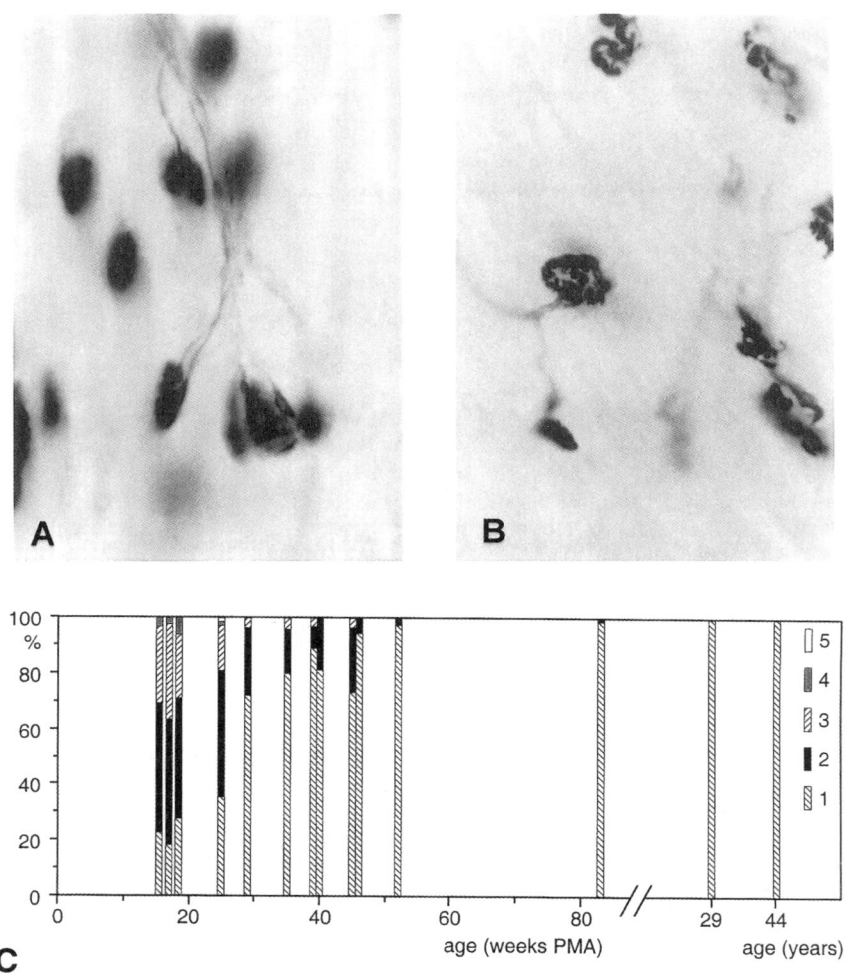

Figure 8 Micrographs of motor endplates with nerve terminals in the psoas muscle of the human. Calibration bar 10 mm. Panel **A**, fetus at 28 weeks PMA (note the polyneural innervation); panel **B**, fetus at 35 weeks PMA (note the increased diameter of the axons and varicosities); panel **C**, regression of polyneural innervation with age; horizontal axis, ages in weeks PMA and years respectively; vertical axis, percentages of enplates with one to five endings. (Reproduced by courtesy of Elsevier).

and increase in their intensity. Gaze control and compensatory eye movements develop from that age, and arm movements aimed at objects gradually emerge. An important prerequisite for these behaviours is adequate postural control (see also Hadders-Algra, this handbook). Prechtl (1989) studied neck muscle activity in babies in prone and supine positions during external perturbations, and he and his co-workers observed that direction-specific activity in the neck muscles occurs from the the 8th to 10th postnatal week onwards. Results on EMG recordings in neck and trunk muscles of babies, while reaching, support this conclusion (Van der Fits & Hadders-Algra, 1998; Van der Fits et al., 1999a).

Successful reaching movements in infants develop around 4–5 months of term age (Bertenthal & Von Hofsten, 1990; Von Hofsten, 1991) and such movements from this age are accompanied by contractions in trunk and neck muscles (Van der Fits et al., 1999a). Only from 15 months does the activity in postural muscles consistently anticipate reaching movements, and this is suggestive for a feed-forward control of activation of postural muscles (Van der Fits et al., 1999b).

In the first weeks after term birth, stepping movements can be elicited in human babies when they are supported (Forssberg, 1985), but this lasts until 7–9 months when rhythmic walking movements reappear (Forssberg, 1985; Forssberg & Dietz, 1997). In human infants standing without support develops from the age of 12 months (Touwen, 1976). Children are able to walk without support from the age of 15 months (Touwen, 1976) but only at the age of 6–8 years has the adult type of walking and postural control developed (Assaiante & Amblard, 1995; Okamoto & Kumamoto, 1972; Massion, 1992; Bril & Breniere, 1993; see Gordon, this handbook). Experiments by Grasso et al. (1998), in which head and trunk movements in infants aged 3.5–8.5 years were recorded when following a curved trajectory, indicated that feed-forward control of locomotor trajectories becomes increasingly important during later stages of walking development. Summarizing these results in the human, postural control gradually extends from rostral to caudal areas and feedback control precedes feed-forward control.

Obviously, motor control in the human differs in many respects from that in rats, but the relation between the development of postural control and movements shows certain parallels. Rats are able to stand with their ventral body surface free from P8, and it was suggested that feedback mechanisms of postural control prevail at this and the following days (Gramsbergen, 1998). The adult-like and fluent walking pattern in rats develops suddenly around the end of the 2nd week. From this age other movements may be superimposed upon ongoing walking, and this strongly suggests that, from that age, posture is controlled in a feed-forward fashion. In a protracted development which lasts until about the end of the 4th week after birth, movement and postural control reach maturity (see above).

Evidence points to the cerebellum being crucial to the coupling of the execution of movements and postural control (e.g. Grillner, 1975 referring to animal species; Forssberg, 1985 referring to human babies). First, in rats, temporal aspects of cerebellar development indicate striking parallels with this feed-forward control becoming operational. The development of the parallel fibre system in the cerebellar cortex is finished at about the 15th day (Puro & Woodward, 1977b) and then the climbing fibres which form the other main afferent system have established their mature terminations with Purkinje cells (Crépel *et al.*, 1976). This is the very age when the immature walking pattern is replaced by the adult-like walking pattern. The second argument is that lesions in the cerebellum performed before P10 (such as a unilateral cerebellar hemispherectomy) only from P14 to P15 lead to abnormalities in motor behaviour (Gramsbergen, 1982, 1984). In developing rats, and other species such as cats, experimental research is now directed towards the involvement of the cerebellum, the precerebellar nuclei and the basal ganglia in the linkage of postural control and the planning and execution of movements (Houk & Wise, 1995). This research will hopefully in the near future lead to an understanding of this important aspect of later motor development. This insight is important not only for understanding normal development but also deviant motor development. Deficient postural control has been claimed to be a major factor in such cases (Aicardi & Bax, 1992).

EPILOGUE

Research into the brain and behaviour of rats from the onset of motility until adult hood has provided us with clues as to their interrelations. In other cases careful observations and meticulously performed experiments have posed new problems. This research was given a new momentum by recent advances in molecular research and genetic engineering, and results and suggestions from this research have led to important readjustments of our insights.

The ever-growing body of knowledge obtained in animal research is the basis from which brain and behaviour relationships in the human can be interpreted. On the other hand, new results from research in the human, e.g. by ultrasound scanning of fetal movements and carefully observing the development of motor behaviour, has prompted new animal experiments. It is hoped that this review helps in planning such new experiments, both in the human and in animals, in order to further advance a comprehensive understanding of brain and behaviour–relationships in the human.

References

Adelman, H.B. (1967). *The embryological treatises of Hieronymus Fabricius of Aquapendente*, a facsimile edition. Ithaca, NY: Cornell University Press.

Agapite, J. & Steller, H. (1997). Neuronal cell death. In W.M. Cowan, T.M. Jessell & S.L. Zipursky (Eds.) *Molecular and cellular approaches to neural development* (pp. 264–289). Oxford: Oxford University Press.

Aicardi, J. & Bax, M. (1992). Cerebral palsy. In J. Aicardi (Ed.) *Diseases of the nervous system in childhood* (pp. 220–374). *Clinics in Developmental Medicine*, 115–118. Oxford: Blackwell.

Almli, C.R. & Fisher, R.S. (1977). Infant rats: sensorimotor ontogeny and effects of substantia nigra destruction. *Brain Research Bulletin*, **2**, 425–459.

Altman, J. & Bayer, S.A. (1982). Development of the cranial nerve ganglia and related nuclei in the rat. In F. Beck, W. Hild, J. van Limborgh, R. Ortmann, J.E. Pauly & T.H. Schiebler (Eds.), *Advances in anatomy, embryology and cell biology* (vol. 74). Berlin: Springer Verlag.

Altman, J. & Bayer, S.A. (1984). The development of the rat spinal cord. In F. Beck, W. Hild, J. van Limborgh, R. Ortmann, J.E. Pauly & T.H. Schiebler (Eds.), *Advances in anatomy, embryology and cell biology* (vol. 85). Berlin: Springer Verlag.

Altman, J. & Bayer, S.A. (1996). *Development of the cerebellar system in relation to its evolution, structure and functions*. Boca Raton, FL: CRC Press.

Altman, J. & Sudarshan, K. (1975). Postnatal development of locomotion in the laboratory rat. *Animal Behaviour*, **23**, 896–920.

Angulo y Gonzalez, A.W. (1932). The prenatal development of behavior in the albino rat. *Journal of Comparative Neurology*, **55**, 395–442.

Anokhin, P.K. (1964). Systemogenesis as a general regulator of brain development *Progress in Brain Research*, **9**, 54–86.

Anochin, P.K. (1967). Das funktionelle System als Grundlage der physiologischen Architektur des Verhaltensaktes. *Brain and behaviour research*. Monograph Series, Band 1. Jena: Gustav Fischer.

Armand, J. (1982). The origin, course and terminations of corticospinal fibers in various mammals. *Progress in Brain Research*, **57**, 327–360.

Arsenio Nunes, M.L. & Sotelo, C. (1985). Development of the spinocerebellar system in the postnatal rat. *Journal of Comparative Neurology*, **237**, 291–306.

Asanuma, C., Ohkawa, R., Stanfield, B.B. & Cowan, W.M. (1988). Observations on the development of certain ascending inputs to the thalamus in rats. I. Postnatal development. *Brain Research*, **469**, 159–170.

Assaiante, C. & Amblard, B. (1995). An ontogenetic model for the sensorimotor organization of balance control in humans. *Human Movement Science*, **14**, 13–43.

Auclair, F., Belanger, M.C. & Marchand, R. (1993). Ontogenetic study of early brain stem projections to the spinal cord in the rat. *Brain Research Bulletin*, **30**, 281–289.

Barcroft, J. & Barron, D.H. (1939). The development of behavior in foetal sheep. *Journal of Comparative Neurology*, **70**, 477–502.

Barron, D.H. (1943). The early development of the motor cells and columns in the spinal cord of the sheep. *Journal of Comparative Neurology*, **78**, 1–26.

Beal J.A. & Bice, T.N. (1994). Neurogenesis of spinothalamic and spinocerebellar tract neurons in the lumbar spinal cord of the rat. *Developmental Brain Research*, **78**, 49–56.

Bekoff, A. & Lau, B. (1980). Interlimb coordination in 20-day-old rat fetuses. *Journal of Experimental Biology*, **214**, 173–175.

Bekoff, A. & Trainer, W. (1979). The development of interlimb co-ordination during swimming in postnatal rats. *Journal of Experimental Biology*, **83**, 1–11.

Bennett, M.R. & Pettigrew, A.G. (1974). The formation of synapses in striated muscle during development. *Journal of Physiology (London)*, **241**, 515–545.

Bennett, M.R., Ho, S. & Lavidis, N. (1986). Competition between segmental nerves at end-plates in rat gastrocnemius muscle during loss of polyneural innervation. *Journal of Physiology (London)*, **381**, 351–376.

Bertenthal, B. & Von Hofsten, C. (1998). Eye, head and trunk control: the foundation for manual development. *Neuroscience and Biobehavioral Reviews*, **22**, 515–520.

Betz, W.J., Caldwell, J.H. & Ribchester, R.R. (1979). The size of motor units during post-natal development of rat lumbrical muscle. *Journal of Physiology (London)*, **297**, 463–478.

Birnholz, J.C., Stephens, J.C. & Faria, M. (1978). Fetal movements patterns: a possible means of defining neurologic developmental milestones *in utero*. *American Journal of Roentgenology*, **130**, 537–540.

Blanck, A., Hard, E. & Larsson, K. (1967). Ontogenetic development of orienting behavior in the rat. *Journal of Comparative Physiology and Psychology*, **63**, 427–441.

Bolles, R.C. & Woods, P.J. (1964). The ontogeny of behavior in the albino rat. *Animal Behavior*, **12**, 427–441.

Brazier, M.A.B. (1988). *A history of neurophysiology in the 19th century*. New York: Raven Press.

Bregman, B.S. (1987). Development of serotonin immunoreactivity in the rat spinal cord and its plasticity after neonatal spinal cord lesions. *Brain Research*, **431**, 245–263.

Bregman, B.S. & Bernstein-Goral, H. (1991). Both regenerating and late-developing pathways contribute to transplant-induced anatomical plasticity after spinal cord lesions at birth. *Experimental Neurology*, **112**, 49–63.

Bril, B. & Breniere, Y. (1993). Posture and independent locomotion in early child hood: learning to walk or learning dynamic postural control? In G. Savelsberg (Ed.) *The development of co-ordination in infancy* (pp. 337–358). Amsterdam: Elsevier.

Brown, M.C., Jansen, J.K.S. & Van Essen, D. (1976). Polyneural innervation of skeletal muscle in new-born rats and its elimination during maturation. *Journal of Physiology (London)*, **261**, 387–422.

Buckingham, M., Houzelstein, D., Lyons, G., Ontell, M., Ott, M.D. & Sassoon, D. (1992). Expression of muscle genes in the mouse embryo. *Symposia of the Society for Experimental Biology*, **46**, 203–217.

Cassidy, G. & Cabana, T. (1993). The development of the long descending projections in the opossum, monodelphis domestica. *Developmental Brain Research*, **72**, 291–299.

Castro, A.J. (1972). Motor performance in rats. The effects of pyramidal tract section. *Brain Research*, **44**, 313–323.

Cazalets, J.R., Borde, M. & Clarac, F. (1995). Localization and organization of the central pattern generator for hindlimb locomotion in newborn rat. *Journal of Neuroscience*, **15**, 4943–4951.

Cholley, B., Wassef, M., Arsénio-Nunes, L., Bréhier, A. & Sotelo, C. (1989). Proximal trajectory of the brachium conjunctivum in rat fetuses and its early association with the parabrachial nucleus. A study combining *in vitro* HRP anterograde axonal tracing and immunocytochemistry. *Developmental Brain Research*, **45**, 185–202.

Clarke, K.A. & Parker, A.J. (1986). A quantitative study of normal locomotion in the rat. *Physiology and Behavior*, **38**, 345–351.

Clarke, K.A. & Williams E (1994). Development of locomotion in the rat, spatiotemporal footfall patterns. *Physiology and Behavior*, **55**, 151–155.

Close, R. (1964). Dynamic properties of fast and slow skeletal muscles of the rat during development. *Journal of Physiology (London)*, **173**, 74–95.

Coggeshall, R.E., Pover, C.M. & Fitzgerald, M. (1994). Dorsal root ganglion cell death and surviving cell numbers in relation to the development of sensory innervation in the rat hindlimb. *Developmental Brain Research*, **82**, 193–212.

Coghill, G.E. (1929). *Anatomy and the problem of behaviour*. Cambridge, UK: Cambridge University Press.

Condon, K., Silberstein, L., Blau, H.M. & Thompson, W.J. (1990). Differentiation of fiber types in aneural musclularture of the prenatal rat hindlimb. *Developmental Biology*, **138**, 275–295.

Corner, M.A. & Crain, S.M. (1964). Spontaneous contractions and bioelectric activity after differentiation in culture of presumptive neuromuscular tissues in the early frog embryo. *Experientia,* **21**, 1–7.

Crain, S.M. (1964). Electrophysiological studies of cord-innervated skeletal muscles in long-term tissue cultures of mouse embryo myotomes. *Anatomical Record*, **148**, 273–274.

Crain, S.M. (1966). Development of "organotypic" bioelectric activities in central nervous tissues during maturation in culture. *International Reviews of Neurobiology*, **9**, 1–43.

Crépel, F., Mariani, J. & Delhaye-Bouchaud, N. (1976). Evidence for a multiple innervation of Purkinje cells by climbing fibres in the immature rat cerebellum. *Journal of Neurobiology*, **7**, 579–582.

Cummings, J.P. & Stelzner, D.J. (1984). Prenatal and postnatal development of lamina IX neurons in the rat thoracic spinal cord. *Experimental Neurology*, **83**, 155–166.

Curfs, M.H.J., Gribnau, A.A.M. & Dederen, P.J.W.C. (1993). Postnatal maturation of the dendritic fields of motoneuron pools supplying flexor and extensor muscles of the distal forelimb in the rat. *Development*, **117**, 535–541.

Dawes, G.S., Fox, H.E., Leduc, B.M., Liggins, G.C. & Richards, R.T. (1972) Respiratory movements and rapid eye movement sleep in the foetal lamb. *Journal of Physiology (London)*, **220**, 199–143.

Dekkers, J., Becker, D.L., Cook, J.E. & Navarrete, R. (1994). Early postnatal changes in the somato-dendritic morphology of ankle flexor motoneurons in the rat. *European Journal of Neuroscience*, **6**, 87–97.

Dennis, M.J., Ziskind-Conhaim, L. & Harris, A.J. (1981). Development of neuromuscular junctions in the rat embryo. *Developmental Biology*, **81**, 266–279.

De Vries, J.I.P., Visser, G.H.A. & Prechtl, H.F.R. (1982). The emergence of fetal behaviour. I. Qualitative aspects. *Early Human Development*, **7**, 301–322.

Diamond, J. & Miledi, R. (1962). A study of foetal and new-born rat muscle fibers. *Journal of Physiology (London)*, **162**, 393–408.

Dobbing, J. (1981). The later development of the brain and its vulnerability. In J.A. Davies & J. Dobbing (Eds.) *Scientific foundations of paediatrics* (pp. 744–759). London: Heinemann.

Dobbing, J. & Sands, J. (1979). Comparative aspects of the brain growth spurt. *Early Human Development*, **3**, 79–83.

Donoghue, J.P. & Wise, S.P. (1982). The motor cortex of the rat: cytoarchitecture and microstimu-lation mapping. *Journal of Comparative Neurology*, **175**, 207–232.

Drachman, D.B. (1968). The role of acetylcholine as a trophic neuromuscular transmitter. In G.E. Wolstenholme & M. O'Connor (Eds.), *Growth of the nervous system* (pp. 251–273). London: Churchill.

Drachman, D.B. & Sokoloff, L. (1966). The role of movement in embryonic joint development. *Developmental Biology*, **14**, 401–420.

Dreyfus-Brisac, C.C., Samson-Dolfuss, D. & Fischgold, H. (1955). L'activité electrique cerébrale du premature et du nouveau-né. *Seminaires de l'Hôpital, Paris*, **31**, 1783–1790 (pp. 135–142).

Edelman, G.M. (1993). A golden age for adhesion. *Cell Adhesion Communications*, **1**, 1–7.

Edwards, D.D. & Edwards, J.S. (1970). Fetal movements; development and time course. *Science*, **169**, 95–97.

Erbkam, K.J. (1837). Lebhafte Bewegung eines viermonatlichen Foetus. *Neue Zeitschrift für Geburtskunde*, **5**, 324–326.

Fitzgerald, J.E. & Windle, W.F. (1942). Some observations on early human fetal activity. *Journal of Comparative Neurology*, **76**, 159–167.

Fitzgerald, M. (1987). Cutaneous primary afferent properties in the hind limb of the neonatal rat. *Journal of Physiology (London)*, **383**, 79–92.

Fitzgerald, M. & Koltzenburg, M. (1986). The functional development of descending inhibitory pathways in the dorsolateral funiculus of the newborn rat spinal cord. *Brain Research*, **389**, 261–270.

Floeter, M.K. & Lev-Tov, A. (1993). Excitation of lumbar motoneurons by the medial longitudinal fasciculus in the *in vitro* brain stem spinal cord preparation of the neonatal rat. *Journal of Neurophysiology*, **70**, 2241–2250.

Florini, J.R., Ewton, D.Z. & Magri, K.A. (1991a). Hormones, growth factors and myogenic differenti-ation. *Annual Review of Physiology*, **53**, 201–216.

Florini, J.R., Ewton, D.Z. & Roof, S.L. (1991b). Insulin-like growth factor-I stimulates myogenic differentiation by induction of myogenic gene expression. *Molecular Endocrinology*, **5**, 718–724.

Florini, J.R., Magri, K.A., Ewton, D.Z., James, P.L., Grindstaff, K. & Rotwein, P.S. (1991c). "Spontaneous" differentiation of skeletal myoblasts is dependent on autocrine secretion of insulin-like growth factor II. *Journal of Biological Chemistry*, **266**, 15917–15923.

Forssberg, H. (1985). Ontogeny of human locomotor control: I. Infant stepping, supported locomotion, and transition to independent locomotion. *Experimental Brain Research*, **57**, 480–493

Forssberg, H., Dietz, V. (1997). Neurobiology of normal and impaired locomotor development. In K.J. Connolly & H. Forssberg (Eds.), *Neurophysiology and neuropsychology of motor development*. (pp. 89–100). London: McKeith Press.

Fulton, B.P. (1995). Gap junctions in the developing nervous system. *Perspectives in Developmental Neurobiology*, **2**, 327–334.

Fulton, B.P. & Walton, K. (1987). Electrophysiological properties of neonatal rat motoneurones studied *in vitro*. *Journal of Physiology (London)*, **382**, 651–678.

Fulton, B.P., Miledi, R. & Takahashi, T. (1980). Electrical synapses between motoneurons in the spinal cord of the newborn rat. *Proceedings of the Royal Society of London, B.* **208**, 115–120.

Geisler, H.C., Westerga, J. & Gramsbergen, A. (1993). Development of posture in the rat. *Acta Neurobiologia Experimentalis*, **53**, 517–523.

Geisler, H.C., Van der Fits, I.B.M. & Gramsbergen, A. (1996a). The effect of vestibular deprivation on motor development in the rat. *Behavioral Brain Research*, **86**, 89–96.

Geisler, H.C. & Gramsbergen, A. (1998). Development of the EMG of the long back muscles after vestibular deprivation. *Journal of Vestibular Research*, **8**, 1–11.

Geisler, H.C., Westerga, J. & Gramsbergen, A. (1996b). The function of the long back muscles; an EMG study in the rat. *Behavioral Brain Research*, **80**, 211–215.

Geisler, H.C., Westerga, J., IJkema-Paassen, J. & Gramsbergen, A. (2000). The development of dendrite bundles in motoneuronal pools of trunk muscles. *Neural Plasticity*, **7**, 193–203.

Goodman, C.S. & Tessier-Lavigne, M. (1997). Molecular mechanisms of axon guidance and target recognition. In W.M. Cowan, T.M. Jessell & S.L. Zipursky (Eds.), *Molecular and cellular approaches to neural development* (pp. 108–178). Oxford: Oxford University Press.

505

Gorgels, T.G.F.M., De Kort, E.J.M., Van Aanholt, H.T.H. & Nieuwenhuys, R. (1989). A quantitative analysis of the development of the pyramidal tract in the cervical spinal cord in the rat. *Anatomy Embryology*, **179**, 377–385.

Gottlieb, G. (1976). Conceptions of prenatal development: behavioural embryology. *Physiological Reviews*, **83**, 215–234.

Gramsbergen, A. (1976). The development of the EEG in the rat. *Developmental Psychobiology*, **9**, 501–515.

Gramsbergen, A. (1982). The effects of cerebellar hemispherectomy in the young rat. I. Behavioral sequelae. *Behavioural Brain Research*, **6**, 85–92.

Gramsbergen, A. (1998). Posture and locomotion in the rat, inter- or independent development. *Neuroscience and BioBehavioral Reviews*, **22**, 547–554.

Gramsbergen, A. (2001a). Locomotion in the rat, behavioural and neurobiological development. In W. Back & G. Clayton (Eds.), *Equine locomotion* (pp. 37–54). London: W.B. Saunders.

Gramsbergen, A. (2001b). The relation between the development of the adult walking pattern and eye-opening in the rat (Submitted).

Gramsbergen, A. & IJkema-Paassen. J. (1984). The effects of early cerebellar hemispherectomy in the rat: behavioral, neuroanatomical and electrophysiological sequelae. In S. Finger & C.R. Almli (Eds.), *Early Brain Damage* (vol. 2, pp. 155–177). London: Academic Press.

Gramsbergen, A. & Mulder, E.J.H. (1998). The influence of betamethason and dexamethason on motor development in the developing rat. *Pediatric Research*, **44**, 1–6.

Gramsbergen, A., IJkema-Paassen, J., Westerga, J. & Geisler, H.C. (1996). Dendrite bundles in moto-neuronal pools of trunk and extremity muscles in the rat. *Experimental Neurology*, **137**, 34–42.

Gramsbergen, A., IJkema-Paassen, J., Nikkels, P.G.J. & Hadders-Algra, M. (1997). Regression of polyneural innervation in the human psoas muscle. *Early Human Development*, **49**, 49–61.

Gramsbergen, A., Schwartze, P. & Prechtl, H.F.R. (1970). The postnatal development of behavioral states in the rat. *Developmental Psychobiology*, **3**, 267–280.

Gramsbergen, A., Van Eykern, L.A., Taekema, H.C. & Geisler, H. C. (1999). The activation of back muscles during locomotion in the developing rat. *Developmental Brain Research*, **112**, 217–228.

Grasso, R., Assaiante, C. Prevost, P. & Berthoz, A. (1998). Development of anticipatory orienting strategies during locomotor tasks in children. *Neuroscience and BioBehavioral Reviews*, **22**, 533–540.

Greensmith, L. & Vrbova, G. (1991). Neuromuscular contacts in the developing rat soleus depend on muscle activity. *Developmental Brain Research*, **62**, 121–129.

Greensmith, L. & Vrbova, G. (1996). Motoneurone survival: a functional approach. *Trends in Neurological Sciences,* **19**, 450–459.

Gribnau, A.A.M., De Kort E.J.M., Dederen, P.J.W.W.C. & Nieuwenhuys R (1986). On the development of the corticospinal tract in the rat. II. An anterograde tracer study of the outgrowth of the corticospinal fibers. *Anatomy and Embryology,* **175**, 101–110.

Grillner, S. (1975). Locomotion in vertebrates: central mechanisms and reflex interaction. *Physiological Reviews*, **55**, 247–304.

Grillner, S. (1981). Control of locomotion in bipeds, tetrapods, and fish. In V.B. Brooks (Ed.), *Handbook of physiology* (pp. 1179–1236). Bethesda, MD: American Physiological Society.

Grillner, S. & Zangger, P. (1975). How detailed is the central pattern generator for locomotion? *Brain Research*, **88**, 367–371.

Grillner, S., Wallén, P., Brodin, L. & Lansner, A. (1991). Neuronal network generating locomotor behaviour in lamprey: circuitry, transmitters, membrane properties, and simulation. *Annual Review of Neuroscience*, **14**, 169–199.

Grishkat, H.L. & Eisenman, L.M. (1995). Development of the spinocerebellar projection in the prenatal mouse. *Journal of Comparative Neurology*, **363**, 93–108.

Hadders-Algra, M. & Forssberg, H. (1996). Ontogeny of postural adjustments during sitting in infancy: variation, selection and modulation. *Journal of Physiology (London)*, **493**, 273–288.

Hadders-Algra, M., Brogren, E. & Forssberg, H. (1998). Development of postural control – differences between ventral and dorsal muscles? *Neuroscience and BioBehavioral Reviews*, **22**, 501–506.

Hall, M. (1850). *Synopsis of the diastaltic nervous system; or the system of the spinal marrow, and its reflex arcs; as the nervous agent in all the functions of ingestion in the animal economy.* London: J. Mallet.

Hall, Z.W. & Sanes, J.R. (1993). Synaptic structure and development: the neuromuscular junction. *Neuron*, **10** (Suppl.), 99–121.

Hamburger, V. (1963). Some aspects of the embryology of behaviour. *Quarterly Review of Biology*, **38**, 342–365.

Hamburger, V. (1973). Anatomical and physiological basis of embryonic motility in birds and mammals. In G. Gottlieb (Ed.), *Studies on the development of behaviour and the nervous system*, Vol. 1: Behavioural embryology (pp. 63–76). New York: Academic Press .

Hamburger, V. (1988). Ontogeny of neuroembryology. *Journal of Neuroscience*, **8**, 3535–3540.

Harrison, R.G. (1904). An experimental study of the relation of the nervous system to the developing musculature of the frog. *American Journal of Anatomy*, **3**, 197–220

Henrikson, C.K. & Vaughn, J.E. (1974). Fine structural relationships between neurites and radial glial processes in developing mouse spinal cord. *Journal of Neurocytology*, **3**, 659–675.

Hobbelen, J. F., Gramsbergen, A. & Van Hof, M.W. (1992). Descending motor pathways and the hopping response in the rabbit. *Behavioural Brain Research*, **51**, 217–221.

Hoh, J.F.Y. (1991). Myogenic regulation of mammalian skeletal muscle fibres. *News in the Physiological Sciences*, **6**, 1–6.

Hollyday, M. (1985). Development of motor innervation in vertebrate limbs. In G.M. Edelman, W.E. Gall, W.M. Cowan (Eds.), *Molecular bases of neural development* (pp. 243–264). Chichester: John Wiley & Sons.

Holstege, G. (1988). Brainstem–spinal cord projections in the cat, related to control of head and axial movements. In J. Buttner-Ennever (Ed.), *Neuroanatomy of the oculomotor system* (pp. 429–468). Amsterdam: Elsevier.

Holstege, G. (1991). Descending motor pathways and the spinal motor system: limbic and non-limbic components. *Progress in Brain Research,* **87**, 307–421.

Holstege, G. (1995). The basic, somatic, and emotional components of the motor system in mammals. In G. Paxinos (Ed.) *The rat nervous system* (2nd ed.; pp. 137–154). San Diego: Academic Press.

Hooker, D. (1930). Spinal cord section in rat fetuses. *Journal of Comparative Neurology*, **50**, 413–459.

Hooker, D. (1952). *The prenatal origin of behaviour*. Lawrence, KA: University of Kansas Press.

Houk, J.C. & Wise, S.P. (1995). Distributed modular architectures linking basal ganglia, cerebellum, and cerebral cortex: their role in planning and controlling action. *Cerebral Cortex*, **2**, 95–110.

Hytten, F.E. (1982). Breech presentation: is it a bad omen? *British Journal of Obstetrics and Gynaecology*, **89**, 879–880.

Ianniruberto, A. & Tajani, E. (1981). Ultrasonograhic study of fetal movements. *Seminars in Perinatology*, **5**, 175–181.

IJkema-Paassen, J. & Gramsbergen, A. (1998). Regression of polyneural innervation in the psoas muscle of the developing rat. *Muscle and Nerve*, **21**, 1058–1063.

Jacobson, M. (1991). *Foundations of neuroscience*. New York: Plenum Press.

Jamon, M. & Clarac, F. (1998). Early walking in the neonatal rat: a kinematic study. *Behavioral Neuroscience,* **112**, 1218–1228.

Jansen, J.K.S. & Fladby, T. (1990). The perinatal reorganization of the innervation of skeletal muscle in mammals. *Progress in Neurobiology*, **34**, 39–90.

Jeannerod, M. (1985). *The brain machine: the development of neurophysiological thought*. Cambridge, MA: Harvard University Press.

Joosten, E.A.J., Gribnau, A.A.M. & Dederen, P.W.J.C. (1987). An anterograde tracer study of the developing corticospinal tract in the rat: three components. *Developmental Brain Research*, **36**, 121–130.

Joosten, E.A.J., Gribnau, A.A.M. & Dederen, P.J.W.C. (1989). Postnatal development of the cortico-spinal tract in the rat. An ultrastructural anterograde tracer study. *Anatomy and Embryology*, **179**, 449–456.

Jouvet-Mounier, D., Astic, L. & Lacote, D. (1970). Ontogenesis of the states of sleep in rat, cat and guinea pig during the first postnatal month. *Developmental Psychobiology*, **2**, 216–239.

Kalverboer, A.F. (1975). *A neurobehavioural study in pre-school children*. Clin. Dev. Med. 54. London: SIMPS, Heinemann.

Kelly, A.M. (1983). Emergence of specialization of skeletal muscle. In L.D. Peachey (Ed.), *Handbook of Physiology, Section 10* (pp. 507–537). Bethesda, MD: American Physiology Society.

Kelly, A.M. & Rubinstein, N.A. (1986). Development of neuromuscular specialization. *Medical Science in Sports Exercise*, **18**, 292–298.

Kelly, A.M. & Zacks, S.I. (1969). The histogenesis of rat intercostal muscle. *Journal of Cell Biology*, **42**, 154–167.

Kernell, D. (1998). The final common pathway in postural control – developmental perspective. *Neuroscience and BioBehavioral Reviews*, **22**, 479–484.

Kirby, M.L., & Holtzman, S.G. (1982). Effects of chronic opiate administration on spontaneous activity of fetal rats. *Pharmacology and Biochemical Behavior*, **16**, 263–269.

Kretschmer, A. & Schwartze, P. (1974). The dependency of sensomotoric reactions on sleep waking behaviour in postnatal growing rats. In L. Jilek & S. Trojan (Eds.) *Ontogenesis of the brain* (vol 2, pp. 327–335). Prague: Charles University Press.

Kudo, N. & Yamada, T. (1985). Development of the monosynaptic stretch reflex in the rat: an *in vitro* study *Journal of Physiology (London)*, **369**, 127–144.

Kudo, N. & Yamada, T. (1987). Morphological and physiological studies of development of the monosynaptic pathway in the rat lumbar spinal cord. *Journal of Physiology (London)*, **389**, 441–459.

Kudo, N., Furukawa F. & Okado, H. (1993). Development of descending fibers to the rat embryonic spinal cord. *Neuroscience Research*, **16**, 131–141.

Kugelberg, E. (1976). Adaptive transformation in rat soleus motor units during growth. *Journal of Neurological Science*, **27**, 269–282.

Kuo, Z-T. (1932). Ontogeny of embryonic behaviour in Aves. V. The reflex concept in the light of embryonic behaviour in birds. *Psychology Reviews*, **39**, 499–515.

Kuo, Z.-T. (1963). Total patterns, local reflexes or gradients of response? *Proceedings, 16th International Congress on Zoology*, **4**, 371–374.

Kuypers, H.G.J.M (1962). Corticospinal connections: postnatal development in the rhesus monkey. *Science*, **138**, 678–680

Kuypers, H.G.J.M. (1982). *A new look at the organization of the motor system*. (Progress in Brain Research, vol. 57, pp. 381–403). Amsterdam: Elsevier.

Lakke, E.A.J.F. (1997). The projections to the spinal cord of the rat during development; a time table of descent. *Advances in Anatomy, Embryology and Cell Biology*, **135**, 1–137.

Lakke, E.A.J.F. & Marani, E. (1991). The prenatal descent of rubro-spinal fibers through the spinal cord of the rat: a study using retrograde transport of (WGA-) HRP. *Journal of Comparative Neurology*, **314**, 67–78.

Lawrence, D.G. & Hopkins, D.A. (1976). The development of motor control in the rhesus monkey: evidence concerning the role of corticomotoneuronal connections. *Brain*, **99**, 235–254.

Lemon, R.N., Armand, J., Olivier, E. & Edgley, S.A. (1997). Skilled action and the development of the corticospinal tract in primates. In K.J. Connolly, & H. Forssberg (Eds.), *Neurophysiology and neuropsychology of motor development* (pp. 78–100). London: MacKeith Press.

Leong, S.K., Shieh, J.Y. & Wong, W.C. (1984). Localizing spinal-cord-projecting neurons in neonatal and immature albino rats. *Journal of Comparative Neurology*, **228**, 18–23.

Liang, F., Moret, V., Wiesendanger M. & Rouiller, E.M. (1991). Corticomotoneuronal connections in the rat: evidence from double-labeling of motoneurons and corticospinal axon arborizations. *Journal of Comparative Neurology*, **308**, 169–179.

Massion, J. (1992). Movements, posture and equilibrium: interaction and coordination *Progress in Neurobiology*, **38**, 35–56.

McLennan, I.S. (1994). Neurogenic and myogenic regulation of skeletal muscle formation: a critical re-evaluation. *Progress in Neurobiology*, **44**, 119–140.

Milburn, A. (1973). The early development of muscle spindles in the rat. *Journal of Cell Science*, **12**, 175–195.

Minkowski, M. (1923). Zur Entwicklunggeschichte, Lokalisation und Klinik des Fuszsohlenreflexes. *Schweizerisches Archiv für Neurologie und Psychiatrie*, **1**, 1–61.

Moessinger, A.C. (1988). Morphological consequences of depressed or impaired fetal activity. In W.P. Smotherman & S.R. Robinson (Eds.), *Behavior of the fetus* (pp. 163–174). Caldwell: The Telford Press.

Moore, S. & Walsh, F. (1985). Specific regulation of N-CAM/D2 CAM cell adhesion molecule during skeletal muscle development. *EMBO Journal*, **4**, 623–630.

Nandi, K.N., Beal, J.A. & Knight, D.S. (1990). Neurogenic period of ascending tract neurons in the upper lumbar spinal cord of the rat. *Experimental Neurology*, **107**, 187–191.

Narayanan, C.H., Fox, M.W. & Hamburger, V. (1971). Prenatal development of spontaneous and evoked activity in the rat. *Behavior,* **40**, 100–134.

Neafsey, E.J., Bold, E.L., Haas, G., Hurley-Gins, K.M., Quirck, G., Sievert, C.F. & Terryberry, R.R. (1986). The organization of the rat motor cortex: a microstimulation mapping study. *Brain Research Reviews,* **11**, 77–96.

Nornes, H.O. & Das, P.D. (1974). Temporal pattern of neurogenesis in the spinal cord of rat. I. An autoradiographic study. Time and origin and migration and settling patterns of neuroblasts. *Brain Research,* **73**, 121–138.

O'Brien, R.A.D., Oestberg, A.J.C. & Vrbova, G. (1978). Observations on the elimination of poly-neural innervation in developing mammalian skeletal muscle. *Journal of Physiology (London),* **282**, 571–582.

O'Hanlon, G.M. & Lowrie, M.B. (1994a). Dendritic development in normal lumbar motoneurons and following neonatal nerve crush in the rat. *Developmental Neuroscience,* **16**, 17–24.

O'Hanlon, G.M. & Lowrie, M.B. (1994b). Both afferent and efferent connections influence postnatal growth of motoneuron dendrites in the rat. *Developmental Neuroscience,* **16**, 100–107.

Okamoto, T. & Kumamoto, M. (1972). Electromyographic study of the learning process of walking in infants. *Electromyography,* **12**, 149–158.

Oppenheim, R.W. (1974). The ontogeny of behaviour in the chick embryo. In, D.S. Lehrman, J Rosenblatt, R.A. Hinde & E.Shaw (Eds.), *Advances in the study of behaviour* (vol. 5, pp. 133–172). New York: Academic Press.

Oppenheim, R.W. (1978). G.E. Coghill (1872–1941) pioneer, neuroembryologist and developmental psychobiologist. *Perspectives in Biological Medicine,* **22**, 45–64.

Oppenheim, R.W. (1981a). Neuronal cell death and some related regressive phenomena during neurogenesis: A selective historical review and progress report. In W.M. Cowan (Ed.), *Studies in developmental neurobiology: essays in honor of Viktor Hamburger* (pp. 74–133). Oxford: Oxford University Press.

Oppenheim, R.W. (1981b). Ontogenetic adaptations and retrogressive processes in the development of the nervous system and behaviour: a neuroembryological perspective. In K.J. Connolly & H.F.R. Prechtl (Eds.), *Maturation and development: biological and psychological perspectives* (pp. 73–109). Clin. Dev. Med., 77/78. London: Heinemann.

Oppenheim, R.W. (1986). The absence of significant postnatal motoneuron death in the brachial and lumbar spinal cord of the rat. *Journal of Comparative Neurology,* **246**, 281–286.

Oppenheim, R.W. (1991). Cell death during development of the nervous system. *Annual Review of Neuroscience,* **14**, 453–501.

Oppenheim, R.W. & Haverkamp, L.J. (1988). Neurotrophic interactions in the development of spinal cord motoneurons. In D. Evered & J. Whelan (Eds.), *Plasticity of the neuromuscular system* (pp. 152–171). Chichester: John Wiley & Sons.

Parmelee, A.H., Schulte, F.J., Akiyama, Y., Wenner, W.H., Schultz, M.A. & Stern, E. (1968). Maturation of EEG activity during sleep in premature infants. *Electroencephalography and Clinical Neurophysiology,* **24**, 319–329.

Pavlov, I.P. (1927). *Conditioned reflexes. An investigation of the physiological activity of the cerebral cortex* (G.V. Anrep, transl.). Oxford: Oxford University Press.

Petrosini, L., Molinari & M., Gremoli, T. (1990). Hemicerebellectomy and motor behaviour in rats. *Experimental Brain Research,* **82**, 472–482.

Pflueger, E.F.W. (1877). Die Lebenszaehigkeit des menschlichen Foetus. *Archiv für die Gesamte Physiologie,* **14**, 628– 629.

Porter, R. & Lemon, R.N. (1993). *Corticospinal function and voluntary movement.* Oxford: Oxford University Press.

Prechtl, H.F.R. (1965) Prognostic value of neurological signs in the newborn infant. *Proceedings of the Royal Society of Medicine,* **58**, 3–4.

Prechtl, H.F.R. (1974). The behavioural states of the newborn infant (a review). *Brain Research,* **76**, 185–212.

Prechtl, H.F.R. (1984). Continuity and change in early neural development. In H.F.R. Prechtl (Ed.), *Continuity of neural functions from prenatal to postnatal life* (pp. 1–15). Oxford: SIMP.

Prechtl, H.F.R. (1989). Development of postural control in infancy. In C. von Euler, H. Forssberg & H. Lagercrantz (Eds.), *Neurobiology of early infant behaviour* (pp. 58–69), London: Macmillan.

509

Prechtl, H.F.R. (1990). Qualitative changes of spontaneous movements in fetus and preterm infant are a marker of neurological dysfunction. *Early Human Development*, **23**, 151–158.

Prechtl, H.F.R. (1997). State of the art of a new functional assessment of the young nervous system. An early predictor of cerebral palsy. *Early Human Development*, **50**, 1–11.

Preyer, W. (1885). *Spezielle Physiologie des Embryo*. Leipzig: Grieben.

Puro, D.G. & Woodward, D.J. (1977a). Maturation of evoked climbing fiber input to rat cerebellar Purkinje cells. *Experimental Brain Research*, **28**, 85–100.

Puro, D.G. & Woodward, D.J. (1977b). Maturation of evoked mossy fiber input to rat cerebellar Purkinje cells. *Experimental Brain Research*, **28**, 427–441.

Rajaofetra, N., Sandillon, F., Geffard, M. & Orivat, A. (1989). Pre- and post-natal ontogeny of serotonergic projections to the rat spinal cord. *Journal of Neuroscience Research*, **22**, 305–321.

Rajaofetra, N., Poulat, P., Marlier, L., Geffard, M. & Privat, A. (1992). Pre- and postnatal development of noradrenergic projections to the rat spinal cord: an immunocytochemical study. *Developmental Brain Research*, **67**, 237–246.

Ramon y Cajal, S. (1909–1911). *Histologie du Systeme Nerveux et des Vertebrés* (2 tomes). Transl., L. Azoulay (reprinted in 1952–1955). Madrid: CSIC.

Ramon y Cajal, S. (1929). *Studies on vertebrate neurogenesis* (Transl. from the french edition [1960] by L. Guth). IL: C.C. Thomas.

Redfern, P.A. (1970). Neuromuscular transmission in new-born rats. *Journal of Physiology (London)*, **209**, 701–709.

Reinold, E. (1976). Ultrasonics in early pregnancy. Diagnostic scanning and fetal motor activity. In P.J. Keller (Ed.), *Contributions to Gynaecology and Obstetrics* (vol. 1, pp. 101–148). Basel: Karger.

Romanes, G.J., (1941). The development and the significance of the cell columns in the ventral horn of the cervical and upper thoracic spinal cord of the rabbit. *Journal of Anatomy (London)*, **76**, 112–130.

Romijn, H.J., Hofman, M.A. & Gramsbergen, A. (1991). At what age is the developing cerebral cortex of the rat comparable to that of the full-term newborn human baby? *Early Human Development*, **26**, 61–67.

Roney, K.J., Scheibel, A.B. & Shaw, G.L. (1979). Dendritic bundles: survey of anatomical experiments and physiological theories. *Brain Research Reviews*, **1**, 225–271.

Roux, W. (1881). *Der Kampf der Theile im Organismus*. Engelmann: Leipzig.

Rubinstein, N.A. & Kelly, A.M. (1981). Development of muscle fiber specialization in the rat hindlimb. *Journal of Cell Biology*, **90**, 128–144

Saito, K. (1979). Development of spinal reflexes in the rat fetus studied *in vitro*. *Journal of Physiology (London)*, **294**, 581–594.

Sanes, J.R. & Scheller, R.H. (1997). Synapse formation: a molecular perspective. In W.M. Cowan, T.M. Jessell & S.L. Zipursky (Eds.), *Molecular and cellular approaches to neural development* (pp. 179–219). Oxford: Oxford University Press.

Sanes, D.H. & Poo, M-M. (1988). *In vitro* analysis of specificity during nerve-muscle synaptogenesis. In D. Evered & J. Whelan (Eds.), *Plasticity of the neuromuscular system* (pp. 116–130). Chichester: Wiley.

Scheibel, M.E. & Scheibel, A.B. (1970). Developmental relationship between spinal motoneuron dendrite bundles and patterned activity in the hind limb of cats. *Experimental Neurology*, **29**, 328–335.

Schoenen, J. (1982). Dendritic organization of the human spinal cord: the motoneurons. *Journal of Comparative Neurology*, **211**, 226–247.

Schreyer, D.J. & Jones, E.G. (1982). Growth and target finding by axons of the corticospinal tract in prenatal and postnatal rats. *Neuroscience*, **7**, 1837–853.

Schreyer, D.J. & Jones, E.G. (1988). Axon elimination in the developing corticospinal tract of the rat. *Developmental Brain Research*, **38**, 89–101.

Shapiro, L.J. & Jungers, W.J. (1994). Electromyography of back muscles during quadrupedal and bipedal walking in primates. *American Journal of Physical Anthropology*, **93**, 491–504.

Sherrington, C.S. (1906). *The integrative action of the nervous system*. New Haven: Yale University Press.

Small, W.S. (1899). Notes on the psychic development of the young white rat. *American Journal of Psychology*, **11**, 80–100.

Smart, J.L. & Dobbing, J. (1971). Vulnerability of developing brain. II. Effects of early nutritional deprivation on reflex ontogeny and development of behaviour in the rat. *Brain Research*, **28**, 85–95.

Smith, C.L (1983). The development and postnatal organization of primary afferent projections to the rat thoracic spinal cord. *Journal of Comparative Neurology*, **220**, 29–43.

Smith, C.L. & Hollyday, M. (1983). The development and postnatal organization of motor nuclei in the rat thoracic spinal cord. *Journal of Comparative Neurology*, **220**, 16–28.

Smotherman, W.P. & Robinson, S.R. (1986). A method for endoscopic visualization of rat fetuses in situ. *Physiology and Behaviour,* **37**, 663–665.

Smotherman, W.P. & Robinson, S.R. (1987). Prenatal expression of species-typical action patterns in the rat fetus. *Journal of Comparative Psychology*, **101**, 190–196.

Smotherman, W.P. & Robinson, S.R. (1988). The uterus as environment: the ecology of fetal behaviour. In E.M. Blass (Ed.), *Handbook of behavioral neurobiology* (vol. 9, pp. 149–196). Developmental psychobiology and behavioral ecology. New York: Plenum.

Smotherman, W.P., Richards, L.S., Robinson, S.R. (1984). Techniques for observing fetal behaviour *in utero*: a comparison of chemomyelotomy and spinal transection. *Developmental Psychobiology*, **17**, 661–674.

Smotherman, W.P., Robinson, S.R. & Miller, B.J. (1986). A reversible preparation for observing the behaviour of fetal rats in utero: spinal anaesthesia with lidocaine. *Physiology and Behaviour*, **37**, 57–60.

Stanfield, B.B. & O'Leary, D.D.M. (1985). The transient corticospinal projection from the occipital cortex during postnatal development of the rat. *Journal of Comparative Neurology*, **238**, 236–248.

Stanfield,B.B., O'Leary, D.D.M. & Fricks, C. (1982). Selective collateral elimination in early postnatal development restricts cortical distribution of rat pyramidal tract neurones. *Nature,* **298**, 371–373.

Stickrod, G., Smotherman W.P. &Kimble, D.P. (1982). *In-utero* taste/odor aversion conditioning in the rat: Effects of nipple preference testing. *Phyiology and Behavior*, **28**, 5–7.

Takeichi, M. (1995). Morphogenetic roles of classic cadherins. *Current Opinion in Biology*, **7,** 619–627.

Taylor, P.M. (1960). Oxygen consumption in newborn rats. *Journal of Physiology*, **154**, 153–168.

Thorstensson, A., Carlson, H., Zomlefer & M., Nilsson, J. (1982). Lumbar back muscle activity in relation to trunk movements during locomotion in man. *Acta Physiologica Scandinavica*, **116**, 13–20.

Tilney, F. (1934). Behaviour in its relation to the development of the brain. *Bulletin of the Neurology Institute (New York)*, **3**, 252–358.

Tompkins, P. (1946). An inquiry into the causes of breech presentation. *American Journal of Obstetrics and Gynaecology*, **51**, 595–606.

Touwen, B.C.L. (1976). *Neurological development in infancy*. Clin Dev. Med., 58 London: Heinemann.

Ugolini, G., Kuypers, H.G.J.M. & Strick, P.L. (1989). Transneuronal transfer of herpes virus from peripheral nerves to cortex and brainstem. *Science* **243**, 89–91.

Van der Fits, I.B.M. & Hadders-Algra, M. (1998). The development of postural response patterns during reaching in healthy infants. *Neuroscience and BioBehavioral. Reviews,* **22**, 521–526.

Van der Fits, I.B.M., Klip, A.W.J., Van Eykern, L.A. & Hadders-Algra, M. (1999a). Postural adjustments during spontaneous and goal directed arm movements in the first half year of life. *Behavioral Brain Research*, **106,** 75–90.

Van der Fits, I.B.M, Otten, E., Klip, A.W.J., Van Eykern, L.A. & Hadders-Algra, M. (1999b). The development of postural adjustments during reaching in 6 - 18 months old infants: evidence for two transitions. *Experimental Brain Research*, **126**, 517–528.

Van der Want, J.J.L., Gramsbergen, A., IJkema-Paassen, J., De Weerd, H. & Liem, R.S.B. (1998). Dendro-dendritic connections between motoneurons in the rat spinal cord. An electronmicroscopic investigation. *Brain Research*, **779**, 342–345.

Vaughn, J.E. & Grieshaber, J.A. (1973). A morphological investigation of an early reflex pathway in developing rat spinal cord. *Journal of Comparative Neurology*, **148**, 177–210.

Vejsada, R., Palecek, J., Hnik, P. & Soukop, T. (1985a). Postnatal development of conduction velocity and fibre size in the rat tibial nerve. *International Journal of Neurosciences*, **3**, 583–595.

Vejsada, R., Hnik, P., Payne R. Ujec, E. & Palecek, J. (1985b). The postnatal functional development of muscle stretch receptors in the rat. *Somatosensory Research*, **2**, 205–222.

Vinay, L. & Clarac, F. (1999). Antidromic discharges of dorsal root afferents and inhibition of the lumbar monosynaptic reflex in the neonatal rat. *Neuroscience*, **90**, 165–176.

Vinay, L., Cazalets, J.-R. & Clarac, F. (1995). Evidence for the existence of a functional polysynaptic pathway from trigeminal afferents to lumbar motoneurons in the neonatal rat. *European Journal of Neuroscience*, **7**, 143–151.

Visser, G.H.A., Laurini, R.N., De Vries, J.I.P., Bekedam, D. & Prechtl, H.F.R. (1985). Abnormal motor behaviour in anencephalic fetuses. *Early Human Development*, **12**, 173–182.

Von Hofsten, C. (1991). Structuring of early reaching movements: a longitudinal study. *Journal of Motor Behavior*, **23**, 280–292.

Wakelam, M. (1985). The fusion of myoblasts. *Biochemistry Journal*, **228**, 1–122.

Wallace, J.A. & Lauder, J.M. (1983). Development of the serotonergic system in the rat embryo an immunocytochemical study. *Brain Research Bulletin*, **10**, 459–479.

Westerga, J. & Gramsbergen, A. (1990). The development of locomotion in the rat. *Developmental Brain Research*, **57**, 163–174.

Westerga, J. & Gramsbergen, A. (1992). Structural changes of the soleus and the tibialis anterior motoneuron pool during development in the rat. *Journal of Comparative Neurology*, **319**, 406–416.

Westerga, J. & Gramsbergen, A. (1993). Changes in the EMG of two major hind limb muscles during locomotor development in the rat. *Experimental Brain Research*, **92**, 479–488.

Westerga, J. & Gramsbergen A. (1994). Development of the EMG of the soleus muscle in the rat. *Developmental Brain Research*, **80**, 233–243.

Wiese, S., Digby, M.R., Gunnersen, J.M., Goetz, R., Pei, G., Holtmann, B., Lowenthal, J. & Sendtner, M. (1999). The anti-apoptotic protein ITA is essential for NGF-mediated survival of embryonic chick neurons. *Nature Neuroscience*, **2**, 978–983.

Wilson, S.J., Ross, J.J. & Harris, A.J. (1988). A critical period for formation of secondary myotubes defined by prenatal undernourishment in rats. *Development*, **102**, 815–821.

Windle, W.F. (1940). *Physiology of the fetus: origin and extent of function in prenatal life*. Philadelphia: Saunders.

Windle, W.F. & Griffin, A.M. (1931). Observations on embryonic and fetal movements of the cat. *Journal of Comparative Neurology*, **52**, 149–188.

Windle, W.F., Minear, W.L., Austin, M.F. & Orr, D.W. (1935). The origin and early development of somatic behavior in the albino rat. *Physiological Zoology*, **8**, 156–185.

Windle, W.F. & Baxter, R.E. (1936). Development of reflex mechanisms in the spinal cord of albino rat embryos. Correlations between structure and function, and comparisons with the cat and the chick. *Journal of Comparative Neurology*, **63**, 189–209.

Woodward, D.J., Hoffer, B.J. & Lapham, L.W. (1969). Postnatal development of electrical and enzyme histochemical activity in Purkinje cells of rat. *Experimental Neurology*, **23**, 120–139.

Yawa, H. (1987). Changes in the dendritic geometry of mouse superior cervical ganglion cells following postganglionic axotomy. *Journal of Neuroscience*, **7**, 3703–3711.

Zomlefer, M.R., Provencher, J., Blanchette, G. & Rossignol, S. (1984). Electromy ographic study of lumbar back muscles during locomotion in acute high decerebrate and in low spinal cats. *Brain Research*, **290**, 249–260.

Further reading

Altman, J. & Bayer, S.A. (1984). The development of the rat spinal cord. In F. Beck *et al.* (Eds.), *Advances in anatomy, embryology and cell biology* (vol. 85). Berlin: Springer Verlag.

Jacobson, M. (1991). *Developmental neurobiology*. New York: Plenum Press.

Prechtl H.F.R. (Ed.) (1984). *Continuity of neural functions from prenatal to postnatal life*. Oxford: Spastics International Medical Publications.

III.6
Development of Hand Motor Control

ANDREW M. GORDON

ABSTRACT

Coordinated movement of the arm and hand is nearly non-existent at birth. In fact, skilled manipulatory behaviours involving precise regulation of individuated finger movements and forces have a surprisingly long and protracted development, lasting many years. This chapter reviews the development of these behaviours. Specifically, the development of reaching, grasping, reach-to-grasp, finger individuation, and complex manual motor skills, as well as the neural mechanisms underlying these behaviours are discussed from a multidisciplinary perspective. Historical perspectives on the topic are compared to more contemporary knowledge. Technological advances in the study of movement and multidisciplinary approaches have afforded an opportunity to re-evaluate our conceptualizations of developmental milestones. Evidence for early reaching behaviours is discussed in relation to the subsequent development of goal-directed reaching, which we now know continues until adolescence. Similarly, the ability to configure the hand during the reach according to the intrinsic properties of the object, which also develops over many years, will be examined. Studies of prehensile forces during object manipulation are also reviewed. These studies have provided insight into the formation of appropriate synergies to simplify motor control, precise sensorimotor integration, and the planning of manipulatory behaviours. Accumulating evidence suggests that there is an important interplay between the developing nervous system and the environment for the development of manual dexterity. Finally, directions for future research on the development of hand motor control will be offered.

INTRODUCTION

Fine manual dexterity is one of the distinguishing features of human motor skills, and has been an essential part of human evolution. The repertoire of complex manual behaviours that are used daily include holding and manipulating objects,

A.F. Kalverboer and A. Gramsbergen (eds.), Handbook of Brain and Behaviour in Human Development, 513–538
© 2001 Kluwer Academic Publishers. Printed in Great Britain.

eating, dressing, writing, gesturing, tool use and, increasingly, interfacing with electronic and computer devices such as cellular telephones and miniature electronic organizers. Unlike the vocal tract, which is not configured anatomically for speech production at birth (see Kent, 1999, for review), the anatomical features of the hand are established at birth. Nevertheless, similar to speech and other coordinated voluntary behaviours such as locomotion, hand motor coordination is nearly non-existent in human newborn infants. In fact, skilled manipulatory behaviours involving precise regulation of individuated finger movements and forces develop over many years. These behaviours require the ability to regulate in an intricate fashion 19 intrinsic and 20 extrinsic hand muscles. The neural substrates underlying this control are not complete at the time of birth and may take several years to develop fully (see Chapter II.10).

In the past decade technological advances and multidisciplinary approaches have provided considerable insight into developmental processes. In this chapter I will review the development of hand motor control. I will start by describing the development of reaching behaviours, followed by the development of grasping, reach-to-grasp, individuated finger movement, and finally the coordination of prehensile forces. Recent advances in our understanding of hand motor skill development will be discussed in relation to historical perspectives. The age at which each skill emerges is plotted in Figure 1 for reference purposes throughout the chapter.

THEORETICAL BACKGROUND

Early scientists studying the development of hand function approached the topic from a biological perspective, with the emergence of skills being defined as a result of genetically determined neuromaturation. Emphasis was placed on the

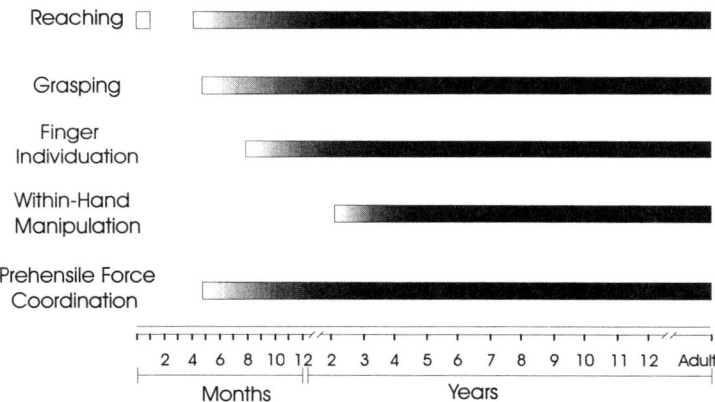

Figure 1 Development of various upper limb skills. Left sides of bars indicate approximate age at which skill emerges. Shading represents development from immature (white) to adult-like (black) coordination. Note that the intermediate shading does not indicate the proficiency of the skill or imply a linear rate of development.

achievement of motor milestones in a fixed order. Though this approach has been attractive because it characterizes development in discrete stages, it treats development as though it were static and does not provide insight into how changes occur. A subsequent shift in thinking occurred, in which emphasis was placed on the environment. In particular, the importance of problem-solving using an interactive search or discovery process in context with the environment has been highlighted. While few scientists would deny some role of the developing nervous system or the impact of the environment, is there clear evidence for the two factors interacting with each other?

There have also been debates regarding when various prehensile behaviours emerge, what the pattern of movement is and when children achieve adult skill levels. For example, when does goal-directed reaching begin? Some claim that voluntary reaching can be observed as early as the first weeks of life, while others believe these are generally random arm movements. Is reaching already adult-like during the second year of life? Some claim that the reaching movements of 2-year-olds are behaviourally nearly indistinguishable from those of adults, while others propose that subtle changes in control mechanisms continue for several years.

When does voluntary grasping emerge? Some have postulated that it can already be seen at the onset of early reaching, whereas others suggest that it occurs considerably later. When does pre-shaping of the hand based on the object's physical characteristics begin? Some have contended that it occurs as soon as voluntary reaching is evident, while others propose that it develops more slowly. Does early grasping really begin with ulnar flexion? Historically this has been thought to be the case, though this has recently been challenged. Does independent finger movement occur because of nature or nurture? How long does it take to develop adult-like hand skills? What is the neural basis of development of reaching and grasping? Does the control of prehensile forces mirror the neural development and maturation? In the following literature review I will provide historical and current perspectives on each of these issues.

CRITICAL REVIEW OF THE LITERATURE

Development of reaching

Before the classic studies of reaching in infants by Bower and associates (Bower et al., 1970; Bower, 1972), the established opinion in the field was that reaching behaviours are absent in the newborn infant (e.g. Gesell & Amatruda, 1964; Piaget, 1953; White et al., 1964). Rather, the arm movements observed at birth were generally considered to be random thrashing. Although random and aimless in appearance, Bower's studies suggested that reaching towards objects can be observed as early as the second postnatal week (Bower et al., 1970), though there is poor directional and distance control. Although initially contested (e.g. Dodwell et al., 1976; Ruff & Halton, 1978), the existence of these "pre-reaching" behaviours during the first month of life was subsequently confirmed by others (e.g. Hofsten, 1982; Trevarthen et al., 1975), and arm movements directed towards the mouth have also been documented at this age (e.g. Butterworth,

1986). Hofsten (1982) speculated that the function of these reaching behaviours appears to be more attentional than manipulative. More recently, van der Meer *et al.* (1995) recorded arm-waving movements while newborn infants lay supine with the head facing to one side (Figure 2). The infants were provided vision of only the limb they were facing, only the contralateral limb (on a monitor) or neither limb. When small forces were applied to each wrist it was found that the infants oppose the forces to keep the limb up and moving normally if they can see it, but not when they cannot see it. These findings suggest that visual control of arm movement begins shortly after birth. The authors speculated that these spontaneous arm movements may aid in the development of a bodily frame of reference, which is important for the later development of reaching skills.

From birth to around 2 months of age these rudimentary "pre-reaching" behaviours diminish (although few studies have systematically tested this U-shaped developmental function, but see Hofsten, 1984). At that time the movements made are staccato and restrained, as if the infant is attempting to get the arm to extend but is unable to do so (Hofsten, personal communication). Furthermore, the hand remains fisted throughout the arm extension (Hofsten, 1984). As the reaching behaviours reappear they are performed in a more functional manner with the hand opening during the forward extension. The diminishment of these early "pre-reaching" behaviours may be analogous to the

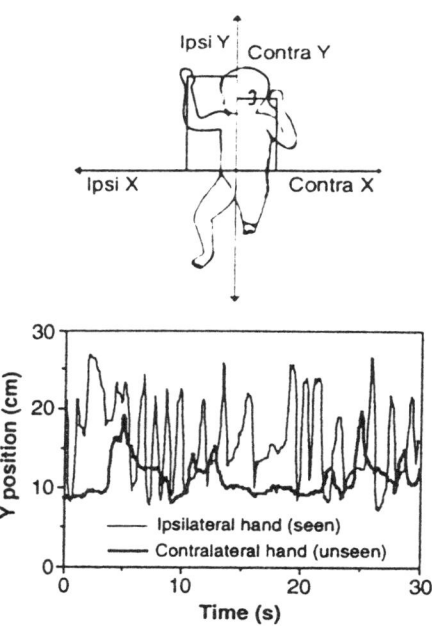

Figure 2 *Top*: schematic of position of infant. *Bottom*: movement in *y*-direction of a newborn infant waving both arms. The thin line represents the seen ipsilateral hand while the bold line represents the unseen contralateral hand. Note that the frequency and amplitude of the movements are considerably greater in the seen hand. (Adapted from van der Meer *et al.*, 1995, with permission.)

common disappearance of the step reflex prior to the emergence of supported locomotion involving voluntary goal-directed infant stepping movements (Peiper, 1963). The disappearance of the step reflex seems to parallel increases in body weight (with disproportional increases in subcutaneous fat), which the weak muscles cannot overcome (Thelen *et al.*, 1984). This could also be the case for arm movements. It has also been hypothesized that further development of the reticulospinal pathways occurs, strengthening connections between brainstem locomotor centres and spinal networks controlling locomotion (e.g. Forssberg, 1985; Forssberg & Dietz, 1997). It is tempting to speculate that a similar reorganization or strengthening of neural pathways controlling reaching occurs, perhaps in order to control the compliance of the rapidly developing upper extremity. It has been suggested that mechanisms required for the accurate movement of the arm in space are similar to, and have evolved from, mechanisms used to accurately place the foot during locomotion (Georgopoulos & Grillner, 1989). Both of these behaviours involve the motor cortex and corticospinal tract (e.g. Liddell & Phillips, 1944), and probably an interneuronal (propriospinal) system at the C3–C4 spinal segments concerned with the neural integration of limb positioning in the spinal cord (e.g. Alstermark & Lundberg, 1992; Pierrot-Deseilligny, 1996).

Regardless of the mechanism, the discrepancy in opinion concerning the emergence of reaching movements seemingly was due to the prevalence of developmental studies beginning at or after the age of 6 weeks; i.e. during the apparent diminishment of "pre-reaching." The transition from the spontaneous movements to reaching occurs when infants are capable of intentionally adjusting the force and compliance of the arm (Thelen *et al.*, 1993).

The precise control of reaching emerges rapidly after this period. Early cinematic studies of infant reaching (e.g. Halverson, 1943; McGraw, 1943) have shown that effective reaching begins around 3–5 months. Subsequently, more detailed kinematic analysis has demonstrated that accurate reaching behaviours can be observed by this age (e.g. Hofsten & Rönnqvist, 1988). At this age children are able to gauge whether an object is within reach (Field, 1977; Gordon & Yonas, 1976) and demonstrate appropriate deceleration prior to contact with the object (Hofsten, 1979). However, the movement trajectories are not linear as in adults; rather they are circuitous (Halverson, 1931, 1943; Fetters & Todd, 1987). The hand path becomes more linear by reduction of the medial deviation prior to lateral deviation (Hofsten, 1979, 1991). Earlier studies (e.g. Halverson, 1931, 1943) suggested that the reaching trajectories became nearly straight by the age of 1 year though, as shown below, subsequent analysis of the movements using modern kinematic recording techniques has shown that this is not the case.

The reaching movements in young infants are also composed of several submovements (Fetters & Todd, 1987), as are the torque profiles (Zaal *et al.*, 1999), indicating that the movements are discontinuous. The lack of unimodal bell-shaped velocity profiles during the reach may suggest an absence of effective anticipatory (prospective) control (see Brooks, 1984), and increased reliance on feedback mechanisms. It has been suggested that proprioception, rather than vision of the limb, guides early grasping (e.g. Clifton *et al.*, 1993). Interestingly, the number of submovements decreases and speed increases during reaching towards objects in the darkness (e.g. Clifton *et al.*, 1994; McCarty & Ashmead,

1999), suggesting that vision of the object influences reaching in young infants. By the age of 9 months fewer submovements are observed (Hofsten, 1979, 1991).

In a recent series of studies Konczak and colleagues have studied the kinematics and dynamics of the reaching movements in children between the age of 4 months and 3 years (Konczak *et al.*, 1995, 1997; Konczak & Dichgans, 1997). Examples of the angular and endpoint kinematics and the total joint torque, muscular torque and rate of muscular torque change are shown in Figure 3 for young infants compared with an adult. As seen in the figure, the hand paths are considerably more curvilinear for the young infants than the adults, as described above. The angular displacement and velocity of the shoulder and elbow become smoother during the course of development, though there is still curvilinearity in the hand path at 64 weeks of age. What is unique about this study is that it provides insight as to why early reaching movements are discontinuous. The multi-segmented trajectories were not due to an inability to generate sufficient peak amplitudes and muscle torque, but were instead due to inadequate temporal force control of the arm segments (Konczak *et al.*, 1995; but cf. Zaal *et al.*, 1999). By 15 months the timing of the muscle torques approximates adult behaviours, though the trajectories are not as smooth as those observed in adults (Konczak *et al.*, 1997). At this age infants still produce the moment using both flexor and extensor muscle torques, while adults produce solely flexor muscle torques, integrating gravitational and reactive torques into their movement production (Konczak *et al.*, 1995). Most kinematic parameters do not achieve adult-like levels before the age of 2 years (Konczak & Dichgans, 1997). While an inter-joint synergy between elbow and shoulder motion emerges between 12 and 15 months, the multijoint coordination continues to improve up to the third year of life. Beyond this age the movement time and velocity do not change, though the hand trajectory continues to straighten and the number of submovements decreases until the age of 12 years (Kuhtz-Buschbeck *et al.*, 1998).

In addition to changes in arm kinematics and dynamics, changes also occur in the strategies employed. In a recent study Berthier *et al.* (1999) studied the proximodistal coordination of reaching in infants. They found that infants largely use shoulder and torso rotation to reach out towards an object. They often begin reaching by bringing their hands to a common starting location and have similar reaching velocities across trials. These findings suggest that early reaching may be characterized by attempts to reduce movement complexity by limiting the degrees of freedom (Bernstein, 1967), primarily by producing movements using trunk flexion. The sitting infant must also coordinate the reaching behaviours with postural control mechanisms in an anticipatory fashion in order to maintain an upright position (see Chapter III.7).

Mirror and non-differentiated movements of both limbs are a part of an infant's motor repertoire early in life (e.g. Connolly & Stratton, 1968; Corbetta & Mounoud, 1990; White *et al.*, 1964; see Fagard, 1998). By about 6 months these bilateral reaches give way to unimanual reaching. The shift could be due to increasing independence between the two hands, or a better mapping of the motor apparati to the intrinsic properties of the object (Fagard & Jacquet, 1996). Despite the existence of ipsilateral corticospinal projections the mirror movements are probably the result of simultaneous activation of the crossed corticospinal projections via the right and left motor cortices, which may diminish during development as a result of increased transcallosal inhibition (Heinen *et al.*, 1998; Mayston *et al.*, 1999; Müller *et al.*, 1997; Reitz & Müller, 1998).

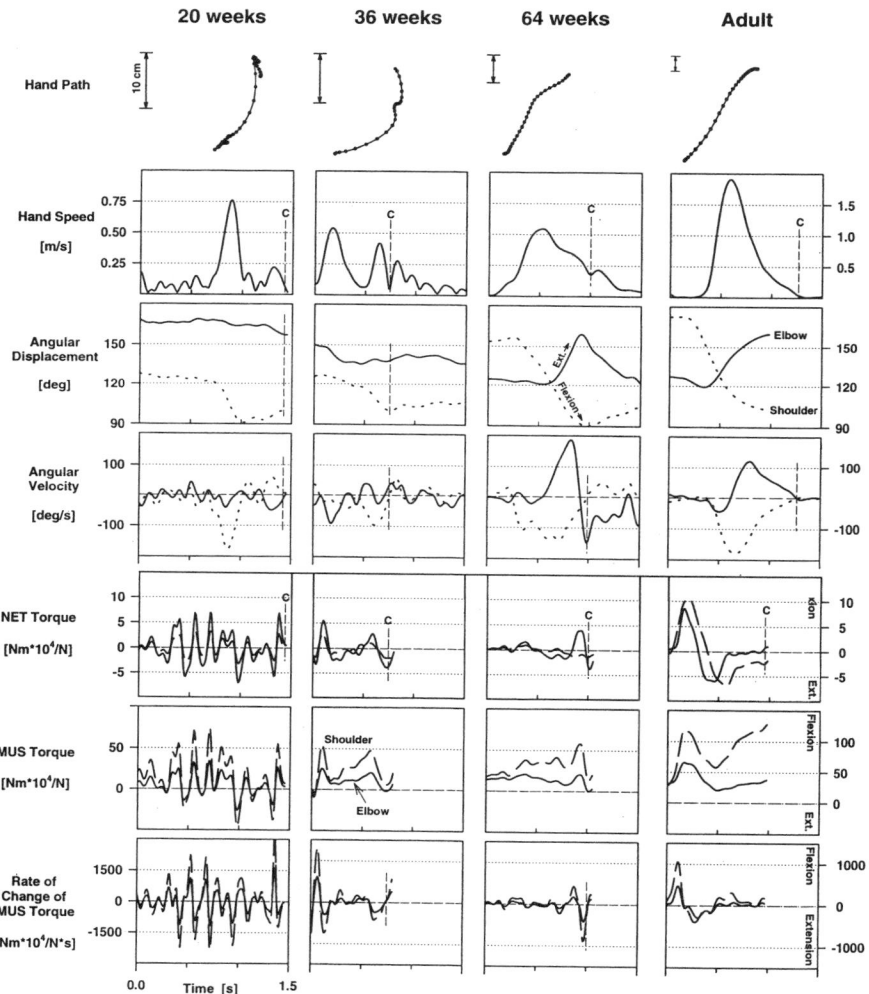

Figure 3 Angular and endpoint kinematics, profiles of total joint torque (NET), muscular torque and muscular torque rate of change for individual reaching movements in four subjects of various ages. The hand path is shown in the sagittal plane, and angular kinematics are based on sagittal projection angles. Reaching contact is indicated by the vertical line (*c*), and torque profiles are cut off after contact. Note that the hand paths become straighter, the movements less segmented, and that less muscular extension torque is employed as development progresses. (Adapted from Konczak *et al.*, 1997, with permission.)

Development of grasping

The newborn infant has a repertoire of hand and finger movements that are endogenously generated or elicited in response to external stimulation. Stereotyped reflexes are the predominant motor patterns observed in the hand. Stimulation of the palm induces hand closure (grasp reflex) (Figure 4), while stimulation of the dorsal portion of the hand may inhibit the grasp reflex or elicit hand opening (McGraw, 1943; Peiper, 1963; Twitchell, 1970). A flexor synergy with flexion of all hand muscles is elicited with traction of the arm. The grasp reflexes gradually diminish, and around 2–3 months postnatally proprioceptive contributions disappear (McGraw, 1943; Touwen, 1976). There is a question as to whether the grasp reflex disappears prior to the emergence of voluntary grasping (Illingworth, 1975), or persists and interacts with the emerging voluntary grasping (Twitchell, 1970). The answer to this question probably lies in between these two extremes, whereby descending control takes advantage of the neural circuitry involved in the grasp reflex. The reflexes become more varied and the contribution of individual fingers can be modified based on the location and characteristics of the tactile stimuli.

Voluntary grasping can be observed around 2–3 months, starting with flexion of all the fingers around the object (Connolly & Elliott, 1972; Halverson, 1931; Twitchell, 1970). It was long thought that prehension gradually develops from ulnar to forefinger grasp contact (Halverson, 1931, 1932, 1937). Halverson (1931) speculated that there might be a late maturation of the neuromusculature of the index finger. However, more recent work suggests that the index finger may play a more leading role in early grasping by contacting and initiating the grasp prior to the other fingers (Lantz *et al.,* 1996). The discrepancy could be due to the orientation of the objects with respect to the hand, since even slight deviations in orientation angles without corresponding adjustments of the wrist would alter the observations. Opposition of the thumb generally occurs around 10–12 months, about the same time at which true individuation of finger movement develops, allowing greater flexibility in the grasps employed (Connolly & Elliott, 1972; Napier, 1956; Newell & McDonald, 1997). A wide repertoire of grasp patterns quickly follows (Connolly, 1973). It should be noted that the types of

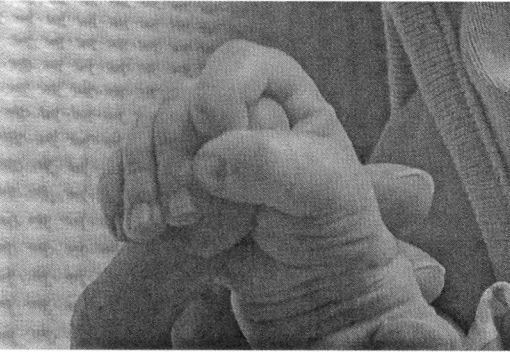

Figure 4 Example of a grasp reflex, in which placement of an object (in this case a finger) in the hand results in flexion of the hand and fingers around it.

grips elicited probably also depend on the size of the object in relation to the hand (see Newell & Cesari, 1998, for review).

Following the emergence of independent finger movement there is a considerable refinement in the ability to perform fine dexterous skills. Tool use (e.g. using a spoon) generally begins during the second year of life and evolves over an extended period (Connolly & Dalgleish, 1989). In a recent study (McCarty *et al.*, 1999), infants were presented with spoons oriented with the handle pointed towards the left or right. The researchers found that younger infants generally disregard the handle orientation and grasp the object with their preferred hand even if this resulted in an "awkward grasp". During the second year they begin to conceptualize the problem and grasp the handle with the appropriate hand using an efficient radial grip. One of the most complex skills, in-hand manipulation, develops over many years (e.g. Exner, 1990; Pehoski *et al.*, 1997a,b). The ability to accomplish various activities of daily living, including writing, buttoning and eating, is dependent on the development of in-hand manipulation. These skills include moving an object within the hand after grasp by rolling it between the fingertips, turning it end over end, moving the fingers linearly over it or moving several objects between the palm and fingertips. Interestingly, while engineers have been able to develop robotic manipulators capable of performing spatial movements and force application far more precise and consistent than the human hand, in-hand manipulation has been a much greater challenge.

Although reaching and grasping clearly develop at different rates and are controlled by functionally distinct anatomical pathways (e.g. Lawrence & Kuypers, 1968b), in reality grasping is not mutually exclusive of reaching. Rather, the hand is typically transported in space in order to bring objects within the aperture of the hand and fingers for subsequent manipulation. In adults the hand is pre-shaped according to the object's size during the deceleration of the hand prior to contact with the object (e.g. Jeannerod, 1981, 1984). There has been considerable controversy over when the aperture of the hand begins to be shaped to the size of the object. Early studies suggested that preshaping of the fingers during the reach may occur as early as the first weeks of life (Bower *et al.*, 1970). However, subsequent work has indicated that the finger shaping is rather imprecise (Newell *et al.*, 1989). There is a sudden transition from reaching without grasp to reaching with grasp, at an individual-specific age, between 14 and 24 weeks (Wimmers *et al.*, 1998). Effective orientation of the wrist during the reach also begins around 4 months of age (Hofsten & Fazel-Zandy, 1984).

Closure of the hand becomes more precise with increasing age (Fagard & Jacquet, 1996; Hofsten & Fazel-Zandy, 1984; Hofsten & Rönnqvist, 1988; Lockman, 1984). Hofsten and Rönnqvist (1988) found that at 5–6 months infants visually mediate closure of the hand in anticipation of contact with an object. However, up until the age of 13 months the hand closure during the reach begins later (i.e. closer to the point of object contact) than in adults (Figure 5). Clear preshaping according to an object's size, and hand orientation with respect to the grip axis of the object is evident by around 9 months (see also Lockman & Ashmead, 1984). The late development of preshaping according to object size may be related to the slow development of independent finger control.

When does grasping in children approximate grasping in adults? In a recent study Kuhtz-Buschbeck *et al.* (1998) examined the development of reach-to-

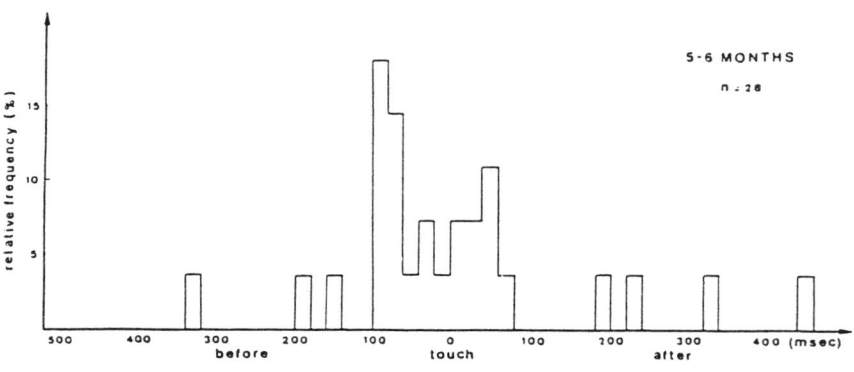

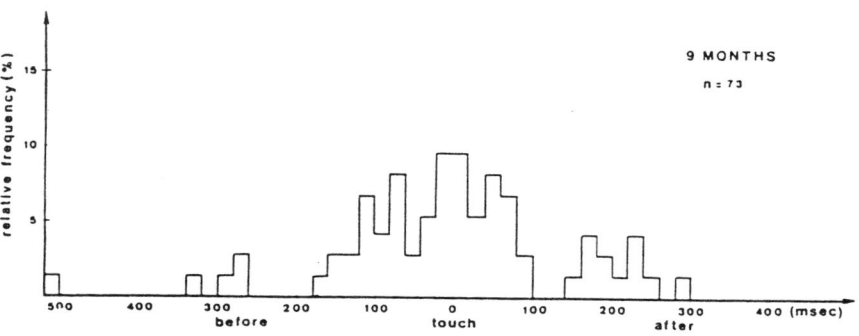

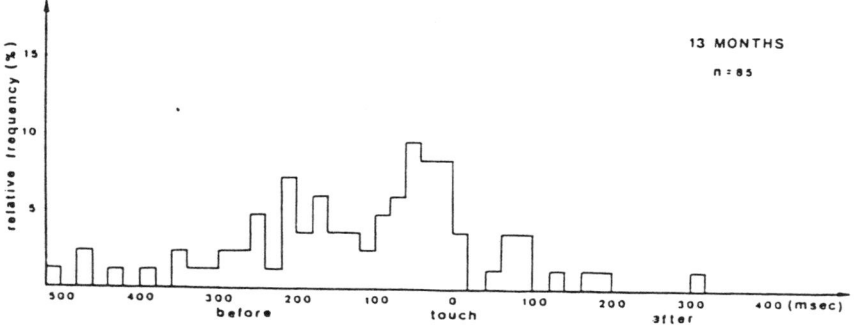

Figure 5 Relative frequency distributions of time intervals between start of hand closure and contact with the object for individual reaches performed by infants 5–6 months old (*top*), 9 months old (*middle*) and 13 months old (*bottom*). Note that hand opening is initiated considerably earlier in the 13-month-old infants. (Adapted from Hofsten & Rönnqvist, 1988, with permission.)

grasp in children age 4–12 years. They found that younger children open the hand wider relative to older children, perhaps allowing more room for error. While the maximum grip aperture is observed during deceleration in all children, it is more synchronous with peak deceleration in the older children, indicating a tighter temporal coupling. The uniformity and regularity of the movements also improves with age. This can be seen in Figure 6, which shows the hand velocity as a function of grip aperture for six superimposed trials in three children of various ages. As shown in the figure, the variability and number of velocity peaks decreases with age. Thus, changes occur in the coordination of reach-to-grasp, with increased stereotypy, until the end of the first decade of life.

Neural substrates for independent finger movement

The motor cortex and corticospinal projections provide the neural substrate for independent finger movements (Tower, 1940; Lawrence & Hopkins, 1976; Lawrence & Kuypers, 1968a; see Lemon, 1999, and Chapter II.10 for review). While corticospinal fibres descend to all levels of the spinal cord by birth, significant cortico-motoneuronal connections to the motor nuclei innervating

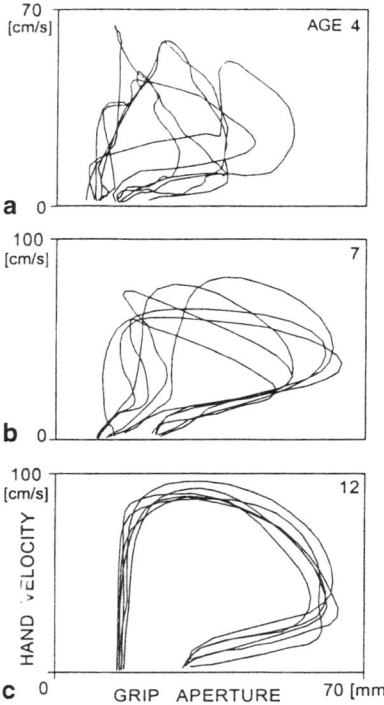

Figure 6 Hand velocity plotted as a function of grip aperture during six superimposed trials of reach-to-grasp for children age 4 years (**a**), 7 years (**b**) and 12 years (**c**). Note the increased uniformity and regularity during development. (Adapted from Kuhtz-Buschbeck *et al.*, 1998, with permission.)

the hand muscles are lacking (Armand *et al.*, 1997, Galea & Darian-Smith, 1995). There is considerable evidence that there are neurophysiological correlates between development and maturation of these projections and the emergence of independent finger movement in both the monkey (Flament *et al.*, 1992) and the human neonate (Eyre *et al.*, 1991; Koh & Eyre, 1988; Müller *et al.*, 1991; Olivier *et al.*, 1997). Furthermore, transcranial magnetic stimulation has shown that conduction time decreases until 10 years of age (Fietzek *et al.*, 2000; Müller *et al.*, 1991), suggesting continued myelination of corticospinal tract fibres.

A key question in understanding the development of the corticospinal projections concerns the issue of nature versus nurture. That is, does the emergence of independent finger motion depend on a genetically determined maturation of the corticospinal projections, or does the maturation of these projections depend on use of the hand and fingers? Recent animal studies suggest that the development of these fine dexterity skills is "epigenetic" in nature (see Sporns & Edelman, 1993). Specifically, the formation of mature corticospinal termination patterns in the cat depends on neural activity in the sensorimotor cortex during early postnatal life (Martin *et al.*, 1999). Furthermore, blockade of neural activity early during development permanently impairs the cat's ability to organize prehensile movements (Martin *et al.*, 2000). These findings may suggest that there is an early critical period, analogous to the critical periods for the development of the acoustic, visual, somatosensory and language systems, in which sensorimotor cortical activity may not only contribute to movement control, but may also shape the anatomical and functional development of corticospinal circuits (see Berardi *et al.*, 2000 for recent discussion of critical periods).

Sensorimotor control of prehensile forces

Dexterous control of the hand and fingers requires not only appropriate movement of the arm and hand during the reaching and grasping of an object, but also the precise coordination of fingertip forces once contact with the object has been made. In one of the most extensive descriptions of prehension in infants of its time, Halverson (1931) anecdotally noted:

> Young infants from birth to 6 months exhibit a grasp of force entirely disproportionate with the pressure necessary to hold and lift the seized object. Whether the object be heavy or light, it is driven hard against the palm in a vice-like palm grip which is purely an expediency for procuring and holding the object. As the age of the infant increases, the force of the grip diminishes, until at 52 weeks he takes the cube with the fingertips in a manner which suggests the presence of some appreciation of the amount of pressure required to lift and hold the cube (Halverson, 1931, p. 258).

The detailed study of prehensile force coordination was not undertaken until more than a half-century later. In an extensive series of studies which began in the 1980s, Johansson & Westling pioneered the study of the fingertip force coordination in adults by employing an intricate instrumented device (Figure 7A) capable of monitoring prehensile forces during the grasping and lifting of objects of various weights and textures (e.g. Johansson & Westling, 1984, 1987, 1988a; see Johansson, 1998, for review).

Figure 7B shows a recording of a grip-lift movement of an adult. While seemingly one simple act, the figure illustrates that the task is actually composed of a series of phases which must be fluently coordinated into one smooth action. During the *contact phase* (T0–T1), there is only a short (~20 ms) delay between contact of each opposing digit with the object since hand closure began during the reach as described above. During the subsequent *preload phase* (T1–T2), the grip (squeeze) force increases slightly until the onset of the vertical lifting drive, indicated by a positive increase in the vertical load force. This delay between contact and load force initiation is typically ~60–90 ms, and the onset of the lifting drive is thought to be triggered by rapidly adapting mechanoreceptive afferents in the fingertips signalling grip establishment as the delay is considerably prolonged following digital anesthesia (Johansson & Westling, 1984) and in individuals with sensory impairments (e.g. Eliasson *et al.*, 1991; Gordon & Duff, 1999a,b). In the following *loading phase* (T2–T3), the grip and load forces increase in parallel (i.e. they are coupled until the load force overcomes the gravitational forces acting on the object and the object is lifted from its support surface). This phase is characterized by rates of force increase (first derivative of the grip and load force) that are mainly single-peaked and bell-shaped, suggesting a well-programmed lift. During the subsequent *transport phase* (T3–T4), the object is repositioned to a desired height above the table surface and then is held stationary in the air in the *static phase*.

Once the basic control mechanisms in adults began to be understood, Forssberg and his colleagues began the comprehensive study of fingertip force coordination during object manipulation in children (Eliasson *et al.*, 1995a;

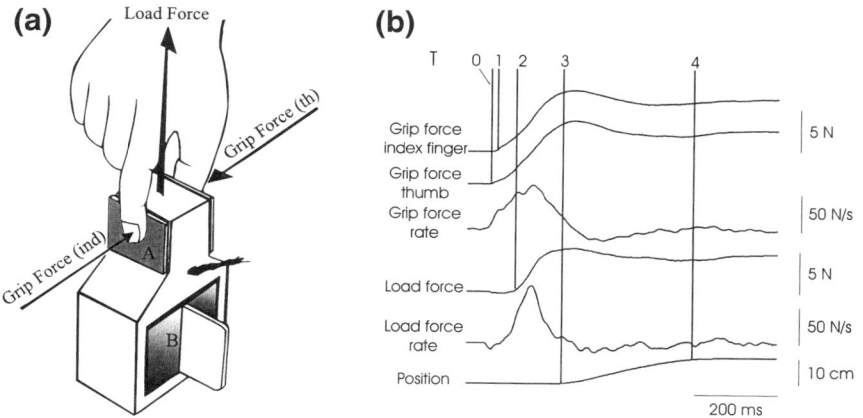

Figure 7 (**a**) Schematic drawing of the grip instrument used during developmental studies. A: Exchangeable grip surfaces covering strain-gauge force transducers at the thumb (th) and index finger (ind), B: exchangeable mass, and infrared light-emitting diode projecting to a photo-electric position-sensing camera (not shown). (**b**) Temporal (T) parameters associated with a lift. The object is contacted first by one finger (T0). Both grip forces begin to increase (T1), before the onset of positive load force (T2). The grip force and load force increase during the load phase until lift-off (T3), and the object is lifted (T4) during the transport phase. (Modified from Forssberg *et al.*, 1991, with permission.)

Forssberg *et al.*, 1991, 1992, 1995; Gordon *et al.*, 1992, 1994; see also Gordon & Forssberg, 1997, for review). As described below, the coordination of this task takes many years to reach adult-like performance. These studies have not only provided a more detailed description of the development of grasping, but have also afforded a window into the mechanisms underlying the neural development and control of movement.

Force coordination

In their first developmental study of this task, Forssberg *et al.* (1991) examined the force coordination in 134 children and adults. Figure 8 shows the coordination of fingertip forces during five superimposed trials in various children between the ages of 1 and 8 years, as well as in an adult. In the adult the force coordination is smooth and rather invariant, reflecting a highly automatized

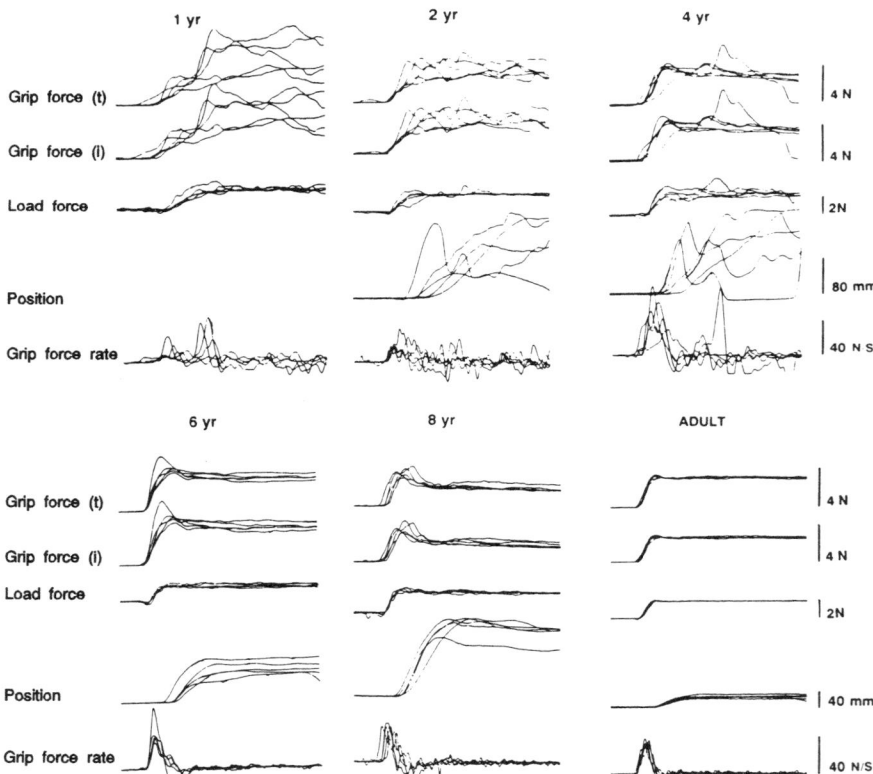

Figure 8 Grip force from the thumb (t) and index finger (i), load force, vertical position, and grip force rate (first derivative of grip force) during several lifts for one child at each age and an adult. Position signals are absent for the 1-year-old since these children often grasped the object from their experimenter's palm. Note the higher and more variable forces in the young children. (Modified from Forssberg *et al.*, 1991, with permission.)

grasping and lifting behaviour. The rates of force increase are unimodal and bell-shaped, indicating a well-programmed force increase. In contrast, the forces are markedly higher in the youngest children, and there is considerably more inter- and intra-trial variability than in the adult. The grip force rates are multi-peaked, reflecting a discontinuous force increase. There is a gradual reduction both in the force and in the variability of force, and the performance approximates that of the adult by 6–8 years of age. At that time children are able to coordinate their force output with the movement kinematics (Paré & Dugas, 1999).

Sequential coordination

The temporal control described above is also prolonged in the children compared with the adults (Forssberg *et al.*, 1991). This is shown in Figure 9, which plots the latencies of the contact phase and preload phase for each age group studied. For the children below 1 year of age the latency between contact of opposing digits is more than three times longer than in the adults. This may reflect less efficient shaping of the hand prior to contact with the object, as described above (Hofsten & Rönnqvist, 1988). Similarly, the preload phase duration is more than four times longer than in the adults, which may represent an inability to efficiently use tactile information to trigger the release of motor commands underlying the lifting drive. These results suggest that young children are unable to form smooth transitions between movement phases. Over the first 3 years of development the durations decrease rapidly, and by the age of 8 years the durations are similar to those in adults.

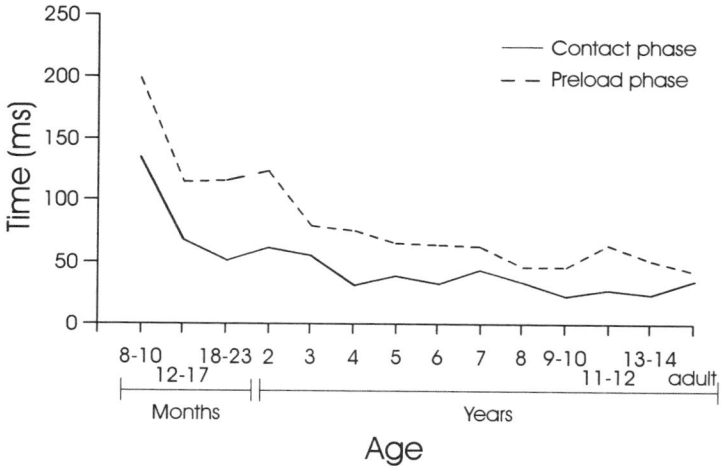

Figure 9 Duration between contact of the thumb and index finger and the preload phase duration for children of various ages and adults. Note the decrease in temporal delays during development. (Plotted from data in Forssberg *et al.*, 1991.)

Force coupling

Figure 10 shows the grip force plotted as a function of the load force during the force increase for children of various ages and an adult (Forssberg *et al.,* 1991). As indicated by the diagonal trajectory of the plots, the forces are tightly coupled in the adult, indicating a functional grip–lift synergy simplifying the control of these two forces (Bernstein, 1967). The grip–lift synergy is not seen in young children. Instead, the grip force initially increases with a negative load force (as seen by forces to the left of the *y*-axis) in which the object is pushed against its surface during the prolonged preload phase. This is in agreement with the observations by Halverson (1931) 70 years ago. Pushing the object down against its support may assist in the stabilization of grasp and/or the achievement of additional tactile information related to the surface texture. After a considerable portion of the final grasp force has been achieved, the load force begins to increase, though still not in parallel with the grip force. Thus, young children exhibit a sequential increase of the grip and load forces. The excessive grasp forces provide an increased safety margin to avoid slips (see below), though it limits the delicate sensorimotor control of the fingertip forces. By the second year the grip and load forces begin to increase more in parallel, and the coupling increases until adolescence. Interestingly, the grip–lift synergy seemingly involves a number of different cortical and subcortical areas since lesions in many CNS sites cause a disruption in the force coupling (Forssberg *et al.,* 1999).

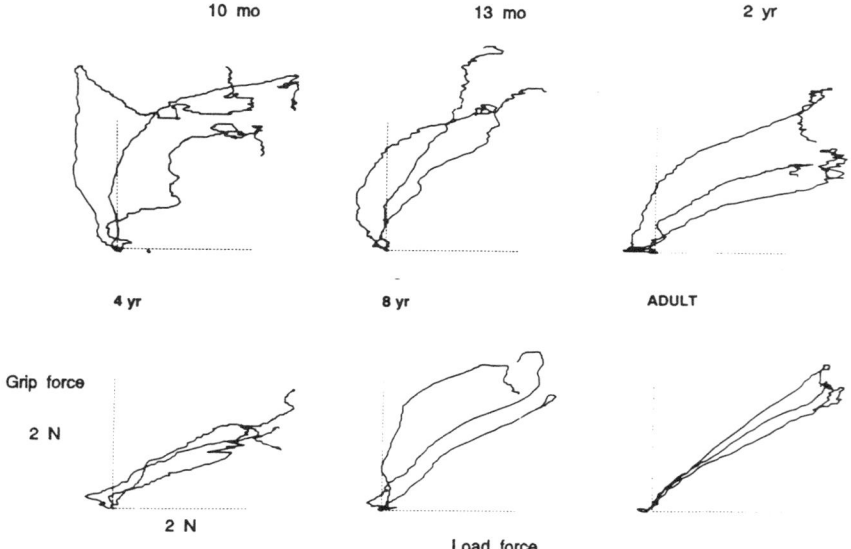

Figure 10 Grip force as a function of load force during the preload and load phase for children of various ages and an adult. Note the increased force coupling during development. (Modified from Forssberg *et al.,* 1991, with permission.)

Sensory control

To efficiently hold an object the grip forces must be adjusted to the weight and texture of the object in order to prevent accidental dropping or fatigue. The forces must exceed the slip ratio (grip force/load force ratio when the object begins to slip from the digits) by creating a small safety margin. They must also be appropriately adjusted to the physical characteristics of the manipulated object; this adjustment is mediated by mechanoreceptive afferents in the skin and by proprioception (Johansson & Westling, 1984; see Johansson, 1998). Following their initial work on the basic coordination of precision grasping in children, Forssberg and colleagues examined the development of sensory (and anticipatory) adaptation of the fingertip force output during grasping (Forssberg *et al.*, 1992, 1995).

Figure 11A shows the safety margin (expressed as a percentage of the minimum grip force/load force ratio necessary to prevent slips) during the static phase for lifts with sandpaper surfaces for children of various ages and adults. The figure shows that the youngest children employ excessive grasping forces (and thus higher safety margins), as described above, which decrease to adult levels by the age of 6 years. As shown in Figure 11B, there is considerably more variation in the static grip force in the young children, which again approximates adult levels by about 6 years. The higher safety margins may be employed to ensure that the grasping variability does not result in grip force/load force ratios fluctuating below the slip ratio.

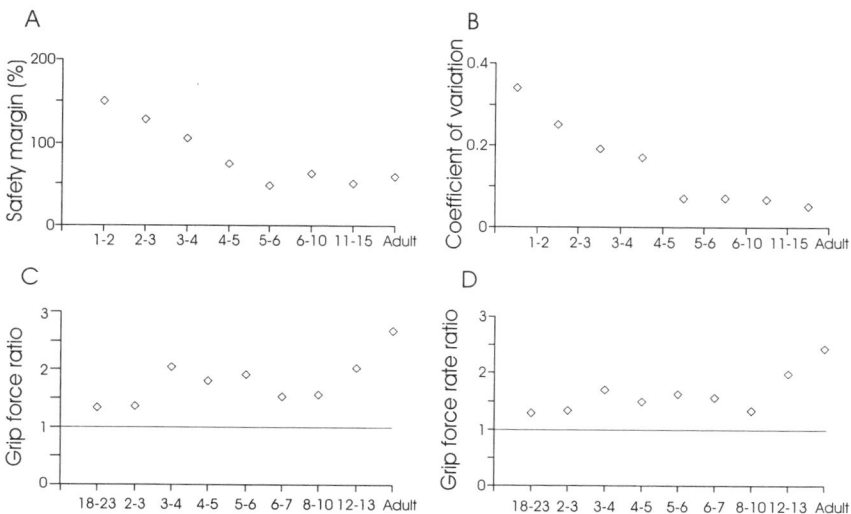

Figure 11 A: Safety margin for lifts with sandpaper contact surfaces expressed as a percentage of the slip ratio; **B**: coefficient of variation (SD/mean) for static grip forces; **C**: ratio between grip forces employed for sandpaper and silk contact surfaces when the object is held in the air; **D**: ratio between grip force rates employed for sandpaper and silk contact surfaces. Note the decrease in force and variability. (Plotted from data in Forssberg *et al.*, 1995.)

Despite the higher and more variable forces the young children typically are able to adapt their grasping forces to the object's surface texture (Figure 11C); i.e. higher forces are used for the more slippery (silk) surfaces than rough ones (sandpaper), as seen by higher ratios between the static grip forces employed for lifts with the sandpaper and silk contact surfaces in the figure. However, the adaptation is not as distinct in the young children as it is in the adults (as seen by higher force ratios between the two grip surfaces). Similar findings have also been observed when the weight of the object, rather than the texture, is manipulated (Forssberg et al., 1992). These findings suggest that children have higher and more variable forces that are not as well adapted to the object's properties using sensory information as in adults.

Anticipatory control

Despite the intricate sensorimotor control mechanisms described above, adjustments in motor output are subject to delays in processing sensory information. Sensory information related to an object's texture takes ~0.1 s to influence the force development in adults, and sensory information signalling an object's weight is not available until after the object is lifted off its support. If the initial grip forces during manipulation are increased too quickly, excessive force may result in the damaging of fragile objects (or hand). On the other hand, if the grip force is developed too slowly in relation to the load force (i.e. an inappropriate ratio between these forces occurs), the object may be dropped. Heavy or slippery objects also require an inordinate amount of time to lift if the forces are developed too slowly. To avoid these problems, and reduce the initial requirement for sensory-based control, the development of isometric forces must be scaled (planned) prior to the initiation of the movement to match the object's expected weight and texture (Gordon et al., 1993; Johansson & Westling, 1988a). The force scaling is based on internal representations of the object gained during previous manipulatory experience. Such "anticipatory control" of the force output is characterized by bell-shaped force rate profiles with the amplitude scaled from the onset towards the target load force (i.e. faster rates of force increase for more slippery or heavier objects). Figure 11D shows the ratio in the grip force rates employed between lifts with the sandpaper and silk contact surfaces for children of various ages and adults. Similar to the sensory adaptation, all age groups displayed anticipatory control, as seen by ratios in the grip force rates above 1. However, the ratios are considerably reduced in young children. Many children below 18 months of age do not show any anticipatory force scaling (Forssberg et al., 1992). As shown above (Figure 7), the force rates are generally multi-peaked (discontinuous), suggesting that they are not as well programmed, and instead employ a cautious "probing" strategy (Gordon et al., 1991). The anticipatory control improved with age. Interestingly, when the abilities to adjust the fingertip forces and scale the force rates are compared, there is a strong relationship between these two parameters across age groups ($r = 0.94$). This highlights the importance of sensory information in both processes (see Gordon & Duff, 1999a,b).

With additional manipulatory experience children begin to be able to use visual geometric size cues to predict the weight of the object in the latter half of the third year (Gachoud *et al.*, 1983; Gordon *et al.*, 1992). Furthermore, children can generalize the internal representations since they can use sensory information obtained during manipulatory experience with one hand to subsequently scale the force increase in advance during lifts with the contralateral hand (Gordon *et al.*, 1994, 1999).

Reflexive control

Despite our best intentions, slips of objects from the fingertips occasionally occur, particularly when lifting objects that are heavier or more slippery than expected, or following an unexpected perturbation (e.g. while holding a dog's leash). Triggered responses following an unexpected loading of an object in adults can be elicited within as little as ~60 ms (Johansson & Westling, 1988b; Winstein *et al.*, 2000). In children the grip response is longer (~100 ms) and weaker (Eliasson *et al.*, 1995a). Young children have a short-latency EMG burst ~20 ms (probably corresponding to the spinal excitatory component of the cutaneomuscular reflex; see Jenner & Stephens, 1982), which is not present in older children. These studies suggest that there is a reorganization of reflexive pathways from more stereotyped patterns organized in the spinal cord to more flexible supraspinal reflexes.

DISCUSSION

The development of manual dexterity is a long process in which complex skills continue to improve until adolescence. This process was traditionally thought to be mediated by genetically determined maturation of the developing nervous and muscular system. However, emerging evidence suggests that the development of arm and hand motor control is more epigenetic in nature; i.e. there is an important interplay between the developing nervous system and the environment (see Sporns & Edelman, 1993).

Technological advances in kinematic and kinetic recording methods are facilitating critical insight into the control processes mediating the development of various motor skills. We now know that infants exhibit a crude form of "pre-reaching" in which they develop visual hand regard that probably aids the development of internal models for the control of subsequent goal-directed reaching at around 4 months. The ability to grasp begins to develop shortly after this. Despite the rapid initial development, both reaching and grasping continue to be refined until adolescence, and the underlying mechanisms are now beginning to be understood. Seemingly, complex skills which have developed relatively late evolutionarily (e.g. in-hand manipulation) have a longer ontogeny.

The ability to record fingertip forces during grasping has also led to considerable advances in our understanding of object manipulation, and provides insights into the control processes underlying movement. Measurement of these forces has dispelled the common belief that, because of immature neuromusculature of the index finger, early grasping is characterized solely by ulnar flexion. We also now know that early grasping lacks many of the features necessary for skilful object manipulation, including the formation of appropriate synergies to simplify motor control, precise sensorimotor integration, and clear anticipatory control. The study of prehensile forces during object manipulation has also provided insights into the nature of developmental neurological disorders such as cerebral palsy (Eliasson *et al.*, 1991, 1992, 1995b; Eliasson & Gordon, 2000; Forssberg *et al.*, 1999; Gordon & Duff, 1999a,b; Gordon *et al.*, 1999).

Though considerable progress has been made in the understanding of the mechanisms underlying the development of manual activity, there are still considerable gaps in our knowledge. For example, the hypothesis that early critical experience-dependent periods exist in which sensorimotor cortical activity shapes the anatomical and functional development of corticospinal circuits needs further exploration. When are these periods and what is the nature of the activity required? Can the developing nervous system underlying grasping be shaped or accelerated by intensive early experience as in locomotion (Vereijken & Thelen, 1997) or postural control (Hadders-Algra *et al.*, 1996; Woollacott, 1994), and does this influence the progression of skills? Further study of the development of finger individuation and its relationship with available grip configuration and complex motor skills is required. Technological limitations have restricted the smallest size in which instrumented devices for the study of prehensile forces can be constructed. Do these measurement limitations influence the observations of grasp control in the same way that object size influences grip configurations? It is also not known how the forces at all five digits are coordinated during early grasping. The study of more complex, non-laboratory-based tasks will also be required to fully comprehend the intricate nature of the development.

In order to truly understand the development of these skills, knowledge from multiple disciplines must be combined, including neurophysiology, paediatrics, developmental psychology, kinesiology and biomechanics, genetics, molecular biology, and engineering and robotics. This will also require that researchers coming from opposite perspectives (e.g. neuromaturation and dynamical systems) work towards forging middle ground. Multidisciplinary collaboration will probably be the key to the complete understanding of the developmental processes underlying motor skill progression.

ACKNOWLEDGEMENTS

This work was supported by a grant from the National Science Foundation (no. 9733679) and the VIDDA foundation. I thank Dr Karen Adolph for helpful comments, and Drs Claes von Hofsten and Ann Gentile for enlightening discussions of this material. I am grateful to Dr Lori Quinn and Eric Dannheim for consent to photograph their daughter's (Annabel) hand, and Dr Ralf Reilmann for expert photography.

References

Alstermark, B. & Lundberg, A. (1992). The C3–C4 propriospinal system: target-reaching and food-taking. In Jami, L., Pierrot-Deseilligny, E., & Zytnicki, D. (Eds.), *Muscle afferents and spinal control of movement* (pp. 327–354). London: Pergamon Press.

Armand, J., Olivier, E., Edgley, S.A. & Lemon, R.N. (1997). Postnatal development of corticospinal projections from motor cortex to the cervical enlargement in the macaque monkey. *Journal of Neuroscience,* **17**, 251–266.

Berardi, N., Pizzorusso, T. & Maffei, L. (2000). Critical periods during sensory development. *Current Opinion in Neurobiology,* **10**, 138–145.

Bernstein, N. (1967). *The coordination and regulation of movements.* Oxford: Pergamon.

Berthier, N.E., Clifton, R.K., McCall, D.D. & Robin, D.J. (1999). Proximodistal structure of early reaching in human infants. *Experimental Brain Research,* **127**, 259–269.

Bower, T. (1972). Object perception in infants. *Perception,* **1**, 15–30.

Bower, T., Broughton, J.M. & Moore, M.K. (1970). Demonstration of intention in the reaching behaviour of neonate humans. *Nature,* **228**, 679–680.

Brooks, V.B. (1984). How are "move" and "hold" programs matched? In Bloedel, J.R., Dichgans, J. & Precht, W. (Eds.), *Cerebellar functions* (pp. 1–23). Berlin: Springer.

Butterworth, G. (1986). Some problems in explaining the origins of movement control. In M.G. Wade & H.T.A. Whiting (Eds.), *Motor development in children: Problems of coordination and control* (pp. 23–32). Dordrecht: Martinus Nijhoff.

Clifton, R.K., Muir, D.W., Ashmead, D.H. & Clarkson, M.G. (1993). Is visually guided reaching in early infancy a myth? *Child Development,* **64**, 1099–1110.

Clifton, R.K., Rochat, P., Robin, D.J. & Berthier, N.E. (1994). Multimodal perception in the control of infant reaching. *Journal of Experimental Psychology, Human Perception and Performance,* **20**, 876–886.

Connolly, K.J. (1973). Factors influencing the learning of manual skills in young children. In R.A. Hinde & J.S. Hind (Eds.), *Constraints on learning* (pp. 337–365). London: Academic Press.

Connolly, K. & Dalgleish, M. (1989). The emergence of a tool-using skill in infancy. *Developmental Psychology,* **25**, 894–912.

Connolly, K.J. & Elliott, J. (1972). The evolution and ontogeny of hand function. In B. Jones (Ed.), *Ethological studies of child behaviour* (pp. 329–383). Cambridge, UK: Cambridge University Press.

Connolly, K.J. & Stratton, P. (1968). Developmental changes in associated movements. *Developmental Medicine and Child Neurology,* **10**, 49–56.

Corbetta, D. & Mounoud, P. (1990). Early development of grasping and manipulation. In C. Bard, M. Fleury & L. Hay (Eds.), *Development of eye-hand coordination across the life span* (pp. 188–213). Columbia: University of South Carolina Press.

Dodwell, P.C., Muir, D. & DiFranco, D. (1976). Responses of infants to visually presented objects. *Science,* **194**, 209–211.

Eliasson, A.C. & Gordon, A.M. (2000). Impaired force coordination during object release in children with hemiplegic cerebral palsy. *Developmental Medicine and Child Neurology,* **42**, 228–234.

Eliasson, A.C., Forssberg, H., Ikuta, K., Apel, I., Westling, G. & Johansson, R.S. (1995a). Development of human precision grip V: Anticipatory and triggered grip actions during sudden loading. *Experimental Brain Research,* **106**, 425–433.

Eliasson, A.C., Gordon, A.M. & Forssberg, H. (1991). Basic coordination of manipulative forces in children with cerebral palsy. *Developmental Medicine and Child Neurology,* **33**, 661–670.

Eliasson, A.C., Gordon, A.M. & Forssberg, H. (1992). Impaired anticipatory control of isometric forces during grasping by children with cerebral palsy. *Developmental Medicine and Child Neurology,* **34**, 216–225.

Eliasson, A.C., Gordon, A.M. & Forssberg, H. (1995b). Tactile control of isometric fingertip forces during grasping in children with cerebral palsy. *Developmental Medicine and Child Neurology,* **37**, 72–84.

Exner, C.E. (1990). The zone of proximal development in in-hand manipulation skills of nondys-functional 3- and 4-year-old children. *American Journal of Occupational Therapy,* **44**, 884–891.

Eyre, J.A., Miller, S. & Ramesh, V. (1991). Constancy of central conduction delays during development in man: investigation of motor and somatosensory pathways. *Journal of Physiology (London),* **434**, 441–452.

Fagard, J. (1998). Changes in grasping skills and the emergence of bimanual coordination during the first year of life. In K.J. Connolly (Ed.), *The psychobiology of the hand* (pp. 123–143). London: MacKeith Press.

Fagard, J. & Jacquet, A.Y. (1996). Changes in reaching and grasping objects of different sizes between 7 and 13 months of age. *British Journal of Developmental Psychology,* **14**, 65–78.

Fetters, L. & Todd, J. (1987). Quantitative assessment of infant reaching movements. *Journal of Motor Behavior,* **19**, 147–166.

Field, J. (1977). Coordination of vision and prehension in young infants. *Child Development,* **48**, 97–103.

Fietzek, U.M., Heinen, F., Berweck, S., Maute, S., Hufschmidt, A., Schulte-Monting, J., Lucking, C.H. & Korinthenberg, R. (2000). Development of the corticospinal system and hand motor function: central conduction times and motor performance tests. *Developmental Medicine and Child Neurology,* **42**, 220–227.

Flament, D., Hall, E.J. & Lemon, R.N. (1992). The development of cortico–motoneuronal projections investigated using magnetic brain stimulation in the infant macaque. *Journal of Physiology (London),* **447**, 755–768.

Forssberg, H. (1985). Ontogeny of human locomotor control I. Infant stepping, supported locomotion and transition to independent locomotion. *Experimental Brain Research,* **57**, 480–493.

Forssberg, H. & Dietz, V. (1997). Neurobiology of normal and impaired locomotor development. In K.J. Connolly & H. Forssberg (Eds.), *Neurophysiology and neuropsychology of motor development* (pp. 78–100). London: MacKeith Press.

Forssberg, H., Eliasson, A.C., Kinoshita, H., Johansson, R.S. & Westling, G. (1991). Development of human precision grip. I. Basic coordination of force. *Experimental Brain Research,* **85**, 451–457.

Forssberg, H., Kinoshita, H., Eliasson, A.C., Johansson, R.S., Westling, G. & Gordon, A.M. (1992). Development of human precision grip. II. Anticipatory control of isometric forces targeted for object's weight. *Experimental Brain Research,* **90**, 393–398.

Forssberg, H., Eliasson, A.C., Kinoshita, H., Johansson, R.S. & Westling, G. (1995). Development of human precision grip IV: Tactlile adaptation of isometric finger forces to the frictional condition. *Experimental Brain Research,* **104**, 323–330.

Forssberg, H., Eliasson, A.C., Redon-Zouitenn, C., Mercuri, E. & Dubowitz, L. (1999). Impaired grip–lift synergy in children with unilateral brain lesions. *Brain,* **122**, 1157–1168.

Gachoud, J.P., Mounoud, P. & Hauert, C.A. (1983). Motor stategies in lifting movements: a comparison of adult and child performance. *Journal of Motor Behavior,* **15**, 202–216.

Galea, M.P. & Darian-Smith, I. (1995). Postnatal maturation of the direct corticospinal projections in the macaque monkey. *Cerebral Cortex,* **5**, 518–540.

Georgopoulos, A.P. & Grillner, S. (1989). Visuomotor coordination in reaching and locomotion. *Science,* **245**, 1209–1210.

Gesell, A. & Amatruda, C.S. (1964). *Developmental diagnosis. Normal and abnormal child development.* New York: Harper and Row.

Gordon, A.M. & Duff, S.V. (1999a). Fingertip forces in children with hemiplegic cerebral palsy. I. anticipatory scaling. *Developmental Medicine and Child Neurology,* **41**, 166–175.

Gordon, A.M. & Duff, S.V. (1999b). Relation between clinical measures and fine manipulative control in children with hemiplegic cerebral palsy. *Developmental Medicine and Child Neurology,* **41**, 586–591.

Gordon, A.M. & Forssberg, H. (1997). The development of neural control mechanisms for grasping in children. In K.J. Connolly & H. Forssberg (Eds.), *Neurophysiology and psychology of motor development* (pp. 214–231). London: MacKeith Press.

Gordon, F.R. & Yonas, A. (1976). Sensitivity to binocular depth information in infants. *Journal of Experimental Child Psychology,* **22**, 413–422.

Gordon, A.M., Forssberg, H., Johansson, R.S. & Westling, G. (1991). The integration of haptically acquired size information in the programming of precision grip. *Experimental Brain Research,* **83**, 483–488.

Gordon, A.M., Forssberg, H., Johansson, R.S., Eliasson, A.C. & Westling, G. (1992). Development of human precision grip. III. Integration of visual size cues during the programming of isometric forces. *Experimental Brain Research,* **90**, 399–403.

Gordon, A.M., Westling, G., Cole, K.J. & Johansson, R.S. (1993). Memor / representations underlying motor commands used during manipulation of common and novel objects. *Journal of Neurophysiology,* **69**, 1789–1796.

Gordon, A.M., Forssberg, H. & Iwasaki, N. (1994). Formation and lateralization of internal representations underlying motor commands during precision grip. *Neuropsychologia,* **32**, 555–568.

Gordon, A.M., Charles, J. & Duff, S.V. (1999). Fingertip forces in children with hemiplegic cerebral palsy. II. Bilateral coordination. *Developmental Medicine and Child Neurology,* **41**, 176–185.

Hadders-Algra, M., Brogren, E. & Forssberg, H. (1996). Training affects the development of postural adjustments in sitting infants. *Journal of Physiology (London)*, **493**, 289–298.

Halverson, H.M. (1931). Study of prehension in infants. *Genetic Society General Psychology Mongraph*, **10**, 110–286.

Halverson, H.M. (1932). A further study of grasping. *Journal Genetic Psychology*, **7**, 34–63.

Halverson, H.M. (1937). Studies of the grasping responses of early infancy. I, II, III. *Journal of Genetic Psychology*, **51**, 371–449.

Halverson, H.M. (1943). *The development of prehension in infants*. J.S. Barker, J.S. Kounin & H.F. Wright (Eds.), London: McGraw-Hill.

Heinen, F., Glocker, F.X., Fietzek, U., Meyer, B.U., Lucking, C.H. & Korinthenberg, R. (1998). Absence of transcallosal inhibition following focal magnetic stimulation in preschool children. *Annals Neurology*, **43**, 608–612.

Hofsten, C. von (1979). Development of visually guided reaching: the approach phase. *Human Movement Studies*, **5**, 160–178.

Hofsten, C. von (1982). Eye–hand coordination in the newborn. *Developmental Psychobiology*, **18**, 450–461.

Hofsten, C. von (1984). Developmental changes in the organization of prereaching movements. *Developmental Psychobiology*, **20**, 378–388.

Hofsten, C.v. (1991). Structuring of early reaching movements: a longitudinal study. *Journal of Motor Behavior*, **23**, 280–292.

Hofsten, C.v. & Fazel-Zandy, S. (1984). Development of visually guided hand orientation in reaching. *Journal of Experimental Child Psychology*, **38**, 208–219.

Hofsten, C.v. & Rönnqvist, L. (1988). Preparation for grasping an object: a developmental study. *Journal of Experimental Psychology, (Human Perception and Performance)*, **4**, 610–621.

Illingworth, R.S. (1975). *The development of the infant and young child. Normal and abnormal.* Edinburgh: Churchill Livingstone.

Jeannerod, M. (1981). Intersegmental coordination during reaching at natural visual objects. In A. Baddeley (Ed.), *Attention and performance* (vol 9, pp. 153–168). Hilsdale, NJ: Erlbaum.

Jeannerod, M. (1984). The timing of natural prehension movements. *Journal of Motor Behavior*, **16**, 235–254.

Jenner, J.R. & Stephens, J.A. (1982). Cutaneous reflex responses and their central nervous pathways studied in man. *Journal of Physiology*, **333**, 405–419.

Johansson, R.S. (1998). Sensory input and control of grip. *Novartis Foundation Symposium*, **218**, 45–59; discussion, 59–63, 45–59.

Johansson, R.S. & Westling, G. (1984). Roles of glabrous skin receptors and sensorimotor memory in automatic control of precision grip when lifting rougher or more slippery objects. *Experimental Brain Research*, **56**, 550–564.

Johansson, R.S. & Westling, G. (1987). Signals in tactile afferents from the fingers eliciting adaptive motor responses during precision grip. *Experimental Brain Research*, **66**, 141–154.

Johansson, R.S. & Westling, G. (1988a). Coordinated isometric muscle commands adequately and erroneously programmed for the weight during lifting task with precision grip. *Experimental Brain Research*, **71**, 59–71.

Johansson, R.S. & Westling, G. (1988b). Programmed and triggered actions to rapid load changes during precision grip. *Experimental Brain Research*, **71**, 72–86.

Kent, R.D. (1999). Motor control: neurophysiology and functional development. In A. Caruso & E. Strand (Eds.), *Clinical management of motor speech disorders in children* (pp. 29–71). New York: Thième.

Koh, T.H.H.G. & Eyre, J.A. (1988). Maturation of corticospinal tracts assessed by electromagnetic stimulation of the motor cortex. *Archives of Diseases in Children*, **63**, 1347–1352.

Konczak, J. & Dichgans, J. (1997). The development toward stereotypic arm kinematics during reaching in the first 3 years of life. *Experimental Brain Research*, **117**, 346–354.

Konczak, J., Borutta, M., Topka, H. & Dichgans, J. (1995). The development of goal-directed reaching in infants: hand trajectory formation and joint torque control. *Experimental Brain Research*, **106**, 156–168.

Konczak, J., Borutta, M. & Dichgans, J. (1997). The development of goal-directed reaching in infants II. Learning to produce task-adequate patterns of joint torque. *Experimental Brain Research*, **113**, 465–474.

Kuhtz-Buschbeck, J.P., Stolze, H., Johnk, K., Boczek-Funcke, A. & Illert, M. (1998). Development of prehension movements in children: a kinematic study. *Experimental Brain Research,* **122**, 424–432.

Lantz, C., Melen, K. & Forssberg, H. (1996). Early infant grasping involves radial fingers. *Developmental Medicine and Child Neurology,* **38**, 668–674.

Lawrence, D.G. & Hopkins, D.A. (1976). The development of motor control in the rhesus monkey: evidence concerning the role of corticomotoneuronal connections. *Brain,* **99**, 235–254.

Lawrence, D.G. & Kuypers, H.G.J.M. (1968a). The functional organization of the motor system in the monkey. I. The effects of bilateral pyramidal lesions. *Brain,* **91**, 1–14.

Lawrence, D.G. & Kuypers, H.G.J.M. (1968b). The functional organization of the motor system in the monkey. II. The effects of lesions of the descending brain-stem pathways. *Brain,* **91**, 15–43.

Lemon, R.N. (1999). Neural control of dexterity: what has been achieved? *Experimental Brain Research,* **128**, 6–12.

Liddell, E.G.T. & Phillips, C.G. (1944). Pyramidal section in the cat. *Brain,* **67**, 1–9.

Lockman, J.J. (1984). Development of detour abilities in infants. *Child Development,* **55**, 482–491.

Lockman, J.J. & Ashmead, D.H.B.E.W. (1984). The development of anticipatory hand orientation during infancy. *Journal of Experimental Child Psychology,* **37**, 176–186.

Martin, J.H., Kably, B. & Hacking, A. (1999). Activity-dependent development of cortical axon terminations in the spinal cord and brain stem. *Experimental Brain Research,* **125**, 184–199.

Martin, J.H., Donarummo, L. & Hacking, A. (2000). Impairments in prehension produced by early postnatal sensory motor cortex activity blockade. *Journal of Neurophysiology,* **83**, 895–906.

Mayston, M.J., Harrison, L.M. & Stephens, J.A. (1999). A neurophysiological study of mirror movements in adults and children. *Annals of Neurology,* **45**, 583–594.

McCarty, M.E. & Ashmead, D.H. (1999). Visual control of reaching and grasping in infants. *Developmental Psychology,* **35**, 620–631.

McCarty, M.E., Clifton, R.K. & Collard, R.R. (1999). Problem solving in infancy: the emergence of an action plan. *Developmental Psychology,* **35**, 1091–1101.

McGraw, M.B. (1943). *The neuromuscular maturation of the human infant* (Reprinted 1990 as *Classics in Developmental Medicine,* No. 4), London: MacKeith Press.

Muller, K., Hömberg, V. & Lenard, H.G. (1991). Magnetic stimulation of motor cortex and nerve roots in children. Maturation of cortico-motoneural projections. *Electroencephalography and Clinical Neurophysiology,* **81**, 63–70.

Muller, K., Kass-Iliyya, F. & Reitz, M. (1997). Ontogeny of ipsilateral corticospinal projections: a developmental study with transcranial magnetic stimulation. *Annals of Neurology,* **42**, 705–711.

Napier, J.R. (1956). The prehensile movements of the human hand. *Journal of Bone and Joint Surgery (America),* **38B**, 902–913.

Newell, K.M. & Cesari, P. (1998). Body scale and the development of hand form and function in prehension. In K.J. Connolly (Ed.), *The psychobiology of the hand* (pp. 162–176), London: MacKeith Press.

Newell, K.M. & Mcdonald, P.V. (1997) The development of grip patterns in infancy. In K.J. Connolly & H. Forssberg (Eds.), *The neurophysiology and neuropsychology of motor development* (pp. 232–256), London: MacKeith Press.

Newell, K.M., Scully, D.M., Mcdonald, P.V. & Baillargeon, R. (1989). Task constraints and infant prehensile grip configurations. *Developmental Psychobiology,* **22**, 817–831.

Olivier, E., Edgley, S.A., Armand, J. & Lemon, R.N. (1997). An electrophysiological study of the postnatal development of the corticospinal system in the macaque monkey. *Journal of Neuroscience,* **17**, 267–276.

Paré, M. & Dugas, C. (1999). Developmental changes in prehension during childhood. *Experimental Brain Research,* **125**, 239–247.

Pehoski, C., Henderson, A. & Tickle-Degnen, L. (1997a). In-hand manipulation in young children: translation movements. *American Journal of Occupational Therapy,* **51**, 719–728.

Pehoski, C., Henderson, A. & Tickle-Degnen, L. (1997b). In-hand manipulation in young children: rotation of an object in the fingers. *American Journal of Occupational Therapy,* **51**, 544–552.

Peiper, A. (1963). *Cerebral function in infancy and childhood.* New York: Consultants Bureau.

Piaget, J. (1953). *The origins in intelligence in children.* London: Routledge & Kegan Paul.

Pierrot-Deseilligny, E. (1996). Transmission of the cortical command for human voluntary movement through cervical propriospinal premotoneurons. *Progress in Neurobiology,* **48**, 489–517.

Reitz, M. & Muller, K. (1998). Differences between 'congenital mirror movements' and 'associated movements' in normal children: a neurophysiological case study. *Neuroscience Letters,* **256**, 69–72.

Ruff, H.A. & Halton, A. (1978). Is there directed reaching in the human neonate. *Developmental Psychology,* **14**, 426–426.

Sporns, O. & Edelman, G.M. (1993). Solving Bernstein's problem: a proposal for the development of coordinated movement by selection. *Child Development,* **64**, 960–981.

Thelen, E., Fisher, D. & Ridley-Johnson, R. (1984). The relationship between physical growth and a newborn reflex. *Infant Behavior and Development,* **7**, 479–493.

Thelen, E., Corbetta, D., Kamm, K., Spencer, J.P., Schneider, K. & Zernicke, R.F. (1993). The transition to reaching: mapping intention and intrinsic dynamics. *Child Development,* **64**, 1058–1098.

Touwen, B. (1976). *Neurological development in infancy. Clinics in developmental medicine,* no. 58, London: Heinemann.

Tower, S.S. (1940). Pyramidal lesion in the monkey. *Brain,* **63**, 36–90.

Trevarthen, C., Hubley, P. & Sheeran, L. (1975). Les activités innée du nourrisson. *La Recherche,* **6**, 447–458.

Twitchell, T.E. (1970). Reflex mechanisms and the development of prehension. In K. Connolly (Ed.), *Mechanisms of motor skill development* (pp. 25–37). New York: Academic Press.

van der Meer, A.L., van der Weel, F.R. & Lee, D.N. (1995). The functional significance of arm movements in neonates. *Science,* **267**, 693–695.

Vereijken, B. & Thelen, E. (1997). Training infant treadmill stepping: the role of individual pattern stability. *Developmental Psychobiology,* **30**, 89–102.

White, B.L., Castle, P. & Held, R. (1964). Observations on the development of visually directed reaching. *Child Development,* **35**, 349–364.

Wimmers, R.H., Savelsbergh, G.J., Beek, P.J. & Hopkins, B. (1998). Evidence for a phase transition in the early development of prehension. *Developmental Psychobiology,* **32**, 235–248.

Winstein, C.J., Horak, F.B. & Fisher, B.E. (2000). Influence of central set on anticipatory and triggered grip-force adjustments. *Experimental Brain Research,* **130**, 298–308.

Woolacott, M.H. (1994). Changes in balance control across the lifespan: can training improve balance efficiency? In K. Taguchi, M. Igarashi & S. Mori (Eds.), *Vestibular and neural front* (pp. 121–129). Amsterdam: Elsevier Science.

Zaal, F.T., Daigle, K., Gottlieb, G.L. & Thelen, E. (1999). An unlearned principle for controlling natural movements. *Journal of Neurophysiology,* **82**, 255–259.

Further Reading

Connolly, K.J. & Forssberg, H. (1997). *Neurophysiology and neuropsychology of motor development.* London: MacKeith Press.

Connolly, K.J. (1998). *The psychobiology of the hand.* London: MacKeith Press.

Porter, R. & Lemon, R.N. (1993). *Corticospinal function and voluntary movement.* Oxford: Oxford University Press.

III.7
Development of Gross Motor Functions

MIJNA HADDERS-ALGRA

ABSTRACT

Edelman's neuronal group selection theory offers an excellent framework for the understanding of motor development. Development starts during fetal life and infancy with the phase of primary variability, which is characterized by abundant variation in motor behaviour. During the last phase of primary variability selection of the most efficient motor behaviour occurs. The selection takes place at function-specific points in time during the first 1½ years after birth. After the process of selection, which is associated with a transient minor reduction in the variation of motor behaviour, variation returns. The phase of secondary or adaptive variability evolves in which children develop the ability to adapt motor behaviour to task-specific conditions.

The development of gross motor functions is primarily dependent on the protracted course of the development of postural control. The latter is characterized by multiple phases of transition occurring at the ages of 2–3 months, 5–6 months, 9–10 months, 15 months, 2–3 years and 6–7 years, and continues until adolescence.

INTRODUCTION

Gross motor development in the human has a protracted course. It takes about a year of postnatal life before a human infant can walk without support (Gesell & Armatruda, 1947; Touwen, 1976). This is remarkably long, even within the group of altricial mammals. For instance, independent, belly-off-the-floor locomotion emerges in the cat around week 4, i.e. after a postnatal period which is equivalent to 40% of the gestational period (Howland *et al.*, 1995), and in the rat at day 11, i.e. after a postnatal period which is equivalent to 50% of the duration of gestation (Westerga & Gramsbergen, 1990). The extended period required for the development of human locomotion is probably related to the

A.F. Kalverboer and A. Gramsbergen (eds.), Handbook of Brain and Behaviour in Human Development, 539–568
© 2001 Kluwer Academic Publishers. Printed in Great Britain.

specific bipedal nature of human gait. Bipedal gait has the great advantage of freeing the hands for fine motor skills. Apparently, during evolution – with its associated neocortical expansion – this advantage prevailed over the disadvantage of a relative postural instability associated with bipedallity (Popper & Eccles, 1977; Rose, 1976).

The development of gross motor functions in children is the theme of this chapter; but before we can start with the gross motor melody, some theoretical tuning has to be performed. The tuning will provide the keys of normal motor development: variation, selection and adaptation.

THEORETICAL BACKGROUND

In the middle of the 20th century motor development was generally regarded as a process based on a gradual unfolding of predetermined patterns in the central nervous system (Illingworth, 1966). The idea that behavioural patterns emerge in an orderly genetic sequence resulted in the distinction of general developmental rules, such as the cephalocaudal and central-to-distal sequences of development (Ames, 1937; Gesell & Amatruda, 1947). Motor development was considered to be the result of an increasing cortical control over lower reflexes (McGraw, 1943; Peiper, 1963). These neural–maturationist ideas left little place for an effect of environmental stimulation and experience on motor development. Still, Myrtle McGraw (1943), a pioneering developmentalist who is usually regarded as a neural–maturationist, acknowledged that motor development can be influenced by environmental stimulation. She concluded, on the basis of the striking developmental differences between Johnny, whose motor development was stimulated excessively, and Jimmy, his "non-trained" twin brother (McGraw, 1939), that experience can enhance motor development, but "that a certain amount of neural maturation must take place before any function can be modified by specific stimulation" (McGraw, 1943).

Others were not satisfied with the neural–maturationist theories: "How can the timetable of motor solutions be encoded in the brain or in the genes?" (Thelen, 1995). Instead, they embraced the ideas of Kugler and co-workers (Kugler et al., 1980; Kugler & Turvey, 1987), at present known as the dynamic systems theory (see Hopkins, this handbook). Thelen and co-workers argue that development can be regarded as a dissipative structure, i.e. as a dynamic system, because patterns of behaviour act as collectives ("attractor states") of the component parts – such as muscle strength, body weight, postural support, the infant's mood, and brain development – within particular environmental and task contexts. The patterns arise through self-organization (Thelen, 1985; Thelen et al., 1993; Ulrich, 1997). According to Thelen (1995, p. 84): "Developmental change can be seen as a series of states of stability, instability, and phase shifts in the attractor landscape, reflecting the probability that a pattern will emerge under particular constraints". Thus, the difference between the neural–maturationist and the dynamic systems theories is that the former considers the (maturational) state of the nervous system as the main constraint for developmental progress, whereas in the latter theory the neural substrate plays only a subordinate role. Or, as Thelen (1995, p. 86) formulates it: "The Dynamic

Systems view differs sharply from the traditional maturational accounts by proposing that even the so-called 'phylogenetic' skills – the universal milestones such as crawling, reaching, and walking – are learned through a process of modulating current dynamics to fit a new task through exploration and selection of a wider space of possible configurations".

Recently Gerald M. Edelman developed a new theoretical concept on neural development: the neuronal group selection theory (Edelman, 1989, 1993; Sporns & Edelman, 1993). This theory could offer the golden mean between the neuro–maturationist and dynamic systems theories. According to Edelman's theory, development starts with so-called primary neuronal repertoires. The cells and the gross connectivity of these primary repertoires are determined by evolution. The repertoires show substantial variation through the dynamic epigenetic regulation of cell division, adhesion, migration, death and neurite extension and retraction (Changeux, 1997; Changeux & Danchin, 1976; Rakic, 1988). Development proceeds with selection on the basis of afferent information produced by behaviour and experience. This experiential selection results in secondary, more adapted repertoires of neuronal groups. The selection process is thought to be mediated by changes in synaptic strength of intra- and intergroup connections, in which the topology of the cells (Nelson *et al.*, 1993) and the presence or absence of coincident electrical activity in pre- and postsynaptic neurons play a role (Changeux & Danchin, 1976; Hebb, 1949). When the selection has just been accomplished, behavioural variation is slightly reduced. Soon, however, abundant variation returns, because the organism and its populations of neurons are constantly exposed to a multitude of experiences. As a result the neural structures of the secondary repertoires are globally mapped with large collections of parallel channels. The coordination between the various maps and channels is coordinated by a higher-order selectional process called re-entry signalling (Edelman, 1993). The presence of the global mapping in the secondary repertoires is the basis of variable behaviour which can be adapted to environmental constraints.

The neuronal group selection theory could end the continuing "nature–nurture" debate, as the theory explicitly states that development is not governed by either a genetically dictated neural substrate or environmental conditions. On the contrary, the theory highlights the notion that development is the result of a complex intertwining of information from genes and environment. Similar ideas – be it without the concept of variation and selection – were already forwarded by Waddington (1962), who portrayed development as an "epigenetic landscape" in which developmental processes were canalized into specific process chains ("chreods"). Waddington's concept of experiential canalization was elaborated by Gottlieb (1991). Gottlieb stressed two points. In the first place he underlined the idea that the various organizational levels of an organism, such as genes, cells, organ systems, organisms and the interactions between organisms, can mutually influence each other. Secondly, he drew the attention to the role of species- and age-specific behaviour, which could play a canalizing role by exposing the individual to specific experiences. The presence of such age-dependent canalizing behaviour had already been noted by McGraw (1943), when she reported that children exhibit an indomitable urge to exercise a function as soon as it emerges. This remarkable drive to exercise a newly developing function underscores the importance of self-produced activity for the creation of optimal neuronal circuitries (cf. Bertenthal *et al.*, 1994; Hadders-Algra *et al.*, 1996b).

The translation of the concepts of the neuronal group selection theory to the domain of motor development results in a developmental progress with two different phases of variability (Figure 1). Motor development starts during early fetal life (see Prechtl, in Chapter III.2 of this handbook) with the phase of "primary variability", a phase which continues during infancy. Motor behaviour at that time is characterized by profuse variation, such as variation in movement trajectories, and variation in temporal and quantitative aspects of motility (Forssberg, 1985; Hadders-Algra *et al.*, 1992, 1996a; Konczak *et al.*, 1995; Touwen, 1976). These variations in motor activity are not neatly tuned to environmental conditions, but the variations themselves constitute a fundamental developmental phenomenon. It is conceivable that the abundant variation in motility is brought about by activity of the epigenetically determined, but rather grossly specified, primary neural repertoires. The neural system of primary repertoires presumably explores by means of self-generated activity, and consequently also by means of self-generated afferent information, all motor possibilities available within the neurobiological and anthropometric constraints set by evolution. The exploration gradually results in selection of the most efficient movement patterns. The timing of the selection process is probably function-specific. For instance, selection of the most efficient sucking movement occurs prior to term age (Hadders-Algra & Dirks, 2000), the selection of a more straight forward-directed arm movement during reaching (Konczak *et al.*, 1995; Thelen *et al.*, 1993), of the most efficient postural adjustments (Hadders-Algra *et al.*, 1996a; Van der Fits *et al.*, 1999b) and of the sounds and syllables occurring in the locally employed language (De Boysson-Bardies, 1993; Vihman, 1993) most probably take place during the second postnatal half-year, and the selection of the heel–toe gait, the most efficient way of placing of the foot, occurs during the second postnatal year (Burnett & Johnson, 1971; Cioni *et al.*, 1993). The process of selection prunes the infant tree of giant motor possibilities, and induces a transient phase with moderate motor variation (Hadders-Algra *et al.*, 1998b). Presumably the timing of the transient phase is – like the timing of selection – function-specific. After the transient phase the phase of 'secondary variability' or 'adaptive variability' starts (Hadders-Algra *et al.*, 1998b; Touwen, 1993). A rich movement repertoire is created with an efficient motor solution for each specific situation. Between the ages of 2 and 3 years motor variation starts to bloom, but it is not until adolescence that the motor repertoire is mature (Gordon & Forssberg, 1997; Hadders-Algra *et al.*, 1998b; Hempel, 1993; Touwen, 1993). The mature situation is characterized by the ability to adapt each movement exactly and efficiently to task-specific conditions or, in the absence of tight constraints, by the ability to generate multiple solutions or strategies for a single motor task (e.g. Diener *et al.*, 1983; Van der Fits *et al.*, 1998).

NON GOAL-DIRECTED MOTOR BEHAVIOUR

At birth, be it term or preterm, the newborn infant has a repertoire of motor behaviours which are necessary for survival (see Prechtl, 1984). To this category

Primary Variability

- Activity of epigenetically determined, grossly specified primary neural repertoires.
- Neural system explores by means of self-generated activity, and consequently by means of self-generated afferent information, all motor possibilities available within the neurobiological and anthropometric constraints set by evolution.
- Abundant variation in motor behaviour.
- Occurring during fetal life and infancy.

Selection

- Experiential selection of the most effective motor patterns and their associated neuronal groups.
- Transient minor reduction in the variation of motor behaviour.
- Occurring during infancy, at function-specific ages.

Secondary or Adaptive Variability

- Creation of secondary neural repertoires with a large collection of parallel channels due to the exposure to the multitide of experiences.
- Mature situation:
  - task constraints: ability to adapt each movement exactly and efficiently to task-specific conditions.
  - no task constraints: multiple motor solutions or strategies for a single motor task.
- Onset: function-specific from mid-infancy onwards. Starting to bloom at 2–3 years; mature in adolescence.

Figure 1 Motor development in the light of Edelman's neuronal group selection theory.

of so-called "innate behaviours" belong feeding (i.e. rooting, sucking, mastication, swallowing), and respiration, as well as various protective reactions such as blinking and coughing. There are also transient movement patterns, such as general movements and infant stepping, which are present during several months after birth.

The motor behaviours, which are present at birth, emerge during early fetal life (see Prechtl, Chapter III.2 of this handbook). From neurobiological research in other species it is known that central pattern generators (CPGs) underlie the generation of this type of movement (e.g. Funk & Feldman, 1995 (respiration); Grillner et al., 1995; Cazalets, this handbook (locomotion); Miller, 1972; Selverston & Moulin, 1987 (feeding). The CPG neurones are coupled in networks which can generate complex basic activation patterns of the muscles without any sensory signals. Yet sensory information of the movement is important in adapting the movement to the environment. The activity of the networks, which are usually located in the spinal cord or brainstem, is controlled from cortical and subcortical centres via descending motor pathways. This control becomes increasingly complex with increasing age.

General movements

The most frequently occurring movement pattern of the human fetus and newborn infant is the general movement (GM). General movements consist of series of gross movements of variable speed and amplitude, which involve all parts of the body but lack a distinctive sequencing of the participating body parts (Prechtl & Nolte, 1984). Remarkably, GMs are among the first movements which the human fetus develops, and they emerge prior to isolated limb movements (De Vries et al., 1982). GMs can already be observed before the completion of the spinal reflex arc, which is accomplished at 8 weeks postmenstrual age (PMA; Okado & Kojima, 1984; Windle & Fitzgerald, 1937). This means that GMs, like other motor behaviours produced by CPG networks, can be generated in the absence of afferent information. This underscores the spontaneous or autogenic nature of the first movements (Hall & Oppenheim, 1987) and refutes the long-held belief that all movements of the fetus and newborn are reflex in character (Feldman, 1920; Humphrey, 1969).

Movements resembling human GMs can also be observed in other species, be it only during prenatal life. For instance, Coghill (1929) described GM-like movements in the embryos of the amphibian *Amblystoma*. During the early phases of development *Amblystoma* exhibits "total behaviour patterns" in which trunk and fore- and hindlimbs participate. Early motor behaviour has especially been studied in the embryonic chick. The basic motility type of the chick embryo is type I motility, which consists of spontaneous, seemingly uncoordinated movements (Hamburger & Oppenheim, 1967). During type I motility all parts of the body can move in any conceivable combination (Hamburger, 1963, 1973). The type I motility disappears when the embryo approaches hatching age, to be absent after hatching. In mammal fetuses (rat: Angulo y Gonzalez, 1932; rabbit and guinea pig: Preyer, 1885) comparable generalized motility can be recognized. In the fetal rat the generalized motility emerges 1 day after the onset of fetal motility, which starts at embryonic day 15 (Angulo y Gonzalez, 1932;

Narayanan *et al.*, 1971). A slight difference between the generalized movements of the rat fetus and those of the chick embryo has been observed. The movements of the rat are in general smoother than those of the chick (Hamburger, 1973). After birth, rats no longer show GM-like movements. Instead they show motility aiming at progression, i.e. weak crawling movements (see Gramsbergen, this handbook) or swimming behaviour (see Cazalets, this handbook). Unfortunately, no detailed reports exist on the various forms of prenatal motility in monkeys. However, like other animals, monkeys do not have GM-like movements after birth (Dunbar & Badam, 1998); in fact the human newborn seems to be the only newborn creature in which generalized movements persist after birth. Possibly the human newborn can afford this type of non-goal-directed motor behaviour, which is especially displayed in the vulnerable supine position (cf. Hamburger *et al.*, 1966), due to the presence of sophisticated parental care (Papoušek & Papoušek, 1984).

Of course, one could query whether the prenatal general movements of the human fetus are identical to those of the chick and rat embryo. The basic description of generalized motility in various species is the same, and includes the notion that generalized movements are movements in which all parts of the body participate in a very variable way. The observation that all parts of the body participate resulted in the term "total" or "mass" movements (rats: Angulo y Gonzalez, 1932; human: Peiper, 1963), and only recently the term "general movements" was introduced (human: Prechtl *et al.*, 1979). The very variable nature in which the various body-parts are coordinated led to the descriptions "impulsive" (various species: Preyer, 1885), "seemingly uncoordinated" (chick: Hamburger, 1973), "uncoordinated" (human: Minkowski, 1938; Prechtl *et al.*, 1979) and more recently to the description 'coordinated' movement pattern (human: Prechtl, 1990). In all studies reported, the generalized motility precedes the emergence of isolated limb movements. Thus, the basic features of generalized motility are shared by all hitherto-studied subjects. A qualitative difference still seems to be present between the generalized movements of the chick and those of the human fetus and infant. The movements of the chick are described as monotonous and lacking rotatory components (Hamburger, 1973), whereas complexity and rich variation in movement trajectory, including rotatory movements, are the hallmark of normal human GMs (Hadders-Algra *et al.*, 1997; Prechtl, 1990). It is conceivable that the rich variation and complexity of human GMs reflect the seemingly aimless and explorative activity of the primary neuronal groups of the relatively large human (sub)cortical areas on the extensive CPG networks of the GMs in the spinal cord and brainstem. This hypothesis is supported by the finding that human GMs, which lack complexity and variation, i.e. GMs which are definitely abnormal, are strong indicators of the development of cerebral palsy (Hadders-Algra *et al.*, 1997; Hadders-Algra & Groothuis, 1999; Prechtl, 1990; Prechtl *et al.*, 1997).

During human development GMs show age-specific characteristics. Little information is known on the qualitative changes of GMs during the first two trimesters of pregnancy. During the third trimester GMs are characterized by an extreme variation and complexity. The movements (in the extrauterine environment described as "preterm" GMs; Hadders-Algra *et al.*, 1997) are wonderfully fluent, and include many movements of the trunk. Around 36–38 weeks PMA a transition in GMs can be observed. The extremely variable "preterm" GMs change into the more slow and forceful "writhing" GMs, in which the trunk participates less than during the previous GM phase (Hadders-Algra et al., 1997; Hopkins & Prechtl, 1984). The "writhing" GMs constitute a temporary form of GMs, as they disappear around 6–8 weeks post-term age (Hadders-Algra & Prechtl, 1992; Hopkins & Prechtl, 1984). EMG recordings of spontaneous motor behaviour (Hadders-Algra et al., 1992, 1997) and H-reflex studies (Hakamada et al., 1988) indicated that the peri-term period, i.e. the period from 36–38 weeks PMA until 6–8 weeks post-term age, is characterized by a temporary increased excitability of the motoneurons. This might explain why previously the motor behaviour around term age was described as the phase of "physiological hypertonia" (Peiper, 1963; Saint-Anne Dargassies, 1974). At the end of the second month post-term the "writhing" GMs are replaced by the final form of GMs, the so-called "fidgety" GMs. The "fidgety" GMs consist of a continuous stream of tiny, elegant movements occurring irregularly all over the body (Hadders-Algra & Prechtl, 1992; Hopkins & Prechtl, 1984). The transition from "writhing" to "fidgety" GMs occurs between 6 and 8 weeks post-term age, thus in a relatively narrow time-window. The finding that this change in GM form is more closely related to postmenstrual age than to postnatal age, suggests that the transition for a major part is based on endogenous maturational processes (Hadders-Algra & Prechtl, 1992). Postnatal experience plays a minor role, as healthy preterm infants in general exhibit their "fidgety" GMs only 1 week earlier than full-term babies do (Cioni & Prechtl, 1990). Surface EMG recordings indicated that the change of "writhing" GMs into "fidgety" GMs is associated with a decrease in the duration and amplitude of the phasic EMG bursts and a decrease in tonic background activity. It was hypothesized that these EMG findings could be explained by changing membrane properties of the motoneurons and interneurons of the participating spinal networks, an increasing effect of Renshaw inhibition, a decrease in motoneuronal excitability, and a regression of polyneural muscle innervation (Hadders-Algra & Prechtl, 1992). Recently the last hypothesis was proven to be correct: the adult type of mononeural muscle innervation develops – at least in the psoas muscle – around 3 months post-term (Gramsbergen et al., 1997). The "fidgety" GMs are present until about 4 months of post-term age. From that age onwards the GMs are gradually replaced by goal-directed movements. This implies that GMs dissolve without progressing to the developmental phases of selection and secondary variability.

Recently, Prechtl (1990) discovered that the quality of GMs is a powerful tool to predict motor outcome. The quality of GMs can be classified as normal, mildly abnormal and definitely abnormal (Hadders-Algra et al., 1997). Normal GMs are characterized by fluency, variation and complexity (Figure 2; Prechtl, 1990). The muscle coordination patterns of these movements are characterized by variation: variation in the muscles which participate, and in the timing and the quantity of the muscle activation (Figure 3; Hadders-Algra et al., 1992, 1997). Despite this variation, muscle activity is not at random. Normal GMs, for instance, during 70–85% of movement-time show a pattern of antagonistic co-activation. Mildly abnormal GMs lack fluency, but do show some movement complexity and variation. The corresponding variable EMG patterns exhibit abnormalities in the temporal and quantitative scaling of phasic muscle activity, suggesting dysfunction of the monoaminergic systems (Figure 3; Hadders-Algra *et al.,* 1997; Hadders-Algra & Groothuis, 1999). Definitely abnormal GMs lack fluency, complexity and variation altogether (Figure 2). This is reflected by the absence of variation in muscle coordination: the patterns consist either of a stereotyped synchronous activation of most participating muscles, or a stereotyped pattern of reciprocal activity (Figure 3; Hadders-Algra *et al.,* 1997). The persistent presence of definitely abnormal GMs despite transitions to following GM phases puts an infant at high risk (70–80%) for the development of cerebral palsy (CP; Ferrari *et al.,* 1990; Prechtl *et al.,* 1993). However, in a considerable number of infants movement quality is not stable over age. The majority of changes occurs at the transition periods at 36–38 weeks PMA and 6–8 weeks post-term. This means, for instance, that definitely abnormal movements occurring before 36–38 weeks can later change into mildly abnormal ones, and mildly abnormal movements into definitely abnormal ones. Consequently, GM quality during the final GM phase – the phase of "fidgety" GMs – has the best predictive power. Definitely abnormal GMs at "fidgety age" are associated with a high risk ($\pm$ 70%) for the development of CP (Hadders-Algra *et al.,* 1997; Prechtl *et al.,* 1997) with the remaining 30% of children developing minor neurological dysfunction (MND), usually in combination with attention problems. Mildly abnormal GMs are associated with a minimal chance for CP, a 50% chance for the development of MND, and – when compared to children with normal "fidgety" GMs – a threefold increase of attention problems and a doubling of aggressive behaviour at school-age. Normal "fidgety" GMs bear a favourable prognosis: no CP and about 10% chance for the development of MND at school-age (Hadders-Algra & Groothuis, 1999). The power of GM assessment to predict CP is higher than that of the traditional neurological examination at early age, due to the fact that some infants with CP do show abnormal GMs, but no traditional neurological signs, such as abnormalities in muscle tone or head balance (Cioni *et al.,* 1997). The amazingly strong relationship between the quality of GMs and motor outcome supports the notion that the neural circuitries underlying GM activity form the neural building blocks for later motor skills.

Postural control

Before birth, little postural control is required. The fetus floats in the amniotic fluid, and the uterine walls provide ample support, especially so during the last phases of pregnancy. Postnatally the situation changes dramatically: the all-round support is missing and the infant is exposed to the forces of gravity. The extrauterine environment induces in preterm infants a change in the posture of the limbs: the intrauterinely preferred flexion posture (Ververs *et al.,* 1998) changes into extension (Saint-Anne Dargassies, 1974). In preterm infants younger than 32 weeks the preferred extension posture is present in both arms and legs. From 32 weeks onwards the extension changes into a preference for flexion, at first in the legs, and from about 36 weeks onwards also in the arms (Amiel-Tison, 1968; Dubowitz & Dubowitz, 1981; Saint-Anne Dargassies, 1974). It should be realized, however, that the age-dependent preference postures can be observed only during the relatively short periods with active wakefulness and not during sleep (Cioni & Prechtl, 1990; Hadders-Algra *et al.,* 1998a; Prechtl *et*

A

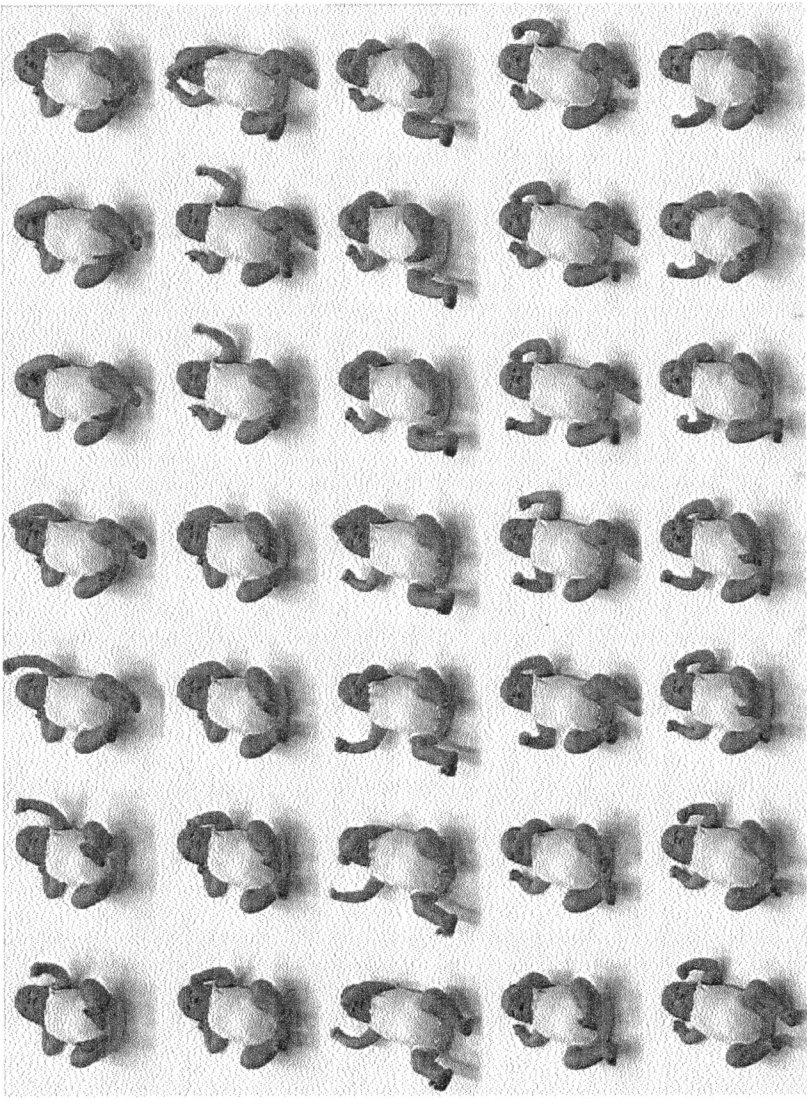

Figure 2 Consecutive frames of video-recordings of GMs of two infants at 3 months post-term. The recordings should be read as a text, starting in the left upper corner of the figure. The interval between the frames is 0.24 s, and the total duration of each video-fragment is 8.16 s. The infant in **A** shows normal "fidgety" GMs. The continuous changes in the position of arms and legs reflect movement variation. Movement complexity is illustrated by the movement of the left leg on line 3: the leg does not show a simple flexion–extension movement, but the flexion–extension is combined with abduction in the hip and endorotation of the foot. The infant in **B** shows definitely abnormal GMs.

B

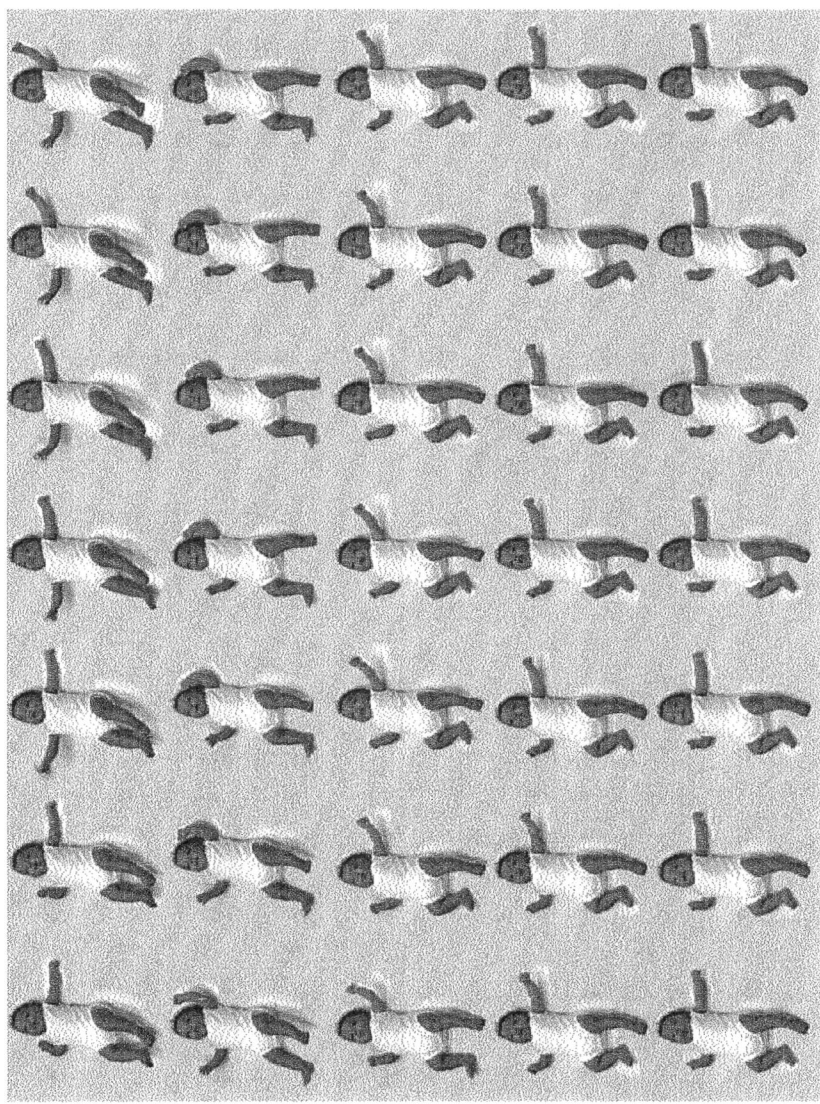

The lack of movement variation is demonstrated by the impression that the infant does not move at all. The infant, however, moved just as much as the other infant. The presence of simple flexion–extension movements illustrates the absence of movement complexity. (The video-recordings form part of a collaborative project with the Department of Developmental and Experimental Clinical Psychology of the University of Groningen. Published with permission of the parents and the *Nederlands Tijdschrift voor Geneeskunde* (Hadders-Algra, 1997)).

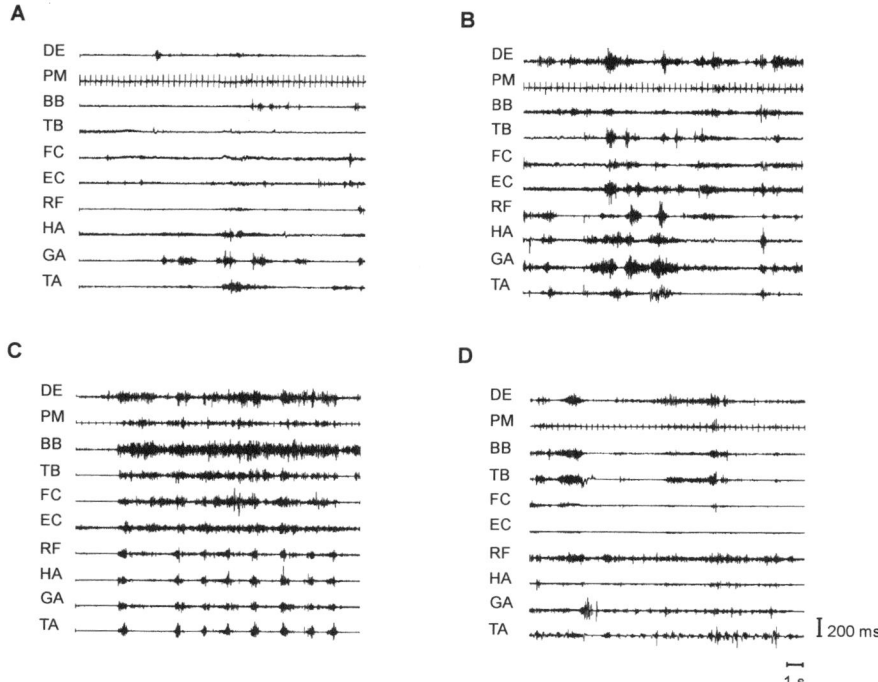

Figure 3 EMG recordings of arm and leg muscles during GMs at 3 months post-term age. **A**: Normal"fidgety" GM; note the variable presence of small phasic bursts of muscle activity. **B**: Mildly abnormal GM: the variation in the spatial distribution of phasic EMG bursts is conserved, but the size (amplitude and duration) of the phasic bursts is disproportionally large. **C** and **D**: Examples of definitely abnormal GMs with stereotyped muscle activation patterns. In **C** the arm and leg muscles are activated in synchrony; in **D** the arm muscles are activated synchronously, but the lower leg muscles (GA and TA) show a stereotyped pattern of reciprocal activity. EMGs are surface EMGs recorded on the right side of the body: BB = m. biceps brachii, DE = deltoid muscle, EC = mm. extensor carpi, FC = mm. flexor carpi, GA = m. gastrocnemius, HA = hamstrings, PM = m. pectoralis major, RF = m. rectus femoris, TA = m. tibialis anterior, TB = m. triceps brachii.

al., 1979; Vles et al., 1989). With respect to the posture of neck and trunk it has been reported that before 32 weeks PMA antigravity postural control of the neck and trunk is entirely lacking (Amiel-Tison, 1968; Saint-Anne Dargassies, 1974). During the following weeks some head control develops, so that at term age low-risk preterm infants, like full-term infants, can keep the head upright for a few seconds while in a sitting position (Prechtl, 1977). Near term, human infants also develop a head-preference, in general to the right side. This preference not only develops in the extrauterine environment (Vles & Van Oostenbrugge, 1988), but also in utero (Ververs et al., 1994).

At term age the dominant posture of arms and legs is strong flexion (Amiel-Tison, 1968). The flexion posture gradually disappears during the following months, with the flexion decreasing somewhat earlier in the arms than in the legs

(McGraw, 1943; Touwen, 1976). At 2–3 month post-term, i.e. concurrent with the development of the "fidgety" GMs, the head can be stabilized on the trunk (Touwen, 1976), vestibular responsiveness has improved (Eviatar *et al.*, 1974), and a steady visual fixation and brisk visual orienting reactions have been developed (Atkinson & Braddick, 1989). Thus, at 2–3 months a general transition in motor behaviour takes place (Hopkins & Prechtl, 1984). Possibly, increasing activities in the basal ganglia, the cerebellum and the parietal, temporal and occipital cortices, which have been demonstrated by functional neuro-imaging studies (Chugani *et al.*, 1987; Rubinstein *et al.*, 1989), play an important role in this transition.

Precursors of goal-directed motility

The newborn infant shows precursors of two types of goal-directed motility: "pre-reaching" movements, which are the precursors of goal-directed reaching and manipulative behaviour, and infant stepping movements, which are the precursors of locomotion.

Newborn infants are unable to voluntarily grasp an object. However, Von Hofsten (1982) showed that neonates move their arms with open hands more often to an attractive object when they visually fixate it, than when they are not paying visual attention to the object. These "pre-reaching" movements (Trevarthen *et al.*, 1981), however, do not result in an actual grasping of the object. When the newborn infant is provided with ample postural support of the neck and shoulder region, however, the "pre-reaching" movements become more goal-oriented and occasionally successful (Amiel-Tison & Grenier, 1983). Thus, lack of sufficient postural control seems to be a major constraint in the development of goal-directed reaching. Indeed, the development of successful reaching around 4 months of age, is associated with an increase in the activity of the trapezius and deltoid muscles, resulting in an improvement of shoulder stabilization (Thelen & Spencer, 1998), and a more efficient organization of neck and trunk muscle activity (Van der Fits *et al.*, 1999a).

Newborn infants – like the fetus (De Vries *et al.*, 1984) – show locomotor-like behaviour in the form of neonatal crawling and stepping movements (McGraw, 1943; Prechtl, 1977). These movements are probably generated by spinal pattern generators analogous to the locomotion in the hindlimbs of kittens after a transection of the thoracic cord (Forssberg *et al.*, 1980) or in spinal lampreys (Grillner *et al.*, 1995). It is still a matter of discussion whether humans have a hierarchical organization of neural networks for locomotion similar to that of quadrupeds, or whether supraspinal circuits have taken over some of the control functions. Recent evidence suggests that monoaminergic drugs can induce locomotion in non-human primates (Hultborn *et al.*, 1993) and several groups have been able to induce locomotion in humans after incomplete spinal lesions (Dietz *et al.*, 1994; Fung *et al.*, 1990). The infant stepping movements are rather primitive in character and differ largely from the flexible plantigrade gait of adulthood (Forssberg, 1985). The non-goal-directed neonatal stepping is characterized by a lack of segment-specific movements, implying that the legs tend to flex and extend as a single unit (Forssberg, 1985; Thelen, 1985), by a lacking adult-type of heel strike, a variable muscle activation with a high degree of antagonistic co-activation, and by short-latency EMG bursts at the foot contact due to segmental reflex activity (Forssberg, 1985). In the absence of specific training the stepping movements can no longer be elicited after the age of 2–3 months (Touwen, 1976). A period of locomotor silence follows, which is succeeded in the third quarter of the first postnatal year by goal-directed progression in the form of crawling and supported locomotion. When the neonatal stepping is trained daily, the stepping response can be elicited until it is replaced by supported locomotion (Yang *et al.*, 1998; Zelazo, 1983). This is perhaps not so surprising in light of the fact that the locomotor pattern of supported locomotion is reminiscent of that of neonatal stepping – both lacking the consistent determinants of plantigrade gait (Forssberg, 1985).

GOAL-DIRECTED MOTOR BEHAVIOUR

Development of goal-directed behaviour during infancy

The major transition of non-goal-directed general movement activity into voluntary, goal-directed motility is a gradual process which lasts several weeks. Goal-directed motility emerges during the final phase of GM activity, the phase of "fidgety" GMs. From about 10 weeks post-term, when the infants show full-blown "fidgety" GM activity, the first goal-directed movements of the arms occur in the form of mutual manipulation of the hands, manipulation of the clothes and reaching movements which do not result in successful grasping (Hopkins & Prechtl, 1984). Initially the general movement activity is interrupted for short periods with goal-directed arm movements. However, when the infants get older, and the reaching movements become successful more often, the reverse occurs: goal-directed arm motility is alternated with short periods of GM activity. In terms of neural networks the gradual change from GM activity into goal-directed behaviour could mean that the generalized networks controlling GM activity are flexibly rearranged into multiple smaller networks, dedicated to the control of specific motor behaviour, such as goal-directed motility of the arms and the legs, and postural control (cf. Simmers *et al.*, 1995). It was recently shown that such a functional shift of neurons in the participation of different neural networks is guided by experience (Nargeot & Moulins, 1997).

The goal-directed motility develops in a craniocaudal sequence, as the reaching movements in the arms develop 1–2 months before the goal-directed movements of the legs during rolling and prone progression (McGraw, 1943; Touwen, 1976). When the upper part of the body is already involved in goal-directed reaching behaviour, the lower part of the body continues with free-wheeling GM-like activity. The goal-directed reaching movements become successful at 4–5 months (Trevarthen *et al.*, 1981; Von Hofsten, 1984), concurrent with the development of stereo-acuity (Granrud, 1986) and substantial improvements in postural control (Van der Fits *et al.*, 1999a).

Motor development during infancy is characterized by intra- and interindividual variation (Touwen, 1976, 1978). The variation occurs, for instance, as variation in the emergence of a function, variation in the performance of a function, and variation in the duration of specific developmental phases. Infancy is the period during which the fetally conceived primary neural repertoires are elaborated further. It is the phase of "primary variability" (Hadders-Algra *et al.,* 1998b; Touwen, 1993) with abundant (cortical) synaptogenesis (Huttenlocher *et al.,* 1982; Rakic *et al.,* 1986). The resulting multifarious primary networks pave the way for the selection of the most appropriate circuitries. The variation in development includes the co-occurrence of different developmental phases. For instance, infants of a certain age can alternate belly crawling with crawling on hands and knees (Figure 6; McGraw, 1943; Touwen, 1976). Healthy infants can also exhibit a temporary regression, an "inconsistency", in the development of a specific function (Touwen, 1976). As long as the regression is restricted to a single function it can be regarded as another expression of developmental variation. The large variation in the attainment of milestones (Figure 4) implies that the assessment of milestones has limited clinical value. A slow development of a single function usually has no clinical significance but, of course, the finding of a general delay is clinically relevant. For the assessment of motor milestones in preterm infants the age of the infant should be corrected for prematurity, i.e. functional age should be calculated from term age onwards (Touwen, 1981). This rule is especially useful during the first half-year post-term, when the variation in the emergence of motor skills lies in the order of about 3 months, which is equivalent to the degree of prematurity. At later ages the variation in motor skill attainment exceeds the degree of prematurity, thereby reducing the significance of age correction in older preterm infants.

Development of postural control during infancy

Postural control is primarily aiming at the maintenance of a vertical posture of head and trunk against the forces of gravity, because a vertical orientation of the proximal parts of the body provides an optimal condition for vision and goal-directed motility (Massion, 1998). The development of postural control is one of the major accomplishments of infant motor development. The postural achievements form the basis for adequate reaching, sitting, standing and walking.

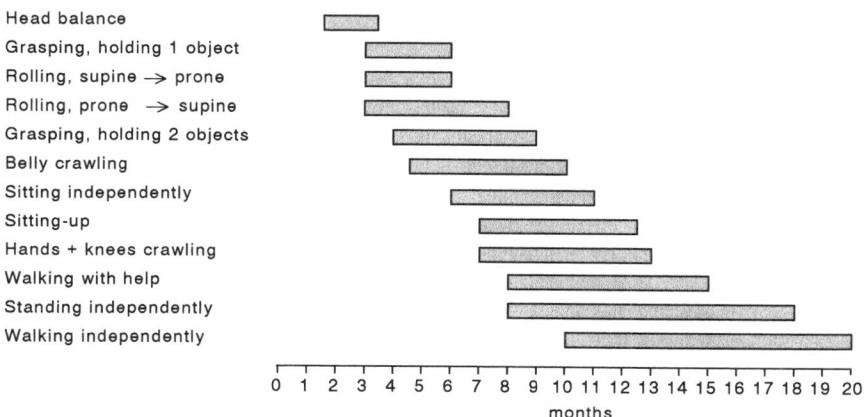

Figure 4 Schematic representation of the ages at which some motor skills emerge during infancy. The lengths of the bars reflect the inter-individual variation. Adapted from Touwen (1976).

At the age of 3 months, when the infant is able to stabilize the head in space, reaching movements emerge (Touwen, 1976). The first reaching movements are accompanied by a relatively unorganized in-concert activation of the neck and trunk muscles. However, at 4–5 months, when the reaching movements result in actual grasping, the variable neck and trunk muscle activity shows a distinct organization with the basic characteristics of adult postural adjustments in non-standing positions. Already at this age postural activity is direction-specific, i.e. the forward sway of the body induced by the reaching movement elicits a primary activation of the dorsal postural muscles, and it shows a top-down recruitment, meaning that the neck muscles are activated prior to the trunk muscles (Van der Fits et al., 1998, 1999a). During the early phases of the development of reaching the quality of the reaching movement shows a strong positive correlation with the quality of postural control (Fallang et al., 2000). This link between reaching and postural control becomes more subtle with increasing age (Fallang et al., 2000; Savelsbergh & Van der Kamp, 1994). When the infant gets older the postural muscles are less often activated during reaching activity, so that at 5–6 months the combination in which the muscles are activated is very variable (Hadders-Algra et al., 1996a; Hirschfeld & Forssberg, 1994; Van der Fits et al., 1999b). The presence of little and very variable postural activity is a transient phenomenon, which coincides with a transient reduction in the variation of postural sway during sitting (Bertenthal et al., 1997). This temporary phase of little postural activity probably serves as a transition phase, as after the transition the process of selection starts, heralding the development of adaptive postural strategies during sitting (Hadders-Algra et al., 1996a; Van der Fits et al., 1999b). The selection, which is dependent on daily life experiences in postural control, and which takes place during the phase when infants learn to sit independently, favours the pattern in which all direction-specific muscles are activated ("complete pattern", Hadders-Algra et al., 1996b, see Figure 5).

Figure 5 (*right*) Effect of daily balance training on the development of postural adjustments. **A**: Schematic presentation of the experimental condition: postural adjustments during sitting are tested on a movable platform by means of multiple surface EMG recordings. On the left: a forward translation of the platform induces a backward sway of the body and results in a (direction-specific) activation of the muscles on the ventral side of the body (NF = neck flexor muscles, RA = m. rectus abdominis, RF = m. rectus femoris). On the right the reverse condition: a backward translation produces a forward sway of the body and causes a response in the muscles on the dorsal side of the body (NE = neck extensor muscles, TE/LE = thoracal and lumbar extensor muscles, HAM = hamstrings). **B**: Developmental changes in postural response patterns during forward translations in trained and non-trained infants at 5–6 and 9–10 months of age. Each horizontal bar represents the response patterns for one subject. The diagram on the right supplies the hatching codes of the response patterns used in the left part of the figure, for instance the 12th pattern, consisting of the "complete" response pattern in which the NF, RA and RF muscles are activated and the dorsal extensor muscles are inhibited (E-INH), is indicated by a black hatching pattern in the left-hand panel. The postural response patterns at 5–6 months show a large intra- and interindividual variation. At 9–10 months the variation is less, especially so in the group of trained infants. The trained infants preponderantly selected the most efficient "complete" pattern in response to the postural perturbation. The training consisted of toy presentation in the border zone of reaching-without-falling, and was carried out by the parents three times a day during a period of 3 months, starting at study entry at 5–6 months. (Adapted from Hadders-Algra et al., 1996b.)

DEVELOPMENT OF GROSS MOTOR FUNCTIONS

A

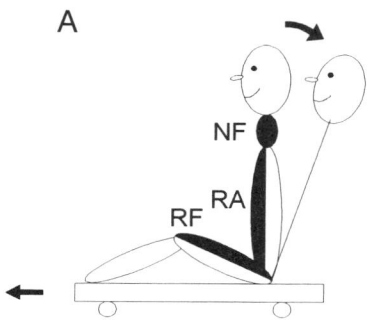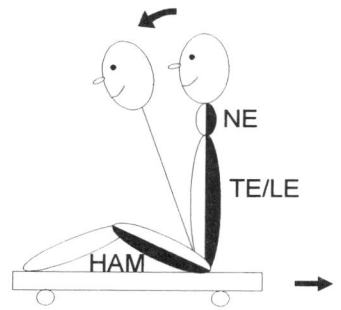

B

NON-TRAINED   TRAINED

5-6 MONTHS

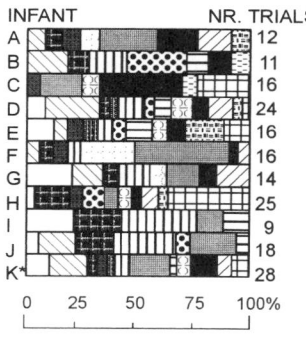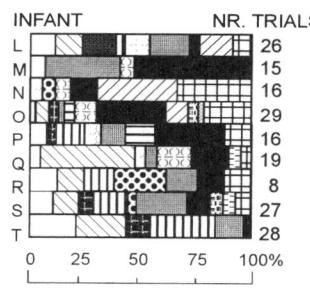

9-10 MONTHS

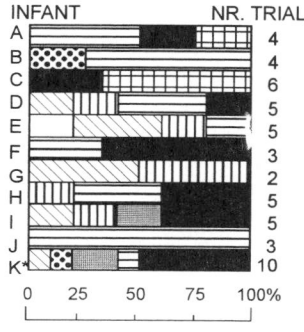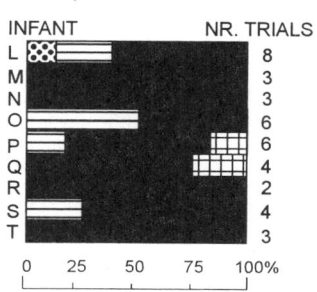

After the selection of the complete pattern, i.e. from 8–9 months onwards, infants obtain the ability to modulate the quantity of the postural muscle activity with respect to specific conditions, such as the pelvis position during sitting. This ability, the capacity to adapt postural muscle activity to environmental conditions, can also be facilitated by training (Hadders-Algra *et al.*, 1996b).

From 8–9 months onwards the ability to stand develops (Figure 4). Relatively little is known on muscle coordination during stance in infants who are not able to stand without help. The information available stems from studies on postural adjustments on a movable platform in which non-standing babies were provided with postural support (Sveistrup & Woollacott, 1996, 1997; Woollacott & Sveistrup, 1992). However, the provision of postural support is known to reduce postural activity to a considerable extent (Cordo & Nashner, 1982; Müller *et al.*, 1992). The studies of Sveistrup & Woollacott indicated that supported infants in the pull-to-stand stage of development seldom show direction-specific activity in the leg and trunk muscles. When pull-to-stand infants were trained intensively for 3 days on the balance platform the prevalence of "complete" direction-specific postural synergies increased (Sveistrup & Woollacott, 1997). Sveistrup & Woollacott also tested postural adjustments during stance in infants who were just able to stand independently, and in infants who had just mastered the skill of independent walking (Sveistrup & Woollacott, 1996; Woollacott & Sveistrup, 1992). Their data revealed that infants who are just able to stand without help often show "complete" postural synergies. The temporal coordination of muscle activity in the synergies is variable, but a preference for the adult "bottom-up" recruitment order can already be distinguished. Newly walking infants invariably select the "complete" postural response patterns with a consistent caudo-to-cranial recruitment.

The study of Van der Fits *et al.* (1999b), on postural adjustments during reaching in a sitting position, indicated that, from about 15 months of age, feed-forward motor control processes become increasingly important. At this age infants show postural muscle activity which precedes the muscle activity in the reaching arm. The development of this type of anticipatory postural control is related to the development of independent walking (Van der Fits *et al.*, 1999b).

Development of progressive behaviour during infancy

When infants develop goal-directed motility in the legs, the leg movements are used to roll out of the supine position, as this position is incompatible with efficient progressive behaviour. Rolling to the side develops at around 3 months of age, and this precursory rolling is soon followed by successful rolling from supine into prone position (Figure 4; Touwen, 1976). Rolling in the reverse direction, i.e. from prone into supine, develops in general several weeks later, but exceptions to this rule – infants who are able to roll from prone into supine prior to being able to roll from supine into prone – are commonly observed (Touwen, 1976).

Progressive behaviour develops in the prone position. It starts when the infant is able to lift the head and upper part of the thorax, using the elbows part of the time for additional support. The first locomotor movements in prone are rather ineffective in terms of progression. They consist of wriggling or pivoting movements, in which the arms and legs are tried out in a seemingly random way

(McGraw, 1943; Touwen, 1976). Next, locomotion in prone progresses from belly crawling with the use of the arms, or the arms and legs, to crawling on hands and knees. Finally, a short-lasting stage of crawling on hands and feet (quadruped or "bear" crawling) can be distinguished (Adolph *et al.*, 1998; Ames, 1937; Largo *et al.*, 1985; McGraw, 1943; Touwen, 1976). Again, however, many exceptions to this standard development of crawling are present in the normal population. The exceptions include children who skip certain stages of crawling, such as belly crawling, or children who never crawl, or children who show totally different patterns of non-walking progression, such as shuffling in sitting position (Adolph *et al.*, 1998; Ames, 1937; Largo *et al.*, 1985). Children can also show several crawling strategies at the same age (Figure 6; Adolph *et al.*, 1998; Touwen, 1976).

The development of crawling is experience-dependent. Babies who sleep consistently in prone position develop the various stages of crawling behaviour significantly earlier than babies sleeping in a supine position (Davis *et al.*, 1998). Moreover, babies who lack prone experience during the daytime, because they spend a substantial part of the day in a so-called baby-walker, show a significant delay in the development of prone locomotion (Crouchman, 1986). Specific training of crawling can accelerate the development of prone progression and is associated with a facilitation of the development of standing and walking (Lagerspetz *et al.*, 1971).

The coordination of arm and leg movements during belly crawling is largely variable. However, when postural control allows for crawling on hands and knees the variation in locomotor coordination drops rapidly, and the efficient pattern of diagonal gait, in which right arm and left leg move together, is selected (Adolph *et al.*, 1998; Burnside, 1927; Freedland & Bertenthal, 1994). The proficiency of crawling increases gradually during the first 10 weeks of crawling on hands and knees. This is reflected by a gradual increase in crawling velocity, and a gradual decrease in the duration of the stance and swing phases of the limbs. Remarkably, crawling proficiency is better in infants who previously were belly crawlers, than in infants who skipped the stage of belly crawling (Adolph *et al.*, 1998). This could imply that experience during belly crawling offers an optimal condition to explore the primary neural repertoires for crawling and select the neural activity patterns resulting in the most efficient crawling solutions.

Development of goal-directed behaviour during childhood

Infancy by definition ends when the child can walk without help. Post-infancy motor development is characterized by a gradual increase in agility, adaptability and the ability to make complex movement sequences. It is the phase of secondary or adaptive variability, during which maturational processes in continuous interaction with experience produces highly adaptive secondary neuronal repertoires (Figure 1; Edelman, 1989; Hadders-Algra *et al.*, 1998b; Touwen, 1993). The creation of the secondary repertoires is associated with extensive synapse rearrangement, the net result of synapse formation and synapse elimination (Huttenlocher *et al.*, 1982; Purves, 1994; Rakic *et al.*, 1986). This is facilitated by increasingly shorter processing times, which can be attributed in part to ongoing myelination (Jernigan *et al.*, 1991; Müller *et al.*, 1994).

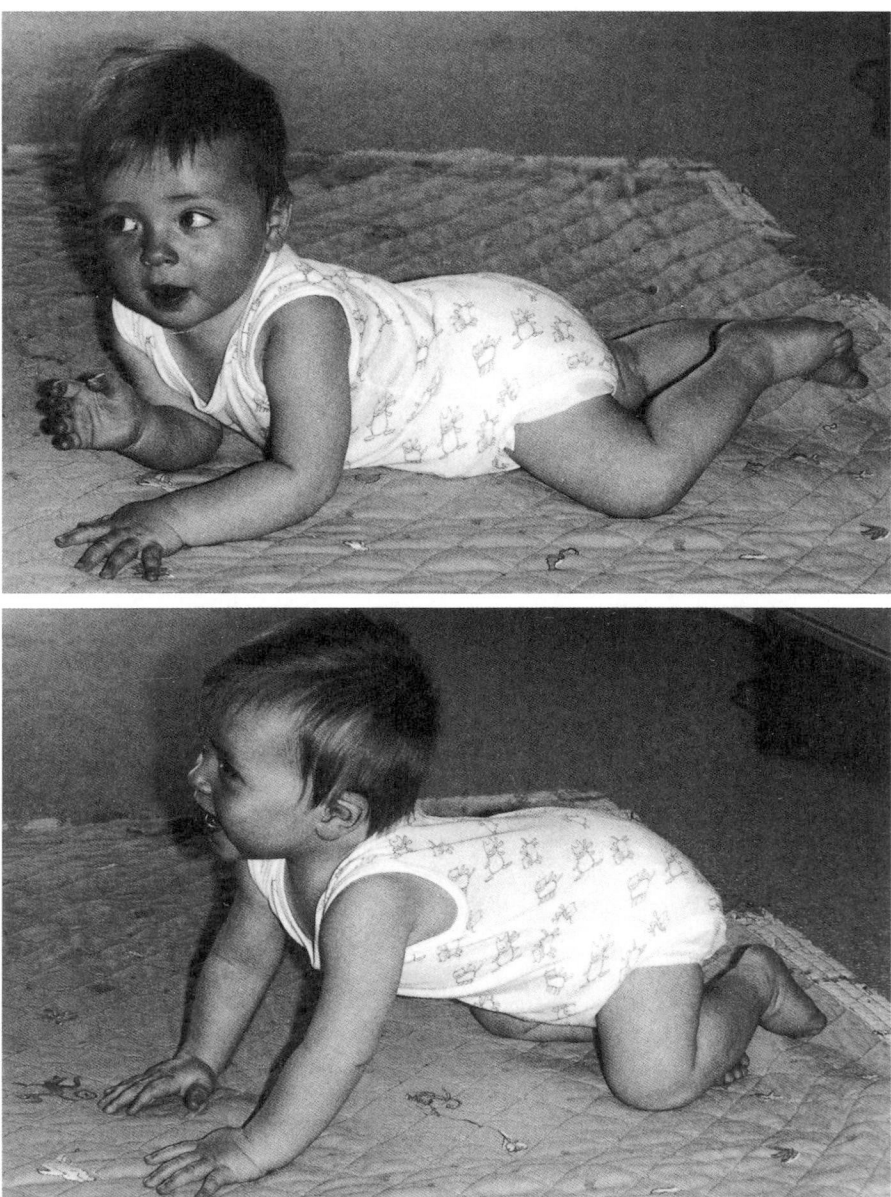

Figure 6 Infant of 10 months who alternates belly crawling (upper panel) with crawling on hands and knees (lower panel). Published with permission of the parents.

Development of postural control during childhood

The early phases of locomotion are dominated by difficulties with the mainte-nance of balance. Postural control at toddler age is achieved by an activation of all direction-specific leg, trunk and neck muscles and a rather large amount of antagonistic co-activation (Hadders-Algra *et al.*, 1998b). This "muscular corset strategy", which explains the stick-like, *en-bloc* locomotor behaviour of the toddler, can be regarded as a smart initial solution of the nervous system to master the large degrees of freedom of the upright moving body (cf. Bernstein, 1935). The *"en-bloc"* postural strategy during early bipedal life is organized with hip stabilization in space as a frame of reference (Assaiante *et al.*, 1993). Vision does not seem to play a large role in toddler postural control, as postural sway of standing toddlers does not increase in the absence of visual input (Ashmead & McCarty, 1991). However, the vestibular otolith system probably contributes substantially to the early mastery of locomotor postural control, as vestibular ocular responses to otolith stimulation change significantly during the emer-gence of independent walking (Wiener-Vacher *et al.*, 1996).

The energy-consuming "muscular corset strategy" is present during the so-called transient toddling phase, which ends at 2½–3 years (Hadders-Algra *et al.*, 1998b). Thereafter, antagonistic co-activation decreases in postural adjustments during sitting and standing (Forssberg & Nashner, 1982; Hadders-Algra *et al.*, 1998b). After 2½–3 years postural adjustments during sitting are characterized by variation in the combination and order in which the direction-specific agonistic muscles are activated (Hadders-Algra *et al.*, 1998b). In the standing condition, which is substantially less stable than the sitting condition, the agonistic muscles continue to be consistently activated, albeit to a less strong extent (Berger *et al.*, 1995; Forssberg & Nashner, 1982; Woollacott *et al.*, 1987). During walking the frame of reference for postural control changes from hip stabilization in space to head stabilization in space (Assaiante, 1998). Anticipatory postural adjustments during gait initiation emerge (Ledebt *et al.*, 1998). However, consistent antici-patory postural behaviour in the bipedal position is first found from 4 years onwards, not only during gait initiation, but also during rising to tiptoe (Haas *et al.*, 1989; Ledebt *et al.*, 1998).

From 2–3 years onwards vision starts to have a distinct effect on postural control during stance. Postural sway increases significantly in the absence of visual control (Hirabayashi & Iwasaki, 1995; Riach & Starkes, 1994), and the absence of visual control during postural adjustments in stance induces monosynaptic reflex activity in the gastrocnemius muscles and an early activation of the neck muscles, phenomena which are absent when vision is not restricted (Woollacott *et al.*, 1987). It should be kept in mind, however, that – at least from pre-school age onwards – the impact of somatosensory information on balance control outweighs that of visual input (Gurfinkel *et al.*, 1995; Shumway-Cook & Woollacott, 1985).

Postural control continues to change with increasing age, a process which is accompanied with increasingly shorter latencies until postural response onset (Haas *et al.*, 1986; Müller *et al.*, 1991). Around 6–7 years another transition in postural control can be distinguished (Assaiante, 1998; Berger *et al.*, 1990; Shumway-Cook & Woollacott, 1985; Woollacott *et al.*, 1987). Postural sway during standing decreases significantly, partially due to increases in body height and body weight, and the size of the feet (Riach & Starkes, 1993, 1994). The decrease in postural sway is associated with a decreasing reliance on visual information (Foudriat et al., 1993; Hytönen et al., 1993; Woollacott et al., 1987), and a significant increase in the stability limits, i.e. the ranges in which the body can sway without losing balance (Riach & Starkes, 1993). The postural muscle responses during perturbations in stance obtain their adult configuration with relatively small response amplitudes when compared to the responses at younger ages (Berger et al., 1995; Woollacott et al., 1987). Postural control after the age of 6 years allows the child to stand on one leg for more than 20 seconds and to tandem walk on a balance beam (Sutherland et al., 1988). Further refinements in postural control continue until adolescence. This results in decreasing postural sway during standing (Hirabayashi & Iwasaki, 1995; Hytönen et al., 1993), the ability to maintain balance during postural perturbations in standing conditions with conflicting somatosensory and visual information (Forssberg & Nashner, 1982), and a fine-tuning of anticipatory postural adjustments during locomotion (Hirschfeld & Forssberg, 1992).

Development of locomotion during childhood

The locomotor pattern of the newly walking child is highly variable and lacks the plantigrade characteristics of adult gait (Burnett & Johnson, 1971; Cioni et al., 1993; Forssberg, 1985; Statham & Murray, 1971). The joints of the legs still tend to be flexed and extended in synchrony, and the adult type of heel-strike in front of the body is missing (Forssberg, 1985). The lack of a consistent heel-strike is associated with calf muscle activity preceding foot contact, monosynaptic stretch reflex activity in the calf muscles at foot contact, and lacking dorsiflexion activity of the tibialis anterior muscle prior to toe-off (Forssberg, 1985; Sutherland et al., 1988). This means that the lower leg muscles are co-activated during a major part of the gait cycle (Berger et al., 1984; Forssberg, 1985). Closer inspection of the early development of the heel-strike reveals that the contact of the foot during landing in nearly walking infants – who still need postural support – and newly walking infants varies between toe-first, flat-foot and heel-first, with a preference for a flat-foot strike (Cioni et al., 1993; Statham & Murray, 1971). After about 4 months of locomotor experience the energetically efficient pattern of heel-strike (Forssberg & Dietz, 1997) is selected significantly more often (Burnett & Johnson, 1971; Cioni et al., 1993), and by 2 years of age all children have developed a consistent heel-strike (Sutherland et al., 1988). Walking proficiency improves considerably during the first 4–5 months of locomotor experience: walking velocity increases, the step frequency and step length increase, the step width decreases and the oscillatory movements of the head and trunk decrease (Bril & Brenière, 1992; Ledebt et al., 1995).

From about 2 years of age the movements of the ankle are out of phase with those of the proximal joints of the leg (Forssberg & Dietz, 1997). Reciprocal activity in the lower leg muscles starts to develop due to a disappearance of calf muscle activity during the swing phase and a change in the timing of the tibialis anterior activity, which now starts prior to toe-off and switches off during mid-stance (Berger et al., 1984; Sutherland et al., 1988). At this age most children are

able to walk on toes and to run (Sutherland *et al.*, 1988). During the following years the locomotor pattern is further refined, reflecting an increasing influence of supraspinal control on the spinal locomotor networks (Forssberg & Dietz, 1997). The monosynaptic reflex activity in the calf muscles disappears around 4 years of age and the size of calf muscle activity during stance increases until at least 7 years of age (Berger *et al.*, 1984). The velocity of the first step during gait initiation increases with increasing age, which in part is related to changes in body size, but also in part by the ability to generate anticipatory postural adjustments. Children are able to produce consistent anticipatory postural behaviour at gait initiation from 4 years onwards, but only at the age of 6 years is the size of the anticipatory backward displacement of the centre of pressure in the feet – like in adults – related to the velocity of the first step (Ledebt *et al.*, 1998). Walking proficiency continues to increase, especially due to body-size related increases in step length and swing time (Wheelwright *et al.*, 1993). The protracted course of adaptive locomotor behaviour and its continuous dependency on postural abilities is illustrated by the finding that 10-year-old children still have more difficulties in the adaptation of gait than adult subjects when tested on a split-belt treadmill imposing different belt speeds to the left and right leg (Zijlstra *et al.*, 1996).

CONCLUDING REMARKS

Gradually it has become clear that variation is a fundamental feature of normal motor development. Development starts with the phase of primary variability, continues with selection, which occurs at function-specific points in time during the first 1½ years after birth, and ends with the phase of secondary or adaptive variability. The development of gross motor functions is primarily dependent on the protracted course of the development of postural control. The latter is characterized by multiple phases of transition occurring at the ages of 2–3 months, 5–6 months, 9–10 months, 15 months, 2–3 years and 6–7 years, and continues until adolescence.

References

Adolph, K.E., Vereijken, B. & Denny, M.A. (1998). Learning to crawl. *Child Development*, **69**, 1299–1312.
Ames, L.B. (1937). The sequential patterning of prone progression in the human infant. *Genetic Psychology Monographs*, **19**, 409–460.
Amiel-Tison, C. (1968). Neurological evaluation of the maturity of newborn infants. *Archives of Disease in Childhood*, **43**, 89–93.
Amiel-Tison, C. & Grenier, A. (1983). N*eurological evaluation of the newborn and infant*. New York: Masson.
Angulo y González, A.W. (1932). The prenatal development of behaviour in the albino rat. *Journal of Comparative Neurology*, **55**, 395–442.
Assaiante, C. (1998). Development of locomotor balance control in healthy children. *Neuroscience and Biobehavioral Reviews*, **22**, 527–532.
Assaiante, C., Thomachot, B. & Aurently, R. (1993). Hip stabilization and lateral balance control in toddlers during the first four months of autonomous walking. *NeuroReport*, **4**, 875–878.
Ashmead, D.H. & McCarty, M.E. (1991). Postural sway of human infants while standing in light and dark. *Child Development*, **62**, 1276–1287.

Atkinson, J. & Braddick, O. (1989). Development of basic visual functions. In A. Slater & G. Bremner (Eds.), *Infant development* (pp. 7–41). London: Lawrence Erlbaum.

Berger, W., Altenmüller, E. & Dietz, V. (1984). Normal and impaired development of children's gait. *Human Neurobiology*, **3**, 163–170.

Berger, W., Horstmann, G.A. & Dietz, V. (1990). Interlimb coordination of stance in children: divergent modulation of spinal reflex responses and cerebral evoked potentials in terms of age. *Neuroscience Letters*, **116**, 118–122.

Berger, W., Trippel, M., Assaiante, C., Zijlstra, W. & Dietz, V. (1995). Developmental aspects of equilibrium control during stance: a kinematic and EMG study. *Gait and Posture*, **3**, 149–155.

Bernstein, N. (1935). The problem of the interrelation of co-ordination and localization. In H.T.A. Whiting (Ed.), *Human motor actions. Bernstein reassessed* (pp. 77–119), Amsterdam: Elsevier.

Bertenthal, B.I., Campos, J.J. & Kermoian, R. (1994). An epigenetic perspective on the development of self-produced locomotion and its consequences. *Current Directions in Psychological Science*, **3**, 140–145.

Bertenthal, B.I., Rose, J.I. & Bai, D.L. (1997). Perception-action coupling in the development of visual control of posture. *Journal of Experimental Psychology, Human Perception and Performance*, **23**, 1631–1643.

Bril, B. & Brenière, Y. (1992). Postural requirements and progression velocity in young walkers. *Journal of Motor Behavior*, **24**, 105–116.

Burnett, C.N. & Johnson, E.W. (1971). Development of gait in childhood: Part II. *Developmental Medicine and Child Neurology* **13**, 207–215.

Burnside, L.H. (1927). Coordination in the locomotion of infants. *Genetic Psychology Monographs*, **5**, 280–372.

Changeux, J-P. (1997). Variation and selection in neural function. *Trends in Neuroscience*, **20**, 291–293.

Changeux, J-P. & Danchin, A. (1976). Selective stabilisation of developing synapses as a mechanism for the specification of neuronal networks. *Nature*, **264**, 705–12.

Chugani, H.T., Phelps, M.E. & Maziotta, J.C. (1987). 18-FDG Positron emission tomography in human brain. Functional development. *Annals of Neurology*, **22**, 487–497.

Cioni, G. & Prechtl, H.F.R. (1990). Preterm and early postterm motor behaviour in low-risk premature infants. *Early Human Development*, **23**, 159–191.

Cioni, G., Duchini, F., Milianti, B., Paolicelli, P.B., Sicola, E., Boldrini, A. & Ferrari, A. (1993). Differences and variations in the patterns of early independent walking. *Early Human Development*, **35**, 193–205.

Cioni, G., Ferrari, F., Einspieler, C., Paolicelli, P, Barbani, M.T. & Prechtl., H.F.R. (1997). Comparison between observation of spontaneous movements and neurological examination in preterm infants. *Journal of Pediatrics*, **130**, 704–711.

Coghill, G.E. (1929). *Anatomy and the problem of behaviour*. Cambridge: Cambridge University Press.

Cordo, P.J. & Nashner, L.M. (1982). Properties of postural adjustments associated with rapid arm movements. *Journal of Neurophysiology,*, **47**, 287–302.

Crouchman, M. (1986). The effects of babywalkers on early locomotor development. *Developmental Medicine and Child Neurology*, **28**, 757–761.

Davis, B.E., Moon, R.Y., Sachs, H.C. & Ottolini, M.C. (1998). Effects of sleep position on infant motor development. *Pediatrics*, **102**, 1135–1140.

De Boysson-Bardies, B. (1993). Ontogeny of language-specific syllabic productions. In B. De Boysson-Bardies, S. De Schonen, P. Jusczyk, P. McNeilage & J. Morton (Eds.), *Developmental neurocognition: speech and face processing in the first year of life* (pp. 353–363). Dordrecht: Kluwer Academic Publishers.

De Vries, J.I.P., Visser, G.H.A. & Prechtl, H.F.R. (1982). The emergence of fetal behaviour. I. Qualitative aspects. *Early Human Development*, **7**, 301–322.

De Vries, J.I.P., Visser, G.H.A. & Prechtl, H.F.R. (1984). Fetal motility in the first half of pregnancy. In H.F.R. Prechtl (Ed.), *Continuity of neural functions form prenatal to postnatal life* (pp. 46–64). Clin. Dev. Med. no. 94, Oxford: Blackwell.

Diener, H.C., Bootz, F., Dichgans, J. & Bruzek, W. (1983). Variability of postural reflexes in humans. *Experimental Brain Research*, **52**, 423–428.

Dietz, V., Colombo, G. & Jensen L. (1994). Locomotor activity in spinal man. *Lancet*, **344**, 1260–1263.

Dubowitz, L.M.S. & Dubowitz, V. (1981). *The neurological assessment of the preterm and full-term newborn infant*. Clin. Dev. Med. no. 79. London: Heinemann.

Dunbar, D.C. & Badam, G.L. (1998). Development of posture and locomotion in free-ranging primates. *Neuroscience Biobehavioral Reviews,* **22**, 541–546.

Edelman, G.M. (1989). *Neural Darwinism. The theory of neuronal group selection*. Oxford: Oxford University Press.

Edelman, G.M. (1993). Neural Darwinism: Selection and reentrant signalling in higher brain function. *Neuron*, **10**, 115–125.

Eviatar, L, Eviatar, A. & Naray, I. (1974). Maturation of neuro-vestibular responses in infants. *Developmental Medicine and Child Neurology,* **16**, 435–446.

Fallang, B, Saugstad, O.D. & Hadders-Algra, M. (2000). Goal directed reaching and postural control in supine position in healthy infants. *Behavioral Brain Research*, **115**, 9–18.

Feldman, W.M. (1920). *The principles of ante-natal and post-natal child physiology, pure and applied*. London: Longmans, Green & Co.

Ferrari, F., Cioni, G. & Prechtl, H.F.R. (1990). Qualitative changes of general movements in preterm infants with brain lesions. *Early Human Development*, **23**, 193–231.

Forssberg, H. (1985). Ontogeny of human locomotor control. I. Infant stepping, supported locomotion and transition to independent locomotion. *Experimental Brain Research,* **57**, 480–493.

Forssberg, H., Dietz, V. (1997). Neurobiology of normal and impaired locomotor development. In K.J. Connolly & H. Forssberg (Eds.), *Neurophysiology and neuropsychology of motor development* (pp. 78–100). Clin. Dev. Med. no. 143/144. London: MacKeith Press.

Forssberg, H. & Nashner, L.M. (1982). Ontogenetic development of postural control in man: adaptation to altered support and visual conditions. *Journal of Neuroscience,* **2**, 545–552.

Forssberg, H., Grillner, S. & Halbertsma, J. (1980). The locomotion of the low spinal cat. I. Coordination within a hindlimb. *Acta Physiologica Scandinavica*, **108**, 269–281.

Foudriat, B.A., Di Fabio, R.P. & Anderson, J.H. (1993). Sensory organization of balance responses in children 3–6 years of age: a normative study with diagnostic implication. *International Journal of Pediatric Otorhinolaryngology*, **27**, 255–271.

Freedland, R.L. & Bertenthal, B.I. (1994). Developmental changes in interlimb coordination: transition to hands-and-knees crawling. *Psychological Science,* **5**, 26–32.

Fung, J., Stewart, J.E. & Barbeau, H. (1990). The combined effects of clonidine and cyproheptadine with interactive training on the modulation of locomotion in spinal cord injured subjects. *Journal of the Neurological Sciences*, **100**, 85–93.

Funk, G.D. & Feldman, J.L. (1995). Generation of respiratory rhythm and pattern in mammals: insights form developmental studies. *Current Opinion in Neurobiology,* **5**, 778–785.

Gesell, A. & Amatruda C.S. (1947). *Developmental diagnosis. Normal and abnormal child development* (2nd. ed.). New York: Harper & Row.

Gordon, A.M. & Forssberg, H. (1997). Development of neural mechanisms underlying grasping in children. In K.J. Connolly & H. Forssberg (Eds.), *Neurophysiology and neuropsychology of motor development* (pp. 214–231). Clin. Dev. Med. no. 143–144, London: MacKeith Press.

Gottlieb, G. (1991). Experiential canalization of behavioural development: theory. *Developmental Psychology,* **27**, 4–13.

Gramsbergen, A., Ijkema-Paassen, J., Nikkels, P.G.J. & Hadders-Algra, M. (1997). Regression of polyneural innervation in the human psoas muscle. *Early Human Development*, **49**, 49–61.

Granrud, C.E. (1986). Binocular vision and spatial perception in 4- and 5-month-old infants. *Journal of Experimental Psychology, Human Perception and Performance,* **12**, 36–49.

Grillner, S., Deliagina, T., Ekeberg, Ö., El Manira, A., Hill, R.H., Lansner, A., Orlovsky, G.N. & Wallén, P. (1995). Neural networks that co-ordinate locomotion and body orientation in lamprey. *Trends in Neuroscience*, **18**, 270–279.

Gurfinkel, V.S., Ivanenko, Yu, P., Levik, S. & Babakova, I.A. (1995). Kinesthetic reference for human orthograde posture. *Neuroscience*, **68**, 229–243.

Haas, G., Diener, H.C., Bacher, M. & Dichgans, J. (1986). Development of postural control in children: short-, medium- and long latency EMG responses of leg muscles after perturbation of stance. *Experimental Brain Research*, **64**, 127–132.

Haas, G., Diener, H.C., Rapp, H. & Dichgans, J. (1989). Development of feedback and feedforward control of upright stance. *Developmental Medicine and Child Neurology,* **31**, 481–488.

Hadders-Algra, M. (1997). De beoordeling van spontane motoriek van jonge baby's: een doeltreffende methode voor de opsporing van hersenfunctiestoornissen. *Nederlandsch Tijdschrift voor Geneeskunde,* **141**, 816–820.

Hadders-Algra, M. & Dirks, T. (2000). *De motorische ontwikkeling van de zuigeling.* Houten: Bohn, Stafleu en Van Loghum.

Hadders-Algra, M. & Groothuis, A.M.C. (1999). Quality of general movements in infancy is related to the development of neurological dysfunction, attention deficit hyperactivity disorder and aggressive behaviour. *Developmental Medicine and Child Neurology,* **41**, 381–391.

Hadders-Algra, M. & Prechtl, H.F.R.(1992). Developmental course of general movements in early infancy. I: Descriptive analysis of change in form. *Early Human Development,* **28**, 201–214.

Hadders-Algra, M., Brogren, E. & Forssberg, H. (1996a). Ontogeny of postural adjustments during sitting in infancy: variation, selection and modulation. *Journal of Physiology,* **493**, 273–288.

Hadders-Algra, M., Brogren, E. & Forssberg, H. (1996b). Training affects the development of postural adjustments in sitting infants. *Journal of Physiology,* **493**, 289–298.

Hadders-Algra, M., Brogren, E. & Forssberg, H. (1998a). Development of postural control – differences between ventral and dorsal muscles? *Neuroscience and Biobehavioral Reviews,* **22**, 501–506.

Hadders-Algra, M., Brogren, E. & Forssberg, H. (1998b). Postural adjustments during sitting at pre-school age: the presence of a transient toddling phase. *Developmental Medicine and Child Neurology,* **40**, 436–447.

Hadders-Algra, M., Klip-Van den Nieuwendijk, A.W.J., Martijn, A. & Van Eykern, L.A. (1997). Assessment of general movements: towards a better understanding of a sensitive method to evaluate brain function in young infants. *Developmental Medicine and Child Neurology,* **39**, 88–98.

Hadders-Algra, M., Van Eykern, L.A., Klip-van den Nieuwendijk, A.W.J. & Prechtl, H.F.R. (1992). Developmental course of general movements in early infancy. II. EMG correlates. *Early Human Development,* **28**, 231–252.

Hakamada, S., Hayakawa, F., Kuno, K. & Tanaka, R. (1988). Development of the monosynaptic reflex pathway in the human spinal cord. *Developmental Brain Research,* **42**, 239–246.

Hall, W.G. & Oppenheim, R.W. (1987). Developmental psychobiology: prenatal, perinatal, and early postnatal aspects of behavioural development. *Annual Review of Psychology,* **38**, 91–128.

Hamburger, V. (1963). Some aspects of the embryology of behaviour. *Quarterly Review of Biology,* **38**, 342–365.

Hamburger, V. (1973). Anatomical and physiological basis of embryonic motility in birds and mammals. In G. Gottlieb (Ed.), *Studies on the development of behaviour and the nervous system.* Vol. 1. *Behavioural embryology* (pp. 52–76). New York: Academic Press.

Hamburger, V. & Oppenheim, R. (1967). Prehatching motility and hatching behaviour in the chick. *Journal of Experimental Zoology,* **166**, 171–204.

Hamburger, V., Wenger, E. & Oppenheim, R. (1966). Motility in the chick embryo in the absence of sensory input. *Journal of Experimental Zoology,* **162**, 133–160.

Hebb, D.O. (1949). *The organization of behaviour.* New York: Wiley.

Hempel, M.S. (1993). Neurological development during toddling age in normal children and children at risk of development disorders. *Early Human Development,* **34**, 47–57.

Hirabayashi, W. & Iwasaki, Y. (1995). Developmental perspective of sensory organization on postural control. *Brain and Development,* **17**, 111–113.

Hirschfeld, H. & Forssberg, H. (1992). Development of anticipatory postural adjustments during locomotion in children. *Journal of Neurophysiology,* **68**, 542–550.

Hirschfeld, H. & Forssberg, H. (1994). Epigenetic development of postural responses for sitting during infancy. *Experimental Brain Research,* **97**, 528–540.

Hopkins, B. & Prechtl, H.F.R. (1984). A qualitative approach to the development of movements during early infancy. In H.F.R. Prechtl (Ed.), *Continuity of neural functions form prenatal to postnatal life* (pp. 179–197). Clin. Dev. Med. no. 94, Oxford: Blackwell.

Howland, D.R., Bregman, B.S. & Goldberger, M.E. (1995). The development of quadrupedal locomotion in the kitten. *Experimental Neurology,* **135**, 93–107.

Hultborn, H., Petersen, N., Brownstone, R. & Nielsen, J. (1993). Evidence of fictive spinal locomotion in the marmoset (*Callitreix jacchus*). *Society of Neuroscience Abstracts,* **19**, 539.

Humphrey, T. (1969). Postnatal repetition of human prenatal activity sequences with some suggestion of their neuroanatomical basis. In R.J. Robinson (Ed.), *Brain and early behaviour* (pp. 43–71). New York: Academic Press.

Huttenlocher, P.R., DeCourten, C., Garey, L.J. & Van der Loos, H. (1982). Synaptogenesis in human visual cortex – evidence for synapse elimination during normal development. *Neuroscience Letters*, **33**, 247–252.

Hytönen, M., Pyykkö, I., Aalto H. & Starck, J. (1993). Postural control and age. *Acta Otolaryngologica*, **113**, 119–122.

Illingworth, R.S. (1966). *The development of the infant and young child: normal and abnormal development* (3rd ed.). Edinburgh: Livingstone.

Jernigan, T.L., Trauner, D.A., Hesselink, J.R. & Talla, P.A. (1991). Maturation of the human cerebrum observed "*in vivo*" during adolescence. *Brain*, **114**, 2037–2049.

Konczak, J., Borutta, M., Topka, H. & Dichgans, J. (1995). The development of goal-directed reaching in infants: hand trajectory formation and joint torque control. *Experimental Brain Research*, **106**, 156–168.

Kugler, P.N. & Turvey, M.T. (1987). *Information, natural law, and the self-assembly of rhythmic movement*. Hillsdale, NJ: Erlbaum.

Kugler, P.N., Kelso, J.A.S. & Turvey, M.T. (1980). On the concept of coordinative structures as dissipative structures. I. Theoretical lines of convergence. In G.E. Stelmach & J. Requin (Eds.), *Tutorials in motor behaviour* (pp. 3–47). New York: North Holland.

Lagerspetz, K., Nygård, M. & Strandvik, C. (1971). The effects of training in crawling on the motor and mental development of infants. *Scandinavian Journal of Psychology*, **12**, 192–197.

Largo, R.H., Molinari, L., Weber, M., Comenale Pinto L. & Duc, G. (1985). Early development of locomotion: significance of prematurity, cerebral palsy and sex. *Developmental Medicine and Child Neurology*, **27**, 183–191.

Ledebt, A., Bril, B. & Brenière, Y. (1998). The build-up of anticipatory behaviour. An analysis of the development of gait initiation in children. *Experimental Brain Research*, **120**, 9–17.

Ledebt, A., Bril, B. & Wiener-Vacher, S. (1995). Trunk and head stabilization during the first months of independent walking. *NeuroReport*, **6**, 1737–1740.

Massion, J. (1998). Postural control systems in a developmental perspective. *Neuroscience and Biobehavioral Reviews*, **22**, 465–472.

McGraw, M. (1939). Later development of children specially trained during infancy, Johnny and Jimmy at school age. *Child Development*, **10**, 1–19.

McGraw, M.B. (1943). *The neuromuscular maturation of the human infant*. Reprinted 1989: *Classics in developmental medicine*, no. 4. London: MacKeith Press.

Miller, A.J. (1972). Significance of sensory inflow to the swallowing reflex. *Brain Research*, **43**, 147–159.

Minkowski, M. (1938). Neurobiologische Studien am menschlichen Foetus. In E. Abderhalden (Ed.), *Handbuch der biologischen Arbeitsmethoden. Abt. V: Methoden zum Studium der Funktionen der einzelne Organe im Tierischen Organismus*. Teil 5B. (pp. 511–619). Berlin: Urban & Schwarzenberg.

Müller, K., Ebner, B. & Hömberg, V. (1994). Maturation of fastest afferent and efferent central and peripheral pathways: no evidence for a constancy of central conduction delay. *Neuroscience Letters*, **166**, 9–12.

Müller, K., Hömberg, V., Coppenrath, P. & Lenard, H.G. (1991). Maturation of lower extremity EMG responses to postural perturbation. Relationship of response-latencies to development of fastest central and peripheral efferents. *Experimental Brain Research*, **84**, 444–452.

Müller, K., Hömberg, V., Coppenrath, P. & Lenard, H.G. (1992). Maturation of set-modulation of lower extremity EMG responses to postural perturbations. *Neuropediatrics*, **23**, 82–94.

Narayanan, C.H., Fox, M.W. & Hamburger, V. (1971). Prenatal development of spontaneous and evoked activity in the rat (*Rattus norwegicus albinus*). *Behaviour*, **40**, 100–134.

Nargeot, R. & Moulins, M. (1997). Sensory-induced plasticity of motor pattern selection in the lobster stomatogastric nervous system. *European Journal of Neuroscience*, **9**, 1636–1645.

Nelson, P.G., Fields R.D., Yu, C. & Liu, Y. (1993). Synapse elimination from the mouse neuromuscular junction *in vitro*: a non-Hebbian activity dependent process. *Journal of Neurobiology*, **24**, 1517–1530.

Okado, N. & Kojima, T. (1984). Ontogeny of the central nervous system: neurogenesis, fibre connection, synaptogenesis and myelination in the spinal cord. In H.F.R. Prechtl (Ed.), *Continuity of neural functions from prenatal to postnatal life* (pp. 31–45). Clin. Dev. Med. no. 94, Oxford: Blackwell.

Papoušek, H., Papoušek, M. (1984). Qualitative transitions in integrative processes during the first trimester of human postpartum life. In H.F.R. Prechtl (Ed.), *Continuity of neural functions from prenatal to postnatal life* (pp. 220–244). Clin.Dev. Med. no. 94, Oxford: Blackwell.

Peiper, A. (1963). *Cerebral function in infancy and childhood* (3rd ed.). New York: Consultants Bureau.

Popper, K.R. & Eccles, J.C. (1977). *The self and its brain*. London: Routledge & Kegan Paul.

Prechtl, H.F.R. (1977). *The neurological examination of the full-term newborn infant* (2nd ed.). Clin. Dev. Med. no. 63. London: Heinemann Medical Books.

Prechtl, H.F.R. (1984). Continuity and change in early neural development. In H.F.R. Prechtl (Ed.), *Continuity of neural functions from prenatal to postnatal life* (pp. 1–15). Clin. Dev. Med. no. 94, Oxford: Blackwell.

Prechtl, H.F.R. (1990). Qualitative changes of spontaneous movements in fetus and preterm infant are a marker of neurological dysfunction. *Early Human Development, 23,* 151–158.

Prechtl, H.F.R. & Nolte, R. (1984). Motor behaviour of preterm infants. In H.F.R. Prechtl (Ed.), *Continuity of neural functions from prenatal to postnatal life* (pp. 79–92). Clin. Dev. Med. no. 94, Oxford: Blackwell.

Prechtl, H.F.R., Einspieler, C., Cioni, G., Bos, A., Ferrari, F. & Sontheimer, D. (1997). An early marker of developing neurological handicap after perinatal brain lesions. *Lancet, 339,* 1361–1363.

Prechtl, H.F.R., Fargel, J.W., Weinmann, H.M. & Bakker, H.H. (1979). Postures, motility and respiration of low-risk pre-term infants. *Developmental Medicine and Child Neurology, 21,* 3–27.

Prechtl, H.F.R., Ferrari, F. & Cioni, G. (1993). Predictive value of general movements in asphyxiated fullterm infants. *Early Human Development, 35,* 91–120.

Preyer, W. (1885). *Specielle Physiologie des Embryo.* Leipzig: Th. Griebens Verlag.

Purves, D. (1994). *Neural activity and the growth of the brain.* Cambridge, UK: Cambridge University Press.

Rakic, P. (1988). Specification of cerebral cortical areas. *Science, 241,* 170–176.

Rakic, P., Bourgeois, J.P., Eckenhoff, M.F., Zecevic, N. & Goldman-Rakic, P.S. (1986). Concurrent overproduction of synapses in diverse regions of the primate cerebral cortex. *Science, 232,* 232–235.

Riach, C.J. & Starkes, J.L. (1993). Stability limits of quiet standing postural control in children and adults. *Gait and Posture, 1,* 105–111.

Riach, C.J. & Starkes, J.L. (1994). Velocity of centre of pressure excursions as an indicator of postural control systems in children. *Gait and Posture, 2,* 167–172

Rose, S. (1976). *The conscious brain* (updated ed.). New York: Vintage Books.

Rubinstein, M., Denays, R., Ham, H.R., Piepsz, A., VanPachterbeke, T., Haumont, D. & Noël, P. (1989). Functional imaging of brain maturation in humans using iodine-123 Idoamphetamine and SPECT. *Journal of Nuclear Medicine, 30,* 1982–1985.

Saint-Anne Dargassies, S. (1974). *Le développement neurologique du nouveau-né à terme et prématuré.* Paris: Masson.

Savelsbergh, G.J.P. & Van der Kamp, J. (1994). The effect of body orientation to gravity on early infant reaching. *Journal of Experimental Child Psychology, 8,* 510–528

Selverston, A.I. & Moulin, M. (1987). *The crustacean stomatogastic system.* Berlin: Springer Verlag.

Shumway-Cook, A. & Woollacott, M.H. (1985). The growth of stability: postural control from a developmental perspective. *Journal of Motor Behavior, 17,* 131–147.

Simmers, J., Meyran, P. & Moulins, M. (1995). Modulation and dynamic specification of motor rhythm-generating circuits in crustacea. *Journal of Physiology, Paris, 89,* 195–208.

Sporns, O. & Edelman, G.M. (1993). Solving Bernstein's problem: a proposal for the development of coordinated movement by selection. *Child Development, 64,* 960–981.

Statham, L. & Murray, M.P. (1971). Early walking patterns of normal children. *Childhood Orthopedics and Related Research, 79,* 8–24.

Sutherland, D.H., Olshen, R.A., Biden, E.N. & Wyatt, M.P. (1988). *The development of mature walking.* Clin. Dev. Med., no. 104/105. Oxford: Blackwell.

Sveistrup, H. & Woollacott, M.H. (1996). Longitudinal development of the automatic postural response in infants. *Journal of Motor Behavior, 28,* 58–70.

Sveistrup, H. & Woollacott, M.H. (1997). Practice modifies the developing automatic postural response. *Experimental Brain Research,* **114,** 33–43.

Thelen, E. (1985). Developmental origins of motor coordination: leg movements in human infants. *Developmental Psychobiology,* **18,** 1–22.

Thelen, E. (1995). Motor development. A new synthesis. *American Psychologist,* **50,** 79–95.

Thelen, E. & Spencer, J.P. (1998). Postural control during reaching in young infants: a dynamic systems approach. *Neuroscience Biobehavioral Reviews,* **22,** 507–514.

Thelen, E., Corbetta, D., Kamm, K., Spencer, J.P., Schneider, K. & Zernicke, R.F. (1993). The transition to reaching: mapping intention and intrinsic dynamics. *Child Development,* **64,** 1058–1098.

Touwen, B.C.L. (1976). *Neurological development in infancy.* Clin. Dev. Med. no. 58. London: Heinemann.

Touwen, B.C.L. (1978). Variability and stereotypy in normal and deviant development. In J. Apley (Ed.), *Care of the handicapped child* (pp. 99–110). Clin. Dev. Med. no. 67, London: Heinemann.

Touwen, B.C.L. (1981). The preterm infant in the extrauterine environment. Implications for neurology. *Early Human Development,* **3/4,** 287–300.

Touwen, B.C.L. (1993). How normal is variable, or now variable is normal? *Early Human Development,* **34,** 1–12.

Trevarthen, C., Murray, L. & Hubley, P. (1981). Psychology of infants. In J.A. Davis & J. Dobbing (Eds.), *Scientific foundations of pediatrics* (2nd ed., pp. 211–274). London: Heinemann.

Ulrich, B.D. (1997). Dynamic systems theory and skill development in infants and children. In K.J. Connolly & H. Forssberg (Eds.), *Neurophysiology and neuropsychology of motor development* (pp. 319–345). Clin. Dev. Med. no. 143–144. London: MacKeith Press.

Van der Fits, I.B.M., Klip, A.W.J., Van Eykern, L.A. & Hadders-Algra, M. (1998). Postural adjustments accompanying fast pointing movements in standing, sitting and lying adults. *Experimental Brain Research,* **120,** 202–216.

Van der Fits, I.B.M., Klip, A.W.J., Van Eykern, L.A. & Hadders-Algra, M. (1999a). Postural adjustments during spontaneous and goal-directed arm movements in the first half year of life. *Behavioral Brain Research,* **106,** 75–90.

Van der Fits, I.B.M., Otten, E., Klip, A.W.J., Van Eykern, L.A. & Hadders-Algra, M. (1999b). The development of postural adjustments during reaching in 6 to 18 months old infants: evidence for two transitions. *Experimental Brain Research,* **126,** 517–528.

Ververs, I.A.P., De Vries, J.I.P., Van Geijn, H.P. & Hopkins, B. (1994). Prenatal head position from 12–38 weeks. I. Developmental aspects. *Early Human Development,* **39,** 83–91.

Ververs, I.A.P., Van Gelder-Hasker, M.R., De Vries, J.I.P., Hopkins, B. & Van Geijn, H.P. (1998). Prenatal development of arm posture. *Early Human Development,* **51,** 61–70.

Vihman, M.M. (1993). Variable paths to early word production. *Journal of Phonetics,* **21,** 61–82.

Vles, J.S.H. & Van Oostenbrugge, R. (1988). Head position in low-risk premature infants: impact of nursing routines. *Biology of the Neonate,* **54,** 307–313.

Vles, J.S.H., Kingma, H., Caberg, H., Daniels, H. & Casaer, P. (1989). Posture of low-risk preterm infants between 32 and 36 weeks postmenstrual age. *Developmental Medicine and Child Neurology,* **31,** 191–195.

Von Hofsten, C. (1982). Eye-hand coordination in the newborn. *Developmental Psychology,* **18,** 450–461.

Von Hofsten, C. (1984). Developmental changes in the organization of pre-reaching movements. *Developmental Psychology,* **20,** 378–388.

Waddington, C.H. (1962). *New patterns in genetics and development.* New York: Columbia University Press.

Westerga, J. & Gramsbergen, A. (1990). Development of locomotion in the rat. *Developmental Brain Research,* **57,** 163–174.

Wheelwright, E.F., Minns, R.A., Law, H.T. & Elton, R.A. (1993). Temporal and spatial parameters of gait in children. I. Normal control data. *Developmental Medicine and Child Neurology,* **35,** 102–113.

Wiener-Vacher, S.R., Toupet, F. & Narcy, P. (1996). Canal and otolith vestibulo-ocular reflexes to vertical and off vertical axis rotations in children learning to walk. *Acta Otolaryngologica (Stockholm),* **116,** 657–665.

Windle, W.F. & Fitzgerald, J.E. (1937). Development of the spinal reflex mechanism in human embryos. *Journal of Comparative Neurology,* **67,** 493–512.

Woollacott, M.H. & Sveistrup, H. (1992). Changes in the sequencing and timing of muscle response coordination associated with developmental transitions in balance abilities. *Human Movement Science,* **11**, 23–36.

Woollacott, M., Debû, B. & Mowatt, M. (1987). Neuromuscular control of posture in the infant and child: is vision dominant? *Journal of Motor Behavior,* **19**, 167–186.

Yang, J.F., Stephens, M.J. & Vishram, R. (1998). Infant stepping: a method to study the sensory control of human walking. *Journal of Physiology,* **507**, 927–937.

Zelazo, P.R. (1983). The development of walking: new findings and old assumptions. *Journal of Motor Behavior*, **2**, 99–137.

Zijlstra, W., Prokop, T. & Berger, W. (1996). Adaptability of leg movements during normal treadmill walking and split-belt walking in children. *Gait and Posture*, **4**, 212–221.

Further reading

Connolly, K.J. & Forssberg, H. (1997). *Neurophysiology and neurophychology of motor development.* Clin. Dev. Med. no. 143–144. London: MacKeith Press.

Edelman, G.M. (1989). *Neural Darwinism. The theory of neuronal group selection.* Oxford: Oxford University Press.

Touwen, B.C.L. (1976). *Neurological development in infancy.* Clin. Dev. Med. no. 58. London: Heinemann Medical Books.

III.8
Neuromotor Development from Kindergarten Age to Adolescence

REMO H. LARGO and JON A. CAFLISCH

ABSTRACT

The following aspects of neuromotor development from kindergarten age to adolescence are described: developmental course and inter-individual variability of timed performance and associated movements (AM), interrelationships between timed performance and AM, development and degree of lateralization, and gender differences. Only motor functions that show age-specific changes and were not – or only mildly – confounded by non-motor variables of perception, memory, training or social rearing conditions are addressed. In the last section of this chapter, methodological issues, the role of maturation rate, genetic endowment and experience, and clinically relevant relationships between motor functions and developmental areas, such as cognition, language and socioemotional behaviour, are discussed.

INTRODUCTION

Children act on their environment and express themselves through their motor system. At school age, motor abilities, such as handwriting, contribute substantially to their academic performance; influence their competence in developmental areas, such as verbal and non-verbal communication; and thus profoundly influence their feelings of well-being and self-esteem (Polatajko, 1999). Mild to moderate motor dysfunction is a significant epidemiological problem, occurring in up to 6% of the children in the general population (American Psychiatric Association, 1994).

Professionals – such as neurologists or paediatricians – who are faced with a presumably "clumsy child" are requested to investigate whether his/her motor performance and quality of movements and posture are still within the normal range, are an expression of a developmental delay, i.e. comparable to the behaviour of a younger child, or indicate frank neurological impairment. Major motor

A.F. Kalverboer and A. Gramsbergen (eds.), Handbook of Brain and Behaviour in Human Development, 569–590
© 2001 Kluwer Academic Publishers. Printed in Great Britain.

impairments, e.g. cerebral palsy, are easily recognized even by a lay person. Less obvious but more frequent are minor motor impairments that can rarely be related to a specific neurological disorder. Clinically these are associated with so-called "soft" or "subtle" neurological signs. Subtle neurological signs are minor findings that are commonly present in young children (Deuel & Robinson, 1987). It is only their persistence into later years that makes them "pathological". "The diagnosis of minimal brain dysfunction is based upon findings that are abnormal only with reference to the child's age... had the child been younger, the findings would have been regarded as normal" (Kinsbourne, 1973). Subtle neurological signs not only serve as markers for mild motor impairment, they have also been related to behavioural disturbances, such as hyperactivity, impulsiveness, learning disabilities, aggressive antisocial conduct disorder, slow cognitive development, and even anxiety and depression.

Solid knowledge of normal neuromotor development is a prerequisite for all professionals dealing with normal and developmentally disturbed children. In recent years the following three major aspects of normal neuromotor development have been studied:

1. *Milestones*, i.e. age at onset of a motor skill. Touwen & Prechtl (1979), e.g. provided age norms for a task in which the child was able to hop on one foot.
2. *Timed performances*, speed at which a motor task was performed. Denckla (1973), and Denckla & Rudel (1974), e.g. published age-specific norms for the time needed to complete a defined number of distinct movements of the fingers, hand or foot.
3. *Quality of movements*, includes parameters of the active extremity, such as smoothness, and of the passive body parts, such as associated movements.

    Schellekens and co-workers (1983) investigated temporal and spatial characteristics of movements in children who reportedly had movement problems in comparison with normal children. The former were slower at tapping, their movements were less smooth and the pattern of acceleration and deceleration exhibited in the movements was different from those observed in the normal group.

    Associated movements (AM), known variously as synkinetic movements, co-movements, or mirror movements, are motor irradiations that are outside the subject's awareness. These AM accompany movements, but are not necessary for the performance of an intended action (Connolly & Stratton, 1968; Woods & Teuber, 1978). For example, Wolff *et al.* (1985) reported on age-related changes of duration of AM in various motor tasks. Duration and degree of AM are the parameters of movement quality most frequently assessed in clinical practice and research.

A number of test instruments are available to assess onset of motor skills, timed performances and movement quality. The most widely used include: Physical and Neurological Examination for Soft Signs (PANESS; Guy, 1976), Examination of the Child with Minor Neurological Dysfunction (Touwen & Prechtl, 1979), and Neurological Examination for Subtle Signs (NESS; Denckla, 1985a). More recently the Zurich Neuromotor Test Assessment (ZNA) has been developed at the Growth and Development Center of the University Children's Hospital Zurich (Largo *et al.*, 2000a–c). The ZNA is a standardized testing procedure in which distinct motor tasks are assessed with regard to timed performance, duration and degree of associated movements of the contralateral and ipsilateral extremity, face, head and body (Table 1).

Most motor tasks have been described previously by Denckla (1973) and Denckla & Rudel (1974), Wolff *et al.* (1985), and Henderson & Sudgen (1992). For the ZNA only tasks of differing complexity which showed sizeable age-specific changes were chosen for evaluation, such as repetitive, alternating or sequential movements tasks. These tasks were modified to reach the highest possible level of standardization for examination procedure, time measurement and scoring of duration and degree of AM of the contralateral and ipsilateral extremity, face, head and body (Table 2). Duration and degree of AM are scored from videotapes for each body part (manual and normative data are available from the first author upon request). Normative data have been established in a cross-sectional study including 644 children ranging in age from 5 to 18 years (Table 3). This chapter concentrates on developmental aspects of timed performances and associated movements (AM), the only two parameters that allow a

Table 1 Motor tasks of the Zurich Neuromotor Assessment (ZNA)

Repetitive movements	Fingers, hand, foot
Alternating movements	Hand (pro-/supination in sitting position), diadochokinesis (pro-/supination in standing position), foot (heel–toe alternation)
Sequential movements	Fingers
Peg board	
Dynamic balance	Sideways, forwards
Static balance	
Stress gaits	Walking on toes, on heels, on outer soles of feet, on inner soles of feet

Table 2 Zurich Neuromotor Assessment (ZNA): definition for timed performance, duration and degree of AM

Timed performance	Time (s) needed to complete a defined number of movements (e.g. 20 repetitive finger movements)
Associated movements (AM)	
Duration	Occurrence of AM during the timed period is noted. AM for contralateral and ipsilateral extremity, face, head and body are scored separately. A score of 0–10 is estimated for each body part (0: no AM; 10: AM throughout the entire timed period).
Total score of duration	Represents the sum of the duration scores of contralateral and ipsilateral AM, face, head and for a motor task.
Degree	Extent to which AM are expressed. A score of 0–3 was estimated for each body part (0: no AM; 3: maximum expression of AM).
Total score of degree	Represents the sum of the degree scores of contralateral and ipsilateral AM, face, head, and body for a motor task.
Handedness	Judged by a hand preference inventory consisting of four unimanual tasks.

Table 3 Zurich cross-sectional study on neuromotor development: Study population ($n = 644$) and mean ages at testing

	Mean testing age/SD (years)	Females	Male
Kindergarten	5.7/0.43	55	52
First grade	7.3/0.35	62	60
Third grade	9.4/0.39	57	57
Sixth grade	12.5/0.44	60	58
Londitudinal study	15.0/0.05	46	44
Londitudinal study	18.0/0.10	45	48
n		325	319

longitudinal description of neuromotor development between kindergarten age and adolescence. The following aspects of neuromotor development are addressed: developmental course and inter-individual variability of timed performance and AM, interrelationships between timed performance and AM, development and degree of lateralization, and gender differences.

Only those motor functions that show age-specific changes and are not – or only mildly – confounded by non-motor variables of perception, memory, training or social rearing conditions (Neuhäuser, 1975) will be discussed. For motor skills, such as drawing, handwriting or sports activities, the reader is referred to authors such as Stott et al. (1984), Wade & Whiting (1986), Gallahue (1989) and Burton & Miller (1998). Current concepts of dynamic system approaches to variability and motor control are discussed in Newell & Corcos (1993). Neurological signs, such as dyskinesia or clonus, which are not part of normal neuromotor development at any age, will therefore not be addressed. The interested reader may consult Gubbay (1975), Touwen & Prechtl (1979), Tupper (1987), Gage (1991) and Connolly & Forssberg (1997).

REVIEW OF THE LITERATURE

Developmental course

Older children obviously perform more efficiently and have less AM than do younger children. Developmental changes have been reported for many motor skills, such as hopping or catching or throwing a ball (Gallahue, 1989; Burton & Miller, 1998). However, the developmental courses of distinct motor functions are less well documented.

With regard to timed performance, Denckla (1973) and Denckla & Rudel (1974) provided some data showing that speed of performance is, as expected, lower in younger children, and reaches a plateau between 8 and 10 years of age. Wolff and his colleagues (1983) essentially confirmed Denckla's findings by reporting age-specific changes in the speed of repetitive, alternating and sequential movements at school age.

In the Zurich study (Largo et al., 2000a), a steady improvement of timed performance throughout the entire prepubertal period has been noted (Figure 1).

Yet the increase in speed differed considerably among the various motor tasks. Repetitive finger and hand movements improved less than did alternating hand and sequential finger movements. The increase of speed appeared to be related to the complexity of movement patterns. This also seemed to apply for the age at which a motor task reached a plateau. Repetitive movements levelled off in performance as early as 12–15 years, while in sequential movements this was not seen until 18 years.

The peg board task reached a plateau at 12 years, while performance in dynamic balance improved up to 15 years and then tended to slow down again by the age of 18 years. The latter observation is probably less a reflection of maturation or training effect than of the large increase, first in muscle strength and then subcutaneous fat tissue, that takes place during puberty (Malina, 1986).

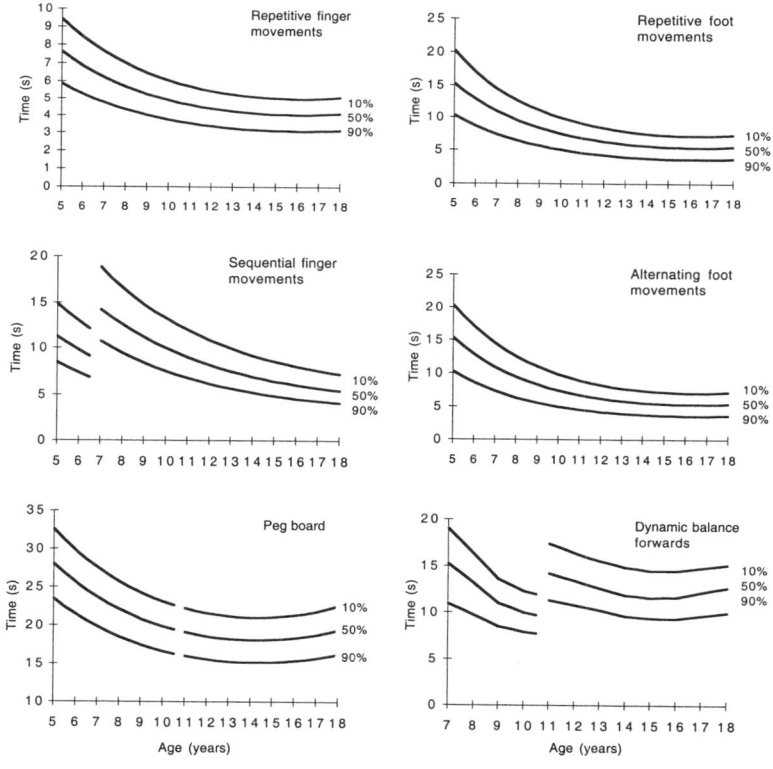

Figure 1 Centile curves of timed performance for six motor tasks in males. Time performance is given for the dominant upper extremity: repetitive finger movements (20 movements between forefinger and thumb), sequential finger movements (age < 7 years 3/ age > 7 years five sequences between thumb and the other four fingers), and peg board; for the right lower extremity: repetitive foot movements (20 taps), alternating foot movements (10 heel–toe alternations), and for dynamic balance (age < 10.5 years 12/ age > 10.5 years 18 side-/forwards hoppings over cord).

Developmental changes of AM among school-age children have been reported by Zazzo (1960), Fog & Fog (1963) and Abercrombie *et al.* (1964). In the finger-lifting method, homolateral and contralateral AM were recorded when the child was asked to move a specified finger. In the clip-pinching method, AM in the contralateral hand were judged when the child exhibited a certain degree of pressure with the thumb. Touwen & Prechtl (1979) provided some semiquantitative data for AM in children performing tasks such as finger opposition, diadochokinesis, walking on heels or hopping on one foot. Vitiello and co-workers (1989) reported a dramatic decrease in the total score of subtle signs at about 6 years of age. They pointed out that this decrease occurs at the very age when children are expected to enter school and learn skills for which a high level of coordination is required.

The Zurich study revealed a variable and rather complex developmental course of contralateral AM (Figure 2). First, timing and extent of reduction of duration were related to the complexity of the motor task. Duration of contra-lateral AM for repetitive hand and finger movements had already decreased during the prepubertal period, while duration of alternating and sequential hand movements were delayed until the onset of puberty. Little developmental change was observed in the peg board task.

Second, duration of contralateral AM did not regularly decrease with age. A non-linear course of duration was observed that was again a function of the complexity of the individual motor task, e.g. duration of contralateral AM decreased much earlier in repetitive than in alternating and sequential finger movements; in the latter there was even a transitory increase of contralateral AM noted. With the peg board there was a decrease of AM in only about half of the children and even a slight re-increase between 15 and 18 years. Comparable observations were made with respect to the degree of contralateral AM as well as to the total score of duration and the total score of degree of AM for both the dominant and the non-dominant extremity (for definition of total scores see Table 2).

Inter-individual variability

In the past, neurological findings, such as tendon reflexes, or simple motor patterns, such as repetitive movements, had been regarded as invariable among healthy children of the same age. However, in recent years there has been a growing awareness that motor functions not only appear and change age-specifically, but they are also highly variable within an age group. Denckla (1973, 1985) and Denckla & Rudel (1974) provided mean values and standard deviations for the speed of repetitive, alternating and sequential movements. Wolff *et al.* (1983, 1985) reported some data on age-specific variability of AM in these motor tasks.

In the Zurich study inter-individual variation of timed performance and contralateral AM was considerable among all motor tasks. At the age of 7 years sequential movements were carried out by males, on average, in 14.2 s; 10% of the males needed more than 18.2 s, and another 10% less than 10.2 s (Figure 1). Inter-individual variations decreased variably with age. Little change was noted in repetitive movements, while alternating and sequential movements showed a major reduction.

Between 5 and 7 years only a few children performing the repetitive finger move-ments showed no contralateral AM, while about 10% of the children displayed

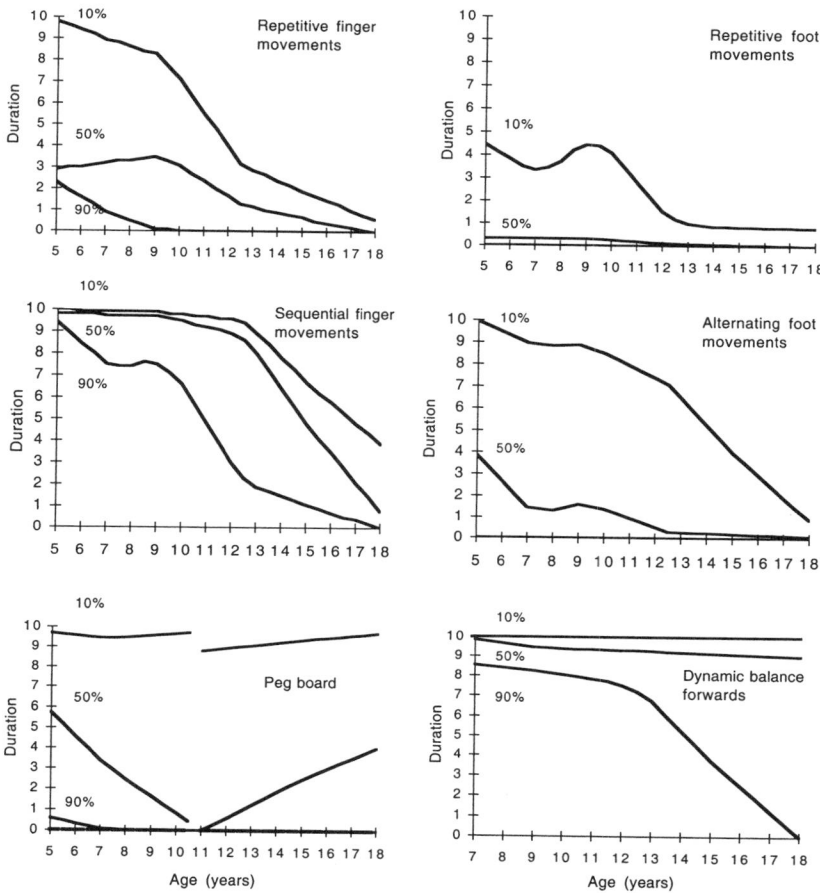

Figure 2 Centile curves of duration for contralateral associated movements (AM) of six motor tasks in males (definition of motor tasks see Figure 1). Scoring: 0 = no AM; 5 = AM during half of the timed period; 10 = AM during the whole timed period.

contralateral AM throughout the entire timed period. By 18 years about 50% of the adolescents had no contralateral AM at all, and an additional 40% showed contralateral AM during less than 10% of the timed period. At kindergarten age most children displayed contralateral AM when performing sequential finger movements. At 18 years no contralateral AM were noted in about 50% of the adolescents and, in an additional 40%, contralateral AM were observed during less than half of the timed period. The extent to which inter-individual variability changed with age varied considerably among the motor tasks. In simple motor tasks, such as repetitive movements, contralateral AM duration was already reduced at an early age, while it remained largely unchanged in the peg board.

A large inter-individual variation was also noted with regard to the degree of contralateral AM, as well as to the total score of AM duration and the total score of AM degree for both the dominant and non-dominant extremity.

Interrelationships between timed performance and AM

Intuitively, performances showing "good" time scores, i.e. high speed, are associated with a short duration and a low degree of AM and vice-versa. To our knowledge there are no data published on the interrelationships between timed performance and AM. In the Zurich study the following observations were made:

1. The dominant hand and the non-dominant hand were significantly correlated with respect to timed performance, as well as duration and degree of AM. For example, a high speed in sequential movements of the dominant hand was accompanied by a corresponding high speed in the non-dominant hand. Short duration and low degree of AM while performing with the dominant hand were associated with comparable duration and degree of AM when the non-dominant hand was active.
2. Timed performances were significantly correlated for the same types of movements. Thus, repetitive movements of fingers, hand and foot were moderately correlated ($r = 0.35$–0.45; $p < 0.05$–0.01). However, there were no significant correlations between different movement patterns at any age, e.g. between repetitive and sequential movements of the fingers or between repetitive and alternating movements of the foot.
3. Durations of AM in different motor tasks were moderately correlated with each other. Thus a child who showed extensive AM in finger repetition also tended to display a long duration of AM in other tasks.
4. Duration and degree of AM noted in a single motor task were significantly correlated with each other. AM occurring extensively during the test period also tended to be expressed more strongly.
5. The clinically most relevant observation was the lack of significant relationships at any age between time measurement on the one hand, and duration and degree of AM on the other. Thus, high speed in a motor task was not significantly correlated with a low duration and a low degree of contralateral AM and vice-versa. These findings might be explained by the fact that duration and degree of AM are a function not of competence, but rather of the stress exerted on the child. A well-coordinated child reaches a higher speed than a less well-coordinated child. However, when making an effort to perform as well as possible, both AM duration and degree may be influenced by the same amount of stress. This assumption is supported by the findings of Wolff *et al.* (1983). To illustrate the links between intended action and unintended motor overflow, they compared each child's timed performance with the frequency of AM produced. The speed of intended actions was inversely related to the frequency of mirror movements in homologous muscles at the opposite side; in other words the effort required to perform an intended action was directly related to the duration of AM.

At this point we would stress that the data from the Zurich study are based on normal children. In developmentally disturbed and neurologically impaired children different and clinically relevant relationships between timed performance and duration and degree of AM may be observed.

Lateralization of motor functions

In right-handed children, "common sense" assumes a somewhat less stringent set of expectations for the left hand than for the right hand. However, is a right-handed child only slightly or very much slower with the left hand? What is a "normal" difference between the dominant and the non-dominant hand? Is there a consistent side difference between the lower extremities?

The clinical relevance of these questions becomes evident in developmentally disturbed children. Children with central nervous system dysfunctions are reported to display late or weak lateralization, which is thought to be a reflection of an anomalous cerebral dominance (Ingram, 1969). Annett (1970) observed an excess of ambidexters among children with low IQ levels. In children with "minimal brain dysfunction", Denckla (1973) and Denckla & Rudel (1974) found a significant percentage who used either hand. Denckla regarded these children as bilaterally clumsy, rather than ambidextrous. Other children showed mixed lateral preference combinations for the hand, foot and eye. She also more frequently noted children who strictly preferred using their right hand, and who were so lateralized that they could not use the left hand in a normal manner, i.e. they were pathologically hyper-lateralized. Reitan (1971) observed an excess of preferred hand superiority in handwriting among brain-damaged children compared with normal children. A lack in the quality of left limb coordination in brain-damaged children has also been observed by Rudel *et al.* (1974).

Side differences, e.g. in children with a hemisyndrome, or differences between the upper and lower extremities, e.g. in children with spastic diplegia, can be reliably assessed only if normative data on lateralization are available for all age levels. Touwen & Prechtl (1979) provided some semiquantitative data on lateralization of gross motor skills, e.g. a normal "right-foot" 4-year-old balances and hops only on the right foot; at 5 years the left foot catches up and strength of lateral preference is less pronounced. Denckla (1973) reported different degrees of lateralization in repetitive and successive finger movements. Right hand superiority was noted for speed of repetitive movements, but the hands were equally proficient in successive finger movements. Denckla's findings are supported by other studies demonstrating right side superiority in tasks requiring repetitive movements of fingers, hands and arms (Wyke, 1967, 1969; Spreen & Gaddes, 1969; Bowen *et al.*, 1972).

Denckla's findings were also essentially confirmed by the Zurich study (Figure 3). Repetitive movements of fingers, hand and foot showed significant lateralization at all ages ($p < 0.01$). Approximately 90% of the subjects performed faster with their dominant hand. Lateralization tended to decrease during puberty.

There was no side preference in sequential finger movements at kindergarten and early school age. During puberty the dominant hand became slightly faster. In alternating hand pro-/supination movements a moderate preference of the dominant hand was noted. With regard to the lower extremities, lateralization in toe tapping – in favour of the dominant side – was moderate but significant at most ages. There were no significant side differences in heel–toe alternations.

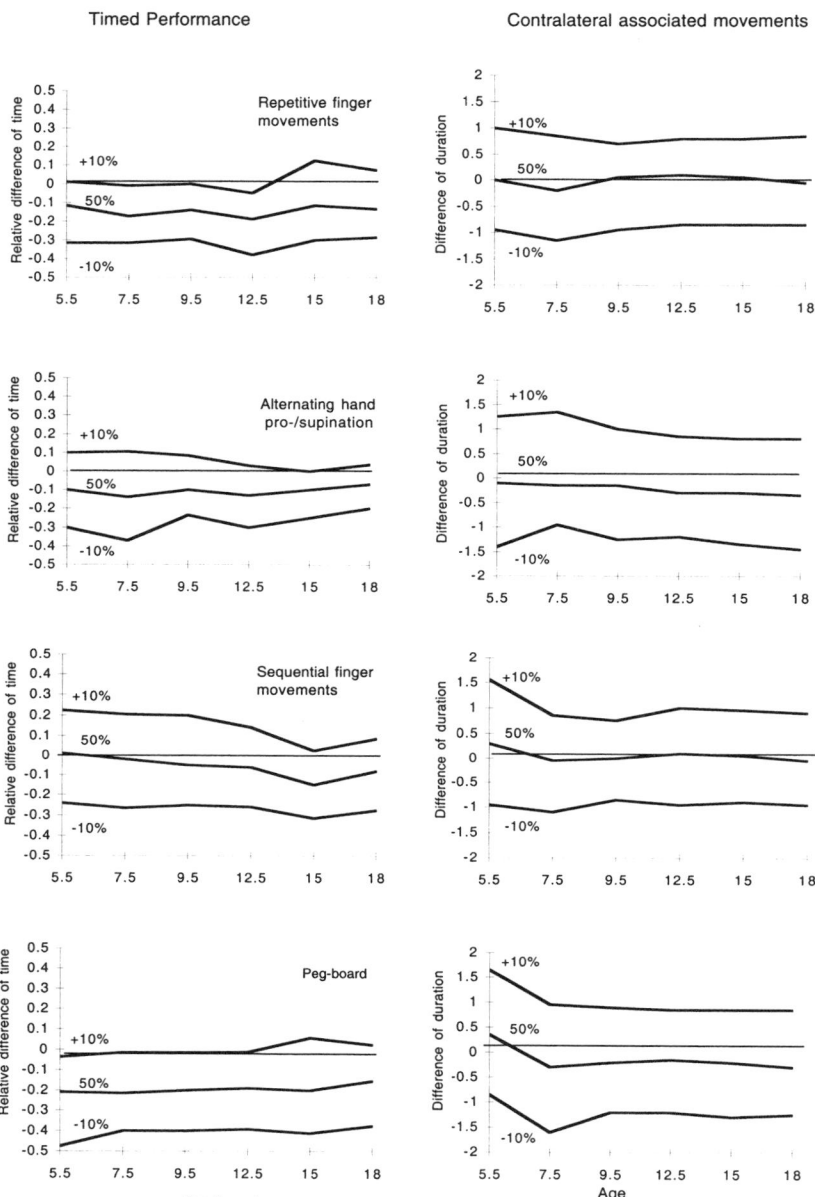

Figure 3 Centile curves of lateralization of timed performance and of duration of contralateral AM for four motor tasks in males. (definition of motor tasks see Figure 1). *Left side*: relative time difference = (time of dominant hand – time of nondominant hand)/((time of dominant hand + time of non-dominant hand)/2). *Right side*: difference of duration of AM = duration of AM dominant hand – duration of AM non-dominant hand.

The speed of repetitive finger movements appears to be dependent on the left hemisphere. Studies in brain-injured adults indicate that bilateral slowing results from left hemisphere damage, whereas contralateral slowing results when there is only right hemisphere damage (Wyke, 1968; Bowen et al., 1972). The equivalence of a child's right and left successive finger speed may be due to a requirement that both hemispheres contribute to each hand.

Right hand superiority is particularly reported in studies of tasks involving the use of instruments, such as handwriting skill (Reitan, 1971), skill in cutting circles with scissors (Zurif & Carson, 1970) and Purdue peg board scores (Rapin et al., 1966). The hemispheric contributions to such tool-using tasks seem to differ from that observed in the repetitive movement studies, as unilateral brain lesions, either right- or left-sided, have been found to impair the Purdue peg board dexterity of both hands (Vaughan & Costa, 1962; Wyke, 1968).

In the Zurich study a highly significant side difference was found in the peg board task. However, the inter-individual variability of lateralization was also remarkable. The children were, on average, 20% faster in their performance time with the dominant hand than with the non-dominant hand. About 10% of the children were even 40% or more faster with their dominant hand, while about 10% performed equally well or even slightly better with their non-dominant hand.

In contrast to timed performance, duration and degree of contralateral AM showed no or only mild side differences (Figure 3). The children tended to display less contralateral AM when they used their dominant hand. This effect was most pronounced in the peg board task. Lateralization of AM was most pronounced, but still moderate, in diadochokinesis. None of these differences reached statistical significance.

Does lateralization change with age? In the Zurich study no significant developmental progression of lateralization was noted. With respect to timed performance the magnitude of the side difference in repetitive and alternating movements remained stable between the ages of 5 and 18 years. It increased slightly in sequential finger movements and decreased somewhat in the peg board, but neither trend was significant. These findings are in agreement with Annett (1970), Denckla (1973) and Denckla & Rudel (1974), who found that the magnitude of right hand superiority did not increase with age.

The degree of functional asymmetry of a motor skill may be explained by a asymmetrical development of the two hemispheres, but it has also been related to two different brainstem systems. Experimental work with monkeys and human clinical observations (Brinkman & Kuypers, 1973) indicate that two brainstem systems, both with cortical inputs, are involved. The lateral system is responsible for independent, individual distal hand and finger movements, and the ventromedial system for proximal and integrated body–limb movements. Each hemisphere is connected to the ventromedial system both contralaterally and ipsilaterally, but only contralaterally to the lateral system.

GENDER DIFFERENCES

There is a general assumption that females perform faster, particularly in fine motor tasks, and are better coordinated than males. Studies on gender differences provided mixed results. In Denckla's studies (1973) and those of Denckla & Rudel (1974), successive finger movements and heel–toe alternation were performed faster by females. However, the males tapped faster, showed a higher percentage who were better with the right hand, and the magnitude of right hand superiority was smaller than in females. In the Zurich study females tended to be faster than males in all motor tasks between 5 and 7 years (Figure 4). Thereafter, in agreement with Denckla, females still performed better in sequential movements while males were slightly faster in repetitive movements. A few significant differences between females and males were noted; however, given the large interindividual variation, these gender differences of timed performance are of no clinical relevance.

In contrast to timed performance, gender differences were quite prominent in duration and degree of contralateral AM. Females displayed less AM than males at most ages and exhibited the decrease in contralateral AM 1–3 years earlier than did males. A comparable shift related to sex has been reported in intellectual development and somatic growth (Waber, 1977). These findings most probably reflect a different maturation rate for females and males.

According to Heap & Wyke (1972), rapid successive movements depend upon sensorimotor programmes requiring intra- and inter-hemispheric co-ordination. Thus, females might develop adequate inter-hemispheric connections at an earlier age than do males. The development of these inter-hemispheric connections may be the relevant feature of earlier cerebral maturation in females, rather than any earlier concentration in one hemisphere of an originally bilaterally subservient function.

Both in hand preference and relative manual skill, females are reported to be more asymmetrical to the right than males at younger ages. In a peg board task, Annett (1970) observed that females were slightly faster than males with the right hand, while males were slightly faster than females with the left hand. Denckla (1973) and Denckla & Rudel (1974) suggested that females may display earlier, but not persistently better, left-hemisphere-dependent skills, whereas males show better, but not earlier, right-hemisphere-dependent skills. In the Zurich study no significant gender differences in lateralization were noted.

The common assumption that females are better coordinated and perform faster than males turned out to be only partially valid. Females appear to be better coordinated because they show AM that are less pronounced and occur less frequently. However, their performances are not significantly superior to those of males.

DISCUSSION

During school age and adolescence, motor functions change rapidly with age. Motor functions begin and complete their maximum development at different chronological ages, and among children of the same age they display a large inter-individual variability and different degrees of lateralization. To improve our understanding of neuromotor development and its disturbances, future research and clinical studies need to consider the following three topics in more depth:

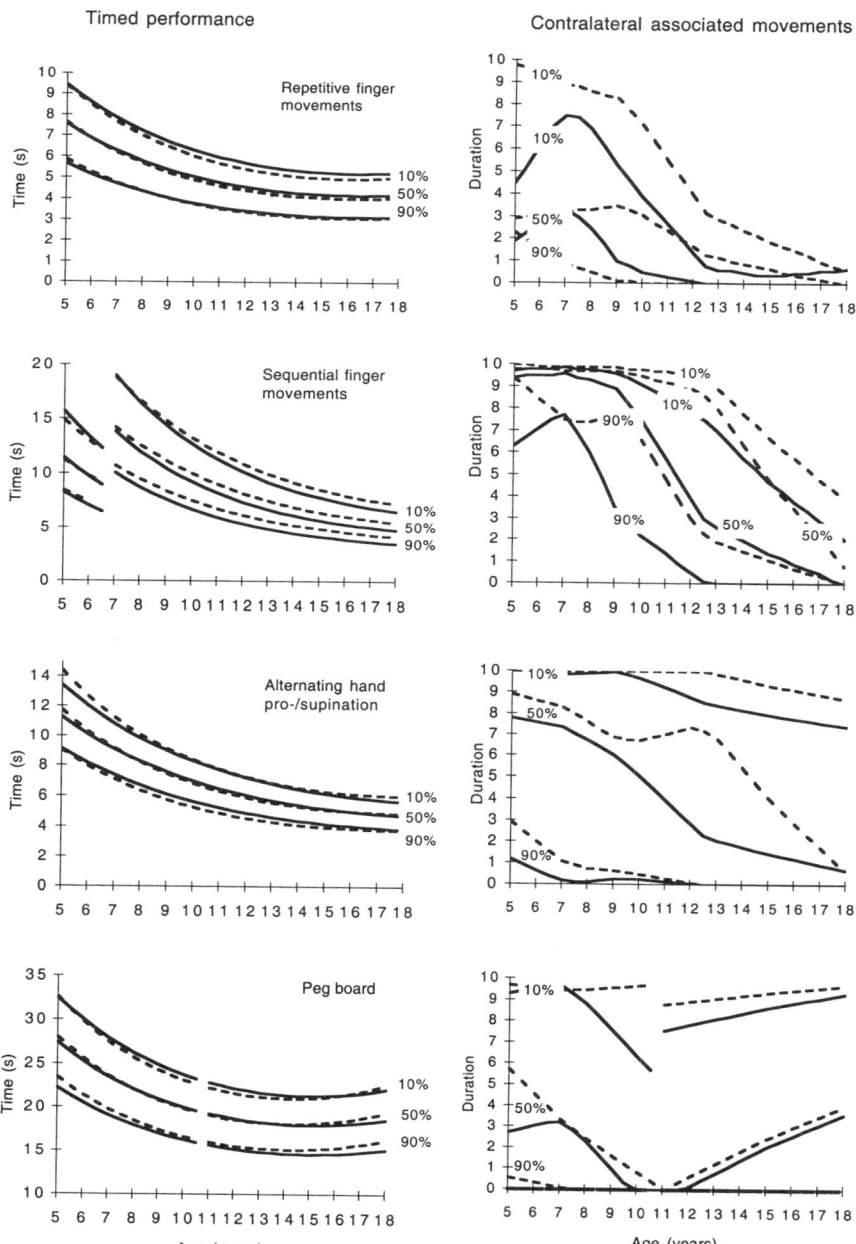

Figure 4 Gender differences. Centile curves of timed performance of the dominant upper extremity and of duration of contralateral AM for four motor tasks in females (___) and males (------) are presented (definition of motor tasks see Figure 1.)

1. methodological standards in neuromotor assessments;
2. role of maturation rate, genetic endowment and experience;
3. relationship between neuromotor functions and other developmental areas.

Methodological standards in neuromotor assessments

Despite the extended use of neurological assessments in research and clinical practice, and the existence of much literature on children with motor dysfunctions, only a few studies have been published on standardization of assessments, their reliability and validity (Neeper & Greenwood, 1987).

Reliability requires a highly standardized examination procedure and well-trained personnel. Kakebeeke *et al.* (1993) investigated the inter-rater and test–retest reliability of the Touwen examination (Touwen & Prechtl, 1979). When the manual was the only reference for instruction, acceptable levels of reliability were not achieved. The reliability estimates for the total test scores were satisfactory, while inter-rater reliability for the nine groups of items and the individual tasks within these groups was poor. When methodology and interpretation of performance was agreed upon among observers, these disagreements diminished. Short-term stability of the total scores was good, but reliability for group and individual item scores remained poor.

Quitkin *et al.* (1976) found that some neurological signs were scored reliably between examiners (kappa coefficient > 0.50), among them, finger–thumb mirror overflow, left-sided pronation–supination, foot-tapping, while others, such as finger–thumb opposition or pronation–supination mirror overflow, were unreliable. In the same study only a very few signs proved stable at re-testing. Vitiello *et al.* (1989) analysed inter-observer and test–retest reliability of neurological subtle signs in 54 psychiatric patients and 25 normal children aged 5–17 years, using the revised version of the Neurological Examination for Subtle Signs (NESS). An acceptable inter-rater reliability (kappa coefficient > 0.50, intraclass correlation coefficient > 0.70) was found for 40 of 64 items tested. This was particularly true for continuous variables, such as time needed to perform 20 consecutive movements. Reliability was lower for overflow movements and dysrhythmias, which are more dependent on subjective interpretation. Test–retest reliability at 2 weeks was unsatisfactory for most of the categorically scored items, including some classic subtle signs, such as overflow or dysrhythmias (kappa and intra-class correlation coefficients < 0.50). Continuous items usually remained stable at retest. The authors recommend that researchers and clinicians should rely more on subtle signs that can be assessed on continuous scales. Denckla (1973) analysed test–retest reliability for timed performances 3 weeks after the initial testing. Correlation coefficients between trials were 0.69 for repetitive and 0.80 for successive movements.

In the Zurich study examinations were videotaped, and subsequently timed performance, duration and degree of AM were scored from the videotapes. Intra-observer and inter-observer reliability for timed performance was found to be very high (Spearman correlations: r: 0.90–0.99 and 0.95–0.99, respectively), while test–retest reliability was lower and more variable (r: 0.56–0.80). With respect to duration and degree of contralateral AM, intra-observer reliability was moderate to high (0.57–0.93 for duration, and 0.55–0.89 for degree, respectively). Inter-observer correlations tended to be lower, but most were comparable to the intra-observer correlations (0.44–0.95 for duration and 0.51–0.98 for degree). Test–retest correlations were moderate to low (0.35–0.69 for duration and 0.35–0.61 for degree).

There are no studies reporting on the validity of neurological signs in children. In the Zurich Cross-sectional Study of 644 school children the agreement between the presence of motor impairment leading to pedagogic and/or therapeutic intervention and the results of the Zurich Neuromotor Assessment (ZNA) was investigated. Some 95% of the children who received any kind of intervention were identified by the assessment (timed performance and duration and degree of AM below the 10th centile), while 25% of the children with a comparably low performance received no therapy. Some 7% of the children received intervention even though their performance and expression of AM was between the 10th and 90th centile.

Summarizing, timed motor performance as well as duration and degree of AM between kindergarten age and adolescence are characterized by long-lasting developmental changes and large inter-individual variations; both change and variation are related to the complexity of the motor task. Thus, for a reliable assessment of motor competence in school-age children, a well-standardized test instrument and age-specific standards for motor performance and movement quality are necessary preconditions.

Numerous studies have demonstrated that the overall frequency of AM is greater among children with developmental disabilities than among age-matched controls (see below for references). However, to really improve the validity of neuromotor testing, further research is needed to answer certain questions, such as: Are contralateral AM more or less discriminating as indicators of motor dysfunction than AM of other body parts, or is a total score of duration of AM the most sensitive measure? Are timed performances more or less sensitive than duration or degree of AM for detecting motor dysfunction in childhood?

Maturation rate, genetic endowment and experience

Maturation rate varies considerably among children of the same age, and therefore affects their development (Wohlwill, 1973). Somatic growth differs by ± 1.5 years at early school age and by ± 3 years during adolescence (Tanner, 1970). Cognitive development is also influenced by maturation rate. Intellectual performance during early adolescence is moderately and consistently correlated with measures in sexual maturation and skeletal age (Kohen-Ratz, 1974; Waber, 1977). Preliminary findings in the Zurich study also indicate a significant and clinically relevant impact of maturation rate on neuromotor development. Moderate but consistently significant correlations were noted between timed performances, duration and degree of AM and skeletal age (Tomaselli et al., 2000). Thus, children who were advanced in their somatic development tended to perform better and to show less AM than did late maturers and vice-versa.

Wolff et al. (1985) suggested that AM may be a reliable measure of developmental age when the assessment is specifically based on motor signs in the age sensitive range. In particular they regarded the neuromotor status as a useful, independent developmental measure for psychological investigations during those periods in development when transient dissociation between chronological and biological age may have important consequences for academic achievement, self-esteem and social adaptation.

There is extensive literature on child–parent resemblance in somatic growth and intellectual development. Child–midparent height correlations are low during the first years of life and, by the age of 4 years, reach a moderate value of about 0.60 (Largo, 1993). The child–parent relationship in intellectual performance during childhood takes a similar course. For the first 2 years correlations are less than 0.20 and increase to about 0.50 by 4 years. Is there also a comparable and thus, for research and clinical work, relevant child–parent resemblance in neuromotor competence? Kovar (1981) provided a few data on this important topic which, however, are not conclusive. There is a great need for research on child–parent resemblance in neuromotor functioning.

Obviously adaptive behaviours, such as catching or throwing a ball, can be improved by exercise. However, to what extent are distinct motor functions, such as repetitive or successive finger movements, susceptible to training? Does the speed of successive finger movements increase and do AM decrease when a child plays the flute? A few studies in children and adults indicate that large numbers of repetitions are necessary in order to significantly improve speed in defined motor activities (Kottke, 1980). More informative data on this therapeutically highly relevant subject are required.

Neuromotor functions and other developmental areas

Performance and quality of motor skills have been related to disturbances in different developmental areas. Subtle signs were reported to be associated with a number of conditions, such as hyperactivity (Lucas et al., 1965; Hertzig et al., 1969; McMahon & Greenberg, 1977; Mikkelsen et al., 1982), impulsivity (Paulsen & O'Donnell, 1979), aggressive antisocial conduct disorder (Wolff & Hurwitz, 1966; Lewis et al., 1979). Gillberg and co-workers extensively studied developmental coordination disorders in children with attention deficit disorder, as well as psychiatric and personality disorders (Landgren et al., 1996, 1998; Gillberg, 1998; Kadesjo & Gillberg, 1998). Presence of subtle neurological signs in childhood appears to be of prognostic significance for affective and anxiety disorders in adolescence (Shaffer et al., 1983, 1986).

Associations between abnormal motor coordination and emotional symptoms have received particular attention in recent years. Patients with various neurological diseases were at increased risk for affective anxiety and obsessive–compulsive disorders. Possible relationships between emotional disturbances, motor dysfunctions, and lesions of the basal ganglia have been reported among both neurological and psychiatric patients (Table 4).

From an evolutionary and ontogenetic perspective, neural systems for the control of motor speech and expressive language may be derivatives of mechanisms for the serial order control of coordinated fine-motor manual skills (MacNeilage, 1970; Liberman, 1974; Wolff et al., 1985, 1990; Kelso & Tuller, 1983; Denckla, 1984). Thus, the development of coordinated action might intersect with language acquisition and, by this pathway, influence the early stages of reading skill acquisition. Annett (1973) and Rudel (1980) found a direct correlation between

Table 4 Emotional, behavioural and motor disturbances related to basal ganglia disorders

Obsessive–compulsive and other affective disorders	Laplane *et al.* (1989), Denkla (1989), Wise & Rapoport (1989), Hollander *et al.* (1990), Insel (1992)
Depression and anxiety disorders	Starkstein *et al.* (1990), Rabins *et al.* (1991)
Wilson's disease	Dening *et al.* (1989)
Parkinson's disease	Stein *et al.* (1990), Cummings (1992)
Sydenham's chorea	Swedo *et al.* (1989)

speed of repetitive fine-motor skills, vocabulary size and language performance among young normal children. Owen *et al.* (1971) and Denckla (1976) reported an inverse correlation between speed of timed manoeuvres and academic achievement in older children with language and learning disabilities.

Much literature has been devoted to the associations between abnormal neurological signs and disturbances in language, cognition, social behaviour and emotional development. However, these findings will be better understood only if we improve our understanding on normal neuromotor development and its relationships to other developmental domains such as language and cognitive functions.

In conclusion, our knowledge of neuromotor development between kindergarten age and adolescence has improved considerably in recent years. There is a growing awareness that motor functions have their specific developmental course, are highly variable among children of the same age, and may or may not vary in lateralization or between genders. This knowledge needs to be implemented in clinical practice and – as indicated above – in the future, major research efforts are required to further improve our understanding of neuromotor development.

ACKNOWLEDGEMENTS

We thank Franziska Hug and Kathrin Muggli for their careful work in data collection, as well as Eva Etzensberger and Ernesto Peter for their contribution to the data analysis. We thank Tanja Kakebeeke, Luciano Molinari, Gerhard Neuhäuser and Kurt von Siebenthal for their useful comments and criticisms that have helped make the content clearer. This work was supported by the Swiss National Science Foundation (Grant No. 3200-045829.95/2).

References

Abercrombie, M.L.J., Lindon, R.L. & Tyson, M.C. (1964). Associated movements in normal and physically handicapped children. *Developmental Medicine and Child Neurology,* **6**, 573–580.
American Psychiatric Association (1994). *Diagnostic and statistical manual of mental disorders* (DSM-IV). Washington, DC: American Psychiatric Association.

Annett, M. (1970). The growth of manual preference and speed. *British Journal of Psychology,* **61**, 545–558.

Annett, M. (1973). Laterality of childhood hemiplegia and the growth of speech and intelligence. *Cortex,* **9**, 4–39.

Bowen, F.P., Hoehn, M.M. & Yahr, M.D. (1972). Cerebral dominance in relation to tracking and tapping performance in patients with parkinsonism. *Neurology,* **22**, 32–39.

Brinkman, J. & Kuypers, H.G. (1973). Cerebral control of contralateral and ipsilateral arm, hand, and finger movements in the split-brain rhesus monkey. *Brain,* **96**, 653–674.

Burton, A.W. & Miller, D.E. (1998). *Movement skill assessment.* Leeds: Human Kinetics Publisher.

Connolly, K.J. & Forssberg, H. (1997). *Neurophysiology and neuropsychology of motor development.* Clin. Dev. Med. no. 143/144. Oxford: MacKeith Press.

Connolly, K.J. & Stratton, P. (1968). Developmental changes in associated movements. *Developmental Medicine and Child Neurology,* **10**, 49–56.

Cummings, J.L. (1992). Depression and Parkinson's disease: a review. *American Journal of Psychiatry,* **149**, 443–454.

Denckla, M.B. (1973). Development of speed in repetitive and successive finger movements in normal children. *Developmental Medicine and Child Neurology,* **15**, 635–645.

Denckla, M.B. (1976). Naming of object-drawings by dyslexics and other learning disabled children. *Brain and Language,* **3**, 1–15.

Denckla, M.B. (1984). Developmental dyspraxia. The clumsy child. In M.D. Levine & P. Satz (Eds.), *Middle childhood: Development and dysfunction* (pp. 251–260). Boston: University Park Press.

Denckla, M.B. (1985). Revised neurological examination for subtle signs. *Psychopharmacology Bulletin,* **21**, 773–800.

Denckla, M.B. & Rudel R.G. (1974). Development of motor co-ordination in normal children. *Developmental Medicine and Child Neurology,* **16**, 729–741.

Denckla, M. (1989). Neurological examination. In J.L. Rapoport (Ed.), *Obsessive compulsive disorder in childhood and adolescence* (pp. 107–118). Washington, DC: American Psychiatric Press.

Dening, T.R. & Berrios, G.E. (1989). Wilson's disease: psychiatric symptoms in 195 cases. *Archives of General Psychiatry,* **46**, 1126–1134.

Deuel, R.K. & Robinson, D.J. (1987). Developmental motor signs. In D.E. Tupper (Ed.), *Soft neurological signs* (pp. 95–129). Orlando, FL: Grune & Stratton.

Fog, E. & Fog, M. (1963). Cerebral inhibition examined by associated movements. In M. Bax & R.C. MacKeith (Eds.), *Minimal cerebral dysfunction* (pp. 52–57). Clin. Dev. Med. no. 10. London: Spastics Society with Heinemann.

Gage, J.R. (1991). *Gait analysis in cerebral palsy.* Clin. Dev. Med. no. 121. Oxford: MacKeith Press.

Gallahue, D.L. (1989). *Motor development. Infants, children, adolescents.* Indianapolis: Benchmark Press.

Gillberg, C. (1998). Hyperactivity, inattention and motor control problems: prevalence, comorbidity and background factors. *Folia Phoniatrica et Logopedica,* **50**, 107–17.

Gubbay, S.S. (1975). *The clumsy child – a study of developmental apraxic and agnosic ataxia.* London: Saunders.

Guy, W. (1976). *PANESS. ECDEU Assessment Manual for Psychopharmacology.* Revised US Department of Health, Education and Welfare, DHEW publication No. (ADM) 76–338 (pp. 383–406).

Heap, M. & Wyke, M. (1972). Learning of a unimanual motor skill by patients with brain lesions: an experimental study. *Cortex,* **8**, 1–18.

Henderson, S.E. & Sugden, D.A. (1992). *Movement assessment battery for children (ABC).* Kent: The Psychological Corporation.

Hertzig, M.E., Bortner, M. & Birch, H.G. (1969). Neurologic findings in children educationally designated as "brain damaged". *American Journal of Orthopsychiatry,* **39**, 437–446.

Hollander, E., Schiffman, E., Cohen, B. Rivera-Stein, M.A., Rosen, W., Gorman, J.M., Fyer, A.J., Papp, L. & Liebowitz, M.R. (1990). Signs of central nervous system dysfunction in obsessive-compulsive disorder. *Archives of General Psychiatry,* **47**, 27–32.

Ingram, T.T. (1969). The development of higher nervous activity in childhood and its disorders. In P.J. Vinken & G.W. Bruyn (Eds.), *Handbook of clinical neurology. Disorders of speech, perception and symbolic behaviour* (vol. 4, pp. 87–101), Amsterdam: North-Holland.

Insel, T. (1992). Toward a neuroanatomy of obsessive-compulsive disorder. *Archives of General Psychiatry,* **49**, 739–744.

Kadesjo, B. & Gillberg, C. (1998). Attention deficits and clumsiness in Swedish 7-year-old children. *Developmental Medicine and Child Neurology*, **40**, 796–804.

Kakebeeke, T.H., Jongmans, M.J., Dubowitz, L.M., Schoemaker, M.M. & Henderson, S.E. (1993). Some aspects of the reliability of Touwens's examination of the child with minor neurological dysfunction. *Developmental Medicine and Child Neurology*, **35**, 1097–1105.

Kelso, J.A.S. & Tuller, B. (1983). A dynamical basis for action systems. In M.S. Gassaniga (Ed.), *Handbook of cognitive neuroscience* (pp. 143–168). New York: Plenum Press.

Kinsbourne, M. (1973). Minimal brain dysfunction as a neurodevelopmental lag. *Annals of the New York Academy of Sciences*, **205**, 268–273.

Kohen-Raz, R. (1974). Physiological maturation and mental growth in pre-adolescence and puberty. *Journal of Child Psychology and Psychiatry*, **15**, 199–213.

Kottke, F.J. (1980). From reflex to skill: the training of coordination. *Archives of Physical Medicine and Rehabiliation*, **61**, 551–561.

Kovar R. (1981). *Human variation in motor abilities and its genetic analysis*. Prague: Charles University.

Landgren, M., Pettersson, R., Kjellman, B. & Gillberg, C. (1996). ADHD, DAMP and other neurodevelopmental/psychiatric disorders in 6-year-old children: epidemiology and co-morbidity. *Developmental Medicine and Child Neurology*, **38**, 891-906.

Landgren, M., Kjellman, B. & Gillberg, C. (1998). Attention deficit disorder with developmental coordination disorders. *Archives of Diseases in Childhood*, **79**, 207–212.

Laplane, D., Levasseur, M., Pillon, B., Dubois, B., Baulac, M., Mazoyer, B., Tran Dinh, S., Sette, G., Danze, F. & Baron, J.C. (1989). Obsessive-compulsive and other behavioural changes with bilateral basal ganglia lesions: a neuropsychological, magnetic resonance imaging and positron tomography study. *Brain*, **112**, 699–725.

Largo, R.H. (1993). Die Regulation des postnatalen Wachstums aus phänomenologischer Sicht. *Nova Acta Leopoldina*, **69**, 245–258

Largo, R.H., Caflisch, J.A., Hug, F. & Muggli, K. (2001a). Neuromotor development from 5 to 18 years: Part 1. Timed performances. *Developmental Medicine and Child Neurology*, **43**, 436–443.

Largo, R.H., Caflisch, J.A., Hug, F. & Muggli, K. (2001b). Neuromotor development from 5 to 18 years: Part 2. Associated movements. *Developmental Medicine and Child Neurology*, **43**, 444–453.

Largo R.H., Caflisch J.A., Hug F. & Muggli K. (2000c). Neuromotor development from 5 to 18 years: Laterality of timed performance and contralateral associated movements. (In preparation.)

Lewis, D.O., Shanok, S.S., Pincus, J.H. & Glaser, G.H. (1979). Violent juvenile delinquents: psychiatric, neurological, psychological, and abuse factors. *Journal of the American Academy of Psychiatry*, **18**, 307–319.

Libermann, A.M. (1974). The specialization of the language hemisphere. In F.O. Schmitt & F.G. Worden (Eds.), *The neurosciences* (pp. 68–87). Third Study Program. Cambridge, MA: MIT Press.

Lucas, A.R., Rodin, E.A. & Simson, C.B. (1965). Neurological assessment of children with early school problems. *Developmental Medicine and Child Neurology*, **7**, 145–156.

MacNeilage, P.F. (1970). Motor control of serial ordering of speech. *Psychology Review*, **77**, 182–196.

McMahon, S.A. & Greenberg, L.M. (1977). Serial neurologic examination of hyperactive children. *Pediatrics*, **59**, 584–587.

Malina, R.M. (1986). Growth of muscle tissue and muscle mass. In F. Falkner & J.M. Tanner (Eds.), *Human growth* (vol. 2, pp. 273–294). New York: Plenum Press.

Mikkelsen, E.J., Brown, G.L., Minichiello, M.D., Millican, F.K. & Rapoport, J.L. (1982). Neurologic status in hyperactive, enuretic, encopretic and normal boys. *Journal of the American Academy of Child Psychiatry*, **21**, 75–81.

Neeper, R. & Greenwood, R.S. (1987). On the psychiatric importance of neurological soft signs. In BB. Lahey & A.E. Kazdin (Eds.), *Advances in cinical child psychology* (vol. 10, pp. 217–258) New York: Plenum Press.

Neuhäuser, G. (1975). Methods of assessing and recording motor skills and movement patterns. *Developmental Medicine and Child Neurology*, **17**, 369–386.

Newell, K.M. & Corcos, D.M. (Eds.) (1993). *Variability and motor control*. Leeds: Human Kinetics.

Owen, F.W., Adams, P.A., Forrest, T., Stolz, L.M. & Fisher, S. (1971). Learning disorders in children: sibling studies. *Monograph on Social Research in Child Development*, **36**, 1–77.

Paulsen, K. & O'Donnell, J.P. (1979). Construct validation of children's behaviour problem dimensions: relationship to activity level, impulsivity, and soft neurological signs. *Journal of Psychology,* **101,** 273–278.

Polatajko, H.J. (1999). Developmental coordination disorder (DCD): Alias the clumsy child syndrome. In K. Whitmore, H. Hart & G. Willems, G. (Eds.), *A neurodevelopmental approach to specific learning disorders* (pp. 119–133). London: MacKeith Press.

Quitkin, F., Rifkin, A. & Klein, D.F. (1976). Neurologic soft signs in schizophrenia and character disorder. Organicity in schizophrenia with premorbid asociality and emotionally unstable character disorders. *Archives of General Psychiatry,* **33,** 845–853.

Rabins, P.V., Pearlson, G.D., Aylward, E., Kumar, A. & Dowell, K. (1991). Cortical magnetic imaging changes in elderly patients with major depression. *American Journal of Psychiatry,* **148,** 617–620.

Rapin, I., Tourke, L.M. & Costa, L.D. (1966). Evaluation of the Purdue Pegboard as a screening test for brain damage. *Developmental Medicine and Child Neurology,* **8,** 45.

Reitan, R.M. (1971). Complex motor functions of the preferred and non-preferred hands in brain-damaged and normal children. *Perceptual and Motor Skills,* **33,** 671–675.

Rudel, R.G. (1980). Learning disability. Diagnosis by exclusion and discrepancy. *Journal of the American Academy of Child Psychiatry,* **19,** 547–569.

Rudel, R.G., Denckla, M.B. & Spalten, E. (1974). The functional asymmetry of Braille letter learning in normal, sighted children. *Neurology,* **24,** 733–738.

Schellekens, J.M., Scholten, C.A. & Kalverboer, A.F. (1983). Visually guided hand movements in children with minor neurological dysfunction: response time and movement organization. *Journal of Psychology and Psychiatry,* **24,** 89–102.

Shaffer, D., O'Connor, P.A., Shafer, S.Q. & Prupis, S. (1983). Neurological soft signs: their origins and significance for behaviour. In M. Rutter (Ed.), *Developmental neuropsychiatry* (pp. 144–163). New York: Guilford Press.

Shaffer, D., Stockman, C.J. & O'Connor, P.A. (1986). Early soft neurological sings and later psychopathology. In N. Erlenmeyer-Kimling & N.E. Miller (Eds.), *Lifespan research on the prediction of psychopathology* (pp. 31–48). Hillsdale, NJ: Erlbaum.

Spreen, O. & Gaddes, W.H. (1969). Developmental norms for 15 neuropsychological tests age 6 to 15. *Cortex,* **5,** 170–191.

Starkstein, S.E., Cohen, B.S., Fedoroff, P., Parikh, R.M., Price, T.R. & Robinson, R.G. (1990). Relationship between anxiety disorders and depressive disorders in patients with cerebrovascular injury. *Archives of General Psychiatry,* **47,** 246–251.

Stein, M.B., Heuser, I.J., Juncos, J.L. & Uhude, T.W. (1990). Anxiety disorders in patients with Parkinson's disease. *American Journal of Psychiatry,* **147,** 217–220.

Stott, H.D., Moyes, F.A. & Henderson, S.E. (1984). The Henderson Revision of the Test of Motor Impairment. San Antonio: Psychological Corporation.

Swedo, S.E., Rapoport, J.L., Cheslow, D.L., Leonard, H.L., Ayoub, E.M., Hosier, D.M. & Wald, E.R. (1989). High prevalence of obsessive-compulsive symptoms in patients with Sydenham's chorea. *American Journal of Psychiatry,* **146,** 246–249.

Tanner, J.M. (1970). Physical growth. In P. Mussen (Ed.), *Carmichael's handbook of child psychology* (pp. 337–372). New York: Wiley.

Tomaselli, G. & Largo R.H. (2002). Early and late maturers in neuromotor development. (In preparation).

Touwen, B.C.L. & Prechtl, H.F.R. (1979). *The neurological examination of the child with minor nervous dysfunction.* Clin. Dev. Med. no. 38. London: Heinemann.

Tupper D.E. (Ed.) (1987). *Soft neurological signs.* New York: Grune & Stratton.

Vaughan, H.G. & Costa, L.D. (1962). Performance of patients with lateralized cerebral lesions. II. Sensory and motor tests. *Journal of Nervous and Mental Disease,* **134,** 237–242.

Vitiello, B., Ricciuti A.J., Stoff, D.M., Behar, D. & Denckla, M.B. (1989). Reliability of subtle (soft) neurological signs in children. *Journal of the American Academy of Child and Adolescent Psychiatry,* **28,** 749–753.

Waber, D.P. (1977). Sex differences in mental abilities, hemispheric lateralization and rate of physical growth at adolescence. *Developmental Psychology,* **13,** 29–38.

Wade, M.G. & Whiting, H.T.A. (1986). *Motor development in children: aspects of coordination and control.* Dordrecht: Martinus Nijhoff.

Wise, S. & Rapoport, J. (1989). OCD: Is it a basal ganglia disorder? In J.L. Rapoport (Ed.), *Obsessive compulsive disorder in childhood and adolescence* (pp. 327–344). Washington, DC: American Psychiatric Press.

Wohlwill, J. (1973). *The study of behavioural development*. New York: Academic Press.

Wolff, P.H. & Hurwitz, I. (1966). The choreiform syndrome. *Developmental Medicine and Child Neurology,* **8**, 160–165.

Wolff, P.H., Gunnoe, C.E. & Cohen, C. (1983). Associated movements as a measure of developmental age. *Developmental Medicine and Child Neurology,* **25**, 417–429.

Wolff, P.H., Gunnoe, C. & Cohen, C. (1985). Neuromotor maturation and psychological performance: a developmental study. *Developmental Medicine and Child Neurology,* **27**, 344-354.

Wolff, P.H., Michel, G.F., Ovrut, M. & Drake, C. (1990). Rate and timing precision of motor coordination in developmental dyslexia. *Developmental Psychology,* **26**, 349–359.

Woods, B.T. & Teuber, H.L. (1978). Mirror movements after childhood hemiparesis. *Neurology*, **28**, 1152–1157.

Wyke, M. (1967). Effect of brain lesions on the rapidity of arm movement. *Neurology*, **17**, 1113–1120.

Wyke, M. (1968). The effect of brain lesions on the performance of an arm-hand precision task. *Neuropsychologia*, **6**, 125–131.

Wyke, M. (1969). Influence of direction on the rapidity of bilateral arm movements. *Neuropsychologia*, **7**, 189–196.

Zazzo, R. (Ed.) (1960). *Manuel pour l'Examen Psychologique de l'Enfant*. Neuchâtel: Delachaux Nestlé.

Zurif, E.B. & Carson, G. (1970). Dyslexia in relation to cerebral dominance and temporal analysis. *Neuropsychologia*, **8**, 351–358.

Further reading

Burton, A.W. & Miller, D.E. (1998). *Movement skill assessment*. Leeds: Human Kinetics Publisher.

Connolly, K.J. & Forssberg, H. (1997). *Neurophysiology and neuropsychology of motor development*. Clin. Dev. Med. no. 143/144. Oxford: MacKeith Press.

Gallahue, D.L. (1989). *Motor development. Infants, children, adolescents*. Indianapolis: Benchmark Press.

Kadesjo, B. & Gillberg, C. (1998). Attention deficits and clumsiness in Swedish 7-year-old children. *Developmental Medicine and Child Neurology*, **40**, 796–804.

Kakebeeke, T.H., Jongmans, M.J., Dubowitz, L.M., Schoemaker, M.M. & Henderson, S.E. (1993). Some aspects of the reliability of Touwens's examination of the child with minor neurological dysfunction. *Developmental Medicine and Child Neurology*, **35**, 1097–1105.

Landgren, M., Pettersson, R., Kjellman, B. & Gillberg, C. (1996). ADHD, DAMP and other neurodevelopmental/psychiatric disorders in 6-year-old children: epidemiology and co-morbidity. *Developmental Medicine and Child Neurology*, **38**, 891-906.

Neuhäuser, G. (1975). Methods of assessing and recording motor skills and movement patterns. *Developmental Medicine and Child Neurology*, **17**, 369–386.

Newell, K.M. & Corcos, D.M. (Eds.) (1993). *Variability and motor control*. Leeds: Human Kinetics Publisher.

Polatajko, H.J. (1999). Developmental coordination disorder (DCD): Alias the clumsy child syndrome. In K. Whitmore, H. Hart & G. Willems (Eds.), *A neurodevelopmental approach to specific learning disorders* (pp. 119–133). London: MacKeith Press.

Schellekens, J.M., Scholten, C.A. & Kalverboer, A.F. (1983). Visually guided hand movements in children with minor neurological dysfunction: response time and movement organization. *Journal of Child Psychology and Psychiatry*, **24**, 89–102.

Shaffer, D., O'Connor, P.A., Shafer, S.Q. & Prupis, S. (1983). Neurological soft signs: their origins and significance for behaviour. In M. Rutter (Ed.), *Developmental neuropsychiatry* (pp. 144–163). New York: Guilford Press.

Wade, M.G. & Whiting, H.T.A. (1986). *Motor development in children: aspects of coordination and control*. Dordrecht: Martinus Nijhoff.

III.9
Understanding Motor Development: Insights from Dynamical Systems Perspectives

BRIAN HOPKINS

ABSTRACT

This chapter provides an introduction to dynamical systems thinking, which in fact consists of three interrelated parts subsumed under the natural–physical approach to complex systems: the dynamical approach, the ecological approach and the perception–action approach. All three approaches share a common motivation derived from the principles of self-organization and all treat the brain, the body and the environment as dynamically coupled systems. Together they offer new insights into motor development as exemplified by studies of infants that have been inspired by one or other approach. The ultimate aim of a natural–physical approach to motor development is the derivation of a law-based theory of qualitative change in the control and coordination of action, which has yet to be achieved by any theory. However, as discussed, the approach offers new ways of thinking about the development of action. In particular, it offers objective guidelines for detecting and modelling developmental transitions and for identifying the mechanisms involved.

INTRODUCTION

Here I am, seated at my desk, trying to write the opening lines of this chapter. The cognitive effort required to get my thoughts on paper is palpable as I fiddle with the pen in my hand. What I am not immediately aware of is the immense complexity of the processes involved in putting pen to paper interspersed with bouts of fiddling with my pen. Let's try to make this complexity more transparent.

My body consists of some 800 muscles, about 200 bones and more than 100 joints. My hand doing the writing has 14 joints and is linked to an arm with another three. Simply initiating a stroke of my pen recruits more than 20 muscles in my

591

A.F. Kalverboer and A. Gramsbergen (eds.), Handbook of Brain and Behaviour in Human Development, 591–620
© 2001 Kluwer Academic Publishers. Printed in Great Britain.

arm and many more in my hand as I perceptually adjust my fingers to the properties of the pen in order to set it in motion. My other arm is resting on the surface of the desk and provides me with the necessary postural stability for me to produce legible writing and to switch from a "penning" mode to "fiddling" and back again. In addition to a complex musculoskeletal system I have an even more complex nervous system with its 10^{14} neurons in 10^3 different varieties. Somehow it is also involved in getting both my thoughts and pen on to paper.

This personalized vignette serves to illustrate some of the key issues that arise when we ask how we control and coordinate natural and everyday actions. To begin with there is the degrees-of-freedom problem: how does the apparent simplicity of such actions arise from an underlying deeper complexity inherent in the nervous and musculoskeletal systems? Then there is the transition problem: how do we change from one mode of action to a qualitatively different one? Yet another is the coupling problem: how are perception, movement and posture integrated in performing an action? Finally, there is the inescapable brain–body problem: what is the role of the nervous system and how does it interact with the musculoskeletal system in the control and coordination of action? The latter problem forms part of the broader issue of embodiment (Clark, 1997): how does the brain in the body interact with the structure of the environment in generating the kinds of experiences that facilitate cognitive processes and other aspects of mental life? This issue stems from the view that the nervous system, the biomechanics of the body and the environment are coupled dynamical systems from whose interactions adaptive behaviours emerge (Chiel & Beer, 1997).

While these sorts of problems have been depicted by means of an action carried out in real time, they are also germane to our understanding of how actions evolve over ontogenetic time. This is not to say that actions considered along both time scales necessarily share the same specific processes. Rather they abide by similar general principles such as those associated with the concept of self-organization (e.g. see Singer, 1986, who proposes on the basis of neurophysiological evidence that adult learning may be seen as a continuation of early forms of self-organization). During development, morphological and neural factors such as limb mass and the ability to regulate muscle stiffness will have markedly different influences on the control and coordination of actions in infants compared to adults. Accounting for the consequences of the changing nature of such influences on early development has been inspired by dynamical systems approaches to perception and action based on physical principles (Kugler & Turvey, 1987). These approaches have their roots in David Hume's (1711–1776) speculation that the laws of physics could explain the workings of the mind. Collectively termed the natural–physical approach to complex systems (Kugler, 1986), they consist of a number of theoretical perspectives united by the principles of self-organization.

UNITY (SELF-ORGANIZATION) IN DIVERSITY (THE NATURAL–PHYSICAL APPROACH)

The natural–physical approach (NPA) consists of three theoretically compatible perspectives (Beek & van Wieringen, 1994): the dynamical approach, the ecological approach and the perception–action approach (see Figure 1). Their

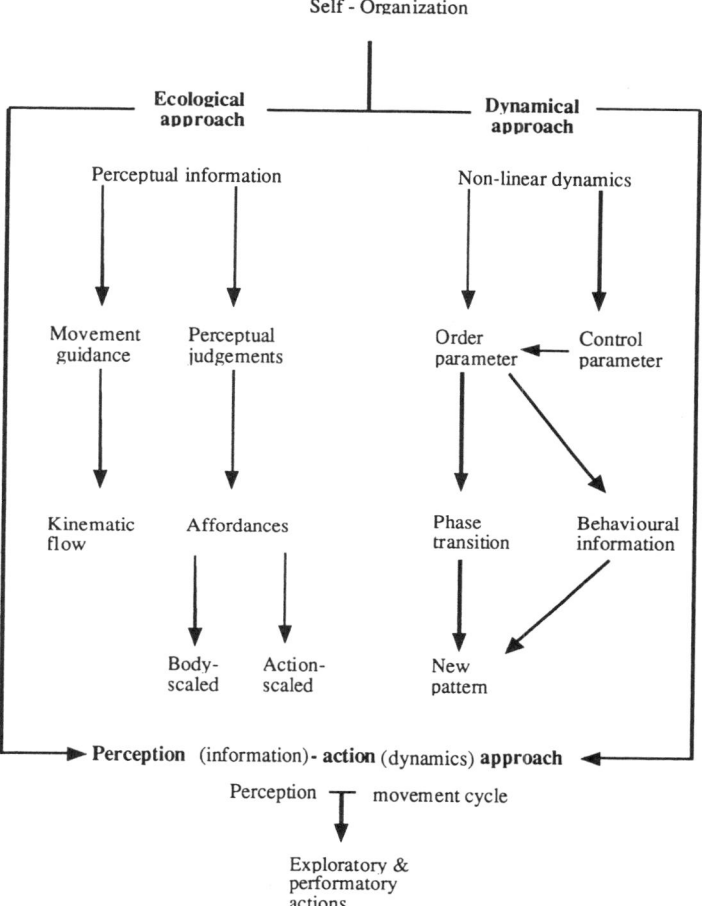

Figure 1 The natural-physical approach to the behaviour of complex systems. A complex system is one whose behaviour is the product of non-linear interactions between its parts and from which new properties can emerge. The starting point is a thermodynamical perspective based on field theory and the science of self-organization in general that provide a theoretical umbrella under which its constituent approaches pursue a common scientific agenda: a natural law-based approach that links information with dynamics in order to explain the organizational principles and mechanisms of pattern formation at the ecological scale of analysis (i.e. the scale at which organisms and their environments are defined). The dynamical approach has two interrelated variants, one addressing movement control and the other movement coordination relative to particular tasks. Quantitative changes in control structures (or symmetry-breaking of constraints) may lead to non-linear reorganizations in coordination corresponding to non-equilibrium phase transitions – all of which highlight principles of self-organization. The ecological approach also has two interrelated variants, one concerned with how body- or action-scaled perception of environmental structures affords potentials for action and the other with how perceptual flow fields guide movement. The theoretical challenge, assumed by the perception–action approach, is to unify the ecological and dynamical approaches into a direct realist theory of perceiving and moving that is general enough to account for stability and change in both real time and ontogenetic time.

common goal to is arrive at a law-based theory of how actions are controlled and coordinated based on self-organizing principles that, for example, "do not appeal to extrinsic executive agencies, motor programs or maturational timetables in order to 'explain' development" (Wolff, 1986, p. 70).

Following Bernstein (1967), coordination is a function constraining the potentially free variables of a system into a behavioural unit that maintains spatial and temporal relationships between movements of segments of the same limb (intralimb coordination) or between movements of two different limbs (interlimb coordination) relative to a particular task. Control is the process by which values or parameters (e.g. force, damping) are assigned to the behavioural unit. Coordination defined in this way is not open to description by Newtonian mechanics, in contrast to control that can be captured by measures such as amplitude, velocity, acceleration and force. Consequently, coordination and control may operate independently in the execution of an action.

Self-organization is a process by which new structures or spatial and temporal patterns spontaneously emerge in open systems without any specification from the outside environment (see also Mareschal & Thomas, this handbook). Thus, new properties can arise as a consequence of some internal regulation in response to changing external circumstances that do not specify what should be changed. This cannot happen in closed systems, but only in complex, open systems (see Table 1 for an overview of the distinctions between closed and open systems).

An important principle of self-organization is symmetry preservation: the maintenance of equilibrium between the components of a system such that inter-actions between them are constrained or limited by certain boundary conditions. Symmetry breaking, another principle of self-organization, can occur when these conditions exceed some critical value or are removed, with the result that the system can make a sudden transition to a new stable state of greater complexity. Often, symmetry breaking is heralded by fluctuations in the behaviour of the system, especially when it is in the process of being displaced to a far-from-equilibrium (i.e. dissipative) state. However, according to irreversible thermody-namics, once the transition is achieved and equilibrium restored, there is no return to the previous state (i.e. the process of change is irreversible in open systems).

A poignant example of self-organization in a biological system is provided by von Holst's (1937/1973) experiment on locomotor movements in the centipede (*Lithobius*). In the intact animal the legs move in a wave-like motion along the whole body. When all legs except two pairs are removed it walks like a quadraped, and like a six-legged insect with a tripod pattern if three pairs are retained. Thus, an aspecific change in boundary conditions results in a system-wide reorgan-ization and a change to a new state of locomotion. Only non- linear systems that can operate in far-from-equilibrium conditions display this form of self-organ-ized behaviour. Such systems are the focus of the NPA, and accounting for their behaviour as a function of time is a primary concern of the dynamical approach.

THE DYNAMICAL APPROACH

This approach bundles together a number of related theories; namely: (1) catas-trophe theory (Thom, 1975), (2) chaos theory (McCauley, 1993) and logistic

Table 1 Distinctions between closed and open systems according to irreversible thermodynamics. In contrast to a closed system, a complex open system can counteract the effects of the Second Law of Thermodynamics through continuously absorbing low-entropy energy from the environment. When in far-from-thermodynamical equilibrium, it generates dissipative structures that may show marked fluctuations and which, in turn, can generate a rapid and irreversible change to a qualitatively different state of greater complexity. Accordingly, development can be depicted as a continuous and irreversible process characterized by fundamental change involving the rearrangement of a system's elements

---

Types of systems

Closed system: one that does not exchange energy or matter with its surround and thus obeys the Second Law of Thermodynamics

Open system: one that exchanges energy with its surround and thus disobeys the Second Law. There are two types of open systems:

(a) *Simple system*: consists of a few elements or degrees of freedom and can only show linear (quantitative) change. There is a 1-to-1 or proportional relationship between input and output in such a system (S)

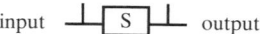

(b) *Complex system*: consists of many elements and can show non-linear (qualitative) change under certain conditions. There is no 1-to-1 relationship between input and output and the system may change abruptly from one state (S_1) to another (S_2)

Complex open system: has two contributions to its total entropy, *viz.* positive entropy (due to irreversible processes within the system) and negative entropy (due to exchange of energy with the surround). If both are in balance the system is stable. When positive entropy exceeds negative entropy the system is irreversibly drawn to the equilibrium point of maximum entropy. With the infusion of more external energy the system may suddenly jump to a new far-from-equilibrium state, which them takes over control of the system's dynamics.

Second Law of Thermodynamics: a closed system evolves to a state of disorder (maximum positive entropy) in which energy is distributed evenly through the system. Thus the system does not show any signs of structural differentiation or complexity.

Entropy: a macroscopical variable or measure of the unavailability of a system's energy to do work. In a broader sense the entropy of a system is a measure of its degree of disorder.

---

growth models (May, 1975), (3) irreversible non-equilibrium thermodynamics (Nicolis & Prigogine, 1989), (4) synergetics (Haken, 1977), and (5) topological dynamics (Bhatias & Szego, 1970). These theories arose as answers to questions about the nature of qualitative change in mathematics (5), physics (4), chemistry (3) and biology (2, 1).

It is important to recognize that the dynamics involved have nothing to do with forces and masses in the conventional sense of Newtonian mechanics. Instead they concern the temporal evolution of a system's behaviour, how it changes and how it seeks stable states over time (Morrison, 1991). Thus the main task of the dynamical approach is to identify the organizational principles, rather than the mechanisms that capture this evolution. The mechanisms may be sought, for example, in the biomechanical properties of limb segments (mass, length), muscles (in terms of the forces they deliver to the limb segments) and tendons (in terms of their elasticity). Manipulation of these properties has been a favoured strategy for those studying motor development from a dynamical approach (see Thelen, 1995). This brings us to the distinction between order and control parameters, that lies at the heart of the dynamical approach.

Order and control parameters: controlling order

An order parameter is a single topological entity that describes the macroscopical behaviour of a complex system in a stable state in terms of low dimensional dynamics. These dynamics, derived from cooperative non-linear couplings between components at more microlevels, are capable of showing multistability, multiple patterns and even deterministic chaos. In the language of synergetics the order parameter's dynamics enslave (or compress or literally give orders to) the enormous dimensionality of processes at the microlevels so that task-specific functions appear at the macroscopical level – something termed the "slaving principle" (Haken, 1977; Wunderlin, 1987). As such they do not refer to the physical mechanisms of pattern formation and may be regarded as analogous, for example, to morphogenetic fields in biological systems.

The mechanisms of pattern formation reside at the level of the control parameter. A control parameter is a boundary condition that acts as a constraint on the collective dynamics of the order parameter (where a constraint can be depicted as a condition that enables as well as restrains, encourages as well as discourages). For action systems there are three classes of constraints (Newell, 1986): organismic (e.g. neural, morphological), environmental (e.g. intrinsic and extrinsic properties of objects) and task-specific (e.g. instructions, goal of task). They are not mutually exclusive classes in terms of their effects, but in fact interact in complex ways as, for example, when locomotor mechanisms (e.g. neural oscillators) interact with the layout of surfaces leading to perceptual constraints or when the musculoskeletal system reacts to environmental forces leading to biomechanical constraints.

When a control parameter is scaled linearly beyond some critical value it can transiently lose its constraining influence on the order parameter, which may then manifest stochastic or even chaotic behaviour before making a sudden transition to a different collective state or mode (see Figure 2). Unlike the role assigned to their symbolic counterparts such as schemata or motor programmes, control parameters do not prescribe variations in the patterning of the order parameter in a strictly deterministic fashion. Rather they control only in the sense of leading the system unspecifically through regions of instabilities or keeping it within a stable operating range.

Developmentally, control parameters may assume the role of rate-limiting factors (Soll, 1979), which differentially constrain the phenotypical expression of relevant processes such that a transition occurs as a function of the process that evolves the slowest. Depending on the age-related transition and the action involved, these factors may be biomechanical, cognitive, hormonal or neural in origin. In the development of reaching, for example, muscle torque (Konczak *et al.*, 1995; Thelen *et al.*, 1993) and postural stability (Out *et al.*, 1998; Rochat, 1992) may be potential rate-limiting factors. This means that both muscle torques

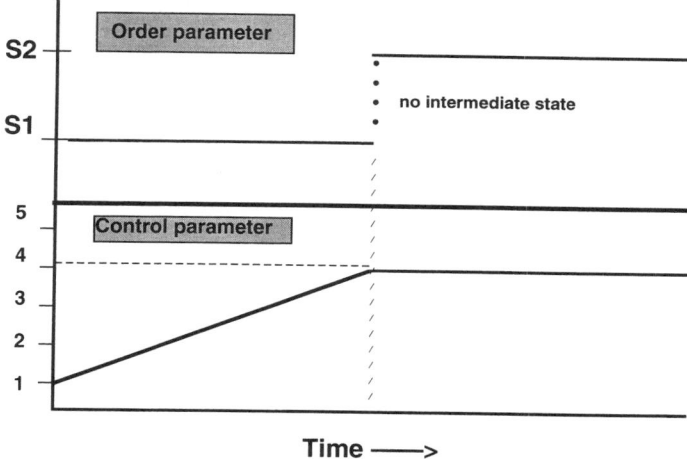

Figure 2 The relationship between order and control parameters. An order parameter or collective variable is a summarizing measure that captures the behavioural state of a complex system with many degrees of freedom. A control parameter is one that moves a system through its collective states. When it increases beyond some critical value it can change the order parameter from one state (S_1) to another (S_2). In non-linear systems this change can occur without any intermediate state, but with fluctuations in the order parameter during the transition between states.

in the shoulders (which counter the postural imbalance induced by unimanual reaching) and postural stability (of the head and trunk) would have to develop before infants can start to make reaching movements.

A much-cited study illustrates not only the distinction between order and control parameters, but also some of the "fingerprints" of self-organization (Kelso, 1984). In this study relative phase constituted the order parameter and movement frequency the control parameter. Relative phase is the angular difference between the motions of two limbs or two segments within a limb that are approximately sinusoidal.

The experimental task was to flex and extend the index fingers in time to an auditory metronome that imposed particular frequencies on the movements. Starting in anti-phase (homologous muscle groups contracting in an alternating manner), there was a *sudden and divergent jump* to the in-phase mode (homologous muscle groups contracting simultaneously) at a critical frequency (about 2 Hz), thus establishing relative phase as an order parameter and frequency as the control parameter. At the transition from the 180° to 0° phase mode, there were *critical fluctuations* around the mean state of the order parameter as indicated by an increase in the standard deviation of the relative phase. Scaling down the frequency after the transition did not result in a switch back to the anti-phase mode, a (weak) indication of *hysteresis* (the stronger form being when there is a return to the original mode at a different value of the control parameter, e.g. a flower opens its petals at one light intensity, but closes them at another). Further "fingerprints" of self-organization were present: *bimodality* in that two stable phase-locking modes were available below (but not above) a critical frequency, and its close relation *inaccessibility* as stable regions between 180° and 0° were not encountered (e.g. a relative phase of 90°). These, and other "fingerprints", referred to as castastrophe flags in castastrophe theory (Gilmore, 1981), are defined in Table 2.

Kelso's (1984) study demonstrated that the discontinuous transition from in-phase to anti-phase was accompanied by a number of phenomena that are hallmarks of a non-equilibrium phase transition – the simplest form of self-organization. The significance of phase transitions is that they provide "windows of opportunity" for revealing the relevant order and control parameters and thus enable description to be turned into prediction (Kelso, 1990). This is because, around the critical values of the control parameter, the number of degrees of freedom in complex systems is reduced to a low-dimensional order parameter description. But first one has to know when such transitions are actually taking place – something that has tended to be treated superficially in developmental psychology (see van der Maas & Hopkins, 1998).

Can the hallmarks of these sorts of transitions be used to detect the occurrence of developmental change? If so, then they offer a means of distinguishing quantitative from qualitative change and continuous from discontinuous change – distinctions that have a long and troublesome history in biology in general (e.g. Nagel, 1961) and developmental psychology in particular (e.g. Brainerd, 1978).

Developmental transitions: from waggling fingers to wobbling infants

A developmental transition has been defined as a change from one stable mode to another in a well-defined system (e.g. reaching) during a restricted period of development (Connell & Furman, 1984). The transitional period, the time taken to change between modes, should be markedly shorter than that spent in the prior and subsequent modes. To give this definition a heuristic function requires a set of rules for detecting the onsets and offsets of a transition. These rules are operationalized in Gilmore's (1981) catastrophe flags.

Some of the flags are illustrated for the case of a developmental change in grasping (i.e. from a scissor to a pincer grasp) by means of a cusp model derived from catastrophe theory (see Figure 3). A cusp, which is the simplest of seven elementary catastrophes that contain a bifurcation mechanism, requires one order parameter and two control parameters to make up the control surface if it is to be modelled (Thom, 1975). For the sake of simplicity, the control parameters are labelled competence (an organismic constraint) and task demands (a task constraint). Combinations of these parameters determine probable behaviour of the order parameter that is represented on a vertical axis. The behaviour surface, which contains a continuum from inefficient to efficient performance, slopes from left to right while its lower part slopes slightly downwa ds from *a* to *b*.

Imagine now an infant who can efficiently perform a scissor grasp, but not yet a pincer grasp. She is presented with a small object (e.g. a pellet), which does not afford being picked up efficiently with a scissor grasp (point *a* on the behaviour surface). Depending on the level of competence (e.g. reflected in the maturation of direct corticospinal pathways), she can now proceed in one of two ways: either

Table 2 Gilmore's (1981) eight catastrophe flags that can be applied to the detection of developmental transitions. Distinguished in terms of whether they occur before or during a transition (i.e. inside or outside the bifurcation set). Flags 1, 2, 3 and 8 observed when Delay function holds (switch between modes before old one becomes completely annihilated), but not when Maxwell function operative (switch between modes before old one becomes unstable). Flags 1, 4, 6 and 7 have featured previously in Piagetian-inspired research on developmental transitions (see van der Maas & Molenaar, 1992). Flags 4, 6 and 7 not sensitive enough to distinguish between qualitative and quantitative change. Flags 1 and 2, and 4 and 5, similar to each other, but differ as in each pair the latter one requires a specific manipulation of a control parameter. Flags 1, 2 and 3 not only important in predicting an upcoming transition but also, for example, in clinical settings: to establish critical values and safety margins when a sudden jump would be literally catastrophic (e.g. testing a new therapy for preterm infants with respiratory distress). When only flags 1 and 3 present this suggests a regression rather than start of a transition.

Occurrence	Flag	Definition	Associated function
Prior to transition	(1) Critical fluctuations	Enhancement of variability in order parameter(s). Associated with marked decline in correlations between order parameter(s).	Delay
	(2) Divergence of linear response*	Pertubation of control parameter(s) close to transition leads to: (i) enhancement of fluctuations in order parameter(s) (if one attractor present) or (ii) switching between behaviours (if more than one attractor present). Gives strong confirmation of which order parameter(s) involved in transition.	Delay
	(3) Critical slowing down*	Time taken to return to stable mode after a small perturbation becomes increasingly longer as transition approached. Concerns speed with which order parameter re-equilibrates, not its magnitude of change.	Delay
During a transition	(4) Sudden jump	Abrupt, non-linear change between two modes of functioning. Does not distinguish between quantitative and qualitative change which needs the other criteria.	Neither
	(5) Divergence*	For one unit of control parameter change, many units of change in order parameter. Two nearby initial values of control parameter may lead to widely divergent values in order parameter (cf. chaotic attractor).	Neither
	(6) Bimodality	Order parameter has two or more distinct modes. It may switch between them if they belong to the same behavioural dimension.	Neither
	(7) Inaccessibility	When bimodality present, intermediate values of order parameter are inaccessible due to presence of repellor in bifurcation set which amplifies bimodal distribution of order parameter values.	Neither
	(8) Hysteresis*	Return to original mode occurs at a different control parameter value following a small perturbation or there is no return at all. Gives strong confirmation of which control parameter(s) involved in transition.	Delay

* Ultimately requires experimental manipulation of control parameter for its detection.

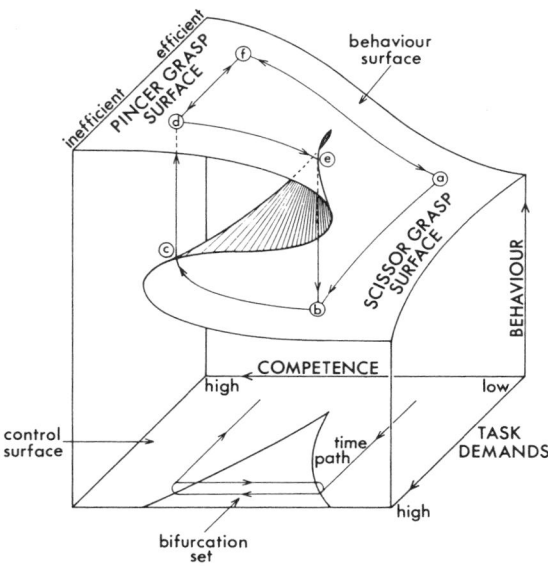

Figure 3 A cusp model depicting developmental change in grasping and some of the catastrophe flags. Smooth continuous change between the two types of grasping shown by path from *a* to *f*, which does not pass through the bifurcation set. If movement along this path was to stop opposite point e, then small differences between the combined values of the two control parameters could be attracted along the downward slope to widely different outcomes as point approached (divergence) – one going along the pincer grasp surface, the other along that for the scissor grasp. A move from *a* to *b*, by altering task demands at a particular competence level, leads to crossing the bifurcation set and therefore discontinuous change (sudden jump, *c* to *d*). Such a change may involve a hysteresis cycle in which the two forms of grasping coexist temporarily (bimodality, *c* and *d*), but with no intermediate values (inaccessibility, between *c* and *d*). Perturbations of task demands relative to competence level prior to entering the bifurcation set could lead to critical fluctuations and slowing down or to one or other type of divergence of linear response.

in the direction of point *f* (a smooth continuous change) or to point *b* (where each point represents an equilibrium state). If to point *b*, she will then travel along the path to point *c* at the edge of the fold (which is projected as a two-dimensional cusp-shaped curve on the control surface). At this point, further progression can be made only by a sudden jump to the upper pincer grasp surface, but at the inefficient end of the continuum. The potential for a hysteresis cycle has now been set up (b→c→d→e→b) so that the infant's behaviour crosses and recrosses the cusp boundary (i.e. bifurcation set) resulting in sudden jumps from one behaviour surface to the other. In doing so, both modes of grasping coexist temporarily (bimodality), but without intermediate forms (inaccessibility). Only when the cycle is broken, by an ongoing competition between the extremes of pincer grasping (indicated by the bidirectional arrow between *d* and *f*), is there a lateral progression to point *f*. When this is achieved, the pincer grasp becomes efficient and dominant (i.e. task-specific).

There is empirical support that grasping develops along the discontinuous pathway from a to f via a hysteresis cycle, rather than the smooth gradual change depicted by the direct link between a and f (Touwen, 1976). This description is just that, a description of qualitative change. To move beyond description requires in the first instance longitudinal data with continuously scaled order and control parameters. They can then be tested for how closely they fit the mathematical requirements of a cusp model, a modelling process facilitated by stochastic catastrophe theory (Cobb & Zachs, 1985; see Hartelman *et al.*, 1998).

All developmental transitions involving qualitative change should be preceded by critical fluctuations. They are, as stressed previously, the "motor" of change and, without the stochastic wobble they introduce, living systems would not be capable of developing to higher ordered states. However, to be able to detect the nature of a transition one needs to test for all flags using a combination of longitudinal and cross-sectional experimental research – no small undertaking. For the development of conservation, van der Maas and Molenaar (1992) suggest how this scenario can be implemented to detect transitions and, through clever experimentation, van der Maas (1993) shows that it can. More recently, preliminary findings on the transitions from no reaching to reaching to reaching with grasping during the first 6 months of life indicate the possibility and fruitfulness of applying this scenario to motor development (Wimmers *et al.*, 1998a,b). In terms of neural ontogeny, it has been suggested that the co-development of frontal (F)- parietal (P) regions of the human cortex as indexed by EEG coherence also exhibit features of a cusp catastrophe (Thatcher, 1998: see Welsh, this handbook). Part of the evidence for this suggestoin is a large jump in the mean left F7–P3 EEG coherence between 5 to 7 years of age within the theta frequency band. Assumed to reflect changes in the coupling between the intracortical systems, there is a shift from in-phase to out-of-phase oscillations between the regions at about 5 years followed by a change back to in-phase activity at around 7 years. Based on a predator-prey model of cortico-cortical connections, Thatcher (1998) hypothesizes that frontal regions regulate the pruning of synapses in posterior areas in a cyclical fashion, thereby giving the development of EEG coherence a stage-like appearance. Labelling the change in mean EEG coherence between 5 and 7 years a 'sudden jump' may seem inappropriate. However, relative to the age range covered (1.5 to 16 years) and given the long cycle times of cerebral dynamics, it does represent a relatively abrupt change. Whether the underlying of intracortical development can be modelled by gradient systems such as a cusp catastrophe is still very much an open question as a rigorous application of the various catastrophe flags to developmental changes in EEG coherence has yet to be carried out.

While all the flags are needed for the convincing detection of qualitative change, two of them are of paramount importance in demonstrating the dynam-ical properties of a developmental transition (i.e. whether it complies with a non-equilibrium phase transition). These are divergence of linear response (observable prior to a transition) and hysteresis (during a transition): the first because it confirms which order parameter is undergoing change and the second the responsible control parameter (see Table 2). Both are difficult to detect, in part because the control parameter has to be known *a priori* and in part because

601

both have to obey the Delay function. If instead the Maxwell function obtains (see Table 2 for the distinction between Delay and Maxwell functions), then both are likely to disappear under stochastic conditions – conditions which are likely to prevail in developmental data that are inherently "noisy" (van der Maas, 1993). Following Gilmore (1981), it should be stressed that the two functions or conventions represent the extremes of a range of possibilities. In between them the nature of a transition becomes fuzzy, such that it is often not possible to define the requisite bifurcation set (i.e. the set of points in the space of control parameters at which a transition occurs from one local minimum to another). Despite this state of affairs, experiments should be designed to probe for the presence of both possibilities by means of incorporating conditions that employ different change times for the control parameters. There are at least two reasons why these functions are of theoretical significance in research on (motor) development.

On the nature of motor development: dynamical functions and time scales

The first reason concerns two opposing views on the nature of motor development. One is that earlier patterns of movement (and posture) have to disappear before new ones can emerge (analogous to the Delay function). The other is that old and new patterns can temporarily coexist before the old one loses its stability (analogous to the Maxwell function). In line with the first view is the claim that the neonatal repertoire of reflexes has to be cortically inhibited before voluntary controlled movements can be become established (McGraw, 1945). Evidence to date does not support this claim. Touwen (1976), for instance, in a longitudinal study of 50 low-risk infants, found no correlation between the disappearance of the palmar grasp and the onset of voluntary radial grasping. What he did find was that the two patterns coexisted for some time, and that during this period the infant switched from one type of grasping to the other when presented with a standard-sized object. In addition, he reported no relationship between the disappearence of radial grasping and the appearance of the pincer grasp.

Such findings seem to suggest that qualitative changes in the development of prehension follow the Maxwell function. We hasten to add that the study in question did not involve any experimental manipulations of control parameters in and around the transitional ages. It was purely correlational in nature and therefore not able to distinguish between the two conventions. In fact, no study to date has expressly addressed this distinction in types of change – an omission requiring rectification if theory building in motor development is to progress.

The second reason is that the Delay and Maxwell functions are intimately tied to differences in the relationships between time scales (see Schöner *et al.*, 1986; Zanone *et al.*, 1993). Space does not permit going into details of the time scales, but the important message is that when their relationships are violated a transition between modes occurs. Basically, if the following set of relationships holds

$$t_{rel} \ll t_{obs} \ll t_{equ} \tag{1}$$

then the system displays local stationarity, is probably in a local attractor and thus not in or near the bifurcation set. That is, local relaxation time (t_{rel}), which is the same as critical slowing down (see Table 2), is considerably shorter than the time scale on which the system is observed (t_{obs}), which in turn is much briefer than the global or equilibrium relaxation time (t_{equ}). The latter scale refers to the time taken to attain a stationary probability distribution from an initial one that is randomly distributed. Put another way, it is the time needed for a system to visit all available states without any change in the value of a control parameter. But as a transition point is approached, the relationships given in (1) are transgressed in that t_{rel} increases and t_{equ} decreases, both assuming the same order as t_{obs}. At the same time a fourth scale has a crucial role in determining the type of transition. This is the time scale along which the control parameter changes (t_{par}). When the following set of relationships obtain

$$t_{rel} \ll t_{par} \ll t_{qu} \tag{2}$$

then the system makes a transition only after the old mode has become unstable (i.e. the Delay function holds). As t_{par} becomes slower, then eventually

$$t_{rell} \ll t_{equl} \ll t_{par} \tag{3}$$

so that the system makes a transition only after the old mode has become unstable and disappears (i.e. the Maxwell function pertains).

These various temporal relationships stress perhaps the major import of the dynamical approach: the necessity of accounting for interactions between processes operating on different time scales in theory and research on developmental change. In doing so, one can begin to answer in a principled manner some long-standing issues about the nature of motor development.

In conclusion, the dynamical approach presents new and largely untested opportunities for studying developmental change. One is that it provides us with objective criteria for pinpointing the onset and offset of developmental transitions. Another is that the castastrophe flags enable one to identify the nature of a transition and whether it has a dynamical basis (i.e. its compliance with the "fingerprints" of a non-equilibrium phase shift). In addition, asynchronies between processes acting on different time scales help us to determine which dynamical properties will be observed and thus how a new behaviour emerges. To avail ourselves of these opportunities requires a carefully thought-out research strategy, the outlines of which have been sketched in by Hopkins *et al.* (1993; see Schmidt & O'Brien, 1998 for a somewhat different strategy). It consists of an interplay between longitudinal and cross-sectional research, the latter involving bifurcation and perturbation experiments around transitional ages identified with non-experimental flags (see Table 2) during the longitudinal phase. The goal of these experiments is to reveal the organizational principles (i.e. the order parameter) and mechanisms (i.e. the control parameters) germane to a particular transition, the first being cast in terms of non-linear dynamics and the second by means of manipulating the couplings between perception, movement and posture.

THE TWO OTHER APPROACHES

The dynamical approach alone, culled either for its metaphors or mathematical formalisms, cannot provide a full account of development. To do so, according to the NPA, requires building a theoretical bridge between dynamics (i.e. action) and information (i.e. perception). The other fundamental building block in this enterprise is the ecological approach to perception. This is founded on Gibson's (1966, 1979) anti-dualistic version of realism concerning how to construe the relationship between animal and environment (animal being preferred to organism by ecological psychologists probably to emphasize the species generality of their ideas). Their rapprochement is seen as a means of eventually achieving a theory of perception–action coupling based on physical principles (i.e. on first principles that do not have recourse to mentalistic concepts). A more apposite label for the ultimate goal is a direct realist theory of perceiving and moving (Hopkins et al., 1993). Relative to motor development, the main question to be tackled by this theory is: "How can information be conceptualized so that it is continuously coordinated with changes in skeletomuscular dynamics that are brought about by changes in skeletomuscular dimensions?" (Kugler et al., 1982, p. 66).

Self-organization revisited

Gibson's theory of direct perception, like Bernstein's (1967) approach to action, is fully in tune with the science of self-organization. Recapitulating, its core feature (the Second Law of Thermodynamics) defines the dissipation of energy towards the symmetrical end state of maximum entropy. Perception provides a means of detecting energy sources (cf. negative entropy) that combat the dissipative tendencies of the Second Law. For example, as Gibson repeatedly pointed out, light contains low, but structured, energy distributions that envelop the animal. They function to specify the location of objects and their properties (e.g. texture) as well as the layout of surfaces. In other words, they contain (or rather specify) information which, when picked up by a perceptual system, allows the animal to use its kinetic properties to generate force fields in an energetically efficient way.

The information specified by the non-inertial properties of the ambient energy distributions may be given in terms of macroscopical descriptions of their kinematics (e.g. time-to-contact; Lee, 1976) or geometry (e.g. surface texture; Gibson, 1966) or both. The important point, in terms of perception–action coupling, is that information does not cause movement (it contains no focal stimuli), but constrains the control and coordination of action. The resultant actions, whether exploratory (i.e. investigative) or performatory (i.e. executive), generate further information to act upon, thus comprising a circular causality between perception and action: actions constrain or enhance what is perceived and what is perceived constrains or guides subsequent actions (see Figure 4). However, this circle is not fully deterministic in that intentions can enter into it as stochastic functions that may change in unpredictable ways how perception feeds off action and vice-versa (Kugler et al., 1990). Further exemplification of perception and of its relationship to action as self-organized processes can be found in Swenson & Turvey (1991).

The principle of mutuality

Both the ecological approach and the perception–action approach are deeply rooted in the principle of mutuality. This principle holds that animal and

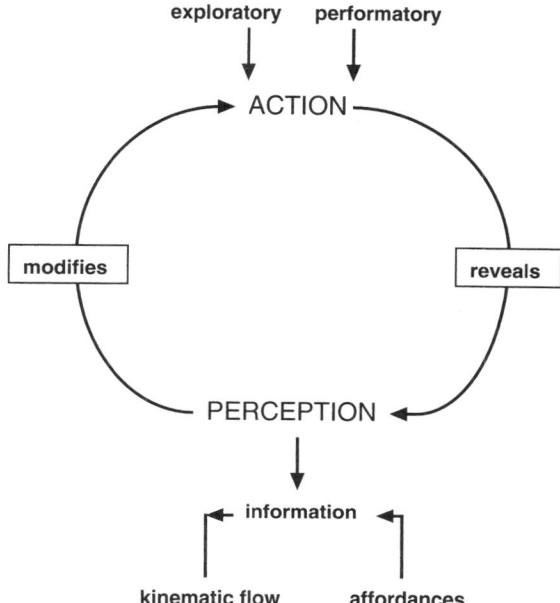

Figure 4 Perception–action coupling and its circular causality. Any action has perceptual consequences, which reveal information about the stability of the observer relative to the environment by means of kinematic (optical, auditory, etc.) flow and about the possibilities for further action by means of affordances. In turn, the perceptual information generated serves to guide or modify the ongoing action. Performatory actions, unlike exploratory actions, are directed towards achieving a specific goal such as grasping an object, and in this sense they are prospectively controlled executive functions.

environment, and thus perception and action, are yoked by the two fundamental reciprocities that function in distinct but complementary ways: a reciprocity between interior and exterior frames of reference (i.e. between internal and external degrees of freedom) and a reciprocity between movement and the detection of the resultant information it generates (i.e. between the generation of force fields and the generation of flow fields).

More controversially, and in contrast to traditional theories of information processing, perceptual systems are active, unbiased samplers of information that can be directly and veridically picked up in the external world. In Sellars's (1973) terminology, it is a distinction between presentationalism (i.e. direct realism in which environmental properties are directly perceived without any psychological mediation) and representationalism (i.e. indirect realism in which order is imposed on sensory inputs by a series of mediating inference-like processes).

Effective affordances

The principle of mutuality is most forcibly expressed in Gibson's (1979) notion of affordances, the points where animal and environment intersect. An affordance, a neologism coined by Gibson, refers to the functional possibilities of an object or environmental layout relative to what an individual can do with respect to them. As such it has both objective and subjective properties – objective in that what something affords depends on its physical characteristics and subjective in that it concerns relevant characteristics of the perceiver, such as body size and action abilities. Seen in this light, affordances define goals that are furnished by the environment and which are realized by affordance-specific control processes termed effectivities (Shaw & Turvey, 1981).

Together, an affordance and its effectivity, define an ecological field. When an animal's effectivities change (e.g. because of changes in the allometric relationships between body parts), then the meaning of an affordance also changes (e.g. an object that afforded standing becomes one that affords sitting on). Put another way, the symmetry-preserving relationship between them is broken and consequently there is some local change in the ecological field or in the animal–environment mutuality (i.e. a transition takes place in the coupling between perception and action).

The ecological approach to development: affording development

Developmental applications of the ecological approach have increased in number over the last few years. A considerable body of research now exists on the development of affordances, some of which has been reviewed by Adolph *et al.* (1993). Much of this (mainly cross-sectional) research concerns age differences in the perception of apertures, surfaces and slopes in relation to locomotion, and of the intrinsic and extrinsic properties of objects relative to reaching and manual or oral exploration. Studies on perceiving the slopes of surfaces and what they afford for locomotion provide a useful illustration of what a developmentally oriented ecological approach has to offer, while at the same time raising an important issue (*viz.* whether the perception of affordances is based on body-scaled or action-scaled metrics during infancy).

In a recent study, crawlers and toddlers were required to ascend or descend four walkways varying in slant from 10° to 40° (Adolph, 1997). There were hardly any differences between the crawling and walking infants on the ascending trials in terms of whether they hesitated or engaged in haptic exploration before attempting to ascend. For the descending trials, there were striking differences between the two groups. On the shallower slopes, the walkers hesitated longer and displayed more haptic exploration. In attempting to descend they tried out a variety of actions (e.g. sitting to slide down or scooting backwards), thereby demonstrating motor equivalence as a means of resolving an uncertain outcome. They clearly perceived the steeper slopes did not afford walking as they promptly slid down or simply avoided descending. The crawlers, in contrast, descended the shallower slopes with little delay or haptic exploration. Unlike the toddlers, they never attempted to descend from a sitting position or by scooting backwards even though they could get to a sitting position from a quadrupedal stance as demonstrated before the experiment.

These findings suggest that toddlers, and to a lesser degree crawlers, may perceive appropriate affordances for locomotion in relation to their own action abilities. This suggestion was investigated by Adolph (1997). Infants with greater crawling experience tended to attempt descents only on the shallower slopes. Among the toddlers, those with longer step lengths tried to walk up the steeper slopes, suggesting that what the slopes afforded were perceived relative to the level of walking ability. However, performance on their descending trials was not accounted for by individual differences over a range of action abilities. Given that perception of optical slant is present in the newborn (Slater & Morrison, 1985), it might be speculated that the requisite perceptual ability is present at birth and that one of its functional expressions is rate-limited by the development of motor systems appropriate to locomoting over slopes. This is not a view shared by Adolph (1997), who contended that what is required is the perception of geographical rather than optical slant – the perception of slope relative to gravity and the body instead of the retina. This contention ignores the possibility that the perception of optical slant may be transformed into that for geographical slant through experiences brought about by dealing with gravitational constraints in the context of locomoting.

Adolph (1997) had only restricted success in demonstrating that action-scaled information contributes to whether or not a slope is perceived as affording traversability. Other studies concerned with infants locomoting up stairs (e.g. Ulrich *et al.,* 1990) and reaching for objects at different distances (e.g. Yonas & Hartman, 1993) have indicated the primacy of action-scaled over body-scaled information in perceiving the boundaries of affordances.

Contrasts in findings between studies could be due to differences in the ages of the infants, in the tasks they were confronted with or interactions between age and task. In general, though, it appears that body-scaled information is less relevant for the perception of affordances during infancy than that which pertains to action abilities (Adolph *et al.,* 1993). A possible explanation for its lack of relevance could reside in the fact that individual differences in the body dimensions of infants are much smaller than at later ages (see Snyder *et al.,* 1977). Certainly at older ages, and particularly with adults, body-sized referenced information has been consistently shown to be a crucial control parameter in perceiving the boundaries of affordances. Expressed in terms of dimensionless ratios between body and environmental variables, examples include the following: leg length to riser height in stair climbing (Warren, 1988), shoulder width to aperture size in passability (Warren & Whang, 1987) and leg length to seat height in sitting (Mark & Vogele, 1988).

There are fewer studies on how and at what ages infants perceive and use kinematically specified information. One opinion is that infants can detect optical flow at an early age, but it is not before 5–10 months that they can use it to control posture (Yonas & Owsely, 1987; Bertenthal & Bai, 1989). This view, based in part on the well-known moving-room experiment (Lee & Aronson, 1974), implies that infants detect information in advance of being able to employ it for guiding action. However, the claim that it is not coupled effectively to action until at least 5 months of age is contradicted by Butterworth (1993), who reports that head control, the first major postural achievement, is sensitive to the direction of optical flow specified by the moving room in 2-month-old infants. Given that around this age infants change from an obligatory-like body-centred posture to a more flexible spatially oriented posture (Geerdink *et al.,* 1994), it would indeed be odd if their head control was impervious to extereoperceptual information that was kinematically specified by the optical flow.

Subsequent research on how optical flow, a term derived by Gibson (1966) to denote change of direction in lines radiating from a moving vantage point, controls the posture and movement of young infants, needs to take account of the variety of meanings it currently embraces. For example, a distinction is now made between radial and lamellar flow – between flow that radiates outwards from a point or singularity in the optical array during forward motion and flow that is restricted to the periphery of the visual system (Koendrink & van Doorn, 1981; Stoffregen, 1985; see Bertenthal & Bai, 1989, who endeavoured to decipher their respective functions in a moving-room experiment with infants who could sit or stand). There are also recent attempts to order and extend the different meanings associated with the tau variable (e.g. Tresilian, 1993; Lee, 1998), which was originally devised as an informational quantity specifying time-to-contact of an object approaching an observer at a constant velocity (Lee, 1976).

Circling perception and action with intent

The perception–action approach, first elucidated by Kugler & Turvey (1987), shares much in common with the ecological approach (e.g. perceiving affordances implies perceiving a coupling between perception and action), but there are differences. In the first instance, it goes beyond the ecological approach in dealing with the principle of mutuality. It treats the mutuality of perception and action as being both closed and open – closed in that flows and forces complement each other, open in that when informational constraints are altered change occurs in the properties of the perception–action circle. The actor's intention is treated as an important informational constraint in this respect (actor being seemingly preferred to animal perhaps because of the greater focus on action in this approach). Here we find a way of treating intention that attempts to explain it without recourse to mentalese or neuronese in the first instance. It is a complex explanation referenced in principles of self-organization, the details of which cannot be given here (see Kugler *et al.*, 1990, for an informative rendering of the details). Its core claim is that intention initiates action, but does not control it.

One feature of the explanation is to treat perception as a remote sensing or early warning device that informs the system of an upcoming goal (e.g. tau specifying time-to-contact). In this way, future states are projected into current perception (intention in perception). When an animal moves, perceptual information flows backwards relative to the flow of action (e.g. locomoting down a sloping surface). Action then has perceptual consequences that specify the goal of action (intention in action). This double-edged treatment of intention suffices to distinguish the main difference between the ecological and perception–action approaches: the former is mainly focused on how perception constrains action and the latter on how changes in action facilitate differentiatial attunement to perceptual information specifying affordances. It is not, of course, a rigid distinction as both approaches contribute to the same scientific agenda, which is to arrive at a theory of informational interactions at the ecological scale of analysis based on natural laws and not just simply metaphors. A number of such laws have been identified (Warren, 1988), which apply to perception and action and to the coupling between them (see Beek & van Wieringen, 1994). The ultimate aim is to arrive at a principled (i.e. mathematical) underpinning of the laws of control – something that has yet to be achieved. It is important to recognize that such laws will only assist in defining what the possibilities are for action in a particular situation. As such, they will not predict which action emerges, as that requires taking into account the actor's intention or goal of the action.

The perception–action approach to development

In its application to development the perception–action approach goes beyond the ecological one in emphasizing that what changes is not just an improving facility of perceptual systems to pick up and differentiate information. Rather, there is change in the organization of action due to successive reorganizations of the functional relationships between perception and action. In terms of the distinction between exploratory and performatory actions (see Figure 4), the young infant has a lot of the former and little of the latter. The developmental conundrum is how spontaneous exploratory actions beget prospectively con- trolled executive actions such as prehension.

Unquestionably, the exploratory actions of the newborn are founded on an intimate coupling between perception and action, as convincingly demonstrated in a clever experiment by van der Meer *et al.* (1995) – a demonstration dia- metrically opposed to Piaget's (1952) depiction of the newborn. To complete the developmental picture, exploratory actions inform infants about the potential properties of their effectivities and from which executive functions are carved out in a process analogous to Edelman's (1989) experiential selection of a secondary repertoire of neural circuits (see Hadders-Algra, this handbook). Thereafter, in a process akin to Edelman's re-entrant mapping between sensory and motor systems, performatory actions assume increasing importance in perceptual and cognitive development. They change the meaning of affordances offered by objects and places and in doing so create new opportunities for extracting information and acquiring knowledge.

There are two points to be stressed about the concept of action. First, actions are composed of task-specific movements. Secondly, there is an intimate relationship between movement and posture. Thus: (1) it is the coupling between perception and movement in performing a particular task that eventuates in action and (2) there is also a coupling between posture and movement in the manufacture of action in that posture imposes boundary conditions on the sorts of movements that are possible while movements constrain the organization of posture. These further specifications to the perception–action approach are depicted in Figure 5. Whilst the first point may seem a question of semantics, the second one is decidedly not. From a developmental perspective, postural achieve- ments (e.g. sitting with support), just like perceptual ones, enable movements to be geared more precisely to the attainment of task-specific actions (e.g. see Fontaine & Pieraut-LeBonniec, 1988). In short, posture exerts important rate- limiting functions during development that have not been systematically investi- gated. Subsequent research needs to address changes in the coupling between posture, movement and perception in the development of action. If carried out within the framework of the perception–action (or ecological) approach, it

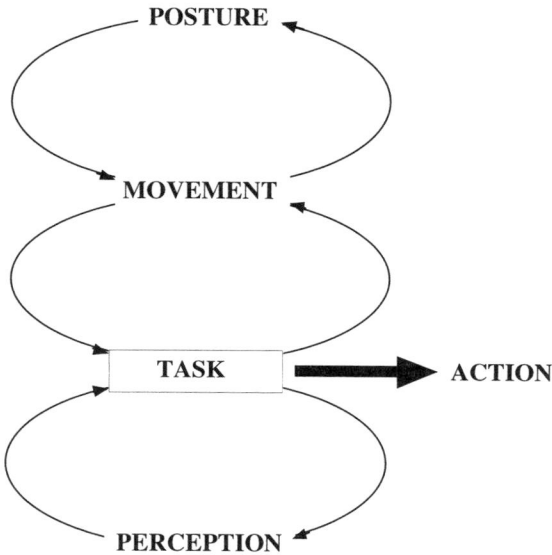

Figure 5 Developmental transformations in an action result from changes in the task-specific couplings between perception, movement and posture. A change in one component will lead to changes in the other components. For example, the acquisition of a new form of postural control will engender change in the sorts of movements that are possible. In turn, the outcomes generated by these new movements will alter what the infant perceives. Change may also be induced by alterations in the nature of the task (e.g. by shifting the infant from a supine to an upright position). Altering the task should result in a reorganization of the couplings between perception, movement and posture and thus to a different form of action. Task-induced changes in action could be particularly evident during developmental transitions when it is assumed that the developing organism has a heightened sensitivity to external perturbations.

should also be capable of contributing to the resolution of a number of problems encountered in the NPA.

CONCLUDING REMARKS: BACK TO THE PROBLEMS

Four issues pertaining to the control and coordination of action were launched in the introduction: the degrees-of freedom problem, the transition problem, the coupling problem and the brain–body problem. Each of them is addressed by the NPA and each of them has deep-seated implications for our understanding of motor development. What are these implications and what, if any, are the further issues they raise?

Degrees-of-freedom problem

Action systems, such as reaching or walking, are high-dimensional systems involving a variety of processes, each of which has potentially many degrees of

Table 3 The degrees-of-freedom problem and Bernstein's solution. The starting point for his solution was to reject there being one-to-one relations between individual muscles and cortical cells. Instead, groups of commonly organized neurons must underlie most actions. Thus, the solution to the degrees-of-freedom problem is that the brain organizes muscles collectively into coordinative structures. As part of his solution, Bernstein also proposed three steps in mastering the degrees of freedom presented by a new action. As it is acquired, gravity-dependent and motion-dependent torques are increasingly exploited, with muscle torques becoming more reserved for initiating and terminating an action (the latter also in conjunction with gravity-dependent torques). With development, less dependence is placed on muscle torques during the formation of an action. A good explanation of the three torques and how they are computed from the total joint torque by means of inverse dynamics can be found in Zernicke & Schneider (1993)

1. A bicycle has 10 degrees of freedom (df) in total (five moving parts, each with its own position and velocity). The human body has many more mechanical degrees of freedom.

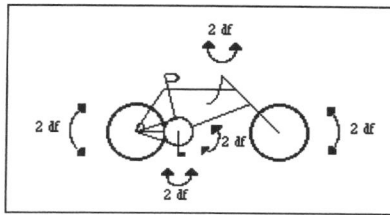

2. Problem: how does the brain control the body's mechanical degrees of freedom as well as those of the bicycle (peripheral indeterminancy problem) in learning to ride it?

Bernstein's solution

Step 1: 'Freezing' degrees of freedom
Active reduction of degrees of freedom that need to be controlled and coordinated within and between limb and torso segments in order to overcome the effects of gravity.

Step 2: 'Unfreezing' degrees of freedom
Release of constrained degrees of freedom that become incorporated into larger functional units (i.e. coordinative structures).

Step 3: 'Exploiting' degrees of freedom
Not just reacting to, but actively exploiting reactive forces (e.g. gravity and motion-dependent torques) arising from the interaction between cyclist and bicycle resulting in the differentiated and energy-efficient use of muscle torques.

Torques: forces acting on a joint causing rotation about the axis of the joint.

Gravitational torques: passives forces arising from gravity acting on the centre of the mass of a limb.

Motion-dependent torques: reactive or inertial forces arising from mechanical interactions between segments of a limb.

Muscle torques: forces arising from active muscle contractions and to a lesser extent from passive deformations of muscle, tendons, ligaments and other tissues around a joint.

freedom. How then are these degrees of freedom constrained so as to achieve such well-ordered, but flexible, actions? A clue is found in a study carried out more than 100 years ago (Ferrier & Yeo, 1881): electrical stimulation of the ventral roots in monkeys resulted not just in contractions of individual muscles, but also in highly coordinated functional synergies.

The notion of functional synergies or coordinative structures, collectives of muscles spanning several joints and constrained to act cooperatively as a task-specific unit, formed the basis of Bernstein's (1967) solution to the degrees-of-freedom problem. In addition, he addressed the associated problem of peripheral indeterminancy (Turvey *et al.,* 1978): how actions are controlled and coordinated relative to the ever-changing properties of muscles, limbs and environment. As part of his solution, Bernstein proposed three descriptive steps in the acquisition of a novel coordinative structure (see Table 3). He implied that this scenario was applicable not only to motor learning, but also to motor development – an implication borne out by research on the development of spontaneous kicking (Thelen, 1985). Interestingly, infants at risk for cerebral palsy do not appear to progress beyond Bernstein's step 1 in that they persist with quite stereotyped joint motions up to the age of 6 months (Vaal *et al.,* 2000). By this age, healthy infants no longer display this stereotypy.

While gaining control over the multiple degrees of freedom in a moving limb is a task to be mastered by the developing infant, they in fact provide a solution to the problem of peripheral indeterminancy. This point is illustrated by a study of kicking movements in 3-month-old infants placed in supine and supported sitting positions (Schneider *et al.,* 1990). In both positions, the infants produced the same kinematic pattern of kicking by means of differentially adjusting their muscle torques, thus demonstrating motor equivalence (the ability to reorganize movements so as to produce the same outcome in the face of changing conditions). Such an ability, therefore, would appear to be the way of resolving the problem of peripheral indeterminancy, and one that captures an important essence of the 2–3-month transformation in neural functions (Prechtl, 1984).

Transition problem

This problem involves both description and explanation. Descriptively, the task is to answer the question "What are the dynamics of change?", and one for which the NPA would appear to be eminently suited. Does motor development consist of a continuous and gradual accumulation of small changes (quantitative change) or of discontinuous reorganizations leading to the emergence of novel patterns of behaviour (qualitative change)? In a sense both these further questions are misleading as, if we follow the NPA, the implication is that a continuous change in a control parameter can give rise to a qualitative change in order parameter. The crux of the matter in understanding the nature of change during motor development resides in the choice of a requisite order parameter for the particular action one wishes to study.

Stemming from von Holst's (1937/1973) work on the swimming movements of the lamprey (*Lambrus*), relative phase has come to be regarded as a cardinal feature of biological coordination (i.e. as an order parameter). This is because it captures the spatiotemporal order of intralimb and interlimb coordination (stability) as well as qualitative changes in these orders (instability). It has been used, for example, in research on the development of intralimb and interlimb coordination during the acquisition of upright locomotion (Clark *et al.,* 1988; Forrester *et al.,* 1993) as well as for the study of age-related changes in clapping (Fitzpatrick *et al.,* 1996).

Relative phase, however, cannot be applied to discrete movements as their motions are not sinusoidal, which has meant that the dynamical approach to biological coordination (rather than control) has been largely restricted to rhythmical movements. The problem is to find an order parameter for discrete movements that is analogous to relative phase and which can be modelled dynamically. One such candidate for intralimb coordination could be relative timing (e.g. between shoulder and elbow), which has the advantage that it remains invariant over changes in movement amplitude and speed, at least for adults (e.g. Soechting & Lacquanti, 1981; Jeannerod, 1984).

Another candidate may involve the spatiotemporal organization of discrete movements such as the speed–curvature relationship (Abend et $al.$, 1982), which indicates where, in a movement trajectory, there is a change in direction. From studies on infant reaching it would appear that this relationship can be interpreted as a developmental invariant (e.g. Fetters & Todd, 1987; von Hofsten, 1991). The intriguing thing about this relational parameter is that it is evident in the arm movements of newborns seated semi-upright, even in the absence of a visual target (von Hofsten & Rönnqvist, 1993). This suggests that the intrinsic dynamics for later goal-directed actions are already present in the newborn. Findings in the same study provide a clue as to an order parameter for interlimb coordination during discrete movements, namely, movement coupling between the two arms in all three dimensions of space (see also Thelen et $al.$, 1993). This also can be considered to be a developmental invariant, but one that undergoes changes in the degree of coupling between the right and left arms during the first year (see Goldfield & Michel, 1986). In addition, it is probably more task-specific than the other potential order parameters for reaching, in that differences in object size may determine whether an infant displays unimanual or bimanual reaching.

Coupling problem

One of the greatest challenges to the provision of a direct realist theory of perceiving and moving is the missing or lost dimension problem (Beek & van Wieringen, 1994). The problem is this: information has the dimensions of length (L) and time (T) from which kinematic parameters such as velocity (LT^{-1}) or acceleration (LT^{-2}) and geometrical quantities such as the shape of a parabolic trajectory (L^2) can be derived, but it has no real mass (M). How then can kinematics specify kinetics (Runeson & Frykholm, 1983)? How can something, which does not contain mass (and thus force, F), specify how an effector system should produce the relevant magnitudes and sequence of forces for a particular task?

The implications of resolving this problem for studying development from the NPA should be made clear. New forms develop through a circular causality between information contained in the low-energy kinematic flow fields and that in the high-energy force fields. Non-linear dynamical entities analogous to coordinative structures such as intentions are formed across these informational fields during development. However, understanding their formation properly cannot be achieved until the missing dimension problem is resolved. As this problem brings with it the problem of intentionality and how it develops, it is effectively taken beyond the purview of Newtonian mechanics. It is the incursion of task-defined intentions into the perception–movement circle that creates the circumstances under which perception benefits from movement and vice-versa.

One solution to the missing dimension problem centres on the concept of behavioural information and the claim that order parameters contain action-specific information (Schöner & Kelso, 1988). According to this solution, order parameters are a veritable storehouse of dynamical behaviours such as metasta-bility and equifinality and also bimodality, critical fluctuations and hysteresis. Each one of these is a potential source of information concerning the stability and loss of stability of the system. As we saw, these macroscopical behaviours are induced by non-specific changes in control parameters at the more microscopical level. However, they may have specific effects on the intrinsic dynamics of the order parameter. Behavioural information has been proposed as such a specific control parameter. It has been defined as a dynamical entity or required behavioural pattern that specifically perturbs the intrinsic dynamics (Schöner & Kelso, 1988). In this sense, it is analogous to Gibson's (1966) notion of perceptual information as specification, but differs from it in that behavioural information is part and parcel of the same coordination dynamics as the order parameter. Thus, there is no need to map kinematics on to kinetics and thus no missing dimension problem.

Kelso (1990) proposed that how behavioural information becomes both specific and meaningful for particular functional tasks is a salient question to pose about infant development. If behavioural information is weak, or even absent, then the infant's behaviour will be determined by intrinsic dynamics. Through discovering the effects of this autonomous behaviour on the environment, behavioural information is generated that may compete with the intrinsic dynamics leading, for example, to fluctuations in the order parameter, and thus the potential for qualitative change. The concept of behavioural information has been successfully applied to adults learning a required non-equilibrium phase transition in a bimanual coordination task paced by a visual metronome (Zanone & Kelso, 1993). It is very much an open question if it can be applied with equal success to developmental transitions.

There is another consideration associated with the coupling problem that concerns the nature of the relationship between movement and posture. The development of actions such as reaching and walking is faced with what is known as the dual state problem (Feld'baum, 1965): discovering the dynamics of coordination while at the same time learning to control the system. Thus, for example, in acquiring the action of reaching, the infant has to find a solution to the task-specific problem of coordinating movements so as to get the hand to the object while at the same time maintaining postural stability. If control and coordination can be derived from different solutions to the same dynamics, then this problem is considerably simplified. Schöner (1990) attempted to do exactly this by means of dynamical modelling. He showed that posture and both rhyth-mical and discrete movements could emerge as different solutions to the same oscillatory (mass-spring) dynamics. It may be possible, therefore, to provide a single dynamical account of movement and posture as both may be part of the same control system – actually not such a new insight as its origins can be traced back to Bernstein (1967). Developmentally, this implies that infants will search for the dynamics that enable them to bring posture into the service of move-ments in achieving task-specific actions – with posture specifically providing stability of segments and of the whole body as a means of making anticipatory or corrective adjustments (see Hadders-Algra, this handbook).

Brain–body problem

The NPA acknowledges that the control and coordination of action ultimately depends on neural mechanisms and that the brain itself is a self-organizing dynamical system (see Haken, 1996). Bernstein (1967) captured how the brain–body problem should be approached in the following way: "the reorganization of the movement begins with its biomechanics...; this biomechanical reorganization sets up new problems for the central nervous system, to which it gradually adapts" (pp. 87–89). One implication of this radical viewpoint is that transformations in the development of action occur not only because of changes in the central nervous system. In addition, there are changes in the biomechanical properties of body segments to which the developing brain adapts. Accordingly, throughout development, there is a continuous and reciprocal interaction between the central nervous system and the musculoskeletal system, and between the latter and the environment. Whether this scenario is applicable to all aspects of motor development is a debatable point. Take, for example, the development of direct corticospinal connections (see Armand, this handbook) and their association with the acquisition of fractionated finger movements (see Gordon, this handbook). In this instance, the requisite distal segments do not undergo any marked changes in their biomechanical properties during the first year of postnatal life as far as we know and as such do not present a 'problem' of adaptation for the central nervous system.

The role of the central nervous system in both control and coordination is explicitly catered for in the NPA as a means of overcoming the storage problem (*viz.* how can the brain store the exact specifications of all movements?). It is treated as a special-purpose device (Runeson, 1977) that transforms dynamical processes (for coordination) and mechanical processes (for control) into task-appropriate actions. In doing so, it does not specify (or store) all the details of control and coordination, an assertion supported by von Holst's (1937/1973) centipede experiment. Rather, it specifies the global dynamics required for maintaining stability against external perturbations or for shifting the system to another coordinative mode (see Kelso *et al.,* 1992). It also complements, but does not control, the mechanical forces generated by the limbs (e.g. as in mass-spring models, which stress that control is shared by the natural resonant properties of the limbs; see Feldman & Levin, 1995). To treat the central nervous system as the only source of control is to commit the Fallacy of Misplaced Concreteness (Bullock & Grossberg, 1991). Developmentally, what is at issue is how changes in the control of movement and posture may engender changes in the coordination of action. Put in a familiar context, the issue is how changes in one or more control parameters result in changes in the dynamics of an order parameter. The task for the developmentalist is to identify not only an appropriate order parameter, but also age-appropriate control parameters that are ultimately open to some form of experimental manipulation.

Now, I have reached the end of the process of putting my thoughts on paper. Being unable to do this without a pen in my hand, the next step is to type them into my word processor. Apparently, in order to facilitate my thought processes, I need to experience a direct and variable interaction between my body and the material world (and which in this specific instance is better afforded by the act of writing than of typing). In this respect at least I am not much different from a developing infant.

References

Abend, W., Bizzi, E. & Morasso, P. (1982). Human arm trajectory formation. *Brain*, **105**, 331–348.

Adolph, K.E. (1997). Learning in the development of infant locomotion. *Monographs of the Society for Research in Child Development*, Serial no. 251, vol. 62, no. 3.

Adolph, K.E., Eppler, M.A. & Gibson, E.J. (1993). Development of perception of affordances. In C. Rovee-Collier & L.P. Lipsitt (Eds.), *Advances in infancy research* (vol. 8, pp. 51–98). Norwood, NJ: Ablex.

Beek, P.J. & Wieringen, P.C.W (1994). Perspectives on the relation between information and dynamics: dialogue. *Human Movement Science*, **13**, 519–533.

Bernstein, N.A. (1967). *The coordination and regulation of movement*. Oxford: Pergamon.

Bertenthal, B.I. & Bai, D.L. (1989). Infant's sensitivity to optical flow for controlling posture. *Developmental Psychology*, **25**, 936–945.

Bhatias, N.P. & Szego, G.P. (1970). *Stability theory of dynamical systems*. New York: Pergamon.

Brainerd, C.J. (1978). The stage question in cognitive developmental theory. *Behavioral and Brain Sciences*, **2**, 173–213.

Bullock, D. & Grossberg, S. (1991). Adaptive neural networks for control of movement trajectories invariant under speed and force rescaling. *Human Movement Science*, **10**, 3–53.

Butterworth, G. (1993). Dynamic approaches to infant perception and action: old and new theories about the origins of knowledge. In L.B. Smith & E. Thelen (Eds.), *A dynamic systems approach to development: applications* (pp. 171–187). Cambridge, MA: MIT Press.

Chiel, H.J. & Beer, R.D. (1997). The brain has a body: adaptive behaviour emerges from interactions of nervous system, body and environment. *Trends in Neuroscience*, **20**, 553–557.

Clark, A. (1997). *Being there: putting the brain, body, and world together again*. Cambridge, MA: MIT Press.

Clark, J.E., Whitall, J. & Phillips, S.J. (1988). Human interlimb coordination: the first six months of independent walking. *Developmental Psychobiology*, **21**, 445–456.

Cobb, L. & Zacks, S. (1985). Applications of catastrophe theory for statistical modeling in the biosciences. *Journal of the American Statistical Association*, **80**, 793–802.

Connell, J.P. & Furman, W. (1984). The study of transitions. In R.J. Emde & R. Harmon (Eds.), *Continuities and discontinuities in development* (pp. 153–173). New York: Plenum Press.

Edelman, G.M. (1989). *The remembered present: a biological theory of consciousness*. New York: Basic Books.

Feld'baum, A.A. (1965). *Optimal control systems*. New York: Academic Press.

Feldman, A.G. & Levin, M.F. (1995). The origin and use of positional frames of reference in motor control. *Behavioral and Brain Sciences*, **18**, 723–806.

Ferrier, D. & Yeo, G.F. (1881). The functional relations of the motor roots of the brachial and lumbosacral plexuses. *Proceedings of the Royal Society of London*, **32**, 12–20.

Fetters, L. & Todd, J. (1987). Quantitative assessment of infant reaching movements. *Journal of Motor Behavior*, **19**, 147–166.

Fitzpatrick, P., Schmidt, R.C. & Lockman, J.J. (1996). Development of clapping behaviour. *Child Development*, **67**, 2691–2708.

Fontaine, R. & Pieraut-LeBonniec, G. (1988). Postural evolution and integration of the prehension gesture in children aged 4 to 10 months. *British Journal of Developmental Psychology*, **6**, 223–233.

Forrester, L.W., Phillips, S.J. & Clark, J.E. (1993). Locomotor coordination in infancy: the transition from walking to running. In G.J.P Savelsbergh (Ed.). *The development of coordination in infancy* (pp. 359–393). Amsterdam: North-Holland.

Geerdink, J.J., Hopkins, B. & Hoeksma, J.B. (1994). The development of head position preference in preterm infants beyond term age. *Developmental Psychobiology*, **27**, 153–168.

Gibson, J.J. (1966). *The senses considered as perceptual systems*. Boston: Houghton-Mifflin.

Gibson, J.J. (1979). *The ecological approach to visual perception*. Boston: Houghton-Mifflin.

Gilmore, R. (1981). *Catastrophe theory for scientists and engineers*. New York: Wiley.

Goldfield, E.C. & Michel, G.F. (1986). Spatio-temporal linkage in infant interlimb coordination. *Developmental Psychobiology*, **19**, 259–264.

Haken, H. (1977). *Synergetics: an introduction*. Berlin: Springer.

Haken, H. (1996). *Principles of brain functioning: a synergetic approach to brain activity, behaviour, and cognition*. Berlin: Springer.

Hartelman, P.A.I., van der Maas, H.L.J. & Molenaar, P.C.M. (1998). Detecting and modelling developmental transitions. *British Journal of Developmental Psychology*, **16**, 97–122.

Hopkins, B., Beek, P.J. & Kalverboer, A.F. (1993). Theoretical issues in the longitudinal study of motor development. In A.F. Kalverboer, B. Hopkins & R. Geuze (Eds.), *Motor development in early and later childhood: longitudinal approaches* (pp. 343-371). Cambridge, UK: Cambridge University Press.

Jeannerod, M. (1984). The timing of natural prehension movements. *Journal of Motor Behavior*, **16**, 235–254.

Kelso, J.A.S. (1984). Phase transitions and critical behaviour in human bimanual coordination. *American Journal of Physiology: Regulative, Integrative and Comparative Physiology*, **246**, R1000–R1004.

Kelso, J.A.S. (1990). Phase transitions: foundations of behaviour. In H. Haken & M. Stadler (Eds.), *Synergetics of cognition* (pp. 249–268). Berlin: Springer.

Kelso, J.A.S. (1994). The informational character of self-organized coordination dynamics. *Human Movement Sciences*, **13**, 393–413.

Kelso, J.A.S., Bressler, S.L., Buchanan, S., De Guzman, G.C., Ding, M., Fuchs, A. & Holroyd, T. (1992). A phase transition in human brain and behaviour. *Physics Letters A*, **169**, 134–144.

Koendrink, J.J. & van Doorn, A.J. (1981). Exterospecific component of the motion parallax field. *Journal of the Optical Society of America*, **71**, 953–957.

Konczak, J., Borutta, M., Topka, H. & Dichgans, J. (1995). The development of goal-directed reaching in infants: hand trajectory formation and joint torque control. *Experimental Brain Research*, **106**, 156–168.

Kugler, P.N. (1986). A morphological perspective on the origin and evolution of movement patterns. In M.G. Wade & H.T.A. Whiting (Eds.), *Motor development in children: aspects of coordination and control* (pp. 77–106). Dordrecht: Martinus Nijhoff.

Kugler, P.N., Kelso, J.A.S. & Turvey, M.T. (1980). On the concept of coordinative structures as dissipative structures: I. Theoretical lines of convergence. In G.E. Stelmach & J. Requin (Eds.), *Tutorials in motor behaviour* (pp. 3–47). Amsterdam: North-Holland.

Kugler, P.N., Kelso, J.A.S. & Turvey, M.T. (1982). On coordination and control in naturally developing systems. In J.A.S. Kelso & J.E. Clark (Eds.), *The development of movement coordination and control* (pp. 5–78). New York: Wiley.

Kugler, P.N., Shaw, R.E, Vincente, K.J. & Kinsella-Shaw, J. (1990). Inquiry into intentional systems I: Issues in ecological physics. *Psychological Research*, **52**, 98–121.

Kugler, P.N. & Turvey, M.T. (1987). *Information, natural law, and the self-assembly of rhythmic movement: theoretical and experimental investigations*. Hillsdale, NJ: Erlbaum.

Lee, D. N. (1976). A theory of visual control of braking based on information about time-to-collision. *Perception*, **5**, 437–459.

Lee, D.N. (1998). Guiding movement by coupling taus. *Ecological Psychology*, **10**, 221–250.

Lee, D.N. & Aronson, E. (1974). Visual proprioceptive control of standing in human infants. *Perception and Psychophysics*, **15**, 529–532.

Mark, L.S. & Vogele, D. (1988). A biodynamical basis for perceiving categories of action: a study of sitting and stairclimbing. *Journal of Motor Behavior*, **19**, 367–384.

May, R. (1976). Simple mathematical models with very complicated dynamics. *Nature*, **261**, 459–467.

McCauley, J.L. (1993). *Chaos, dynamics and fractals: an algorithmic approach to deterministic chaos*. Cambridge, UK: Cambridge University Press.

McGraw, M. (1945). *The neuromuscular maturation of the human infant*. New York: Columbia University Press.

Morrison, F. (1991). *The art of modeling dynamic systems*. New York: Wiley.

Nagel, E. (1961). *The structure of science*. London: Routledge & Kegan Paul.

Newell, K.M. (1986). Constraints on the development of coordination. In M.G. Wade & H.T.A. Whiting (Eds.), *Motor development in children: aspects of coordination and control* (pp. 341–360). Dordrecht: Martinus Nijhoff.

Nicolis, G. & Prigogine, I. (1989). *Exploring complexity: an introduction*. New York: Freeman.

Out, L., van Soest, A. J., Savelsbergh, G.J.P. & Hopkins, B. (1998). The effect of posture on early reaching movements. *Journal of Motor Behavior*, **30**, 260–272

Piaget, J. (1952). *The origins of intelligence in children*. New York: Norton.

Prechtl, H.F.R. (Ed.) (1984). *Continuity of neural functions from prenatal to postnatal life*. Oxford: Blackwell.

Rochat, P. (1992). Self-sitting and reaching in 5-to-8 month-old infants: the impact of posture and its development on early eye hand coordination. *Journal of Motor Behavior*, **24**, 210–220.

Runeson, S. (1977). On the possibility of "smart" perceptual mechanisms. *Scandinavian Journal of Psychology*, **18**, 172–180.

Runeson, S. & Frykolm, G. (1983). Kinematic specification of dynamics as an informational basis for person and ation perception: expectation, gender, recognition, and deceptive intention. *Journal of Experimental Psychology: Human Perception and Performance*, **16**, 227–247.

Schmidt, R.C. & O'Brien, B. (1998). Modeling interpersonal coordination dynamics: implications for a dynamical theory of developing systems. In K.M. Newell & P.C.M. Molenaar (Eds.), *Applications of nonlinear dynamics to developmental process modeling* (pp. 221–240). Mahwah, NJ: Erlbaum.

Schneider, K., Zernicke, R.F. Ulrich, B.D. & Jensen, J.L. (1990). Understanding movement control in infants through the analysis of limb intersegmental dynamics. *Journal of Motor Behavior*, **22**, 493–520.

Schöner, G. (1990). A dynamic theory of coordination of discrete movements. *Biological Cybernetics*, **63**, 257–270.

Schöner, G., Haken, H. & Kelso, J.A.S. (1986). A stochastic theory of phase transitions in human hand movement. *Biological Cybernetics*, **53**, 247–257.

Schöner G. & Kelso, J.A.S. (1988). A dynamic pattern theory of behavioural change. *Journal of Theoretical Biology*, **135**, 501–524.

Sellars, R.W. (1973). *Neglected alternatives*. Lewisburg, PA: Bucknell University Press.

Shaw, R.E. & Turvey, M.T. (1981). Coalitions as models for ecosystems: a realist perspective on perceptual organization. In M. Kubovy & J.R. Pomerantz (Eds.), *Perceptual organization* (pp. 343–408). Hillsdale, NJ: Erlbaum.

Singer, W. (1986). The brain as a self-organizing system. *European Archives of Psychiatry and Neurological Sciences*, **236**, 4–9.

Slater, A. & Morrison, V. (1985). Shape constancy and slant perception at birth. *Perception*, **14**, 337–344.

Snyder, R.G., Schneider, L.W., Owings, C.L., Reynolds, H.M., Golumb, D.H. & Schork, M.A. (1977). *Anthropometry of infants, children and youths to age 18 for product safety design, SP-450*. Warrendal, PA: Society of Automotive Engineers.

Soechting, J.E. & Lacquanti, F. (1981). Invariant characteristics of a pointing movement in man. *Journal of Neuroscience*, **1**, 710–720.

Soll, D. (1979). Timers in developmental systems. *Science*, **203**, 841–849.

Stoffregen, T.A. (1985). Flow structure versus retinal location in the optical control of stance. *Journal of Experimental Psychology: Human Perception and Performance*, **11**, 554–565.

Swenson, R. & Turvey, M.T. (1991). Thermodynamic reasons for perception-action cycles. *Ecological Psychology*, **3**, 317–348.

Thatcher, R.W. (1998). A predator-prey model of human cerebral development. In K.M. Newell & P.C.M. Molenaar (Eds.), *Applications of nonlinear dynamics to developmental process modeling* (pp. 67–128). Mahwah, NJ: Erlbaum.

Thelen, E. (1985). Developmental origins of motor coordination: leg movements in human infants. *Developmental Psychobiology*, **18**, 1–22

Thelen, E. (1995). Motor development: a new synthesis. *American Psychologist*, **50**, 79–95.

Thelen, E., Corbetta, D., Kamm, K., Spencer, J.P., Schneider, K. & Zernicke, R.F. (1993). The transition to reaching: mapping intention and intrinsic dynamics. *Child Development*, **64**, 1058–1098.

Thom, R. (1975). *Structural stability and morphogenesis: an outline of a general theory of models*. (D.H. Fowler, trans). Reading, MA: Benjamin.

Tresilian, J.R. (1993). Four questions of time to contact: a critical examination of research on interceptive timing. *Perception,* **22**, 653–680.

Touwen, B.C.L. (1976). *Neurological development in infancy*. London: Heinemann.

Turvey, M.T., Shaw, R.E. & Mace, W. (1978). Issues in theory of action: degrees of freedom, coordinative structures and coalitions. In J. Requin (Ed.), *Attention and performance. VII* (pp. 557–598). Hillsdale, NJ: Erlbaum.

Ulrich, B., Thelen, E. & Niles, D. (1990). Perceptual determinants of action: stair-climbing choices in infants and toddlers. In J.E. Clark & J. Humphrey (Eds.), *Advances in motor development research*. vol. 3. New York: AMS Publishers.

Vaal, J., van Soest, A.J., Hopkins, B., Sie, L.T.L. & van der Knaap, M.S. (2000). Development of spontaneous leg movements in infants with and without periventricular leukomalacia. *Experimental Brain Research*, **135**, 94–105.

van der Maas, H.L.J. (1993). Catastrophe analysis of stagewise cognitive development. Academic thesis, Universiteit van Amsterdam.

van der Maas, H.L.J. & Hopkins, B. (1998). Dynamical systems theory: so what's new? *British Journal of Developmental Psychology*, **16**, 1–13.

van der Maas, H.L.J. & Molenaar, P.C.M. (1992). Cognitive development: an application of catastrophe theory. *Psychological Review*, **99**, 395–417

van der Meer, A.L.H., van der Weel, F.R. & Lee, D. N. (1995). The functional significance of arm movements in neonates. *Science*, **267**, 693–695.

von Hofsten, C. (1991). Structuring of early reaching movements: a longitudinal study. *Journal of Motor Behavior,* **23**, 280–292.

von Hofsten, C. & Rönnqvist, L. (1993). The structuring of neonatal arm movements. *Child Development*, **64**, 1046–1057.

von Holst, E. (1937/1973). *The behavioural physiology of animals and man: the collected papers of Erich von Holst,* Vol. 1 (R. Martin, ed. and trans.). Coral Gables, FL: University of Miami Press.

Warren, W. H. (1988). Action modes and laws of control for the visual guidance of action. In O.G. Meijer & K. Roth (Eds), *Complex movement behaviour: the 'action-motor' controversy* (pp. 339–380). Amsterdam: North-Holland.

Warren, W.H. & Whang, S. (1987). Visual guidance of walking through apertures: body-scaled information for affordances. *Journal of Experimental Psychology: Human Perception and Performance*, **13**, 371–383.

Wimmers, R.H., Savelsbergh, G.J.P., van der Kamp, J. & Hartelman, P (1998a). A developmental transition in prehension modeled as a cusp castastrophe. *Developmental Psychobiology*, **32**, 23–35.

Wimmers, R.H, Savelsbergh, G.J.P, Beek, P.J. & Hopkins, B. (1998b). Evidence for a phase transition in the early development of prehension. *Developmental Psychobiology,* **32**, 235–248.

Wolff, P.H. (1986). The maturation and development of fetal motor patterns. In M.G. Wade & H.T.A. Whiting (Eds.), Motor development in children: aspects of coordination and control (pp. 65–74). Dordrecht: Martinus Nijhoff.

Wunderlin, A. (1987). On the slaving principle. *Springer Proceedings in Physics*, **19**, 140–147.

Yonas, A. & Granrud, C.E. (1985). Reaching as a measure of infants' spatial perception. In G. Gottlieb & N. Krasnegor (Eds.), *Measurement of audition and vision in the first year of life: a methodological overview* (pp. 301–322). Norwood, NJ: Ablex.

Yonas, A. & Hartman, B. (1993). Perceiving the affordance of contact in four- and five-month-old infants. *Child Developmen ,* **64**, 298–308.

Yonas, A. & Owsley, C. (198). Development of visual space perception. In P. Salapatek & L. Cohen (Eds.), *Handbook of infant perception,* vol. 2: *From perception to cognition*. Orlando, FL: Academic Press.

Zanone, P.G., Kelso, J.A.S. & Jeka, J.J. (1993). Concepts and methods for a dynamical approach to behavioural coordination and change. In G.J.P. Savelsbergh (Ed.), *The development of coordination in infancy* (pp. 89–135). Amsterdam: North-Holland.

Zernicke, R.F. & Schneider, K. (1993). Biomechanics and developmental neuromotor control. *Child Development*, **64**, 982–1004.

Further reading

Brown, C. (1995). *Chaos and castastrophe theories*. London: Sage.

Coveney, P. & Highfield, R. (1990). *The arrow of time: the quest to solve science's greatest mystery*. London: Allen.

Freeman, W.J. (1999). *How brains make up their minds*. London: Weidenfeld & Nicolson.

Goldfield, E.C. (1995). *Emergent forms: origins and early development of human action and perception*. Oxford: Oxford Univeristy Press.

Kaplan, D. & Glass, L. (1995). *Understanding nonlinear dynamics*. New York: Springer.

Kelso, J.A.S. (1995). *Dynamic patterns: the self-organization of brain and behaviour*. Cambridge, MA: MIT Press.

Latash, M.L. & Turvey, M.T. (Eds.) (1996). *Dexterity and its development*. Mahwah, NJ: Erlbaum.

Thelen, E., Schöner, G., Scheier, C. & Smith, L.B. (2000). The dynamics of embodiment: a field theory of infant perseverative reaching. *Behavioral and Brain Sciences*, (In press).

IV. DEVELOPMENT OF PERCEPTION AND COGNITION

IV.1
Introduction: Development of Perception and Action

BRIAN HOPKINS

Distinctions between perception and cognition are difficult to make, which is probably the reason why there have been few attempts to do so (however, see Hopkins & Kalverboer, 1983). At the risk of oversimplification, perception can be regarded as an active process of discriminating, recognizing and identifying something, and cognition as psychological processes for acquiring knowledge and understanding of the world. Processes typically ascribed to cognition are, for example, remembering, imagining, conceiving, thinking and reasoning. Attention, however, is a construct catered for in various ways by theories of perception and cognition as is the concept of state. Thus, not surprisingly, we find (visual) attention and (infant) state as recurring themes in this section.

Another expectable theme running through the chapters is the relationship between brain and behaviour mainly, but not only, during infant development. Each chapter testifies to the progress that has been made in understanding this relationship while at the same time acknowledging that there are still fundamental issues to be resolved (e.g. the binding problem). Much of the progress in recent years has been due to technological advances in registering brain activity in human and animal children while engaged in tasks designed to assess perceptual and cognitive functions. The potential danger of such advances can be a (benign) neglect of the essential enterprise of theory-building about the development of brain–behaviour relationships. To some extent this enterprise is currently a salient feature of the burgeoning field of developmental cognitive neuroscience, which despite its name also addresses the development of perceptual functions. Its impact is readily discernible in most of the ensuing chapters.

The chapter by **Sireteanu** is devoted to the structural and functional development of the visual system during the period of infancy. It deals with the development of low-level visual functions (e.g. visual acuity) and high-level visual functions (e.g. segmentation and integration of visual information). The latter set of functions raises the binding problem, namely the mechanisms by which features of objects such as colour, location and shape are integrated to form coherent wholes. The notion of a "grandmother cell" in which sensory signals

A.F. Kalverboer and A. Gramsbergen (eds.), Handbook of Brain and Behaviour in Human Development, 623–628
© 2001 Kluwer Academic Publishers. Printed in Great Britain.

converge on to a neuron dedicated to the job of identifying their combinations so as to transform them into percepts is no longer valid. Most solutions now involve some form of temporal synchrony between neurons in a network that is context-dependent (e.g. see Roelfsema *et al.,* 1997). Nevertheless, we are far from a generally accepted solution to the binding problem and especially one that addresses it within a developmental framework. In terms of structural development, research on the pathways from the magnocelluar (M) and parvocelluar (P) layers of the lateral geniculate nucleus to the primary visual cortex are related to the distinction between the ventral ("vision-for-perception" via the P pathway) and dorsal ("vision-for-action" via the M pathway) visual streams. Once again we are confronted with the binding problem, but now between rather than within systems. How does a system specialized for the identification of objects (*viz.* the ventral stream) cooperate with one involved in controlling actions to objects arranged in an egocentric spatial framework (*viz.* the dorsal stream)? Evidence is cited from a PET scan study with macaques that the dorsal M pathway develops earlier than the ventral P pathway, which suggests that goal-directed actions may be functional before the ability to integrate complex visual information. That, of course, depends on what one means by "actions" and "perceptual integration" within a developmental perspective. Citing other, behaviourally based, evidence, the next chapter arrives at the opposite conclusion, namely, that the P pathway develops more rapidly than the magnocelluar stream.

The ventral–dorsal stream distinction also figures in the chapter by **Johnson**, as does that between low- and high-level visual functions. While there is some overlap in content with the previous chapter, it differs in a number of ways. One difference is a critical consideration of the methods available for appraising both sets of functions in young infants. Another is a coverage of topics that range from oculomotor functions (e.g. saccadic eye movements and those concerned with the maintenance of visual stability) through attentional mechanisms to abilities that straddle the border between perception and cognition (e.g. the perception of object unity). In dealing with each of these topics, the question is asked whether a particular function or ability is present in the newborn or at some later age. Clearly, many of the low-level attributes are evident in a rudimentary form at birth, but show great improvements during the subsequent 4 months. How they contribute to, or rather constrain, the acquisition of higher-level abilities such as entailed in object perception remains a somewhat neglected issue in research on infant visual development. Inevitably, the binding problem rears its head again when considering age-related differences in how infants respond to the object unity task. Within-system solutions along the lines of the temporal synchrony hypothesis are reviewed, and hints given as to the developmental mechanisms that might be involved. The chapter concludes by identifying the need for investigating this hypothesis in terms of the development of synchro-nized activity across cortical areas, Other recommendations for future research include accounting for individual differences in the functional development of the dorsal and ventral streams as well as with regard to the role of temperament and state when infants are confronted with attentionally demanding tasks.

A between-systems binding problem lies at the heart of the chapter by **King, Schnupp, Doubell** and **Baron**. Specifically, how are vision (sensitive to spatiotemporal change) and audition (sensitive to spectrotemporal change) combined in the guidance of actions such as the localization of objects in space and how does such cross-modal integration develop? The model system for answering such questions is the superior colliculus (SC) in mammals and its homologue in avian species, the optic tectum. However, neural substrates implicated in the inter-action of auditory and visual information are widely distributed. They include not only the SC, but also the inferior colliculus, premotor neurons in the brainstem and spinal cord that receive inputs from the SC, the pulvinar (visual) nuclei with indirect connections to the SC, the basal ganglia and the frontal eye fields as well as the posterior parietal and prefrontal cortices. The SC has the advantage of being a laminated structure in which there is a rather precise topographical mapping between its superficial and deeper layers. The superficial layers receive only visual information and contain a map of visual space, while neurons in the deep layers are the recipients of converging visual, auditory and somatosensory inputs and form a motor map for the control of gaze direction. Developmentally, the superficial layers assume an adult-like map, seemingly without visual experience, before the deeper layers – at least as far as the ferret and rat are concerned. Human infants, however, appear to be able to integrate visual and auditory information at birth under certain conditions, which would suggest the neural foundations for this ability are developed prenatally. Subsequent development is complicated by a rapid growth in head size in the human infant. Having a mixture of altricial and precocial traits at the birth, the human infant can be expected to follow a different developmental course in the ability to integrate information across the auditory and visual modalities compared to other species. This is substantiated by research in which the animal model concerns the cat, the ferret the rat or the barn owl.

The theme of species differences in development should be borne in mind when reading the chapter by **Bolhuis**. Here, a wealth of material is presented on filial imprinting and song learning in avian species, both of which raise the ever-green problem of sensitive periods. Parallels are drawn between filial imprinting in the chick and face discrimination in human newborns as shown by their preference to track moving face-like patterns rather than ones that do not resemble a face. However, this preference diminishes some 2–3 months later, a finding that signals circumspection in drawing parallels with the development of filial imprinting. Song learning in birds is dealt with in terms of its potential neural substrates and once again parallels are made with human development – this time with regard to the development of speech perception and preverbal vocalizations, and the lateralization of these functions. As for bird song, recent evidence suggests dissociable brain regions for perception and production as revealed by the protein products of so-called immediate early genes (given the remarkable claims made by Pepperberg, 1999, for the speech abilities of the African grey parrot, it is a mute point as to whether song birds are the best source of comparison in these respects). The pitfalls associated with the use of the concept of sensitive periods are captured in two ways: first, by pointing out that they concern two types of explanation at three different levels of analysis and secondly, through examples drawn from research on filial imprinting and bird song learning. At a more general level, timely reminders are issued about the stultifying effects on the study of development of continuing to adhere to such outdated concepts as innateness and dichotomies that adhere to variants of the distinction between nature and nurture.

Salzarulo, Giganti, Fagioli and **Ficca** in their chapter address a neglected topic, but one central to our understanding of perceptual and cognitive development. This is how sleep and wake states develop during the period of infancy. States have been identified in terms of biophysical measures, especially using EEG, and by means of behavioural indicators. Each state is taken to represent a qualitatively different mode of nervous activity as each one manifests different input–output relationships. EEG states and behavioural states differ as to the labels applied. For example, the behavioural states 1 and 2 are classified as quiet sleep and paradoxical sleep, respectively, using criteria derived from EEG measures. This difference sometimes leads to confusion as to what is being described and how it is to be explained, and the problem becomes particularly pertinent when studying states in preterm infants before term age. Another source of confusion is drowsiness: does it constitute a state or period of transition between states? Given its relatively short duration and lack of stability, it is better treated as a transitional phenomenon. Turning from sleep to wake states, we find a lack of empirical studies dealing with how they develop and whether new ones emerge after the newborn period. The problem here is that present behavioural classifications of wake states are really applicable only to the first 2–3 months after birth. Beyond this age, state and attention become inextricably intertwined, as discussed in the present chapter. Also discussed are the age-dependent effects of wakefulness on sleep and age-related changes in the effectiveness of external influences on (mainly sleep) state regulation. The latter include the effects of ambient temperature, nutrition, parental behaviour and co-sleeping, pacifier use and vestibular and visual stimulation. What they show is that, when states are perturbed, they have the capacity to maintain their stability or to self-organize into a new state.

The concept of self-organization forms the theoretical heart of the chapter by **Mareschal** and **Thomas**. With this concept as a reference point, the strengths and weaknesses of connectionist modelling in simulating general and more specific features of cognitive development are reviewed. An important claim is that such models allow one to address a central developmental issue in a principled manner, namely, the emergence of new properties. This issue embraces the cognitive or learning paradox, a persistent problem in accounting for transformations in cognitive development (*viz.* how can new and more powerful structures emerge in a system that does not contain the information to describe them?). Representational re-description is one solution to this problem and is portrayed as a principle of self-organization that "drives" cognitive development. Examples of applying connectionist models to specific aspects of cognitive development involve category learning and object permanence in infants, and weight conservation and seriation in older children. Specific applications to language development (e.g. phonology; vocabulary) and to developmental disorders (e.g. autism; Williams syndrome) are also considered. Missing are connectionist models applied to the development of interpersonal communication (see Olthof *et al.*, 2000). If we are to better understand how the concept of self-organization can account for developmental change, then the sorts of simulation models discussed can provide a useful first step in this endeavour. A crucial test of their relevance is whether they can account for the acquisition of new abilities without any *a-priori* instructions.

The next chapter, by **Welsh**, attempts the formidable task of relating what is known about the structural development of the prefrontal cortex (PFC) to the development of executive functions (EFs). A major obstacle in this respect is the lack of a commonly accepted definition of EFs and appropriate tests for evaluating them in young children. These functions are assumed to consist of a number of component abilities, chief among which is working memory. A further complication is that there are contrasting views on how the PFC develops, due possibly to the fact that one is based on human data and the other on findings with monkeys. In humans, synaptogenesis in the PFC appears to be gradual while for the rhesus monkey this process occurs simultaneously across the PFC and other cortical areas. A model more in keeping with human evidence is adopted, which has been applied to the development of EEG coherence. In this model cortical development undergoes cyclical reorganizations that are to some extent controlled by the PFC. Data from five studies on the development of EEG coherence show some conformity with the multistage pattern of development described by the model. Achieving a closer correspondence between this model and developmental changes in cortical synaptogenesis will benefit from suggestions made in this chapter. The same applies to the problems of defining EFs and deriving age-appropriate tasks that capture developmental changes in terms of their essential features. What also needs to be done is to identify cognitive abilities during infancy that constitute precursors of EFs appearing later in development, As yet, we lack a testable theory for tackling this particular problem.

In the final chapter, by **Kalverboer**, we are confronted with the precursor problem in another guise. What are the early antecedents of developmental disorders and how best can they be identified and measured? Answers to these questions are exemplified with reference to children who manifest that well-researched, but ill-defined, clinical entity known as the attention deficit hyperactivity disorder (ADHD). One set of antecedents is sought in a three-component model of visual attention (see Berger & Posner, this handbook). That component referred to as the posterior attention system is singled out for special consideration. Consisting of two attentional mechanisms, labelled "disengagement" and "inhibition of return", research conducted by the author and colleagues suggests that they do not develop synchronously in healthy infants between the ages of 6 weeks and 6 months. The implications of this finding for infants at risk for developing attentional deficits are presently unclear, but worthy of further investigation if only because of the failure of stages-of-information processing approaches in pinpointing the nature of such deficits. Another set of antecedents is couched in terms of "state regulation deficiencies" in such processes as arousal, activation and evaluation. This is quite a different approach to the concept of state espoused by Salzarulo *et al.* A combination of this approach to state with that described for visual attention is recommended as one of the goals to be achieved in future research on the antecedents of deviant developmental outcomes such as ADHD. Moreover, the dearth of longitudinal studies in this area of research is badly in need of rectification if we are to gain any relevant insights into the risk mechanisms or causal pathways that eventuate in such outcomes.

In conclusion, the chapters in this section point to certain problems and themes that will probably assume growing importance in the study of perceptual and cognitive development. These include the binding problem, how the distal and ventral streams co-develop, the theoretical integration of concepts relating attention and state, the role of environmentally induced gene activity in promoting plasticity in relevant neural structures, and the provision of further clarifications about how self-organization can engender qualitative change for particular functions during restricted periods of development. Unfortunately, making meaningful and useful distinctions between perception and cognition will probably not figure in anybody's future agenda.

References

Hopkins, B. & Kalverboer, A.F. (1983). Perception and cognition. *Journal of Child Psychology and Psychiatry,* **24**, 37–38.

Olthof, T., Kunnen, E.S. & Boom, J. (2000). Simulating mother-child interactions: exploring the varieties of a non-linear dynamic systems approach. *Infant and Child Development,* **9**, 33–60.

Pepperberg, I.M. (1999). *The Alex studies: cognitive and communicative abilities of grey parrots.* Cambridge, MA: Harvard University Press.

Roelfsema, P.R., Engel, A.K., Konig, P. & Singer, W. (1997). Visuomotor integration is associated with zero-time lag synchronization among cortical areas. *Nature,* **385**, 157–161.

IV.2
Development of the Visual System in the Human Infant

RUXANDRA SIRETEANU

This chapter is dedicated to the memory of Gesine Mohn (1956–1992): an excellent scientist and a dear friend.

ABSTRACT

The visual system is immature in the human newborn. Key structures, such as the fovea centralis of the eye, or the microcircuitry of the primary visual centres in the brain, require several years to evolve and reach the adult level. Functionally the human infant is also visually immature: the visual acuity in the newborn is about 1/50 of adult acuity; the visual field of each eye barely extends from the midline to about 20 degrees in the periphery, thus providing little or no overlap of the images of the two eyes; as a consequence binocular interaction, including stereopsis, is absent at birth. It takes several years for these functions to develop. Higher-order visual functions, e.g. segmentation of oriented textures, preference for the least salient object in a visual scene, and the complementary function of binding distant items into a coherent spatial representation, require even longer developmental periods, extending well into late adolescence. The progressive maturation of visual functions in childhood reflects the asynchronous development of the neural structures involved in the process of seeing.

INTRODUCTION

In the adult human brain, $3–5 \times 10^{12}$ nerve cells are connected by 5×10^{14} synapses. According to a long-held dogma all nerve cells are present and in place at the time of birth; yet the newborn brain weighs only 300–350 grams – about one-fourth of the weight of the adult human brain. Adult brain weight is reached

A.F. Kalverboer and A. Gramsbergen (eds.), Handbook of Brain and Behaviour in Human Development, 629–652
© 2001 Kluwer Academic Publishers. Printed in Great Britain.

at about 12.5 years of age; about 80% of the postnatal weight gain is completed by the age of 4 years (Himwich, 1973). This tremendous increase in brain weight during the first years of postnatal life must be due to the increase in size and complexity of interneural connections, as well as the increase in number, size and complexity of the non-neural brain cells. Indeed, glial cells still divide and proliferate after birth; some of them, the oligodendrocytes, provide the myelin sheath, a lipid and protein coating which wraps around the young axons. This coating emerges after birth and continues to develop postnatally (Yakovlev & Lecours, 1967). As a result the conduction velocity of the newly myelinated axons increases, giving the young organism the ability to receive, process and integrate incoming sensory signals, and to orchestrate and implement an appropriate behavioural response.

Cortical cells also change their morphology after birth. Dendrites and spines sprout, elongate and become exquisitely ramified, providing a million-fold increase in postsynaptic surface; this process enormously increases the chance of nerve cells to communicate with each other (Conel, 1939–67; see Figure 1). The number of synaptic contacts literally explodes between birth and the 8th postnatal month (Huttenlocher, 1974; Garey, 1984). Many of these connections are not destined to survive. In the kitten, transitory projections exist from the auditory to the visual cortex in the juvenile brain (Innocenti & Clarke, 1984) and even directly from the frontoparietal and temporal cortices to the primary visual areas (Dehay et al., 1984).

The period of exuberant development during the first postnatal months, during which neural processes expand and multiply, is followed by a period dominated by the elimination of the superfluous: programmed cell death (apoptosis) eliminates excedentary neurons; synapse elimination reduces the number of interneuronal contacts (Sidman & Rakic, 1973; Rakic et al., 1986; Huttenlocher & de Courten, 1987). These processes continue well into adolescence, and are guided by experience; the adult human brain is thus "sculpted" according to the needs of a changing environment (for a comprehensive review see Schlote, 1983).

According to a more recent view cell proliferation and cell death occur simultaneously, but at different rates, throughout the developmental period. Thus, at any given age, a dynamic equilibrium exists between cell generation and cell elimination (Shankle et al., 1998a; Landing et al., 1998a). At prenatal stages cell proliferation prevails; during the early postnatal period programmed cell death predominates, such that a minimum cell count is reached around 15 months postnatally. From this age the total number of cortical neurons increases slowly, but steadily, until it approximately doubles at around 6 years of age (Shankle et al., 1998b; 1999).

Not all regions of the infant brain develop at the same pace; we speak of a "heterochrony" of brain development (Spatz, 1966; Kahle, 1969; see Figure 2). Thus, while axons in the spinal cord acquire their myelin shea h around the fourth prenatal month, cortical association fibres, and especialiy those subserving higher cognitive functions, still evolve until puberty. Long-range association fibres, particularly fibres interconnecting the two cerebral hemispheres, are still in the process of becoming myelinated around the end of the second decade of life (Yakovlev & Lecours, 1967). Consequently, functional maturation in the

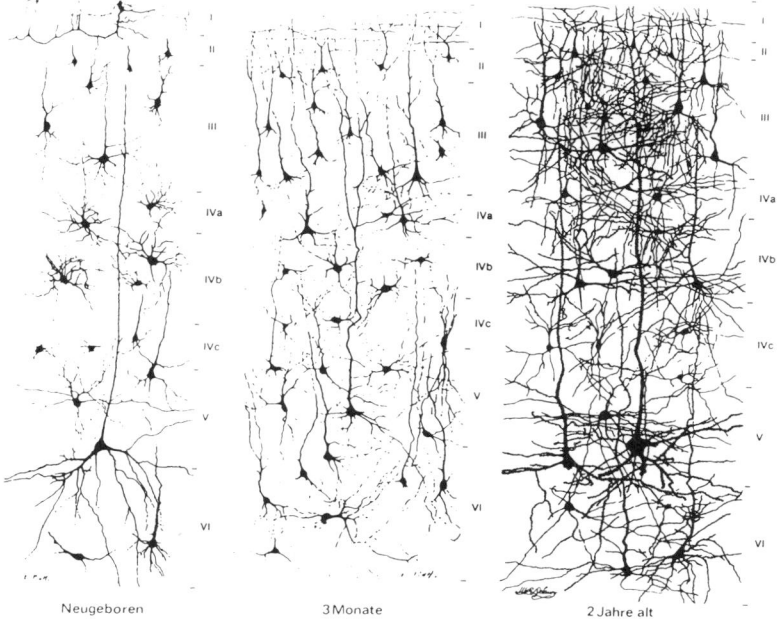

Figure 1 Structural development of the visual cortex of human infants from birth (left) to 3 months (middle) and 2 years of age, assessed using the Golgi method. The different width of the cortex was compensated for by changing the scale. (Conel, 1939–67.)

human newborn proceeds asynchronously. Sensory functions tend to mature before motor functions; the systems processing tactile and auditory information tend to mature earlier than the visual system; and even within a single sensory modality different submodalities develop at different rates (Kahle, 1969; reviewed in Schlote, 1983).

In this chapter I shall present a selective review on the development of the visual system in the first years of life of the human being. Selected anatomical data are presented alongside psychophysical developmental data, including data from my own laboratory. Whenever possible I shall include correlative data obtained from animal studies.

THE VISUAL PATHWAY

The Eye

The retina of the newborn eye contains all the receptive cells needed to transform electromagnetic energy from the outside world into neural signals, which are subsequently transported to the visual centres in the brain. The gradient of retinal maturation is inside-out: the ganglion cells mature earlier than the

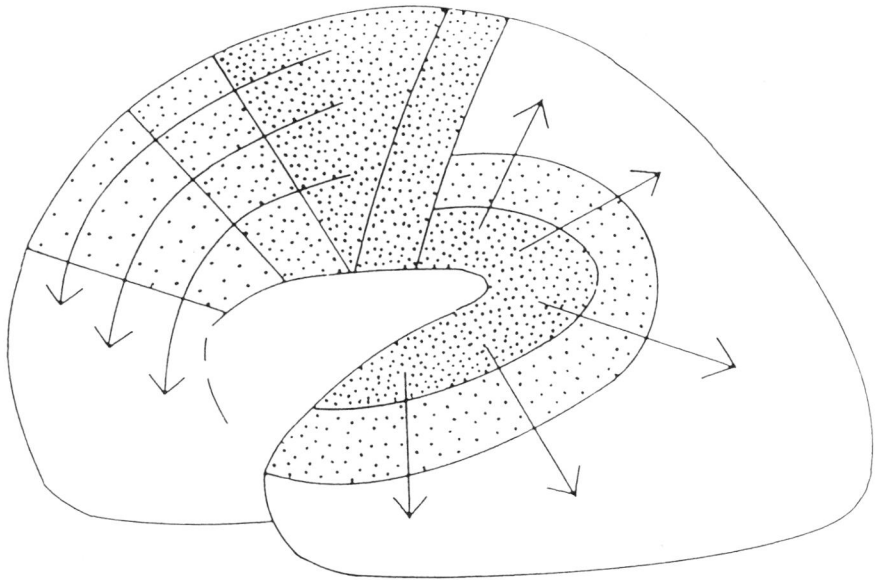

Figure 2 Developmental gradients in the cerebral cortex of the human. Maturation proceeds from the early differentiated precentral region to the frontobasal cortex. In the central region differentiation proceeds from the insula to the adjacent areas. (Kahle, 1969.)

bipolar, amacrine and horizontal cells, which in turn mature before the photo-receptor cells. Peripheral parts of the retinal layers acquire an adult-like appearance later than more central parts (Mann, 1964; Provis *et al.*, 1985). The myelination of the visual pathway occurs mainly postnatally (Nakayama, 1967; Magoon & Robb, 1981; Garey, 1984).

The spatial configuration of the photoreceptor cell mosaic undergoes considerable postnatal transformations: retinal cones are still migrating towards the fovea centralis after birth, thereby increasing their packing density and hence the resolving power of the eye; concomitantly, the cells forming the inner layers of the retina migrate outwards, leaving the foveal cones with a direct, unhindered view of the outer world. As a consequence of this double migration the foveal cones become more slender and the fibres of Henle elongate. This process takes several years to complete: the fovea centralis acquires its adult-like appearance at around 4–5 years of age (Hendrickson & Yuodelis, 1984: Yuodelis & Hendrickson, 1986; see Figure 3).

In contrast to the immaturity of the central part of the retina in the newborn human, the retinal mid-periphery appears rather mature, and its development is completed much earlier; at 4 months postnatally the morphology of the peripheral retina appears adult-like (Hendrickson & Drucker, 1992). A similar developmental sequence was described in the macaque monkey (Hendrickson & Kupfer, 1976; Packer *et al.*, 1989, 1990).

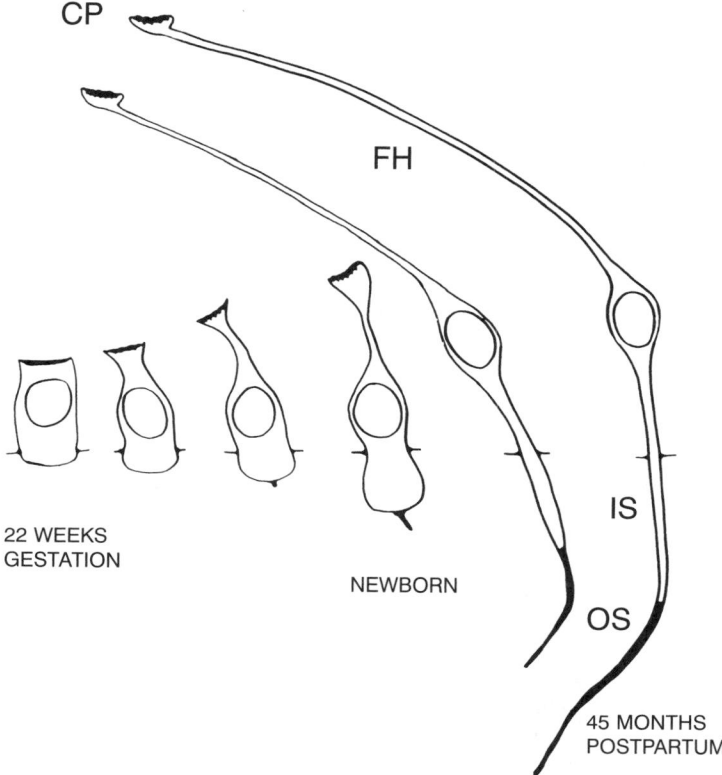

Figure 3 Developmental changes of the foveal cones in the human retina from 22, 24, 26, 34 and 36 weeks of gestation, at birth and at 15 and 45 months postnatally. The inner segment (IS) develops before birth while most of the outer segment (OS) develops after birth. Both the fibre of Henle (FH) and cone pedicle (CP) are present before birth. All four of these structures show a striking thinning and elongation over this developmental span, especially at postnatal ages (drawn to comparative scale). (Hendrickson & Yuodelis, 1984.)

The lateral geniculate nucleus

Visual information coming from the two retinae is projected onto separate layers of a diencephalic structure, the lateral geniculate nucleus of the thalamus. Here, monocular information is still preserved, in the form of separate, alternating retinotopic maps of the contralateral visual hemifield (see Figure 4). Information concerning coarse, transient, colourless, dynamic events of the outer world is mainly processed by two layers in the lateral geniculate nucleus containing larger cells with thicker axons (the "magnocellular" layers, belonging to the "M pathway"), while the processing of information about finely structured, detailed, coloured, but mainly slowly changing visual events is carried out in the four layers containing cells with smaller somata and thinner axons (the "parvocellular" layers, belonging to the "P pathway"; Livingstone & Hubel, 1988; see also Schiller *et al.*, 1990).

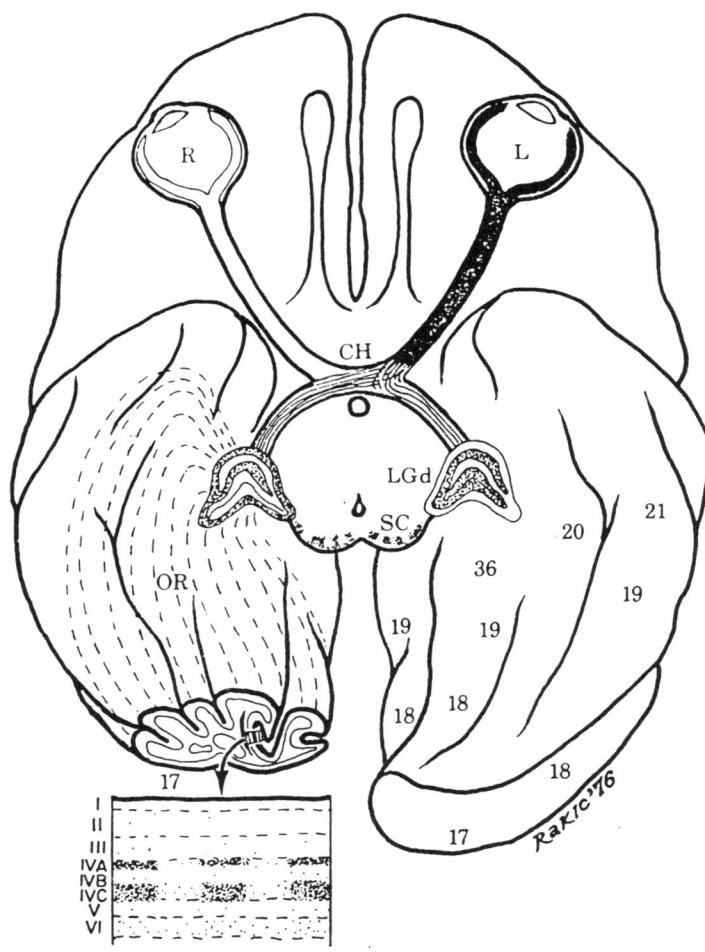

Figure 4 The visual pathway in the primate (from the eye to the primary visual cortex). The projection of the left eye is labelled dark. *Inset*: ocular dominance columns in the striate cortex. CH: optic chiasma; OR: optic radiation. (Rakic, 1976.)

The division of the lateral geniculate nucleus into two magnocellular and four parvocellular layers, with each division subserving distinct functional roles, occurs prenatally. There are hints that the layers receiving inputs from the contralateral eye might become functional before those receiving inputs from the ipsilateral eye. Cells in the magnocellular division, presumably concerned with the processing of information concerning visual motion and depth, mature before cells in the parvocellular division, deemed to be involved in the processing of information concerning the fine-grained analysis of visual scenes and the processing of spectral information finally leading to the sensation of colour (Garey & Saini, 1981; but see Hickey, 1977).

The cortical microcircuitry

The primary visual cortex of the primate brain (area V1, or Brodmann's area 17) consists of modules in which eye-dominance information and information about the orientation of visual contours is processed in an orderly fashion (Hubel & Wiesel, 1977; see Figure 4). The segregation of the cortical surface into territories subserving the right and the left eye emerges during prenatal development and is practically completed at 4–6 months after birth (Hickey & Peduzzi, 1987; Horton & Hedley-Whyte, 1984), when binocular function starts to emerge (Braddick *et al.*, 1980; Held, 1985).

Similar to the other cortical areas, the complexity of the intracortical wiring dramatically increases during the first postnatal years (Conel, 1939–67). The number of synapses in the primary visual cortex multiplies rapidly until the eighth postnatal month, then decreases slowly over several years, to stabilize at around 10 years of age (Garey & de Courten, 1983; Huttenlocher & de Courten, 1987). The intracortical myelinization process continues well into late childhood (Yakovlev & Lecours, 1967). The number of anatomically defined cortical columns increases until at least 6 years postnatally (Landing *et al.*, 1998b).

Recent studies involving injection of fluorescent crystals into slices of fixed human material, and their transportation along intracortical fibres, have revealed that the long-range, tangential intracortical connections are rudimentary at birth; they evolve slowly, and it is not before 6 years of age that they reach the adult state (Burkhalter & Bernardo, 1989). These connections appear ideally suited to provide a substrate for the segmentation of visual scenes and the complementary function of connecting distant visual elements into coherent visual objects (Burkhalter *et al.*, 1993).

Thus the cortical machinery involved in the process of seeing is present, but immature, at birth. Visual processing is sluggish, due to the low conduction velocity of the still-unmyelinated axons; the incomplete segregation of the ocular dominance fails to provide the anatomical substrate necessary for the computing of stereoscopic depth; and the lacking tangential connections fail to provide the prerequisite for the segmentation of visual scenes.

Extrastriate visual pathways

Information from the primary visual area is fanned out to a belt of extrastriate areas surrounding the occipital pole of the brain and including areas located in the temporal lobe (areas V2, V3, VP, V4, TEO, and areas in the inferotemporal cortex) and in the parietal lobe (areas V2, V3a, MT, MST, FST, VIP, LIP; for a review see Goodale, 1993; see Figure 5). There is anatomical evidence that the higher-order visual areas might achieve maturity earlier than the primary visual cortex. The gradient of maturation proceeds from the insula (a region buried deep in the lateral fissure) caudally towards the occipital cortex (Kahle, 1969; see Figure 2). Thus, the primary visual cortex (visual area V1, or Brodmann's area 17), morphologically definitely more sophisticated than hierarchically subsequent

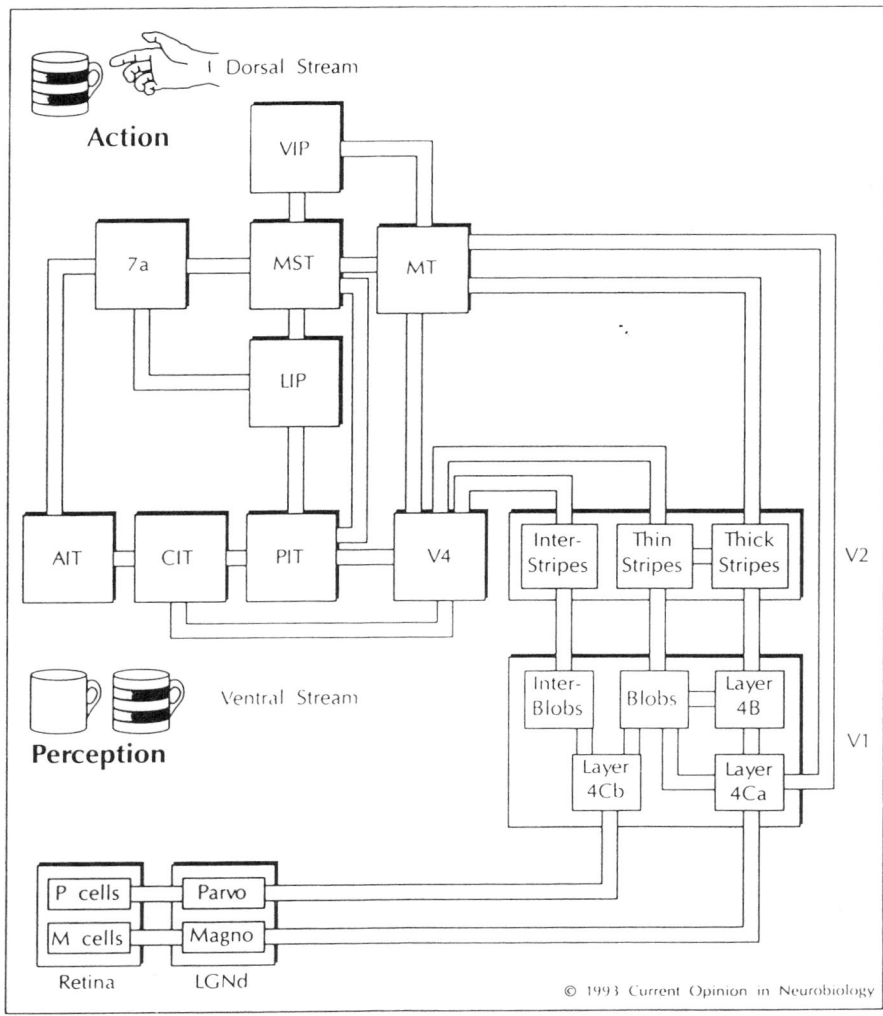

Figure 5 Schematic diagram illustrating the main pathways of the dorsal and ventral visual stream in the primate. Notice that while the P and M pathways are well segregated until they reach area V1, they become heavily intermingled after that. Notice also that, even though two main streams of processing can be identified, they are heavily interconnected. (AIT: anterior inferotemporal cortex; CIT: central inferotemporal cortex; PIT: posterior inferotemporal cortex; VIP: ventral intraparietal sulcus). (Goodale, 1993.)

areas in the temporal and the parietal lobe, is the last to achieve maturity (Pandya *et al.*, 1988).

The apparently counterintuitive finding of a developmental gradient from higher to lower hierarchical areas was confirmed by studies in macaque monkeys:

indeed, single cells in area V2 reach their final position before cells in area V1 (Sidman & Rakic, 1973; Rakic, 1976).

Feed-forward connections from lower to hierarchically higher areas tend to develop before the reciprocal, feed-back connections (Schlote, 1983). Thus, feed-back connections from area V2 to area V1 develop later than the forward projection connecting area V1 to area V2 (Burkhalter, 1993).

M/P pathways

The division of labour into a P pathway, primarily concerned with the analysis of detailed spatial information and colour, and an M pathway, devoted to the analysis of motion and spatial localization of visual objects, is carried on at the level of the primary visual cortex. "P information" is processed in the so-called "parvo-blob" pathway, which passes through layer IVCb in area V1, to the cytochrome oxidase-rich blobs in layers 2 and 3 of area V1, then to the thin stripes in area V2, through portions of area V4 in the temporal cortex and finally reaching visual areas processing complex visual patterns such as individual faces in the inferotemporal cortex (IT). This pathway (the so-called "ventral pathway") was suggested to be involved in the processing of form and colour information of visual objects (Hubel & Livingstone, 1987). "M information" proceeds from the layer IVB in area V1, through the thick stripes in area V2, to the visual motion-sensitive areas in the middle temporal lobe (MT, MST, FST), to reach intermodal association areas such as area 7a in the posterior parietal cortex. This so-called "dorsal pathway" was suggested to be involved in information concerning the localization of visual objects in the extrapersonal space (Hubel & Livingstone, 1987; for a critical review see Goodale, 1993).

Recent studies suggest that this division of labour is not concerned with the processing of "what" versus "where" information, as proposed by Ungerleider & Mishkin (1982), but rather with a dichotomy between "what" versus "how" information: the ventral pathway was suggested to mediate aspects of visual information concerning the *perception* of visual objects and their encoding in visual memory, while the dorsal pathway was believed to be primarily involved in mediating of goal-directed, visually controlled *action*, e.g. prehension (Goodale, 1993; Haffenden & Goodale, 1998; Goodale & Haffenden, 1998; see Figure 5).

There are anatomical indications that the dorsal, M pathway matures earlier than the ventral, P pathway (Distler *et al.*, 1996). This would suggest that goal-directed actions might reach maturity earlier than the perceptual integration of complex visual information.

Higher-order cortical areas

A further gradient of cortical maturation proceeds from the central sulcus towards the prefrontal cortical regions (Kahle, 1969; see Figure 2). Indeed, areas in the prefrontal cortex, which are involved in visual recognition and working memory tasks, develop late in both human and monkey infants (Diamond, 1985; Diamond & Goldman-Rakic, 1989). A functional dichotomy was suggested for prefrontal areas, with more dorsal areas being involved in visuomotor coordination, and areas in more ventral regions being involved in short-term memory tasks

(Goldman-Rakic, 1998). If the developmental dichotomy suggested for the parietal and temporal cortical divisions is carried on in the prefrontal regions, a later development of cognitive visual tasks involving memory components, as in delayed non-matching to sample (DNMS) tasks, might be expected (Overman, 1990).

VISUAL FUNCTIONS

Basic visual functions

Visual acuity and contrast sensitivity

The morphological immaturity of the human visual system at birth suggests that vision in the human newborn is severely limited. Indeed, basic visual functions such as visual acuity, contrast sensitivity and even the extent of the functional visual field are restricted. Visual acuity was assessed in human infants with a multitude of behavioural and electrophysiological methods (Teller *et al.*, 1974; Gwiazda *et al.*, 1978; Mayer & Dobson, 1982; Fiorentini *et al.*, 1983, 1984; Norcia & Tyler, 1985; Courage & Adams, 1990; Sireteanu *et al.*, 1994). In spite of a large variability between the results obtained with the different techniques there is general agreement that visual acuity is about 1/50 of the adult value at birth; acuity develops rapidly until about 6 months, then more slowly until the end of the first year, to reach the adult level around 4–6 years of age. Normative data using the Teller Acuity Cards (McDonald *et al.*, 1985; see Figure 6) show a very good agreement across different laboratories (Mayer *et al.*, 1995; Salomao & Ventura, 1995; Neu & Sireteanu, 1997; see Figure 7).

Parallel to the development of acuity, contrast sensitivity improves during early childhood; with increasing age, infants' peak spatial frequency and maximum contrast sensitivity move towards progressively higher values (Atkinson *et al.*, 1977; Pirchio *et al.*, 1978; Allen *et al.*, 1989; Fiorentini, 1992).

This dramatic increase in acuity and contrast sensitivity is most likely to be due to the postnatal transformation of retinal morphology. Physiologically, the size of the retinal receptive fields decreases, in spite of a postnatal enlargement of the eyeball and of the dendritic fields of the retinal ganglion cells; this decrease is probably due to a delayed development of inhibitory connections (Russoff & Dubin, 1977).

Peripheral vision

The non-uniform development of the retina, with a very late emergence and maturation of the fovea (Hendrickson & Yuodelis, 1984; Yuodelis & Hendrickson, 1986), predicts a functional immaturity in the central part of the visual field in the human newborn, accompanied by a relative functional maturity in the visual mid-periphery. Indeed, it appears that the rapid postnatal improvement in overall visual acuity is closely related to the development of the fovea; the development of acuity in the peripheral visual field is much more limited and it reaches a plateau at an earlier age (Spinelli *et al.*, 1983; Sireteanu *et al.*, 1984, 1994; see Figure 8). Infants also show a slightly, but consistently better, visual acuity in their temporal than the nasal visual field (Sireteanu *et al.*, 1994).

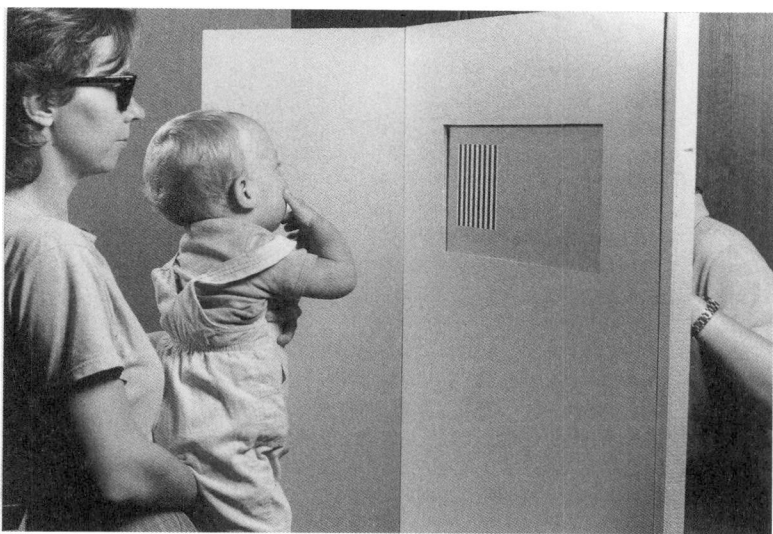

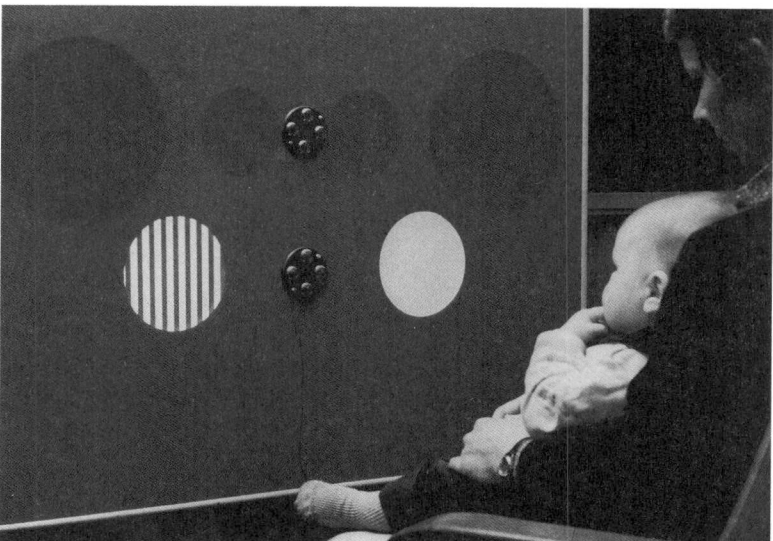

Figure 6 Experimental set-up for testing visual acuity in human infants. *Upper panel*: Teller acuity cards. The attention of the infant is drawn to the rectangular opening. The infant is presented with a grey card containing a highly contrasted, rectangular, vertical, square-wave grating of increasing spatial frequencies. The infant's reaction is watched through a small peep-hole in the centre of the card. (Katz & Sireteanu, 1990.) *Lower panel*: forced-choice preferential looking. The infant is facing a screen containing two circular apertures onto which the stimuli can be projected. A naive observer decides, on the basis of the infant's reaction, on which side the stimulus (a highly contrasted, vertical square-wave grating) was presented. Visual acuity corresponds to the spatial frequency which evokes 75% correct responses. (Sireteanu *et al.*, 1994.)

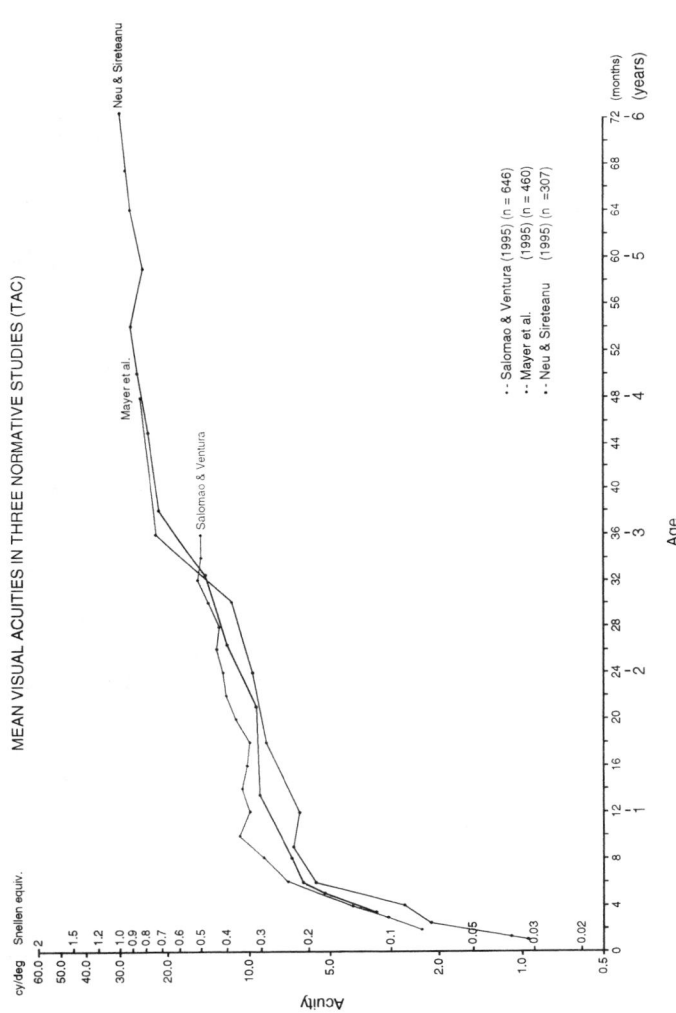

Figure 7 Mean visual acuities obtained with the Teller acuity cards in three normative studies. (Neu & Sireteanu, 1997.)

BINOCULAR ACUITY

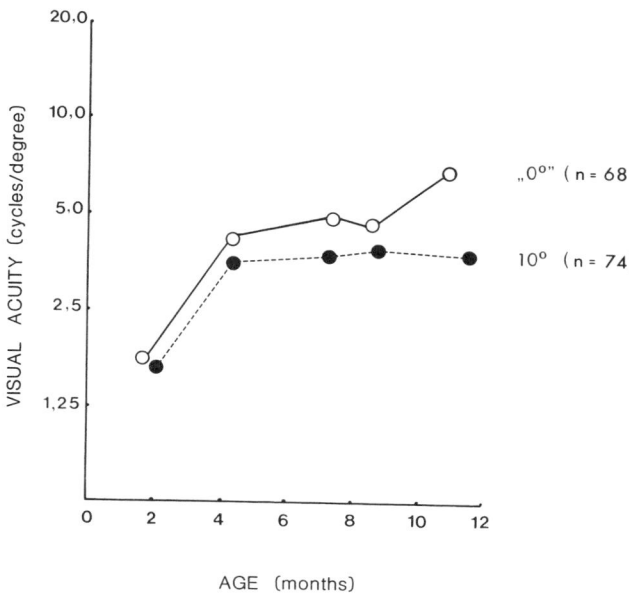

Figure 8 Grating acuity in the central and peripheral visual field. To test acuity in the central visual field a forced-choice procedure was used. Acuity in the peripheral visual field was assessed by monitoring the direction of the first fixation of the infant away from the central fixation stimulus. (Sireteanu *et al.*, 1984.)

The transient nasotemporal asymmetry of visual acuity parallels the asymmetric enlargement of the visual fields in human infants. Indeed, the monocular visual fields in the newborn are extremely restricted: each eye perceives a narrow strip of the outer world, extending from the midline to about 20 degrees temporally, with little or no overlap of the fields of the two eyes; the fields expand progressively, to reach the adult extent at around 1 year of age (Mohn & van Hof-van Duin, 1986; Schwartz *et al.*, 1987; Lewis & Maurer, 1992).

Animal studies

Behavioural studies in kittens and juvenile macaque monkeys, using methods designed for human infants, show striking parallels to human development. The development of the visual fields in young kittens is very similar to the development of the visual fields in human babies (Sireteanu & Maurer, 1982; Sireteanu, 1995).

Visual acuity, tested with operant methods, comes close to the best acuity of single cells in the lateral geniculate nucleus and the visual cortex (monkeys: Teller *et al.*, 1978; Boothe, *et al.*, 1988; Blakemore & Vital-Durand, 1986; Jacobs

& Blakemore, 1988; kittens: Freeman & Marg, 1975; Mitchell *et al.*, 1976; Derrington, 1977; Fiorentini *et al.*, 1983). When tested with unreinforced preferential-looking methods or using the Teller acuity cards, visual acuity falls short of the acuity obtained in monkeys with an operant conditioning method (Sireteanu *et al.*, 1992) or, in kittens, with the jumping stand procedure (Sireteanu, 1985; Katz & Sireteanu, 1990; see Figures 9 and 10).

Electrophysiological studies in the lateral geniculate nucleus of macaque monkeys reveal that the superiority of acuity in the foveal region over the visual periphery, seen in adult animals, is reflected in the acuity of single cells. In contrast, newborn monkeys show a relatively "flat" distribution of acuity. As a consequence an enormous increase in acuity takes place for single cells receiving inputs from the central visual field, with little or no increase of acuity for cells receiving information from the peripheral visual field (Blakemore & Vital-Durand, 1979).

Higher visual functions

Texture segmentation, spatial integration

The segmentation of visual scenes based on the coherence of visual features such as local luminance or line orientation requires the existence and functioning of intracortical connections between neural structures encoding these particular features. It is therefore not surprising that the segmentation of textures based on contour orientation is late to develop, while segmentation based on local luminance can be demonstrated relatively early (Sireteanu & Rieth, 1992; see Figures 11 and 12).

While the exact age of onset of this function depends on the constellation of parameters used in the experimental situation (Atkinson & Braddick, 1992; Sireteanu & Rieth, 1992; Rieth & Sireteanu, 1994a; Sireteanu & Rettenbach, 1996; Rettenbach *et al.*, 1999), there is consensus that the extraction of figures on the basis of the coherence of local contour orientation depends on the existence of long-range connections between single cells with oriented receptive fields in the primary visual cortex; these connections are absent in the human newborn and it is not before 5–6 years of age that they reach an adult appearance (Burkhalter *et al.*, 1993).

A strikingly similar developmental pattern is followed by a function requiring the long-range integration of oriented contours (Kovacs & Julesz, 1993; Field *et al.*, 1993): children aged less than 5–6 years encounter difficulties when asked to locate a contour defined by the alignment of oriented Gabor patches (Kovacs *et al.*, 1999). This developmental similarity suggests that seemingly diverse functions such as the segmentation of oriented textures and the binding of distant oriented patches into a coherent figure might share a common neural substrate.

Visual saliency

Young infants orient spontaneously towards items differing from their surround by their increased saliency. Attentive adults perform equally well if the item to

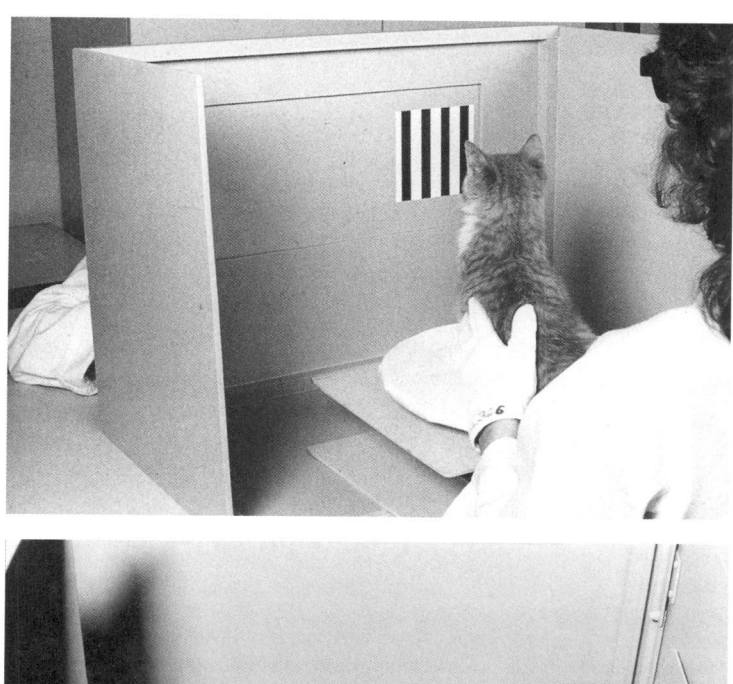

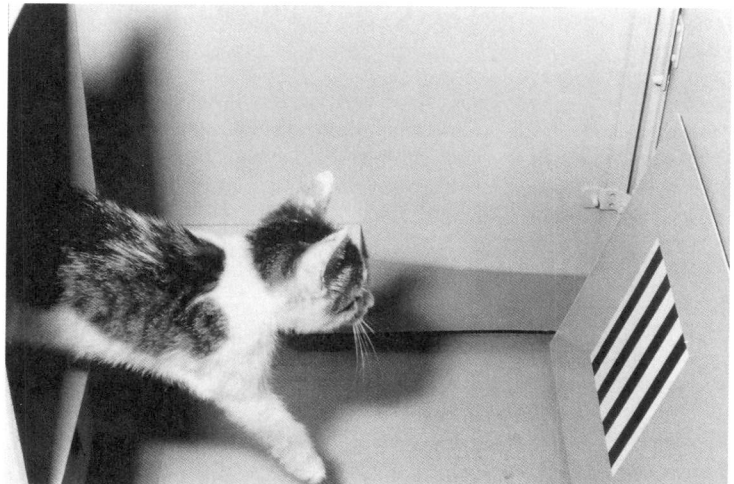

Figure 9 Experimental set-up for testing visual acuity in kittens. *Upper panel*: the kitten is facing a grey card on which a highly contrasted, vertical square-wave grating of different spatial frequencies was presented (Teller acuity cards). The kitten's reaction was monitored by a naive observer through a small peep-hole in the centre of the card (forced-choice preferential looking). *Lower panel*: the kitten is trained to jump on a platform in front of a grey card containing a highly contrasted, square-wave grating on one side (Teller acuity cards of different spatial frequencies). Following a correct jump the kitten is rewarded by food. An incorrect jump is not punished. Visual acuity corresponds to the spatial frequency which evokes 75% correct responses. (Katz & Sireteanu, 1992.)

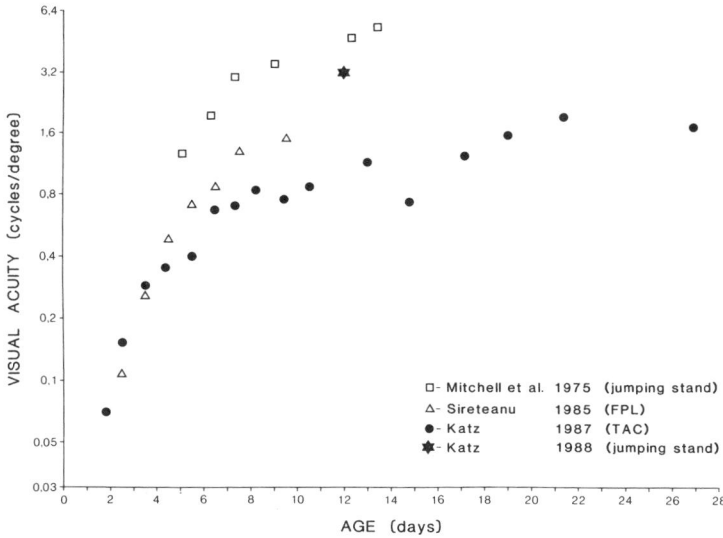

BINOCULAR ACUITY
(normal kittens)

Figure 10 Mean visual acuity obtained with the preferential looking technique (filled circles) and with the jumping stand procedure (star). For a comparison previous data obtained with a forced-choice preferential looking method (open triangles) and with the jumping stand procedure (open squares). (Katz & Sireteanu, 1992.)

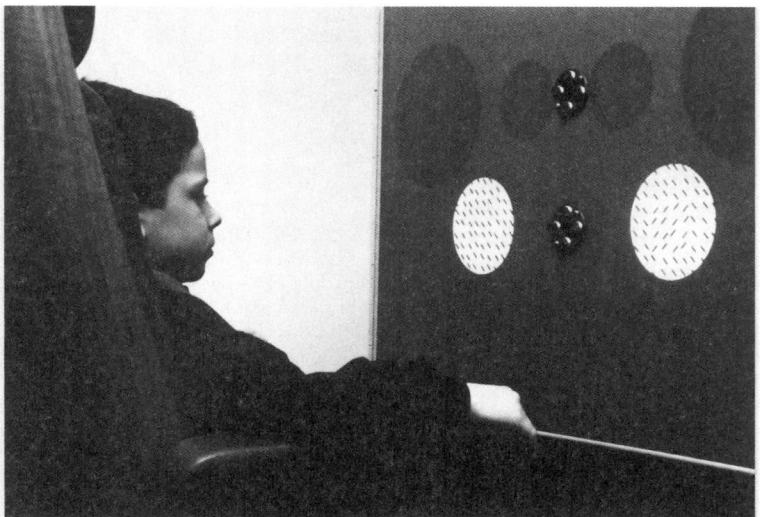

Figure 11 Experimental set-up for assessing the ability of infants and children to segment visual textures defined by orientation contrast. A surface containing a target defined by a group of 16 discrepant elements was paired with a texture not containing the target. (Sireteanu & Rieth, 1992.)

644

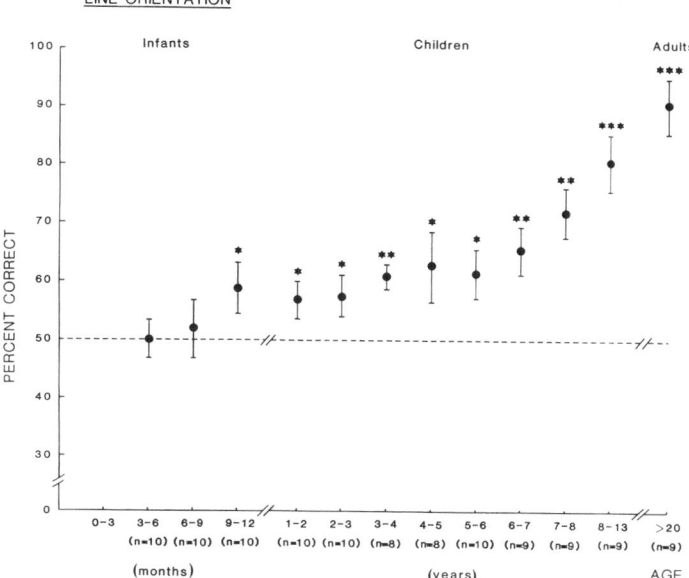

Figure 12 Mean preference for the target containing the discrepant oriented texture for subjects ranging from 3 months to > 20 years of age. Notice the relatively late emergence of statistically reliable preference for targets defined by orientation contrast (around the end of the first year of life). (Sireteanu & Rieth, 1992.)

be detected differs from the surround by being "more salient" or "less salient" than the items of the surround, while adults whose attention is engaged in a concurrent task are selectively impaired in the detection of the "least salient" item (Braun, 1994). This selective impairment mimics a lesion in primate area V4 (Schiller & Lee, 1991).

The preference for the "least salient" item in a visual scene is not functional before 1 year of age, and can first be demonstrated in 3-year-old children (Sireteanu, 2000; see Figures 13 and 14). Obviously the neural mechanisms necessary for the extraction of a "least salient" item in a visual scene are not functional in the young infant. As in the case of the segmentation and integration of oriented contours, this function is likely to involve higher-order visual structures situated on the ventral pathway, which are known to require a lengthy developmental period.

Context-dependence (the Ebbinghaus illusion)

There is accumulating evidence that a whole range of visual functions requiring the extraction of context-dependent information also involve a long developmental period. The Ebbinghaus illusion, in which a circle appears larger or smaller depending upon the size of surrounding elements, is the prototype of an

SLIDE STIMULI

("most salient" vs. "least salient" stimulus)

Figure 13 Stimuli used for assessing the ability of infants and children to orient towards stimuli of differing visual saliency. *Upper panel*: the target is more salient than the surround. *Lower panel*: the target is less salient than the surround. (Sireteanu & Encke, in preparation; Sireteanu, 2000.)

illusion based on size contrast. This illusion was shown to affect the eye but not the hand: normal adult subjects perceive the illusory size distortion, but the aperture of their grasp corresponds to the veridical size of the circle (Agliotti *et al.,* 1995; Haffenden & Goodale, 1998), thus confirming that visual perception and the mechanisms involved in the visual guidance of actions such as prehension involve distinct neural mechanisms. These functions have been shown to be dissociable in neurological patients (Goodale *et al.*, 1991).

In young children, perception of the Ebbinghaus illusion is much reduced before the age of 6 years (Weintraub, 1979; Zanuttini, 1996). The influence of semantic factors on the illusion also increases between 4 and 8 years of age (Zanuttini, 1996). This illusion is probably based on the long-range, lateral interactions involving the ventral visual pathway (Goodale & Haffenden, 1998).

Further evidence for a developmental dissociation comes from studies on object recognition versus object discrimination: in human infants as in infant monkeys, visual discrimination tasks which are based on a non-cognitive, habit formation, corticostriatal system, are relatively quick to develop, but a lengthy developmental period is required for the acquisition of delayed non-matching-to-sample tasks (DNMS), which presumably involve the functioning of a corticolimbic, cognitive memory system (Bachevalier & Mishkin, 1984; Overman *et al.*, 1992).

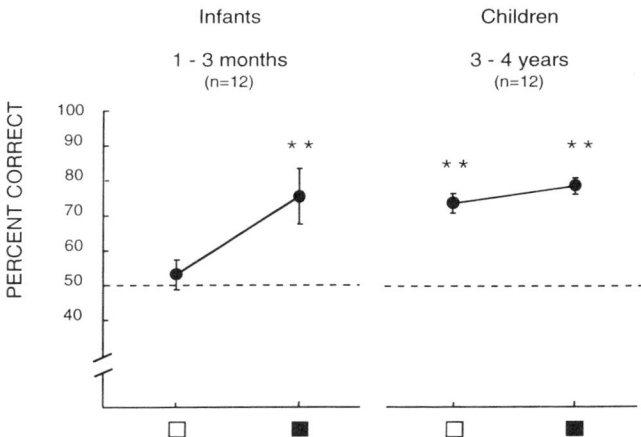

Figure 14 Infants aged 1–3 months strongly prefer the "most salient" stimulus, but show no evidence for preference for the "least salient stimulus". Children 3–4 years of age show a strong preference for both stimuli. (Sireteanu & Encke, in preparation; Sireteanu, 2000.)

CONCLUDING REMARKS

This chapter has reviewed evidence on the early development of the visual system and visual functioning in human infants. The studies presented here suggest that human newborns enter the world with a rudimentary visual apparel, which enables them to obtain only a sketchy impression of the outer world: the visual field is restricted, binocular function is absent, the visual contours lack sharpness, and the ability to extract figures out of a cluttered background and to bind disparate visual items into coherent figures is nearly absent. Some of these functions require a short time to evolve and mature. However, many visual functions, especially those implying the functioning of the ventral visual pathway, are still developing into the teen years, while the child and later adolescent is already using vision to act upon and interact with the environment. Knowledge concerning the early limitations of vision might help to provide adequate means in educational programmes.

ACKNOWLEDGEMENTS

I am indebted to my colleagues Regina Kellerer, Klaus-Peter Boergen, Maria Fronius, Dan H. Constantinescu, Christiane Rieth, Regina Rettenbach, Birgit Katz, Beate Neu, Irmgard Encke, Manuela Wagner and Iris Bachert for their cooperation in the studies mentioned in this chapter. Thanks are due to Wolf Singer for continuing support and encouragement. Part of the experimental work was financially supported by grants from the Deutsche Forschungsgemeinschaft.

References

Agliotti, S., de Souza, J.F.X. & Goodale, M.A. (1995). Size-contrast illusions deceive the eye but not the hand. *Current Biology*, **5,** 679–685.

Allen, D., Tyler, C.W. & Norcia, A.M. (1989). Development of grating acuity and contrast sensitivity in the central and peripheral visual field of the human infant. *Investigative Ophthalmology and Visual Science (Supplement)*, **30**, 311.

Atkinson, J. & Braddick, O. (1992). Visual segmentation of oriented textures by infants. *Behavioural Brain Research*, **49**, 123–131.

Atkinson, J., Braddick, O. & Moar, K. (1977). Development of contrast sensitivity over the first 3 months of life of the human infant. *Vision Research*, **17**, 1037–1044.

Bachevalier, J. & Mishkin, M. (1984). An early and late developing system for learning and retention in infant monkeys. *Behavioural Neuroscience*, **98**, 770–778.

Blakemore, C. & Vital-Durand, F. (1979). Development of the neural basis of visual acuity in monkeys: speculation on the origin of deprivation amblyopia. *Transactions of the Ophthalmological Society, UK*, **99**, 363–368.

Blakemore, C. & Vital-Durand, F. (1986). Organization and post-natal development of the monkey's lateral geniculate nucleus. *Journal of Physiology*, **380**, 453–491.

Boothe, R.G., Kiorpes, L., Williams, R.A. & Teller, D.Y. (1988). Operant measurements of contrast sensitivity in infant macaque monkeys during normal development. *Vision Research*, **28**, 387–396.

Braddick, O.J., Atkinson, J., Julesz, B., Kropfl, W., Bodis-Wolner, I. & Raab, E. (1980). Cortical binocularity in infants. *Nature*, **288**, 363–365.

Braun, J. (1994). Visual search among items of different salience: Removal of visual attention mimics a lesion in area V4. *Journal of Neuroscience*, **14**, 554–567.

Burkhalter, A. (1993). Development of forward and feedback connections between areas V1 and V2 of human visual cortex. *Cerebral Cortex*, **3**, 476–487.

Burkhalter, A. & Bernardo, K.L. (1989). Organization of cortico-cortical connections in the human visual cortex. *Proceedings of the National Academy of Sciences, USA*, **86**, 1071–1075.

Burkhalter, A., Bernardo, K.L. & Charles, V. (1993). Development of local circuits in human visual cortex. *Journal of Neuroscience*, **13**, 1916–1931.

Conel, J.L. (1939–67). *The postnatal development of the human cerebral cortex*, vols 1–8. Cambridge, MA: Harvard University Press.

Courage, M.L. & Adams, R.J. (1990). The early development of visual acuity in the binocular and monocular visual fields. *Infant Behaviour and Development*, **13**, 123–128.

Dehay, C., Bullier, J. & Kennedy, H. (1984). Transient projections from the fronto-parietal and temporal cortex to areas 17, 18 and 19 in the kitten. *Experimental Brain Research*, **57**, 208–212.

Derrington, A. (1977). Development of selectivity in kitten striate cortex. *Journal of Physiology*, **276**, 46–47.

Diamond, A. (1985). The development of the ability to use recall to guide action, as indicated by infants' performance of A–nonB. *Child Development*, **56**, 868–883.

Diamond, A. & Goldman-Rakic, P.S. (1989). Comparison of human infants and rhesus monkeys on Piaget's A–nonB task: evidence for dependence on dorsolateral prefrontal cortex. *Experimental Brain Research*, **74**, 24–40.

Distler, C., Bachevalier, J. Kennedy, C., Mishkin, M. & Ungerleider, L.G. (1996). Functional development of the corticocortical pathway for motion analysis in the macaque monkey: a 14C-2-deoxyglucose study. *Cerebral Cortex*, **6**, 184–195.

Field, D., Hayes, A. & Hess, R. (1993). Contour integration by the human visual system: Evidence for a local "association field". *Vision Research*, **33**, 173–193.

Fiorentini, A. (1992). Parallel processes in human visual development. In I. Brannan (Ed.), *Applications of parallel processing of vision* (pp. 81–118). Amsterdam: Elsevier.

Fiorentini, A., Pirchio, M. & Sandini, G. (1984). Development of retinal acuity in infants evaluated with pattern electroretinogram. *Human Neurobiology*, **3**, 93–95.

Fiorentini, A., Pirchio, M. & Spinelli, D. (1983). Development of retinal and cortical responses to pattern reversal in infants: a selective review. *Behavioural Brain Research*, **10**, 99–106.

Freeman D.N. & Marg, E. (1975). Visual acuity development coincides with the sensitive period in kittens. *Nature*, **254**, 614–615.

Garey, L.J. (1984). Structural development of the visual system of man. *Human Neurobiology*, **3**, 75–80.

Garey, L.J. & de Courten, C. (1983). Structural development of the visual cortex in monkey and man. *Behavioural Brain Research*, **10**, 3–13.

Garey, L.J. & Saini, K.D. (1981). Golgi studies on the normal development of neurons in the lateral geniculate nucleus of the monkey. *Experimental Brain Research*, **44**, 117–128.

Goldman-Rakic, P.S. (1998). The prefrontal landscape: implications of functional architecture for understanding human mentation and the central executive. In A.C. Roberts, T.W. Robbins *et al.* (Eds.), *The prefrontal cortex: Executive and cognitive functions* (pp. 87–102). New York: Oxford University Press.

Goodale, M.A. (1993). Visual pathways supporting perception and action in the primate cerebral cortex. *Current Opinion in Neurobiology*, **3**, 578–585.

Goodale, M.A. & Haffenden, A. (1998). Frames of reference for perception and action in the human visual system. *Neuroscience and Biobehavioural Reviews*, **22**, 161–172.

Goodale, M.A. & Milner, A.D. (1992). Separate visual pathways for perception and action. *Trends in Neuroscience*, **15**, 20–25.

Goodale, M.A., Milner, A.D., Jakobson, L.S. & Carey, D.P. (1991). A neurological dissociation between perceiving objects and grasping them. *Nature*, **349**, 154–156.

Gwiazda, J. Brill, S., Mohindra, I. & Held, R. (1978). Infant visual acuity and its meridional variation. *Vision Research*, **18**, 1557–1564.

Haffenden, A. & Goodale, M.A. (1998). The effect of pictorial illusion on prehension and perception. *Journal of Cognitive Neuroscience*, **10**, 122–136.

Held, R. (1985). Binocular vision – behavioural and neural development. In J. Mehler & R. Fox (Eds.), *Neonate cognition: beyond the blooming and buzzing cconfusion* (pp. 37–44). Hillsdale, NJ: Erlbaum.

Hendrickson, A. & Drucker, D. (1992). The development of parafoveal and mid-peripheral human retina. *Behavioural Brain Research*, **49**, 21–31.

Hendrickson, A. & Kupfer, C. (1976). The histogenesis of the fovea in the macaque monkey. *Investigative Ophthalmology and Visual Science*, **15**, 746–756.

Hendrickson, A. & Yuodelis, C. (1984). The morphological development of the human fovea. *Ophthalmology*, **91**, 603–612.

Hickey, J.L. (1977). Postnatal development of the human lateral geniculate nucleus: Relationship to a critical period for the visual system. *Science*, **198**, 836–838.

Hickey, J.L. & Peduzzi, J.D. (1987). Structure and development of the visual system. In P. Salaptek & L. Cohen (Eds.), *Handbook of Infant Perception* (pp. 1–42). New York: Academic Press.

Himwich, W.A. (1973). Problems in interpreting neuronal changes occuring in developing and aging animals. In D.H. Ford (Ed.), *Neurobiological aspects of maturation and aging*. Amsterdam: Elsevier.

Horton, J.C. & Hedley-Whyte, E.T. (1984). Mapping of cytochrome oxydase patches and ocular dominance columns in human visual cortex. *Philosophical Transactions of the Royal Society of London, B,* **278**, 377–409.

Hubel, D.N. & Livingstone, M.S. (1987). Segregation of form, color, and stereopsis in primate area 18. *Journal of Neuroscience*, **7**, 3378–3415.

Hubel, D.N. & Wiesel, T.N. (1977). Functional architecture of macaque monkey visual cortex. *Proceedings of the Royal Society, London, B, Biological Sciences*, **198**, 1–59.

Huttenlocher, P.R. (1974). Dendritic development in neocortex of children with mental defect and with infantile spasms. *Neurology*, **24**, 507–511.

Huttenlocher, P.R. & de Courten, C.H. (1987). The development of synapses in striate cortex of man. *Human Neurobiology*, 6, 1–9.

Innocenti, G.M. & Clarke, S. (1984). Bilateral transitory projection to visual areas from auditory cortex in kittens. *Developmental Brain Research*, **14**, 143–148.

Jacobs, D.S. & Blakemore, C. (1988). Factors limiting the postnatal development of visual acuity in the monkey. *Vision Research*, **28**, 947–959.

Kahle, W. (1969). *Die Entwicklung der Menschlichen Großhirnhemisphäre*. Berlin: Springer Verlag.

Katz, B. & Sireteanu, R. (1990). The Teller acuity card test: a useful method for the clinical routine? *Clinical Vision Sciences*, **5**, 307–323.

Katz, B. & Sireteanu, R. (1992). Development of visual acuity in kittens: a comparison between jumping stand and Teller acuity card test. *Clinical Vision Sciences*, **7**, 219–224.

Kovacs, I. & Julesz, B. (1993). A closed curve is much more than an incomplete one: Effect of closure in figure-ground discrimination. *Proceedings of the National Academy of Sciences, USA*, **90**, 7495–7497.

Kovacs, I., Kozma, P., Feher, A. & Benedek, G. (1999). Late maturation of visual spatial integration in humans. *Proceedings of the National Academy of Sciences, USA*, **96**, 12204–12209.

Landing, B.H., Shankle, W.R. & Boyd, J.P. (1998a). Quantitative microscopic anatomy, illustrated by its potential role in furthering understanding of the processes of structuring the developing human cerebral cortex. *Acta Pediatrica Japonica*, **40**, 400–418.

Landing, B.H., Shankle, W.R. & Hara, J. (1998). Constructing the human cerebral cortex during infancy and childhood: Types and numbers of cortical columns and numbers of neurons in such columns at different age-points. *Acta Pediatrica Japonica*, **40**, 530–543.

Lewis, T.L. & Maurer, D. (1992). The development of the temporal and nasal visual fields during infancy. *Vision Research*, **32**, 903–911.

Livingstone, M.S. & Hubel, D.N. (1988). Segregation of form, color, movement and depth: Anatomy, physiology, and perception. *Science*, **240**, 740–749.

Magoon, E.H. & Robb, R.M. (1981). Development of myelin in human optic nerve and tract. *Archives of Ophthalmology*, **99**, 655–659.

Mann, I. (1964). *The development of the human eye*. London: British Medical Association.

Mayer, D.L & Dobson, V. (1982). Visual acuity development in infants and young children, as assessed by operant preferential looking. *Vision Research*, **22**, 1141–1151.

Mayer, D.L., Beiser, A.S., Warner, A.F., Pratt, E.M., Raye, K.N. & Lang J.M. (1995). Monocular acuity norms for the Teller acuity cards between ages of one month and four years. *Investigative Ophthalmology and Visual Science*, **36**, 671–685.

McDonald, M., Dobson, V., Sebris, S.L., Baitch, L., Varner, D. & Teller, D.Y. (1985). The acuity card procedure: a rapid test of visual acuity. *Investigative Ophthalmology and Visual Science*, **26**, 1158–1162.

Mitchell, D.E., Giffin, F., Wilkinson, F., Anderson, P. & Smith, M.L. (1976). Visual resolution in young kittens. *Vision Research*, **16**, 363–366.

Mohn, G. & van Hof-van Duin, J. (1986). Development of the binocular and monocular visual fields of human infants during the first year of life. *Clinical Vision Sciences*, **1**, 51–64.

Nakayama, K. (1967). Studies on the myelination of the optic nerve. *Japanese Journal of Ophthalmology*, **11**, 132–140.

Neu, B. & Sireteanu, R. (1997). Monocular acuity in preschool children: assessment with the Teller and Keeler acuity cards in comparison to the C-test. *Strabismus*, **5**, 185–201.

Norcia, A.M. & Tyler, C.W. (1985). Spatial frequency sweep VEP: visual acuity during the first year of life. *Vision Research*, **25**, 1399–1408.

Overman, W.H. (1990). Performance on traditional match-to-sample, non-match-to-sample, and object discrimination tasks by 12 to 32 month-old children: A developmental progression. In A. Diamond (Ed.), *The development and neural basis of higher cognitive functions* (pp. 365–383). New York: New York Academy of Sciences Press.

Overman, W., Bachevalier, J., Turnen, M & Peuster, A. (1992). Object recognition versus object discrimination: Comparison between human infants and infant monkeys. *Behavioural Neuroscience*, **106**, 15–29.

Packer, O., Hendrickson, A.E. & Curcio, C.A. (1989). Photoreceptor topography of the retina in the adult pigtail macaque (*Macaca nemestrina*). *Journal of Comparative Neurology*, **288**, 165–183.

Packer, O., Hendrickson, A.E. & Curcio, C.A. (1990). Development and redistribution of photoreceptors across the *Macaca nemestrina* (pigtail macaque) retina. *Journal of Comparative Neurology*, **298**, 472–493.

Pandya, D.N., Seltzer, B. & Barbas, H. (1988). Input–output organization of the primate cerebral cortex. *Comparative Primate Biology*, **4**, 39–80.

Pirchio, M., Spinelli, D., Fiorentini, A. & Maffei, L. (1978). Infant contrast sensitivity evaluated by evoked potentials. *Brain Research*, **141**, 179–184.

Provis, J.M., von Driel, Billson, F.A. & Russell, P. (1985). Development of the human retina: Patterns of cell distribution and redistribution in the ganglion cell layer. *Journal of Comparative Neurology*, **233**, 429–451.

Rakic, P. (1976). Differences in time of origin and in eventual distribution of neurons in areas 17 and 18 of visual cortex in rhesus monkeys. *Experimental Brain Research (Supplement)*, **1**, 244–248.

Rakic, P., Bourgeois, J.-P., Eckenhof, E.F., Zevcevic, N. & Goldman-Rakic, P. (1986). Concurrent overproduction of synapses in diverse regions of the primate cortex. *Science*, **232**, 232–235.

Rettenbach, R., Diller, G. & Sireteanu, R. (1999). Do deaf people see better? Texture segmentation and visual search compensate in adult but not in juvenile subjects. *Cognitive Neuroscience*, **11**, 560–583.

Rieth, C. & Sireteanu, R. (1994a). Texture segmentation and "pop-out" in infants and children: the effect of test field size. *Spatial Vision*, **8**, 173–191.

Rieth, C. & Sireteanu, R. (1994b). Texture segmentation and visual search based on orientation contrast: an infant study with the familiarization–novelty preference method. *Infant Behavior and Development*, **17**, 359–369.

Russoff, A. & Dubin, M.W. (1977). Development of receptive field properties of retinal ganglion cells in kittens. *Journal of Neurophysiology*, **40**, 1188–1198.

Salomao, S.R. & Ventura, D.F. (1995). Large sample population for visual acuities obtained with Vistech–Teller cards. *Investigative Ophthalmology and Visual Science*, **36**, 657–670.

Schiller, P.H. & Lee, K. (1991). The role of primate extrastriate area V4 in vision. *Science*, **251**, 1251–1253.

Schiller, P.H., Logothetis, N. & Charles, E.R. (1990). Functions of the colour-opponent and broad-band channels of the visual system. *Nature*, **343**, 68–70.

Schlote, W. (1983). Entwicklung des Nervensystems. In W. Doerr & G. Seifert (Eds.), *Spezielle Pathologische Anatomie* (vol. 13, pp. 1–171). Berlin: Springer Verlag.

Schwartz, T.L., Dobson, V., Sandstrom, D.J. & van Hof-van Duin, J. (1987). Kinetic perimetry assessment of binocular visual field shape and size in young infants. *Vision Research*, **27**, 2163–2175.

Shankle, W.R., Landing, B.H., Rafil, M.S., Schianos, A., Chen, J.M. & Hara, J. (1998). Evidence for a postnatal doubling of neuron number in the developing human cerebral cortex between 15 months and 6 years. *Journal of Theoretical Biology*, **191**, 115–140.

Shankle, W.R., Rafil, M.S., Landing, B.H. & Fallon, J.H. (1999). Approximate doubling of numbers of neurons in postnatal human cerebral cortex and in 35 specific cytoarchitectural areas from birth to 72 months. *Pediatric and Developmental Pathology*, **2**, 144–159.

Shankle, W.R., Romney, A.K., Landing, B.H. & Hara, J. (1998). Developmental patterns in the cytoarchitecture of the human cerebral cortex from birth to 6 years examined by correspondence analysis. *Proceedings of the National Academy of Sciences, USA*, **95**, 4023–4028.

Sidman, R.L. & Rakic, P. (1973). Neuronal migration, with special reference to the human brain: a review. *Brain Research*, **62**, 1–35.

Sireteanu, R. (1985). The development of visual acuity in very young kittens. A study with forced-choice preferential looking. *Vision Research*, **25**, 781–788.

Sireteanu, R. (1995). Development of the visual field: Results from human and animal studies. In F. Vital-Durand, J. Atkinson & O. Braddick (Eds.), *Infant Visual Development* (pp. 17–31). Oxford University Press.

Sireteanu, R. (2000). Texture segmentation, "pop-out" and feature binding in infants and children. In C. Rovee-Collier, L.P. Lipsitt & H. Hayne (Eds.), *Progress in infancy research* (vol. 1, pp. 183–249). Mahwah, NJ: Lawrence Erlbaum.

Sireteanu, R. & Maurer, D. (1982). The development of the kitten's visual field. *Vision Research*, **22**, 1105–1111.

Sireteanu, R. & Rettenbach, R. (1996). Textursegmentierung und visuelle Suche: Entwicklung, Lernen und Plastizität. *Klinische Monatsblätter der Augenheilkunde*, **208**, 3–10.

Sireteanu, R. & Rieth, C. (1992). Texture segregation in infants and children. *Behavioural Brain Research*, **49**, 133–139.

Sireteanu, R., Fronius, M. & Constantinescu, D.H. (1994). The development of visual acuity in the peripheral visual field of human infants: binocular and monocular measurements. *Vision Research*, **34**, 1659–1671.

Sireteanu, R., Katz, B., Mohn, G. & Vital-Durand, F. (1992). Teller acuity cards for testing visual development and the effects of experimental manipulation in macaques. *Clinical Vision Sciences*, **7**, 107–117.

Sireteanu, R., Kellerer, R. & Boergen, K.-P. (1984). The development of peripheral visual acuity in human infants. A preliminary study. *Human Neurobiology*, **3**, 81–85.

Spatz, H. (1966). Gehirnentwicklung (Introversion–Promination) und Endocranialausguß. In R. Hassler & H. Stephan (Eds.), *Evolution of the forebrain* (pp. 136–152). Stuttgart: Thieme.

Spinelli, D., Pirchio, M. & Sandini, G. (1983). Visual acuity in the young infant is highest in a small area. *Vision Research*, **23**, 1133–1136.

Teller, D.Y., Morse, R., Borton, R. & Regal, D. (1974). Visual acuity for vertical and diagonal gratings in human infants. *Vision Research*, **14**, 1433–1439.

Teller, D.Y., Regal, D.M., Videen, T.O. & Pulos, E. (1978). Development of visual acuity in infant monkeys (Macaca nemestrina) during the early postnatal weeks. *Vision Research*, **18**, 561–566.

Ungerleider, L.G. & Mishkin, M. (1982). Two cortical visual systems. In D.J. Ingle, M.A. Goodale & R.J.W. Mansfield (Eds.), *Analysis of visual behaviour* (pp. 549–586). Cambridge, MA: MIT Press.

Weintraub, D.J. (1979). Ebbinghaus illusion: Context, contour, and age influence the judged size of a circle amidst circles. *Journal of Experimental Psychology: Human Perception and Performance*, **5**, 353–364.

Yakovlev, P.I. & Lecours, A.R. (1967). The myelogenetic cycles of regional maturation of the brain. In A. Miniakowski (Ed.), *Regional Development of the Brain in Early Life* (pp. 3–70). Oxford: Blackwell.

Yuodelis, C. & Hendrickson, A. (1986). A qualitative and quantitative analysis of the human fovea during development. *Vision Research*, 26, 847–855.

Zanuttini, L. (1996). Figural and semantic factors in change in the Ebbinghaus illusion across four age groups of children. *Perceptual and Motor Skills*, **82**, 15–18.

IV.3
Neurophysiological and Psychophysical Approaches to Infant Visual Development

SCOTT P. JOHNSON

ABSTRACT

Infants are born with some visual skills, but in general vision is rather poor in neonates. However, perceptual development is rapid, and by 6 months of age the visual system is nearly adult-like in function. Evidence from three domains is presented, basic visual function; eye movements, attention, and visual stability; and perception of object unity, and the possible neurophysiological underpinnings of development in each of these areas are discussed. A common developmental pattern is evident across these domains: most perceptual behaviours in neonates appear to be mediated primarily by subcortical mechanisms. Across the first several months after birth, cortical structures begin to contribute to improvements in sensory, perceptual, and cognitive function. Important progress is being made towards the goal of developing a unified account of the neurophysiology of visual development, although much remains to be explored. The chapter concludes with a discussion of potentially fruitful future research directions. In particular a critical need exists for direct evidence linking cortical and cognitive development.

INTRODUCTION

Consider the state of visual perception in the human neonate. At birth she is suddenly confronted with the optic array, an amalgam of colours, shapes, and motion that shifts with every eye movement and the movement of every object. Making sense of this perceptual environment might seem a daunting task, and some traditional views of visual development echoed this viewpoint. For example, James (1890/1950) famously summarized what he believed to be the infant's perceptual experience:

A.F. Kalverboer and A. Gramsbergen (eds.), Handbook of Brain and Behaviour in Human Development, 653–676
© 2001 Kluwer Academic Publishers. Printed in Great Britain.

The baby, assailed by eyes, ears, nose, skin, and entrails at once, feels it all as one great blooming, buzzing confusion [and] any number of impressions, from any number of sensory sources, falling simultaneously on a mind which has not yet experienced them separately, will fuse into a single undivided object for that mind (p. 488).

This position was reiterated by Piaget (1952, 1954), who proposed that, at birth, percepts across modalities are not unified and actions are not coordinated, and that the neonate's visual world consists of a patchwork or "tableaux" of moving colours and shapes (rather than segregated, coherent objects). Casual observation of the young infant confirms these suggestions: Prior to the second half-year after birth most motor behaviours appear to be disorganized, and coordination of vision and reaching seems beyond the infant's capacity. Carefully controlled research over the past several decades, however, has revealed that James's and Piaget's positions (as well as "common-sense" views of the infant as disorganized) are overly pessimistic with respect to the state of infant perception: many visual functions are in place at birth or soon after. A strong base of descriptive evidence has resulted, some of which is described in this chapter. Nevertheless, important questions remain concerning the origins of visual skills in infancy, and the mechanisms of development of these skills.

The function of visual perception is to provide information to the observer in order to facilitate subsequent action on the environment. The importance of veridical perception of our surroundings is attested by the allotment of cortical tissue devoted to vision: by some estimates over 50% of the cortex of the macaque monkey (a phylogenetically close cousin to *Homo sapiens*) is involved in visual perception, and there are perhaps 30 distinct cortical areas that participate in visual or visuomotor processing (Felleman & Van Essen, 1991; for general reviews see DeYoe & Van Essen, 1988; Van Essen *et al.*, 1992; Zeki & Shipp, 1988). Visual attributes are processed in parallel throughout much of the mature visual system, by distinct neural pathways: the *parvocellular* system is relatively more sensitive to fine detail and colour, whereas the *magnocellular* system is more sensitive to global form, motion, and binocular depth information. These two processing streams originate in retinal ganglion cells, proceed through the lateral geniculate nucleus in the thalamus, and then to the cortex, remaining largely segregated, but partly interconnected (Schiller, 1996). In the adult, area V1, through which visual information is first routed in the cortex, contains cells that are selectively sensitive to luminance, colour, motion, orientation, and depth (Hubel & Wiesel, 1979; Movshon & Newsome, 1998; Zeki, 1993). V1 is a laminated structure, with several layers parallel to the surface of the cortex. The layers have different functions: input from the lateral geniculate nucleus, output to subcortical and higher cortical areas, communicative networks of neurons within a layer, and so on. V1 acts as the cortex's "first-pass" processor of visual information, and distributes signals to higher cortical areas for further, more specialized analysis (Livingstone & Hubel, 1983). From V1 the parvocellular processing stream continues to the temporal lobe, which participates in object identification and recognition (the so-called "what" system), whereas the magnocellular stream continues to the parietal lobe, which participates in determining spatial location and action (the "where" or "how" system) (Milner & Goodale, 1995, Mishkin *et al.*, 1983).

Much of the structure of V1 and other visual areas arises in the absence of visual experience, but complete visual development is dependent on some amount of exposure to the environment (see Daw, 1995, for review). The extent to which normal development depends on or is independent of experience has long been a source of contention among scientists and philosophers (see, e.g. Elman *et al.*, 1996; Greenough & Alcantara, 1993; Spelke & Newport, 1998). Certainly this chapter will not answer this question definitively, but significant insights on the issue, provided by investigations of visual function in infants, will be discussed.

Visual functions do not mature in parallel. Some appear to be available at birth (e.g. some eye movement systems), whereas others require several months or longer to become fully operational (e.g. some object perception skills). One general theme that pervades contemporary theorizing is that visual development is best accounted for by divergent rates of maturation of various processing streams. For example, Bronson (1974) proposed that visual processing is limited to subcortical structures in the neonate, and that cortical functions emerge gradually over the first few months. More sophisticated models have recently emerged, however (some of which are discussed in subsequent sections of this chapter), that acknowledge some cortical function at birth, and specify more precise roles for cortical and subcortical visual areas. Human infants are born with a full complement of cortical neurons, and the details of several kinds of morphological change that occur across infancy are known (e.g. pruning of synaptic connections, selected as a function of experience; increased myelination and resultant improvements in neural conduction; growth of patterns of connectivity within V1 and other visual areas). In general, however, the best way to characterize the observable changes in behaviour that may result from neuro-physiological development remains a source of controversy among researchers and theoreticians.

The rest of the chapter is divided into five sections. The first three are concerned with basic visual function; eye movements, attention, and visual stability; and perception of object unity; each of which is organized into subsections on definitions, methods, and evidence and theory. A "conclusions" section then presents a brief summary, and is followed by some suggestions for future research directions. It will become clear that research on individual topics in infant visual development is informed by disparate standpoints, reflecting what is known about each topic (with respect to human development) and the neural loci of the functions within the visual system (which is often inferred from research with animals and human adults with brain lesions).

BASIC VISUAL FUNCTION

Definitions

Basic visual function can be defined in terms of the ability to discern detail in the scene, as well as the luminance, colour, motion, orientation, and depth of object surfaces. These low-level attributes are used to determine surface and object boundaries and their distance from the observer. *Visual acuity* is an index of the observer's ability to perceive fine detail. Discrimination of variance in luminance is quantified as *contrast sensitivity*, the minimum detectable luminance contrast, which is often measured with striped gratings that vary in spatial frequency (stripe width) and contrast (the luminance difference between light and dark stripes). The ability to detect chromatic (colour) variations is quantified as *wavelength sensitivity*, a measure of the minimum detectable colour contrast. Sensitivity to motion is somewhat difficult to quantify, because infants at birth are able to follow object movement with their eyes, but exhibit limitations in *direction sensitivity* (e.g. distinguishing coherent from incoherent motion, or categorizing leftward versus rightward motion), as indexed by other methods, for several months after birth. *Orientation sensitivity* is the ability to distinguish edges of varying orientations. Finally, *depth perception* is also difficult to quantify, because there are at least 15 sources of information specifying the relative and absolute distances of surfaces from the observer (Cutting & Vishton, 1995). Yonas and Granrud (1984) suggested that infants are first sensitive to kinetic (motion-based) depth information (at around 3 weeks), followed by binocular information (i.e. sensitivity to binocular disparity or stereopsis, the difference between the inputs to the two retinae at around 4–5 months), and finally pictorial information (i.e. information that can be depicted in a two-dimensional scene at around 6 months).

Methods

A method that has proven to be particularly fruitful in determining infant visual capacity is the *preferential looking* technique. This method capitalizes on the tendency of infants to look longer at a visual stimulus that is more interesting or novel, relative to a second, less interesting stimulus. A popular variant of this technique is forced-choice preferential looking, or FPL, in which an infant is presented with two stimuli, placed side-by-side (Teller, 1977). The infant is observed, and looking time to each side is recorded. FPL has been used, for example, to assess the development of contrast sensitivity by placing a uniform grey stimulus alongside gratings of varying contrasts and spatial frequencies, while recording infants' responses (Atkinson *et al.*, 1977). The assumption is that the infants will look more at the striped stimulus if the stripes are discriminable. If contrast sensitivity is insufficient to resolve the stripes, looking at the two sides will be more or less equivalent. A variant of FPL is the *habituation* method, in which the infant is presented with a single stimulus repeatedly, as looking times are recorded. After many presentations looking typically wanes, which is assumed to reflect an increase in familiarity and a decline in interest on the infants' part. After looking declines to a preset criterion (e.g. half the initial looking times, measured across blocks of three trials), the infant is presented with a pair of new stimuli, either sequentially or side-by-side. Given the tendency of infants to look longer at relatively novel stimuli after habituation (Bornstein, 1985), a preference for one versus the other post-habituation stimulus is thought to reflect a novelty preference. The researcher can determine which aspects of a visual display are salient to the infant by manipulation of stimulus characteristics.

An alternative method for exploring the development of some basic visual functions is the recording of *visual-evoked potentials*, or VEP, in which electrodes are placed against the back of the skull in order to record activity in the visual cortex. VEP has been used, for example, to assess the development of visual acuity by presenting infants with grating stimuli that reverse phase (i.e. black stripes change to white, and vice-versa) in a particular temporal pattern (Norcia & Tyler, 1985). If changes in the VEP are time-locked to changes in the stimulus, it is assumed that there are cortical responses to the information in the stimulus. By making the stripes progressively smaller, therefore, the lower limits of visual acuity can be assessed.

Evidence and theory

Vision is poor in the neonate, and it is estimated that visual acuity is about 20/600 (Brown et al., 1987). Acuity, contrast and wavelength sensitivity, motion and orientation sensitivity, and depth perception all undergo improvement over the first few months after birth. Questions of mechanisms of development have centred around whether these improvements are best characterized as sub-cortical (exclusive of cortical structures) or cortical, or some combination of the two (cf. the chapter by Sireteanu in this handbook). The effective field of view in infants expands considerably after birth, as indexed with preferential looking methods when a stimulus is introduced into the periphery (Hood, 1995; Maurer & Lewis, 1998).

The development of contrast and wavelength sensitivity and acuity can be ascribed in part to changes in the eye itself, such as the size and shape of rods and cones, and their migration towards the fovea (the retina's centre, containing a high concentration of cones, where acuity is maximized) (Abramov et al., 1982; Yuodelis & Hendrickson, 1986). There are also improvements in the eye's optics, and growth of the eyeball (such that the retinal image becomes larger and falls on more receptors). Banks & Shannon (1993) incorporated information from retinal and optical immaturity into an "ideal observer", a computer model of the visual system whose performance on various visual tasks was compared with expected performance based on subcortical limitations. Banks and Shannon determined that limitations in contrast sensitivity, wavelength sensitivity, and acuity cannot be entirely explained by subcortical development, thus implicating such post-retinal factors as increased neural efficiency (increasing the signal-to-noise ratio throughout the visual system) and the development of subcortical–cortical and cortico–cortical pathways. At present little is known about the precise contributions of development in humans of these post-retinal factors to improvements in visual acuity after birth.

Motion sensitivity is present in rudimentary form at birth, indicated by the following of moving stimuli with the eyes. However, VEPs do not reveal cortical responses to motion until some time after 8 weeks of age, suggesting that neonates' looking responses to moving stimuli are mediated by subcortical mechanisms (Wattam-Bell, 1991). Direction sensitivity as revealed by preferential looking or habituation paradigms is poor in infants younger than 6–8 weeks (e.g. Wattam-Bell, 1996; but see Laplante et al., 1996), confirming the suggestion that cortical processing of motion is unlikely to be robust in very young infants (see Banton & Bertenthal, 1997). The emergence of cortical motion sensitivity might be a function of the development of pathways that include a motion processing area, which has been called MT in the monkey (M.H. Johnson, 1990; Movshon &

Newsome, 1992). These pathways are thought to be driven primarily by the magnocellular system.

Similarly, orientation discrimination is present at birth: after a period of familiarization to a single orientation, neonates often will look longer at a novel orientation (Slater *et al.*, 1988). Orientation discrimination is not reflected in the VEP, and undergoes a gradual strengthening over the first several postnatal months in response to changes in orientation (Braddick *et al.*, 1986; Hood *et al.*, 1992). In contrast to motion sensitivity, however, the evidence is stronger for cortical involvement in orientation sensitivity to birth. This is because neonates will exhibit habituation to one orientation and recover interest to a different post-habituation orientation (Slater *et al.*, 1983a), and habituation and recovery of interest are most likely cortical phenomena (see Slater, 1995), although the precise neural mechanisms underlying habituation have not yet been identified. Neonates have also been found to discriminate among different shapes with this technique (e.g. Slater *et al.*, 1983b). Shape identification is associated with the inferotemporal cortex, which is driven largely by the parvocellular system.

Depth perception, as discussed previously, is composed of sensitivity to and utilization of three classes of visual information: kinetic information, binocular disparity, and pictorial information. Sensitivity to kinetic information was discussed in the previous paragraph. Little is known about the neuro-physiological basis of sensitivity to pictorial depth information, although its developmental timing has received extensive investigation: infants appear to begin utilizing this information to guide reaching some time between 5 and 7 months (see Yonas & Granrud, 1984). However, the onset of stereopsis has been well investigated with respect to its neurophysiological origins (Held, 1993). Functional stereopsis depends on the appropriate identification of "eye-of-origin" information; that is, whether input to a particular cortical visual neuron is from the left or the right eye. Some neurons in adult visual cortex respond primarily to left eye input and others primarily to right eye input. Later processing stages combine this information. Stereopsis is thought to be a function of the segregation of *ocular dominance columns* in the input layer (layer IV) of area V1. Prior to the onset of stereopsis, inputs from the left and right eyes synapse onto the same neurons in V1, precluding the provision of eye-of-origin information to later processing stages. After about 4 months in humans (earlier in other primates), inputs from the two eyes segregate such that they synapse largely onto separate neurons in layer IV. From there, signals carry eye-of-origin information to cells that process disparity (DeAngelis *et al.*, 1998; Poggio *et al.*, 1988), but developmental aspects of this latter process remain to be investigated in humans.

Several researchers have proposed that, given the delay in emergence of visual functions thought to be supported by the magnocellular system (e.g. direction sensitivity, stereopsis) relative to parvocellular functions (e.g. orientation sensitivity, shape discrimination), the parvocellular processing stream matures more rapidly than the magnocellular stream (e.g. Atkinson, 1992). This proposal enjoys additional support from research on individual differences in global (i.e. large-scale) versus local (i.e. small-scale) processing of stimulus elements (Colombo, 1995) and eye movements in infancy, some of which is presented in the next section.

EYE MOVEMENTS, ATTENTION, AND VISUAL STABILITY

Definitions

James (1890/1950) defined *attention* as "the taking possession by the mind, in clear and vivid form, of one out of what seem several simultaneously possible objects or trains of thought" (pp. 403–404). This definition considers stimulus selection and deselection as hallmarks of attentional engagement and disengagement, implying "withdrawal from some things in order to deal effectively with others" (p. 404). Recent theorizing has centred around eye movements as indices of infant attention, on the basis that infants, like adults, are motivated to foveate stimuli of interest. Eye movements are produced by six extraocular muscles connected to the eyeball, and these are controlled directly by neurons in the brainstem. Several subcortical and cortical structures provide signals to the brainstem in the adult, and thereby influence eye movements indirectly (Büttner & Büttner-Ennever, 1988; Schiller, 1998).

Recently, models of infant neurophysiological development have been proposed that are based on changes in eye movement patterns over the first few postnatal months (e.g. M.H. Johnson, 1990). There are four primary eye movement systems, each of which is thought to be produced by separate circuitry (although there is some disagreement about overlap in the functions of some circuits; see von Hofsten & Rosander, 1998), and all channelled through the brainstem. *Saccades* are the most common form of eye movement, consisting of quick changes of fixation whose function is to place the retinal image of an object of interest on the fovea for closer inspection. *Smooth pursuit* refers to the tracking of small targets moving against a background, in order to maintain foveation of the target. The *optokinetic response* (also known as *optokinetic nystagmus*, or OKN) is a semi-reflexive (i.e. it can be suppressed consciously) eye movement pattern driven by a large visual array that moves with respect to a stationary observer. This can be understood by imagining oneself in a car looking out the window at the passing terrain. The eyes follow the terrain to the right and then quickly "snap" back left, following right, snap left, and so on in a repetitive fashion. The *vestibulo-ocular response* (or VOR) is a semi-reflexive eye movement pattern driven by a stationary large visual array placed in front of, or surrounding, a moving observer. This can be understood by imagining oneself in a swivel chair, rotating anticlockwise, in an otherwise stationary room. The eyes fixate some object in the room and hold fixation briefly (moving to the right with respect to the head, and thereby counter-rotating with respect to the body) and then quickly snap back (moving left with respect to the head), fixate some other object, snap back, and so on in a repetitive fashion. The relation of eye movements to the head is similar in both OKN and VOR: a relatively slow, smooth track followed by a quick saccade-like movement. However, the stimulus conditions and the underlying neural circuitry controlling each are different.

Visual stability refers to the ability to stabilize the retinal image despite perturbations due to eye, head, and body movements, and motion in the observer's surroundings. Visual stability is vital for the effective extraction of detailed information about the visual environment, because the observer is rarely stationary. Retinal-image stabilization is achieved by a combination of three oculomotor systems: OKN, VOR, and smooth pursuit. With appropriate control of all three systems the observer can maintain fixation on targets even if he or she is moving, if the background or target is moving, or some combination of these situations. For example, suppose an observer is looking out of the window of a train, and a fly lands on the window. In order to fixate the fly the tendency towards OKN (driven by large-field motion, such as the landscape seen out the window) must be cancelled, by initiating smooth pursuit in the opposite direction relative to OKN, resulting in a stable gaze. Stereopsis and vergence eye movements also contribute to visual stability, because fixed targets and background are often at different distances from the observer.

Methods

Saccades can be recorded with a *corneal reflection* system, which incorporates a camera that obtains an image of the pupil and an infrared reflection of the cornea (the outside surface of the eye). The centres of the pupil and cornea shift relative to one another as the observer looks around, and a comparison of these centres across time reveals where the observer is looking with great temporal and spatial precision (although questions have been raised concerning the accuracy of corneal reflection with infants; see Slater & Findlay, 1972). Saccades can also be recorded with videotape, which is later coded off-line (frame-by-frame) with respect to changes in fixation. Some temporal resolution is sacrificed with videotape, and spatial resolution is crude, but sufficient for some purposes (e.g. if one only wishes to know whether the infant looks towards the right or left side of a display). Smooth pursuit can be recorded with corneal reflection, but many researchers prefer *electro-oculography* (EOG), consisting of recordings made from small electrodes placed near the outer canthi (the corner of the eye next to the template). The eye is asymmetrical with respect to polarity (the front of the eye is more positive than the rear), and the electrodes measure the difference in polarity at the skin on the left and right sides of the face as the eyes move. EOG offers high temporal resolution and avoids many of the technical difficulties imposed by corneal reflection, such as the necessity of keeping the infant's head steady (to maintain the pupil within camera range). A limitation of EOG is that it works best only when recording horizontal eye movements. Finally, eye movements have been recorded in fetuses via *ultrasonography* (Birnholz, 1981; Bots *et al.*, 1981), but little is known about the functional significance of these movements, other than the implication that some of the neural circuitry for saccade generation is in place before birth, and that the rapid eye movements characteristic of "active sleep" (i.e. dreaming) are functional by 36 weeks after conception (as well as eye inactivity, associated with deep sleep).

Contributions of OKN, VOR, and smooth pursuit to visual stability have been recorded with EOG while the infant is either stationary and views a large display (Aslin & Johnson, 1996), or is rotated or moved laterally in front of the display (Daniel & Lee, 1990; Aslin & Johnson, 1994). Infants have also been placed inside a large drum that either rotates as the infant remains stationary (von Hofsten & Rosander, 1996), or remains stationary as the infant is rotated inside it (von Hofsten & Rosander, 1997). The stimulus consists of moving texture that fills most of the visual field; this surface can also be applied to the inside of the drum. It is also possible to record head movements with a motion-registration system concurrently with eye movements; eye and head movements are then combined to indicate point of gaze (Aslin & Johnson, 1994; Daniel & Lee, 1990; von Hofsten & Rosander, 1998). Visual stability under conditions that would normally elicit OKN or VOR can be assessed by the extent to which the infant looks at a small target placed in front of the background, and thereby suppresses the tendency to engage in OKN or VOR (Aslin & Johnson, 1994, 1996).

Evidence and theory

Neonates, if awake and alert, examine their surroundings with a series of fixations. Young infants' fixations, however, often do not extent beyond areas of high contrast, such as edges (Bronson, 1994), or remain centred around a limited set of stimulus features (Bronson, 1991; S.P. Johnson & Johnson, 2001). In addition, several investigators have identified an "externality effect" that seems to bias young infants' responses to patterns in favour of outer contours, rather than internal detail (Bushnell et al., 1983; Milewski, 1976). The externality effect appears to diminish after around 1 month of age (Bushnell, 1979), and is often overcome when there is motion of the internal elements (Girton, 1979). Infants older than 3 months, in contrast, will more often scan in what appears to be an exploratory fashion. Older infants will also scan between individual stimuli more readily than will younger infants (Bronson, 1997).

This pattern of development has been interpreted as a shift from reflexive to more purposive scanning (e.g. M.H. Johnson, 1990). Further evidence of greater control over saccades with development comes from the visual expectation paradigm, in which an infant views a sequence of individually presented stimuli, one on either side of a display, that are shown individually according to some regular (or irregular) schedule (see Haith, 1993). The infant's eye movements are monitored for evidence that he or she anticipates the onset of a stimulus by directing a saccade towards the appropriate location before its appearance. Such a pattern has been reported in response to simple stimulus sequences in infants as young as 2 months (e.g. Wentworth & Haith, 1992), and improves rapidly in terms of anticipations to more complex sequences (Canfield & Haith, 1991). Reflexive saccades have been linked to a subcortical pathway from retina, through the thalamus, to V1 and then to the superior colliculus, a midbrain structure with inputs to the brainstem (see Schiller, 1998). The later emergence of purposive saccades is consistent with the maturation of a cortical pathway linking V1, parietal area LIP, the frontal eye fields in the frontal lobe, and the superior colliculus (Canfield & Smith, 1999; M.H. Johnson, 1990, 1995; Schiller, 1998). The frontal lobe has been implicated in other tasks involving planning, such as reaching to an appropriate, hidden location (Diamond, 1991).

In contrast to saccades, smooth pursuit appears to be limited in very young infants. When presented with a small, moving target, infants younger than 2 months will often attempt to track it with a series of "catch-up" saccades, rather than smooth eye movements (Aslin, 1981). After 2 months, infants are better able to stay on a target as it moves. There is some evidence that younger infants engage in short bouts of smooth pursuit if the target speed is not too high, but smooth pursuit is not robust (Kremenitzer et al., 1979; Roucoux, et al., 1983).

The limitation in pursuit is not an inability to move the eyes smoothly: neonates will readily engage in OKN, which contains a slow-movement component. Rather, very young infants may be incapable of engaging in predictive eye movements, such that the future location of a moving target cannot be computed (a function of the frontal eye fields; see Lynch, 1987), or they may be unable to track due to limitations in motion processing (a function of area MT; see Komatsu & Wurtz, 1989). Alternatively, it might be that immaturity of retinal photoreceptors prohibits firm registration of the target on the fovea, such that a series of saccades is necessary to recentre gaze and maintain fixation.

Several researchers have proved "covert" shifts of attention in infants, which are made in the absence of eye movements but are revealed by a facilitation in reaction time to a precued target. Such effects have been reported in 4-month-olds, but not younger infants (see M.H. Johnson, 1995). A related phenomenon is "inhibition of return" (IOR), which describes a *delay* in eye movements towards a previously cued location (the difference between facilitative and inhibitory effects of precuing centres around its timing; see Posner & Peterson, 1990). IOR has been demonstrated in 6-month-olds, but until recently was not reported in younger infants (see Clohessey *et al.*, 1991). Covert attention is thought to rely on a cortical circuit involving the parietal lobe, which is involved in processing of temporal information, and the superior colliculus (Posner & Dehaene, 1994). IOR has been linked to saccade planning, which presupposes involvement of the frontal eye fields (M.H. Johnson, 1995). One difficulty for this account is a pair of recent reports of IOR in neonates (Simion *et al.*, 1995; Valenza *et al.*, 1994), suggesting that, in its initial ontogenetic forms, IOR is mediated by subcortical mechanisms (as of yet unspecified).

Relative to eye movements, attentional disengagement has received little formal investigation. It has been reported that 1- and 2-month-olds will often exhibit much longer looking times than either neonates or 4-month-olds during habituation experiments (Hood *et al.*, 1996; S.P. Johnson, 1996; Slater *et al.*, 1996) and other looking time paradigms (e.g. M.H. Johnson *et al.*, 1991; Mohn & Van Hoff-van Duin, 1986). This so-called "sticky fixation" has been tied to tonic inhibition of the superior colliculus by the substantia nigra and basal ganglia, which is later released by cortical mechanisms subserving peripheral expansion of the visual field and attendant improvements in stimulus selection (M.H. Johnson, 1990; see also Hood *et al.*, 1998; Maurer & Lewis, 1998). However, there are other, more cognitive explanations for longer looking times in 1- and 2-month-olds (S.P. Johnson, 1996).

The final aspect of attention to be discussed is that of vigilance, or maintaining an alert state in readiness to inspect stimuli. Richards and colleagues have documented changes in oculomotor behaviour as a function of attentional state, which is indexed by heart rate. During periods of *sustained attention* (characterized by a deceleration in heart rate), infants are less easily distracted by peripheral stimuli if fixated on a central stimulus, and show enhanced refixation in the absence of a central stimulus, relative to other periods (Richards & Hunter, 1997). Sustained attention is also associated with improvements in smooth pursuit, an effect that increases between 8 and 26 weeks (Richards & Holley, 1999). The link between distractibility and heart rate deceleration has been found in infants as young as 8 weeks, but is not as robust as that observed in older infants (Richards, 1989). These results are consistent with the research, discussed previously, demonstrating gains in purposive eye movements (saccades and smooth pursuit) some time after 2 months. Richards & Hunter (1998) described these findings in terms of a general "arousal" system that enhances performance on a variety of visual tasks, such as form and colour discrimination, motion detection, and eye movements, by heightening the general responsiveness of the visual system and other sensory systems. The arousal system, which incorporates areas in frontal cortex, the limbic system, and the reticular formation, has widespread effects across the cortex and subcortical structures, and some neurotransmitter systems (see Richards & Hunter, 1998).

As mentioned previously, visual stability under conditions in which objects or the observer moves requires coordination between OKN, VOR, and smooth pursuit eye movement systems. Neonates have been found to exhibit OKN (Shupert & Fuchs, 1988). Before 4 months of age an intriguing asymmetry is evident when OKN is recorded monocularly: OKN is nearly absent when the stimulus elements move nasally (towards the nose), but is robust when stimulus elements move temporally (away from the nose). After 4 months this asymmetry disappears (Atkinson & Braddick, 1981). This developmental pattern has been taken as further evidence for a shift from subcortical to cortical control over visual function in the first few months after birth. This is because temporal OKN (but not nasal OKN) is found in many non-human species and has been linked to subcortical pathways (e.g. in cats, from retina through the pretectum, a midbrain structure, to the brainstem). In humans temporal OKN is also thought to rely on subcortical structures. The appearance of nasal OKN is hypothesized to be due to the influence of visual cortex (Atkinson & Braddick, 1981; Braddick *et al.*, 1996).

VOR, like OKN, is functional in neonates (Shupert & Fuchs, 1988). Unlike OKN, however, VOR appears in mature form in very young infants, implying that the structures responsible for VOR are in place at birth (e.g. von Hofsten & Rosander, 1996). VOR is generated by signals arising in the semicircular canals in the inner ear, which are activated by translatory or rotational acceleration of the head. These signals proceed directly to the brainstem, imparting a very high degree of spatial and temporal precision in controlling compensatory eye movements (Büttner & Büttner-Ennever, 1988). Suppression of OKN and VOR was not observed in 1-month-olds, but was found to be reliable in 2- and 4-month-olds (Aslin & Johnson, 1994, 1996). This timing is consistent with the increased control in smooth pursuit noted at 2 months (Aslin, 1981), and suggests that the onset of smooth pursuit provides significant contributions to the development of visual stability (i.e. in cancelling OKN and VOR).

PERCEPTION OF OBJECT UNITY

Definitions

Approaches to the development of object perception have focused on changes in infants' ability to perceive bounded, coherent entities that are separate from one another and from the background (e.g. S.P. Johnson, 1997; Needham *et al.*, 1997; Spelke, 1990). At present little is known about the development of the higher-level neural mechanisms subserving object perception, which are exceedingly complex (e.g. Felleman & Van Essen, 1991). One noteworthy approach to object perception is the representational framework for the extraction of shape from two-dimensional images described by Marr (1982). Marr's approach has been criticized for a lack of biological plausibility (e.g. Churchland, 1986), a concern I will not deal with here; the important point for the present discussion is that his approach to object perception makes the question seem more tractable. Marr parsed the problem into four components: (1) derivation of intensity values for each point in the image; (2) organization of these values into a set of primitive tokens that are grouped according to similarity, movement, and so on (the "primal sketch"); (3) arrangement of the primitives into a set of coherent surfaces, preserving their relative distances from the observer (the "2.5-D sketch"); and (4) finalization of a description of volumes in space that represent segregated objects (the "3-D model"). The neural computations that achieve each of the four steps in this order might be expected to be progressively more complex, and indeed the first two are realized largely in the initial stages of visual processing (e.g. from the retina through V1; see Zeki, 1993). Surface representation, or the 2.5-D sketch, is thought to be accomplished by "mid-level" cortical mechanisms, perhaps beginning in V1 (Nakayama *et al.*, 1996), and it is on this problem that this section of the chapter will concentrate.

The development of surface segregation and representation has been investigated with the *object unity* task. Infants are presented with a display that to adults appears to consist of a partly occluded object (e.g. a rod) whose centre is partly occluded by another object (e.g. a box). The question is whether infants respond to this display as if it consists of two objects (as an adult would report), or three objects (a top rod part, a centre box, and a bottom rod part), or perhaps some indeterminate percept. That is, the researcher attempts to distinguish between responses to the distal stimulus (what is "understood" by the observer) and the proximal stimulus (what is literally seen).

The object unity task recalls what has been termed the *binding problem*, which refers to the fact that an object is characterized by a specific set of physical properties, such as colour, size, reflectance, shape, and spatial location (S.P. Johnson, 2001). To appropriately segment individual objects, the brain must somehow bind together their individual characteristics to form an impression of separate, coherent entities, while avoiding mix-ups between object features (e.g. if presented with a green circle and a red square, binding green and round rather than green and square). In the object unity task the binding problem is evident in the fact that the observer must note that the top and bottom rod parts belong together, and that they share similar attributes (i.e. a particular colour, size, orientation, and motion, and a roughly similar spatial location). Moreover, the rod parts must be distinguished from both the background and the occluder. A neurophysiological model of the binding problem is discussed subsequently, along with consideration of possible developmental mechanisms.

Methods

Kellman & Spelke (1983) developed an elegant habituation method to explore 4-month-olds' perception of partly occluded objects. In one variant of this task the infants were habituated to a rod-and-box display, after which they viewed two test displays, presented singly and in alternation: a complete rod, and a "broken" rod, consisting of two aligned rod parts with a gap in the centre. Both test displays were consistent with the visible portions of the rod in the original (rod-and-box) display. Kellman and Spelke reported that, when the two rod parts were aligned and underwent common lateral motion above and below the box, the infants looked significantly longer during test at the broken rod, relative to the complete rod (infants in a control group exhibited no consistent test display preference). Given that infants generally look longer at novel displays after habituation, these results were taken to indicate that the infants perceived the rod's unity, despite incomplete information for connectedness.

Evidence and theory

In addition to lateral motion, 4-month-olds appear to perceive object unity in displays in which the rod moves vertically, or back and forth in depth (Kellman et al., 1986). This effect obtains even if the top rod part is paired with a highly dissimilar surface, if the two surfaces move together (Kellman & Spelke, 1983). In contrast, young infants do not perceive unity if the display is stationary, or if the box moves relative to a stationary rod, or if the rod and box move together (Kellman & Spelke, 1983; cf. Craton, 1996). To account for these results, Kellman and Spelke proposed that some aspects of object perception are available at birth: "Humans may begin life with the notion that the environment is composed of things that are coherent, that move as units independently of one another, and that tend to persist, maintaining their coherence and boundaries as they move" (1983, p. 521). A modified form of this position was offered by Kellman (1993), who suggested that, before 6 months, infants are "edge-insensitive", meaning that edge orientation (e.g. alignment) does not enter into the unit formation process. Rather, motion alone dictates perception of object unity. Some time after 6 months edge sensitivity emerges, such that perception of object unity can be based on non-motion information, such as alignment and surface appearance, in addition to kinetic information.

These views have not fared well in light of recent evidence. First, perception of object unity does not appear to be available to infants at birth: after habituation to a rod-and-box stimulus, neonates look longer at a complete rod test display, relative to a broken rod (the opposite pattern to that observed in 4-month-olds). This implies perception of disjoint objects in the rod-and-box display (Slater et al., 1996). Two-month-olds have been found to perceive object unity (S.P. Johnson & Aslin, 1995), but their performance is not as robust as that of older infants (S.P. Johnson & Náñez, 1995). Second, perception of object unity in 4-month-olds is strongly dependent on edge alignment: if the rod parts are not aligned, unit formation appears to be blocked, despite common motion (S.P. Johnson & Aslin, 1996). In addition to edge alignment and motion, 4-month-olds have been found sensitive to other kinds of visual information in surface segregation tasks: local and global configuration, or "good form" (S.P. Johnson et al., 2000), depth information (S.P. Johnson & Aslin, 1996; Smith et al., 2001), colour (S.P. Johnson & Aslin, 2000), illusory contours (S.P. Johnson & Aslin, 1998), and others (see S.P. Johnson, 2000). To account for these results, S.P. Johnson and

665

Aslin (1996; S.P. Johnson, 1997, 2000) proposed a *threshold model*: veridical surface segregation depends on both the visual information available in a particular display, and the readiness of the infant to attend to this information (which may vary, depending on age). That is, veridical object perception occurs when *sufficiency* of visual information is met with *efficiency* of perceptual and cognitive skills. There does not seem to be a single cue that uniquely supports unit formation. Rather, object unity and other surface attributes appear to be multiply specified to young infants, as they are to adults. In infants, however, performance is highly age-dependent: there is no available evidence of veridical surface segregation or perception of object unity in neonates, but these skills emerge rapidly such that, by 2 months, infants perceive unity under some circumstances, and these abilities become much more robust by 4 months (S.P. Johnson, 1997).

How might infants' perception of object unity offer insights into possible developmental mechanisms involved in the brain's solution to the binding problem? Visual attributes are coded in parallel across the cortex, and the visual system is organized hierarchically: early stages (V1) contain cells that are individually tuned to a variety of aspects of the visual scene (orientation, colour, motion, and so on) whereas cells in areas that are higher in the hierarchy are tuned to more specific features. It has been speculated that the binding problem is solved with patterns of synchronous firing of assemblies, or networks, of cells across the hierarchy (Crick & Koch, 1990; Engel *et al.*, 1992; Singer, 1993, 1994). On this *temporal coding* view a visual stimulus activates a constellation of feature detectors throughout the visual system. Individual stimuli will tend to activate unique cell assemblies, the global activity of assemblies' subcomponents binding together stimulus features to impart a stable percept of distinct, coherent objects. The "glue" that binds together unique object representations is the synchrony of neuronal discharges, at around 40 Hz, in the various visual areas. This synchrony is governed by attention to individual objects, and is transient, lasting only a few hundred ms (Revonsuo *et al.*, 1997).

Much of the evidence in favour of temporal coding comes from demonstrations of synchronized responses in spatially separate regions of the cat and monkey visual system (in both subcortical and cortical areas), obtained from single-cell recordings (e.g. Eckhorn *et al.*, 1988; Gray *et al.*, 1989; Roelfsema *et al.*, 1997; Kreiter & Singer, 1996). In humans, evidence for temporal coding in perceptual and learning tasks has been obtained with EEG (Miltner *et al.*, 1999; Rodriguez *et al.*, 1999) and with stimulus-timing methods (Sohmiya *et al.*, 1998).

The neural mechanisms subserving perception of object unity are unclear, but the temporal coding hypothesis suggests that the rod parts are perceptually bound due in part to the synchronized responses of cells across active visual areas. In rod-and-box displays with fewer cues (e.g. a stationary rod), fewer visual areas may be directly involved, yet adults may still readily perceive unity. However, individual areas in the infant's visual system may not yet be as finely tuned to surface characteristics as they are in adults. Moreover, there is probably a higher degree of neural "noise" (and a resultant reduction in signal/noise ratio) imposing limitations on efficient neural transduction in the young infant's cortex, which may restrict, for example, contrast sensitivity and other low-level functions (Skoczenski & Aslin, 1995; Skoczenski & Norcia, 1998; cf. Banks & Shannon,

1993). Therefore, cell circuitries in infants analogous to those in adults may be engaged to some extent by a particular display, yet the totality of their responses may be insufficient to activate a percept of a partly occluded surface. In this sense, then, the threshold model may bear important implications to a viable theory of the developmental neurobiology of object perception: more visual information is more likely to activate synchronization mechanisms, which are necessary to impart veridical percepts. It is also possible that synchronization provides important inputs into cortical development, by guiding certain experience-dependent processes such as patterns of neural connectivity and differentiation of neurotransmitters (see Greenough *et al.*, 1987; S.P. Johnson, 2000, 2001; Rodman, 1994; Singer, 1995, for discussion).

CONCLUSIONS

Clearly, infants are not born as "blank slates". Neonates are endowed with a complement of reflexive eye movements, and exhibit some rudimentary visual skills, such as sensitivity to variations in contrast and motion. Visual skills are limited, however, and undergo rapid advances over the ensuing few months. Most can be accounted for by the operations of subcortical structures. A general pattern across the four domains of research explored in this chapter is that the period from birth to 4 months sees a remarkable rate of development of the cortical functions subserving more sophisticated visual proficiency, such as visual stability and object perception. The strong parallels in timing across domains are suggestive of common developmental mechanisms underlying these changes, and comprehensive theories of what these mechanisms might be are beginning to emerge. For example, M.H. Johnson (1997) proposed an account of the development of experience-dependent and experience-independent cortical architectures and representations based on evidence from infant development, neurophysiology, and connectionist modelling. This account has elicited controversy (see S.P. Johnson, 1998), but it is very well argued, highly provocate, and thoughtful.

FUTURE RESEARCH DIRECTIONS

The development of cortical functionality

The question pervading each of the preceding sections in this chapter concerns the extent to which a functioning visual cortex supports sensory, perceptual, and cognitive abilities in the infant, or whether these abilities rely more on subcortical functions (cf. Atkinson, 1992; M.H. Johnson, 1990; Maurer & Lewis, 1979; Slater, 1995; Sokol & Jones, 1979). To date the majority of evidence bearing on this question is indirect, coming from behavioural methods. More direct evidence is needed. The use of electroencephalography (EEG) to measure electrical activity across the scalp is becoming more common (e.g. in the work of D. and V. Molfese, M.H. Johnson, and J. Richards), suggesting that important advances in answering questions of developments in cortical function across infancy are on the horizon.

Attentional disengagement

Recent reports of IOR in neonates (Simion *et al.*, 1995; Valenza *et al.*, 1994) suggest that this behaviour is mediated by subcortical systems, but it is not yet clear whether other kinds of disengagement are independent of cortical mechanisms. For example, sticky fixation has been described as difficulty disengaging attention, due to inhibition of eye movement signals produced by the superior colliculus, which peaks at 1–2 months (M.H. Johnson, 1990). However, S.P. Johnson (1996) suggested that this behaviour might actually reflect the tendency of young infants to look longer at stimuli that they are actively processing, such that they exhibit extended looking when confronted with a stimulus they are on the cusp of understanding. This latter, more cognitive account of disengagement is supported by evidence of differences in looking times in 4-month-olds as a function of stimulus characteristics, depending on whether the infants appear to perceive occlusion in the displays (S.P. Johnson & Aslin, 2000). Clearly, additional evidence is needed to distinguish the contributions of cortical and subcortical mechanisms to attentional disengagement.

Individual differences: cortical processing streams

Colombo (1995) recently presented a neurophysiological account of individual differences in attention and visual processing in infancy, based on possible idiosyncratic dissimilarities in the developmental trajectories of magnocellular and parvocellular channels. Although largely based on perceptual paradigms, this approach has important implications for advancing our understanding of cognitive development, because of the continuity of some information processing mechanisms from infancy to childhood (e.g. global versus local processing, post-habituation recovery to novelty), and would profit from additional, more direct evidence of individual differences in cortical function.

Individual differences: temperament

Individual differences have been observed in the effectiveness with which infants are able to disengage from a visual stimulus (e.g. M.H. Johnson *et al.*, 1991). That is, some infants appear more capable of self-regulation (in this case, regulating attention) than others, and self-regulation is strongly linked to temperament (see Ruff & Rothbart, 1996). Individual differences in temperament are observed in neonates, and there is evidence for continuity over development (Kagan, 1991). There are several theoretical approaches to the biological correlates of temperament (see Gunnar, 1990), but the possibility of common neurophysiological mechanisms between temperament and attention remains to be systematically investigated.

Infant state

Infant attention appears to be closely tied to their internal state; specifically, the activity of an arousal system whose engagement can be indexed by measuring heart rate (Richards & Hunter, 1998). Currently, few infant vision researchers take heart-rate measures during testing sessions, recording oculomotor behaviours exclusively. This may be largely because of the technical demands imposed by this semi-invasive procedure, and perhaps a lack of awareness of its utility. However, such independent information on the infant's internal state might prove invaluable to building more comprehensive models of developmental neurophysiological function.

Neurophysiological basis of habituation and novelty preferences

A number of experiments have documented that neonates will habituate and subsequently exhibit novelty preferences (see Slater, 1995). The logic of this method dictates that the infant forms some sort of representation of the habituated stimulus, and then compares it to the test stimuli, directing more attention to novel display. On the surface this behaviour would necessarily seem supported by cortical mechanisms, but this view was challenged with a sensory-adaptation model that posits only limited cortical involvement (Dannemiller & Banks, 1983). This challenge was refuted (Slater & Morison, 1985), and the refutation challenged (Dannemiller & Banks, 1986). At present the reliability of novelty preferences at birth to a wide variety of stimuli tends to support a view of more extensive cortical involvement (Slater, 1995). However, further research is necessary to clarify the mechanisms involved.

Synchronized cortical activity

The account of a neurophysiological solution to the binding problem, presented in the section on Perception of Object Unity, is plausible but remains to be tested empirically. The ontogeny of synchronization across cortical areas also remains to be investigated. Recent advances in technology to record EEG in young infants offer great promise toward achieving these goals (e.g. Csibra et al., 2000; M.H. Johnson et al., 1998), given that recording of coherent EEG activity in adults is possible with such systems (Miltner et al., 1999; Rodriguez et al., 1999; Tallon-Baudry & Bertrand, 1999).

References

Abramov, I., Gordon, J., Hendrickson, A., Hainline, L., Dobson, V. & La Bossiere, E. (1982). The retina of the newborn infant. *Science*, **217**, 265–267.

Aslin, R.N. (1981). Development of smooth pursuit in human infants. In D.F. Fisher, R.A. Monty & J.W. Senders (Eds.), *Eye movements: cognition and visual perception* (pp. 31–51). Hillsdale, NJ: Erlbaum.

Aslin, R.N. & Johnson, S.P. (1994, May). Suppression of VOR by human infants. Poster presented at the Association for Research in Vision and Ophthalmology conference, Sarasota, FL.

Aslin, R.N. & Johnson, S.P. (1996). Suppression of the optokinetic reflex in human infants: implications for stable fixation and shifts of attention. *Infant Behavior and Development*, **19**, 233–240.

Atkinson, J. (1992). Early visual development: differential functioning of parvocellular and magnocellular pathways. *Eye*, **6**, 129–135.

Atkinson, J. & Braddick, O. (1981). Development of optokinetic nystagmus in infants: an indicator of cortical binocularity? In D.F. Fisher, R.A. Monty & J.W. Senders (Eds.), *Eye movements: cognition and visual perception* (pp. 53–64). Hillsdale, NJ: Erlbaum.

Atkinson, J., Braddick, O. & Moar, K. (1977). Development of contrast sensitivity over the first 3 months of life in the human infant. *Vision Research*, **17**, 1037–1044.

Banks, M.S. & Shannon, E. (1993). Spatial and chromatic visual efficiency in human neonates. In C.E. Granrud (Ed.), *Visual perception and cognition in infancy* (pp. 1–46). Hillsdale, NJ: Erlbaum.

Banton, T. & Bertenthal, B.I. (1997). Multiple developmental pathways for motion processing. *Optometry and Vision Science*, **74**, 751–760.

Birnholz, J.C. (1981). The development of human fetal eye movement patterns. *Science*, **213**, 679–681.

Bornstein, M.H. (1985). Habituation of attention as a measure of visual information processing in human infants: summary, systematization, and synthesis. In G. Gottlieb & N.A. Krasnegor (Eds.), *Measurement of audition and vision in the first year of postnatal life: a methodological overview* (pp. 253–300). Norwood, NJ: Ablex.

Bots, R.S.G.M., Nijhius, J.G., Martin, C.B. Jr & Precht, H.F.R. (1981). Human fetal eye movements: detection *in utero* by ultrasonography. *Early Human Development*, **5**, 87–94.

Braddick, O.J., Atkinson, J. & Hood, B. (1996). Striate cortex, extrastriate cortex, and colliculus: some new approaches. In F. Vital-Durand, J. Atkinson & O.J. Braddick (Eds.), *Infant vision* (pp. 203–220). Oxford: Oxford University Press.

Braddick, O.J., Wattam-Bell, J. & Atkinson, J. (1986). Orientation-specific cortical responses develop in early infancy. *Nature*, **320**, 617–619.

Bronson, G.W. (1974). The postnatal growth of visual capacity. *Child Development*, **57**, 251–274.

Bronson, G.W. (1991). Changes in infants' visual scanning across the 2- to 14-week age period. *Journal of Experimental Child Psychology*, **49**, 101–125.

Bronson, G.W. (1994). Infants' transitions toward adult-like scanning. *Child Development*, **65**, 1243–1261.

Bronson, G.W. (1997). The growth of visual capacity: evidence from infant scanning patterns. In C. Rovee-Collier & L.P. Lipsitt (Eds.), *Advances in infancy research* (vol. 11, pp. 109–141). Greenwich, CN: Ablex.

Brown, A.M., Dobson, V. & Maier, J. (1987). Visual acuity of human infants at scotopic, mesopic, and photopic luminances. *Vision Research*, **27**, 1845–1858.

Bushnell, I.W. (1979). Modification of the externality effect in young infants. *Journal of Experimental Child Psychology*, **28**, 211–229.

Bushnell, I.W., Gerry, G. & Burt, K. (1983). The externality effect in neonates. *Infant Behavior and Development*, **6**, 151–156.

Büttner, U. & Büttner-Ennever, J.A. (1988). Present concepts of oculomotor organization. In J.A. Büttner-Ennever (Ed.), *Neuroanatomy of the oculomotor system* (pp. 3–32). New York: Elsevier.

Canfield, R.L. & Haith, M.M. (1991). Young infants' visual expectations for symmetric and asymmetric stimulus sequences. *Developmental Psychology*, **27**, 198–208.

Canfield, R.L. & Smith, E.G. (1999, April). Making connections: visual expectations and the functional development of cortical visual pathways. Paper presented at the Society for Research in Child Development conference, Albuquerque, NM.

Churchland, P.S. (1986). *Neurophilosophy: toward a unified science of the mind–brain*. Cambridge, MA: MIT Press.

Clohessy, A.B., Posner, M.I., Rothbart, M.K. & Vecera, S.P. (1991). The development of inhibition of return in early infancy. *Journal of Cognitive Neuroscience*, **3**, 345–350.

Colombo, J. (1995). On the neural mechanisms underlying developmental and individual differences in visual fixation in infancy: two hypotheses. *Developmental Review*, **13**, 97–135.

Craton, L.G. (1996). The development of perceptual completion abilities: infants;' perception of stationary, partially occluded objects. *Child Development*, **67**, 890–904.

Crick, F. & Koch, C. (1990). Towards a neurobiological theory of consciousness. *Seminars in the Neurosciences*, **2**, 263–275.

Csibra, G., Davis, G., Spratling, M.W. & Johnson, M.H. (2000). Gamma oscillations and object processing in the infant brain. *Science*, **290**, 1582–1585.

Cutting, J.E. & Vishton, P.M. (1995). Perceiving layout: the integration, relative potency, and contextual use of different information about depth. In W. Epstein & S. Rogers (Eds.), *Handbook of perception and cognition, vol. 5: Perception of space and motion* (pp. 69–107). San Diego, CA: Academic Press.

Daniel, B.M. & Lee, D.N. (1990). Development of looking with the head and eyes. *Journal of Experimental Child Psychology*, **50**, 200–216.

Dannemiller, J.L. & Banks, M.S. (1983). Can selective adaptation account for early infant habituation? *Merrill-Palmer Quarterly*, **29**, 151–158.

Dannemiller, J.L. & Banks, M.S. (1986). Testing models of infant habituation: a reply to Slater and Morison. *Merrill-Palmer Quarterly*, **32**, 87–91.

Daw, N.W. (1995). *Visual development*. New York: Plenum.

DeAngelis, G.C., Cumming, B.G. & Newsome, W.T. (1998). Cortical area MT and the perception of stereoscopic depth. *Nature*, **394**, 677–680.

DeYoe, E.A. & Van Essen, D.C. (1988). Concurrent processing streams in monkey visual cortex. *Trends in Neurosciences*, **11**, 219–226.

Diamond, A. (1991). Neuropsychological insights into the meaning of object concept development. In S. Carey & R. Gelman (Eds.), *The epigenesis of mind: essays on biology and cognition* (pp. 67–110). Hillsdale, NJ: Erlbaum.

Eckhorn, R., Bauer, R., Jordan, W., Brosch, M., Kruse, W., Munk, M. & Reitboeck, H.J. (1988). Coherent oscillations: a mechanism of feature linking in the visual cortex? *Biological Cybernetics*, **60**, 121–130.

Elman, J.L., Bates, E.A., Johnson, M.H., Karmiloff-Smith, A., Parisi, D. & Plunkett, K. (1996). *Rethinking innateness: a connectionist perspective on development*. Cambridge, MA: MIT Press.

Engel, A.K., König, P., Kreiter, A.K., Schillen, T.B. & Singer, W. (1992). Temporal coding in the visual cortex: new vistas on integration in the nervous system. *Trends in Neurosciences*, **15**, 216–226.

Felleman, D.J. & Van Essen, D.C. (1991). Distributed hierarchical processing in the primate cerebral cortex. *Cerebral Cortex*, **1**, 1–47.

Girton, M.R. (1979). Infants' attention to intrastimulus motion. *Journal of Experimental Child Psychology*, **28**, 416–423.

Gray, C.M., König, P., Engel, A.K. & Singer, W. (1989). Oscillatory responses in the cat visual cortex exhibit inter-columnar synchronization which reflects global stimulus properties. *Nature*, **338**, 334–337.

Greenough, W.T. & Alcantara, A. (1993). The roles of experience in different developmental information stage processes. In B. de Boysson-Bardies, S. de Schonen, P., Jusczyk, P. MacNeilage & J. Morton (Eds.), *Changes in speech and face processing in infancy: a glimpse at developmental mechanisms of cognition* (pp. 3–16). Dordrecht: Kluwer.

Greenough, W.T., Black, J.W. & Wallace, C.S. (1987). Experience and brain development. *Child Development*, **58**, 539–559.

Gunnar, M.R. (1990). The psychobiology of infant temperament. In J. Colombo & J. Fagen (Eds.), *Individual differences in infancy: reliability, stability, and prediction* (pp. 387–409). Hillsdale, NJ: Erlbaum.

Haith, M.M. (1993). Future-oriented processes in infancy: the case of visual expectations. In C.E. Granrud (Ed.), *Visual perception and cognition in infancy* (pp. 235–264). Hillsdale, NJ: Erlbaum.

Held, R. (1993). Two stages in the development of binocular vision and eye alignment. In K. Simons (Ed.), *Early visual development: normal and abnormal* (pp. 250–257). New York: Oxford University Press.

Hood, B.M. (1995). Shifts of visual attention in the human infant: a neuroscientific approach. In C. Rovee-Collier & L.P. Lipsitt (Eds.), *Advances in infancy research* (vol. 9, pp. 163–216). Norwood, NJ: Ablex.

Hood, B.M., Atkinson, J. & Braddick, O. (1998). Selection-for-action and the development of orienting and visual attention. In J.E. Richards (Ed.), *Cognitive neuroscience of attention: a developmental perspective* (pp. 219–250). Mahwah, NJ: Erlbaum.

Hood, B.M., Atkinson, J., Braddick, O. & Wattam-Bell, J. (1992). Orientation selectivity in infancy: behavioural evidence for temporal sensitivity. *Perception*, **21**, 351–354.

Hood, B.M., Murray, L., King, F., Hooper, R., Atkinson, J. & Braddick, O. (1996). Habituation changes in early infancy: longitudinal measures from birth to 6 months. *Journal of Reproductive and Infant Psychology*, **14**, 177–185.

Hubel, D.H. & Wiesel, T.N. (1979). Brain mechanisms of vision. *Scientific American*, **241**, 45–53.

James, W. (1950). *The principles of psychology*. New York: Dover (original work published 1890).

Johnson, M.H. (1990). Cortical maturation and the development of visual attention in early infancy. *Journal of Cognitive Neuroscience*, **2**, 81–95.

Johnson, M.H. (1995). The development of visual attention: a cognitive neuroscience perspective. In M.S. Gazzaniga (Ed.), *The cognitive neurosciences* (pp. 735–747). Cambridge, MA: MIT Press.

Johnson, M.H. (1997). *Developmental cognitive neuroscience*. Cambridge, MA: Blackwell.

Johnson, M.H., Csibra, G. & Davis, G. (2000, July). Event-related potential responses to Kanizsa squares in infancy. Poster presented at the International Conference on Infant Studies, Brighton, UK.

Johnson, M.H., Gilmore, R.O. & Czibra, G. (1998). Toward a computational model of the development of saccade planning. In J.E. Richards (Ed.), *Cognitive neuroscience of attention: a developmental perspective* (pp. 103–130). Mahwah, NJ: Erlbaum.

Johnson, M.H., Posner, M.I. & Rothbart, M.K. (1991). Components of visual orienting in early infancy: contingency learning, anticipatory looking, and disengaging. *Journal of Cognitive Neuroscience*, **3**, 335–344.

Johnson, S.P. (1996). Habituation patterns and object perception in young infants. *Journal of Reproductive and Infant Psychology*, **14**, 207–218.

Johnson, S.P. (1997). Young infants' perception of object unity: implications for development of attentional and cognitive skills. *Current Directions in Psychological Science*, **6**, 5–11.

Johnson, S.P. (Ed.). (1998). Mark H. Johnson's *Developmental cognitive neuroscience* [Special issue]. *Early Development and Parenting*, **7** (3).

Johnson, S.P. (2000). The development of visual surface perception: insights into the ontogeny of knowledge. In C. Rovee-Collier, L. Lipsitt & H. Hayne (Eds.), *Progress in infancy research*, (vol. 1, pp. 113–154). Mahwah, NJ: Erlbaum.

Johnson, S.P. (2001). Visual development in human infants: binding features, surfaces, and objects. *Visual Cognition* (In press).

Johnson, S.P. & Aslin, R.N. (1995). Perception of object unity in 2-month-old infants. *Developmental Psychology*, **31**, 739–745.

Johnson, S.P. & Aslin, R.N. (1996). Perception of object unity in young infants: the roles of motion, depth, and orientation. *Cognitive Development*, **11**, 161–180.

Johnson, S.P. & Aslin, R.N. (1998). Young infants' perception of illusory contours in dynamic displays. *Perception*, **27**, 341–353.

Johnson, S.P. & Aslin, R.N. (2000). Young infants' perception of transparency. *Developmental Psychology*, **36**, 808–816.

Johnson, S.P. & Johnson, K.L. (2001). Young infants' perception of partly occluded objects: evidence from scanning patterns. *Infant Behavior and Development*, (In Press).

Johnson, S.P. & Náñez, J.E. (1995). Young infants' perception of object unity in two-dimensional displays. *Infant Behavior and Development*, **18**, 133–143.

Johnson, S.P., Bremner, J.G., Slater, A. & Mason, U. (2000). The role of good form in young infants' perception of partly occluded objects. *Journal of Experimental Child Psychology*, **76**, 1–25.

Kagan, J. (1991). Continuity and discontinuity in development. In S.E. Brauth, W.S. Hall & R.J. Dooling (Eds.), *Plasticity of development* (pp. 11–26). Cambridge, MA: MIT Press.

Kellman, P.J. (1993). Kinematic foundations of infant visual perception. In C.E. Granrud (Ed.), *Visual perception and cognition in infancy* (pp. 121–173). Hillsdale, NJ: Erlbaum.

Kellman, P.J. & Spelke, E.S. (1983). Perception of partly occluded objects in infancy. *Cognitive Psychology*, **15**, 483–524.

Kellman, P.J., Spelke, E.S. & Short, K.R. (1986). Infant perception of object unity from translatory motion in depth and vertical translation. *Child Development*, **57**, 72–86.

Komatsu, H. & Wurtz, R.H. (1989). Relation of cortical areas MT and MST to smooth pursuit eye movements. In M. Ito (Ed.), *Neural programming* (pp. 137–148). Tokyo: Japan Scientific Societies Press.

Kreiter, A.K. & Singer, W. (1996). Stimulus-dependent synchronization of neuronal responses in the visual cortex of the awake macaque monkey. *Journal of Neuroscience*, **16**, 2381–2396.

Kremenitzer, J.P., Vaughan, H.G. Jr, Kurtzberg, D. & Dowling, K. (1979). Smooth-pursuit eye movements in the newborn infant. *Child Development*, **50**, 442–448.

Laplante, D.P., Orr, R.R., Neville, K., Vorkapich, L. & Sasso, D. (1996). Discrimination of stimulus rotation by newborns. *Infant Behavior and Development*, **19**, 271–279.

Livingstone, M.S. & Hubel, D.H. (1983). Specificity of cortico-cortical connections in monkey visual system. *Nature*, **304**, 531–534.

Lynch, J.C. (1987). Frontal eye field lesions in monkeys disrupt visual pursuit. *Experimental Brain Research*, **68**, 437–441.

Marr, D. (1982). *Vision*. San Francisco, CA: Freeman.

Maurer, D. & Lewis, T.L. (1979). A physiological explanation of infants' early visual development. *Canadian Journal of Psychology*, **33**, 232–252.

Maurer, D. & Lewis, T.L. (1998). Overt orienting toward peripheral stimuli: normal development and underlying mechanisms. In J.E. Richards (Ed.), *Cognitive neuroscience of attention: a developmental perspective* (pp. 51–102). Mahwah, NJ: Erlbaum.

Milewski, A.E. (1976). Infants' discrimination of internal and external pattern elements. *Journal of Experimental Child Psychology*, **22**, 229–246.

Milner, A.D. & Goodale, M.A. (1995). *The visual brain in action*. Oxford: Oxford University Press.

Miltner, W.H.R., Braun, C., Arnold, M., Witte, H. & Taub, E. (1999). Coherence of gamma-band EEG activity as a basis for associative learning. *Nature*, **397**, 434–436.

Mishkin, M., Ungerleider, L.G. & Macko, K.A. (1983). Object vision and spatial vision: two cortical pathways. *Trends in Neurosciences*, **6**, 414–417.

Mohn, G. & Van Hof-van Duin, J. (1986). Development of the binocular and monocular fields of human infants during the first year of life. *Clinical Vision Science*, **1**, 51–64.

Movshon, J.A. & Newsome, W.T. (1992). Neural foundations of visual motion perception. *Current Directions in Psychological Science*, **1**, 35–39.

Movshon, J.A. & Newsome, W.T. (1996). Visual response properties of striate cortical neurons projecting to area MT in macaque monkeys. *Journal of Neuroscience*, **16**, 7733–7741.

Nakayama, K., He, Z.J. & Shimojo, S. (1996). Visual surface representation: a critical link between lower-level and higher-level vision. In D.N. Osherson (series Ed.) and S.M. Kosslyn & D.N. Osherson (vol. Eds.), *Visual Cognition, vol. 2: an invitation to cognitive science* (2nd ed., pp. 1–70). Cambridge, MA: MIT Press.

Needham, A., Baillargeon, R. & Kaufman, L. (1997). Object segregation in infancy. In C. Rovee-Collier & L.P. Lipsitt (Eds.), *Advances in infancy research* (vol. 11, pp. 1–44). Norwood, NJ: Ablex.

Norcia, A.M. & Tyler, C.W. (1985). Spatial frequency sweep VEP: visual acuity during the first year of life. *Vision Research*, **25**, 1399–1408.

Piaget, J. (1952). *The origins of intelligence in children*. New York: International Universities Press.

Piaget, J. (1954). *The construction of reality in the child*. New York: Basic Books.

Poggio, G.F., Gonzalez, F. & Krauser, F. (1988). Stereoscopic mechanisms in monkey visual cortex: binocular correlation and disparity selectivity. *Journal of Neuroscience*, **8**, 4531–4550.

Posner, M.I. & Dehaene, S. (1994). Attentional networks. *Trends in Neurosciences*, **17**, 75–79.

Posner, M.I. & Peterson, S.E. (1990). The attention system of the human brain. *Annual Review of Neuroscience*, **13**, 25–42.

Revonsuo, A., Wilenius-Emet, M., Kuusela, J. & Lehto, M. (1997). The neural generation of a unified illusion in human vision. *Neuroreport*, **8**, 3867–3870.

Richards, J.E. (1989). Sustained visual attention in 8-week-old infants. *Infant Behavior and Development*, **12**, 426–436.

Richards, J.E. & Holley, F.B. (1999). Infant attention and the development of smooth pursuit tracking. *Developmental Psychology*, **35**, 856–867.

Richards, J.E. & Hunter, S.K. (1997). Peripheral stimulus localization by infants with eye and head movements during visual attention. *Vision Research*, **37**, 3021–3035.

Richards, J.E. & Hunter, S.K. (1998). Attention and eye movements in young infants: Neural control and development. In J.E. Richards (Ed.), *Cognitive neuroscience of attention: a developmental perspective* (pp. 131–162). Mahwah, NJ: Erlbaum.

Rodman, H.R. (1994). Development of inferior temporal cortex in the monkey. *Cerebral Cortex*, **4**, 484–498.

Rodriguez, E., George, N., Lachaux, J., Martinerie, J., Renault, B. & Varela, F.J. (1999). Perception's shadow: long-distance synchronization of human brain activity. *Nature*, **397**, 430–433.

Roelfsema, P.R., Engel, A.K., König, P. & Singer, W. (1997). Visuomotor integration is associated with zero time-lag synchronization among cortical areas. *Nature*, **385**, 157–161.

Roucoux, A., Culee, C. & Roucoux, M. (1983). Development of fixation and pursuit eye movements in human infants. *Behavioural Brain Research*, **10**, 133–139.

Ruff, H.A. & Rothbart, M.K. (1996). *Attention in early development*. New York: Oxford University Press.

Schiller, P.H. (1996). On the specificity of neurons and visual areas. *Behavioural Brain Research*, **76**, 21–35.

Schiller, P.H. (1998). The neural control of visually guided eye movements. In J.E. Richards (Ed.), *Cognitive neuroscience of attention: a developmental perspective* (pp. 3–50). Mahwah, NJ: Erlbaum.

Shupert, C. & Fuchs, A.F. (1988). Development of conjugate eye movements. *Vision Research*, **30**, 1077–1092.

Simion, F., Valenza, E., Umiltà, C. & Barba, B.D. (1995). Inhibition of return in newborns is temporo-nasal asymmetrical. *Infant Behavior and Development*, **18**, 189–194.

Singer, W. (1993). Synchronization of cortical activity and its putative role in information processing and learning. *Annual Review of Physiology*, **55**, 349–374.

Singer, W. (1994). Putative functions of temporal correlations in neocortical processing. In C. Koch & J.L. Davis (Eds.), *Large-scale neuronal theories of the brain* (pp. 201–237). Cambridge, MA: MIT Press.

Singer, W. (1995). Development and plasticity of cortical processing architectures. *Science*, **270**, 758–764.

Skoczenzki, A.M. & Aslin, R.N. (1995). Assessment of vernier acuity development using the "equivalent intrinsic blue" paradigm. *Vision Research*, **35**, 1879–1887.

Skoczenski, A.M. & Norcia, A.M. (1998). Neural noise limitations on infant visual sensitivity. *Nature*, **391**, 697.

Slater, A. (1995). Visual perception and memory at birth. In C. Rovee-Collier & L.P. Lipsitt (Eds.), *Advances in infancy research* (vol. 9, pp. 107–162). Norwood, NJ: Ablex.

Slater, A.M. & Findlay, J.M. (1972). The measurement of fixation position in the newborn baby. *Journal of Experimental Child Psychology*, **14**, 349–364.

Slater, A. & Morison, V. (1985). Selective adaptation cannot account for early infant habituation: a response to Dannemiller and Banks (1983). *Merrill-Palmer Quarterly*, **31**, 99–103.

Slater, A.M., Brown, E., Mattock, A. & Bornstein, M. (1996). Continuity and change in habituation in the first 4 months from birth. *Journal of Reproductive and Infant Psychology*, **14**, 187–194.

Slater, A., Johnson, S.P., Brown, E. & Badenoch, M. (1996). Newborn infants' perception of partly occluded objects. *Infant Behavior and Development*, **19**, 145–148.

Slater, A., Morison, V. & Rose, D. (1983a). Locus of habituation in the human newborn. *Perception*, **12**, 593–598.

Slater, A., Morison, V. & Rose, D. (1983b). Perception of shape by the newborn baby. *British Journal of Developmental Psychology*, **1**, 135–142.

Slater, A., Morison, V. & Somers, M. (1988). Orientation discrimination and cortical function in the human newborn. *Perception*, **17**, 597–602.

Smith, W.C., Johnson, S.P. & Spelke, E.S. (2000). Motion and edge sensitivity in perception of object unity (Manuscript submitted for publication).

Sohmiya, S., Sohmiya, K. & Sohmiya, T. (1998). Connection between synchronization of oscillatory activities at early stages and a final stage in the visual system. *Perceptual and Motor Skills*, **86**, 1107–1116.

Sokol, S. & Jones, K. (1979). Implicit time of pattern evoked potentials in infants: an index of maturation of spatial vision. *Vision Research*, **19**, 747–755.

Spelke, E.S. (1990). Principles of object perception. *Cognitive Science*, **14**, 29–56.

Spelke, E.S. & Newport, E.L. (1998). Nativism, empiricism, and the development of knowledge. In W. Damon (series Ed.) & R.M. Lerner (Ed.), *Handbook of Child Psychology, vol. 1: Theoretical models of human development* (5th ed., pp. 275–340). New York: Wiley.

Tallon-Baudry, C. & Baudry, O. (1999). Oscillatory gamma activity in humans and its role in object representation. *Trends in Cognitive Sciences*, **3**, 151–162.

Teller, D.Y. (1977). The forced-choice preferential looking procedure: a psychophysical technique for use with human infants. *Infant Behavior and Development*, **2**, 135–153.

Valenza, E., Simion, F. & Umiltà, C. (1994). Inhibition of return in newborn infants. *Infant Behavior and Development*, **17**, 293–302.

Van Essen, D.C., Anderson, C.H. & Felleman, D.J. (1992). Information processing in the primate visual system: an integrated systems perspective. *Science*, **255**, 419–423.

van Hofsten, C. & Rosander, K. (1996). The development of gaze control and predictive tracking in young infants. *Vision Research*, **36**, 81–96.

von Hofsten, C. & Rosander, K. (1997). Development of smooth pursuit tracking in young infants. *Vision Research*, **37**, 1799–1810.

von Hofsten, C. & Rosander, K. (1998). The establishment of gaze control in early infancy. In F. Simion & G. Butterworth (Eds.), *The development of sensory, motor and cognitive capacities in early infancy: from perception to cognition* (pp. 49–66). Hove, East Sussex, UK: Psychology Press.

Wattam-Bell, J. (1991). Development of motion-specific cortical responses in infancy. *Vision Research*, **31**, 287–297.

Wattam-Bell, J. (1996). Visual motion processing in 1-month-old infants: habituation experiments. *Vision Research*, **36**, 1679–1685.

Wentworth, N. & Haith, M.M. (1992). Event-specific expectations of 2- and 3-month-old infants. *Developmental Psychology*, **28**, 842–850.

Yonas, A. & Granrud, C.E. (1984). The development of sensitivity to kinetic, binocular, and pictorial depth information in human infants. In D. Engle, D. Lee & M. Jennerod (Eds.), *Brain mechanisms and spatial vision* (pp. 113–145). Dordrecht: Martinus Nijhoff.

Yuodelis, C. & Hendrickson, A. (1986). A qualitative and quantitative analysis of the human fovea during development. *Vision Research*, **26**, 847–855.

Zeki, S. (1993). *A vision of the brain*. Oxford: Blackwell Science.

Zeki, S. & Shipp, S. (1988). The functional logic of cortical connections. *Nature*, **335**, 311–317.

Suggestions for further reading

Büttner-Ennever, J.A. (Ed.) (1988). *Neuroanatomy of the oculomotor system*. New York: Elsevier.

Daw, N.W. (1995). *Visual development*. New York: Plenum.

Johnson, M.H. (1997). *Developmental cognitive neuroscience*. Cambridge, MA: Blackwell.

Johnson, S.P. (2000). The development of visual surface perception: insights into the ontogeny of knowledge. In C. Rovee-Collier, L., Lipsitt & H. Hayne (Eds.), *Progress in infancy research* (vol. 1, pp. 113–154). Mahwah, NJ: Erlbaum.

Richards, J.E. (Ed.) (1998). *Cognitive neuroscience of attention: a developmental perspective*. Mahwah, NJ: Erlbaum.

Singer, W. (1993). Synchronization of cortical activity and its putative role in information processing and learning. *Annual Review of Physiology*, **55**, 349–374.

Singer, W. (1995). Development and plasticity of cortical processing architectures. *Science*, **270**, 758–764.

Vital-Durand, F., Atkinson, J. & Braddick, O.J. (Eds.) (1996). *Infant vision*. Oxford: Oxford University Press.

IV.4
Neuronal Mechanisms for Integrating Visual and Auditory Space During Development

ANDREW J. KING, JAN W.H. SCHNUPP,
TIMOTHY P. DOUBELL and JÉRÔME BARON

ABSTRACT

Sensory information is initially processed in the brain in separate, modality-specific streams and is then brought together and synthesized to form a coherent multimodal representation of the environment. The integration of information from different sensory channels brings significant advantages, as it leads to improvements in the perception of events that activate several sensory systems concurrently, such as objects that can be seen as well as heard or felt. Multisensory interactions can also be advantageous during development, because information from one sensory modality can be used to calibrate emerging sensory faculties in another modality. This is highlighted by neurophysiological studies of the superior colliculus, a midbrain nucleus that coordinates spatial information from multiple sensory modalities so that individual neurons can process multisensory cues. The steps involved in merging different sensory representations in the superior colliculus rely on correlated visual and auditory experience to calibrate the developing representation of auditory space. Although these neurophysiological mechanisms are specifically concerned with the integration of multisensory spatial information, similar principles may underlie other aspects of perception and behaviour that depend on interactions between different sensory inputs.

INTRODUCTION

Each of our senses depends on specialized receptor cells that respond to particular aspects of our environment, such as light, sound or heat, together with

A.F. Kalverboer and A. Gramsbergen (eds.), Handbook of Brain and Behaviour in Human Development, 677–696
© 2001 Kluwer Academic Publishers. Printed in Great Britain.

well-defined regions of the central nervous system that process and represent sensory information in a modality-specific way. Scientific investigations of the senses, particularly at the level of individual neurons and their circuits, have tended to treat each system in isolation. However, the receptor cells of more than one modality, for example, in both the eyes and the ears, often register the same external event. Consequently, the integration and coordination of information derived from different sensory systems plays an important role in stimulus perception, attention and the control of movement (Calvert *et al.*, 1998; Driver & Spence, 1998; Stein & Meredith, 1993).

The most comprehensive evidence for crossmodal effects on perception and behaviour has come from psychophysical studies carried on adult human subjects (reviewed by King, 1990; Stein & Meredith, 1993; Welch & Warren, 1986). Thus, combining information across the senses has been shown to lead to improvements in stimulus detection, localization and identification and to faster reactions made following stimulus presentation. Some of the most compelling demonstrations of the influence of one sensory modality on the responses made to another come from situations in which different sensory cues provide conflicting information. For example, our ability to understand speech can be significantly enhanced, particularly in noisy situations, by the visual cues produced by the synchronous movements of the speaker's lips and face (Sumby & Pollack, 1954; Massaro, 1987). But if the visual and auditory cues become mismatched, by dubbing a speech sound on to the soundtrack of a film showing a person mouthing a different but related syllable, the listener can hear a different sound that represents a combination of the audible and visual signals (McGurk & MacDonald, 1976). Vision can also influence the perceived location of a sound source (Bertelson & Radeau, 1981; Jack & Thurlow, 1973), as, for example, in the case of a ventriloquist's dummy. In this instance subjects tend to mislocalize the speaker's voice in the direction of the spatially discordant visual cues produced by the dummy's moving lips. These studies highlight the propensity of the central nervous system to coordinate and synthesize information derived from different sensory modalities so that, within certain limits, conflicting cues may be perceived as if they originate from the same event. Although visual perception can be influenced by other sensory cues (e.g. Sekuler *et al.*, 1997; Welch & Warren, 1986), vision tends to dominate over the other modalities, particularly in localization tasks.

How infants learn to perceive and integrate the different modality signals associated with multisensory events is a fundamental issue in human development. Some authors (e.g. Piaget, 1954) have proposed that sensory systems initially function independently and that equivalences between them have to be learned, whereas others (e.g. Bower, 1971; Gibson, 1969) have argued that infants show an innate predisposition to unify the senses. One way of integrating multisensory cues originating from a common source is to pick out amodal attributes, such as the temporal synchrony of the visual and auditory signals produced by an object repeatedly striking a surface. Within a few months from birth, human infants appear to be able to utilize temporal synchrony to link auditory and visual inputs (Bahrick, 1992; Lewkowicz, 1992; Spelke, 1979), but they require a larger inter-sensory interval than adults before the different modality signals are perceived as separate events (Lewkowicz, 1996). Prelinguistic infants can also associate the visual and auditory components of speech, an ability that depends on the spectral composition of the sounds rather than on crossmodal timing (Kuhl & Meltzoff, 1982). In contrast to adults, however, infants are unable to match non-speech sounds to visually presented speech (Kuhl et al., 1991).

Attempts to examine whether auditory and visual space are coordinated during early infancy have produced conflicting results (Aronson & Rosenbloom, 1971; McGurk & Lewis, 1974). Nevertheless, some studies have reported that human infants can orient towards sound sources more accurately in the presence of visual cues (Morrongiello & Rocca, 1987) and, even within hours of birth, show signs of intersensory bias towards spatially discordant visual targets (Castillo & Butterworth, 1981).

Taken together, these studies indicate that humans can match crossmodal information in early infancy, although this ability is limited in comparison with that exhibited by adults. It seems likely that the subsequent maturation of intersensory processing depends, at least in part, on experience with multi-sensory stimuli, as shown by studies in other species (King, 1999; Sleigh et al., 1998; Stein et al., 2000).

The use of modern functional imaging techniques and evoked potential measurements, together with studies of neurological patients, can provide valuable insights into the neural basis for multisensory interactions in humans. However, an understanding of the mechanisms by which individual neurons synthesize information that originates from multiple, independent sensory channels can be obtained only from electrophysiological studies in non-human animals. Neurons that receive converging inputs from different sensory modalities have been described in many different regions of the brain, including both specialized multisensory areas and even, to some extent, the primary sensory pathways (Calvert et al., 1998; King & Hartline, 1999; Stein & Meredith, 1993). The most extensively studied area is the superior colliculus (SC) in the midbrain, which provides an excellent model system for investigating both the principles underlying multisensory integration and the factors responsible for merging the neural representations of visual and auditory space during development (King, 1999; Stein et al., 2000).

THEORETICAL BACKGROUND: SENSORIMOTOR ORGANIZATION OF THE SUPERIOR COLLICULUS

The SC forms part of the roof of the midbrain and one of its main functions is the mediation of rapid, reflexive orienting movements of the eyes and head towards novel targets (Stein & Meredith, 1993). The SC is a laminated nucleus which, anatomically and functionally, can be broadly subdivided into superficial and deep layers. Microstimulation studies have shown that the deep layers contain a "motor map" for controlling the direction of gaze (Sparks, 1999). Thus, stimulation of the rostral SC will either maintain fixation or result in small saccadic eye movements, whereas stimulation of caudal regions can produce large contralateral gaze shifts, involving movements of the eyes and head. Similarly, stimulation of the medial or lateral SC will result in orientation movements directed up or down, respectively. In mammals, neurons in the superficial layers are exclusively visual, whereas the deep layers receive visual, auditory and somatosensory inputs. Each of these sensory representations is topographically arranged to form a neural map of space (Figure 1), which matches the organization of the motor map. For example, neurons in the medial SC respond optimally to auditory or visual stimuli placed above the animal or to stimulation of the upper body surface. These sensory signals are then used to elicit orienting movements towards the stimulus.

Some of the deep SC neurons receive converging multisensory inputs and therefore possess overlapping receptive fields for different modalities. This gives rise to some interesting phenomena, including "multisensory facilitation", where a response evoked by multimodal (e.g. combined visual and auditory) stimulation may be greater than the sum of the responses to the different stimuli presented in isolation. These response enhancements tend to be particularly marked when the individual sensory stimuli are weak (Stein & Meredith, 1993). Functionally, multisensory facilitation is likely to be extremely useful. First, biologically important events, such as potential predators or prey, may try very hard to avoid detection by using camouflage and stealth to become "minimally effective stimuli". Secondly, individual sensory neurons in the SC are spontaneously active and their responses tend to be "noisy", i.e. the number of additional discharges caused by a stimulus can vary widely from presentation to presentation. Under these conditions, detecting the presence of a stimulus and pinpointing its position accurately on the basis of the discharges of SC neurons can become an exceedingly difficult task, but one where combining information from separate sensory input channels can make a great difference.

If the noisy responses to weak unimodal stimulation are treated as random variables, then the optimal strategy for combining information across modalities would be to apply Bayes' rule of conditional probabilities[1]. Anastasio *et al.* (2000) have suggested that the discharge rates of multisensory neurons in the SC may indeed be estimates of the conditional probabilities for the presence of a stimulus, given particular discharge rates at each of the sensory input channels. In order to compute these conditional probabilities, SC neurons would have to possess sigmoidal input–output functions, which appears to be borne out by the experimental data available so far (King & Schnupp, 2000). However, in order to produce accurate estimates of these Bayesian probabilities a neuron may have to tune its input–output functions according to the prior probabilities for the presence of a target and the statistics of the spontaneous and driven discharges reaching the SC. Whether developing SC neurons are capable of this is presently unknown.

Of course, inputs from different sensory streams should be interpreted as signalling high conditional probabilities for the presence of a target only if they are temporally more or less synchronous and originate from the same region of

---

[1]Bayes' rule is a widely applied mathematical theorem that defines the probability of a particular event given that another event has previously occurred. See Applebaum (1996) for further details.

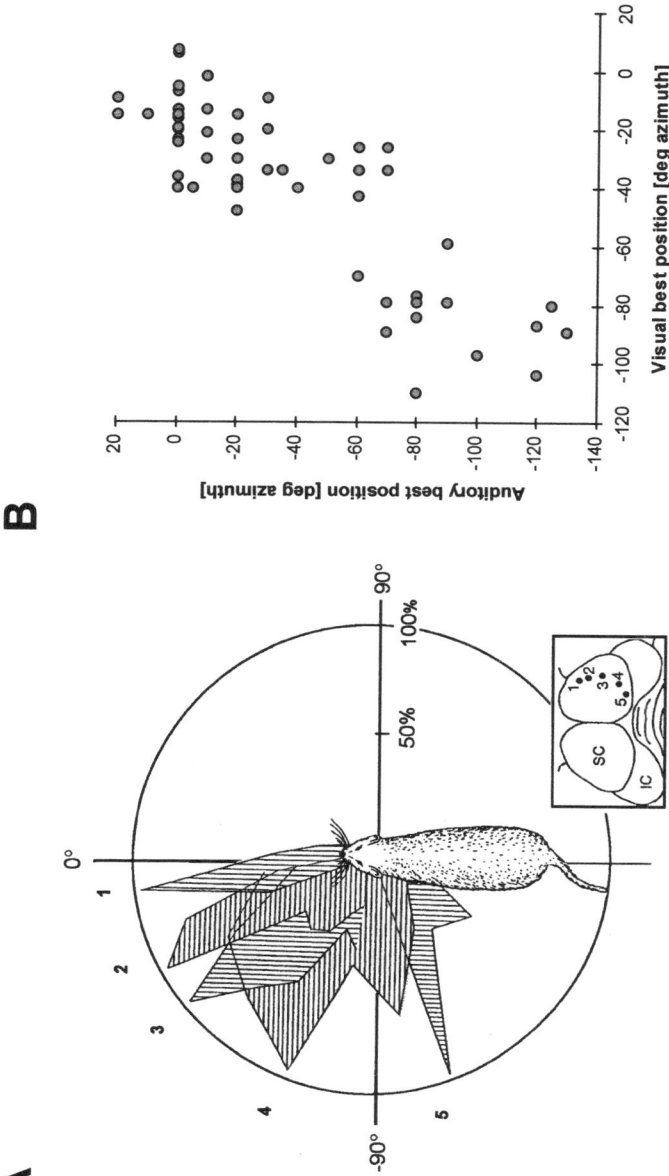

Figure 1 **A**: Representation of auditory space by sensory neurons in the SC of the ferret. The hatched areas are "polar plots" of receptive field profiles recorded from single auditory neurons in the SC following presentation of noise bursts from different loudspeaker directions in the horizontal plane (azimuth). The receptive field profiles typically display a single peak, corresponding to the direction of maximal response or "best position". The receptive field profiles shown were recorded at various locations along the rostrocaudal extent of the SC, as indicated on the inset at the bottom right. The auditory best positions vary systematically from in front of the animal round into the contralateral hemifield as a function of recording site. **B**: Alignment of sensory maps. Visual neurons in the superficial layers of the SC show the same topographic arrangement of receptive field best positions as deep-layer auditory neurons. Consequently, visual and auditory best positions recorded in the same vertical electrode penetration through the SC are highly correlated.

space, as would be the case if they arise from a single target object (e.g. an event that can be seen as well as heard). This too appears to be borne out by experimental data. Thus, multisensory signals that are widely disparate in space (e.g. a visual stimulus presented at one position paired with a sound from a very different location) or time tend not to produce response enhancements and may even depress or eliminate the responses observed with unimodal stimulation (Kadunce *et al.*, 1997; King & Palmer, 1985; Meredith & Stein, 1986).

The organization of the SC, with multiple sensory maps that are topographically aligned both with each other and with a common motor map, provides an efficient means by which both unimodal and multimodal stimuli can evoke orienting movements that redirect attention towards the target. Map registration would also appear to be essential if the integrative properties of deep SC neurons are to synthesize the different modality cues associated with individual multisensory events. However, the different sensory systems derive and represent spatial information in different ways. For example, a two-dimensional map of the visual world is projected onto the retina and can therefore be reconstructed within the brain as a result of topographically ordered afferent projections. Consequently, the coordinates of visual space are centred on the retina. Similarly, a somatosensory map is body-centred because this arises as a result of spatially organized projections from primary afferents innervating receptors in the skin. Formation of a map of auditory space is more problematic, however, because, unlike the retina or the skin, the auditory receptor surface in the cochlea encodes the frequency components of a sound rather than its location in space. Sound source locations must therefore be derived from acoustical cues produced by interactions with the head and ears (Figure 2). These cues comprise differences in sound level and arrival time between the two ears, or changes in the sound spectrum due to direction-dependent filtering of the sound as it passes through the outer ear (King & Carlile, 1995). The coordinates of auditory space are therefore centred on the ears and head. Because the synthesis of a neural map of auditory space involves tuning neurons to particular values of these acoustic localization cues, this sensory representation is often referred to as a "computational map", to distinguish it from the "projectional" maps formed by the visual and somatosensory systems.

Coordinating the spatial representations for different sensory modalities involves establishing a correspondence between specific values of the auditory localization cues and positions on the retina and the body surface. Map registration, apparently so important for the multisensory integration performed by SC neurons, should therefore be degraded whenever the sense organs move independently. For example, the retinocentric reference frame for the visual map will change as the direction of gaze changes. Of course, the acoustic cues for sound location remain unaffected by a change in eye position, so we might expect a misalignment of the visual and auditory maps in the SC whenever gaze is directed anywhere but straight ahead. However, at least a proportion of both auditory (Hartline *et al.*, 1995; Jay & Sparks, 1984; Populin & Yin, 1998) and somatosensory (Groh & Sparks, 1996) neurons in the SC respond to changes in eye position in a way that suggests a rapid, adaptive reorganization of their spatial receptive fields.

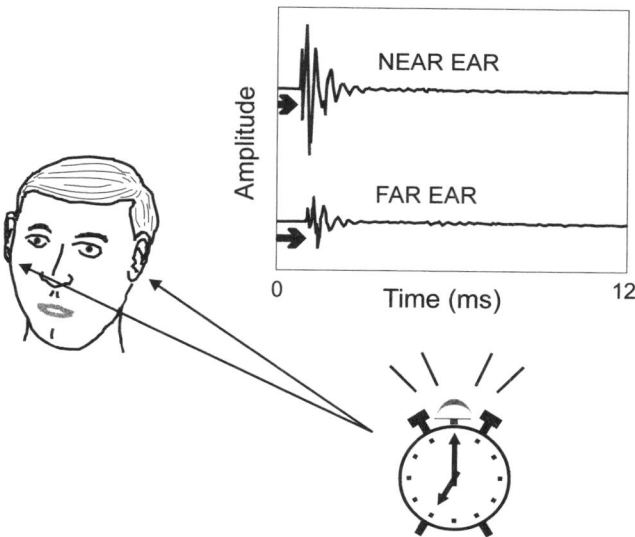

Figure 2 Acoustic cues for sound localization. Sounds originating from one side of the head will arrive first at the ear closer to the source, giving rise to an interaural difference in time of arrival. In addition, the directional filtering properties of the outer ears and the shadowing effect of the head combine to produce an interaural level difference for high-frequency sounds. The "binaural" cues for sound location are illustrated by the waveform of the sound, which is both delayed and reduced in amplitude at the listener's far ear. "Monaural" spectral cues generated by the outer ears also allow sounds to be localized using one ear alone.

Establishing and maintaining map alignment during development is also a complicated process. This particularly affects the emerging auditory representation because, as the head and outer ears grow, the values of the various acoustic localization cues corresponding to a particular sound direction will change (Clifton *et al.*, 1988; King, 1999). Consequently there would appear to be a need for the midbrain circuitry responsible for the auditory spatial tuning of SC neurons to be "calibrated" on the basis of experience, in order to set up and preserve the correspondence with other sensory representations. Much of the remainder of this chapter will look at how this calibration might be achieved in the developing animal.

DEVELOPMENT OF SENSORY MAPS AND MULTISENSORY INTEGRATION IN THE SUPERIOR COLLICULUS

The sensory maps in the superficial and deep layers of the SC differ considerably in the time courses of their maturation. In most mammals, axons growing out from the retina reach the superficial layers and establish a basic retinotopic projection before birth, with adult-like topographic order being present in the map of visual space by the time of eye opening (Kao et al., 1994; King & Carlile, 1995; Wallace et al., 1997). The development of this map therefore does not depend on visual experience, but seems to be guided by a combination of molecular guidance cues and spontaneous retinal activity patterns (Constantine-Paton et al., 1990; Goodhill & Richards, 1999). By contrast, the various sensory representations in the deeper layers mature much more gradually over the course of several weeks of postnatal development (King & Carlile, 1995; Wallace & Stein, 1997; Withington-Wray et al., 1990a). In the case of the auditory representation, the gradual emergence of topographic order may, at least in part, reflect developmental changes in the auditory localization cues that accompany growth of the head and ears (King, 1999).

In very young kittens, deep SC neurons respond only to tactile stimuli, with sensitivity to auditory and then to visual stimulation emerging at a later stage (Stein et al., 1973). Multisensory neurons initially lack the marked enhancement of responses to combined-modality stimulation seen in mature animals. The number of cells exhibiting these response characteristics gradually increases with age, possibly due to the maturation of descending inputs from "association" areas of the cortex, as deep SC neurons become progressively more selective for stimulus location (Wallace & Stein, 1997). On the basis of these neuro-physiological observations it would appear that, in contrast to humans, infant cats may be unable to synthesize multisensory inputs to guide orientation behaviour.

ROLE OF SENSORY EXPERIENCE IN THE DEVELOPMENT OF SENSORY MAP ALIGNMENT

In humans, crossmodal illusions, such as the ventriloquism effect, which result from the introduction of a mismatch between the cues available through different sensory modalities, provide a powerful illustration of how one sensory system can influence another. Similarly, manipulation of sensory cues available to young animals has been used to demonstrate the role of experience in aligning the different sensory maps in the SC during development. These studies (reviewed in King, 1999; Stein et al., 2000) have concentrated mainly on the relationship between the visual and auditory representations. Perhaps not surprisingly, given its computational nature, the map of auditory space is particularly dependent on sensory experience, both auditory and visual, for its normal development.

As an animal grows, changes in the size, shape and separation of the ears progressively alter the monaural and binaural localization cue values corresponding to particular sound directions (King & Carlile, 1995). One approach to the study of auditory map plasticity has been to observe the neurophysiological consequences of changing these relationships experi-mentally. For example, rearing guinea-pigs in an environment of intense, constant, omnidirectional noise, which presumably masks the experience of individual, localizable auditory stimuli, has been shown to disrupt the development of the auditory map in the SC (Withington-Wray et al., 1990b). Similarly, removing the pinna and concha of the outer ear will degrade the spectral localization cues available and, in juvenile ferrets, this procedure also impairs the emergence of topographic order in the auditory space map (Schnupp *et al.*, 1998). However, while some manipulations are highly disruptive, others lead to compensatory changes in the developing space map. Inserting an ear plug into one ear canal will attenuate and slightly delay sounds to that ear, thereby dramatically altering interaural level and time difference cue values. Nevertheless, adult barn owls (Knudsen, 1985) and ferrets (King *et al.*, 1988) raised with one ear plugged have largely normal auditory maps, in register with the visual map, but only if the recordings are made with the ear plug in place. However, visually deprived owls fail to adapt completely to a period of monaural occlusion during infancy, indicating that these changes are at least partially guided by visual experience (Knudsen & Mogdans, 1992).

This guiding influence of vision on the developing auditory space map is also borne out by a number of other experimental findings. For example, deprivation of patterned visual experience in young barn owls (Knudsen *et al.*, 1991), ferrets (King & Carlile, 1993) or guinea-pigs (Withington-Wray *et al.*, 1990c) prevents the auditory space map from developing normally and therefore disrupts intersensory map alignment. Displacing an animal's visual field with respect to the head, either through a surgically induced strabismus (King *et al.*, 1988; Figure 3) or with prism lenses (Knudsen & Brainard, 1991), induces a corresponding systematic shift in the auditory representation, so that the visual and auditory maps remain aligned. However, the auditory map appears unable to adapt to a complete inversion of the visual field by surgical rotation of the eye – this procedure leads to a degradation of the auditory map, rather than to adaptive changes (King & Carlile, 1995).

These experiments indicate that correlated visual and auditory experience arising from objects that can be both seen and heard is used to refine the maturing auditory representation. As a result it shares the same coordinates as the visual map, which remains unchanged by any of the manipulations described above. Perhaps surprisingly, recent evidence suggests that the guiding visual signals used to calibrate the developing representation of auditory space may arise from the exclusively visual neurons in the superficial layers, rather than the visual component of the multisensory receptive fields of deep-layer neurons. As described in the previous section, the more precise map in the superficial layers matures before any of the deep-layer maps. Superficial-layer neurons project topographically to the deep layers (Lee *et al.*, 1997) and, in neonatal ferrets, partial lesions of the superficial layers lead to a failure of auditory map development that is restricted to the region directly below the lesion (King *et al.*, 1998). Equivalent lesions of the superficial layers of the SC in adult ferrets, in contrast, do not affect the underlying auditory map (Figure 4).

In order to relate experience-induced changes in neuronal response properties to the development of intersensory functions in humans, it is, of course, necessary to examine whether similar principles apply at a behavioural level in the species concerned. Barn owls have relatively fixed eyes and ears and exhibit a particularly close correspondence between the neural representations of visual and auditory space in the optic tectum, the avian homologue of the SC. Manipulation of either visual (Brainard & Knudsen, 1998; Knudsen *et al.*, 1991) or auditory (Knudsen, 1985; Knudsen *et al.*, 1984) experience in this species has

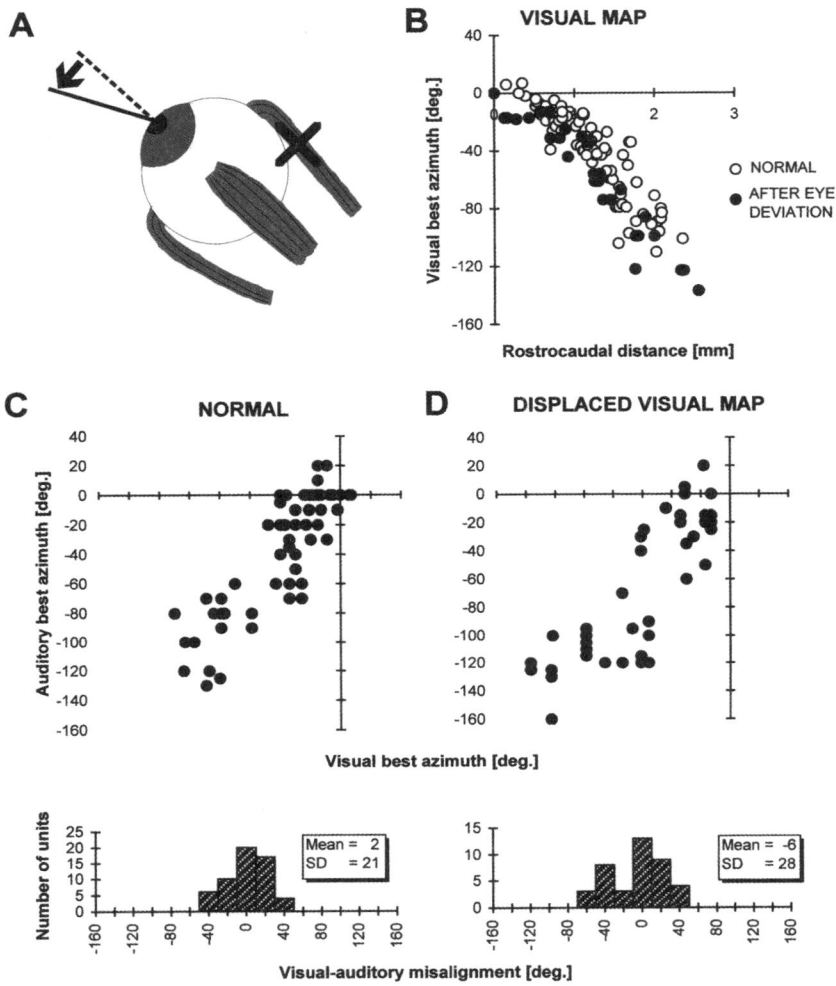

Figure 3 Visual calibration of the developing auditory space map in the SC. **A**: Removal of the medial rectus muscle just before natural eye opening induces a small, outward deviation of the eye. **B**: Recordings made from adult ferrets show that, as expected, the visual best azimuths of superficial-layer neurons are shifted laterally relative to the head by a corresponding amount. **C, D**: These data are plotted against the auditory best azimuths of neurons recorded in the deeper layers of the same vertical electrode penetrations. The histograms plot the angular difference between the visual and auditory best azimuths. Neither the mean nor the variance of the visual–auditory misalignments differ significantly between the normal (**C**) and visually-displaced (**D**) groups, suggesting that a corresponding shift in the representation of auditory space has taken place. (Adapted from King *et al.*, 1988).

remarkably similar effects on the auditory spatial tuning of these neurons and on auditory localization behaviour. In mammals, which have less precise maps of auditory space, the situation is potentially more complex because information about the position of the eyes (and perhaps other sense organs too) appears to be incorporated when coordinating the sensory input with the motor output. Nevertheless, equivalent changes have been observed in sound localization behaviour to those induced in the SC by raising ferrets with altered acoustic cues (King et al., 2000; Parsons et al., 1999).

Because physiological and behavioural studies highlight the predominant role of vision in coordinating spatial information provided by the different senses, it is important to consider the effects of blindness on audition and the other senses. In visually deprived owls the map of auditory space in the optic tectum and the precision of sound-evoked orienting responses are degraded in a similar manner (Knudsen et al., 1991). In contrast, behavioural studies in humans (Lessard et al., 1998; Röder et al., 1999), cats (Rauschecker & Kniepert, 1994) and ferrets (King & Parsons, 1999) have reported improved auditory localization following early loss of vision, a finding that is presumably related to the expansion of auditory representations in several brain regions, including areas that are normally visual in function (reviewed by Rauschecker, 1995). This illustrates another example of intersensory plasticity, in which the central nervous system undergoes compensatory changes following altered sensory inputs.

BASIS FOR MERGING VISUAL AND AUDITORY REPRESENTATIONS OF SPACE DURING DEVELOPMENT

The facilitatory effects of vision on sound localization by humans have been interpreted by some authors in terms of visual signals providing a frame of reference and by others as a consequence of information about target-directed movements of the eyes and head (King, 1990). In a similar vein the mechanisms responsible for visual calibration of the developing auditory space map include visual feedback about the accuracy of sound-evoked orienting movements, direct template matching between the two sensory representations, and detection of visual–auditory misalignment by bimodal neurons, which is then fed back to auditory neurons at an earlier stage in the pathway.

Progress in determining which of these mechanisms may be involved clearly requires that the site of auditory plasticity be identified. In mammals this is not known. In barn owls, however, prism rearing systematically shifts the tuning of neurons to interaural time differences, one of the binaural localization cues, both in the optic tectum and at a lower level in the auditory pathway, the external nucleus of the inferior colliculus (ICX; Brainard & Knudsen, 1993). These changes, which most likely result from signals fed back from the tectum, are caused by the formation of novel, excitatory connections to the ICX from the central nucleus of the inferior colliculus (Feldman & Knudsen, 1997), in conjunction with inhibition of existing connections that would otherwise convey sensitivity to the normal range of interaural time differences (Zheng & Knudsen, 1999). Interestingly, the continued existence of subthreshold inputs may explain why auditory neurons in older birds can readjust their spatial tuning to normal

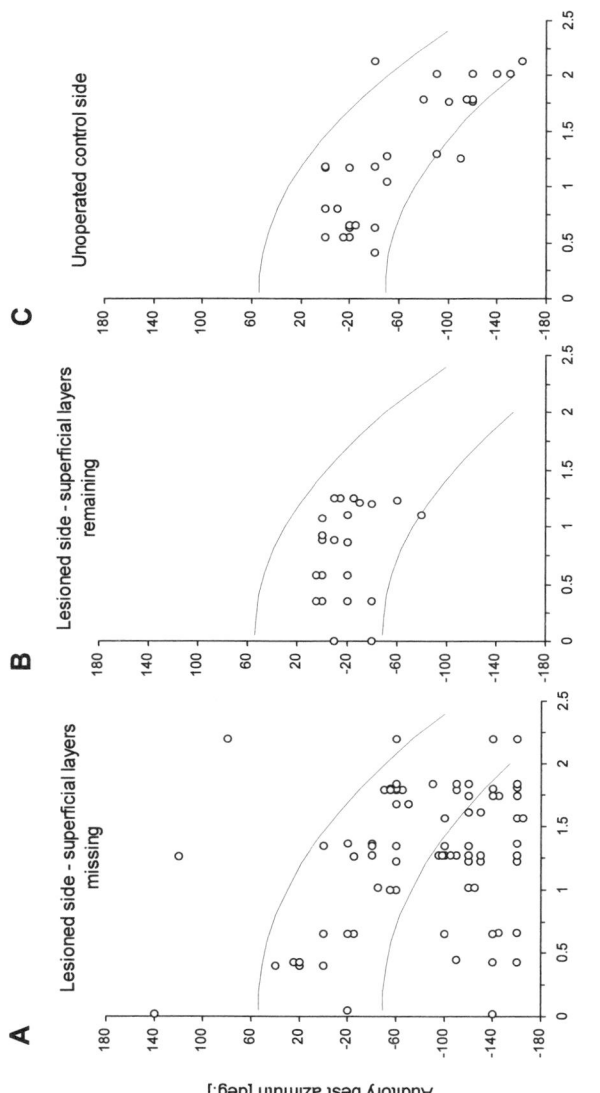

Figure 4 Impact of superficial layer lesions on auditory map development. The data points show the auditory best positions plotted against the rostrocaudal location of the recorded neurons in the deep SC layers of adult ferrets that had undergone partial unilateral removal of the superficial, visual layers in infancy. The continuous lines delineate the 95% confidence interval for the scatter in the auditory representation in a group of normal control animals. Where the superficial layers are missing (**A**) the auditory topography is highly disorganized; this is indicated by the large number of best positions falling outside the normal range. In contrast, auditory best positions recorded in regions where the superficial layers were still intact (**B**) or in the contralateral, unoperated SC (**C**) fall within the normal range. (Adapted from King *et al.*, 1998).

values following removal of the prisms, and even adapt again in adulthood to a similar optical displacement of the visual field, a capacity that is largely missing in naïve adults (Knudsen, 1998). Whether equivalent intersensory adjustments in auditory responses are possible in adult mammals has yet to be investigated, although it is tempting to speculate that similar unmasking effects may be related to the visual capture of sound localization and its after-effects.

The cellular mechanisms that underlie the adaptive matching of acoustic cues and visual signals, and particularly the process by which vision guides the refinement of auditory spatial tuning in the midbrain, have yet to be clarified. One popular idea is that the activity-dependent steps that lead to a reorganization of auditory circuits may conform to a "Hebbian" mechanism of synaptic plasticity, named after the Canadian psychologist who originally set out the foundations for this scheme. This form of associative synaptic plasticity requires coincident pre- and postsynaptic activity, which may lead to long-term potentiation of synaptic responses (Konnerth et al., 1996). The probability of coordinating pre- and postsynaptic activity, and therefore of stabilizing particular synapses, is thought to increase if converging signals, in this case from the eyes and the ears, are temporally correlated, as may be the case when modality-specific cues arise from a common source (Figure 5A). If this process were weighted in an experience-dependent manner, temporally correlated signals would arise most often from spatially congruent stimuli, thereby matching neuronal sensitivity to acoustic localization cues to a previously formed visual map.

The NMDA class of glutamate receptors appears to constitute an ideal molecular device for producing Hebbian synapses, and the presence of high levels of these receptors is thought to be a necessary factor for the activity-dependent plasticity of neural circuits (Constantine-Paton et al., 1990; Fox, 1995). Consistent with this idea, we found that high levels of NMDA–receptor binding sites are found in the SC soon after the functional onset of hearing and vision and that, in the deep layers, these levels decline most dramatically during the period of auditory map formation (Figure 6). Moreover, chronic application of NMDA–receptor antagonists to the superficial layers disrupts the emergence of topographic order in this representation (Schnupp et al., 1995). However, these studies fall well short of establishing a causal role for NMDA receptors in auditory map plasticity and other biochemical factors are also likely to be involved. Indeed, despite the attractiveness of the Hebbian scheme, an alternative possibility is that non-Hebbian mechanisms of synaptic enhancement, not requiring correlated pre- and postsynaptic activity, may also contribute. This could be based on presynaptic interactions between converging inputs (Figure 5B) of the sort implicated in classical conditioning in *Aplysia* or in NMDA-independent long-term potentiation in the hippocampus (Lechner & Byrne, 1998).

A potential problem with mechanisms of synaptic plasticity based on the detection of temporally correlated sensory signals is that only a proportion of the deep SC neurons seem to be overtly multimodal and, in kittens, these appear relatively late in development (Wallace & Stein, 1997). Moreover, if, as suggested by the barn owl data, the site of visual calibration is actually before the SC in a purely auditory area of the midbrain (Brainard & Knudsen, 1993), it would appear to be necessary to assume the existence of subthreshold visual influences

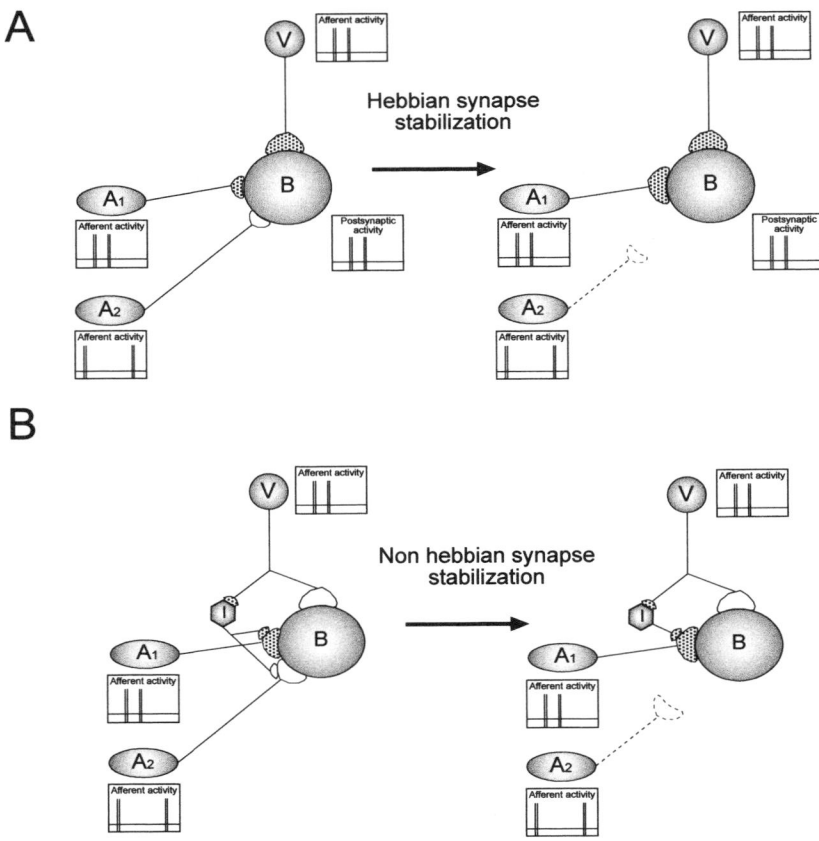

Figure 5 A: Hebbian scheme of synaptic plasticity whereby a bimodal cell (B) receives a strong input from a visual (V) neuron and weak inputs from two auditory neurons (A1 and A2). When V and A1 are coactivated simultaneously the postsynaptic cell B becomes sufficiently depolarized to activate signalling pathways that lead to selective strengthening of the input from neuron A1. The input from cell A2, which is not coactive with the visual neuron, becomes weaker and is eliminated. **B**: An alternative non-Hebbian scheme showing that, in addition to its direct inputs onto neuron B, neuron V can indirectly modulate the A1 synapse via a presynaptic interneuron (I). In this scheme the presynaptic terminal of A1 is the site of coincidence detection. Action potentials generated in A1 invade its own synaptic terminal, activating a second-messenger cascade, which can be further enhanced by coincident activation of prejunctional synapses from interneuron I, leading to long-lasting enhancement of transmitter release.

on the tuning of these neurons to auditory localization cues. In this regard, demonstrations in other systems that long-term potentiation can reinforce the synapses of other, nearby, cells – most likely through the release of diffusible biochemical messengers, may be particularly relevant (e.g. Kossel *et al.*, 1990; Schuman & Madison, 1994).

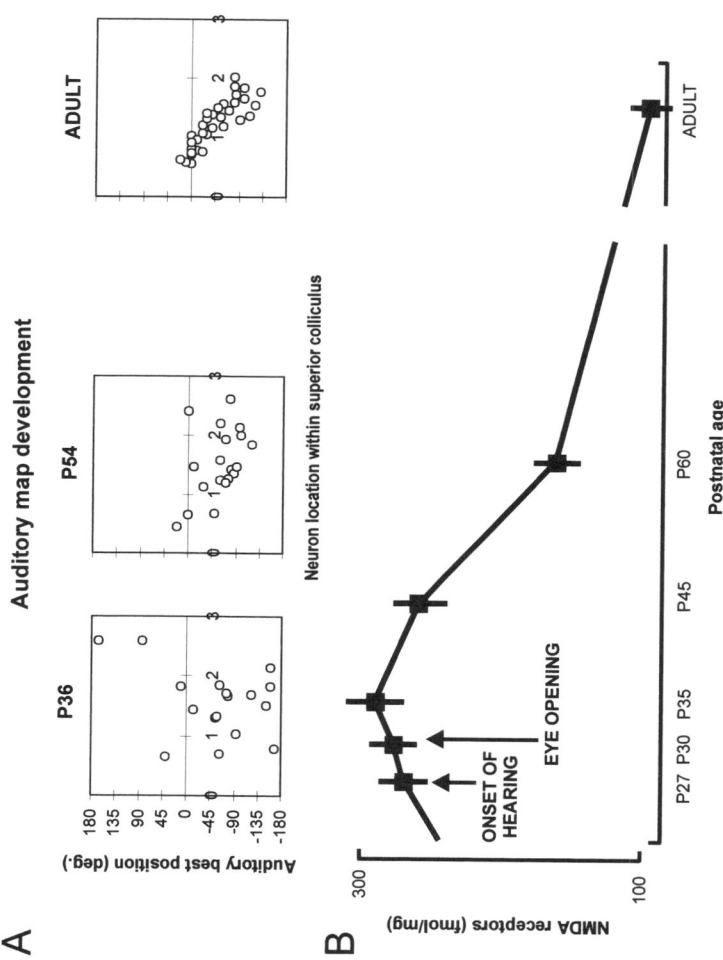

Figure 6 Relationship between the maturation of the auditory space map and the developmental expression of NMDA receptors in the ferret SC. **A:** Auditory best positions are plotted against the rostrocaudal location of the recording electrode in the deep SC. Ferrets begin to hear at about 28 days after birth. The topographic order in the auditory representation is initially poor, but gradually improves over the next few weeks. **B:** *In-vitro* quantitative autoradiography reveals that the level of MK801-specific binding to NMDA receptors in the deeper layers of the SC peaks around the time of eye opening and then declines gradually with a time-course comparable to that of the maturation of the auditory space map.

DISCUSSION

Many aspects of perception and behaviour rely on the cooperative interactions that take place between different sensory systems. Our knowledge of the neural basis for the coordination and integration of modality-specific information is still limited, but studies of the SC have provided many important insights into these processes. The registration of visual, auditory and tactile maps, which appears to be maintained in spite of both the independent mobility of the sense organs and of growth-related changes in their relative geometry, provides a simple and efficient way of synthesizing multisensory signals to guide reflexive orienting responses.

Studies of the maturation of intersensory registration in the SC have provided one of the clearest examples of the functional significance of experience-mediated plasticity in the brain. As a result of their influence on the developing auditory responses, the more accurate and reliable spatial cues provided by the visual system play the dominant role in aligning the different sensory maps. Substantial progress has been made in this area in the last few years, and we are now in a position to ask questions about the molecular and cellular mechanisms that underlie both the initial matching of visual and auditory signals and the experience-dependent adjustment of auditory spatial tuning that can be induced at later stages of development.

Localization behaviours that are more complex than stimulus-evoked orienting responses involve various regions of the cortex. For example, following unilateral lesions of the auditory cortex, cats and other species will no longer perform a learned approach task to sound sources located in the contralateral hemifield (Jenkins & Merzenich, 1984). In contrast to the SC, however, there is no evidence for a map of auditory space in the cortex (Middlebrooks, 1999). In addition, posterior parietal and prefrontal areas of the cortex appear to play a role in both visual and auditory space perception (Bushara *et al.*, 1999). The contribution of these areas to specific multisensory phenomena is poorly understood, and we know practically nothing about the development of cortical neurons that participate in these tasks. Nevertheless, based on the properties of multisensory neurons in the cortex that have so far been described, it seems reasonable to anticipate that the principles established at the level of the SC in a variety of species will also be applicable to more complex aspects of perception.

In attempting to draw parallels with crossmodal phenomena in humans it is interesting to note that a similar intersensory bias characterizes both stimulus localization and identification. There are some key differences between the capacity of human infants to match sensory information provided by different modalities and the multisensory properties of SC neurons that have been reported in the early postnatal stages of other species. However, neuro-physiological studies of development most often employ altricial species – usually cats or ferrets for studies in mammals – which have limited sensory capacities at birth. The sensory systems of neonatal primates, including humans, are more mature, but still undergo protracted periods of postnatal development during which experience will influence the circuitry of the brain.

References

Anastasio, T.J., Patton, P.E. & Belkacem-Boussaid, K. (2000). Using Bayes' rule to model multisensory enhancement in the superior colliculus. *Neural Computation*, **12**, 1165–1187.

Applebaum, D. (1996). *Probability and information: an integrated approach.* Cambridge, UK: Cambridge University Press.

Aronson, E. & Rosenbloom, S. (1971). Space perception in early infancy: perception within a common auditory–visual space. *Science,* **172**, 1161–1163.

Bahrick, L.E. (1992). Infants' perceptual differentiation of amodal and modality-specific audiovisual relations. *Journal of Experimental Child Psychology*, **53**, 180–199.

Bertelson, P. & Radeau, M. (1981). Cross-modal bias and perceptual fusion with auditory-visual spatial discordance. *Perception and Psychophysics*, **29**, 578–584.

Bower, T.G.R. (1971). The object in the world of the infant. *Scientific American,* **225**, 30–38.

Brainard, M.S. & Knudsen, E.I. (1993). Experience-dependent plasticity in the inferior colliculus: a site for visual calibration of the neural representation of auditory space in the barn owl. *Journal of Neuroscience,* **13**, 4589–4608.

Brainard, M.S. & Knudsen, E.I. (1998). Sensitive periods for visual calibration of the auditory space map in the barn owl optic tectum. *Journal of Neuroscience,* **18**, 3929–3942.

Bushara, K.O., Weeks, R.A., Ishii, K., Catalan, M.J., Tian, B., Rauschecker, J.P. & Hallett, M. (1999) Modality-specific frontal and parietal areas for auditory and visual spatial localization in humans. *Nature Neuroscience,* **2**, 759–766.

Calvert, G.A., Brammer, M.J. & Iversen, S.D. (1998). Crossmodal identification. *Trends in Cognitive Sciences*, **2**, 247–253.

Castillo, M. & Butterworth, G. (1981). Neonatal localisation of a sound in visual space. *Perception*, **10**, 331–338.

Clifton, R.K., Gwiazda, J., Bauer, J.A., Clarkson, M.G. & Held, R.M. (1988). Growth in head size during infancy: implications for sound localization. *Developmental Psychology,* **24**, 477–483.

Constantine-Paton, M., Cline, H.T. & Debski, E. (1990). Patterned activity, synaptic convergence, and the NMDA-receptor in developing visual pathways. *Annual Review of Neuroscience,* **13**, 129–154.

Driver, J. & Spence, C. (1998). Attention and the crossmodal construction of space. *Trends in Cognitive Sciences*, **2**, 254–262.

Feldman, D.E. & Knudsen, E.I. (1997). An anatomical basis for visual calibration of the auditory space map in the barn owl's midbrain. *Journal of Neuroscience,* **17**, 6820–6837.

Fox, K. (1995). The critical period for long-term potentiation in primary sensory cortex. *Neuron*, **15**, 485–488.

Gibson, J.J. (1969). *Principles of perceptual learning and development.* New York: Appleton-Century-Crofts.

Goodhill, G.J. & Richards, L.J. (1999). Retinotectal maps: molecules, models and misplaced data. *Trends in Neuroscience*, **22**, 529–534.

Groh, J.M. & Sparks, D.L. (1996). Saccades to somatosensory targets. III. Eye-position-dependent somatosensory activity in primate superior colliculus. *Journal of Neurophysiology,* **75**, 439–453.

Hartline, P.H., Pandey Vimal, R.L., King, A.J., Kurylo, D.D. & Northmore, D.P.M. (1995). Effects of eye position on auditory localization and neural representation of space in superior colliculus of cats. *Experimental Brain Research,* **104**, 402–408.

Jack, C.E. & Thurlow, W.R. (1973). Effects of degree of visual association and angle of displacement on the "ventriloquism" effect. *Perceptual and Motor Skills*, **37**, 967–979.

Jay, M.F. & Sparks, D.L. (1984). Auditory receptive fields in primate superior colliculus shift with changes in eye position. *Nature,* **309**, 345–347.

Jenkins, W.M. & Merzenich, M.M. (1984). Role of cat primary auditory cortex for sound-localization behavior. *Journal of Neurophysiology,* **52**, 819–847.

Kadunce, D.C, Vaughan, J.W., Wallace, M.T., Benedek G. & Stein, B.E. (1997). Mechanisms of within- and cross-modality suppression in the superior colliculus. *Journal of Neurophysiology,* **78**, 2834–2847.

Kao, C.-Q., McHaffie, J.G., Meredith, M.A. & Stein, B.E. (1994). Functional development of a central visual map in cat. *Journal of Neurophysiology,* **72**, 266–272.

King, A.J. (1990). The integration of visual and auditory spatial information in the brain. In D.M. Guthrie (Ed.), *Higher order sensory processing* (pp. 75–113). Manchester: Manchester University Press.

King, A.J. (1999). Sensory experience and the formation of a computational map of auditory space. *Bioessays*, **21**, 900–911.

King, A.J. & Carlile, S. (1993). Changes induced in the representation of auditory space in the superior colliculus by rearing ferrets with binocular eyelid suture. *Experimental Brain Research*, **94**, 444–455.

King, A.J. & Carlile, S. (1995). Neural coding for auditory space. In M.S. Gazzaniga (Ed.), *The cognitive neurosciences* (pp. 279–293). Cambridge, MA: MIT Press.

King, A.J. & Hartline, P.H. (1999). Multisensory convergence. In G. Adelman & B.H. Smith (Eds.), *Encyclopedia of neuroscience* (pp. 1236–1240). Amsterdam: Elsevier Science.

King, A.J. & Palmer, A.R. (1985). Integration of visual and auditory information in bimodal neurones in the guinea-pig superior colliculus. *Experimental Brain Research*, **60**, 492–500.

King, A.J. & Parsons, C.H. (1999). Improved auditory spatial acuity in visually deprived ferrets. *European Journal of Neuroscience*, **11**, 3945–3956.

King, A.J. & Schnupp, J.W.H. (2000). Sensory convergence in neural function and development. In M.S. Gazzaniga (Ed.), *The new cognitive neurosciences* (2nd ed., pp. 437–450). Cambridge, MA: MIT Press.

King, A.J., Parsons, C.H. & Moore, D.R. (2000). Plasticity in the neural coding of auditory space in the mammalian brain. *Proceedings of the National Academy of Sciences, USA*, **97**, 11821–11828.

King, A.J., Schnupp, J.W.H. & Thompson, I.D. (1998). Signals from the superficial layers of the superior colliculus enable the development of the auditory space map in the deeper layers. *Journal of Neuroscience*, **18**, 9394–9408.

King, A.J., Hutchings, M.E., Moore, D.R. & Blakemore, C. (1988). Developmental plasticity in the visual and auditory representations in the mammalian superior colliculus. *Nature*, **332**, 73–76.

Knudsen, E.I. (1985). Experience alters the spatial tuning of auditory units in the optic tectum during a sensitive period in the barn owl. *Journal of Neuroscience*, **5**, 3094–3109.

Knudsen, E.I. (1998). Capacity for plasticity in the adult owl auditory system expanded by juvenile experience. *Science*, **279**, 1531–1533.

Knudsen, E.I. & Brainard, M.S. (1991). Visual instruction of the neural map of auditory space in the developing optic tectum. *Science*, **253**, 85–87.

Knudsen, E.I. & Mogdans, J. (1992). Vision-independent adjustment of unit tuning to sound localization cues in response to monaural occlusion in developing owl optic tectum. *Journal of Neuroscience*, **12**, 3485–3493.

Knudsen, E.I., Esterly, S.D. & du Lac, S. (1991). Stretched and upside-down maps of auditory space in the optic tectum of blind-reared owls; acoustic basis and behavioral correlates. *Journal of Neuroscience*, **11**, 1727–1747.

Knudsen, E.I., Esterly, S.D. & Knudsen, P.F. (1984). Monaural occlusion alters sound localization during a sensitive period in the barn owl. *Journal of Neuroscience*, **4**, 1001–1011.

Konnerth, A., Tsien, R.Y., Mikoshiba, K. & Altman, J. (Eds.) (1996). *Coincidence detection in the nervous system*. Strasbourg: HFSP.

Kossel, A., Bonhoeffer, T. & Bolz, J., (1990). Non-Hebbian synapses in rat visual cortex. *Neuroreport*, **1**, 115–118.

Kuhl, P.K. & Meltzoff, A.N. (1982). The bimodal perception of speech in infancy. *Science*, **218**, 1138–1141.

Kuhl, P.K., Williams, K.A. & Meltzoff, A.N. (1991). Cross-modal speech perception in adults and infants using nonspeech auditory stimuli. *Journal of Experimental Psychology*, **17**, 829–840.

Lechner, H.A. & Byrne, J.H. (1998). New perspectives on classical conditioning: a synthesis of hebbian and non-hebbian mechanisms. *Neuron*, **20**, 355–358.

Lee, P.H., Helms, M.C., Augustine, G.J. & Hall, W.C. (1997). Role of intrinsic synaptic circuitry in collicular sensorimotor integration. *Proceedings of the National Academy of Sciences, USA*, **94**, 13299–13304.

Lessard, N., Paré, M., Lepore, F. & Lassonde, M. (1998). Early-blind human subjects localize sound sources better than sighted subjects. *Nature*, **395**, 278–280.

Lewkowicz, D.J. (1992). Infants' response to temporally based intersensory equivalence: the effect of synchronous sounds on visual preferences for moving stimuli. *Infant Behavior and Development*, **15**, 297–324.

Lewkowicz, D.J. (1996). Perception of auditory-visual temporal synchrony in human infants. *Journal of Experimental Psychology*, **22**, 1094–1106.

McGurk, H. & Lewis, M. (1974). Space perception in early infancy: perception within a common auditory–visual space? *Science,* **186**, 649–650.

McGurk, H. & MacDonald, J. (1976). Hearing lips and seeing voices. *Nature,* **264**, 746–748.

Massaro, D.W. (1987) *Speech perception by ear and by eye: a paradigm for psychological inquiry.* Hillsdale, NJ: Erlbaum.

Meredith, M.A. & Stein, B.E. (1996). Spatial determinants of multisensory integration in cat superior colliculus neurons. *Journal of Neurophysiology,* **75**, 1843–1857.

Middlebrooks, J.C. (1999). Cortical representations of auditory space. In M.S. Gazzaniga (Ed.), *The new cognitive neurosciences* (2nd ed., pp. 425–436). Cambridge, MA: MIT Press.

Morrongiello, B.A. & Rocca, P.T. (1987). Infants' localization of sounds in the horizontal plane: effects of auditory and visual cues. *Child Development,* **58**, 918–927.

Parsons, C.H., Lanyon, R.G., Schnupp, J.W.H. & King, A.J. (1999). Effects of altering spectral cues in infancy on horizontal and vertical sound localization by adult ferrets. *Journal of Neurophysiology,* **82**, 2294–2309.

Piaget, J. (1954). *The construction of reality in the child.* New York: Basic Books.

Populin, L.C. & Yin, T.C.T. (1998). Sensitivity of auditory cells in the superior colliculus to eye position in the behaving cat. In A.R. Palmer, A. Rees, A.Q. Summerfield & R. Meddis, (Eds.), *Psychophysical and physiological advances in hearing* (pp. 441–448). London: Whurr.

Rauschecker, J.P. (1995). Compensatory plasticity and sensory substitution in the cerebral cortex. *Trends in Neuroscience,* **18**, 36–43.

Rauschecker, J.P., Kniepert, U. (1994) Auditory localization behaviour in visually deprived cats. *European Journal of Neuroscience,* **6**, 149–160.

Röder, B., Teder-Sälejärvi, W., Sterr, A., Rösler, F., Hillyard, S.A. & Neville, H.J. (1999). Improved auditory spatial tuning in blind humans. *Nature,* **400**, 162–166.

Schnupp, J.W.H., King, A.J. & Carlile, S. (1998). Altered spectral localization cues disrupt the development of the auditory space map in the superior colliculus of the ferret. *Journal of Neurophysiology,* **79**, 1053–1069.

Schnupp, J.W.H., King, A.J., Smith, A.L. & Thompson, I.D. (1995). NMDA–receptor antagonists disrupt the formation of the auditory space map in the mammalian superior colliculus. *Journal of Neuroscience,* **15**, 1516–1531.

Schuman, E.M. & Madison, D.V. (1994). Locally distributed synaptic potentiation in the hippocampus. *Science,* **263**, 532–536.

Sekuler, R., Sekuler, A.B. & Lau, R. (1997). Sound alters visual motion perception. *Nature,* **385**, 308.

Sleigh, M.J., Columbus, R.F. & Lickliter, R. (1998). Intersensory experience and early perceptual development: postnatal experience with multimodal maternal cues affects intersensory responsiveness in bobwhite quail chicks. *Developmental Psychology,* **34**, 215–223.

Sparks, D.L. (1999). Conceptual issues related to the role of the superior colliculus in the control of gaze. *Current Opinion in Neurobiology,* **9**, 698–707.

Spelke, E.S. (1979). Perceiving bimodally specified events in infancy. *Developmental Psychology,* **15,** 626–636.

Stein, B.E. & Meredith, M.A. (1993). *The merging of the senses.* Cambridge, MA: MIT Press.

Stein, B.E., Labos, E. & Kruger, L. (1973). Sequence of changes in properties of neurons of superior colliculus of the kitten during maturation. *Journal of Neurophysiology,* **36**, 667–679.

Stein, B.E., Wallace, M.T. & Stanford, T.R. (2000). Merging sensory signals in the brain: the development of multisensory integration in the superior colliculus. In M.S. Gazzaniga (Ed.), *The new cognitive neurosciences* (2nd ed., pp. 55–71). Cambridge, MA: MIT Press.

Sumby, W.H. & Pollack, I. (1954). Visual contribution to speech intelligibility in noise. *Journal of the Acoustical Society of America,* **26**, 212–215.

Wallace, M.T. & Stein, B.E. (1997). Development of multisensory neurons and multisensory integration in cat superior colliculus. *Journal of Neuroscience,* **17**, 2429–2444.

Wallace, M.T., McHaffie, J.G. & Stein, B.E. (1997). Visual response properties and visuotopic representation in the newborn monkey superior colliculus. *Journal of Neurophysiology,* **78**, 2732–2741.

Welch, R.B. & Warren, D.H. (1986). Intersensory interactions. In K.R. Boff, L. Kaufman & J.P. Thomas (Eds.), *Handbook of perception and human performance,* vol. 1: *Sensory processes and perception* (pp. 1–36). New York: Wiley.

Withington-Wray, D.J., Binns, K.E. & Keating, M.J. (1990a). The developmental emergence of a map of auditory space in the superior colliculus of the guinea pig. *Developmental Brain Research,* **51**, 225–236.

Withington-Wray, D.J., Binns, K.E., Dhanjal, S.S., Brickley, S.G. & Keating, M.J. (1990b). The maturation of the superior collicular map of auditory space in the guinea pig is disrupted by developmental auditory deprivation. *European Journal of Neuroscience,* **2**, 693–703.

Withington-Wray, D.J., Binns, K.E. & Keating, M.J. (1990c). The maturation of the superior collicular map of auditory space in the guinea pig is disrupted by developmental visual deprivation. *European Journal of Neuroscience,* **2**, 682–692.

Zheng, W. & Knudsen, E.I. (1999). Functional selection of adaptive auditory space map by $GABA_A$-mediated inhibition. *Science*, **284**, 962–965.

Further reading

Calvert, G.A., Brammer, M.J. & Iversen, S.D. (1998). Crossmodal identification. *Trends in Cognitive Sciences*, **2**, 247–253.

Driver, J. & Spence, C. (1998). Attention and the crossmodal construction of space. *Trends in Cognitive Sciences*, **2**, 254–262.

King, A.J. (1999). Sensory experience and the formation of a computational map of auditory space. *Bioessays*, **21**, 900–911.

King, A.J. & Hartline, P.H. (1999). Multisensory convergence. In G. Adelman & B.H. Smith (Eds.), *Encyclopedia of neuroscience* (2nd ed., pp. 1236–1240). Amsterdam: Elsevier Science.

Stein, B.E., Wallace, M.T. & Stanford, T.R. (2000). Merging sensory signals in the brain: the development of multisensory integration in the superior colliculus. In M.S. Gazzaniga (Ed.), *The new cognitive neurosciences* (2nd ed., pp. 55–71). Cambridge, MA: MIT Press.

IV.5
Biological Approaches to the Study of Perceptual and Cognitive Development

JOHAN J. BOLHUIS

ABSTRACT

In the biological analysis of perceptual and cognitive development, the distinction between innate and acquired factors is increasingly abandoned in favour of interactionist interpretations. Sensitive periods in bird song learning and imprinting can be a result of external experience, or they can have an endogenous cause, but even then they can be altered through physiological manipulations. There are parallels between the development of filial preferences in birds and face recognition in humans, as well as between bird song learning and language development. Aspects of early learning in birds are similar to those in mammals, including humans.

INTRODUCTION: BIOLOGY AND DEVELOPMENT

The study of the development of human brain and behaviour can benefit greatly from empirical evidence from biological disciplines such as ethology and behavioural neuroscience. A biological approach, involving mostly research on animals, can be fruitful for the study of human develoment for a number of reasons. First, there is likely to be a degree of continuity between humans and animals when it comes to basic principles of the ontogeny of perceptual and motor abilities. Second, experiments on animals often allow us to manipulate their central nervous systems or their behaviour in ways that would be impossible in humans, other than in pathologies provided by Nature, or in the effects of neurosurgery. Apart from these prudent considerations, of course, there is a great deal of interest in the development of brain and behaviour in animals for its own sake, and the study of human development can benefit from the knowledge

697

A.F. Kalverboer and A. Gramsbergen (eds.), Handbook of Brain and Behaviour in Human Development, 697–724
© 2001 Kluwer Academic Publishers. Printed in Great Britain.

acquired in this domain. Several chapters in this handbook show that a lot of the knowledge pertaining to basic principles of the development of the brain is derived from animal research. Furthermore, many of the concepts that are important in the novel discipline of developmental cognitive neuroscience have been derived from work on animals (e.g. Johnson & Morton, 1991; Johnson, 1993).

A biological approach can often offer new perspectives on developmental phenomena, compared with a "cognitivist" approach. An example is the analysis of Piaget's (1954) "A-not-B" object permanence task in human infants. In this task infants can watch a reward being hidden in one of two locations. After a delay of several seconds the infant is encouraged to find the reward. From an early age infants will successfully reach to the appropriate location and retrieve the reward. However, if, after a number of such trials, the experimenter switches the side on which the reward is hidden in full view of the infant, it will continue to reach to the location where it originally saw the reward being hidden. Only when infants are about 8–10 months old do they begin to be able to perform this task successfully. Michel & Moore (1995) proposed an explanation for this problem at a cognitive level, suggesting the inadequacy of a neural explanation for cognitive change. In contrast, Johnson & Morton (1991) argued for the limitations of a cognitive explanation for the phenomenon and advocated a biological approach to interpret these findings, referring to the work of Diamond & Goldman-Rakic (1989). On the basis of evidence obtained from experiments with humans and monkeys, these authors suggested that the advance in infant behaviour in Piaget's object permance task is mediated by developments in the dorsolateral prefrontal cortex (cf. Johnson, 1994).

The cognitive and the biological domain are two different levels of analysis (Churchland & Sejnowski, 1988), which would ideally come together in the study of development. Developmental problems can be investigated at both levels, and the analyses can be autonomous within each discipline. However, as the "A-not-B" example shows, information from one domain can affect the analysis in the other in important ways. In this present chapter I will provide further examples illustrating that a cognitive analysis of development can be supported by a biological interpretation, using findings from ethological or neurobiological research. For instance, it will become clear that the analysis of the development of face recognition in human infants has benefited greatly from the study of filial predispositions in the domestic chick (see also de Haan & Halit, this volume). In addition, the findings in the chick resulted from a continual interplay between a behavioural and neurobiological approach (Bolhuis & Honey, 1998; Horn, 1998).

THEORETICAL BACKGROUND: BEYOND INNATENESS

The early days of ethology witnessed a battle of ideas between those who thought that behavioural development could be analysed in terms of innate and acquired components, and those who thought that development involved a more complex dynamic. The early debate involved Lorenz (1965), Lehrman (1953) and others. This is not the occasion to dwell upon this controversy in detail (Bolhuis, 1999b). Recently we have collected some of the key papers on the debate (Bolhuis & Hogan, 1999), while Elman et al. (1996) have put forward some contemporary interpretations of an interactionist view of development in a volume they entitled *Rethinking Innateness*. Briefly, an important aspect of the debate boils down to what Lehrman (1970) called "semantics". That is, Lehrman (1970) thought that Lorenz and he were really interested in two different problems. Lehrman was interested in studying the effects of all types of experience on all types of behaviour at all stages of development, very much from a causal perspective, whereas Lorenz was interested only in studying the effects of functional experience on behaviour mechanisms at the stage of development at which they begin to function as modes of adaptation to the environment (Hogan, 1988). Hogan (1988) argued that both these viewpoints are equally legitimate, but that Lorenz's functional criterion corresponds to the way most people think about development. Certainly, many authors still find it difficult to abandon Lorenzian terminology (and thus presumably Lorenzian concepts), even when they aspire to be "interactionists" (cf. Bolhuis, 1999b). That is, one often finds that the old Lorenzian dichotomy is merely rephrased in interactionist terms, e.g. when development is seen as a continuous interaction between "innate predispositions" and "environmental factors". It is important to realize that there is no direct, one-to-one, effect of genes on behaviour. Genes are sections of DNA that code for proteins, not for behaviour patterns (Keverne, 1997). Furthermore, it is impossible to rear an individual in complete isolation from the environment; it simply wouldn't survive, as "structures and activity patterns" (Lehrman, 1953) at any time during development need to interact with the internal and external environment of the organism. Lehrman (1953, 1970) – and Oyama (1985) after him – argued that terms such as "innate" or "genetically fixed" obscure the importance of investigating *processes* in order to understand mechanisms of behavioural development. The best way to summarize Lehrman's dynamic view of development is to quote from his 1953 paper:

> The problem of development is the problem of the development of new structures and activity patterns from the resolution of the interaction of *existing* structures and patterns, within the organism and its internal environment, and between the organism and its outer environment. At any stage of development, the new features emerge from the interactions within the *current* stage and between the *current* stage and the environment. The interaction out of which the organism develops is *not* one, as is so often said, between heredity and environment. It is between *organism* and environment! And the organism is different at each different stage of its development (p. 345; italics by Lehrman).

In the next section of this chapter some examples of contemporary evidence concerning "predispositions" may illustrate the way in which the concept of "innate" has been left behind to make way for a more subtle, interactionist interpretation of development.

PREDISPOSITIONS AND PERCEPTUAL PREFERENCES

Predispositions have been found to play a role in perceptual mechanisms that are involved in bird song learning (Marler, 1991) and auditory preferences in ducklings (Gottlieb, 1980). Further, there is an important influence of predispositions in the perception of faces in neonatal human infants (Johnson & Morton, 1991; see de Haan & Halit, this volume), and in the development of filial preferences in chicks (Bolhuis, 1996; Johnson *et al.*, 1985). In the study of the development of filial preferences in chicks we have used the term filial predispositions (Bolhuis, 1996). These were defined as perceptual preferences that develop in young animals without experience with the particular stimuli involved (e.g. Bolhuis, 1999a; Bolhuis & Honey, 1998).

A predisposition for species-specific sounds has been demonstrated in song learning in certain avian species (Marler, 1987, 1991). Song bird species need to learn their song from a tutor male (see below and DeVoogd, 1994; Marler, 1976). Under certain circumstances young males of some species can learn their songs, or at least part of their songs, from tape recordings of tutor songs. When fledgling male song sparrows (*Melospiza melodia*) and swamp sparrows (*Melospiza georgiana*) were exposed to taped songs that consisted of equal numbers of songs of both species, they preferentially learnt the songs of their own species. Males of both species are able to sing the songs of the other species, thus it appears that perceptual predispositions are involved in what Marler (1991) called "the sensitization of young sparrows to conspecific song" (p. 200).

In an extensive series of experiments, Gottlieb (e.g. 1971, 1980, 1982, 1988) investigated the mechanisms underlying the preferences that young ducklings of a number of species show for the maternal call of their own species over that of other species. Gottlieb (1971) found that differential behaviour towards the species-specific call could already be observed in an early embryonic stage, before the animal started to vocalize itself. However, a post-hatching preference for the conspecific maternal call was found only when the animals received exposure to embryonic contact-contentment calls, played back at the right speed (Gottlieb, 1980) and with a natural variation (Gottlieb, 1982), within a certain period in development (Gottlieb, 1985). Thus, the expression of the species-specific predisposition in ducklings is dependent on particular experience earlier in development (see below, and Bolhuis, 1996, Gottlieb, 1980).

The development of filial behaviour in the chick involves two systems that are neurally and behaviourally dissociable (Bolhuis, 1996; Bolhuis & Honey, 1998; Horn, 1985, 1998). Filial imprinting is the process through which social preferences of young animals (mainly precocial birds) become restricted to a particular stimulus as a result of exposure to that stimulus (see Bolhuis, 1991, for review). Horn and his collaborators found that, in domestic chicks, a restricted region of the forebrain (the intermediate and medial hyperstriatum ventrale or IMHV) is crucially involved in imprinting (for reviews see Horn, 1985, 1998). Neural evidence that distinguished two different systems came from a study (Horn & McCabe, 1984) showing that the effects of bilateral lesions to the IMHV were much greater in chicks that were trained by exposing them to a rotating red box than with chicks that received imprinting training with a rotating stuffed jungle fowl (*Gallus gallus spadiceus*). Behavioural evidence for the existence of a predisposition was provided by a study in which day-old dark-reared chicks received imprinting training by exposing them to either a rotating red box or to a rotating stuffed jungle fowl hen (Johnson *et al.*, 1985; see Figure 1). Chicks in a control group were exposed to white overhead light for the same amount of time. The approach preferences of the chicks were measured in a subsequent test in which the two training stimuli were presented simultaneously. Preferences were tested at either 2 h (Test 1) or 24 h (Test 2) after the end of training. At Test 1 the chicks preferred the object to which they had been exposed previously. At Test 2 there was a significantly greater preference for the fowl in both experimental groups as well as in the light-exposed control group. Thus, the preference for the jungle fowl increased from the 2 h to the 24 h test, and did so regardless of the stimulus with which the chicks had been trained.

These results suggest that the preferences of trained chicks are influenced by at least two different systems. On the one hand there is an effect of experience with particular stimuli (reflected in the differences k1–k4 in Figure 1), that is, filial imprinting. On the other hand there is an emerging predisposition to approach stimuli resembling conspecifics (reflected in Figure 1 as Δy for the control group). The results in Figure 1 suggest that training with a particular stimulus is not necessary for the predisposition to emerge. In subsequent studies it was found that visual experience was not necessary to "trigger" or induce (Bolhuis, 1996; Gottlieb, 1976) the predisposition. The predisposition can emerge in dark-reared chicks, provided that they receive a certain amount of non-specific stimulation within a certain period in development (Johnson et al., 1989).

The stimulus characteristics that are important for the filial predisposition to be expressed were investigated in chicks that had developed the predisposition, in tests involving an intact stuffed jungle fowl versus one in a series of increasingly

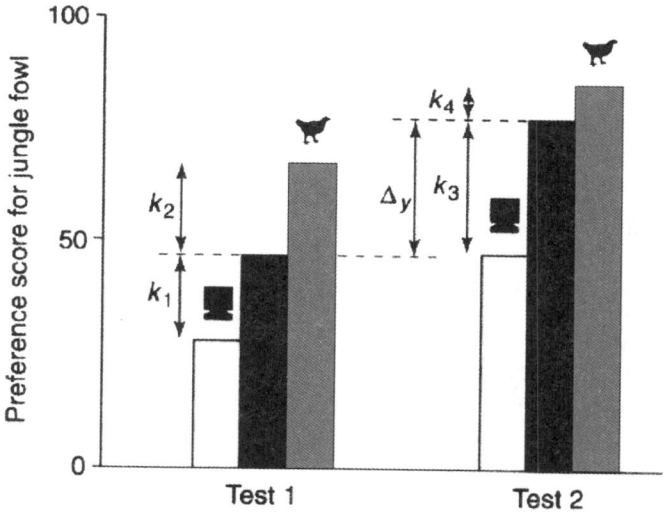

Figure 1 Mean preference scores, expressed as a preference for the stuffed fowl, of chicks previously trained by exposure to a rotating stuffed jungle fowl, a rotating red box, or exposed to white light. Preference scores are defined as: activity when attempting to approach the stuffed jungle fowl divided by total approach activity during the test. Preferences were measured in a simultaneous test either 2 h (Test 1) or 24 h (Test 2) after the end of training; $k1$–$k4$ represent the differences between the preferences of the trained chicks and the controls; Δy represents the difference in preference between the control chicks at Test 2 and at Test 1. See text for further explanation. (Adapted from Horn (1985), by permission of the Oxford University Press, after Johnson et al. (1985)).

degraded versions of a stuffed jungle fowl (Johnson & Horn, 1988). The degraded versions ranged from one in which different parts of the model (wings, head, torso, legs) were reassembled in an unnatural way, to one in which the pelt of a jungle fowl had been cut into small pieces that were stuck onto a rotating box. The intact model was preferred only when the degraded object possessed no distinguishable jungle fowl features. In addition chicks did not prefer an intact jungle fowl model over an alternative object that contained only the head and neck of a stuffed fowl. Thus, the head and neck region contains stimuli that are relevant for the predisposition. In subsequent experiments it was found that the chicks did not prefer a stuffed jungle fowl hen over a stuffed Gadwall duck (*Anas strepera*) or even a stuffed pole cat (*Mustela putorius*). Thus, the predisposition is not species- or even class-specific. Subsequent studies showed that eyes are an important stimulus, but that other aspects of the stimulus are also sufficient for the expression of a predisposition (Bolhuis, 1996). There are interesting similarities between the development of face recognition in human infants, and the development of filial preferences in chicks (Johnson & Morton, 1991). Newly born infants have been shown to track a moving face-like stimulus more than a stimulus that lacks these features, or in which these features have been jumbled up (Johnson *et al.*, 1991; Johnson & Morton, 1991; Johnson, 1994). It is not known whether the development of face preferences in infants is dependent on previous experience, but the parallel with the emergence of perceptual predispositions in birds is striking. Similarly, in both human infants and in young precocial birds the features of individual stimuli need to be learned. De Haan (this handbook) discusses the similarities between face recognition and avian predispositions in detail.

SENSITIVE PERIODS

The concept of sensitive periods has had an important influence in the study of development of animals and humans (see reviews in Bornstein, 1987; Rauschecker & Marler, 1987). Here we shall adopt the definition given by Bateson & Hinde (1987): "The sensitive period concept implies a phase of great susceptibility preceded and followed by lower sensitivity, with relatively gradual transitions" (p. 20). Early authors (e.g. Lorenz, 1937) used the term "critical period", borrowed from embryology. However, on the basis of subsequent evidence, Bateson & Hinde (1987; cf. Bateson, 1979) argue that the periods of increased sensitivity are not sharply defined, and consequently they suggested the use of the term "sensitive period", which is now widely (but by no means universally) used. In the analysis of mammalian visual development the term "critical period" is still used frequently (e.g. Hirsch *et al.*, 1987; Wiesel, 1982), although it is clear that the period of time during which neural plasticity in the visual cortex is possible is not narrowly defined, but that sensitivity to external experience changes gradually (Timney, 1987; Wiesel, 1982).

Bateson & Hinde (1987) distinguished between two types of explanation and three different levels of analysis, leading to six possible ways to explain evidence for a sensitive period. The two types of explanation were called immediate and preceding, respectively, depending on whether the causal factors accompany or immediately precede a change in sensitivity or whether these causes lie further back in time. The three levels of analysis distinguished were called organismic, physiological and molecular; we shall be concerned only with the first two levels in this chapter. The authors emphasize that the different types of explanation need not be incompatible with each other, but they argue that in a discussion of sensitive periods it should be clear which type of explanation is used. We shall discuss recent evidence concerning sensitive periods in three developmental paradigms, namely imprinting, the development of perceptual predispositions, and song learning.

Sensitive periods in imprinting

Filial imprinting is the process through which early social preferences become restricted to a particular stimulus or class of stimuli, as a result of exposure to that stimulus (Bolhuis, 1991). This early learning phenomenon is often regarded as a classic example of a developmental process involving sensitive periods. In fact, Lorenz (1937) originally suggested that occurrence of imprinting within a sensitive period (he called it a "critical period") was one of the criteria that distinguished the process from conventional learning. Early imprinting research led to the suggestion that there was a narrow sensitive period within which imprinting could occur. This sensitive period was thought to occur within the first 24 h after hatching in ducklings and chicks, and last for not more than a few hours. Subsequent research has demonstrated that there is not a strict sensitive period for filial imprinting (for reviews see Bateson, 1979; Bateson & Hinde, 1987; Bolhuis, 1991). In the analysis of sensitive periods it is important to distinguish between the onset and the decline of increased sensitivity – the beginning and the end of the sensitive period, so to speak – as there are often different causal factors for these two events. In filial imprinting it is likely that the onset of the sensitive period can be explained in terms of "immediate physiological" factors, such as an increase in visual effiency and an increase in motor ability in precocial birds some time after hatching (Bateson & Hinde, 1987). Different causal factors are thought to be involved in the end of the sensitive period for imprinting. It has been suggested that the ability to imprint comes to an end after the animal has developed a social preference for a certain stimulus as a result of exposure to that stimulus. The animal will stay close to the familiar object and avoid novel ones; it will thus receive very little exposure to a novel stimulus and there will be little opportunity for further imprinting (Bateson, 1979; Bolhuis, 1991; Sluckin & Salzen, 1961). This interpretation implies that imprinting will remain possible if the appropriate stimulus is not presented. Indeed, chicks that are reared in isolation retain the ability to imprint for longer than socially reared chicks (see Bateson, 1979, for review). An apparent decline in sensitivity in isolated chicks might be explained as resulting from imprinting of the animals to stationary visual aspects of the rearing environment, impairing subsequent imprinting to imprinting stimuli (Bateson, 1964). Thus, a "preceding organismic" explanation of the end of the sensitive period postulates that it is the imprinting process itself that brings the sensitive period to an end.

Bateson & Hinde (1987) suggested that competing models to explain sensitive periods should come from the same class of explanations. These authors

BOLHUIS

contrasted two "preceding physiological" explanations for the end of the sensitive period for imprinting; namely a "physiological clock" model and a "competitive exclusion" model. In the former, some kind of endogenous physiological mechanism would cause the onset and end of sensitivity. This is in fact the conventional view of the causes of sensitive periods, as it has been used in a number of disciplines (see reviews in Bornstein, 1987; Rauschecker & Marler, 1987). In the competitive exclusion model (Bateson, 1987), neural growth is associated with a particular sensory input from the environment. The model assumes that there is a limited capacity for this neural growth to impinge upon the systems that are responsible for the execution of the behaviour involved (e.g. approach in the case of imprinting). Input from different stimuli is "competing" for access to these executive systems. Once neural growth associated with a certain stimulus has passed beyond the halfway point of capacity of the executive systems, subsequent stimuli will be less able to gain access to these systems. This model expresses at the "preceding physiological" level the same principles as the "preceding organismic" does for the end of the sensitive period for imprinting. The essential difference between the two models is that one assumes that there is an internal clock mechanism while the other emphasizes the role of external experience (Bateson, 1987; ten Cate, 1989). The evidence from filial imprinting studies is not consistent with an internal clock model, but supports a model in which external experience, specifically experience with the imprinting stimulus, is the causal factor for the end of the sensitive period (see Bateson, 1987; Bolhuis, 1991). Recent studies show that social preferences acquired through sexual imprinting can be altered by experience in adulthood, suggesting that here too the evidence is consistent with an experience-dependent end to the sensitive period (see Bischof, 1994, for discussion).

Predispositions and sensitive periods

Sensitive periods have also been found to play a role in the development of perceptual predispositions. Gottlieb (1985) observed that the expression of a species-specific auditory predisposition in hatchling ducklings required specific embryonic experience during a sensitive period. Similarly, it was found that there is a sensitive period for the induction of a perceptual predisposition in the chick. Johnson et al. (1989) first reported that the emergence of the predisposition could be induced in dark-reared chicks at 24 h or 36 h after hatching, but not at 12 h or 42 h after hatching. A control experiment showed that the time of induction (i.e. exposure to the non-specific stimulation) was important, not the time of testing. As the induction of the predisposition is not dependent upon specific visual experience, it is unlikely that this sensitive period is dependent on experience in the way that the sensitive period for imprinting is. Rather, it seems that the sensitive period for the predisposition is endogenous, and is best explained as resulting from an internal clock mechanism. However, subsequent studies suggested that the internal clock can be "reset", as it was found that the sensitive period is not fixed in developmental time, but that it can be altered by certain manipulations. The onset of the sensitive period can be delayed by early application of either the catecholaminergic neurotoxin DSP4 (Davies et al., 1992) or the anaesthetic equithesin (Bolhuis & Horn, 1997). In the study by

Davies *et al.* (1992) the whole of the sensitive period (which appeared to have the same duration as before) was delayed by application of the neurotoxin. These results show that sensitive periods with endogenous causes, that are best explained by an internal clock model, still exhibit an amount of flexibility. It is possible that the causal factors underlying the shift in sensitive periods may also play a role later in life and lead to renewed sensitivity to external experience (Bateson & Hinde, 1987). Such renewed plasticity has been suggested for the effects of visual experience on ocular dominance columns (Wiesel, 1982) in the visual cortex of the cat (Kasamatsu *et al.*, 1979). There is a sensitive period during which monocular deprivation affects the formation of ocular dominance columns in the visual cortex of the cat, the rhesus monkey and the human infant (Wiesel, 1982). It was reported that local infusion of noradrenaline (norepi-nephrine) led to renewed plasticity in the visual cortex of adult cats (Kasamatsu *et al.*, 1979). There have been failures to replicate these results, however, and there is a controversy in the literature as to the validity of these findings (Hirsch *et al.*, 1987; Sillito, 1983; Timney, 1987).

Sensitive periods in bird song learning

Vocalizations have an important function in communication between conspecifics in many species of both songbirds and non-songbirds. Songbirds (oscines) need to learn their song from a tutor. The developmental time at which this learning takes place varies widely between species: in some species learning occurs early in development, while in other species learning takes place when the birds have dispersed to another area than where they were reared, usually when they are in their first breeding season (see Snowdon & Hausberger, 1997, for reviews). In addition, a distinction is made between "open-ended learners" (such as the canary) that can learn new songs throughout life, and species in which one song is learned early in life that remains unaltered (Marler, 1987). We will discuss these "age-limited learners" (Marler & Peters, 1987) only (for reviews see DeVoogd, 1994; Marler, 1976). The classic work of Marler and his co-workers has led to the distinction of two main phases in the song copying process: a memorization phase and a motor learning phase (see Figure 2). During the memorization phase (when the bird does not yet sing itself) the young male is thought to form a "template" or memory of the paternal song. Later, during the motor learning phase, the young will start to vocalize, gradually matching his own vocalizations with the information stored in the template. Song learning in "age-limited learners" thus goes through various stages, called subsong, plastic song and crystallized or full song, respectively. During subsong the animal produces highly variable and unstructured song patterns. This phase has been likened to "babbling" in human infants (Snowdon & Hausberger, 1997; see below).

During the period of plastic song the male produces song patterns that it has learned earlier, but there is still a degree of plasticity in song output. This is the phase when the animal is thought to "match" its own song output with the stored template. Eventually, there is crystallization of vocal output into full song, which is fixed for life (Marler, 1976, 1987). Marler argued that sensory templates are similar to Lorenzian "innate releasing mechanisms". However, now that we have seen that the term "innate" is meaningless for an explanation of behavioural

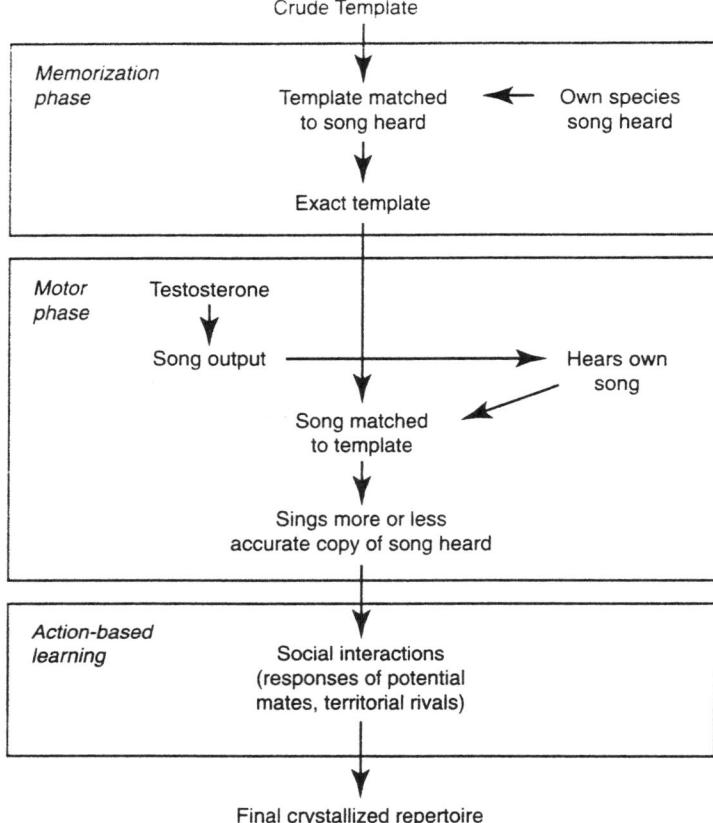

Figure 2 Stages in song development in some songbird species, and their presumed underlying mechanisms. See text for futher explananation. (From Shettleworth (1998), adapted from Slater (1983), with permission.)

development, it would be better to adopt the terminology of subsequent authors who have argued that templates are essentially the same as "perceptual mechanisms" or "cognitive structures" (e.g. Hogan, 1988), terms which are neutral in that respect.

Sensitive periods are an important feature of song learning in some species of song birds. (Marler, 1987, 1991). For instance, in the zebra finch (*Taeniopygia guttata castanotis*), song learning takes place until approximately 65 days after hatching (Slater *et al.*, 1988). Baptista & Petrinovich (1984; Petrinovich & Baptista, 1987) were the first to claim that tape tutoring leads to an artificially short sensitive period, and that exposure to live, interactive song tutors yields a much longer sensitive period for song learning. In one of their experiments (Petrinovich & Baptista, 1987), they exposed young white-crowned sparrows (*Zonotrichia leucophrys*) to either a taped song or to a live tutor until day 50.

Subsequently, all birds were exposed to a novel live tutor until their song had crystallized, around day 200. There was considerable learning of songs presented to the young birds after day 50, which was supposed to be the end of the sensitive period for song learning in this species (but see Nelson, 1998, for a discussion of the different claims for a sensitive period). Recently, Nelson (1998) reviewed the literature concerning live versus tape tutoring and the sensitive period for song learning. He suggested that the findings reported by Petrinovich and Baptista are the result of confounding two kinds of learning. Nelson distinguishes between learning by memorization, during the sensitive phase, and learning by selection of previously memorized songs, in a process termed "action-based learning" (Marler, 1991) or "selection-based learning" (Nelson & Marler, 1994). Essentially, Nelson and Marler suggest that the young male learns several different songs, which are then produced later in life. On the basis of social interactions with other adult birds, most songs disappear from the animal's repertoire and only one song remains (see Figure 2). Nelson (1998) suggests that the birds in the study of Petrinovich & Baptista (1987) had the opportunity to interact with the live tutors, and not with the birds that had produced the taped songs. Thus, according to Nelson, in both means of tutoring there is the same sensitive period, and the apparent lengthening of the sensive period is a result of the differential tutoring procedure used for live and tape tutors.

Marler (1987) evaluated the influence of social experience on the length of the sensitive period for bird song learning. Song sparrows exposed to a continuously changing programme of tape-recorded songs, learned most of their songs between about 20 and 50 days after hatching (Marler & Peters, 1987). However, there was considerable variation in that some birds learned songs heard until 200 days after hatching. These birds also learned from songs heard earlier in the exposure period, indicating that the sensitive period was extended rather than delayed. Thus, there is variation in the length of the sensitive period for song learning from tapes. Subsequently, Marler & Peters (1988) directly compared the sensitive period for tape tutoring and live tutoring in the swamp sparrow, by giving young males a programme of tape- or live tutoring during the first year of life. The sensitive period was similar between these two conditions, with again considerable inter-individual variation in sensitive period length (within conditions), as in the song sparrow. Thus, social tutoring does not extend the sensitive period in the swamp sparrow. However, on the basis of work on other species, such as that of Baptista and Petrinovich discussed above, Marler (1987) suggested that sensitive periods are not fixed but labile. This was shown particularly in experiments in which young birds were first exposed to a taped song, followed by exposure to a live tutor. In several cases exposure to the live tutor resulted in learning of the song of that tutor, some considerable time after the closure of the sensitive period during which tape tutoring is still effective. Marler (1987) argued that, when birds are deprived of the appropriate stimulation at a certain stage in development, the sensitive period can be extended considerably. Findings in work on the zebra finch are consistent with this suggestion. Morrison & Nottebohm (1992) and Slater et al. (1993) found that the sensitive period for song learning could be extended when zebra finch males were isolated visually or auditorily after day 35. Isolated males developed stable songs, but these could be modified when they were exposed to novel songs as adults. These results suggest a "self-terminating" sensitive period, not dissimilar to suggestions made for filial imprinting, discussed earlier. Thus, the song learning results suggest a role for external experience, and they do not support an "internal clock" model of sensitive periods (Bateson & Hinde, 1987; ten Cate, 1989; but see Nelson, 1998, for a different view). As yet it is not known what the underlying mechanisms are. At the "preceding physiological" level (Bateson & Hinde, 1987), it is possible that external stimulation results in changes in hormone levels (Korsia & Bottjer, 1991; Whaling *et al.*, 1998), which somehow could prevent further learning via a "competitive exclusion" mechanism (Bateson, 1987). At the "preceding organismic" level it could be that, once the animal has memorized a particular song, it attends less to novel songs, or it may avoid individuals that produce novel songs.

Conclusions

These results from a number of different developmental paradigms do not support the concept of a narrowly defined sensitive period; rather they suggest that sensitive periods come in various shapes and sizes. There is inter-individual variation in the length of sensitive periods. The beginning of a sensitive period may have different causal factors than the decline of sensitivity. In song learning, sensitive periods may be extended when there is social interaction with the song tutor, but it is still controversial whether this phenomenon is an artefact of the experimental design. Alternatively, the sensitive period may be extended when the appropriate stimulus is withheld, similar to the situation in filial imprinting. In both song learning and imprinting, sensitive periods are influenced by external experience, and they are not compatible with an interpretation in terms of an internal clock mechanism. The sensitive period for the induction of a filial predisposition in the chick has endogenous causes – as has the sensitive period for the effects of visual deprivation on the mammalian visual cortex – but there the sensitive period can be delayed as a result of certain physiological manipulations early after hatching.

AVIAN VOCAL LEARNING AND HUMAN LANGUAGE DEVELOPMENT

There are similarities between the way speech develops in human infants, and vocal learning in birds (Doupe & Kuhl, 1999; Marler, 1976; Snowdon & Hausberger, 1997). Human newborns prefer their mother's voice over that of other adult females, as a result of intrauterine auditory learning (DeCasper & Fifer, 1980; Spence & DeCasper, 1987; Locke, 1993, 1994). In addition, at 2–4 days after birth, infants prefer the language spoken by the mother prenatally (Moon *et al.*, 1993). Locke (1994) concluded that "All of these findings suggest that the human newborn begins extrauterine life with an auditory pre-adaptation, a strong inclination to track conspecific voice." In addition, there is rapid postnatal learning of characteristics of the mother's voice (Locke, 1994). Marler (1976) suggested that these perceptual preferences before speech has developed indicate that human infants have auditory templates for speech sounds, analogous to sensory templates in song birds.

A second aspect of human speech development that has a strong similarity with bird song learning is babbling (Koopmans-van Beinum & van der Stelt, 1986). According to Locke (1994), "No other primate does anything quite like babbling" (p. 308). However, he also suggests that song bird subsong (see Figure 1) is quite analogous to babbling. Marler (1976) too has pointed out the similarity between subsong and human babbling (see also Doupe & Kuhl, 1999; Snowdon & Hausberger, 1997). He suggested that improvement in a child's babbling may reflect "a process of matching vocal output to sensory templates by auditory feedback" (Marler, 1976, p. 327), similar to template matching in songbirds such as the white-crowned sparrow.

Social interaction with the parent is considered to be important for language development (Locke, 1994; Snowdon & Hausberger, 1997). Locke (1994) has argued that visual stimuli emanating from the parent or caretaker, especially facial expressions, play an important role in the development of human language (see also contributions in Section V in this volume, for similar views). One function of this visual stimulation may be that it focuses the infant's attention on the vocal utterances of the parent. Similar processes may occur during avian vocal development. For instance, social interactions between the "pupil" and the song tutor enhance song learning (see above and Snowdon & Hausberger, 1997), although it is not clear through what mechanism. It could be that visual stimuli draw the young animal's attention to the vocal example provided by the song tutor (Hultsch et al., 1997; ten Cate, 1994; ten Cate & Houx, 1999). There are examples of audiovisual interactions during filial imprinting that have been examined in some more detail. Van Kampen & Bolhuis (1991) found that auditory imprinting is enhanced by simultaneous exposure to an auditory and a visual imprinting stimulus (see Bolhuis & Van Kampen, 1992, for a review of auditory learning in imprinting). The results of subsequent studies have implications for the mechanisms of early learning, and will be discussed below.

MULTIPLE NEURAL SUBSTRATES UNDERLYING THE PERCEPTION AND PRODUCTION OF BIRD SONG AND HUMAN LANGUAGE

This is not the place to discuss neuropsychological theories of language development (Doupe & Kuhl, 1999). However, given the parallels between human speech development and bird song learning (see Marler, 1976; Doupe & Kuhl, 1999), I will discuss some recent evidence concerning the neural mechanisms of bird song learning. This evidence also suggests a dissociation between perception and production of vocalizations.

Figure 3A illustrates the so-called "song control nuclei" that were identified on the basis of an extensive series of studies of bird song, using neuroanatomical methods, lesions and electrophysiology (for reviews see DeVoogd, 1994; Nottebohm, 1991). Work from a number of different laboratories led to the suggestion of two forebrain song pathways, the rostral and the caudal pathway. The rostral pathway includes the high vocal centre (HVC), area X, and LMAN (see Figure 3 for names of brain regions). Lesions to LMAN or area X disrupted song development in juvenile zebra finch males, but similar lesions placed in these nuclei in adult birds did not affect song output. These results suggested

A

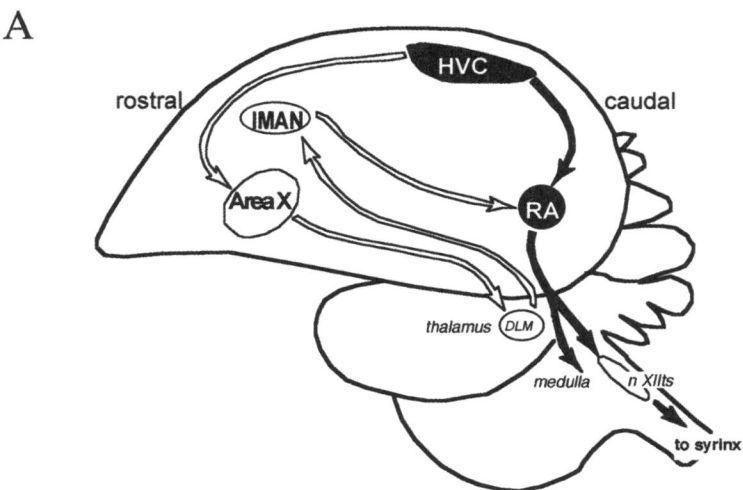

B

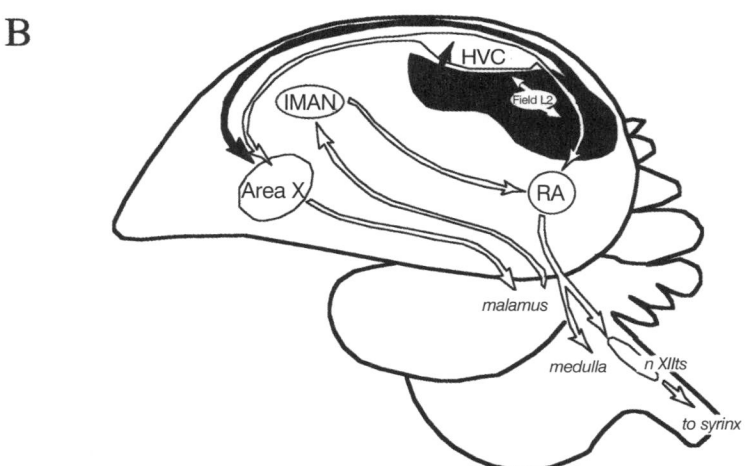

Figure 3 Schematic diagrams of the brain of a songbird (medial parasagittal plane), with various regions that play a role in bird song. **A**: The rostral and caudal pathways. The caudal pathway is indicated with black lines; the rostral pathway is indicated with white lines. **B**: Some of the nuclei shown in Figure 3A, together with other regions that have been shown to be involved in the perception of song. The black area in **B** represents the caudo-medial neostriatum (NCM) and the caudal part of the hyperstriatum ventrale (CHV). (From Clayton, 2000, with permission). Abbreviations: HVC: high vocal center; RA: nucleus robustus archistriatalis; nXIIts: nucleus hypoglossus; DLM: medial part of the dorsolateral thalamic nucleus; lMAN: lateral part of the magnocellular nucleus of the anterior neostriatum.

710

that LMAN and area X may be involved in the acquisition of song. In contrast, lesions to nuclei in the caudal pathway (including HVC and RA), or to any of its connections result in immediate, profound and irreversible deficits in song in adult birds. This suggests that the motor pathway for song begins with HVC and consists of the direct projection to RA, and its projections to nXIIts, the hypoglossal nucleus in the brainstem that projects to the syrinx, where the sound is produced. Electrophysiological recordings from units in the HVC in adult white-crowned sparrows and zebra finches revealed that most of the responsive units responded preferentially to the bird's own song, even when compared to the song of its tutor. However, when zebra finches can choose between exposure to their own song and that of their tutor, they prefer the latter (Adret, 1993). This suggests that tutor songs as well as own songs are memorized and stored some-where in the brain, possibly in different locations. In spite of considerable progress made in this field, Nottebohm, one of the pioneers of the neurobiology of bird song, stated. "We do not know how and where learned sounds are stored" (Nottebohm *et al.*, 1990).

Immediate early genes and bird song: A dissociation between production and perception

Immediate early genes (IEGs) may become expressed when a cell is activated (Dragunow & Faull, 1989). The expression of IEGs occurs much faster than that of "normal" or "late" genes. IEGs code for proteins that, once they have been synthesized in the cytoplasm, return to the cell nucleus to affect the transcription of "late" genes. In turn, these late genes may code for proteins that can modify the properties of cells, such as certain enzymes or receptor molecules. Thus, IEGs can play an important role in the chain of events that leads to plastic changes in cells such as neurons. The expression of IEGs can also be used as a marker for cell activation. McCabe & Horn (1998) were the first to use the expression of IEGs to successfully study the neural substrate of recognition memory. They investigated the expression of Fos, the protein product of the immediate early gene c-fos, in the chick forebrain in filial imprinting. These authors found a significant positive correlation between imprinting preference and the number of Fos-positive cell nuclei in the IMHV, but not in another part of the brain, the hippocampus.

Mello, Clayton and co-workers (Mello *et al.*, 1992; Mello & Clayton, 1994; Clayton, 2000) were the first to apply the IEG method to the study of bird song. They studied the expression of the immediate early gene ZENK in adult canaries and zebra finches after exposure to taped conspecific song, heterospecific song, or tone bursts. Exposure to conspecific song led to a significant increase in the expression of ZENK in the medial caudal neostriatum (NCM) and in the caudo-medial part of the hyperstriatum ventrale (CHV), compared to the two control groups (see Figure 3B). The result was surprising because there was no significant effect of exposure to song on IEG expression in the traditional "song control nuclei", such as the high vocal centre (HVC), area X, LMAN or RA (see Figure 3 for nomenclature), nor was there an effect in field L, a primary auditory projection area. These stimulus-specific and area-specific effects make it unlikely that the IEG result reflects increased levels of arousal. The regions showing significant IEG expression have not traditionally been implicated in the production or perception of song. More recent research (Jarvis & Nottebohm, 1997) shows that there is a dissociation between brain regions (NCM and CHV) involved in the perception of song, and the traditional "song control nuclei" that are active when song is produced (see Figure 3B). That is, Jarvis & Nottebohm (1997) measured IEG expression in zebra finches that could hear a playback of an unfamiliar song. Some of the birds would "countersing" to the taped song, while others would remain silent. In birds that were hearing only, there was a significant increase in IEG expression in the NCM and CHV, compared to controls that did not hear a song. Birds that sang while the taped song was played showed additional gene expression in the traditional "song control nuclei". Furthermore, deafened birds that sang showed increased IEG expression in the "song control nuclei", but not in NCM or CHV. The authors dubbed this phenomenon "motor-driven gene expression".

Despite these dramatic advances in the neurobiology of bird song, we still do not know much more about the neural substrates of song learning than when Nottebohm *et al.* (1990) professed our ignorance in the quote at the end of the previous section. We do know that repeated exposure to a song (in adult zebra finches) leads to a waning of the IEG reponse in the NCM to that song (Mello *et al.*, 1995), which suggests a possible role for this structure in the detection of familiarity and novelty. Also, a significant IEG response to song in the NCM does not occur before 20 days after hatching in zebra finches, which suggests a possible role for this structure in the acquisition of songs. A recent study from our own laboratory provides firmer evidence for a role for the NCM in song learning and memory (Bolhuis *et al.*, 2000). We found that, in zebra finch males that had been tape-tutored with a particular song, subsequent exposure to that song led to a significant increase in expression of the IEG protein products Fos and ZENK in the NCM, compared to birds that were not exposed to song. Furthermore, there was a significant positive correlation between the expression of both these IEG products and the number of song elements that the birds had copied from their tutor songs. There were no such IEG responses in the traditional "song control nuclei". This is the first evidence suggesting that a restricted region of the songbird forebrain may be the neural substrate for song learning (cf. Marler & Doupe, 2000).

Lateralization in bird song and human language

In the canary and the white-crowned sparrow, lesions of the left hypoglossal nerve or the left HVC cause more disruption of song than similar lesions to the right side of the brain (Nottebohm & Nottebohm, 1976; Nottebohm, 1980). These early results suggested an interesting parallel in left hemisphere dominance between human speech and bird song (Studdert-Kennedy, 1981). However, in canaries the musculature of the left side of the syrinx is much larger than that of the right side (Nottebohm & Nottebohm, 1976; Nottebohm, 1980). In addition, the volume of the HVC (which correlates with song repertoire) does not differ between the two hemispheres. Each hemisphere innervates only the ipsilateral syrinx muscles. Thus, a lesion to the left side of the brain would be likely to have a greater disruptive effect on song, as the left brain innervates the larger left side musculature. Subsequent electrophysiological studies involving simultaneous recording from the left and right HVCs of the canary and the zebra finch (McCasland, 1987) confirmed the suggestion that there may not be lateralization in the control of bird song. The temporal relationship between neuronal activity patterns and singing was identical for both HVCs. Recent work by Goller & Suthers (1995) confirmed the dissociation between central and more peripheral mechanisms in the lateralization of bird song. These authors made simultaneous recordings of sound, air flow, and activation of syringeal and respiratory muscles during singing in brown thrashers (*Toxostoma rufum*). It was found that peripheral lateralization could be achieved through closure of a syringeal valve by means of activation of a subset of ipsilateral syringeal muscles. Other syringeal muscles, primarily involved in the phonetic structure of the bird's song, were active bilaterally. This leads to what the authors termed "unilateral gating of bilateral [central] motor patterns". Such a mechanism is useful for the enormous acoustic flexibility and variability that is seen in bird song. However, it is clear that the nature of lateralization is quite different between humans and songbirds.

There is evidence for hemispheric differences in the perception of bird song in the zebra finch. Cynx *et al.* (1992) lesioned the left or right auditory relay nucleus of the thalamus in adult male zebra finches, thus disrupting auditory input to the ipsilateral forebrain. The birds were then trained in two different auditory discrimination tasks, in which they had to discriminate (i) between their own song and the song of a cage mate; (ii) between two versions of a novel zebra finch song that differed only in the harmonic structure of the final element. Birds with lesions on the right side performed better than their left side-lesioned counterparts on the former task, while left side-lesioned birds performed better on the latter task. The authors concluded that the two sides of the zebra finch brain process conspecific sounds differently. Nottebohm *et al.* (1990) interpreted these results as an indication that in zebra finches "the left hemisphere is better at discriminating between stimuli that differ in many ways ('holistic perception') and the right hemisphere is better at discriminations requiring a more 'analytic' processing of input" (p. 121). Studdert-Kennedy (1981) concluded, on the basis of work on human language perception (Studdert-Kennedy & Shankweiler, 1970), that the left hemisphere is specialized for phonological analysis of spoken language, while the right hemisphere "perceives language holistically, seizing meaning from the 'auditory contours' of words rather than by phonological analysis" (Studdert-Kennedy, 1981, p. 546). Thus there is an interesting parallel between zebra finches and humans in the lateralization of function with regard to vocal perception. The side of the brain that is thought to be involved in the "holistic" versus the "analytic" function differs between the two species; Cynx *et al.* (1992) suggested that, unlike the canary, the zebra finch may have right hemisphere dominance for song production. If that were the case, it would seem that the hemisphere that is involved in the "analytic" perception of vocalizations is also the dominant hemisphere for vocal production in both species.

In humans, evidence for a left hemisphere dominance for speech perception was found immediately after birth (Segalowitz & Chapman, 1980; see also Locke, 1993). In zebra finches the immediate early gene response to the presentation of song appears some time between day 20 and day 30 of development (Jin & Clayton, 1997). We do not know whether these responses are lateralized.

AVIAN PERCEPTUAL DEVELOPMENT: REPRESENTATION FORMATION DURING EARLY LEARNING

Lorenz (1937), regarded imprinting as a process that has nothing to do with learning. If imprinting were wholly different from other forms of learning, it would not be a good model for the study of human development. More recently, the special character of imprinting has been called into question by a number of authors (e.g. Hoffman & Ratner, 1973; Bateson, 1990; Bolhuis *et al.*, 1990; see Bolhuis, 1996, for discussion). Hoffman & Ratner (1973) interpreted filial imprinting as a form of classical (Pavlovian) conditioning, with movement of the imprinting stimulus as the unconditioned stimulus (US) and other, neutral aspects of the stimulus as conditioned stimuli (CSs). Imprinting may proceed without stimulus movement, so movement cannot be a crucial US (see Bolhuis *et al.*, 1990, for further discussion). It has been known for some time that imprinting stimuli have reinforcing qualities (see Bolhuis, 1991, for review). Bolhuis *et al.* (1990) reviewed the evidence for an associative learning intereptation of imprinting, and suggested that exposure to an imprinting object could lead to some kind of reinforcing event (an affective state) which might lead to conditioned responding to neutral aspects of the stimulus. As Shettleworth (1998) pointed out, this interpretation in itself is a tautology and says little more than "objects that support imprinting support imprinting". However, if such a mechanism would underlie imprinting, phenomena that are typical of Pavlovian conditioning, such as blocking and overshadowing, would have to occur in imprinting. There is now evidence that Pavlovian conditioning does indeed play a role in filial imprinting (De Vos & Bolhuis, 1990; Van Kampen & De Vos, 1995).

Van Kampen (1996) suggested that imprinting involves both conditioning and perceptual learning. Perceptual learning involves learning of the characteristics of stimuli, rather than their relationship to other stimuli, as in conditioning (cf. Shettleworth, 1998). Similarly, Bateson (1990) pointed out that imprinting, like other forms of learning, involves the formation of representations of stimuli. The dissociation between conditioning and perceptual learning may be nominal. Current theories of perceptual learning suggest that the underlying mechanisms may involve the formation of associations between stimuli or stimulus elements (Bateson & Horn, 1994; McLaren *et al.*, 1989; Pearce, 1994). We will see that the imprinting evidence supports such an interpretation.

Recent findings concerning imprinting with audiovisual compound stimuli shed light on the mechanisms involved in the development of filial preferences in the chick (for recent reviews see Bolhuis & Honey, 1998; Bolhuis, 1999). Some of these mechanisms were revealed when it was taken into account that in the natural situation the mother hen not only has a particular size, shape and coloration but also emits a specific call. Similarly, during imprinting training in the laboratory, the young birds are often exposed to a visual stimulus accompanied by a recording of the maternal call of a hen. In a number of studies chicks were exposed to either a visual stimulus together with a maternal call, or to the visual stimulus alone. Subsequently, both groups received a preference test in which the visual training stimulus and a novel visual stimulus were presented without a maternal call. Chicks given the maternal call during training showed a significantly greater preference for the training stimulus than those that had not. Thus, the presence of a call during visual imprinting training can enhance the preference acquired for the visual stimulus. A separate study demonstrated that this effect is reciprocal: auditory imprinting was enhanced as a result of audiovisual compound training. Furthermore, a similar enhancement effect was found in an experiment involving learning the different visual aspects of an imprinting stimulus. An interesting interpretation of these findings is that a similar mechanism might underlie this facilitation of imprinting and the poten-tiation of learning observed when animals receive conditioning with a simultaneous compound (Rescorla & Durlach, 1981). For instance, in rats, aversive conditioning of an odour is enhanced when that odour is paired with a taste during the conditioning session. Such potentiation may be a result of within-event learning. That is, during conditioning associations may form between the representations of the two elements of the compound stimulus. Similarly, during imprinting training with an audiovisual compound, representations of the auditory and visual components of the compound may become associated. The formation of such an association will mean that during a test in which only the visual component is presented there may be two sources of preferential responding – that acquired by the visual component and that due to the associatively activated representation of the auditory component of the compound.

Some of the predictions of such a within-event learning account of poten-tiation of imprinting were tested in an experiment by Bolhuis & Honey (1994) in which chicks were exposed to a compound stimulus containing an auditory element (a taped maternal call) and a visual element (a moving two-dimensional image). The potentiation effect was replicated in subsequent preference tests with the training visual stimulus and a novel visual stimulus. That is, chicks that had been exposed to the compound stimulus showed a significantly greater preference for the training (visual) stimulus than chicks that had been exposed to the visual stimulus only. Similarly, chicks that had been exposed to the compound stimulus showed a greater preference for the visual stimulus than chicks that had been exposed to the visual and the auditory stimuli in separate sessions. Furthermore, a session of imprinting training with the call *in addition to* training with the compound stimulus, reduced the preference for the visual stimulus, whether exposure to the call occurred *after* or *before* exposure to the audiovisual compound. The temporal relationship between the visual and auditory components of an imprinting stimulus was varied in a series of experiments involving an extended series of trials of brief exposure to the stimulus elements with relatively long inter-stimulus intervals. During imprinting training, presentations of a visual stimulus either preceded, followed or were separated by 30 seconds from presentations of a simultaneous compound comprising another visual stimulus and an auditory stimulus. In subsequent preference tests, chicks in each condition showed an equivalent preference to approach the visual stimulus that had been presented in compound with the auditory stimulus, rather than the other visual stimulus. In contrast, in a separate study, Bolhuis and Honey (unpublished) found that if the auditory stimulus was replaced with a heat reinforcer (and the chicks were trained in a cooled cabinet), imprinting proceeded better if presentation of the heat reinforcer followed presentation of the visual stimulus, compared to when it was presented simultaneously.

These results suggest that filial imprinting and Pavlovian conditioning proceed in different ways. In the case of imprinting with an audiovisual compound stimulus, an alternative interpretation to that of within-event learning would assume that the auditory stimulus acts as a kind of US to the initially neutral visual stimulus (the CS). Imprinting would then be a kind of Pavlovian conditioning. However, in conventional Pavlovian conditioning sequential presentation of the CS and the US is far superior to simultaneous presentation of these two stimuli. As we have seen, in imprinting with an audiovisual compound it is the other way around: simultaneous presentation of the auditory and the visual stimulus is superior to sequential presentation. The results with audiovisual compound imprinting are analogous to findings in similar ("sensory preconditioning") paradigms in rats (Rescorla & Durlach, 1981; see Dickinson & Burke, 1996, for human parallels). They support the suggestion that the potentiation of imprinting is a product of within-event learning, that is undermined when one element of the compound is presented in isolation. Furthermore, the findings demonstrate that in filial imprinting, unlike Pavlovian conditioning, learning proceeds most readily when training involves the simultaneous presentation of stimuli.

As indicated above, bird song learning is enhanced by social interactions with the tutor. Recent studies have attempted to investigate the mechanisms underlying this enhancement. It is possible that song learning is enhanced as a result of simultaneous exposure to a visual stimulus, analogous to within-event learning in filial imprinting. Houx & ten Cate (1999) found that zebra finch song learning was not enhanced when exposure to taped songs was either followed or preceded by presentation of a visual stimulus (a stuffed adult zebra finch male). Subsequently, Bolhuis et al. (1999) did not find enhanced song learning when the taped songs and the visual stimulus were presented simultaneously, compared to sequential presentation. In a separate experiment these authors showed that the visual stimulus was noticed by the experimental males, and a potent reinforcer in an instrumental task. In both these studies there was significant song learning from tapes. Elucidation of the mechanisms underlying the enhancement of song learning by social interactions awaits further experiments (Snowdon & Hausberger, 1997).

DISCUSSION: DYNAMIC INTERACTIONS DURING DEVELOPMENT

In this present chapter I have discussed biological approaches to the development of brain and behaviour. Most of my examples were taken from animal paradigms of behavioural and neural development. Work in this field has led to novel insights into developmental processes. Dichotomous views of "nature" versus "nurture" have made way for a much more subtle interpretation, in which development is seen as a continual interaction between the individual and its internal and external environment. Good examples of an interactionist approach can be found in the study of perceptual predispositions, which have been found, for example, in the development of bird song and of visual and auditory filial preferences.

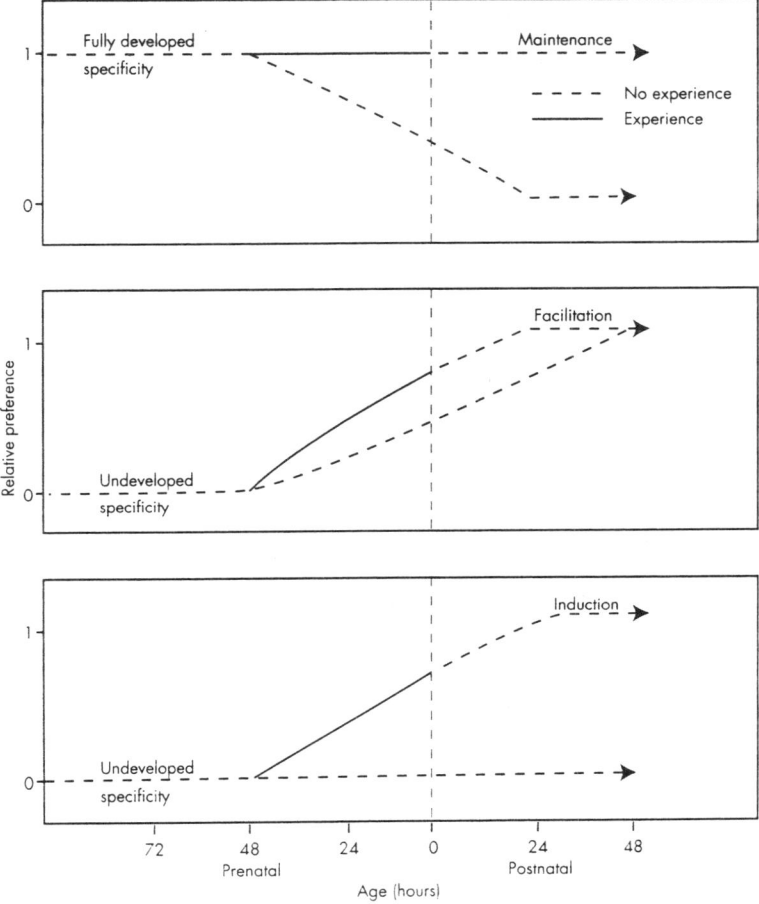

Figure 4 Three ways in which experience can influence the development of perceptual preferences: maintenance, facilitation and induction. In this diagram the behaviour of interest is shown on the *y*-axis as a preference for the maternal call of the bird's own species. On the *x*-axis 0 is the time of hatching; in other behaviours there may be another time scale. Maintenance and facilitation were demonstrated in some of Gottlieb's own work on the development of auditory preferences in ducklings (see Gottlieb, 1980), and in other paradigms (see text). Induction was demonstrated e.g. in the development of a filial predisposition in domestic chicks (e.g. Bolhuis, 1996). (After Gottlieb (1980), with permission.)

Gottlieb (1976, 1980) introduced the terms *maintenance, facilitation,* and *induction* to distinguish some of the ways in which experience can affect the development of perceptual mechanisms (Figure 4). All these three possibilities have been shown to occur in the development of filial preferences (Bolhuis, 1991, 1996). For instance, Gottlieb's (1980, 1982) demonstration, discussed above, that ducklings' preference for maternal calls of their own species is dependent upon particular embryonic experience is an example of maintenance. The effects of non-specific experience on the development of the filial predisposition in chicks (Bolhuis, 1986; Horn, 1985; Johnson *et al.*, 1985) may be categorized as a case of induction; the predisposition is not expressed initially, and emerges only after non-specific experience. Similar results have been reported for instance in the development of food recognition in chicks (Hogan, 1988). Examples of facilitation and maintenance have also been found in the development of perceptual preferences in birds (e.g. Bolhuis, 1996; Gottlieb, 1980). Maintenance of sexual preferences was found in investigations of sexual imprinting in zebra finches (Bischof, 1994). Recent reviews of developmental paradigms suggest that the (inducing, maintaining or facilitating) effect of non-specific, seemingly unrelated, experience on specific perceptual mechanisms is a fundamental principle of behavioural development (Bolhuis, 1996, 1999b; Hogan, 1988, 1994). Gottlieb's terms are of course descriptive, and do not directly address the underlying mechanisms. However, as Shettleworth (1994) asserts, Gottlieb's classification may be a useful way of categorizing effects of experience on behavioural development. Neurobiological investigations may shed light on the mechanisms underlying the emergence of predispositions. It is known that the development of the filial predisposition in chicks is not dependent on the integrity of the IMHV, the forebrain region that is crucial for filial imprinting (Johnson & Horn, 1986; see above). Thus it appears that the neural substrate for the predisposition is outside the IMHV, but it is not yet known whether it is localized to a particular region of the chick brain. Identification of the neural substate of predispositions may be the key to understanding the effects of experience on their development.

This review shows that sensitive periods play an important role in the development of brain and behaviour. However, it is often difficult to establish the existence of a sensitive period in the development of a particular behaviour, and it is even more difficult to analyse the underlying mechanisms. What we can conclude is that sensitive periods come in different shapes and sizes, and that the underlying mechanisms can also vary widely. Particularly the end of apparent sensitivity may be the result of an endogenous mechanism, or, alternatively, it may be caused by the specific experience of the individual. These multiple causes of sensitive periods imply a considerable flexibility. Behaviour whose development is affected by a sensitive period may be altered in the adult stage (e.g. Bischof, 1994). Even when sensitive periods are likely to have endogenous causes, their timing can be affected by environmental or physiological factors (Bolhuis & Horn, 1997; Davies *et al.*, 1992; Marler, 1987).

Developmental processes involved in imprinting and bird song acquisition have often been used as a model for human development. We have seen that there are parallels between learning and predispositions in the perceptual development of birds and human infants. These parallels exist at both the behavioural and the neural level. There has been a reciprocal application of developmental concepts between the animal and human domain. In addition, behavioural and neural analyses of early learning in animals have become prominent models of learning and memory processes in developing as well as adult humans. The success of these animal paradigms will continue to stimulate biological approaches to the study of perceptual and cognitive development.

References

Adret, P. (1993). Operant conditioning, song learning and imprinting to taped song in the zebra finch. *Animal Behaviour*, **46**, 149–159.

Baptista, L.F. & Petrinovich, L. (1984). Social interaction, sensitive phases and the song template hypothesis in the white-crowned sparrow. *Animal Behaviour*, **34**, 1359–1371.

Bateson, P.P.G. (1964). Relation between conspicuousness of stimuli and their effectiveness in the imprinting situation. *Journal of Comparative and Physiological Psychology*, **58**, 407–411.

Bateson, P. (1979). How do sensitive periods arise and what are they for? *Animal Behaviour,* **27**, 470–486.

Bateson, P. (1987). Imprinting as a process of competitive exclusion. In J.P. Rauschecker & P. Marler, (Eds.), *Imprinting and cortical plasticity. Comparative aspects of sensitive periods* (pp. 151–168). New York: John Wiley.

Bateson, P. (1990). Is imprinting such a special case? *Philosophical Transactions of the Royal Society, London, B*, **329**, 125–131.

Bateson, P. & Hinde, R.A. (1987). Developmental changes in sensitivity to experience. In M.H. Bornstein (Ed.), *Sensitive periods in development* (pp. 19–34). Hillsdale, NJ: Lawrence Erlbaum Associates.

Bateson, P. & Horn, G. (1994). Imprinting and recognition memory: a neural net model. *Animal Behaviour*, **48**, 695–715.

Bischof, H.-J. (1994). Sexual imprinting as a two-stage process. In J.A. Hogan & J.J. Bolhuis (Eds.), *Causal mechanisms of behavioural development* (pp. 82–97). Cambridge, UK: Cambridge University Press.

Bolhuis J.J. (1991). Mechanisms of avian imprinting: a review. *Biological Reviews*, **66**, 303–345.

Bolhuis J.J. (1994). Neurobiological analyses of behavioural mechanisms in development. In J.A. Hogan & J.J. Bolhuis (Eds.), *Causal mechanisms of behavioural development* (pp. 16–46). Cambridge, UK: Cambridge University Press.

Bolhuis J.J. (1996). Development of perceptual mechanisms in birds: predispositions and imprinting. In C.F. Moss & S.J. Shettleworth (Esd.), *Neuroethological studies of cognitive and perceptual processes* (pp. 158–164). Boulder, CO: Westview Press.

Bolhuis, J.J. (1999a). Early learning and the development of filial preferences in the chick. *Behavioural Brain Research*, **98**, 245–252.

Bolhuis, J.J. (1999b). The development of animal behavior. From Lorenz to neural nets. *Naturwissenschaften*, **86**, 101–111.

Bolhuis J.J. & Hogan J.A. (1999). *The development of animal behavior. A reader*. Oxford: Blackwell.

Bolhuis J.J., & Honey R.C. (1994). Within-event learning during filial imprinting. *Journal of Experimental Psychology, Animal Behavior Processes*, **20**, 240–248.

Bolhuis J.J. & Honey R.C. (1998). Imprinting, learning and development: from behaviour to brain and back. *Trends in Neurosciences*, **21**, 306–311.

Bolhuis, J.J. & Horn, G. (1997). Delayed induction of a filial predisposition in the chick after anaesthesia. *Physiology and Behavior,* **62**, 1235–1239.

Bolhuis, J.J. & Van Kampen, H.S. (1992). An evaluation of auditory learning in filial imprinting. *Behaviour*, **122**, 195–230.

Bolhuis, J.J., Johnson, M.H. & Horn, G. (1989). Interacting mechanisms during the formation of filial preferences: the development of a predisposition does not prevent learning. *Journal of Experimental Psychology: Animal Behavior Proceedings*, **15**, 376–382.

Boluis, J.J., de Vos, G.J. & Kruijt, J.P. (1990). Filial imprinting and associative learning. *Quarterly Journal of Experimental Psychology*, **42B**, 313–329.

Bolhuis, J.J., Van Mil, D.P. & Houx, B.B. (1990). Song learning with audio-visual compound stimuli in zebra finches. *Animal Behaviour*, **58**, 1285–1292.

Bolhuis, J.J., Zijlstra, G.G.O., Den Boer-Visser, A.M. & Van der Zee, E.A. (2000). Localized neuronal activation in the zebra finch brain is related to the strength of song learning. *Proceedings of the National Academy of Sciences, USA,* **97**, 2282–2285.

Bornstein, M.H. (Ed.) (1987). *Sensitive periods in development*. Hillsdale, NJ: Lawrence Erlbaum Associates.

Churchland, P.S. & Sejnowski, T.J. (1988). Perspectives on cognitive neuroscience. *Science*, **242**, 741–745.

Clayton, D.F. (2000). The neural basis of avian song learning and perception. In J.J. Bolhuis (Ed.), *Brain mechanisms of perception, learning and memory* (pp. 113–125). Oxford: Oxford University Press.

Cynx, J., Williams, H. & Nottebohm, F. (1992) Hemispheric differences in avian song discrimination. *Proceedings of the National Academy of Sciences, USA,* **89**, 1372–1375.

Davies, D.C., Johnson, M.H. & Horn, G. (1992). The effect of the neurotoxin DSP4 on the development of a predisposition in the domestic chick. *Developmental Psychobiology*, **25**, 251–259.

DeCasper, A. & Fifer, W.P. (1980). On human bonding: newborns prefer their mothers' voices. *Science*, **208**, 1174–1176.

DeVoogd, T.J. (1994). The neural basis for the acquisition and production of bird song. In J.A. Hogan & J.J. Bolhuis (Eds.), *Causal mechanisms of behavioural development* (pp. 49–81). Cambridge, UK: Cambridge University Press.

De Vos, G.J. & Bolhuis, J.J. (1990). An investigation into blocking of filial imprinting in chicks during exposure to a compound stimulus. *Quarterly Journal of Experimental Psychology*, **42B**, 289–312.

Diamond, A. & Goldman-Rakic, P.S. (1989). Comparison of human infants and rhesus monkeys on Piaget's AB task: evidence for dependence on dorsolateral prefrontal cortex. *Experimental Brain Research,* **74**, 24–40.

Dickinson, A. & Burke, J. (1996). Within-compound associations mediate the retrospective revaluation of causality judgements. *Quarterly Journal of Experimental Psychology,* **49B**, 60–80.

Doupe, A.J. & Kuhl, P.K. (1999). Birdsong and human speech: common themes and mechanisms. *Annual Review of Neuroscience,* **22**, 567–631.

Dragunow, M. & Faull, R. (1989). The use of *c-fos* as a metabolic marker in neuronal pathway tracing. *Journal of Neuroscience Methods,* **29**, 261–265.

Durlach, P.J. & Rescorla, R.A. (1980). Potentiation rather than overshadowing in flavor-aversion learning: an analysis in terms of within-compound associations. *Journal of Experimental Psychology: Animal Behavior Processes,* **6**, 175–187.

Elman, J.L., Bates, E.A., Johnson, M.H., Karmiloff-Smith, A., Parisi, D. & Plunkett, K. (1996). *Rethinking innateness*. Cambridge, MA: MIT Press.

Goller, F. & Suthers, R.A. (1995). Implication for lateralization of bird song from unilateral gating of bilateral motor patterns. *Nature*, **373**, 63–66.

Gottlieb, G. (1971). *Development of species identification in birds*. Chicago, IL: University of Chicago Press.

Gottlieb, G. (1976). The roles of experience in the development of behavior and the nervous system. In G. Gottlieb (Ed.), *Neural and behavioral specificity: studies in the development of behavior and the nervous system* (pp. 237–280). New York: Academic Press.

Gottlieb, G. (1980). Development of species identification in ducklings: VI. Specific embryonic experience required to maintain species-typical perception in Peking ducklings. *Journal of Comparative and Physiological Psychology*, **94**, 579–587.

Gottlieb, G. (1982). Development of species identification in ducklings: IX. The necessity of experiencing normal variations in embryonic auditory stimulation. *Developmental Psychobiology,* **15**, 507–517.

Gottlieb, G. (1985). Development of species identification in ducklings: XI. Embryonic critical period for species-typical perception in the hatchling. *Animal Behaviour*, **33**, 225–233.

Gottlieb, G. (1988). Development of species identification in ducklings: XV. Individual auditory recognition. *Developmental Psychobiology,* **21**, 509–522.

Hebb, D.O. (1949). *The organization of behavior*. New York: John Wiley & Sons.

Hirsch, H.V.B., Tieman, D.G., Tieman, S.B. & Tumosa, N. (1987). Unequal alternating exposure: effects during and after the classical critical period. In J.P. Rauschecker & P. Marler (Eds.), *Imprinting and cortical plasticity. Comparative aspects of sensitive periods* (pp. 287–320). New York: John Wiley & Sons.

Hoffman, H.S. & Ratner, A.M. (1973). A reinforcement model of imprinting: implications for socialization in monkeys and men. *Psychological Review*, **80**, 527–544.

Hogan, J.A. (1988). Cause and function in the development of behavior systems. In E.M. Blass (Ed.), *Handbook of behavioral neurobiology* (vol. 9. pp. 63–106). New York: Plenum Press.

Hogan, J.A. (1994). Development of behavior systems. In: J.A. Hogan & J.J. Bolhuis (Eds.), *Causal mechanisms of behavioural development*. (pp. 242–264). Cambridge, UK: Cambridge University Press.

Honey, R.C. & Bolhuis, J.J. (1997). Imprinting, conditioning, and within-event learning. *Quarterly Journal of Experimental Psychology,* **50B**, 97–110.

Honey, R.C., Horn, G. & Bateson, P. (1993). Perceptual learning during filial imprinting: evidence from transfer of training studies. *Quarterly Journal of Experimental Psychology*, **46B**, 253-269.

Horn, G. (1985). *Memory, imprinting, and the brain*. Oxford: Clarendon Press.

Horn, G. (1998). Visual imprinting and the neural mechanisms of recognition memory. *Trends in Neurosciences*, **21**, 300–305.

Horn, G. & McCabe, B.J. (1984). Predispositions and preferences. Effects on imprinting of lesions to the chick brain. *Animal Behaviour*, **32**, 288–292.

Houx, B.B. & ten Cate, C. (1999). The influence of classical contingencies on song learning in zebra finches. *Journal of Comparative Psychology*, **113**, 235–242.

Hultsch, H., Schleuss, F. & Todt, D. (1997). Auditory-visual stimulus pairing enhances perceptual learning in a songbird. *Animal Behaviour*, **58**, 143–149.

Jarvis, E.D. & Nottebohm, F. (1997). Motor-driven gene expression. *Proceedings of the National Academy of Sciences, USA*, **94**, 4097–4102.

Jin, H. & Clayton, D.F. (1997). Localized changes in immediate-early gene regulation during sensory and motor learning in zebra finches. *Neuron*, **19**, 1049–1059.

Johnson, M.H. (1993). *Brain development and cognition: a reader*. Oxford: Blackwell.

Johnson, M.H. (1994). Cortical mechanisms of cognitive development. In J.A. Hogan & J.J. Bolhuis (Eds.), *Causal mechanisms of behavioural development* (pp. 267–288). Cambridge, UK: Cambridge University Press.

Johnson, M.H. & Horn, G. (1986). Dissociation between recognition memory and associative learning by a restricted lesion to the chick forebrain. *Neuropsychologia*, **24**, 329–340.

Johnson, M.H. & Horn, G. (1988). Development of filial preferences in dark-reared chicks. *Animal Behaviour*, **36**, 675–683.

Johnson, M.H. & Morton, J. (1991). *Biology and cognitive development: the case of face recognition*. Oxford: Blackwell Publishers.

Johnson, M.H., Bolhuis, J.J. & Horn, G. (1985). Interaction between acquired preferences and developing predispositions during imprinting. *Animal Behaviour*, **33**, 1000–1006.

Johnson, M.H., Davies, D.C. & Horn, G. (1989). A sensitive period for the development of a predisposition in dark-reared chicks. *Animal Behaviour*, **37**, 1044–1046.

Johnson, M.H., Dziurawiec, S., Ellis, H.D. & Morton, J. (1991). Newborns' preferential tracking of face-like stimuli and its subsequent decline. *Cognition*, **40**, 1–19.

Kasamatsu, T., Pettigrew, J.D. & Ary, M. (1979). Restoration of visual cortical plasticity by local microperfusion of norepinephrine. *Journal of Comparative Neurology*, **185**, 163–182.

Keverne, E.B. (1997). An evaluation of what the mouse knockout experiments are telling us about mammalian behaviour. *BioEssays*, **19**, 1091–1098.

Koopmans-van Beinum, F.J. & van der Stelt, J.M. (1986). Early stages in the development of speech movements. In B. Lindblom & R. Zetterstrom (Eds.), *Precursors of early speech* (pp. 37–50). New York: Stockton Press.

Korsia, S. & Bottjer, S.W. (1991). Chronic testosterone treatment impairs vocal learning in male zebra finches during a restricted period of development. *Journal of Neuroscience*, **11**, 2362–23271.

Lehrman, D.S. (1953). A critique of Konrad Lorenz's theory of instinctive behavior. *Quarterly Review of Biology*, **28**, 337–363.

Lehrman, D.S. (1970). Semantic and conceptual issues in the nature-nurture problem. In L.R. Aronson, E. Tobach, D.S. Lehrman & J.S. Rosenblatt (Eds.), *Development and evolution of behavior* (pp. 17–52). San Francisco, CA: Freeman.

Locke, J.L. (1993). *The child's path to spoken language*. Cambridge, MA: Harvard University Press.

Locke, J.L. (1994). The biological building blocks of spoken language. In J.A. Hogan & J.J. Bolhuis (Eds.), *Causal mechanisms of behavioural development* (pp. 300–324). Cambridge, UK: Cambridge University Press.

Lorenz, K. (1937). The companion in the bird's world. *Auk*, **54**, 245–273.

Lorenz, K. (1965). *Evolution and modification of behavior*. Chicago, IL: University of Chicago Press.

Marler, P. (1976). Sensory templates in species-specific behavior. In J.Fentress (Ed.), *Simpler networks and behavior* (pp. 314–329). Sunderland, MA: Sinauer.

Marler, P. (1987). Sensitive periods and the roles of specific and general sensory stimulation in birdsong learning. In J.P. Rauschecker & P. Marler (Eds.), *Imprinting and cortical plasticity. Comparative aspects of sensitive periods* (pp. 99–135). New York: John Wiley & Sons.

Marler, P. (1991). Song-learning behavior: the interface with neuroethology. *Trends in Neurosciences*, **14**, 199–206.

Marler, P. & Doupe, A.J. (2000). Singing in the brain. *Proceedings of the National Academy of Sciences, USA*, **97**, 2965–2967.

Marler, P. & Peters, S. (1987). A sensitive period for song acquisition in the song sparrow, *Melospiza melodia*: a case of age-limited learning. *Ethology*, **76**, 89–100.

Marler, P. & Peters, S. (1988). Sensitive periods for song acquisition from tape recordings and live tutors in the swamp sparrow, *Melospiza georgiana*. *Developmental Psychobiology*, **15**, 369–378.

McCabe, B.J. & Horn, G. (1994). Learning-related changes in Fos-like immunoreactivity in the chick forebrain after imprinting. *Proceedings of the National Academy of Sciences of the USA*, **91**, 11417–11421.

McCasland, J.S. (1987). Neuronal control of bird song production. *Journal of Neuroscience*, **7**, 23–39.

McLaren, I.P.L., Kaye, H., & Mackintosh, N.J. (1989). An associative theory of the representation of stimuli: applications to perceptual learning and latent inhibition. In R.G.M. Morris (Ed.), *Parallel distributed processing: implications for psychology and neurobiology* (pp. 102–130). Oxford: Oxford University Press.

Mello, C.V. & Clayton, D.F. (1994). Song-induced ZENK gene expresion in auditory pathways of songbird brain and its relation to the song control system. *Journal of Neuroscience*, **14**, 6652–6666.

Mello, C.V. & Ribeiro, S. (1998). Zenk protein regulation by song in the brain of songbirds. *Journal of Comparative Neurology*, **393**, 426–438.

Mello, C.V., Vicario, D.S. & Clayton, D.F. (1992). Song presentation induces gene-expression in the songbird forebrain. *Proceedings of the National Academy of Sciences of the USA*, **89**, 6818–6822.

Mello, C.V., Nottebohm, F. & Clayton, D.F. (1995). Repeated exposure to one song leads to a rapid and persistent decline in an immediate early gene's response to that song in zebra finch telencephalon. *Journal of Neuroscience*, **15**, 6919–6925.

Michel, G.F. & Moore, C.L. (1995). *Developmental psychobiology. An interdisciplinary science*. Cambridge, MA: MIT Press.

Moon, C., Cooper, R.P. & Fifer, W.P. (1993). Two-day olds prefer their native language. *Infant Behaviour and Development*, **16**, 495–500.

Morrison, R.G. & Nottebohm, F. (1992). Role of a telencephalic nucleus in the delayed song learning of socially isolated zebra finches. *Journal of Neurobiology*, **24**, 1045–1064.

Nelson, D.A. (1998). External validity and experimental design: the sensitive phase for song learning. *Animal Behaviour*, **56**, 487–491.

Nelson, D.A. & Marler, P. (1994). Selection-based learning in bird song development. *Proceedings of the National Academy of Sciences, USA*, **91**, 10498–10501.

Nottebohm, F. (1980). Brain pathways for vocal learning in birds: a review of the first 10 years. *Progress in Psychobiology, Physiology and Psychology*, **9**, 85–124.

Nottebohm, F. (1991). Reassessing the mechanisms and origins of vocal learning in birds. *Trends in Neurosciences*, **14**, 206–211.

Nottebohm, F. & Nottebohm, M.E. (1976). Left hypoglossal dominance in the control of canary and white-crowned sparrow song. *Journal of Comparative Physiology, A*, **108**, 171–192.

Nottebohm, F., Alvarez-Buylla, A., Cynx, J., Kirn, J. Ling, C-Y., Nottebohm, M, Suter, R. Tolles, A. & Williams, H. (1990). Song learning in birds: the relation between perception and production. *Philosopical Transactions of the Royal Society, London, B*, **329**, 115–124.

Oyama, S. (1985). *The ontogeny of information*. Cambridge: Cambridge University Press.

Pearce, J.M. (1994). Similarity and discrimination: a selective review and a connectionist model. *Psychological Review*, **101**, 587–607.

Petrinovich, L. & Baptista, L.F. (1987). Song development in the White-crowend Sparrow: modification of learned song. *Animal Behaviour*, **35**, 961–974.

Piaget, J. (1954). *The construction of reality in the child*. New York: Basic Books.

Rauschecker, J.P. & Marler, P. (Eds.) (1987). *Imprinting and cortical plasticity: comparative aspects of sensitive periods*. New York: John Wiley & Sons.

Rescorla, R.A. & Durlach P.J. (1981). Within-event learning in Pavlovian conditioning. In N.E. Spear & R.R. Miller (Eds.), *Information processing in animals: memory mechanisms* (pp. 81–111). Hillsdale, NJ: Lawrence Erlbaum.

Segalowitz, S.J. & Chapman, J.S. (1980). Cerebral asymmetry for speech in neonates: a behavioral measure. *Brain and Language*, **9**, 281–288.

Shettleworth, S.J. (1994). The varieties of learning in development: toward a common framework. In J.A. Hogan & J.J. Bolhuis (Eds.), *Causal mechanisms of behavioural development* (pp. 358–376). Cambridge: Cambridge University Press.

Shettleworth, S.J. (1998). *Cognition, evolution and behavior*. Oxford: Oxford University Press.

Sillito, A.M. (1983). Plasticity in the visual cortex. *Nature*, **303**, 477–478.

Slater, P.J.B. (1983). The development of individual behaviour. In T.R. Halliday & P.J.B. Slater (Eds.), *Animal behaviour*, vol. 3: *Genes, development and learning* (pp. 82–113). Oxford: Blackwell.

Slater, P.J.B., Eales, L.A. & Clayton, N.S. (1988). Song learning in zebra finches (*Taeniopygia guttata*): progress and prospects. *Advances in the Study of Behavior*, **18**, 1–34.

Slater, P.J.B., Jones, A. & ten Cate, C. (1993). Can lack of experience delay the end of the sensitive phase for song learning? *Netherlands Journal of Zoology*, **43**, 80–90.

Sluckin, W. & Salzen, E.A. (1961). Imprinting and perceptual learning. *Quarterly Journal of Experimental Psychology*, **8**, 65–77.

Snowdon, C.T. & Hausberger, M. (Eds.) (1997). *Social influences on vocal development*. Cambridge, UK: Cambridge University Press.

Spence, M.J. & DeCasper, A.J. (1987). Prenatal experience with low-frequency maternal-voice sounds influence neonatal perception of maternal voice samples. *Infant Behavior and Development*, **10**, 133–142.

Stent, G.S. (1973). A physiological mechanism for Hebb's postulate of learning. *Proceedings of the National Academy of Sciences, USA*, **70**, 997–1001.

Studdert-Kennedy, M. (1981). The beginnings of speech. In K. Immelmann, G.W. Barlow, L. Petrinovich & M. Main (Eds.), *Behavioral Development* (pp. 533–561). Cambridge, UK: Cambridge University Press.

Studdert-Kennedy, M. & Shankweiler, D. (1970). Hemispheric specialization for speech perception. *Journal of the Acoustic Society of America*, **48**, 579–594.

ten Cate, C. & Houx, B. (1999). Social interactions and song learning: are behavioural contingencies important? *Proceedings of the 22nd International Ornithological Congress*, 156–164.

ten Cate, C. (1994). Perceptual mechanisms in imprinting and song learning In J.A. Hogan & J.J. Bolhuis (Eds.), *Causal mechanisms of behavioural development* (pp. 116–146). Cambridge, UK: Cambridge University Press.

ten Cate, C. & Houx, B. (1999). Social interactions and song learning: are behavioural contingencies important? *Proceedings of the 22nd International Ornithological Congress*, 156–164.

Timney, B. (1987). Dark rearing and the sensitive period for monocular deprivation. In J.P. Rauschecker & J.P. Marler. (Eds.), *Imprinting and cortical plasticity: comparative aspects of sensitive periods* (pp. 321–345). New York: John Wiley & Sons.

Van Kampen, H.S. (1996). A framework for the study of filial imprinting and the development of attachment. *Psychonomic Bulletin and Review*, **3**, 3–20.

Van Kampen, H.S. & Bolhuis, J.J. (1991). Auditory learning and filial imprinting in the chick. *Behaviour*, **117**, 303–319.

Van Kampen, H.S. & Bolhuis, J.J. (1993). Interaction between auditory and visual learning during imprinting. *Animal Behaviour*, **45**, 623–625.

Van Kampen, H.S. & De Vos, G.J. (1994). Potentiation in learning about the visual features of an imprinting stimulus. *Animal Behaviour*, **47**, 1468–1470.

Van Kampen, H.S. & De Vos, G.J. (1995). A study of blocking and overshadowing in filial imprinting, *Quarterly Journal of Experimental Psychology*, **48B**, 346–356.

Whaling, C.S., Soha, J.A., Nelson, D.A., Lasley, B. & Marler, P. (1998). Photoperiod and tutor access affect the process of vocal learning. *Animal Behaviour*, **56**, 1075–1082.

Wiesel, T.N. (1982). Postnatal development of the visual cortex and the influence of environment. *Nature*, **299**, 583–591.

Zann, R.A. (1996). *The zebra finch. A synthesis of field and laboratory studies*. Oxford: Oxford University Press.

FURTHER READING

Bolhuis J.J. & Hogan J.A. (1999). *The development of animal behavior. A reader*. Oxford: Blackwell.

Hogan, J.A. & Bolhuis, J.J. (1994). *Causal mechanisms of behavioural development*. Cambridge: Cambridge University Press.

Johnson, M.H. (1993). *Brain development and cognition: a reader.* Oxford: Blackwell.

Michel, G.F. & Moore, C.L. (1995). *Developmental psychobiology. An interdisciplinary science*. Cambridge, MA: MIT Press.

IV.6
Development of State Regulation

PIERO SALZARULO, FIORENZA GIGANTI, IGINO FAGIOLI AND
GIANLUCA FICCA

ABSTRACT

Sleep states are constellations of variables that are stable over time,
emerging and developing throughout the early epochs of both prenatal
and postnatal life. If an evident link exists between the development of
states and the maturation of the central nervous system, many other
factors, either physical or related to processes of interaction, may
interfere with the mechanisms implied in state regulation.

Here we focus on several of these factors, including experimental
manipulation, parental care and nutrition, in an attempt to assess their
role in state regulation and to show how they can contribute to producing changes in the development of states.

INTRODUCTION

The concept of state was introduced in the field of development mainly by the
work of Wolff and of Prechtl; its importance is now widely acknowledged. Sleep
states are constellations of physiological and behavioural variables that are stable
over time and repeat themselves, which emerge and then develop throughout the
early epochs of both prenatal and postnatal life (Prechtl & O'Brien, 1982; Wolff,
1987). The link with the maturation of the central nervous system (CNS) is
evident, and has been underscored by several investigators (Curzi-Dascalova &
Mirmiran, 1996; Prechtl & O'Brien, 1982).

The different criteria used to identify "states" may justify some differences
obtained by investigators on the amount of each state and above all on the time
of life at which they emerge. However, we would like to stress that, on the whole,
the *developmental trend* of each state is sufficiently generalizable, whatever the
criteria or the species.

The regulation of states implies neurophysiological and neurochemical mechanisms, which express what the brain can perform at each step (age). Factors
from outside, both physical and related to human interaction, can modulate for

A.F. Kalverboer and A. Gramsbergen (eds.), Handbook of Brain and Behaviour in Human Development, 725–742
© 2001 Kluwer Academic Publishers. Printed in Great Britain.

a short or long time the state characteristics, sometimes leading to a change in the developmental trend of states; this is the case for some pathological conditions (e.g. malnutrition, see Salzarulo et al., 1982).

ANIMAL STUDIES OF THE MECHANISMS INVOLVED IN STATES

Studies performed in animals (rat, cat, guinea-pig) in the early 1970s (Jouvet-Mounier et al., 1970; Adrien, 1976) showed the time of emergence of behavioural states and also suggested mechanisms which are possibly involved.

Changes with age, while sharing the same sequence, are different according to the species. At birth, animals born mature (e.g. the precocial guinea-pig) show a lower amount of paradoxical sleep than those who are born immature (the altricial rat and cat) as emphasized by Verley & Garma (1975).

Studies (Jouvet-Mounier et al., 1970; Verley & Garma, 1975) performed in altricial animals, both intact and lesioned, showed that there are profoundly different events according to age.

In the development of the rat, paradoxical sleep is typified by low-voltage fast waves in the ECoG (electrocorticogram), loss of neck muscle tone, muscular twitches accompanied by rapid eye movements, and irregular respiratory and heart rates (Jouvet-Mounier et al., 1970). This is preceded by a nearly continuous motor activity characterized by more or less violent jerks of limbs, head and body in general, with a complete disappearance of nuchal muscular tone (Adrien, 1976), which has been called "seismic sleep". While the term "sleep" has been used, physiological and chemical regulations seem radically different from those involved later in paradoxical sleep (Adrien, 1976) and led some authors to suggest that mechanisms underlying "seismic sleep" and paradoxical sleep are different, the first being generated by the spinal cord (Adrien, 1976; Jouvet-Mounier et al., 1970). Others suggested that the embryonic motility of seismic sleep "is one of the building blocks of the adult paradoxical sleep" (Verley & Garma, 1975, p. 119).

Gramsbergen et al. (1970) illustrated in the rat the progressive changes with age in the global amount and episode duration of each of five behavioural states, including waking. It is worth pointing out that quiet sleep is the state which shows an increase with age of mean duration, a finding that is completely in agreement with human infant development (Fagioli & Salzarulo, 1982a).

HUMAN STUDIES

In humans, as far as basic criteria (use of more than one parameter, window length, etc.) for defining states are concerned, many contributions have reported distinct behavioural states from very early stages of development. Studies performed in utero by real-time ultrasonic imaging, and by cardiotacograph, showed the emergence of behavioural states at about 36–38 weeks gestational age (Nijhuis et al., 1982) based on three parameters: eye movements, body motility and cardiac rhythm. On the other hand, studies performed in infants born before term give non-uniform results and raise interesting problems – mainly, but not only, concerning states of wakefulness.

The existence of behavioural states in preterm infants was ascertained by Curzi-Dascalova *et al.* (1988) using two parameters as criteria for states: EEG and eye movements. Using these criteria, states were identified as early as 31 weeks gestational age, which is an earlier age than the one for behavioural states *in utero*. This discrepancy is possibly related more to methodological differences between the two studies than to the uterine or extrauterine conditions. In the latter study (Curzi-Dascalova *et al.*, 1988), as well in other studies by the same group (Curzi-Dascalova & Mirmiran, 1996) and by others, waking states do not appear to be clearly distinguished from sleep states.

Behavioural criteria, including different motility and EEG patterns, have been considered useful in distinguishing sleep states in preterm infants between 30 and 39 weeks postmenstrual age (Shani *et al.*, 1995).

Parmelee & Stern (1972) also found states relatively early in preterm infants, taking into account three behavioural parameters (respiratory pattern, eye movements and above all body movements). It is interesting to compare the developmental trend found in that study with that reported for rats (Adrien, 1976). The similarity between the two is striking (Figure 1). Each state, paradoxical sleep (PS) and slow-wave sleep (SWS) in rats, active sleep (AS) and quiet sleep (QS) in humans (for a summary of current terminology used for humans see Salzarulo & Fagioli, 1999, p. 63), follows the same trend in both species: increasing and then decreasing PS and AS, and increasing QS. Even more impressive is the time-course of "seismic" sleep in rats and transitional sleep in humans, despite the fact that criteria for seismic sleep (continuous motility) and transitional sleep (mixed elements of AS and QS) are different. This leads to the hypothesis that both represent a "primitive" form of activity of the immature brain, involving the expression of a low level of inhibition (Sterman, 1979; Ktonas *et al.*, 1990).

Criteria used for defining wakefulness are apparently lacking in preterm infants. Does this mean that wakefulness does not exist, or that the preterm characteristics, being particular, are not captured by the usual criteria? Wolff (1984) attempted to define the characteristics of wakefulness in full-term newborns and infants up to 3 months of age. He not only showed that the global amount of wakefulness increases with age, he also showed change of a qualitative nature. Wolff (1966, 1973a) proposed a taxonomy of waking states, which includes two different types of waking: alert inactivity and waking activity. These states can be recognized as early as the first weeks after birth. Moreover, there is a third category of waking, alert activity, that does not emerge clearly until the second month. This was defined as "the spontaneous condition in which the eyes are open, the infant makes continuous or intermittent movements of the limbs, and at the same time can pursue a moving visual target with the head and eyes" (Wolff, 1984, p. 151). The mean duration of alert activity increases rapidly and discontinuously at around 6–8 weeks, while at about the same time the mean duration of alert inactivity strongly decreases (Wolff, 1984).

It is worthwhile noting that many investigators have shown behavioural changes towards the end of the second month (Dittrichova & Lapakova, 1964; Papoušek, 1969; Emde, 1980). Prechtl (1986) stressed the fact that the CNS undergoes a

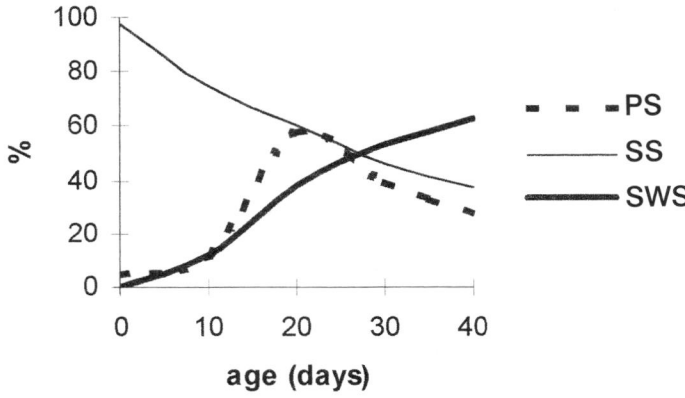

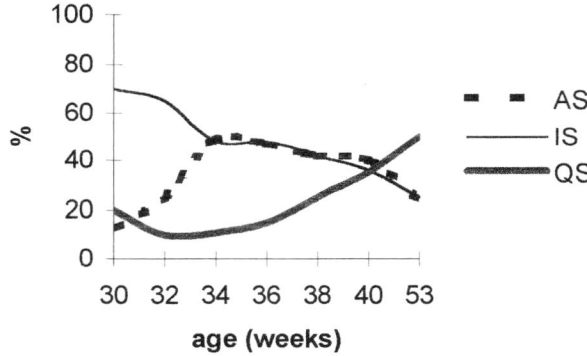

Figure 1 *Top*: state development in rats (modified from Adrien, 1976). Abscissa: days after birth (= 0); ordinate: percentage of states. PS = mechanisms of paradoxical sleep; SS = seismic sleep; SWS = slow-wave sleep. *Below*: state development in human from preterm to full-term infants (redrawn from data by Parmelee & Stern, 1972). Abscissa: conceptional age (weeks); ordinate: percentage of states. AS = active sleep; QS = quiet sleep; IS = indeterminate (or transitional) sleep.

major reorganization at that age. These developmental changes are shown, by, among other things, a reduction of non-specific motor activity during sleep, the decrease of daytime fussing, the ability to learn in a conditioning situation, the onset of a "social smile" and the beginning of non-cry vocalizations.

SLEEP STATES

After term, sleep states undergo profound developmental changes, particularly in the first year of life. However, important changes are also observed later, in particular concerning rhythmicity and EEG characteristics. The changes in the amount of sleep states have been stressed by many researchers. Both polygraphic and video studies (Roffwarg *et al.*, 1964; Navelet *et al.*, 1982; Louis *et al.*, 1997; Anders & Keener, 1985) showed a decrease in the amount of active sleep (REM or paradoxical sleep) and an increase of quiet sleep (QS). It has also been shown that those changes, reflecting maturational properties of CNS, concern episode duration (QS) and number (PS) (Fagioli & Salzarulo, 1982a). The increase in QS duration is related to improvements in the ability to sustain a state. We will see that waking episodes demonstrate a similar trend, slighty delayed with respect to that of sleep.

Further research showed that the increase of NREM sleep and the decrease of REM sleep and ambiguous sleep (including some of the characteristics of both REM and NREM and corresponding to other terminologies such as transitional, indeterminated, etc.) are not evenly distributed in the 24-hour periods. In fact, the amount of NREM sleep increases mainly during the night-time, whereas both REM and ambiguous sleep decrease during the daytime (Fagioli & Salzarulo, 1982a). However, several authors have focused on another aspect of state change – namely, organization.

The decrease of ambiguous sleep indirectly reflects changes in state organization: the less ambiguous sleep, the less disorganized states. In conjunction with the above there is an increasing ability to build up regular sequences of state cycles, and to sustain each state. NREM–REM sleep cycles undergo several changes with age. Although the mean duration of all the cycles during the whole 24-hour period does not change significantly over the first 6 months (Fagioli & Salzarulo, 1982b), the mean duration of the nocturnal cycles shows a significant increase across the whole first year of life (Ficca *et al.*, 2000). Although a reciprocal modification with age in the proportion of the two sleep states is observed, the NREM increase overcomes the relative REM decrease, and accounts for the lengthening of the mean duration of the nocturnal cycles (Ficca *et al.*, 2000a). Moreover, a better sleep organization with age, related to the cyclic occurrence of the two states, is suggested by the increase in the proportion of total time spent in cycle on the total sleep time (Ficca *et al.*, 2000a). This indicator has been related in different studies to biological factors such as weight change and protein anabolism in both infants and children (Fagioli *et al.*, 1981; Salzarulo & Fagioli, 1995) and to psychological functions such as long-term memory in young adults (Ficca *et al.*, 2000b) and in the elderly (Mazzoni *et al.*, 1999).

TRANSITION BETWEEN STATES

The presence of several states within a sleep episode makes it interesting to look at the modalities of passage from one to the next.

The shift from state to state diminishes across age in number for 24 hours, following the increased duration of QS (Louis *et al.*, 1997; Fagioli & Salzarulo, 1982a). Older infants show fewer shifts with increased sleep state duration and with similar total sleep duration.

Several authors (Parmelee & Stern, 1972; Ellingson, 1975; Fagioli & Salzarulo, 1982a) have recognized sleep bouts without the characteristic of either PS or QS and termed them transitional, ambiguous or indeterminate sleep. All these terminologies indicate a poorly organized sleep, which diminishes in both amount (Fagioli & Salzarulo, 1982a; Louis et al., 1997) and bout duration (Fagioli & Salzarulo, 1982a) as a function of age. Moreover, it is interesting to look at the behaviour of individual variables. Some years ago, Monod & Curzi-Dascalova (1973) showed sequences of several physiological variables at the transition between QS and PS, and between PS and QS. It is impressive that this order is quite fixed and can be found in most individuals of similar age, a finding partially confirmed by Shirataki & Prechtl (1977). Furthermore, even when the transition is directly from one state to another (without ambiguous or indeterminate sleep in between), some physiological activities show a change not always synchronized with state change. Recently we described EEG activity level at the transition between QS and PS and vice-versa (Salzarulo et al., 1991). In very young infants (1–3 weeks old), the activity level increases in QS well before the beginning of PS. From 4–5 week onwards, until the end of the first year of life, the change in EEG activity level is quite abrupt and corresponds closely to the beginning of PS.

A different developmental trend is observed for the PS–QS transition. While the youngest infants show a progressive decrease of EEG activity level beginning well before QS onset (a trend mirroring that of the QS–PS shift), the older ones manifest a change beginning just at QS onset (and continuing thereafter, except in infants 4–11 weeks old). In addition, EEG level does not change between PS and a peculiar type of QS (without slow-wave sleep) in infants older than 5 months (Bes et al., 1991).

A trend similar to that of the EEG was observed for EMG phasic activity (Liefting et al., 1994) in very young infants: the number of phasic events starts to increase several minutes before PS in the QS–PS transition (Figure 2, top). The older infants (19–20 weeks) show a similar trend (Figure 2, foot), while in the same age range EEG activity level changes abruptly. In the opposite direction (PS–QS), the decrease of EMG phasic activity starts early and becomes stabilized before QS onset in young as well as in older infants (Figure 3). In both cases (QS–PS and PS–QS transitions), the state is "announced" well before its "real" onset. This confirms what has been reported by Monod & Curzi-Dascalova (1973) with regard to some phasic events (e.g. eye movements).

FACTORS INVOLVED IN STATE REGULATION

We have shown that development implies changes in both the quantitative and qualitative aspects of states. It is well recognized, as we stressed at the beginning, that major changes are supported by CNS maturation, which gives to those changes a universal developmental trend. The contribution of specific factors is less easily ascertained. It is also important to separate short-term effects (i.e., the immediate effects of external events), from effects due to, or related to, a long-lasting presence of a factor which could change the steps and/or the amount/characteristics of the state developmental trend.

Naturalistic studies have shown a relationship between some spontaneous events (waking) and subsequent sleep states. The kind of state at sleep onset is related to the duration of prior wakefulness and changes with age (Fagioli et al.,

Transition QS-PS (2-3 weeks)

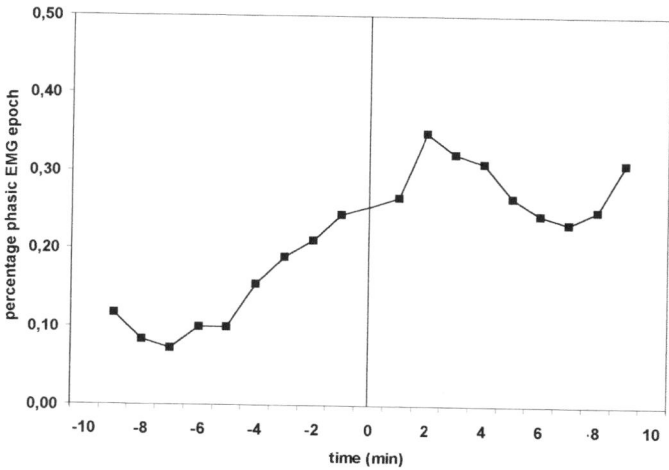

Transition QS-PS (19-20 weeks)

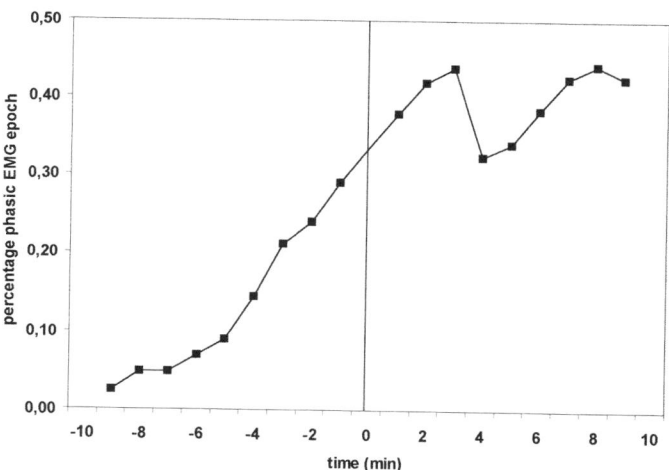

Figure 2 EMG phasic events before and after the transition from QS to PS in two groups of infants. Abscissa: time (min); 0 corresponds to transition between the two states; ordinate: percentage of phasic events calculated each minute.

1981). In infants less than 5 months old, short periods of waking are frequently followed by PS, and long periods by QS. In infants older than 6 months, long periods of waking are again followed mainly by a QS onset, while short periods of waking are mainly followed by ambiguous sleep. In other words, QS onset is linked from the very beginning to long prior waking episodes. These results give

Transition PS-QS (2-3 weeks)

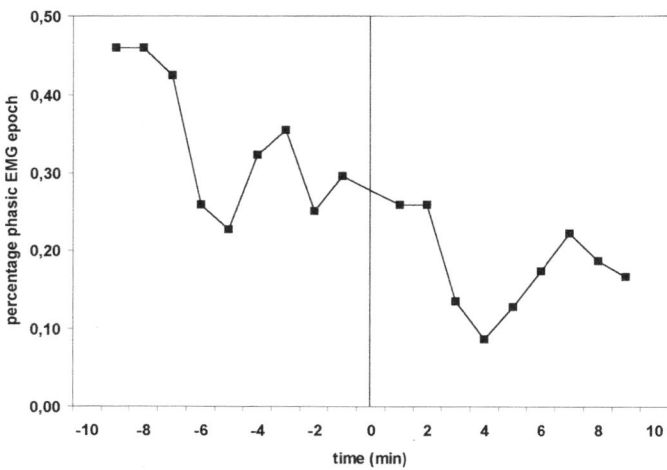

Transition PS-QS (19-20 weeks)

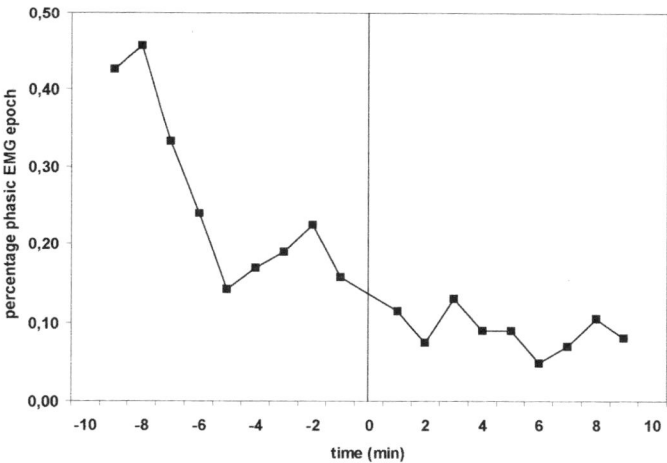

Figure 3 EMG phasic events before and after the transition from PS to QS in two groups of infants. For legend see Figure 2.

support to the hypothesis that state at sleep onset is regulated by an interaction between age and waking duration.

EEG background activity also, and, in particular, the amount of slow-wave activity (SWA) during NREM periods, seems to be regulated by prior waking duration in infants (Salzarulo & Fagioli, 1992), as previously shown in adults (Webb & Agnew, 1971; Feinberg, 1974; Borbély, 1982). In fact, the duration of the previous waking has recently been shown to be correlated with the amplitude of the increase of SWA at the beginning of the first NREM sleep period of the subsequent sleep, in a group of infants from 2 months to 1 year old (Fagioli & Salzarulo, 1998), showing the effect of wakefulness already in infants. These data fit very well with the predictions derived from the mathematical models concerning sleep regulation elaborated in adults (Daan et al., 1984; Achermann et al., 1993). These models posit that during the waking period a sort of brain "fatigue" increases progressively, and that during sleep it is eliminated. However, the relationships between waking duration (and therefore of the amount of "fatigue") and the sleep measures are not easy to show. In fact, the waking duration is less correlated with sleep duration or with duration of a given sleep state or stage, while it is correlated with SWA measures of NREM sleep. In addition, SWA can be detected only through EEG recordings, and need computerized technology to be quantified.

Experimental manipulations of sleep and wakefulness

Can sleep states be changed by external events? To explore changes in state, stimuli can be introduced when babies are either awake or asleep. In this perspective we are greatly indebted to Wolff (1959, 1963, 1987) who made extensive and controlled observations in newborns and infants. Effects are considerably different according to the state. Generally speaking, it is much easier to modify state 2, in Prechtl's (1977) and Wolff's (1987) terminology, than state 1. The latter is characterized by "strong homeostatic constraints" (Wolff, 1987, p. 44), which tend to re-equilibrate the state under external or internal influences within a limited pre-established time. Nevertheless, it is possible to prolong state 1 provided that the stimulus (monotonous sound) is introduced early after state onset and then continued. Effects on state 2 are more variable (Wolff, 1987). To reduce the duration of state, tickling was used. It is much easier to shorten state 2 than state 1, provided that the stimulus is sufficiently repeated.

These results show the response of the organism, according to the state, to a well-defined physical stimulus. The respective role of other kinds of stimuli should be ascertained in controlled experiments. Anedoctal observations generally show that is easy to stop state 2, whatever the stimulus characteristics. It is quite easy to prolong wakefulness. The mean duration of wakefulness can be prolonged with various "interesting spectacles" (Wolff, 1987). For example, during the first month a large red pencil oscillating back and forth rapidly in the baby's visual field maintained the infant's interest. By the third month this stimulus has no effect. What is needed is a human face moving back and forth across the visual field to prolong the alert state (Wolff, 1987). In contrast, when rhythmical rocking was started within the first 3 minutes of spontaneous wakefulness, and prolonged until the infant fell asleep, the mean duration of sustained waking could be reduced significantly. All the results, on the whole, provide an important message: state duration can be experimentally modified *within certain limits*.

Temperature could be a powerful factor inducing changes in states. Coolness has been found to be more effective than warmth (Brück et al., 1962).

Diminution of ambient temperature in neonates (Bach et al., 1994; Fleming et al., 1988) and in infants (Azaz et al., 1992) is followed by an increase in active sleep and a reduction in QS, and overall by a greater discontinuity of sleep, with intrusions of several bouts of waking with increased body movements (Azaz et al., 1992). The mechanisms responsible for these changes are still obscure. However, the interaction between metabolism, thermoregulation and sleep states should be implicated.

Nutrition and sleep states

The role of nutrition on behavioural states has deserved special attention for a long time. When speaking about nutrition it is necessary to separate the amount of food intake from feeding rhythms effects (Salzarulo, 1984). While there is agreement about the effects on states in the first instance, the role of the second one is much more uncertain.

The effect of the feeding rhythms on sleep – wake patterns in infants has been examined by several authors (Hellbrugge et al., 1964; Mills, 1974; Kripke, 1974). According to Ashton (1971), food intake appears to induce sleep in newborns and is considered a synchronizer of the sleep–wake alternation. Nevertheless, it has been emphasized that infants fall asleep more frequently when they are fed on demand than when they have scheduled meals (Gaensbauer & Emde, 1973). This result suggests that both are regulated by a common basic rhythm. The feeling of hunger, which is a function of the time elapsed from the last meal, could also cause the increase in motility observed at the end of the night (Hellbrugge, 1968; Irwin, 1932).

However, a series of studies performed by our group (Salzarulo et al., 1979, 1980; Fagioli et al., 1981, 1988) indicated that the alternation of behavioural states in the 24-hour period is similar in orally fed and continuously fed infants. The increase of motility at the end of the night has been observed in both groups. These studies show that waking–sleep cycles develop without a link to feeding rhythms, and suggest that both could be linked to a common "generator". Similar conclusions were reached by McMillen et al. (1991) in a study on preterm and full-term infants.

The influence of meals on sleep states was reported by some authors (Ashton, 1971; Harper et al., 1977) supporting the notion that meal timing could regulate state sequences. For example, Ashton (1973) points out that in neonates motor reactivity in the first epoch of active sleep is significantly lower than in the first epoch following feeding. Ashton (1971) and Harper et al. (1977) showed that active sleep is more prevalent immediately after a waking period with feeding than after a such period without food intake, emphasizing that the first REM period following feeding is shorter than subsequent REM episodes. We found no differences between orally-fed and continuously-fed infants for the time of sleep onset and of awakening (Fagioli et al., 1981), and for REM latency (Schulz et al., 1983).

Parenting and its role on state regulation

While the role of parenting on infant development has received much attention, both in clinical and experimental studies, its role on state regulation is still little known. Recently its potential interest has been underscored (Mosko *et al.*, 1996) and discussed at meetings (Messer & Parker, 1997; Salzarulo & Toselli, 1997). The debate concerns the extent to which sleep–wake rhythms and states can differ according to several kinds of parenting and, above all, can be permanently influenced by them.

Sleep parenting can be understood as any effective parental behaviour addressed directly or indirectly to infant sleep, or creating particular environmental conditions which could influence infant sleep. Parents can focus attention on infant sleep and then intervene, while in other cases attention to the infant is directed when he/she is awake (Anders, 1979).

Complete and coherent information on the effects of different parenting behaviour on infant sleep is lacking. Some insight comes from studies investigating wake–sleep diurnal distribution. Sander (1969), in a controlled study, found patterns of wake–sleep rhythm in infants living in a nursery to be different from those of infants living with a foster mother.

Night-waking, mainly the number of awakenings, has been mentioned by several investigators in a clinical context (Salzarulo & Chevalier, 1983; Blurton-Jones *et al.*, 1978). The differences according to parental behaviour (parents near the bed at sleep onset or leaving the baby to fall asleep alone) have been stressed by recent research (Sadeh *et al.*, 1991; Sadeh, 1994; Rikert & Johnson, 1988). The presence of parents at bedtime seems to predispose to night awakenings and night waking duration (Adair *et al.*, 1991).

Rikert & Johnson (1988) found that both "systematic ignoring" and "scheduled awakening" by parents are effective in decreasing spontaneous awakenings and crying episodes. In particular the authors reported that systematic ignoring was more effective than scheduled awakening practice only during the first week of treatment. However, in a controlled study, using objective measures of the level of motility, Sadeh (1994) found no changes in the number of previously existing awakenings when different parental behaviours are installed.

Co-sleeping, as part of parenting, deserves special attention for several reasons. Recent reviews (Lozoff *et al.*, 1985; Anders & Taylor, 1994; Mosko & McKenna, 1997) mentioned worldwide habits concerning the place and human context where infants sleep. Surprisingly enough, in many societies (about 50%) co-sleeping is a customary sleeping habit; and even in Western countries co-sleeping is relatively diffuse (Wolf *et al.*, 1996; Toselli *et al.*, 1998). Sleep continuity seems to be modified by co-sleeping. A study performed by Lozoff *et al.* (1996) showed that night-waking is always more frequent in co-sleeping infants, whatever the ethnic origin and sociocultural status. Elias *et al.* (1986) reported that the duration of sleep episodes is shorter in co-sleeping infants than in infants sleeping alone; adding breast-feeding leads to an even shorter duration.

Factors involved in co-sleeping which are responsible for the diminution of sleep are difficult to identify. What is included in co-sleeping that could be involved in the modulation of sleep architecture (patterns)? We could refer first to environmental temperature and then to body contact. Maternal proximity with interacting behaviours seems to be the most probable (see Mosko *et al.*, 1996).

There is little knowledge regarding the effects of co-sleeping on the components of sleep, e.g. states and stages. Alterations of sleep states, a topic approached only by Mosko *et al.* (1996), seem to be restricted to a diminution of SWS in co-sleeping infants. To ascertain the real effectiveness of co-sleeping on state organization, additional investigations seem necessary using further parameters and criteria, such as the duration of states and the amount of ambiguous sleep. It should be remarked that, if the diminution of SWS is replicated, this should not be in favour of co-sleeping, due to the importance of this kind of sleep for biological processes (Feinberg & March, 1995; Horne, 1988; Borbély, 1982).

Among various practices involved in parenting, pacifier use is a typical one. The role of the pacifier, mainly its quieting effect, has been well ascertained both in full-term and in preterm infants (Kessen & Leutzendorff, 1963; Woodson *et al.*, 1985). While being obviously linked to parental practices, the pacifier effect was studied mainly as a physical agent. Non-nutritive sucking (i.e. sucking activity which is not followed by food ingestion) can be induced by the pacifier and, besides a soothing effect, if used during sleep it decreases response thresholds in young infants (Wolff & Simmons, 1967). We still do not know the consequences of pacifier use on states at sleep onset.

Body contact and vestibular–proprioceptive stimulation also have a soothing effect in neonates (Korner & Thoman, 1972). In particular the vestibular–proprioceptive stimulation effect induced by a water bed on sleep states has been highlighted by DeRiggi (1990). The water bed produces an increase in total sleep time and in the amount of quiet sleep.

As a final comment on the topic of parenting, we wish to point out that careful consideration should be given to the various aspects involved in parenting (in particular sociocultural ones). The duration of practices and beliefs (Toselli *et al.*, 1998) should also be taken into account, together with the long-term effects (i.e. effects on the developmental trend of states).

ATTENTION AND ITS RELATIONSHIP WITH STATE

During early development many variables may affect attentional processes – that is the way infants perceive, elaborate and respond to discrete environmental stimuli. Beyond age and experience, one main factor is their current state.

Seminal studies (e.g. Wolff, 1966) have enlightened us regarding the importance of states, as early as the first weeks of life, as an interface expression of CNS activity between perception and its output, in terms of behavioural phenomena. As a suggestive example, Wolff reminds us how physically identical stimuli (e.g. a jar to the crib of the infant), may provoke different responses in term of motor outputs whether the infant is in "deep" sleep, "light" sleep or fully awake (Wolff, 1984).

Far from being always reputed to be the necessary assumption for research on attention in infants, this evidence has been often neglected, introducing a biasing factor in many studies addressed to comprehension of the time patterning of the main indicators of attention. In a review of the literature on the orienting response (OR) and heart rate deceleration folllowing stimuli presentation, Graham & Jackson (1970) observed that very few studies succeeded in showing an OR in the newborn. According to these authors, most of these studies had not controlled for babies' state, so that the absence of response could correspond to those periods when the baby was sleeping.

In infants, attention may be inferred by means of direct observation, from peculiar behavioural patterns which can either consolidate across time or be progressively replaced by other features. The basic attention behaviour is eye scanning, indexed by the number of saccadic movements produced by the subject during exploration of the environment.

Within behavioural states, two conditions have been described when the infant appears to be especially reactive to stimuli from the environment. During *"alert inactivity"* (Wolff, 1984, 1987) there are very few body movements, with a relaxed face, the respiration is regular and the eyes are open; during *"alert activity"* (Wolff, 1984), controlled movements of limited intensity may be seen.

From 8 weeks onwards, seemingly, the baby becomes able to "engage in two or more acquired goal-directed actions at once" (Wolff, 1987, p. 25). Thus, the occurrence of alert activity marks a turning point in the development of the sensorimotor processes of attention. The increase in the ability to sustain prolonged periods of wakefulness could be also partly due to the increase in the number of goal-oriented activities.

Beyond eye movements, other phenomena have been proposed as expressions of information perception and processing. One line of research has used non-nutritive sucking (NNS) as an index of attention. At present no definitive experimental evidence has been obtained that NNS may imply attention. Perhaps NNS may indirectly intervene on the readiness to respond to stimuli by remarkably affecting arousal levels, and specifically by inhibiting it and therefore acting as a "tranquillizer" (Woodson & Hamilton, 1988).

Beyond being identified through the above indices, attention in infants may be addressed as a dynamic process, during which the baby has to gather attention and direct it selectively to a given event, maintain the focus on the stimulus and finally disengage from it. All these events require the functioning of certain substrates and mechanisms, and correspond to changes in physiological variables. Since these changes may steadily persist for appreciable amounts of time, as long as a given phase of the attentional process persists, Ruff & Rothbart (1996) propose that their *ensembles* might be considered as proper states themselves. However, we should remember that for some authors (Wolff, 1987; Prechtl, 1974), the definition of state implies longer time intervals (> 1–3 minutes) than those involved in attention.

Attention can be initiated only if a proper level of autonomic and cortical activation is reached. The consequent "aroused" state would have different features in newborns than in older infants (Kahneman, 1973). During the first year of life, there should be a pattern of marked reduction in motor activity, accompanied by heart rate deceleration. After the first year of life a second, more developed, attention system replaces the former. This system is mainly operating during problem-solving processes and is characterized by a generalized pattern of sympathetic dominance, whose main feature is an increase in heart rate (Kahneman, 1973). Although this pattern still includes a clear coupling between the functioning of the attentional system, with the related neurovegetative changes (increase of arousal), and reduction of body motility, this coupling is less marked than at previous ages.

737

CONCLUSIONS

At the end of this chapter we wish to stress the concept of developmental change supported by CNS maturation and modulated by external events/factors. To reach a "state condition", clearly distinct from another "state", a long developmental course should be considered, mainly in prenatal life. A state is the result of the integrative capabilities of the CNS, and can be modulated, or disrupted, by environmental or pathological events.

Two qualifications should be re-iterated. Some physiological activities are not included in the "package" regulated synchronously; they shift more or less independently of the "state variables". Their functional significance should be ascertained. Several "natural" external factors have been shown to influence waking–sleep rhythms. However, their effect on states and, above all, on state regulation, has not been reported. This should be one of the important issues in the study of states during development.

The role of states in exchanges between subjects and the external world makes their future investigation an up-to-date research topic for psychologists and neurobiologists.

ACKNOWLEDGEMENTS

We thank P. Ktonas for revising the English of a previous version of the chapter, and C. Chiorri for help in preparing some figures and references. Partially supported by CNR and MURST grants.

References

Achermann, P., Dijk, D.J., Brunner, D.P. & Borbely, A.A. (1993). A model of human sleep homeostasis based on EEG slow-wave activity: quantitative comparison of data and simulations. *Brain Research Bulletin*, **31**, 97–113.

Adair, R., Baucher, H., Philipp, B., Levenson, S. & Zuckerman, B. (1991). Night waking during infancy: role of parental presence at bedtime. *Pediatrics*, **87**, 500–504.

Adrien, J. (1976). Le sommeil du nouveau-né. *La Recherche*, **63**, 74–76.

Anders, T.F. (1979). Night-waking in infants during the first year of life. *Pediatrics*, **63**, 860–864.

Anders, T.F., Keener, M.A. (1985) Developmental course of nighttime sleep–wake patterns in full-term and premature infants during the first year of life. *Sleep*, **8**, 173–192.

Anders, T.F. & Taylor, T.R. (1994). Babies and their sleep environment. *Children's Environments*, **11**, 123–134.

Ashton, R. (1971). Behavioural sleep cycles in the human newborns. *Child Development*, **42**, 2098–2100.

Ashton, R. (1973). The influence of state and prandial condition upon the reactivity of the newborn to auditory stimulation. *Journal of Experimental Child Psychology*, **15**, 315–327.

Azaz, Y., Fleming, P.J., Levine, M., McCabe, R., Stewart, A. & Johnson, P. (1992). The relationship between environmental temperature, metabolic rate, sleep state and evaporative water loss in infants from birth to three months. *Pediatrics Research*, **32**, 417–423.

Bach, V., Bouferrache, B., Kremp, O., Maingourd, Y. & Libert, J.P. (1994). Regulation of sleep and body temperature in response to exposure to cool and warm environments in neonates. *Pediatrics*, **93**, 789–796.

Bes, F., Schulz, H., Navelet, Y. & Salzarulo, P. (1991). The distribution of slow wave sleep across the night: a comparison for infants, children and adults. *Sleep*, **14**, 5–12.

Blurton-Jones, N., Ferreira, M.C. & Brown, M.F. (1978). The association between perinatal factors and later night waking. *Developmental Medicine and Child Neurology,* **20**, 427–434.

Borbély, A.A. (1982). A two process model of sleep regulation. *Human Neurobiology,* **1**, 195–204.

Brück, K., Parmelee, A.H. & Brück, M. (1962). Neutral temperature range and range of thermal comfort in premature infants. *Biology of the Neonate,* **4**, 32–51.

Curzi-Dascalova, L. & Mirmiran, M. (1996). *Manuel des techniques d'enregistrement et d'analyse des stades de sommeil et de veille chez le prématuré et le nouveau-né à terme.* Paris: Inserm.

Curzi-Dascalova, L., Peirano, P. & Morel-Kahn, F. (1988). Development of sleep states in normal premature and full-term newborns. *Developmental Psychobiology,* **21**, 431–444.

Daan, S., Beersma, D.G.M. & Borbely, A. (1984). Timing of human sleep: recovery processes gated by a circadian pacemaker. *American Journal of Physiology,* **246**, R161–R178.

De Riggi, P. (1990). The effect of waterbeds on preterm infants' sleep. In S.G. Funk, E.M. Tornquist, M.T. Champagne, L.A. Copp & R.A. Wiese (Eds.), *Key aspects of recovery. Improving nutrition, rest and mobility* (pp. 98–110). New York: Springer.

Dittrichova, J. & Lapakova, V. (1964). Development of the waking state in young infants. *Child Development,* **35**, 365–370.

Elias, M.F., Nicholson, N.A., Bora, C. & Johnston, J. (1986). Sleep/wake patterns of breast-fed infants in the first 2 years of life. *Pediatrics,* **77**, 322–329.

Ellingson, R.J. (1975). Ontogenesis of sleep in the human. In G.C. Lairy and P. Salzarulo (Eds.), *The experimental study of human sleep: methodological problems* (pp. 129–156). Amsterdam: Elsevier.

Emde, R. (1980) Toward a psychoanalitic theory of affect. In S.I. Greenspan & G.H. Pollok (Eds.), *The course of life: psychoanalitic contributions toward understanding personality development. Infancy and early childhood* (vol. 1, pp. 72–90). Washington, DC: NIMH.

Fagioli, I. & Salzarulo, P. (1982a). Sleep states development in the first year of life assessed trough 24-hour recordings. *Early Human Development,* **6**, 215–228.

Fagioli, I. & Salzarulo, P. (1982b). Organisation temporelle des cycles de sommeil dans le 24 heures chez le nourrisson. *Revues de EEG et Neurophysiologie,* **12**, 344–348.

Fagioli, I. & Salzarulo, P. (1998). Prior spontaneous nocturnal waking duration and EEG during quiet sleep in infants: an automatic analysis approach *Behavioral Brain Research,* **91**, 23–28.

Fagioli, I., Salomon, F. & Salzarulo, P. (1981). L'endormissement chez l'enfant en nutrition continue depuis la naissance: enregistrements de vingt-quatre heures. *Revues de EEG et Neurophysiologie,* **11**, 37–44.

Fagioli, I., Bes, F. & Salzarulo, P. (1988). 24-hour behavioural states distribution in continuously fed infants. *Early Human Development,* **18**, 151–156.

Feinberg, I. (1974). Changes in sleep cycle patterns with age. *Journal of Psychiatric Research,* **10**, 283–306.

Feinberg, I. & March, J.D. (1995). Observations on delta homeostasis, the one model of NREM-REM alternation and the neurologic implications of experimental dream. *Behavioral Brain Research,* **69**, 97–108.

Ficca, G., Fagioli, I. & Salzarulo, P. (2000a). Sleep organization in the first year of life: developmental trends of the quiet sleep-paradoxical sleep cycle. *Journal of Sleep Research,* **9**, 1–4.

Ficca, G., Lombardo, P., Rossi, L. & Salzarulo, P. (2000b). Morning recall of verbal material depends on prior sleep organization. *Behavioral Brain Research,* **112**, 159–163.

Fleming, P.J., Levine, M.R., Azaz, Y. & Johnson P. (1988). The effect of sleep state on the metabolic response to cold stress in newborn infants. In C.T. Jones (Ed.), *Fetal and neonatal development* (pp. 635–639). Ithaca, NY: Perinatology Press.

Gaensbauer, T.J. & Emde, R. (1973). Wakefulness and feeding in human newborns. *Archives of General Psychiatry,* **28**, 894.

Graham, F.K. & Jackson, J.C. (1970). Arousal systems and infant heart rate responses. *Advances in Child Developmental Behavior,* **5**, 59–117.

Gramsbergen, A., Schwartze, P. & Prechtl, H.R.F. (1970). The post natal development of behavioural states in the rat. *Developmental Psychobiology,* **3**, 267–280.

Harper, R.M., Hoppenbrouwers, T., Bannett, D., Hodgman, J., Sterman, M.B. & McGinty, D.J. (1977). Effects of feeding on state and cardiac regulation in the infant. *Developmental Psychobiology,* **10**, 507–517.

Hellbrügge, T. (1968). Ontogénese des rythmes circadiaires chez l'enfant. In J. Ajuriaguerra (Ed.), *Cycles biologiques et psychiatrie* (pp. 159–183). Paris: Masson.

Hellbrügge, T., Ehrengut Lange, J., Rutenfranz, J. & Stehr, K. (1964). Circadian periodicity of physiological functions in different stages of infancy and childhood. *Annals of the New York Academy of Sciences*, **117**, 361–373.

Horne, J.A. (1988). *Why we sleep. The functions of sleep in humans and other mammals*. New York: Oxford University Press.

Irwin, O.C. (1932). The amount of motility of seventy-three newborn infants. *Journal of Applied Physiology*, **14**, 415.

Jouvet-Mounier, D., Astic, L. & Lacote, D. (1970). Ontogenesis of the states of sleep in rat, cat, and guinea pig during the first postnatal month. *Developmental Psychobiology*, **2**, 216–239.

Kahneman, D. (1973). *Attention and effort*. Englewood. Cliffs, NJ: Prentice-Hall.

Kessen, W. & Leutzendorff, A.M. (1963). The effect of nonnutritive sucking on movement in the human newborn. *Journal of Comparative and Physiological Psychology*, **56**, 69–72.

Korner, A.F. & Thoman, E.B. (1972). The relative efficacy of contact and vestibular-proprioceptive stimulation in soothing neonates. *Child Development*, **43**, 443–453.

Kripke, F. (1974). Ultradian rhythms and sleep. In L.E. Sceving, F. Halberg & J.E. Pauly (Ed.), *Chronobiology* (p. 475). Stuttgart: Thieme.

Ktonas, P.Y., Bes, F., Rigoard, M.T., Wong, C., Mallart, R. & Salzarulo, P. (1990). Developmental changes in the clustering pattern of sleep rapid eye movement activity during the first year of life: a Markov-process approach. *Electroencephalography and Clinical Neurophysiology*, **75**, 136–140.

Liefting, B., Bes, F., Fagioli, I. & Salzarulo, P. (1994). Electromyographic activity and sleep states in infants. *Sleep*, **17**, 718–722.

Louis, J., Cannard, C., Bastuji, H. & Challamel, M.J. (1997). Sleep ontogenesis revisited: a longitudinal 24-hour home polygraphic study on 15 normal infants during the first two years of life. *Sleep*, **20**, 323–333.

Lozoff, B., Wolf, A.W. & Davis, N.S. (1985). Sleep problems seen in pediatric practice. *Pediatrics*, **75**, 477–483.

Lozoff, B., Askew, G.L. & Wolf, A.W. (1996). Cosleeping and early childhood sleep problems: effects of ethnicity and socioeconomic status. *Developmental and Behavioural Pediatrics*, **17**, 9–15.

Mazzoni, G., Gori, S., Formicola, G., Gneri, C., Massetani, R., Murri, L. & Salzarulo, P. (1999). Word recall correlates with sleep cycles in elderly subjects. *Journal of Sleep Research*, **8**, 185–188.

McMillen, I.C., Kok, J.S.M., Adamson, T.M., Deayton, J.M. & Nowak, R. (1991). Development of circadian sleep–wake rhythms in preterm and full-term infants. *Pediatric Research*, **29**, 381–384.

Messer, D. & Parker, C. (1997). Maternal expectations, behaviour and children's sleeping. Workshop "Parenting and sleep" Seventh meeting of the European Sleep Club, Lyon, 12–13 September.

Mills, J.N. (1974). Development of circadian rhythms in infancy In J.A. Davis & J. Dobbing (Eds.), *Scientific foundation of paediatrics* (p. 758). London: Heinemann.

Monod, N. & Curzi-Dascalova, L. (1973). Les états transitionnels de sommeil chez le nouveau-né à terme. *Revues de EEG et Neurophysiologie*, **3**, 87–96.

Mosko, S. & McKenna, C.R.J. (1997). Maternal sleep and arousal during bedsharing with infants. *Sleep*, **20**, 142–150.

Mosko, S., McKenna, C.R.J. & Drummond, S. (1996). Infant sleep architecture during bedsharing and possible implications for SIDS. *Sleep*, **19**, 677–684.

Navelet, Y., Benoit, O. & Bouard, G. (1982). Nocturnal sleep organization during the first months of life. *Electroencephalography and Clinical Neurophysiology*, **54**, 71–78.

Nijhuis, J.G., Prechtl, H.F.R., Martin, C., Jr. & Bots, R.S.G.M. (1982). Are there behavioural states in the human fetus? *Early Human Development*, **6**, 177–195.

Papoušek, H. (1969). Individual variability in learned responses. In R.J. Robinson (Ed.), *Brain and early behaviour* (pp. 251–262). London: Academic Press.

Parmelee, A.H. & Stern, E. (1972). Development of states in infants. In C.D. Clemente, D.P. Purpura & F.E.Meyer (Eds.), *Sleep and maturing nervous system* (pp. 199–228). New York: Academic Press.

Prechtl, H.F.R. (1974). The behavioural states of the newborn infant (a review). *Brain Research*, **76**, 185–212.

Prechtl, H.R.F. (1977). *The neurological examination of the fullterm newborn infant. Clinics in Developmental Medicine*, vol. 63. London: Heinemann.

Prechtl, H.R.F. (1986). New persepctives in early human development. *European Journal of Obstetrics, Gynecology and Reproductive Biology,* **21**, 347–355.

Prechtl, H.F.R. & O'Brien, M.J. (1982). Behavioural states of the full-term newborn. The emergence of a concept. In P. Stratton (Ed.), *Psychobiology of the human newborn* (pp. 53–73). New York: Wiley.

Rikert, V.I. & Johnson, M. (1988). Reducing nocturnal awakening and crying episodes in infants and young children: a comparison between scheduled awakenings and systematic ignoring. *Pediatrics,* **81**, 203–212.

Roffwarg, H.P., Dement, W.C. & Fischer, C. (1964). Preliminary observations of the sleep-dream pattern in neonates, infants, children, adults. In E. Harms (Ed.), *Problems of sleep, dreams in children* (pp. 60–72). New York: MacMillan.

Ruff, H.A. & Rothbart, M.K. (1996). Development of attention as a state. In H.A. Ruff & M.K. Rothbart (Eds.), *Attention in early development* (pp. 93–109). New York: Oxford Unversity Press.

Sadeh, A. (1994). assesment of intervention for infant night waking: parental reports and activity-based home monitoring. *Journal of Consulting and Clinical Psychology,* **62**, 63–68.

Sadeh, A., Lavie P., Scher A., Tirosh E. & Epstein R. (1991). Actigraphic home-monitoring of sleep-disturbed and control infants and young children: a new method for pediatric assesment of sleep-wake patterns. *Pediatrics,* **87**, 494–499.

Salzarulo, P. (1984). Sleep nutrition and metabolism during development In Pollitt & Amante (Eds.), *Energy intake and activity* (pp. 323–327). New York: A. Liss.

Salzarulo, P. & Chevalier, A. (1983). A.Sleep problems in chlidren and their relationship with early disturbances of the waking-sleeping rhythm. *Sleep,* **6**, 47–51.

Salzarulo, P. & Fagioli, I. (1992). Post-natal development of sleep organization in humans: speculations on the emergence of the "S process". *Neurophysiologie Clinique,* **22**, 107–115.

Salzarulo, P. & Fagioli, I. (1995). Sleep for development or development for waking?: some speculations from a human perspective. *Behavioral Brain Research,* **69**, 23–29.

Salzarulo, P. & Fagioli, I. (1999). Changes of sleep states and physiological activities across the first year of life. In A. Kalverboer, M.L. Genta & B. Hopkins (Eds.), *Current issues in developmental psychology. Biopsychological perspectives* (pp. 53–74). Dordrecht: Kluwer.

Salzarulo, P. & Toselli, M. (1997). Infant's sleep parenting: between ideas and cares. Workshop "Parenting and sleep" Seventh meeting of the European Sleep Club, Lyon, 12–13 September.

Salzarulo, P., Fagioli, I, Salomon, F., Duhamel, J-F. & Ricour, R. (1979). Alimentation continue et rythme veille-sommeil chez l'enfant. *Archives Française de Pediatrie,* (Suppl.), **36**, 26–32.

Salzarulo, P., Fagioli, I., Salomon, F., Ricour, C., Raimbault, G., Ambrosi, S., Cicchi, O., Duhamel, J.F. & Rigoard, M.T (1980). Sleep patterns in infants under continuous feeding from birth. *Electroencephalography and Clinical Neurophysiology,* **49**, 330–336.

Salzarulo, P., Fagioli, I., Salomon, F. & Ricour, C. (1982). Developmental trend of quiet sleep is altered by early human malnutrition and recovered by nutritional rehabilitation. *Early Human Development,* **7**, 257–264.

Salzarulo, P., Fagioli, I., Peirano, P., Bes, F. & Schulz, H. (1991). Levels of EEG background activity and sleep states in the first year of life. In G.M. Terzano, P.L. Halasz & A.C. Declerck (Eds.), *Phasic events and dynamic organization of sleep* (pp. 53–63). New York: Raven Press.

Sander, L.W. (1969). Regulation and organization in the early infant-caretaker system. In R.J. Robinson (Ed.), *Brain and early behaviour* (p. 311). London: Academic Press.

Schulz, H., Salzarulo, P., Fagioli, I. & Massetani, R. (1983). REM latency: development in the first year of life. *Electroencephalography and Clinical Neurophysiology,* **56**, 316–322.

Shani, R., Schulze, F.K., Stefanski, M., Myers, M.M. & Fifer, W.P. (1995). Methodological issues in coding sleep states in immature infants. *Developmental Psychobiology,* **28**, 85–101.

Shirataki, S. & Prechtl, H.R.F. (1977). Sleep transitions in newborn infants: preliminary study. *Developmental Medicine and Child Neurology,* **19**, 316–325.

Sterman, M.B. (1979). Ontogeny of sleep: implications for functions. In R. Drucker-Colin, M. Shkurovich & M.B. Sterman (Eds.), *The function of sleep.* (pp. 207–231). New York: Academic Press.

Toselli, M., Farneti, P. & Salzarulo, P. (1998). Maternal representation and care of infant sleep. *Early Development and Parenting,* **7**, 73–78.

Verley, R. & Garma, L. (1975). The criteria of sleep stages during ontogeny in different animal species. In G.C. Lairy & P. Salzarulo (Eds.), *The experimental study of human sleep: methodological problems* (pp. 109–128). Amsterdam: Elsevier.

Webb, W.B. & Agnew, H.W. Jr (1971). Stage 4 sleep: influence of time-course variables. *Science*, **174**, 1354–1356.

Wolf, A.W., Lozoff, B., Latz, S. & Paludetto, R. (1996). Parental theories in the manegement of young children's sleep in Japan, Italy and United States. In S. Harkeness & C.M. Super (Eds.), *Parents' cultural belief systems. Their origins, expressions and consequences* (pp. 364–384). London: Guilford.

Wolff, P.H. (1959). Observations on newborn infants. *Psychosomatic Medicine*, **221**, 110–118.

Wolff, P.H. (1963). Observation of behaviour in early infancy. Paper presented to the New Orleans Psychoanalytic Society, New Orleans, June.

Wolff, P.H. (1966). The causes, controls and organization of behaviour in the neonate. *Psychological Issues*, 5, Monograph 17. New York: International University Press.

Wolff, P.H. (1973a). The organization of behaviour in the first three months of life. *Nervous and Mental Disease*, **51**, 132–153.

Wolff, P.H. (1984). Discontinuous changes in human wakefullness around the end of the second month of life; a developmental perspective. In H.F.R Prechtl (Ed.), *Continuity of neural functions from pre-natal to post-natal life* (pp. 144–158). Oxford: Blackwell.

Wolff, P.H. (1987). *The development of behavioural states and the expression of emotions in early infancy.* Chicago, IL: University of Chicago Press.

Wolff, P.H. & Simmons, M.A. (1967). Nonnutritive sucking and response thresholds in young infants *Child Development*, **38**, 631–638.

Woodson, R. & Hamilton, C. (1988). The effect of nonnutritive sucking on heart rate in preterm infants. *Developmental Psychobiology*, **21**, 207–213.

Woodson, R., Drinkwin, J. & Hamilton, C. (1985). Effects of nonnutritive sucking on state and activity: term–preterm comparison. *Infant Behaviour and Development*, **8**, 435–441.

Further reading

Kalverboer, A., Genta, M.L. & Hopkins, B. (1999). *Current issues in developmental psychology. Biopsychological perspectives*. Dordrecht: Kluwer.

Ruff, H.A. & Rothbart, M.K. (1996). *Attention in early development*. Oxford: Oxford Unversity Press.

Wolff, P.H. (1987). *The development of behavioural states and the expression of emotions in early infancy.* Chicago, IL: University of Chicago Press.

IV.7
Self-organization in Normal and Abnormal Cognitive Development

DENIS MARESCHAL and MICHAEL S.C. THOMAS

ABSTRACT

This chapter discusses self-organization as a motor for cognitive develop-ment. Self-organization occurs in systems with many degrees of freedom and is ubiquitous in the brain. The principal means of investigating the role of self-organization in cognitive development is through connec-tionist computational modelling. Connectionist models are computer models loosely based on neural information processing. We survey a range of models of cognitive development in infants and children and identify the constraints on self-organization that lead to the emergence of target behaviours. A survey of connectionist models of abnormal cognitive development illustrates how deviations in these constraints can lead to the development of abnormal behaviours. Special attention is paid to models of development in autistic children.

INTRODUCTION

We have come a long way in understanding the processes that underlie brain development since the days of Piaget's attempts to relate cognitive development to an unfolding biological substrate (e.g. Piaget, 1971, 1980). *Developmental cognitive neuroscience* is a new field of research that addresses that very issue. The aim of this field is to bridge the gap between children's cognitive develop-ment (as assessed by behavioural studies) and the underlying development of the brain (Johnson, 1997).

Although the age-old debate concerning the relative importance of nature and nurture in determining development rages on, it has recently taken a new twist. Few people now claim that innate knowledge is hard-wired in *a-priori* neural connections (representational innateness). Rather, it is generally accepted that both nature and nurture play a role in children's cognitive development. The pivotal question that remains is the extent to which plasticity dominates

A.F. Kalverboer and A. Gramsbergen (eds.), Handbook of Brain and Behaviour in Human Development, 743–766
© *2001 Kluwer Academic Publishers. Printed in Great Britain.*

development and the extent to which structural constraints are genetically determined such that experience plays only a limited role in fine-tuning these structures (Elman *et al.*, 1996).

This chapter will explore how the concept of self-organization can provide an account of behavioural development in infants and children. Along the way it will explore how constraints (or boundary conditions) guide self-organization. An important tool for exploring self-organization in cognitive development is connectionist computational modelling. Connectionist network models (or artificial neural network models) are computer models loosely based on neural information processing. These models allow us to explore how different system constraints interact with an environment to give rise to observed system behaviours. They also provide a means of exploring how deviations in self-organization (due to a shift in boundary conditions) can result in the emergence of abnormal behaviours.

In the rest of this chapter we begin by discussing self-organization in the brain. We then turn to discussing self-organization in cognitive development. Following this, connectionist modelling is introduced as a means of investigating self-organization in development. The two subsequent sections review connectionist models of normal cognitive development and abnormal cognitive development.

SELF-ORGANIZATION IN THE BRAIN

Many chapters in this handbook provide examples of self-organization. Self-organization occurs when structure emerges in response to a system's dynamic interactions with an environment. Self-organization is a fundamental characteristic of the brain (Willshaw & von der Malsburg, 1976; Grossberg, 1982; Changeux *et al.*, 1984; Edelman & Finkel, 1985; von der Malsburg, 1995; Keslo, 1995). It can occur at several time scales: a learning time scale of hours and days, a developmental time scale of months and years, but also on the functional time scale of seconds and minutes. All stages of brain organization involve an element of self-organization (Keslo, 1995; Johnson, 1997). It is unlikely that the genes can (in any direct way) encode the full information necessary to describe the brain (Elman *et al.*, 1996). Given that the cerebral cortex alone contains some 10^{14} synapses, and given the variability of vertebrate brain structure, it is difficult to see how individual wiring diagrams could be encoded within the limited coding space of the genome (von der Malsburg, 1995).

There are many well-studied examples of self-organization in the physical and biological sciences (Prigogine & Stengers, 1986). Self-organization occurs in systems with a large number of degrees of freedom (e.g. synapses in the brain). Initially the system is undifferentiated (randomly organized) but, as a result of small adaptive changes, an order begins to emerge among the elements of the system. These changes can self-amplify, resulting in a form of positive feedback. If there is a limitation in resources this limitation can lead to competition and selection among the changes. Finally, changes can cooperate, enhancing the "fitness" of some changes over the others in spite of competition. With respect to self-organization in the brain, synaptic adjustment rules (such as the Hebbian learning rule; Hebb, 1949) can lead to ordered connection patterns that in turn lead to structured behaviours.

A fundamental characteristic of self-organizing systems is that global order can arise from local interactions. This is extremely important in the brain where local interactions between cellular neighbours create states of global order and ultimately generate coherent behaviour. However, we should be careful in what we understand by local here, because nerve cells are connected by long axons. Local neural interactions are not necessarily topologically arranged (von der Malsburg, 1995). Connected cells can be neighbours although they are physically located at other ends of the brain. One implication of this is that some ordered structures in the brain may not initially "look" ordered to our eyes, because we rely heavily on spatial contiguity to perceive patterns.

There are two relevant parameters in network self-organization. The first is the information or activation transmitted through the network (cf. action potentials). The second is the connection strength between successive units (cf. synaptic strength). Connections control neural interactions and are characterized by a continuous weight variable. These connections reflect the size of the effect exerted on one unit by another unit. Organization in the brain can therefore take place at two levels: activity and connectivity. Changes in activation levels reflect self-organization at the instantaneous, functional level whereas changes in connectivity correspond to self-organization on a learning or developmental time scale.

SELF-ORGANIZATION IN COGNITIVE DEVELOPMENT

The fundamental question of cognitive development is where new behaviours come from. Traditionally, developmentalists have looked for the source of these behaviours either in the organism or in the environment. Perhaps new structures arise as a result of instructions stored beforehand in some code (e.g. the genes); or perhaps new behaviours are acquired by absorbing the structures and patterns of the environment directly. Oyama (1985) has suggested that both of these accounts are fundamentally preformationist in that it is either the genes or the environment that determines the nature of the structures that are developed. She argues that attributing the origin of structure to the genes or the environment simply pushes back to another level questions concerning the causal origin of these structures. Therefore, it fails to answer the fundamental question of cognitive development. Oyama further suggests that it is the concept of self-organization that rescues developmentalists from this logical hole of infinite regress. In biological systems pattern and order can emerge from the process of interaction without the need for explicit instructions.

There is now ample evidence of self-organization occurring during development in both linguistic and cognitive domains. Specific examples include children's understanding of balancing relations (Karmiloff-Smith & Inhelder, 1971), children's drawing abilities (Karmiloff-Smith, 1990), and language acquisition (Karmiloff-Smith, 1985). Karmiloff-Smith (1992) suggests that what drives development is the endogenous principle of "representational re-description". Even when performance seems to be adequate, there are pressures arising from within the cognitive system to re-describe existing knowledge in more abstract and accessible forms. These pressures arise from the need to make information in one functional module accessible to another functional module.

745

There have been several attempts to explain the apparent stage-like growth of competence in children in terms of self-organizing principles. Van Geert (1991) outlined a framework for discussing language and cognitive development as growth under limited resources. He formulated a dynamic systems model of development in terms of logistic growth equations. This model describes development as a result of supportive and competitive interactions between "cognitive growers". Similarly, Van der Maas & Molenaar (1992) presented an account of stage transitions on conservation tasks in terms of catastrophe theory. According to this account, discrete and qualitative shifts in behaviour arise as a result in continuous changes in the underlying parameters of a system.

Although these models provide a good account of how new behaviours can emerge through the continuous adjustment of abstract parameters, and as a result of endogenous pressures, they rarely relate those parameters to any measurable cognitive quantities. Nor do they relate them to some underlying neurological substrate. Explicit accounts of self-organization in behaviour have been limited to describing how actions are elicited in infancy, or to how separate motor systems become coupled to induce higher levels of motor action (e.g. Goldfield, 1995). A more effective means of exploring self-organization in cognitive development (and to relate development to neural information processing) is to construct neurally based computer simulations of cognitive development.

CONNECTIONIST COMPUTATIONAL MODELLING

Connectionist models are computer models loosely based on the principles of neural information processing (Rumelhart & McClelland, 1986; Elman *et al.*, 1996; McLeod *et al.*, 1998). They are information processing models and are not intended to be neural models. They embody general principles such as inhibition and excitation within a distributed, parallel processing system. They attempt to strike the balance between importing some of the key ideas from the neuro-sciences while maintaining sufficiently discrete and definable components to allow questions about behaviour to be formulated in terms of a high-level cognitive computational framework.

From a developmental perspective, connectionist networks are ideal for modelling because they develop their own internal representations as a result of interacting with an environment (Plunkett & Sinha, 1992). However, these networks are not simply *tabula rasa* empirical learning machines. The representations they develop can be strongly determined by initial constraints (or boundary conditions). These constraints can take the form of different associative learning mechanisms attuned to specific information in the environment (e.g. temporal correlation or spatial correlation), or they can take the form of architectural constraints that guide the flow of information in the system. Although connectionist modelling has its roots in associationist learning paradigms, it has inherited the Hebbian rather than the Hullian tradition. That is, what goes on inside the box (inside the network) is as important in determining the overall behaviour of the networks as is the correlation between the inputs (stimuli) and the outputs (responses).

Connectionist networks are made up of simple processing units (idealized neurons) interconnected via weighted communication lines. Units are often represented as circles and the weighted communication lines (the idealized synapses) as lines between these circles. Activation flows from unit to unit via these connection weights. Figure 1a shows a generic connectionist network in which activation can flow in any direction. However, most applications of connectionist networks impose constraints on the way activation can flow. These constraints are embodied by the pattern of connections between units.

Figure 1b shows a typical *feed-forward* network. Activation (information) is constrained to move in one direction only. Some units (those units through which information enters the network) are called *input units*. Other units (those units through which information leaves the network) are called *output units*. All other units are called *hidden units*. In a feed-forward network, information is first encoded as a pattern of activation across the bank of input units. That activation then filters up through a first layer of weights until it produces a pattern of activation across the band of hidden units. The pattern of activation produced across the hidden units constitutes an *internal re-representation* of the information originally presented to the network. The activation at the hidden units continues to flow through the network until it reaches the output unit. The pattern of activation produced at the output units is taken as the network's response to the initial input.

Each unit in the network is a very simple processor that mimics the functioning of an idealized neuron. The unit sums the weighted activation arriving into it. It then sets its own level of activation according to some non-linear function of that weighted sum. The non-linearity allows the unit to respond differentially to different ranges of input values. The key idea of connectionist modelling is that of collective computation. That is, although the behaviour of the individual components in the network is simple, the behaviour of the network as a whole can be very complex. It is the behaviour of the network as a whole that is taken to model different aspects of infant development.

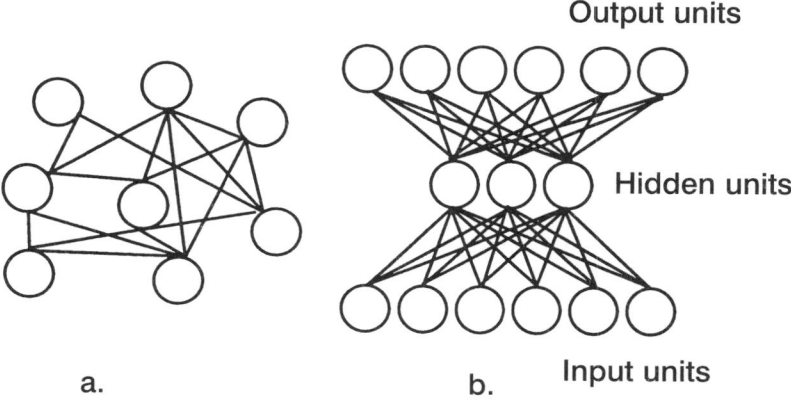

Figure 1 Schema of (**a**) a generic and (**b**) a feed-forward connectionist network.

Given the units' response functions, the network's behaviour is determined by the connection weights. As activation flows through the network it is transformed by the set of connection weights between successive layers in the network. Thus, learning (i.e. adapting one's behaviour) is accomplished by tuning the connection weights until some stable state of behaviour is obtained. *Supervised* networks adjust their weights until the output response (for a given input) matches a target response. The target can be obtained from an explicit teacher, or it can arise from the environment, but it must come from outside the system. *Unsupervised* networks adjust their weights until some internal constraint is satisfied (e.g. maximally different inputs must have maximally different internal representations). *Backpropagation* (Rumelhart *et al.*, 1986) is a popular training algorithm for supervised connectionist networks that incrementally updates the network weights so as to minimize the difference between the network's output and some desired target output. These networks self-organize in such a way as to internalize structures in the environment.

Through adaptation the connection weights come to encode regularities about the network's environment that are relevant to a task the network must solve. Networks are very sensitive to the distribution statistics of relevant features in their environment. A feed-forward network with a single layer of hidden units can approximate arbitrarily well any finite output response function, given enough hidden units (Cybenko, 1989). Further details of the similarities between connectionist network learning and statistical learning procedures can be found elsewhere (e.g. Hertz *et al.*, 1991).

There are two levels of knowledge in these networks. The connection weights encode generalities about the problem that have been accumulated over repeated encounters with the environment. One can think of this as a form of long-term memory or category-specific knowledge as opposed to knowledge about a particular task or object. In contrast, the pattern of activation that arises in response to inputs, encodes information about the current state of the world. Internal representations are determined by an interaction between the current input (activation across the input units) and previous experience as encoded in the connection weights.

Many connectionist networks are very simple. They may contain some 100 units or so. This is not to suggest that the part of the brain solving the corresponding task only has 100 neurons. It is important to understand that most connectionist models are not intended as neural models, but rather as information-processing models of behaviour. The models constitute examples of how systems with similar computational properties to the brain can give rise to a set of observed behaviours. Sometimes, individual units are taken to represent pools of neurons or cell assemblies. According to this interpretation the activation level of the units corresponds to the proportion of neurons firing in the pool (e.g. Changeux & Dehaene, 1989).

CONNECTIONIST MODELS OF COGNITIVE DEVELOPMENT

Infancy provides an excellent opportunity to model self-organizing processes because behaviour is closely tied to perceptual-motor skills. We begin this section by describing two models of infant cognitive development and then turn to describing self-organizing models of children's cognitive development.

Many infant categorization tasks rely on preferential looking or habituation techniques, based on the finding that infants direct more attention to unfamiliar or unexpected stimuli (Reznick & Kagan, 1984). Connectionist autoencoder networks have been used to model the relation between sustained attention and representation construction (Mareschal & French, 1997, 2000; Mareschal *et al.*, 2001 Schafer & Mareschal, 2001). An autoencoder is a feed-forward connectionist network with a single layer of hidden units (Figure 1b). It is called an autoencoder because it associates an input with itself. The network learns to reproduce on the output units the pattern of activation across the input units. The successive cycles of training in the autoencoder are an iterative process by which a reliable internal representation of the input is developed.

This approach to modelling novelty preference assumes that infant looking times are positively correlated with the network error. The greater the error, the longer the looking time, because it takes more training cycles to reduce the error. The degree to which error (looking time) increases on presentation of a novel object depends on the similarity between the novel object and the familiar object. Presenting a series of similar objects leads to a progressive error drop on future similar objects.

An unusual asymmetry has been observed in natural category formation in infants (Quinn *et al.*, 1993; Quinn & Eimas, 1996). When 3–4-month-olds are initially exposed to a series of pictures of cats, they will form a category of cat that excludes dogs. We can tell this because, after habituating to a series of pictures of cats, they nevertheless show a novelty response to a picture of a dog. However, when they are exposed to a series of pictures of dogs, they will form a category of dog that does include cats. Thus, there would be no novelty response to a picture of a cat after a series of dogs. We used the autoencoder network above to explain this behaviour. The original cat and dog pictures were measured along 10 dimensions and presented to the networks for categorization. The same presentation procedure was used as with the infants. These networks developed CAT and DOG categories with the same exclusivity asymmetry as the 3–4-month-olds. Moreover, the model predicted that learning DOG after CAT would disrupt the prior learning of CAT whereas learning CAT after DOG would not disrupt the prior learning of CATS. A subsequent study with 3–4-month-olds found this also to be true of infants (Mareschal *et al.*, 2001).

The asymmetry was explained in terms of the distribution of cat and dog feature values in the stimuli presented to the infants. Most cat values fell within the range of dog values but the converse was not true. Thus for a system that processes the statistical distribution of features of a stimulus, the cats would appear as a subset of the dog category. Further analyses revealed that these networks could parse the world into distinct categories according to the correlation of feature values in the same way as 10-month-olds have been shown to do (Younger, 1985; Mareschal & French, 2000). In short, this model demonstrates that categorical representations can self-organize in a neural system as a result of exposure to familiarization exemplars.

We now turn to discussing object-directed behaviours in infancy. Newborns possess sophisticated object-oriented perceptual skills (Slater, 1995) but the age at which infants are able to reason about hidden objects remains unclear. Using *manual search* to test infants' understanding of hidden objects, Piaget concluded it was not until 7.5–9 months that infants can understand that an object continues to exist beyond direct perception. He concluded this because infants younger than this age fail to reach successfully for an object hidden behind an occluding screen (Piaget, 1952, 1954). More recent studies using a violation of expectancy paradigm have suggested that infants as young as 3.5 months do have some understanding of hidden objects. These studies rely on non-search indices such as surprise instead of manual retrieval to assess infant knowledge (e.g. Baillargeon *et al.*, 1985; Baillargeon, 1993). Infants *watch* an event in which some physical property of a hidden object is violated (e.g. solidity). Surprise at this violation (as measured by increased visual inspection of the event) is interpreted as showing that the infants know: (a) that the hidden object still exists, and (b) that the hidden object maintains the physical property that was violated (Baillargeon, 1993). The nature and origins of this developmental lag between understanding the continued existence of a hidden object and searching for it remains a central question of infant cognitive development.

An initial attempt to account for this behaviour in terms of the principles of self-organization was put forward by Munakata *et al.* (1997). These authors reported on a connectionist model that learned to keep track of the potential reappearance of a hidden object. The model experienced a number of events in which a screen would move past a stationary object, thereby hiding the object. With experience the model learned to predict when the object would reappear. Knowledge about the hidden object was gradually encoded in the network connection weights in response to experience with the environment. Similar accounts of infant performance on delayed response tasks that involved reaching for a hidden object, accounts also based on the interaction and competition between neural systems, have been proposed (Dehaene & Changeux, 1989; Munakata, 1998).

Mareschal *et al.* (1999) describe a model also designed to address this question (Figure 2). This model embodies more neurophysiological constraints than the previous models. Anatomical, neurophysiological, and psychophysical evidence points to the existence of two processing routes for visual object information in the cortex (Ungerleider & Mishkin,1982; Van Essen *et al.*, 1992; Goodale, 1993; Milner & Goodale, 1995). Although the exact functionality of the two routes remains a hotly debated question, it is generally accepted that they contain radically different kinds of representations. The dorsal (or parietal) route processes spatial–temporal object information, whereas the ventral (or temporal) route processes object feature information.

The Mareschal *et al.* model is more complex than the simple autoencoder networks described above. Rather than drawing individual units (as in Figure 1), each box represents a layer of units and each arrow represents a full set of connections between successive layers. The dotted lines delimit separate modules. Information enters the networks via a simplified retina. The object recognition network develops a spatially invariant feature-based representation of the object (cf. the functions of the ventral cortical route) whereas the trajectory prediction network develops a spatial temporal representation of the object (cf. the functions of the dorsal cortical route). The response integration network recruits and coordinates these representations as and when required by an active, voluntary response.

Like infants, the model showed a developmental lag between expectancy and retrieval. Active tasks such as retrieval of a desired hidden object required the integration of information across the multiple object representations, whereas surprise or dishabituation tasks may require access to only one of the representations separately. A developmental lag appeared between retrieval and surprise-based tasks because of the added cognitive demands of accessing two object representations simultaneously in an active response task. The model predicted that dishabituation tasks requiring infants to access cortically separable representations would also show a development lag, as compared to tasks that required access to only one cortical representation.

We now turn to considering models of cognitive development in children rather than infants. The recent revival in connectionist modelling of cognitive development was triggered by the need to provide a mechanistic account of Piagetian cognitive development. Piaget described the motors of development in terms of assimilation (that new information is changed to match existing knowledge better), accommodation (that existing knowledge is changed to match new information better), and equilibration (a combination of assimilation and accommodation, Piaget, 1977). However, it was never entirely clear what these terms referred to (Boden, 1994). The first connectionist models of children's cognitive development attempted to couch connectionist modelling in terms of assimilation and accommodations (Plunkett & Sinha, 1992). So, for example, the pattern of activation produced by the presentation of an input to a network was described as assimilation, while weight adjustment in response to an error signal was seen as accommodation (McClelland, 1989). In an alternative account, weight adjustments were seen as assimilative learning and the addition of new

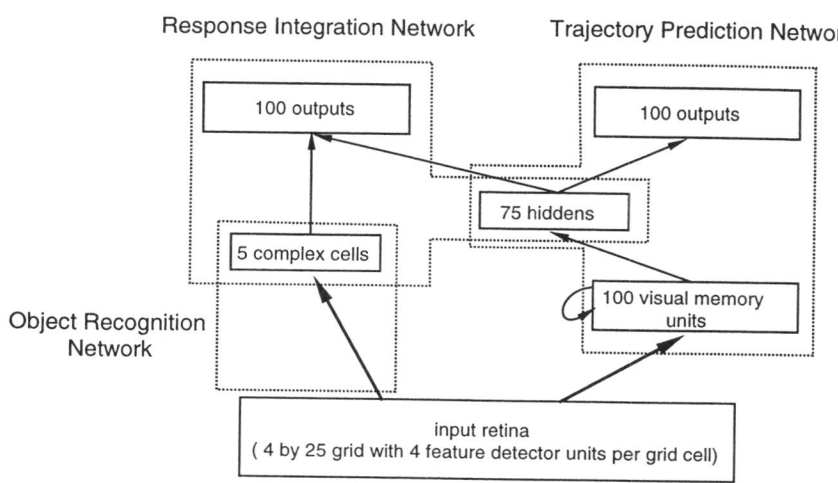

Figure 2 Schema of Mareschal, Plunkett, & Harris, (1999) object processing model.

units (the growth of the networks) was seen as accommodation (Mareschal, 1991; Shultz *et al.*, 1995). As a result of the focus on Piagetian development many of the early models tried to capture the trademark of Piagetian theory: stage-like development.

Perhaps the first connectionist model of cognitive development was Papert's (1963) Genetron model. Papert used a series of simple neural networks (perceptrons) to model how children learn to assess which of two sticks is the longest. He found that, like the children, perceptrons would focus on one end of the stick before considering both ends in conjunction. In his model, development was viewed as the hierarchical integration of successive perceptron modules.

A classic post-Piagetian problem-solving task on which children show stage-like development is Siegler's (1976) balance-scale task. Siegler examined children's developing abilities to predict which side of a balance scale would tip when a set of weights was positioned at varying distances on either side of the fulcrum. Detailed analyses of behaviours across a battery of balance-scale problems revealed that children progressed through four stages in the development of balance scale understanding. In stage 1 children always chose the side with the most weights as the one that would go down. In stage 2 children would attend predominantly to the weight dimension, but if there were equal weights on both sides they would chose the side with the greatest distance as the one to go down. In stage 3 children would succeed on weight problems, but would guess in conflict problems where there were more weights on one side but greater distance on the other side. Finally, in stage 4 children would solve all problems by implicitly computing the torque on each side of the fulcrum.

Performance on this task has been modelled by presenting feed-forward connectionist networks with a series of balance scale problems (McClelland & Jenkins, 1991; Shultz *et al.*, 1995). Networks were presented with weight and distance information for the left and right sides of the fulcrum, and had to learn to predict which way the balance scale would tip. The networks developed through a series of stages analogous to the children. Development was accounted for in terms of the self-organizing properties of weight adjustment and the possible addition of new units to the network. The macroscopic behaviour of stage development emerged as a result of fine-grained, local adjustments to the weights. An important assumption of these models was that the children's learning environment was biased towards learning from problems where weight was the important dimension. That is, the dimension of weight was made more important in the training set. Thus, self-organization in these networks reflected the distribution of problems that children learn from.

Mareschal & Shultz (1999) describe a connectionist model of the development of children's seriation (or sorting) abilities. Seriation is a task originally developed by Piaget (1965) to probe children's developing transitive reasoning skills. Piaget identified four stages of development. In stage 1 young children made no effort to sort a series of sticks when asked to do so. In stage 2 they would group sticks according to local size comparisons, but were unable to extend this order to the entire array. Their sorted series consisted of small groups of consistently sorted sticks but would not extend over the entire set of sticks. In stage 3 children successfully sorted the entire array of sticks, but used a slow and fastidious empirical trial-and-error strategy. Finally, in stage 4, children quickly and efficiently sorted the entire series by systematically choosing the smallest un-ordered stick at each move.

Mareschal and Shultz modelled this task by decomposing the seriation task into the two subtasks. These were: (1) identifying *where* a stick should be moved to and (2) identifying *which* stick need to be selected. Each task was learnt by a distinct connectionist module. There was no initial architectural difference between modules that biased a network to learn one task better than the other task. However, one module received feedback about the correct stick to move and the other module received feedback about the correct location that a stick needed to be moved to. In response to this feedback the two modules self-organized differently, and in such a way as to encode the order relationship that ultimately enabled them to accomplish their individual subtasks. During this process the whole network progressed through the same behavioural stages (described above) as do the children. Again, the macroscopic stage-wise developmental profiles could be attributed to fine-grained changes in the connection weights between units in the sub-networks. Also of interest was the fact that the observed level of competence of a given network resulted from the interacting competence of the two modules developing at their own distinct rates. Observed sorting errors could be explained in terms of an interaction between errors on selecting a stick and errors in placing a stick in a partially ordered series.

In continuing with the general trend to model Piagetian tasks, Shultz (1998) described a connectionist model of conservation behaviours in children. Conservation is the understanding that some quantities are preserved during some perceptual transformations (e.g. the total amount of clay is preserved when a lump of clay is stretched out). The model shows a shift from attending to perceptual information to transformation information as a result of self-organization in the network.

Tasks on which children's development appears not to show stages have also been modelled. Sirois & Shultz (1998) described a model of discrimination shift learning in children. Discrimination shift tasks represent a basic form of concept learning in which the subjects must shift the criteria for categorization during the task. Manipulations of the amount of experience the networks had with the task simulated the developmental phenomena. This suggests that human developmental differences in shift learning arise from spontaneous over-adaptation in older participants.

Connectionist models have also been used to investigate processes of early language development (see Plunkett, 1997, for a review). These include models of early phonological development in learning to categorize speech sounds (Nakisa & Plunkett, 1998; Schafer & Mareschal, 2001), learning to segment the speech stream into discrete words (Christiansen *et al.*, 1998), vocabulary development (Gasser & Smith, 1998; Plunkett *et al.*, 1992), the acquisition of inflectional morphology, for instance in forming the past tense of verbs (Daugherty & Seidenberg, 1992; Plunkett & Juola, 1999; MacWhinney & Leinbach, 1991; Plunkett & Marchman, 1991, 1993, 1996; Rumelhart & McClelland, 1986) and in forming the plural of nouns (Forrester & Plunkett, 1994; Juola & Plunkett, 1999; Plunkett & Nakisa, 1997), and models of the acquisition of syntax (Elman, 1990, 1992, 1993).

In all of these models an initial network structure is postulated by the modeller, but the connection weights are randomized, so that the model has no initial content. The model is then exposed to language input of some sort, and required to adjust its weights, either to form concise representations with an unsupervised algorithm (e.g. Nakisa & Plunkett's (1998) model of early phonological development) or to learn a mapping between two domains using a supervised algorithm (e.g. in Christiansen *et al.*'s (1998) model the network had to learn a mapping between the current phoneme and the next phoneme in the speech stream; in Rumelhart & McClelland's (1986) model the network had to learn a mapping between the stem and the past tense form of verbs). During training the network weights self-organize to reflect the structure of the language task. Thus in the lexical segmentation model, Christiansen *et al.* showed how the model could learn to combine probabilistic cues about phonemes, relative lexical stress, and boundaries between utterances, to predict word boundaries. Each of these cues on its own is an unreliable indicator, yet after training on a large corpus of child-directed speech the model was able to use these cues to reliably identify word boundaries.

These models have been successful in showing how distributional information in the child's language input can be very useful in helping the child to learn language. It is the latent structure in the input that allows these self-organizing models to form structured representations appropriate to the linguistic task (Redington & Chater, 1998). The fact that such information can be extracted from the linguistic environment has led to a reconsideration of the degree to which aspects of language must be innately coded into the developing language system (Elman *et al.*, 1996). However, the difficulty of extracting information which is not readily apparent in the language input, such as underlying grammatical structure, continues to drive claims that significant aspects of our knowledge of the structure of language are innately specified (see e.g. Chomsky, 1986, 1988, 1995; Pinker, 1994, 1999). This remains a very active area of research (see e.g. Henderson & Lane, 1998, for recent work on the acquisition of grammatical structures in connectionist networks).

In summary connectionist models provide concrete examples of how the processes of self-organization can account for many aspects of behavioural change during cognitive development. Moreover, connectionist models are loosely based on the principles of neural information processing and thereby provide a means of linking behavioural development with neural or brain development.

CONNECTIONIST MODELS OF DEVELOPMENT DISORDERS

The ability of self-organizing connectionist models to capture changes during development also presents us with the opportunity to investigate *disorders* of development. Connectionist models have been used to explore a range of developmental disorders, including autism (Cohen, 1994, 1998; Gustafsson, 1997), developmental dyslexia (Harm & Seidenberg, 1999; Manis *et al.*, 1996; Plaut *et al.*, 1996; Seidenberg & McClelland, 1989; Zorzi *et al.*, 1998), Specific Language Impairment (Hoeffner & McClelland, 1993), the development of morphology in a damaged language system (Marchman, 1993), and language processing in Williams syndrome (Thomas & Karmiloff-Smith, 2001a).

Connectionist models of normal development carry with them a number of assumptions about constraints (boundary conditions). These include: (1) the initial state of the network, in terms of the number of units, layers, connections, and the pattern of connectivity (collectively known as the network architecture); (2) the way a particular cognitive problem is presented to the network, in terms of the input and output representations; (3) the dynamics of activation changes in the network; (4) the learning algorithm that the network will use to change its connection weights (and potentially, its architecture); and (5) the regime of training which the network will undergo.

Models of developmental disorders have assumed that one or more of these elements is atypical in the disordered system. In particular the following initial changes have been proposed to simulate each disorder: "too few" hidden units (developmental dyslexia, autism), "too many" hidden units (autism, Williams syndrome), "too much" lateral inhibition in the output layers of self-organizing feature maps (autism), alterations in input and output representations (developmental dyslexia, Williams syndrome, specific language impairment), and elimination of intermediate layers of units (developmental dyslexia, Williams syndrome). We will return later to consider what constitutes "too few" "too many", or "too much". For the time being the key aspect to note with these models is that, with the exception of Marchman (1993), who explored the ability of a network to recover from damage at different points in development, these models all postulate differences in the system *prior to training*. That is, the system undergoes a process of development under a set of constraints that is different right from the start. As a consequence the developmental profile that we find in these systems may be qualitatively different from the normal case, rather than simply being a delayed version, or a version of development which is normal but terminates at an earlier stage. As we shall see, connectionist models provide us with a new type of vocabulary to consider the nature of this disordered development.

The application of connectionist modelling methods to developmental disorders is still very new. However, enough work has been done to outline the potential contribution of this approach, as well as its possible weaknesses. In the following section we will consider one example in more detail, that of autism.

Autism is a developmental disorder characterized by a central triad of deficits. These deficits are in social interaction, communication, and imagination (Wing & Gould, 1979). In addition, other features have been associated with the syndrome. These include a restricted repertoire of interests, an obsessive desire for sameness, savant abilities, excellent rote memory, a preoccupation with parts of objects, improved perceptual discrimination, and an impaired ability to form abstractions or generalize knowledge to new situations (see Happé, 1994, for a review). Autism has multiple biological causes (Gillberg & Coleman, 1992), and a disposition to the disorder is probably inherited (Simonoff *et al.*, 1998). Autism appears to have multiple behavioural subgroups, so that it is perhaps best characterized as a spectrum of disorders rather than a single disorder (Wing, 1988).

Cohen (1994) focused on the pattern of improved perceptual discrimination and impaired generalization found in the syndrome. He concluded that evidence from neuropathological investigations of the brains of affected individuals was suggestive of abnormal wiring patterns in various brain regions, perhaps caused by deficits in neuronal migration during fetal development (Piven *et al.*, 1990), by

755

curtailment of normal neuronal growth (Bauman, 1991), and/or by aberrant development (Courchesne *et al.*, 1993). In comparison with the normal brain, in Cohen's view, the structural deficits were consistent with too few neurons in some areas, such as the cerebellum, and too many neurons in other areas, such as the amygdala and hippocampus (see also Bauman, 1999). Cohen took a simple three-layer feed-forward network and trained it on a classification task using *Backpropagation*. The network had two output units for the binary classification task, and 11 inputs units on which to present exemplars of the two classes. Cohen explored the effect of varying the number of units in the hidden layer between 1 and 13, first on the ability of the network to learn the classification in the training set, and secondly on its ability to generalize this classification to a new test set.

Cohen found that, with one hidden unit, performance on both the training set (examples previously presented to the network) and generalization set (novel examples) was poor. With 13 hidden units the network quickly and accurately learnt the classification for the training set. Early in training generalization was also good, but with further training generalization then declined, as the network was subject to overlearning. In overlearning the network's representations become increasingly focused on details of the training set. Cohen suggested that his networks captured different characteristics found in autistic subjects. In some cases autistic children have trouble acquiring simple discriminations and attend to a restricted range of stimuli (corresponding to too few neurons). In others, autistic children have good discrimination, and indeed very good memory, but this relies on representing too many unique details of stimuli, details that subsequently interfere with generalization (corresponding to too many neurons).

This study illustrates some of the advantages and disadvantages of using connectionist networks to model disordered development. Let us begin with the disadvantages. Using a modeler's rule of thumb, Cohen decides that a network with three hidden units will be the "normal" case (p. 12). Less than this and learning suffers, greater than this and generalization suffers. But where does the rule of thumb come from? Well, connectionist modellers have traditionally attempted to set the number of hidden units precisely to achieve optimal learning on the training set *and* optimal generalization. Cohen can provide us with no independent justification for the "normal" set of parameters in the network; and this holds true for all of the models mentioned above. Thus far, we simply do not have solid grounds for saying why some set-up of architecture, representations, and training regime should represent the "normal" case for development – other than that it seems to capture the normal pattern of development. On the other hand it is worthwhile to show that a single model can account for normal development using one set of parameters, and atypical development using a second set. This suggests we may be on the right track. Nevertheless, work remains to be done to ground these parameter settings. In Cohen's case this might be done by pursuing the hypotheses that certain patterns of neural connectivity are a sufficient condition for autism, and that variation in this connectivity can account for the various subgroups of autism.

On the plus side, Cohen's model does generate some interesting new hypotheses. One of these is that the apparent regression in functioning that some autistic children show between 18 and 30 months of age could be explained by over-learning in systems which have too many neurons. The second is that behavioural therapy may be effective because its training regimes weaken previously established, idiosyncratic connections in the networks which have too many neurons, and strengthen connections for knowledge that forms a better basis for generalization. These are useful unanticipated hypotheses that have arisen from the modelling process.

Gustafsson (1997) followed a similar line of thought in proposing that the capacity of autistic individuals to form representations of sensory experience may be impaired. Gustafsson suggested that their cognitive impairment arose in the development of "cortical feature maps". These maps are thought to develop in systems using unsupervised learning, where the aim is for a set of processing units to self-organize such that they form a concise representation of the information they receive. This concise representation may then be used to drive higher-order processes. In Gustafsson's view the formation of such feature maps is damaged in autism due to too much lateral inhibition in this layer of units. The idea can be explained as follows. A simple, unsupervised network has two connected layers of units, the input units and the output units. On a given learning trial the output units compete with each other to become the one that will represent the pattern of activation arriving from the input units. Through exposure to many different patterns, areas of neighbouring output units come to specialize in representing certain patterns.

Lateral inhibition is the process that mediates the competition between the output units. Now, if there is "too much" competition between these units, the output units may come to represent very precise details of particular input patterns. For example, instead of a unit coming to represent a book, it might represent the book on a shelf at a certain angle with another book next to it. While such fine-grained information would be useful in classification tasks (such as spotting the book on the shelf), this context-bound information is a poor basis for generalization (e.g. recognizing the book over a range of situations). In relation to autism, fine-grained information would explain good perceptual discrimination abilities. The inability to form abstractions of sensory data might explain impairments to higher-order processes, since such processes would have to rely on "raw data" or root memory (Hermelin & O'Connor, 1970).

Although Gustafsson did not run any computer simulations, Oliver et al. (2000) have examined the ways in which just such a process of feature map formation can be disrupted by changes in the initial properties of a self-organizing connectionist network. Moreover, these authors did so in an attempt to develop a more explicit conceptual framework in which to consider developmental disorders. Oliver et al. employed a model based on that proposed by Miller et al. (1989) to explain the emergence of ocular dominance columns in the primary visual cortex of mammals. Ocular dominance columns are sets of neurons tuned to respond to information coming from only one eye. Initially, neurons in this area of visual cortex are not specialized, responding to both eyes. However, through a process of self-organization, the sheet of neurons segregates itself into strips that respond only to one eye or the other. These strips of neurons supply information that can be used in subsequent computations involved in stereovision. Miller et al.'s neural network model had an output layer which received input from two retinas. In the output layer, output neurons competed with each other to respond to input patterns. Miller et al. showed how this

757

network, constructed using biologically plausible principles, could self-organize its output layer to segregate into areas responding only to input from one or other of the retinas.

Similarly, Oliver *et al.* employed a two-layer network, whereby an output layer of 30×30 units received information from a single input retina of 30×30 units. The network was shown a set of four stimuli, in the form of bars lying across the input retina. The bars had a certain similarity structure, whereby A was similar to B, B was similar to C, and C was similar to D. Oliver *et al.* showed that, using their initial parameter set, the output layer formed a topographic map of the possible inputs. That is, certain areas of the output layer specialized in responding to input A, other areas specialized in responding to input B, other areas to C, and so on. Moreover, the similarity structure of the input was reflected in the topographic map: the area responding to A was next to that responding to B, the area for B was next to that for C, and so on.

Oliver *et al.* then went on to disrupt the network in various different ways. In each case they performed the manipulation *prior to* exposing the network to the training stimuli. They varied the threshold of the output units, disrupted the connectivity between the input and output layers, disrupted the connectivity responsible for lateral inhibition in output layer, and changed the similarity of the input stimuli to each other. These manipulations demonstrated that small initial differences in the constraints under which the model developed could have significant effects on the outcome of development. The resulting topographic map suffered a range of disruptions, including output units failing to specialize at all or simply turning off, specialization emerging but not in organized areas, and organized areas emerging but without adjacent areas representing similar-looking bars.

Three points are of particular note here. First, loss of function in the final performance of the network was not due to removing specific parts of the network responsible for that function. Final performance was the outcome of a developmental process. Second, changes in the initial network conditions altered the trajectory of development that the network followed. Disordered development was *qualitatively* different.

Oliver *et al.*'s model demonstrates that Gustafsson's conjecture is a plausible one. Changes in factors such as the initial levels of lateral inhibition in the output layer may well disrupt the featural representations subsequently formed by the network, and indirectly, any processes that use these representations. This leads us to the third point. The outcome of this differential development may not simply be a generalized impairment, but a pattern of *strengths and weaknesses*. The representations may be changed in ways that are good for some tasks but bad for others. Thus, context-bound features will support superior discrimination, but will lead to inferior generalization. Even abilities that seem unimpaired (sometimes referred to as "intact" or "spared" abilities) may, as a consequence of disordered development, be achieved by qualitatively different cognitive processes (Karmiloff-Smith, 1998).

This explanation of autism is exciting in that it attempts to build bridges between levels of description, from its basis in biological principles through computational modelling to behavioural outcome. However, the explanation is somewhat limited. Both Cohen and Gustafsson's models tackle associated perceptual impairments of the disorder; but neither addresses the central triad of features that is generally taken to characterize the syndrome – deficits in socialization, communication, and imagination. It is far from clear that their biologically constrained network models can be straightforwardly extended to account for deficits in, say, play behaviour (Thomas, 2000). Currently, when we extend connectionist principles to investigating high-level behaviour of this sort, these principles serve less as biologically motivated constraints and more as an extended metaphor, a set of conceptual tools to formulate new ideas that themselves must be empirically tested (Thomas & Stone, 1998; see Cohen, 1998, for an example of such conceptualization). Perhaps *it is* the case that context-bound sensory feature maps impair imagination, socialization, and so on. If so, we must derive testable hypotheses and collect empirical support for this proposal. For example, we might identify a particular social task, hypothesize how such a task might be achieved using "social" features, and then predict how context-bound social features would lead to certain kinds of deficits in this task. Support for a hypothesis of this type would strengthen Cohen and Gustafsson's respective claims that something akin to learning with too many neurons or learning with unusual levels of lateral inhibition could account for the range of deficits associated with autism.

It is likely that the application of connectionist modelling to developmental disorders will have at least two implications for this field. The first will be to favour dynamic rather than static accounts of selective deficits. The second will be to force clarification of the relationship of individual differences within normal development to atypical development.

With regard to the first point it is important to note that, in recent times, it has been common to characterize developmental disorders as if they were equivalent to cases of adult brain damage – as if they represented a normal cognitive system largely "intact" but for the loss of some specialized functional module. This is nicely illustrated by one of the leading explanations of the triad of deficits in autism, the "theory of mind" hypothesis (Baron-Cohen et al., 1985). Under this hypothesis autistic people lack the specific ability to understand and reason about mental states. For instance, they would not be able to understand other people's behaviours in terms of their (potentially false) beliefs and desires. This inability to "think about thoughts" leads to specific impairments in social skills, communication, and imagination. Leslie (1987) has suggested that a particular, discrete functional component of the cognitive system might be responsible for this ability (the "Theory of Mind Module"). Moreover, Baron-Cohen suggests that this module is innately built into the human mind (e.g. Baron-Cohen, 1998). Autism, in this view, is the consequence of highly selective damage to one component of an otherwise normally developing cognitive system. Similar explanations of behavioural deficits in developmental disorders can also be found for Specific Language Impairment and for Williams syndrome (Clahsen & Almazan, 1998; Pinker, 1991, 1994, 1999).

As discussion of Oliver et al.'s (2000) model demonstrates, connectionist accounts of developmental disorders place much greater emphasis on the process of development itself. That is, the selective behavioural deficits found in the end-state of an atypically developing system are the outcome of a developmental process which, due to differing initial constraints, has followed an alternative trajectory. The system is qualitatively different, not a normal system with some elements missing. This view is consistent with recent criticisms of the static adult brain damage account of developmental disorders (Bishop, 1997a; Karmiloff-Smith, 1998). The adult brain damage account is feasible only if one of two rather unlikely assumptions turns out to be true. These are either that the computational modules comprising the cognitive system develop independently, or that the content of these modules is static and innately specified (as implied by Baron-Cohen, 1998). However, Bishop (1997b) has argued that interactivity rather than independence is the hallmark of early development; and a number of researchers have argued that functional modules are the outcome of a developmental process, rather than being a precursor to it (Elman *et al.*, 1996; Karmiloff-Smith, 1992).

Indeed connectionist modelling thus far tends to lend support to the idea that damage to a system prior to development has a quite different effect from damage of the end-state system (Elman *et al.*, 1996; Thomas, 2000; Thomas & Karmiloff-Smith, 2001b). The same kind of pre-training network manipulations that were used to model developmental disorders have also been used *after network training is complete*, but now to capture a range of *acquired* disorders, including Alzheimer's disease, Parkinson's disease, epilepsy, phantom limbs, stroke, frontal lobe damage, prosopagnosia, semantic memory deficits, acquired dyslexia, alexia, schizophrenia, and unipolar depression (see Reggia *et al.*, 1996, for a selection of such models. See Marchman, 1993, for an investigation of the effect of lesioning a network while development is still under way.)

The second implication of connectionist models for the study of developmental disorders is with regard to the relation of atypical development to individual differences. McLeod *et al.* (1998) have suggested that connectionist models may be able to capture individual differences between people's cognitive processes by varying network parameters such as learning rate, number of hidden units, and initial weight configurations. Given that the same types of manipulations have been used to capture disordered development, this suggests that (albeit implicitly) connectionist theorists support the view that developmental disorders and individual differences lie on a continuum. However, this view has been strongly contested. Developmental disorders are frequently thought to involve organic damage, and there is evidence that atypical development is qualitatively different from the development found at the lower end of the normal distribution of typical development (Bennett-Gates & Zigler, 1998). Work remains to be done to explore the extent to which initial differences in constraints overlap for typical and atypical groups in development, and the extent to which these constraints are different.

DISCUSSION AND FUTURE DIRECTIONS

This chapter has illustrated how models can be used to investigate self-organization in cognitive development. When artificial neural network models are used, developmental phenomena can be related to processes of self-organization in the brain. This offers one way of addressing the central question of developmental cognitive neuroscience (Johnson, 1997); namely, the relation between brain development and cognitive development. The models also provide a means of exploring the cause of abnormal development in terms of divergent self-organization. A number of outstanding questions still need to be addressed.

The biological plausibility of connectionist models has been questioned (O'Reilly, 1998). Although these models are functionally equivalent to some of the features of neural information processing, many other processes exist in the brain. These other processes (e.g. diffusion of neural modulators) may radically affect the self-organizing properties of these systems. Such processes must be incorporated in any future modelling work. One example is synaptogenesis, the creation of new synapses. Until recently, synaptogenesis was not thought to play an important role in later brain development. However, Quartz & Sejnowski (1997) have argued that synaptogenesis and arborization (the development of new dendritic trees on which new synapses may be formed) play an important part in increasing the computational power of natural neural networks. Although these ideas have been incorporated in some models of cognitive development (see Mareschal & Shultz, 1996, for a review) they still remain relatively rare. The realization that synaptogenesis may play an important role in some aspects of cognitive development begs the question of which processes are best accounted for in terms of synaptogenesis and which are best accounted for in terms of self-organization in systems with fixed numbers of synapses. Shultz & Mareschal (1997) provide one tentative answer by suggesting that abilities that emerge early in development, and that are culturally invariant, should be modelled with static network models, whereas later developing and culturally specific abilities should be modelled with a synaptic growth process.

Many of the models described above speak of cognitive modules. These are components dedicated to processing certain types of information. While there is no denying that modularization exists both in the brain and in cognition (Karmiloff-Smith, 1992), the origin of these modules is still unresolved. Again, the modules may be pre-wired, or they may emerge though self-organizing processes such as competition (e.g. Jacobs, 1999; Jacobs *et al.*, 1991).

Throughout this chapter we have restricted our review to a single class of connectionist models (see Grossberg, 1982, for an example of another class of models). The appropriateness of this class of model for capturing self-organization in cognitive development has been brought into question. Raijmaker *et al.* (1996) have suggested that existing connectionist models do not capture dynamic aspects of self-organization sufficiently well. They suggested that temporal oscillations in fully connected recurrent networks are the appropriate markers of self-organization in neural systems (see also Molenaar, 1986).

Finally, the models described in this chapter make no attempt to account for explicit or metacognitive knowledge; that is, knowledge that can be verbalized and that can be subjected to syntactic logical operations. Because of this shortcoming it has been suggested that connectionist networks cannot reveal anything of interest about cognitive development (Marcus, 1998). What this issue really raises is the degree to which explicit and metacognitive knowledge is important in the observed development of behaviour. It also suggests that modelling work needs to be done in order to understand how the brain might instantiate these abilities, and how they might emerge as self-organizing processes.

References

Baillargeon, R. (1993). The object concept revisited: new directions in the investigation of infants' physical knowledge. In C.E. Granrud (Ed.), *Visual perception and cognition in infancy* (pp. 265–315). London: LEA.

Baillargeon, R., Spelke, E.S. & Wasserman, S. (1985). Object permanence in 5-month-old infants. *Cognition*, **20**, 191–208.

Baron-Cohen, S. (1998). Modularity in developmental cognitive neuropsychology: evidence from autism and Gilles de la Tourette syndrome. In J.A. Burack, R.M. Hodapp & E. Zigler (Eds.), *Handbook of mental retardation and development* (pp. 334–348). Cambridge, UK: Cambridge University Press.

Baron-Cohen, S. (2001). Does the study of autism justify minimalist innate modularity? *Learning and Individual Differences (In press)*.

Baron-Cohen, S., Leslie, A.M. & Frith, U. (1985). Does the autistic child have a "theory of mind"? *Cognition*, **21**, 37–46.

Bauman, M. (1991). Microscopic neuroanatomic abnormalities in autism. *Pediatrics*, **87**, 791–796.

Bauman, M. (1999). Autism: clinical features and neurobiological observations. In H. Tager-Flusberg (Ed.), *Neurodevelopmental disorders* (pp. 383–399). Cambridge, MA: MIT Press.

Bennett-Gates, H. & Zigler, E. (1998). Resolving the development-difference debate: an evaluation of the triarchic and systems theory models. In J.A. Burack, R.M. Hodapp & E. Zigler (Eds.), *Handbook of mental retardation and development* (pp. 115–131). Cambridge, UK: Cambridge University Press.

Bishop, D.V.M. (1997a). Cognitive neuropsychology and developmental disorders: uncomfortable bedfellows. *Quarterly Journal of Experimental Psychology*, **50A**, 899–923.

Bishop, D.V.M. (1997b). *Uncommon understanding: development and disorders of language comprehension in children*. Hove, UK: Psychology Press.

Boden, M.A. (1982). Is equilibration important? A view from artificial intelligence. *British Journal of Psychology*, **73**, 65–173.

Boden, M.A. (1994). *Piaget*. London, UK: Fontana Press.

Chomsky, N. (1986). *Knowledge of language: its nature, origin and use*. New York: Praeger.

Chomsky, N. (1988). *Language and problems of knowledge: the Managua Lectures*. Cambridge, MA: MIT Press.

Chomsky, N. (1995). *The minimalist program*. Cambridge, MA: MIT Press.

Changeux, J.P. & Dehaene, S. (1989). Neuronal models of cognitive function. *Cognition*, **33**, 63–109.

Changeux, J.P., Heidman, T. & Patte, P. (1984). Learning by selection. In P. Marler & H.S. Terrace (Eds.), *The biology of learning* (pp. 115–133). Berlin: Springer-Verlag.

Christiansen, M.H., Allen, J. & Seidenberg, M.S. (1998). Learning to segment speech using multiple cues: a connectionist model. *Language and Cognitive Processes*, **13**, 221–268.

Clahsen, H. & Almazan, M. (1998). Syntax and morphology in Williams syndrome. *Cognition*, **68**, 167–198.

Cohen, I.L. (1994). An artificial neural network analogue of learning in autism. *Biological Psychiatry*, **36**, 5–20.

Cohen, I.L. (1998). Neural network analysis of learning in autism. In D.J. Stein & J. Ludik (Eds.), *Neural networks and psychopathology* (pp. 274–315). Cambridge: Cambridge University Press.

Courchesne, E., Press, G.A. & Yeung-Courchesne, R. (1993). Parietal lobe abnormalities detected with MR in patients with infantile autism. *American Journal of Roentgenology*, **160**, 387–393.

Cybenko, G. (1989). Approximation by superpositions of a sigmoidal function. *Mathematics of Control, Signals, and Systems*, **2**, 303–314.

Daugherty, K. & Seidenberg, M.S. (1992). Rules or connections? The past tense revisited. In *Proceedings of the Fourteenth Annual Conference of the Cognitive Science Society* (pp. 259–264). Hillsdale, NJ: Lawrence Erlbaum.

Dehaene, S. & Changeux, J.P. (1989). A simple model of prefrontal cortex function in delayed-response tasks. *Journal of Cognitive Neuroscience*, **1**, 244–261.

Edelman, G.M. & Finkel, L.H. (1985). Neuronal group selection in the cerebral cortex. In G.M. Edelman, W.E. Gall & W.M. Cowan (Eds.), *Dynamic aspects of neocortical function* (pp. 635–695). New York: Wiley-Interscience.

Elman, J.L. (1990). Finding structure in time. *Cognitive Science*, **14**, 179–211.

Elman, J.L. (1992). Grammatical structure and distributed representations. In S. Davis (Ed.), *Connectionism theory and practice* (pp. 139–173). Oxford, UK: Oxford University Press.

Elman, J.L. (1993). Learning and development in neural networks: the importance of starting small. *Cognition*, **48**, 71–99.

Elman, J.L., Bates, E.A., Johnson, M.H., Karmiloff-Smith, A., Parisi, D. & Plunkett, K. (1996). *Rethinking innateness: a connectionist perspective on development*. Cambridge, MA: MIT Press.

Forrester, N. & Plunkett, K. (1994). Learning the Arabic plural: the case for minority default mappings in connectionist networks. In A. Ramand & K. Eiselt (Eds.), *Proceedings of the Sixteenth Annual Conference of the Cognitive Science Society* (pp. 319–323). Mahwah, NJ: Erlbaum.

Gasser, M. & Smith, L.B. (1998). Learning nouns and adjectives: a connectionist account. *Language and Cognitive Processes*, **13**, 269–306.

Gillberg, C. & Coleman, M. (1992). *The biology of the autistic syndromes* (2nd ed.). London: MacKeith.

Goldfield, E.C. (1995). *Emergent forms. Origins and early development of human action and perception*. Oxford: Oxford University Press.

Goodale, M.A. (1993). Visual pathways supporting perception and action in the primate cerebral cortex. *Current Opinion in Neurobiology*, **3**, 578–585.

Grossberg, S. (1982). How does the brain build a cognitive code? *Psychological Review*, **87**, 1–51.

Gustafsson, L. (1997). Inadequate cortical feature maps: a neural circuit theory of autism. *Biological Psychiatry*, **42**, 1138–1147.

Happé, F. G. E. (1994). *Autism: an introduction to psychological theory*. London: UCL Press.

Harm, M.W. & Seidenberg, M.S. (1999). Phonology, reading acquisition, and dyslexia: insights from connectionist models. *Psychological Review*, **106**, 491–528.

Hebb. D. (1949). *The organization of behaviour*. New York: Wiley.

Henderson, J. & Lane, P. (1998). A connectionist architecture for learning to parse. In *Proceedings of COLING-ACL'98*, pp. 531–537.

Hertz, J., Krogh, A. & Palmer, R.G. (1991). *Introduction to the theory of neural computation*. Redwood City, CA: Addison-Wesley.

Hermelin, B. & O'Connor, N. (1970). *Psychological experiments with autistic children*. Oxford: Pergamon Press.

Hoeffner, J.H. & McClelland, J.L. (1993). Can a perceptual processing deficit explain the impairment of inflectional morphology in developmental dysphasia? A computational investigation. In E.V. Clark (Ed.), *Proceedings of the 25th Child Language Research Forum* (pp. 38–49). Stanford, CA: Stanford University Press.

Jacobs, R.A. (1999). Computational studies of the development of functionally specialized neural modules. *Trends in Cognitive Science*, **3**, 31–38.

Jacobs, R.A., Jordan, M.I. & Barto, A.G. (1991). Task decomposition through competition in a modular connectionist architecture: the what and where vision tasks. *Cognitive Science*, **15**, 219–250.

Johnson, M.H. (1997). *Developmental cognitive neuroscience*. Oxford, UK: Blackwell.

Karmiloff-Smith, A. (1985). Language and cognitive processes from a developmental perspective. *Language and Cognitive Processes*, **1**, 60–85.

Karmiloff-Smith, A. (1990). Constraints on representational change: evidence from children's drawing. *Cognition*, **34**, 57–83.

Karmiloff-Smith, A. (1992). *Beyond modularity: a developmental perspective on cognitive science*. Cambridge, MA: MIT Press.

Karmiloff-Smith, A. (1998). Development itself is the key to understanding developmental disorders. *Trends in Cognitive Sciences*, **2**, 389–398.

Karmiloff-Smith, A. & Inhelder, B. (1971). If you want to get ahead, get a theory. *Cognition*, **3**, 195–212.

Karmiloff-Smith, A. & Thomas, M.S.C. (2000). Developmental disorders. In M.A. Arbib & P.H. Arbib (Eds.), *Handbook of brain theory and neural networks* (2nd ed.). Cambridge, MA: MIT Press (In press).

Keslo, S. (1995). *Dynamic pattern: the self-organization of brain and behaviour*. Cambridge, MA: MIT Press.

Leslie, A.M. (1987). Pretence and representation: the origins of "theory of mind". *Psychological Review*, **94**, 412–426.

MacWhinney, B. & Leinbach, J. (1991). Implementations are not conceptualizations: revising the verb learning model. *Cognition*, **40**, 121–157.

763

Manis, F.R., Seidenberg, M.S., Doi, L.M., McBride-Chang, C. & Peterson, A. (1996). On the bases of two subtypes of developmental dyslexia. *Cognition*, **58**, 157–195.

Marchman, V. (1993). Constraints on plasticity in a connectionist model of the English past tense. *Journal of Cognitive Neuroscience*, **5**, 215–234.

Marcus, G. (1998). Rethinking eliminative connectionism. *Cognitive Psychology*, **37**, 243–282.

Mareschal, D. (1991). Cascade-correlation and the Genetron: possible implementations of equilibration. Technical Report 91-10-17. McGill Cognitive Science Center, McGill University, Montreal, Canada.

Mareschal, D. & French, R. M. (1997). A connectionist account of interference effects in early infant memory and categorization. In M.G. Shafto & P. Langley (Eds.), *Proceedings of the 19th annual conference of the Cognitive Science Society* (pp. 484–489). Mahwah, NJ: Erlbaum, Mahwah.

Mareschal, D. & French, R.M. (2000). Mechanisms of categorization in infancy. *Infancy*, **1**, 111–122.

Mareschal, D. & Shultz, T. R. (1999). Development of children's seriation: a connectionist approach. *Connection Science*, **11**, 153–188

Mareschal, D. & Shultz, T.R. (1996). Generative connectionist architectures and constructivist cognitive development. *Cognitive Development*, **11**, 59–88.

Mareschal, D., Plunkett, K. & Harris, P. (1999). A computational and neuropsychological account of object-oriented behaviours in infancy. *Developmental Science*, **2**, 306–317.

Mareschal, D., French, R.M. & Quinn, P.C. (2001). A connectionist account of asymmetric category learning in early infancy. (Submitted).

McClelland, J.L. (1989). Parallel distributed processing: implications for cognition and development. In R.G.M. Morris (Ed.), *Parallel distributed processing: implications for psychology and neurobiology* (pp. 9–45). Oxford, UK: Oxford University Press.

McClelland, J.L. & Jenkins, E. (1991). Nature, nurture, and connections: Implications of connectionist models for cognitive development. In K. van Lehn (Ed.), *Architectures for intelligence* (pp. 41–73). Hillsdale, NJ: LEA.

McLeod, P., Plunkett, K. & Rolls, E.T. (1998). *Introduction to connectionist modeling of cognitive processes*. Oxford: Oxford University Press.

Miller, K.D., Keller, J.B. & Stryker, M.P. (1989). Ocular dominance column development: analysis and simulation. *Science*, **245**, 605–615.

Milner, A.D. & Goodale, M.A. (1995). *The visual brain in action*. Oxford: Oxford University Press.

Molenaar, P.C.M. (1986). On the impossibility of acquiring more powerful structures: a neglected alternative. *Human Development*, **29**, 245–251.

Munakata, Y. (1998). Infants perseveration and implication for object permanence theories: a PDP model of the AB task. *Developmental Science*, **1**, 161–211.

Munakata, Y., McClelland, J.L., Johnson, M.N. & Siegler, R.S. (1997). Rethinking infant knowledge: towards an adaptive process account of successes and failures in object permanence tasks. *Psychological Review*, **104**, 686–713.

Nakisa, R.C. & Plunkett, K. (1998). Innately guided learning by a neural network: the case of featural representation of speech language. *Language and Cognitive Processes*, **13**, 105–128.

Oliver, A., Johnson, M.H., Karmiloff-Smith, A. & Pennington, B. (2000). Deviations in the emergence of representations: a neuroconstructivist framework for analysing developmental disorders. *Developmental Science*, **3**, 1–23.

O'Reilly, R.C. (1998). Six principles for biologically based computational models of cortical cognition. *Trends in Cognitive Sciences*, **2**, 455–462.

Oyama, S (1985). *The ontogeny of information*. Cambridge, UK: Cambridge University Press.

Papert, S. (1963). Intelligence chez l'enfant et chez le robot. *Etudes D'Epistemologie Génétiques*, **15**, 131–194.

Piaget, J. (1952). *The origins of intelligence in the child*. New York: International Universities Press.

Piaget, J. (1954). *The construction of reality in the child*. New York: Basic Books.

Piaget, J. (1965). *The child's concept of number*. New York: Norton Library.

Piaget, J. (1971). *Biology and knowledge*. Edinburgh: Edinburgh University Press.

Piaget, J. (1977). *The development of thought: equilibration of cognitive structures*. Oxford: Blackwell.

Piaget, J. (1980). *Adaptation and intelligence. Organic selection and phenocopy*. Chicago, IL: Chicago University Press.

Pinker, S. (1991). Rules of language. *Science*, **253**, 530–535.

Pinker, S. (1994). *The language instinct: the new science of language and mind*. New York: Penguin.

Pinker, S. (1999). *Words and rules*. London: Weidenfeld & Nicolson.

Piven, J., Berthier, M.L., Starkstein, S.E., Nehme, E., Pearlson, G. & Folstein, S. (1990). Magnetic resonance imaging evidence for a defect of cerebral cortical development in autism. *American Journal of Psychiatry*, **147**, 734–739.

Plaut, D.C., McClelland, J.L., Seidenberg, M.S. & Patterson, K.E. (1996). Understanding normal and impaired word reading: computational principles in quasi-regular domains. *Psychological Review*, **103**, 56–115.

Plunkett, K. (1997). Theories of early language acquisition. *Trends in Cognitive Sciences*, **1**, 146–153.

Plunkett, K. & Juola, P. (1999). A connectionist model of English past tense and plural morphology. *Cognitive Science*, **23**, 463–490.

Plunkett, K. & Marchman, V. (1991). U-shaped learning and frequency effects in a multi-layered perceptron: implications for child language acquisition. *Cognition*, **38**, 1–60.

Plunkett, K. & Marchman, V. (1993). From rote learning to system building: acquiring verb morphology in children and connectionist nets. *Cognition*, **48**, 21–69.

Plunkett, K. & Marchman, V. (1996). Learning from a connectionist model of the acquisition of the English past tense. *Cognition*, **61**, 299–308.

Plunkett, K. & Nakisa, R.C. (1997). A connectionist model of the Arabic plural system. *Language and Cognitive Processes*, **12**, 807–836.

Plunkett, K. & Sinha, C. (1992). Connectionism and developmental theory. *British Journal of Developmental Psychology*, **10**, 209–254.

Plunkett, K., Sinha, C., Moeller, M.F. & Strandsby, O. (1992). Vocabulary growth in children and a connectionist net. *Connection Science*, **4**, 293–312.

Prigogine, I. & Stengers, I. (1986). *Order out of chaos*. London: Flamingo Press.

Quartz, S.R. & Sejnowski, T.J. (1997). The neural basis of cognitive development: a constructivist manifesto. *Behavioural and Brain Sciences*, **20**, 537–596.

Quinn, P.C. & Eimas, P.D. (1996). Perceptual organization and categorization in young infants. *Advances in Infancy Research*, **10**, 1–36.

Quinn, P.C., Eimas, P.D. & Rosenkrantz, S.L. (1993). Evidence for representations of perceptually similar natural categories by 3-month-old and 4-month-old infants. *Perception*, **22**, 463–475.

Raijmaker, M.E.J., van Koten, S. & Molenaar, P.C.M. (1996). On the validity of simulating stagewise development by means of PDP networks: application of catastrophe analysis and an experimental test of rule-like network performance. *Cognitive Science*, **20**, 101–136.

Redington, M. & Chater, N. (1998). Connectionist and statistical approaches to language acquisition: a distributional perspective. *Language and Cognitive Processes*, **13**, 129–192.

Reggia, J.A., Ruppin, E. & Berndt, R.S. (Eds.) (1996). *Neural modeling of brain and cognitive disorders*. London: World Scientific.

Reznick, J.S. & Kagan, J. (1983). Category detection in infancy. *Advances in Infancy Research*, **2**, 79–111.

Rumelhart, D.E. & McClelland, J.L. (1986). *Parallel Distributed Processing*, vol. 1. Cambridge, MA: MIT Press.

Rumelhart, D.E. & McClelland, J.L. (1986). On learning the past tense of English verbs. In J.L. McClelland, D.E. Rumelhart & the PDP Research Group (Eds.), *Parallel distributed processing: explorations in the microstructure of cognition*, vol. 2: *Psychological and Biological Models* (pp. 216–271). Cambridge, MA: MIT Press.

Rumelhart, D.E., Hinton, G.E. & Williams, R.J. (1986). Learning representations by back-propagating errors. *Nature*, **323**, 533–536.

Schafer, G. & Mareschal, D. (2001). Qualitative shifts in behaviour without qualitative shifts in processing: the case of speech sound discrimination and word learning in infancy. *Infancy*, **1** (In press).

Seidenberg, M.S. & McClelland, J.L. (1989). A distributed, developmental model of word recognition and naming. *Psychological Review*, **96**, 452–477.

Shultz, T.R. (1998). A computational analysis of conservation. *Developmental Science*, **1**, 103–126.

Shultz, T. R. & Mareschal, D. (1997). Rethinking innateness, learning, and constructivism. *Cognitive Development*, **12**, 563–586.

Shultz, T.R., Mareschal, D. & Schmidt, W.C. (1994). Modeling cognitive development on balance scale phenomena. *Machine Learning*, **16**, 57–86.

Shultz, T.R., Schmidt, W.C., Buckingham, D. & Mareschal, D. (1995). Modeling cognitive development with a generative connectionist algorithm. In T.J. Simon & G.S. Halford (Eds.), *Developing cognitive competence: new approaches to process modeling* (pp. 205–261). Hillsdale, NJ: Erlbaum.

Siegler, R.S. (1976). Three aspects of cognitive development. *Cognitive Psychology*, **8**, 481–520.

Simonoff, E., Bolton, P. & Rutter, M. (1998). Genetic perspectives on mental retardation. In J.A. Burack, R.M. Hodapp & E. Zigler (Eds.), *Handbook of mental retardation and development* (pp. 41–79). Cambridge, UK: Cambridge University Press.

Sirois, S. & Shultz, T.R. (1998). Neural network modeling of developmental effects in discrimination shifts. *Journal of Experimental Child Psychology*, **71**, 235–274.

Slater, A. (1995). Visual perception and memory at birth. *Advances in Infancy Research*, **9**, 107–162.

Thomas, M.S.C. (2000). Neuroconstructivism's promise. Commentary on Oliver *et al. Developmental Science*, **3**, 35–37.

Thomas, M.S.C. & Karmiloff-Smith, A. (2001a). Modelling language acquisition in atypical phenotypes. Manuscript submitted for publication.

Thomas, M.S.C. & Karmiloff-Smith, A. (2001b). Are developmental disorders like cases of adult brain damage? Implications from connectionist modelling. Manuscript submitted for publication.

Thomas, M.S.C. & Stone, A. (1998). "Cognitive" connectionist models are just models, and connectionism is a progressive research programme. Commentary on Green on Connectionist Explanation. *Psycoloquy*, **9**, http://www.princeton.edu/sharnad/psyc.html.

Ungerleider, L.G. & Mishkin, M. (1982). Two cortical visual systems. In D.J. Ingle, M.A. Goodale & R.J.W. Mansfield (Eds.), *Analysis of visual behaviour* (pp. 549–586). Cambridge, MA: MIT Press.

van der Maas, H.L.C. & Molenaar, P.C.M. (1992). Stagewise cognitive development: an application of catastrophe theory. *Psychological Review*, **99**, 395–417.

van Essen, D.C., Anderson, C.H. & Felleman, D.J. (1992). Information processing in the primate visual system: an integrated systems perspective. *Science*, **255**, 419–423.

van Geert, P. (1991). A dynamic system model of cognitive and language growth. *Psychological Review*, **98**, 3–53.

von der Malsburg, C. (1995). Self-organization in the brain. In M.A. Arbib (Ed.), *The handbook of brain theory and neural networks* (pp. 375–381). Cambridge, MA: MIT Press.

Willshaw, D.J. & von der Malsburg, C. (1976). How patterned neural connections can be set up by self-organization. *Proceedings of the Royal Society of London, B*, **194**, 431–445.

Wing, L. (1988). The continuum of autistic characteristics. In E. Schopler & G.B. Mesibov (Eds.), *Diagnosis and assessment in autism* (pp. 91–110). New York: Plenum Press.

Wing, L. & Gould, J. (1979). Severe impairments of social interaction and associated abnormalities in children: epidemiology and classification. *Journal of Autism and Developmental Disorders*, **9**, 11–29.

Younger, B.A. (1985). The segregation of items into categories by ten-month-old infants. *Child Development*, **56**, 1574–1583.

Zorzi, M., Houghton, G. & Butterworth, B. (1998). Two routes or one in reading aloud? A connectionist dual-process model. *Journal of Experimental Psychology: Human Perception and Performance*, **24**, 1131–1161.

Further reading

Elman, J.L., Bates, E.A., Johnson, M.H., Karmiloff-Smith, A., Parisi, D. & Plunkett, K. (1996). *Rethinking innateness: a connectionist perspective on development*. Cambridge, MA: MIT Press.

Karmiloff-Smith, A. (1998). Development itself is the key to understanding developmental disorders. *Trends in Cognitive Science*, **2**, 389–398.

Keslo, S. (1995). *Dynamic pattern: the self-organization of brain and behaviour*. Cambridge, MA: MIT Press.

Nelson, C.A. & Luciana, M. (2001). *Handbook of developmental cognitive neuroscience*. Cambridge, MA: MIT Press.

Siegler, R.S. (1997). *Children's thinking* (3rd ed.). London: Prentice Hall.

Thelen, E. & Smith, L. (1994). *A dynamic systems approach to the development of cognition and action*. Cambridge, MA: MIT Press.

IV.8
The Prefrontal Cortex and the Development of Executive Function in Childhood

MARILYN C. WELSH

ABSTRACT

The purpose of this chapter is to draw parallels between the maturation of the prefrontal cortex (PFC) and the development of a cluster of behaviours known as executive function (EF). Although several indices of neurobiological development (e.g. myelination, synaptogenesis) suggest a relatively late and protracted course of PFC development, there exists substantial debate regarding the precise nature of the developmental trajectory due to the variety of methodologies employed. A theory of cyclic cortical reorganization proposed by Thatcher (e.g. 1992, 1994a, 1997) posits that the frontal cortex is instrumental in guiding successive "waves" of synapse formation and elimination throughout the cortex to progressively shape the brain to meet changing environmental demands. Thatcher's analysis of EEG coherence data gathered from children and adolescents indicate three main cycles of cortical development and reorganization from 18 months to 5 years, 5 years to 10 years, and 10 years to 14 years. For the purposes of this chapter the conceptualization of EF includes the components of planning, working memory, inhibition, and flexible set shifting, and the degree to which the normal development of EF conforms to Thatcher's three cycles was explored. Again, the variety of operational definitions of EF and the diversity of methodologies complicate the integration of this literature; however, some interesting quantitative and qualitative changes in EF were identified that are consistent with the three cycles of cortical reorganization.

INTRODUCTION

The objective of this chapter is to review research that provides the most current understanding of the development of the prefrontal cortex (PFC), and the

A.F. Kalverboer and A. Gramsbergen (eds.), Handbook of Brain and Behaviour in Human Development, 767–790
© 2001 Kluwer Academic Publishers. Printed in Great Britain.

potential connection of this neurobiological process to behavioural development during childhood. Such an enterprise is complicated by a number of factors. Research exploring the development of prefrontal cortical structure and function has utilized a range of populations (e.g. human and non-human primates), diverse indices of development (e.g. synaptogenesis, myelination, EEG coherence), and a variety of methodologies (e.g. different types of histological, morphological, electrophysiolgical, and neuroradiological techniques). Tracking the development of the behaviours mediated by the PFC is challenging given the lack of consensus regarding how to operationally define, and therefore measure, these behaviours. Even if these empirical literatures were consistent, clear, and convergent; drawing parallels between the neurobiological and developmental literatures can be problematic. As Huttenlocher and Dabholker (1997) pointed out, "cortical anatomy *per se* reveals little concerning function" (p. 69).

For the purposes of this chapter the working definition of those behaviours controlled by the PFC of developing children will be the construct of "executive function (EF)". Most researchers and theorists have referred to prefrontal cortical functions by the umbrella term EF (e.g. Denckla, 1996; Welsh & Pennington, 1988); however, the exact nature of this cognitive construct in terms of component processes and the relations among such processes is the subject of much debate (e.g. Eslinger *et al.*, 1997). This chapter will open with a discussion of current definitions of EF, which will be integrated into a framework for the component skills that can be explored developmentally. In the second section, the development of the prefrontal region of the cortex will be described, following the authoritative work of Fuster (1997). Particular attention will be given to the debate regarding whether the development of this cortical region is or is not protracted over the childhood years, as compared to other cortical areas. A potential means of reconciling these two positions on the neurobiological development of the PFC is to consider the possibility of cycles of cortical development, and the model proposed by Thatcher (1994a,b, 1997) will be reviewed in the third section. Thatcher's model of cyclic cortical reorganization has identified three main cycles or stages of cortical development in which the frontal region plays a prominent role, and these stages roughly conform to early childhood, middle childhood, and late childhood/early adolescence. The fourth section of this chapter explores whether the current empirical research suggests quantitative and/or qualitative changes in EFs that would be consistent with Thatcher's putative stages of prefrontal development. Finally, in the fifth section, conclusions that can be preliminarily drawn from this integration of diverse neurobiological and behavioural literatures will be discussed, as well as directions for future research.

DESCRIPTION OF THE EF DOMAIN

EF is a domain of processing that is required for future-oriented, goal-directed behaviour; it is generally assumed to include such processes as set maintenance, planning, inhibition, and flexibility (Welsh & Pennington, 1988) presumably mediated by the PFC (Stuss & Benson, 1986). There appears to be no argument with the position that EF is necessary (but not sufficient) for successful, adaptive behaviour, whether you speak to a parent of a 2-year-old, a teacher struggling to convey the process of coherent composition, or a neurologist treating a patient with recent frontal cortical damage. Ironically, as important as EF appears to be to adaptive behaviour, there is still substantial debate regarding the most basic issue: exactly what *is* EF, and how can one best measure the behaviours reflecting this construct across the lifespan? At present there is no commonly accepted definition of EF with regard to its core, underlying processes and the covariation among these processes. This dilemma is reflected in the following statement, "The lack of progress in research on EF may well be due to the failure to establish shared meaning in psychological, neuropsychological, and educational realms of inquiry" (Borkowski & Burke, 1996, p. 224).

As Denckla (1996) pointed out, if one reads the current neuropsychological literature it is virtually impossible to decouple the term EF from its putative physical substrate, the PFC of the brain. Tranel *et al.* (1995) wrote: "it must be acknowledged that the capacities subsumed by executive functions have been linked to the prefrontal region throughout the entire history of neuroscience, and to some extent, the psychology and anatomy are inseparable". It is important to note here that the frontal cortex comprises over one-third of human cortical tissue, and is composed of a variety of anatomical components, each with specialized cortical and subcortical connections (Damasio & Anderson, 1993). Rather than linking EF to the frontal lobe *en masse*, neuroscientific research suggests that it is more accurate to attribute executive processes to the PFC, and even more specifically to two regions within the PFC: dorsolateral and orbital–frontal (Iversen & Dunnett, 1990; cited in Barkley, 1997).

The clinical neuropsychological literature is replete with detailed case histories of individuals who suffered focal frontal damage and who subsequently were impaired in their future-oriented, goal-directed behaviour. For example, Luria (1966) described frontal-damaged patients who lacked self-initiated, purposeful behaviour that would necessitate the skills of anticipation, planning, and monitoring. More recently, Stuss & Benson (1986) described the cognitive sequelae following damage to the PFC, including: distraction by irrelevant stimuli; inability to flexibly switch mental sets (i.e. perseveration); failure to initiate appropriate activity; problems maintaining effort over time; inability to use feedback; and failure to plan and organize activity to attain goals. Other processes attributed to the PFC based on human lesion studies include: appreciation of context (Fuster, 1989), novel problem solving (Duncan, 1995; Duncan *et al.*, 1995), and "memory for the future" (Ingvar, 1985). Fuster (1985, 1997) has developed a model of PFC function based on his studies of single-cell recording of electrical activity during performance on a delayed-response task, a classic frontal cortical measure. He has attributed three functions to the PFC: (1) a temporally retrospective memory function of working memory, (2) a temporally prospective function of anticipatory set or planning, and (3) an interference-control mechanism that suppresses behaviour that is incompatible with the current goal. Seminal clinical observations by Lezak (1983, 1995), Luria (1973), and Stuss & Benson (1986) have led to frameworks of EF that converge on four similar "factors", recently pointed out by Goldstein & Green (1995): (1) motivation, initiation, anticipation or volition; (2) task analysis, goal formation, and planning; (3) purposive action, implementing or carrying out the plan; and (4) self-monitoring and evaluating the effectiveness of one's performance. It is important to note that, for the most part, these clinical descriptions that formed the basis of our current definitions of EF have concerned adult patients. It is still very much an empirical question whether such characterizations of frontal or executive processes are appropriate for children of different ages.

Denckla (1996) recently argued for a shift in focus from EF as "supraordinate", "higher-order", or "meta" processes such as task analysis (Borkowski & Burke, 1996) to a characterization of the core, cognitive processes involved. One potential difficulty with defining EF in terms of complex cognitive processes is that this operationalization will be less appropriate for the investigation of EF processes in very young children and infants. To illustrate this point, for decades frontal lobe function was defined as performance on the Wisconsin Card Sorting Test (WCST; e.g. Milner, 1964) in the field of clinical neuropsychology. This task presumably requires EF skills such as inhibition and flexibility; however, additional complex cognitive processes, including concept formation and hypothesis testing, are also relevant. Poor performance on the WCST led to the misguided conclusion often cited in the neuropsychological literature that the PFC is nonfunctional until age 10–12 years (e.g. Golden, 1981). In light of new information a more accurate conclusion appears to be that the WCST, and other similar neuropsychological measures of frontal function, were inappropriate for assessing EF in young children because of the additional cognitive demands. Denckla's description of EF involves central control processes that allow one to manipulate mental representations of relevant information during the delay between the stimulus and the response, which necessitates control of incoming interference and modulation of outgoing responses. A dual focus on working memory and inhibition as the key processes mediated by the PFC has been articulated recently by Pennington and colleagues (Pennington, 1994; Pennington *et al.*, 1995; Roberts & Pennington, 1996), and this appears to be a reasonable way to explore this domain of functioning in the youngest individuals. For example, Diamond and colleagues (e.g. Diamond *et al.*, 1994) assess EF in infants by means of Piaget's A-not-B object permanence task, a close analogue to the delayed-response tasks found to be sensitive to prefrontal function in non-human primates (Jacobsen, 1935). Performance on this task requires the maintenance of information across the delay and inhibition of interfering stimuli and maladaptive responses. One contemporary view of EF that has enjoyed recent empirical support is the perspective that essentially all of the cognitive and behavioural sequelae of PFC damage can be explained as a dysfunction of the working memory system. Consistent with the (Kimberg & Farah, 1993; Kimberg, D'Esposito & Farah, 1998) view of prefrontal cortical function, Goldman-Rakic and colleagues (see Funashi *et al.*, 1989) have identified a specific class of neurons in the principal sulcus region of the PFC that fire selectively during the delay period in a delayed-response task; that is, these neuronal networks appear to be mediating the mental representation of the stimulus in working memory.

The preceding review should make it clear that there exists no universally accepted definition of EFs. The theories and models described above have evolved from research with individuals of various ages, and it is not clear whether a somewhat different definition of EF is needed depending on the developmental period under investigation. In order to be able to trace the nature of prefrontally mediated behaviours across development, one would want to identify some core, cognitive "primitives" that would be as relevant to the behaviour of a toddler as they would be to the behaviour of an adolescent. Whether a single cognitive component, such as working memory (e.g. Kimberg & Farah, 1993), is sufficient to completely describe the nature of prefrontal function awaits further investigation. A broader view of PFC functions offered by Grafman (1999) arose from his synthesis of a variety of conceptual frameworks from Luria to Stuss & Benson, and such functions include supervisory attention, contextual memory, planning, reasoning, social cognition, and decision making. Similarly, Eslinger (1996; Eslinger *et al.*, 1997) defined EF as the mechanisms that provide control, organization, and direction to behaviour for the purposes of achieving either short-term or long-term goals. Although working memory would be one necessary mechanism for providing such control, organization, and direction, under certain complex and demanding circumstances working memory alone would not be sufficient to accomplish the future-oriented task (Welsh, 2001). Given that a purpose of this chapter is to track the development of EF across the childhood years, the definition of this domain will be expanded to include the following component processes: working memory, inhibition, planning and flexibility. In the next section the development of the cortical region presumed to mediate these EF processes will be described.

DEVELOPMENT OF THE PFC

Our current understanding of the development of the PFC, both prenatally and postnatally, will be discussed in order to provide a framework for exploring the development of EF, those behaviours presumed to be controlled by this brain region. It is axiomatic that anatomical development must precede functional development. Following from this assumption, neurodevelopmental events, such as dendritic arborization and synaptogenesis, should mark the "minimal age of onset of function in a given neural system" (Huttenlocher & Dabholker, 1997, p. 70). Therefore, it seems reasonable to explore what is known about the unfolding of these neurodevelopmental events in the PFC across childhood and to examine the question, "Do changes in the behaviours assumed to be linked to this cortical region follow a similar developmental progression?" As the following review will illustrate, there is unfortunately a lack of agreement across researchers regarding the time-course of PFC development in the childhood years.

Fuster's seminal volume on the PFC provides the overarching framework for describing the neuroanatomical development of this cortical region (Fuster, 1997). During gestation the PFC develops according to a precisely timed sequence of expansion, cell migration, and lamination, as is the case for the rest of the neo-cortex (e.g. Mrzljak *et al.*, 1998; Rakic, 1974; Sidman, 1974; Uylings *et al.*, 1990; as cited in Fuster, 1997). Cortical cell migration and differentiation occur simultaneously with the arrival of thalamocortical fibres (Marrin-Padilla, 1970; Sidman & Rakic, 1973; as cited in Fuster, 1997). Radial glial cells guide the migration of cells from the germinal zones to their final destinations in the characteristic six cortical layers. In humans the mature laminar structure of the cortex appears to be in place by 7 months gestation and basically complete by birth (Mrzljak *et al.*, 1990; cited in Fuster , 1997). Juraska and Fifkoya (1979) found that the neurons extend their dendritic processes upon their arrival at their respective cortical layers (see Fuster, 1997). Even at birth the dendritic arbors of the neurons comprising the PFC are relatively immature; dendritic branching, and therefore density, increases rapidly from birth through 2 years of age (Fuster, 1997).

The basic cytoarchitectural composition of the PFC appears to be "pre-established" at birth (Fuster, 1997); however, it is important to note that the PFC is one of several large cortical regions without clear anatomical boundaries (Huttenlocher & Dabholker, 1997). It is characterized by a granular layer IV, which indicates that this region has been specified for associative or sensory functions, as opposed to motor behaviours (Huttenlocher & Dabholker, 1997). A histological examination of the entire cerebral cortex reveals an identical vertical columnar organization (Huttenlocher & Dabholker, 1997); however, currently it is believed that differential patterns of postnatal environmental and intracortical inputs result in a fine-tuning and sculpting of the synaptic organization, and perhaps other characteristics, of this brain region (e.g. Goldman-Rakic *et al.*, 1997; Fuster, 1997; Huttenlocher & Dabholker, 1997).

It is the precise nature of this sculpting process, particularly the timing of these neurodevelopmental events across the childhood years, that is the subject of current investigation and debate. For example, Fuster (1997) cites research that indicates the fine-tuning and differentiation of the pyramidal neurons of layer III of the PFC proceeds until puberty (Mrzljak *et al.*, 1990). According to Fuster (1995), cortical layer III appears to be the site of the origin and termination of an extensive network of corticocortical connections, perhaps underlying the formation of associative memories. Fuster (1997) suggests that such a prolonged period of neuronal development in this cortical layer would make sense in light of the profound changes occurring in the realms of cognitive and social development from birth to puberty. Such developmental changes in behaviour would undoubtedly involve the associative processing and sharing of information that demands communication among diverse cortical systems.

Neurodevelopmental events such as myelination and dendritic arborization provide evidence for a protracted period of development of the PFC across childhood. Flechsig's pioneering work in the early 1900s (Flechsig, 1901, 1920; as cited in Fuster, 1997) established a chronological series of myelogenetic stages in which the last brain regions to myelinate are cortical association areas (including the PFC), and specifically layers II and III. Flechsig's early observations have been supported by more recent histological analyses of postmortem brain tissue by Yakovlev & Lecours (1967). As technology has improved, techniques such as magnetic resonance imaging (MRI) have also yielded evidence of later myelination of the prefrontal region of the cortex. In several studies (e.g. Barkovich, 1990; Salamon, 1990; Wolpar & Barnes, 1992; cited in Huttenlocher & Dabholker, 1997), MRI has revealed myelination of the prefrontal white matter occurring last among cortical areas and becoming complete around age 1. Whereas age 1 does not appear to reflect "protracted" development of the PFC, Huttenlocher & Dabholker (1997) suggest that the resolution level of routine MRI may be insufficient to reveal the subtle differences in myelination across cortical areas that might exist across childhood. However, in a more recent study by Paus *et al.* (1999), researchers were able to identify, via MRI, a gradual increase in white matter density in cortical fibre tracts with age. Of particular relevance to this chapter, the maturation of the frontal–temporal pathway showed a protracted course of development up to age 17 years (the oldest age group studied) and primarily in the left hemisphere. It is important to note here that this pathway, the arcuate fasciculus, contains the fibres connecting Wernicke's area of the temporal lobe to Broca's area of the frontal lobe; thus these findings may be more relevant to the development of speech functions than to the development of EF processes mediated by the PFC.

The issue of whether one can assume that the localization of such neurodevelopmental events, such as myelination, can be related directly to the development of specific behaviours (e.g. speech, EF, etc.) is the central question of this review and others like it. Flechsig's early proposals that his myelogenetic stages can be linked to behavioural developments observed from infancy through adolescence were widely criticized in the 1900s (Fuster, 1997). This assumption that the two correlated processes, myelination of the brain and behavioural development, reflect a causal connection remains controversial today (Huttenlocher & Dabholker, 1997). Of course, one cannot conclude a causal relationship between any neurobiological process (e.g. synpatogenesis, dendritic development) and function. One way to explore the potential connection between brain and behaviour development is to examine cases in which neurobiological development is delayed or deviant. For example, Van der Knaap *et al.* (1991; cited in Huttenlocher & Dabholker, 1997) found that delayed myelination in infants was related to developmental delays in behaviour, suggesting a causal relationship between this index of brain development and behaviour.

Another commonly used measure of brain development is dendritic branching. Extensive, albeit early, data collected by Conel (1939–1963; cited in Huttenlocher & Dabholker, 1997) demonstrated later dendritic development on cortical pyramidal neurons in the prefrontal cortex, than in posterior cortical regions. At birth, dendritic arborization in the middle frontal gyrus is very primitive compared with that of the superior temporal gyrus of the auditory cortex. Visual inspection of dendritic branching in these Golgi-stained samples indicated that the developmental lag in the prefrontal cortical neurons was apparent only from 1 to 3 months of age. Newer, quantitative analyses of dendritic development in the PFC have found both progressive (i.e. growth) and regressive (i.e. loss) events to occur beyond this age range (e.g. Becker *et al.*, 1984; as cited in Huttenlocher & Dabholker, 1997) and even up until puberty (Mrzlajak *et al.*, 1990; as cited in Fuster, 1997).

By far the most widely discussed measure of ontogenetic brain changes is synaptogenesis and synapse elimination. It is this index of brain development that has spurred the most controversy regarding the timing of PFC development, as compared to other cortical areas. Research by Huttenlocher in the 1970s examined synaptogenesis and synaptic elimination in layer III of the middle frontal gyrus of human PFC. Huttenlocher (1979) found a rapid increase in synaptic density up to age 1, and a plateau in this density, substantially greater that adult levels, from age 1 to 7 years. Finally, there is a gradual decrease in synaptic density from age 7 to adolescence, at which time adult levels are reached. More recent research in the Huttenlocher laboratory (cited in Huttenlocher & Dabholker,

1997) has compared the time-course of synaptogenesis in primary visual cortex, primary auditory cortex, and PFC (layer II of anterior middle frontal gyrus). Consistent with earlier findings the PFC region exhibited a later peak in synaptic density, at about 3 years, than did the primary sensory areas. There is a gradual decrease in synaptic density after this point; however, an excess number of synaptic contacts (as compared to adults) is maintained throughout the childhood years. Huttenlocher & Dabholker (1997) point out that in some ways this synaptic development is counterintuitive in light of behavioural development. The sharp increase and high point in synaptic contacts occurs at an age when EF processes are only just beginning to manifest in a child's behaviour. In contrast, the protracted period of improvement in executive processes that is apparent across the school age years actually parallels a gradual *decline* in synaptic density in the PFC. It has been proposed by Changeux and colleagues (Changeux & Danchin, 1976; Changeux *et al.*, 1984) that it is precisely this *decrease* in synaptic number that reflects the stabilization of the functional networks underlying advances in cognitive processing, such as EF. In a computer simulation of Changeux's theory of selective stabilization, Edelman (1987) demonstrated that an initially random set of excitatory and inhibitory connections could evolve into a sytematic network of strong connectivity after receiving patterns of repetitive stimulation. Interestingly, these and other "selectionist" theories of neurodevelopment (Quartz & Sejnowski, 1997) focus on the *loss* of non-critical, redundant connections as the key mechanism underlying the development of the functional networks associated with important behavioural developments.

In contrast to Huttenlocher's empirical evidence of later and more protracted synaptic development in the PFC of human cerebral cortex, Goldman-Rakic and colleagues (Goldman-Rakic *et al.*, 1997; Rakic *et al.*, 1986) have found *concurrent* synaptogenesis across a range of cortical areas (including the PFC) in the rhesus monkey. These researchers found a period of rapid increase in synaptic density perinatally, which peaks at about 2 months of age; this high level of synaptic contacts is maintained until about age 3 years (sexual maturation). This pattern of sharp increase and plateau of excess synaptic contacts between 2 months and 3 years for rhesus monkeys conforms to the pattern seen by Huttenlocher in the PFC of humans between 1 year and puberty. However, Goldman-Rakic and colleagues observed this pattern of synaptogenesis in *all* areas of the cortex examined. They have acknowledged that their finding of contemporaneous synaptogenesis across several cortical regions stands in stark contrast to the data from the Huttenlocher laboratory. These contradictory findings could reflect a species difference in cortical development, although Goldman-Rakic *et al.* (1997) suggest that it is unlikely that there would be substantial differences between two closely related primate species. There are also several methodological differences that distinguish the work in the two different laboratories with regard to how the cortical tissue was prepared, synaptic contacts counted, and the data expressed. When Rakic *et al.* (1994) reanalysed and plotted the data from human visual cortex and PFC collected by Huttenlocher and colleagues, they replicated the finding of *concurrent* synaptogenesis observed in the rhesus monkey.

It is clear that the question of the developmental timetable of synaptogenesis in the PFC, as compared with other cortical regions, is far from resolved. Both research groups suggest that their neurodevelopmental data are consistent with evidence of behavioural development of those processes presumed to be mediated by the PFC. Huttenlocher & Dabholker (1997) contend that earlier synaptogenesis in primary visual cortex than in the PFC is consistent with the relatively early emergence of visual abilities (e.g. stereopsis) as compared to the gradual emergence of EF across childhood and well into adolescence. In contrast, Goldman-Rakic *et al.* (1997) maintain that early and concurrent synaptogenesis across the cortex makes sense given behavioural data indicating integrative processing within a functioning working memory system as early as infancy (e.g. Diamond *et al.*, 1994). These discrepant positions are at least partly reflective of the wide range of perspectives on those behaviours controlled by the PFC. If one defines EF in terms of the core, cognitive processes (e.g. working memory and inhibition), then one could argue that these processes can be observed as early in life as some emerging visual skills and develop in parallel with many other behavioural processes in the perceptual, motor, and language domains (Goldman-Rakic *et al.*, 1997). However, if one defines EF as the recruit-ment of a range of cognitive processes (e.g. planning, task analysis, hypothesis testing) to subserve future-oriented, goal-directed behaviour, then a later emergence and more protracted development of EF, as compared to more "basic" cognitive functions, also rings true (e.g. Huttenlocher & Dabholker, 1997). To fan the flames of this debate further, Fuster (1987) has suggested that, even if there is not a developmental lag in synaptogenesis in the PFC, this does not preclude the possibility of a variety of qualitative functional changes in this brain region via electrochemical facilitation of existing synaptic contacts. Therefore, the question of whether synaptogenesis *per se* is the most appropriate marker of the *functional* development of the PFC is still very much open to empirical investigation.

In summary, the picture that emerges regarding the pattern of neurobiological development of the PFC varies depending on the particular index of develop-ment selected, the methodology employed, and the species studied. Whereas the indices of myelination and dendritic branching have shown a later develop-mental "peak" in the PFC, compared with other cortical regions, sometimes this regional difference is observed earlier in ontogeny (e.g. under 1 year of age) than one would suspect if this "later development" was reflective of the emergence of EF. It remains to be seen if more sophisticated methodologies in the future will be effective in revealing subtle regional differences in neurodevelopment across the childhood years (e.g. Paus *et al.*, 1999). With regard to the index of synaptic development and elimination, the question of whether the developmental trajec-tory of the PFC is different (and later) than other regions of human neocortex is unresolved. Goldman-Rakic, Bourgeois and Rakic (1997) has suggested that, as opposed to viewing cortical development as a hierarchical temporal sequence in which "ontogeny recapitulates phylogeny", it is more appropriate to conceptualize the cortex as an interconnected structure that develops as a "whole cloth". As Goldman-Rakic *et al.* (1997) write, "The whole-cloth view of the cortex as a woven tapestry in which the entire piece emerges by progressive addition of threads to all portions simultaneously derives from consideration of the comparative time course of synapse formation and synaptic density in diverse regions of the primate cortex" (pp. 33–34). The next section will explore whether it may be useful to conceptualize the neurobiological development of the PF in terms of *cycles* of development, instead of imposing the assumption of a strict linear progression. In Thatcher's model of cyclic reorganization of cortical development he describes how frontal cortical development may be linked to the structural and functional development of posterior cortical regions, consistent with the "whole-cloth" view and concurrent synaptogenesis proposed by Goldman-Rakic and colleagues. However, given the prominent role given to the frontal cortex, in terms of the "driving force" behind these cyclic patterns of development, Thatcher's model may also be consistent with the Huttenlocher view of the more protracted course of frontal development across the childhood years.

THATCHER'S MODEL OF CYCLIC CORTICAL REORGANIZATION

Thatcher (1991, 1994a,b, 1997) has proposed a theory of cortical development as a series of cycles in which functional subsystems are both differentiated and integrated, and the frontal cortex plays a prominent role in this development. The theory is based on analyses of the development of human electroencephalographic (EEG) coherence, which can be understood as the degree to which there is "phase synchrony" or "shared activity" among spatially separated generators of electrical activity in the brain (Thatcher, 1997). EEG coherence between two cortical regions, for example the frontal cortex and posterior cortex, implies that the two spatially distant regions are both structurally and functionally connected by means of corticocortical connections between neuronal assemblies (Thatcher, 1997). An increase in coherence is presumed to reflect an increase in the number and/or strength of these intracortical connections; whereas a decrease in coherence is thought to be due to a decrease in the number and/or strength of such connections (Thatcher, 1994a). Across child development Thatcher has identified oscillations, or an "ebb and flow", in the pattern of EEG coherence that he has suggested reflect a repetitive pattern of synaptic overproduction followed by synaptic elimination. However, it is important to note that Thatcher has acknowledged a current lack of understanding regarding the precise neurophysiological mechanisms underlying these oscillations in EEG coherence. These mechanisms could include any one or more of the following: synaptogenesis, myelination, axonal sprouting, enlargement of existing synaptic terminals, pruning of synaptic contacts, changes in the amount of presynaptic neurotransmitter and/or in the postsynaptic response (Thatcher, 1994a,b).

Thatcher's model of cyclical cortical development has been based on computerized EEG data collected from 436 children ranging in age from 6 months to 16 years (Thatcher *et al.*, 1987). He has argued that this analysis of EEG coherence data provides a window on the functional differentiation and integration of cortical regions via a repetitive pattern of "growth spurts" embedded within 4-year anatomical cycles that are characterized by phase transitions (Thatcher, 1994b). Cycle I occurs from age 1.5 years to approximately 5 years, cycle II is from approximately 5 years to 10 years, and cycle III lasts from approximately 10 years to 14 years. Each cycle involves an iterative pattern of synaptic overproduction and elimination. A sequential lengthening of intracortical connections between frontal and posterior areas observed in the left hemisphere suggests an integration of previously differentiated subsystems;

whereas a sequential contraction to shorter distance local cortical connections in the right hemisphere implies a differentiation of previously integrated subsystems (Thatcher, 1994a). As Thatcher (1997) writes, these cycles represent "a developmental spiral staircase in which brain structures are periodically revisited, resulting in a stepwise increase in differentiation and integration throughout the postnatal period" (p. 85). The ultimate goal of such cyclic reorganization of the human cortex across development is to gradually sculpt the microanatomy of the brain to progressively adapt the functional neural architecture to the dynamic nature of the demands imposed by a changing environment. In this way Thatcher's view of cortical development could be viewed as consistent with a neural constructivist position (Quartz & Sejnowski, 1997), in which the brain changes both by subtraction (i.e. selecting out redundant connections) and by addition (i.e. building new connections) in order to generate increasingly effective mental representations needed by the developing individual.

An important point for the current chapter is that, in Thatcher's model, the frontal cortex assumes the critical role of controlling, to a great extent, the postnatal selection and pruning of synaptic connections (Thatcher, 1997). Data from Thatcher's laboratory (Thatcher, 1991, 1992) indicate that there is a competitive dynamic between the long-distance frontal connections and the short-distance posterior connections, with the frontal connections essentially "winning the battle". The mean EEG coherence data from Thatcher (1992) has been mathematically modelled and found to best fit a "predator–prey" model of cycles of synaptic overabundance and pruning in which the frontal regions act as "predators" and the posterior cortical areas behave as synaptic "prey" (Thatcher, 1997). According to this model there is a continuous cycling of synaptic overproduction followed by synpatic elimination in both the frontal and posterior cortical regions. However, the nature of the EEG coherence data suggests that the frontal regions are directly in control of the organization and reorganization of posterior cortical regions through the process of selection and pruning of synpatic contacts to meet current environmental demands and contingencies.

The prominence of the frontal cortex in shaping the microanatomy of the brain can be illustrated by Thatcher's description of the microcycles of development that occur within each of the three main cycles (Thatcher, 1994a). For example, cycle I begins at age 1.5 years with no frontal control of posterior cortical regions, but with synaptic number and/or strength of short-distance fibres of the posterior cortex at a peak (i.e. differentiation phase). At age 2.5 years there appears to be emerging frontal control over posterior cortex, as indicated by a decrease in short-distance posterior connections and an increase in long-distance frontal-posterior connections. At about 3.5 years this frontal influence peaks (i.e. integration phase), and subsequently declines. The assumption is that frontal control involves the selection and elimination of superfluous posterior synaptic contacts; i.e. those that are inactive with respect to environmental demands (Changeux et al., 1984; Edelman, 1987). As the supply of local posterior synaptic connections diminishes, so does the need for frontal control or predation. Finally, at the end of this 4-year cycle (about 5.5 years), long-distance frontal–posterior influence reaches a low point; whereas local, short-distance posterior activity increases again (the beginning of a new differentiation cycle). This increase is assumed to reflect a "restocking" of posterior cortical synapses that the frontal cortex can replace or reorganize during the next cycle of development. Therefore one can view each 4-year growth cycle in Thatcher's model as involving four microcycles or stages: (1) development of short-distance synaptic connections in the posterior cortex, (2) emerging frontal cortical influence over these posterior synapses, (3) a peak of frontal influence in which posterior synpatic contacts are selectively pruned, and (4) decreasing frontal control concurrent with increasing development of new posterior connections.

How does Thatcher's model of cyclic cortical reorganization, and the unique role of the frontal cortex in this process, fit with our earlier discussion of the development of the PFC with respect to synaptogenesis (i.e. Goldman-Rakic *et al.*, 1997; Huttenlocher & Dabholker, 1997)? Before addressing this question two important caveats are in order. First, Thatcher refers to the controlling influence of the frontal cortex in these cycles of synaptic overproduction and elimination. Strictly speaking, "frontal cortex" refers to all cortical tissue anterior to the central sulcus (e.g. Damasio & Anderson, 1993), which includes regions such as the primary and supplementary motor strips, Broca's area, and frontal eye fields. Given the nature of Thatcher's index of brain development (EEG coherence), localization to brain regions more precise than "frontal cortex" and "posterior cortex" may be problematic. Whether this predation by the frontal cortex during the brain growth cycles involves the entire frontal cortex or specific regions of the frontal cortex (e.g. PFC) has not been addressed in the theory. Secondly, although the oscillations of EEG coherence appear to reflect the changes in the number and strength of synaptic contacts, the degree to which other neurobiological processes in addition to synaptogenesis (e.g. myelination, neurochemical alterations) contribute to these changes is unknown.

One major difference between both the Goldman-Rakic and Huttenlocher views of frontal cortical development and Thatcher's model of cyclical cortical development relates to the assumed underlying trajectory of neurobiological development. Both Goldman-Rakic and Huttenlocher assume the synaptic overproduction phase involves a more or less linear pattern of increasing number of synaptic contacts (or density), followed by a plateau in which this high number of synapses is maintained, and finally a linearly decreasing number or density of synapses until adult levels are reached in adolescence. Where the positions of Goldman-Rakic and Huttenlocher diverge concerns whether this general linear pattern of synaptogenesis and synaptic elimination occurs concurrently for all cortical areas (Goldman-Rakic) or exhibits different time-courses across child development depending on the cortical region of interest (Huttenlocher). Goldman-Rakic and colleagues have suggested that their data indicating concurrent synpatogenesis are consistent with a "whole-cloth" view of cortical development in which various cortical areas are inextricably interconnected, and thus should develop simultaneously. In contrast, Huttenlocher and colleagues report data that are consistent with the view that many cortical areas follow the characteristic "rise and fall" in synaptic density; however, certain cortical regions (e.g. primary visual cortex) exhibit these phases of synaptic development substantially earlier in development than the prefrontal cortex. A very different view of cortical (and by implication frontal cortical) development is proposed in Thatcher's cyclical reorganization model, antithetical to the position that a single linear trajectory in which the number of synaptic contacts rise and fall best characterizes synaptic development. The analytical modelling of the EEG coherence data collected in Thatcher's laboratory suggests three main cycles of cortical development, and embedded within each cycle are microcycles involving frontal integration of differentiated neuronal systems within posterior cortex. Therefore this model implies frontal cortical influence on brain function and behaviour relatively early in development (i.e., from 18 months) and throughout childhood and adolescence. However, this frontal influence is also likely to manifest differently in behaviour across the childhood years as the frontal cortical connections serve to integrate progressively more differentiated sensory systems.

This distinction between the more "linear" view of cortical development espoused by the Huttenlocher and Goldman-Rakic research groups and the "cyclic" view proposed by Thatcher can be considered within the context of the debate between two perspectives on neural development: selectionism and neural constructivism (Quartz & Sejnowski, 1997). Quartz and Sejnowski argue that selectionist theories focus on the *loss* of redundant connections as the critical mechanism underlying the functional development of the brain; and in this light the protracted, linearly decreasing synaptic density during the formative years of cognitive development makes sense. However, neural constructivist theories suggest both regressive and progressive neurobiological events underlie the functional development of the brain; and in this context a cyclical, progressively increasing "spiral staircase" analogy proposed by Thatcher is appropriate. The debate between the selectionist and neural constructivist views of brain development rages on, and it is far from resolved with respect to a range of behavioural developments. The next section will explore the development of a particular set of behaviours, those that are putatively mediated by the prefrontal cortical region from toddlerhood through early adolescence.

DEVELOPMENT OF EXECUTIVE FUNCTIONS

As previously described, those behaviours presumed to reflect PFC function involve planning, working memory, inhibition, and flexibility; and these are referred to collectively as EF. In light of the current debate regarding the structural and functional development of the PFC, and given Thatcher's provocative model proposing three developmental cycles of cortical development, it begs the question: to what degree, and in what ways, do EF processes change across the childhood years? Is there empirical evidence for "stage-like" development of EF, and do these findings conform to Thatcher's three cycles of development? Are there dramatic shifts (in a quantitative sense) in EF performance from one cycle to the next? What is the nature of EF during these three developmental periods and are there qualitative differences? In the following review the development of EF will be traced across Thatcher's three developmental cycles: cycle I (1.5–5 years), cycle II (5–10 years), and cycle III (10–14 years). Research specifically designed to explore the development of EF across a wide age range is relatively new and, for the most part, has focused on school-aged children. For this reason research that has examined the normal development of component processes of EF, working memory and inhibition/flexibility, will also be discussed. It is important to note that these studies are wide-ranging in terms of the conceptualization

779

of the cognitive process (EF, working memory, inhibition), the experimental tasks used to measure the process, and the age range examined, all of which makes comparisons problematic. However, despite the innumerable conceptual and methodological differences across research groups exploring this phenomenon, some interesting developmental trends emerge (see also, Welsh, 2001).

Studies specifically designed to track the normal development of cognitive processes attributed to PFC function have been a relatively recent phenomenon. The studies generally adopt a common approach: children are tested on tasks selected from either neuropsychology or cognitive psychology because they are presumed to measure one or more of the EF processes described earlier. The authors of these studies begin with a conception of frontal lobe function that derives from the adult clinical neuropsychological literature, and then select or create age-appropriate tasks that might reveal the "child versions" of the same EF skills. Five studies will be discussed that illustrate the above strategy, and that were designed specifically to address the question: what is the normal developmental pattern exhibited by behaviours attributed to the PFC? For the purpose of this chapter particular attention will be paid to whether the developmental transitions in performance found in these studies conform to the three cycles of development proposed by Thatcher.

Passler et al. (1985), and a follow-up study by Becker, et al. (1987), explored the normal development of EF from age 6 through 12 years; EF was operationalized as performance on a battery of tasks that assessed inhibition and flexible set shifting of motor responses and mnemonic representations utilizing both verbal and non-verbal stimuli. The results of these two studies were very consistent, demonstrating that the most striking developmental advances in inhibition and flexibility occurred between 6 and 8 years. Moreover, inhibitory control with respect to mnemonic representations achieved maturity relatively late in development, at about 12 years. Although both studies focused on inhibition, the performance on the recency memory task could be interpreted as reflecting the interaction of working memory and inhibition. That is, the children had to actively maintain temporal-order information in working memory, as well as inhibit inappropriate responses during the recency judgement trials.

Fiducia & O'Leary (1990) examined the normal development of EF in children from 7 to 13 years of age by administering two tasks that required the generation of a working memory representation consisting of task-relevant information, while inhibiting the tendency to include irrelevant, distractor stimuli. Fiducia & O'Leary found that performance on one experimental task improved significantly at two points in development, 7–10 years and 10–13 years, while performance on the other task improved only between the ages of 10 and 13. This evidence of two developmental shifts, one during middle childhood and one during preadolescence, is consistent with the findings of Passler et al. (1985) and Becker et al. (1987). Fiducia and O'Leary note that a significant correlation between the two tasks emerged at age 13, suggesting that a common ability to inhibit the processing of irrelevant stimuli in complex working memory representations may mature at this age. Again, this interpretation is consistent with the Passler et al. (1985) and Becker et al. (1987) finding that performance on tasks placing heavy demands on both working memory and inhibition reached adult-level performance in early adolescence.

In a fourth study designed to explore the normal development of frontal lobe behaviours, Welsh *et al.* (1991) utilized a more wide-ranging battery of EF tasks. Tests were selected from both the neuropsychology and developmental literatures in order to measure EF processes such as planning, organized search, and inhibition. Performance of 100 children ranging in age from 3 to 12 years, and a sample of 10 adults, indicated a multistage pattern of development as observed by the Passler and Becker research groups. The age at which adult-level performance was achieved depended on the nature of the task, in terms of the cognitive processes required and the difficulty level. Tasks that tapped simple planning and organized search of an external stimulus display exhibited adult-level performance as early as age 6. The two tasks that primarily demanded inhibition skills matured by age 10. However, tasks that required more complex planning and the organized search of long-term memory information did not reach adult-level performance by age 12. A recent study by Luciana & Nelson (1999) explored the development of EF in young, normal children ranging in age from 4 to 8 years, comparing their performance on tests of planning, working memory, inhibition and flexibility to that of young adults. Consistent with the findings of Welsh *et al.* (1991), these authors found simple planning and working memory processes emerged by 5 years of age. Similar to the Passler *et al.* (1985) and Becker *et al.* (1987) results, a significant improvement was seen at age 8 years on more complex tasks that increased the demands for planning, inhibition, and flexibility. Moreover, 8-year-olds did not exhibit an adult-level of performance on these more difficult EF tasks, suggesting a later developmental shift or shifts in performance.

Other studies have explored the development of EF skills by examining developmental trends on established neuropsychological measures of frontal lobe function, such as the Wisconsin Card Sorting Test (WCST), the Tower of London (TOL), and the Tower of Hanoi (TOH). The WCST presumably demands EF skills of working memory, inhibition, and flexibility (e.g. Roberts & Pennington, 1996), and several studies have converged on the finding that mature performance, especially with regard to perseveration, is observed at about age 10 years (Chelune & Baer, 1986; Chelune & Thompson, 1987; Levin *et al.*, 1991; Welsh *et al.*, 1991). Disc-transfer problem solving tasks, such as the TOL and TOH, require that a start state is transformed to a goal state via a correct sequence of moves. These tasks typically have been considered to be measures of planning (e.g. Lezak, 1995); however, there is recent evidence that the tasks also demand working memory and inhibition/flexibility processes (Goel & Grafman, 1985; Roberts & Pennington, 1996, Welsh *et al.*, 1999). Research exploring the normal development of performance on the TOL has identified three main developmental shifts: early emergence of skill by about 5 years of age (Luciana & Nelson, 1999), a second spurt in performance between ages 7 and 9 (Anderson *et al.*, 1996; Kepler, 1998; Luciano & Nelson, 1999), and finally a later improvement in skill between 12 and 13 years (Anderson *et al.*, 1996; Levin *et al.*, 1991). The TOH task, though similar in structure to the TOL, correlates only moderately with the latter task (Welsh *et al.*, 1999, 2000) and may demand a somewhat different combination of EF skills. However, like the TOL, Welsh (Welsh *et al.*, 1991; Welsh, 1991) found that skilled performance emerged on easier problems by age 6 years; whereas performance on more difficult problems had not attained adult-level performance by age 12 years.

What common developmental patterns emerge from this review of the literature on EF in children and how do these patterns fit with Thatcher's three cycles of frontal cortical development? One approach to answering this question is to examine the evidence for consistent developmental improvements at the ages that demarcate Thatcher's cycles I and II (about 5 years of age) and cycles II and III (about 10 years of age). Taken together, these studies do indicate a multistage pattern of development with some EF skills displayed by relatively young children and other skills not maturing until 6 or more years later. For example, some simple planning, organized search, spatial working memory and inhibitory control are seen in young children between the ages of 4 and 6 years old (Luciana & Nelson, 1999; Passler *et al.*, 1985; Welsh *et al.*, 1991) which would conform to the transition between Thatcher's cycles I and II. There is substantial development of inhibition and flexible set shifting abilities between 6 and 8 years (Becker *et al.*, 1987; Luciana & Nelson, 1999; Passler *et al.*, 1985) which would be occurring within Thatcher's cycle II. Given that there are "microcycles" of development along both left to right and anterior to posterior gradients within each of Thatcher's main cycles of development, developmental spurts in behaviour might be expected *within* main cycles as well as *between* them. The improvements at age 8 years are followed by a consolidation of sorts in which inhibitory control and flexibility skills on a range of tasks (WCST, MFFT, conflict, perseveration, etc.) attain maturity by age 10 (Anderson *et al.*, 1996; Passler *et al.*, 1985; Welsh *et al.*, 1991); this would mark the end of cycle II and the beginning of cycle III. Finally, tasks that require more complex planning strategies and the manipulation of mnemonic information appear to show an early developmental shift between approximately 7 and 10 years of age (during cycle II), as well as a more protracted course of development extending into adolescence, or cycle III (Anderson *et al.*, 1996; Fiducia & O'Leary, 1990; Kepler, 1998; Levin *et al.*, 1991; Welsh *et al.*, 1991).

A related, but somewhat different, question regarding the development of EF pertains to whether there is evidence for qualitative changes in EF behaviour across the three cycles of development. Because the traditional EF literature typically does not provide adequate information regarding EF skills during the age period characteristic of Thatcher's cycle I (1.5–5 years), it is useful to examine the literature concerning the normal development of working memory and inhibitory processes. With regard to both visual and verbal working memory, research has identified very early emergence of these skills. Children as young as 18–24 months have shown surprisingly good ability on both spatial working memory tasks (Foreman *et al.*, 1990) and verbal working memory (Adams & Gathercole, 1995). Performance on visual working memory span tasks continues to develop during the preschool years, as indicated by developmental improvements observed from 3 to 4 years and from 4 to 5 years by Mutter *et al.* (1999). However, both visual and verbal working memory tasks that involve interference control and inhibition mediated by the central executive (Baddeley, 1992) exhibit developmental improvements in performance across childhood and into adolescence (Case, 1992); that is, development is continuing well into Thatcher's cycles II and III.

As in the case of working memory, an emergence of inhibitory skills also is observed during Thatcher's cycle I, from 1.5 to 5 years of age. It appears that children as young as 18 months show a primitive form of inhibition in self-control tasks (e.g. Kopp, 1982); however, flexible strategies to aid the inhibition of responses are more likely to emerge in the 3–4 year-old range. Using very different experimental paradigms several researchers have observed a developmental shift in inhibitory control from age 3 to age 4 or 5 (Cuneo & Welsh, 1990; Diamond & Taylor, 1996; Zelazo & Frye, 1998). Mutter *et al.* (1999) recently replicated the developmental improvement found from 3 to 4 years of age on the visual search task used by Cuneo & Welsh (1990), a task that demands not only inhibition but also flexible set shifting. The findings of all of these studies dovetail nicely with the seminal observations of Luria (1966) that verbal control over behaviour begins to emerge at about 4 years of age; such control or mediation of behaviour most likely involves working memory, inhibition, and flexible set shifting. The development of inhibition skills on tasks that require the additional demand of cognitive flexibility may be domain- and task-specific, with motor inhibition developing during Thatcher's cycle I and early cycle II and inhibition of irrel-evant linguistic and mnemonic information exhibiting most of its maturation during the cycle II age period (e.g. Bjorklund & Harnishfeger, 1990).

Combining the EF, working memory, and inhibition literatures, what tentative conclusions can be drawn regarding the nature of EF behaviour during Thatcher's three cycles of frontal cortical development? As presented in Table 1, during cycle I, from approximately 18 months to 5 years of age, there is evidence for emerging, and sometimes proficient, working memory, inhibition, and simple flexibility, especially when the tasks rely heavily on motor responses for perfor-mance. Cycle II, 5–10 years of age, is perhaps the most dynamic period for EF development. Depending on the task, developmental improvements in planning, working memory, inhibition, and flexibility have been observed as early as age 6 years and as late as age 10 years. Looking across a wide range of studies one can no doubt find significant developmental improvements in performance at *each age* within cycle II, and this clearly does not lead to the identification of a single,

Table 1 Thatcher's cycles of cortical reorganization and the development of EF skills

Thatcher Cycle	Age Range	Relevant EF skills
Cycle I	18 months–5 years	Emerging working memory, inhibitory control, simple flexible set shifting. Executive control of motor behaviour most proficient.
Cycle II	5–10 years	Perhaps the most dynamic period of EF development. Improvements in planning, working memory, inhibition, and flexible set shifting. Depending on the task and domain, significant developmental improvements have been observed at ages 6, 8, and 10 years.
Cycle III	10–14 years	Continued development of EF skills throughout, and beyond, this age period. Adult-level performance on some tests of verbal working memory, inhibition, and flexibility emerge between 10 and 12 years of age. Protracted development on tasks that appear to require the coordination of many EF processes, such as working memory, inhibition, and planning (e.g. Tower of Hanoi, Tower of London).

convergent, developmental "shift" in behaviour. However, any one of a number of methodological differences across studies (e.g. tasks, age ranges and age groupings, analytic techniques), as well as simple measurement error, could easily be responsible for the lack of convergence across these studies. Finally, it appears that performance on many EF tasks does not reach adult-level performance by the end of cycle II; instead it continues to develop during cycle II (10–14 years) and probably beyond this age period. There are consistent findings that the EF skills of verbal working memory, inhibition and flexibility mature between the ages of 10 and 12 years. Interestingly, those skills that appear to show a protracted period of development beyond age 12, visual working memory and the coordination of working memory and inhibition, may be just the processes underlying performance on many planning tasks. Performance on tasks presumed to require planning skills, such as the TOH and TOL, also show a similarly protracted period of development.

CONCLUSIONS AND FUTURE DIRECTIONS

It should be clear from the preceding review that there remain many unanswered questions regarding the course of neurobiological development of the prefrontal cortical region of the brain. If one focuses attention on the process of synaptogenesis as the index of the functional development of the PFC, evidence does appear to converge on a rapid increase in synaptic density in the first few years of life, followed by relative stability of this density at a level much higher than the mature brain, and finally a gradual pruning of these excess synaptic connections until adult levels are reached in early adolescence. The findings of two major laboratories exploring this phenomenon diverge on the question of whether the PFC alone exhibits this relatively protracted course of synaptic development (Huttenlocher and colleagues) or whether this developmental pattern is characteristic of a wide range of cortical regions (Goldman-Rakic and colleagues). It is important to note that the concurrent synaptogenesis hypothesis proposed by the Goldman-Rakic group is derived from research on non-human primates, whereas, the Huttenlocher view evolved from years of research examining postmortem human brain tissue. As technology improves the analytical techniques available for research in neuroscience, investigations of the pattern of synaptogenesis in the PFC will, to a greater extent, utilize consistent experimental procedures, facilitating comparison across studies.

Interestingly, the cyclic reorganization model of cortical development proposed by Thatcher suggests three cycles of development that roughly conform to the three phases of synaptogenesis observed by both the Huttenlocher and Goldman-Rakic research groups: cycle I (1.5–5 years) would be consistent with the early sharp rise in synaptic density, cycle II (5–10 years) would relate to the plateau period of high synaptic density, and cycle III (10–14 years) conforms to the slow pruning phase. However, Thatcher's model is based on in-vivo examinations of brain function in developing humans, specifically EEG coherence, and it is unclear how to relate his findings to the observations of synaptic density in human or non-human primate brains made by Goldman-Rakic and by Huttenlocher. EEG coherence, as studied by Thatcher, presumably reflects the process of synaptogenesis and synapse elimination; however, these synaptic connections are thought to be either short-distance connections within the posterior cortex or long-distance intracortical connections between the frontal and posterior cortex, depending on the timing within the developmental cycle. In the future, brain research utilizing varying techniques and different levels of analysis (histological, morphological, electrophysiological, metabolic, etc.) will need to be to integrated to assess whether a singular view of the structural and functional development of the human brain emerges. It would be particularly interesting to examine whether Thatcher's cycles of development, based on EEG coherence data, prove to be convergent with data germane to PFC development yielded by the histological research, such as that conducted in the Goldman-Rakic and the Huttenlocher laboratories.

With regard to research on the normal development of cognitive processes attributed to the PFC, there are some consistent trends emerging from the literature exploring EF in children. These developmental patterns conform, to some degree, to Thatcher's three cycles of cortical development: there is evidence for the emergence and sometimes maturation of certain EF processes during the toddler and preschool period, substantial quantitative and qualitative change in these skills during the school-age period, and continued development of some of the more sophisticated EF abilities into adolescence. These findings concerning the normal growth of EF are also consistent with the linear progression in synaptic density observed by both Goldman-Rakic and Huttenlocher across childhood: emergence of many skills coincides with the rapid increase in synaptic density, dramatic changes in EF occur during the plateau period of "exuberant connectivity" (Chugani, 1994), and maturation of a sophisticated form of EF are observed as synaptic contacts are efficiently pruned and the brain sculpted to acquire its adult morphology. It is important to note that growth spurts in EF are not always observed at the ages predicted by Thatcher's stage theory (i.e. age 5 and 10), and that other theories of cortical development (e.g. Hudspeth & Pribram, 1992) might predict performance changes at somewhat different ages. Although some tentative conclusions can be drawn regarding the normal development of EF, in order to accurately compare studies across laboratories exploring this phenomenon a consistent and widely accepted definition of the construct is needed. Such agreement across researchers would ultimately lead to the use of common experimental tasks and methods of assessment, which should be examined for reliability and validity across a wide age range.

In closing, there is intriguing evidence from recent research that the PFC mediates certain cognitive behaviours relatively early in development, and this influence increases and changes across the childhood and adolescent years. Unfortunately, because of the wide variety of theoretical perspectives and experimental methodologies characteristic of the studies in the disciplines of neuroscience, neuropsychology, and cognitive psychology, there are many conflicting views on these developmental issues. Developmental neurobiologists do not agree on the precise developmental course of the structure and function of the PFC; neuropsychologists and cognitive psychologists hold differing perspectives on the development of EF processes. Given the ambiguity on both of these fronts, how can one hope to link the neurobiological and behavioural developmental progressions? On one point most researchers can agree: the behaviours putatively mediated by the PFC are complex, fascinating, and inherent to the unfolding of normal cognitive and social development. A clearer understanding of brain and behaviour development, as it concerns the PFC and EF processes, will continue to challenge researchers for many years to come.

References

Adams, A.M. & Gathercole, S.E. (1995). Phonological working memory and speech production in preschool children. *Journal of Speech and Hearing Research*, **38**, 403–414.

Anderson, P., Anderson, V. & Lajoie, G. (1996). The Tower of London test: validation and standardization for pediatric populations. *Clinical Neuropsychology*, **10**, 54–56.

Baddeley, A.D. (1992). Working memory. *Science,* **255**, 556–559.

Barkley, R.A. (1997). Behavioural inhibition, sustained attention, and executive functions: constructing a unifying theory of ADHD. *Psychological Bulletin*, **121**, 65–94.

Becker, M.G., Isaac, W. & Hynd, G.W. (1987). Neuropsychological development of nonverbal behaviours attributed to "frontal lobe" functioning. *Developmental Neuropsychology*, **3**, 275–298.

Bjorklund, D.F. & Harnishfeger, K.K. (1990). The resources construct in cognitive development: diverse sources of evidence and a theory of inefficient inhibition. *Developmental Review*, **10**, 48–71.

Borkowski, J.G. & Burke, J.E. (1996). Theories, models, and measurements of executive functioning: An information processing perspective. In L.G. Reid & N.A. Krasnegor (Eds.), *Attention, memory, and executive function* (pp. 235–261). Baltimore, MD: P.H. Brookes.

Case, R. (1992). The role of the frontal lobes in the regulation of cognitive development. *Brain and Cognition,* **20**, 51–73.

Changeux, J.-P. & Danchin, A. (1976). Selective stabilization of developing synapses as a mechanism for the specification of neuronal networks. *Nature*, **264**, 705–712.

Changeux, J -P., Heidmann, T. & Patte, P. (1984). Learning by selection. In P. Marler & H.S. Terrace (Eds.), *The biology of learning* (pp. 115–137). New York: Springer-Verlag.

Chelune, G.J. & Baer, R.L. (1986). Developmental norms for the Wisconsin Card Sorting Test. *Journal of Clinical and Experimental Neuropsychology*, **8**, 219–228.

Chelune, G.J. & Thompson, L.L. (1987). Evaluation of the general sensitivity of the Wisconsin Card Sorting Test among young and older children. *Developmental Neuropsychology*, **3**, 81–90.

Chugani, H.T. (1994). Development of regional brain glucose metabolism in relation to behaviour and plasticity. In G. Dawson & K.W. Fischer (Eds.), *Human behaviour and the developing brain* (pp. 153–175). New York: Guilford Press.

Cuneo, K.M. & Welsh, M.C. (1990). Perseveration in young children: developmental and neuropsychological perspectives. *Child Study Journal*, **22**, 73–91.

Damasio, A.R. & Anderson, S.W. (1993). The frontal lobes. In K.M. Heilman & E. Valenstein (Eds.), *Clinical neuropsychology* (3rd ed., pp. 409–460). New York: Oxford University Press.

Denckla, M.B. (1996). Biological correlates of learning and attention: what is relevant to learning disability and attention deficit hyperactivity disorder?. *Journal of Developmental and Behavioural Pediatrics*, **17**, 114–119.

Diamond, A. & Taylor, C. (1996). Development of an aspect of executive control: development of the abilities to remember what I said and to "Do as I say, not as I do". *Developmental Psychobiology*, **29**, 315–334.

Diamond, A., Werker, J.F. & Lalonde, C. (1994). Toward understanding commonalities in the development of object search, detour navigation, categorization, and speech perception. In G. Dawson & K.W. Fischer (Eds.), *Human behaviour and the developing brain* (pp. 380–426). New York: Guilford Press.

Duncan, J. (1995). Attention, intelligence, and the frontal lobes. In M.S. Gazzaniga (Ed.), *The cognitive neurosciences* (pp. 721–733). Cambridge, MA: MIT Press.

Duncan, J., Burgess, P. & Emslie, H. (1995). Fluid intelligence after frontal lobe lesions. *Neuropsychologia*, **33**, 261–268.

Edelman, G.M. (1987). *Neural Darwinism: the theory of neuronal group selection*. New York: Basic Books.

Eslinger, P.J. (1996). Conceptualizing, describing, and measuring components of executive function: a summary. In G.R. Lyon & N.A. Krasnagor (Eds.), *Attention, memory, and executive function* (pp. 367–395). Baltimore, MD: Paul H. Brookes.

Eslinger, P.J., Biddle, K.R. & Grattan, L.M. (1997). Cognitive and social development in children with prefrontal cortex lesions. In N. Krasnegor, G.R. Lyon,& P.S. Goldman-Rakic (Eds.), *Development of the prefrontal cortex: evolution, neurobiology, and behaviour* (pp. 295–353). Baltimore,MD: Paul H. Brookes.

Fiducia, D. & O'Leary, D.S. (1990). Development of a behaviour attributed to the frontal lobes and the relationship to other cognitive functions. *Developmental Neuropsychology*, **6**, 85–94.

Foreman, N., Warry, R. & Murray, P. (1990). Development of reference and working spatial memory in preschool children. *Journal of General Psychology*, **117**, 267–276.

Funahashi, S., Bruce, C.J. & Goldman-Rakic, P.S. (1989). Mnemonic coding of visual space in the monkey's dorsolateral prefrontal cortex. *Journal of Neurophysiology*, **61**, 331–349.

Fuster, J.M. (1985). The prefrontal cortex, mediator of cross-temporal contingencies. *Human Neurobiology*, **4**, 169–179.

Fuster, J.M. (1989). *The prefrontal cortex* (2nd ed.). New York: Raven Press.

Fuster, J.M. (1997). *The prefrontal cortex* (3rd ed.). New York: Raven Press.

Goel, V. & Grafman, J. (1995). Are the frontal lobes implicated in "planning" functions? Interpreting data from the Tower of Hanoi. *Neuropsychologia*, **33**, 623–642.

Golden, C.J. (1981). The Luria–Nebraska Children's Battery: theory and formulation. In G.W. Hynd & J.E. Obrzut (Eds.). *Neuropsychological assessment and the school-aged child* (pp. 277–302). New York: Grune & Stratton.

Goldman-Rakic, P.S., Bourgeois, J.-P. & Rakic, P. (1997). Synaptic substrate of cognitive development: life-span analysis of synaptogenesis in the prefrontal cortex of the nonhuman primate. In N. Krasnegor, G.R. Lyon & P.S. Goldman-Rakic (Eds.), *Development of the prefrontal cortex: evolution, neurobiology, and behaviour* (pp. 27–47). Baltimore, MD: Paul H. Brookes.

Goldstein, F.C. & Green, R.C. (1995). Assessment of problem solving and executive functions. In R.L. Mapou & J. Spector (Eds.), *Clinical neuropsychological assessment: a cognitive approach* (pp. 49–81). New York: Plenum Press.

Grafman, J. (1999). Experimental assessment of adult frontal lobe function. In B.L. Miller & J.L. Cummings (Eds.), *The Human Frontal Lobes: Functions and Disorders* (pp. 321–344).

Hudspeth, W.J. & Pribram, K.H. (1992). Psychophysiological indices of cognitive maturation. *International Journal of Comparative Neurology*, **52**, 19–29.

Huttenlocher, P.R. & Dabholker, A.S. (1997). Developmental anatomy of prefrontal cortex. In N. Krasnegor, G.R. Lyon & P.S. Goldman-Rakic (Eds.), *Development of the prefrontal cortex: evolution, neurobiology, and behaviour* (pp. 69–83). Baltimore, MD: Paul H. Brookes.

Ingvar, D.H. (1985). "Memory of the future": an essay on the temporal organization of conscious awareness. *Human Neurobiology*, **4**, 127–136.

Jacobsen, C.F. (1935). Functions of frontal association area in primates. *Archives of Neurology and Psychiatry*, **33**, 558–569.

Kepler, M.D. (1998). Assessing executive functions using the Stroop, Tower of London, Children's Behaviour Questionnaire, and Behaviour Rating Inventory of Executive Functions. Unpublished master's thesis, University of Northern Colorado, Greeley, Co, USA.

Kimberg, D.Y., D'Esposito, M. & Farah, M.J. (1998). Cognitive functions in the prefrontal cortex: working memory and executive control. *Current Directions in Psychological Science*, **6**, 185–192.

Kimberg, D.Y. & Farah, M.J. (1993). A unified account of cognitive impairments following frontal lobe damage: the role of working memory in complex, organized behaviour. *Journal of Experimental Psychology: General*, **122**, 411–428.

Kopp, C.B. (1982). Antecedents of self-regulation: a developmental perspective. *Developmental Psychology*, **18**, 199–214.

Levin, H.S., Culhane, K.A., Hartmann, J., Erankovich, K., Mattson, A.J., Haward, H., Ringolz, G., Ewing-Cobbs, L. & Fletcher, J.M. (1991). Developmental changes in performance on tests of purported frontal lobe functioning. *Developmental Neuropsychology*, **7**, 377–395.

Levin, H.S., Mendelsohn, D., Lilly, M.J., Fletcher, J.M., Culhane, K.A., Chapman, S.B., Harward, H., Kusnerick, L., Bruce, D. & Eisenberg, H.M. (1994). Tower of London performance in relation to magnetic resonance imaging following closed head injury to children. *Neuropsychologia*, **28**, 126–140.

787

Lezak, M.D. (1983). *Neuropsychological assessment* (2nd ed.). New York: Oxford University Press.

Lezak, M.D. (1995). *Neuropsychological assessment* (3rd ed.). New York: Oxford University Press.

Luciana, M. & Nelson, C.A. (1999). The functional emergence of prefrontally-guided working memory systems in four- to eight-year-old children. *Neuropsychologia*, **36**, 273–293.

Luria, A.R. (1966). *Higher cortical functions in man*. New York: Basic Books.

Luria, A.R. (1973). *The working brain*. New York: Basic Books.

Milner, B. (1964). Some effects of frontal lobectomy in man. In J.M. Warren & K. Akert (Eds.), *The frontal granular cortec and behaviour* (pp. 313–334). New York: McGraw-Hill.

Mutter, B., Alcorn, M.B., Welsh, M.C. & Granrud, C. (June, 1999). Working memory and inhibitory control as predictors of false-belief task performance. Poster presented at the annual meeting of American Psychological Society, Denver, CO, USA, June.

Passler, M.A., Isaac, W. & Hynd, G.W. (1985). Neuropsychological development of behaviour attributed to frontal lobe functioning in children. *Developmental Neuropsychology*, **1**, 349–370.

Paus, T., Zijdenbos, A., Worsley, K., Collins, D.L., Blumenthal, J., Gledd, J.N., Rapoport, J.L. & Evans, A.C. (1999). Structural maturation of neural pathways in children and adolescents: *in vivo* study. *Science*, **283**, 1908–1910.

Pennington, B.F. (1994). The working memory function of the prefrontal cortices: implications for developmental and individual differences in cognition. In M.M. Haith & J.B. Benson (Eds.), *The J.D. and C.T. MacArthur Foundation series on mental health and development: the development of future-oriented processes* (pp. 243–289). Chicago, IL: University of Chicago Press.

Pennington, B.F., Benneto, L., McAleer, O.K. & Roberts, R.J. (1995). Executive functions and working memory: Theoretical and measurement issues. In G.R. Lyon & N.A. Krasnegor (Eds.), *Attention, memory, and executive function* (pp. 327–348). Baltimore, MD: Brookes.

Rakic, P., Bourgeois, J.-P., Eckenhoff, M.F., Zecevic, N. & Goldman-Rakic, P.S. (1986). Concurrent overproduction of synapses in diverse regions of the primate cerebral cortex. *Science*, **232**, 232–235.

Roberts, R.J. & Pennington, B.F. (1996). An interactive framework for examining prefrontal cognitive processes. *Developmental Neuropsychology*, **12**, 105–126.

Stuss, D.T. & Benson, D.F. (1986). *The frontal lobes*. New York: Raven Press.

Thatcher, R.W. (1991). Maturation of the human frontal lobes: physiological evidence for staging. *Developmental Neuropsychology*, **7**, 397–419.

Thatcher, R.W. (1992). Cyclic cortical reorganization during early childhood. *Brain and Cognition*, **20**, 24–50.

Thatcher, R.W. (1994a). Cyclic cortical reorganization: origins of cognition. In G. Dawson & K. Fischer (Eds.), *Human behaviour and the developing brain* (pp. 232–268). New York: Guilford Press.

Thatcher, R.W. (1994b). Psychopathology of early frontal lobe damage: dependence on cycles of development. *Development and Psychopathology*, **6**, 565–596.

Thatcher, R.W. (1997). Human frontal lobe development: a theory of cyclical cortical reorganization. In N. Krasnegor, G.R. Lyon & P.S. Goldman-Rakic (Eds.), *Development of the prefrontal cortex: evolution, neurobiology, and behaviour* (pp. 85–113). Baltimore, MD: Paul H. Brookes.

Thatcher, R.W., Walker, R.A. & Giudice, S. (1987). Human cerebral hemispheres develop at different rates and ages. *Science*, **236**, 1110–1113.

Tranel, D., Anderson, S.W. & Benton, A.L. (1995). Development of the concept of Executive Function and its relationship to the frontal lobes. In F. Boller & J. Grafman (Eds.), *Handbook of neuropsychology* (vol. 9, pp. 126–148). Amsterdam: Elsevier.

Welsh, M.C. (1991). Rule-guided behavior and self-monitoring on the Tower of Hanoi disk-transfer task. *Cognitive Development*, **6**, 59–76.

Welsh, M.C. (2001). Developmental and clinical variations in executive functions. In U. Kirk & D. Molfese (Eds.), *Developmental variations in language and learning*. Hillsdale, NJ: LEA (In Press).

Welsh, M.C. & Pennington, B.F. (1988). Assessing frontal lobe functioning in children: views from developmental psychology. *Developmental Neuropsychology*, **4**, 199–230.

Welsh, M.C., Pennington, B.F. & Groisser, D.B. (1991). A normative-developmental study of executive function: a window on prefrontal function in children. *Developmental Neuropsychology*, **7**, 131–149.

Welsh, M.C., Satterlee-Cartmell, T. & Stine, S. (1999). Towers of Hanoi and London: contribution to working memory and inhibition to performance. *Brain and Cognition*, **41**, 231–242.

Welsh, M.C., Huizinga, M., McGarraugh, M., Devine, S., Beatty, K., Adams, C. & Ryan, C. (2000). Reliability and validity of the Tower of Hanoi – revised. Poster presented at the annual meeting of the International Neuropsychological Society, Denver, CO, USA, February.

Quartz, S.R. & Sejnowski, T.J. (1997). The neural basis of cognitive development: a constructivist manifesto. *Behavioural and Brain Sciences*, **20**, 537–596.

Yakovlev, P.I. & Lecours, A.-R. (1967). The myelogenetic cycles of regional maturation of the brain. In A. Minowski (Ed.)., *Regional development of the brain in early life* (pp. 3–64). Oxford: Blackwell.

Zelazo, P.D. & Frye, D. (1998). Cognitive complexity and control: II. The development of executive function in children. *Current Directions in Psychological Science*, **7**, 121–126.

IV.9
Approaches to the Study of Early Risk for Neurobehavioural Disorders: Exemplified by the Early Visual Attention/ADHD Issue

ALEX F. KALVERBOER

ABSTRACT

The explanatory power of studies on early risk for neurobehavioural disorders has improved considerably during the past two decades. Insights into the dynamics of early organism–environment interaction have grown, not the least due to careful observational and (semi-)experimental studies, in which approaches from developmental neurology, developmental psychology and ethology are linked. Biological and environmental constraints have become clearer, as well as infants' and caretakers' abilities for learning and communication. Insight has grown into the infant's information-processing capacities in relation to the regulation of its behavioural state, as well as into the caretaker's ability to modulate the infant's behaviour, relative to the signals emitted. Careful experimental studies focusing on core attentional processes can be executed, guided by models on information processing and underlying brain mechanisms. Techniques for brain scanning and for refined behavioural registrations allow us to obtain direct information on structure–function relationships from the earliest phases of development onwards. However, notwithstanding this progress in theory and technique, there is still a lack of longitudinal studies with respect to the early roots of neurobehavioural disorders at later age, which give a real insight into how a risk condition develops into a real defect/disorder. Why is this so, and how can the situation be improved? This chapter comments on this issue, exemplified by the study of early visual attention and the case of early risk for attention deficit hyperactivity disorder (ADHD).

A.F. Kalverboer and A. Gramsbergen (eds.), Handbook of Brain and Behaviour in Human Development, 791–814
© 2001 Kluwer Academic Publishers. Printed in Great Britain.

INTRODUCTION

Large improvements have been made during recent decades in the study of early risk for neurobehavioural disorders. From descriptions of risk factors in terms of often poorly specified medical conditions (e.g. anoxia, hypoxia, length of pregnancy) with an unclear relationship to their possible effects on neurobehavioural development, we have reached the stage of refined measurements of brain and behavioural processes.

Brain processes can be measured directly and in real time concomitant with the organism's reactions. Structural characteristics can be meaningfully linked to processes in the brain. Insight is rapidly growing into how complex functions are "represented" in the brain. However, how these connections develop, and how the nervous system is able to functionally compensate for structural deficiencies, is still largely unknown.

Psychological functions can also be described in great detail, guided by models which allow for experimental manipulations that are theoretically based. Models become increasingly refined and are able to account for the complex interactions between the reactions of the organism and the regulation of state. In the past decade a wealth of data has been collected that give a refined picture of complex mental processes such as memory, attention, perceptions and emotions and – at a high level of integration – so-called executive functions, which are very much the topic of the present cognitive neurosciences.

Together these converging developments in brain and behaviour research may allow for the advanced study of early risk factors for a deviant development, bringing this to a level of refinedness that was unthinkable some decades ago. From the behavioural side information-processing theory in particular may contribute to this development. Also important are new developments which lay a strong focus on the dynamics of complex interactions between emotion and cognition and away from a too one-sided cognitive approach. However, to what extent are we obtaining a better understanding of normal and deviant brain–behaviour development in the human? How may approaches (models) at different levels of explanation help to achieve a more integrated view of developmental processes and risk for deviance? In this chapter some aspects of this problem will be discussed.

After a brief history I will sketch the present state of the art, taking as an example the approach in our laboratories and elsewhere to the study of early visual attention, aiming at a better understanding of early roots of normal and deviant neurobehavioural development. This will be related to present-day approaches to the study of attention deficit hyperactivity disorder (ADHD), which is thought to be one of the probable outcomes of early risk conditions in which attentional mechanisms play a role.

A BRIEF HISTORY

Since approximately 1950 medical and behavioural sciences have made more systematic attempts to distinguish the (high) risk organisms from others at an early stage of development. Triggers were clinical entities, which implied complex behavioural and learning disorders, e.g. minimal brain dysfunction (MBD), now attention deficit hyperactivity disorder (ADHD); clumsiness, now developmental coordination disorder (DCD); dyslexia, etc. Such syndromes were (and still are) thought to be at least to some extent eventually associated with a structural (including biochemical) defect in the brain due to a less than optimal functioning of the CNS.

Initially, in such studies, groups to be compared were distinguished exclusively on the basis of early medical (prenatal and perinatal) conditions, such as preterm birth, intrauterine growth retardation, anoxia/hypoxia, prolonged labour, etc. Such conditions were thought to contribute to the risk for later neurobehavioural deviance (see Kopp, 1983, for a review) Recent examples of such approaches can be found in Robson & Cline (1998) and Clerkes-Julkowski (1998). Such risk conditions are usually related to outcomes of clinical assessments and impressions, psychological tests and behavioural questionnaires, filled out by parents or teachers. As a result an enormous number of studies has been generated, which usually report statistical differences between groups at risk and control groups, but are almost impossible to be interpreted in terms of "risk mechanisms". The New York collaborative project was an improvement in that newborn neurological examinations were included in the longitudinal study (Laufer & Denhoff, 1957). Unfortunately the early neurological examinations were far from satisfying, so that ultimately no further insight was gained into the roots of later clinical conditions such as MBD.

A clear improvement was the application, from about 1955 onwards, of developmental neurological assessments which allowed us to obtain an age-adequate picture of the optimality of the infant's and child's neurological functions in the course of ontogeny (Prechtl & Beintema, 1964; Touwen, 1979). This improvement in neurological assessment, combined with careful behavioural observations, strongly based on insights derived from ethology, allowed for the development of neurobehavioural profiles of children at risk and for relating early neurology and behaviour to eventual problems at later ages (Kalverboer, 1975). However, the examination procedures were too time-consuming to be generally introduced into clinical practice. These assessment procedures, however, are still valuable tools in brain–behaviour research.

A next important step was the development and improvement of precise experimental and observational procedures to study children's psychological functions; first in school-aged children, but gradually more and more suited for precise experimental study at younger ages, down until directly after birth. Information-processing models, initially developed for the study of adult cognitive processes, became applicable to research on cognitive functions from the earliest phases of extrauterine life onward.

Parallel improvements were made in the methodology of longitudinal studies and the statistical treatment of data, obtained in follow-up studies. With a growing insight into the complicated dynamics of developmental processes, more refined procedures for the study of behavioural and neural functions and their structural concomitants were developed.

Particularly in the past two decades the explanatory power of studies on early risk for neurobehavioural deviance has considerably increased due to the possibility of directly and experimentally studying the psychological and biological processes involved in the development of problem behaviour from the earliest phases of ontogeny onwards. This improvement is due mainly to two factors: first the development of refined techniques for the experimental study of biological and cognitive functions and the analysis of data, and second theoretical developments, on the basis of which models can be composed which give the opportunity to meaningfully relate data on brain–behaviour relationships obtained over a relatively wide age range from very early age onwards and, to quote Mareschal (1998), "to take seriously constraints on how the brain functions and develops" (p. 147).

To illustrate this I now will focus on a domain of study important in present-day early-risk research, namely the development of attention and in particular visual attention. This research topic is considered in the perspective of early risk for problem behaviour in ADHD, as described in the *Diagnostic and Statistical Manual of Mental Disorder*, 4th ed. (DSM-IV, 1994). I will try to illustrate how dense follow-up studies may contribute to our insight into the risk mechanisms during ontogeny, and indicate how initially remote lines of research may converge (e.g. the study of attentional processes in relation to state organization) and become part of a more integrated approach to the study of the early roots of developmental deviance. However, the limitations should also be clear. Until now only first steps have been taken.

ADHD AS A TRIGGER FOR STUDIES ON EARLY RISK

Children with ADHD are characterized by distractibility, a lack of sustained attention, and an insufficient impulse control. However, it is a heterogeneous category, due to the fact that the label ADHD is applied for children with very different profiles of behavioural problems. DSM-VI, the international standard for clinical nomenclature, includes children in this category if they present either six signs or more of "inattention", or six signs or more signs of "hyperactivity/impulsivity". As a consequence, studies on risk mechanisms related to ADHD have to do with ambiguous outcome criteria. More exact outcome definitions, based on operationalizations in terms of coherent subcategories, are a prerequisite for improving the theoretical power of early risk studies.

In order to better understand the processes that may contribute to a clinical condition, such as ADHD, traditionally two sorts of approach have been applied in prospective studies:

1. the study of groups thought to have been exposed to a factor that has impaired the condition of the brain, such as pre- and dysmaturity, pre- and perinatal complications, such as asphyxia, or metabolic disorders, e.g. phenylketonuria (PKU) or congenital hypothyroidism (CH) and
2. the study of relatively complex early childhood characteristics, such as difficult temperament.

Almost inevitably such studies cannot provide basic insights into mechanisms playing a role in the development of later ADHD. The impact of medical complications for the integrity of structure and functions of the nervous system may be very varied, and is often unclear. Attempts to identify possible behavioural precurcors of ADHD (previously labelled MBD, minimal brain dysfunction), such as temperamental characteristics, for example overactivity, hyperexcitability or lack of rhythmicity, were also not very fruitful. Such behaviours were generally selected on the basis of phenomenological similarities with the clinical features of ADHD, based on the conviction of an almost linear connection between the early and later behavioural repertoire (e.g. see Thomas & Chess, 1977). However, this was too simple a conception, as it failed to account for the complex transactions taking place between organism and environment during ontogeny. Concepts were too diverse and, especially in infant studies, too diffuse, whereas adequate theory was lacking.

INTERACTIVE DESIGNS

Attempts were then made to develop paradigms based on theoretical considerations, for example on how the infant's behavioural organization (e.g. postural control, state regulation, orienting behaviour) may affect its ability to experience and perceive contingencies in social and specific (contingency learning) task conditions.

Problems in later social and cognitive development might be partly mediated by the caretaker's ability to refinedly adapt to children's intentions. With knowledge of the constraints, set by the maturation of the nervous system and by particular impairments that might affect the child's adaptive capacities, further insight into the development of neurobehavioural problems could be obtained. A nice example of such an approach is Wijnroks' longitudinal study on the development of preterms and dysmaturily born children between 6 months and 2 years of age (Wijnroks, 1994). He made a detailed study of interactions between the infants' behaviour, with a special focus on state regulation and postural control, and the caretaker's sensitive responsiveness, hypothesizing that such interactions may have a direct impact for later cognitive competence and quality of attachment. He connected notions derived from developmental neurology and developmental psychology in an attempt to identify particular "mechanisms" that might contribute to later behavioural and learning problems.

An example of such a mechanism in his study is a lack of contingency between the infant's behaviour and its social partner's reactions, due to problems in early postural control (mainly indicated by signs of hyperextension), which are frequently observed in preterms and dysmaturily born infants, in combination with a lack of sensitive responsiveness by the caretaking environment. It was expected that such a condition might negatively affect later cognitive and social functioning, in particular contingency learning (Brighi, 1997; Wijnroks, 1997) and attachment quality. Results were varied and not easy to interpret: early problems in postural control were negatively related to competence in problem solving, whereas high-intensity stimulation by the mother went together with more negative behaviour in the infant.

Undoubtedly there remains some doubt as to how interactions may work, and on how relationships can be explained, e.g. between quality of postural control and cognitive competence. Conclusions necessarily remain at a global level. Nevertheless, this approach is a significant next step in unravelling the complex determinants of later neurobehavioural disorders, being an attempt to take account of the continuous transactions between the child's and the environment's actions, which have been so nicely demonstrated in Papoušek's refined studies on "early signalling" and "intuitive parenting" (Papoušek & Papoušek, 1983, 1993; Papoušek *et al.*, 1999). Two characteristics of such studies strongly contribute to the power of research on early risk for deviance in the domain of "developmental cognitive neuroscience" (Johnson, 1997), namely the linking of insights, derived from different disciplines, in particular developmental neurology, developmental psychology, and ethology, and the association of experimental approaches and refined observations in laboratory and real-life conditions.

THE DIRECT STUDY OF EARLY ATTENTION

A recent development on which I focus in this chapter, and which may significantly increase the explanatory power of early risk research with respect to ADHD, is the study of basic constituents of attentive behaviour from an early age onwards. Such studies should be guided by notions about their relationships to the development of structures and processes in the brain, on the one hand, and about their relevance for the organism's adaptive behaviour in ecological conditions, on the other hand. Researchers such as Johnson *et al.*, 1991, have paved the way to the precise measurement of processes, such as "disengagement" and "inhibition of return", which are considered as markers for the posterior attention system (PAS). They presented thoughtful discussions on how such processes might relate to nervous system development, an underlying network in the brain has been hypothesized, and how they may affect daily life adaptations, and have an impact on temperament and personality development (e.g. in Rothbart's soothability studies, see also Derryberry & Rothbart, Chapter V.8, this handbook). Furthermore, such basic attentional processes seem to be directly related to characteristic features of the ADHD pattern, such as dreaminess, distractibility and lack of sustained attention. However, to date, dense prospective studies on such basic processes in children thought to be at risk for neurobehavioural disorders such as ADHD, parallel to "normal" controls, are not available.

A first step is the direct study of basic attentional processes, in particular early visual attention, in children at risk as compared to normal controls. This is the topic of the subsequent paragraphs.

THE SIGNIFICANCE OF EARLY VISUAL ATTENTION FOR THE CHILD'S DEVELOPMENT

Almost no infant behaviour plays such a crucial role in early communication and exploration as visual orientation. Mothers expend much effort in trying to catch an infant's gaze and orient the baby to her face. When babies fail to do so this may upset the mother, and even lead to rejection or aggressive emotional reactions. A clinical case of child abuse was directly related to an infant's disorder in directing its gaze (Kalverboer *et al.*, 1970). Orienting reactions indicate children's focus of interest: mothers put much effort into attracting children's attention, and in later phases of development the process of "mutual gazing" indicates how attention is shifting from the nearby environment (in particular the mother's face) to more remote aspects of the environment. Exploratory activity develops out of face-to-face contact (Schaffer, 1971). Impairments in orienting behaviour, due either to a lack of postural control or to impairments in oculomotor functions, or in visual acuity, or to insufficiencies in attentional capacities, may have strong impacts on the quality of the interaction process and negatively affect the caretaker's emotional reactions. Studying the determinants of such "visual attention" impairments may shed light on possible implications for later attention deficits, such as in children with ADHD.

In developmental studies, visual orientation has been used as the main index for early learning. How the child starts to distinguish between novel and familiar, how it habituates to new conditions, is studied in the context of the visual orientation paradigm. Differences in reaction times regarding different categories of visually presented stimuli are used as indications of learning capacities and of the maturation of brain functions. So far, however, a too-simple view has been held on early visual orientation and how it may indicate the emotional and cognitive status of the organism. First, almost exclusively the direction of gaze has been applied as the behaviour, which is indicative for a change in cognitive activity. Second, it has been largely overlooked that visual behaviour not only reflects the infant's capacity to process information, but also depends on the behavioural state of the infant.

In search of the roots of later adaptive problems in children, especially in the domain of disorders in information processing as ADHD, interest in early attentive processes has greatly increased in the past decade. This was strongly triggered by the development of new insights in the field of neuropsychology of selective attention.

PRESENT-DAY VIEWS ON "ATTENTION SYSTEMS"

Triggered by pioneering work by Posner's group (Posner, 1990, 1995; Posner & Petersen, 1990; Posner & Rothbart, 1991), the early development of visual attention attracted a strong research interest in the past decade. Posner postulated three attention systems in the brain, which are thought to play a crucial role in the organism's adaptive behaviour during ontogeny. First, there is the posterior attention system, the PAS (Posner & Petersen, 1990), which is involved in the selection of relevant information by shifts in visual orientation. Brain structures considered to be involved in the PAS are the parietal lobe, the superior colliculus and the pulvinar nucleus of the thalamus. Second, there is the anterior attention system, the AAS, which is considered to be actively processing relevant information and involved in making more complex comparisons. This is supposed to exercise executive control over voluntary behaviour. The anterior cingulated gyrus (mid-frontal) and the prefrontal cortex are thought to be core structures in the AAS network. The third attention system is the vigilance system, which has a function in maintaining a sustained level of wakefulness (alertness). Figure 1, after Webster & Ungerleider (1998), indicates how these

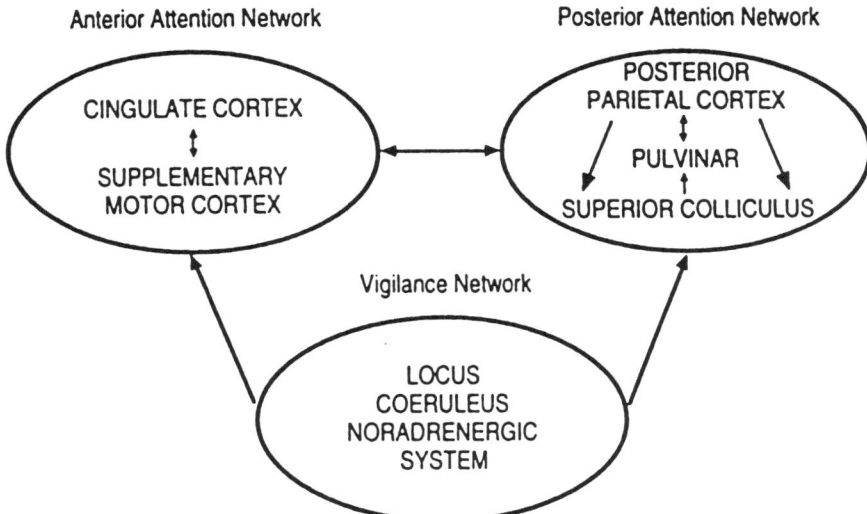

Anterior Attention Network   Posterior Attention Network

Vigilance Network

Figure 1 Schematic representation of the neural network model of attention described by Posner and colleagues in Webster & Ungerleider, 1998 and after Posner, 1990, 1995.

systems may relate to various structures of the brain. It also demonstrates how the vigilance network influences both the posterior and anterior system. However, evidence for brain involvement is mainly based on PET scan studies in human adults and in primates, not on data derived directly from the human developing brain.

BACKGROUND OF OUR STUDIES: THE PAS, OVERT AND COVERT ATTENTION

In particular the PAS, as postulated by Posner and co-workers, has in the past decade in various centres been the basis for experimental research on the early development of visual attention in the human. In our laboratory, studies on the development of the PAS have been largely triggered by the search for early risk mechanisms that may play a role in the development of later problem behaviour in children with ADHD. The PAS is thought to play a crucial role in shifting visual attention (see Figure 2, after Posner *et al.*, 1987).

This scheme represents the successive phases in the process of a shift of attention from one location to another. Successive mental operations are thought to begin with the presentation of a cue. To begin with the system has to be alerted and the ongoing activity interrupted. The new cue is localized and the gaze has to be disengaged from the previous focus of attention (in this process the parietal lobe is thought to play a crucial role). Now, eye movements can take place to the newly cued location (involvement of the superior colliculus of the midbrain). Engagement occurs to the new location, whereas the tendency to move back is

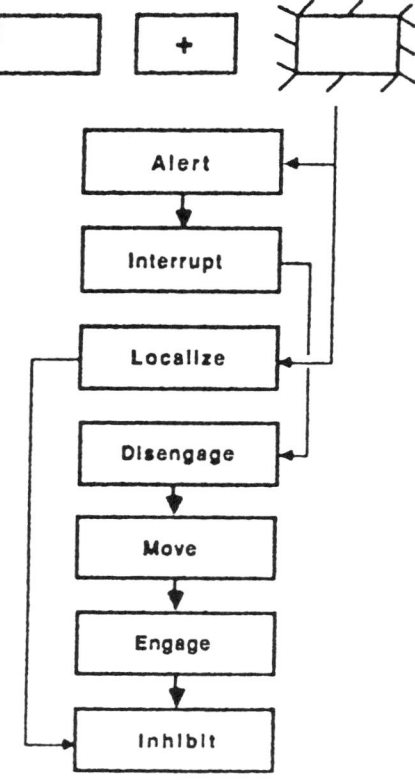

Figure 2 Schematic representation of the mental operations that play a role when visual attention shifts from a fixated target to a new cued location. The four last operations involve the posterior attention system (Posner *et al.*, 1987).

inhibited (involvement of the pulvinar nucleus of the thalamus). The last four operations involve the PAS. Disengagement and inhibition of return are considered to be crucial indexes of the functioning of this system.

Core concepts are covert and overt attention. By covert attention is meant the internal preparation which precedes a shift of visual orientation and allows gaze to be detached from the original location (disengagement). Overt attention is the condition of being attentive to a particular target, usually indicated by the direction of gaze. Developmental researchers usually point to overt changes in visual orientation as an index of changes in attention. However, there is evidence for the existence of the process of covert visual attention, which implies shifts in "inner orientation", which are not directly discernible in the overt behaviour of the individual. They precede changes in overt attention and function to modulate the activity of the visual system, and to prepare it in order to meet the specific demands of the task at hand. They have to be inferred from the enhancement of processes at the location at which attention becomes focused. Targets at that location are detected faster and at lower thresholds than targets at other locations (Posner, 1990). Research in adults has provided evidence for the existence of covert visual attention systems, anatomically distinct from the visual system, with the function of modulating its activity (Posner, 1988; Klein, 1979).

Can the PAS be seen as a functionally homogeneous system, of which the processes of "disengagement" and "inhibition of return" are markers, or is it much more loosely structured than suggested? Direct evidence is scarce. Insight into the functional significance during early ontogeny of the various brain structures considered to be involved in the PAS is very limited, and not obtained in direct longitudinal studies on the developing infant. The same is true for the developmental course of the so-called markers of the PAS. Consequently, notions about the possible links between these brain structures and these "markers" during development are also still highly speculative.

Given this state-of-the-art, the best strategy might be to focus directly on the development of crucial processes in the early development of attention, instead of considering these processes as "markers" of an attention system, the structure and developmental course of which are still largely unknown (see the criticism on the emphasis upon marker tasks by Richards, 1997).

DISENGAGEMENT AND INHIBITION OF RETURN (IOR) IN INFANCY RESEARCH

In our laboratories of experimental clinical psychology the development of visual attention in the first year of life is directly studied in normal newborns as well as in children at risk, in collaboration with developmental neurology and paediatrics (Butcher & Kalverboer, 1997; Butcher, 2000). Focus is on the processes "disengagement" and "inhibition of return", which are of particular significance in the early development of visual attention and plausibly related to attention deficits at later age. "Disengagement" concerns the ability to shift gaze and attention away from a persisting fixation point; "inhibition of return" the ability to inhibit the tendency to return to a previously cued location. Functions of these two processes are different but related, both being prerequisites for efficient exploratory behaviour. Disengagement is required preceding the execution of a movement, away from a previous target, while IOR allows for a sufficient inspection of a chosen target.

In a series of studies the developmental course of these two processes was closely followed: data were collected at 2-week intervals from 6 weeks to 6 months of age, the period during which the PAS is thought to come into operation. The study concerned a group of full-term healthy infants (Butcher et al., 1999, 2000) and two groups of preterms, with a varying degree of transient periventricular echogenicity (PVE) (Butcher et al., 2001a,b). PVE indicates a frequently occurring form of perinatal brain injury which is associated with damage to areas of the brain that mediate visual attentional processing (Fawer & Calame, 1991). Butcher's findings are intriguing. Preliminary results suggest that disengagement and inhibition of return have different developmental courses. Disengagement shows a rapid increase from 6 to 16 weeks, and then slows down. IOR shows the most rapid development around week 14, and has become established at about 25 weeks of age. These first results do not give support to the notion of a tightly integrated attentional system, as previously postulated, but suggest a much looser structure of such a system and underline Richards' doubts with respect to a "marker" approach. They are illustrative for our present-day lack of insight in the developmental course of basic functions in the early development of information processing in the human.

The development of disengagement is impaired in a substantial number of preterm infants with transient periventricular echogenicity. Although differences are small they can, according to Butcher *et al.* (1999), be expected to affect the amount and efficiency of visual exploration. Difficulties in IOR, also found in many preterm infants with PVE, suggest that these children have difficulties in resisting capture by intense stimuli. Differences between full-term and preterm infants are small, but they may have a significant effect on children's daily life adaptation. This poses the question of the ecological implications of experimental findings.

IMPACT FOR DAILY LIFE ADAPTATIONS

Evidently, to obtain further insight into the ecological relevance, experiments should be complemented by careful observations in so-called free-field/real-life conditions. How do caretakers adapt their behaviour to the infant, and vice-versa? What may these developments imply for the infant's behaviour in the natural habitat? As yet, insight into the ecological significance of the development of the PAS is almost lacking. Some first results of two pilot studies will be reported here, based on observations in a pseudo-experimental set-up in the infant's home situation. The crucial question is whether and how caretakers adapt to the infant's signals, and how basic attention processes may affect the infant's real-life interactive and explorative behaviour.

EARLY ATTENTION AND MOTHER–INFANT INTERACTION

Attention-demanding behaviours in mothers

In how far may the development of these visual attention processes possibly be affected by the infant's interactions with the social environment?

Young infants differ considerably in the amount of attention they pay to the face of the mother in caretaker–infant interactions. This was shown in a small longitudinal study of infants from 2 to 7 months of age, in which children were observed in their homes in loosely standardized interactive conditions (Fijn van Draat, 1995). The study suggested that the mother's behaviour could be an important determinant of the baby's visual attention. Mothers strongly differ in the patterns of vocalization and expressive behaviour with respect to their babies, and these differences may affect the amount of looking of baby at the mother's face. In a further experimental study attention-demanding behaviour of the mother was systematically manipulated (Cordes & Venekamp, 1996). Mothers were required during three successive interaction epochs (each lasting 4 minutes) to pay attention to their babies in different manners. In the first epoch they were asked to behave in the way they usually do, in the second epoch to show a minimum of attention-attracting behaviour and in the third to try to maximally attract the attention of the infant. Two samples of mother–infant dyads were cross-sectionally studied, namely 19 dyads in which the baby was 2 months old and 19 dyads of a mother with a 4-month-old baby. Each epoch was divided in two 2-minute sessions, namely a face-to-face session (without toys) and a toy session in which the mother could visually present toys to the baby, either in the periphery or frontally. The main results were as follows:

1. Mothers show a large variety of attention-demanding behaviours in epoch 3. Such behaviours are vocalizations and verbalizations, e.g. much and affectionate talking, making sounds of varying pitch and queer sounds, using the name of the baby, laughing and singing, and movements and gestures, e.g. changing the distance between her and the baby, touching the baby in a playish manner, playing "hand-games", showing clear and varied facial expressions. Mothers largely differ in the richness of their behavioural repertoire. Interestingly enough, the usual behaviour, shown in epoch 1, is for some mothers very similar to epoch 3 (the "maximum" instruction), for others it is similar to epoch 2 (the "minimum" condition). This confirms the preliminary guess that mothers largely differ in their attempts to attract babies' attention in their usual daily-life conditions (see also Papoušek & Papoušek, 1983).

2. In all face-to-face epochs mother's attention-demanding behaviours have a clear positive effect on the baby's gazing. In epoch 2 we see a lot of short fixations in different directions: there is continuous attention to the environment, but seemingly not structured by the mother's behaviour. In the toy condition there is a lot of fixation to the object, but still also, in an alternating fashion, the baby looks at the mother's face, as if she is a continuous base for new orientations in other directions.

3. Two-month-old babies look more at their mothers than do four-month-olds. Older babies are more easily attracted by the toys presented, and remain longer at peripheral targets (the processes disengagement and IOR are evidently more developed). Finally, we see more mutuality in the older dyads than in the younger ones as the behaviours of partners are more smoothly adapted to each other, as could be shown in refined sequence analyses.

Attentional development and soothability

Rothbart *et al.* (1990) have suggested that there might be a direct relationship between the development of the PAS on the one hand and soothability as an aspect of temperament on the other (see also Derryberry & Rothbart, Chapter V.8, this handbook). This is feasible, as attention for the external world may distract the infant from an unpleasant state. Consequently, when the infant is able to disengage and to remain oriented at a new stimulus this may have a soothing effect.

In a detailed study on the development of soothability in infants from 6 to 26 weeks of age we expected a transition around the third or fourth month, as around that period infants can actively change their focus of attention during longer periods, due to the development of the PAS, in particular the ability to inhibit return to an initial focus of attention. However, we failed to find such a clear linear relationship between attentional processes and soothability. Although, around the fourth month, mothers tended more and more to use active strategies to distract the infant (by talking, offering toys, pointing at an object and in the meantime trying to attract the infant's attention by turning it in the direction of that object), we could find no evidence for a direct relationship with IOR (which we measured in the home situation by presenting attractive toys, in a systematic manner, in central and peripheral positions to the infant, scoring fixation times). Mothers have a variety of strategies to sooth their infants. Soothability is embedded in the complex concept of temperament, and mothers strongly adapt their behaviour to the infant's behavioural tendencies, in which attentional processes play only a limited role. Research concerning these relationships is difficult. Parallel developments do not yet imply causal connections. We found, for example, that around the 14th week of life most infants enter a more stable period with respect to their soothability, in that they develop a sort of characteristic pattern. Around the same time the most important markers of PAS are observable. However, no meaningful relationship between these two phenomena could be established.

In conclusion, undoubtedly infants as well as mothers vary greatly in their attentive and attention-demanding behaviours. The developmental course of both sorts of processes is far less consistent than was anticipated by those who believed in clear-cut transitions in the PAS around the fourth month of life. In some infants attention development is almost linear; in others we see sudden changes. How basic attentional processes may affect complex interactive and cognitive processes is still far from clear.

Apart from deficiencies in early attention, state regulation deficiencies may contribute to the risk for ADHD. Some remarks on this issue follow in subsequent paragraphs.

Attention and state-regulation

Attentional processes are largely affected by the regulation of behavioural states. There is, for example, growing evidence for state regulation problems contributing to sustained attention deficits in ADHD (van der Meere, 1996). Therefore, attention–state interactions should be an important focus of early risk research.

Since the early 1960s the concept of state has played a crucial role in studies on neurological development during the early phases of ontogeny, not least due to the work of Wolff (1987) and Prechtl and co-workers (Prechtl, 1974, see also Salzarulo et al.'s contribution to this handbook). In developmental neurology a series of thorough observational and experimental studies on the early development of state regulation have been executed, indicating the impact that variations and deficiencies in state regulation may have for the infant's reactivity, and consequently for its adaptation to the social and physical environment. However, developmental psychology largely failed, at least until the mid-1990s to pay any serious attention to the possible role of the regulation of state in learning and adaptive behaviour. For example, habituation studies, a wealth of which have been executed, have completely neglected how the infant's reactivity and consequently habituation may be affected by fluctuations in the infant's/child's behavioural state. Wolff (1984) states, in an overview on studies on early human wakefulness: "visual attentional processes of young infants have been investigated in considerable detail, but such studies usually take for granted the general disposition of waking" (p. 144). Apart from a few exceptions this statement is still valid. Wolff stresses the importance of wakefulness for any coherent theory of development (see also Wolff, 1987). This remark can be generalized in that variations/deficiencies in state regulation may affect attentional processes and learning as well as social interaction (see also Papoušek, Chapter V.4, this handbook).

EARLY ATTENTION AND LATER ADHD

How can early infancy studies, as discussed above, contribute to our insight in ADHD problems at later age? Whether results of such studies can be meaningfully related to indications of a clinical condition, such as ADHD, very much depends on the conceptual and methodological basis of approaches to the study of aspects of children's information processing at early ages as compared to later phases in development. First some remarks on the study of ADHD.

Information processing in ADHD

Ideally, the study of possible risk mechanisms as components of a later clinical condition, such as ADHD, should have its starting point in a theoretically based relationship, postulated between such early processes and core aspects of the latter condition. In reality there is a large gap between approaches to early risk attentional processes and to later ADHD. To date, studies on ADHD are mainly experimental, based on information-processing models, using the reaction-time paradigm, with a lack of studies on the ecological relevance of findings. Initially, such experimental studies were strongly based on Sternberg's linear stage model of human information processing (Sternberg, 1969). In this model four sequential stages are postulated in the transition from stimulus input to response output; namely: encoding, memory search, decision making and response organization. Interest in autonomic versus controlled information processing came up on the basis of the Shiffrin & Schneider's capacity-limited information-processing model (Shiffrin & Schneider, 1977; Schneider & Shiffrin, 1977). This model focuses on the clinically and developmentally important distinction between attention-demanding (capacity-limited) processes and processes which occur automatically and (almost) effortlessly. Almost all studies fail to find evidence for a limitation in information-processing capacity in ADHD. In general no indications for inabilities in orienting, reorienting, encoding, focused and divided attention are found in ADHD (see Van der Meere, 1996, for a review). This is after more than two decades of studies on possible defects in the various linearly defined stages of information processing.

Studies on processes such as disengagement and IOR are almost lacking in ADHD, with a few exceptions, such as the studies by McDonald *et al.* (1999) and Nigg *et al.* (1997). On the other hand no early attention studies are available which are inspired by the information-processing models applied in ADHD studies. A prerequisite for the improvement of the study of early risk for deviance is that research efforts on risk mechanisms and their possible clinical effects become more theoretically and conceptually related.

State regulation in ADHD research

State regulation is a core topic in the present-day experimental study of information processing in ADHD. This line of research is strongly linked to attempts to gain further insight into the way in which so-called energetic factors might play a role in children's attentive behaviour, indicated by the effects of psychopharmacological events, in particular stimulant drugs, on these children's functioning. Sternberg's additive factor model and Shiffrin and Schneider's two-process theory do not account for the effect of the behavioural state on task performance. This was done by Sanders (1983), who integrated in his hierarchical model of processing sources the previous two models with three energetic systems, which were previously described by Pribram & McGuinness (1975). These are respectively an arousal system, underlying a phasic reaction to input, an activation system which regulates the tonic readiness to respond, and an effort system that is thought to coordinate arousal and activation, but also to be active in controlled information processing. Based on Sanders' model, experimental ADHD research has, during the past decade, focused strongly on how deficiencies in a child's state regulation might interfere with optimal information processing. State regulation problems in ADHD/hyperactivity (a non-optimal arousal state) were already indicated in the early 1980s by Douglas and others (Douglas, 1983, 1988). Currently, a more differentiated state concept is used, in which the processes of arousal and activation are distinguished, related respectively to the alerting effect of sensory activity and the control of motor readiness. Furthermore, an effort system, controlled by an evaluation system and affected by motivational factors, is defined, which can compensate for a suboptimal state. In ADHD a deficit in the effort/activation system of the subject is postulated, leading to sustained attention problems. However, such problems are found only in prolonged task conditions, when stimuli are presented at a high rate of sequence, or when the experimenter is absent (van der Meere, 1996). In terms of Sanders' model this is indicative for a non-optimal activation state. An observational study (Alberts & van der Meere, 1992) suggests self-stimulating activities in ADHD, that increase in the course of a task, such as looking away, and are meant to compensate for a non-optimal activation state. Psychophysiological studies, using evoked potential and heart rate indices for state, also confirm the activation/effort hypothesis in ADHD.

In studies such as these state is experimentally manipulated, usually by changes in the rate of presented stimuli. How deficiencies in attention and state regulation may relate to these children's clinical problems or daily life behaviour is still unclear. To clarify such a relationship a plea has been made for a chain of approaches, in which behavioural assessments, ethological observations, and reaction time studies, focusing on behavioural and psychophysiological variables, are combined (Kalverboer, 1977; van der Meere, 1996). Clinical features as well as experimental findings turn out to be largely situationally dependent, comparable to what was found in an observational study on a large group of preschool children with minor neurological dysfunctions (Kalverboer, 1975). In ADHD, problem behaviours are typically reported in boring conditions, whereas experimental findings depend strongly on the testing conditions and the child's motivation. Especially in this category of children there is a large discrepancy between what these children are optimally able to do in highly structured laboratory conditions and what they actually do in the real-life situation.

How can the quality of research on early risk for ADHD be improved? The next section will discuss some of the main points to be considered for a fruitful approach to the study on risk mechanisms over a larger age range.

FROM EARLY INFANCY TO ADHD

In clinical descriptions of problem behaviour in children, labelled ADHD, attentional deficits evidently play a crucial role. How these problems can be traced back to deficiencies in the basic components of attention – namely selectivity, control, and vigilance, as distinguished in the information processing models – is still a question of debate. Present-day experimental evidence strongly emphasizes these children's state regulation problems interfering with their information-processing efficiency. Experimental studies have not provided clear evidence for selective attention deficiencies in terms of the stage models of information processing. Therefore, deficiencies in components of the PAS, involved in the selection of information by visual orientation, such as disengagement and OIR, may play only a minor role as possible precursors of ADHD symptomatology. In any case such processes concern only a few of the possible determinants of later attentional deficits. Our studies on early visual attention are presented here as an example of how systematic research on the roots of later neurobehavioural disorders can be designed. Such studies can help to throw new light on the background of ADHD, in that they relate attentional processes to the organism's sensory-motor functions and their neuroanatomical basis. Follow-up studies, which also take account of the development of the systems involved in control and vigilance, may improve our insight in the ontogenetic roots of complex clinical entities. Especially, attention–state regulation interactions may be an interesting focus in future longitudinal studies on early risk for neurobehavioural disorders, e.g. ADHD. Prerequisite is a thorough consideration of concepts and models, to be applied in longitudinal research from early infancy onwards in order to develop coherent approaches to the study of early risk for later deviance.

THE OUTCOME CRITERIA: THE CLINICAL CONCEPT ADHD OR MEANINGFUL SUBCATEGORIES

ADHD, as a clinical entity, is not at all a simple phenomenon. The label is in fact an umbrella term for a large variety of problem children, some of whom are overactive, others hypoactive, some dreamy, some distractible, some both, and some having emotional problems. They may be aggressive in the group (many of them are described as bullies), or tend to isolate themselves (being the victims, not the bullies). Evidently such a condition cannot simply be explained in terms of one or a few underlying processes.

Based on present-day insights Swanson *et al.* (1998) suggested a "tentative alignment of symptom domains of ADHD, cognitive processes of attention, and the neural networks of attention" (p. 455). They rightly consider ADHD as a complex "polytypic syndrome" with multiple biological bases. This is evidently the case, a situation comparable to PDD-NOS (pervasive developmental disorder – not otherwise specified), DCD (developmental coordination disorder), autism, and many other DSM categories, which have been clinically and not theoretically defined and are not suited for the selection of groups to detect basic brain–behaviour relationships (see also Kalverboer, 1996). Swanson *et al.* suggest splitting the ADHD symptoms up into three subgroups, corresponding to the three aspects in the ADHD symptomatology, namely, the alerting deficit (indicated by statements on these children's behaviour, such as: "has difficulty sustaining attention" and "fails to finish"), thought to be related to right frontal brain function, a selective attention deficit ("distracted by extraneous stimuli" and "does not seem to listen"), related to an orienting deficit and linked to a network centred bilaterally in the parietal lobe and an executive function deficit (signs of impulsivity, such as "difficulty awaiting turn", and "interrupts or intrudes on others"), thought to be related to a network centred in the anterior cingulated brain region. They suggest a research programme based on these distinctions. Apart from the still-existing lack of clearness on the localization and homogeneity of neural networks (see Posner & DiGirolamo, 1980; Johnson, 1998) and the lack of direct human developmental evidence, this seems a reasonable proposal, not least in view of the controversial reports on brain involvement in ADHD until now. For example, Swanson *et al.* report how Heilman *et al.* (1991) propose in ADHD a right hemisphere (norepinephrine deficit) theory, whereas Malone *et al.* (1994) suggest a left hemisphere (dopamine deficit) theory, evidently emphasizing different ADHD symptoms.

WEAKNESSES IN METHODOLOGY: HOW VALID ARE CLINICAL DESCRIPTIONS?

Much emphasis should be laid on obtaining really valid and reliable data on the clinical phenomena (the behavioural repertoire of the child), an as yet almost completely neglected issue in experimental (cognitively and pharmacologically oriented) studies on ADHD. Usually the only available clinical information (even in the most sophisticated experimental studies) is derived from non-standardized interviews and from behavioural questionnaires, filled out by parents, teachers and/or clinicians and generally of dubious validity and reliability. Undoubtedly, knowledge of parents' and teachers' perceptions of children's behaviour may be important, as their views largely structure a child's environment (Kalverboer *et al.*, 1994). However, such information should be collected with the application of reliable and well-validated instruments.

ECOLOGICAL VALIDITY: A PLEA FOR THE ETHOLOGICAL APPROACH

To really gain more insight in a clinical entity, such as ADHD, detailed and standardized observations of children's behaviour in various ecologically relevant conditions would be required (see Kalverboer, 1975, as an example of such an approach), and ethology provides the basic approaches to obtaining such information. Clinical researchers such as Ounsted (1955) and Hutt *et al.* (1965) have done pioneering work in humans with their observational studies on hyperactive, brain-damaged and autistic children. Later, detailed follow-up studies on the development of children with neurological deficiencies have elaborated these observations on the children's behavioural repertoire, so that clinically meaningful information could be obtained, expressed in terms of neurobehavioural profiles (Kalverboer, 1975). In these studies careful observations revealed as early as the early 1970s how situationally dependent neurobehavioural relationships were in children with minor neurological deficiencies. Their problems were typically manifested in boring conditions with a lack of attractive play materials. In more stimulating play conditions they functioned as well as controls, or even better. Free-field observations revealed these situation–behaviour interactions, which did not show up in more rigidly structured experimental conditions and as a result had also been overlooked in experimental studies on ADHD for a long time.

Unfortunately, after Tinbergen & Tinbergen (1972), ethology, the approach "par excellence" for the study of the organism in its "natural" condition, has failed to maintain interest in the direct study of human natural history, although some of its impact can be traced back in developmental psychology. In ADHD there is a lack of natural history studies and systematic observations in daily-life conditions. If only a fraction of the methodological and theoretical efforts presently invested in the experimental model-driven approach to ADHD was invested in obtaining valid accounts of children's daily-life adaptations the quality of brain–behaviour studies in ADHD would improve considerably. Then, close follow-up studies on carefully diagnosed newborns and young children at risk, both neurologically and behaviourally, would really provide an insight into the mechanisms underlying these children's problem behaviours.

COLLABORATION BETWEEN DISCIPLINES

In trying to understand the developmental course of neurobehavioural patterns and their underlying mechanisms, insights derived from developmental neurology are of the utmost importance. They prevent researchers from too-simple conceptualizations as they show that there are large fluctuations and relapses in the development of (functional) patterns (Touwen, 1976), that forms are not simply developing in a linear sequence (e.g. the palmar grasp is not simply a precursor of voluntary, radial grasping), but that patterns may exist in parallel. Behavioural patterns, related to the maturation of the nervous system, can be considered as ontogenetic adaptations, having a function in a particular phase of development. Such notions are indispensable for obtaining insight into how developmental transitions come about.

Such an approach should be complemented by refined experimental and observational studies of cognitive processes. Laboratory studies on basic mechanisms, such as overt and covert attention, should be paralleled by detailed observations on daily-life adaptations, in the best tradition of work as done by ethologists and developmental psychologists such as MacGraw, Tinbergen and Papoušek.

A PLEA FOR LONGITUDINAL STUDIES

There is a lack of longitudinal studies over a larger age range (e.g. from birth onwards until middle school age) on the development of attentional mechanisms in relation to neurological development. Data available until the present have been obtained mainly in studies of a limited scope, mostly cross-sectional, in infants and young children. What is required are precise analyses of the developmental pathways of these processes in relation to the development of the visual system and brain structures, to be complemented by careful observations of the complex patterns of interactive and exploratory behaviour in the child's natural condition. Such studies are of particular importance as deficiencies in basic mechanisms of early attention may interfere with early communication and learning (e.g. habituation) and an efficient exploration of the environment. They may help to better understand how early deviances may contribute to later clinical conditions and are a prerequisite for getting insight in how "the self-organizing and adaptive nature of brain development can channel (also deviant) developmental trajectories to one or a small number of adaptive outcomes" (Johnson, 1997, p. 201).

FINAL REMARKS

What is the state-of-the-art in present-day research on the sequelae of early biological risk factors? Earlier I indicated four key problems in such research, namely (Kalverboer, 1988):

1. the long time lapse between the first manifestations of risk and their possible effects requiring longitudinal studies over a long time period;
2. the difficulty in defining behavioural disorders in various stages of development in a theoretically consistent way;
3. difficulties in the early assessment of the central nervous system;
4. the continuous reorganization of behaviour during development.

My conclusion at that time was that long-term predictions of developmental outcomes on the basis of the child's prenatal and perinatal history were not possible. This statement still holds, supported by latest insights into the dynamics of development, although evidently great improvements have been made, especially technically (e.g. in brain-imaging techniques). It has become clear in the past decade that one should not overestimate the predictive value of early signs of risk. To date too little is understood about neural plasticity and the dynamics of human development to know what, in the long run, the predictive value of early risk factors will be (Prechtl, 1981, p. 211). Conceptually and methodologically, so far only the first steps have been taken in coming to a more integrated approach to the study of neurobehavioural development in the human, notwithstanding the invention of a new term 'developmental cognitive neuroscience'.

Evidently the longitudinal study of basic attentional mechanisms can pay a significant contribution to our insight into complex cognitive and interactional processes and their developmental course. However, this is only one set of determinants from a large number of influencing factors in the emotional and cognitive domain. How deficiencies may contribute to later problem behaviour such as ADHD is a topic for further research. Research is time-consuming; it demands building up good cohorts of groups at-risk from a very early age onwards and a most precise methodology and, not unimportant, is not easily funded. Clearly enough there is still a long way to go, notwithstanding the promising techniques which have become available (such as brain imaging, and refined experimental procedures for the study of cognitive and behavioural functions from birth onwards) and the growing insight into how the brain might organize complex attentional and cognitive functions.

References

Alberts, E. & van der Meere, J.J. (1992). Observations of hyperactive behaviour during vigilance. *Journal of Child Psychology and Psychiatry*, **33**, 1355–1364.

Brighi, A. (1997). Mother–infant interaction and contingency learning in full-term infants. *Early Development and Parenting,* **6**, 37–45.

Butcher, P.R. (2000). Longitudinal studies of visual attention in infants: the early development of disengagement and inhibition of return. Meppel, the Netherlands: Aton.

Butcher, P.R. & Kalverboer, A.F. (1997). The early development of covert visuo-spatial attention and its impact on social looking. *Early Development and Parenting*, **6**, 15–26.

Butcher, P.R., Kalverboer, A.F. & Geuze R.H. (1999). Inhibition of return in very young infants: a longitudinal study. *Infant Behaviour and Development*, **22**, 303–319.

Butcher, P.R., Kalverboer, A.F., Geuze, R.H. & Stremmelaar, E.F. (2001a). A longitudinal study of the development of shifts of gaze to a peripheral stimulus in preterm infants with transient periventricular echogenicity. *Journal of Experimental Child Psychology*, (In Press).

Butcher, P.R., Kalverboer, A.F., Geuze, R.H. & Stremmelaar, E.F. (2000) The development of IOR in preterm infants with transient periventricular echogenicity: a longitudinal study. In Butcher, P.R. *Longitudinal studies of visual attention in infants: the early development of disengagement and inhibition of return.* (Chapter 6, pp. 93–111). Meppel, the Netherlands: Aton.

Butcher, P.R., Kalverboer, A.F. & Geuze, R.H. (2001b). Infants' shifts of gaze from a central to a peripheral stimulus: a longitudinal study of development between 6 and 26 weeks. *Infant Behaviour and Development*, **23**, 3–21.

Clerkes-Julkowski, M. (1998). Learning disability, attention-deficit disorder and language impairmen as outcomes of prematurity: a longitudinal descriptive study. *Journal of Learning Disabilities,* **31**, 294–306.

Cordes, K.E. & Venekamp, I. (1996). On the relationship between mother's behaviour and visual behaviour of the baby. Research report, Laboratory of Experimental Clinical Psychology, University of Groningen, the Netherlands.

Diagnostic and Statistical Manual of Mental Disorders, (4th ed.), (*DSM-IV*) (1994). Washington, DC: American Psychiatric Association.

Douglas, V.I. (1983). Attentional and cognitive problems. In M. Rutter (Ed.), *Developmental neuropsychiatry* (pp. 280–328). New York: Guilford Press.

Douglas, V.I. (1988). Cognitive deficits in children with attention disorder with hyperactivity. In L.M. Bloomingdale & J.A. Sergeant (Eds.), *Attention deficit disorder: criteria, cognition, intervention* (pp. 65–83). Oxford: Pergamon Press.

Fawer, C.L. & Calame, A. (1991). Significance of ultrasound appearances in the neurological development and cognitive abilities of preterm infants at 5 years. *European Journal of Pediatrics*, **150**, 515–520.

Fijn van Draat, R.E. (1995). The early development of attentional processes: the Posterior Attention System. Research report, Laboratory of Experimental Clinical Psychology, University Groningen, the Netherlands.

Heilmann, K.M., Voellar, K.S. & Nadeau, S.E. (1991). A possible pathophysiological substrate of ADHD. *Journal of Child Neurology*, **6**, S76–S81.

Hutt, C., Hutt, S.J. & Ounsted, C.A. (1965). The behaviour of children with and without upper C.N.S. lesions. *Behaviour,* **24**, 246–268.

Johnson, M.H. (1997). *Developmental cognitive neuroscience: an introduction*. Oxford: Blackwell.

Johnson, M.H. (1998). Developing an attentive brain. In R. Parasuraman (Ed.), *The attentive brain* (pp. 4276–443). Cambridge, MA: The MIT Press.

Johnson, M.H., Posner M.I. & Rothbart, M.K. (1991). Components of visual orienting in early infancy: contingency learning, anticipatory looking and disengaging. *Journal of Cognitive Neuroscience*, **3**, 336–343

Kalverboer, A.F. (1975). A neurobehavioural study in preschool children. *Clinics in Developmental Medicine*, no. 54. London: Heinemann/Spastics Society.

Kalverboer, A.F. (1977). Human ethology: the possible use of direct systematic observation in human psychopharmacology. In H.M. van Praag & J. Bruynfels (Eds.), *Neurotransmission and disturbed behaviour (*pp. 280–293). Haarlem: Erven Boon.

Kalverboer, A.F. (1988). Follow-up of biological high-risk groups. In M. Rutter (Ed.), *Studies of psycho-social risk: the power of longitudinal data* (pp. 114–137). Cambridge. UK: Cambridge University Press.

Kalverboer, A.F. (1996). Developmental psychopathology. In: Husén, T. & Postlethwaite, T.N. (Eds.), *The international encyclopedia of education* (pp. 67–73). London: Pergamon Press.

Kalverboer, A.F., Genta, M.L. & Hopkins, J.B. (1999). *Current issues in developmental psychology: biopsychological perspectives*. Dordrecht: Kluwer Academic Publishers.

Kalverboer, A.F., Le Coultre, R. & Casaer, P. (1970). Implications of congenital ophthalmoplegia for the development of visuo-motor functions (illustrated by a case with the Moebius syndrome). *Developmental Medicine and Child Neurology*, **12**, 642–654.

Kalverboer, A.F., van der Schot, L.W.A., Hendrikx, M.M.T., Huisman, J., Slijper, F.M.E. & Stemerdink, B.A. (1994). Social behavior and task orientation in early-treated PKU. *Acta Paediatrica,* Supplement, **407**, 1104–1105.

Klein, R.M. (1979). Does oculomotor readiness mediate cognitive control of visual attention? In R.S. Nickerson (Ed.), *Attention and performance, VIII* (pp. 259–288). Hillsdale, NJ: Erlbaum.

Kopp, C.B. (1983) Risk factors in development. In M.M. Haith & J.J. Campos (Eds.), *Infancy and developmental psychobiology* (pp. 1081–1088). *Handbook of child psychology,* vol. 2 (P. Mussen, ed.). New York: Wiley.

Laufer, M.W. & Denhoff, E. (1957). Hyperkinetic behavior syndrome in children. *Journal of Pediatrics,* **5550**, 463–474.

Malone, M.A., Kershner, J.R. & Swanson, J.M. (1994). Hemispheric processing and methylphenidate effects in ADHD. *Journal of Child Neurology,* **9**, 181–189.

Mareschal, D. (1998). Developental cognitive neuroscience and connectionist models of infancy. *Early Development and Parenting,* **7**, 147–152.

Mc Donald, S., Bennett, K.M.B., Chambers, H. & Castello, U. (1999). Covert orienting and focussing of attention in children with attention deficit hyperactivity disorder. *Neuropsychologia*, **37**, 345–353.

Nigg, J.T., Swanson, J.M. & Hinshaw, S.P. (1997). Covert visual attention in boys with attention deficit hyperactivity disorder: lateral effects, methylphenidate response and results for parents. *Neuropsychologia*, **35**, 165–176.

Ounsted, C.A. (1955). The h /perkinetic syndrome in epileptic children. *Lancet*, **2**, 303–311.

Papoušek, H. & Papoušek, м. (1983). Biological basis of social interactions: implications of research for an understanding of behavioural deviance. *Journal of Child Psychology and Psychiatry,* **24**, 117–129.

Papoušek, H. & Papoušek, M. (1993). Early interactional signalling: the role of facial movements. In A.F. Kalverboer, J.B. Hopkins & R.H. Geuze (Eds.), *Motor development in early and later childhood: longitudinal approaches* (pp. 136–152). Cambridge, UK: Cambridge University Press.

Papoušek, H., Papoušek, M. & Koester, L.S. (1999). Early integration of experience: the interplay of nature and culture. In A.F. Kalverboer, M.L. Genta & J.B. Hopkins (Eds.), *Current issues in developmental psychology: biopsychological perspectives* (pp. 27–51). Dordrecht: Kluwer.

Parasuraman, R. (Ed.) (1998). *The attentive brain*. Cambridga, MA: The MIT Press.

Posner, M.I. (1988). Structures and functions of selective attention. In T. Boll & B. Bryant (Eds.), *Clinical neuropsychology and brain function* (pp. 173–202). Washington, DC: APA.

Posner, M.I. (1990). Hierarchical distributed networks in the neuropsychology of selective attention, In A. Caramazza (Ed.), *Cognitive neuropsychology and neurolinguistics* (pp. 187–210). Hillsdale, NJ: Erlbaum.

Posner, M.I. (1995). Attention in cognitive neuroscience: an overview. In M.S. Gazzaniga (Ed.), *Handbook of cognitive neuroscience* (pp. 615–624). Cambridge, MA: MIT Press.

Posner, M.I. & DiGirolamo G.J. (1980). Executive Attention: Conflict, target Detection, and Cognitive Control. In R. Parasuraman (Ed.), *The attentive brain* (pp. 401–423). Cambridge, MA: MIT Press.

Posner, M.I. & Petersen, S.E. (1990). The attention system of the human brain. *Annual Review of Neuroscience, 13*, 25–42.

Posner, M.I., Rothbart, M.K. (1991) Attentional mechanisms and conscious experience. In A.D. Milner & M.D. Rugg (Eds.), *The neuropsychology of consciousness* (pp. 91–112). London: Academic Press.

Posner, M.I., Inhof, A.W., Friedrich, F.J. & Cohen, A. (1987). Isolating attention systems: a cognitive anatomical analysis. *Psychobiology, 15*, 107–121.

Posner, M.J., Inhof, A.W., Friedrich, F.J. & Cohen, A. (1987). Isolating attention systems: a cognitive anatomical analysis. *Psychobiology, 15*, 107–121.

Prechtl, H.F.R. (1974). The behavioural states of the newborn infant (a review). *Brain Research, 76*, 185–212.

Prechtl, H.F.R. (1981). The study of neural development as a perspective of clinical problems. In K.J. Connolly & H.F.R. Prechtl (Eds.), *Maturation and development: biological and psychological perspectives* (pp. 198–215). London: SIMP/Heinemann.

Prechtl, H.F.R. & Beintema, D.J. (1964). *The neurological examination of the full-term newborn infant*. Clin. Dev. Med. no.12. London: SIMP/Heinemann.

Pribram, K.H. & McGuinness, D. (1975). Arousal, activation and effort in the control of attention. *Psychological Review, 82*, 116–149.

Richards, J.E. (1997). Focussing on visual attention. *Early Development and Parenting, 7*, 153–158.

Richards, J.E. (Ed.) (1998). *Cognitive neuroscience of attention: a developmental perspective*. Mahwah, NJ: Erlbaum.

Robson, A. & Cline B. (1998). Developmental consequences of intra-uterine growth retardation. *Infant Behavior and Development, 21*, 331–344.

Rothbart, M.K., Posner, M.J. & Boylan, A. (1990). Regulatory mechanisms in infant temperament. In Enns, J. (Ed.), *The development of attention: research and theory* (pp. 47–66). Amsterdam: North Holland.

Sanders, A.F. (1983). Towards a model of stress and human performance. *Acta Psychologica, 53*, 61–97.

Schaffer, H.R. (Ed.) (1971). *The origins of human social relations*. New York: Academic Press.

Schneider, W. & Shriffrin, R.M. (1977). Controlled and automatic human information processing; detection, search and attention. *Psychological Review, 84*, 433–466.

Shriffrin, R.M. & Schneider, W. (1977). Controlled and automatic human information processing: perceptual learning, automatic attending, and a general theory. *Psychological Review, 84*, 127–190.

Sternberg, S. (1969). Discovery of processing stages: extensions of Donders' method. In W.G. Koster (Ed.), *Attention and performance* (vol. 2, pp. 276–315). Amsterdam: North Holland.

Swanson, J., Posner, M.I., Cantwell, D., Wigal, S., Crinella, F., Filipek, P., Emerson, J.M., Tucker, D. & Nalcioglu, O. (1998). Attention-deficit/hyperactivity disorder: symptom domains, cognitive processes, and neural networks. In R. Parasuraman (Ed.), *The attentive brain*, (pp. 445–460). Cambridge, MA: MIT Press.

Thomas, A. & Chess, S. (1977). *Temperament and development*. New York: Brunner/Mazel.

Tinbergen, N. & Tinbergen, E.R. (1972). Early childhood autism – an ethological approach. *Zeitschrift für Tierpsychologie, 10*, 1–53.

Touwen, B.C.L. (1976). *Neurological development in infancy*. Clin. Dev. Med. no. 58. London: SIMP/Heinemann.

Touwen, B.C.L. (1979). *The examination of the child with minor neurological dysfunction*. Clin. Dev. Med. no. 71. London: SIMP/Heinemann.

Van der Meere, J.J. (1996). The role of attention. In S. Sandberg (Ed.), *Monographs on child and adolescent psychiatry: hyperactivity disorders of childhood* (pp. 111–148). Cambridge, UK: Cambridge University Press.

Webster, M.J. & Ungerleider, L.G. (1998). Neuroanatomy of visual attention. In R. Parasuraman (Ed.), *The attentive brain* (pp. 19–34). Cambridge, MA: MIT Press.

Wijnroks, L. (1994). *Dimensions of mother–infant interactions and the development of social and cognitive competence in preterm infants*. Doctoral thesis, State University Groningen, the Netherlands.

Wijnroks, L. (1997). Mother–infant interaction and contingency learning in pre-term infants. *Early Development and Parenting*, **6**, 27–36.

Wolff, P.H. (1984). Discontinuous changes in human wakefulness around the end of the second month of life: a developmental perspective. In H.F.R. Prechtl (Ed.), *Continuity of neural functions from prenatal to postnatal life* (pp. 144–158). Clin. Dev. Med. no. 94. Oxford: SIMP.

Wolff, P.H. (1987). *The development of behavioural states and the expression of emotions in early infancy*. Chicago, IL: University of Chicago Press.

Further reading

Johnson, M.H. (1997) *Developmental cognitive neuroscience: an introduction*. Oxford: Blackwell.

Prechtl, H.F.R. (Ed.) (1984). *Continuity of neural functions*, Oxford: SIMP/Heinemann.

Richards, J.E. (Ed.) (1998). *Cognitive neuroscience of attention: a developmental perspective*. Mahwah, NJ: Erlbaum.

813

V. DEVELOPMENT OF COMMUNICATION AND EMOTION

V.1
Introduction

ALEX F. KALVERBOER

All contributions to this section concern the strongly related issues of communication and emotion in relation to structure, physiology and function of the brain. This is a growing field of interest in brain–behaviour research. In this section (psycho-) physiological, neurobiological, ethological, psychological and psycholinguistic approaches are represented.

The chapter by **Hofer** (V.2), concerns biological developmental events that may have important regulatory effects on particular physiological systems and related interactive behaviours. The author makes clear how animal research has revealed a network of simple behavioural and biological processes that may underlie the initial formation of attachment in early human development. He discusses how, within the observable parent–offspring interactions, a number of sensorimotor, thermal and nutrient-based events have regulatory effects on particular physiological systems and (interactive) behaviours in the infant. Separation effects in infant rats can be traced back to the abrupt withdrawal of all these regulatory processes, whereas such processes may have long-term effects on the shaping of adult physiology and behaviour. Such findings are a most inspiring basis for the direct study of regulatory processes in early human development and provide a bridge between biological and psychological processes.

The chapter by **Trevarthen** (V.3) is on the neurobiology of early communication. Its core theme is that innate motives and emotions anticipate the development of intersubjective contacts and relations. They have their basis in an extensive "preadapted" neural system, which develops prenatally and concerns the whole brain, including the subcortical areas. This system is especially elaborate in humans, and is concerned with linking emotions and cognitions. Trevarthen gives an extensive review of prenatal brain development, underlining the primacy of "body mapping cell arrays" and the formation of environment-expectant cerebral tissues, which are adapted for cognitive assimilation of socioemotional information. Clearly enough infants' abilities are not just "emerging constructions of experience". The chapter is thought-provoking and evidently contrasts with views in current cognitive neuroscience, which are considered by the author as individualistic, rational/constructivist and corticocentric.

A.F. Kalverboer and A. Gramsbergen (eds.), Handbook of Brain and Behaviour in Human Development, 817–820
© 2001 Kluwer Academic Publishers. Printed in Great Britain.

Papoušek's chapter on "Signalling in early social interaction and communication" (V.4) contains an eloquent review of a series of studies on the roots of early human communication, to which the author has paid a major contribution. Individual behaviour patterns are seen as a rich source of information about the individual's internal state and its intentions, and become a means of dialogic exchange with the social partner. In humans new dimensions of communicative capacities have evolved resulting from a species-specific ability to create new representative symbols and use them for purposes of verbal communication, and internal representation in rational thinking. Fundamental steps of this process can be detected in early phases of human ontogeny. Contributions of microanalytic, experimental, interdisciplinary, and comparative approaches to the study of preverbal forms of communication are discussed in this chapter. There is strong evidence for innate determinants of infant's signalling patterns, indicating an "intrinsic motivation" for social communication, and which, if unfulfilled, may be a strong pathogenetic factor. There is a universality in rhythmic patterns, due to human timing tendencies and with a variability that reflects the ongoing dynamics of the interaction. Caretaker's reactions, described in terms of "intuitive parental didactics" are the result of a coevolution (early emergence, universality, minimal awareness) of predispositions for signalling in the infants' social environment. Even deaf mothers apply such didactical principles, using sign language. The caregiver tends to interpret the infant's behaviour as intentional and communicative

Pantoja, Nelson-Goens, and **Fogel** (V.5) discuss a dynamic systems approach to the study of early emotional development in the context of mother–infant communication. This approach is particularly useful for the study of how emotions self-organize and change over time. It is contrasted with the differential emotions theory (which considers basic emotions as biologically inherited programmes; a limited number of basic emotions is associated with different cognitive processes across time) and the functionalist theory (which envisages emotions as "a state of action readiness", evolving from the interplay between the organism and the "appraised" environment). Dynamic systems theory, with as a core concept "self-organization", also considers emotions as relational processes, not as hardwired states. Their review of the neurophysiological concomitants (mainly focusing on brain lateralization) and behavioural components of emotions indicates the large gap which still exists between brain and behaviour studies in this domain. Therefore, their plea for the integration of detailed observations over time of infants with their primary caregivers in studies on neurological mechanisms is well taken. The authors stress the heuristic value of their qualitative approach, which they want to combine with quantitative methods, and which are illustrated by excerpts from materials of an ongoing study.

De Haan and **Halit** (V.6) discuss the developmentally and theoretically intriguing issue of the development of face recognition during infancy and its neural basis. The core question is how the constraints imposed by the structure of the brain interact with input from the environment during ontogeny to create the mature pattern of specialization. Present-day evidence suggests that there is a subcortical mechanism, present very early in postnatal life, that acts to preferentially direct young infants' attention to face-like patterns and which does not require visual experience. Infants seem to encode and recognize the features that are most salient in a display. By the second month of life a transition in face processing occurs due to cortical development. How specific is this cortical system for face processing? Evidence to date is most consistent with the view that there is a general-purpose recognition system, which specializes during early ontogeny. Shifts in sensitivity during development, documented to date, are not specific to faces (e.g. shifts in sensitivity from external to internal features of a display). Early cortical processing is less face-specific than in later life. Experience plays an important role in the development of specialization; however, whether there is a critical period has yet to be determined.

The chapter by **Dehaene-Lambertz, Mehler**, and **Pena** (V.7), is a psycholinguistic contribution which discusses how different fields of cognitive neuroscience (developmental studies, studies on patients with lesions in the brain, and brain imaging studies), psychology (left/right hemisphere involvement), and psycholinguistics (acquisition of a second language) may contribute to the clarification of the issue of how language arises in the infant's brain. Although there is a wealth of material available from such studies, insight into how the human brain is uniquely able to acquire language is still very limited. The authors argue rightly enough that there is a lack of direct developmental studies, and of specific instruments for the study of linguistic capacities (the frequently used verbal IQs are inadequate), whereas too many studies merely compare groups, while a single-subject approach would be required.

The final chapter, by **Derryberry** and **Rothbart** (V.8), discusses the temperamental organization of reactive processes related to positive and negative emotions and regulatory processes related to the development of attention, in particular to the (reactive) posterior and the (voluntary) anterior attention system. More flexible and voluntary means of control are afforded by the development of attentional systems which regulate reactive approach and avoidance tendencies. These processes are discussed in terms of underlying neural systems, a reward-responsive circuitry, an approach-avoidance related motor system, and a dopaminergic facilitatory mechanism. On the basis of longitudinal findings on the stability of temperament from infancy to childhood, it is argued that differences in temperament and in emotional and behavioural reactions may provide a basis for later personality development. This chapter attempts to link knowledge and approaches from different disciplines (evolution theory-based ethology, personality theory, cognitive neuropsychology, and psychopharmacology) in the study of emotional development. Of course there are methodological complications, e.g. when data from parent-questionnaires and home and laboratory observations are related to psychophysiological measures. However, despite limitations the temperamental approach can make a valuable contribution to the interdisciplinary study of emotional development, especially when the study of emotions and neural mechanisms (e.g. in relation to frontal, executive, functions) becomes more refined.

V.2
Origins of Attachment and Regulators of Development within Early Social Interactions: From Animal to Human

In recent years animal research has revealed a network of simple behavioural and biological processes that underlie the psychological constructs we use to describe early social relationships. The developmental events responsible for the initial formation of attachment in the fetal and newborn periods, and the origins of our first vocal communication system, are beginning to be elucidated at the level of specific behavioural and neural processes. In addition we have found, hidden within the observable interactions of parent and offspring, a number of sensorimotor, thermal and nutrient-based events that each have unexpected regulatory effects on particular infant behaviours and physiological systems. The complex pattern of responses to early maternal separation in infant rats can be traced to the abrupt withdrawal of all these various independent regulatory processes that had been acting on individual components of the infant's physiology and behaviour. These regulatory processes also appear to mediate long-term shaping effects exerted by early relationships on adult physiology and behaviour. In human development early regulatory interactions may provide a bridge between biological and psychological processes in the development of our earliest emotional states and mental representations.

INTRODUCTION

Growing evidence suggests that the basic functional building blocks for much of the mammalian infants' affective and communicative development are laid down in the early postnatal period, primarily through interaction with its mother (Hann *et al.,* 1998; Panksepp, 1998; Carter *et al.,* 1997; Goldberg *et al.,* 1995; Schore,

A.F. Kalverboer and A. Gramsbergen (eds.), Handbook of Brain and Behaviour in Human Development, 821–840
© 2001 Kluwer Academic Publishers. Printed in Great Britain.

1994). Yet the first behaviours that human infants direct towards caretakers do not resemble those that we associate with emotional expression, and the few early emotional expressions that do occur – such as smiling, crying and grimacing – are often disconnected from other behaviours and from appropriate eliciting events. They appear "spontaneously" when we see no reason for them, and often fail to appear when anticipated (for example, the social smile does not appear for 2–3 months). Only the defensive responses to pain and cold are clearly present soon after birth. We have learned, through bitter experience in Romania (Earls & Carlson, 1998), that if infants are fed, but have little or no interaction with a caretaker, their affect expression and capacity for social communication will remain essentially in this primitive state. Thus, interactions with a caretaker appear to play the same role for an infant's affective and communicative development as early visual input has for its ability to see. It is remarkable that the effects of early deprivation on the development of emotion and communication were not firmly established until the latter half of this century, although anecdotal experience with abandoned and wild children in the 19th century suggested major effects (Malson, 1972; Singh & Zingg 1966).

The general term used for the processes by which the infant's first social relationship is formed, regulated and expressed is "attachment" or, colloquially, "bonding". Unfortunately these words convey the idea of the emergence of a unitary functional system, whereas attachment is not unitary, but made up of a number of different processes, much as is the case for appetite. Thus, there is now evidence that the affects generally associated with attachment, such as the joy of reunion, the comfort of sustained contact, the anxiety of separation and the despair of loss, emerge out of the sustained activity of a number of different processes that take place within the ongoing parent–infant interaction, as do the patterns of social communication that are associated with these affect states.

Thus, in order to understand the development of attachment and the responses to separation, we need to look closely within the observable parent–infant interaction beginning prior to birth. Through the identification and analysis of the component processes underlying these interactions, we will find clues to the brain substrates for early affects and communication patterns.

However, the early parent–infant interaction also has long-term shaping effects on the physiology and behaviour of adult offspring as described in this chapter. Indeed, experimental studies in laboratory animals have confirmed clinical observations of early experience effects that reach beyond adulthood, to the next generation (Denenberg & Whimbey, 1963; Skolnick et al., 1980). Since an experimental analysis of the mother–infant interaction in humans is not possible, and long-term observational studies from infancy to adulthood are so very difficult and take so long, investigators have turned to rapidly developing laboratory species such as the rat.

This chapter will describe what we have learned from experimental studies in laboratory animals about the developmental processes that underlie: (1) the early formation of an enduring relationship between mother and infant, (2) the elicitation and organization of the emotional states that follow separation and reunion, and (3) the capacity of different mother–infant interaction patterns to shape the development of offspring brain and behaviour. At each point in the story I will attempt to relate the conclusions of the animal research to human development.

THEORETICAL BACKGROUND

Until recently research and theory in the field of the development of social communication and emotion had proceeded in isolation from the fields of cognitive
and brain development. Yet at the centre and origin of the first relationship lies
a daunting cognitive task for the newborn mammal: learning to identify, remember
and prefer its own mother, learning to approach, to remain close and to orient
itself to her so that suckling can occur. Once thought to be limited to simple
reflex acts, the learning and memory capacities of newborn and infant mammals
have been the subject of considerable research over the past two decades, but
this work has generally not been incorporated into the attachment literature. In
this early period of life we can see how emotion and cognition arise together
from the same interactive experiences. In newly hatched birds the phenomenon
of "imprinting" played an important role in the early development of attachment
theory, and has been the subject of elegant analyses at the level of neural
processes of learning as well as at the behavioural level (reviewed in Rauschecker
& Marler, 1987). However, nothing similar has been found in slower-developing
mammals such as rodents or primates.

Bowlby, and the attachment field that grew from his landmark synthesis of
ethology and psychoanalysis, theorized that some form of imprinting-like process
must underlie the formation of attachment in slower-developing mammals such
as humans (Bowlby, 1969). Attachment theory holds that the early responses of
newborns consist primarily of species-typical reflex acts. These are thought to
become slowly organized into a behavioural system that maintains proximity to
the mother. Through repeated use and maturation, Bowlby theorized, this
simple system gradually develops into a full-fledged attachment system (at about
7–8 months in the human) as it becomes imbued with affect and organized around
an "internal working model" (mental representation), specified by the infant's
particular experiences with its mother up to that point in time. The mother goes
through a similar process, strongly influenced by (unconscious) traces of her own
experiences with her mother. The affective "bond" thus formed is expressed in
the infant's early positive emotional states of joy, comfort, security in the
presence of the mother and the negative states of protest, anxiety, anger and
despair elicited by separation from her.

In addition to the uncertainty as to how attachment is initiated, the theory has
difficulty in explaining the slower-developing physiological and behavioural effects
of early maternal separation and falls prey to circular reasoning in that the best
evidence for the existence of attachment is the immediate (protest) response to
separation, yet attachment is then used as an explanation for the very protest
response from which it was inferred. Despite these unresolved issues, current
attachment research is focused on the delineation of different patterns of attachment, their correlation with subsequent child development and the generational
transmission of attachment patterns, as if the basic processes underlying attachment formation and the response to separation were settled.

The other theoretical framework for early social bond formation derives from
the European ethologists, whose emphasis on fixed action patterns and innate
releasing mechanisms has gradually incorporated more flexibility in its formu-
lations, based on the mounting neuroscience evidence for activity-dependent

developmental brain processes. Emotion theorists too are shifting away from an emphasis on distinct types of "universal" emotions with fixed, species-typical expressions. Instead, a "transactional approach" is being used in which an emotion is thought to constitute a stage in an organism's interaction with its environment during which responses are organized adaptively in relation to the demands of the particular situation eliciting the emotional state (Campos *et al.,* 1994).

These shifts in theoretical orientation have set the stage for a new view of attachment as deriving from specifiable regulatory processes recently found to be embedded in the interactions between infant and mother (Hofer, 1995). These hidden regulatory processes do not constitute a single motivational system, but rather represent a network in which different physiological and behavioural systems of the infant are regulated by different components of the interaction between mother and infant. The mother–infant dyad can thus be viewed as a self-regulating system, an extension of the infant's own homeostatic system. Its developmental role is not only to supply the environment that is necessary for the infant to express its genetic potential but also to contribute the specific experiences that modulate genetic expression and thus shape the infant's development.

This view of the mother–infant dyad as a self-regulating biological system provides us with an entirely different understanding of the response to separation, in which individual systems respond, each in its own way, to withdrawal of the component of the interaction that had been its regulator during the pre-existing relationship. The complex pattern of biobehavioural responses to maternal separation early in the postnatal period is thus the result of withdrawal of all these regulatory interactions at once. In this view, as development proceeds, the regulatory interactions become more complex. At first these consist of simple patterns in the frequency and intensity of stimulation related to the nursing cycle. Soon contingent and rhythmic events, intersensory stimulation and learned expectancies characterize the interaction. Emotional states are now regulated rather than individual systems. In humans, complex traces of these experiences are laid down in memory as mental representations, and eventually become the basis for the child's behaviour, and for the adult's responses to separation and loss.

REVIEW OF LITERATURE

There are several recent reviews of attachment covering a broad range of animal and human research (Cassidy & Shaver, 1999; Goldberg *et al.,* 1995; Kraemer, 1992; Nelson & Panskepp, 1998; Panskepp, 1998). Therefore I will focus on two generally neglected areas, the initial development of attachment and the mechanisms for its long-term effects on development. To do this I will concentrate on the single species in which we have gained the most new understanding: the laboratory rat. Studies in animals can never in themselves answer questions about human nature, but they can generate new ideas, methods, and concepts that allow us to see human development in a new light and provide research questions that we would never otherwise have thought to ask. I will attempt to suggest some of these implications as I tell the story of attachment from its origins in the fetus to its lasting effects in the next generation.

Prenatal origins

Interest in behaviour during the prenatal period was late to develop, because the fetus was so hard to study and its behaviour was thought to be made up of little more than rudimentary movements and primitive reflexes. With the advent of ultrasound imaging techniques and a method for exteriorizing the rat fetus for experimentation, however, we have begun to learn something about this forgotten period in life and to realize how much behaviour, including learning, begins *in utero*. The earliest edited book on fetal behaviour appeared in 1988 (Smotherman & Robinson).

The first strong evidence for fetal learning came from studies on early voice recognition in humans, in which it was found that babies recognize and prefer their own mother's voice, even when tested within hours after birth (de Casper & Fifer, 1980). Bill Fifer continued these studies in our department using an ingenious device through which newborns can choose between two tape-recorded voices by sucking at different rates on a pacifier rigged to control an audiotape player (reviewed in Fifer & Moon, 1995). He has found that newborn infants, in the first hours after birth, prefer human voices to silence, female voices to males, their native language to another language and their own mother to another mother reading the same Dr Seuss story. In order to obtain more direct evidence for the prenatal origins of these preferences (rather than very rapid postnatal learning), Fifer filtered the high-frequency components from the tapes to make the mother's voice resemble recordings of maternal voice by hydrophone placed within the amniotic space of pregnant women. This altered recording, in which the words were virtually unrecognizable to adults, was preferred to the standard mothers' voice by newborns in the first hours after birth, a preference that tended to wane in the second and third postnatal days. Furthermore, there is now evidence that newborns prefer familiar speech and music sequences to which they have been repeatedly exposed prenatally.

In a striking inter-species similarity, rat pups were shown to discriminate and prefer their own dam's amniotic fluid in preference to that of another dam (Hepper, 1987). Newborn pups were also shown to require amniotic fluid on a teat in order to find and attach to it for their first nursing attempt (Blass, 1990). Robinson & Smotherman (1995) have directly tested the hypothesis that pups begin to learn about their mother's scent *in utero* and have begun to explore neural substrates for this very early form of plasticity. They have been able to demonstrate one trial taste aversion learning and classical conditioning in late-term rat fetuses, using intraoral cannula infusions and perioral stimulation. Taste aversions learned *in utero* were expressed in the free feeding responses of weanling rats nearly 3 weeks later. They went on to determine that aversive responses to vibrissa stimulation were attenuated or blocked by intraoral milk infusion, a prenatal "comfort" effect they found to be mediated by mu-opioid receptors.

These forms of fetal learning, involving maternal voice in humans and amniotic fluid in rodents, appear to play an adaptive role in preparing the infant for its first extrauterine encounter with its mother. They are thus the earliest origins we have yet found for attachment to the mother.

First approach behaviours after birth

In the late prenatal period, rat pups engage in a number of spontaneous behaviours *in utero*, including curls, stretches, trunk and limb movements. These acts were observed to increase markedly in frequency with progressive removal of intrauterine space constraints, as pups were observed first through the uterine wall, then through the thin amniotic sac and finally in a warm saline bath (Smotherman & Robinson, 1986). When newborn pups are observed on the first day after birth they show few spontaneous behaviours either when alone on the nest floor or in a litter group. However, Jon Polan and I have recently found that when 1-day-old pups, after as few as two or three nursing bouts, are stimulated gently by soft surfaces from above, as when the mother hovers over them, they show a surprisingly vigorous repertoire of behaviours (Polan & Hofer, 1999a). These include the curling and stretching seen prenatally, but now also include movement towards the suspended surface, wriggling, audible vocalizations and, most strikingly, turning upside-down towards the surface above them. Evidently these behaviours propel the pup into close contact with the ventrum, maintain it in proximity and keep it oriented towards the surface. They thus appear to be very early attachment behaviours. In a series of experiments we found that these are not stereotyped reflex acts, but organized responses that are graded according to the number of maternal modalities present on the surface (e.g. texture, warmth, odour). Furthermore, they are enhanced by periods of prior maternal deprivation, suggesting a motivational component. By 2 days of age we found that pups discriminate their own mother's odour in preference to equally familiar nest odours (Polan & Hofer, 1998) and by the first week postnatal, Hepper (1986) has shown that pups discriminate and prefer their own mother, father and siblings to other lactating females, males or age-mates.

These results show that a "behavioural attachment system" capable of approach and proximity maintenance to the mother, and motivated by brief periods of separation from her, may occur much earlier in development than previously supposed. The remarkable specificity of the approach response of the infant rat to individual family members within the first few postnatal days demonstrates that specificity of attachment does not require long experience or advanced cognitive and emotional capabilities. Olfaction in the rat and vision in the human provide the necessary basis for approach responses that are specific to a single individual. However, this remarkable capability can develop independently of the specificity of its comfort response; for even a 2-week-old pup will show an equal comfort response with any female that is available. This non-specificity is limited, however, for 2–3-week-old pups clearly avoid the odour of unfamiliar males (but not of familiar males) and show immobility when exposed to them (Takahashi, 1992).

Recent work in humans, inspired by these findings in lower animals, have shown that human newborns too are capable of slowly locomoting across the bare surface of the mother's abdomen and locating the breast scented with amniotic fluid in preference to the untreated breast (Varendi *et al.*, 1996). Apparently, human newborns are not as helpless as previously thought, and possess approach and orienting behaviours that anticipate the recognized onset of maternal attachment at 6–8 months.

Early postnatal learning

Although specific olfactory and/or auditory predispositions towards the infant's own mother may be acquired prenatally, after birth the newborn enters a new world where contingent events, so important for more advanced forms of learning, are now occurring with great frequency. Sullivan *et al.* (1986a) showed that associating a novel odour with simulated licking of the pup, after just a few repetitions, resulted in the pup learning to select, approach and remain close to that odour. Several different kinds of tactile stimulation, even tail pinch and mild electric shock, also induced preferences for the associated odour during the first week of postnatal life. Such strong and clearly aversive tactile stimulation ceased to induce preferences after the first postnatal week, however, and then began to induce avoidance responses during the second week. These results defined a sensitive period for the formation of positive associations reinforced by intense tactile stimulation. We next found that odours conditioned in this way not only produced olfactory preferences in a choice test, but when an inert littermate was scented with the odor, they elicited increased active huddling behaviour, probing and pawing, and also increased the time the trained pup spent in contact with the target animal (Sullivan *et al.*, 1986b). Thus the odour came to arouse the same behaviours normally elicited during interaction with the mother.

This rapid learning process resembles imprinting in birds and, because of the effectiveness of even aversive stimulation at an early age, it reminds us of the clinical observation that strong attachments can occur in children of abusive parents. This learning is limited to an early sensitive period, does not require standard reinforcing events, and accommodates even intense levels of stimulation as reinforcing. Cues learned in this way can be highly specific to an identifying maternal feature, they acquire the capacity to elicit states of increased arousal and operate, at a distance, as incentive cues in a motivational system that ensures close proximity of the infant to the mother (Rosenblatt, 1983).

This form of early olfactory learning has become a model system for neurochemical and neuroanatomical studies that have established the existence of a distributed memory system involving the amygdala, the hippocampus and thalamocortical systems, as well as the olfactory bulb and cortex (Wilson & Sullivan, 1994). Norepinephrine appears to play a dual role in this learning, enhancing olfactory system responsiveness during training and permitting later consolidation. Dopaminergic, serotonergic, glutamatergic and GABA receptors have also been implicated.

Bowlby proposed that early specific proximity-seeking behaviour in mammals, including humans, would eventually be explained by the discovery of an imprinting-like process. It would appear that these results in rat pups describe just such a process.

Maternal separation responses and regulation of infant systems

Soon after birth, prenatally acquired perceptual biases, stimulus-guided tactile responses and associative learning create a powerful behavioural control system through which the infant maintains close proximity to its mother. However, there is another important attribute of attachment, by which the emotional tie of the infant to its mother has been inferred: the response to separation. This has been supposed to be an integral part of the proximity-maintenance system, one that represents the affective expression of its motivational nature. Thus, the degree or strength of attachment is thought to be responsible for the intensity of the response to separation, and the separation response itself is taken to represent a full expression of the attachment behaviours in the absence of their "goal object".

Experiments in our laboratory led us to a very different view, in which the processes underlying attachment and the responses to separation are seen as separate and distinct early in life (Hofer, 1975b, 1983). The response of infant rats, and primates, to maternal separation has been found to involve a complex pattern of changes in a number of different behavioural and physiological systems (Kraemer, 1992; Hofer, 1996b). We found that this pattern was not an integrated psychophysiological response, as had been supposed, but the result of a novel mechanism. During separation, each of the individual systems of the infant rat responded to the absence of one or another component of the infants' previous interaction with its mother. Providing one of these components to a separated pup, for example maternal warmth, maintained the level of brain biogenic amine function underlying the pups' general activity level (Stone et al., 1976; Hofer, 1980) but had no effect on other systems, for example the pups' cardiac rate. Heart rate fell 40% after 18 hours of separation, regardless of whether supplemental heat was provided (Hofer, 1971). The heart rate, normally maintained by sympathetic tone, we found was regulated by maternal provision of milk to neural receptors in the lining of the pup's stomach (Hofer & Weiner, 1975). By studying a number of other systems, such as those controlling sleep–wake states (Hofer, 1976a), activity level (Hofer, 1975a), sucking pattern (Brake et al., 1982) vocalization (Hofer & Shair, 1980) and blood pressure (Shear et al., 1983), we concluded that in maternal separation all the regulatory components of the mother–infant interaction were withdrawn at once, yielding a pattern of increases or decreases in level of function of the infant's systems, depending upon whether the particular system had been up- or down-regulated by the previous mother–infant interaction. Other investigators, using this approach, have discovered other regulating systems of this sort.

For example, removal of the dam from rat pups was found to produce a rapid (30 min) fall in the pup's growth hormone (GH) levels, and vigorous tactile stroking of maternally separated pups (mimicking maternal licking) prevented the fall in GH (reviewed by Kuhn & Schanberg, 1991). Brain substrates for this effect were then investigated, and it appears that GH levels are normally maintained by maternal licking, acting through serotonin (5-HT) 2A and 2C receptor modulation of the balance between growth hormone-releasing factor (GRH) and somatostatin (SS), that together act on the anterior pituitary release of GH (Katz et al., 1996). The withdrawal of maternal licking by separation allows GRH to fall and SS to rise, resulting in a precipitate fall in GH.

There are several biological similarities between this maternal deprivation effect in rats and the reactive attachment disorders of infancy. Applying this knowledge to low birthweight prematurely born babies, Tiffany Field and co-workers joined the Schanberg group. They used a combination of stroking and limb movement, administered three times a day for 15 minutes each time, and continued throughout their 2 weeks hospitalization. This intervention increased weight gain, head circumference and behaviour development test scores in relation to a randomly chosen control group, with beneficial effects discernible many months later (Field *et al.*, 1986).

The separation cry and the origins of vocal communication

One of the best-known responses to maternal separation is the infant's isolation call, a behaviour that occurs in a wide variety of species (Newman, 1988; Lester & Boukydis, 1985). In the rat this call is in the ultrasonic range (40 kHz) and appears on the first or second postnatal day (Noirot, 1968). Pharmacological studies show that the ultrasonic vocalization (USV) response to isolation is attenuated or blocked in a dose-dependent manner by clinically effective anxiolytics that act at benzodiazepine and serotonin receptors, and conversely USV rates are increased by compounds known to be anxiogenic in humans, such as benzo-diazepine receptor inverse agonists (beta-carboline, FG 1742) and GABAa receptor ligands such as pentylenetetrazol (Miczek *et al.*, 1991; Hofer, 1996a). Within serotonin and opioid systems, receptor subtypes known to have opposing effects on experimental anxiety in adult rats also have opposing effects on infant calling rates. Neuroanatomical studies in infant rats show that stimulation of the periaqueductal grey area produces USV, and chemical lesions of this area prevent calling (Goodwin & Barr, 1998). The more distal motor pathway is through nucleus ambiguus and both laryngeal branches of the vagus nerve (Wetzel *et al.*, 1980). Higher centres known to be involved in cats and primates suggest a neural substrate for isolation calls involving primarily the hypothal-amus, amygdala, thalamus, and hippocampus (Jurgens, 1998), brain areas known to be involved in adult human and adult animal anxiety and defensive responses.

This evidence strongly suggests that separation produces an early affective state in rat pups, that is expressed by the rates of infant calling. How does this calling behaviour, and its inferred underlying affective state, develop as a com-munication system between mother and pup? Infant rat USVs are a powerful stimulus for the lactating rat, capable of causing her to interrupt an ongoing nursing bout, initiate searching outside the nest and direct her search towards the source of the calls (Smotherman *et al.*, 1978). The mother's retrieval response to the pup's vocal signals then results in renewed contact between pup and mother. This contact in turn quietens the pup. The isolation and comfort responses in attachment theory are described as expressions of interruption and re-establishment of a social bond. Such a formulation would predict that, since the pup recognizes its mother by her scent (as described above), pups made acutely anosmic would fail to show a comfort response. However, anosmic pups show comfort responses that are virtually unaffected by loss of their capacity to recognize their mother in this way (Hofer & Shair, 1991). Instead, we and others have found multiple regulators of infant USV within the contact between mother and pup: warmth, tactile stimuli, and milk as well as her scent (Hofer, 1996b). Provision of these modalities separately, by experimental design, and then in combination elicits a graded response, with the maximum isolation calling rates occurring when all are withdrawn at once and with the full comfort response being elicited only when all are present at once.

829

After the first week we have recently found that another more complex vocal response begins to emerge (Hofer *et al.,* 1998). The pup now begins to regulate its isolation calling response in relation to social cues that were present *prior* to isolation. If the 2-week-old pup has been with its mother even briefly (1 min) prior to separation, it calls at three or four times the rate typically found after having been in the company of its littermates. If it has been with a virgin female, or if its mother has been entirely passive (anaesthetized), the pup calls at an intermediate rate (Hofer *et al.,* 1999). If, on the other hand, it has encountered the smell of an unfamiliar adult male (a potential predator) prior to isolation, it will vocalize little, if at all (Shair *et al.,* 1997). We called the maternal effect "potentiation", and have found that this capacity can be acquired by virgin females, and even by adult males, if the pups have been reared with either one, in addition to their dam, prior to testing at 12 days postnatal age (Brunelli *et al.,* 1998).

We hypothesize that this is a form of affect regulation analogous to that found when a human mother returns briefly to the day-care centre for some reason, only to experience on her second departure an unexpected storm of vocal protest from her toddler. In learning experiments in rat pups it has been found that trained 12-day-olds vocalize at greatly increased rates when their learned expectancy of reward (by maternal contact) is first violated in early extinction (Amsel *et al.,* 1977). The adaptive value of this form of regulation may be that it enables isolated pups to call at maximum rates when the mother has recently been with them and is likely still to be nearby, and to be silent when predator cues are present prior to isolation.

Thus, in the development of interactions between the infant and its mother, a vocal communication system becomes established. The infant separation call rate appears to be controlled by the same neural systems that mediate anxiety in adult animals and humans, apparently a strongly conserved response system. Early in development the system is regulated by the multiple components of the infants' immediate social interaction, or by their withdrawal in separation. Later in development the system comes to be regulated also by traces of the infants' recent past experience with specific social cues that may predict the risk:benefit ratio of the calling response attracting predators, or maternal responses.

Effects of maternal regulatory interactions on infant development

The actions of maternal regulators of infant biology and behaviour are not limited to the mediation of responses to maternal separation discussed above. They exert their regulatory effects continuously, throughout the preweaning period and even beyond. A good illustration is the recent discovery of a major role for the mother–infant interaction in the development of the hypothalamic–adreno-cortical axis (HPA). In mid-infancy, from postnatal days 4 to 14, the rat pups' HPA response to isolation, and to mild stressors such as saline injections, is known to be less intense than in the newborn or weaning periods, a stage known as the "stress-hyporesponsive period". Surprisingly, it was recently found that this species-typical developmental stage is not the product of an intrinsic developmental programme, but the result of hidden regulators at work within the ongoing mother–infant interaction. First, it was found that 9–12-day-old pups' basal corticosterone (CORT) level and the magnitude of the adrenocorticotrophic hormone (ACTH) and CORT response to isolation were increased five-fold after 24 hours of maternal separation (Stanton *et al.*, 1988). Next, by utilizing our concept of hidden regulators, Sucheki *et al.* (1993) attempted to prevent these separation-induced changes by supplying various components of the mother–infant interaction. They found that repeated stroking of the separated infants for as little as three 1-min periods prevented the increase in ACTH response, while providing milk by cheek cannula during separation prevented the separation-induced blunting of the adrenal CORT response to ACTH. Tracing these regulatory effects back to brain systems, Levine's group has recently found that stroking (representing maternal licking) regulates the intensity of the cfos messenger RNA (mRNA) response in the paraventricular nucleus of the hypo-thalamus (PVN) and the corticotropin releasing hormone (CRH) receptor mRNA expression as well, in PVN, the amygdala and other limbic system sites (Van Oers *et al.*, 1998). It is intaorally administered milk, however, that regulates glucocorticoid receptor mRNA in the CA1 region of the hippocampus and CORT release from the adrenal in response to ACTH.

Through this anatomical neuromodulator analysis Levine and colleagues discovered that maternal licking and milk delivery during suckling exerted an unexpected and prolonged attenuating effect on the responsiveness of the HPA. This maternal regulatory effect, once established in the first few postnatal days, continues throughout most of the nursing period, finally declining as weaning occurs from day 15 to 21. These regulatory interactions achieve this effect by increasing the inhibitory feedback from hippocampal glucocorticoid receptors and by decreasing the hypothalamic stimulation of CRF and ACTH output. These regulatory effects on the pup's brain can be rapidly reversed by maternal separation.

Based on this evolving story, Plotsky & Meaney (1993) explored the possibility that repeated shorter (3 h) maternal separations during this developmental period might have effects on the HPA that persisted beyond the preweaning period into adulthood. They found elevated basal ACTH levels, an increased ACTH response to mild stress, and elevated CRH mRNA expression in the amygdala in adults that had been repeatedly separated as infants. Plotsky (personal communication) has recently found that it was not the repeated brief separations themselves, but their effects on maternal behaviour after reunion that exerted the long-range shaping effect on HPA development. For if the dams were provided with foster pups during the 3-h separation period, the long-term effects on her own pups were eliminated.

These findings suggest that qualitative differences in the mother–infant interaction can act to shape development through altered regulation of infant systems. Meaney and his colleagues used a maternal behaviour observational approach developed previously in our laboratory (Myers *et al.*, 1989) to directly test this implication of the concept of maternal regulators. They found that dams in their colony that were observed naturally to have high levels of licking, grooming, and of the high-arched nursing position, reared pups that were later found to have lower than normal HPA axis responsivity to restraint stress as adults. In contrast, the offspring of dams that naturally showed the lowest levels of these interactions resembled the adults with a history of repeated early maternal separations described above (Liu *et al.*, 1997). The group was led to this comparison of different mothering patterns by their studies on the classic "handling" effect, in which pups are briefly removed from the home cage and then returned after a few minutes, each day for the first 1–2 weeks of postnatal life. This manipulation they found reduced HPA reactivity in adult offspring by causing increased glucocorticoid receptor levels and decreased levels of CRH synthesis in the PVN of the adult hypothalamus. When mother–infant interactions of handled litters were observed, the dams showed increases in licking and grooming and in high-arched nursing position in comparison to control litters, measures Liu *et al.* then used in the natural variation studies described above.

Maternal effects in genetic models

In 1974 two Czech investigators reported the results of a study in which they cross- fostered at birth the offspring of two different strains of rats that they had selectively bred for high and low aggression towards mice (Flandera & Novakova, 1974). On cross-fostering pups between mothers of the two different strains they found that the traits that emerged in the offspring were those typical of their foster mothers rather than of their own genotype. Since the pups never observed adults interacting with mice, the different traits were transmitted in some other way by the mothers during interaction with their foster pups between birth and weaning. Low-aggression strain pups fostered to high-aggression strain dams showed high levels of aggression (65% of them killed mice) both at 30 and 90 days of age. Conversely, pups from the high-aggression strain, when cross-fostered to low-aggression strain mothers, showed minimal levels of mouse killing (12.5%) as adolescents (30 days). However, their genotype gradually emerged as the offspring developed increased levels of aggression by 90 days of age (50%). This study is a good example of how selective breeding can exert its effect indirectly, through altering the behaviour of the mother towards the pups, as well as through more direct effect on the phenotype of the offspring. Unfortunately these findings were not further explored to determine the behavioural and brain mechanisms for postnatal transmission of the traits.

A more recent example, in which postnatal mechanisms have been explored, involved spontaneously hypertensive rats (SHR) and their Wistar Kyoto (WKY) progenitor control strain. It has been found that cross-fostering of the young of the SHR strain at birth to WKY mothers normalizes their adult blood pressure (BP), but the low BP of WKY rats was not increased by cross-fostering them at birth to SHR dams (McCarty *et al.*, 1992). The BP of both strains was unaffected by the fostering manipulation itself within strains, and cross-fostering at various ages within the 3 weeks following birth showed that the sensitive period for the normalization of BP lay within the first or second postnatal week.

We have explored the possible mechanisms for this maternal effect in a series of studies (reviewed in Myers *et al.*, 1992). First, we developed methods for the rapid survey of ongoing mother–infant interactions naturally occurring in the maternity cages of a colony, methods later used by Meaney and his group in the HPA studies described above. In our study we learned that, although members of each of these inbred strains were genetically identical, there was significant inter-litter variability both in adult BP and in maternal behaviour within each strain, as well as between the two strains. Three maternal behaviours accounted for most of the variability in adult BP both between litters and between strains. Pups of mothers that showed more of these behaviours had higher BP as adults. By looking carefully at the components of maternal behaviour that were most highly correlated with adult BP of the offspring (high-arched nursing position, contact time and licking), we then made an educated guess that led us to the next chapter in the story. First, we examined the milk let-down that occurred during the high-arched maternal nursing position. We found a sudden transient rise in blood pressure of 50% during natural and oxytocin-induced milk ejection, a greater increase than during any other pup activity (Shair *et al.*, 1986). This surge in blood pressure was found to be caused by a major increase in neural activity in the adrenergic vasoconstrictor system that controls blood vessel tone throughout the body. Shair & Hofer (1993) then found that this surge was triggered by contact of the milk with sensory nerve endings on the tongue and in the throat. Myers has gone on to show that the rate of weight gain during a critical period of nursing is a powerful correlate of adult BP, and that experimentally increasing the level of milk let-down events, during this 4-day period, by a temporary reduction in litter size, significantly increased adult BP of offspring (Myers *et al.*, 1992). The importance of these studies was further increased when Myers and his colleagues found that human infants also show major increases in BP in response to maternal milk let-down (Cohen *et al.*, 1992).

In these studies the differing adult phenotypic traits of genetically selected lines of laboratory animals have been powerfully shaped (or even determined in the case of mouse-killing behaviour) by the particular maternal environment in which they were reared as pups. We are beginning to analyze the behavioural mechanisms by which specific components of the mother–infant interaction exert these shaping effects on the expression of the genotype through development.

DISCUSSION

The research approaches described in this chapter have begun to answer the three questions posed in the Introduction: how the infant first acquires the propensity to maintain close contact with its caretaker, why interruption of this contact produces such extensive biological and behavioural responses in the offspring, and how seemingly minor variations in the ongoing patterns of the mother–infant interaction become translated into long-term differences between individuals in adulthood. Until recently we have attempted to answer these questions primarily in terms of inferred motivational and emotional states maturing in the infant and child. Attachment has been viewed as a unified system, separate from other motivational systems, the complex response to separation has been seen as an emotional response expressing high levels of this motivational state, and the long-term effects that follow variations in the quality of the early mother–infant relationship as reflecting persistent differences induced in emotion regulation within the maturing attachment system.

The new work outlined in this chapter reveals a level of developmental process that underlies these inferred emotional responses, and that provides different explanations for the observed responses of the infant. These processes act at a sensorimotor and physiological level that translates readily to the level of neural events, and opens up the potential for us to understand how these emotional states become organized in the developing brain of the infant.

In Section II of this book, the neural events responsible for the early development of brain structure were described in terms of their cell and molecular mechanisms. Recent discoveries of the extent of plasticity in the formation of brain structure and its cellular mechanisms (Fox *et al.,* 1999; Jessell & Kandel, 1993) provide a set of processes that can be linked directly to behavioural events within the early mother–infant interaction. These mechanisms include a number of processes that depend upon neural activity: stimulating the release of neurotrophins; directive cell–cell interactions during cell migration; prevention of cell death; receptor organization; and the formation, stabilization and elimination of synapses. In all these instances the presence of or absence of neural activity in particular pathways provides an organizational mechanism that both defines subsequent behavioural interactions with the environment, and is defined by them. This cascade of events eventually serves to fit the individual brain to the environment it inhabits. Many synapses form early on, but those that are activated repeatedly or simultaneously, or in close sequence, are selectively preserved, providing the basis for differential responding to particular patterns and combinations of stimuli. The major sources of sensory stimulation and elicitation of neural activity, on which these early brain changes depend, derives from the infant's interactions with its mother. This unitary source for these early activity-dependent processes, the mother, ensures that the infant's behaviour is organized around its interactions with her, and forms the basis for what we refer to as a highly specific attachment bond.

The special characteristics of early learning are likely also to depend on the unique properties of these early brain processes. As Spear and his colleagues have shown (Spear *et al.,* 1988; Spear & McKinsie, 1994), pre-weanling rat pups will form conditioned responses if two stimuli are simultaneously presented, as well as if one stimulus closely precedes (and predicts) the next, whereas adults will learn only from the predictive form of contingency. Further, rat pups will learn to respond more intensely if two different modalities of independently conditioned stimuli are subsequently combined, whereas adults respond less well ("overshadowing"). These differences are thought to reflect the adults' tendency to differentiate between sensory modalities and the infants' tendency to combine or "unitize". Finally, young infants form positive associations even with aversive stimuli (see above). These special characteristics of early learning would appear to strongly predispose the infant to form a specific attachment to its mother. Eventually they will be traced to differences from the adult in the cellular and molecular conditions for strengthening and inhibiting the specific synaptic connections upon which learning is based.

From physiological regulation to mental representation

The early learning and widespread regulatory interactions described above are not involved solely in biological and behavioural development, but are also the first experiences out of which mental representations and their associated emotions arise in human development. So far as we understand the process, experiences made up of the infant's individual acts, parental responses, sensory impressions and associated affects are laid down in memory during and after early parent–infant interactions. These individual units of experience are integrated into something like a network of attributes in memory, invested with associated affect, and resulting in the formation of an internal working model of the relationship. In this way the learning and regulatory interactions we have described in infancy become ingredients for the mental representations and related affective qualities that form the inner experience of older children and adults (Stern, 1985). Once formed, it seems likely that these organized mental structures come to act as superordinate regulators of behavioural and biological systems underlying motivation and affect, gradually supplanting the sensori-motor, thermal and nutrient regulatory systems found in younger infants. This would link biological systems with internal object representations in humans, and would account for the remarkable upheavals of biological as well as psycho-logical systems that take place in adults in response to cues signalling impending separation, or in response to losses established simply upon hearing of a death, for example, by telephone (Hofer, 1984, 1996b).

Mother–infant relationships which differ in quality, and which necessarily involve different levels and patterns of regulation in a variety of systems, will be reflected in the nature of the mental representations present in different children as they grow up. The emotions aroused during early crying responses to separation, during the profound state changes associated with the prolonged loss of all maternal regulators, and during the reunion of a separated infant with its mother, are clearly intense. These emotional states have commanded our attention, they are what everyone intuitively recognizes about attachment and separation, and what we feel about the people we are close to. These inner experiences occur at a different level of psychobiological organization than the changes in autonomic, endocrine and neurophysiological systems we have been able to study in rats and monkeys, as well as in younger human infants.

The discovery of regulatory interactions and the effects of their withdrawal allow us to understand not only the responses to separation in young organisms of limited cognitive–emotional capacity, but also the familiar experienced emotions and memories that can be verbally described to us by older children and adults. It is not that rat pups respond to loss of regulatory processes, while human infants respond to emotions of love, sadness, anger and grief. Human infants, as they mature, can respond at the symbolic level *as well as* at the level of the behavioural and physiological processes of the regulatory interactions. The two levels appear to be organized as parallel and complementary response systems. Even *adult* humans continue to respond in important ways at the sensorimotor-physiological level in their social interactions, separations and losses, continuing a process begun in infancy. A good example of this is the mutual regulation of menstrual synchrony among close female friends, an effect that takes place out of conscious awareness and has recently been found to be mediated at least in part by a pheromonal cue (Stern & McClintock, 1998). Other examples may well include the role of social interactions in entraining circadian physiological rhythms and the remarkable effects of social support on the course of medical illness (Hofer, 1984). In this way adult love, grief and bereavement may well contain elements of the simpler regulatory processes that we can clearly see in the attachment responses of infant animals to separation from their social companions.

This is perhaps the most challenging area for future research, to find out how to apply what we have learned in basic brain and behaviour studies, to the human condition. At present there is widespread use of the concept of regulation as inherent in the mother–infant interaction in humans. This is generally used in two ways: first, to refer to the graded effects of different patterns of interaction on the emotional responses of the infant, the so-called "regulation of affect" (Schore, 1994); second, it is used to refer to how the behaviours initiated by the infant or mother and/or their responses to each other act to regulate the interaction itself, its tempo or rhythm (hence its "quality") or the distance (both psychological and physical) between the members of the dyad. The word "regulation" is probably overused at present, and inappropriately applied to any influence of one member of the dyad upon the other.

A role for non-verbal features of the mother–infant interaction in the specification of lasting mental representations is a central hypothesis of clinical attachment theory. It would be difficult to confirm this clinically useful idea with certainty, but the long-term effects described in this chapter lend some support, by analogy. Prospective studies from infancy to childhood would be most interesting, and could reveal which residues of particular early interactions can be related to which later characteristics of the stories, play or social relationships of older children, and eventually of adults.

ACKNOWLEDGEMENTS

This work was supported by a Research Scientist Award (5 K05 MH38632) and project grant (5 R01 MH40430) from the National Institute of Mental Health and by The Sackler Institute for Developmental Psychobiology, Columbia University College of Physicians and Surgeons.

References

Amsel, A., Radek, C.C., Graham, M. & Letz, R. (1977). Ultrasound emission in infant rats as an indicant of arousal during appetitive learning and extinction. *Science*, **197**, 786–788.

Blass, E.M. (1990). Suckling: determinants, changes, mechanisms, and lasting impressions. *Developmental Psychology*, **26**, 520–533.

Bowlby, J. (1969). *Attachment and loss*, vol. 1: *Attachment*. New York: Basic Books.

Brake, S.C., Sager, D.J., Sullivan, R. & Hofer, M.A. (1982). The role of intraoral and gastrointestinal cues in the control of sucking and milk consumption in rat pups. *Developmental Psychobiology*, **15**, 529–541.

Brunelli, S.A., Masmela, J.R., Shair, H.N. & Hofer M.A. (1998). Effects of biparental rearing on ultrasonic vocalization (USV) responses of rat pups (*Rattus norvegicus*). *Journal of Comparative Psychology*, **112**, 331–343.

Campos, J.J., Mumme, D.L., Kermoian, R. & Campos, R.G. (1994). A functionalist perspective on the nature of emotion. In N.B. Fox (Ed.), *Monographs of the Society for Research in Child Development,* serial no. 240, vol. 59, nos. 2–3.

Carter, C.S., Lederhandler, I.I. & Kirkpatrick, B. (1997). *The integrative neurobiology of affiliation*. New York: New York Academy of Sciences.

Cassidy, J. & Shaver, P.R. (1999). (Eds.), *Handbook of attachment theory and research*. New York: Guilford Press.

Cohen, M., Witherspoon, M., Brown, D.R. & Myers, M.M. (1992). Blood pressure increases in response to feeding in the term neonate. *Developmental Psychobiology*, **25**, 291–298.

DeCasper, A.J. & Fifer, W.P. (1980). Of human bonding; newborns prefer their mothers' voices. *Science*, **208**, 1174–1176.

Denenberg V.H. & Whimbey, A.E. (1963). Behaviour of adult rats is modified by the experiences their mothers had as infants. *Science*, **142**, 1192–1193.

Earls, F. & Carlson, M. (1998). Recovery from profound early social deprivation. In D.M. Hann, L.C. Huffman, I.I. Lederhendler. & D. Meinecke, (Eds.), *Advancing research ondevelopmental plasticity: integrating the behavioural science and neuroscience of mental health* (pp. 185–195). National Institutes of Mental Health, National Institutes of Health, Rockville, Md.

Field, T.M., Schanberg, S.M., Scafidid, F., Bauer, C.R., Vega-Lahr, N., Garcia, R., Nystrom, J. & Kuhn, C.M. (1986). Tactile/kinesthetic stimulation effects on preterm neonates. *Pediatrics*, **77**, 654–658.

Fifer, W.P. & Moon, C.M. (1995) The effects of fetal experience with sound. In J.-P. Lecanuet, W.P. Fifer, N.A., Krasnegor & W.P. Smotherman (Eds.), *Fetal development – a psychobiological perspective* (pp. 351–368). Hillsdale, NJ: Lawrence Erlbaum.

Flandera V. & Novakova V. (1974). Effect of mother on the development of aggressive behaviour in rats. *Developmental Psychobiology*, **8**, 49–54.

Fox, K., Henley, J. & Issac, J. (1999). Experience-dependent development of NMDA receptor transmission. *Nature Neuroscience*, **2**, 297–299.

Goldberg, S., Muir, R. & Kerr, J. (1995). *Attachment theory: social, developmental, and clinical perspectives*. Hillsdale, NJ: Analytic Press.

Goodwin, G.A. & Barr, G.A. (1998). Behavioral and heart rate effects of infusing of kainic acid into the dorsal midbrain during early development in the rat. *Developmental Brain Research*, **107**, 11–20.

Hann, D.M., Huffman, L.C., Lederhendler, I.I. & Meinecke, D. (Eds). (1998). *Advancing research on developmental plasticity: integrating the behavioural science and neuroscience of mental health*. National Institutes of Mental Health, National Institutes of Health, Rockville, Md.

Hepper, P.G. (1986). Parental recognition in the rat. *Quarterly Journal of Experimental Psychology*, **38B**, 151–160.

Hepper, P.G. (1987). The amniotic fluid: an important priming role in kin recognition. *Animal Behavior*, **35**, 1343–1346.

Hofer, M.A. (1971). Cardiac rate regulated by nutritional factor in young rats. *Science*, **172**, 1039–1041.

Hofer, M.A. (1975a). Studies on how early maternal separation produces behavioural change in young rats. *Psychosomatic Medicine*, **37**, 245–264.

Hofer, M.A. (1975b). Infant separation responses and the maternal role. *Biological Psychiatry*, **10**, 149–153.

Hofer, M.A. (1976a). The organization of sleep and wakefulness after maternal separation in young rats. *Developmental Psychobiology*, **9**, 189–205.

Hofer, M.A. (1976b). Olfactory denervation: Its biological and behavioural effects in infant rats. *Journal of Comparative Physiology and Psychology*, **90**, 829–838.

Hofer, M.A. (1980). The effects of reserpine and amphetamine on the development of hyperactivity in maternally deprived rat pups. *Psychosomatic Medicine*, **42**, 513–520.

Hofer, M.A. (1983). On the relationship between attachment and separation processes in infancy. In R. Plutchik & H. Kellerman (Eds.), *Emotions in early development* (pp. 199–216). New York: Academic Press.

Hofer, M.A. (1984). Relationships as regulators: a psychobiological perspective on bereavement. *Psychosomatic Medicine*, **46**, 183–197.

Hofer, M.A. (1994). Early relationships as regulators of infant physiology and behaviour. *Acta Paediatrica*, Supplement **397**, 9–18.

Hofer, M.A. (1995). Hidden regulators: implications for a new understanding of attachment, separation and loss. In S. Goldberg, R. Muir & J. Kerr (Eds.), *Attachment theory: social, developmental and clinical perspectives* (pp. 203–230). Hillsdale, NJ: Analytic Press.

Hofer, M.A. (1996b). On the nature and consequences of early loss. *Psychosomatic Medicine*, **58**, 570–581.

Hofer, M.A. (1996a). Multiple regulators of ultrasonic vocalization in the infant rat. *Psychoneuroendocrinology*, **21**, 203–217.

Hofer, M.A. & Shair, H.N. (1980). Sensory processes in the control of isolation-induced ultrasonic vocalization by 2-week-old rats. *Journal of Comparative Physiology and Psychology,* **94**, 271–299.

Hofer, M.A. & Shair, H.N. (1991). Trigeminal and olfactory pathways mediating isolation distress and companion comfort responses in rat pups. *Behavioral Neuroscience,* **105**, 699–706.

Hofer, M.A. & Weiner, H. (1975). Physiological mechanisms for cardiac control by nutritional intake after early maternal separation in the young rat. *Psychosomatic Medicine,* **37**, 8–24.

Hofer, M.A., Masmela, J.R., Brunelli, S.A. & Shair, H.N. (1998). The ontogeny of maternal potentiation of the infant rats' isolation call. *Developmental Psychobiology,* **33**, 189–201.

Hofer, M.A., Masmela, J.R., Brunelli, S.A. & Shair, H.N. (1999). Behavioral mechanisms for active maternal potentiation of isolation calling in rat pups. *Behavioral Neuroscience,* **113**, 51–61.

Jessell, T.M. & Kandel, E.R. (1993). Synaptic transmission: a bidirectional and self-modifiable form of cell-cell communication. *Cell, 72/Neuron,* **10** (Suppl.), 1–30 January.

Jürgens, U. (1998). Neuronal control of mammalian vocalization. *Naturwissenschaften,* **85**, 367–388.

Katz, L.M., Nathan, L., Kuhn, C.M. & Schanberg, S.M. (1996). Inhibition of GH in maternal separation may be mediated through altered serotonergic activity at $5HT_{2A}$ and $5HT_{2C}$ receptors. *Psychoneuroendocrinology,* **21**, 219–235.

Kraemer, G.W. (1992). A psychobiological theory of attachment. *Behavioural and Brain Sciences,* **15**, 493–511.

Kuhn, C.M. & Schanberg, S.M. (1991). Stimulation in infancy and brain development. In B.J. Carroll (Ed.), *Psychopathology and the brain* (pp. 97–112). New York: Raven Press.

Lester, B.M. & Boukydis, C.F. (Eds.) (1985). *Infant crying: theoretical and research perspectives.* New York: Plenum Press.

Liu, D., Diorio, J., Tannenbaum, B., Caldji, C., Francis, D., Freedman, A., Sharma, S., Pearson, D., Plotsky, P.M. & Meaney, M.J. (1997). Maternal care, hippocampal glucocorticoid receptors, and hypothalamic–pituitary–adrenal responses to stress. *Science,* **277**, 1659–1661.

Malson, L. (1972). *Wolf children and the problem of human nature.* New York: Monthly Review Press.

McCarty, R., Cierpial M.A., Murphy, C.A., Lee, J.H. & Field-Okotcha, C. (1992). Maternal involvement in the development of cardiovascular phenotype. *Experientia,* **48**, 315–322.

Meaney, M.J., Bhatnagar, S., Diorio, J., Larocque, S., Francis, D., O'Donnell, D., Shanks, N., Sharma, S., Smythe, J. & Viau, V. (1993). Molecular basis for the development of individual differences in the hypothalamic-pituitary-adrenal stress response. *Cellular and Molecular Neurobiology,* **13**, 321–347.

Miczek, K.A., Tornatsky, W. & Vivian, J. (1991). Ethology and neuropharmacology: rodent ultrasounds. In *Advances in pharmacologic sciences* (pp. 409–429). Basel: Birkhauser.

Myers, M.M., Brunelli, S.A., Squire, J.M., Shindledecker, R. & Hofer, M.A. (1989). Maternal behavior of SHR rats and its relationship to offspring blood pressure. *Developmental Psychobiology,* **22**, 29–53.

Myers, M.M., Shair, H.N. & Hofer, M.A. (1992). Feeding in infancy: short- and long-term effects on cardiovascular function. *Experientia,* **48**, 322–333.

Nelson, E.E. & Panksepp, J. (1998). Brain substrates of infant–mother attachment: contributions of opioids, oxytocin, and norepinephrine. *Neuroscience and Biobehavioural Reviews,* **22**, 437–452.

Newman, J.D. (Ed.) (1988). *The physiological control of mammalian vocalization.* New York: Plenum Press.

Noirot, E. (1968). Ultrasounds in young rodents. II. Changes with age in albino rats. *Animal Behaviour,* **16**, 129–134.

Panksepp, J. (1998). *Affective neuroscience: the foundations of human and animal emotions.* New York: Oxford University Press.

Plotsky, P.M. & Meaney, M.J. (1993). Early postnatal experience alters hypothalamic corticotropin-releasing factor (CRF) mRNA. Median eminence CRF content and stress-induced release in rats. *Molecular Brain Research,* **18**, 195–200.

Polan, H.J. & Hofer, M.A. (1998). Olfactory preference for mother over home nest shavings by newborn rats. *Developmental Psychobiology,* **33**, 5–20.

Polan, H.J. & Hofer, M.A. (1999a). Maternally directed orienting behaviours of newborn rats. *Developmental Psychobiology,* **34**, 269–280.

Polan, H.J. & Hofer, M.A. (1999b). Psychobiological origins of infant attachment and separation responses. In J. Cassidy & P.R. Shaver (Eds.), *Handbook of attachment theory and research* (pp. 174–194). New York: The Guilford Press.

Rauschecker, J.P. & Marler, P. (1987). *Imprinting and cortical plasticity: comparative aspects of sensitive periods.* New York: John Wiley.

Robinson, S.R. & Smotherman, W.P. (1995). Habituation and classical conditioning in the rat fetus: opioid involvements. In J.-P. Lecanuet, W.P. Fifer, N.A. Krasnegor & W.P. Smotherman (Eds.), *Fetal development -- a psychobiolotgical perspective* (pp. 295–314). Hillsdale, NJ: LawrenceErlbaum.

Rosenblatt, J.S. (1983). Olfaction mediates develomental transition in the altricial newborn of selected species of mammals. *Developmental Psychobiology,* **16**, 347–375.

Schore, A.N. (1994). *Affect regulation and the origin of the self.* Hillsdale, NJ: Erlbaum.

Shair, H.N. & Hofer, M.A. (1993). Afferent control of pressor responses to feeding in young rats. *Physiology and Behavior,* **53**, 565–576.

Shair, H.N., Masmela, J.R., Brunelli, S.A. & Hofer, M.A. (1997). Potentiation and inhibition of ultrasonic vocalization of rat pups: regulation by social cues. *Developmental Psychobiology,* **30**, 195–200.

Shair, H.N., Brake, S.C., Hofer, M.A. & Myers, M.M. (1986). Blood pressure responses to milk ejection in the young rat. *Physiology and Behavior,* **37**, 171–176.

Shear, M.K., Brunelli, S.A. & Hofer, M.A. (1983). The effects of maternal deprivation and of refeeding on the blood pressure of infant rats. *Psychosomatic Medicine,* **45**, 3–9.

Singh, J.A.L. & Zingg, R.M. (1966). *Wolf-children and feral man.* Archon Books. The Shoestring Press, Hamden, CT.

Skolnick, N.J., Ackerman, S.H., Hofer, M.A. & Weiner, H. (1980). Vertical transmission of acquired ulcer susceptibility in the rat. *Science,* **208**, 1161–1163.

Smotherman, W.P. & Robinson, S.R. (1986). Environmental determinants of behaviour in the rat fetus. *Animal Behaviour,* **34**, 1859–1873.

Smotherman, W.P. & Robinson, S.R. (Eds.) (1988). *Behaviour of the fetus.* Caldwell, NJ: Telford Press.

Smotherman, W.P., Bell, R.W., Hershberger, W.A. & Coover, G.D. (1978). Orientation to rat pup cues: effects of maternal experiential history. *Animal Behaviour,* **26**, 265–273.

Spear, N.E. and McKinzie, D.L. (1994) Intersensory integration in the infant rat. In D.J. Lefkowicz & R. Lickliter (Eds.), *The development of intersensory perception* (pp. 133–161). Hillsdale, NJ: Erlbaum.

Spear, N.E., Kraemer, P.J., Molina, J.C. & Smoller, D.E. (1988). Developmental change in learning and memory: infantile disposition for unitization. In J. Delacour, & J.C.S. Levy (Eds.), *Systems with learning and memory abilities* (pp. 27–52). Amsterdam: Elsevier.

Stanton, M.D., Gutierrez, Y.R. & Levine, S. (1988). Maternal deprivation potentiates pituitary-adrenal stress responses in infant rats. *Behavioral Neuroscience,* **102**, 692–700.

Stern, D.N. (1985). *The interpersonal world of the infant: a view from psychoanalysis and developmental psychology.* New York: Basic Books.

Stern, K. & McClintock, M.K. (1998). Regulation of ovulation by human pheromones. *Nature,* **392**, 177–179.

Stone, E., Bonnet, K. & Hofer, M.A. (1976). Survival and development of maternally deprived rats: role of body temperature. *Psychosomatic Medicine,* **33**, 242–249.

Suchecki, D., Rosenfeld, P. & Levine, S. (1993). Maternal regulation of the hypothalamic-pituitary-adrenal axis in the infant rat: the roles of feeding and stroking. *Developmental Brain Research,* **75**, 185–192.

Sullivan, R.M., Hofer, M.A. & Brake, S.C. (1986a). Olfactory-guided orientation in neonatal rats is enhanced by a conditioned change in behavioural state. *Developmental Psychobiology,* **19**, 615–623.

Sullivan, R.M., Brake, S.C., Hofer, M.A. & Williams, C.L. (1986b). Huddling and independent feeding of neonatal rats can be facilitated by a conditioned change in behavioural state. *Developmental Psychobiology,* **19**, 625–635.

Takahashi L.K. (1992). Developmental expression of defensive responses during exposure to conspecific adults in preweanling rats (*Rattus norvegicus*). *Journal of Comparative Psychology,* **106**, 69–77.

van Oers, H.J.J., de Kloet, E.R., Whelan, T. & Levine, S. (1998). Maternal deprivation effect on the infant's neural stress markers is reversed by tactile stimulation and feeding but not by suppressing corticosterone. *Journal of Neuroscience,* **18**, 10171–10179.

Varendi, H., Porter, R.H. & Winberg, J. (1996). Attractiveness of amniotic fluid odor: evidence of prenatal olfactory learning? *Acta Paediatrica,* **85**, 1223–1227.

Wetzel, D.M., Kelley, D.B. & Campbell, B.A. (1980). Central control of ultrasonic vocalizations in neonatal rats: I. Brain stem motor nuclei. *Journal of Comparative Physiology and Psychology,* **94**, 596–605.

Wilson, D.A. & Sullivan, R.M. (1994). Neurobiology of associative learning in the neonate: early olfactory learning. *Behavioral Neural Biology,* **61**, 1–18.

V.3
The Neurobiology of Early Communication: Intersubjective Regulations in Human Brain Development

COLWYN TREVARTHEN

ABSTRACT

Infant communication research has brought evidence, both behavioural and neurobiological, of innate motives and emotions that anticipate development of intersubjective contacts and relations. An account of the adaptations in the child's brain for mirroring expressions before language, and of developmental changes of brain structure and function in early years, entails a review of prenatal brain development. Body-mapping cell arrays emerge autonomously in the embryo CNS, before the peripheral sensory and motor nerves, and embryogenic principles exercise a primary influence over the formation of the environment-expectant cerebral tissues in the fetus. Aminergic neural regulators of internal physiological state and of brain system formation are identical with the intrinsic motive formation that governs both expressions of communication and elaboration of cognitive systems postnatally. So the neural mechanism adapted for cognitive assimilation of socioemotional information "brain-to-brain", as well as for transmitting knowledge and skills in a cultural context, is laid down prenatally – by intercellular communication regulated by gene expression, and by the formation of body-movement-representing systems in the fetal brain that will "mirror" other persons' actions.

A psychobiology of childhood that takes account of the development of the whole brain, and of the motives of the whole child in communication, contrasts with current developmental cognitive neuroscience theory, which is individualistic, rational/constructivist and corticocentric.

A.F. Kalverboer and A. Gramsbergen (eds.), Handbook of Brain and Behaviour in Human Development, 841–882
© 2001 Kluwer Academic Publishers. Printed in Great Britain.

INTRODUCTION: INFANT COMMUNICATION AND COGNITIVE NEUROSCIENCE

The knowledge we now possess of how infants communicate their coherent intentions and feelings raises questions for cognitive neuroscience. Cognitive theory explains the performance of adults in perception, problem-solving and language, taking evidence from procedures developed to assess academic skills. Cognitive neuroscience focuses on the "acquisition, storage and use of information by the nervous system" (Wilson & Keil, 1999, p. lvii). In cognitive psychology of infants the findings of experiments measuring the development of perception and problem-solving abilities and preferences are conceived as emergent constructions of experience, driven at first by reflex actions and "biological" needs, and bolstered by intelligent parenting. However, communicating infants actively adapt to human movements in regulated rhythmic patterns, sympathetically mirroring or complementing expressions that indicate motives of interpersonal relating. How these motives can be interpreted in terms of information processing, emerging perception of objects and discriminatory attention is not clear. Nor are they fully explained by a "dynamic systems theory" of events emerging in a complex biomechanical and sensory-motor system of the moving body (Fogel & Thelen, 1987).

The findings of infant research also challenge a neuroscience that focuses on cognitive processing in the neuronal net, conceived as a dynamic interconnected system of cells that reacts with plastic reorganization to competing environmental and embodied inputs. The "initial state" of the spontaneously active infant brain evidently has adaptive functional "constraints" that anticipate and motivate future learning and system formation. The motives and emotions for intersubjective relating to a parent's intuitive sympathetic communications are powerful anticipatory forces in mental growth. They influence cognition and learning in human company before infants can explore impersonal physical objects by grasping them in their hands and manipulating them, testing their affordances. Intersubjective motives underpin the development of thinking and language.

The comparative neurological approach of "affective neuroscience" (Panksepp, 1998) shows that the integrative mechanism for *emotions* is subcortical. Its output is an evolutionary elaboration of processes that mediate their effects through the special visceral efferents of the cranial nerve nuclei of the brainstem, which are coupled to outputs of the autonomic nervous system in regulation of the animal's internal state. In contrast, a subject's *purposes and attentions* for gaining and choosing experience of the world outside the body can be communicated only by mirroring of the control of somatic muscles that guide body movements and focus receptors on goal objects. For control of the latter goal-directed behaviours a foundation may again be found in brainstem structures. The combined operation of visceral and somatic mirror reactions gives the infant access to the other subject's anticipatory "motor images", and permits direct motive-to-motive engagement with a companion, and common interest in a shared environment. This is the physiological foundation for the psychological phenomenon of *intersubjectivity*, which couples brains of a socially cooperative species in joint cognition, and which is specially elaborated in humans (Trevarthen, 1986, 1998a,b).

Developments in infancy confirm that there is an extensive central neural system in the human brain concerned with joining emotions and cognitions in the service of preverbal communication. The toddler's mastery of language rests upon the ability of the infant to engage in joint attention and joint purposes with other human beings, and to sympathize with their emotions. These functions do not depend on words or symbolic reasoning. They are founded on innate motivating principles of the whole human brain, not just a neocortical capacity to a assimilate a wider experience.

THE FUNCTIONS OF EARLY COMMUNICATION

First steps to thought-sharing

A healthy child of 3 years is expected to be alert to the intelligence and emotions in other persons, and to have, besides a lively well-coordinated body, a capacity for understanding imagined meanings, and for creative expression of these in mimesis (narrative body expression), and in language. How this consciousness and expressive communication with others is generated remains a mystery, in spite of several decades of intense research. Constructivist theories to explain how intermental abilities are assembled from supposed primary sensory-motor and cognitive mechanisms have repeatedly failed.

An awake newborn is immediately capable of coordinated whole-body movement guided by an integrated awareness. The baby can visually fixate an "out-of-the-body" object and can modify the synchronized head–eye–hand coordination to track a moving exterior target with a series of oculomotor saccades and pulsing arm extensions, performing fluid, rate-controlled "'prereaching" – joint rotations of arm and hand aimed to a point outside the body (Trevarthen *et al.*, 1981; von Hofsten, 1984; Trevarthen, 1984a). These coherent actions can occur when there is no object to be perceived – generated within a spontaneous cerebral "motor image" of the inherent temporospatial field of activity of the whole body.

Piaget, who assumed the newborn to be incoherent in action and experience, to have to "construct" consciousness of its body and of the outside world, presents visual tracking of a moving object by the "sensory-motor" infant as the paradigm of "imitation", an accommodative cognitive activity that is coupled with the expression of a *self-regulatory* emotion of "serious intent". This function he contrasted to "play", which is triumphantly assimilative and performed with *self-satisfying* "pleasure in mastery" (Piaget, 1962). Piaget disregarded the potential communicative function of these object-related behaviours, as he overlooked the infant's social awareness.

Before a baby is strong enough to handle and explore non-living objects effectively, he or she shows preferential responses to the motive states and feelings of other persons (Aitken & Trevarthen, 1997; Fogel, 1993; Papoušek & Papoušek, 1987; Stern, 1985, 1993; Stern *et al.*, 1985; Trevarthen, 1979, 1987a, 1993a,b, 1998a; Trevarthen *et al.*, 1981; Weinberg & Tronick, 1994). Infants are born sensitive to the emotions of vocal expression and touch, and can express clear emotions – in smiles and coos of recognition, frowns of annoyance, and hand movements that signal changing states of alertness, distress or interest, and

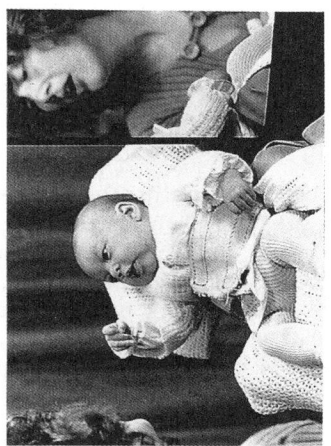

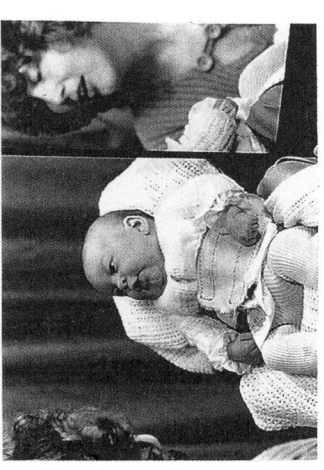

Figure 1 (*also see opposite*) Protoconversation. *Above*: A 6-week-old girl coos while mother listens, and raises her right hand in an expressive gesture as her mother speaks to her. *Middle*: Thirty seconds of the vocal exchange between this girl and her mother showing, from top to bottom, the rhythmic structure of sounds in a spectrograph, pitch plots in sones derived from the spectrograph showing how the voices of mother and infant begin around middle C (C4), rising to fill the octave above C4, and fall back to C4. The infant's vocalisations are enclosed in boxes in the spectrograph and pitch plot. The words uttered by the mother are shown underneath, and infant vocalisations are marked by asterisks. *Below*: The timing of syllables, bars and phrases is indicated. The 30 second episode may be described as a 'narrative'; it shows a rise and fall of excitement or emotion, which the two subjects share.

844

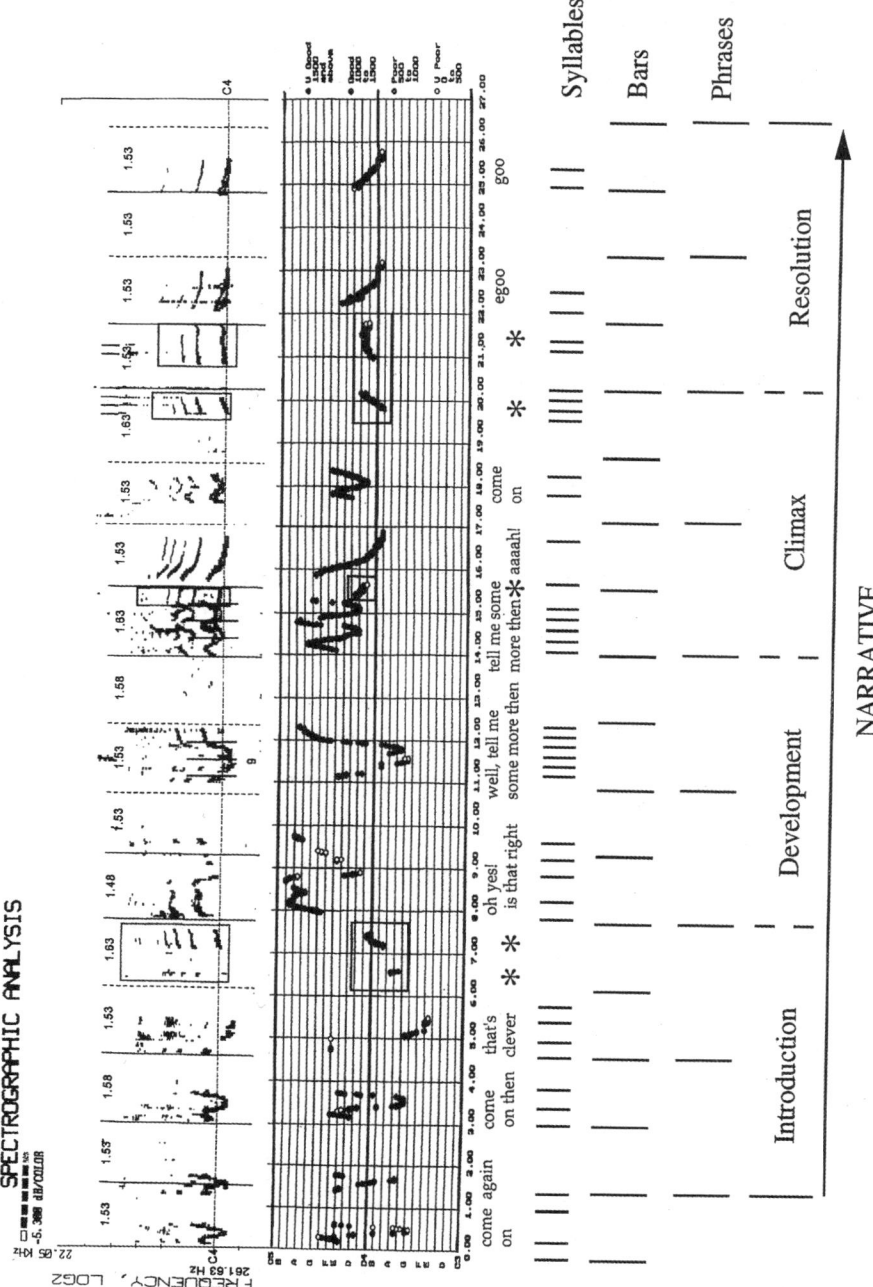

readiness for making communication. Newborns also see human faces, and have preferential curiosity for their individual form and expressions.

While some expressions of a young infant are clearly adapted to elicit parental care for internal physiological homeostasis – states of behavioural arousal and sleep, comfort, feeding and protection from environmental dangers – other expressions (smiling, gaze approach/avoidance, coo vocalizations and certain hand gestures) transmit "relational" affects, and are clearly adapted to regulate intersubjective contact. Features of a mother's talking are learned from experience *in utero* (DeCasper & Spence, 1986; Fifer & Moon, 1995). Her speech can be identified by her newborn at once, and recognition of the visible appearance of her face is acquired within hours, aided by the newborn's capacity for interest in imitation of, and interaction with, facial expressions (Field *et al.*, 1982; Heimann, 1989; Kugiumutzakis, 1999; Meltzoff & Moore, 1977, 1992, 1994; Nagy & Molnár, 1994; Reissland, 1988; Zeifman *et al.*, 1996; Trevarthen *et al.*, 1999; Uzgiris, 1991).

Neonatal imitation, a rather peculiar form of sympathetic behaviour elicited by exaggerated "modelling" behaviour of an adult interrupted by waiting for a reaction from the infant, exhibits the cardinal features of elementary conscious intentional behaviour – flexibility of sensory confirmation (intermodal equivalence of the senses) and rate-specified rhythmic motor execution (motor equivalence for different parts of the body) (Trevarthen *et al.*, 1999). It involves apprehension of the intrinsic *motive* that generates both the movement form and its perceptual validation (Trevarthen, 1998b, 1999a). It functions as reciprocal communication from the start, operating with emotions of pleasure, interest, surprise, etc. (Trevarthen *et al.*, 1999), anticipating reactions from the partner. Newborns may also exhibit recognition of the other by non-imitative signs of sympathetic intersubjective response: smiles to a parent's voice, gestures that convey emotions of pleasure or annoyance, and transformations of state of arousal. In quiet alert state, alert newborns show phasic social expectancies that may assist the development of emotionally satisfying exchanges with an adult.

The inborn communicative impulses develop rapidly. Within 2 months an infant will turn promptly to a mother's voice and look at her eyes, then participate in a rhythmic face-to-face *protoconversation* engaging simultaneously sight, hearing and touch (Bateson, 1979; Fogel & Hannan, 1985; Stern, 1985; Trevarthen, 1979, 1993b) (Figure 1). When the mother speaks affectionately, in prosodically clear and expressive "intuitive motherese", the infant becomes animated to produce cycles of expressiveness, with smiles, vocalizations, lip and tongue movements of "prespeech", movements of the limbs and gestures of the hands. These expressions are clustered to produce emissions resembling "utterances" and phrases of adult conversation. Cycles of *assertive expression,* alternating with *apprehensive watching*, are generated as mother and baby enter into a pattern of turn-taking (Trevarthen *et al.*, 1999). The hand movements of the infant are complex gestures, distinguishable from pointing or grasping actions made in orientation to objects that the infant is looking at or tracking. This difference, indicating a distinct motive process for manual expression of social feelings, can be seen in newborns (Rönnqvist & von Hofsten, 1994). The infant's behaviours demonstrate, from birth, the cognitive and performative skills of *primary intersubjectivity*, called "primary" because it serves as the foundation for the elaboration of future more complex voluntary cooperations in awareness and intentionality.

In the next few months the infant gains gravity-exploiting postural regulation of the head and limbs, sits up and begins to crawl. Interest in surroundings and in tracking and gaining possession of objects increases (Figure 2). At the same time ritualized games develop in which patterned and repeating moves are negotiated with humour and teasing between infant and parent (Trevarthen & Hubley, 1978; Pecheux *et al.*, 2000). The infant, while attending critically to the consequences of his or her investigative moves to track and focus on objects seen, heard, touched, smelled and tasted, shows *self–other awareness* (Bråten, 1988; Aitken & Trevarthen, 1997), "showing off" and making emotional reactions to an observer's attention, not only in game routines with other persons, but also to a mirror self-image (Reddy, 1991; Reddy *et al.*, 1997; Trevarthen, 1990a, 1998a; Trevarthen *et al.*, 1981, 1999). By 6 months the baby is supremely sensitive to the timing of experiences contingent upon his or her own agency and anticipation, and is especially alert to the moves of playmates. Responses to rhythms in human body movement, including the polyrhythms of music, become stronger and more exuberantly emotional (Trevarthen, 1999a; Mazakopaki, 2000).

This playfulness and imitation with the dynamics of the body as an expressive instrument foreshadows the fanciful mimesis that is so characteristic of toddlers' play. It conveys a desire to participate in imaginative "narratives" or "stories" of action and emotion – the framework of conscious and emotionally charged understanding of purposes on which all cultural intelligence, including language, is built. By 1 year a baby can, without language, share arbitrary experiences with familiar persons, displaying an individual socially adapted personality while pointing to objects of joint interest and responding to pointing, using a protolanguage combining vocalizations and gestures, and attending to and imitating conventional expressions and actions (Adamson & Bakeman, 1991; Butterworth & Grover, 1988; Halliday, 1975; Hubley & Trevarthen, 1979; Meltzoff, 1995; Stern *et al.*, 1985; Tomasello, 1988; Trevarthen, 1990a, 1998b; Trevarthen & Hubley, 1978; Trevarthen *et al.*, 1981; Uzgiris, 1991). These behaviours may be described as manifestations of a need for *companionship* in understanding (Trevarthen, 2000) (Figure 3).

Conversational synchrony and reciprocity of motor impulses

"Conversational analysis" of natural interactions and turn-taking between communicating adults have demonstrated synchrony between movements of expression, both within subjects (intra-synchrony) and between them (inter-synchrony). Application of microanalytic techniques has shown that mother–infant interactions achieve comparable levels of efficiency in coordination and precision of timing (Beebe *et al.*, 1985; Feldstein *et al.*, 1993; Jaffe *et al.*, 2000). Central coherence of timing of movements and the processing of sensory data is essential to biomechanical efficiency and unity of conscious action (Bernstein, 1967; Trevarthen, 1984b, 1999a). This internal integration of "plans" for movement becomes the origin of intersubjective sympathy and cooperative intentions.

Piaget (1953) explained how the infant acquires ("constructs") cognitive schemata from "circular reactions" – "primary", representing only the body; "secondary" conceiving manipulated objects; 'tertiary', involving combinatory

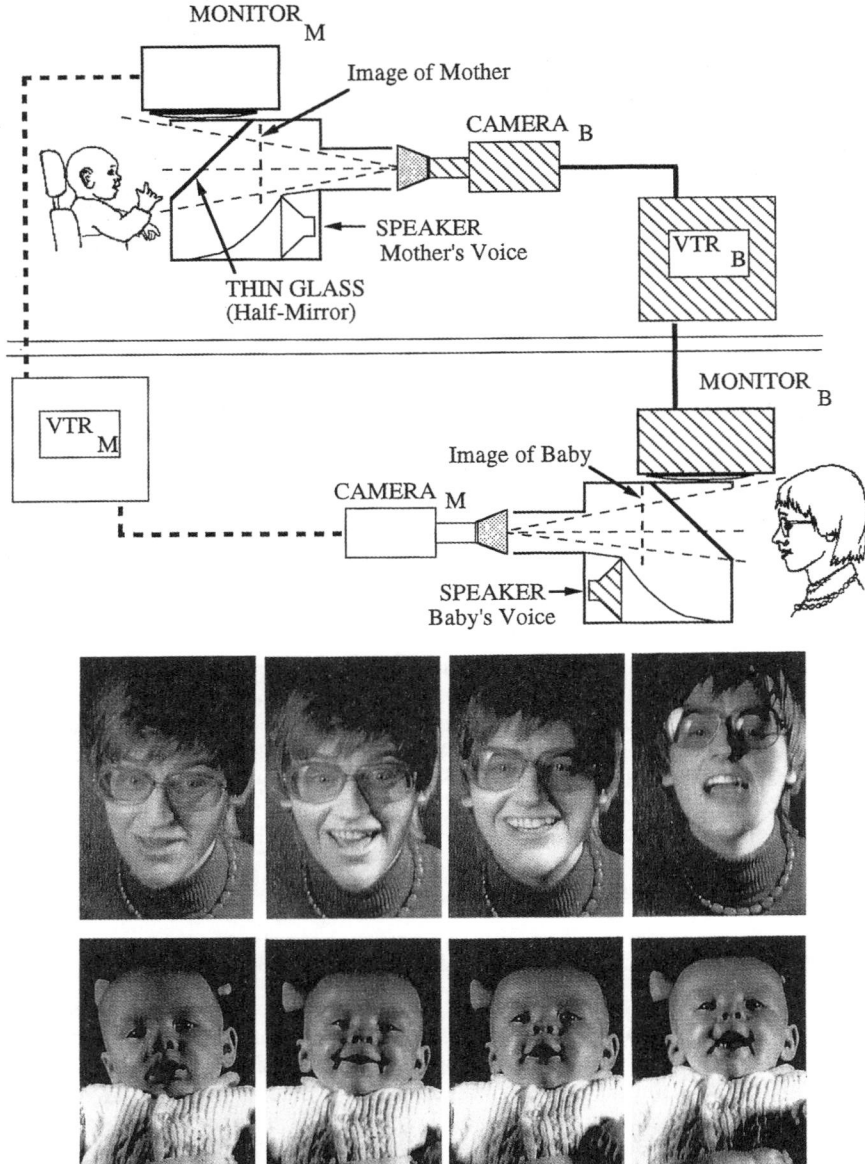

Figure 2 The Double Television method of testing an infant's sensitivity to the contingency of the mother's communicative responses. An 8-week-old infant girl interacts 'on line' with the TV image and recorded sound of her mother's talk. *Above*: Two concealed cameras, one facing each subject, transmit the images between the subjects in two separate rooms. They see each other on monitors by reflection in sloping plates of glass. *Below*: Images of moments in their 'live' engagement. Note: When 90 seconds of the mother's happy conversation was replayed to the baby, the baby became disturbed by the non contingent behaviour (see text).

and symbolic (i.e. social or interpersonal) representations. He acknowledged the place of rhythm, "at the junction of organic and mental life", and observed that, "rhythm ... involves a way of linking elements together which already heralds an elementary form of what appears as the reversibility characteristic of higher mental processes" (Piaget, 1966). Donald (1991, p. 186) claims that rhythm is a special feature of *expressive* behaviour, and central to human mimesis. He says:

> "Rhythm is an integrative mimetic skill relating to both verbal and visuomotor mimesis. Rhythm is a uniquely human attribute; no other creature spontaneously tracks and imitates rhythms in the way humans do, without training. Rhythmic ability is supramodal,... it may be played out with any motor modality, including the hands, feet, head, mouth or the whole body.... Rhythm is therefore evidence of a central mimetic controller that can track various movement modalities simultaneously and in parallel."

Stern has underlined the importance of dynamic "intermodal fluency" in the establishment of rhythmically coordinated "affect attunement" between infant and adult (Stern, 1985; Stern *et al.*, 1985).

Rhythmic cycles and phases of motivation that generate and regulate purposeful movement are given a primary regularity by motive states in the brain in three fundamental dimensions: *morphology* (structure), *intensity* (energy) and *timing* (process) (Trevarthen, 1986). The communicative interactions of young infants show regulation in all three of these dimensions.

Table 1 Developmental changes in the first 18 months of infancy (see Figure 3)

Cognitive and somatic developments	Developments in communication
A: Regulation of sleep, feeding and breathing. Innate "pre-reaching".	Imitation of expressions. Smiles to voice.
B: Pre-reaching declines. Swipes and grabs.	Fixates eyes with smiling. Protoconversations. Mouth and tongue imitations. Distressed by "still-face" test.
C: Smooth visual tracking, with strong head support. Reaching and catching.	"Person–person" games, mirror recognition.
D: Interest in surroundings increases. Accurate reach and grasp. Binocular stereopsis. Manipulative play with objects.	Imitation of clapping and pointing. "Person–person–object" games.
E: Babbling. Persistent manipulation, rhythmic banging of objects. Crawling and sitting, pulling up to stand.	Playful, self-aware imitating. Showing off. Stranger fear.
F: Combines objects, "executive thinking". Categorizes experiences. Walking.	Cooperation in tasks; follows pointing. Declarations with "joint attention". Protolanguage. Clowning.
G: Self-feeding with hand.	Mimesis of purposeful actions, uses of "tools" and cultural learning. May use first words.

Infants' emotions, and their function in communication

Infants demonstrate the subjective regulation of changing communicative purposes with emotions. Even self-motivated "instrumental" actions, aimed to control events in the outside world, give signs of emotional evaluation, the expressions of which can only affect another human being (Papoušek, 1967). The motives and emotions in protoconversation have been tested by experiments that interrupt communication between infant and adult. These amplify both positive and negative emotional regulations.

In the *still- or blank-face test* (Tronick et al., 1978; Murray & Trevarthen, 1985) a mother who has been communicating happily with her 2-month-old stops on a signal from the experimenter, and simply looks at the infant for about 1 minute without any reaction to what the infant does. The infant shows "appeals" for communication by smiling, vocalizing and gesturing, punctuated by sober staring at the mother, then avoidance of eye-contact and distress. The expression of sad self-conscious confusion would, in an older person, be seen as embarrassment or "shame".

A second *video interaction replay* experiment was designed to deal with the objection that, in the still-face test, the infant was reacting to loss of stimulation, not perceiving loss of communication (Murray & Trevarthen, 1985). A double video (DTV) or "videophone" link enables infant and mother to communicate "live" by hearing vocalizations and seeing one another's face expressions on a television monitor directly in front (Figure 2). Once happy communication was obtained, an animated and playful portion of the recording of the mother approximately 1 minute in length is replayed to the infant. The projection of the mother's behaviour is exactly as before, but the physical recording does not react contingently to what the baby is expressing at any moment. Infants show occasional accidental interaction with the taped behaviour of the mother, confusion when it fails to respond in time and appropriately, then distress and avoidance, as in the "still-face" experiment. In both tests it takes time for the infant to recover when the mother resumes normal communication, or is "on line" again (Wienberg & Tronick, 1996).

In a converse experiment, replay of the infant's behaviour to the mother in the DTV apparatus causes her to feel "something is wrong". Different mothers experience different emotions and make different verbal evaluations when the infant appears not to "connect", some criticizing the baby, some blaming themselves (Murray & Trevarthen, 1986).

It has been claimed that infants under 2 months lack (have not yet "constructed") a coherent, intentional "self", and therefore do not anticipate agency in another person, and cannot be sensitive to the contingency of their responses (Rochat et al., 1998; Rochat & Striano, 1999). However, replication of this DTV replay experiment with improvements, confirming the infant's responses in every detail, proves that 2-month-olds can predict the timing and emotion of a mother's expressions in communication (Nadel et al., 1999). Indeed, awareness of the timing of other persons' responses, and anticipation of an appropriate response in time, can be demonstrated for a premature newborn (Trevarthen, 1999a). The mental clock, by which another's sympathy can be judged, is innate, not assimilated from experience of a moving body.

Older infants' reactions to the still-face and DTV replay tests show that a capacity to withstand disengagement without distress increases with the infant's growing interest in the environment at large. Infants over 4 months actively investigate surroundings, and they can use this to escape an unresponsive mother (Trevarthen, 1990a, 1998a; Hains & Muir, 1996; Muir & Hains, 1999). When they are older than 6 months, "secure" infants respond to the still-face test as an entertaining game (Trevarthen, 1990a, 1998a, p. 40).

The perturbation tests show how sensitive a young infant is to affectionate parenting, and they explain how failure of caregiver support is potentially harmful (Field, 1992; Cohn & Tronick, 1983; Murray & Cooper, 1997; Tronick & Weinberg, 1997). It has been found that the first 3 months is a period of high susceptibility for lasting effects of maternal postnatal depression, especially in boys (Murray, 1992; Murray *et al.*, 1993; Trevarthen & Aitken, 2000).

"Non-basic" emotions

In a recent critical review of theories of emotion in infancy, Draghi-Loren *et al.* (2001) show how academic psychology has accepted the Cartesian distinction between emotions and thinking. In the constructivist theory of social development the motivating function of emotions in communication remains problematic (Rothbart, 1994; Tronick, 1989). Emotions are assumed to arise from "biological" regulations inside the body, and to be elaborated by thought or reason, which requires the acquisition of representations and logical operations describing and explaining the outside world (Lewis, 1999; Sroufe, 1996). The self-consciousness of these representations is verbal and reflective rather than intuitive and enactive, and it is instilled by parental rewards and sanctions, or persuaded by rational arguments.

The evidence from early infancy suggests, on the contrary, that the *relational emotions* (Stern, 1993), which are specifically adapted to real-time regulation of the balance of initiatives and reactions between one human being and another, and which contribute to relationships of affectionate attachment, trust and companionship, and to defence against abuse, mistrust and disregard, are fundamental to the ecology of emerging human consciousness. It follows that emotions described as "complex", "non-basic" and "acquired" are, in fact, *primary and necessary* to the child's entry into the social/cultural world. The theory that cognitions construct the classical 7 reactive emotions – "interest, joy, surprise, fear, anger, sadness, disgust" – with perhaps two more, "contempt" and "shame" (Izard, 1994), is not compatible with how infants respond to persons. Emotions, not just *in*, but *between* persons, include "pride, jealousy, coyness, shame, resentment, rage", and the lasting evaluations and empathy of "admiration, love, hate, contempt" that we can develop for particular individuals *as persons*. They have foundations in dynamic reactions of even young infants to the feel of "being present" with another, and receptive to their changing motive expressions (Aitken & Trevarthen, 1997; Fogel, 1993; Stern, 1993; Trevarthen, 1993a; Tronick & Weinberg, 1997).

Intuitive parenting and the musicality of "motherese"

When a parent talks to a baby the voice has characteristic rhythmic and melodic features and specific vocal qualities of pitch and timbre. *Motherese* or *infant-directed speech (IDS)* is organized in repeated phrases, and creates animating incidents and slowly changing, cyclic "narratives" of emotion (see Figure 1).

The *dynamic narrative envelopes* of a mother's utterances have been identified as necessary training for the infant's developing self-awareness and consciousness of agency (Papoušek & Papoušek, 1987; Stern, 1999). On the other hand, the precision of the infant's mirroring, even with a restricted vocal repertoire, has been taken as evidence for the interpersonal nature of the communication (Beebe *et al.*, 1985; Stern, 1985; Stern *et al.*, 1985; Trevarthen, 1998a, 1999a; Trevarthen *et al.*, 1999). Newborns can synchronize with salient moments in the adult's message by gesture or utterance, and "coo" sounds can be matched in pitch and quality (timbre) between them. Adults speaking to infants tend to approach the pitch and quality of the infant's preceding utterance (Malloch, 1999). There is a sensitive two-way mirroring of the emotional values of expression that overrides the great difference in maturity of the participants.

IDS has exaggerated *expressivity*. It projects the feelings, interests and intentions of the speaker. As Mary Catherine Bateson pointed out, protoconversation is a form of human communication that is related to education in the forms and meanings of language, and also to the rhythms and melodies of religious ritual and communion, and traditional healing practices (Bateson, 1979, pp. 74–76).

Speech to infants in different languages demonstrates *universal rhythmic and prosodic features* (Fernald, 1992; Grieser & Kuhl, 1988; Papoušek *et al.*, 1991) (see Figure 1). Expressions used by parents to elicit infant attention are similar in English and Mandarin, rising contours elicit and maintain infant attention more than falling pitch (Papoušek *et al.*, 1991). Infants are more interactive, interested and emotionally positive to IDS. Three different functions have been attributed to IDS: this speech *engages attention* (Papoušek *et al.*, 1991), *communicates affect*, facilitating social interaction (Fernald, 1992; Kitamura & Burnham, 1998), and *helps language acquisition* (Papoušek, 1994). All three are consequences of the infant's innate motives for communicating the primary impulses of their agency.

The similarity of IDS and vocal play to universal features of music, and to the rhythmic and rhyming forms, phrases and verses of poetry, has led to a concept of preverbal or subverbal *communicative musicality* as a fundamental basis for sharing of motives and feelings. Infants discriminate subtle features of musical sounds and melodic forms and rhythmic phrases, especially as these are represented in a mother's voice (Fassbender, 1996; Jusczyk & Krumhansl, 1993; Malloch, 1999; Trainor, 1996; Trehub, 1990; Trehub *et al.*, 1997; Trehub *et al.*, 1997; Trevarthen, 1999a). These are manifestations of a fundamentally innate process by which mutual emotional "attunement" is established between human subjects (Stern *et al.*, 1985; Tønsberg & Hauge, 1996).

With older infants, parents' songs and games become more lively, and ritual forms often repeated give great satisfaction to both (Stern, 1990). The songs modulate the infant's emotion, and reflect the extent to which he or she engages in communication (Papoušek *et al.*, 1991; Trevarthen, 1999a). In laboratory discrimination tests with infants about 6 months old, play songs evoke increased alerting to the external world, and joint attention, while lullabies encourage more self-focused infant behaviours (Rock *et al.*, 1999).

The *intrinsic motive pulse* (IMP) of musicality (Trevarthen, 1999a) is active in infants' play with adults, and when infants respond to musical sound in laboratory tests, making infants move in rhythm and register interest and happiness (Fassbender, 1996; Jusczyk & Krumhansl, 1993; Papoušek, 1994; Trehub, 1990; Trehub *et al.*, 1997; Zentner & Kagan, 1996). Protoconversations with 6-week-olds are on a slow *adagio* (one beat in 900 milliseconds; 70/minute) (see Figure 1). In animated games the beat accelerates to *andante* (1/700 milliseconds; 90/minute) or *moderato* (1/500 milliseconds; 120/minute). Stern describes "feeling qualities" transmitted with distinctive "activation contours", "captured by such kinetic terms as 'crescendo, decrescendo, fading, exploding, bursting, elongated, fleeting, pulsing, wavering, effortful, easy', and so on" (Stern, 1993, p. 206).

Trehub concludes that "infants' representation of melodies is abstract and adult-like" (Trehub, 1990, p. 437), and that "the design features of infant music should embody pitch levels in the vicinity of the octave beginning with middle C (262 Hz), simple contours that are unidirectional or that have few changes in pitch direction (e.g. rise–fall), slow tempos (approximately 2.5 notes/sec), and simple rhythms" (Trehub, 1990, p. 443). These predictions match baby songs and the prosodic patterns parents use to excite or calm their infants (Papoušek, 1994; Papoušek *et al.*, 1991; Fernald, 1992; Trainor, 1996; Trehub *et al.*, 1997b). Computer-assisted acoustic analysis of mother–infant vocal play (Malloch, 1999) confirms a musician's perception that parameters of *timing* (rhythm), *quality* (pitch, and vocal spectrum or timbre) and *narrative* (the melodic/emotional development over time) are all controlled in speech or song with an infant.

Rhythmic movements of infants' hands may synchronize with adult speech (Condon & Sander, 1974). A remarkable example is offered by the case of a 5-month-old baby, born totally blind, who is "conducting" with appropriately paced and lifted hand gestures the rhythm and melodic line of her mother's singing of a song well known to the baby (Trevarthen, 1999a).

Development of cooperative awareness

Longitudinal studies document age-related transformation of the infant's motives through the first year, which influence the mother's expressions in play (Figure 3; Table 1). The baby's growing interest in manipulation and visual inspection and tracking leads the mother to play games with objects. At the end of the first year, *joint interest* of mother and infant in their surroundings is triggered by the infant's emerging curiosity about what the mother is doing (Trevarthen & Hubley, 1978). This change has momentous consequences in learning, and affects the ways mothers act with and speak to their infants.

Cooperative manipulation of objects, with orientation to observe and cooperate with the probable intentions of companions in an arbitrary task, appear months before the first word is uttered (Adamson & Bakeman, 1991; Halliday, 1975; Trevarthen & Hubley, 1978). At 10 months an infant understands meaning in "co-operative awareness" with "joint purpose" (Hubley & Trevarthen, 1979;

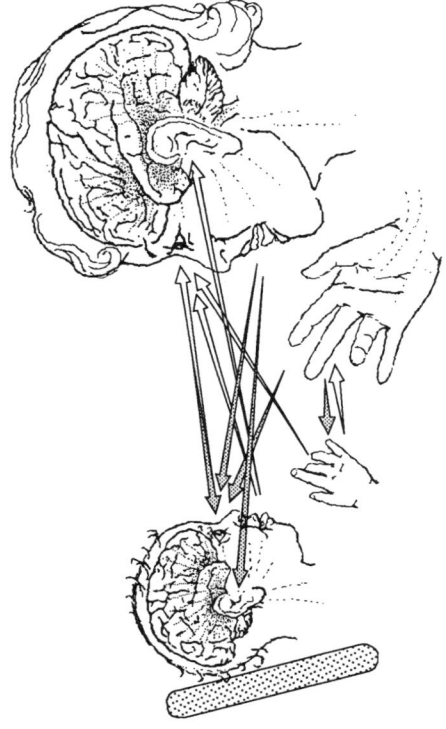

Figure 3 (*also see opposite*) Development of communication and cooperation in infancy. *Above*: Touch, hearing and sight are used concurrently to coordinate the many channels of expression in protoconversation with a young infant. *Middle*: Age-related phases in the development of intersubjectivity, self-consciousness and cooperative awareness in the first 18 months after birth. *Below*: Major growth indicators in the cerebral cortex of an infant.

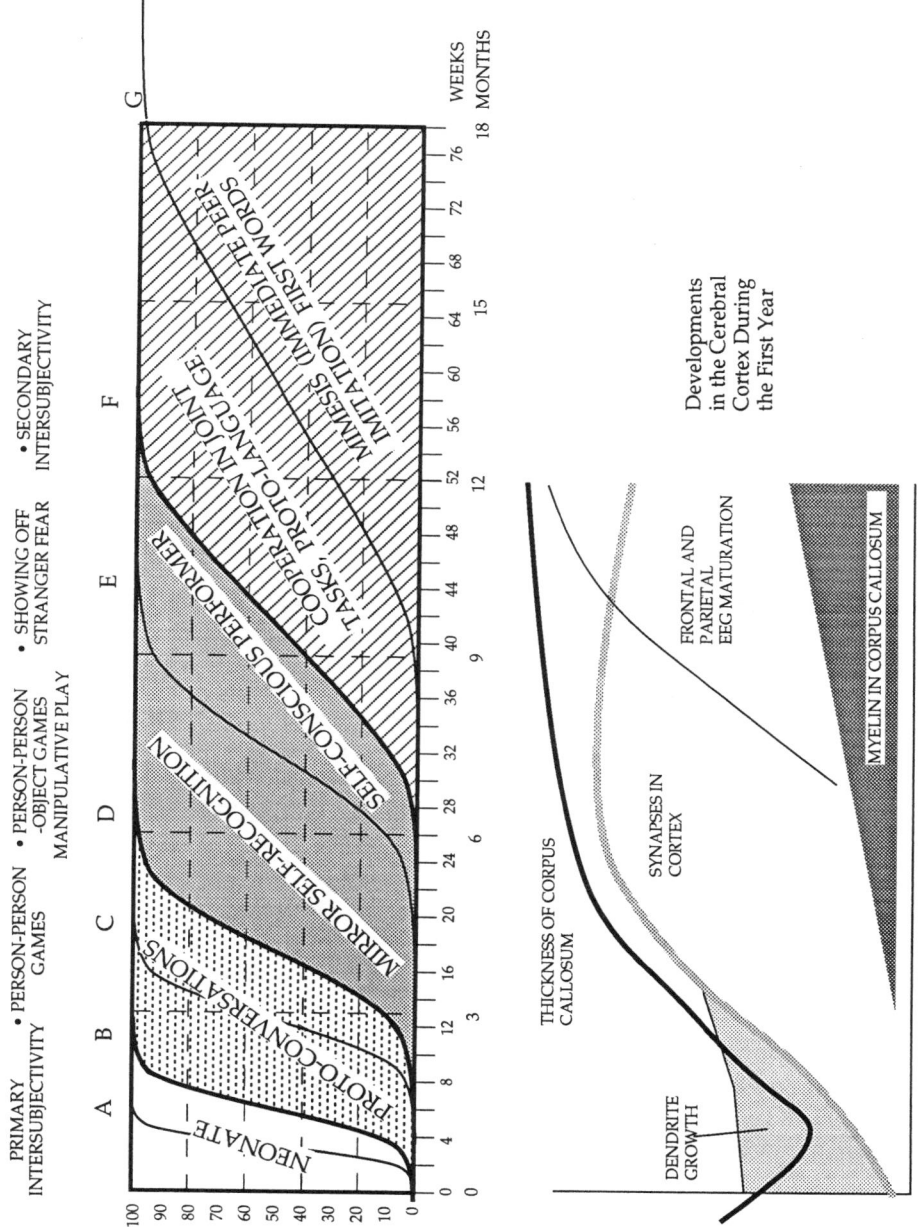

Developments in the Cerebral Cortex During the First Year

Tomasello, 1988; Tomasello *et al.*, 1993). This is described as the development of *secondary intersubjectivity* (Trevarthen & Hubley, 1978).

In the second year a baby shares *arbitrary experiences and meanings*, displaying to familiar persons an individual, socially adapted personality while attending to and imitating conventional vocalizations and gestures, as well as noticing objects that other persons use, imitating their actions and understanding their purposes (Adamson & Bakeman, 1991; Butterworth & Grover, 1988; Locke, 1993; Meltzoff, 1995).

Hearing language; catching the "point" of words in narratives of intention

A child picks up language by doing things with it, and by noticing what other persons do with it (Bruner, 1983; Locke, 1993; Ninio & Snow, 1996). The acts negotiating social participation with emotion come earlier in development than intention-directing "proto-imperatives" (Ninio & Snow, 1996; Snow *et al.*, 1996), just as "person–person games" come before "person–person–object games" in the middle of the first year (Trevarthen & Hubley, 1978).

The early stage of grammar learning is not simply a matter of coordinating vocalizations with *intentions and attentions* – requests, pointing, showing, giving. It has concern for human *feelings and sensitivities* which form the backing texture of all live communication and "experiencing together". Joint attention, strongly associated with the picking up of words in the second and third years (Rollins & Snow, 1998; Tomasello, 1988), is not just a convergence of lines of sight and directions of instrumental action. Nadel has demonstrated the importance of immediate imitation of actions and utterances among toddlers for sharing meaning, and she underlines the pleasure and humour of the sharing, which is expressed by gesture and vocal prosody (Nadel & Pezé, 1993). Children and adults alike are easily caught in dramatic make-believe (Harris, 1998), identifying the roles of "characters", and infants play with *emotional narratives* long before they can talk (Trevarthen, 1999a).

An "experience-expectant" speech awareness allows a newborn to begin learning phonological and prosodic features of the mother's language immediately (Locke, 1993). Kitamura & Burnham (1998) conclude that the affectionate tone and exaggerated intonation of IDS is "the pivotal quality that attracts infant attention". Evidently, "maternal affective expression provides a primitive method of conveying meaning to the prelingual infant, an essential first step in developing a facility for processing language." (*loc. cit.*, p. 235).

A rapid "explosion" of vocabulary starts about 20–30 months after birth, when the child has a new compulsion to share items and narratives of experience, as well as a desire to identify categories that "need" to be named (Tomasello, 1988). There are large individual differences in this learning, as there are in temperament at this "difficult" age (Locke, 1993). The first "telegraphic" productions of two- or three-word sentences do not appear to be linked to awareness of phonological, grammatical and syntactic rules before about 30 months – "formulaic" or mimetic stereotypes are part of rituals of shared expressiveness and joint interest.

Left and right asymmetries of gestures and motive states

Asymmetries observed in expressive gestures of young infants lead to the conclusion that they reflect left–right differences in "motor images" (Jeannerod, 1994) for producing and comprehending communication. Neonates show a right ear preference for speech, but a left ear preference for heartbeat or music. Microanalyses of films of protoconversations and photographic records showed that, at 2 months, protoconversational utterances of cooing and/or "prespeech" (articulations of lips and tongue) were most often accompanied by raised right hand gestures, and extending the hand and pointing with the fingers was frequently precisely synchronous with the oral gesture. When they were distressed and in more reflective and observant phases, infants held their left hands towards their bodies and sometimes touched or fingered their clothes. They appear to be more "assertive" with their right hands, and more receptive or "apprehensive" when moving their left hands (Trevarthen, 1996).

THE HUMAN BRAIN AT BIRTH AND DEVELOPMENTS IN CHILDREN

The initial human state: body-related and environment-related motive regulators in the neonatal brain

A newborn infant's cerebral hemispheres are one-third the size of the adult hemispheres, and the cerebellum is even less developed (Figure 4). The frontal, parietal and temporal neocortices are disproportionately small, the temporal lobe being about half the length of the adult lobe, showing that the hippocampus as well as the temporal neocortex is undeveloped (Trevarthen, 1987b).

Primary sensory and motor cortices are the most mature, and subcortical components are well formed, particularly the highly differentiated aminergic regulatory systems of the brainstem that innervate all cortical regions and remain as motivation regulators in the brain throughout life (Trevarthen, 1985). Rapid branching of dendrites on cortical cells around term coincides with intense synaptogenesis. After 3 months dendritic growth slows, but synapse density continues to increase until 6 months (Huttenlocher, 1994). The first 3 months is a time when sleep, respiration and circulatory control undergo extensive consolidation, linked with developments in the body's motor capacity.

Myelinization studies show that, while some cerebral tissues are well differentiated at birth, others must undergo extensive development. In the first 3 months the greatest change is in motor pathways, sensory roots of the spinal cord, and in visual projections to the midbrain tectum, thalamus and cortex (Yakovlev & Lecours, 1967; Gilles et al., 1983; Trevarthen, 1985). Subthalamic (brainstem) auditory pathways are more mature than visual ones, but post-thalamic (cortical) tactile and auditory projections develop over the first few years, much more slowly than those for vision. Synaptic mechanisms of layer 4 neurons in the visual cortex appear to be completed rapidly around 5 months, when binocular stereopsis develops (Held, 1985).

Neurobiology and attachment for protection versus companionship in learning

Bowlby (1958) formulated attachment theory on ethological evidence of the effects of maternal deprivation in rhesus monkeys (Higley & Suomi, 1986). Early body contact played a critical role in development of appropriate social behaviour, and disruption to early contact resulted in stress and later pathological behaviour. Research on rodents, cats and primates shows that structure, neurochemistry and function of a mammal brain change greatly around birth, and these changes critically depend on parental support (Blass, 1999; McKenna & Mosko, 1994; Panksepp, 1998; Rosenblatt, 1994; Schore, 1994; Zeifman *et al.*, 1996).

Transformations in the *ergotropic/trophotropic balance*, i.e. between energy expending action on the environment and nurturing and protective self-maintenance (Hess, 1954), are regulated inside the infant human and young child by the same genetic factors as those that instructed the differentiation of somatic and visceral sensory and motor systems of the brain in the embryo and fetus. Intrinsic regulations of brain activities that direct movements and attention become the predictors and creators of experience (Panksepp, 1998; Trevarthen & Aitken, 1994; Tucker, 1992). Developments in perceptuomotor skills and cognition depend on changes in communication, changes that are linked to reorganizations in autonomic and neurohumoral state and dependent on the quality of responses from a caregiver.

Human infants are born more immature than other primates and, like altricial young of other mammals, have strong adaptations for integrating their state regulation with maternal care (Carter *et al.*, 1997). Young infants respond to the touch, movement, smell, temperature, etc. of a mother, and sleeping with the mother may support the development of cardiac and respiratory self-regulations before the development of cerebral mechanisms of breath control for speech (McKenna & Mosko, 1994), which involves integration between forebrain and brainstem centres (Ploog, 1992). Newborns gain regulation from the rhythms of maternal breathing and heart beat, and fetuses are sensitive to maternal vocal patterns (DeCasper & Spence, 1986; Fifer & Moon, 1995). Infants' arousal and expressions of distress are immediately responsive to stimulation from breastfeeding, their state regulation benefiting from the sugar and fat content of breast milk (Blass, 1996). These interactions are not merely physiological. For example, the response to breast milk is facilitated if the newborn has sight of the mother's eyes (Zeifman *et al.*, 1996).

The communicative precocity of human newborns indicates that emotional responses to caregivers must play a crucial role in early brain development (Schore, 1994). They are likely to influence differentiation of perceptual discrimination, cognitive processing, memory, voluntary deployment of attention to environmental objects, and executive functioning or problem solving (Trevarthen, 1989, 1990b; Schore, 1994). The regulatory mechanisms of the infant brain are subject to changes by endocrine steroids and other hormones (McEwen, 1997; Suomi, 1997). The infant's endocrine status, neurochemistry and brain growth respond to maternal stimulation (Hofer, 1990; Kraemer, 1997; Schanberg & Field, 1987; Schore, 1994). Modern perinatal medicine finds evidence that this human regulation may be only partly substituted by artificial clinical procedures (Als, 1995).

Developments in the meso-limbic cortices of the temporal and frontal lobes in infants and toddlers transform autonomic self-regulation, emotions in communication, and the motives for action and experience (Schore, 1994), but at all stages these later-maturing limbic and neocortical circuits emerge in recip- rocal involvement with the multi-modal regulatory systems of the intrinsic motive formation (IMF) in the brainstem core (Trevarthen & Aitken, 1994). Expressions of pleasure and displeasure are formulated subcortically (Steiner, 1979), and vocal expression of other emotions, and the later-developing neocor- tical mechanisms of speech, are organized around nuclei of the brainstem, the basal ganglia, thalamus and limbic cortex (Ploog, 1992; Trevarthen, 1998c). The adaptive structure of a newborn human brain, and especially those parts that serve social or interpersonal perception and expression, set directions and limits to any future acquisition of skills or knowledge by a child (Figure 5).

There is physiological evidence that, in normal motor activity, internal gener- ation of time and distance parameters for an animal's movements are set up in the brain before reafferent information is received from the senses. The brain generates "motor images" in anticipation of perception (Jeannerod, 1994). Psychophysical studies show that "time in the mind" defines moments and "phrases" of conscious perception, as of motor actions (Pöppel, 1994).

Coordinated, rhythmically regulated action can be observed in movements of a human fetus by mid gestation (Cioni & Castellaci, 1990; de Vries *et al.*, 1984) and the quality of fetal movements displays temperamental differences, even between identical twins (Piontelli, 1992). Cerebral representations of movement in time can explain the abilities of newborn infants to produce rhythmic and melodic forms of movement that match intrinsic kinematic features of adult movements, and with these to enter into synchronized, imitative or comple- mentary interactions with the adult partner (Jaffe *et al.*, 2000; Trevarthen, 1986, 1999a). They also explain the sensitivity that young infants have to contingent stimulus effects of their own actions (Gergeley & Watson, 1999) and the distress that is caused by making a partner's movements non-contingent (Murray & Trevarthen, 1985; Nadel *et al.*, 1999).

From the start of life the mechanisms of emotion and cognition are interde- pendent, as they clearly are in the neuropsychology of adults (Damasio, 1999; Tucker *et al.*, 2000). Special importance is given to groups of bioaminergic neurons of the reticular formation that innervate most areas of the growing brain of embryo and fetus (Holstege *et al.*, 1997). These regulate the global state of brain functions, initiate purposeful movement and focus awareness, and are them- selves regulated by caregiver–infant interaction. Emotions serve as adaptive guides for channelling the energy of action (Panksepp, 1992, 1998). In infancy the frontal lobes appear to have a major role in communication of emotions and regulation of the relationship to a parent (Davidson & Fox, 1982; Field *et al.*, 1995). Consolidation of a "working model" of mutual regulation of emotions with a care- giver in both protective "attachment" and adventurous "companionship", gives the child increased power to gain advantages from collaborative response to the environment (Reddy *et al.*, 1997; Schore, 1994; Stern, 1985, 1999; Trevarthen, 2000) (Figure 3).

The brain mechanisms of imitation

Neural "imitator" elements that translate from the observed action of another subject to matching motor image for hand use by the self have been identified by electrophysiological recording of nerve cell activity in area F5 of the inferior premotor cortex of monkeys (Di Pelligrino *et al.*, 1992). The "mirror" neurons are active when the monkey makes a movement, say to grasp a piece of food, and also when similar movement, to the same object, is made by the experimenter. A comparable substrate for recognition of facial expressions has been found in the temporal cortex (Rolls, 1992), and systems that mediate mirroring of vocalizations are also likely to be located in the temporal lobe.

Prefrontal "mirror neurons" have been taken as possible candidates for the mechanism that enables infants to imitate, to associate emotion with imitation of other persons' expressions and gestures, and to develop language (Rizzolatti & Arbib, 1998). However, given the relative immaturity of the frontal cortex in infants, and the extensive representation of expressive organs in the brainstem, it is more likely that unidentified subcortical components of a "mirror system" are responsible at least for the many imitations that neonates perform. Mirroring actions involves "multimodal", or "transmodal" sensory correspondence, and there are many multimodal neural populations in the brainstem, for example in the superior colliculus (Sparks & Groh, 1995). These convergent sensory fields are integrated with extensive systems that formulate motor images for action and expression (Damasio, 1999; Holstege *et al.*, 1997; Panksepp, 1998).

The maturational sequence of brain functions in the child

Infants' perceptual, cognitive, motoric and communicative functions exhibit "non-linear", age-related transformations in emotion, interest and ability (Rijt-Plooij & Plooij, 1993), and also in the motives of intersubjectivity (Figure 2). These may be described as stages in the emergence of cognitive mechanisms and behavioural skills (Fischer & Rose, 1997), but the intrinsic regulations of the developing brain of the subject are *motivational*, searching for experience and causing changes in cognition and skill. Measurements of head circumference show *brain growth spurts* at 3–4 weeks, 7–8 weeks and 10–11 weeks (Fischer & Rose, 1994).

General cell and circuit differentiation, indicated by glucose metabolic rate, increases through infancy to around 5–7 years, falling thereafter (Chugani, 1994). In the newborn highest rates of metabolism are found in the primary sensory and motor cortices, thalamus, brainstem and the vermis of the cerebellum, but the basal ganglia, hippocampus and cingulate gyrus are also active. Parietal, temporal and primary visual cortices, basal ganglia and the cerebellar hemispheres are increasingly active during the first 3 months. Lateral and inferior frontal cortex shows increase of glucose consumption after 6–8 months and the dorsal and medial frontal cortices show comparable increase only between 8 and 12 months. At 1 year the infant's pattern of glucose utilization resembles that of an adult.

In the last 3 months of the first year the ratio of EEG alpha activity to beta activity in the prefrontal cortex increases with the developments in intelligence and locomotion (Bell & Fox, 1996). Brain mechanisms for cognitive and motor maturation may be regulated by the same intrinsic factors. EEG coherence studies indicate that cycles of development swing between left and right cortex throughout childhood (Thatcher, 1994). In summarizing his findings, Thatcher describes cycles of hemispheric development: "One interpretation is that the frontal regions control or significantly influence the cycles of synaptic influence in posterior cortical regions ... . This process is nonlinear in both space and time and is manifested behaviorally by relatively sudden changes in cognitive competence" (Thatcher, 1994, p. 588).

Cycles of differentiation and reorganization manifested as "growth surges" in EEG, and in head circumference (Dawson & Fischer, 1994) show three accelerations of cortical development later in childhood – at 3–5 years, 6–7 years and around puberty.

Executive and linguistic intelligence and growth of frontal lobes

The prefrontal cortex (PFC), constituting 30% of the total human adult neocortex, is said to perform "complex cognitive tasks involving active memory, abstract reasoning and judgement" (Lewis, 1997). It is attributed an essential role in the integration of information in the temporal domain (Fuster, 1989; Goldman-Rakic, 1987), a function that would explain the close reciprocal anatomical relationship between the PFC and the cerebellum. The delayed response task which depends upon an intact PFC develops at about 1 year in humans (Diamond, 1985), but the adult functions of the PFC do not mature until after puberty (Lewis, 1997).

The conventional picture of PFC functions gives first place to experience-processing cognitive activities, rather than motivational or emotional functions. However, both phylogenetic and ontogenetic considerations would indicate that this part of the brain has evolved to integrate intrinsic visceral regulations with environmental information (Damasio, 1999). The PFC has importance in expressive communication, and in the multisensory integrations required to recognize underlying intentions and feelings in behaviours of other subjects (Goldman-Rakic, 1987; Diamond, 1985; Schore, 1994). The rostroventral premotor area of monkey frontal cortex (area F5), representing movements of hands, face and larynx, is probably homologous with Broca's speech area in humans. In both species this region has a key role in the development of imitated skills. In humans it mediates in the acquisition of all productive aspects of language, including speech (Rizzolatti & Arbib, 1998). Lateral prefrontal cortex is implicated in speech production and the dynamics of language, while understanding of speech involves dorsolateral temporal cortex and areas above the Sylvian sulcus (Damasio & Damasio, 1992).

The orbitofrontal cortex, linked to the mediolateral temporal cortex, undergoes important elaboration in infancy (Schore, 1994). Other structures that have a comparatively long development in humans, elaborating of new functions from fetal stages to adulthood, include the reticular formation, hippocampus, dorsomedial temporal lobe and parts of the cerebellum (Yakovlev & Lecours, 1967; Gilles *et al.*, 1983). These all mediate collaboration between intrinsic motivating activities and acquisition of new organization under environmental instruction. All must be important in human cultural learning

Early development of language in the brain

Preverbal communication will involve both limbic and neocortical mechanisms in temporal and prefrontal lobes, as well as many subcortical structures of forebrain, diencephalon, midbrain and hindbrain (Damasio & Damasio, 1992; Trevarthen, 1998c).

Perisylvian cortex of the left hemisphere is active in the perception and production of speech or sign (Damasio & Damasio, 1992), and left-hemisphere lesions interfere with language expression and word and sentence comprehension, leaving prosody and simple word understanding intact (Trevarthen, 1984c). Right-hemisphere lesions interfere with prosody, melodic expression, and poetic or metaphoric awareness, as well as perception of emotion in the expression of others. Activity in the right hemisphere is correlated more with reception of the prosodic features important in recognition of language users and transmission of affect. In infancy, communication depends more on these emotive functions of communication, which correlates with evidence that the infant's right hemisphere is more developed and more active than the left. A conspicuous increase in left hemisphere temporofrontal associative links between 2 and 5 years relates to the differentiation of syntactic functions in language. Before that, area 44 of the left frontal lobe (Broca's area) develops more basal dendrites between 18 and 36 months as syllable and word combinations are learned, and as the child stabilizes hand preference for meaningful or conventional acts (Pulvermüller & Schumann, 1994).

The cognitive and executive systems in the neocortex are activated, regulated and intercoordinated by neurochemical output from cells of the brainstem reticular formation and the limbic system, and these generate different emotions in left and right of the brain, predisposing the left brain to dopaminergic initiative and action, and the right to noradrenergic awareness and observation (Davidson & Hugdahl, 1995; Trevarthen, 1989, 1990c, 1996; Trevarthen & Aitken, 1994). Knowing how to act and what to perceive, as well as how to predict events and causes, will involve parts of the brain that organize experience in strategic ways evaluated with relation to the self. These parts are of two main kinds, located in prefrontal cortex for "procedural" (or "executive") and propositional functions, and in the medial temporal cortex connected with the hippocampus for conceptual and "declarative" aspects. Clearly, these different functions of motives will have an important role in organizing "language" in the developing brain (Trevarthen, 1998c).

PRENATAL BRAIN DEVELOPMENT

The fertile "chaos" that a vast overproduction of neuroblasts, axon outgrowths, dendrites and synapses creates in the developing cortex of an infant is deeply and productively constrained by structures and processes that form in the embryo brainstem before the nerve net is electrically active or receptive to the environment (O'Rahilly & Müller, 1994). Interneuron systems of the autonomic nervous system, hypothalamus, reticular formation, basal ganglia and limbic system which mediate intermodal sensory integration, the coordination of motor patterns and

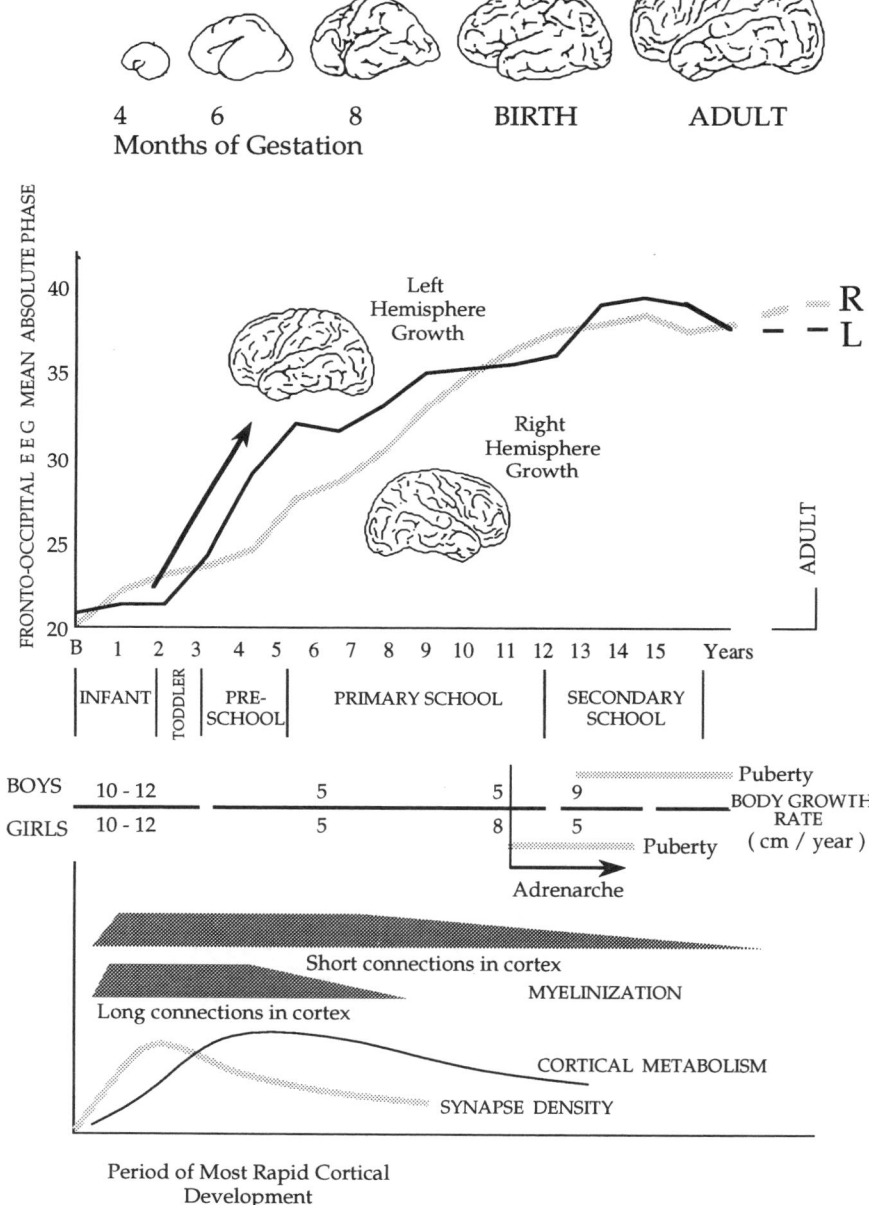

Figure 4 Development of the two cerebral hemispheres in childhood.

motive states, appear in the first month of gestation (Figure 6). These structures, constituting the intrinsic motive formation (IMF), are well developed before cortical neurons are generated (Trevarthen & Aitken, 1994).

Determination of embryo brain organization

To explain the initial mapping of prospective action space we need to know how genes and other factors affect the multiplication and survival of populations of undifferentiated cells in the embryo. Brain cells become organized in systems in the inert body of the embryo, out of contact with any sensory receptors.

The polarity and symmetry of the human organism is determined when cells of body and brain (neural plate) are still indistinguishable (Ruiz i Altaba, 1994). Regulator genes encode protein transcription factors that turn genes on and off, and the regional expression of these inductive proteins in a temporal sequence down the embryo body reflects cascades of homeobox (Hox) gene activations that define pattern formation in segments of the embryo and its brain (Ruiz i Altaba, 1994; Krumlauf, 1994; Vicente & Kennedy, 1997). Division of the embryo body into segments in the third and fourth weeks triggers segmentation of the neural tube into neuromeres. Inside each neuromere the proliferation, differentiation and grouping of neural cells are regulated by Hox genes. The hypothalamus, top-level coordinator between the autonomic nervous system and somatic and visceral systems of the CNS, develops early, in response to signals from the head mesoderm (Rubenstein *et al.*, 1998)

The differentiation of face, cerebral hemispheres and diencephalon (thalamus, subthalamus, hypothalamus and optic cup) is determined by Pax genes and cell adhesion molecules (CAMs) that segregate cell types, map the body in the brain and determine future affinities of neurons and the formation of nuclei, axonal tracts and synaptic arrays (Davidson, 1993; Edelman, 1988).

How neurons form functional systems

The cells that differentiate in the embryo CNS gain distinctive affinities for molecular elements in the extracellular matrix, on the surface of glia cells, and on other neurons (Goodman & Shatz, 1993). Forces of attraction and repulsion direct cell migration and the paths that axons and dendrites follow in the late embryo (weeks 5 and 6) (O'Rahilly & Müller, 1994). Neuroblasts divide around the ventricle of the neural tube, migrate out along radial glia fibres, and form layered arrangements in the hippocampus, cerebral cortex and cerebellum (Hynes & Lander, 1992). The tips of axons and dendrites are guided by "growth cones" (Suter & Forscher, 1998), sensory-motor structures that explore the intercellular terrain, sensing different molecular environments and laying down the path of the nerve cell extension (Vicente & Kennedy, 1997; Haydon & Drapeau, 1995; Magge & Madsen, 1997). Finally, nerve terminals form intricate arrays of synaptic contacts on other nerve cells, with hundreds of contacts per cell. The formation and maintenance of effective synapses is affected by the excitations received from other neurons, and by growth factors and steroid hormones produced by endocrine glands (McEwen, 1989, 1997; Xiao & Link, 1998). Building of brain systems involves enormous programmed loss of elements, survival of nerve cells depending on genes that promote or inhibit cell death, and on protein growth factors (Cameron *et al.*, 1998; Davies, 1994).

Psychological functions of the mature brain depend on "activity-dependent patterning" or "plasticity" of synapses, shaped by stimuli from the world and from the body (Hebb, 1949; Singer, 1986), and the greatest plasticity will be in early childhood, and even in the fetus, electrical impulse traffic triggered by excitation of receptors, modifies the structures formed by spontaneous developmental processes. Nevertheless, the brain is never a passive receiver of stimulus input, and major structural responses to epigenetic factors or the environment are regulated to occur at certain ages in critical periods (Cynader *et al.*, 1990).

Brainstem state-regulating and motive systems

In the mid-embryo stage (week 4), the nervous system is electrically inactive. The first generalized movements occur in week 8 (de Vries *et al.*, 1984), but already in week 5 monoamine transmission pathways grow from the brainstem into the primordial cerebral hemispheres. Key components of the emotional motor system (hypothalamus, basal ganglia and amygdala) are in place when the neocortex is unformed (Holstege *et al.*, 1997; O'Rahilly & Müller, 1994; Rodier, 2000) (Figure 4). By the end of the second month the main components of the brain are in place, and eyes, vestibular canals and cochlea, hands, nose and mouth, and the hands are rapidly differentiating their distinctive forms, each dedicated to the picking up of a particular forms of physical information from the environment. Core regulatory mechanisms of the central nervous system – the peri-aqueductal grey of the midbrain, the hypothalamus, reticular formation, basal ganglia and limbic system – are laid down in the first trimester, but the cognitive systems of the cerebral cortex do not appear until the second trimester. The motivating and life-maintaining structures form a link between regulation of gene instructions in prenatal brain morphogenesis and the acquired adaptations of the developing mind (Trevarthen & Aitken, 1994). Defects in this link are implicated in disorders of empathy and cognition, including autism and schizophrenia (Aitken & Trevarthen, 1997; Keshavan & Murray, 1997; Tager-Flusberg, 1999; Trevarthen *et al.*, 1998).

The fetal cerebral cortex

A rudimentary cerebral cortex is visible by week 6. Cortical cells proliferate and migrate in waves from week 7 radially along glia strands, to form layers VI to II in inside-out order (Rakic, 1991). Migrating cortical cells pass through the "subplate", cells of which receive initial synaptic contacts of afferents growing from brainstem, basal forebrain, thalamus and ipsilateral and contralateral cortex. Subplate cells project axons into the cortical plate, pioneer paths to thalamus and other subcortical structures, and then die after the cortex is mature (Lewis, 1997). The subplate of humans is the largest and most persistent, and abnormal genetic instructions there may lead to schizophrenia and other psychiatric disorders.

Brainstem monoaminergic neurons penetrate the cortical plate at 13 weeks, and sensory thalamic axons arrive in subsequent weeks (Kostovic & Goldman-Rakic, 1983; Lund, 1997). Thus neural motive regulators of cortical activity

(Singer, 1986) mature ahead of those that are destined to bring environmental information from special receptors. At the mid-fetal stage (week 24) dopamine axons, important in coordination of innate motor patterns, are concentrated in the deep parts of the cortical plate and the upper subplate, at the time when thalamic sensory afferents end their 'waiting' in the subplate (Verney et al., 1993; Zecevic & Verney, 1995). Earlier formed, deeper neuron layers in the cortex linked with limbic structures mediate in the development of cognitive structures, which take inputs from sensory and corticocortical connections (Tucker, 1992; Benes, 1994). Experiments with monkeys prove that the relative size of cortical areas depends on input from subcortical centres (Rakic, 1991).

Fetal movements, the earliest intentions and their asymmetries

Research on the behaviour, psychology and physiology of the fetus indicates that in the last trimester functions are established in anticipation of an active post-natal life, and especially for assimilating maternal care (Lecanuet et al., 1995). Motor coordinations exist that are adapted for visual exploration, reaching and grasping, walking, and for expressive communication, by facial expression and gesture (de Vries et al., 1984; Prechtl, 1984). Respiratory movements and amniotic breathing appear several weeks before birth. It is of particular interest that heart rate changes are integrated with phases of motor activity from 24 weeks (James et al., 1995). This is indicative of the formation of a prospective control of autonomic state coupled to readiness for muscular activity on the environment, a feature of brain function, which Jeannerod (1994) has cited as evidence for the formation of cerebral "motor images" underlying conscious awareness and purposeful movement.

Asymmetries of psychological function and behaviour that are elaborated through childhood (Trevarthen, 1990b) may be induced prenatally by anatomical and neurochemical asymmetries to be found in all levels of the human brain (Cynader et al., 1981; Huttenlocher, 1994; Trevarthen, 1996). Manual preference observed in thumb-sucking in individual fetuses of 15 weeks gestational age correlates with hand preferences seen in the second year after birth (Hepper, 1995). The human neocortex is visibly asymmetric in fetal stages, a feature which originates in the embryo (Rakic, 1991). The right hemisphere is more advanced than the left in surface features from about the 25th week (Rosen & Galaburda, 1985) and this bias persists until the left hemisphere shows a postnatal growth spurt starting in the second year (Thatcher et al., 1987). The asymmetric areas around the Sylvian sulcus are crucial in culture-related processes of the adult human brain, including language.

NOTES ON THE PSYCHOBIOLOGY OF CEREBRAL MOTIVES AND EMOTIONS, AND THEIR ROLE IN COGNITIVE GROWTH

Key functions of the human head and brain: self-regulations and social signalling

When primitive vertebrates evolved as active predators, presegmental tissues at the anterior end became an independently mobile head, with receptors for sensing the world, jaws for controlling food intake and gill arches for gas exchange. Simultaneously a forebrain developed to coordinate a more vigorous, energy-hungry life with more foresightful intelligence (Gans & Northcutt, 1983). In protochordates, surviving ancestors of vertebrates, the environment-sensing brain rests above the primordial hypothalamus, which is a centre of the neuro-hormonal system for self-regulation.

In socially intelligent higher vertebrates the head additions, with the autonomic system, have become the expressive/receptive systems of the face, throat, eyes and ears, to which, in humans, the hands have been recruited. The communication of psychological expectations between subjects by expressive movements constitutes a function which integrates internal organismic or visceral regulations with environment-directed or somatic ones (Trevarthen, 1987b, 1989; Porges, 1997). Brain mechanisms of ancient phylogenetic origin (MacLean, 1990; Porges, 1997; Panksepp, 1998), which evolved in reptiles, birds and mammals to signal bodily needs and to attract parental care, have been augmented by, or transformed into, mechanisms of sympathetic mind-engagement, with great creative potential.

In the adult human, core regulatory systems of interneurons in the brainstem direct attentional orientations, coordinate purposeful movements of the body and its parts, and mediate the equilibria between autonomic and exploratory or executive states. The same systems emerge in the embryo brain as regulators of morphogenesis in cognitive systems (Zecevic & Verney, 1995). An important output from this intrinsic motive formation (IMF; Trevarthen & Aitken, 1994) controls the "sensory-accessory motor systems" of the special receptors of the head and hands. The eyes, the ears and cochlea, the lips and tongue and the palms and fingers are separately aimable and tunable. Movements of these structures rhythmically "aim" and "censor" the uptake of perceptual information in different modalities of high sensitivity and resolution, and these motor adjustments occur in the exploratory and focusing phases of attending to the outside world, before the final commitment of a "consummatory act". They therefore exhibit predictive information to an observer about emerging impulses of the subject, and they signal awareness and intentions (Panksepp, 1998; Trevarthen, 1993b, 1997, 1998b). All the organs of human linguistic expression are recruited from this accessory motor set (Jürgens, 1979; Ploog, 1992).

Directed attention links brainstem and cortex (Morecraft *et al.*, 1993), and emotions arise as centrally generated states involving all levels of the brain (Heilman & Satz, 1983; Panksepp, 1992; Gainott *et al.*, 1993). Neocortical areas are intimately coupled to the basal ganglia, which communicate directly with hindbrain and midbrain motor structures, and to the limbic system, which has close anatomical connections with visceral, autonomic and endocrine parts of the brainstem reticular formation and hypothalamus (Hess, 1954; MacLean, 1990; Ploog, 1979; Nauta & Domesick, 1982). The cerebellum, too, has massive reciprocal connections with the neocortex, notably between the prefrontal cortex and the neocerebellar vermis. All these are implicated in the interpersonal communication of dynamic emotions.

Two centres have been identified as essential for relating emotions to social experience and for the development of a reflective (cognitive) understanding and voluntary control of emotions; in the amygdala and temporal mesocortex

Table 2 Human cranial nerves; visceral and somatic functions, and adaptations for intersubjective communication

Communication movements	Visceral and somatic movements	Cranial nerves	Visceral and somatic senses	Communication senses
		1: Olfactory	Odour, taste	Smelling/tasting; other
		2: Optic	Light sense, vision	Seeing other
Looking at other Direction of gaze Pupil changes	Eye rotation Lens and pupil Movements	3: Oculomotor	Eye-muscle sense	
Looking at other Direction of gaze	Eye rotation	4: Trochlear	Eye-muscle sense	
Face expressions Vocalizing Sucking, kissing	Mastication	5: Trigeminal	Facial feelings	Feeling other's touch; feeling own face
Looking at other Eye expressions Crying	Eye rotation lifting Eyelids, tears	6: Abducent	Eye-muscle sense	
Face expressions Speech Listening to speech	Eating Middle ear muscles	7: Facial	Taste, tongue, mouth	Tasting/feeling other
		8: Auditory	Hearing, balance	Hearing other; hearing self
Vocalizing, laughing Expression in voice Coughing	Coughing, biting Salivating Swallowing	9: Glossopharyngeal	Taste	Tasting/feeling other
Vocal expression Circulatory signs Panting, gasping	Heart and gut activity Vomiting, fainting Respiration	10: Vagus	Taste, heart, lungs, gut	Feeling own emotion
Head expressions Vocalizing Laughing, coughing	Head movements Shoulder movements Swallowing	11: Accessory		
Vocalizing, speaking Licking, sucking	Tongue	12: Hypoglossal		

The frontal cortex has descending control of expressions and of attention to persons (which is blocked by fear, anxiety and stress)

(Aggleton, 1993; Bachevalier, 1994; LeDoux, 1993; Rolls, 1990) and in the orbitofrontal cortex (Cummings, 1993; Dawson, 1994; Goldman-Rakic, 1992). Orbitofrontal and temporal pole cortex develop extensively after birth and are presumed to modulate the activities of the infant's brainstem emotional system (Schore, 1994). Both regions develop asymmetries (Ross, 1996).

Self–other intelligence and cultural learning

It is now recognized that the principal selective pressure for evolution of larger brains in the line leading to humans comes from the advantages of social intelligence, and social learning (Donald, 1991). Cultural learning (Tomasello *et al.*, 1993), the cumulative passing on of artificial knowledge, techniques and beliefs from generation to generation puts wholly new demands on intersubjective representations. All human communities and societies, however different they may be in size, structure and technical sophistication, and in whatever form the history of their knowledge and beliefs are coded, stored and transmitted, are animated in mutual interest. Social knowledge produced by interaction requires a means of representing knowledge in sociable forms, and even infants have the rudiments of this ability. In childhood, intersubjective processes and "narratives of collaborative awareness" are an inextricable part of self-organization by education (Bruner, 1996).

Human brains, then, are adapted for intersubjective growth and cultural learning. The psychological evidence from infancy implies that the neural arrays for self-maintenance and self-coordination, and those for regulating transactions with objects, have evolved with a third system, which regulates self–other inter-coordination and "altero-ception" (Bråten, 1988; Trevarthen, 1999b; Aitken & Trevarthen, 1997; Trevarthen *et al.*, 1999). More than in any other species, human brain and behaviour development does not make sense if the individual is considered in isolation, and if the impress of information on receptive neural circuitry is taken as the principal process driving development.

Sociocultural, "motivative" neuroscience

Two core factors of brain function and anatomical organization, which are generally neglected in cognitive science, the *timing and modulation of motor representations and the relation of body form to regulation and guidance of action in space*, are crucial to human communication at all stages of its development.

Piaget's genetic epistemology charted the unfolding regulation of cognitive schemes in an infant as an individual observer/actor (Piaget, 1953). Application of a similar epigenetic model to inter-subjective psychology leads to enquiry about the initial state of the infant's motives and awareness in interaction with parents. In attachment theory the adaptations of this initial psychosocial state to companionship and cultural learning is not well comprehended. According to the current formulation the infant is born needing maternal care to regulate endogenous physiological "tension", but has no specific need for organizing psychological structures and processes in cooperation with a human companion (Sroufe, 1996).

Edelman's (1987) metaphor of "neural Darwinism" for within-brain epigenesis of psychological processes applies the theory of selection at the level of competing neural components in the cerebral cortex, which is conceived as a non-linear dynamic system, in which only the most ecologically responsive cell groups in the neocortical networks survive. Edelman's theory has encouraged confidence in the adaptive plasticity of human cognitive growth. It takes support from evidence that exuberantly over-abundant components in the developing

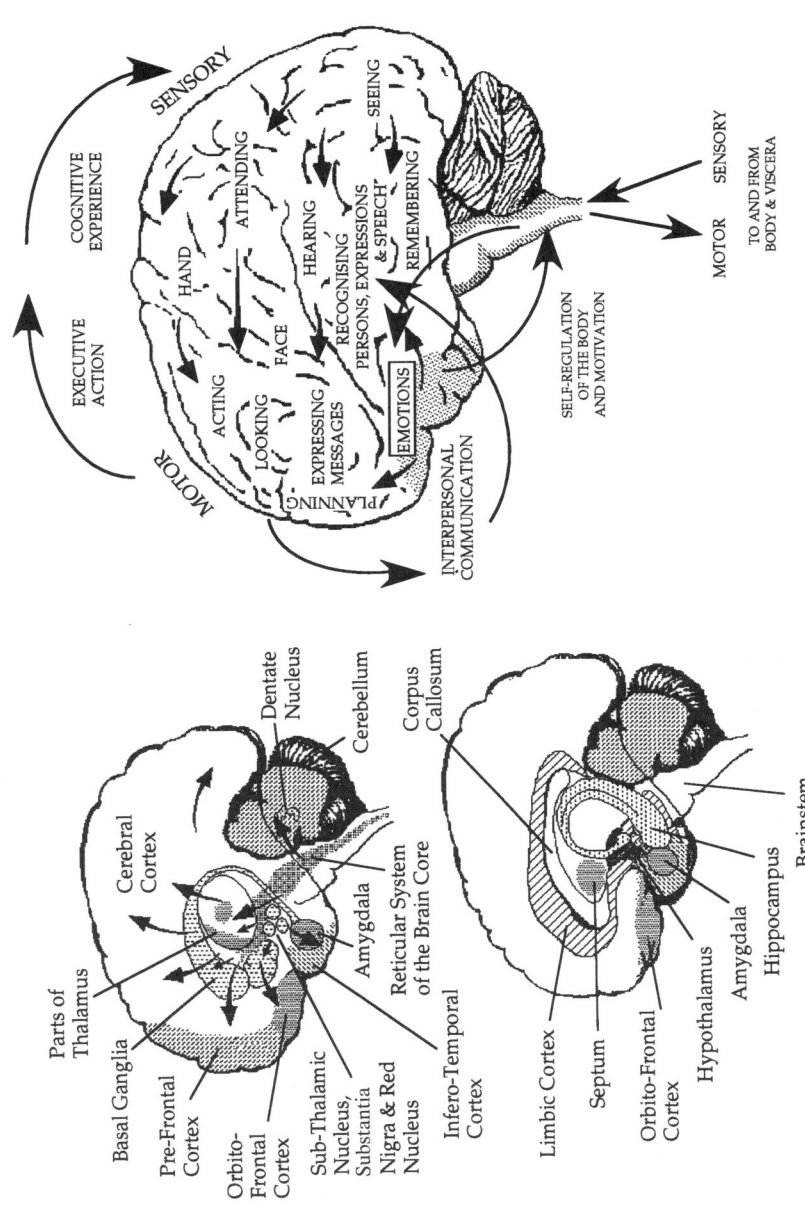

Figure 5 Cortical and subcortical levels of the brain are reciprocally connected in the functions of motivation, emotion, intentional action, perception and learning, and ascending projections from the reticular core of the brainstem modulate cortical functions. Communication involves all levels of the brain, and couples self-regulatory functions and social signalling of motives and emotions through mediation of the brainstem core, limbic cortex, amygdala, infero-temporal cortex and orbitofrontal cortex.

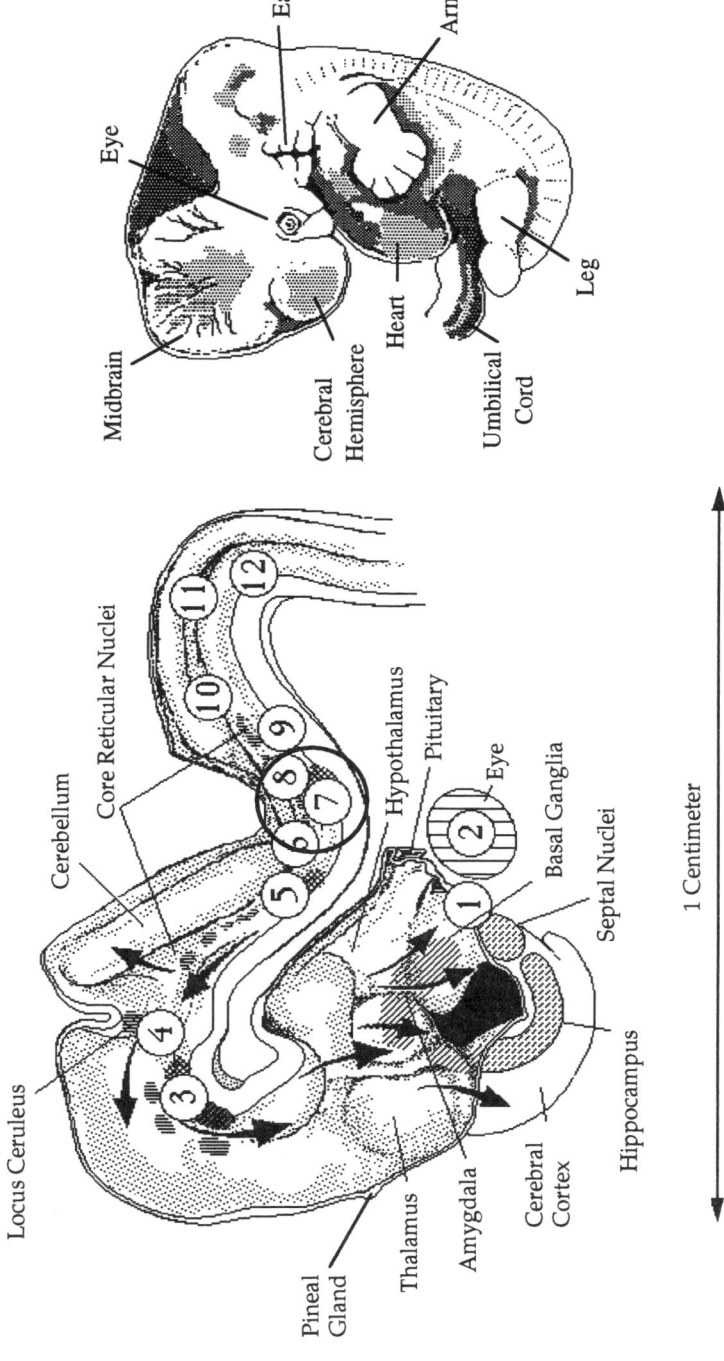

Figure 6 Human embryo at 7 weeks after conception. The cranial nuclei and reticular system of the brainstem have developed, and they project into the forebrain hemispheres before the neurons of the cerebral cortex are differentiated. The numbers for the cranial nuclei correspond to those in Table 2.

nervous system are subject to draconian selection, through synaptic pruning and other mechanisms of elimination thought to increase functional specificity (Changeux, 1985). Postnatal selection theory offers no satisfactory explanation for pan-specific features of human psychological function and brain organization; such as those that have an active part in the determination of social and cultural adaptation in all human groups, and which are displayed by infants from the first weeks.

It would appear that in future neuroscience will find effective application in developmental psychology by adopting a whole-brain approach, integrating interest in embodied cognition with acceptance of the regulatory role of intrinsic human motives and emotions for communication. It will accept that no human brain can develop effective volitional and cognitive "belief" structures if it is unable to exercise an innate motivation to interact sympathetically with the effects expressed by other human brains.

References

Adamson, L.B. & Bakeman, R. (1991). The development of shared attention during infancy. In R. Vasta (Ed.), *Annals of child development* (vol. 8, pp. 1–41). London: Kingsley.

Aggleton J.P. (1993). The contribution of the amygdala to normal and abnormal emotional states. *Trends in Neurosciences*, **16**, 328–333.

Aitken, K.J. & Trevarthen, C. (1997). Self–other organization in human psychological development. *Development and Psychopathology*, **9**, 651–675.

Als, H. (1995). The preterm infant: a model for study of fetal brain expectation. In J.-P. Lecanuet, W.P. Fifer, N.A. Krasnegor & W.P. Smotherman (Eds.), *Fetal development: a psychobiological perspective* (pp. 439–471). Hillsdale, NJ: Erlbaum.

Bachevalier, J. (1994). Medial temporal lobe structures and autism: a review of clinical and experimental findings. *Neuropsychologia*, **32**, 627–648.

Bateson, M.C. (1979). "The epigenesis of conversational interaction": a personal account of research development. In M. Bullowa (Ed.), *Before speech: the beginning of human communication* (pp. 63–77). London: Cambridge University Press.

Beebe, B., Jaffe, J., Feldstein, S., Mays, K. & Alson, D. (1985). Inter-personal timing: the application of an adult dialogue model to mother-infant vocal and kinesic interactions. In F.M, Field & N. Fox (Eds.), *Social perception in infants* (pp. 217–248). Norwood, NJ: Ablex.

Bell, M.A. & Fox, N.A. (1996). Crawling experience is related to changes in cortical organization during infancy: evidence from EEG coherence. *Developmental Psychobiology*, **29**, 551–561.

Benes, F.M. (1994). Development of the corticolimbic system. In G. Dawson & K.W. Fischer (Eds.), *Human behavior and the developing brain* (pp. 176–206). New York: Guilford Press.

Bernstein, N. (1967). *Coordination and regulation of movements*. New York: Pergamon.

Blass, E.M. (1996). Mothers and their infants: peptide-mediated physiological, behavioral and affective changes during suckling. *Regulatory Peptides*, **66**, 109–112.

Blass, E.M. (1999). The ontogeny of human face recognition: orogustatory, visual and social influences. In P. Rochat (Ed.), *Early social cognition: understanding others in the first months of life* (pp. 35–65). Mahwah, NJ: Erlbaum.

Bowlby J. (1958). The nature of the child's tie to his mother. *International Journal of Psychoanalysis*, **39**, 1–23

Bråten, S. (1988). Between dialogical mind and monological reason: postulating the virtual other. In M. Campanella (Ed.), *Between rationality and cognition* (pp. 205–235). Turin: Albert Meynier.

Bruner, J.S. (1983). *Child's talk. Learning to use language*. New York: Norton.

Bruner, J.S. (1996). *The culture of education*. Cambridge, MA: Harvard University Press.

Butterworth, G. & Grover, L. (1988). The origins of referential communication in human infancy. In L. Weiskrantz (Ed.), *Thought without language* (pp. 5–24). Oxford: Clarendon.

Cameron, H.A., Hazel, T.H. & McKey, R.D.G. (1998). Regulation of neurogenesis by growth factors and neurotransmitters. *Journal of Neurobiology*, **36**, 287–306.

Carter, C.S., Lederhendler, I.I. & Kirkpatrick, B. (1997). *The integrative neurobiology of affiliation*. *Annals of the New York Academy of Sciences*, **807**. New York: New York Academy of Sciences.

Changeux, J.-P. (1985) *Neuronal man: the biology of mind*. New York: Pantheon.

Chugani, H.T. (1994). Development of regional brain glucose metabolism in relation to behavior and plasticity. In G. Dawson & K.W. Fischer (Eds.), *Human behavior and the developing brain* (pp. 153–175). New York: The Guilford Press.

Cioni, G. & Castellaci, A.M. (1990). Development of fetal and neonatal motor activity: implications for neurology. In H. Bloch & B. Bertenthal (Eds.), *Sensory-motor organization and development in infancy and early childhood* (pp. 135–144). Dordrecht: Kluwer.

Cohn, J.F. & Tronick, E.Z. (1983). Three-month-old infants' reaction to simulated maternal depression. *Child Development*, **54**, 185–193.

Condon, W.S. & Sander, L.W. (1974). Neonate movement is synchronized with adult speech: interactional participation and language acquisition. *Science*, **183**, 99–101.

Cummings, J.L. (1993). Frontal-subcortical circuits and human behavior. *Archives of Neurology*, **50**, 873–880.

Cynader, M., Lepore, F. & Guillemot, J.P. (1981). Inter-hemispheric competition during postnatal development. *Nature*, **290**, 139–140.

Cynader, M., Shaw, C., Prusky, G. & van Huizen, F. (1990). Neural mechanisms underlying modifiability of response properties in developing cat visual cortex. *Research Publications – Association for Research in Nervous & Mental Disease,* **67**, 85–108.

Damasio, A.R. (1999). *The feeling of what happens: body, emotion and the making of consciousness*. London: Heinemann.

Damasio, A.R. & Damasio, H. (1992). Brain and language. *Scientific American*, **267**, 88–95.

Davidson, E. (1993). Molecular biology of embryonic development: how far have we come in the last ten years? *Bio Essays*, **16**, 603–615.

Davidson, R.J. & Fox, N.A. (1982). Asymmetric brain activity discriminates between positive and negative affective stimuli in human infants. *Science*, **218**, 1235–1237.

Davidson, R.J. & Hugdahl, K. (Eds.) (1995). *Brain asymmetry*. Cambridge, MA: MIT Press.

Davies, A. (1994). Intrinsic programmes of growth and survival in developing vertebrate neurons. *Trends in the Neurosciences*, **17**, 195–199.

Dawson, G. (1994). Development of emotional expression and emotion regulation in infancy: contributions of the frontal lobes. In G. Dawson & K.W. Fischer (Eds.), *Human behavior and the developing brain* (pp. 346–379). New York: Guilford Press.

Dawson, G. & Fischer, K.W. (1994). *Human behavior and the developing brain*. New York: Guilford Press.

de Vries, J.I.P., Visser, G.H.A. & Prechtl, H.F.R. (1984). Fetal motility in the first half of pregnancy. In H.F.R. Prechtl (Ed.), *Continuity of neural functions from prenatal to postnatal life* (pp. 46–64). Oxford: Blackwell.

DeCasper, A.J. & Spence, M.J. (1986). Prenatal maternal speech influences newborns' perception of speech sounds. *Infant Behavior and Development*, **9**, 133–150.

Di Pelligrino, G., Fadiga, L., Fogassi, L., Gallese, V. & Rizzolatti, G. (1992). Understanding motor events: a neurophysiological study. *Experimental Brain Research*, **91**, 176–180.

Diamond, A. (1985). The development of the ability to use recall to guide action, as indicated by infants' performances on A-not B. *Child Development*, **56**, 868–883.

Donald, M. (1991). *Origins of the modern mind*. Cambridge, MA: Harvard University Press.

Draghi-Lorenz, R., Reddy, V. & Costall, A. (2001). Rethinking the development of "non-basic" emotions: a critical review of existing theories. *Developmental Review*, **21**(3), (In Press).

Edelman, G. (1987). *Neural Darwinism: the theory of neuronal group selection*. New York: Basic Books.

Edelman, G.M. (1988). *Topobiology: an introduction to molecular embryology*. New York: Basic Books.

Fassbender, C. (1996). Infants' auditory sensitivity towards acoustic parameters of speech and music. In I. Deliege & J. Sloboda (Eds.), *Musical beginnings: origins and development of musical competence* (pp. 56–87). Oxford: Oxford University Press.

Feldstein, S., Jaffe, J., Beebe, B., Crown, C.L., Jasnow, M., Fox, H. & Gordon, S. (1993). Co-ordinated interpersonal timing in adult–infant vocal interactions: a cross site replication. *Infant Behavior and Development*, **16**, 455–470.

Fernald, A. (1992). Human maternal vocalizations to infants as biologically relevant signals: an evolutionary perspective. In J.H. Barkow, L. Cosmides, & J. Tooby (Eds.), The *adapted mind: evolutionary psychology and the genration of culture* (pp. 345–382). Oxford: Oxford University Press.

Field, T. (1992). Infants of depressed mothers. *Development and Psychopathology*, **4**, 49–66.

Field, T., Fox, N.A., Pickens, J. & Nawrocki, T. (1995). Right frontal EEG activation in 3-to-6 month old infants of depressed mothers. *Developmental Psychology*, **31**, 358–363.

Field, T.N., Woodson, R., Greenberg, R. & Cohen, D. (1982). Discrimination and imitation of facial expressions by neonates. *Science*, **218**, 179–181.

Fifer, W.P. & Moon, C.M. (1995). The effects of fetal experience with sound. In J.-P. Lecanuet, W.P. Fifer, N.A. Krasnegor & W.P. Smotherman (Eds.), *Fetal development: a psychobiological perspective* (pp. 351–366). Hillsdale, NJ: Erlbaum.

Fischer, K.W. & Rose, S.P. (1994). Dynamic development of coordination of components in brain and behaviour: a framework for theory and research. In G. Dawson & K.W. Fischer (Eds.), *Human behavior and the developing brain* (pp. 3–66). New York: Guilford Press.

Fischer, K.W. & Rose, S.P. (1997). Dynamic growth cycles of brain and cognitive development. In R.W. Thatcher, G.R. Lyon, J. Rumsey & N. Krasnegor (Eds.), *Developmental neuroimaging, mapping the development and behavior* (pp. 263–279). San Diego, CA: Academic Press.

Fogel, A. (1993). *Developing through relationships*. Chicago, IL: University of Chicago Press.

Fogel, A. & Hannan, T.E. (1985). Manual actions of nine- to fifteen-week-old human infants during face-to-face interaction with their mothers. *Child Development*, **56**, 1271–1279.

Fogel, A. & Thelen, E. (1987). Development of early expressive action from a dynamic systems approach. *Developmental Psychology*, **23**, 747–761.

Fuster, J.M. (1989). *The prefrontal cortex: anatomy, physiology and neuropsychology of the frontal lobe*. New York: Raven Press.

Gainotti G., Caltagirone C. & Zoccolotti P. (1993). Left/right and cortical/subcortical dichotomies in the neuropsychological study of human emotions. *Cognition and Emotion*, **7**, 71–93.

Gans, C. & Northcutt, R.G. (1983). Neural crest and the origin of vertebrates: a new head. *Science*, **220**, 268–274

Gergeley, G. & Watson, J. (1999). Early social development: contingency perception and the social bio-feedback model. In P. Rochat (Ed.), *Early social cognition: understanding others in the first months of life* (pp. 101–136). Mahwah, NJ: Erlbaum.

Gilles, F.H., Shankle, W. & Dooling, E.G. (1983). Myelinated tracts: growth patterns. In F.H. Gilles, A. Leviton & E.G. Dooling (Eds.), *The developing brain: growth and epidemiological neuropathology* (pp. 117–183). Boston, MA: John Wright.

Goldman-Rakic, P.S. (1987). Development of cortical circuitry and cognitive function. *Child Development*, **58**, 601–622.

Goldman-Rakic, P. (1992). Dopamine mediated mechanisms of the prefrontal cortex. *Seminars in the Neurosciences*, **4**, 149–160.

Goodman, C. & Shatz, C. (1993). Developmental mechanisms that generate precise patterns of neural connectivity. *Cell*, **72**/*Neuron*, **10** (Supplement), 77–98.

Grieser, D.L. & Kuhl, P.K. (1988). Maternal speech to infants in a tonal language: support for universal prosodic features in motherese. *Developmental Psychology*, **24**, 14–20.

Hains, S.M.J. & Muir, D.W. (1996). Effects stimulus contingency in infant–adult interactions. *Infant Behavior and Development*, **19**, 49–61.

Halliday, M.A.K. (1975). *Learning how to mean: explorations in the development of language*. London: Edward Arnold.

Harris, P.L. (1998). Fictional absorbtion: emotional responses to make-believe. In S. Bråten (Ed.), *Intersubjective communication and emotion in early ontogeny* (pp. 336–353). Cambridge: Cambridge University Press.

Haydon, P. & Drapeau, P. (1995). From contact to connection: early events during synaptogenesis. *Trends in the Neurosciences*, **18**, 196–201.

Hebb, D.O. (1949). *The organization of behavior*. New York: Wiley.

Heilman, K.M. & Satz, P. (Eds.) (1983). *Neuropsychology of human emotion*. London: Guildford Press.

Heimann, M. (1989). Neonatal imitation, gaze aversion, and mother–infant interaction. *Infant Behaviour and Development*, **12**, 495–505.

Held, R. (1985). Binocular vision: behavioral and neuronal development. In J. Mehler & R. Fox (Eds.), *Neonate cognition: beyond the blooming buzzing confusion* (pp. 37–44). Hillsdale, NJ: Erlbaum.

Hepper, P.G. (1995). The behavior of the fetus as an indicator of neural functioning. In J.-P. Lecanuet, W.P. Fifer, N.A. Krasnegor & W.P. Smotherman (Eds.), *Fetal development: a psychobiological perspective* (pp. 405–417). Hillsdale, NJ: Erlbaum.

Hess, W.R. (1954). *Diencephalon: autonomic and extrapyramidal functions.* Orlando, FL: Grune & Stratton.

Higley, J.D. & Suomi, S.J. (1986). Parental behaviour in non-human primates. In W. Sluckin & M. Herbert (Eds.), *Parental behaviour* (pp. 152–207). Oxford: Basil Blackwell.

Hofer, M.A. (1990). Early symbiotic processes: hard evidence from a soft place. In R.A. Glick & S. Bone (Eds.), *Pleasure beyond the pleasure principle* (pp. 55–78). Newhaven, CT: Yale University Press.

Holstege, G., Bandler, R. & Saper, C. B. (Eds.) (1997). *The emotional motor system.* Amsterdam: Elsevier.

Hubley, P. & Trevarthen C. (1979). Sharing a task in infancy. In I. Uzgiris (Ed.), *Social interaction during infancy* (pp. 57–80). *New Directions for Child Development,* (vol. 4, pp. 57–80). San Francisco, CA: Jossey-Bass.

Huttenlocher, P.R. (1994). Synaptogenesis in human cerebral cortex. In G. Dawson & K.W. Fischer (Eds.), *Human behavior and the developing brain* (pp. 137–152). New York: Guilford Press.

Hynes, R. & Lander, A. (1992). Contact and adhesive specifications in the associations, migrations and targeting of cells and axons. *Cell,* **68,** 303–322.

Izard, C.E. (1994). What develops in emotional development? Intersystems Connections. In P. Ekman & J. Davidson (Eds.), *The nature of emotions* (pp. 356–361). Oxford: Oxford University Press.

Jaffe, J., Beebe, B., Felstein, S., Crown, C.L. & Jasnow, M.D. (2000). Rhythms of dialogue in infancy: coordinated timing and social development. (Submitted to SRCD Monographs, April).

James, D., Pillai, M. & Smoleniec, J. (1995). Neurobehavioral development in the human fetus. In J.-P. Lecanuet, W.P. Fifer, N.A. Krasnegor & W.P. Smotherman (Eds.), *Fetal development: a psychobiological perspective* (pp. 101–128). Hillsdale, NJ: Erlbaum.

Jeannerod, M. (1994). The representing brain: neural correlates of motor intention and imagery. *Behavioral and Brain Sciences,* **17,** 187–245.

Jürgens, U. (1979). Neural control of vocalization in non-human primates. In H.D. Steklis, M.J. Raleigh (Eds.), *Neurobiology of social communication in primates: an evolutionary perspective* (pp. 1–44). New York: Academic Press.

Jusczyk, P.W. & Krumhansl, C.L. (1993). Pitch and rhythmic patterns affecting infants' sensitivity to musical phrase structure. *Journal of Experimental Psychology: Human Perception and Performance,* **19,** 627–640.

Keshavan, M.S. & Murray, R. (Eds.) (1997). *Neurodevelopment and adult psychopathology.* Cambridge, UK: Cambridge University Press.

Kitamura, C. & Burnham D. (1998). The infant's response to maternal vocal affect In C. Rovee-Collier, L.P. Lipsitt & H. Hayne (Eds.), *Advances in infancy research* (vol. 12, pp. 221–236). Stamford, CT: Ablex.

Kostovic, I. & Goldman-Rakic, P.S. (1983). Transient cholinesterase staining in the mediodorsal nucleus of the thalamus and its connections in the developing human and monkey brain. *Journal of Comparative Neurology,* **219,** 431–447

Kraemer, G.K. (1997). Psychobiology of early social attachment in rhesus monkeys. Clinical applications. In Carter, C.S., Lederhendler, I.I. & Kirkpatrick, B (Eds.), *The integrative neuro-biology of affiliation. Annals of New York Academy of Sciences,* **807,** 401–418. New York: New York Academy of Sciences.

Krumlauf, R. (1994). Hox genes in vertebrate development. *Cell,* **78,** 191–201.

Kugiumutzakis, G. (1999). Genesis and development of early infant mimesis to facial and vocal models. In J. Nadel & G. Butterworth (Eds.), *Imitation in infancy* (pp. 36–59). Cambridge, UK: Cambridge University Press.

Lecanuet, J.-P., Fifer, W.P., Krasnegor, N.A. & Smotherman, W. P. (Eds.) (1995). *Foetal development: a psychobiological perspective.* Hillsdale, NJ: Erlbaum.

LeDoux, J.E. (1993). Emotional systems in the brain. *Behavior and Brain Research,* **58,** 69–79.

Lewis, D.A. (1997). Development of the primate frontal cortex. In M.S. Keshavan & R.M. Murray (Eds.), *Neurodevelopment and adult psychopathology* (pp. 12–30). Cambridge, UK: Cambridge University Press.

Lewis, M. (1999). Social cognition and the self. In P. Rochat (Ed.), *Early social cognition: understanding others in the first months of life* (pp. 81–98). Mahwah, NJ: Erlbaum.

Locke, J.L. (1993). *The child's path to spoken language.* Cambridge, MA: Harvard University Press.

Lund, J.S. (1997). Development of the cerebral cortex: an overview. In M.S. Keshavan & R.M. Murray (Eds.), *Neurodevelopment and adult psychopathology* (pp. 3–11). Cambridge, UK: Cambridge University Press.

MacLean, P. (1990). *The triune brain in evolution.* New York: Plenum Press.

Magge, S. & Madsen, J.R. (1997). Molecular biology of axonal outgrowth. *Pediatric Neurosurgery,* **27,** 168–175.

Malloch, S. (1999). Mother and infants and communicative musicality. In 'Rhythms, musical narrative, and the origins of human communication'. *Musicae Scientiae,* Special Issue, 1999–2000 (pp. 13–28). Liège: European Society for the Cognitive Sciences of Music.

Mazakopaki, K. (2000). Subjective and intersubjective appreciation of music by infants. Poster, International Conference on Infant Studies, Brighton, 16–19 July.

McEwen, B.S. (1989). Endocrine effects on the brain and their relationship to behavior. In G. Siegel, B. Agranoff, R.W. Albers & P. Molinoff (Eds.), *Basic neurochemistry* (4th edn, pp. 893–913). New York: Raven Press.

McEwen, B.S. (1997). Hormones as regulators of brain development: life-long effects related to health and disease. *Acta Paediatrica Supplement,* **422,** 41–44.

McKenna, J.J. & Mosko, S. (1994). Sleep and arousal, synchrony and independence among mothers and infants sleeping apart and together (same bed): an experiment in evolutionary medicine. *Acta Paediatrica Supplement,* **397,** 94–102.

Meltzoff, A.N. (1995). Understanding the intentions of others: re-enactment of intended acts by 18-month-old children. *Developmental Psychology,* **31,** 838–850.

Meltzoff, A.N. & Moore, M.H. (1977). Imitation of facial and manual gestures by human neonates. *Science,* **198,** 75–78.

Meltzoff, A.N. & Moore, M.K. (1992). Early imitation within a functional framework: the importance of personal identity, movement and development. *Infant Behavior and Development,* **15,** 479–505.

Meltzoff, A.N. & Moore, M.K. (1994). Imitation, memory, and the representation of persons. *Infant Behavior and Development,* **17,** 83–99.

Morecraft, R.J., Geula, C. & Mesulam, M-M. (1993). Architecture of connectivity within a cingulo-fronto-paraietal neurocognitive network for directed attention. *Archives of Neurology,* **50,** 279–284.

Muir, D. & Hains, S. (1999). Young infants' perception of adult intentionality: adult contingency and eye direction. In P. Rochat (Ed.), *Early social cognition: understanding others in the first months of life* (pp. 155–187), Mahwah, NJ: Erlbaum.

Murray, L. (1992). The impact of postnatal depression on infant development. *Journal of Child Psychology and Psychiatry,* **33,** 543–561.

Murray, L. & Cooper, P.J. (Eds.) (1997). *Postpartum depression and child development.* New York: Guilford Press.

Murray, L. & Trevarthen C. (1985). Emotional regulation of interactions between two-month-olds and their mothers. In T. Field & N. Fox (Eds.), *Social perception in infants* (pp. 177–197). Norwood, NJ: Ablex.

Murray, L. & Trevarthen C. (1986). The infant's role in mother–infant communication. *Journal of Child Language,* **13,** 15–29.

Murray, L., Kempton, C., Woolgar, M. & Hooper, R. (1993). Depressed mothers' speech to their infants and its relation to infant gender and cognitive development. *Journal of Child Psychology and Psychiatry,* **34,** 1083–1101.

Nadel, J. & Pezé, A. (1993). Immediate imitation as a basis for primary communication in toddlers and autistic children. In J. Nadel & L. Camioni (Eds.), *New perspectives in early communicative development* (pp. 139–156). London: Routledge.

Nadel, J., Carchon, I., Kervella, C., Marcelli, D. & Réserbat-Plantey, D. (1999). Expectancies for social contingency in 2-month-olds. *Developmental Science,* **2,** 164–173

Nagy, E. & Molnár, P. (1994). *Homo imitans or Homo provocans?* Abstract, *International Journal of Psychophysiology,* **18**, 128.

Nauta, W.J.H. & Domesick, V.B. (1982). Neural associations of the limbic system. In A.L. Beckman (Ed.), *The neural basis of behavior* (pp. 175–206). New York: SP Medical and Scientific Books.

Ninio, A. & Snow, C.E. (1996). *Pragmatic development*. Boulder, CO: Westview.

O'Rahilly, R. & Müller, F. (1994). *The embryonic human brain: an atlas of developmental stages.* New York: Wiley-Liss.

Panksepp, J. (1992). A critical role for "affective neuroscience" in resolving what is basic about basic emotions. *Psychological Review,* **99**, 554–560.

Panksepp, J. (1998). *Affective neuroscience: the foundations of human and animal emotions*. New York: Oxford University Press.

Papoušek, H. (1967). Experimental studies of appetitional behaviour in human newborns and infants. In H.W. Stevenson, E.H. Hess & H.L. Rheingold (Eds.), *Early behaviour: comparative and developmental approaches* (pp. 249–277). New York: Wiley.

Papoušek, H. & Papoušek, M. (1987). Intuitive parenting: a dialectic counterpart to the infant's integrative competence. In J.D. Osofsk (Ed.), *Handbook of infant development* (2nd edn, pp. 669–720). New York: Wiley.

Papoušek, M. (1994). Melodies in caregivers' speech: a species specific guidance towards language. *Early Development and Parenting,* **3**, 5–17.

Papoušek, M., Papoušek, H. & Symmes, D. (1991). The meanings and melodies in motherese in tone and stress languages. *Infant Behavior and Development,* **14**, 415–440.

Pecheux, M.G., Ruel, J. & Findji, F. (2000). Exploration des objects dans l'interaction mère-bébé: etude longitudinale. *Enfance* (In press).

Piaget, J. (1953). *The origins of intelligence in children*. London: Routledge & Kegan Paul.

Piaget, J. (1962). *Play, dreams and imitation in childhood*. London: Routledge & Kegan Paul.

Piaget, J. (1966). *The psychology of intelligence*. Totowa, NJ: Littlefield Adams.

Piontelli, A. (1992). *From fetus to child*. London: Routledge.

Ploog, D. (1979). Phonation, emotion, cognition, with reference to the brain mechanisms involved. In G. Wolstenholme & M. O'Connor (Eds.), *Brain and mind* (pp. 79–98). (Ciba Foundation Symposium, 69, New Series). Amsterdam: Excerpta Medica.

Ploog, D. (1992). Neuroethological perspectives on the human brain: from the expression of emotions to intentional signing and speech. In A. Harrington (Ed.), *So human a brain* (pp. 3–13). Boston, MA: Birkhäuser.

Pöppel, E. (1994). Temporal mechanisms in perception. *International Review of Neurobiology,* **37**, 185–202.

Porges, S.W. (1997). Emotion: an evolutionary by-product of the neural regulation of the autonomic nervous system. In C.S. Carter, I.I. Lederhendler & B. Kirkpatrick (Eds.), *The integrative neurobiology of affiliation* (pp. 62–78). *Annals of the New York Academy of Sciences*, vol. 807. New York: New York Academy of Sciences.

Prechtl, H.F.R. (1984). Continuity and change in early human development. In H.F.R. Prechtl (Ed.), *Continuity of neural functions from prenatal to postnatal life* (pp. 1–13). (Clin. Dev. Med. no. 94). Oxford: Blackwell.

Pulvermüller, F. & Schumann, J.H. (1994). Neurobiological mechanisms of language acquisition. *Language Learning,* **44**, 681–734.

Rakic, P. (1991). Development of the primate cerebral cortex. In M. Lewis (Ed.), *Child and adolescent psychiatry: a comprehensive textbook* (pp. 11–28). Baltimore, MD: Williams and Wilkins.

Reddy, V. (1991). Playing with others' expectations; teasing and mucking about in the first year. In A. Whiten (Ed.), *Natural theories of mind* (pp. 143–158). Oxford: Blackwell.

Reddy, V., Hay, D., Murray, L. & Trevarthen, C. (1997). Communication in infancy: Mutual regulation of affect and attention. In G. Bremner, A. Slater & G. Butterworth (Eds.), *Infant development: recent advances* (pp. 247–274). Hove: Psychology Press.

Reissland, N. (1988). Neonatal imitation in the first hour of life: observations in rural Nepal. *Developmental Psychology,* **24**, 464–469.

Rijt-Plooij, H.H.C. van de & Plooij, F.X. (1993). Distinct periods of mother-infant conflict in normal development: sources of progress and germs of pathology. *Journal of Child Psychology and Psychiatry,* **34**, 229–245.

Rizzolatti, G. & Arbib, M.A. (1998). Language within our grasp. *Trends in the Neurosciences*, **21**, 188–194.

Rochat, P. & Striano, T. (1999). Social-cognitve development in the first year. In P. Rochat (Ed.), *Early social cognition: understanding others in the first months of life* (pp. 3–34). Mahwah, NJ: Erlbaum.

Rochat, P. Neisser, U. & Marian, V. (1998). Are young infants sensitive to interpersonal contingency? *Infant Behaviour and Development*, **21**, 355–366.

Rock, A.M.L., Trainor, L.J. & Addison, T.L. (1999). Distinctive messages in infant-directed lullabies and play songs. *Developmental Psychology*, **35**, 527–534.

Rodier, P.M. (2000). The early signs of autism. *Scientific American*, **282**, 38–45.

Rollins, P.R. & Snow, C.E. (1998). Shared attention and grammatical development in typical children and children with autism. *Journal of Child Language*, **25**, 653–673.

Rolls, E.T. (1990). A theory of emotion, and its application to understandng the neural basis of emotion. *Cognition and Emotion*, **4**, 161–190.

Rolls, E.T. (1992). Neurophysiological mechanisms underlying face processing within and beyond the temporal visual cortical areas. *Philosophical Transactions of the Royal Society of London, Ser. B*, **335**, 11–21.

Rönnqvist, L. & Hofsten, C. von (1994). Neonatal finger and arm movements as determined by a social and an object context. *Early Development and Parenting*, **3**, 81–94.

Rosen, G.D. & Galaburda, A.M. (1985). Development of language: a question of asymmetry and deviation. In J. Mehler & R. Fox (Eds.), *Neonate cognition: beyond the blooming, buzzing confusion* (pp. 307–326). Hillsdale, NJ: Erlbaum.

Rosenblatt, J.S. (1994). Psychobiology of maternal behaviour: contribution to the clinical understanding of maternal behaviour among humans. *Acta Paediatrica*, Supplement, **397**, 3–8.

Ross, E.D. (1996). Hemispheric specialization for emotions, affective aspects of language and communication and the cognitive control of display behaviours in humans. In G. Holstege, R. Bandler & C.B. Saper (Eds.), *The emotional motor system* (pp. 583–594). Amsterdam: Elsevier.

Rothbart, M.K. (1994). What develops in emotional development? Emotional Development: Changes in Reactivity and Self-regulation. In P. Ekman & J. Davidson (Eds.), *The nature of emotions* (pp. 369–372). Oxford: Oxford University Press.

Rubenstein, J.L.R., Shamamura, K., Martinez, S. & Puelles, L. (1998). Regionalization of the prosencephalic neural plate. *Annual Review of Neuroscience*, **21**, 445–477.

Ruiz i Altaba, A. (1994). Pattern formation in the vertebrate neural plate. *Trends in the Neurosciences*, **17**, 233–243.

Schanberg, S.M. & Field, T.M. (1987). Sensory deprivation, stress and supplemental stimulation in the rat pup and preterm human neonate. *Child Development*, **58**, 1431–1447.

Schore, A.N. (1994). *Affect regulation and the origin of the self: the neurobiology of emotional development*. Hillsdale, NJ: Erlbaum.

Singer, W. (1986). The brain as a self-organizing system. *European Archives of Psychiatry and Neurological Sciences*, **236**, 4–9.

Snow, C.E., Pan, B., Imbens-Bailey, A. & Herman, J. (1996). Learning how to say what one means: a longitudinal study of children's speech act use. *Social Development*, **5**, 56–84.

Sparks, D.L. & Groh, J.M. (1995). The superior colliculus: A window for viewing issues in integrative neuroscience. In M.S. Gazzaniga (Ed.), *The cognitive neurosciences* (pp. 565–584). Cambridge, MA: MIT Press.

Sroufe, L.A. (1996). *Emotional development: the organisation of emotional life in the early years*. Cambridge: Cambridge University Press.

Steiner, J.E. (1979). Human facial expressions in response to taste and smell stimulation. *Advances in Child Development and Behavior*, **13**, 257–295.

Stern, D.N. (1985). *The interpersonal world of the infant: a view from psychoanalysis and development psychology*. New York: Basic Books. (Second Edition, 2000, with new Introduction).

Stern, D.N. (1990). Joy and satisfaction in infancy. In R.A. Glick & S. Bone (Eds.), *Pleasure beyond the pleasure principle* (pp. 13–25). Newhaven, CT: Yale University Press.

Stern, D.N. (1993). The role of feelings for an interpersonal self. In U. Neisser (Ed.), *The perceived self: ecological and interpersonal sources of the self-knowledge* (pp. 205–215). New York: Cambridge University Press.

Stern, D.N. (1999). Vitality contours: the temporal contour of feelings as a basic unit for constructing the infant's social experience. In P. Rochat (Ed.), *Early social cognition: understanding others in the first months of life* (pp. 67–80). Mahwah, NJ: Erlbaum.

Stern, D.N., Hofer, L., Haft, W. & Dore, J. (1985). Affect attunement: the sharing of feeling states between mother and infant by means of inter-modal fluency. In T.M. Field & N.A. Fox (Eds.), *Social perception in infants* (pp. 249–268). Norwood, NJ: Ablex.

Suomi, S.J. (1997). Long-term effects of different early rearing experiences on social, emotional, and physiological development in nonhuman primates. In M.S. Keshavan & R.M. Murray (Eds.), *Neurodevelopment and adult psychopathology* (pp. 104–116). Cambridge: Cambridge University Press.

Suter, D.M. & Forscher, P. (1998). An emerging link between cytoskeletal dynamics and cell adhesion molecules in growth cone guidance. *Current Opinion in Neurobiology*, **8**, 106–116.

Tager-Flusberg, H. (1999). *Neurodevelopmental disorders*. Cambridge, MA: MIT Press.

Thatcher, R.W. (1994). Psychopathology of early frontal lobe damage: Dependence on cycles of development. *Development and Psychopathology*, **6**, 565–596.

Thatcher, R.W., Walker, R.A. & Giudice, S. (1987). Human cerebral hemispheres develop at different rates and ages. *Science*, **236**, 1110–1113.

Tomasello, M. (1988). The role of joint attentional processes in early language development. *Language Sciences*, **10**, 69–88.

Tomasello, M., Kruger, A.C. & Ratner, H.H. (1993). Cultural learning. *Behavioral and Brain Sciences*, **16**, 495–552.

Tønsberg, G.H. & Hauge, T.S. (1996). The musical nature of prelinguistic interaction. The temporal structure and organisation in co-created interaction with congenital deaf-blinds. *Nordic Journal of Music Therapy*, **5**, 63–75.

Trainor, L.J. (1996). Infant preferences for infant-directed versus non-infant-directed play songs and lullabies. *Infant Behavior and Development*, **19**, 83–92.

Trehub, S.E. (1990). The perception of musical patterns by human infants: the provision of similar patterns by their parents. In M.A. Berkley & W.C. Stebbins (Eds.), *Comparative perception*, vol 1, *Mechanisms* (pp. 429–459). New York: Wiley.

Trehub, S.E., Schellenberg, G. & Hill, D. (1997). The origins of music perception and cognition: a developmental perspective. In I. Deliège & J.A. Sloboda (Eds.), *Perception and cognition of music* (pp. 103–128). Hove: Psychology Press.

Trehub, S.E., Unyk, A.M., Kamenetsky, S.B., Hill, D.S., Trainor, L.J., Henderson, J.L. & Saraza M. (1997). Mothers' and fathers' singing to infants. *Developmental Psychology*, **33**, 500–507.

Trevarthen, C. (1979). Communication and cooperation in early infancy. A description of primary intersubjectivity. In M. Bullowa (Ed.), *Before speech: the beginning of human communication* (pp. 321–347). London: Cambridge University Press.

Trevarthen, C. (1984a). How control of movements develops. In H.T.A. Whiting (Ed.), *Human motor actions: Bernstein reassessed* (pp. 223–261). Amsterdam: Elsevier North-Holland.

Trevarthen, C. (1984b). Biodynamic structures, cognitive correlates of motive sets and development of motives in infants. In W. Prinz & A.F. Saunders (Eds.), *Cognition and motor processes* (pp. 3327–350). Berlin: Springer Verlag.

Trevarthen, C. (1984c). Hemispheric specialization. In S.R. Geiger *et al.* (Eds.), *Handbook of physiology;* (Section 1: *The nervous system*, vol. 2: *Sensory processes* (pp. 1129–1190). (Section editor, I. Darian-Smith). Washington: American Physiological Society.

Trevarthen, C. (1985). Neuroembryology and the development of perceptual mechanisms. In F. Falkner & J.M. Tanner (Eds.), *Human growth* (pp. 301–383). New York: Plenum.

Trevarthen, C. (1986). Development of intersubjective motor control in infants. In M.G. Wade & H.T.A. Whiting (Eds.), *Motor development in children: aspects of coordination and control* (pp. 209–261). Dordrecht: Martinus Nijhof.

Trevarthen, C. (1987a). Infancy, mind in. In R.L. Gregory & O.L. Zangwill (Eds.), *Oxford companion to the mind* (pp. 362–368). Oxford: Oxford University Press.

Trevarthen, C. (1987b). Brain development. In R.L. Gregory & O.L. Zangwill (Eds.), *Oxford companion to the mind* (pp. 101–110). Oxford: Oxford University Press.

Trevarthen, C. (1989). Development of early social interactions and the affective regulation of brain growth. In C. von Euler, H. Forssberg & H. Lagercrantz (Eds.), *Neurobiology of early infant behaviour* (pp. 191–216). (Wenner-Gren Center International Symposium Series, vol. 55) Macmillan; New York: Stockton Press.

Trevarthen, C. (1990a). Signs before speech. In T.A. Sebeok & J. Umiker-Sebeok (Eds.), *The semiotic web, 1989* (pp. 689–755). Berlin: Mouton de Gruyter.

Trevarthen, C. (1990b). Growth and education of the hemispheres. In C. Trevarthen (Ed.), *Brain circuits and functions of the mind: essays in honour of Roger W. Sperry* (pp. 334–363). New York: Cambridge University Press.

Trevarthen, C. (1990c). Integrative functions of the cerebral commissures. In F. Boller & J. Grafman (Eds.), *Handbook of neuropsychology* (vol. 4, pp. 49–83). Amsterdam: Elsevier.

Trevarthen, C. (1993a). The function of emotions in early infant communication and development. In J. Nadel & L. Camaioni (Eds.), *New perspectives in early communicative development* (pp. 48–81). London: Routledge.

Trevarthen, C. (1993b). The self born in intersubjectivity: an infant communicating. In U. Neisser (Ed.), *The perceived self: ecological and interpersonal sources of self-knowledge* (pp. 121-173). New York: Cambridge University Press.

Trevarthen, C. (1996). Lateral asymmetries in infancy: Iimplications for the development of the hemispheres. *Neuroscience and Biobehavioral Reviews*, **20**, 571–586.

Trevarthen, C. (1997). Foetal and neonatal psychology: Intrinsic motives and learning behaviour. In F. Cockburn (Ed.), *Advances in perinatal medicine* (pp. 282–291). (Proceedings of the XV European Congress of Perinatal Medicine, Glasgow, 10–13 September 1996). New York: Parthenon.

Trevarthen, C. (1998a). The concept and foundations of infant intersubjectivity. In S. Bråten (Ed.), *Intersubjective communication and emotion in early ontogeny* (pp. 15–46). Cambridge, UK: Cambridge University Press.

Trevarthen, C. (1998b). The nature of motives for human consciousness. *Psychology: The Journal of the Hellenic Psychological Society* (Special Issue: "The place of psychology in contemporary sciences", Part 2. Guest Editor, T. Velli), **4**, 187–221.

Trevarthen, C. (1998c). Language development: mechanisms in the Brain. In G. Adelman & B. Smith (Eds.), *The encyclopedia of neuroscience* (pp. 1018–1026); second edition on CD-ROM Amsterdam: Elsevier.

Trevarthen, C. (1999a). Musicality and the intrinsic motive pulse: evidence from human psycho-biology and infant communication. In "Rhythms, Musical Narrative, and the Origins of Human Communication". *Musicae Scientiae*, Special Issue, 1999-2000 (pp. 157–213). Liège: European Society for the Cognitive Sciences of Music.

Trevarthen, C. (1999b). Intersubjectivity *The MIT encyclopedia of cognitive sciences* (pp. 413–416). (General editors: R. Wilson & F. Keil). Cambridge, MA: MIT Press.

Trevarthen, C. (2000). Intrinsic motives for companionship in understanding: their origin, development and significance for infant mental health. *Infant Mental Health Journal* (In press).

Trevarthen, C. & Aitken, K. (2000). Intersubjective foundations in human psychological development. *Annual Research Review. Journal of Child Psychology and Psychiatry and Allied Disciplines*. (Accepted for publication in 2001).

Trevarthen, C. & Aitken, K.J. (1994). Brain development, infant communication, and empathy disorders: intrinsic factors in child mental health. *Development and Psychopathology*, **6**, 599–635.

Trevarthen, C., Aitken, K.J., Papoudi, D. & Robarts, J.Z. (1998). *Children with autism: Diagnosis and interventions to meet their needs*, (2nd edn.). London: Jessica Kingsley.

Trevarthen, C. & Hubley, P. (1978). Secondary intersubjectivity: confidence confiding and acts of meaning in the first year. In A. Lock (Ed.), *Action, gesture and symbol: the emergence of language* (pp. 183–229). London: Academic Press.

Trevarthen, C., Kokkinaki, T. & Fiamenghi, G.A. Jr (1999). What infants' imitations communicate: With mothers, with fathers and with peers. In J. Nadel & G. Butterworth (Eds.), *Imitation in infancy* (pp. 127–185). Cambridge, UK: Cambridge University Press.

Trevarthen, C., Murray, L. & Hubley, P.A. (1981). Psychology of infants. In J. Davis & J. Dobbing (Eds.), *Scientific foundations of clinical paediatrics* (2nd ed., pp. 211–274). London: Heinemann.

Tronick, E.Z. (1989). Emotions and emotional communication in infants. *American Psychologist*, **44**, 112–126.

Tronick, E.Z. & Weinberg, M.K. (1997). Depressed mothers and infants: failure to form dyadic states of consciousness. In L. Murray & P.J. Cooper (Eds.), *Postpartum depression and child development* (pp. 54–81). New York: Guildford Press.

Tronick, E.Z., Als H., Adamson, L., Wise, S. & Brazelton, T.B. (1978). The infant's response to entrapment between contradictory messages in face-to face interaction. *Journal of the American Academy of Child Psychiatry*, **17**, 1–13

Tucker, D.M. (1992). Developing emotions and cortical networks. In M.R. Gunnar & C.A. Nelson (Eds.), *Minnesota symposium on child psychology*, vol. 24: *Developmental behavioural neuroscience* (pp. 75–128). Hillsdale, NJ: Erlbaum.

Tucker, D.M, Derryberry, D. & Luu, P. (2000). Anatomy and physiology of human emotion: vertical integration of brainstem, limbic, and cortical systems. In J. Borod (Ed.), *Handbook of Neuropsychology of emotion* (pp. 56–79). New York: Oxford.

Uzgiris, I.C. (1991). The social context of infant imitation.' In M. Lewis & S. Feinman (Eds.), *Social influences and socialization in infancy* (pp. 215–251). New York: Plenum Press.

Verney, C., Milosevic, A., Alvarez, C. & Berger, B. (1993). Immunocytochemical evidence of well-developed dopaminergic and noradrenergic innervations in the frontal cerebral cortex of human fetuses at midgestation. *Journal of Comparative Neurology*, **336**, 331–344.

Vicente, A.M. & Kennedy, J.L. (1997). The genetics of neurodevelopment and schizophrenia, In M.S. Keshavan & R.M. Murray (Eds.), *Neurodevelopment and adult psychopathology* (pp. 31–56). Cambridge, UK: Cambridge University Press.

Von Hofsten, C. (1984). Developmental changes in the organization of pre-reaching movements. *Developmental Psychology*, **20**, 378–388.

Weinberg, M.K. & Tronick, E.Z. (1994). Beyond the face: an empirical study of infant affective configurations of facial, vocal, gestural and regulatory behaviors. *Child Development*, **65**, 1503–1515.

Weinberg, M.K. & Tronick, E.Z. (1996). Infant affective reactions to the resumption of maternal interaction after the still-face. *Child Development*, **67**, 905–914.

Wilson, R. & Keil, F. (1999) (Eds.). *The MIT encyclopedia of cognitive sciences*. Cambridge, MA: MIT Press.

Xiao, B.-G. & Link, H. (1998). Immune regulation within the central nervous system. *Journal of Neurological Sciences*, **157**, 1–12.

Yakovlev, P.I. & Lecours, A.-R. (1967). The myelogenetic cycles of regional maturation of the brain. In A. Minkowski (Ed.), *Regional development of the brain in early life* (pp. 3–70). Edinburgh: Blackwell.

Zecevic, N. & Verney, C. (1995). Development of the catecholamine neurons in human embryos and fetuses, with special emphasis on the innervation of the cerebral cortex. *Journal of Comparative Neurology*, **351**, 509–535.

Zeifman, D., Delaney, S. & Blass, E. (1996). Sweet taste, looking, and calm in 2- and 4-week-old infants: the eyes have it. *Developmental Psychology*, **32**, 1090–1099.

Zentner, M.R. & Kagan, J. (1996). Perception of music by infants. *Nature*, **383**, 29.

V.4
Signalling in Early Social Interaction and Communication

HANUŠ PAPOUŠEK

ABSTRACT

Observable patterns of an individual's behaviour may carry various sorts of information: expression of the individual's internal state, representation of the corresponding context, or an appeal if the social environment perceives and interprets it as such. Elicited social consequences may increase the probability that the individual will learn using the eliciting behavioural pattern as an intentional social signal and as a means of dialogic exchange of information with social partners. Such a communicative capacity hides a significant adaptive potential and has achieved a remarkable complexity during the evolution of primates with their high forms of social organization. In humans new dimensions of communicative capacities have evolved resulting from species-specific abilities to freely create new representative symbols and use them for purposes of verbal communication, internal mental representation in rational thinking, establishment of unique cultures, and the construction of conscious self.

Fundamental steps in this species-specific evolutionary progress can be detected in early phases of human ontogeny. Their analyses have profited from microanalytical, experimental, interdisciplinary, and comparative approaches to preverbal forms of communication and have provided new arguments for the interpretation of both the phylogeny and ontogeny of human communication. The contributions are discussed with respect to the adaptive relevance of human communication in the evolution of speech and in the genesis of culture.

INTRODUCTION: QUESTIONS TO ASK ABOUT EARLY SIGNALLING

Living systems have evolved so as to detect biologically relevant changes in both non-living and living environments, answer them with innate sets of specific

A.F. Kalverboer and A. Gramsbergen (eds.), Handbook of Brain and Behaviour in Human Development, 883–900
© 2001 Kluwer Academic Publishers. Printed in Great Britain.

behaviours, and learn how to improve either reading and processing of incoming signals, or responding to such signals in the most adaptive ways. Moreover, living systems are genetically predisposed to use certain types of behaviour as signals and to learn how to adaptively vary the emission of their own signals. Particularly important to living systems is the use of signals for the purpose of communication among conspecific or heterospecific members of living populations where the sending member intentionally uses certain signals or combinations of them, while the receiving member intentionally modifies responses. Such an advanced level of signalling in the form of social communication serves in the animal world for many adaptively relevant purposes, from the search for sexual partners, care for progeny as well as social organization up to exchange and accumulation of experience.

To deal with signalling in living systems means to consider a chain of possibilities of uneven complexity reaching from unintentional to intentional signals on the one hand, and from innate levels of signalling to learned or cultivated levels on the other. Simple speculations on this chain of variations in signalling raise questions such as whether the chain has something in common with the evolution of communication, which level of complexity human communication has reached, whether human communication bears a character of a species-specific communication and, if so, how this specificity develops during ontogeny.

For obvious reasons both partners emitting signals and partners receiving these signals have to be considered as a unit if signalling is to be discussed in relation to human social communication. This assumption is of particular significance for the analyses of the early ontogeny of signalling. Not only is the postpartum dependence of infants on caregivers ultimately of paramount importance during early ontogeny, but so is the probability of detecting the basis of innate predispositions for signalling.

Direct experimental methods for the verification of innateness are known to represent an almost insoluble problem in the case of human mental capacities. However, viewed from comparative perspectives, data on early human development may provide indirect pieces of evidence of innate determinants (Papoušek & Papoušek, 1987) such as:

1. Functional involvement of the concerned behavioural patterns in species-specific means of evolutionary adaptation.
2. High probability of occurrence in response to cue elicitors or within given interactional contexts.
3. Universality of the pattern across sex, age, culture, or species.
4. Early emergence during ontogeny.
5. Minimal awareness and rational control of the given behavioural pattern in subjects carrying out the pattern.
6. Coevolution of counterparts to the pattern in conspecifics.
7. Presence of the pattern in cases where environmental influence was for some reason seriously restricted or eliminated.
8. Conceptual integrity in best-fitting interpretations of the determinants of a given pattern with respect to other, functionally interrelated, behavioural patterns.

The deeper the evolutionary roots of such capacities, the more the powerful forms of intrinsic motivation may be related to them (see also Trevarthen, this handbook). Moreover, a strong intrinsic motivation behind a need may also be interrelated to particularly effective pathogenetic factors of early behavioural disorders if the need remains insufficiently fulfilled. Interestingly, infants' needs of informational enrichment, cognitive integration of experience, social communication, play, and mastery or symbolic competence have seldom appeared in lists of such needs in contrast to bodily, nutritional, and emotional needs that are universal among mammals. Consequently, insufficient satiation of these needs only gradually finds recognition in the list of factors causing deviations in infant mental development and health.

During early ontogeny a poorly fulfilled need may lead to special deviations in developmental processes. It may cause stressful frustration in an infant and thus disturb communicative and emotional interchanges with a caregiver. At another time, however, it may lead to a functional constraint on the development of a partial capacity that, in further developmental steps, should serve as a crucial base for the development of successive capacities. In the latter case a defect with long-term consequences can occur during infancy and disturb the acquisition of further capacities at later ages. This danger has been stressed, and perhaps overexaggerated, in relation to emotional bonding in attachment theories. However, it has gained insufficient attention in relation to the development of early signalling, social communication, playfulness, and other assumptions for later capacities. For example, public media have recently, in 1999, drawn attention to a barely believable number of 4 million functional analphabets in Germany who completed fundamental education based on good German standards and remained unable to acquire such fundamental abilities for sociocultural integration as reading and writing. Previous reports 3 years earlier mentioned only 3 million analphabets. However, data on developmental trends are as insufficient as is the scientific attention paid to the question of whether this problem might be related to early defects in ontogeny of signalling or social communication.

COEVOLUTION OF PREDISPOSITIONS FOR SIGNALLING IN PREVERBAL INFANTS

Infancy researchers have for a long time noted that human infants are not incompetent, altricial beings. At the very beginning of postpartum life, newborn infants overcome burdens of physiological adjustments in respiration, blood circulation, and nutrition, mount an immunological defence against massive microbial invasions, and reorganize their biorhythms in accordance with day/night cycles. Thus, newborns master their developmental chances with specific competences that allow them to cope with burdens that might surpass adaptive potentials in adults.

Some neonatal competences disappear later, such as the ability to coordinate sucking, swallowing, and breathing so as to avoid inhaling (breast) milk. This coordination partly depends on a relatively short distance between the entrance to the air passage and the cranial base. Conversely, a new competence emerges while profiting from the increasing resonant space between larynx and cranial base, and from a simultaneous neuronal transformation allowing prolonged vowel-like sounds during periods of comfort in infants. This example demonstrates phasic changes in the development of infant capacities that, at the same time, are relevant to social signalling in forms of cooing – sounds that are evidently pleasant to caregivers' ears and seldom remain disregarded. The new infant competence undergoes repeated transformations leading to the acquisition of the first words in about 1 year, thus revealing precocity in those means of evolutionary adaptation that differentiate humans from other living organisms (Papoušek & Papoušek, 1997).

The history of infancy research documents that humanogenetic aspects of evolution – the ontogeny of symbolic capacities above all – have been taken into consideration only gradually. The first discoveries of infant competences in learning, fundamental cognitive operations, and emotional bonding were astonishing; yet concerned developmental processes that are common to all mammalian species. In the meantime, infancy researchers have found conceptual guidelines and methodological tools for the detection of specific pathways in human evolution. These pathways may be marked with inconspicuous forms of signalling during early interactions. However, these elementary forms may have coevolved in both infants and their caregivers as predispositions for unique capacities characterizing adult humans such as complex forms of social communication, rational thinking, cultural integration, and self-consciousness.

Consider, for example, babyishness – the special morphological characteristics of the head and face in young infants. This is a non-intentional and non-behavioural item, yet it functions as a cue-signal eliciting altruistic and protective attitudes towards infants in their social environment. Human caregivers respond to babyishness even if they have not yet been informed about this phenomenon, and they do so with evident universality across age, sex, and cultures. Moreover, the phenomenon is universally distributed among all mammalian species and birds (Lorenz, 1943). According to these criteria the innateness of babyishness is beyond any doubt, and the innateness of social responses to it is also highly probable.

The case of infant crying

The neonatal cry may represent a different example. At the beginning, crying can be elicited by hunger, pain or other unpleasant stimuli (Wolff, 1969) very much like an innate reflex, and may universally elicit feelings of alarm and activate physiological and mental readiness in caregivers to help and soothe the infant. Sooner or later, however, the infant becomes aware of the contingency of social interventions on his or her own cry signals, and may use crying intentionally.

Early intentionality is no easy topic for experimental studies. As long as it is interpreted as an act based on neuronal regulation including both feed-back and feed-forward loops (Bruner, 1975; Dore, 1975), crying might become intentional very early, as indicated by experiments in human newborns who consistently redirect head turns in order to reach a milk formula corresponding to their preference (Papoušek. 1967). Although neonatal crying has simiʾar acoustic qualities in many mammalian species (Newman, 1988), a specialist – with the help of appropriate equipment – may detect additional information on present or imminent deviations in air passages, resonant cavities, vocal cords, or regulation of breathing. In doing so the dependence of emitted information on the capacity – and technical equipment – of the receiver is confirmed (Hirschberg & Szende, 1985).

Yet another type of insight may be gained in analyses of the dynamics of crying (Papoušek *et al.,* 1986). Neonatal crying mostly corresponds to tonic responses with a gradual build-up and decline, lasting several seconds, and separated by several minutes from signs of comfort that may follow. In infants above 7 months of age crying appears with increasing frequency in phasic forms that characterize communicative signals. It shows steep build-ups and declines, and may be followed by opposite phasic signals such as laughter within 1 or 2 seconds. Typically, such phasic forms of cry appear in contexts where the infant, as a little behaviourist, tries to manipulate the caregiver with punishing or rewarding emotional signals.

The case of infant imitation

The first postpartum steps in learning how to communicate are facilitated by an innate predisposition to imitation that human infants share with the young of primates. As evident in newborns, imitation first concerns oral and facial gestures (Meltzoff & Moore, 1977, 1983; Field *et al.,* 1982), then – depending on a neural transformation around the end of the second month – hand gestures, vocal sounds, absolute pitch or melodies (Kessen *et al.,* 1979; Legerstee, 1990), and other types of procedural activities relevant to both non-vocal and vocal forms of communication. Eventually, objects of imitation change with the infant's age. Observations in infants above 7 months of age indicate that they imitate only those actions that they can cognitively integrate, and that are meaningful to them (Masur & Ritz, 1984; Uzgiris, 1984).

After another neural transformation including hemispheric differentiation in the brain around 7–8 months of age, infants increasingly imitate declarative elements of communication (i.e. naming). Procedural imitation and learning precedes declarative imitation and learning during both phylogeny and ontogeny, is more resistant to amnesia than declarative learning (Squire, 1987), and may lead to permanent results even in infants. Kuhl (1983) assumes that, in this way, infants set the mind for the prototypical sounds of their culture, often for the rest of their lives. Kuhl & Meltzoff (1996) found supporting evidence in the characteristics of the vowels "a", "i" and "u" that infants acquired by imitation at the age of 5 months.

Prior to the appearance of declarative symbols the infant can demonstrate procedural precursors in goal-directed movements. Consider, for example, the infant's capacity to detect and conceptualize the rules of experimental rewards (Papoušek & Bernstein, 1969). The experimenter introduces a rule, according to which a 4-month-old can turn on an attractive visual display through three consecutive head-turns to the left, in one experiment, or four consecutive head-turns to the right, in another. The infant copes with the problem – as documented in polygraphic records – and learns not only to carry out four consecutive head movements in the slow tempo of exploratory movements, but patterns the movements within one session into fast, rhythmical triads or tetrads, respectively, that are followed by motionless waiting for the reward. Such a response elicits the impression of a symbol for a numeric concept but can only be a quasi-symbol as long as the infant cannot yet count and use true declarative symbols. Parents would tend to interpret such performances as counting, since they take even less conspicuous procedural performances for communicative messages with a declarative content.

The preverbal origins of language and symbolic representation

With the progress in the infant's integrative capacities originally non-intentional signals may be used intentionally. Conversely, the caregiver may intentionally decide either to answer the infant's signals, or to ignore them, according to consciously selected educational principles. As explained in the next section, even educational principles may be a part of innate behavioural patterns, although typically they are determined by integration of life experience and by the influence of cultural institutions.

The contribution of culture to variability and extent of human signalling or social communication is enormous and very often crucial. For example, the caregiver may know nothing about infrared irradiation that indicates the course of some autonomic processes in infants, and certainly cannot perceive it. However, the caregiver may find adequate information in libraries or other cultural facilities and learn to decode infant infrared signals with the help of special devices.

The relevance of words is another example deserving attention. Speech can result from a genetically determined biological predisposition. However, due to their use in a culture, words acquire new properties and can function as a force of their own. The integrative power of words can serve as an example. Human capacity of abstract representation allows us to use words either as concrete names for individual objects or events, or as abstract names for entire categories. Consequently, abstract names can be used for hierarchical classifications and their availability can increase the tendency to use them as more people become aware of advantages of classifications and learn to use them. Simple classifications, such as finch–songbird–bird–vertebrate–animal have an impact on the ways of thinking about, for instance, how far a classification can go in both directions: in the one, such thinking can occasionally lead to the idea of gods, in the other to the idea of atoms. The significance of culture and communication is mutually interdependent. Human culture has sometimes been viewed as the opposite to nature and a human-specific phenomenon. However, simple precursors of cultures have been described in other species, especially primates (e.g. Bard & Gardner, 1996; Kawai, 1963). Similarly, the use of words in social communication had been considered as human-specific, until categorical proto-words were discovered in primate communication (Seyfarth *et al.,* 1980; Winter *et al.,* 1966). Thus both culture and the use of speech have their biological precursors in non-human animals and can be viewed as products of natural evolution. Consequently, the main predicate differentiating human and non-human forms of culture seems to be the level of variability and complexity.

Attempts to reveal the crucial parameter responsible for the distinct categorical increase in the complexity of human versus animal forms of culture and social communication have led to similar conclusions in the neurosciences, philosophy, and theories of biological systems. The Russian neurophysiologist Pavlov (1932) saw the crucial criterion in the human-specific capacity for using learned signals of higher orders. He envisaged specialized centres for those capacities in speech-related areas of human cortex and called them the second signalling system. Von Bertalanffy (1968), in a pioneering attempt to apply system theories in interpretation of anthropogenesis, came to the conclusion that the fundamental process leading to human-specific features is the evolution of the capacity for using freely created and fully representational symbols. His view was influenced by the philosophical position of Cassirer (1953).

In Bertalanffy's and Cassirer's concepts, symbols are not identical with words; they may have an iconic character, consist of mathematical formulas or other codes, but must stand for things or events for which they have been consciously and freely chosen. As internal representations they may be used for the individual's mental operations independently of physical laws that may restrict their use in reality. This form of symbolic capacity has led to the origins of languages, rational thinking, emergence of unprecedented levels in human cultures, and human self-consciousness. For this reason Cassirer suggested we scientifically classify Man as *Homo symbolicum*, rather than *Homo sapiens*.

Although, logically, the significance of arguments on freely and consciously created symbols is undeniable, neuroscientists have not yet been able to identify brain structures or humoral factors responsible for the difference between such symbols and other forms of signals (Gazzaniga, 1995). By exclusion, then, we may accept the argument of systems theoreticians assuming that, in dynamic systems, the complexity of multimodal and multidimensional experience may *per se* lead to the emergence of new qualities through the determinants of self-regulation.

In summary

Generally speaking, the repertoire of signals relevant to living organisms is extensive and categorizable in various directions. During the evolution of mammals this repertoire has increasingly functioned in the service of dialogical and reciprocal exchanges of information that are characteristic of social communication. Evolving capacities of reading and processing signals from conspecific environment on the one hand, and improving modifications of one's own signals on the other, have enabled a progressive accumulation of experience. In the case of primates the participation of visible gestures in face or hands, and audible expressions, have conspicuously dominated social communication and led to the use of precursors of words. The new level of mammalian communication in primates coincided with the advent of other significant progress; namely, the propagation of newly learned skills as a part of the process of cultural enrichment. Yet another conspicuous gradation in the complexity of primate social communication is apparent in the human-specific capacity for symbol formation.

Obviously, from the point of view of social communication, the repertoire of signals can be aligned with evolutionary trends whereby, tacitly, biological circumstances, such as ecological factors, genetic predispositions, and the degree of brain evolution, are taken into consideration as crucial determinants of signal processing. Disclosure of parallel trends in both phylogenetic and ontogenetic developments of communicative signalling raises the question of parallelism and functional interrelationships between predispositions for communicative signalling that may have coevolved in infants and caregivers.

COEVOLUTION OF PREDISPOSITIONS FOR SIGNALLING IN THE INFANT'S SOCIAL ENVIRONMENT

As explained in the preceding section, human infants are precocious among mammals in communicative development, and yet they have to learn a lot to bring their innate predispositions to the phylogenetically advanced level of social verbal dialogues within their specific culture in just 1 or 2 years. Does this achievement depend entirely on infant capacities, or have supportive strategies evolved on the side of the infant's social environment to facilitate either the development of general prerequisites for communication, or its culture-specific forms, or both?

Facilitating infant learning and the role of intuitive parenting

Experimental analyses have shown that infant learning improves with age not only due to maturation but also under the influence of experimental modifications comparable to didactic interventions, such as arrangements in which easy tasks precede more difficult tasks, or in which repeated experience allows infants to learn how to learn (Papoušek, 1977; Papoušek & Papoušek, 1984).

It has long been unclear whether the social environment can facilitate infant learning that is crucial to the acquisition of language and, if so, how this would occur. Interestingly, in contrast to the existence of sophisticated didactic methods and cultural institutions for teaching individuals second or third languages, no methods seemed to have existed – either in cultural institutions or among parents – for teaching the first language. The impression of nonexistent didactics related to the acquisition of mother tongue was corrected when interviews and questionnaires used in studies of parental capacities were replaced with audio-visual techniques. The new methods allowed microanalytical evaluations and led to the detection of didactic interventions that parents or other social partners carry out unknowingly and intuitively (Papoušek & Papoušek, 1978, 1987).

Neurophysiologically, intuitive behaviours are faster and less strenuous than are rationally controlled behaviours. Adults often need less than 500 ms to intuitively respond to infant signals, although they need a minimum of 500–600 ms for a mere preceptual awareness of stimulation, even if their cortex is stimulated directly (Vander et al., 1990). Needless to say, rational decisions concerning didactic interventions may last more than seconds.

A closer look at the infants' naturalistic settings reveals a wealth of interactional episodes giving the infant opportunities for learning, prediction, anticipatory acting, or concept formation and transformation. Such opportunities seldom result from interactions with the physical environment (Newson, 1979; Papoušek & Papoušek, 1984), but plenty of them occur during interactions with the social environment. Every minute of an alert parent–infant interaction may include one or several episodes of multimodal and reciprocal interchanges that are so favourable for improvement of the infant's integrative and communicative competencies that they elicit an impression of didactic interventions (Papoušek & Papoušek, 1984, 1987).

Parents are typically unaware of carrying out any didactically relevant interventions, and often deny or misinterpret interventions that have been convincingly documented in videorecords. For example, mothers will observe a sleeping newborn from their "reading distance" of 40–50 cm but halve the distance (as appropriate for the newborn's vision) when the newborn looks at them; they will place themselves in the centre of the newborn's field of vision, try to attract his/her visual attention in various ways, and reward the newborn with expressive greeting responses for achievement of an eye-to-eye contact. However, mothers are not aware of these behaviours, and exhibit them even if they strongly believe that the newborn cannot yet see anything (Schoetzau & Papoušek, 1977).

Studies involving newborns and young infants have shown that learning requires an alert waking state in the infant, simple structures of stimuli and learning trials, a large number of repetitions of trials, gradual ordering of tasks in terms of complexity, use of adequate rewards, and sensitivity to feedback signals indicating the limits of the infant's tolerance (Papoušek & Papoušek, 1987, 1989a). Somewhat analogous assumptions have been stressed for the development of early communication; namely, consistency in the performance of simple and repetitive communicative elements (Messer, 1980), fostering visual contact (Fogel, 1977; Papoušek & Papoušek, 1977), and dialogic engagement of several modalities (Turkewitz & McGuire, 1978). Analyses of parental interventions subsumed under the Papoušeks' concept of "intuitive parenting" (Papoušek & Papoušek, 1987) document the existence of a systematic form of support serving both the infant's improvement of integrative capacities and the acquisition of the mother-tongue. The support corresponds to principles of formal didactics, although it is based upon biologically determined predispositions. The repertoire of such support is rather universal across sex, age, and culture of caregivers, according to available evidence (Papoušek & Papoušek, 1987). Thus, the environmental guidance that has been acknowledged by linguists in relation to speaking children functions, as well as with regard to the behaviour of preverbal infants, although, in the case of preverbal infants, it is based on biological and non-conscious rather than cultural and conscious determinants.

Typical examples of the parents' intuitive adjustments in their own signalling concern the prosody of infant-directed speech. The acoustic properties of infant-directed melodic contours and the modes in which they are displayed (repetitiveness, slow tempo, contingency on infant behaviour) serve a number of functions: to draw attention to parental speech; to affect arousal and attention; to help the infant to detect, categorize, and abstract elementary holistic units from the flow of speech; to provide the first categorical messages; to segment the flow of speech into linguistically meaningful units; and to mediate phonological and linguistic information (Papoušek et al., 1991; 1995). They also serve as models for those features with which infants first learn to control their vocal sounds; namely, the modulation of melodic contour and vowel-like resonance (Papoušek, 1994).

In one respect, human infants differ categorically from all other animals: even if human infants were absolutely unable to communicate, they have parents who have evolved as experts in both communication and metacommunication, and become capable of communicating even with strange species, or find enough information on it in books. What the newborn intends to express in not as important as the caregiver's capacity to perceive and detect information in neonatal behaviour. The newborn responds to various environmental changes with innate or prenatally acquired patterns of behaviour that lack the intention to communicate, while the caregiver makes a supreme effort to establish a dialogue, and tends to interpret even non-vocal forms of the infant's behaviour as intentional communicative feedback (Papoušek & Papoušek, 1987). The non-cry, quiet vocal sound, that will become the main vehicle of future words and representative symbols, develops only gradually during the preverbal age and is still too sparsely differentiated in newborns to provide the parent with reliable information about the infant's motivation and needs. However, the parent is reliably predisposed to processing combinations of minimal vocal utterances with facial, gestural, or bodily displays as means of infantile communication.

The infant's signalling capacity develops gradually in competition with other adaptive processes and depends, among others, on transformations in the central nervous system. Around the end of the second month the learning of goal-directed movements distinctly accelerates (Papoušek, 1977) and affective facial displays accompanying the course of learning or problem solving become more expressive (Papoušek, 1967). Caregivers can more easily read cues in facial expressions and hand gestures – the two areas predisposed in primates for non-vocal communication – which indicate how far the infant processes interactional experience. The content of caregivers' comments in infant-directed speech indicates that they start viewing the infant as a competent social partner. Infants who had been able to imitate only oral activities and facial expressions after the birth start imitating vocal sounds around this age. Interactions between caregivers and infants increasingly acquire a dialogic character (Papoušek, 1994).

Caregiver–infant interaction, often viewed as a mere expression of emotional bonding, obviously means more – a sequence of episodes activating integrative processes (learning, imitation, cognition, and play) in infants, and eliciting emotional expressions in response to successful integrations. In fact, parent–infant dyads represent prototypes of didactic systems opposite in the amount of experience, with infants predisposed for learning, parents predisposed for teaching, and acquisition of speech in infants set as the main target of "intuitive parental didactics". Some aspects of vocal communication, such as some specific modifications of prosody in infant-directed speech, have already been demonstrated in comparative studies to be universal across sex, age, and cultures so far investigated (Ferguson, 1964; Fernald, 1992; Fernald et al., 1989; Papoušek et al., 1991). Their universality indirectly proves their innateness.

"Experiments in nature": deaf infants and twins

Studies with deaf infants and deaf mothers throw additional light upon the problem of innateness. Deaf infants cannot learn the language of their cultural environment without special help. However, the development of their preverbal vocalization is parallel to the development in hearing infants, but distinctly slower (Oller & Eilers, 1988). Obviously, vocal development depends on innate regulations during an earlier age. However, audiovocal experience is soon necessary for further progress. Deaf mothers cannot modify infant-directed speech in didactic ways; thus they cannot support speech acquisition in deaf infants. However, if deaf mothers teach their deaf infants the American Sign Language they unknowingly modify signing according to the same didactic principles as hearing mothers do so in the audiovocal modality (Erting et al., 1990). Deaf

infants develop signing in steps that are similar to the steps in vocal development, including babbling in the manual mode (Petitto & Matentette, 1991).

Yet another "experiment in nature" has been studied by Bornstein (1985) in
twins. In comparison with singletons, each twin is exposed to less than half of
maternal didactic interventions, as measured at the age of 5 months. At the age
of 15 months twins are already significantly delayed in cognitive and verbal competencies. Thus, various findings help in elucidating the evolution and preverbal
ontogeny of infantile and parental predispositions related to the acquisition of
speech, the participation of both genetically transmitted programmes and
environmental impacts, and the amodal character of innate predispositions.

The role of parental imitation and sensitivity to infant signals

Unlike other mammals, including non-human primates, human parents strongly
tend to imitate infants' facial and vocal displays. They are the first to imitate, as
evident in parent–newborn interactions, they encourage infants in various ways
to imitate and reward imitations with expressions of pleasure. Unlike imitations
of facial or manual behaviours, vocal imitations concern more complex processing, as they give infants a chance to compare the product of imitation with the
model sound. Parents modify infant-directed speech in a way that diminishes the
original disparities between adult speech and infant sounds, and thus increases
the overall likelihood of imitation (Papoušek et al., 1987a). Thus, vocal imitation
supports the development of perceptual, integrative, and communicative processes concurrently, and mostly in combination with pleasant, mutually rewarding
emotional experience (Papoušek & Papoušek, 1984; Papoušek, 1996). Between
3 and 6 months of infant age, vocal imitation is increasingly incorporated in
nursery games, initiated by mothers, and in infants' vocal plays. Such complex
engagement has much in common with interpretations of the roles of imitation
and play in cognitive development (Harkins & Uzgiris, 1991; Papoušek &
Papoušek, 1989b; Piaget, 1962; Uzgiris, 1989).

In general, preverbal communication encompasses behaviours in more than
one modality, requires bodily contact or close proximity, and concerns three main
communicative functions: delivery of fundamental, biologically relevant
information on *current* conditions; assessment of the infant's momentary motivation and needs, and supportive guidance in relation to affective, integrative, and
communicative self-regulation. In this sense preverbal communication represents a fundamental part of prosocial behaviour in humans.

Before infant vocalization becomes sufficiently modifiable to enable delivery
of differential messages, caregivers seem to utilize all observable paralinguistic
patterns in infant behaviour as potential carriers of dialogic messages. However,
this assumption is difficult to verify statistically in interactional analyses. Too many
elements of communication occur simultaneously and complicate identification
of individual cue-signals, such as infant hand signals. Kestermann (1982)
experimentally explored the effectiveness of hand signals displayed as the only
variable in slides showing infants in specifically prepared drawings (Papoušek et
al., 2000). Participants were asked to respond with non-verbal acts – offering a
milk bottle or a toy, or turning off a light as if to allow the infant to fall asleep.
Non-verbal responses were chosen to reduce the role of rational decisions and to

increase the probability of intuitive responses.

Hand signals of hunger, readiness to interact, fatigue, and falling asleep appeared to function as cues for adequate caregiving in samples of school-girls, women, and men. Additional interviews revealed that most participants were unaware of the reasons for their decisions and assumed that the crucial cues came from infants' facial expressions. Kestermann's data also demonstrated that the level of appropriate intuitive responses depended not only on biological predispositions but also on the amount of previous experience with infants. For instance, mothers of infants performed better in experiments than school-girls or childless women. Fathers who had regularly been involved in care of infants did not significantly differ from experienced mothers.

Rhythms and games

Parental vocalizations do not occur in a vacuum, but are usually accompanied by various forms of non-vocal stimulation. During the fifth month after birth, the first precursors of consonants appear in the infant's vocalization, and the parent responds in two distinct ways: first, with displaying adequate models of consonants, alluring the infant into their imitations, and rewarding the imitations; and secondly, with introduction of rhythmical games into the repertoire of interchanges. Such games can doubtless be categorized as play. However, they also bear the character of a didactic intervention insofar as they support rhythmical repetitions and segmentations of movements which, among others, are necessary for the production of several consecutive syllables within one breath. According to Kelso *et al.,* (1983; see also Oller & Eilers, 1992), the "minimal rhythmic units" in nursery games, finger tapping or syllables in various languages, are remarkably similar, apparently due to general tendencies in human timing (Turner, 1985).

Timing control and temporal organization seem to be important both in the internal organization of infant experience (Lewkowicz, 1989), and in the coordination of infant motor behaviour (Dent *et al.,* 1991; Wolff, 1991). According to Korner (1979), the mutuality achieved by a given parent–infant pair depends on the parents' ability to tune their temporal patterning to the infant's individuality. A detailed analysis of rhythms occurring in maternal interactions with 3-month-old infants (Koester *et al.,* 1989) has revealed a high frequency of rhythmical patterns (occupying 48.3% of 3-minute interactions) and a variability of rhythms that is independent of any inherent pacemaker but reflects the ongoing dynamics of the interaction and the type of movements involved.

Interactional rhythmical games have been shown to emerge in the repertoire of parental interventions at 5 months of infant age, peak at 7 months, and then gradually decrease towards the end of the first year (Papoušek, 1994). The peak overlaps with the onset of reduplicated canonical syllables, as if the use of rhythmical games were to pave the way for the infant's ability to segment expiration for the purpose of rhythmically chaining repetitive syllables in one utterance. Reduplicated syllables, bisyllables such as "gaga" in particular, mark a turning point around the age of 7 months, heralding the emergence of a new capacity for processing declarative information and using representative symbols. This striking new element in infant vocal expertise sets the stage for new parental strategies to come: during dyadic interactions and play they foster joint cooperative action, instances of joint attention to objects and events in the environment and of joint reference; moreover, they utilize infant bisyllables as potential protowords and assign them lexical meaning while associating them with the corresponding focus of attention (Papoušek, 1994). Thus, the milestone of canonical syllables marks the transition from mere procedural learning to processing of declarative information, from procedural refinements of signalling to the emergence of symbolic signalling and the use of conventional representative symbols. Similar developmental transitions in infant signalling and parental interventions are reflected in the development of both solitary and collaborative forms of play.

From games to play

At the beginning of the third month learning becomes much faster (Papoušek, 1967) the coordination of orofacial and hand movements improves so as to allow more expressive displays of facial, gestural, and vocal expressions of mood, integrative processes, or attempts to establish social contacts. Some of the new activity patterns elicit impressions of playful activities, although they are difficult to discriminate from exploration in the service of cognitive needs. Infancy researchers have focused attention to two forms of play – vocal play as particularly suitable for investigation of the initial period of postpartum development (Papoušek & Papoušek, 1981, 1989b; Papoušek et al., 1987b), and solitary or collaborative symbolic play for the following periods (Bornstein & Tamis-LeMonda, 1995).

In general the analysis of vocal play gives a valuable insight into the ontogenetic origins of playfulness and confirms that parents as caregivers support the occurrence of play in various indirect and direct ways (Papoušek & Papoušek, 1989b; Papoušek et al., 1987b). Some of their supportive tendencies, such as vocal matching or a differential use of melodic contours in baby talk, function from the moment of the infant's birth and obviously fulfil some other fundamental didactic tasks before they become regular parts of playful interchanges accompanied by reciprocal expressions of pleasure. The developmental progress in infants' vocal play, evident at 4 months, opens further avenues towards more complex forms of play and simultaneously towards a more differentiated utilization of potentials included in vocal play. This development may also take other directions, such as the introduction of musicality, use of musical toys, further ritualized games, and speech exercises, to name only a few. This increasing complexity quickly exceeds the limits within which it is possible to verify one's interpretations objectively. Solitary and interactive vocal play alike contribute to procedural refinement of infants' vocal control and enrichment of vocal repertoire with both phonological elements of the mother-tongue and creative inventions of new vocal patterns.

The next forum in which children may advance their cognitions about objects, people, and actions, and construct increasingly sophisticated representations of the world and relations between symbols and their external referents, is symbolic play observable after 12 months of age (Bornstein & Lamb, 1992; Piaget, 1962; Vygotsky, 1978). Play develops in sophistication as it moves from sensorimotor exploration through concrete and functional activities with objects eventually to include generative expressions of pretence. In the latter form of play, children enact experience and events through symbolic gestures, and they engage in actions that are detached from real objects (Bornstein & Tamis LeMonda, 1995). Symbolic play can include simple scenarios about self (pretending to drink from empty toy cups) and about other things (putting dolls to sleep), sequences of pretence (pouring into empty cups and then pretending to drink), and even substitutions (pretending to talk into an object as though it were a telephone) (Bornstein & Tamis LeMonda, 1995). In short, play has been viewed as manifesting the child's growing representational competence or capacity for expressing symbolic thought.

During collaborative play with parents, children are guided in the re-creation, expression, and elaboration of symbolic themes. Adults play in ways which children observe and learn from, they induce children to play, and they provide objects for play. Children may initiate pretend-play sequences, but they also presumably learn from and imitate the pretence they see (Uzgiris, 1984). Such collaborative play interactions are thought to foster more sophisticated expressions of play in immediate exchanges, and potentially advance children's cognitive competence over more extended periods. Parental participation in collaborative play is believed to raise the level of expression of symbolic sophistication in the child's play, make it richer or more diverse, or sustain it (Dunn & Wooding, 1977; Slade, 1987).

Bornstein *et al.* (1994) reported that the degree to which mothers participate in and guide their children's symbolic play affects the level of their children's symbolic play. Children engage in symbolic play more, and for longer periods, when in collaboration than when alone. Similar to the aforementioned didactic tendency reported in vocal interchanges with infants, mothers respond to child play most frequently at levels equal or just above the preceding level of play and so channel child play towards greater sophistication (Damast *et al.*, 1996).

Evidence of maternal influence on child play during collaborative play raised questions as to whether the beneficial effects carry over to children's symbolic play when the child plays alone and whether they last over a longer time frame. As reviewed by Fein & Fryer (1995), few empirical studies actually support the view that the level of maternal play sophistication influences the level of child play sophistication beyond interactive circumstances. In contrast, some concurrent studies point to a potential influence of parental behaviour on child play. Belsky *et al.* (1980) hypothesized that, to at least some degree, parents directly influence their children to explore the environment and play at higher levels. They observed mothers and infants at home at ages of 9, 12, 15 and 18 months, and mothers who stimulated their babies more had infants who explored their environments more competently. The data of Hoppe-Graff & Engel (1996) also showed a long-term construction of symbolic competence that was learned in play with parents in 9 months (from approximately 9 to 18 months).

CONCLUDING COMMENTS

Developmental neurologists and neurophysiologists have come to the conclusion that even the first year of infancy should be viewed as a phasic developmental process, characterized by periods of distinct transformations in the modularity of the central nervous system and by corresponding shifts in the development of observable behaviours or experimentally detectable integrative capacities. Such a theoretical position challenges infancy researchers to pay greater attention to early communicative processes. The increased interest of neuroscientists in the conscious regulation of human behaviour has also brought with it evidence that a great deal of everyday decisions in humans occur at the non-conscious level rather than on the basis of rational, fully conscious mental processes (Gazzaniga, 1995).

To explain the processes underlying symbolic representation in the brain promises a key solution to the main facets in the puzzle of human mind – verbal communication, mental representation, rational cognition, cultural integration, and consciousness. It serves as a challenge to scientists in various fields, from the neurosciences and psychology to biology and physics.

References

Bard, K.A. & Gardner, K.H. (1996). Influences on development in infant chimpanzees: Enculturation, temperament, and cognition. In A. Russon, K.A. Bard & S.T. Parker (Eds.), *Reaching into thought: the minds of great apes* (pp. 235–256). Cambridge, UK: Cambridge University Press.

Belsky, J., Goode, M.K. & Most, R.K. (1980). Maternal stimulation and infant exploratory competence: cross-sectional, correlational, and experimental analyses. *Child Development, 51*, 1163–1178.

Bertalanffy, L. von (1968). *Organismic psychology theory.* Barre, MA: Clark University with Barre Publishers.

Bornstein, M.H. (1985). How infant and mother jointly contribute to developing cognitive competence in the child. *Proceedings of the National Academy of Sciences, USA, 82*, 7470–7473.

Bornstein, M.H. & Lamb, M.E. (1992). *Development in infancy: an introduction* (3rd ed.) New York: McGraw-Hill.

Bornstein, M.H. & Tamis-LeMonda, C.S. (1995). Parent–child symbolic play: three theories in search of an affect. *Developmental Review, 15*, 382–400.

Bornstein, M.H., Haynes, O.M., O'Reilly, A.W. & Painter, K.M. (1994). Solitary and collaborative pretense play in early childhood: sources of individual variation in the development of representational competence. Unpublished manuscript. National Institute of Child Health and Human Development, Bethesda, MD.

Bruner, J.S. (1975). The ontogenesis of speech acts. *Journal of Child Language, 2*, 1–9.

Cassirer, E. (1953). *The philosophy of symbolic forms*, vol. 1. New Haven: Yale University.

Damast, A.M., Tamis-LeMonda, C.S. & Bornstein, M.H. (1996). Mother–child play: sequential interactions and the relation between maternal beliefs and behaviors. *Child Development, 67*, 1752–1766.

Dent, R.D., Mitchell, P.R. & Sancier, M. (1991). Evidence and role of rhythmic organization in early vocal development in human infants. In J. Fagard & P.H. Wolff (Eds.), *The development of timing control and temporal organization in coordinated action: invariant relative timing, rhythms and coordination* (pp. 135–149). Advances in Psychology, 81. Amsterdam: North-Holland.

Dore, J. (1975). Holophrases, speech acts and language universals. *Journal of Child Language, 2*, 21–40.

Dunn, J. & Wooding, C. (1977). Play in the home and its implications for learning. In B. Tizard & D. Harvey (Eds.), *Biology of play* (pp. 45–58). Spastic International Medical Publications. Philadelphia: Lippincott.

Erting, C.J., Prezioso, C. & Hynes, M.O. (1990). The interactional context of deaf mother–infant communication. In V. Volterra & C. Erting (Eds.), *From gesture to language in hearing and deaf children* (pp. 97–106). Berlin: Springer-Verlag.

Fein, G.G. & Fryer, M.G. (1995). Maternal contributions to early symbolic play competence. *Developmental Review, 15*, 367–381.

Ferguson, C.A. (1964). Babytalk in six languages. *American Anthropologist,* **66**, 103–114.

Fernald, A. (1992). Meaningful melodies in mothers' speech to infants. In H. Papoušek, U. Jürgens & M. Papoušek (Eds.), *Nonverbal vocal communication: comparative and developmental approaches* (pp. 262–282). New York: Cambridge University Press.

Fernald, A., Taeschner, T., Dunn, J., Papoušek, M., Boysson-Bardies, B. & Fukui, I. (1989). A cross-language study of prosodic modifications in mothers' and fathers' speech to preverbal infants. *Journal of Child Language,* **16**, 977–1001.

Field, T., Woodson, R., Greenberg, R. & Cohen, D. (1982). Discrimination and imitation of facial expressions by neonates. *Science,* **218**, 179–181.

Fogel, A. (1977). Temporal organization in mother–infant face-to-face interaction. In H.R. Schaffer (Ed.), *Studies in mother–infant interaction* (pp. 119–151). London: Academic Press.

Fogel, A. (1990). The process of developmental change in infant communicative action: using dynamic systems theory to study individual ontogenies. In J. Colombo & J. Fagen (Eds.), *Individual differences in infancy: reliability, stability, prediction* (pp. 341–358). Hillsdale, NJ: Erlbaum.

Gazzaniga, M.S. (1995). Consciousness and the cerebral hemispheres. In M.S. Gazzaniga (Ed.), *The cognitive neurosciences* (pp. 1391–1400). Cambridge, MA: MIT Press.

Harkins, D.A. & Uzgiris, I.C. (1991). Hand-use matching between mothers and infants during the first year. *Infant Behavior & Development,* **14**, 289–298.

Hirschberg, J. & Szende. T. (1985). *Pathologische Schreistimme, Stridor und Hustenton im Säuglingsalter.* Budapest: Akadémiai Kiadó.

Hoppe-Graff, S. & Engel, I. (1996). Entwicklungsmuster und Erwerbsprozesse früher Symbolkompetenzen. Abschlußbericht des DFG-Projektes Ho 922/4-2. Universität Leipzig, August.

Kawai, M. (1963). On the newly acquired behaviors of the natural troop of Japanese monkeys on Koshima Island. *Primates,* **4**, 113–115.

Kelso, J.A.S., Tuller, B. & Harris, K.S. (1983). A dynamic pattern perspective on the control and coordination of movement. In P.F. MacNeilage (Ed.), *The production of speech* (pp. 137–173). New York: Springer.

Kessen, W., Levine, J. & Wendrich, K. (1979). The imitation of pitch in infants. *Infant Behavior and Development,* **2**, 93–99.

Kestermann, G. (1982). Gestik von Säuglingen: Ihre kommunikative Bedeutung für erfahrene und unerfahrene Bezugspersonen (Infants' gestures: their significance for experienced and inexperienced caregivers). Unpublished doctoral dissertation, University of Bielefeld, Germany.

Koester, L.S., Papoušek, H. & Papoušek, M. (1989). Patterns of rhythmic stimulation by mothers with three-month-olds: a cross-modal comparison. *International Journal of Behavioural Development,* **12**, 143–154.

Korner, A.F. (1979). Maternal rhythms and waterbeds: a form of intervention with premature infants. In E.B. Thoman (Ed.), *Origins of the infant's social responsiveness* (pp. 95–124). Hillsdale, NJ: Erlbaum.

Kuhl, P.K. (1983). Perception of auditory equivalence classes for speech in early infancy. *Infant Behavior and Development,* **6**, 263–285.

Kuhl, P.K. & Meltzoff, A.N. (1996). Infant vocalizations in response to speech: vocal imitation and development change. *Journal of the Acoustical Society of America,* **100**, 2425–2438.

Legerstee, M. (1990). Infants use multimodal information to imitate speech sounds. *Infant Behavior and Development,* **13**, 343–354.

Lewkowicz, D.J. (1989). The role of temporal factors in infant behavior and development. In I. Levin & D. Zakay (Eds.), *Time and human cognition: a life-span perspective* (pp. 9–62). Advances in Psychology, vol. 59. Amsterdam: North-Holland.

Lorenz, K.Z. (1943). Die angeborenen Formen möglicher Erfahrung. (In German). *Zeitschrift der Tierpsychologie,* **5**, 235–409.

MacLean, P.D. (1990). *The triune brain in evolution. Role in paleocerebral functions.* New York: Plenum Press.

Masur, E.S. & Ritz, E.G. (1984). Patterns of gestural, vocal, and verbal imitation performance in infancy. *Merrill-Palmer Quarterly,* **30**, 359–392.

Meltzoff, A.N. & Moore, M.K. (1977). Imitation of facial and manual gestures by human neonates. *Science,* **198**, 75–78.

Meltzoff, A.N. & Moore, M.K. (1983). Newborn infants imitate adult facial gestures. *Child Development,* **54**, 702–709.

Messer, D.J. (1980). The episodic structure of maternal speech to young children. *Journal of Child Language*, **7**, 29–40.

Newman, J.D. (1988). *Physiological control of mammalian vocalization*. New York: Plenum Press.

Newson, J. (1979). Intentional behaviour in the young infant. In D. Schaffer & J. Dunn (Eds.), *The first year of life. Psychological and medical implications of early experience* (pp. 91–96). Chichester: Wiley.

Oller, D.K. & Eilers, R.E. (1988). The role of audition in infant babbling. *Child Development*, **59**, 441–449.

Oller, D.K. & Eilers, R.E. (1992). Development of vocal signalling in human infants; Toward a methodology for cross-species vocalization comparisons. In H. Papoušek, U. Jürgens & M. Papoušek (Eds.), *Nonverbal vocal communication: comparative and developmental approaches.* (pp. 174–191). Cambridge, UK: Cambridge University Press.

Papoušek, H. (1967). Experimental studies of appetitional behavior in human newborns and infants. In H.W. Stevenson, E.H. Hess & H.L. Rheingold (Eds.), *Early behavior: comparative and developmental approaches* (pp. 249–277). New York: Wiley.

Papoušek, H. (1977). Entwicklung der Lernfähigkeit im Säuglingsalter (Development of learning ability during infancy). In G. Nissen (Ed.), *Intelligenz, Lernen und Lernstörungen: theorie, praxis und therapie* (pp. 75–93). (Intelligence, learning, and learning disorders: theory, practice, and therapy). Berlin: Springer-Verlag.

Papoušek, H. & Bernstein, P. (1969). The functions of conditioning stimulation in human neonates and infants. In A. Ambrose (Ed.), *Stimulation in early infancy* (pp. 229–252). London: Academic Press.

Papoušek, H. & Papoušek, M. (1977). Mothering and the cognitive headstart: Psychobiological considerations. In H.R. Schaffer (Ed.), *Studies in mother–infant interaction* (pp. 63–85). London: Academic Press.

Papoušek, H. & Papoušek, M. (1978). Interdisciplinary parallels in studies of early human behavior: from physical to cognitive needs, from attachment to dyadic education. *International Journal of Behavioral Development*, **1**, 37–49.

Papoušek, H. & Papoušek, M. (1984). Learning and cognition in the everyday life of human infants. In J.S. Rosenblatt, C. Beer, M.-C. Busnel & P.J.B. Slater (Eds.), *Advances in the study of behavior*, vol. 14 (pp. 127–163). New York: Academic Press.

Papoušek, H. & Papoušek, M. (1987). Intuitive parenting: a dialectic counterpart to the infant's integrative competence. In J.D. Osofsky (Ed.), *Handbook of infant development* (2nd ed., pp. 669–720). New York: Wiley.

Papoušek, H. & Papoušek, M. (1989a). Intuitive parenting: aspects related to educational psychology. In B. Hopkins, M.-G. Pecheux & H. Papoušek (Eds.), *Infancy and education: psychological considerations. European Journal of Psychology of Education*, **4** (2, Special Issue), 201–210.

Papoušek, H. & Papoušek, M. (1997). Preverbal communication in humans and the genesis of culture. In U. Segerstrale & P. Molnár (Eds.), *Nonverbal communication: where nature meets culture* (pp. 87–107). Mahwah, NJ: Erlbaum.

Papoušek, H., Papoušek, M. & Bornstein, M.H. (1985). The naturalistic vocal environment of young infants: on the significance of homogeneity and variability in parental speech. In T. Field & N. Fox (Eds.), *Social perception in infants* (pp. 269–297). Norwood, NJ: Ablex.

Papoušek, H., Papoušek, M. & Kestermann, G. (2000). Preverbal communication: emergence of representative symbols. In N. Budwig, I.C. Uzgiris & J.W. Wertsch (Eds.), *Communication: an arena of development* (pp. 81–107). Norwood, NJ: Ablex Publishing Corporation.

Papoušek, M. (1994). *Vom ersten Schrei zum ersten Wort: Anfänge der Sprachentwicklung in der vorsprachlichen Kommunikation.* (From the first cry to the first word: The beginnings of speech development in preverbal communication). Bern: Huber.

Papoušek, M. & Papoušek, H. (1981). Musical elements in the infant's vocalizations: their significance for communication, cognition and creativity. In L.P. Lipsitt (Ed.), *Advances in infancy research* (vol. 1, pp. 163–224). Norwood, NJ: Ablex.

Papoušek, M. & Papoušek, H. (1989b). Forms and functions of vocal matching in precanonical mother–infant interactions. *First Language*, **9**, 137–158 (Special Issue on "Precursors to speech").

Papoušek, M., Papoušek, H. & Haekel, M. (1987a). Didactic adjustments in fathers' and mothers' speech to their three-month-old infants. *Journal of Psycholinguistic Research*, **16**, 491–516.

Papoušek, M., Papoušek, H. & Harris, B.J. (1987b). The emergence of play in parent–infant inter-actions. In D. Görlitz & J.F. Wohlwill (Eds.), *Curiosity, imaginations, and play on the development of spontaneous cognitive and motivational processes* (pp. 214–246). Hillsdale, NJ: Erlbaum.

Papoušek, H., Papoušek, M. & Koesler, L. (1986). Sharing emotionality and sharing knowledge: A microanalytic approach to parent–infant communication. In C.E. Izard & P. Read (Eds.), *Measuring emotions in infants and children* (pp. 93–123), Cambridge, England: Cambridge University Press.

Papoušek, M., Papoušek, H. & Symmes, D. (1991). The meanings of melodies in motherese in tone and stress languages. *Infant Behavior and Development*, **14**, 414–440.

Pavlov, I.P. (1932). The reply of a physiologist to psychologists. *Psychological Review*, **39**, 81–127.

Petitto, L.A. & Marentette, P.F. (1991). Babbling in the manual mode: evidence for the ontogeny of language. *Science*, **251**, 1493–1496.

Piaget, J. (1962). *Play, dreams, and imitation in childhood*. New York: Norton.

Schoetzau, A. & Papoušek, H. (1977). Mütterliches Verhalten bei der Aufnahme von Blickkontakt mit dem Neugeborenen (Maternal behavior associated with establishment of visual contact with newborns). *Zeitschrift für Entwicklungspsychologie und pädagogische Psychologie*, **9**, 231–239.

Seyfarth, R.M., Cheney, D.L. & Marler, P. (1980). Monkey responses to three different alarm calls: evidence of predator classification and semantic communication. *Science*, **210**, 801–803.

Slade, A. (1987). A longitudinal study of maternal involvement and symbolic play during the toddler period. *Child Development*, **58**, 367–375.

Squire, L.R. (1987). *Memory and brain*. Oxford: Oxford University Press.

Turkewitz, G. & McGuire, I. (1978). Intersensory functioning during early development. *Journal of Mental Health*, **7**, 165–182.

Turner, F. (1985). *Natural classicism*. New York: Paragon.

Uzgiris, I.C. (1989). Infants in relation: performers, pupils, and partners. In W. Damon (Ed.), *Child development today and tomorrow* (pp. 288–311). San Francisco, CAREGIVERS: Joseey-Bass.

Uzgiris, I.C. (1984). Imitation in infancy: its interpersonal aspects. In M. Perlmutter (Ed.), *The Minnesota Symposia on Child Psychology* (vol. 17, pp. 1–32). Hillsdale, NJ: Erlbaum.

Vander A.J., Sherman, J.H. & Luciano, D.S. (1990). *Human physiology. The mechanisms of body functions* (5the ed.). New York: McGraw-Hill.

Vygotsky, L.S. (1978). *Mind in society: the development of higher psychological processes*. Cambridge, MA: Harvard University Press.

Winter, P., Ploog, D. & Latta, J. (1966). Vocal repertoire of the squirrel monkey (*Saimiri sciureus*), its analysis and significance. *Experimental Brain Research*, **1**, 359–384.

Wolff, P.H. (1969). The natural history of crying and other vocalizations in early infancy. In B. Foss (Ed.), *Determinants of infant behavior* (vol. 4, pp. 81–109). London: Methuen.

Wolff, P.H. (1991). Endogenous motor rhythms in young infants. In J. Faard & P.H. Wolff (Eds.), *The development of timing control and temporal organization in coordinated action: invariant relative timing, rhythms and coordination* (pp. 119–133). Advances in Psychology, vol. 81. Amsterdam: North-Holland.

Further reading

Budwig, N., Uzgiris, I.C. & Wertsch, J.V. (Eds.), *Communication: An arena of development*. Norwood, NJ: Ablex Publishing Company.

Papoušek, M. (1994). Melodies in Caregivers' speech: a species-specific guidance towards language. *Early Development and Parenting*, **3**, 5–17.

Papoušek, H. & Papoušek, M. (1995). Intuitive parenting. In M.H. Bornstein (Ed.), *Handbook of parenting*, vol. 2. *Biology and ecology of parenting* (pp. 117–136). Mahwah, NJ: Lawrence Erlbaum Ass.

Papoušek, H., Jürgens, U. & Papoušek, M. (Eds.) (1992). *Nonverbal vocal communication: compar-ative & developmental approaches*. Cambridge: Cambridge University Press.

V.5
A Dynamical Systems Approach to the Study of Early Emotional Development in the Context of Mother–Infant Communication

ANDRÉA P.F. PANTOJA, G. CHRISTINA NELSON-GOENS
and ALAN FOGEL

ABSTRACT

In this chapter we discuss a dynamical systems approach to the study of early emotional development in the context of mother–infant communication. We begin by introducing the problem of how one can examine emotional development, followed by an introduction of two dominant theories of emotions: the differential emotions theory and the functionalist theory. We then present a dynamical systems approach to emotions. We also review the literature on both the behavioural and neurophysiological components of emotions. Finally, using excerpts from an ongoing longitudinal study, we illustrate the theoretical and methodological innovations using a dynamical systems approach. We conclude encouraging an avenue that embodies the use of qualitative methods.

INTRODUCTION

Traditionally the study of emotions has been primarily focused on facial expressions and other singular components of emotions. A dynamical systems

This work was in part supported by a grant to Andréa P. F. Pantoja from the National Science Foundation of Brazil-CNPq (200828/94) and a grant to Alan Fogel from the National Institute of Mental Health (R01 MH48680).

A.F. Kalverboer and A. Gramsbergen (eds.), Handbook of Brain and Behaviour in Human Development, 901–920
© *2001 Kluwer Academic Publishers. Printed in Great Britain.*

approach to emotions, however, emphasizes the importance of examining how various emotion components such as facial actions, body movements, vocalizations, gestures, and actions of the brain constitute different emotions over time. The goal then is to explore how emotions can transcend individual modes of expression (e.g. Camras, 1991; Fogel *et al*., 1992; Wolff, 1987). In this section we use two predominant theories of emotion to provide a foundation for discussing a dynamical systems approach.

Differential emotions theory

Based on Darwin's earlier work, (Tomkins & McCarter, 1995) conceptualized emotions as biologically inherited programmes controlling autonomic, blood-flow, respiratory, vocal, and facial muscles responses. Different correlated sets of responses constitute eight specific basic emotions; thus the name differential emotions theory (DET). Tomkins claimed that the face was the primary location of emotion expression and, for him, the face takes the lead in establishing and creating an awareness of an emotional state, with the other constituents coming into play more slowly (Demos, 1988).

Tomkins' emphasis on the face was followed up by Ekman (1972, 1977; as cited in Demos, 1988) and Izard (1968, 1971; as cited in Demos, 1988) who independently explored the different basic facial patterns proposed by Tomkins. The investigation of the different facial patterns led Ekman & Friesen to develop a very exhaustive and precise measure of facial actions (1975, 1978) called the facial action coding system (FACS). FACS is a detailed anatomically based coding system which allows investigators to identify many distinct facial movements. The main reason for focusing on such analysis of the facial muscles rests on the premise that the face is a valid and primary indicator of an internal emotional state (Demos, 1988). Perhaps because this theory is dominant in the field of developmental psychology, the face has been so intensively examined by emotion theorists. Demos (1988) asserts:

> In adopting this particular theoretical framework (I refer here to the differential affect theory set forth by Silvan Tomkins [1962, 1963] and its subsequent elaboration by Paul Ekman and Carrol Izard), we are making the assumption that facial expressions in young children represent the primary, the most precise, and the clearest indicators of affective states (p. 128).

Differential emotions theorists (e.g. Ekman & Friesen, 1975; Izard, 1997) propose that there are only a limited number of basic emotions. They argue that subtle nuances in the basic emotions due to the association with various cognitive processes are not accurate evidence towards the emergence of a larger repertoire of emotions. For example, when one thinks about an absent loved one, one may feel nostalgic. When one thinks about the death of a loved one, one may experience a profound feeling of grief. For differential emotions theorists those experiences are illustrations of the basic emotion sadness. These different emotional experiences (i.e. nostalgia and grief) are interpreted as variations in intensity of the same basic emotion when associated with distinct cognitive processes.

Thus, different people with different personal and cultural histories may experience a variety of distinct emotions due to the multiplicity of links between emotion and cognition; but the basic emotions remain the same across time, individuals, and cultures. "DET has always argued that a given discrete emotion state, though often occurring with a *pattern* of other discrete emotion states, retains its unique motivational properties" (Izard, 1997, p. 68). This premise allows differential emotions theorists to dissect human experience into autonomous units as in the case of pride. Pride is conceived as another illustration of the link between a basic emotion (i.e. joy) and a specific cognitive process (i.e. mastery or achievement).

Therefore, differential emotions theorists maintain that emotional development can be explained by the change in the expression of an innate set of emotions (i.e. basic emotions) with each of these emotions endowed with a distinct motivational property (Izard, 1997). That is, these basic emotions are present from birth and their "core" does not change across time, but rather their mode of expression. This theoretical postulate is based on Tomkins' view of emotions as biologically inherited programmes with the face as having priority over visceral, behavioural, experiential, and relational changes (Ekman, 1995).

The notion that there exist a number of basic emotions that remain relatively unchanged across time has been challenged. Some emotion theorists (e.g. Barrett, 1998) would argue that, as emotions become associated with different cognitive processes across time, a change in the repertoire of emotions occurs. The functionalist theory, described next, questions the assumption of a core set of innate emotions.

Functionalist theory

Another predominant developmental view of emotion is the functionalist theory (e.g. Barrett, 1993, 1998; Campos *et al*., 1994; Frijda, 1993). According to the functionalist theory, emotions are not biologically inherited programmes, nor are they entities that reside in the brain or face. Instead, emotions are processes that evolve from the interaction between the organism and the environment as appraised by the organism. Frijda (1993) describes emotions in the following manner:

> Emotional experience, at its most prototypical, is not subjective experience. It is in part a perception and in part a felt interaction with the environment, or felt inclination or disinclination thereto. It is something between him and me, or between her and me, or it and me. ... Most emotions, being interactions, are events over time and are felt as events over time (p. 249).

Frijda & Tcherkassof (1997) emphasized that individuals rarely use emotion labels to describe emotions. Instead they make use of narrative descriptions of emotionally charged situations in which the protagonist is often in an interaction with the environment. Emotions are not internal discrete states that are universally labelled; nor are they expressed primarily in the face. Emotions are the individual's relational activity to the environment; emotions are states of action readiness (Frijda, 1993; Frijda & Tcherkassof, 1997).

For example, one may cry over happiness. In this case crying cannot be considered as evidence for sadness. For Frijda these "deviations" from what differential emotions theory would expect are very frequent in everyday life. Thus, the importance of studying emotions as part of the individual's relationship to the environment (Frijda & Tcherkassof, 1997). Various types of action readiness correspond to different types of organism–environment inter-action or "interactional goals". States of action readiness will vary as a function of their goal (e.g. obtain proximity, avoid contact, etc.) and their level of activation (e.g. hyperactive, neutral, etc.).

In contrast to the differential emotions theory these states of action readiness do not reflect hard-wired programmes, they are not a product of discrete internal emotional states. Emotions as states of action readiness reflect an individual's relation to the environment. Therefore, emotions evolve as communicative devices produced in the service of social motives and intentions (of which the organism may or may not be aware).

When conceiving of emotions as communicative devices, other functionalist theorists (e.g. Barrett, 1993, 1998; Campos *et al.*, 1994) emphasize the process emotions serve for the organism in its immediate context. For Barrett (1993, 1998) and Campos *et al.* (1994), different emotions have specific functions that emerge from the distinct individual–environment relationships. Emotions that comprise similar functions constitute an emotion family. A particular emotion family is defined by three adaptive functions it promotes: behaviour-regulatory functions; social-regulatory functions; and internal-regulatory functions. The concept of emotion families reflects a significant shift away from what was previously held by differential emotions theorists. Emotions are processes that have adaptive functions for the organism in relation to the environment. What defines a certain emotion is not a particular facial display that "expresses" an internal discrete state of the organism (Barrett, 1998). Instead, emotions can be identified only in terms of their function for the organism, and these functions do not present a one-to-one correspondence to specific behaviours (including facial behaviours).

The methodological implications of this theoretical shift are clear. Emotions should now be understood in terms of how the behaviour is organized. A particular facial display does not express a clear-cut internal state of the organism (Campos *et al.*, 1994). Facial movements, when studied, are not divorced from context, since the context will provide information about at least one of the three possible regulatory functions of the emotion. Placing an emphasis on the functions of emotions for the organism–environment interaction, functionalist theorists make an indirect call for new interpretative methods of analysis. Barrett (1993) suggests that the appropriateness of a coding system is relative to the purpose and theoretical grounding of the investigation.

> It will depend on the research question, however, what facial coding system, if any, should be used and how it should be used. If the research question involved only negative versus positive emotion, then the discrete emotion templates (e.g. Izard, 1979) may serve as a *guide* (but not a mandate) for determination of which movements should be classified as negative and which should be positive. If what is important

is whether the movements specify approach versus withdraw, the template system is not appropriate, and a functional analysis of the movements is needed (p. 165).

Emotions as functional processes will change across time as the individual's relationship with the environment changes. Thus, emotional development from a functionalist perspective entails changes in the emotional experience for the individual involved, the manner in which emotions are instigated, and the observable features of the emotions. In contrast to differential emotions theory, the emotion of pride would not merely be an illustration of the basic emotion of joy, linked with achievement. For functionalist theorists pride is a qualitatively different emotion because it involves at least three critical differences. First, joy involves having something, whereas pride encompasses meeting standards and/or goals. Second, joy does not require evaluation of oneself, whereas pride does. Finally, joy does not necessarily consist of approaching somebody, whereas pride includes an inclination to approach the other (Barrett, 1998). Although joy and pride may be part of related emotion families they promote different regulatory functions, thereby constituting different emotions. Functionalist theory and its contemplation of emotions as an individual's socially adaptive process of relating to the world (as opposed to innate basic discrete emotions) represent a critical shift in the conceptualization of emotion and its development.

In the next paragraphs we turn our discussion to our dynamical systems approach to emotional development. We aim to continue synthesizing some of the potential controversies currently found in the field of developmental psychology.

DYNAMICAL SYSTEMS APPROACH TO EMOTIONAL DEVELOPMENT

Similar to the functionalist theory, in the dynamical systems approach, emotions are envisioned as relational processes, not hard-wired internal states expressed outwards. In our opinion, conceptualizing emotions as relational processes is a major theoretical contribution to the field of developmental psychology. The dynamical systems approach also places emphasis on the emergent properties of emotional development. A critical research question becomes: What are the change processes involved in the development of emotions? We find the dynamical systems approach conceptually useful because it begins to provide heuristic tools to examine the change processes implicated in the unfolding of different emotions over time. This approach strongly relies on the principle of self-organization (Fogel et al., 1992; Messinger et al., 1999).

Self-organization refers to the continuous process of interaction among the constituents of a given system. Of particular note, constituents are theoretical notions that aim to actualize the different parts entailed in the composition of a system. In this chapter the emotions of human beings are the system of interest. The various muscles of the face examined by differential emotions theorists, for example, can be conceptualized as constituents of the emotion system. From our dynamical systems approach the face is one among many equally important constituents of emotions, including differing body postures, vocalizations, activities of the brain, and the interactive contexts in which human beings are engaged (Fogel, 1993).

Through self-organization these emotion constituents form patterns of co-activity which provide an individual's emotion system with its particular character over time. Emotions are a unique phenomenon because they inform human beings about the significance of their relationships with the environment, including their relationships with others. The emphasis must be on how different emotions emerge out of the interaction among the various emotion constituents.

There are many implications of dynamical systems theory for the study of emotion that are beyond the scope of this chapter (see however, Camras, 1991; Fogel *et al.*, 1992, 1997; and Lewis, 1995, for more details). Here we focus on self-organization in our developmental investigations of emotions. We stress the emergence of emotion patterns over time rather than configurations of emotions on faces (differential emotions theory) or families of emotions united by a common function (functionalist theory). The focus is then on how different emotions self-organize.

Furthermore, a dynamical systems approach to emotions is inherently developmental because self-organization is a process simultaneously contemplated at two distinct time-scales. There is the self-organizing process occurring at the level of the moment-by-moment change (referred to herein as real-time change). Examples are the minute transformations observed second after second, as individuals interact with one another. During a pleasant conversation they may gradually lean towards each other, relaxing their bodies, slightly tilting their heads, turning their gazes to one another, producing a smile on their faces, and gently raising the intonation of their voices, thereby producing melodic sounds. The behavioural variations that occur when individuals interact are illustrations of self-organization at the level of real-time changes.

At the same time, individuals organize and transform their lives at the developmental level (referred to herein as developmental-time changes). For example, in the multiple encounters described above, individuals may co-create a pleasant inclination towards one another, thereby facilitating their engagement in mutually gratifying and creative experiences. A developmental-time change gradually emerged from the details that constitute individuals' day-to-day life; that is, the real-time changes observed in the lives of individuals (for a more detailed account on time-scales see Lewis, 1995; Lyra, 2001; and Thelen & Ulrich, 1991). Our contention is that additional emphasis on self-organizing processes, that integrate real-time and developmental-time changes, should be more systematically introduced in developmental investigations of emotions (Fogel *et al.*, 1992).

In this chapter we intend to offer a scientific approach that attempts to capture self-organizing processes within two time-scales. Thus we suggest that emotions be studied over time while considering both time-scales simultaneously. The implications of taking this approach are evident. Descriptions of microscopic details that humans experience day to day and over time are at the core of our developmental analysis. Based on the principle of self-organization we contend that emotions develop out of the day-to-day changes (real-time changes) observed throughout an individual's life (developmental-time changes). Later we will present a narrative–qualitative analytical technique in which both the real-time and the developmental-time changes are dynamically encompassed in our developmental investigations of emotions.

In sum this chapter is based in part on the idea that emotions can be thought of as emergent patterns of activity integral in the creation of meaningful relationships. Note that the term "relationships" is not limited to interpersonal relationships. Emotions are also relational in that they can occur in the context of live relationships with inanimate and animate things, with natural and cultural objects, and with the self (Fogel *et al.*, 1997). In the case of infants, interpersonal relationships are particularly relevant because it is with significant others that infants spend a great deal of their waking time creating primary relationships.

In the next section we present a critical review of the relevant literature on emotions and emotional development. We begin by discussing the behavioural components of emotions, followed by a description of neurophysiological indices of emotions. The goal is to illustrate how these different emotion constituents may act together, forming patterns of co-activities we often call emotions.

BEHAVIOURAL AND NEUROPHYSIOLOGICAL COMPONENTS OF EMOTIONS

As mentioned earlier, we argued that emotions emerge from the dynamic inter-play among emotion constituents such as CNS activation, ANS arousal, actions of the face, body and voice, psychological processes, and processes related to the relationship between the individual and the environment. Although our investigations focus on the behavioural components of emotions in the context of early communication, we postulate that all emotion constituents, both at the behavioural and neurophysiological levels, are crucial in the development of emotions.

Behavioural components of infant emotions

Research on early development reveals various changes in infants' emotions as indicated by the changes in their behaviours across the first 7 months of life (e.g. Camras, 1991; Fogel, 1997; Messinger *et al.*, 1997, 1999; Weinberg & Tronick, 1994; Wolff, 1987). This body of research supports the notion that emotions are an integral part of communication processes, even during early infancy. At 1 month of age, for instance, infants present a relatively small repertoire of emotional behaviours, and in most cases they are related to the emotional experience of distress. Over time, as new emotional experiences emerge (as indicated by the emergence of smiling and cooing, for example), the duration of distress decreases (Wo'ff, 1987).

As a way of investigating the development of an infant's positive emotion, Wolff (1987) described developmental changes in infants' smiling across the first 3 months of life. Changes in infants' smiling were related to changes in the infants' relationship with their environment. In the first week of life infants did not smile reliably to sounds, visual stimuli, tactile stimulation, bouncing, or the like. A change in infants' smiling is observed towards the end of the second week. Approximately half of the sample smiled regularly to the human voice when the infants were awake. By the fourth week almost all the infants (20 of the 22 infants) smiled consistently to both human and non-human sounds, although the human voice was the most critical attraction to the infants at this age. At 5 weeks of age, infants smiled more often to the combination of voice and face than to either of the two alone. By 8 weeks the human voice alone made the infants rotate their heads and eyes as if searching for the source of the sound, and frequently the infants smiled only after making visual contact with their mothers' faces. At the same time Wolff found that infants became more selective to different speech sounds; that is, maternal baby talk and natural human voice were more effective to capture infants' visual attention than the investigator's high-pitched voice. Finally, by the eleventh or twelfth week (around 3 months), infants equally smiled at all familiar faces, and any familiar face combined with familiar voice was more effective than either of the two alone. These results suggest that the infants' positive emotion, as indicated by smiling, gradually self-organizes as the infants and their caregivers co-create specific forms of social communication, specifically face-to-face communication.

After 4 months, infants begin to experience an expanded variety of positive emotions as illustrated by the increased complexity in infants' smiling (Fogel, 1997). By this age differing smiles are observed in the context of changing mother–infant communication (Messinger *et al.*, 1997). Cheek-raise smiles that result in a narrowing of the eye aperture (called a Duchenne smile) are observed in the context of social communication, when infants are held upright, face-to-face with a reciprocating mother. Another smile (called play smile) that involved a dropping of the jaw, is observed when infants are held close to their mothers while being kissed or tickled. Messinger *et al.* (1997) suggest that Duchenne smiles might be related to an emotional experience of positive, visually mediated, connectedness between the infant and mother, while play smiles might be related to more tactile and physically arousing events.

More recently, Messinger *et al.* (1999) compared Duchenne and non-Duchenne smiles in the context of mother–infant face-to-face interaction using a real-time-based analysis. The results indicated that Duchenne and non-Duchenne smiles are not only correlated within interactive episodes but have similar developmental trajectories. Indeed, Duchenne smiles are often preceded by non-Duchenne smiles. This study suggests that, in infancy, those different types of smiles may not be indicative of qualitatively distinct emotional meanings. The authors propose that Duchenne and non-Duchenne smiles may constitute different temporal phases of a continuous emotional process since they often occur in similar situations during infancy.

The development of an infant's emotional experience of surprise has been investigated by Camras (1991). In the second month of life surprise involves raised eyebrows, eye blink, a distinct pattern of soft breathing and limb mobility suggesting arousal and excited attention. When the visually guided reaching develops in the first 4 months, surprise is often observed in the specific context of the infant attempting to reach for an object. By 5 months, surprise occurs during a sequence composed of reaching, grasping, and mouthing of an object.

Moving beyond the face, Weinberg & Tronick (1994) found four distinct emotional patterns at 6 months of age. In the first emotional pattern, infants present positive facial actions (such as smiling) accompanied by positive vocalizations,

mouthing of body parts, and gazing at mother in the context of social engagement. During object engagement, infants display facial actions of interest (i.e.
neutral eyebrows and eyes track target) accompanied by mouthing of and gazing
at objects. Facial actions of sadness (i.e. arched eyebrows, mouth down, and
tears), accompanied by fussy vocalizations, composes a third emotional pattern
called passive withdrawal. Finally, the emotional pattern called active protest is
defined by facial actions of anger (i.e. frown, clenched jaw, eyes narrowed, and
red face), fussy and cry vocalizations, pick-up requests, and escape movements.

The findings briefly discussed in this section illustrate how infants' emotions
are flexible, dynamic, defined by a multiplicity of components (e.g. body, vocal,
and facial behaviours), and change in relation to changes in specific communicative contexts. According to a dynamical systems approach to emotions, it
would be inexact to conclude that the communication process elicits the emotion
in a causal–linear manner. Instead, we argue in this chapter that infants'
emotions and the different communication contexts in which they are involved
unfold together in time.

Neurophysiological components of infant emotions

Different brain structures (such as thalamus, hippocampus, and the cingulate
gyrus) and neurochemical processes (such as norepinephrine and serotonin)
function along with the autonomic nervous system and play a large part in the
experience of emotion. The most relevant findings to be highlighted in this
chapter are those centered on brain lateralization of emotions. Specifically,
researchers have consistently found that the left frontal area is associated with
approach-related emotions (or positive emotions) and the right frontal area with
withdrawal-related emotions (or negative emotion) (e.g. Davidson *et al.*, 1990;
Ekman *et al.*, 1990; Wheeler *et al.*, 1993).

The primary neurophysiological method used to measure lateral activation in
the cortex is the electroencephalogram (EEG). EEG recordings represent the
summation of neuronal activity within the cortex. These recordings also give
information about the functioning of the cortex under specific conditions. The
focus of the study of lateralization of emotion is on the alpha bands recorded
from the cortex (Kinsbourne & Hiscock, 1983; Parmelee & Sigman, 1983).

In studies looking at the lateralization of different positive and negative
emotions, investigators have focused on the shift between left and right
hemisphere alpha power. Increased activation of one hemisphere of the brain is
signalled by a decrease in power from one side of the cortex in relation to the
other side (Kinsbourne & Hiscock, 1983).

Individual differences in emotionality have prompted Wheeler *et al.* (1993) to
hypothesize that individuals with stronger left frontal activation will respond
more intensely to elicitors of positive emotion and less intensely to elicitors of
negative emotion. These investigations assess the associations between negative
and positive facial expressions and particular patterns of frontal cerebral activation using EEG.

Following up this hypothesis, Fox & Davidson (1986, 1987, 1988) measured
infants' left- versus right-side cerebral activation. In one of their studies (Fox &
Davidson, 1986), 2- and 3-day-old infants were administered distilled water, a

sucrose solution, and then a citric acid solution. Recordings of EEG were taken at baseline and for 30 seconds immediately following the presentation of the liquid. Thirty-three male and female newborns with two right-handed parents were initially recruited, and data analysis was performed on the 11 usable subjects. The newborns' faces were videotaped, and the facial expressions of disgust and interest were coded for the 30 seconds period of EEG recording. Results indicated that the three taste conditions elicited differential cerebral activation. Surprisingly, the water elicited comparable duration of disgust expressions to the citric acid solution and caused increased activation in the right frontal and parietal areas. The sucrose solution elicited the opposite cerebral activation in both the frontal and parietal regions.

In another study, 35 females 10 months old were tested in an approach paradigm with mother and stranger. EEG and video-recordings were taken during all phases of the approach by stranger and parent. With regard to cerebral activation, greater left-sided frontal activation occurred during mother reaching for infant (Fox & Davidson, 1987).

In their research with 10-month-olds, Fox & Davidson (1988) also found that coding infants' facial expressions with a more fine-grained system during artifact-free EEG recordings produced interesting findings. These investigators used an approach paradigm to elicit facial expressions in the infants. Mothers approached their infants with a smiling face and a stranger approached the infants with a neutral face. The investigators recorded brain activity from left and right frontal and parietal regions. A significant difference was found in EEG activity with greater left frontal activation during smiles that involved cheek raising. Fox & Davidson (1988) cautiously suggest that the increased left frontal activity during "felt" smiles (smiles with cheek raising) (Ekman & Friesen, 1982) could be associated with the infant's approach towards mother. These investigators surmised that a smile without raising the cheek constituted withdrawal or wariness of the stranger.

In sum, a body of literature exists offering insight into the neurophysiological basis of emotion in young human beings. Exposure to this literature makes it evident that investigations into the laterality of emotions are only one aspect of this broad topic of the neurophysiological aspects of emotions. We recognize that the use of electroencephalogram recordings of cerebral activation give only part of the picture of the effect of hemispheric activation on emotions. Research in this area has improved upon itself by tightening controls surrounding the testing of infants, implementing checks to make sure that a particular emotion is elicited, and by using only artifact-free data.

These findings are interpreted herein as illustrations of how different emotions self-organize from the joint cooperation of CNS activation (e.g. right versus left hemispheres), facial expressions (e.g. smiles with or without cheek raise), and communication contexts (e.g. mother versus stranger conditions). Certainly these findings can be interpreted as evidence for differential emotions theory. In our opinion, in order to better support either theory, research on the joint cooperation of CNS activation and facial expressions needs to be conducted at the levels of real-time and developmental-time changes simultaneously. Nevertheless, the studies reviewed in this section point out the complexity of the phenomenon of emotional development. Our sensitivity to this complexity is reflected, in part, by our appreciation for applying a dynamical systems approach to our developmental investigations of emotions.

Although we propose in this chapter that emotions are emergent from the relationship among different emotion components (including the brain), our contribution to this handbook centers on the examination of early emotional development at the behavioural level. Specifically, we focus on the changes in both the mother and the infant's emotional experiences as indicated by their faces, vocalizations, body postures, gestures, breathing activities, and the communication contexts in which these experiences are embedded. We do not intend to suggest, however, that there might be a behavioural primacy over neurophysiological components for understanding emotional development. We simply propose that examining how infants' emotions emerge in the context of mother–infant communication constitutes an important step for the advancement of the early emotional development literature. In the next paragraphs we illustrate our dynamical systems approach for studying emotions.

EMOTIONS AND MOTHER–INFANT COMMUNICATION

The existing literature on early emotional development has certainly provided great advances in our understanding of emotions and emotional development. However, the detailed analysis of the change processes involved in early emotional development has been relatively neglected. This chapter is an attempt to address this question using a dynamical systems approach to the investigation of emotional development. In order to present the theoretical application and methodological ramifications of this approach, we will offer excerpts from an ongoing longitudinal study.

We present a case from a larger sample of mother–infant dyads living in the western region of the United States. The mother was asked to visit a laboratory setting with her infant three times a week for a period of 16 consecutive weeks, starting when the infant was 10 weeks old. The laboratory playroom was furnished with carpeting, a sofa, blankets, and a toy-crate with age-appropriate toys. This intensive longitudinal data set allowed for the examination of the change processes implicated in emotional development, both at the levels of real time and developmental time. Eleven segments from the illustrative dyad are given. These segments (presented below in italics) were selected because they contain behavioural components (e.g. smiles and vocalizations) believed to be indicative of positive emotions.

This chapter focuses on how the different behavioural components of emotions self-organize into coherent patterns suggestive of emotional positivity. The purpose is to demonstrate how emotions gradually emerge as mother–infant dyads develop various communicative routines over time. In so doing we aim to encourage the use of systematic approaches to study developmental processes at the scales of real time and developmental time.

Nathan and Linda: An illustrative case

From the first visit we will present four segments in which Nathan, a 10-week-old infant, and Linda, the mother, appeared to be engaged in mutually gratifying experiences of being together. Within each of these first four segments a positive connection appears to be gradually co-created by this dyad. Smiles, positive vocalizations, mutual gazing, relaxed body postures cohere over time into an identifiable emotion pattern we refer to as "mutually gratifying connection".

Segment 1, Visit 1 (03:43–03:46)
Nathan has a pacifier in his mouth. Linda and Nathan are looking at one another, while she holds his arms, bringing them towards her, and kissing them. As Nathan drops his pacifier, which falls into Linda's lap, Linda stops kissing Nathan's hands, bringing his arms out, looking directly into his eyes by slightly leaning her head forward. Nathan protrudes his lips towards Linda, vocalizing, raising his eyebrows and moving his head up and down while Linda, raising the corners of her lips and the intonation of her voice, says "Hello!". Nathan continues protruding his lips towards Linda, vocalizing, raising his eyebrows and moving his head up and down as if talking to Linda. Meanwhile, Linda repeats Nathan's sounds, smiling at him.

Note in this first segment how Nathan and Linda gradually introduce behavioural components into their ongoing communication in relation to one another. For example, as Nathan drops his pacifier, Linda facilitates eye contact; as Nathan protrudes his lips towards Linda, Linda begins raising her lip corners and intonation of her voice. Within a period of 3 seconds the dyad begins to build up a mutually gratifying connection as their behaviours coalesce into a short-lived sequence that will become part of a pattern of positive emotion to emerge in later segments. In the next minute the following happens:

Segment 2, Visit 1 (04:48–05:16)
Nathan and Linda are again looking at one another. Nathan has his pacifier in his mouth. Linda has her eyebrows raised and softly talks to Nathan, rubbing her right hand on Nathan's stomach. As Linda continues rubbing her hand on Nathan's stomach, Nathan jerks his body, moving his left arm abruptly. At this point Linda exaggerates her facial expression, opening her mouth, raising her eyebrows even more and saying "Oh! Does that tickle?", while Nathan stares at Linda. Linda begins to gently tickle Nathan, raising the corners of her lips and continuing saying "Does that tickle?". Nathan begins vocalizing and grabbing his T-shirt while both maintain eye contact. Linda removes Nathan's pacifier from his mouth, smiling, opening her eyes wide, and whispering. At this point Nathan begins making cooing mouth movements, sometimes vocalizing, moving his head up and down, waving his left arm and stretching his trunk with his arms, while Linda continues smiling, whispering, and gently tickling Nathan.

This second segment lasts approximately 28 seconds. The dyad appears to begin re-configuring their mutually gratifying connection when Linda accidentally tickles Nathan while she rubs his stomach. Note how Linda makes a positive mock-surprise face (open mouth and raised eyebrows) as if punctuating the onset of the changes that might potentially emerge in their ongoing communication. Indeed, following this non-verbal punctuation a particular series of real-time changes partially identified in the previous segment is actualized by Nathan and Linda. We can observe mutual gazing, Nathan's vocal and mouth movements as well as Linda's smiling and widening of her eyes. Their mutually gratifying moments appear to constitute seeds of a pattern of positive emotion that is gradually coalescing and becoming part of their relationship.

During this second segment Nathan and Linda also innovate these moments of mutually gratifying connection with the addition of new behavioural components into their ongoing flow of communication, such as Linda's tickling and Nathan's stretching of his trunk with his arms. We interpret the enactment of these behaviours as indication that Nathan and Linda are beginning to develop and recognize a specific form of positive communication.

Forty seconds later another episode of this mutually gratifying connection is observed:

Segment 3, Visit 1 (05:47–05:52)
As Nathan begins making cooing mouth movements, without vocalizing and while looking at Linda intently, Linda looks at Nathan smiling (lip corners up). Nathan begins raising his lip corners, continuing to make cooing mouth movements and looking at Linda. Whenever Nathan smiles (lip corners up), Linda replies "Yeeeeeah!", raising her cheeks as well as the intonation of her voice, thereby forming a bigger smile (lip corners up and cheeks raised).

Although brief in duration (5 seconds), this third segment displays part of the behavioural components that constitute the moments of mutually gratifying connection previously experienced by Nathan and Linda. In fact this segment begins with Nathan making cooing mouth movements and Linda smiling while they both maintain eye contact. Shortly after, an innovation is incorporated by Nathan into their ongoing communication: Nathan also begins smiling. Linda appears to go along with Nathan: she innovates their mutually gratifying connection by raising her cheeks while smiling. We propose that the real-time changes observed in this third segment are reiterations, while at the same time transformations, of the previous moments of mutually gratifying connection Nathan and Linda develop together. They re-enact the mouth movements, mutual gazing, postural orientation, and intonation of voice previously observed as part of their mutually gratifying connection. Simultaneously, innovations are introduced into this quasi-stable pattern of behaviours, thereby transforming and reiterating it.

The last segment of mutually gratifying connection observed during the first visit occurs almost 2 minutes following the third segment:

Segment 4, Visit 1 (07:48–08:12)
Linda begins kissing Nathan's hands again, talking to him, raising her eyebrows and lip corners, while Nathan sucks on his pacifier, looking at Linda. Nathan begins producing long and loud sounds (as if giggling), as Linda begins gently stretching his arms, smiling and repeating his sounds back. As this stretching game goes on, Nathan drops his pacifier as he laughs with Linda, looking at her.

A similar set of behaviours observed in the first three segments recurs during this fourth segment. For example, Linda raises her eyebrows and lip corners while talking to Nathan and stretching his arms, and Nathan vocalizes while looking at Linda. This behavioural recurrence suggests the relative stability of these moments of mutually gratifying connection co-created by Nathan and Linda. On the other hand, more innovations of these moments are also observed. Particularly, Nathan produces long and loud sounds and begins laughing with Linda during the real-time emergence of this stretching game.

Two days later another three segments of the mutually gratifying connection are observed. These moments are shorter in duration as if indicating a change in the dyad's level of familiarity with these routines (see concept of abbreviation in Fogel & Lyra, 1997; Lyra & Rossetti-Ferreira, 1995; Lyra & Winegar, 1997). Nathan and Linda maintain their gaze with bodies oriented to one another. Their smiles become bigger (lip corners up and mouth open). In fact they open their mouths together as if recognizing a communicative routine. Linda maintains physical proximity throughout these moments, often touching and kissing Nathan. As illustrated in the next three segments, Nathan and Linda continue composing positive emotion patterns.

Segment 5, Visit 2 (03:51–03:52)
Nathan is sitting on Linda's lap, intently looking at Linda's shirt, while Linda vocalizes "tsh tsh tsh", raising her lip corners and moving her legs up and down as if trying to initiate a "horse" game. As Nathan shifts his gaze towards Linda's face, Linda exaggerates her smile and vocalizations, opening her eyes wide. Nathan begins opening his mouth with Linda, but then he immediately looks down, back to her shirt, with the same intent face as before. As Nathan shifts his gaze back to Linda's shirt, Linda lowers her lip corners and eyes while maintaining a subtle smile and vocalizations.

After about 6 minutes another short-lived episode of the mutually gratifying connection is observed:

Segment 6, Visit 2 (10:29–10:31)
Nathan is lying on Linda's lap, tonguing his lips with his eyes closed, while Linda observes him quietly. Linda begins kissing Nathan's hands, looking at him as Nathan continues tonguing his lips with his eyes closed. When Nathan opens his eyes he finds Linda directly in front of him, kissing his hands, leaning over him and looking at him. When Linda realizes that Nathan has opened his eyes, she abruptly stops kissing Nathan, moving his arms apart, raising her lip corners, and saying: "Ah!". In unison, Nathan opens his mouth and begins vocalizing while continuing to look at Linda.

Another brief segment recurs within the next minute. This segment is particularly unique because it emerges from a moment of negative emotion in which Nathan is distressed.

Segment 7, Visit 2 (11:35–11:36)
Nathan is looking at Linda with a relaxed body and face, holding his hands together, while Linda vocalizes at him "oh oh oh", resting her hands on his hands. Linda begins saying "It's ok!" several times, while patting Nathan's arms. Meanwhile, Nathan keeps looking at her, opening his mouth, raising his lip corners, and relaxing his arms.

Note how mutual gazing, postural orientation, change in the voice, vocalizations, and mutual smiling – behavioural components co-enacted during the dyad's first visit – quickly and smoothly become part of Linda and Nathan's ongoing communication in this second visit. Further, mutual mouth opening becomes integrated into their communication. We interpret these recurrences (i.e. the behavioural components observed in the first visit) and innovations (i.e. the additional components observed in the second visit) of the moments of mutually gratifying connection as indicative that Linda and Nathan recognize the emergence and development of an emotionally positive and meaningful communicative routine. Through these narrative descriptions of the real-time changes in Nathan and Linda's communication we illustrate the building up of a particular emotion over developmental time. We do so by emphasizing the movement through which the dyad's vocalizations, postures, touches, smiles and laughter start to coalesce into a recognizable pattern of positive emotion we refer to as mutually gratifying connection.

During their third visit, four segments of the mutually gratifying connection are observed. Yet Linda and Nathan continue transforming these moments. Mutual laughter becomes more systematically incorporated into their communicative routines. The positivity experienced by Nathan and Linda appears to intensify. Dynamic stability is an integral part of the process of emotional development of this dyad.

Note how Nathan begins initiating their moments of mutually gratifying connection more consistently and explicitly: he gazes at Linda, protruding his tongue (see segment 8); or he begins vocalizing, raising his lip corners while looking at Linda (see segment 9). By introducing mutually identifiable behavioural components of positive emotion into their ongoing communication, Nathan not only appears to recognize the meaningful routines co-created with Linda, he also explicitly contributes to the re-occurrence of these routines.

Segment 8, Visit 3 (00:32–00:37)
While Linda finishes adjusting Nathan's posture into his usual supine position for facilitating eye-contact, Nathan looks at Linda, sticking his tongue out. As Nathan begins looking at Linda, Linda holds his arms, slightly leaning over him, raising her lip corners and says "Boysenberries" [referring to the drawing on her shirt that Nathan was looking at seconds before]. Nathan continues looking at Linda, raising his lip corners, while Linda continues looking at Nathan, repeating "Boysenberries", moving his arms while raising the intonation of her voice. When Linda finishes saying "berries", both Nathan and Linda begin opening their mouths together while maintaining eye-contact and their lip corners raised, thereby forming a bigger smile. While Nathan holds his big smile on his face, looking at Linda, Linda closes her mouth, maintaining her lip corners raised and moving Nathan's arms as she says again "Boysenberries" with the same melodic voice. Nathan produces a long and loud sound, opening his mouth even more and moving his tongue inside of it, while his lip corners remain raised. At this point Linda watches Nathan, raising her eyebrows and protruding her lips as if cooing at him.

Within the next 20 seconds, a second segment is briefly observed:

Segment 9, Visit 3 (00:54–00:55)
As Nathan begins vocalizing again, raising his lip corners and looking at Linda, Linda raises her eyebrows and lip corners, slightly tilting her head to her right side, touching Nathan's hands and saying "It is!".

Another 2½ minutes have passed. A third and short-lived segment recurs. Of particular note, this segment emerges as part of a moment in which Linda attempts to help Nathan stay calm and relaxed.

Segment 10, Visit 3 (02:26–02:27)
Nathan begins vocalizing while sucking on his pacifier, looking at Linda, raising his inner eyebrows and slightly raising his lip corners. At the same time Linda also raises her eyebrows and lip corners, continuing to hold the pacifier in Nathan's mouth with her index finger, whispering.

And finally, a minute later, a relatively longer segment is observed:

Segment 11, Visit 3 (03:14–03:24)
Nathan is looking at Linda, sucking on his pacifier, while Linda gently plays with Nathan's feet, rubbing them together. As Linda begins rubbing Nathan's feet more vigorously, she begins making synchronized sounds "tsh tsh tsh" with her movements, looking at Nathan, while Nathan continues looking at Linda with a relaxed face and body, sucking on his pacifier. Nathan produces a long sound, looking at Linda with a relaxed face, while Linda continues rubbing his feet together and moving his legs up and down, saying "tsh tsh tsh", raising her lip corners and showing her teeth. At this point Nathan gradually becomes more engaged in this rubbing the feet game, to the point of dropping his pacifier which falls into Linda's lap, as he vocalizes. Linda continues rubbing his feet together, smiling and softly talking to Nathan, as Nathan begins raising his lip corners while looking at Linda with relaxed body, vocalizing, and tonguing.

In sum, both Nathan and Linda changed their behaviours in relation to one another, thereby forming and transforming their moments of "mutually gratifying connection" at the scale of real time. We propose that, at the same time, Nathan and Linda developed a communicative pattern composed of reciprocal gazing, different forms of smiling (lip corner raise and lip corner raise accompanied by mouth opening or cheek raise), vocalizations, postural orientation, and physical proximity. This specific patterning of behaviours informed the dyad of their mutually positive experience. From this process the dyad was exhibiting what we refer to as developmental-time changes in their emotions.

With these detailed narratives of the mutually gratifying moments between Nathan and Linda, we attempted to illustrate our application of a dynamical systems perspective to examine how emotions (specifically, variations of positive emotion) self-organize over the course of real time and developmental time. It is worth repeating that emotions were conceptualized as patterns of activity providing the mother–infant dyad with information about the meaning of their relationship.

CONCLUSION

We contend that our use of the dynamical systems approach has strong potential to continue yielding a comprehensive understanding of emotional development. The analysis presented above focused on positive emotions in the social contexts co-created by a mother and her young infant. The data seemed to support our view of emotions as emergent and relational processes. The patterns of positive emotions described in our narrative data encompassed infant smiles that involve lip corner raise (referred to as non-Duchenne smile) as well as smiles composed of lip corner raise and mouth opening (referred to as play smile). The data also provided evidence that different behaviours, including the mother's, emerged in conjunction with the infant smiles. Reciprocal gaze, cooing-like mouth movements, postural orientation, vocalizations, laughter, fluctuations in the intonation of the voice, and physical proximity were integral components of the pattern of positive emotion. The mother also displayed play smiles as well as smiles involving lip corner raise and the contraction of the cheeks around the eyes (referred to as Duchenne smiles) as part of the emotion pattern under investigation.

We claim that important contributions to the field have been provided by differential emotions theory and functionalist theory. Various studies on adults and infants sustain the notion that discretely distinct configurations on the face are observed, as predicted by differential emotions theory (e.g. Ekman & Friesen, 1982). The work of other scholars (e.g. Barrett, 1998) has also provided supporting evidence towards the view of emotions as communicative devices united by a common function, as maintained by functionalist theory. Our dynamical systems approach emphasizes that emotions and mother–infant communication are inseparable processes that flow together in the day-to-day occurrences of human beings. When examined through continuous real-time and microscopic analysis, our investigations reveal the dynamic nature of the phenomenon of emotional development.

For those developmental researchers concerned with how emotions self-organize and change over time, the utilization of the dynamical systems approach may be helpful. From this handbook, it becomes apparent that many theoretical variations and methodological applications of dynamical systems exist. Our approach is most unique in its commitment to describing in great detail the processes of change that occur in emotional development and in its utilization of qualitative–narrative methods. Our viewpoint is that a fuller understanding of human development is derived from an openness to the contributions of different scientific approaches. We encourage an avenue that embodies the use of qualitative and quantitative methods. We also invite those interested in neurological measures of emotions, such as the work on brain lateralization, to integrate into their work repeated observations of infants with their primary caregivers in more naturalistic social contexts. This, we hope, will bring the field to a novel understanding of development.

References

Barrett, K.C. (1993). The development of nonverbal communication of emotion: a functionalist perspective. *Journal of Nonverbal Behavior*, **17**, 145–169.

Barrett, K.C. (1998). A functionalist approach to the development of emotion. In M.F. Mascolo & S. Griffin (Eds.), *What develops in emotional development?* (pp. 109–133). New York: Plenum.

Campos, J.J., Mumme, D.L., Kermoian, R. & Campos, R.G. (1994). A functionalist perspective on the nature of emotion. Commentary on The development of emotion regulation: biological and behavioral considerations. *Monographs of the Society for Research in Child Development*, **59**, 284–303.

Camras, L.A. (1991). Conceptualizing early infant affect. In K. Strongman (Ed.), *International review of studies on emotion* (pp. 16–28). New York: Wiley.

Davidson, R.J., Ekman, P., Saron, C., Senulis, J. & Friesen, W. (1990). Approach–withdrawal and cerebral asymmetry: emotional expression and brain physiology I. *Journal of Personality and Social Psychology*, **58**, 330–341.

Demos, E.V. (1982). Facial expressions of infants and toddlers: a descriptive analysis. In T. Field & A. Fogel (Eds.), *Emotion and early interaction* (pp. 127–160). Hillsdale, NJ: LEA.

Demos, E.V. (1988). Affect and the development of the self: a new frontier. In A. Goldberg (Ed.), *Frontiers in self psychology progress in self psychology* (vol. 3, pp. 27–53). Hillsdale, NJ: Analytic Press.

Ekman, P. (1992). An argument for basic emotions. *Cognition and Emotion*, **6**, 169–200.

Ekman, P. (1994). All emotions are basic. In P. Ekman and R.J. Davidson (Eds.), *The nature of emotion: fundamental questions* (pp. 15–19). New York: Oxford University Press.

Ekman, P. (1995). Silvan Tomkins and facial expression. In V. Demos (Ed.), *Exploring affect: the selected writings of Silvan S. Tomkins* (pp. 209–214). New York: Cambridge University Press.

Ekman, P. & Friesen, W. (1975). *Unmasking the Face*. Englewood Cliffs, NJ: Prentice-Hall.

Ekman, P. & Friesen, W. (1978). *Manual for the facial affect coding system*. Palo Alto, CA: Consulting Psychologists Press.

Ekman, P. & Friesen, W.V. (1982). Felt, false, and miserable smiles. *Journal of Non-verbal Behavior*, **6**, 238–252.

Ekman, P., Davidson, R. J. & Friesen, W. (1990). The Duchenne smile: emotional expression and brain physiology. II. *Journal of Personality and Social Psychology*, **58**, 342–353.

Fogel, A. (1993). *Developing through relationships*. Chicago, IL: University of Chicago Press.

Fogel, A. (1997). *Infancy: infant, family, and society* (3rd ed.). St. Paul, MN: West Publishing Co.

Fogel, A. & Lyra, M.C.D.P. (1997). Dynamics of development in relationships. In F. Masterpasqua & P. Perna (Eds.), *The psychological meaning of chaos: translating theory into practice* (pp. 75–94).

Fogel, A., Dickson, K.L., Hsu, H., Messinger, D.S., Nelson-Goens, G.C. & Nwokah, E. (1997). Communication of smiling and laughter in mother–infant play: research on emotion from a dynamic systems perspective. In K. Barrett (Ed.), *Emotion and communication* (pp. 5–24). San Francisco, CA: Jossey-Bass.

Fogel, A., Nwokah, E., Dedo, J., Messinger, D.S., Dickson, K. L., Matusov, E. & Holt, S. A. (1992). Social process theory of emotion: a dynamic systems perspective. *Social Development*, **1**, 122–142.

Fox, N.A. & Davidson, R.J. (1986). Taste-elicited changes in facial signs of emotion and the asymmetry of brain electrical activity in human newborns. *Neuropsychologia*, **24**, 417–422.

Fox, N.A. & Davidson, R.J. (1987). Electroencephalogram asymmetry in response to the approach of a stranger and maternal separation in 10 month old infants. *Developmental Psychology*, **23**, 233–240.

Fox, N.A. & Davidson, R.J. (1988). Patterns of brain electrical activity during facial signs of emotion in 10 month old infants. *Developmental Psychology*, **24**, 230–236.

Izard, C.E. (1997). Emotions and facial expressions: a perspective from differential emotions theory. In J.A. Russel & J.M. Fernandez-Dols (Eds.), *The psychology of facial expression* (pp. 57–77). New York: Cambridge University Press.

Kinsbourne, M. & Hiscock, M. (1983). The normal and deviant development of functional lateralization of the brain. In P. Mussen (series Ed.), and M.M. Haith & J.J. Campos (vol. Eds.), *Handbook of child psychology*: vol 2: *Infancy and developmental psychobiology* (pp. 157–280). New York: Wiley.

Lewis, M.D. (1995). Cognition–emotion feedback and the self-organization of developmental paths. *Human Development*, **38**, 71–102.

Lyra, M.C.D.P. (2001). Desenvolvimento de um sistema de relações historicamente construído: Contribuições da comunicação no início da vida. *Psicologia: Reflexão e Crítica* (In press).

Lyra, M.C.D.P & Rossetti-Ferreira, M.C. (1995). Transformation and construction in social interaction: a new perspective of analysis of the mother–infant dyad. In J. Valsiner (Ed.), *Child development within culturally structured environment: comparative–cultural and constructivist perspectives* (pp. 51–77). Hillsdale, NJ: LEA.

Lyra, M.C.D.P. & Winegar, L.T. (1997). Processual dynamics of interaction through time: Adult-child interactions and process of development. In A. Fogel, M.C.D.P. Lyra & J. Valsiner (Eds.), *Dynamics and indeterminism in developmental and social processes* (pp. 93–109). Hillsdale, NJ: LEA.

Messinger, D.S., Fogel, A. & Dickson, K.L. (1997). A dynamic systems approach to infant facial action. In J.A. Russell & J.M. Fernandez-Dols (Eds.), *The psychology of facial expression* (pp. 205–226). New York: Cambridge University Press.

Messinger, D.S., Fogel, A. & Dickson, K. L. (1999). What's in a smile? *Developmental Psychology*, **35**, 701–708.

Parmelee, A. H. & Sigman, M. (1983). Perinatal brain development and behavior. In P. Mussen (series Ed.), and M.M. Haith & J.J. Campos (vol. Eds.), *Handbook of child psychology*: vol 2: *Infancy and developmental psychobiology* (pp. 95–155). New York: Wiley.

Thelen, E. & Ulrich, B.D. (1991). Hidden skills: a dynamic systems analysis of treadmill stepping during the first year. *Monographs of the Society for Research in Child Development*, **56**.

Tomkins, S.S. & McCarter, R. (1995). What and where are the primary effects? Some evidence for a theory. In V. Demos (Ed.), *Exploring effect: The selected writings of Silvan S. Tomkins* (pp. 217–262). New York: Cambridge University Press.

Weinberg, M.K. & Tronick, E.Z. (1994). Beyond the face: an empirical study of infant affective configurations of facial, vocal, gestural, and regulatory behaviors. *Child Development*, **65**, 1503–1515.

Wheeler, R.E., Davidson, R.J. & Tomarken, A.J. (1993). Frontal brain asymmetry and emotional reactivity: a biological substrate of affective style. *Psychophysiology*, **30**, 82–89.

Wolff, P.H. (1987). *The development of behavioral states and the expression of emotions in early infancy*. Chicago, IL: University of Chicago Press.

Further reading

Camras, L.A. (1991). Conceptualizing early infant affect. In K. Strongman (Ed.), *International review of studies on emotion* (pp. 16–28). New York: Wiley.

Dickson, K.L., Fogel, A. & Messinger, D. (1997). The social context of smiling and laughter in infants. In M. Mascolo & S. Griffen (Eds.), *The development of emotion* (pp. 253–273). New York: Plenum Press.

Fogel, A. (1993). *Developing through relationships*. Chicago, IL: University of Chicago Press.

Fogel, A. & Lyra, M.C.D.P. (1997). Dynamics of development in relationships. In F. Masterpasqua & P. Perna (Eds.), *The psychological meaning of chaos: translating theory into practice* (pp. 75–94). Washington, DC: APA.

Fogel, A., Dickson, K.L., Hsu, H., Messinger, D., Nelson-Goens, C. & Nwokah, E. (1997). Communication of smiling and laughter in mother–infant play: research on emotion from a dynamic systems perspective. In K. Caplovitz Barrett (Ed.), *New Directions in Child Development: The communication of emotion: Current research from diverse perspectives*, **77** (pp. 5–24). San Francisco, CA: Jossey-Bass.

Fogel, A., Nwokah, E., Dedo, J., Messinger, D.S., Dickson, K.L., Matusov, E. & Holt, S. A. (1992). Social process theory of emotion: a dynamic systems perspective. *Social Development*, **1**, 122–142.

Frijda, N.H. (1993). Moods, emotion episodes, and emotions. In M. Lewis & J.M. Haviland (Eds.), *Handbook of Emotions* (pp. 381–403). New York: The Guilford Press.

Frijda, N.H. & Tcherkassof, A. (1997). Facial expressions as modes of action readiness. In J.A. Russell & J.M. Ferandez-Dols (Eds.), *The psychology of facial expression* (pp. 78–98). New York: Cambridge University Press.

Gottman, J.M. (1993). Studying emotion in social interaction. In M. Lewis & J.M. Haviland (Eds.), *Handbook of Emotions* (pp. 475–487). New York: The Guilford Press.

Haviland, J.M. & Kahlbaugh, P. (1993). Emotion and identity. In M. Lewis & J.M. Haviland (Eds.), *Handbook of Emotions* (pp. 327–339). New York: The Guilford Press.

Kellert, S.H. (1993). *In the wake of chaos: unpredictable order in dynamical systems*. Chicago, IL: University of Chicago Press.

Lewis, M.D. (1995). Cognition–emotion feedback and the self-organization of developmental paths. *Human Development*, **38**, 71–102.

Lewis, M.D. (1997). Personality self-organization: cascading constraints on cognition-emotion interaction. In A. Fogel, M.C.D.P. Lyra, & J. Valsiner (Eds.), *Dynamics and indeterminism in developmental and social processes* (pp. 193–216). Hillsdale, NJ: LEA.

Lyra, M.C.D.P. (2001). Desenvolvimento de um sistema de relações historicamente construído: Contribuições da comunicação no início da vida. *Psicologia: Reflexão e Crítica* (In press).

Lyra, M.C.D.P & Rossetti-Ferreira, M.C. (1995). Transformation and construction in social interaction: a new perspective of analysis of the mother–infant dyad. In J. Valsiner (Ed.), *Child development within culturally structured environment: comparative–cultural and constructivist perspectives* (pp. 51–77). Hillsdale, NJ: LEA.

Lyra, M.C.D.P. & Winegar, L.T. (1997). Processual dynamics of interaction through time: Adult-child interactions and process of development. In A. Fogel, M.C.D.P. Lyra & J. Valsiner (Eds.), *Dynamics and indeterminism in developmental and social processes* (pp. 93–109). Hillsdale, NJ: LEA.

Messinger, D.S., Fogel, A. & Dickson, K.L. (1997). A dynamic systems approach to infant facial action. In J.A. Russell & J.M. Fernandez-Dols (Eds.), *The psychology of facial expression* (pp. 205–226). New York: Cambridge University Press.

Messinger, D. S., Fogel, A. & Dickson, K. L. (1999). What's in a smile? *Developmental Psychology, 35*, 701–708.

Thelen, E. & Ulrich, B.D. (1991). Hidden skills: a dynamic systems analysis of treadmill stepping during the first year. *Monographs of the Society for Research in Child Development, 56*.

White, G.M. (1993). Emotions inside out: the anthropology of effect. In M. Lewis & J.M. Haviland (Eds.), *Handbook of Emotions* (pp. 29–39). New York: The Guilford Press.

V.6
Neural Bases and Development of Face Recognition During Infancy

MICHELLE DE HAAN and HANIFE HALIT

ABSTRACT

The human face is a complex visual stimulus with clear biological significance. Visual processing of faces in adults appears to be mediated
by specialized systems that are distinct from those used to process
most other objects. A question of fundamental interest to developmentalists is how the constraints imposed by the structure of the brain
interact with input from the environment during ontogeny to create the
pattern of specialization observed in the mature system. In this chapter
we review several hypotheses of the mechanisms of development of
face-processing systems during infancy and evaluate evidence relevant
to these hypotheses. We conclude that the development of face-processing during infancy is best characterized as a gradual specialization
of an initially more general-purpose processing system.

INTRODUCTION

The human face is typically one of a baby's first sights following birth, and from
this time babies show an interest in orienting to face-like patterns (Goren *et al.*,
1975; Johnson *et al.*, 1991; Maurer & Young, 1983; Valenza *et al.*, 1996). This
precocious ability seems adaptive for young infants who rely on caretakers for
their survival (Bowlby, 1958). However, in spite of this impressive early
competency, face processing follows a protracted developmental course before
becoming adult-like (Campbell & Tuck, 1995; Johnston & Ellis, 1995).
Developmental theories of face processing must be able to explain both the
infant's early sophistication in face processing and how time and experience
moulds these processing mechanisms into the specialized, localized ones
observed in the adult. Since adult-like abilities do not appear all at once fully
mature, the developmental approach can give insight into what abilities are most
fundamental and allow us to study certain components of the system in isolation

A.F. Kalverboer and A. Gramsbergen (eds.), Handbook of Brain and Behaviour in Human Development, 921–938
© *2001 Kluwer Academic Publishers. Printed in Great Britain.*

in a way that is not possible with adults. This is important both to understand the assembly of the face-processing system itself and to develop a more general model of the development of specialization of function in the brain. In this chapter we outline current views on the development of face processing during the first year of life, and examine how infants' skills begin to be transformed into the expert ones of the adult.

THEORETICAL BACKGROUND: NEWBORNS

Most theorists would agree on at least two aspects of the development of face processing: (1) in the first months of life face processing is mediated primarily subcortically, and (2) at approximately 6–8 weeks of age or later there is a major transition in face processing attributable to the functional emergence of cortical mechanisms. These are points of agreement for both theories of the development of infants' recognition of facedness (i.e. that a stimulus is a face and not another type of object), facial identity (de Schonen & Mathivet, 1989; Morton & Johnson, 1991) and facial emotion (Nelson & de Haan, 1997). The main points of disagreement are: (1) the degree to which the systems involved are specific to faces, and (2) the role of experience in shaping their development.

Sensory hypothesis

According to one view, newborns' responses to visual stimuli are based on the visibility of the stimuli to the infant. In this hypothesis there is no aspect of the system that is responding specifically to faces; instead the preferential orienting of the newborn to faces is merely a consequence of more general mechanisms guiding visual attention. For example, in the linear systems model of infant visual attention, visual preferences in the first months of life are based on the amplitude spectra of the stimuli (the amount of energy in a pattern, defined by the amplitudes and orientations of the component spatial frequencies; Banks & Ginsburg, 1985; Banks & Salapatek, 1981). Infants will prefer a pattern that has more energy at the spatial frequencies to which they are more sensitive. This model is quite good at predicting many of an infant's visual preferences, especially for high-contrast stimuli, in the first months of life (Banks & Ginsburg, 1985; Banks & Salapatek, 1981). With respect to faces, the model predicts that the newborn will prefer to look at a face than at another pattern if the face has an amplitude spectrum that makes it the more visible pattern of the two (Kleiner & Banks, 1987). In the infant's natural visual environment the human face has properties (e.g. large, high-contrast features) that tend to make it a relatively highly visible stimulus (Slater, 1998).

Social hypothesis

According to a different view infants have an innate preference for face-like stimuli that is based not only on the visibility of the stimuli but on a more specific knowledge of the configuration of the face. For example, in the Conspec/Conlern hypothesis an innate, subcortical mechanism, Conspec, causes newborns to orient to patterns with elements in a face-like arrangement (Morton & Johnson, 1991). Empirical studies suggest that the Conspec mechanism is active and leads to preferential orienting to faces when stimuli are presented in the peripheral vision or are moving, but not when stimuli are presented in central vision (Morton & Johnson, 1991; Valenza *et al.*, 1996).

REVIEW OF THE LITERATURE: NEWBORNS

Since visual experience begins at birth, studying the newborn's ability to perceive and remember faces is of considerable interest because it allows us to assess what representational biases the infant is born with that might guide subsequent learning/experience. Newborn babies will move their eyes, and sometimes their heads, longer to keep a moving face-like pattern in view than several other non-face patterns (Goren *et al.*, 1975; Johnson *et al.*, 1991; Maurer & Young, 1983). It seems that all that is needed to elicit this response is a very schematic version of the face: merely a triangular arrangement of three blobs for eyes and a mouth is sufficient (Johnson & Morton, 1991). This result alone is compatible both with the sensory hypothesis and the social hypothesis of newborn face processing. However, the observations that babies: (1) will orient more towards a face than a non-face stimulus that has the optimal spatial frequency for the newborn visual system (Valenza *et al.*, 1996) and (2) will orient more towards a stimulus with phase of face than non-face even when the amplitude information is equal (Kleiner & Banks, 1987; Mondloch *et al.*, 2001) cannot be explained by the sensory hypothesis. Since infants show this response as little as hours after birth it is either a very quickly learned or a congenital representational bias. The possibility that experience plays some role in this response cannot be ruled out, since the face is one of the first sights of the typical newborn (the only study that did not allow any exposure had other methodological probems such as the experimenter who held the infants not being blind to the stimuli; Goren *et al.*, 1975). One test to assess whether learning is involved in the preferential orienting to faces would be to examine whether infants would show a similar preferential orienting to moving or peripheral familiar stimuli compared to unfamiliar stimuli as they do for the face over other patterns. If they did not, it would support the view that the orienting is not only a general orienting to a familiar visual pattern, but does represent a more specific congenital bias for the face.

Consistent with both the sensory and social hypotheses, the orienting of newborns to face-like stimuli in the visual environment appears to be subcortically mediated. One indication of this is that newborns show the preference only in conditions to which the subcortical systems are sensitive (when stimuli are moving or are in the peripheral visual field but not when they are in the central visual field). Further experimental evidence in support of this view is that infants orient more towards face-like patterns than inverted face patterns in the temporal, but not the nasal, visual field (Simion *et al.*, 1998). Since the retinotectal (subcortical) pathway is thought to have greater input from the temporal hemifield and less input from the nasal hemifield than the geniculostriate (cortical) pathway, this asymmetry in the preferential orienting to faces is consistent with subcortical, but not cortical, involvement. The lack of preferential orienting to faces in the nasal field is not simply due to a general lack of sensitivity, since in this same visual field there was increased responding to an optimal spatial frequency non-face stimulus over another non-face stimulus (Simion *et al.,* 1998).

Not only do newborns preferentially orient to faces, but they also look longer at the mother's face than a stranger's face as early as hours to days after birth (even when cues from her smell and her voice are eliminated; Bushnell *et al.,* 1989; Field *et al.,* 1984; Pascalis *et al.,* 1995). This observation represents a difficulty both for the sensory and for the social hypothesis, since it demonstrates that from the beginning there is a mechanism involved in face-processing that is sensitive to experience with specific faces. To account for the preference for the mother's face, the sensory hypothesis would have to postulate that each mother's face was more visible than the stranger's face, which is highly unlikely, if not impossible. The social hypothesis cannot account for this at all, since it does not predict that one face should be preferred over another if both have a face-like configuration. Thus, both views must postulate the existence of an additional mechanism capable of learning about individual faces based on experience. This early learning may be based on a hippocampal system (Johnson & de Haan, 2001; Nelson, 1995), since the hippocampus is known be involved in memory, to functionally mature early relative to memory-related neocortical areas, and to mediate memory in visual paired comparison tasks in infant monkeys (the same type of task used to assess memory in human infants; Webster *et al.,* 1995).

While newborns are able to learn to recognize individual faces, their representations of faces may rely less, or perhaps not at all, on the internal facial features, and instead be based on the external features of the face such as the contour and headline. This was demonstrated in a study showing that newborns look longer at the mother's face than a stranger's face when her full face is presented, but not when only her internal facial features are presented (Pascalis *et al.,* 1995). This pattern is opposite to that of adults, for whom the internal features are more important for recognition of familiar faces (Ellis *et al.,* 1979). It may be that the external regions of the face are high-contrast regions of the face which are more visible to the newborn than are the internal facial features. This interpretation would be consistent with the "externality effect" reported in other studies of infant scanning of patterns at this age. These studies show that before 2 months of age infants tend to scan a frame that surrounds an inner shape rather than the shape itself (Ganon & Swartz, 1980; Maurer & Salapatek, 1976). This effect extends to faces, in that newborns tend to scan the external contour, rather than the inner features, of a static face (Maurer & Salapatek, 1976; but see Maurer, 1983). However, there is some evidence that newborns are also sensitive to the internal features of the face. In two studies (Field *et al.,* 1982, 1983), 36-hour-old infants were presented with a happy, sad or surprised expression posed by a live female model until they looked for less than 2 seconds, and then saw the other two expressions presented in the same way. Infant's looking times increased when the expression changed, suggesting that they could tell the expressions apart. While this study suffers from some methodological problems (e.g. there was no comparison group tested with the same procedure without changing the expressions to see if the increased looking was due to the change in expression or only to movement of the face, etc.), if replicated with a better-controlled procedure it would suggest that, at least in live faces, some information from the internal features is encoded. Moreover, it is known that the externality effect can be overcome by making the internal features of a pattern more salient – for example, by increasing the contrast of the features or moving them (Bushnell, 1979). In real-life learning situations it is possible that infants do in fact encode internal facial features because they are made more salient through such factors as the motion that occurs while a person speaks or expresses emotion. Thus, it might be only under certain laboratory conditions that newborns show a lack of recognition of the internal facial features.

In summary, the available studies suggest that there is a subcortical mechanism present early in life that has an innate or quickly learned knowledge about the structure of the face and acts to preferentially direct young infants' attention to face-like patterns. In addition, there is a mechanism that mediates learning of individual faces. There is conflicting evidence as to the type of facial information that is encoded. It seems most parsimonious to conclude that the infant is most likely to encode and recognize the features that are most salient in a given display (which might vary from experiment to experiment with differences in procedures, e.g. live model versus photos; static versus moving, etc.).

THEORETICAL BACKGROUND: TRANSITION AT 2 MONTHS

According to both the sensory and the social hypotheses, by the second month of life a transition in infant face processing occurs due to cortical development. The specificity of the emerging cortical system for face processing and the role of experience in shaping its development are matters of debate. Three possible scenarios of development of the cortical system are discussed below.

Experience-independent hypothesis

According to one view, face processing is mediated by an innate, domain-specific cortical system (Fodor, 1983; Farah et al., 2000). In this view the cortical system is specific for faces from the time it emerges, and there is no major role for visual experience in its assembly. The only role allowed for experience is that it may act to trigger or "switch on" the domain-specific system. In this view, changes in face-processing skills with age are explained by more general improvements in visual inputs, increases in processing speed, or emergence of new input/output connections of the system.

Experience-expectant hypothesis

According to a different view development can be explained in terms of a process in which there is a neurological preparation for a species-typical experience (Greenough & Black, 1992). In this hypothesis experience plays a critical role in the assembly of the cortical system for face processing but the timing of the experience is critical. The neural system is thought to remain open to environmental input for only a limited period, and if the necessary inputs fail to occur, or are atypical, the "window of opportunity" is closed and development follows an abnormal trajectory. Thus, in this view structure within the environment that is typically common to all members of the species (e.g. seeing patterned light, seeing faces) is allowed to shape development so that not all aspects must be pre-specified and regulated by the genome. Three hypotheses about face recognition, which are not mutually exclusive, fall into this category. What is common among the hypotheses is that they all state that visual experience must occur at particular times in development in order for face processing to develop normally.

In the Conspec/Conlern hypothesis it is argued that one purpose of the subcortical Conspec mechanism may be to provide a "face-biased" input into the developing cortical systems over the first months of life (Morton & Johnson, 1991). In this way cortical systems become specialized for face processing mainly because faces are a dominant input in the first weeks of development. In his perceptual narrowing hypothesis, Nelson (1995) has argued that the mechanism of specialization of the cortical face-processing system may be analogous to that of cortical speech-processing systems. In development of speech processing, infants initially are tuned to a rather broad range of speech contrasts, but with time and experience this window narrows. For example, 6-month-old infants can discriminate speech sounds from both native and non-native languages; 12-month-olds and adults can discriminate only the native language contrast (Werker & Lalonde, 1988; Werker & Tees, 1984). In the same way the useful cues for human face recognition may initially be part of a larger class of stimuli/cues (e.g. there may be a very broad prototype or wide face space of "visual object" or "any face"). With time and experience the category might be narrowed to include only human faces, and then narrowed even further to include only cues that are relevant to specific aspects of face processing (e.g. identity, emotion). One strength of this view is that it is the only one to explicitly address the question of how different aspects of face processing (e.g. identity, emotion, facial speech) come to be processed by different neural systems in adults (Campbell *et al.*, 1996).

While the Conspec/Conlern and perceptual narrowing hypotheses provide an explanation for why face processing might come to differ from object processing, these models do not provide an explanation for some of the specific character-istics of the adult face-processing system. In their differential rates of development hypothesis of de Schonen and colleagues (de Schonen & Mathivet, 1989; de Schonen *et al.*, 1986) provide the most specific developmental account for two characteristics observed in the adult face-processing system: the right hemisphere advantage for faces (reviewed in Rhodes, 1985), and the bias for configural, rather than feature-based, encoding of faces (Farah, 1990; Rhodes *et al.*, 1998). De Schonen argues that these aspects of specialization arise because of three things that coincide during development in the first months of life: (a) babies are first learning about faces, (b) babies' visual systems are more sensitive to low than high spatial frequency visual input, and (c) the right hemisphere is developing more quickly than the left. As a consequence of these factors occurring simultaneously, the right hemisphere becomes specialized for representing faces configurally.

Experience-dependent hypothesis

According to another view experience plays an important role in shaping the cortical system for face recognition, but the mechanism is viewed as a general learning process which can occur in the same way at any time in development and for non-face classes of visual stimuli . For example, Diamond & Carey (1986; see also Gauthier & Tarr, 1997) have argued that the development of expertise in face processing is no different from development of expertise in processing other classes of visual stimuli. The reason why face processing appears "special" is only because in everyday life there may be few other classes of stimuli for which, as for faces, we are required to discriminate among and recognize such a large number of highly similar exemplars. However, under certain circumstances individuals may acquire a similar expertise for a non-human face class of stimuli, such as dogs, cows, or even artifical patterns. What differentiates this view from the experience-expectant view is that it views the system as equally sensitive to input throughout the lifespan.

REVIEW OF THE LITERATURE: TRANSITION AT 2 MONTHS

There is ample evidence that there is a major change in infant visual processing at approximately 2 months of age, involving functional emergence of cortical visual systems (Johnson, 1990). For example, under monocular viewing conditions newborns more frequently orient towards stimuli in the temporal than in the nasal hemifield (Lewis & Maurer, 1992), and they show optokinetic nystagmus more readily to stimuli moving in the temporal to nasal than in the nasal to temporal direction (Teller *et al.*, 1983). Since the retinotectal (subcortical) pathway is thought to have greater input from the temporal hemifield and less input from the nasal hemifield than the geniculostriate (cortical pathway), these asymmetries indicate greater subcortical than cortical control of vision in the newborn. However, by around 2 months of age infants are more dominated by cortical visual processing; therefore their responses are more frequently determined by geniculocortical (nasal hemifield) input such that nasal–temporal asymmetries diminish (Johnson, 1990). The newborn preferential tracking to faces appears to decline around 6 weeks of age (Johnson *et al.*, 1991) and by 2–3 months of age babies begin to show a preference for faces with features naturally arranged over faces with features scrambled, and for a face with normal contrast over a contrast-reversed face (Danemiller & Stephens, 1988) in the central visual field (Maurer & Barerra, 1981).

Early specialization for face processing

As part of the numerous changes that occur in infant vision and memory at 2 months, are there specific changes in face processing? The experience-independent hypothesis predicts that specialization of face processing should be present from the time the cortical face-processing system emerges; in contrast, the experience-expectant and experience-dependent hypotheses would predict a more gradual emergence of specialization. In this section we will examine four experimental approaches to investigating the specialization of early face processing.

Changes in function

One approach to evaluating the specificity of the infant's cortical face processing system is to determine whether the changes in face-processing that occur during development are specific to faces, or whether they also occur for other classes of stimuli. One change in face processing that occurs around 2 months of age is that infants become more sensitive to the internal facial features of static faces. For example, after 2 months of age infants are able to remember faces from only the internal features both after a 2- and a 24-hour delay (Pascalis *et al.*, 1998), they show preferential looking to the mother even if these features are covered (Morton, 1993), and they can recognize a familiar face from a novel viewpoint based only on internal features (Pascalis *et al.*, 1998). Studies of infants' scanning patterns also show that by this age infants scan the internal features of the face more, and there is an especially noticeable increase in the scanning around the eyes (Maurer & Salapatek, 1976). Several prominent theories of adult face processing postulated that adults encode facial identity using information about the spatial relations among the features of the face (e.g. Diamond & Carey, 1986; Rhodes *et al.*, 1998). This enhanced sensitivity to internal facial features might allow infants to encode faces in a manner more similar to the way in which adults encode faces. However, this increased sensitivity to internal features is not specific for faces and also occurs for other, non-face patterns (Hainline, 1978; Milewski, 1976).

A second change in face processing that occurs at this time is that infants begin to relate information between individual faces. This ability is important because it allows the infant to process facial information in a more adult-like manner. For adults one important aspect of encoding facial identity is thought to be information regarding how an individual face differs from a protoypic or average representation (Benson & Perrett, 1991; Mauro & Kubovy, 1992; Rhodes, 1993; Rhodes *et al.*, 1993). By this view adults have formed a prototypic representation of facial features in their relative positions, and one way they can encode a new face is in terms of how it deviates from this prototype. A recent study tested infants' abilities to recognize individual faces and "average" faces made by a computer averaging of a set of faces. Following a familiarization to four individual faces, 3-month-olds were able to recognize both the individual faces and an average of the four, while 1-month-olds can recognize only the individual face (de Haan *et al.*, in press; see also Langlois *et al.*, 1995 for similar results with older infants). A similar change appears to occur in infants' processing of emotional expressions in the face, as only by 4–7 months of age can infants form categories of facial expression (i.e. recognize an expression as the same in spite of differences in the models' gender, identity, intensity of expression, etc.). Thus, it appears that only after 2 months does infants' previous experience with faces affect how they encode a new face. However, this ability is not specific to faces but occurs more generally for pattern processing at this time. For example, infants are able to form perceptual categories of a variety of complex, natural categories (see Quinn, 1998, for review). While there have been no studies to carefully compare the formation and retention of perceptual categories of faces versus other stimuli that might reveal subtle differences, at present there is no evidence to suggest that this effect is special to faces or facial expressions (see de Haan & Nelson, 1998).

In summary, there are at least two changes that occur in face processing at the time when the cortical system is functionally emerging that probably enable the infant to encode faces in a more adult-like manner. However, these changes appear to be general changes in visual pattern processing rather than specific changes in face processing. This does not support the view that emergence of a domain-specific face-processing system mediates these changes.

Neural correlates

A second approach to investigating the development of cortical specialization for face processing is to determine how early face-specific cortical activation is observable. In adults, event-related potential (ERP) studies of electrocortical activation during face processing show that there are face-selective brain potentials. The N170 is a negative deflection peaking approximately 170 ms after stimulus onset that is most prominent over occipitotemporal areas. This potential shows sensitivity to faces in that it tends to be of larger amplitude and shorter latency for faces than other objects (Bentin *et al.*, 1996). It is also influenced by factors such as stimulus inversion, that affect behavioural measures of recognition of faces. The N170 is of larger amplitude and longer latency for inverted compared to upright human faces, an effect that is specific to human faces and does not occur for animal faces (de Haan, Johnson & Pascalis, submitted) or objects (Rossion *et al.*, 2000). Recent studies suggest that an N170-like component is observable in infants as young as 6 months of age, and is larger for human than for animal faces (de Haan, Johnson & Pascalis, submitted). However the "infant N170" component differs from the adult N170 in that: (a) it is of longer latency and (b) is not influenced by face inversion until 12 months of age (Halit, de Haan & Johnson, in preparation). This result supports the perceptual narrowing hypothesis in that it suggests that at 6 months of age the infant's face-processing system may be more broad and less specifically tuned to upright, human faces than is the adult's face-processing system.

Hemispheric differences

A third approach to investigating the specialization of the infant face-processing system is to determine whether infants, like adults, show a right hemisphere differences for face processing. One way to assess the development of hemispheric differences in face processing is with a divided visual field procedure. Because of the nature of the neural projections from the retina to the primary visual cortex, information concerning stimuli presented to the left of the centre of gaze (left visual field; LVF) will reach the right hemisphere before the left hemisphere, while information presented to the right of the centre of gaze (right visual field; RVF) will reach the left hemisphere before the right hemisphere. In adults, faces are typically detected more quickly and accurately in the LVF than in the RVF, and this the LVF/right hemisphere is said to have an "advantage" for face processing. This procedure can be used with infants if latency to eye movements to the peripherally presented patterns, rather than error rates or reaction times, is used as the dependent measure. Four to 9-month-olds move their eyes more quickly to the mother's face than to a stranger's face if the faces are presented in the LVF but not if they are presented in the RVF (de Schonen *et al.*, 1986; de Schonen & Mathivet, 1990). In contrast, simple geometrical shapes are discriminated equally well in either visual field (if anything, better in the RVF; de Schonen *et al.*, 1986). These results suggest that by 4–9 months of age the right hemisphere may be more proficient than the left in recognizing faces.

A second way to assess hemispheric differences is by investigating whether ERP indices of face processing differ over right- versus left-sided electrodes. ERP studies show that by 6 months of age the negative central (Nc) potential is larger for familiar (mother) than unfamiliar faces over the right, but not the left, anterior temporal recordings (de Haan & Nelson, 1997, 1999). In contrast, the Nc is larger for familiar than unfamiliar objects bilaterally (de Haan & Nelson, 1999).

Together, the divided visual field results and the ERP results suggest that a right hemisphere advantage for face processing is present by 4–9 months of age. However, the hemispheric difference may not reflect a difference in face processing *per se* but a more general pattern of hemispheric difference in face processing. Deruelle & de Schonen (1995) have shown that the right hemisphere is superior at detecting changes in spatial arrangement of features and the left is superior at detecting differences in the detail of the features. Thus, the hemispheric differences observed for face processing may be a manifestation of these more general hemispheric differences in pattern processing.

Early lesions of putative face-processing areas

A fourth way of assessing the specificity of the early face-processing system is by observing the consequences of early damage to the cortical areas that in adults are thought critical for face processing. If the face-processing system is domain-specific and anatomically localized from very early in life, then damage to all or parts of it should produce major impairments in face processing. In adults a syndrome known as prosopagnosia has been described, which is an impairment in face recognition which occurs in spite of relatively or apparently completely intact basic visual processing and object recognition. There appear to be sub-categories of prosopagnosia showing different patterns of face-processing deficit, but the damage in many cases is in the occipital and/or temporal cortical areas and is either bilateral or right-sided (De Renzi 1986; De Renzi *et al.*, 1994; Landis *et al.*, 1986). Do lesions early in life have similar effects? Using the prospective approach, Mancini *et al.* (1998) have studied the effects of perinatal unilateral lesions on later face-processing abilities. Their data show that when children are tested at 5–14 years of age the effects of perinatal lesions are fairly mild: less than half the children showed impaired performance relative to controls on tests of face or object identity recognition. Furthermore: (a) face-processing deficits were no more common than object-processing deficits following a right hemisphere lesion, (b) face-processing deficits were no more common after right-sided than left-sided damage, and (c) a face-processing deficit never occurred in absence of an object-processing deficit. This general pattern of results is similar to that reported in other studies (Ballantyne & Trauner, 1999; Mancini *et al.*, 1994) and suggests that the infant's system is more plastic following damage than is the adult's system.

However, a few retrospective reports in which prosopagnosia in children or adults has been linked to earlier brain damage suggest a different conclusion. In four reported cases the authors concluded that the face-processing deficits documented were due to more general visual processing deficits or agnosia (Ariel & Sadeh, 1996; de Haan & Campbell, 1991; Temple, 1992; Young & Ellis, 1989). It should be noted that in these cases the deficits in face processing were quite general and included recognition of identity, emotion, and other facial information. This is in contrast to the pattern in adult prosopagnosics, who not infrequently show a more selective deficit in a particular aspect of face processing (e.g. Campbell *et al.*, 1996). Even in cases in which more selective deficits in face processing are claimed (Bentin *et al.*, 1999; Farah *et al.*, 2001), one difficulty in interpreting cases identified as adults is that it is difficult to precisely determine *when* the neural damage relevant to the prosopagnosia occurred.

In summary, there is at present no compelling evidence that specific impairments in face processing can occur as a consequence of infant brain lesions. Such cases are reported, and presumably occur, less frequently following early than late neural damage. In general the evidence is more consistent with the view that there is a general-purpose recognition system operating early in life, and that it shows some degree of plasticity following early damage.

Role of visual experience

The four lines of evidence outlined above suggest that the infant's face-processing system is not fully specialized early in life. This leads to the question of the role experience might play in the further development of the face-processing system. Below we outline two approaches to investigating the role of experience in the development of the cortical face-processing system.

Deprivation of visual experience

If visual experience plays a major role in organizing the face-recognition system, then there should be observable effects of depriving visual experience on development of the system. One study to address this question examined the development of face processing in a small number of patients treated for congenital, bilateral cataracts (a cataract is an opacity of the lens that, when dense and central, prevents patterned input from reaching the retina). These patients had their cataracts removed and were fitted with contact lenses after 6 weeks of age. Would these patients behave more like newborns, who have no visual experience, or like age-mates? The results showed that a newborn-like preference for orienting to moving faces was still present after 6 weeks of age in the cataract-treated patients but not the normally sighted controls (Mondloch *et al.*, 1998). Moreover, on further tests that differentiate between normally developing newborns and 2-month-olds, the patients typically behaved like newborns. These data show that experience plays at least the minimal role of triggering a domain-specific system, and leaves open the possibility that it plays a more substantial role via experience-expectant or experience-dependent mechanisms. Further studies are needed to determine whether development is delayed but eventually "catches up" (as might be predicted in an experience-dependent view) or whether there are more fundamental changes in the nature of the system as a result of this atypical pattern of experience (which might be predicted from an experience-expectant view).

Early-learned and later-learned expertise

One way to test the experience-dependent hypothesis is to determine whether training and development of expertise in processing a non-face category of stimuli leads to activation of "face-processing" systems. One approach has been to look at the effects of training on the inversion effect. The inversion effect refers to the finding that inversion impairs recognition of faces more than recognition of objects (review in Valentine, 1991). Two studies show that experts in naturally learned (dogs; Diamond & Carey, 1986) or experimentally learned ('greebles'; Gauthier et al., 1997) categories also show inversion effects that differ from those shown by novices. Moreover, a recent functional magnetic resonance imaging study demonstrated that training in greeble recognition caused an increase in activation of regions of the fusiform gyrus that had been identified in prior work as "face-specific" because they were more activated by faces than objects (Gauthier et al., 1999). This area was also more activated in one expert in greeble recognition than in novices during passive viewing of greebles. These results support the experience-dependent view in that they show that some characteristics of adult face processing that are often considered "special" to faces can be acquired later in life for a non-face category. It may be that expertise in discriminating and remembering exemplars from a particular class leads to their being processed at a subordinate or "individual" level more automatically (Tanaka & Taylor, 1991). For example, we may more automatically encode a face as an individual person, rather than just "a face", while for other, non-expert, categories we may typically encode only at the more basic level of "chair", "cat", etc.

SUMMARY: TRANSITION AT 2 MONTHS

The results reviewed above provide a portrait of the early cortical face-processing system. First, studies of changes in the cognitive representations of faces suggest that, while important changes occur at the time the cortical system emerges, the changes documented to date are not specific to faces. Second, studies of neural activity following face processing suggest that the adult-like pattern of sensitivity to upright, human faces emerges only gradually over the first year of life and beyond. Third, studies of hemispheric differences suggest that there is an adult-like right hemisphere bias in face processing early in life but, as biases also occur more generally for processing different aspects of pattern information in non-face patterns, this bias may represent more general hemispheric biases in pattern processing rather than a quality specific to faces. Fourth, studies of developmental prosopagnosia suggest that a face-specific recognition deficit is less often observed following perinatal damage than it is following brain damage sustained during adulthood, suggesting that the early face-processing system may be more widely distributed, more general-purpose, and/or more plastic following damage than the adult one. Together, these lines of evidence suggest that the early cortical face-processing system is less face-specific than it will be later in life, and only gradually exhibits adult-like face specificity. The studies of the effects of visual experience support the view that experience plays an important role in development of specialization of the cortical face-processing system. First, studies of visual deprivation suggest that some experience is needed at least to trigger the cortical system; whether there is a critical period has yet to be determined. Second, studies comparing face processing with later-learned expertise suggest that some characteristics of face processing that appear unique to face can occur for processing of other patterns with sufficient learning experience. These results show that experience and learning may be necessary for normal development of face processing and may be sufficient to engage a "unique" type of processing.

CONCLUSIONS

Young infants express a preference for orienting to faces and an ability to learn about individual faces shortly after birth. The mechanism mediating preferential orienting to faces in the newborn may represent an ability that does not require visual experience to emerge. These abilities may well represent visual experience-independent abilities. In spite of these impressive abilities the newborn's processing of faces appears quite different from adults' processing of faces in terms of the nature of information processed in the face. Major transitions in infants' visual attention and processing of faces occur at about 2 months of age. The available evidence is consistent with the view that this is due to emergence of cortical influence in face processing. However, the available evidence suggests that the cortical system is not specific to faces at this time – neither the functional changes that occur nor the neural activity related to face processing appear specific to faces at this time. What may make faces processed differently from some other objects at this time might be the global encoding of the right hemisphere. Further experience-dependent learning probably leads to development of expertise in processing faces that may result in processing effects (inversion effects) and neural activation (fusiform areas) that are different from many other objects, but these systems probably remain open to further experience and development of expertise in other categories of visual stimuli. This specialization may involve mechanisms that detect the expert class early, and more quickly relegate it to appropriate processing (process just the key features most useful) while non-expert stimuli might be subject to a slower, less "stimulus-tailored" kind of encoding.

References

Ariel, R. & Sadeh, M. (1996). Congenital visual agnosia and prosopagnosia in a child: a case report. *Cortex*, **32**, 221–240.

Ballantyne, A.O. & Trauner, D.A. (1999). Facial recognition in children after perinatal stroke. *Neuropsychiatry Neuropsychology and Behavioral Neurology*, **12**, 82–87.

Banks, M. & Ginsburg, A.P. (1985). Infant visual preferences: a review and new theoretical treatment. *Advances in Child Development and Behavior*, **19**, 207–246.

Banks, M. & Salapatek, P. (1981). Infant pattern vision: a new approach based on the contrast sensitivity function. *Journal of Experimental Child Psychology*, **31**, 1–45.

Benson, P.J. & Perrett, D.I. (1991). Perception and recognition of photographic quality facial caricatures: implications for recognition of natural images. *European Journal of Cognitive Psychology*, **3**, 105–135.

Bentin, S., Allison, T., Puce, A., Perez, E. & McCarthy, G. (1996). Electrophyisological studies of face perception in humans. *Journal of Cognitive Neuroscience*, **8**, 551–565.

Bentin, S., Deouell, L.Y. & Soroker, N. (1999). Selective visual streaming in face recognition: evidence from developmental prosopagnosia. *NeuroReport*, **10**, 823-827.

Bowlby, J. (1958). The nature of the child's tie to his mother. *International Journal of Psychoanalysis*, **39**, 350–373.

Bushnell, I.W. (1979). Modification of the externality effect in young infants. *Journal of Experimental Child Psychology*, **28**, 211–229.

Bushnell, I.W.R., Sai, F. & Mullin, J.T. (1989). Neonatal recognition of the mother's face. *British Journal of Developmental Psychology*, **7**, 3–15.

Campbell, R. & Tuck, M. (1995). Recognition of parts of famous-face photographs by children: an experimental note. *Perception*, **24**, 451–456.

Campbell, R., Brooks, B., de Haan, E. & Roberts, T. (1996). Dissociating face processing skills: decision about lip-read speech, emotion and identity. *Quarterly Journal of Experimental Psychology A*, **49**, 295–314.

Campbell, R., Pascalis, O., Coleman, M., Wallace, S.B. & Benson, P.J. (1997). Are faces of different species perceived categorically? *Proceedings of the Royal Society London B: Biological Sciences*, **264**, 1429–1434.

Clarke, S., Riahi-Ayra, S., Tardif, E., Eskenasy, A.C. & Probst, A. (1999). Thalamic projections to the fusiform gyrus in man. *European Journal of Neuroscience*, **11**, 1835–1838.

Dannemiller, J.L. & Stephens, B.R. (1988). A critical test of infant pattern preference models. *Child Development*, **59**, 210–216.

de Haan, E.H. & Campbell, R. (1991). A fifteen-year follow-up of a case of developmental prosopagnosia. *Cortex*, **27**, 489–509.

de Haan, M., Johnson, M.H., Maurer, D. & Perrett, D.I. (In Press). Recognition of individual faces and average face prototypes by one- and three-month-old infants. *Cognitive Development*, **00**, 000–000.

de Haan. M. & Nelson, C.A. (1997). Recognition of the mother's face by six-month-old infants: a neurobehavioral study. *Child Development*, **68**, 187–210.

de Haan, M. & Nelson, C.A. (1998). Discrimination and categorisation of facial expressions of emotion during infancy. In A. Slater (Ed.), *Perceptual development* (pp. 287–309). Hove: Psychology Press.

de Haan, M. & Nelson, C.A. (1999). Brain activity differentiates face and object processing by 6-month-old infants. *Developmental Psychology*, **34**, 1114–1121.

de Haan, M., Pascalis, O. & Johnson, M.H. (under review). Specialization of neural mechanisms underlying face recognition in human infants. *Journal of Cognitive Neuroscience*, **00**, 000–000.

de Renzi, E. (1986). Prosopagnosia in two patients with CT scan evidence of damage confined to the right hemisphere. *Neuropsychologia*, **24**, 385–389.

de Renzi, E., Perani, D., Carlesimo, G.A., Silveri, M.C. & Fazio. F. (1994). Prosopagnosia can be associated with damage confined to the right hemisphere – an MRI and PET study and review of the literature. *Neuropsychologia*, **32**, 893–902.

de Schonen, S. & Mathivet, E. (1989). First come, first served: a scenario about the development of hemispheric specialization in face recognition during infancy. *European Bulletin of Cognitive Psychology (CPC)*, **9**, 3–44.

de Schonen, S. & Mathivet, E. (1990). Hemispheric asymmetry in a face discrimination task in infants. *Child Development*, **61**, 1192–1205.

de Schonen, S., Gil de Diaz, M. & Mathivet, E. (1986). Hemispheric asymmetry in face processing in infancy. In H.D. Ellis, M.A. Jeeves, F. Newcombe & A. Young (Eds.), *Aspects of face processing* (pp. 199–208). Dordrecht: Martinus Nijhoff.

Deruelle, C. & de Schonen, S. (1995). Pattern processing in infancy: hemispheric differences in the processing of shape and location of visual components. *Infant Behavior and Development*, **18**, 123–132.

Diamond, R. & Carey, S. (1986). Why faces are and are not special: an effect of expertise. *Journal of Experimental Child Psychology: General*, **115**, 107–117.

Farah, M.J. (1990). *Visual agnosia*. Cambridge, MA: MIT Press.

Farah, M.J., Rabinowitz, C., Quinn, G.E. & Liu, G.T. (2000). Early commitment of neural substrates for face recognition. *Cognitive Neuropsychology*, **17**, 117–123.

Field, T.M., Cohen, D., Garcia, R. & Collins, R. (1983). Discrimination and imitation of facial expressions by term and preterm neonates. *Infant Behavior and Development*, **13**, 497–511.

Field, T.M., Cohen, D., Garcia, R. & Greenberg, R. (1984). Mother–stranger face discrimination by the newborn. *Infant Behavior and Development*, **7**, 19–25.

Field, T.M., Woodson, R., Greenberg, R. & Cohen, D. (1982). Discrimination and imitation of facial expressions by neonates. *Science*, **218**, 179–181.

Fodor, J.A. (1983). *The modularity of mind*. Cambridge, MA: MIT Press.

Ganon, E.C. & Swartz, K.B. (1980). Perception of internal elements of compound figures by one-month-old infants. *Journal of Experimental Child Psychology*, **30**, 159–170.

Gauthier, I. & Tarr, M.J. (1997). Becoming a "Greeble" expert: exploring mechanisms for face recognition. *Vision Research*, **37**, 1673–1682.

Gauthier, I., Tarr, M.J., Anderson, A.W., Skudlarski, P. & Gore, J.C. (1999). Activation of the middle fusiform 'face area' increases with expertise in recognizing novel objects. *Nature Neuroscience*, **2**, 568–573.

Goren, C.C., Sarty, M. & Wu, P.K.Y. (1975). Visual following and pattern discrimination of face-like stimuli by newborn infants. *Pediatrics*, **56**, 544–549.

Greenough, W.T. & Black, J.E. (1992). Induction of brain structure by experience. In M.R. Gunnar & C.A. Nelson (Eds.), Minnesota symposia on child psychology. vol. 24. *Developmental behavioral neuroscience* (pp. 155–200). Hillsdale, NJ: Lawrence Erlbaum Associates.

Hainline, L. (1978). Developmental changes in visual scanning of face and nonface patterns by infants. *Journal of Experimental Child Psychology*, **25**, 90–115.

Johnson, M.H. (1990). Cortical maturation and the development of visual attention in early infancy. *Journal of Cognitive Neuroscience*, **2**(2), 81–95.

Johnson, M.H., Dziurawiec, S., Ellis, H. & Morton, J. (1991). Newborns' preferential tracking of face-like stimuli and its subsequent decline. *Cognition*, **40**, 1–19.

Johnson, M.H. & de Haan, M. (2001). Developing cortical specialisation for visual-cognitive function: the case of face recognition. In J. McClelland and R.S. Seigler (Eds.), *Mechanisms of cognitive development*. (pp. 253–270). Mahwah, NJ: Lawrence Erlbaum Associates.

Johnson, M.J. & Morton, J. (1991). *Biology and cognitive development. The case of face recognition*. Oxford, UK: Blackwell.

Johnson, R.A., & Ellis, H.D. (1995). Age effects in the processing of typical and distinctive faces. *Quarterly Journal of Experimental Psychology A*, **48**, 447–465.

Kleiner, K.A. (1987). Amplitude and phase spectra as indices of infants' pattern preferences. *Infant Behavior and Development*, **10**, 45–59.

Kleiner, K.A., & Banks, M.S. (1987). Stimulus energy does not account for 2-month-olds' face preferences. *Journal of Experimental Psychology: Human Perception and Performance*, **13**, 594–600.

Landis, T., Cummings, J.L., Christen, L., Bogen, J.E. & Imhof, H.G. (1986). Are unilateral right posterior cerebral lesions sufficient to cause prosopagnosia? Clinical and radiological findings in six additional patients. *Cortex*, **22**, 243–252.

Langlois, J.H., Musselman, L.E., Rubenstein, A.J., Smoot, M.T., Hallam, M.J. & Oakes, L.M. (1995). *Infants average faces: a basis for attractiveness preferences*. Poster presented at the Biennial meeting of the Society for Research in Child Development, Indianapolis, IN.

Lewis, T.L. & Maurer, D. (1992). The development of the temporal and nasal visual fields during infancy. *Vision Research*, **32**, 903–911.

Mancini, J., de Schonen, S., Deruelle, C. & Massoulier, A. (1994). Face recognition in children with early right or left brain damage. *Developmental Medicine and Child Neurology*, **36**, 156–166.

Mancini, J., Casse-Perrot, C., Giusiano, B., Girard, N., Camps, R., Deruelle, C. & de Schonen, S. (1998). *Face processing development after a perinatal unilateral brain lesion*. Human Frontiers Science Foundation Developmental Cognitive Neuroscience Technical Report Series No. 98.6.

Maurer, D. (1983). The scanning of compound figures by young infants. *Journal of Experimental Child Psychology*, **35**, 437–448.

Maurer, D. & Barrera, M. (1981). Infants' perception of natural and distorted arrangements of a schematic face. *Child Development*, **52**, 196–202.

Maurer, D. & Barrera, M. (1981). Infants' perception of natural and distorted arrangements of a schematic face. *Child Development*, **47**, 523–527.

Maurer, D. & Salapatek, S. (1976). Developmental changes in the scanning of faces by young infants. *Child Development*, **47**, 523–527.

Maurer, D. & Young, R.E. (1983). Newborns' following of natural and distorted arrangements of facial features. *Infant Behavior and Development*, **6**, 127–131.

Mauro, R. & Kubovy, M. (1992). Caricature and face recognition. *Memory and Cognition*, **20**, 433–440.

Milewski, A.E. (1976). Infants' discrimination of internal and external pattern elements. *Journal of Experimental Child Psychology*, **22**, 229–246.

Mondloch, C.J., Lewis, T.L., Budreau, D.R., Maurer, D., Dannemiller, J.L., Stephens, B.R. & Kleiner-Gathercoal, K.A. (2001). Face perception during early infancy. *Psychological Science* (In press).

Mondloch, C.J., Lewis, T.L., Maurer, D. & Levin, A.V. (1998). The effects of visual experience on face preferences during infancy. *Human Frontiers Scientific Foundation Developmental Cognitive Neuroscience Technical Report 98.4.* McMaster University, Canada.

Morton, J. (1993). Mechanisms of infant face processing. In B. de Boysson-Bardies, S. de Schonen, P. Jusczyk, P. McNeilage & J. Morton (Eds.), *Developmental neurocognition: speech and face processing in the first year of life* (pp. 93–102). London: Kluwer.

Morton, J. & Johnson, M.H. (1991). CONSPEC and CONLERN: a two-process theory of infant face recognition. *Psychological Review*, **2**, 164–181.

Nelson, C.A. (1995). The ontogeny of human memory: a cognitive neuroscience perspective. *Developmental Psychology*, **31**, 723–738.

Nelson, C.A. & de Haan, M. (1997). A neurobehavioral approach to the recognition of facial expressions in infancy. In J. A. Russell & J.M. Fernandez-Dols (Eds.), *The psychology of facial expression* (pp. 176–204). Cambridge, UK: Cambridge University Press.

Pascalis, O. & de Schonen, S. (1994). Recognition memory in 3- to 4-day-old human neonates. *Neuroreport*, **8**, 1721–1724.

Pascalis, O., de Haan, M., Nelson, C.A. & de Schonen, S. (1998). Long-term recognition memory for faces assessed by visual paired comparison in 3- and 6-month-old infants. *Journal of Experimental Psychology: Learning, Memory and Cognition*, **24**, 249–260.

Pascalis, O., de Schonen, S., Morton, J. Deruelle, C. & Fabre-Grenet, M. (1995). Mother's face recognition in neonates: a replication and an extension. *Infant Behavior and Development*, **18**, 79–86.

Quinn, P.C. (1998). Object and spatial categorisation in young infants: "what" and "where" in early visual perception. In A. Slater (Ed.), *Perceptual development: visual, auditory and speech perception in infancy* (pp. 131–165). Hove, UK: Psychology Press.

Rhodes, G. (1985). Lateralized processes in face recognition. *British Journal of Psychology*, **76**, 249–271.

Rhodes, G. (1993). Configural coding, expertise, and the right hemisphere advantage for face recognition. *Brain and Cognition*, **22**, 19–41.

Rhodes, G., Brake, S. & Atkinson, A.P. (1993). What's lost in inverted faces? *Cognition*, **47**, 25–57.

Rhodes, G., Carey, S., Byatt, G. & Proffitt, F. (1998). Coding spatial variations in faces and simple shapes: a test of two models. *Vision Research*, **38**, 2307–2321.

Rossion, B., Gauthier, I., Tarr, M.J., Despland, P., Bruyer, R., Linotte, S. & Crommelinck, M. (2000). The N170 occipito-temporal component is delayed and enhanced to inverted faces but not to inverted objects: an electrophysiological account of face-specific processes in the human brain. *NeuroReport*, **11**, 69–74.

Simion, F., Valenza, E., Umilta, C. & Dalla Barba, B. (1998). Preferential orienting to faces in newborns: a temporal–nasal asymmetry. *Journal of Experimental Psychology: Human Perception and Performance*, **24**, 1399–1405.

Slater, A. (1998). The competent infant: innate organisation and early learning in infant visual perception. In A. Slater (Ed.), *Perceptual development*. Hove, UK: Psychology Press.

Slater, A., Morison, V. & Rose, D. (1983). Perception of shape by the new-born baby. *British Journal of Developmental Psychology*, **1**, 135–142.

Tanaka, J.W. & Taylor, M. (1991). Object categories and expertise: is the basic level in the eye of the beholder? *Cognitive Psychology*, **23**, 457–482.

Teller, D.Y., Succop, A. & Mar, C. (1993). Infant eye movement asymmetries: stationary counterphase gratings elicit temporal-to-nasal optokinetic nystagmus in two-month-old infants under monocular test conditions. *Vision Research*, **33**, 2139–2152.

Valentine, T. (1989). Upside down faces: a review of the effect of inversion upon face recognition. *British Journal of Psychology*, **79**, 471–491.

Valenza, E., Simion, F., Cassia, V. M. & Umilta, C. (1996). Face preference at birth. *Journal of Experimental Psychology: Human Perception and Performance*, **22**, 892–903.

Webser, M.J., Ungerleider, L.G. & Bachevalier, J. (1995). Development and plasticity of the neural circuitry underlying visual recognition memory. *Canadian Journal of Physiology and Pharmacology*, **73**, 1364–1371.

Werker, J.F. & Lalonde, C.E. (1988). Cross-language speech perception: initial capabilities and developmental change. *Developmental Psychology*, **24**, 672–683.

Werker, J.F. & Tees, R.C. (1984). Phonemic and phonetic factors in adult cross-language speech perception. *Journal of Acoustical Society of America*, **75**, 1866–1878.

Further reading

de Schonen, S. & Mathivet, E. (1990). Hemispheric asymmetry in a face discrimination task in infants. *Child Development*, **61**, 1192–1205.

Johnson, M.J. & Morton, J. (1991). *Biology and cognitive development. The case of face recognition*. Oxford, UK: Blackwell.

Nelson, C.A. & de Haan, M. (1997). A neurobehavioral approach to the recognition of facial expressions in infancy. In J.A. Russell & J.M. Fernandez-Dols (Eds.), *The psychology of facial expression* (pp. 176-204). Cambridge, UK: Cambridge University Press.

V.7
Cerebral Bases of Language Acquisition

G. DEHAENE-LAMBERTZ, J. MEHLER and M. PENA

ABSTRACT

Humans have tried to understand the origin of their own mental abilities since the beginnings of science. They have asked: Why is it that only humans have a language? Why can they acquire grammar that makes it possible to generate utterances and encode infinite meanings? What makes it possible for speakers to convey true and also false assertions? And why is it that only humans create recursive systems such as music, mathematics and science? Two major accounts have been proposed to explain how the human brain attains language. One account assumes that it is the larger brain of humans that allows them to take advantage of their environment through associative mechanisms. The second one postulates modules dedicated to compute certain types of linguistic information. In this chapter we explore how the different fields of cognitive neuroscience (developmental studies, brain lesions and brain imaging studies), psychology and linguistics contribute to the clarification of the issue of how language arises in the infant's brain.

INTRODUCTION

The development of language and cognition is a matter of great importance both for theoreticians and clinicians. Understanding development makes it possible to characterize human nature; namely, the components that distinguish the human mind from the mind of other vertebrates. For clinicians, understanding the laws of development helps them to discover abnormalities and establish whether the child is in need of assistance. These concerns are complementary since understanding clinical deficits requires adequate theories of cognitive development in normal children.

939

A.F. Kalverboer and A. Gramsbergen (eds.), Handbook of Brain and Behaviour in Human Development, 939–966
© 2001 Kluwer Academic Publishers. Printed in Great Britain.

Two major accounts are proposed to explain how the human mind attains language and cognitive abilities. In the one it is assumed that the brains of vertebrates, even though differing in size, are similar, but it is differences in sensory systems which make different sorts of information available to the species. From these broad generalizations, theorists working within this tradition assume that organisms learn all they know from the environment through associative mechanisms; larger brains have greater associative networks and thus can learn more things, but always through this unique mechanism. Basically, aside from the ability to associate, brains are blank slates at birth.

The other account assumes that organisms have specialized *modules* or mental organs to compute certain types of information. Moreover, some of these modules implement unique computational algorithms. For example, humans have a device that modular theorists usually describe as uniquely devoted to process faces and not other visual objects (Kanwisher *et al.*, 1997). Likewise, there is also some evidence that there is a module in the left insula where utterances are compiled and sent to the articulator (Dronkers, 1996). Many other modules have been described and their loci have been established. According to some theorists these modules reflect the evolutionary history of the species. For instance, some claim that language has evolved because of the advantages derived from improved communication among conspecifics (Nowak *et al.*, 2000), though other scenarios may explain its emergence. Regardless of the evolutionary perspective one adopts, modular theorists assume that different species are equipped with innate domain-specific learning skills. In more modern terms they propose different computational abilities for different domains. In particular, theorists assume that the language acquisition device, or LAD (Chomsky, 1965) is possibly the single most important module that humans have acquired. Indeed, it is the one that enables the acquisition of syntax, the essence of any grammatical system.

Those who assume that humans learn language because only they have the brain size to store all the necessary associations also predict that more intelligent children must acquire language faster and better than less intelligent ones. *A fortiori*, they also predict that the more evolved apes acquire "more" language than the more primitive ones. Theorists who uphold these views believe that all species have some language, and that as one ascends the phylogenetic ladder one finds richer languages. In contrast, modular theorists postulate that LAD is a specific device that appeared only in human brains, and they do not expect to find language in other species. Likewise, modular theorists predict that a lesion may attain the language faculty without affecting other intellectual functions. The theoretical position which is taken in this field obviously determines how development is studied. This is something to bear in mind when reading the rest of this chapter.

Below we present a review of the major landmarks that characterize language acquisition during the first few months of life. Then we concentrate on the effect that brain lesions and/or unfavourable surroundings can have for language acquisition. We also review the impediments that arise in children who do not receive the necessary input to learn language. We ask, at the end of this chapter, how age of acquisition can affect language learning.

EARLY STUDIES OF LANGUAGE ACQUISITION: LEARNING OR BIOLOGY?

Historical perspective

The characteristic landmarks of language acquisition can be found in several classical treatises of paediatric science. Gesell and co-workers, in a book entitled *The first five years of life* (Gesell *et al.*, 1940) described the stages of language development in some detail. Gesell focuses on the audible part of the language faculty; namely speech. Gesell acknowledges that the age when infants acquire language varies greatly. Language begins to emerge by the end of the first year of life. That is when parents first notice playful behaviour suggesting the child's desire to communicate. However, the first recognizable words and short phrases arise by the end of the second year of life. Gesell completes his survey with the assertion that it takes roughly 4 years for language to be mastered.

It is unclear whether Gesell's landmarks reflect the fact that most children have parents who speak to them and spend time teaching them about language, or whether the regularity of language development comes about because of a mature nervous system. These issues being addressed had to await Lenneberg's *Biological foundations of language*, a remarkable book, that appeared in 1967. Lenneberg became the first psychologist to study language acquisition in relation to biology, and he reaches the conclusion that language arises in humans because they, and only they, have the necessary biological endowment, a conclusion that is quite similar to Chomsky's. Lenneberg was familiar with Chomsky's views (Chomsky, 1957, 1959) yet he complements the formal and logical proposals of the latter with a biological perspective.

Lenneberg states that, by 12 weeks, babies cry less when someone talks to them than when they hear other noises, suggesting that speech stimuli have a special meaning to them. Babies also display a pitch-modulated vowel-like sound that is called *cooing*. Although it does not sound like something they have heard from adults, cooing is universal in infants. By 16 weeks infants clearly respond to speech by turning the head towards the speaker. By 20 weeks vowel-like patterns are combined with consonantal sounds to form patterns that are quite different from those of mature speakers. These landmarks suggest that the first stages of language acquisition are under strong maturational control.

By 6 months cooing changes to become *babbling*, a pattern that is characterized by single, mostly simple syllables, for example ma, da, mu, da. Two months later infants begin to reduplicate syllables such as baba, or mama, and intonation patterns can be recognized. By 10 months infants try, albeit unsuccessfully, to imitate speech. This is also when infants first indicate that they can distinguish the meaning of some words. The first words, e.g., *mama* or *dada*, tend to emerge at about 12 months, although in some babies these may arise a few months later. When the child starts to speak late, the first production tends to be short sentences rather than words. Some time between 1 year and 1½ years the infant recognizes words and adjusts to commands. At this age the infant has a small vocabulary and babbles with elaborated intonation patterns. Generally a few short sentences are produced, but little if any communicative intent is apparent. Six months later one notices a considerable expansion of the vocabulary and short sentences are frequently produced. The miracle of what is often called the lexical explosion takes place just a few months later.[1] Soon thereafter the language of the child begins to approximate that of the adults in their surround (Lenneberg, 1967). The lawful regularity of language acquisition, given the huge variability of the infants' environment, is taken as an indication that there is a strong underlying genetic component.

Chomsky and Lenneberg rekindled the interest for the study of language acquisition changing the almost universal belief among psychologists that language is just learning. The changing *zeitgeist* raised awareness as to the colossal achievement underlying language acquisition. Today psychologists and paediatricians are aware that the average high-school student has a vast knowledge of words, and Miller (1991) estimated this to be around 60,000 word roots. This implies that, on average, humans learn more than 10 words a day from birth until the age of 16, and it should be realized that, in addition, they have to learn the grammar in a record time. What explains the human brain having such a unique competence? This is, in part, the topic that will concern us below. Before doing this, however, we need to complete the description of the landmarks established by Gesell and Lenneberg. Both observed language production in the child, and from these observations they drew inferences about language acquisition. As we will see below, most contemporary research on language acquisition focuses on speech perception and comprehension. Observers had the intuition that comprehension precedes production, and that Lenneberg's landmarks had to be complemented.

Contemporary studies of speech perception

Languages of the world use a limited number of distinctive sounds or phonemes. These are usually categorized as vowels or consonants, a universal contrast that appears to be part of the production system (Caramazza *et al.*, 2000). When learning the language the child has to discover which phonemes are relevant to this specific language. Likewise, the child has to figure out the rules that govern the combination of segments into words. How and when does the infant begin to learn about speech sounds? Is the infant unconcerned until it has learned some of the words in the language, or does learning start before that stage? A major outcome of recent research is that infants learn a lot about the phonological structure of the language even before they rely on meaning.

Studies have shown that 2-month-olds have already acquired some information about the language which is spoken in their surround. Indeed, at birth, infants may discriminate sentences of two different languages. This assertion is valid only in case the languages belonging to different rhythmic classes (Mehler *et al.*, 1988; Ramus *et al.*, 1999). For example, 4-day-old French infants recognized a switch from Russian to French sentences (Mehler, 1988; see also Moon *et al.*, 1993). Two-month-old British and French babies (Christophe & Morton, 1998)

---

[1]That is, quite abruptly infants learn many new words every day. A few months later it becomes impossible to say how many lexical items the child knows; but see Bloom for a more tempered view.

were able to discriminate between English and French; moreover, the infants oriented faster towards the source of their native language being spoken (Dehaene-Lambertz & Houston, 1998). The infants' behaviour remains unaltered when the utterances are low-pass filtered (Mehler *et al.*, 1988; Dehaene-Lambertz & Houston, 1998) suggesting that this initial representation is based on global properties. The best cues to explain the behaviour therefore seem to be rhythm and/or prosody.

Initially it was assumed by many investigators that the ability to discriminate between spoken languages was a specific and innate property of the language acquisition device. This, however, appeared a simplification. Today we know that pairs of languages are discriminated at birth only if they belong to different rhythmic classes. Indeed, French newborns discriminate English from Italian or Japanese from English, but they are not able to distinguish English from Dutch (Nazzi *et al.*, 1998). Of course, English and Dutch are stress-timed languages while Italian, French and Spanish are syllable-timed and Japanese is mora-timed[2]. Moreover, French newborns distinguish a medley of Spanish and Italian sentences from one with Dutch and English sentences (or vice-versa), but fail to react to heterogeneous combinations, i.e. Spanish and Dutch sentences contrasted with a combination of Italian and English sentences (Nazzi *et al.*, 1998). Ramus & Mehler (1999; see also Ramus *et al.*, 1999) proposed that infants use the duration of intervals between vocals and the variability of the consonantal intervals to compute rhythm. Thus, in a language such as Japanese that licenses

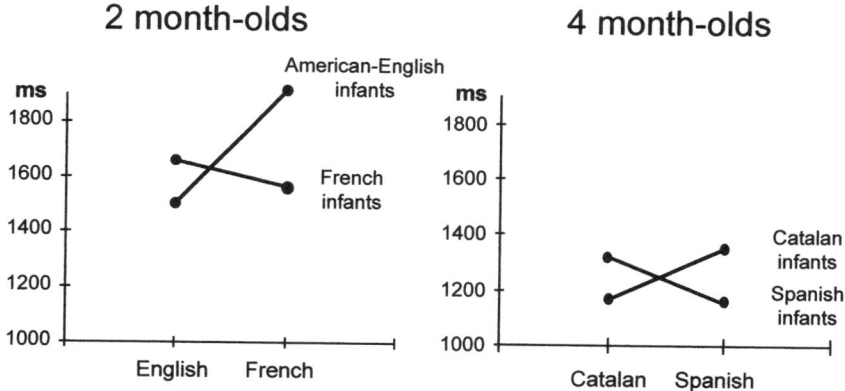

Figure 1 Eye orientation latencies towards the native language and a foreign one in 2- and 4-month-old infants. Infants orient faster towards their native language. (From Dehaene-Lambertz & Houston, 1998; Bosch & Sebastin-Gallés, 1997.)

---

[2]We use the terms syllable times and stress times as classificatory terms without espousing the notion that the former have syllable isochrony while the latter have stress interval isochrony.

only one syllabic structure (consisting of a consonant [C] and a vowel [V]), one expects to find little variation in consonantal duration, as compared to a language that allows structures such as CCCVCCC (e.g. springs, splits, etc.). Moreover, more of the overall duration ought to be devoted to the production of vowels than in languages with more varied syllable structures, e.g. English. After testing his ideas with utterances from eight languages, Ramus showed that the behaviour of neonates could be explained by assuming that they identify rhythm, as illustrated above. In more recent research Ramus used a new technique in which speech was artificially re-synthesized and utterances were de-lexicalized. He showed that infants still discriminate pairs of languages that differ in rhythm even when the utterances have been delexicalized. In contrast, when rhythm is equalized the infants are no longer able to discriminate the languages, even though other parameters, such as intonation, remain unchanged.

Ramus *et al.* (1999) plotted utterances of several languages pitting the vowel intervals against the variability of the consonantal intervals. They found that there is good correspondence between the classes of language, as conjectured by phonologists (Abercrombie, 1967), and the notion of rhythm as defined here. These studies have allowed us to model the behaviour of infants and also to conjecture that, through rhythm, the infant might discover the syllable complexity of the language in the environment. These studies illustrate the importance of the more global properties of speech signals when infants make the first adjustments to language. However, in the next few months the infant begins to pay attention to segmental properties.

By 6 months infants have already identified the vowels of their native language (Kuhl *et al.*, 1992). Kuhl *et al.* have shown that American and Swedish infants react differentially to the prototypical vowels used in English and Swedish respectively. Polka & Werker have shown that, also at 6 months, infants show a decrease of sensitivity to non-native vowels. Consonants are characterized a few months later. By 10–12 months infants stop reacting to non-native consonant contrasts . This is not observed when the tested contrasts cannot be assimilated to acquired native categories (Best *et al.*, 1988). The above results illustrate the importance of learning during the first 12 months of life, something most parents are not even aware of. Some investigators have argued that this kind of learning is also present in other species trained to respond to speech stimuli. Kluender *et al.* (1998) trained European starlings with vowels and found that they behave as the infants studied by Kuhl. More research, however, will be necessary to understand whether learning is all it takes to converge to the phonology of the maternal language, or if specific constraints have to be added to simulate acquisition.

Infants who approach the end of the first year of life have already discovered many properties of the phonetic repertoire of their language and start paying attention to some frequent words. At 9 months, but not at 6 months, infants prefer to listen to a list of frequent words in their native language rather than a list of words from another language (Jusczyk *et al.*, 1993). At the same age infants also prefer non-words with phoneme arrangements (phonotactic sequences) that are similar to the patterns in the native language (Jusczyk *et al.*, 1994). Moreover, 9-month-olds prefer lists of words with metrical properties that correspond to their maternal language. For instance, American infants prefer a list of bisyllabic strong–weak (SW) words[3] (the typical pattern of words in English) over a list of weak–strong words (a pattern that is less frequent in English) (Jusczyk *et al.*, 1993b). Before 9-months most infants have learned only a few words, which licenses the assumption that many phonological properties are determined before learning a lexicon. However, one wonders how the relevant parameters are learned. How, for instance, does the infant discover that in English, but not in some other language, the metrical pattern is SW? Or that in a language such as Japanese a syllable cannot have a coda? Many of these regularities hold for segmented speech, e.g. word-initial syllables tend to be strong. How can these regularities be learned before the child acquires a lexicon that contains that data that seem essential to establish such inferences in the first place? Knowledge of the prosodic and phonotactic properties of the native language may help the child extract words from the speech stream without having access to a memorized lexicon. Indeed, imagine that by some mechanism or another the infant figures out that, in English, SW is the metrical structure of words. If so the child might use that knowledge and postulate that the beginning of words coincides with strong syllables. This may very well be the case. However, we still have to ask how the child learns these properties.

Recently Saffran *et al.* proposed that the statistical regularities in the data can provide the child with a powerful mechanism to segment speech pre-lexically. They showed that 8-month-old infants can use transition probabilities between syllables to segment continuous monotonous synthetic speech. The authors synthesized syllables with equal pitch, energy and duration. With those syllables they constructed a sequence containing four trisyllabic (CV-CV-CV) words. The speech stream was continuous and there was no clue to indicate where one word stops and the next begins. The only cue to words boundaries are probabilistic; namely, the CV segments within the words are more predictable than the CVs between words. The words were concatenated at random. After 2 minutes of exposure to such a speech sequence infants react as if they distinguish the words in the tape from part words, e.g. items with the two last syllables of one word and the first syllable of another word. Saffran has shown that 8-month-old infants use statistics to segment monotonous speech sequences. It is not possible to say whether this behaviour emerges only at 8 months old or whether it is already present in younger infants. Regardless of the exact age at which infants become sensitive to transition probabilities this result shows that infants can segment speech without having to pair sounds with meanings. Once speech streams are segmented the child has the database to discover some phonological properties of the language before acquiring a lexicon.

Another mechanism the infant might use to segment speech streams would be to use auditory cues if they exist in natural speech. Several investigators have shown that some boundary cues are present in connected speech. This induced some investigators to explore whether infants are sensitive to such cues. In a series of experiments Christophe *et al.* (1994) have shown that neonates are

---

[3]A strong syllable is one whose vowel is not reduced and which often bears stress. A weak syllable has a vowel that is reduced to schwa.

sensitive to cues that indicate boundaries between words or phrases. However, it is not known whether these cues are used to parse continuous speech. They have shown that infants distinguish lists of disyllabic words if one list has items that contain a boundary while the other list consists of items spliced from within a word. The items of both lists sound as /MATI/. The items from the first list are spliced from passages such as *panoraMA TYpique*, or *laMA Tibetain* and those of the other list from words such as *matheMATIcien* or *autoMATIquement*. The infants readily distinguish the lists, which confirms that acoustic cues signalling boundaries are present in connected speech and also that infants are sensitive to them. Yet we ignore whether infants spontaneously use these cues to segment continuous speech during language acquisition.

Marcus *et al.* (1999) described another mechanism that infants might use to learn the maternal language. Indeed, while statistics may be a very powerful mechanism to segment continuous speech it is clear that this might not be sufficient to discover more abstract properties that natural language incorporates. The notion that surface regularities are all it takes to learn language collapsed with Chomsky's (Chomsky, 1957, 1965) demonstrations. Marcus *et al.* (1999) have established that infants of 7 months old can compute abstract underlying regularities in a list of "sentences". After a familiarization period with "sentences" that conform to the same structure (for example AAB: "gagati") 7-month-old infants expect novel sentences will have the same structure (e.g., "wowofe" for the AAB structure) rather than another structure, (e.g., "wofewo", an ABA structure). Notice, however, that the novel sentences have "words" that have never been heard before. Marcus *et al.* argue that, since the infant has never heard these words, what she finds familiar can only be the structure of the "sentence" which can also be called an abstract regularity or "rule".

There are other experiments that ask questions about segmentation in another fashion. For instance Jusczyk & Hohne (1997) exposed 8-month-old babies to a story over the course of a few weeks. One week later the infants are tested with words that appeared in the stories or with similar words that were not parts of the story. The infants recognize the familiar words, which means that they are able to parse continuous speech, extract words and commit them to memory for at least 1 week. Two months later, infants become sensitive to function words, e.g. articles, prepositions or auxiliaries. Using event-related potentials (ERPs), Shafer *et al.* studied 10- and 11-month-olds with normal passages or with passages whose function words were replaced by nonsense syllables. The authors report that babies expect words such as "the", "is", "a" at 11 but not at 10 months. We do not know, however, whether infants understand that close class words have a special status in the language or whether they have just noticed that syllables that denote them tend to occur with greater frequency.

Summarizing, we can say that the infant's brain forms a representation of the language spoken in the surrounds before the end of the first year of life. This representation is initially influenced by the prosodic properties of the language. After the second semester of life the representation integrates segmental information, i.e. the inventory of phonemes used in the language and the combinatorial regularities displayed by the segments. Word shapes and the frequency of word forms become available at the same time. An issue we want to stress is that these stages are not connected to the meaning of the utterances to which the infant is exposed. We conjecture that, initially, language learning is predominantly driven by form rather than by content. Paying attention to the form of the speech signal may also be essential to make inroads into learning the syntax of one's language. For example, some languages put complements to the right of the head while others languages put them to the left of the head. In English or French complements follow the head, but in Turkish or Japanese complements precede the head (compare "Boys read books" in English to "KitabI yazdim", that is "books boys read" in Turkish). Imagine the difficulties of an infant who is trying to learn the meaning of the words. Must she conjecture that *books* means *books* and *read* means *read* or does she have to postulate that *books* means *boys* and *read* means *books*? This is a quandary, and nobody really knows how the child will manage to solve this predicament. However, some colleagues have imagined that the prosody of the language might be of some help. This hypothesis is currently being investigated by the psycholinguists.

Prosody might help to discover abstract properties of a language, a hypothesis called *phonological bootstrapping* by Gleitman & Wanner. They proposed it to explain how infants might distinguish open- from closed-class items. Phonological bootstrapping may be extended to explain other aspects of language acquisition. Nespor *et al.* suppose that there are prosodic correlates in utterances with the head direction (parameter) in the language. They argue that in languages in which the head has the complement to its right, prominence falls at the end of phonological phrases, whereas in languages in which the head has the complement to its left, prominence falls at the beginning of phonological phrases. If infants perceive and encode the prominence of phonological phrases in the language they might be able to discover whether their language is a member of the head–complement or the complement–head variety. If so, one of the major problems we pointed out above, as to the difficulty of learning the meaning of words, disappears. For prosodic bootstrapping to be effective, infants must be preprogrammed to acquire language and to have the knowledge that a particular prosodic property elucidates issues of syntax. Prosodic bootstrapping entails that the form of speech determines abstract syntactic properties of the language.

Most of the studies reviewed above bolster the hypothesis that the meaning of words is learned after the infant has determined the phonological properties of their native language. Many months go by before infants start to pay attention to words, *cum* words. Several more months are still necessary before they experience the lexical explosion when the size of their lexicon increases almost exponentially. This nearly universal linguistic calendar holds across many different languages, a fact that is crying out for an explanation. Are these mechanisms specific to language acquisition or are they merely the reflection of general learning mechanisms? Are there specific brain areas involved in speech processing while infants are trying to acquire their first language and, if so, do they correspond to the classical language areas in adults? Are the language areas of adult brains the consequence of learning the native language or the successors of the structures that make such learning possible? Language is mainly represented in the left hemisphere (LH) of adults. We can now ask whether the LH makes it possible to acquire language or whether the LH becomes involved with language as a consequence of its acquisition? A related issue is why children have greater facilities to learn than adults? Is there a specific time window for learning a language? Are young brains more plastic or is it that young children are less affected by interference by the maternal language?

To examine these questions we review three lines of research. First, we present studies using functional imaging with infants to determine the cerebral correlates of language acquisition. Next, we present studies that have tried to determine how brain lesions affect language learning and performance in young babies and children. Finally, we present studies in bilingual adults to determine the influence of age of acquisition on linguistic representations.

CEREBRAL ORGANIZATION OF LANGUAGE IN INFANTS: PRECURSORS AND DEVELOPMENT

In most adults the LH is involved in language; a fact that, in the field of neuro-psychology, has been acknowledged for well over a century. Yet not much is known about the cortical organization and its relation with function in the very young infant. The anatomical asymmetries observed in the classical language areas are already present in the fetal brain after 31 weeks of gestation; e.g. the left planum temporale is larger than its right counterpart (Chi *et al.*, 1977; Witelson & Pallie, 1973). Is this neonatal anatomical asymmetry correlated with a functional asymmetry or is it caused by a privileged association of language with the LH. Three types of studies have tried to disentangle this issue: first, behavioural studies with dichotic listening in normal children; secondly, ERP studies that attempt to establish the brain regions involved with linguistic processing in infants; and thirdly, studies that explore the impact of a brain lesion on language acquisition in children.

Behavioural studies

Dichotic listening[4] is used to determine whether linguistic stimuli are processed by the left hemisphere. When syllables or words are presented dichotically, right-handed adults show a right ear advantage (REA) which is usually attributed to the involvement of the LH in speech processing. The effect is reversed when subjects process other types of acoustic stimuli, as for example timbre or melodic contours. The magnitude of the REA for speech differs across studies from 60 to 80% in right-handed adults (Ahoniska *et al.*, 1993). Dichotic listening is considered a valid method to estimate the association of the LH with speech. Therefore we can use it to assess how infants process speech stimuli; in particular we can ask whether there is a REA for dichotically presented syllables in infants. Entus was the first to use dichotic presentation coupled with high-amplitude sucking , and Glanville *et al.* (1977) used a similar method coupled with the orienting response. Both groups reported that very young infants display a REA for dichotically presented syllables and a left-ear advantage (LEA) for musical tones. Bertoncini

---

[4]Dichotic presentation refers to the simultaneous stimulation of the right and left ear to establish whether one ear is better at processing its stimulus. In adults one observes a REA for language but not for many other auditory stimuli.

et al. (1989) reported similar results in neonates and in adults. Indeed, 55% of the neonates display a REA for syllables, 32.5% a LEA and in 12.5% no ear advantage was found. The reader should remember that Bertoncini *et al.* were unable to select the infants on the basis of handedness. Two other studies failed to find a REA for speech in very young infants (Best *et al.*, 1982; Vargha-Khadem & Corballis, 1979). The first of these studies, however, used poorly matched stimuli while the second study did find a trend in the same direction as in the studies presented above. In conclusion, although a REA seems to be present in very young infants, as in adults, the findings are not as conclusive as one would like them to be. This might in part be due to the experimental paradigms used to test infants. On the other hand, however, we see no persuasive argument to reject the notion that even in the very young infant the LH is already disposed to process speech stimuli.

Functional imaging in infants

Functional imaging techniques such as PETscan and fMRI have helped us to uncover cortical networks underlying cognitive processing. In the past decade a fairly good picture has emerged of the cortical structures that sustain language production and language comprehension. Imaging research has explored these issues with adult volunteers, but infants have been little studied. There are several reasons for this situation, which are mainly due to ethical and technical considerations. For instance, positron emission tomography, PETscan, is made possible by the administration of a radioactive tracer, which cannot be envisioned with healthy infants. fMRI, a technique in constant improvement, is difficult to apply in non-sedated healthy infants. The hope, however, is that new and improved methods will make it easier to study infants by improving the sound-shielding of the machines, by making the techniques non-invasive and less sensitive to movement. Currently, evoked potentials are successfully used in very young infants. Optical topography (Meek *et al.*, 1998) is a newly developed method which currently is being tested and used in several laboratories. The evoked related potentials (ERPs) in response to a change of stimulus that occurs after hearing a stimulus which is repeated several times (either of an acoustic or a phonetic nature) are characterized by a negativity recorded over the frontal areas, and this is called a mismatch negativity (MMN). This response is observed 100–200 ms after stimulus onset and its main generators are located in the temporal lobe. Giard and co-workers (Giard *et al.*, 1995) have shown that a change in one of the features of a sound, e.g. intensity, duration, or frequency, induces a MMN that is different for each of these features. Thus, MMN can be used to study the way auditory stimuli are coded. Using a similar procedure Dehaene-Lambertz & Dehaene (1994) also observed a mismatch response in infants. As in adult subjects the MMN in children differs as a function of the feature that is changed. Indeed, the topography of the MMN response in infants is different for a voice change (male versus female) and a syllable change (/ba/ vs /da/) (Dehaene-Lambertz, 2000). This suggests that, as in adults, the infants' auditory system has several dedicated networks, each one processing a specific property of the stimulus. Furthermore, one of these neuronal networks is specifically involved in the coding of a phonetic change. Dehaene-Lambertz &

Baillet (1998) used three syllables (S1, S2, and S3), synthesized along a /ba/ /da/ continuum so that the same acoustical difference exists between S1 and S2 and between S2 and S3. However, S1 and S2 are perceived as /ba/ and S3 as /da/, a phonetic boundary being located between S2 and S3. The mismatch response when the babies listen to S3 followed by S2 (/da/ /ba/) is significantly stronger than the one recorded when they listen to S1 followed by S2 (/ba/ /ba/), even though the acoustical distance is the same in both cases. Moreover, the MMM responses can be distinguished since crossing the phonetic boundary yields a dorsal and posterior activation compared to an acoustical change within the category (Figure 2).

A question that has been frequently asked in this section concerns the involvement of the LH in speech processing. In this context we can ask whether the network described above is lateralized towards the LH. In adults, Näätänen *et al.* (1997), using magneto-encephalography (MEG), have shown that the neural networks responsible for phonetic processing are asymmetric, involving predominantly the left temporal lobe. Investigations in infants (Dehaene-Lambertz & Dehaene, 1994; Dehaene-Lambertz & Baillet, 1998) indicated that the evoked potentials to syllables have higher voltages over the LH and they proposed the hypothesis of an asymmetry favouring the LH to process syllables. In more recent research, Dehaene-Lambertz (2000) compared the evoked response to tones and to syllables. In both cases the amplitude was significantly higher above the LH and the observed asymmetry was not greater for syllables than for tones. In contrast with the behavioural experiments reported above, Dehaene-Lambertz presented the same stimulus to both ears while the babies were distracted by a very interesting visual presentation[6]. REA in adults is observed only when dichotic presentations are used to induce competition between stimuli (O'Leary *et al.*, 1996). Moreover, an active task increases hemispheric asymmetry in adults (Imaizumi *et al.*, 1997). Both facts may explain the contrasting results found in behavioural studies and some of the ERP experiments.

Further support of a bilateral network involved in phonological processing is coming from the study of some very young patients. An infant who contracted a vascular lesion at birth in the left peri-sylvian areas was tested when she was 2 weeks old using ERPs. A mismatch response was present for a change in timbre as for a phonetic change (Dehaene-Lambertz *et al.*, 2001). In adults a comparable lesion yields a deficit in phoneme processing but leaves timbre processing intact. The infant patient, 2 weeks after the lesion, seems to have none of these deficits. Several reasons may explain this. Possibly the functional involvement of the LH with phonology arises in the adults because the initial abilities of the RH are pruned by the LH superiority in this domain. Another possibility might be that the RH does not have initial abilities to do phonological processing but learns quite quickly if the LH does not participate. Other possibilities are that the infant can relearn much faster than an adult, or that

---

[6]The presentation of interesting visual stimuli is done to keep the infants quiet and still during ERP recording. In order to obtain data in ERP it is necessary to repeat trials many times, making ERP experiments very annoying. The visual stimuli are not synchronized with the auditory stimuli to avoid any contamination of the auditory ERP by visual ERP.

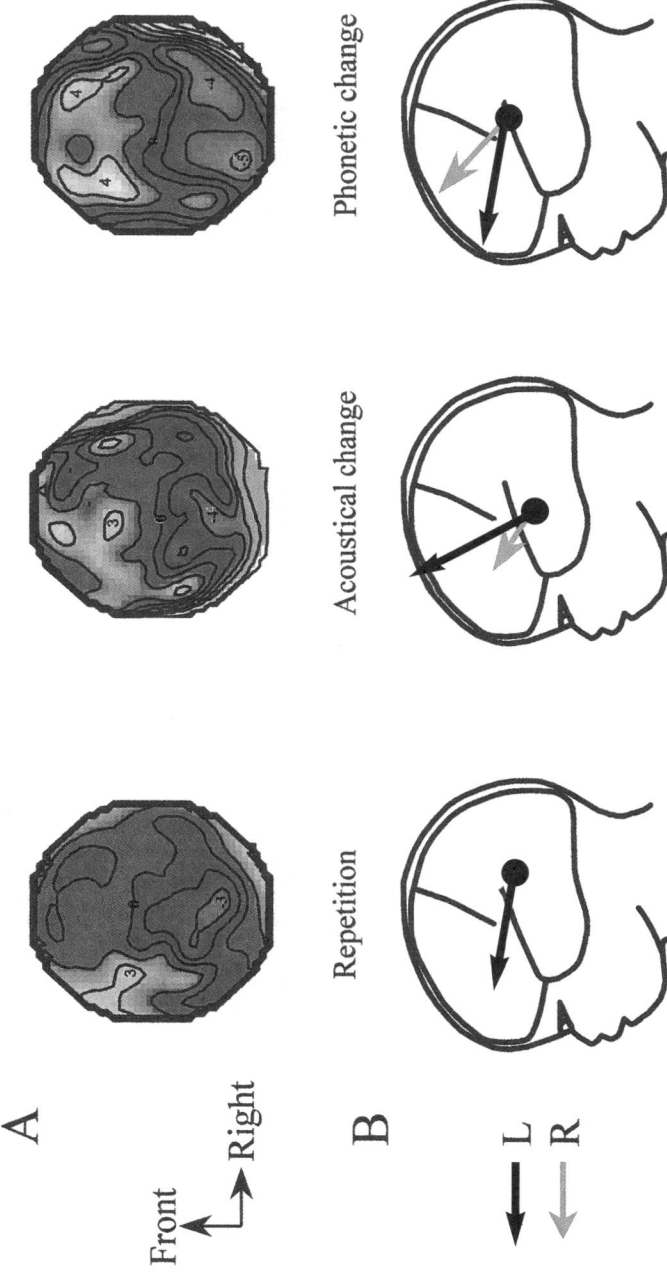

Figure 2 Infants' evoked responses to syllable change reflect both auditory and phonetic processes of mismatch detection. **A**: Topography of evoked responses in 3-month-old infants at the time of the maximal response to novelty (around 400 ms after syllable onset). **B**: Dipole model of the evoked response to a syllable when the syllable is repeated, when an acoustical and a phonetic change are detected. (From Dehaene-Lambertz & Baillet, 1998.)

what the infant has learned is identical to what the non-lesion infant has learned, or that the infant has regained some phonological abilities but not anything like that of the non-injured infant. At any rate, all these questions illustrate why it is so important to intensify neuropsychological research.

Brain organization has also been studied in older infants. Mills *et al*. (1993, 1997) explored the evoked responses to known and unknown words which were equalized for number of syllables and duration with infants from 13 to 20 months. As a control they used the same stimuli but played backwards. A negative response at 200–350 ms is observed for both known and unknown words. This negativity is attenuated for backward words, suggesting that language-like stimuli are processed differently than non-speech stimuli (even though they have a similar acoustical spectrum). This result is reminiscent of that reported in Perani's PETscan study. In that study an increased LH activation was demonstrated for a foreign language played forward compared to the same played backwards (Perani *et al*., 1996)[7].

Using ERPs it was thus possible to demonstrate that dedicated neural networks are involved in speech processing from an early age onwards. These studies did not provide decisive data on early lateralization of function. This may be due to the poor spatial resolution of the method. We expect, however, important progress in the near future on brain imaging studies in developing infants. Coupling ERP with other techniques that provide better spatial resolution, as for instance, fMRI and optical topography, seems a particularly promising direction in this research.

Neuropsychology: Impact of left and right brain lesions on language learning

In our quest to understand what enables the human brain to acquire language we have to study developmental clinical neuropsychology. Investigations in children with brain lesions offer another way to identify the areas mediating language acquisition. Obviously, anatomical information would be most useful in those cases in which the patients were also studied behaviourally. This, however, is seldom the case, which explains some of the misunderstandings and thus ferments many polemics among scientists. Scientists tend to espouse either a view of the organization of the infant's brain being structured and specified, or there are others who stress the brain as being un-structured at birth. Both of these theories often emerge from the same clinical observations. The explanation for this, as we point out above, is that the behavioural data are sufficiently ambiguous to leave open many alternative interpretations. We will come back to this point below.

As a starting point in our brief review we need to tease out some issues that could otherwise become a source of confusion. First, it is necessary to recall that aphasia presents itself differently in children and in adults. The most common symptom is the reduction in oral productions that can turn into complete mutism during the acute period. In contrast, common adult symptoms, e.g. logorrhoea or

---

[7]In this section we have not reviewed all ERP studies but only those that are informative concerning the cerebral bases of linguistic processes.

jargon aphasia, are almost never observed (Hécaen, 1976). This observation can lead to a sampling phenomenon in children's studies. Possibly, when spontaneous speech is impoverished, rather than strongly deviant, parents ignore this, or say that the child has regressed to an earlier stage of development. Unfortunately this means that infant patients often fail to go through a thorough linguistic evaluation following trauma. Second, most cerebral lesions (tumours or vascular accidents) in adults tend to be unilateral, but in children and infants diffuse injuries (anoxia, infection or brain traumatism) are the most frequent condition. Until recently it was difficult to be sure whether a lesion was unilateral. Generally, the lesion site was determined by the side of the hemiplegia which, however, is often a misleading indicator. Furthermore, in order to obtain large groups of patients, numerous studies have included children in whom the age at onset of the lesion was different, lesions with various aetiologies, various causes of brain lesion, different ages at testing, and co-occurrence of other pathologies. Conflating all these factors that might influence language development and brain organization has made it very difficult to draw any conclusions (to understand the danger of such a neglect, read Carter *et al*., 1982; Woods & Teuber, 1978).

If language arises because of properties that are inherent to the LH, then early lesions ought to give rise to severe language deficits[8]. Comparable deficits ought to be rare following a RH lesion. A perfunctory review of the literature illustrates the failure to study single patients; most data come from studies of large groups. A similar situation pervaded adults' clinical neuropsychology in the years before the advent of cognitive neuropsychology. In fact cognitive neuropsychology changed the focus from the study of large groups to that of single patients. Caramazza (1986, 1991) became the advocate for this change. He argued that only the single-subject approach makes it possible to understand the functional difficulties that patients encounter. A proper description of these can help establish a model of the normal cognitive architecture underlying performance. In turn this model helps us to understand the nature of patients' problems. Unfortunately this approach is often still ignored in developmental neuropsychology, a field that is badly in need of change.

Vargha-Khadem and her colleagues have studied a group of 196 hemiplegic children. Of these, 106 had a LH lesion and 90 a RH lesion (Vargha-Khadem *et al*., 2001). The patients were aged 5 years or older at the time of testing. The authors report that they were unable to observe a relation between the side of the lesion and verbal or performance IQ (VIQ, PIQ). Regardless of the side of the lesion, the IQ for the group was under 100 and the verbal IQ was slightly higher than the performance IQ. However, following a LH lesion, expression was generally more impaired than comprehension. Aram and Eisele (1994) found similar results in a study of a group of patients who had suffered a focal vascular lesion. Children with LH lesions had lower VIQ and PIQ than controls, while children with RH lesions had only a reduced PIQ relative to controls. Interestingly,

---

[8]The reason we do not say that subsequent to early LH lesions no language ought to arise is that some language may still be attained through the general intelligence of humans. However, even then one expects to find many problems with grammatical aspects of the competence.

there was no significant interaction between the side of lesion and the verbal or performance IQ. Similar findings have been reported by many other authors (Aram & Whitacker, 1988; Bates *et al.*, 1999) and these are in apparent contradiction to what is found in adult aphasics (Warrington *et al.*, 1986). This, as we will argue later, may be due to the way in which the IQ tests are constructed.

Yet Vargha-Khadem *et al.*'s study suggests that children who acquire a lesion between the first months of life and up to 5 years have a poorer performance than children born with a congenital lesion, or children who contract a lesion after 5 years of life. This pattern is similar for verbal and performance IQ. Woods & Carey (1979) and Riva & Cazzaniga (1986) reached a similar conclusion: lesions before the age of 1 year have a larger impact on IQ scores than lesions occurring later. However, in this last study anoxia, which can cause diffuse damage, was the main pathology in younger babies, whereas focal damage, such as tumours or vascular accidents, were predominant in the older group. Because the predominant type of brain lesion changes across ages, it is difficult to disentangle these two factors.

Vargha-Khadem *et al.* excluded patients with a history of uncontrolled seizures from the study. However, they point out the deleterious effect of seizures even if these are well controlled by therapy. The patients with a history of seizures scored 8.2 IQ points lower than patients without seizures. Across several other studies, epilepsy appears as an important factor that can affect verbal and performance IQ (Ballantyne & Trauner, 1999; Isaacs *et al.*, 1996; Muter *et al.*, 1997; Vargha-Khadem *et al.*, 2001). This might be due to the fact that chaotic brain activity prevents the functional reorganization of healthy tissue and/or to the side-effects of anticonvulsant drugs (Vargha-Khadem *et al.*, 1997). The deleterious impact of seizures may explain the contradicting findings in a more recent study by Vargha-Khadem and collaborators in comparison with those in a previous study with a smaller group of patients. They reported that LH lesions have a greater impact on linguistic development as evaluated by the token test. This smaller group comprised more children with a history of seizures (personal communication). Once again, this underscores the need for a careful selection of children before drawing conclusions from studies in groups of children.

A group of young patients who help us understand issues of brain specialization are children who have lost a hemisphere after surgical removal or by disconnection. Children with hemispherectomy teach us about the residual capacities in a single hemisphere. Fortunately, such patients are quite rare. Hemispherectomies are performed when symptoms, especially epileptic seizures, are so severe that the removal of one half of the brain is thought to have the potential to improve the clinical condition of the patient. Such patients are bound to differ in age of onset of the symptoms, duration and abruptness of the initial disease and age of surgery. Each one of these factors in themselves may have a large impact on cerebral reorganization. When groups of hemispherectomized patients are assessed, results pattern like those described above. Language development appears to be globally like that of control children matched for IQ regardless of the ablation of one or the other hemisphere (Vargha-Khadem, 1998).

Undeniably, group studies may sometimes be useful to gauge parameters that influence cognitive development. However, such global studies are not likely to further our understanding of how the brain of infants and young children is organized. Consider verbal IQ, a measure that is often used in studies of patients to measure language development. In fact, verbal IQ at best is a measure of the knowledge that has been acquired through verbal communication, or that requires understanding of verbal instructions. It is not clear what exactly is being measured. We are, however, certain that VIQ is not a good index to evaluate the linguistic capacities of the child. For example, giving the correct answer to questions as "Who is Christophe Colomb" or "What is the capital of Greece", (questions taken from the French version of the WISC III) are unrelated to linguistic development. The child might fail to answer correctly because he/she does not know Colomb or because he/she thinks that "who" means "what". We just raise these two sources of failure out of many more that we might conceive. Notice, however, that answering correctly is also related to the education that the patient has received and not to the language acquired. Children increase their language by being exposed to it, but encyclopaedic knowledge does not grow like that. The reader may be under the impression that we selected two atypical items to make our point. This is not so; all the items of the verbal IQ test tend to rely on acquired knowledge about the world, not knowledge of language. Thus, verbal IQ cannot ascertain whether a child has language problems or a deficit of attention, memory or encyclopaedic knowledge. In our opinion the field needs to design better ways to assess linguistic abilities in the growing child. Indeed, we are aware that therapists have no good instruments to evaluate patients at different ages. Psycholinguists have not yet created reliable instruments to determine how language acquisition proceeds across different social groups and languages. Being aware of this situation, we propose a very demanding solution. Child neuropsychology, like adult cognitive neuropsychology, ought to adopt the single-subject methodology. This implies that each child should be considered as an experiment of nature. For each child one ought to establish the linguistic (and non-linguistic) domains that are within the patient's competence and then to single out the specific problem. Neuropsychologists ought to obtain help from psycholinguists to evaluate whether matched controls encounter the same problems. To evaluate each child properly one has to test the child repeatedly during many sessions. This means that the child has to be willing to collaborate with the therapist. This arduous path may make it possible to understand the effects of focal lesions on language at different points in development. The available data do not yet give access to a similar understanding.

A few studies with children patients have tried to overcome some of the problems we have singled out. Aram and collaborators (1987) carefully selected a number of patients to evaluate aspects of linguistic residual syntactic and lexical capacities associated with brain lesions. Each patient was matched to his own control subject. The authors illustrate how deleterious brain lesions can be. The linguistic performance of children with either RH or LH damage is inferior to that of controls. Both groups make more errors than controls in naming tasks, for example. Interestingly, LH patients are slower in naming than controls, while the RH patients are faster, suggesting that the deficits underlying naming may be different in both groups. Lexical access appears most impaired in LH patients while attention problems and impulsiveness are mostly apparent in RH patients (Aram *et al.*, 1987). Both groups produce shorter and less complex sentences than the controls. Moreover, the performance of children with LH damage is worse than that of children with RH lesions. The mean length of utterances produced by the LH patients is shorter; a lower percentage of their sentences are correct and the sentences are of a lower complexity than those in RH patients (Aram & Ekelman, 1986). Similar results are reported by Eisele & Aram (1994) in comprehension and imitation of complex sentences: children with LH lesions are more impaired than those with RH lesions, although both groups are worse than their controls. Reilly *et al.* (1998) have studied the impairment of the narrative discourse in brain-damaged children. Contrary to Aram *et al.*, they do not find that the LH patients perform worse than the RH ones. However, their groups are not as homogeneous as the ones in Aram's study. Moreover, factors such as incidence of premature birth may have biased the results[9].

Single-case studies are the norm when reporting on hemispherectomized children. Vargha-Khadem *et al.* (1997) studied Alex, a patient whose LH was surgically removed when he was 9 years old. Alex had been diagnosed as suffering from a Sturge–Weber syndrome affecting his LH. This gave rise precociously to a severe case of epilepsy with no language development. Vargha-Khadem *et al.* state that Alex's comprehension was like that of a normal 3–4-year-old[10] with production limited to babbling. Following removal of the LH and the cessation of anticonvulsant drugs Alex rapidly began to acquire language. Two years after surgery, Alex's performance was greatly improved and he was able to produce sentences such as *"that wind-up toy will slide because it is on a slippery surface"*. He continued to improve for several years after surgery. However, Alex's language has quite a few problems. Not only has he difficulties in repeating non-words but he also has considerable difficulties with some aspects of phonology and morphology. Likewise, when asked to apply a morphological rule to a word he tends to fail. This suggests that Alex may have difficulties with parts of English grammar when he cannot use past memories or context as help.

Vargha-Khadem *et al.* (1991) have also studied a group of six hemispherectomized children (three left and three right). They report that the left-hemispherectomized patients perform less well compared to right-

---

[9]Although the aetiologies of the brain lesion are not given in this paper, some children are described with subcortical lesions only. This type of lesion is often found as a consequence of anoxia in preterm babies. As we have noted above, anoxia often has bilateral consequences even if the main lesion could appear unilateral on brain imagery. Furthermore, preterm babies have very specific brain lesions, that are of subcortical origin (periventricular) but could disorganize the last migration waves within the superficial layers of the cortex. This type of injury could affect much larger cortex areas than the focal porencephaly (Inder *et al.*, 1999).

[10]The authors' evaluation of Alex's comprehension level is problematic. Three-year-olds have been known to vary a great deal in language comprehension. Some children understand what they are told almost as well as an adult, barring some exceptionally complex constructions and infrequent words. Other children understand very little. So, stating that Alex's comprehension is like that of a 3–4-year-old is literally non-interpretable. Even if we are told that Alex could hardly understand any commands we are not told how this was assessed.

hemispherectomized and control subjects in production and comprehension tests that involve morphological items. Moreover, similar to Alex, they failed badly when they were tested with non-words. Stark & McGregor tested two children, one with a left and the other with a right hemispherectomy, both of whom displayed some linguistic difficulties. They have impoverished spontaneous speech but the left hemispherectomized child had more difficulties with phonetic tasks and with tasks that require the use of non-words. The child with a right hemispherectomy was able to acquire a better language postsurgically, and used longer and more complex sentences. Yet she clearly had some limitations. In comprehension she makes errors with the more complex syntactical sentences. Day & Ulatowska (1979) studied an 11-year-old girl whose LH was removed when she was 4 years old. Testing revealed that she had considerable difficulties producing irregular verbs, comparatives and irregular plurals.

The above review shows that our knowledge on how the different aspects of language processing are organized in the infants' brain still falls short. In particular one can assert that, in most adults, language is lateralized to the LH, but we cannot be certain about this in infants. The case studies reviewed above suggest that production difficulties are frequent in children with a malfunctioning LH and worse than in those with RH problems. In contrast, it is harder to assess comprehension difficulties. Boatman *et al.* (1999) followed children with left hemispherectomy longitudinally. The six patients were right-handed children (7–14 years) who had suffered from a Rasmussen encephalitis syndrome that affected their left hemisphere. These children were tested before surgery, 1 week, 6 months and 1 year after the surgery. Before surgery, all patients were impaired compared to controls in syllable discrimination tasks, in the token test and in a word-matching test. Word repetition was correct whereas naming was impaired. One week after surgery, five of the six patients were mute and all were unable to perform in a picture–word-matching-test. In contrast, four of the five patients showed an improved ability to discriminate syllables. Six months later the children were still severely aphasic but, except for one patient, phoneme discrimination was still improving. One year later the patients' performance was approaching normal levels and similar to that of controls. The naming scores remained below that of controls but it was still improving. Their spontaneous speech was largely telegraphic and limited to single words. These results indicate that the RH can process phonemes and discriminate one from another. Moreover, the RH progressively becomes more apt to comprehend speech inputs. None of the above studies, however, demonstrates that the rewired RH can perform as well as the LH. The studies underscore that simple and elementary syntax and phonology can, at least in part, be taken over by the RH.

In conclusion, wherever a brain lesion is located, it may impoverish speech. In general brain-damaged children do not master their native language as well as normal controls. Several studies point to the fact that left lesions are more damaging than right lesions, especially for syntax and morphology. We are fully aware that this tenuous conclusion is more a reflection of the poor state of knowledge in our field than of anything else. The years to come will have to focus on issues of plasticity of cortical structures and many other issues that are necessary to gain a better understanding of how and why humans, and only humans, gain access with such an incredible facility to a complex grammatical system like the one which underlies all natural languages.

Most developmental neuropsychologists study children older than 4 or 5 years; that is, after they have mastered the basic properties of language. A recent exception to this trend was reported by Bates *et al.* (1997). They studied a group of 53 children (36 left lesions and 17 right lesions) most of them with lesions contracted before they were 6 months old. Cerebral imagery was used to determine the loci of the lesion. The authors correlated the subjects' linguistic performances with lesion site. A parental questionnaire was used to try to assess the patients' linguistic performance at different ages. Before the patients were 1 year old no differences were found between patients with a left and a right lesion. When the patients were 19–30 months old those with a left temporal lesion or with a right or left frontal lesion produced fewer words and shorter sentences that other patients. The proportion of closed-class words relative to the total word production, however, was similar to what is observed in a normal population, suggesting a delay in acquisition rather than a deviant pattern of acquisition. The delay in production of the children with left temporal lesions is observed until 5 years. After age 5 the correlation between word production and comprehension and lesion site is no longer significant. The mean performance of children with brain lesions is like the poorest performance of controls. Yet some patients can reach excellent levels of verbal performances (Bates, 1998). Using tasks that rely upon narratives Reilly *et al.* (1998) found that left temporal lesions have an effect on linguistic capacities. Until the age of 7, narrative skills are delayed. The use of morphology and of complex sentences is poor in these patients. For these authors, rather than a qualitative deviation from the normal pattern, brain-damaged children are slowed down for language acquisition (Stiles *et al.*, 1998). However, it should be noted that, in general, these patients rarely catch up with normals. The examination of older patients has uncovered differences between RH and LH patients suggesting that there may be intrinsic differences between the tasks that the hemispheres accomplish.

In summary, young children's language is less impaired after a brain lesion; possibly because it is more rudimentary to start with. Young patients generally make progress and acquire a fairly adequate language, even if their language is at the lower values of the scale when compared to intact children. Moreover, language acquisition, mis-measured by verbal IQ, is similar regardless of the lesion side. Indeed, RH and LH lesions affect VIQ and PIQ in similar ways. Some observers take this result to argue that the cerebral cortex is highly plastic and that there is no reason to think that language is a property that is closely linked to the LH. Yet, as we mentioned above, there are too many problems with the studies that investigate linguistic capacities in young patients with the depth that is necessary to understand initial dispositions and plasticity issues with lack of distortion. Until better data become available we ought to be very prudent drawing conclusions.

LANGUAGE LEARNING ENHANCED IN AN AGE WINDOW?

A property of language which strongly points to its biological basis is the relation between age and the learning performance in language acquisition. Several studies have pointed towards an effect of age on the consequences of brain lesion, suggesting that the brain capacity to learn a language may change with maturation. Could this be related to the difficulties that are encountered by normal people when they want to learn a second language? We will now examine how age can affect language learning.

There are two kinds of polyglots relevant to this problem; first, those who learn two languages, so to say, while in the crib, can be seen as taking advantage of the brain's capacity to learn language with great ease in the first years of life; secondly, those who learn the second language with increased difficulty beyond puberty, that is, when the brain has become less flexible. Is there a clear correlation of proficiency of the second language with the age of acquisition? Is the speed and facility with which individuals learn a second language correlated with age? These are questions that we will address below. These issues are related, though in a different way, to those on the organization of the cortical language areas which were discussed above. Tentatively, one can draw the conclusion that the phonology and morphology of the second language (L2) are more difficult to learn than the lexical and semantic components.

It is a common observation that, even after many years in a country of adoption, one does not usually speak the L2 with the same proficiency as do the natives. Of course, some people can reach a very satisfactory performance in a L2 although late bilinguals usually do not perform as well as natives. This has been shown for segment representations (Cutler *et al.*, 1992), phonetic perception (Pallier *et al.*, 1997), grammaticality judgements (Johnson & Newport, 1989; Weber-Fox & Neville, 1996) and speech production (Flege *et al.*, 1997). Lenneberg wrote that puberty marks the end of the critical period after which there is a dramatic decrease in the capacity to learn a language. Since that statement, numerous studies have argued that, rather than an abrupt break at puberty, there is a slow decrease in the ability to learn that starts at birth and continues until adulthood.

Flege *et al.* (1997) have shown that experts can detect a non-native English spoken by English–Italian bilinguals even though they learned English as early as 3 years of age. Pallier *et al.* (1997) show that Catalan–Spanish adult bilinguals, who have practised both languages since the age of 6, perceive vowels in terms of a model that is adequate to their dominant language rather than for each one of their languages. Spanish dominants do not discriminate the open and closed /e/ vowels (non-existent in Spanish), whereas the Catalan-dominant speakers have no difficulty with these particular vowels. Maybe we need to add that these bilinguals are as proficient as one can hope to find.

Johnson & Newport (1989) have also reported difficulties for bilinguals with their L2 (e.g. Chinese immigrants in the US). The syntactic proficiency, the authors discovered, was correlated with the age of acquisition rather than with the duration of exposure to the L2. Weber-Fox & Neville (1996) have performed a similar behavioural study that they complemented by using evoked potentials. Their results also show that, even though some of the bilinguals have spoken the L2 more frequently than the L1 since the age of 3, they do not perform as mono-linguals in a variety of syntactic tests. These results suggest that exposure to a new language over many years is not sufficient to ensure its complete mastery. Or are there circumstances in which the L2 can be mastered without the slightest hitch? How can one reconcile the difficulty in learning some aspects of the L2 despite massive exposure, training and motivation, with the recent demonstrations that brains remain plastic and that cortical areas can be re-mapped even in old organisms? Why are the cerebral areas that deal with L1 so efficiently incapable to represent L2 equally well?

The two following hypotheses can be put forward. First, age of acquisition determines performance in the L2 because a critical period, or a window of opportunity, makes it difficult to learn language afterwards. Second, once one has learned a L1 this interferes with the learning and performance of L2. Several observations favour the first hypothesis. Newport studied congenital deaf people, who were of different ages when first exposed to sign language and attained a competence in sign language that was inversely related to the age of acquisition. Furthermore, the type of syntactical errors of late learners was different from the ones made by early learners. To a large extent, late learners appear as being similar to tardy bilinguals (Mayberry & Fischer, 1989). This result goes against the interference hypothesis mentioned above.

Recently, Grimshaw *et al.* (1998) studied a deaf child, E.M., who during infancy had had no contact with either oral language or signed language. This child was fitted with a hearing aid at the age of 15 and was able to learn Spanish. However, despite his normal intelligence, his Spanish was deficient when tested 4 years after he was given the hearing aid. The authors report some striking deficiencies. For example, the boy was unable to understand simple negation, a structure that is usually acquired by 2½ years (Brown, 1973). The boy appeared to have mastered the singular versus plural contrast, yet his performance was not flawless. His language production consisted of one word associated with a home-sign gesture. Even compared with a 3- or 4-year-old child his language appeared very deficient. Hopefully this boy will have improved with continuing exposure, but it appears that language will remain a difficult challenge to overcome in such cases. This single-case study indicates that interference may not be the critical issue with learning a language at a later age. A critical period, or a closing window of opportunities, seems to be a more adequate analogue to explain the facts. Studies with feral children also tend to go in this direction although the problems with language acquisition in this population are mostly difficult to interpret due to the intense emotional disorder induced by neglect or to pre-existing deficits that caused the neglect in the first place.

The case study reported by Grimshaw *et al.* (1998) appears to go against the conclusions that can be drawn from Alex, the child who was hemispherectomized at the age of 9 years (Vargha-Khadem *et al.*, 1997). Vargha-Khadem and co-workers claim that Alex's rapid acquisition of speech after the age of 9 challenges the critical period hypothesis. However, notice that, contrary to E.M., Alex had been exposed to language and, according to presurgical evaluation, his comprehension was like that of a 4-year-old. Possibly Alex had acquired some language processing, at any rate enough to become attuned to the native language even in the absence of production. Various explanations may explain the absence of production. One must agree, however, that Alex's performance after such a protracted period of mutism is quite spectacular. However, his case is not a strong argument against a critical period hypothesis. It is, as we have stated above, more of a challenge for the notion that there is little plasticity of brain areas dedicated to language processing. Again we must come to the conclusion that we are in dire need of more data. Hopefully the development of a more systematic study of single cases, such as E.M. and Alex, will clarify the issues we are exploring. Let us now turn to the evidence related to the manner in which L1 and L2 are represented in the brain of bilinguals.

How is L2 represented in the brain of bilinguals? Are the same areas involved for L1 and L2? One would have hoped that some of these questions would have been addressed in patient studies. Several authors have tried to assess the outcome of stroke in bilinguals. Initially, many reports suggested that L1 and L2 were represented in different brain areas. Further studies did not confirm these initial observations. All combinations were described from the loss of both languages following stroke to the preservation of a single language, either L1 or L2 (Paradis, 1995). Functional cerebral imagery (fMRI and PETscan) has helped us to understand the origin of such diverging clinical descriptions. Our laboratory has collaborated with the imaging laboratories in Milan and in Orsay to study comprehension of L1 and L2 in late and early French–English bilinguals using fMRI. In all late bilingual subjects listening to L1 always activates a set of left perisylvian areas, mainly the middle and superior temporal gyri clustering along the superior temporal sulcus. Listening to L2 activates a more variable network of left and right temporal and frontal areas. These results corroborate the earlier PETscan results suggesting that the representation of L1 is more predictable than that of L2 in late bilinguals. For the authors these results support the hypothesis that the acquisition of L1 relies on a dedicated cerebral network, mainly in the LH, while later acquisition of L2 is not necessarily associated with a reproducible biological substrate. These results have shed some light on explaining the problems in the variability of the deficits observed in neuropsychology.

The dispersion of the areas activated when late bilinguals listen to L2 may arise because different individuals utilize different strategies to comprehend the stories, or to the fact that during learning of L2 they used different methods, ranging from conversation and grammar books to merely listening to tapes. Perani et al. (1998) tested two groups of highly proficient bilinguals. The first group was composed of Catalan–Spanish bilinguals who learned L2 before the age of 6 and generally before the age of 4. The second group was composed of Italian–Engish subjects who learned English after puberty but used L2 daily, thereafter attaining a high level of proficiency. In this study with high-proficiency bilinguals no major differences were found between L1 and L2, suggesting that the diffuse representation of L2 in the experiment described above may hold only for low-proficiency bilinguals. In this study it was shown that the level of proficiency is a much better predictor of the cortical representation of L2 than age of acquisition. If volunteers have acquired a very high proficiency level, L1 and L2 will have a similar representation regardless of the age when L2 was acquired. This is an important finding that might have to be explored somewhat further when other studies with similar populations are carried out with a larger variety of tasks than the ones used by Perani et al. One such study, carried out by Kim et al. (1997), already provides us with some additional information. They use fMRI to study implicit production in both languages of bilingual volunteers. They report that in late bilinguals one finds that the two languages use different loci in Broca's area, whereas the representations overlap in early learners. In

Wernicke's area activations for both languages overlap regardless of the age of acquisition. The contrast between this study and that of Perani *et al.* shows that the task that volunteers perform is an essential aspect of the results one obtains. Indeed, while in comprehension Perani *et al.* did not observe any difference in the representation of L1 and L2, in production Kim did not observe differences in Wernicke's area. Notice that this was not the case for the low-proficiency bilinguals we tested. Neville *et al.* studied Chinese–English bilinguals who had arrived in United States at various ages, using ERPs. ERP differences between bilinguals and native English speakers in syntactic tasks are correlated with the age of arrival in the US, with a left–right asymmetry that decreases for late-learners. Similar results are described for ASL–English bilinguals (Bavelier *et al.*, 1998; Neville *et al.*, 1997). Obviously the studies reviewed above give us an indication as to the extraordinary potential which imaging has to clarify the function–structure relationship in bilingual speakers. However, for the time being very few studies have been conducted, and we must wait for more studies in order to begin understanding the way in which brain structure guides early and late acquisition of L2. Yet some generalities seem to be emerging. L1 comprehension involves mainly the perisylvian areas of the LH. These areas are the same in most subjects across a sample of different native languages, including sign languages. In contrast, a L2, even when learned very early on, is still not like the language representation in monolinguals as indicated in a large number of behavioural tests. However, when highly proficient bilinguals are tested the representation of L1 and L2 are quite similar to one another. These conclusions are valid for bilinguals who have mastered two oral languages, while the situation is more confusing for sign and oral languages (Neville *et al.*, 1997; Paulesu & Mehler, 1998).

EPILOGUE

Through this chapter we have reviewed the data coming from the different fields of cognitive neuroscience: developmental studies, studies into brain lesions and brain imaging studies, in order to understand what makes the human brain uniquely able to acquire language. We take it that the existing data do not support one or the other of the accounts of how humans acquire language. In particular, brain lesion studies are still inconclusive, due to the preference given to studies of groups of young patients, often without giving adequate descriptions of the individual symptoms. Although neuropsychological studies are generally more difficult in children than in adults, we think that the careful study of single cases is necessary and unavoidable if we want to attain gains similar to those made by cognitive neuropsychology in adults. We also hope that the development of cerebral brain imaging techniques to be used safely with infants can provide new data and hence more sophisticated theories. Last but not least, we think that the success of the enterprise will be based on a rich interaction of neuroscience, psychology and linguistics.

ACKNOWLEDGEMENT

G.D. and M.P. are supported by the McDonnell Foundation.

References

Abercromie, D. (1967). *Elements of general phonetics*. Edinburgh: Edinburgh University Press.

Ahoniska, J., Cantell, N., Tolvanen, A. & Lyytenen, H. (1993). Speech perception and brain laterality: the effect of ear advantage on auditory event-related potential. *Brain and Language*, **45**, 127–146.

Aram, D.M. & Eisele, J.A. (1994). Intellectual stability in children with unilateral brain lesions. *Neuropsychologia*, **32**, 85–95.

Aram, D.M. & Ekelman, B.L. (1986). Spoken syntax in children with acquired unilateral hemisphere lesions. *Brain and Language*, **27**, 75–100.

Aram, D.M. & Whitacker, H.A. (1988). Cognitive sequelae of unilateral lesions acquired in early childhood. In D.L. Molfese & S.J. Segalowitz (Eds.), *Developmental implications of brain lateralization* (pp. 171–i97). New York: Guilford Press.

Aram, D.M., Ekelman, B.L. & Whitacker, H.A. (1987). Lexical retrieval in left and right brain lesioned children. *Brain and Language*, **31**, 61–87.

Ballantyne, A.O. & Trauner, D.A. (1999). Stability of IQ with age in children after perinatal stroke. Paper presented at the Cognitive Neuroscience Meeting, Washington.

Bates, E. (1998). Brain and language in children and adults. Paper presented at the Brain Development and Cognition in Human Infants: Development and Functional Specialization of the Cortex Conference, San Feliu de Guixols, Spain, 23–28 September.

Bates, E., Thal, D., Trauner, D., Fenson, J., Aram, D., Eisele, J. & Nass, R. (1997). From first words to grammar in children with focal brain injury. *Developmental Neuropsychology*, Special issue on Origins of Language Disorders 13, 275–344.

Bates, E., Vicari, S. & Trauner, D. (1999). Neural mediation of language development: perspectives from lesion studies of infants and children. In H. Tager-Flusberg (Ed.), *Neurodevelopmental disorders* (pp. 533–581). Cambridge, MA: MIT Press.

Bavelier, D., Corina, D., Jezzard, P., Clark, V., Karni, A., Lalwani, A., Rauschecker, J.P., Braun, A., Turner, R. & Neville, H.J. (1998). Hemispheric specialization for English and ASL: left invariance–right variability. *NeuroReport*, **9**, 1537–1542.

Bertoncini, J., Morais, J., Bijeljac-Babic, R., MacAdams, S., Peretz, I. & Mehler, J. (1989). Dichotic perception and laterality in neonates. *Brain and Cognition*, **37**, 591–605.

Best, C.T., Hoffman, H. & Glanville, B.B. (1982). Development of infant ear asymmetries for speech and music. *Perception and Psychophysics*, **31**, 75–85.

Best, C.T., McRoberts, G.W. & Sithole, N.M. (1988). Examination of the perceptual reorganization for speech contrasts: Zulu chick discrimination by English-speaking adults and infants. *Journal of Experimental Psychology: Human Perception and Performance*, **3**, 345–360.

Boatman, D., Freeman, J., Vining, E., Pulsifer, M., Miglioretti, D., Minahan, R., Carson, B., Brandt, J. & McKhann, G. (1999). Language recovery after left hemispherectomy in children with late-onset seizures. *Annals of Neurology*, **46**, 579–586.

Bosch, L. & Sebastian-Gallés, N. (1997). Native-language recognition abilities in 4-month-old infants from monolingual and bilingual environments. *Cognition*, **65**, 33–69.

Brown, R. (1973). *A first language: the early stages*. Cambridge, MA: Harvard University Press.

Caramazza, A. (1986). On drawing inferences about the structure of normal cognitive processes from patterns of impaired performance: the case for single-patient studies. *Brain and Cognition*, **5**, 41–66.

Caramazza, A. (1991). Data, statistics, and theory: a comment on Bates, McDonald, MacWhinney, and Applebaum's "A maximum likelihood procedure for the analysis of group and individual data in aphasia research". *Brain and Language*, **41**, 43–51.

Caramazza, A., Chialant, D., Capasso, R. & Miceli, G. (2000). Separable processing of consonants and vowels. *Nature*, **403**, 428–430.

Carter, R.L., Hohenegger, M.K. & Satz, P. (1982). Aphasia and speech organisation in children. *Science*, **218**, 797–799.

Chi, J.G., Dooling, E.C. & Gilles, F.H. (1977). Gyral development of the human brain. *Annals of Neurology*, **1**, 86–93.

Chomsky, N. (1957). *Syntactic structure*. The Hague: Mouton.

Chomsky, N. (1959). A review of Skinner's. *Verbal Behavior*, **35**, 26–58.

Chomsky, N. (1965). *Aspects of a theory of syntax*. Cambridge, MA: MIT Press.

Christophe, A., Dupoux, E., Bertoncini, J. & Mehler, J. (1994). Do infants perceive word boundaries? An empirical study of the bootstrapping of lexical acquisition. *Journal of the Acoustical Society of America*, **95**, 1570–1580.

Christophe, A. & Morton, J. (1998). Is Dutch native English? Linguistic analysis by 2-month-olds. *Developmental Science*, **1**, 215–219.

Cutler, A., Mehler, J., Norris, D.G. & Segui, J. (1992). The monolingual nature of speech segmentation by bilinguals. *Cognitive Psychology*, **24**, 381–410.

Day, P.S. & Ulatowska, H.K. (1979). Perceptual, cognitive and linguistic development after early hemispherectomy: two case studies. *Brain and Language*, **7**, 17–33.

Dehaene-Lambertz, G. (2000). Cerebral specialization for speech and non-speech stimuli in infants. *Journal of Cognitive Neuroscience*, **12**, 449–460.

Dehaene-Lambertz, G. & Baillet, S. (1998). A phonological representation in the infant brain. *NeuroReport*, **9**, 1885–1888.

Dehaene-Lambertz, G. & Dehaene, S. (1994). Speed and cerebral correlates of syllable discrimination in infants. *Nature*, **370**, 292–295.

Dehaene-Lambertz, G. & Houston, D. (1998). Faster orientation latency toward native language in two-month-old infants. *Language and Speech*, **41**, 21–43.

Dehaene-Lambertz, G., Pena, M., Charolais, A. & Landrieu, P. (2001). Is the left hemisphere useful for phoneme discrimination in infants? (In preparation).

Dronkers, N.F. (1996). A new brain region for coordinating speech articulation. *Nature*, **384**, 159–161.

Eisele, J.A. & Aram, D.M. (1994). Comprehension and imitation of syntax following early hemisphere damage. *Brain and Language*, **46**, 212–231.

Entus, A.K. (1977). Hemispheric asymmetry in processing of dichotically presented speech and nonspeech stimuli by infants. In S.J. Segalowitz & F.A. Gruber (Eds.), *Language development and neurological theory* (pp. 63–73). New York: Academic Press.

Flege, J.E., Frieda, E.M. & Nozawa, T. (1997). Amount of native-language (L1) use affects the pronunciation of an L2. *Journal of Phonetics*, **25**, 169–186.

Gesell, A., Halverson, H.M., Thompson, H., Ilg, F.L., Castner, B.M., Ames, L.B. & Amatruda, C.S. (1940). *The first five years of life*. New York: Harper & Row.

Giard, M.H., Lavikainen, J., Reinikainen, R., Perrin, F., Bertrand, O., Pernier, J. & Näätänen, R. (1995). Separate representations of stimulus frequency, intensity, and duration in auditory sensory memory: an event-related potential and dipole-model analysis. *Journal of Cognitive Neuroscience*, **7**, 133–143.

Glanville, B.B., Best, C.T. & Levenson, R. (1977). A cardiac measure of cerebral asymmetries in infant auditory perception. *Developmental Psychology*, **13**, 54–59.

Gleitman, L.R. & Wanner, E. (1982). Language acquisition: the state of the art. *Language acquisition: the state of the art*, 3–48.

Grimshaw, G.M., Adelstein, A., Bryden, M.P. & MacKinnon, G.E. (1998). First-language acquisition in adolescence: evidence for a critical period for verbal language development. *Brain and Language*, **63**, 237–255.

Hécaen, H. (1976). Acquired aphasia in children and the ontogenesis of hemispheric functional specialization. *Brain and language*, **3**, 114–1134.

Imaizumi, S., Mori, K., Kiritani, S., Kawashima, R., Sugiura, M., Fukuda, H., Itoh, K., Kato, T., Nakamura, A., Hatano, K., Kojima, S. & Nakamura, K. (1997). Vocal identification of speaker and emotion activates different brain regions. *NeuroReport*, **8**, 2809–2812.

Inder, T.E., Hüppi, P.S., Warfield, S., Kikinis, R., Zientara, G.P., Barnes, P.D., Jolesz, F. & Volpe, J.J. (1999). Periventricular white matter injury in the premature infant is followed by reduced cerebral cortical gray matter volume at term. *Annals of Neurology*, **46**, 755–760.

Isaacs, E., Christie, D., Vargha-Khadem, F. & Mishkin, M. (1996). Effects of hemispheric side of injury, age at injury, and presence of seizure disorder on functional ear and hand asymmetries in hemiplegic children. *Neuropsychologia*, **34**, 127–137.

Johnson, J.S. & Newport, E.L. (1989). Critical period effects in second language learning: the influence of maturational state on the acquisition of English as a second language. *Cognitive Psychology*, **21**, 60–99.

Jusczyk, P.W. & Hohne, E.A. (1997). Infants memory for spoken words. *Science*, **277**, 1984–1985.

Jusczyk, P.W., Friederici, A., Wessels, J., Svenkerud, V. & Jusczyk, A. (1993a). Infants' sensitivity to the sound pattern of native language words. *Journal of Memory and Language*, **32**, 402–420.

Jusczyk, P.W., Cutler, A. & Redanz, N.J. (1993b). Infants' preference for the predominant stress patterns of English words. *Child Development*, **64**, 675–687.

Jusczyk, P.W., Luce, P.A. & Charles-Luce, J. (1994). Infants' sensitivity to phonotactic patterns in the native language. *Journal of Memory and Language*, **33**, 630–645.

Kanwisher, N., McDermott, J. & Chun, M.M. (1997). The fusiform face area: a module in human extrastriate cortex specialized for face perception. *Journal of Neurosciences*, **17**, 4302–4311.

Kim, K.H.S., Relkin, N.R., Lee, K.N. & Hirsch, J. (1997). Distinct cortical areas associated with native and second languages. *Nature*, **338**, 171–174.

Kluender, K.R., Lotto, A.J., Holt, L.L. & Bloedel, S.L. (1998). Role of experience for language-specific functional mappings of vowel sounds. *Journal of the Acoustical Society of America*, **104**, 3568–3582.

Kuhl, P.K., Williams, K.A., Lacerda, F., Stevens, K.N. & Lindblom, B. (1992). Linguistic experiences alter phonetic perception in infants by 6 months of age. *Science*, **255**, 606–608.

Lenneberg, E.H. (1967). *Biological Foundations of Language*. New York: John Wiley & Sons.

Marcus, G.F., Vijayan, S., Rao, S.B. & Vishton, P.M. (1999). Rule learning by seven-month-old infants. *Science*, **283**, 77–80.

Mayberry, R.I. & Fischer, S.D. (1989). Looking through phonological shape to lexical meaning: The bottleneck of non-native sign language processing. *Memory & Cognition*, **17**, 740–754.

Meek, J.H., Firbank, M., Elwell, C.E., Atkinson, J., Braddick, O. & Wyatt, J.S. (1998). Regional hemodynamic responses to visual stimulation in awake infants. *Pediatric Research*, **43**, 840–843.

Mehler, J., Jusczyk, P., Lambertz, G., Halsted, N., Bertoncini, J. & Amiel-Tison, C. (1988). A precursor of language acquisition in young infants. *Cognition*, **29**, 143–178.

Miller, G.A. (1991). *The science of words*. New York: Scientific American Library.

Mills, D.L., Coffey-Corina, S.A. & Neville, H.J. (1993). Language acquisition and cerebral specialization in 20-month-old infants. *Journal of Cognitive Neuroscience*, **5**, 317–334.

Mills, D.L., Coffey-Corina, S.A. & Neville, H.J. (1997). Language comprehension and cerebral specialization from 13 to 20 months. *Developmental Neuropsychology*, **13**, 397–445.

Moon, C., Cooper, R.P. & Fifer, W. (1993). Two-day-olds prefer their native language. *Infant Behavior and Development*, **16**, 495–500.

Muter, V., Taylor, S. & Vargha-Khadem, F. (1997). A longitudinal study of early intellectual development in hemiplegic children. *Neuropsychologia*, **35**, 289–298.

Näätänen, R., Lehtokovski, A., Lennes, M., Cheour, M., Huotilainen, M., Iivonen, A., Vainio, M., Alku, P., Ilmoniemi, R.J., Luuk, A., Allik, J., Sinkkonen, J. & Alho, K. (1997). Language-specific phoneme representations revealed by electric and magnetic brain responses. *Nature*, **385**, 432–434.

Nazzi, T., Bertoncini, J. & Mehler, J. (1998). Language discrimination by newborns: towards an understanding of the role of rythm. *Journal of Experimental Psychology: Human Perception and Performance*, **24**, 1–11.

Nespor, M., Guasti, M.T. & Christophe, A. (1996). Selecting word order: the rhythmic activation principle. In U. Kleinhenz (Ed.), *Interfaces in phonology* (pp. 1–26). Berlin: Akademie Verlag.

Neville, H.J., Coffey, S.A., Lawson, D.S., Fisher, A., Emmorey, K. & Bellugi, U. (1997). Neural systems mediating American Sign Language: effects of sensory experience and age of acquisition. *Brain and Language*, **57**, 285–308.

Newport, E.L. (1990). Maturational constraints on language learning. *Cognitive Science*, **14**, 11–28.

Nowak, M.A., Plotkin, J.B. & Jansen, V.A.A. (2000). The evolution of syntactic communication. *Nature*, **404**, 495–498.

O'Leary, D.S., Andreasen, N.C., Hurtig, R.R., Hichiwa, R.D., Watkins, G.L., Ponto, L.L.B., Rogers, M. & Kirchner, P.T. (1996). A positron emission tomography study of binaurally and dichotically presented stimuli: effects of level of language and directed attention. *Brain and Language*, **53**, 20–39.

Pallier, C., Bosch, L. & Sebastian, N. (1997). A limit on behavioral plasticity in speech perception. *Cognition*, **64**, B9–B17.

Paradis, M. (1995). *Aspects of bilingual aphasia*. Oxford: Pergamon Press.

Paulesu, E. & Mehler, J. (1998). Right on in sign language. *Nature*, **392**, 233–234.

Perani, D., Dehaene, S., Grassi, F., Cohen, L., Cappa, S.F., Dupoux, E., Fazio, F. & Mehler, J. (1996). Brain processing of native and foreign languages. *NeuroReport*, **7**, 2439–2444.

Perani, D., Paulesu, E., Sebastian-Gallés, N., Dupoux, E., Dehaene, S., Bettinardi, V., Cappa, S.F., Mehler, J. & Fazio, F. (1998). The bilingual brain: proficiency and age of acquisition of the second language. *Brain*, **121**, 1841–1852.

Polka, L. & Werker, J.F. (1994). Developmental changes in perception of non-native vowel contrasts. *Journal of Experimental Psychology: Human Perception and Performance*, **20**, 421–435.

Ramus, F. & Mehler, J. (1999). Language identification with suprasegmental cues: a study based on speech resynthesis. *Journal of the Acoustical Society of America*, **105**, 512–521.

Ramus, F., Nespor, M. & Mehler, J. (1999). Correlates of the linguistic rhythm in the speech signal. *Cognition*, **73**, 265–292.

Reilly, J.S., Bates, E.A. & Marchman, V.A. (1998). Narrative discourse in children with early focal brain injury. *Brain and Language*, **61**, 335–375.

Riva, D. & Cazzaniga, L. (1986). Late effects of unilateral brain lesions sustained before and after age one. *Neuropsychologia*, **24**, 423–428.

Saffran, J.R., Aslin, R.N. & Newport, E.L. (1996). Statistical learning by 8-month-old infants. *Science*, **274**, 1926–1928.

Shafer, V.L., Shucard, D.W., Shucard, J.L. & Gerken, L.A. (1998). An electrophysiological study of infants' sensitivity to the sound patterns of English speech. *Journal of Speech, Language and Hearing Research*, **41**, 874–886.

Stark, R.E. & McGregor, K.K. (1997). Follow-up study of a right and a left-hemispherectomized child: Implications for localization and impairment of language in children. *Brain and Language*, **60**, 222–242.

Stiles, J., Bates, E.A., Thal, D., Trauner, D. & Reilly, J. (1998). Linguistic, cognitive, and affective development in children with pre- and perinatal focal brain injury: a ten year overview from the San Diego longitudinal project. In C. Rovee-Collier, L. Lipsitt & H. Hayne (Eds.), *Advances in infancy research* (vol. 12, pp. 131–163). Stanford, CA: Ablex.

Vargha-Khadem, F. (1998). Effects of unilateral vs bilateral pathology on development of speech and language functions. Paper presented at the Brain Development and Cognition in Human Infants: Development and Functional Specialization of the Cortex Conference, San Feliu de Guixols, Spain, 23–28 September.

Vargha-Khadem, F. & Corballis, M. (1979). Cerebral assymetry in infants. *Brain and Language*, **8**, 1–9.

Vargha-Khadem, F., Isaacs, E.B., Papaleloudi, H., Polkey, C.E. & Wilson, J. (1991). Development of language in six hemispherectomized patients. *Brain*, **114**, 473–495.

Vargha-Khadem, F., Carr, L.J., Isaacs, E., Brett, E., Adams, C. & Mishkin, M. (1997). Onset of speech after left hemispherectomy in a nine-year-old boy. *Brain*, **120**, 159–182.

Vargha-Khadem, F., Isaacs, E., Watkins, K. & Mishkin, M. (2001). Ontogenic specialization of hemispheric function. In J. Oxbury, C.E. Polkey & M. Duchowny (Eds,), *Intractable focal epilepsy: medical and surgical treatment* (In Press). London: Saunders.

Warrington, E.K., James, M. & Maciejewski, C. (1986). The WAIS as a lateralizing and localizing diagnostic instrument: a study of 656 patients with unilateral cerebral lesion. *Neuropsychologia*, **24**, 223–239.

Weber-Fox, C.M. & Neville, H.J. (1996). Maturational constraints on functional specialization for language processing: ERP and behavioral evidence in bilingual speakers. *Journal of Cognitive Neuroscience*, **8**, 231–256.

Werker, J.F. & Tees, R.C. (1984). Cross-language speech perception: evidence for perceptual reorganisation during the first year of life. *Infant Behavior and Development*, **7**, 49–63.

Witelson, S.F. & Pallie, W. (1973). Left hemisphere specialization for language in the newborn: neuroanatomical evidence for asymmetry. *Brain*, **96**, 641–646.

Woods, B.T. & Teuber, H.L. (1978). Changing patterns of childhood aphasia. *Annals of Neurology*, **3**, 273–280.

Woods, M.T. & Carey, S. (1979). Language deficits after apparent clinical recovery from childhood aphasia. *Annals of Neurology*, **3**, 273–280.

V.8
Early Temperament and Emotional Development

DOUGLAS DERRYBERRY and MARY K. ROTHBART

ABSTRACT

This chapter discusses the temperamental organization of reactive processes related to positive emotions (smiling and laughter, approach) and
negative emotions (frustration, fear), as well as regulatory processes
related to attention. Evidence is reviewed suggesting that reactive
approach and frustration tendencies are associated between infancy
and middle childhood. Fear becomes dissociated from frustration, and
serves as a reactive inhibitory control upon approach tendencies. More
flexible and voluntary means of control are afforded by the development
of attentional systems which regulate reactive approach, frustration,
and fear tendencies. These processes are discussed in terms of their
underlying neural systems, stability during infancy and early childhood,
and relation to adult personality.

INTRODUCTION

From even the earliest months of life infants differ dramatically in their
developing emotional and behavioural reactions. These differences are
intriguing in that they provide a relatively simple model for linking initial
expressions of personality to the development of underlying brain systems.
Perhaps even more intriguing is the possibility that these early differences in
temperament provide a foundation for later emotional and personality
development.

Many of the infant's reactions are highly reactive in nature. They are elicited
by relatively simple external or internal stimuli, and follow a fairly automatic
course across time. Examples include the smiling and approach reactions, fearful
and irritable forms of distress, and orienting of the head and eyes to novelty
shown during the first year of life. These early reactions are thought to reflect

967

A.F. Kalverboer and A. Gramsbergen (eds.), Handbook of Brain and Behaviour in Human Development, 967–988
© 2001 Kluwer Academic Publishers. Printed in Great Britain.

development within subcortical neural systems related to basic emotional and motivational tendencies (Rothbart *et al.*, 1994a; Derryberry & Rothbart, 1997). In characterizing these reactive processes, temperament approaches emphasize not only their qualitative nature, but also their intensity and fluctuations across time. Given a fearful response, for example, infants may differ in how rapidly they demonstrate the reaction, the fear's peak intensity, how quickly it builds to this peak, how long the peak intensity is maintained, and how quickly it recovers to a non-fearful state. During the early months the caregiver provides the major regulation of these reactions, applying soothing techniques such as holding and rocking to dampen the rising distress and promote a rapid recovery (Rothbart & Derryberry, 1981).

As development proceeds, however, the infant becomes more skilled at regulating his or her own reactive emotionality. This increasing regulatory capacity is thought to reflect development within frontal cortical systems related to attention and voluntary control (Berger & Posner, this handbook; Rothbart *et al.*, 1994a). In the case of a fear reaction, infants may deploy their attention towards a safe or comforting object in the environment, such as the caregiver or a favourite blanket, thereby attenuating their rising fear. At later ages the child may be motivated by fear to resist temptation and to delay gratification by directing attention away from the attractive properties of a stimulus. Although such regulatory control is viewed as a basic capacity of the nervous system, temperament approaches emphasize the pronounced differences in self-regulation evident across individual children. Taken together with individual differences in reactive emotional processes, differences in self-regulation allow many paths of emotional development (Rothbart & Bates, 1998).

In this chapter we review recent findings on the reactive and regulatory components of temperament. The first two sections focus on more reactive processes in the positive and negative emotions, and the third section focuses on the regulation of positive and negative emotions. Within each section we begin with a theoretical overview and a brief description of the underlying neural systems. We then discuss the emotional processes as expressed during infancy, followed by a consideration of their stability during childhood. Finally, we conclude the chapter with a discussion of current gaps in our understanding and possible directions for future research.

Our exposition will be framed around three very general issues. The first involves the adequacy of theoretical approaches proposing that the emotional elements of personality are best characterized as general dimensions of positive emotionality and negative emotionality. Specifically, we discuss research suggesting that a more differentiated framework, involving multiple forms of positive and negative emotion, may prove more useful. The second issue concerns the nature of regulatory processes at work within temperament. Here we distinguish two different forms of self-regulatory control: a more reactive inhibitory regulation related to fear and a more voluntary self regulation related to effortful control. The third issue concerns the stability of temperament during childhood. We review some of our recent longitudinal findings between infancy and 7 years of age, and discuss potential links between infant and childhood temperament.

POSITIVE EMOTIONALITY

Recent approaches to adult personality suggest that different types of positive and negative emotions tend to cohere within individuals to form two general dimensions. Positive emotionality reflects variability in emotions such as happiness and self-assurance, whereas negative emotionality reflects tendencies such as fear, frustration, and guilt. Several theorists have proposed that these dimensions of emotionality may be related to the major dimensions of personality. Within the two-dimensional space defined by extraversion and neuroticism, positive emotionality has been suggested to align directly with extraversion, whereas negative emotionality underlies the dimension of neuroticism (Eysenck & Eysenck, 1985; Tellegen, 1985; Watson & Clark, 1992). Alternatively, a diagonal alignment has been proposed in which positive emotionality increases in strength as one moves from the stable introvert to the neurotic extrovert quadrant, whereas negative emotionality increases in strength from stable extraversion to neurotic introversion (Gray, 1987; Wallace *et al.*, 1991). In either case these emotionality dimensions provide an attractive framework for temperament theorizing by suggesting that variability in emotional processes is fundamental to the major dimensions of personality.

Neural systems of positive emotionality

Within this general framework temperament theorists have proposed that emotionality arises from individual differences in distinct neural systems related to positive and negative emotions. The neural systems are commonly thought to have their headquarters in limbic structures such as the amygdala, hippocampus, and hypothalamus. Most models assume that information is initially processed within cortical regions devoted to sensory and associative processing of spatial, object, and motion attributes. From the cortex this information converges upon the limbic circuitry, where its emotional and motivational relevance is evaluated. The limbic emotional circuits then project to brainstem effector circuits that adjust autonomic and motor activity, as well as to brainstem arousal systems that project back upon the cortex (Derryberry & Tucker, 1992).

Theorizing about underlying systems has been influenced by neuro-physiological and animal studies, so that emotions tend to be framed in terms of relatively simple stimuli and behaviours. In the case of positive emotions, theorists have emphasized sets of circuits that function to promote approach behaviour in response to rewarding or appetitive stimuli. For example, Gray's (1987, 1994) "behavioural activation system" responds to signals predicting reward and non-punishment to produce approach behaviour and emotions of hope and relief, respectively. Depue's "behavioural facilitation system" also produces hope and relief, as well as irritative aggression when an appetitive goal is blocked (Depue & Collins, 1999; Depue & Iacono, 1989). Panksepp's (1982, 1998) "expectancy-foraging" or "seeking" system responds to regulatory homeostatic imbalances such as hunger to generate appetitive search behaviour and an emotional state of "desire". These approach-based reward systems are attractive in that they emphasize an evolved capacity likely to be conserved across species, with behavioural manifestations clearly relevant to early human development.

Although the precise circuitry underlying these functions remains unclear, the models feature reward-responsive circuits, approach-related motor systems, and a dopaminergic facilitatory mechanism. Reward-responsive neurons have been found within the orbital frontal cortex, amygdala, and hypothalamus (e.g. Cador *et al.*, 1989; Rolls, 1987). The basolateral nucleus of the amygdala, for example, mobilizes approach behaviour by means of projections to brainstem regions (nucleus accumbens and pendunculopontine nucleus) and cortical motor systems. The amygdaloid neurons also activate dopaminergic neurons within the brainstem's ventral tegmental area, which in turn exert a facilitatory influence upon processing of goal-related information within the nucleus accumbens, limbic system, and frontal cortex. According to Gray's model of the "behavioural activation system", reward-related projections from the amygdala to the nucleus accumbens activate the next step in a motor programme that maximally increases proximity to the incentive stimulus. Converging dopaminergic projections to the accumbens enable the switch to this next neuronal set, thereby facilitating goal-oriented behaviour (Gray, 1994; Gray & McNaughton, 1996). In terms of a broader "behavioural facilitation system", Depue & Collins (1999) propose that a circuit involving the nucleus accumbens, ventral pallidum, and dopaminergic neurons codes the intensity of the rewarding stimuli, while related circuits involving the medial orbital cortex, amygdala, and hippocampus integrate the "salient incentive context". Individual differences in the functioning of this network are thought to arise from functional variation in the dopaminergic projections, which encode the intensity of the incentive motivation and facilitate contextual processing across multiple limbic and frontal sites. As development proceeds, dopaminergic facilitation helps to stabilize synaptic contacts within the medial orbital and limbic circuits, and thus to enhance responsivity to positive incentive stimuli (Depue & Collins, 1999).

Positive emotionality during infancy

Studies of early temperament have relied on parent-report questionnaires, observational visits to the home, laboratory measures of behavioural and psychophysiological responses, and behaviour genetic techniques (Rothbart & Bates, 1998). In our research we have used parent-report and laboratory measures in a longitudinal study of infants at the ages of 3, 6.5, 10, and 13.5 months. Parents filled out the Infant Behavior Questionnaire (IBQ), which includes scales assessing activity level, smiling and laughter, fear, distress to limitations, duration of orienting, and soothability. Reliability of these scales is demonstrated through Cronbach's alphas ranging from 0.72 to 0.85 (Rothbart, 1981), and their correlations with home observation measures demonstrates convergent validity (Rothbart, 1986). Scales relevant to positive emotionality include smiling and laughter and activity level (motor reactivity). In the laboratory, infants were videotaped during the presentation of non-social stimuli (e.g. small squeezable toys, a mechanical dog, a rapidly opening parasol) and social stimuli (e.g. experimenter's speech, a peek-a-boo game). Smiling and laughter to these stimuli were coded in terms of their latency, intensity, and duration, and approach behaviour was assessed through the latency to grasp low-intensity toys. We also measured activity level in terms of the 13.5-month-olds' movement among toys distributed across a grid-lined floor.

Individual differences in positive emotionality first appear at the age of 2–3 months, forming a cluster of reactions including smiling and laughter, vocal reactivity, and activity level (Rothbart, 1989). When assessed across different episodes the infant's tendency to express these positive reactions also appears to be independent of their negative reactions (Goldsmith & Campos, 1986). Our laboratory and parent-report measures indicate that the probability of smiling and laughter increases across the ages of 3, 6.5, 10, and 13.5 months. The form of these expressions also changes. For example, smiling and laughter to the rapidly opening parasol increases in intensity and duration across the four ages, but decreases in latency. As expected, given models relating positive emotion to approach behaviour, smiling and laughter in infancy predict concurrent (Rothbart, 1988) and 6–7-year-old (see below) approach tendencies. Moreover, significant stability was found across all ages for laboratory smiling and laughter, with even stronger stability evident in the parent reports (Rothbart et al., 2001b). Similar evidence of stability has recently been reported by Lemery et al. (1999), who found a composite measure of positive emotionality (based on questionnaire measures of smiling and laughter, pleasure, and sociability) to show stability between the ages of 3 and 18 months.

As the infant's motor control develops, individual differences in approach behaviour appear. Several researchers, relying on Thomas & Chess' (1977) approach–withdrawal dimension, have reported stability of approach–withdrawal from 6 to 12 months (Peters-Martin & Wachs, 1984) and from 6 months to 2 years (Persson-Blennow & McNeil, 1980). In our laboratory studies we have attempted to distinguish approach from withdrawal tendencies by measuring infants' latency to grasp familiar, low-intensity toys. Rapid approach of these low-intensity toys is positively related to smiling and laughter, and shows stability from 6.5 to 13.5 months. When faced with a novel or intense toy, however, latencies to approach increase from 6.5 to 10 months, and do not show stability from 6.5 to 13 months (Rothbart, 1988). As discussed in more detail below, these delays and instability may reflect the development of fear reactions that inhibit approach to novelty and intensity late in the first year.

Additional measures relevant to positive emotionality include the infant's motor reactivity and activity level. When assessed in terms of motor reactivity (movement of arms and legs in response to stimuli), evidence of stability during infancy tends to be mixed, although our IBQ findings indicate some stability of motor reactivity from 3 to 13.5 months (Rothbart et al., 2001b), and a composite measure of activity level has been found to be stable from 3 to 18 months (Lemery et al., 1999). However, it is important to distinguish motor reactivity coupled with positive affect from that accompanying distress. When assessed at 4 months, motor reactivity coupled with positive affect predicts approach-oriented behaviour at 4 months (Calkins & Fox, 1994), while motor reactivity coupled with distress predicts later behavioural inhibition (Calkins et al., 1996; Kagan, 1998). Our laboratory research assessed activity level through more intentional movements among toys spread out on the floor. This measure was not correlated with the parent-report measure, but was positively correlated with parent-reported smiling and laughter, negatively correlated with fear, and related to more rapid laboratory approach. It also made interesting predictions into childhood (see below).

971

Positive emotionality during childhood

Temperament research in toddlers and older children has relied primarily on parent-report instruments, such as the Toddler Behavior Assessment Questionnaire (Goldsmith, 1996) and the Children's Behavior Questionnaire (CBQ; Rothbart et al., 1998). The CBQ contains multiple scales for 3–7-year-olds. Factor analyses of these scales show three broad factors, surgency or extraversion, negative affectivity, and effortful control (Ahadi et al., 1993). The surgency factor is primarily defined by scales assessing positive emotionality and approach in terms of positive anticipation, high-intensity pleasure (sensation-seeking), activity level, impulsivity, smiling and laughter, and a negative loading from shyness. The negative affectivity factor involves positive loadings for shyness, discomfort, fear, anger/frustration, and sadness, and a negative loading from soothability-falling reactivity. The effortful control factor is defined by positive loadings from inhibitory control, attentional focusing, low-intensity pleasure (non-risk taking pleasure), and perceptual sensitivity. Effortful control correlates negatively with both surgency and negative affectivity (Rothbart et al., 1994b). These factors map fairly well conceptually on the extraversion/positive emotionality, neuroticism/negative emotionality, and conscientiousness/constraint dimensions found in adult personality (Ahadi & Rothbart, 1994).

In early analyses with 6- and 7-year-olds we examined contributions of these three factors to socioemotional processes assessed through scales measuring aggression, empathy, guilt/shame, help-seeking, and negativity (Rothbart et al., 1994b). The surgency factor was positively related to aggression and negatively related to guilt/shame. As discussed in more detail below, aggression was also positively related to negative affectivity and negatively related to effortful control, while guilt/shame was positively related to both negative affectivity and effortful control. For now, the relation between the surgency factor and aggression confirms earlier predictions that aggressiveness is related to more reactive approach tendencies (Quay, 1993).

More recently we examined the stability of positive emotionality from infancy to the age of 7 years in a small ($n = 29$) longitudinal sample (Rothbart et al., 2001b). Because the sample is small the results must be viewed tentatively. Nevertheless, the results suggest some stability in both parent-report and laboratory measures, and appear helpful in tracing the contributions of positive emotionality to later individual differences. Infant smiling and laughter as measured by the IBQ and in the laboratory showed little stability between infancy and 7 years of age. However, IBQ smiling and laughter from 3, 6.5, and 10 months predicted 7-year positive anticipation and impulsivity. Laboratory smiling and laughter at 10 and 13.5 months also predicted later positive anticipation, and at 13.5 months predicted later impulsivity. The IBQ measure of activity level (non-intentional motor reactivity) was not systematically related to later measures. The laboratory measure (intentional movement among toys at 13.5 months) did not significantly predict later activity level, but was related to high positive anticipation, impulsivity, motor activation, and low sadness at 7 years. Similarly, children showing rapid approach (short latencies to grasp low-intensity toys) at 6.5, 10, and 13.5 months showed high positive anticipation, high impulsivity, high motor activation, and low sadness at 7 years.

These findings indicate that, even though the smiling and laughter measures are not stable, smiling and laughter, activity level, and approach latency all predict approach-related elements of surgency/positive emotionality at 7 years. Indications of stability in aspects of positive emotionality have also been reported from other laboratories. For example, Buss & Plomin (1975) concluded that activity level is not stable from infancy, but that, from 12 months, activity level shows moderate stability. Questionnaire measures of activity level have been found to be stable from 14 to 20 months (Plomin *et al.*, 1993) and from 2 to 4 years (Lemery *et al.*, 1999). An approach factor has also been found stable during toddlerhood and early childhood (Pedlow *et al.*, 1993), and both approach and activity level were stable from 2 to 12 years (Guerin & Gottfried, 1994). Finally, Caspi & Silva (1995) found that children who were high on approach or confidence at age 3–4 were high on social potency and low on self-reported control (i.e. more impulsive) at age 18.

Although these findings suggest an underlying stability for positive emotionality from infancy to childhood, some interesting dissociations are also apparent. In our longitudinal study (Rothbart *et al.*, 2001b), infant activity level predicted not only positive emotionality at age seven, but also high anger/frustration and low soothability–falling reactivity. Together with the earlier finding relating 7-year surgency to aggression (Rothbart *et al.*, 1994b), this suggests that strong approach tendencies may contribute to negative emotionality as well as positive emotionality (Derryberry & Reed, 1994; Rothbart *et al.*, 1994b). A more complex pattern emerges with the infant's approach behaviour. Children showing short-latency grasps at 6.5, 10, and 13.5 months showed high levels of positive anticipation and impulsivity, along with high anger–frustration and aggression at 7 years. This again suggests that strong approach tendencies contribute to later negative as well as positive emotionality. In addition, children showing rapid approach as infants tended to be low in attentional control and inhibitory control at age 7. This is consistent with our finding of a negative relation between surgency and effortful control factors (Rothbart *et al.*, 1994b), and suggests that strong approach tendencies may constrain the development of voluntary self-control. If approach tendencies are viewed as the "accelerator" towards action, and inhibitory tendencies as the "brakes", it is not surprising that stronger accelerative tendencies may weaken the braking influence of inhibitory control (Rothbart *et al.*, 2001b).

In general these findings indicate that the underlying emotional systems related to positive approach behaviour tend to show stability during infancy and early childhood. However, it appears that the positive components are not entirely isolated from other aspects of temperament. So far we have seen that strong approach tendencies may facilitate the frustrative and aggressive aspects of negative emotionality and constrain the regulatory processes of effortful control. To examine these interactions in more detail we turn to these other components of temperament.

NEGATIVE EMOTIONALITY

Negative emotionality is often viewed as a general dimension that subsumes emotions such as fear, anticipatory anxiety, sadness, frustration/anger, guilt, and so on. For example, the "five-factor" model of adult personality includes these negative emotions as components of the neuroticism superfactor. Thus, neuroticism/negative emotionality is often viewed as orthogonal to extraversion/positive emotionality (Eysenck & Eysenck, 1985; Tellegen, 1985; Watson & Clark, 1992). As evident in our discussion above, however, the relationship between anger/frustration and strong approach tendencies suggests that a more differentiated model may be necessary. The need for distinct systems related to negative emotionality is also evident in recent physiological models.

Neural systems of negative emotionality

Negative emotionality appears to arise from at least three sets of neural circuits related to defensive functions. The first involves primitive circuits related to fear, referred to as a "fear" system by Panksepp (1998) and as part of a general "fight/flight" system by Gray (1994; Gray & McNaughton, 1996). Relevant circuits descend from the central amygdala, medial and anterior hypothalamus, and periaqueductal grey to various motor and autonomic effectors of the lower brainstem. The periaqueductal grey appears to mediate explosive forms of escape given unconditioned pain and imminent threat, whereas the hypothalamic circuits provide more directed forms of escape given more distant threats. The higher-level amygdaloid projections allow for more coordinated escape behaviour based on conditioned as well as unconditioned signals. Activity within the periaqueductal and hypothalamic regions is thought to be related to intense states of fear, with panic reactions serving as an extreme example (Barlow *et al.*, 1996; Gray, 1994; Gray & McNaughton, 1996).

A second set of circuits involves angry and frustrative behaviours. Although Gray combines fearful and aggressive behaviours within the same "fight–flight" system, distinct pathways mediate these two forms of defence (Bandler & Keay, 1996; Bandler & Shipley, 1994; Risold *et al.*, 1994). Thus, Panksepp (1982, 1998) has argued for separable "fear" and "rage" systems. The rage system underlies emotional reactions involving frustration and anger, and is thought to have evolved from the invigorated reactions that serve to dissuade predators when an animal has been captured. In combination with the fear system, the rage system orchestrates defensive aggression in the face of threat. In addition, Panksepp suggests that the rage system is closely related to the "seeking" system involved in positive expectations and approach behaviour. One function of the seeking system involves the tracking of reward contingencies within the frontal cortex. If expected rewards are not registered, the higher-level circuits activate the circuitry of the rage system. In a related proposal, Depue and Iacono (1989) suggest that the reward-related "behavioural facilitation system" promotes irritative aggression when rewards are blocked. These models assume that the behavioural invigoration related to anger is adaptive in response to blocked or absent rewards.

A third set of circuits is located higher in the limbic system and responds primarily to conditioned signals to produce anticipatory anxiety. While many researchers focus on the amygdala's role in conditioned emotion (Davis, 1992; LeDoux, 1996), Gray and McNaughton (1996) emphasize the hippocampus as a pivotal structure in the "behavioural inhibition system" (BIS). This complex set of circuits responds to novel signals and to signals predicting punishment or non-reward. Upon detecting such an input the BIS inhibits approach behaviour by means of projections from the subiculum to the nucleus accumbens and cingulate cortex motor areas. The subiculum also projects to the thalamic reticular nucleus and entorhinal cortex, through which it increases attention to relevant environmental stimuli. The BIS is thought to be particularly important in conflict situations, such as instances when a dangerous environment must be approached. Under these circumstances the animal may proceed through a relatively long period of anxious "risk assessment", attempting to localize potential threats within the environment and to resolve conflicts between competing avoidance and approach tendencies.

More generally the BIS can be viewed as a control system capable of inhibiting the approach functions of the reward-related behavioural activation system and the escape functions of the fight–flight system. In addition to such inhibition the BIS serves an important role in directing attention and facilitating the processing of important environmental information. Individual differences in the reactivity of the BIS are thought to underlie an "anxiety" dimension that increases in strength as one moves from stable extraversion to neurotic introversion. High BIS reactivity appears related to clinical problems involving anticipatory anxiety, as opposed to the panic disorder related to the fight–flight system (Gray & McNaughton, 1996).

Negative emotionality during infancy

Initial forms of negative emotionality include early irritable forms of distress, followed by more organized states related to frustration and fear. Irritability during the first several months may arise from both internal and external sources, and is related to both later fear and frustration (Rothbart & Bates, 1998; Rothbart *et al.*, 2001a). As discussed below, this instability may in part reflect the development of attentional systems that allow a young infant to disengage from the irritating source of stimulation (Johnson *et al.*, 1991; Berger & Posner, this handbook).

As the infant begins to actively engage the environment, emotions of frustration and fear begin to appear. In our longitudinal research, parent-report measures included IBQ scales assessing fear, distress to limitations (frustration), and soothability. Laboratory measures were based on the infant's distress reactions to situational elicitors of fear (novel, intense, unpredictable stimuli) and frustration (the placement of attractive toys out of reach or behind a Plexiglas barrier). We also assessed soothability through the average duration of fear and frustration measures, and of distress assessed in non-fear and frustration episodes. The laboratory measures show the intensity of fear reactions to decrease with age, while the intensity and duration of frustration increases (Rothbart *et al.*, 2001b).

The IBQ measures of fear and frustration tend to be positively correlated, and to show modest stability across 3-month assessment intervals during the first year of life (Rothbart, 1989). Similar stability is seen for soothability. For the laboratory measures, modest stability in predicting fear is found across the four age intervals, but in the case of frustration, stability is evident only from 3 to 6.5 months and from 10 to 13.5 months. Fear and frustration are increasingly uncorrelated in the laboratory. Overall distress is stable from 6.5 months on. Stability for anger and sadness during infancy has also been found in analyses based on facial expressions (Malatesta *et al.*, 1989; Sullivan *et al.*, 1992). Although our infant data show no clear differentiation of fear and frustration, Fox (1989) has found that frustration to arm restraint at 5 months is related more to approach of strangers and novel events at 14 months than to negative withdrawal. In addition, Lemery *et al.* (1999) have found parent-report measures to load on separate (but correlated) factors representing fear and anger–distress. Both the fear and anger–distress composites demonstrate stability from 3 to 18 months.

Late in the first year some infants begin to show inhibited approach behaviour to unfamiliar and intense stimuli. Such inhibition can actually be predicted by a measure of combined crying and motor reactivity taken at 4 months, and is accompanied by enhanced sympathetic and adrenal reactivity (Calkins *et al.*, 1996; Kagan *et al.*, 1988, 1992). It is likely that fear-related inhibitory control reduces the stability of approach during infancy. Once inhibition is established, individual differences in the relative strength of approach versus inhibition appear to be relatively enduring aspects of temperament in novel or intense situations (Rothbart & Bates, 1998). In addition, the incorporation of behavioural inhibition into the fear reaction may promote a differentiation of fear and frustration. While frustration may at times invigorate approach tendencies (Newman, 1987), fear promotes a more consistent inhibition of approach. When viewed in light of this inhibitory function, fear can be seen to serve an important, yet relatively reactive, form of regulatory control (Derryberry & Rothbart, 1997; Rothbart, 1989; Rothbart & Bates, 1998). Such inhibition is clearly adaptive, for it helps protect the infant from inadvertent approach responses. As can be seen in later development, fear also makes important contributions to socialization.

Negative emotionality in childhood

As mentioned above, our early research with the CBQ found fear at 6–7 years to combine with discomfort, anger/frustration, sadness, and low soothability/falling reactivity to form a general factor of negative affectivity. In predicting socio-emotional measures negative affectivity was positively related to aggression, empathy, guilt/shame, help-seeking, and negativity (Rothbart *et al.*, 1994b). The relations with empathy and guilt/shame are of special interest in suggesting a link between negative emotionality and conscience-related emotions. We will discuss these in more detail in light of our longitudinal study, focusing separately on the contributions of infant fear and frustration.

As with approach tendencies, fear-related inhibition shows considerable stability across childhood and even into adolescence (Kagan, 1998). Longitudinal research indicates stability of fearful inhibition from 2 to 4 years (Lemery *et al.*, 1999), from 2 to 8 years (Kagan *et al.*, 1988), and from preschool to the age of 18 (Caspi & Silva, 1995). In our longitudinal sample (Rothbart *et al.*, 2001b), laboratory fear at 13.5 months and early distress at 3 months predicted fear at 7 years, as did the parent-report measure of fear at 6.5, 10, and 13.5 months. Parent-report and laboratory measures of infant fear at these three ages also predicted shyness. In addition, IBQ fear at the three oldest ages predicted later sadness, as did laboratory fear at 13.5 months. Laboratory fear at 13.5 months also predicted low-intensity (non-risk-taking) pleasure at 7 years. Neither IBQ nor laboratory fear predicted frustration/anger in childhood.

While these findings illustrate fear's stability and relation to internalizing tendencies, other relations are indicative of its regulatory capacity. Infants showing high laboratory feat at 13.5 months showed low positive anticipation, impulsivity, activity level, and aggression at age 7. These negative relations are consistent with models such as Gray's (1996), where the anxiety-related behavioural inhibition system inhibits the approach-related behavioural activation system. Related evidence can be found in children with co-morbid ADHD and anxiety, who show reduced impulsivity relative to those with ADHD alone (Pliszka, 1989). In addition, aggressiveness appears to decrease between kindergarten and first grade in children who show internalizing patterns (Bates et al., 1995).

Also important are relations between infant laboratory fear at 13.5 months and measures of empathy and guilt/shame at 7 years. Parent-reports indicated that infants with greater fear at the three younger ages showed more empathy and guilt/shame during childhood. These relations suggest a role for fear in the development of early conscience, and converge with the work of Kochanska (1991, 1995). Kochanska has found fearful preschool-aged children to show better internalization of moral principles, with this relationship heightened when mothers used gentle, non-power-oriented discipline. These relations are important in suggesting one pathway through which early temperament can influence the development of high-level social–cognitive processes.

As expected, given our initial evidence of dissociated fear and frustration, infant frustration shows a different pattern across age. Only the 3 months IBQ distress to limitations measure predicted anger/frustration at 7 years. Laboratory frustration, however, predicted 7-year anger/frustration at 6.5 and 10 months. Laboratory frustration at 10 months also predicted additional components of negative affectivity, including high discomfort, high guilt/shame, and low soothability. The only relation between infant frustration and childhood fear was a negative correlation between IBQ distress to limitations and 7-year fear (Rothbart et al., 2001b).

Complementing the finding that early positive emotionality contributes to later frustration, early frustration appears to be related to later positive emotionality/ approach. Frustration shown in the laboratory at 10 months was related to high anger/frustration, activity level, positive anticipation, impulsivity, aggression, and high-intensity pleasure at 7 years. IBQ distress to limitations predicted later positive anticipation at all ages, high-intensity pleasure at 3 and 10.5 months, impulsivity at 3, 10, and 13.5 months, and low sadness at 10 and 13.5 months. Again these relations suggest a link between anger/frustration and strong approach tendencies. Further evidence of such a link appears in the relations of Lemery et al.'s (1999) distress/anger and fear factors: distress/anger was positively related to activity level, while fear and activity level were independent.

In general the findings regarding negative emotionality point towards two tentative conclusions. First, they suggest that models emphasizing general factors of positive and negative emotionality are limited in characterizing temperament development. Although they remain correlated, measures of the negative emotions of fear and frustration diverge in several ways. While infant fear is related to relatively weak approach behaviour and internalizing tendencies during childhood, infant frustration is related to strong approach behaviour and to externalizing as well as internalizing tendencies. The relation between approach and frustration is generally consistent with Panksepp's (1998) suggestion that unsuccessful reward-related activity of the seeking system may activate the anger/frustration functions of the rage system. Strong approach tendencies may heighten the value of positive expectations, as well as the frustrative feelings when those expectations are not met.

Second, fear can be seen to take on an important inhibitory role in early development. It appears to constrain impulsive approach and aggressive behaviours, and may even contribute to the development of conscience. This may seem somewhat paradoxical, for fear is often viewed as a maladaptive emotion. From an evolutionary point of view, however, it is easy to see how fearful inhibition would help protect the individual from potentially harmful behaviours within the physical or social environment. This adaptive potential is well represented in Gray's (1996) "behavioural inhibition system", which functions to inhibit approach behaviour while enhancing attention to important sources of information. What needs to be emphasized is that this fearful form of inhibitory control, especially during childhood, remains a relatively reactive process that can be easily elicited by situational cues. In some individuals this system may also have a very low threshold, resulting in rigid, over-controlled inhibited patterns across a range of situations. This may protect the child from harm, but it is likely to limit the child's approach and resulting experience with the world. Fortunately, temperament involves additional forms of control that provide greater efficiency and flexibility than that afforded by fear. We turn next to these higher forms of regulation.

REGULATORY PROCESSES

Beyond the inhibitory control provided by fear, additional neural systems related to attention make a crucial contribution to temperament. In general the idea is that individuals can voluntarily deploy their attention, allowing them to regulate their more reactive tendencies. In situations where immediate approach is less than optimal, for example, the child can limit his or her attention to the rewarding properties of the stimulus, and thereby resist temptation and delay gratification. Similarly, when faced with a threatening stimulus, children can constrain their fear by flexibly attending to environmental sources of safety as well as threat. In both of these examples individual differences in attention will influence the child's capacity to suppress his/her more reactive tendencies, to take in additional sources of information, and to plan more efficient strategies for coping.

We have referred to this capacity as "effortful control". As mentioned above, our factor analyses of the CBQ identify a general factor of effortful control (attentional shifting, attentional focusing, inhibitory control, perceptual sensitivity) that is distinct from the surgency and negative emotionality factors (Ahadi et al., 1993). Adult research has employed scales assessing individual differences in attentional focusing, attentional shifting, and inhibitory control (Derryberry & Rothbart, 1988). These scales are correlated with one another, consistent with a general capacity of "effortful control". They are not related to measures of positive emotionality (low-intensity pleasure, high-intensity pleasure, and relief), but are inversely related to negative emotionality (fear, frustration, discomfort, and sadness). More recent factor analyses suggest that effortful control is most closely related to the conscientiousness factor of adult personality (Rothbart *et al.*, 2001a). This makes sense in that efficient use of attention fits well with the orderly, planful, and dutiful characteristics of conscientious individuals (Ahadi & Rothbart, 1994).

Neural systems of effortful control

Recent progress in cognitive neuroscience has identified several attentional systems important to early temperament (Posner & Petersen, 1990; Posner & Raichle, 1994). The "posterior attentional system", which involves a network interconnecting the parietal cortex, thalamus, and superior colliculus, is a more reactive system involved in orienting attention from one location to another. Its component operations, including the ability to disengage, move, and engage attention, develop rapidly between the ages of 3 and 6 months (Ruff & Rothbart, 1996).

The more voluntary "anterior attentional system" begins to develop late in the first year, and is thought to underlie effortful control. Focused within the anterior cingulate region of the frontal lobe, the anterior system receives converging inputs from other systems that process motivational (frontal cortex, amygdala, hippocampus), spatial (parietal cortex), object (temporal cortex), semantic (temporal and frontal cortex), and response-related information (basal ganglia, supplementary motor area) information. This massive input allows the anterior system to carry out "executive" functions based on a wide range of relevant information. These functions are accomplished by cingulate projections to temporal, parietal, and frontal cortices, as well as to multiple limbic and brainstem structures. One crucial function involves regulating the more reactive posterior system, allowing the individual to voluntarily disengage and move attention to selected sources of information. In addition, the anterior system inhibits prepotent or dominant responses, allowing more flexible adjustments in behaviour. At a more cognitive level the anterior system is involved in directing attention to semantic information, in inhibiting dominant conceptual associations, and in interacting with other frontal areas to sustaining working memory. Finally, the anterior system appears involved in the detection of errors, a function crucial to effective self-regulation. These cognitive regulations allow the individual more flexibility in restructuring thought and planning future courses of action (Posner & Petersen, 1990; Posner & Raichle, 1994; Posner & Rothbart, 1998; Berger & Posner, this handbook).

Effortful control during infancy

Because the anterior attentional system begins to develop late in infancy, we begin with reactive processes related to the posterior system. In the early months of life the infant's attention is often driven by stimuli that are intense or novel. In some instances infants have difficulty disengaging from a stimulus, and such "obligatory attention" may lead to a state of rising distress. With the development of the posterior system between 3 and 6 months, the infant becomes better able to disengage and attenuate his or her distress. Development within the posterior system may in part account for limited stability in distress from 3 to later months (Johnson et al., 1991). More generally, early development of the posterior system, even though it tends to be stimulus-driven during this period, may allow some infants to attenuate their distressful states. We have found that babies who can easily disengage attention from an arousing stimulus are reported to be more soothable and less subject to negative affect by their mothers (Rothbart et al., 1992). Providing an object for orienting has the effect of soothing distress in young infants, although distress may return if the object is removed (Harman et al., 1997). Parents often use distraction as a soothing technique.

Our longitudinal measures of infant's attention include parent-reports of duration of orienting as well as laboratory measures of fixation time and sustained play. The IBQ duration of orienting measure shows stability across the four measurement periods from 3 to 13.5 months. This measure, which might be seen as a duration of interest measure, was related, at least during the younger ages, to measures of positive emotionality (Rothbart et al., 1994c). The laboratory fixation time measure showed stability from 10 to 13.5 months. Although these findings indicate some stability of attention during infancy, such consistency has not always been found by other researchers (Ruff & Rothbart, 1996). This is not too surprising because the development of the anterior system late in the first year, allowing increased executive control and planning, may change the meaning of individual differences in orienting. However, we did find that long fixation times in the laboratory were related to high fear, high sadness, high shyness, and low high-intensity (risk-taking) pleasures at the age of 7 (Rothbart et al., 2001b). In earlier analyses we found that children whose mothers described them on the IBQ as more fearful, looked longer on average and looked away from objects less than did less fearful infants. Infants whose mothers reported higher smiling and laughter looked away from presented objects more frequently (Rothbart et al., 1994c).

Effortful control during childhood

Initial signs of anterior development appear late in the first year. These include alterations in fixation times and the ability to inhibit and correct the course of a movement, followed by the appearance of spontaneous alternation at around 18 months (Ruff & Rothbart, 1996; Berger & Posner, this handbook). The anterior system continues to develop during childhood and into adolescence, allowing more sophisticated forms of self-regulation based on verbal information, representations of the self, and projections concerning the future. Because these functions depend on a new and separable attentional system, there is little reason to expect stability from infant measures such as fixation to later effortful control. We have found no such relations in our longitudinal study.

Nevertheless, other research suggests some stability during childhood. In Mischel's work, for example, the number of seconds delayed by preschool children while waiting for physically present rewards predicted parent-reported attentiveness and ability to concentrate as adolescents (Mischel, 1984; Shoda et *al.*, 1990). In addition, questionnaire measures of attention, which may tap reactive as well as effortful processes, are suggestive of stability during childhood. These include stability of "attention-persistence" from 14 to 20 months (Plomin *et al.*, 1993), and of "distractibility" from 18 to 24 months (Wilson & Matheny, 1986) and from 2 to 12 years (Guerin & Gottfried, 1994). Finally, a measure of "cooperation-manageability", which may approximate some aspects of effortful control, was found stable throughout toddlerhood and early childhood (Pedlow *et al.*, 1993).

We have recently used a marker task for anterior cingulate function in which the child must respond to a spatially conflicting stimulus by inhibiting the dominant response and executing a subordinate response. Performance on this task improves considerably between 27 and 36 months, with children showing less preservation of the previous response. Children who perform well are described by their parents as more skilled at attentional shifting and focusing, less impulsive, and less prone to frustration reactions (Gerardi *et al.*, 1996). Using a very similar task with adults, individuals who perform poorly tend to be high in anxiety and low on self-reported attentional control (Derryberry & Reed, 2001). These findings are consistent with the notion that attentional skill, measured through questionnaire or laboratory methods, may help individuals constrain negative forms of emotion.

Our studies with 6–7-year-olds have found effortful control to be defined in terms of scales measuring attentional focusing, inhibitory control, low-intensity pleasure, and perceptual sensitivity. Effortful control is negatively related to both the surgency and negative affectivity factors. These negative relations are in keeping with the notion that attentional skill may help attenuate negative affect, while also serving to constrain impulsive approach tendencies. An interesting example involves the negative relation between effortful control and aggression. As mentioned earlier, aggression also relates positively to surgency and negative affectivity (especially anger; Rothbart *et al.*, 1994b). Since effortful control makes no unique contribution to aggression, effortful control may regulate aggression indirectly by controlling reactive tendencies underlying surgency and negative affectivity. For example, children high in effortful control may be able to direct attention away from the rewarding aspects of aggression, or to decrease the influence of negative affectivity by shifting attention away from the negative cues related to anger. Eisenberg and her colleagues found that 4–6-year-old boys with good attentional control tend to deal with anger by using non-hostile verbal methods rather than overt aggressive methods (Eisenberg *et al.*, 1994).

As mentioned earlier, empathy in 6–7-year-olds is positively related to negative affectivity (but negatively related to anger), and can be predicted by high levels of fear during infancy (Rothbart *et al.*, 1994b, 2001b). However, empathy is even more strongly related to effortful control, with children high in effortful control showing greater empathy. In a study of elderly hospital volunteers, Eisenberg & Okun (1996) found attentional control to be positively related to sympathy and perspective taking, and negatively related to personal distress. In contrast, negative emotional intensity was positively related to sympathy and personal distress. Effortful control may support empathy by allowing the individual to attend to the thoughts and feelings of another without becoming overwhelmed by their own distress. Similarly, guilt/shame in 6–7-year-olds is positively related to effortful control and negative affectivity (Rothbart *et al.*, 1994b). Negative affectivity may contribute to guilt by providing the individual with strong internal cues of discomfort, thereby increasing the probability that the cause of these feelings is attributed to an internal rather than external cause (Dienstbier, 1984). Effortful control may contribute further by providing the flexibility needed to relate these negative feelings of responsibility to one's own specific actions and to the negative consequences for another (Derryberry & Reed, 1994, 1996).

Consistent with these influences on empathy and guilt, effortful control also appears to play a role in the development of conscience. As mentioned above, the internalization of moral principles appears to be facilitated in fearful preschool-aged children, especially when their mothers use gentle discipline (Kochanska, 1991, 1995). In addition, internalized control is facilitated in children high in effortful control (Kochanska *et al.*, 1996). Again we see the influence of two separable control systems regulating the development of conscience. While fear may provide reactive inhibition and strong negative affect for association with moral principles, effortful control provides the attentional flexibility required to link negative affect, action outcomes, and moral principles.

When viewed as a whole these findings begin to illustrate the importance of temperament to the child's emotional, cognitive, and social development. We do not mean to underestimate environmental influences, but the underlying temperament systems may serve central roles in the self-organization of personality. This is particularly evident in the functions of attention, which serves to select and coordinate the most important information, and contributes to the storage of this information in memory. While much theorizing emphasizes behaviour and the immediate environment, children are also highly thoughtful, often employing attention to "replay" their positive and negative experiences. Across development one would expect these emotional and attentional processes to progressively stabilize certain forms of information, and thus to shape the child's representation of the self and world (Derryberry & Reed, 1994, 1996; Rothbart *et al.*, 1994b).

SUMMARY AND CONCLUSIONS

In conclusion, temperament approaches appear to be making progress in tracing the early roots of emotional development. We have seen that elements of positive emotionality (e.g. smiling and laughter, approach tendencies, and activity level) and negative emotionality (e.g. fear and frustration) appear in relatively undifferentiated forms during infancy. Although these general positive and negative tendencies remain influential as the child develops, we also see a differentiation within both domains, with the more surgent aspects of positive emotionality (approach and activity level) relating to the frustrative aspects of negative emotionality (anger and aggression). At the same time the negative emotion of fear assumes an inhibitory role, providing a reactive form of regulation over approach tendencies. With the progressive development of effortful control the child becomes able to regulate both the reactive approach and fearful tendencies, and thus accomplishes more sophisticated and planful forms of self-regulation. Individual differences in these reactive and regulatory processes appear to provide a foundation for personality development in emotional, cognitive, and social realms.

Nevertheless, it can be seen that temperament approaches have a long way to go, with many gaps in our knowledge that remain to be filled. As discussed above, greater resolution is required in discussing positive and negative forms of emotionality. At this point there is good evidence for decomposing negative emotionality along the lines of fear and frustration/anger, with frustration sharing some of its variability with positive emotionality (approach). However, other negative emotions, such as sadness with its relation to depression, may also share aspects of positive and negative emotionality (Chorpita *et al.*, 1998). In addition there may be other types of positive emotions beyond the surgent, sensation-seeking forms of approach. MacDonald (1992) has proposed that a specialized reward system (an "affectional system") has evolved to facilitate close family relationships by promoting feelings of warmth and empathy. Exploring such emotions would enrich temperament theories and provide links to adult dimensions such as agreeableness.

Similarly, our knowledge of regulatory functions remains quite limited. Although progress has been made in distinguishing voluntarily anterior and reactive posterior attentional processes, the anterior system performs multiple functions that have yet to be studied. These frontal functions may constitute separable skills that contribute to temperament in different ways. In studies of adults, for example, Matthews (1997) has proposed that the dimension of extraversion reflects a pattern of cognitive skills: extroverts are better at divided attention, resistance to distraction, retrieval from memory and short-term memory, whereas introverts are better at vigilance, reflective problem-solving, and long-term memory. At this point we can only speculate about these executive functions in children.

Finally, large gaps remain in our understanding of the relation between emotional and attentional processes. Our examples above suggest that effortful attention can regulate positive (approach) and negative (fear) emotions. However, there is a large adult literature indicating that emotions such as fear can regulate attention, biaising attention in favour of threatening information (Eysenck, 1992; Mathews & MacLeod, 1994). Such negative biases may result from a reactive fear influence on the posterior attentional system. These influences are likely to be extremely important to the child's experience, and yet they have received little attention in the developmental literature (Derryberry & Reed, 1996; Vasey & Daleiden, 1996). Furthermore, anatomical studies indicate that the anterior system receives massive input from limbic emotional systems, and it makes sense that emotional states may at times recruit effortful functions. In this case we so far have little guidance from either the developmental or the adult literature.

In spite of these limitations we believe that temperament approaches have much to offer the study of emotional development. Considerable progress has been made across the past 20 years, and we are beginning to see links beyond emotion to include the cognitive and social aspects of personality. Perhaps most important, the study of temperament benefits from the convergence of research from many areas beyond developmental psychology. This chapter may not have done justice to the interdisciplinary foundations of temperament research, but we hope that it conveys some of its promise for the future.

References

Ahadi, S.A. & Rothbart, M.K. (1994). Temperament, development, and the Big Five. In C.F. Halverson Jr, G.A. Kohnstamm & R.P. Martin (Eds.), *The developing structure of temperament and personality from infancy to adulthood* (pp. 189–207). Hillsdale, NJ: Erlbaum.

Ahadi, S.A., Rothbart, M.K. & Ye, R.M. (1993). Children's temperament in the US and China: similarities and differences. *European Journal of Psychology*, **7**, 359–377.

Bandler, R. & Keay, K.A. (1996). Columnar organization in the midbrain periaqueductal gray and the integration of emotional expression. In G. Holstege, R. Bandler & C.B. Saper (Eds.), *Progress in brain research*, vol. 107: *The emotional motor system* (pp. 285–300). Amsterdam: Elsevier.

Bandler, R. & Shipley, M.T. (1994). Columnar organization in the midbrain periaqueductal gray: modules for emotional expression? *Trends in Neurosciences*, **17**, 379–389.

Barlow, D.H., Chorpita, B.F. & Turovsky, J. (1996). Fear, panic, anxiety, and disorders of emotion. In D.A. Hope (Ed.), *Nebraska symposium on motivation*, vol. 43: *Perspectives on anxiety, panic, and fear* (pp. 251–328). Lincoln, NA: University of Nebraska Press.

Bates, J.E., Pettit, G.S. & Dodge, K.A. (1995). Family and child factors in stability and change in children's aggressiveness in elementary school. In J. McCord (Ed.), *Coercion and punishment in long-term perspectives* (pp. 124–138). New York: Cambridge University Press.

Buss, A.H. & Plomin, R. (1975). *A temperament theory of personality development*. New York: Wiley.

Cador, M., Robbins, T.W. & Everitt, B.J. (1989). Involvement of the amygdala in stimulus–reward associations: interaction with the ventral striatum. *Neuroscience*, **30**, 77–86.

Calkins, S.D. & Fox, N.A. (1994). Individual differences in the biological aspects of temperament. In J.E. Bates & T.D. Wachs (Eds.), *Temperament: individual differences at the interface of biology and behavior* (pp. 199–217). Washington, DC: American Psychological Association.

Calkins, S.D., Fox, N.A. & Marshall, T.R. (1996). Behavioral and physiological antecedents of inhibition in infancy. *Child Development*, **67**, 523–540.

Caspi, A. & Silva, P.A. (1995). Temperamental qualities at age three predict personality traits in young adulthood: longitudinal evidence from a birth cohort. *Child Development*, **66**, 486–498.

Chorpita, B.F., Albano, A.M. & Barlow, D.H. (1998). The structure of negative emotions in a clinical sample of children and adolescents. *Journal of Abnormal Psychology*, **107**, 74–85.

Davis, M. (1992). The role of the amygdala in fear and anxiety. *Annual Review of Neuroscience, 1992*, **15**, 353–375.

Depue, R.A. & Collins, P.F. (1999). Neurobiology of the structure of personality: dopamine, facilitation of incentive motivation, and extraversion. *Behavioral and Brain Sciences*, **22**, 491–555.

Depue, R.A. & Iacono, W.G. (1989). Neurobehavioral aspects of affective disorders. *Annual Review of Psychology*, **40**, 457–492.

Derryberry, D. & Reed, M.A. (1994). Temperament and the self-organization of personality. *Development and Psychopathology*, **6**, 653–676.

Derryberry, D. & Reed, M.A. (1996). Regulatory processes and the development of cognitive representations. *Development and Psychopathology*, **8**, 215–234.

Derryberry, D. & Reed, M.A. (2001). Individual differences in attentional control: adaptive regulation of response interference. (Manuscript submitted for publication.)

Derryberry, D. & Rothbart, M.K. (1988). Affect, arousal, and attention as components of temperament. *Journal of Personality and Social Psychology*, **55**, 958–966.

Derryberry, D. & Rothbart, M.K. (1997). Reactive and effortful processes in the organization of temperament. *Development and Psychopathology*, **9**, 633–652.

Derryberry, D. & Tucker, D.M. (1992). Neural mechanisms of emotion. *Journal of Consulting and Clinical Psychology*, **60**, 329–338.

Dienstbier, R.A. (1984). The role of emotion in moral socialization. In C.E. Izard, J. Kagan & R.B. Zajonc (Eds.), *Emotions, cognition, and behavior* (pp. 484–514). Cambridge, UK: Cambridge University Press.

Eisenberg, N. & Okun, M.A. (1996). The relations of dispositional regulation and emotionality to elders' empathy-related responding and affect while volunteering. *Journal of Personality*, **64**, 157–183.

Eisenberg, N., Fabes, R.A., Nyman, M., Bernzweig, J. & Pinulas, A. (1994). The relations of emotionality and regulation to children's anger-related reactions. *Child Development*, **65**, 109–128.

Eysenck, H.J. & Eysenck, M.W. (1985). *Personality and individual differences: a natural science approach*. New York: Plenum.

Eysenck, M.W. (1992). *Anxiety: the cognitive perspective*. Hillsdale, NJ: Erlbaum.

Fox, N.A. (1989). Psychophysical correlates of emotional reactivity during the first year of life. *Developmental Psychology*, **25**, 364–372.

Gerardi, G., Rothbart, M.K., Posner, M.I. & Kepler, S. (1996). The development of attentional control: performance on a spatial Stroop-like task at 24, 30 and 36–38 months of age. Poster session presented at the annual meeting of the International Society for Infant Studies, Providence, Rhode Island.

Goldsmith, H.H. (1996). Studying temperament via construction of the Toddler Behavior Assessment Questionnaire. *Child Development*, **67**, 218–235.

Goldsmith, H.H. & Campos, J.J. (1986). Fundamental issues in the study of early temperament: the Denver Twin Temperament Study. In M.H. Lamb & A. Brown (Eds.), *Advances in developmental psychology* (pp. 231–283). Hillsdale, NJ: Erlbaum.

Gray, J.A. (1987). Perspectives on anxiety and impulsivity: a commentary. *Journal of Research in Personality*, **21**, 493–509.

Gray, J.A. (1994). Framework for a taxonomy of psychiatric disorder. In S.H.M. van Goozen, N.E. Van de Poll & J.A. Sergeant (Eds.), *Emotions: essays on emotion theory* (pp. 29–60). Hillsdale, NJ: Erlbaum.

Gray, J.A. & McNaughton, N. (1996). The neuropsychology of anxiety: reprise. In D.A. Hope (Ed.), *Nebraska symposium on motivation: perspectives on anxiety, panic, and fear* (vol. 43, pp. 61–134). Lincoln, NA: University of Nebraska Press.

Guerin, D.W. & Gottfried, A.W. (1994). Developmental stability and change in parent reports of temperament: a ten-year longitudinal investigation from infancy through preadolescence. *Merrill-Palmer Quarterly*, **40**, 334–355.

Harman, C., Rothbart, M.K. & Posner, M.I. (1997). Distress and attention interactions in early infancy. *Motivation and Emotion*, **21**, 27–43.

Johnson, M.H., Posner, M.I. & Rothbart, M.K. (1991). Components of visual orienting in early infancy: contingency learning, anticipatory looking and disengaging. *Journal of Cognitive Neuroscience*, **3**, 335–344.

Kagan, J. (1998). Biology and the child. In W.S.E. Damon & N.V.E. Eisenberg (Eds.), *Handbook of child psychology*, vol. 3: *Social, emotional and personality development* (5th ed., pp. 177–235). New York: Wiley.

Kagan, J., Reznick, J.S. & Snidman, N. (1988). Biological bases of childhood shyness. *Science*, **240**, 167–173.

Kagan, J., Snidman, N. & Arcus, D.M. (1992). Initial reactions to unfamiliarity. *Current Directions in Psychological Science*, **1**, 171–173.

Kochanska, G. (1991). Socialization and temperament in the development of guilt and conscience. *Child Development*, **62**, 1379–1392.

Kochanska, G. (1995). Children's temperament, mothers' discipline, and security of attachment: multiple pathways to emerging internalization. *Child Development*, **66**, 597–615.

Kochanska, G., Murray, K., Jacques, T.Y., Koenig, A.L. & Vandegeest, K.A. (1996). Inhibitory control in young children and its role in emerging internalization. *Child Development*, **67**, 490–507.

LeDoux, J. (1996). *The emotional brain*. New York: Simon & Schuster.

MacDonald, K. (1992). Warmth as a developmental construct: an evolutionary analysis. *Child Development*, **63**, 753–773.

Lemery, K.S., Goldsmith, H.H., Linnert, M.D. & Mrazek, D.A. (1999). Developmental models of infant and childhood temperament. *Developmental Psychology*, **35**, 189–204.

MacDonald, K. (1995). Evolution, the Five-Factor Model, and levels of personality. *Journal of Personality*, **63**, 525–567.

Malatesta, C.A., Culver, C., Tesman, J.R. & Shepard, B. (1989). The development of emotion expression during the first two years of life. *Monographs of the Society for Research in Child Development*, **54** (1–2 serial no. 219).

Mathews, A.M. & MacLeod, C. (1994). Cognitive approaches to emotion and emotional disorders. *Annual Review of Psychology*, **45**, 25–50.

Matthews, G. (1997). Extraversion, emotion and performance: a cognitive–adaptive model. In G. Matthews (Ed.), *Cognitive science perspectives on personality and emotion* (pp. 399–442). Amsterdam: Elsevier.

Mischel, W. (1983). Delay of gratification as process and as person variable in development. In D. Magnusson & V.P. Allen (Eds.), *Human development: an interactional perspective* (pp. 149–165). New York: Academic Press.

Newman, J.P. (1987). Reaction to punishment in extroverts and psychopaths: implications for the impulsive behavior of disinhibited individuals. *Journal of Research in Personality*, **21**, 464–480.

Panksepp, J. (1982). Toward a general psychobiological theory of emotions. *Behavioral and Brain Sciences*, **5**, 407–467.

Panksepp, J. (1998). *Affective neuroscience*. New York: Oxford University Press.

Pedlow, R., Sanson, A., Prior, M. & Oberklaid, F. (1993). Stability of maternally reported temperament from infancy to 8 years. *Developmental Psychology*, **29**, 998–1007.

Persson-Blennow, I. & McNeil, T.F. (1980). Questionnaires for measurement of temperament in one- and two-year-old children: development and standardization. *Journal of Child Psychology and Psychiatry*, **21**, 37–46.

Peters-Martin, P. & Wachs, T. (1984). A longitudinal study of temperament and its correlates in the first 12 months. *Infant Behavior and Development*, **7**, 285–298.

Pliszka, S.R. (1989). Effect of anxiety on cognition, behavior, and stimulant response in ADHD. *Journal of the American Academy of Child and Adolescent Psychiatry*, **28**, 882–887.

Plomin, R., Emde, R.N., Braungart, J.N., Campos, J., Corley, R., Fulker, D.W., Kagan, J., Reznick, J.S., Robinson, J., Zahn-Waxler, C. & DeFries, J.C. (1993). Genetic change and continuity from fourteen to twenty months: the MacArthur Longitudinal Twin Study. *Child Development*, **64**, 1354–1376.

Posner, M.I. & Petersen, S.E. (1990). The attention system of the human brain. *Annual Review of Neuroscience*, **13**, 25–42.

Posner, M.I. & Raichle, M.E. (1994). *Images of mind*. New York: Scientific American Library.

Posner, M.I. & Rothbart, M.K. (1998). Attention, self-regulation and consciousness. *Philosophical Transactions of the Royal Society of London B*, **353**, 1915–1927.

Quay, H.C. (1993). The psychology of undersocialized aggressive conduct disorder: a theoretical perspective. *Development and Psychopathology*, **5**, 165–180.

Risold, P.Y., Canteras, N.S. & Swanson, L.W. (1994). Organization of projections from the anterior hypothalamic nucleus: a *Phaseolus vulgaris*–leucoagglutinin study in the rat. *Journal of Comparative Neurology*, **348**, 1–40.

Rolls, E.T. (1987). Information representation, processing, and storage in the brain: analysis at the single neuron level. In J.P. Changeaux & M. Konishi (Eds.), *The neural and molecular bases of learning* (pp. 503–540). New York: Wiley.

Rothbart, M.K. (1981). Measurement of temperament in infancy. *Child Development*, **52**, 569–578.

Rothbart, M.K. (1986). Longitudinal observation of infant temperament. *Developmental Psychology*, **22**, 356–365.

Rothbart, M.K. (1988). Temperament and the development of inhibited approach. *Child Development*, **59**, 1241–1250.

Rothbart, M.K. (1989). Temperament and development. In G.A. Kohnstamm, J.E. Bates & M.K. Rothbart (Eds.), *Temperament in childhood* (pp. 187–247). New York: John Wiley & Sons.

Rothbart, M.K. & Bates, J.E. (1998). Temperament. In W.S.E. Damon & N.V.E. Eisenberg (Eds.), *Handbook of child psychology*, vol. 3: *Social, emotional and personality development* (5th ed., pp. 105–176). New York: Wiley.

Rothbart, M.K. & Derryberry, D. (1981). Development of individual differences in temperament. In M.E. Lamb & A.L. Brown (Eds.), *Advances in developmental psychology* (vol. 1, pp. 37–86). Hillsdale, NJ: Erlbaum.

Rothbart, M.K., Ahadi, S.A. & Evans, D. (2001a). Temperament and personality: origins and outcomes. *Journal of Personality and Social Psychology* (Submitted).

Rothbart, M.K., Ahadi, S.A., Hershey, K. & Fisher, P. (1998). Investigations of temperament at 3–7 years: the Children's Behavior Questionnaire (Submitted).

Rothbart, M.K., Ahadi, S.A. & Hershey, K.L. (1994b). Temperament and social behavior in childhood. *Merrill-Palmer Quarterly*, **40**, 21–39.

Rothbart, M.K., Derryberry, D. & Hershey, K. (2001b). Stability of temperament in childhood: laboratory infant assessment to parent report at seven years. In V.J. Molfese & D.L. Molfese (Eds.), *Temperament and personality development across the life span* (pp. 85–119). Hillsdale, NJ: Erlbaum.

Rothbart, M.K., Derryberry, D. & Posner, M.I. (1994a). A psychobiological approach to the development of temperament. In J.E. Bates & T.D. Wachs (Eds.), *Temperament: Individual differences at the interface of biology and behavior* (pp. 83–116). Washington, DC: American Psychological Association.

Rothbart, M.K., Posner, M.I. & Rosicky, J. (1994c). Orienting in normal and pathological development. *Development and Psychopathology*, **6**, 635–652.

Rothbart, M.K., Ziaie, H. & O'Boyle, C. (1992). Self-regulation and emotion in infancy. In N. Eisenberg & R.A. Fabes (Eds.), *Emotion and self-regulation in early development: New directions in child development* (pp. 7–24). San Francisco, CA: Jossey-Bass.

Ruff, H.A. & Rothbart, M.K. (1996). *Attention in early development: themes and variations*. New York: Oxford University Press.

Shoda, Y., Mischel, W. & Peake, P.K. (1990). Predicting adolescent cognitive and self-regulatory competencies from preschool delay of gratification: identifying diagnostic conditions. *Developmental Psychology*, **26**, 978–986.

Sullivan, M.W., Lewis, M. & Alessandri, S.M. (1992). Cross-age stability in emotional expressions during learning and extinction. *Developmental Psychology*, **28**, 58–63.

Tellegen, A. (1985). Structures of mood and personality and their relevance to assessing anxiety, with an emphasis on self-report. In A.H. Tuma & J.D. Maser (Eds.), *Anxiety and the anxiety disorders* (pp. 681–706). Hillsdale, NJ: Erlbaum.

Thomas, A. & Chess, S. (1977). *Temperament and development*. New York: Brunner/Mazel.

Vasey, M.W. & Daleiden, E.L. (1996). Information-processing pathways to cognitive interference in childhood. In I.G. Sarason, G.R. Pierce & B.R. Sarason (Eds.), *Cognitive interference: theories, methods, and findings* (pp. 117–138). Malwah, NJ: Erlbaum.

Wallace, J.F., Newman, J.P. & Bachorowski, J. (1991). Failures of response modulation: impulsive behavior in anxious and impulsive individuals. *Journal of Research in Personality*, **25**, 23–44.

Watson, D. & Clark, L.A. (1992). On traits and temperament: general and specific factors of emotional experience and their relation to the five-factor model. *Journal of Personality*, **60**, 441–476.

Wilson, R.S. & Matheny, A.P.J. (1986). Behavior-genetics research in infant temperament: the Louisville Twin Study. In R. Plomin & J. Dunn (Eds.), *The study of temperament: changes, continuities, and challenges* (pp. 81–98). Hillsdale, NJ: Erlbaum.

Further reading

Derryberry, D. & Rothbart, M.K. (1997). Reactive and effortful processes in the organization of temperament. *Development and Psychopathology*, **9**, 633–652.

Posner, M.I. & Raichle, M.E. (1994). *Images of mind*. New York: Scientific American Library.

Rothbart, M.K. & Bates, J.E. (1998). Temperament. In W.S.E. Damon & N.V.E. Eisenberg (Eds.), *Handbook of child psychology* (vol. 3, pp. 105–176). *Social, emotional and personality development* (5th ed.). New York: Wiley.

VI.   EPILOGUE

VI.1
Concluding Comments: Research on Brain–Behaviour Development in the 21st Century

ALBERT GRAMSBERGEN, BRIAN HOPKINS and ALEX F. KALVERBOER

SOME PRELIMINARY COMMENTS: RECAPITULATION AND QUESTIONS OF TERMINOLOGY

This handbook contains contributions which concern the development of brain and behaviour with a special focus on the human. Contributions on brain development have included not only neuroanatomical and neurophysiological descriptions and explanations, but also changes in hormonal activities and metabolic processes that are crucial in growth and neurotransmission. By "brain", we mean not only the neocortex, but also subcortical structures and the spinal cord. "Behaviour" is in our connotation an umbrella term, which roughly covers the domains of motor functions, perception, cognition and emotion, and how such functions may interrelate and play a role in complex adaptations to the external world. The focus has been on how the various aspects of brain and behaviour develop in the human and how they interrelate and affect each other across developmental time. This is an important field of study, not only to better understand mechanisms in normal and deviant neurobehavioural development, but also to gain improved insights into adult brain–behaviour relationships.

Needless to say it was never our aim to cover or even review such a broad field of scientific interest. Some domains of study have not been included, such as the development of olfaction and taste, and of tactile functions. Within domains of study we made selections based on our aim to represent each one by a limited number of contributions, which together are indicative for present-day thinking on neurobehavioural development. We realize that only the first steps have been made in producing a well-integrated body of knowledge on brain–behaviour development in the human. Consequently, the function of this book depends to a large extent on how the reader uses it, and thus in how far its various parts may fall into place so that the broader picture of neurobehavioural development we have tried to achieve can be appreciated.

A.F. Kalverboer and A. Gramsbergen (eds.), Handbook of Brain and Behaviour in Human Development, 991–1000
© 2001 Kluwer Academic Publishers. Printed in Great Britain.

It is almost impossible to find a suitable term to cover this broader picture, and attempts to do so have so far not been very convincing. For example, the label "developmental cognitive neuroscience", coined by Johnson (1997) is not entirely satisfactory. It suggests an exclusive interest in cognitive processes, a one-sidedness that has for a long time been the overriding tendency in developmental and experimental psychology, and omits consideration of spinal mechanisms (e.g. in terms of proprioneurons) that play such important roles in the control and coordination of behaviour as well as of subcortical structures crucial in emotional development (see Panksepp, 1998). Hopefully our connotations of the terms "brain" and "behaviour" are sufficiently explicit to avoid misunderstanding about the aim and focus of the handbook.

WHY RESEARCH INTO THE DEVELOPMENT OF BRAIN AND BEHAVIOUR?

From the early stages of development the nervous system is involved in regulating bodily functions in such a way as to optimize both present and future states, the abilities required for perceiving and interpreting consistency and change in the environment, and for marshalling task-appropriate behaviours. The formation of the brain has an undeniable genetical basis. Its structure rapidly increases in complexity, but secondarily it is shaped by regressive processes that are epigenetically determined. As the size and the metabolic demands of the growing brain change during development, and as the organism is confronted with changing environmental demands, both the brain and its behaviour have to adjust to and anticipate these changes, as well as ensuring that the prerequisites for later functioning can be met. Not only the processing of information from internal sources but also from the outside world by means of perceptual, emotional and cognitive processes have to develop. While these processes have their origins in prenatal life, they undergo their most major developmental changes after birth.

In order to understand brain functioning at adult age it is therefore of paramount importance to understand its developmental history, a statement that holds for the species as a whole as well as for individual members of that species. This handbook provides an overview of recent advances in our knowledge until the year 2000 across a number of cognate areas that seem to be important for a broad understanding of brain–behaviour development, and also as a basis for further research in the coming years.

THE GROWING INFLUENCE OF MOLECULAR BIOLOGY: SOME CAUTIONARY REMARKS

The past couple of decades have seen an enormous explosion of knowledge in the field of molecular biology, especially with regard to genome sequencing. Research involving animal preparations such as the roundworm *Caenorhabditis elegans* and the fruitfly *Drosophila melanogaster*, with apparently simple and uniform nervous systems and relatively small numbers of coding genes, and more recently mammals such as knockout and mutant mice, have given us new perspectives and many details about how various developmental processes are regulated. The identification of alterations in one gene or a few genes that lead to neurological abnormalities and illnesses (e.g. Huntington's disease) or which contribute to the development of abnormalities (e.g. Alzheimer's disease) are important and clinically relevant outcomes of systematic research into the human genome. However, such a bottom-up interpretation of how genes relate to behaviour is overly narrow, as there is increasing evidence that environmental factors can regulate genetic activity. One convincing example is light-induced changes in the gene expression of proteins thought to be involved in regulating the circadian clock of fruitflies (Myers et al., 1996)

Unravelling the genetical basis of development, however, seems several orders more complex as genes come to expression in intricate cascades with mutual interactions, which both in time and along diverse topographical axes ultimately lead to the global structure of the nervous system (e.g. see Harris & Hartenstein, 1998). These aspects of dynamical interactions between genetical expressions can only be studied in developing organisms. However, given findings concerning the so-called sensory-evoked early gene expression, which appears to be a life-long propensity (Mack & Mack, 1992), such interactions should not be considered to be only genetically determined.

THE CONTRIBUTIONS OF NEUROANATOMY

Investigations into the development of the brain and behaviour or into the development of relationships between both, are driven by motives of a diverse nature. Hamburger (1988) identified two main reasons for such research interests. For neuroanatomists in particular, comparing the early stages of neuro-ontogeny (or what Ramon y Cajal, 1937, in his autobiography insightfully referred to as the "nursery stage") with the mature state offers possibilities for gaining insights into the intricate and highly complex neuroanatomy of neural circuitries in adulthood. There are advantages and disadvantages associated with this comparative exercise. An advantage is that new discoveries in the brain functioning of adults (e.g. the involvement of the cerebellum in non-motor functions such as language comprehension) can bring with them a host of developmental questions that had not been previously considered. A disadvantage is that questions about development are framed in terms of what a child cannot do relative to an adult. Accordingly, age-specific abilities and ontogenetic adaptations in general are ignored, with a consequent impoverishment of developmental theory.

Another motivation stems from an interest in developmental processes in their own right. From antiquity, debates between advocates of preformationism and epigenesis have strongly influenced this particular interest. The former holds that the fertilized egg essentially contains all the organs in invisible form, while the latter adheres to the theory that organs and functions differentiate from the fertilized egg, (see Oppenheim, 1982, for the history of these debates). Almost 20 years after Hamburger it seems appropriate to consider where research into the development of brain and behaviour has brought us in our understanding of their mutual relationships.

In neuroanatomy, comparisons of structures at early and at later stages, and interpolations between them, have given us crucial insights into the architecture of the adult brain. The deduction of the concept of the neuron by Cajal (see Hamburger, 1988) and its alliance with Sherrington's equally intuitive notion of the synapse to form the neuron doctrine is one of the persisting legacies of neuroanatomy. Neuroanatomists now have at their disposal sensitive tracing methods, antibodies against transmitters and neuroactive proteins, each of which has allowed many aspects of neural circuitry at adult age to be revealed in great detail by means of both light and electron microscopy. Antibody staining has shown how connections are established and how transmitter systems develop. Currently there is a growing interest in the role of nitric oxide in neural trans-mission. Not only that, but it has been implicated in generating synaptic plasticity and in controlling the balance between cell proliferation and differentiation. However, we are now in a position to conclude that morphological studies on the development of the nervous system have torpedoed the optimism which many neuroanatomists have fostered of clear-cut relationships between brain structures and behaviour.

THE PUZZLE OF TRANSIENT STRUCTURES AND PROCESSES

As to morphological development itself, it has been shown that transient structures sometimes develop that later disappear, projections may grow that are then retracted altogether, and other projections grow speedily to near their destination and then either wait for some time or make provisional contacts until the ultimate connections are established. The significance of many of these pro-cesses is still not understood. Temporary projections or structures might be a necessary stage in normal morphological development, or remnants of older phylogenetical stages that do not interfere with further development. Together with waiting periods they could also play a role in permitting or inhibiting a behaviour, which in turn is important for survival.

When considering our present knowledge, the morphological development of the CNS appears to be by no means a simple and straightforward process. Moreover, the task of relating the above-mentioned transient events to behavioural development has still to be achieved. Thus, increases in our knowledge of morphological development have not always been paralleled by a concomitant increase in understanding.

BEHAVIOURAL DEVELOPMENT INVOLVES MANY UNRESOLVED PROBLEMS

Muscular contractions or motility can be observed as soon as the motoneuronal axons innervate the primitive muscle mass, and this seems to imply a straightforward cause-and-effect relationship. What has been demonstrated is that early movements are essential for the proper development of muscles, the skeleton, lungs and the activity of the nervous system at later stages. One aspect which seems puzzling, however, is that, soon after the first neuromuscular connections have been formed, a wide variety of complex and seemingly coordinated movement patterns emerge that resemble those after birth. This seems to indicate that the coupling and the temporal order in which muscle groups are activated must be determined from the outset.

Perhaps even more complex and puzzling are the specific behavioural patterns that emerge shortly after birth. The rhythmical axial turnings of the head in a to-and-fro fashion that can be observed in newborns may largely have lost its function in the human (see Prechtl, 1958). However, nutritive sucking movements and swallowing are vital to survival. In the human these movements are embedded in an elaborate behavioural pattern involving simultaneous grasping of the hands, and this attains a high degree of proficiency a few days after birth. The amazing speed with which this pattern develops after birth seems to point to the activation of a circuitry already developed in the fetal period. At later stages, and after having fulfilled their vital role, these patterns wane, and remarkably they cannot be elicited in later life.

Other examples are the feeding pattern in newborn rats consisting of sucking and licking while lying on their backs for prolonged periods, and hatching behaviour in the chick (see Oppenheim, 1973). It has been shown that hatching behaviour can still be elicited after birth by the forced anteflexion of the neck, but normally it is only expressed for a limited period of time (e.g. see Bekoff, this handbook). As these behaviours cannot be elicited in the adult animal, this indicates that some time after they have fulfilled an adaptive role, the neural circuits disappear altogether, or they may be rebuilt as elements of circuits subserving other functions (Prechtl, 1981). Still other behaviours may be expressed only for a short period, and these may be the remnants of patterns that might have had adaptive significance earlier in evolution, the Moro reflex being a striking example (Prechtl, 1981). The functions of these behaviours in the human are not clear. Such behavioural patterns and their circuitry obviously have withstood evolutionary changes because they did not interfere with functioning. Results from investigations into brain development and from ethological and comparative studies on behavioural development therefore point towards neuro-ontogeny as being a complicated interaction between processes aiming at future functioning, processes essential for survival in early environments and with alternative capabilities and with the heritage of a long and capricious evolutionary history.

After birth, mature neuromuscular innervation patterns are gradually established, the force production of muscles increases together with the possibility for subtly recruiting their power. Corticospinal projections develop along with central areas involved in governing and regulating these movements. Postural control seems to be a limiting factor for head control, trunk control and the standing position to develop, and for these functions to be smoothly regulated. Patterns of movement coordination (e.g. during locomotion and prehension) at these early stages differ from those later, and it is sometimes difficult to conclude whether this is due to immature circuitries and effectors or different biomechanical constraints that serve to minimize energy expenditure. An additional task for the central and peripheral nervous systems, as well as for the musculoskeletal system, is to account adequately for changes in body dimensions and increases in mass, which might push the adaptive capabilities of these systems to their limits, especially during growth spurts.

Gradually the specific adaptations needed for survival wane and the development of the still-immature infant is buffered and facilitated by appropriate forms of parental care. Sense organs and central areas involved in perception and cognition gradually develop. Neural projections are refined, the speed and efficiency of information transmission increases, and a rearrangement in the allocation of brain functions may occur in order to increase further the efficiency of neural processing, lateralization being one of the most obvious examples of this reorganization.

GUIDELINES FOR THE FUTURE

The time is now ripe to promote the further integration of approaches to the study of neurobehavioural development. There is a clear scientific and political interest in the study of brain and mind in all their complexities, the topic of the neural basis of consciousness being a point in case (see Damasio, 1999). In part, further integration rests on technical and methodological advancements, which allow for refined measurements on the dynamics of structure–function relationships. Guided by sophisticated brain-based models, complex adaptive behaviours can be analysed in terms of aspects of information processing. The dynamics of interpersonal communication involving preverbal infants can be studied in great detail and related with some degree of circumspection to developing brain processes. Furthermore, advances in non-linear mathematical models and related methodologies are beginning to reveal the details of developmental change as exemplified by recent studies on developmental transitions in domains such as motor and language development. Nevertheless, we are still many steps removed from a well-integrated approach to the study of brain–behaviour development.

What are the main prerequisites for achieving a better-integrated approach to the study of brain–behaviour development? The following may serve as general guidelines:

1. *To focus on the dynamics of developmental processes instead of considering development as a succession of "fixed stages", the characteristics of which should be identified*

An important issue in this context, which should be addressed in the future, is the problem of how the tempo of development is regulated, and how changes in more or less stable states during development are induced (e.g. by processes of self-organization). According to developmental geneticists it is feasible that, as soon as thresholds or levels in concentrations of amino acids or proteins are reached or surpassed, other genes come to expression, but then the problem still remains of how the timing of the particular metabolic processes is regulated. In developmental neurobiology it is known that the proliferation of neuroblasts and glioblasts (at least at later stages) is influenced by growth hormones, but how they are in turn regulated is not known. Another issue is how the establishment of synaptic interactions is governed and whether or not the formation of dendritic spines is dependent on such interactions. Sometimes neuroanatomical connections are established that can be related to new behavioural abilities (e.g.

the transition of palmar to pincer grasping with the establishment of direct corticospinal connections). Although such action-based transitions have been studied systematically only on an incidental basis, they offer the best opportunities for detecting meaningful relationships between processes in brain and behaviour development.

2. *To aim at a problem-centred approach in which the contribution of each discipline is formulated in direct interaction with and complementary to other disciplines*

Still too often separate bodies of knowledge obtained in particular fields of study are just loosely and superficially linked. This is especially true for studies on higher cognitive functions in relation to the CNS, e.g. concerning lateralization in the brain on the one hand and language development on the other, or on structural differentiation in the brain and information processing. Modern techniques of brain imaging already to a certain extent allow the study of the dynamics of processes in various parts of the brain during complex cognitive tasks such as reading and problem solving. However, the applicability of such techniques in developmental studies is still limited. More importantly, a further understanding at the functional level of processes such as language production and comprehension is a prerequisite for goal-directed research into their underlying circuitries. The inventarization of all the areas of the brain involved in one way or the other in such complex functions (e.g. by means of fMRI or PETscanning) should still be regarded as a preliminary step for an in-depth analysis.

On the other hand, attempts at relating descriptions of developmental processes across each of these research domains have been disappointing. Partly this is due to seemingly insurmountable differences among complex systems with many dimensions and by the temporal resolution of the processes involved. Stereological aspects of neurotransmitter molecules and their receptors are several orders of magnitude smaller than the dimensions of neurons and their dendrites and axons. The same holds for the time domain: neurochemical processes take place in pico- or micro-seconds, neurophysiological processes in milliseconds and changes in connectivity probably in minutes to days. The problem is how to relate these processes with vastly different time scales and properties.

On a more positive note, molecular biologists and geneticists are beginning to identify the determinants of differentiation, migration and cell aggregation (e.g. one promising candidate appears to be a family of insulin and insulin-like growth factors termed cytokines). For their part neurobiologists have uncovered some of the salient developmental changes in muscles, transmitters, fibre connections, and CNS regions, each of which makes only partial contributions to behavioural development. Selectionist theories provide explanations at a more encompassing level in terms of how changes might be achieved and then become stabilized. For some problems neuronal modelling might help to promote initial insights into particular developmental processes, and even suggest possibilities that had not been previously considered. The basis of all these efforts to understand development has to remain a meticulous description of behavioural development and the provision of a fruitful environment for an interdisciplinary team of scientists who understand each other's language and methodologies.

3. *The application of procedures for the experimental study, observation, and analysis of data, which are of comparable levels of refinedness in the various approaches, and which contribute to the interdisciplinary study of structural and functional characteristics of the nervous system, and their behavioural concomitants*

In studies on children's development in general, and especially in such studies of children at risk for clinical problems, there is a gap between the results obtained from laboratory studies and those available from observations of functioning in daily life conditions (already indicated by Ounsted in 1955 as "the problem of the cage") Generally, data on their behaviour in daily-life conditions are obtained with questionnaires of insufficient reliability and validity. Relating such dubious and very global clinical information to detailed experimental measurements (e.g. on cognitive processes) is meaningless, but still common practice in particular in the clinical area of research. Refined observations of children's behaviour in well-selected natural conditions is essential for improving the external validity and theoretical power of such studies.

4. *To carefully consider how various theories and models, which are applied to different phases of human development and with respect to different functional domains, may relate to each other. Allied to this endeavour is the need for a more coherent theoretical framework that promotes the longitudinal study of developmental processes over a broader age range than has hitherto been the case*

There has been a long tradition in developmental research to focus on the characteristics of particular stages of development as exemplified in the Piagetian and neo-Piagetian approaches to cognitive development. In recent years there is an increasing focus on developmental transitions and their underlying mechanisms. In short, the study of development has become decidedly dynamical.

Dynamical systems approaches rooted in the principles of self-organization offer some promising guidelines for rejuvenating theory-building in the developmental sciences and for challenging some of our most cherished notions about the nature of ontogenetic development. Being in essence metatheories they do not provide guidelines for what should constitute the specific contents of a particular domain of development. Rather, the guidelines on offer are of a more generic nature. One set of guidelines directs to accounting for both stabilities and instabilities in development regardless of whether we are dealing with the brain, behaviour or both. Another is to acknowledge that development is not only deterministic, but also highly stochastic, and in this respect providing the tools for delineating the two processes. Further guidelines are provided for teasing out the mechanisms that break age-specific constraints on development by means of perturbation and bifurcation experiments in and around periods of previously identified developmental transitions. As such, the application of these approaches to particular domains of development requires not only the use of longitudinal designs, but also their combination with cross-sectional studies at appropriate transitional ages previously identified by mathematical models designed to capture the salient features of self-organizing systems when they are in a far-from-equilibrium state. To date there are few studies that have adopted this formidable, but much-needed, enterprise.

As for studying relationships across different functional domains, there are some recent illuminating examples, especially within developmental psychology. One such example can be found in Campos *et al.* (2000), who attempted to show that the onset of locomotion brings with it a pervasive set of changes in perception, spatial cognition, and social and emotional development (see also the associated set of equally illuminating commentaries).

As a final remark, we cite a strongly worded comment made by a molecular biologist:

> Control of the genome means control of development, and control of development means control of behaviors (Miklos, 1993, p. 851).

Such a statement appears to represent a rampant belief in genetic determinism. Fortunately sufficient evidence has accrued to counteract it, some of which was mentioned above. Substitute "brain" for "genome" and we have a neural determinism that flies in the face of even more compelling evidence. No, the complexities of development cannot be reduced to such simplifications and thus distinctions between "experience-independent" and "experience-dependent" developmental processes serve only to resurrect the ghost of the purportedly long-dead nature–nurture debate.

References

Campos, J.J., Anderson, D.I., Barbu-Roth, M.A., Hubbard, E.M, Hertenstein, M.J. & Witherington, D. (2000). Travel broadens the mind. *Infancy*, **1**, 149–219.

Damasio, A.R. (1999). *The feeling of what happens: body and emotion in the making of consciousness*. New York: Harcourt Brace.

Hamburger, V. (1988). Ontogeny of neuroembryology. *Journal of Neuroscience*, **8**, 3535–3540.

Harris, W.A. & Hartenstein, V. (1998). Cellular determination. In M.J. Zigmond, F.E. Bloom, S.C. Landis, J.L. Roberts & L.R. Squire (Eds.), *Fundamental neuroscience* (Chapter 17). San Diego, CA: Academic Press.

Johnson, M.H. *Developmental Cognitive Neuroscience*. Oxford, UK: Blackwell.

Mack, K.J. & Mack, P.A. (1992). Induction of transcription factors in somatosensory cortex after tactile stimulation. *Molecular Brain Research*, **12**, 141–147.

Miklos, G. (1993). Molecules and cognition: the latterday lessons of levels, language, and lac. Evolutionary overview of brain structure and function in some vertebrates and invertebrates. *Journal of Neurobiology*, **24**, 842–890.

Myers, M.M., Wager-Smith, K., Rothenfluh-Hilfiker, A. & Young, M.W. (1996). Light-induced degradation of TIMELESS and entrainment of the *Drosophila* circadian clock. *Science*, **271**, 736–740.

Oppenheim, R.W. (1973). Prehatching and hatching behaviour: a comparative and physiological consideration. In G. Gottlieb (Ed.), *Studies on development of behavior and the nervous system*, vol I: *Behavioral embryology* (pp. 163–244). New York: Academic Press.

Oppenheim, R.W. (1982). Preformation and epigenesis in the origins of the nervous system and behavior: issues, concepts, and their history. In P.P.G. Bateson & P.H. Klopfer (Eds.), *Perspectives in ethology*, vol 5, *Ontogeny* (pp. 1–100). New York: Plenum Press.

Ounsted, C.A. (1955). The hyperkinetic syndrome in epileptic children. *Lancet*, **2**: 303–311.

Panksepp, J. (1998). *Affective neuroscience: the foundations of human and animal emotions*. New York: Oxford University Press.

Prechtl, H.F.R. (1958). The directed head turning response and allied movements of the human baby. *Behaviour*, **13**, 212–242.

Prechtl, H.F.R. (1981). The study of neural development as a perspective of clinical problems. In K.J. Connolly & H.F.R. Prechtl (Eds.), *Maturation and development: biological and psychological perspectives* (pp. 198–215). Clin. Dev. Med. no. 77/78, London: Heinemann/SIMP.

Ramon y Cajal, S. (1937). *Recollections of my life* (transl. E. Horne Craigie). Cambridge, MA: MIT Press.

Index